AF743966

SYSTEM STRUCTURE AND CONTROL 1995

A Proceedings volume from the IFAC Conference, Nantes, France, 5 - 7 July 1995

Edited by

M. GUGLIELMI
LAN-ECN, Nantes, France

Published for the

INTERNATIONAL FEDERATION OF AUTOMATIC CONTROL

by

PERGAMON
An Imprint of Elsevier Science

UK	Elsevier Science Ltd, The Boulevard, Langford Lane, Kidlington, Oxford, OX5 1GB, UK
USA	Elsevier Science Inc., 660 White Plains Road, Tarrytown, New York 10591-5153, USA
JAPAN	Elsevier Science Japan, Tsunashima Building Annex, 3-20-12 Yushima, Bunkyo-ku, Tokyo 113, Japan

First edition 1996

Library of Congress Cataloging in Publication Data

A catalogue record for this book is available from the Library of Congress

British Library Cataloguing in Publication Data

A catalogue record for this book is available from the British Library

ISBN 0-08-042356 6

This volume was reproduced by means of the photo-offset process using the manuscripts supplied by the authors of the different papers. The manuscripts have been typed using different typewriters and typefaces. The lay-out, figures and tables of some papers did not agree completely with the standard requirements: consequently the reproduction does not display complete uniformity. To ensure rapid publication this discrepancy could not be changed: nor could the English be checked completely. Therefore, the readers are asked to excuse any deficiencies of this publication which may be due to the above mentioned reasons.

The Editor

Transferred to Digital Printing 2009

IFAC CONFERENCE ON SYSTEM STRUCTURE AND CONTROL 1995

Sponsored by
International Federation of Automatic Control (IFAC)
Technical Committee on Theory

Organized by
Laboratoire d'Automatique de Nantes (LAN)
Ecole Centrale de Nantes (ECN)
Université de Nantes, France
Centre National de la Recherche Scientifique
on behalf of
Association Francaise des Sciences et Technologie de l'Information et des Systemes

PREFACE

These proceedings are the final results of the third workshop on the System Structure and Control held in Nantes in July 1995. During the three years between the second workshop held in Prague and the third one, much research has been done on various aspects of system analysis and applications. This conference has provided a forum for the researchers in automatics control around the world to present their results and exchange their ideas and experience. Based on these discussion, we hope that we will identify future trends of the research in this area and promote applications of the structural approach of control theory.

M. GUGLIELMI
Directeur du LAN

CONTENTS

H_2, H_∞ AND OPTIMAL CONTROL

CONTROL OF LINEAR SYSTEMS

TIME-DELAY SYSTEMS I

SINGULAR SYSTEMS

TIME-DELAY SYSTEMS II

CONTROL OF ROBOTS AND MANIPULATORS

ROBUST CONTROL

STRUCTURE AND CONTROL II

MULTIDIMENSIONAL SYSTEMS

STABILITY OF NONLINEAR SYSTEMS

APPLICATIONS II

PLENARY PAPER II

STRUCTURAL PROPERTIES OF LINEAR SYSTEMS

NONLINEAR SYSTEMS II

ON SOME USES OF FACTORIZATIONS IN LINEAR SYSTEM THEORY

P.A. Fuhrmann*

Earl Katz Family Chair in Algebraic System Theory, Department of Mathematics. Ben-Gurion University of the Negev, Beer Sheva, Israel

Abstract. We present a short, and far from exhaustive, survey of some of the uses of factorization theory in the study of linear systems. The whole development is based on model theory, be it in the context of vectorial polynomial, rational or Hardy spaces. Invariant subspaces of operators are related to factorizations of matrix functions. The connection to system theory is via realizations based on various coprime factorizations. Among topics touched upon are the study of factorizations of rational functions and in particular the important cases of inner and all-pass functions. We survey the subject of coprime factorizations and their connection, via spectral factorizations, to various Riccati equations. In this connection we describe some geometric aspects of Wiener-Hopf factorizations. We proceed to describe connections with geometric control theory and give a Hilbert space characterization of stabilizability and detectability subspaces. Finally we describe a complete parametrization of all minimal spectral factors in the case of a coercive spectral density function.

Key Words. factorization, spectral factorization, coprime factorizations, invariant subspaces, realization theory, inner functions, Wiener-Hopf factorizations, geometric control theory, parametrization of spectral factors.

1. Introduction

Factorization theory has its roots in ancient mathematics, motivated by the study of primes and the factorization of integers into the product of primes. Rational numbers were known to the Greeks though their properties were studied mainly from the geometric viewpoint of proportions. The existence of irrational numbers stems from this source as do the Euclidean algorithm and the Bezout equation. Modular arithmetic was known in the East, e.g. the Chinese remainder theorem.

With the shift of emphasis from numbers to functions came the study zeros of functions, starting with the study of zeros of polynomials. The analogies between the ring of integers and polynomials became apparent and led to certain abstractions. The introduction of the notions of groups, rings, fields, ideals and modules led to the establishment of abstract algebra.

In this short paper we will discuss the interrelation between factorization theory and linear system theory. This connection has been a first rate example of the crossfertlization between pure and applied mathematics. Factorization results and related concepts such as coprimeness have had a great impact in the study of linear systems. However, system theory in turn has proved a great source for research problems in the area of factorization theory. Moreover, it is hard to imagine now a study of factorizations without some of the tools provided by system theory.

It is natural to seek for the underlying reasons behind this symbiotic relation. In my opinion, the reasons for this are to be found in the multifaceted way of representing linear systems. Such systems can be described externally by input/output maps, impulse responses, transferfunctions or behaviours. But they also have internal descriptions in terms of differential or difference equations, and state space models. The link connecting external and internal descriptions of a linear system is provided by realization theory. Thus external properties of a linear system are reflected by corresponding, internal, properties of realizations. Coprime factorizations are directly connected to realizations. Other properties related to factorizations have internal characterizations. Thus external properties have their counterpart in internal properties, like signature or Hamiltonian symmetry. Input/output stability is reflected in the stability of generators. The whole area of dynamic stabilization, including robustness issues, is best approached via doubly coprime factorizations.

The structure of the paper is as follows. We begin by studying ideals and invariant subspaces in spaces of functions and their representations. Set operations, which are geometric in nature are related to the arithmetic of certain factorizations. Model theory, as related to invariant subspaces and as a tool in the study of operators, is briefly described. A special case, where stability is assumed, is studied more closely. This case provides a point of contact between algebraic and analytic models.

We proceed to introduce several types of Hankel operators. The representations of their kernels and images are directly related to certain coprime factorizations, which in turn provide a key to the construction of realizations. A short study of doubly coprime factorizations (DCF) is done on two levels. We also indicate how several other classes of rational functions admit appropriate coprime factorizations. This is based on Fuhrmann and Ober [1993].

Next we describe one of the most widely applicable results on factorization of rational functions namely that of Sakhnovich [1976], which was popularized by Bart, Gohberg, Kaashoek and Van Dooren [1979]. This is a state space characterization of the existence of a factorization of a biproper rational function. The assumption is quite restrictive, but it is still the best of its kind currently available.

We mention a few results on Wiener-Hopf factorizations and the connection to invertibility of Toeplitz operators. Following that we proceed to introduce controlled and conditioned invariant subspaces, focusing on internally and externably stabilizable and detectable subspaces. We first derive an algebraic characterization of these spaces. Following that, using the fact that stability is built in, we proceed to give also an analytic characterization in terms of Hardy spaces and inner functions. These results are taken from Fuhrmann and Gombani [1995].

Inner functions are the basic component in thew description of invariant subspaces of Hardy spaces. State space characterizations of inner functions are well known, e.g. Genin et al [1983], and are given in terms of a particular solution of a homogeneous Riccati equation. We show how polynomial spectral factorization leads to an analogous characterization of inner functions. Factorizations of inner functions are related to general solutions of the same Riccati equation. This leads to the study of spectral factorizations for rational functions, positive on the imaginary axis. Extrmal spctral factors ar special casees of Wiener-Hopf factorizations. We will use these as a starting point for the parametrization of the set of all minimal spectral factors.

Due to limitations of space, this short survey is far from exhaustive, with the most notable omission being the approach to optimization problems via spectral and J-spectral factorizations. For some other aspects of factorization theory, in the spirit of this paper we refer to Fuhrmann [1989].

2. Polynomial and rational models

We begin by introducing polynomial models. In this paper, given a matrix function, it can be viewed alternatively as acting on column vectors by multiplication on the left or on row vectors on the right. We can pass from one case to the other by transposition, however we will find it useful to distinguish the two cases. Thus, generally, all function spaces as well as operators will have a sub(super)script, using the letters r or c, depending on whether the the space consists of row vectors or column vectors. For operators that act in a self evident way, the subscripts will be dropped.

We will denote by $\mathbf{R}^m$ the space of all m-vectors with coordinates in $\mathbf{R}$; $\mathbf{R}^m$ is similarly defined. Let π_+ and π_- denote the projections of $\mathbf{R}^m((z^{-1}))$ the space of truncated Laurent series on $\mathbf{R}^m[z]$ and $z^{-1}\mathbf{R}^m[[z^{-1}]]$, the space of formal power series vanishing at infinity, respectively. Since

$$\mathbf{R}^m((z^{-1})) = \mathbf{R}^m[z] \oplus z^{-1}\mathbf{R}^m[[z^{-1}]]$$

π_+ and π_- are complementary projections. In $\mathbf{R}^m[z]$ we define the shift operator by $(S_+p)(z) = zp(z)$, whereas S_- acting in $z^{-1}\mathbf{R}^m[[z^{-1}]]$ is defined by $S_-h = \pi_-h$.

Given a nonsingular polynomial matrix D in $\mathbf{R}^{m\times m}[z]$ we define two projections π_D^c in $\mathbf{R}^m[z]$ and π^D in $z^{-1}\mathbf{R}^m[[z^{-1}]]$ by

$$\pi_D^c f = D\pi_- D^{-1} f \qquad for\ f \in \mathbf{R}^m[z]$$

$$\pi_c^D h = \pi_- D^{-1}\pi_+ Dh \qquad for\ h \in z^{-1}\mathbf{R}^m[[z^{-1}]]$$

and define two linear subspaces of $\mathbf{R}^m[z]$ and $z^{-1}\mathbf{R}^m[[z^{-1}]]$ by

$$X_D^c = Im\pi_D^c,$$

and

$$X_c^D = Im\pi^D.$$

An element f of $\mathbf{R}^m[z]$ belongs to X_D^c if and only if $\pi_+D^{-1}f = 0$, i.e. if and only if $D^{-1}f$ is a strictly proper rational vector function. The space X_D^r is analogously defined. In particular $f \in X_D^r$ if and only if fD^{-1} is strictly proper. We refer to X_D^r and X_D^c as **polynomial models** whereas to X^D

and X^D as **rational models**.

We turn X_D^c into an $\mathbf{R}[z]$-module by defining

$$p \cdot f = \pi_D^c pf \qquad for\ p \in \mathbf{R}[z],\ f \in X_D^c.$$

Since $Ker\pi_D^c = D\mathbf{R}^m[z]$ it follows that X_D^c is isomorphic to the quotient module $\mathbf{R}^m[z]/D\mathbf{R}^m[z]$. Similarly, we introduce in X_c^D a module structure by

$$p \cdot h = \pi_- ph \qquad for\ p \in \mathbf{R}[z],\ h \in X_c^D.$$

In X_D^c we will focus on a special map S_D^c, a generalization of the classical companion matrix, which corresponds to the action of the identity polynomial z, i.e.,

$$S_D^c f = \pi_D^c sf \qquad for \qquad f \in X_D^c.$$

Thus the module structure in X_D^c is identical to the module structure induced by S_D^c through $p \cdot f = p(S_D^c)f$. With this definition the study of S_D^c is identical to the study of the module structure of X_D^c. In particular the invariant subspaces of S_D^c are just the submodules of X_D^c which are characterized next. They are related to factorization of polynomial matrices.

Similarly, the module structure introduced in X_c^D is the same one induced by the shift map

$$S_c^D h = S_- h = \pi_- sh \text{ for } h \in X_c^D.$$

Polynomial and rational models are closely related. Thus we have.

Proposition 1. The polynomial model X_D^c and the rational model X_c^D are isomorphic, with the isomorphism $\rho_D^c : X_c^D \longrightarrow X_D^c$ given by $f \mapsto D^{-1}f$. Moreover we have

$$S_D^c \rho_D^c = \rho_D^c S_c^D.$$

The class of polynomial and rational models is rich enough to model, up to similarity, all linear transformations acting in finite dimensional vector spaces. In fact, if A is a square matrix then S_{zI-A} is isomorphic to A. In particular this gives an easy approach to the study of canonical forms for A, via the study of the pencil $zI - A$.

The following is a key theorem in factorization theory. It connects S_D^c-invariant subspaces of the polynomial model X_D^c, or alternatively, S_c^D-invariant subspaces of the rational model X_c^D, with factorizations of the nonsingular polynomial matrix D. The underlying idea goes back to the work of Brodskii on factorizations of the Livsic characteristic function.

Theorem 2.

1. A subset M of X_D^c is a submodule, or equivalently an S_D^c invariant subspace, if and only if $M = D_1 X_{D_2}^c$ for some factorization $D = D_1 D_2$ with $D_i \in \mathbf{R}^{m\times m}[z]$.
2. A subset M of X_c^D is a submodule, or equivalently an S_c^D invariant subspace, if and only if $M = X_c^{D_2}$ for some factorization $D = D_1 D_2$ with $D_i \in \mathbf{R}^{m\times m}[z]$.

We summarize now the connection between the geometry of invariant subspaces and the arithmetic of polynomial matrices.

Theorem 3. Let M_i, $i = 1, \ldots, s$ be submodules of X_D^c, having the representations $M_i = E_i X_{F_i}$, that correspond to the factorizations

$$D = E_i F_i.$$

Then the following statements are true.

(i) $M_1 \subset M_2$ if and only if $E_1 = E_2 R$, i.e. if and only if E_2 is a left factor of E_1.

(ii) $\cap_{i=1}^s M_i$ has the representation $E_\nu X_{F_\nu}^c$ with E_ν the l.c.r.m. of the E_i and F_ν the g.c.r.d. of the F_i.

(iii) $M_1 + \cdots + M_s$ has the representation $E_\mu X_{F_\mu}^c$ with E_μ the g.c.l.d. of the E_i and F_μ the l.c.l.m. of all the F_i.

Corollary 4. Let $D = E_i F_i$, for $i = 1, \ldots, s$. Then

(i) We have

$$X_D^c = E_1 X_{F_1}^c + \cdots + E_s X_{F_s}^c$$

if and only if the E_i are left coprime.

(ii) We have $\cap_{i=1}^s E_i X_{F_i}^c = 0$ if and only if the F_i are right coprime.

(iii) The decomposition

$$X_D^c = E_1 X_{F_1}^c \oplus \cdots \oplus E_s X_{F_s}^c$$

is a direct sum if and only if $D = E_i F_i$ for all i, the E_i are left coprime and the F_i are right coprime.

The next result summarizes the relation between factorization and the spectral decomposition of linear maps.

Theorem 5. Let $D(z) \in \mathbf{R}^{n\times n}[z]$ be nonsingular and let $d(z) = det D(z)$ be its characteristic polynomial. Suppose d has a factorization $d = e_1 \cdots e_s$ with the e_i pairwise coprime. Then D admits factorizations

$$D = D_i E_i$$

with $det D_i = d_i = \Pi_{j \neq i} e_j$, $det E_i = e_i$ and such that

$$X_D^c = D_1 X_{E_1}^c \oplus \cdots \oplus D_s X_{E_s}^c$$

Moreover

$$det(S_D | D_i X_{E_i}) = e_i.$$

Theorem 3. and Corollary 4. connect coprimeness conditions to the geometry of invariant subspaces. The next result connects coprimeness to invertibility properties of intertwining maps.

Theorem 6. Let D and $\overline{D}$ be nonsingular polynomial matrices. Then

1. A map $Z : X_D^c \longrightarrow X_{\overline{D}}^c$ satisfies $ZS_D^c = S_{\overline{D}}^c Z$ if and only if there exist polynomial matrices E and $\overline{E}$ for which
$$\overline{E}D = \overline{D}E \tag{1}$$
and in terms of which
$$Zf = \pi_{\overline{D}} \overline{E} f \qquad \text{for } f \in X_D^c. \tag{2}$$
2. The map Z defined by (2) is injective if and only if D and E are right coprime.
3. The map Z defined by (2) is surjective if and only if $\overline{D}$ and $\overline{E}$ are left coprime.

Assuming the coprimeness conditions of the previous theorem, there exist polynomial matrices satisfying $\overline{X}D - \overline{Y}E = I$ and $\overline{D}X - \overline{E}Y = I$, or in matrix form

$$\begin{pmatrix} \overline{X} & -\overline{Y} \\ -\overline{E} & \overline{D} \end{pmatrix} \begin{pmatrix} D & Y \\ E & X \end{pmatrix} = \begin{pmatrix} I & \overline{X}Y - \overline{Y}X \\ 0 & I \end{pmatrix}.$$

Multiplying on the left by the inverse of the matrix on the right and modifying accordingly the definition of the matrices $\overline{X}, \overline{Y}$, we get the polynomial doubly coprime factorization (DCF)

$$\begin{pmatrix} \overline{X} & -\overline{Y} \\ -\overline{E} & \overline{D} \end{pmatrix} \begin{pmatrix} D & Y \\ E & X \end{pmatrix} = \begin{pmatrix} I & 0 \\ 0 & I \end{pmatrix}.$$

Doubly coprime factorizations are extremely useful. We give an example. Reversing the order of multiplication in the previous DCF, we obtain $D\overline{Y} = Y\overline{D}$, which indicates that the inverse of the map Z of (2) is given by $Z^{-1}g = \pi_D Y g$. That this is indeed true can be directly verified by a simple computation, using the Bezout equations arising from the second DCF. In the sequel we will see how DCF arise in the context of Hardy spaces.

Note that (1) can be written as $\overline{D}^{-1}\overline{E} = ED^{-1}$. This opens up the possibility of applying this theorem to the analysis of realizations based on polynomial models. This is taken up next.

3. Realization theory

3.1. *Shift realizations*

A $p \times m$ rational matrix function can be considered as a transfer function in two ways.

We will be interested in realizations associated with rational functions having the following representations

$$G = VT^{-1}U + W. \tag{3}$$

where T, U, V, W are appropriately sized polynomial matrices. Each representation of the form (3), is a basis for a state space realization and corealization. This is summed up in the following theorem.

Theorem 7. Let G be a proper $p \times m$ rational function. Assume $G = VT^{-1}U + W$, where T, U, V, W are appropriately sized polynomial matrices.

1. Define, in the state space X_T^c, a quadruple of maps A, B, C, D, with $A : X_T^c \longrightarrow X_T^c$, $B : \mathbf{R}^m \longrightarrow X_T^c$, $C : X_T^c \longrightarrow \mathbf{R}^p$ and $D : \mathbf{R}^m \longrightarrow \mathbf{R}^p$ by
$$\begin{cases} A = S_T \\ B\xi = \pi_T U \xi, \\ Cf = (VT^{-1}f)_{-1} \\ D = G(\infty). \end{cases} \tag{4}$$
Then $G = \left(\begin{array}{c|c} A & B \\ \hline C & D \end{array}\right)$. The realization is reachable if and only if T and U are left coprime and observable if and only if T and V are right coprime.
2. Define, in the state space X_T^r, a system by
$$\begin{cases} A = S_T^r \\ \xi B = \xi \pi_T V, \\ Cf = (fT^{-1}U)_{-1} \\ D = G(\infty). \end{cases} \tag{5}$$
Then $G = \left(\begin{array}{c|c} D & C \\ \hline B & A \end{array}\right)$. This corealization is reachable if and only if T and V are right coprime and observable if and only if T and U are left coprime.

Using the realization (4) and the corealization (5), as well as the isomorphism between polynomial and rational models, one can transfer these realizations to the context of rational models. We will find ample use for both representations when studying problems of geometric control theory. It should be noted that the advantage of basing a realization procedure on representations of the form (3), in contrast to one sided matrix fractions, is that it allows us to study from a functional point of view realizations that are neither reachable nor observable.

Theorem 8. Let

$$G = ED^{-1} = \overline{D}^{-1}\overline{E}$$

be a matrix fraction representation of a proper, $p \times m$ rational function.

1. In the state space X^c_D a system is defined by

$$\begin{cases} A = S^c_D \\ B\xi = \pi_D \xi, \\ Cp = (ED^{-1}p)_{-1} \\ D = G(\infty). \end{cases} \tag{6}$$

   Then $G = \left(\begin{array}{c|c} A & B \\ \hline C & D \end{array}\right)$; this realization is reachable and it is observable if and only if E and D are right coprime.

2. In the state space $X^c_{\overline{D}}$ a system is defined by

$$\begin{cases} A = S^c_{\overline{D}} \\ B\xi = \pi_{\overline{D}}\overline{E}\xi, \\ Ch = (\overline{D}^{-1}h)_{-1} \\ D = G(\infty). \end{cases} \tag{7}$$

   Then $G = \left(\begin{array}{c|c} A & B \\ \hline C & D \end{array}\right)$; this realization is observable and it is reachable if and only if $\overline{E}$ and $\overline{D}$ are left coprime.

3. In the state space X^D_c a system is defined by

$$\begin{cases} A = S^D_c \\ B\xi = \pi_- D^{-1}\xi, \\ Ch = (Eh)_{-1} \\ D = G(\infty). \end{cases} \tag{8}$$

   Then $G = \left(\begin{array}{c|c} A & B \\ \hline C & D \end{array}\right)$; this realization is reachable and it is observable if and only if E and D are right coprime.

4. In the state space $X^{\overline{D}}_c$ a system is defined by

$$\begin{cases} A = S^{\overline{D}}_c \\ B\xi = \pi_- \overline{D}^{-1}\overline{E}\xi, \\ Ch = (h)_{-1} \\ D = G(\infty). \end{cases} \tag{9}$$

   Then $G = \left(\begin{array}{c|c} A & B \\ \hline C & D \end{array}\right)$; this realization is observable and it is reachable if and only if $\overline{E}$ and $\overline{D}$ are left coprime.

5. Under the coprimeness assumptions the realizations (6) and (7) are isomorphic. The isomorphism $Z : X^c_D \longrightarrow X^c_{\overline{D}}$ being given by

$$Zf = \pi_{\overline{D}}\overline{E}f. \tag{10}$$

   Similarly, the realizations (8) and (9) are isomorphic. The isomorphism $\overline{Z} : X^D_c \longrightarrow X^{\overline{D}}_c$ being given by

$$\overline{Z}h = \pi_- Eh. \tag{11}$$

The polynomial coprime factorizations appearing in Theorems 6. and 8. arise most naturally in the study of Hankel operators. In fact, if G is a $p \times m$ rational matrix function, the Hankel operator $H_G : F^m[z] \longrightarrow z^{-1}F^p[[z^{-1}]]$ is defined by

$$H_G p = \pi_- Gp. \tag{12}$$

It is easy to check that $S_- H_G = H_G S_+$, which shows that $\operatorname{Ker} H_G$ is a submodule of $F^m[z]$ which as a linear space is of finite codimension, whereas $\operatorname{Im} H_G$ is a finite dimensional submodule of $z^{-1}F^p[[z^{-1}]]$. Thus there exist polynomial matrices D and $\overline{D}$ for which

$$\begin{cases} \operatorname{Ker} H_G &= DF^m[z] \\ \operatorname{Im} H_G &= X^{\overline{D}} \end{cases}$$

This leads immediately to the coprime factorizations $G = \overline{D}^{-1}\overline{E} = ED^{-1}$. Thus a link has been established between the study of Hankel operators and that of intertwining maps. The same connection can be made also in the analytic context and is basic to the study of the Nehari problem and the commutant lifting theorem. These are central tools in H^∞-control.

From any minimal realization we can easily construct a basis for the model space associated with a right coprime factorization. This construction goes back to Hautus and Heymann [1978] and is extremely useful.

Theorem 9. Let G be a proper rational function of McMillan degree n. Let

$$G = \left(\begin{array}{c|c} A & B \\ \hline C & D_\infty \end{array}\right) \tag{13}$$

be a minimal realization. Then $C(sI - A)^{-1} = \overline{D}^{-1}K(s)$ for some polynomial matrices K and $\overline{D}$. Defining $\overline{N}(s) = K(s)B$ we have

$$G = D_\infty + \overline{D}^{-1}\overline{N}. \tag{14}$$

A basis for $X^c_{\overline{D}}$ is given by the columns k_i of K and a basis for $X^{\overline{D}}_c$ is given by the columns of

$C(sI-A)^{-1}$. In particular

$$X_c^{\overline{D}} = \{C(sI-A)^{-1}\xi \mid \xi \in R_c^n\}. \quad (15)$$

Moreover, the matrix representation of the shift realization corresponding to (14) with respect to the basis $\mathcal{B} = \{h_1, \ldots, h_n\}$ is given by (13).

As a result we conclude that, given any polynomial matrix Q, $\overline{D}^{-1}Q$ is strictly proper if and only if there exists a constant matrix K for which $Q(z) = \Phi(z)K$. Similarly PD^{-1} is strictly proper if and only if there exists a constant matrix L for which $P(z) = L\Psi(z)$.

Theorem 10. Let G be a proper rational matrix having the minimal realization $G = \left(\begin{array}{c|c} A & B \\ \hline C & D \end{array}\right)$. Then

1. If G_1 is proper and $\operatorname{Ker} H_{G_1} \supset \operatorname{Ker} H_G$ then G_1 admits a realization of the form

$$G_1 = \left(\begin{array}{c|c} A & B \\ \hline C_0 & D_0 \end{array}\right).$$

2. If G_1 is proper and $\operatorname{Im} H_{G_1} \subset \operatorname{Im} H_G$ then G_1 admits a realization of the form

$$G_1 = \left(\begin{array}{c|c} A & B_0 \\ \hline C & D_0 \end{array}\right).$$

Corollary. Let $G = ND^{-1}$ have the reachable realization (A, B, C) and let $G' = MD^{-1}$. Then G' has a realization (A, B, C_0) for some C_0.

3.2. *Hardy spaces*

The theme of realization theory can be developed also in the analytic context of Hardy spaces. This has some disadvantages, inasmuch as stability is generally assumed, but it has the great advantage of a Hilbert space structure which is ideally suited to the study of optimization poroblems. Also it has the technical advantage that backward invariant subspaces are, simultaneously, the counterparts of both polynomial and rational models, and thus they provide a more symmetric setting.

Let L^∞ denote the space of measurable, essentially bounded, matrix valued functions on the imaginary axis. We denote by H_+^∞ the Banach space of all bounded, matrix valued, analytic functions in the open right half plane. In order not to encumber the notation, we omit the reference to the size of the matrices as this should always be clear from the context. In the case of square matrices H_+^∞ is a Banach algebra. By a theorem of Fatou, these functions have nontangential boundary values a.e. on the imaginary axis. Thus H_+^∞ can be identified with a closed subspace of L^∞. H_+^2 will denote both the Hardy space of vector valued, analytic functions in the open right half plane which satisfy

$$\|f\|^2 = \sup_{x>0} \int_{-\infty}^{\infty} \|f(x+iy)\|^2 dy < \infty,$$

as well as the subspace of L^2 of their boundary values. Similarly we define H_-^2. As a cosequence of the Fourier-Plancherel and the Paley-Wiener theorems we have the orthogonal direct sum decomposition

$$L^2 = H_+^2 \oplus H_-^2.$$

We denote by P_+ and P_- the orthogonal projections of L^2 onto H_+^2 and H_-^2 respectively.

We will use the same notation for row and column vector spaces. Usually it will be clear from the context which space is considered.

Before the introduction of Hankel operators we digress a bit on invariant subspaces of H_+^2. Since, with an eye towards continuous time problems, we are using the half planes for our definition of the H^2 spaces, we do not have the shift operators coveniently at our disposal, and in order to keep the scope of the paper within reasonable limits we avoid introducing the translation semigroups and their Fourier transforms. This forces us to a slight departure from the usual convention.

The algebra H_+^∞ can be made an algebra of operators acting on H_+^2 by letting, for $\psi \in H_+^\infty$, induce a multiplication map $M_\psi : H_+^2 \longrightarrow H_+^2$ which is defined by

$$M_\psi f = \psi f, \qquad f \in H_+^2. \quad (16)$$

In algebraic language we have introduced an H_+^∞ module structure in H_+^2. The adjoint of M_ψ is given by

$$M_\psi^* f = P_+ \psi^* f, \qquad f \in H_+^2. \quad (17)$$

Both M_ψ and M_ψ^* are special cases of Toeplitz operators.

A subspace $\mathcal{M} \subset H_+^2$ is called an **invariant subspace** if, for each $\psi \in H_+^\infty$ we have $M_\psi \mathcal{M} \subset \mathcal{M}$. Similarly, a subspace $\mathcal{M} \subset H_+^2$ is called a **backward invariant subspace** if, for each $\psi \in H_+^\infty$ we have $M_\psi^* \mathcal{M} \subset \mathcal{M}$. Clearly backward invariant subspaces are just orthogonal complements of invariant subspaces. Invariant subspaces have been characterized by Beurling in terms of inner functions. We recall that an $m \times n$ matrix function $M \in H_+^\infty$ is called **inner** if $\|M\|_\infty \leq 1$ and its boundary values on the imaginary axis are isometric a.e., i.e. $M(it)^* M(it) = I$.

Theorem 11. [Beurling-Lax-Halmos] A nontrivial subspace $\mathcal{M} \subset H_+^2$ is an invariant subspace if and only if $\mathcal{M} = MH_+^2$ for some inner function M.

It can be shown that a nontrivial subspace $\mathcal{M} \subset H_+^2$ is an invariant subspace with a finite dimensional orthogonal complement if and only if $\mathcal{M} = MH_+^2$ for some rational inner function M.

The availability of Beurling's theorem allows the arithmetization of the geometry of invariant subspaces.

Proposition 12. Let H_+^2 be the Hardy space of column vector functions. Then

1. Let $\mathcal{M}, \mathcal{N}$ be full invariant subspaces having the representations $\mathcal{M} = QH_+^2$ and $\mathcal{N} = SH_+^2$ respectively. Then $\mathcal{M} \subset \mathcal{N}$ if and only if $Q = ST$ for some inner function T.
2. Let $\mathcal{M}_i = Q_i H_+^2, i = 1, \ldots, s$ be full invariant subspaces. Then
$$\mathcal{M}_1 + \cdots + \mathcal{M}_s = QH_+^2$$
where Q is the greatest common left inner factor of all Q_i.
3. Let $\mathcal{M}_i = Q_i H_+^2, i = 1, \ldots, s$ be full invariant subspaces. Then
$$\cap \mathcal{M}_i = RH_+^2$$
where R is the least common right inner multiple of all Q_i.

The class of subspaces $H_c(Q)$, with Q inner, is important for modelling and for the development of realization theory. Since elements $\psi \in H_+^\infty$ act in H_+^2 by multiplication and leave QH_+^2 invariant, they induce linear operators acting on $H(Q)$ which are defined by $T_{\psi,Q}f = P_{H(Q)}\psi f$. The adjoints are given by $T_{\psi,Q}^* f = P_+\psi^* f$. Clearly, if $\psi \in H_+^\infty$, we have $||T_{\psi,Q}|| \leq ||\psi||_\infty$.

We note that, for $\tau \leq 0$, the functions $exp_\tau(s) = e^{-\tau s}$ are all in H_+^∞. The operators T_{exp_τ} form a strongly continuous semigroup of operators on $\{MH_+^2\}^\perp$.

The results of Proposition 12. can be adapted to the context of the coinvariant subspace $H_c(M)$. This provides the counterpart to Theorems 2. and 3.

Theorem 13. Let M be an inner function and $H_c(M) = \{MH_+^2\}^\perp$. Then

1. A subspace $\mathcal{V}$ of $H_c(M)$ is invariant under all operators $T_{\psi,M}$ if and only if $\mathcal{V} = M_1 H_c(M_2)$, where $M = M_1 M_2$ is a factorization of M into the product of inner functions.
2. Let $\mathcal{V}_i = M_i H_c(N_i)$, $i = 1, \ldots, s$ be invariant subspaces of $H_c(M)$ corresponding to the factorizations $M = M_i N_i$. Then $\mathcal{V}_1 \subset \mathcal{V}_2$ if and only if $M_1 = M_2 R$ for some inner function R. Alternatively Then $\mathcal{V}_1 \subset \mathcal{V}_2$ if and only if $N_2 = RN_1$ for some inner function R.
3. We have
$$\mathcal{V}_1 + \cdots + \mathcal{V}_s = M_\gamma H_c(N_\gamma)$$
where M_γ is the greatest common left inner factor of all M_i.
4. We have
$$\cap_{i=1}^s \mathcal{V}_i = M_\lambda H_c(N_\lambda)$$
where M_λ is the least common right inner multiple of all M_i, and N_λ is the greatest common right inner divisor of all N_i.

We say that inner functions N_i, $i = 1, \ldots, s$ are **mutually right coprime** if, for each i, N_i is right coprime with the least common left inner multiple of the N_j, $j \neq i$.

Proposition 14. The inner functions N_i are mutually right coprime if and only if for the inner function N defined by
$$H_+^2 N = \cap_{i=1}^s H_+^2 N_i$$
we have
$$\det N = \Pi_{i=1}^s \det N_i.$$

Corollary 15. Given the factorizations $M = M_i N_i$, we have the algebraic direct sum decomposition
$$H_c(M) = M_1 H_c(N_1) \dot{+} \cdots \dot{+} M_s H_c(N_s)$$
if and only if M_i are mutually left coprime and N_i are mutually right coprime.

The following theorem contains a continuous time version of the celebrated commutant lifting theorem. In this connection see Sarason [1968] and Sz.-Nagy and Foias [1970]. The invertibility conditions are from Fuhrmann [1968].

Theorem 16. Given two square inner functions Q and $\overline{Q}$ in H_+^∞. Then

1. A map $Z : H(Q) \longrightarrow H(\overline{Q})$ is an H_+^∞-module homomorphism. i.e. satisfies
$$T_{\psi,\overline{Q}} Z = Z T_{\psi,Q} \qquad \text{for } \psi \in H_+^\infty$$
if and only if there exist $H, \overline{H} \in H_+^\infty$ satisfying
$$\overline{H} Q = \overline{Q} H \tag{18}$$

and in terms of which

$$Zf = P_{H(Q)}\overline{H}f \qquad f \in H(Q). \qquad (19)$$

2. The map Z defined by (19) is injective if and only if Q and H are right coprime.
3. The map Z defined by (19) is surjective if and only if $\overline{Q}$ and $\overline{H}$ are left coprime.
4. The functions $H, \overline{H}$ can be taken to satisfy $\|H\|_\infty = \|\overline{H}\|_\infty = \|Z\|$.

Clearly, equation (18) can be rewritten as

$$G = \overline{Q}^{-1}\overline{H} = HQ^{-1}$$

which provides a factorization, referred to as the Douglas-Shapiro-Shields (DSS) factorization, of the L^∞ function G. Not every L^∞ function admits such a factorization, and for more on this see Fuhrmann [1981].

Given a function $G \in L^\infty$ the Hankel operator $H_G : H^2_+ \longrightarrow H^2_-$ is defined by

$$H_G f = P_- G f \qquad f \in H^2_+.$$

The following result on range inclusion of Hankel operators is of use. It is taken from Fuhrmann [1975,1981].

Proposition 17. Let $G, G_1 \in H^\infty_-$. Then we have

$$\operatorname{Im} H_{G_1} \subset \operatorname{Im} H_G$$

if and only if $G_1 = GF + H$ for some $F, H \in H^\infty_+$.

Theorem 18. Let $\Gamma(t)$ be a $p \times m$ matrix function, defined on $[0, \infty)$, whose Laplace transform, G, is strictly proper, stable and rational, i.e. in H^2_+. Then

1. A time domain minimal realization of G is given in the state space

$$X = span\{\Gamma(t+\tau)\xi | \tau \geq 0, \xi \in \mathbf{R}^n\} \quad (20)$$

by

$$\begin{cases} Af &= f' \\ B\xi &= \Gamma(t)\xi \\ Cf &= f(0) \end{cases}$$

2. Let the space X be defined by (20) and let $\mathcal{X} = \mathcal{L}X$ be its Laplace transform. Then $\mathcal{X}^\perp$ is an invariant subspace of H^2_+ having the representation

$$\mathcal{X}^\perp = KH^2_+$$

for some rational, normalized inner function K.

3. Let the normalized, rational inner function K have the polynomial coprime factorizations

$$K = D^{-1}\overline{D} = \overline{E}E^{-1}$$

then

$$H_c(K) = \{KH^2_+\}^\perp = X^D_c,$$

and

$$H_r(K) = \{H^2_+ K\}^\perp = X^E_r.$$

4. With respect to the corealization (5) of K in the state space X^E_r we have

$$\begin{cases} (Af)(s) = sf(s) - \lim_{s\to\infty} sf(s) \\ \quad = S^E_r f \\ \xi B = \xi K \pi_- = \xi(K(s) - I). \end{cases}$$

5. The Hilbert space adjoints of the maps (A, B) defined above are, for $f \in H_r(K)$, given by

$$\begin{cases} (A^* f)(s) \\ \quad = -sf(s) + \lim_{\sigma\to\infty} \sigma f(\sigma) K(s) \\ B^*\xi = -\lim_{s\to\infty} sf(s). \end{cases}$$

6. A frequency domain minimal corealization of K is given in the state space $\mathcal{X} = \mathcal{L}X$ by

$$\begin{cases} (Af)(s) &= sf(s) - \lim_{s\to\infty} sf(s) \\ \xi B &= \xi(K - I) \\ fC &= \lim_{s\to\infty} sf(s) \\ D &= I. \end{cases} \qquad (21)$$

Part 3. of Theorem 18. sheds light on one of the intersection points between the algebraic and analytic theories. It allows the simultaneous interpretation of the same object both as a rational model space as well as a backward invariant subspace.

Suppose now that the inner function has a factorization of the form $K = K_1K_2$, with K_i inner and $K_i(\infty) = I$. This of course implies the inclusions $K_1K_2H^2_+ \subset K_1H^2_+$ and, equivalently, $\{K_1H^2_+\}^\perp \subset \{K_1K_2H^2_+\}^\perp$. This can be seen also from a different perspective. Let $K_1 = E_1\overline{D_1}^{-1}$ and $K_2 = \overline{E}_2D_2^{-1}$ and let $E_2D_1^{-1} = \overline{D_1}^{-1}\overline{E}_2$. So

$$\begin{aligned} K &= E_1\overline{D_1}^{-1}\overline{E}_2D_2^{-1} = E_1E_2D_1^{-1}D_2^{-1} \\ &= (E_1E_2)(D_2D_1)^{-1} \end{aligned}$$

Using $H_c(K) = X^D_c$, we have the inclusion $X^{D_2D_1}_c \supset X^{D_1}_c$. This of course is consistent with the characterization of rational model inclusions given in Theorem 2.

4. Transfer function factorization

4.1. *Factorization of rational functions*

Given (A_i, B_i, C_i), $i = 1, 2$, minimal realizations of G_i with $\delta(G_i) = dim X_i$ the McMillan degree of G_i then we have a realization of $G_2 G_1$ in the state space $X_1 \oplus X_2$. This realization may not be canonical, but it gives an upper bound on the McMillan degree of a product of rational matrices. Specifically we have the inequality $\delta(G_2 G_1) \leq \delta(G_1) + \delta(G_2)$.

We tackle now the inverse problem. Namely given a transfer function, when can it be factorred into the product of two transfer functions. We will assume throughout this section that the transfer function is **biproper**, that is proper with a proper inverse, which is equivalent to the costant term equal being invertible. If the constant term is actually equal to the identity, we will call it a *normalized biproper function*. In particular the inverse of a normalized biproper function is also normalized biproper.

We pass now to the basic theorem concerning factorization of transfer function due to Sakhnovich [1976], see also Bart, Gohberg, Kaashoek and van Dooren [1979].

Theorem 19. **[Sakhnovich]** Let $G = \left(\begin{array}{c|c} A & B \\ \hline C & I \end{array}\right)$ be a biproper rational function with the realization given in the state space X. Then a necessary and sufficient condition for G to admit a factorization $G = G_2 G_1$ with G_i normalized biproper rational functions is that $X = M_1 \oplus M_2$ with M_1 an $A-$invariant subspace, M_2 an $A^{\times}-$invariant, where $A^{\times}$ is defined by $A^{\times} = A - BC$.

4.2. *Factorization of inner functions*

We proceed now to study square inner functions and their factorization properties. However in this case we are, given a normalized rational inner function K, we are interested in factorizations of the form $K = K_2 K_1$, where the K_i are also normalized inner functions. It is a consequence of Beurling's theorem that factorizations of inner functions are related to invariant subspace inclusions. However here we focus on the induced action in the model space, or backward invariant subspace $H_c(K)$. To make the connection with the Sakhnovich factorization theorem we need to establish the invariance with respect to a single operator. Contrary to the case of $\psi \in H_+^\infty$ where the induced map on $H_c(K)$ was defined by $T_{\psi,K} f = P_{H_c(K)} \psi f$, the "multiplication by s" operator is unbounded in H_+^2 and hence we cannot apply the orthogonal projection $P_{H_c(K)}$ to $sf(s)$. We circumvent this difficulty, using the identification $H_c(K) = X_c^D$, where $K = D^{-1}\overline{D}$ is a polynomial coprime factorization of K, to define the operator $A : H_c(K) \longrightarrow H_c(K)$ given by $Af = sf(s) - \lim_{s \to \infty} sf(s)$. In fact, it was shown in Fuhrmann [1981] that $H_c(K) = X_c^D$ is coinvariant, i.e. invariant under all operators M_ψ^* if and only if it is invariant under all the maps $f \mapsto \pi_- pf$, for an arbitrary polynomial. Obviously the last condition is equivalent to the invariance under the map $S_c^D f = \pi_- sf = sf(s) - \lim_{s \to \infty} sf(s)$. It is certainly not an accident that this map appears as the generator in the shift realization.

Suppose now that the inner function K has the minimal realization (21), in the state space $H_r(K) = X_c^E$. Since $H_r(K)$ has a Hilbert space structure, we can compute the Hilbert space adjoints of all maps appearing in this realization. This was done in Theorem 18. But the innerness of K implies $K^* = K^{-1}$ and hence we get

$$\begin{aligned} fA^* &= -sf(s) + \lim_{s \to \infty} sf(s) \cdot K(s) \\ &= -(sf(s) - \lim_{s \to \infty} sf(s) \\ &\quad + \lim_{s \to \infty} sf(s)(K(s) - I) \\ &= -f(A - CB) \end{aligned}$$

So, if $\mathcal{M} \subset H_r(K)$ is a backward invariant subspace, it has to be of the form $\mathcal{M} = H_r(K_1)$ for a factorization $K = K_2 K_1$. Clearly we have the orthogonal direct sum decomposition

$$H_r(K) = H_r(K_1) \oplus H_r(K_2) K_1$$

and $\mathcal{M}^\perp = H_r(K_2) K_1$ is an A^*-invariant subspace, hence also $A - CB$-invariant. Note that, as we deal here with row vector Hardy spaces, we have $A^{\times} = A - CB$, differing from the definition of $A^{\times}$ in Sakhnovich's theorem.

Before proceeding with the analysis of a state space characterization of the factorization of inner functions, we digress a bit on how spectral factorization techniques enter into the construction of rational inner functions. Note that if p is a Hurwitz polynomial and p^* is defined by $p^*(s) = \overline{p(-\overline{s})}$, then $q(s) = \frac{p^*}{p}$ is an inner function in the right half plane. Thus the pole structure of q determines q uniquely, up to a constant factor of modulus 1. We find it convenient to normalize the inner function by requiring that its value at infinity is equal to 1. There are other inner functions derived in the same way but of lower McMillan degree if we allow pole-zero cancellations.

In the multivariable case, the pole structure can be prescribed from either side. Moreover it can be done in two distinct but related ways. We consider the prescription of left poles. One way is to specify a stable, nonsingular, polynomial matrix to act as left denominator. The other way depends on the fact that, given such a D, there exists a unique, up to isomorphism, observable pair (A, C), with A stable, such that

$$X_c^D = \left\{C(sI - A)^{-1}\xi | \xi \in \mathbf{C}^n\right\}.$$

Thus it suffices to prescribe such a pair. Naturally, in the functional form the construction depends on polynomial spectral factorization, see in this connection Callier [1985] and the further references therein. This translates to a homogeneous Riccati equation in the state space setting. This is the content of the next theorem.

Theorem 20.

1. Given an observable pair (A, C), with A stable. Then $\left(\begin{array}{c|c} A & B \\ \hline C & I \end{array}\right)$ is a normalized inner function if and only if $B = -YC^*$ with Y a nonnegative solution of the homogeneous Riccati equation

$$AY + YA^* + YC^*CY = 0. \qquad (22)$$

The realization $\left(\begin{array}{c|c} A & B \\ \hline C & I \end{array}\right)$ is observable. The reachable subspace is given by ImY.

2. Let D be a square, nonsingular, stable polynomial matrix. Let $(A.C)$ be determined Then $U = D^{-1}N$ is an inner function if and only if N is a solution of the spectral factorization problem

$$NN^* = DD^*.$$

The maximal McMillan degree inner function is given if we take N to be the antistable spectral factor.

As a corollary to the previous theorem we obtain the following result, giving a well known characterization of rational inner functions. See Genin, Van Dooren, Kailath, Delosme and Morf [1983], Finesso and Picci [1982] and also Fuhrmann and Ober [1993] where a slightly more general result is derived. Of course one can give a direct proof of this result based on the state space isomorphism theorem. This is the approach taken in the previously mentioned papers.

Theorem 21.

1. A function U is rational and inner in H_+^∞ if and only if it has a minimal state space realization of the form

$$U = \left(\begin{array}{c|c} A & B \\ \hline -B^*X & I \end{array}\right)$$

with A stable and X the unique, positive definite solution of the homogeneous ARE

$$A^*X + XA + XBB^*X = 0$$

2. A function U is rational and inner in H_+^∞ if and only if it has a minimal state space realization of the form

$$U = \left(\begin{array}{c|c} A & -YC^* \\ \hline C & I \end{array}\right)$$

with A stable and Y the unique, positive definite solution of the homogeneous ARE

$$AY + YA^* + YC^*CY = 0$$

Theorem 21 establishes a connection between inner functions and the Riccati equation. On the other hand we know already the connection between factorizations of inner functions and invariant subspaces. The direct link between solutions of Riccati equations and invariant subspaces goes back to the classic paper Willems [1971]. In this paper Willems parametrizes the set of solutions of an ARE via the set of invariant subspaces of a certain related operator. For more on this theme, as well as connections to factorization theory, we refer to Finesso and Picci [1982] and Fuhrmann [1985,1989].

In view of the preceeding remarks, it is natural to develop a factorization theory for inner functions that would unify arithmetic (i.e. factorization), geometric (i.e. invariant subspaces) and dynamic (i.e. Riccati equations oriented) approaches. This leads to the following.

Theorem 22.

1. Let U be a square inner function in the right half plane, normalized so that $U(\infty) = I$. Let

$$U = \left(\begin{array}{c|c} A & -Y_+C^* \\ \hline C & I \end{array}\right)$$

be a minimal realization with A stable and Y_+ the unique, positive definite solution of the homogeneous ARE

$$AY + YA^* + YC^*CY = 0 \qquad (23)$$

Then there exists a bijective correspondence between

(a) Normalized left inner factors of U.
(b) Invariant subspaces of A.
(c) Nonnegative definite solution of the homogeneous ARE (23).

The correspondence is given by

$$U_1 = \left(\begin{array}{c|c} A & -YC^* \\ \hline C & I \end{array}\right)$$

where Y is a nonnegative definite solution of the homogeneous ARE (23), and $\mathcal{V} = ImY \subset \mathbf{R}^n$ is A invariant.

2. Let U be a square inner function in the right half plane, normalized so that $U(\infty) = I$. Let

$$U = \left(\begin{array}{c|c} A & B \\ \hline -B^*X & I \end{array}\right)$$

be a minimal realization with A stable and X the unique, positive definite solution of the homogeneous ARE

$$A^*X + XA + XBB^*X = 0. \qquad (24)$$

Then there exists a bijective correspondence between

(a) Normalized right inner factors of U.

(b) Nonnegative definite solution of the homogeneous ARE (24).

(c) Invariant subspaces of A.

The correspondence is given by

$$U_2 = \left(\begin{array}{c|c} A & B \\ \hline -B^*X_2 & I \end{array}\right)$$

where X_2 is a nonnegative definite solution of the homogeneous ARE (24). and $\mathcal{V} = KerX \subset \mathbf{R}^n$ is A invariant.

There is a natural partial ordering of left inner factors of a given inner function. There is another natural partial order in the set of symmetric solutions of the Riccati equation (22). It is not surprising that these partial orders are related through the following theorem. The proof is based on previous results and hence omitted.

Theorem 23.

1. Let $U(s)$ be inner, and let $U(s) = I - C(sI - A)^{-1}Y_+C^*$ with $Y_+ > 0$ the unique, positive definite solution of the ARE

$$AY + YA^* + YC^*CY = 0. \qquad (25)$$

Let $U_i(s) = I - C_i(sI - A)^{-1}Y_iC_i^*, \quad i = 1, 2$ be two left factors of U with Y_i nonnegative definite solutions of the Riccati equation (25). Then U_2 is a left inner factor of U_1 if and only if $Y_2 \leq Y_1$.

2. Let $U(s)$ be inner, and let $U(s) = I - B^*X(sI - A)^{-1}B$ with $X > 0$ the unique, positive definite solution of the ARE

$$A^*X + XA + XBB^*X = 0. \qquad (26)$$

Let $U_i(s) = I - B_i^*X_i(sI - A)^{-1}B_i, \quad i = 1, 2$ be two left factors of U with X_i nonnegative definite solutions of the Riccati equation (26). Then U_2 is a right inner factor of U_1 if and only if $X_2 \leq X_1$.

This result has important applications in stochastic realization theory.

4.3. *Coprime Factorizations Over H_+^∞*

We continue our study by focusing on coprime factorizations over H_+^∞. THe motivation for this is twofold. On the one hand was the study of analytic, or rather meromorphic, continuation properties in terms of Hankel operator ranges. This was initiated by Douglas, Shapiro and Shields [1971] and carried over to the multivariable case in Fuhrmann [1975] and led to the DSS factorizations. The associated DCF are instrumental in the spectral analysis of intertwining maps, see Fuhrmann [1981]. The other driving force was stabilization theory. It turns out that stabilizing a linear system by dynamic feedback is equivalent to the solvability of a Bezout equation for the coprime factors over H_+^∞. We refer to Vidyasagar [1985] for an extensive study of the coprime factorization approach.

Two elements $M, N \in H_+^\infty$ are called (strongly) right coprime if there exists an H_+^∞ solution to the Bezout equation $\overline{V}M - \overline{U}N = I$. Similarly, $\overline{M}, \overline{N} \in H_+^\infty$ are called (strongly) left coprime if there exists an H_+^∞ solution to the Bezout equation $\overline{M}V - \overline{N}U = I$. Note that M, N need not have the same size, only the same number of columns. There is also a weaker notion of coprimeness in H_+^∞, see Fuhrmann [1981] for the details, but we will not discuss it here as we focus on rational functions and in this case the two notions of coprimeness coincide. In the sequel we will take coprimeness to mean the strong one.

Let G be a proper rational matrix-valued function. Then the factorization $G = NM^{-1}$ is called a *right factorization* of G, if N, M are stable rational functions and M is invertible with proper inverse. If N, M are right coprime, then the factorization is called a *right coprime factorization.* Similarly we define a left coprime factorization $G = \overline{M}^{-1}\overline{N}$. The two proper stable rational block matrices

$$\begin{pmatrix} M & U \\ N & V \end{pmatrix}, \quad \begin{pmatrix} \overline{V} & -\overline{U} \\ -\overline{N} & \overline{M} \end{pmatrix},$$

with M ($\overline{M}$) having a proper inverse, form a *doubly coprime factorization* of the proper rational function G, if

$$\begin{pmatrix} \overline{V} & -\overline{U} \\ -\overline{N} & \overline{M} \end{pmatrix}\begin{pmatrix} M & U \\ N & V \end{pmatrix} = \begin{pmatrix} I & 0 \\ 0 & I \end{pmatrix}$$

and

$$G = NM^{-1} = \overline{M}^{-1}\overline{N}.$$

Doubly coprime factorizations are related to the Youla-Kucera parametrization of all stabilizing controllers. This accounts for the importance of coprime factorizations over H_{+}^{∞}.

Now, clearly, every proper rational transfer function G has a coprime factorization over H_{+}^{∞}. To see this, assume $G = ED^{-1}$ is a polynomial right coprime factorization. Choose a stable, nonsingular polynomial matrix T such that DT^{-1} is biproper, with D, T right coprime. This can easily be done using Rosenbrock's theorem. Thus we can write $G = (ET^{-1})(DT^{-1})^{-1} = NM^{-1}$, with $N, M \in H_{+}^{\infty}$. This is the approach taken in Fuhrmann and Ober [1993].

Note that in the previous process we had great freedom in the choice of the denominator matrix T. This choice is severely restricted if we make additional requirements on the coprime factorizations. In fact if $G \in H_{-}^{\infty}$ and we require the denominators to be inner functions, we recover the DSS factorizations. If we make no restriction on G other that that it is proper and rational, but require the coprime factorizations to be normalized (NCF), i.e., with $G = \overline{M}^{-1}\overline{N} = NM^{-1}$, that $M^*M + N^*N = I$ and $\overline{M}\overline{M}^* + \overline{N}\overline{N}^* = I$ hold, the choice of T becomes essentially unique and has to be obtained via a polynomial spectral factorization procedure. To see this consider the normalized left coprime factorization case. We have $N = ET^{-1}$ and $M = DT^{-1}$ and the normalization condition translates into $E^*E + D^*D = T^*T$. So T is obtained by polynomial spectral factorization, under the additional requirement that $DT^{-1} = I$. The importance of normalized coprime factorizations became clear in the work of McFarlane and Glover [1989] on robustness issues in stabilization theory. For more on this see Ober and Sefton [1991] and Georgiou and Smith [1990]. Other classes of functions that admit specially normalized coprime factorizations are the class of bounded real functions and the related class of positive real functions. All the normalizations can be given in the form

$$\begin{pmatrix} M^* & N^* \end{pmatrix} J \begin{pmatrix} M \\ N \end{pmatrix} = I,$$

where J is an appropriate metric. For H_{-}^{∞} functions the metric is the degenerate one $J_S = \begin{pmatrix} I & 0 \\ 0 & 0 \end{pmatrix}$. For the standard NCF we have $J_L = \begin{pmatrix} I & 0 \\ 0 & I \end{pmatrix}$. For the bounded real case we take $J_B = \begin{pmatrix} I & 0 \\ 0 & -I \end{pmatrix}$, and finally for the positive real case the metric is chosen to be $J_P = \begin{pmatrix} 0 & I \\ I & 0 \end{pmatrix}$.

From a computational point of view, it is desirable to have state space formulas available for the various (normalized) coprime factorizations. A general approach to such a derivation was initiated in Khargonekar and Sontag [1982]. For normalized coprime factorizations the modified formulas were derived by Meyer and Franklin [1987] and Vidyasagar [1988]. A comprehensive study is given in Fuhrmann and Ober [1993] and the following theorem is taken from there. Some simplifying assumptions have been made in order to fit the format of this paper. Spectral factorizations are central to these derivations and are translated into the solution of various Riccati equations. Usually one Riccati equation arises for the left factorization and the other for the right one.

Theorem 24. Let G be proper and have the minimal realization $G = \left(\begin{array}{c|c} A & B \\ \hline C & D \end{array}\right)$.

1. If $G \in H_{-}^{\infty}$ is strictly proper, then J_S-normalized coprime factorizations are given by

$$\begin{pmatrix} M_S \\ N_S \end{pmatrix} = \left(\begin{array}{c|c} A - BB^*X & B \\ \hline -B^*X & I \\ C & 0 \end{array}\right)$$

$$\begin{pmatrix} \overline{N}_S & \overline{M}_S \end{pmatrix} = \left(\begin{array}{c|cc} A - ZC^*C & -B & ZC^* \\ \hline -C & 0 & I \end{array}\right)$$

   where X, Z are, respectively, solutions of the homogeneous Riccati equations

$$A^*X + XA - XBB^*X = 0$$
$$AZ + ZA^* - ZC^*CZ = 0$$

2. For any strictly proper G, J_L-normalized coprime factorizations are given by

$$\begin{pmatrix} M_L \\ N_L \end{pmatrix} = \left(\begin{array}{c|c} A - BB^*X & B \\ \hline -B^*X & I \\ C & 0 \end{array}\right)$$
$$\begin{pmatrix} -\overline{N}_L & \overline{M}_L \end{pmatrix} = \left(\begin{array}{c|cc} A - ZC^*C & B & ZC^* \\ \hline -C & 0 & I \end{array}\right)$$

   where X, Z are, respectively, solutions of the Riccati equations

$$A^*X + XA + C^*C - XBB^*X = 0$$
$$AZ + ZA^* + BB^* - ZC^*CZ = 0$$

3. For any strictly proper, bounded real G, J_B-normalized coprime factorizations are given

by

$$\begin{pmatrix} M_B \\ N_B \end{pmatrix} = \left(\begin{array}{c|c} A+BB^*X & B \\ \hline B^*X & I \\ C & D \end{array}\right)$$

$$\begin{pmatrix} -\overline{N}_B & \overline{M}_B \end{pmatrix} = \left(\begin{array}{c|cc} A+ZC^*C & B & ZC^* \\ \hline C & 0 & I \end{array}\right)$$

where X, Z are, respectively, solutions of the Riccati equations

$$\begin{aligned} A^*X + XA + C^*C + XBB^*X &= 0 \\ AZ + ZA^* + BB^* + ZC^*CZ &= 0 \end{aligned}$$

4. For any proper positive real G normalized so that $D+D^* = I$, J_P-normalized coprime factorizations are given by

$$\begin{pmatrix} M_P \\ N_P \end{pmatrix} = \left(\begin{array}{c|c} A-B(C-B^*X) & B \\ \hline -(C-B^*X) & I \\ D^*C+DB^*X & D \end{array}\right)$$

$$\begin{pmatrix} -\overline{N}_P & \overline{M}_P \end{pmatrix} =$$

$$\left(\begin{array}{c|cc} A-(B-ZC^*)C & -[BD^*+ZC^*D] & (B-ZC^*) \\ \hline C & -D & I \end{array}\right)$$

where X, Z are, respectively, solutions of the Riccati equations

$$\begin{aligned} &(A^* - C^*B^*)X + X(A-BC) \\ &\quad + C^*C + XBB^*X = 0 \end{aligned}$$

$$\begin{aligned} &(A-BC)Z + Z(A^* - C^*B^*) \\ &\quad + BB^* + ZC^*CZ = 0. \end{aligned}$$

The derivation of doubly coprime factorizations is now easy. For the details we refer to Fuhrmann and Ober [1993].

5. Wiener-Hopf factorizations

In the analysis of stationary stochastic processes the phase function associated with a rational spectral density function is an important tool. The phase function T_0 is defined by $T_0 = \overline{W_+}^{-1}W_-$, where W_- is the stable, minimum phase spectral factor and W_+ is the antistable, maximum phase spectral factor. The phase function itself is an all-pass function and the previous factorization is a special case of a right Wiener-Hopf factorization, to be defined next. However, any rational all-pass function has also left and right DSS factorizations. Thus

$$T_0 = K_+Q_+^* = Q_-^*K_-,$$

with $Q_\pm, K_\pm$ all inner functions. The next two theorems analyse this situation and give also characterizations of invertibility based on considerations of geometry as well based on properties of Hankel operators. The content of this section is based on Fuhrmann and Gombani [1995]. Theorem 26 is standard, see Gohberg and Feldman [1971]. Some of the results of Theorem 27 can be found in Nikolskii [1985].

We recall the definition of Toeplitz operators. As we are working with both row and column spaces we must distinguish between two types of Toeplitz operators.

Definition 25. Let G be an $m \times m$ matrix function in L^∞. We define the Toeplitz operator T_G^c acting on column Hardy space by

$$\begin{aligned} T_G^c : H_+^2 &\longrightarrow H_+^2 \\ f &\mapsto P_+Gf. \end{aligned}$$

Similarly, we define the Toeplitz operator T_G^r acting on row Hardy space by

$$\begin{aligned} T_G^r : H_+^2 &\longrightarrow H_+^2 \\ f &\mapsto P_+fG. \end{aligned}$$

In both cases G is called the **symbol** of the corresponding Toeplitz operator.

We say that

$$G = G_-\Delta_r G_+$$

is a **right Wiener-Hopf factorization** if $G_-^{\pm 1} \in H_-^\infty$, $G_+^{\pm 1} \in H_+^\infty$ and

$$\Delta_r(s) = diag\left(\left(\frac{s-1}{s+1}\right)^{\kappa_1}, \ldots, \left(\frac{s-1}{s+1}\right)^{\kappa_m}\right)$$

with $\kappa_1 \geq \cdots \geq \kappa_m$. The indices $\kappa_1, \cdots, \kappa_m$ are called the **right Wiener-Hopf factorization indices**. Left factorizations and left factorization indices are similarly defined.

It is well known that if $G \in L^\infty$ is continuous, and in particular if it is rational, then Wiener-Hopf factorization exist and the factorization indices, though not necessarily the factorizations, are uniquely defined.

Theorem 26. Let $G \in L^\infty$ and $G = G_-\Delta_r G_+$ its right Wiener-Hopf factorization. Then

1. The following statements are equivalent.
   (a) The Toeplitz operator T_G^c is injective.
   (b) The Toeplitz operator $T_{\Delta_r}^c$ is injective.
   (c) All right Wiener-Hopf factorization indices are nonnegative.
2. The following statements are equivalent.
   (a) The Toeplitz operator T_G^c is surjective.
   (b) The Toeplitz operator $T_{\Delta_r}^c$ is surjective.
   (c) All right Wiener-Hopf factorization indices are nonpositive.

3(a) The Toeplitz operator T_G^c is invertible if and only if all right Wiener-Hopf factorization indices are trivial.

(b) The Toeplitz operator T_G^r is invertible if and only if all left Wiener-Hopf factorization indices are trivial.

In the following theorem the inner functions can be taken to arise from a DSS factorization of an all-pass function, namely $T_0 = Q_-^* K_- = K_+ Q_+^*$.

Theorem 27. Let Q_-, Q_+, K_-, K_+ be inner functions satisfying

$$Q_- K_+ = K_- Q_+.$$

with the coprimeness conditions

$$\begin{aligned} K_- \wedge_L Q_- &= I \\ K_+ \wedge_R Q_+ &= I \end{aligned}$$

satisfied. Then

1. The following statements are equivalent
   (a) The Toeplitz operator $T^c_{Q_+K_+^*} = T^c_{K_-^*Q_-}$ is injective.
   (b) The Toeplitz operator $T^c_{K_+Q_+^*} = T^c_{K_-^*Q_-}$ is surjective.
   (c) All right Wiener-Hopf factorization indices of $Q_+K_+^* = K_-^*Q_-$ are nonnegative.
   (d) We have
   $$H_c(K_-) \cap Q_- H_+^2 = \{0\}$$
   (e) We have
   $$K_- H_+^2 + Q_- H_-^2 = L^2.$$
   (f) We have
   $$K_- H_+^2 + H_c(Q_-) = H_+^2.$$
   (g) We have
   $$P_{H_c(K_-)} H_c(Q_-) = H_c(K_-).$$
   (h) All the singular values of the Hankel operator $H^c_{K_+Q_+^*} = H^c_{K_-^*Q_-}$ are < 1.
2. The following statements are equivalent
   (a) The Toeplitz operator $T^r_{Q_+K_+^*} = T^r_{K_-^*Q_-}$ is injective.
   (b) The Toeplitz operator $T^r_{K_+Q_+^*} = T^r_{K_-^*Q_-}$ is surjective.
   (c) All left Wiener-Hopf factorization indices of $Q_+K_+^* = K_-^*Q_-$ are nonnegative.
   (d) We have
   $$H_r(K_+) \cap H_+^2 Q_+ = \{0\}$$
   (e) We have
   $$H_+^2 K_+ + H_-^2 Q_+ = L^2.$$
   (f) We have
   $$H_+^2 K_+ + H_r(Q_+) = H_+^2.$$
   (g) We have
   $$P_{H_r(K_+)} H_r(Q_+) = H_r(K_+).$$
   (h) We have
3. The following statements are equivalent
   (a) The Toeplitz operators $T^c_{Q_+K_+^*} = T^c_{K_-^*Q_-}$ and $T^r_{Q_+K_+^*} = T^r_{K_-^*Q_-}$ are both invertible.
   (b) All, left and right, Wiener-Hopf factorization indices of $Q_+K_+^* = K_-^*Q_-$ are trivial.
   (c) We have
   $$\begin{cases} H_c(K_-) \cap Q_- H_+^2 &= \{0\} \\ H_c(K_-) + Q_- H_+^2 &= H_+^2 \end{cases}$$
   (d) We have
   $$\begin{cases} P_{H_r(K_-)} H_r(Q_-) &= H_r(K_-) \\ P_{H_c(Q_-)} H_c(K_-) &= H_c(Q_-) \end{cases}$$
   (e) We have
   $$\begin{cases} K_- H_-^2 \cap Q_- H_+^2 &= \{0\} \\ K_- H_-^2 + Q_- H_+^2 &= L^2 \end{cases}$$
   (f) All singular values of the Hankel operators $H^c_{Q_+K_+^*}$ are less than 1.

$$\dim \operatorname{Ker} T^c_{Q_+K_+^*} = \dim\{f \mid \|H^c_{Q_+K_+^*} f\| = \|f\|\}.$$

6. Geometric control theory

In the analysis and synthesis of control systems, especially when using geometric methods as in Wonham [1979], it is necessary to extend the notion of an invariant subspace. Thus, given a transfer function G with realization $G = \left(\begin{array}{c|c} A & B \\ \hline C & D \end{array}\right)$ in a state space X, we say that a subspace $V \subset X$ is a **controlled invariant** subspace if given any initial condition $x_0 \in V$ there exists a control function u for which the solution to $\dot{x}(t) = Ax(t) + Bu(t)$ remains in V. It is easy to check that V is controlled invariant if and only if there exists a feedback map F such that $(A+BF)V \subset V$. Thus we can introduce the dual notion. We say that a subspace $W \subset X$ is **conditioned invariant** if there exists an output injection map H such that $(A + HC)W \subset W$. The concepts of controlled and conditioned invariant subspaces were first introduced by Basile and Marro [1969]. The following is a standard characterization. For a realization $G = \left(\begin{array}{c|c} A & B \\ \hline C & D \end{array}\right)$ in the state space X a subspace $\mathcal{V} \subset X$ is controlled invariant if and only if $A\mathcal{V} \subset \mathcal{V} + ImB$. Similarly, a subspace $\mathcal{V} \subset X$ is conditioned invariant if and only if $A(\mathcal{V} \cap \operatorname{Ker} C) \subset \mathcal{V}$.

It is of interest to the notions of controlled and conditioned invariant subspaces directly to the transfer function itself. This can be done using the shift realizations. The following proposition is taken from Fuhrmann and Willems [1980] and Fuhrmann [1981]. Since invariant subspaces of the shift operator are in correspondence with factorizations, it is only natural to expect that factorizations play a major role in the characterization of controlled and conditioned invariant subspaces. This is indeed the case, although the factorizations leading to the following result have been eliminated from the statement.

Proposition 28. Let G be a $p \times m$ proper rational function having the polynomial coprime factorization $G = ED^{-1} = \overline{D}^{-1}\overline{E}$. Then

1. With respect to the realization (8) in the state space X_c^D, a subspace $V \subset X_c^D$ is controlled invariant if and only if $V = \pi^D L$ for some submodule $L \subset z^{-1}F^m[[z^{-1}]]$.
2. With respect to the realization (7) in the state space $X_{\overline{D}}^c$, a subspace $V \subset X_{\overline{D}}^c$ is conditioned invariant if and only if $V = X_{\overline{D}}^c \cap M$ for some submodule $M \subset F^p[z]$.

Given a linear transformation A in X and an A-invariant subspace V, we denote by $A|_V$ the restriction of A to V. By a slight abuse of notation we will denote by $A|_{X/V}$ the **induced map**, i.e. the map induced by A in the quotient space X/V. This notation extends to conditioned and controlled invariant subspaces. Thus if $V \subset X$ is a controlled invariant subspace for the pair (A, B) and if F is a feedback such that $(A + BF)V \subset V$, then we use the notation $A+BF|_V$ and $A+BF|_{X/V}$ for the restricted and induced maps respectively.

Following Schumacher [1981], we say that a controlled invariant subspace V for the pair (A, B) is stabilizable, or **inner stabilizable** if there exists a feedback F such that V is $A+BF$-invariant and $A+BF|_V$ is stable. Analogously, we say that a controlled invariant subspace V is **outer stabilizable** if there exists a feedback F such that V is $A + BF$-invariant and $A + BF|_{X/V}$ is stable. Similarly we define **inner antistabilizable** subspaces.

There are natural dual concepts. Thus a conditioned invariant subspace V is **inner detectable** if there exists an injection map H such that V is $A + HC$-invariant and $A + HC|_V$ is stable. A subspace V is **outer detectable** if there exists an injection map H such that V is $A+HC$-invariant and $A + HC|_{X/V}$ is stable. Again, the concepts of inner and outer antidetectability are naturally defined.

A simple application of duality considerations in an inner product space X implies that $\mathcal{V} \subset \mathcal{X}$ is an outer detectable subspace with respect to the pair (C, A) if and only if $\mathcal{V}^\perp$ is inner stabilizable with respect to the pair (A^*, C^*). With the previous definitions, the characterizations given in Proposition 28 can be made more specific to take into account stabilizability and detectability. To this end we define a submodule $M \subset F^p[s]$ to be a stable submodule if $M = F^p[s]E$ and E is a stable polynomial matrix. Analogously we define antistable submodules. In a similar fashion we define stable, finite dimensional, submodules of $s^{-1}F^m[[s^{-1}]]$ if they are of the form $L = X_r^D$, with D a stable polynomial matrix. Antistable submodules require D to be antistable. Clearly stable submodules coincide with finite dimensional coinvariant subspaces of H_+^2.

The proof of the following proposition is based on results in Fuhrmann and Willems [1980] and Fuhrmann [1981].

Proposition 29. Let G be a $p \times m$ transfer function having the polynomial coprime factorizations

$$G = E_r D_r^{-1} = D_l^{-1} E_l.$$

Then

1. With respect to the corealization

$$\begin{cases} A = S_{D_r}^r \\ \xi B = \xi E_r \pi_{D_r}, \\ pC = (pD_r^{-1})_{-1} \\ D = G(\infty). \end{cases} \tag{27}$$

   in the state space $X_{D_r}^r$, a subspace $\mathcal{V}$ is outer antidetectable if and only if

$$\mathcal{V} = X_{D_r}^r \cap F^m[s]E_+$$

   for some antistable polynomial matrix E_+.
2. With respect to the corealization (27) in the state space $X_{D_r}^r$, a subspace $\mathcal{V}$ is outer detectable if and only if

$$\mathcal{V} = X_{D_r}^r \cap F^m[s]F_-$$

   for some stable polynomial matrix F_-.

3. With respect to the corealization

$$\begin{cases} A = S_r^{D_l} \\ \xi B = \xi D_l^{-1}\pi_-, \\ fC = (fE_l)_{-1} \\ D = G(\infty). \end{cases} \tag{28}$$

in the state space $X_r^{D_l}$, a subspace $\mathcal{V} \subset X_r^{D_l}$ is inner stabilizable if and only if there exists a stable polynomial matrix E_- such that

$$\mathcal{V} = X_r^{E_-}\pi_r^{D_l}.$$

4. With respect to the corealization (28) in the state space $X_r^{D_l}$, a subspace $\mathcal{V} \subset X_r^{D_l}$ is inner antistabilizable if and only if there exists an antistable polynomial matrix E_+ such that

$$\mathcal{V} = X_r^{E_+}\pi_r^{D_l}.$$

We are ready now to state the analytic counterpart to Proposition 29.

Proposition 30. Let K be an inner function.

1. With respect to the pair A, C, defined in the state space $H_r(K)$ by

$$\begin{cases} (Af)(s) &= sf(s) - \lim_{s\to\infty} sf(s) \\ fC &= \lim_{s\to\infty} sf(s) \end{cases} \tag{29}$$

the subspace $\mathcal{V} \subset H_r(K)$ is outer antidetectable if and only if

$$\mathcal{V} = H_r(K) \cap H_+^2 Q'$$

for some inner function Q'.

2. Let K, Q'' be skewprime inner functions and let $KQ'' = \overline{Q}''K_+$. With respect to the pair A, C, defined as in (29), the subspace $\mathcal{V} \subset H_r(K)Q''$ is outer detectable if and only if

$$\begin{aligned} \mathcal{V} &= H_r(K)Q'' \cap H_r(K_+) \\ &= H_r(K)Q'' \cap H_-^2 K_+. \end{aligned}$$

3. We have

$$\begin{aligned} &([H_r(K) \cap H_+^2 Q']Q'')^\perp \\ &\quad = [P_{H_r(K)} H_r(Q')]Q'' \end{aligned}$$

$$\begin{aligned} &(H_r(K)Q'' \cap H_r(K_+))^\perp \\ &\quad = P_{H_r(K)Q''}(H_r(\overline{Q}'')K_+) \end{aligned}$$

4. With the pair A, B, defined in the state space $H_r(K)$ by

$$\begin{cases} (Af)(s) &= sf(s) - \lim_{s\to\infty} sf(s) \\ \xi B &= \xi(K - I) \end{cases} \tag{30}$$

the subspace $\mathcal{V} \subset H_r(K)$ is inner antistabilizable if and only if there exists an inner function Q' such that

$$\mathcal{V} = P_{H_r(K)} H_r(Q')$$

5. With respect to the pair A, B, defined as in (30), the subspace $\mathcal{V} \subset H_r(K)Q''$ is inner stabilizable if and only if there exists an inner function Q' such that

$$\mathcal{V} = P_{H_r(K)Q''} H_r(\overline{Q}'')K_+.$$

7. Parametrization of the set of minimal spectral factors

Assume we are given a coercive, rational spectral density matrix, i.e. it is positive definite on the extended imaginary axis. Moreover, for simplicity, we assume it is normalized by $\Phi(\infty) = I$. Our aim is to parametrize the set of all minimal McMillan degree spectral factors, that is square rational matrix functions W satisfying $WW^* = \Phi$. This is best done by considering the four extremal spectral factors, $W_-, W_+, \overline{W}_-, \overline{W}_+$, the minimum phase stable, the maximum phase stable, the minimum phase antistable and the maximum phase antistable factors respectively. This implies the existence of four inner functions $Q_\pm, K_\pm$ for which $W_+ = W_-Q_+$, $\overline{W}_+ = \overline{W}_-Q_-$, $W_- = \overline{W}_-K_-$ and $W_+ = \overline{W}_+K_+$. Any minimal stable spectral factor is of the form $W = W_-Q'$, where Q' is a left factor of Q_+ and any minimal, minimum phase spectral factor W is of the form $W = \overline{W}_-K'$, where K' is a left factor of K_-. It can be shown that any minimal spectral factor W is uniquely determined by such factorizations of K_- and Q_+. This is given schematically by the following diagram.

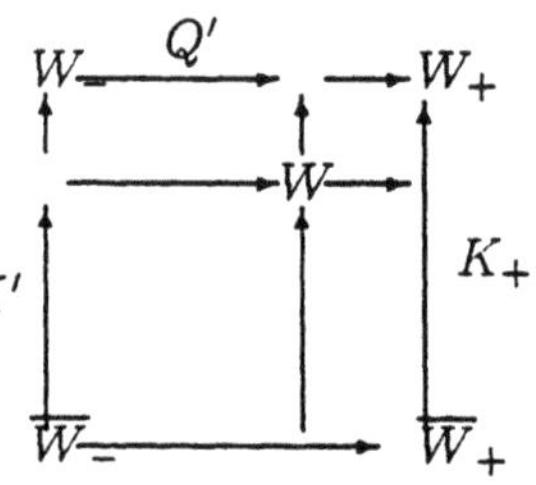

Now factorizations of inner functions are in a bijective correspondence with the set of all nonnegative solutions of a corresponding homogeneous Riccati equation. Leaving out all the details, which can be found in Fuhrmann [1995], this leads to the next theorem which describes the parametrization in state space terms.

Theorem 31. Let $W_- = \left(\begin{array}{c|c} A & B \\ \hline C & I \end{array}\right)$ be a minimal realization of the stable, minimum phase spectral factor of a normalized spectral function Φ. Let X be a nonnegative definite solution of the Riccati equation

$$X(A^* - C^*B^*) + (A - BC)X + XC^*CX = 0,$$

and let Z be a nonnegative definite solution of the Riccati equation

$$A^*Z + ZA + ZBB^*Z = 0.$$

Then each of the following realizations

$$W_1 = \left(\begin{array}{c|c} A + BB^*Z & B - (I + XZ)^{-1}X(C^* + ZB) \\ \hline C + B^*Z & I \end{array} \right)$$

and

$$W_1 = \left(\begin{array}{c|c} A + (B - XC^*)(B^* - CX)Z(I + XZ)^{-1} & B - XC^* \\ \hline C + (B^* - CX)Z(I + XZ)^{-1} & I \end{array} \right)$$

gives a complete parametrization of all minimal spectral factors of Φ.

The two realizations above are isomorphic and the intertwining isomorphism is given by the map $I + XZ$.

Acknowledgement

The author wishes to gratefully acknowledge the support of GIF under Grant No. I 184.

References

H. Bart, I. Gohberg, M.A. Kaashoek and P. Van Dooren [1979], "Factorizations of transfer functions", .

G. Basile and G. Marro [1969], "Controlled and conditioned invariant subspaces in linear system theory", J. Optim. Th. & Appl., 3, 306-315.

F.M. Callier [1985], "On polynomial matrix spectral factorization by symmetric extraction", *IEEE Trans. on Auto. Contr.*, 30, 453-464.

R.G. Douglas, H.S. Shapiro & A.L. Shields [1971], "Cyclic vectors and invariant subspaces for the backward shift", *Ann. Inst. Fourier, Grenoble* 20,1, 37-76.

L. Finesso and G. Picci [1982], "A characterization of minimal spectral factors", *IEEE Trans. Autom. Contr.* AC-27, 122-127.

P.A. Fuhrmann [1968], "On the corona problem and its application to spectral problems in Hilbert space", *Trans. Amer. Math. Soc.*, 132, 55-67.

P.A. Fuhrmann [1968], "A functional calculus in Hilbert space based on operator valued analytic functions", *Israel J. Math.*, 6, 267-278.

P.A. Fuhrmann [1976], "Algebraic system theory: An analyst's point of view", *J. Franklin Inst.*, 301, 521-540.

P.A. Fuhrmann [1975], "On Hankel operator ranges, meromorphic pseudo-continuation and factorization of operator valued analytic functions", *J. Lon. Math. Soc.* (2) 13, 323-327.

P.A. Fuhrmann [1981], "Duality in polynomial models with some applications to geometric control theory," *Trans. Aut. Control*, AC-26, 284-295.

P.A. Fuhrmann [1981], *Linear Operators and Systems in Hilbert Space*, McGraw-Hill.

P.A. Fuhrmann [1984], "The algebraic Riccati equation - a polynomial approach", *Systems and Control Letters*, 5, 369-376.

P.A. Fuhrmann [1989], "Elements of Factorization Theory", in H. Nijmeijer, J. M. S. Schumacher (eds.), *Three Decades of Mathematical System Theory. A Collection of Surveys at the Occasion of the Fiftieth Birthday of Jan C. Willems.* Lecture Notes in Control and Information Sciences Vol. 135. Springer Verlag, Berlin.

P.A. Fuhrmann [1994], "A duality theory for robust control and model reduction", *Lin. Alg. Appl.*, vols. 203-204, 471-578.

P.A. Fuhrmann [1995], "On the characterization and parametrization of minimal spectral factors", to appear, *Journal of Mathematical Systems, Estimation and Control.*

P.A. Fuhrmann and A. Gombani [1995], "A study of rectangular spectral factors", preprint.

P.A. Fuhrmann and R. Ober [1993], "On coprime factorizations", the T. Ando Anniversary Volume, in *Operator Theory: Advances and Applications*, vol. 62, 39-75. Birkhäuser Verlag.

P.A. Fuhrmann and J.C. Willems [1979], "Factorization indices at infinity for rational matrix functions", *Integral Equat. and Oper. Theory*, 2, 287-301.

P.A. Fuhrmann and J.C. Willems [1980], "A study of (A,B)-invariant subspaces via polynomial models", *Int. J. Contr.* 31, 467-494.

Y. Genin, P. Van Dooren, T. Kailath, J.M. Delosme and M. Morf [1983], "On Σ-lossless transfer functions and related questions, *Lin. Alg. Appl.*, 50, 251-275.

T.T. Georgiou and M.C. Smith [1990], "Optimal robustness in the gap metric", *IEEE Trans. on Auto. Contr.*, 35, 673-686.

I.C. Gohberg and I.A. Feldman [1971], *Convolution Equations and Projection Methods for their Solution*, Amer. Math. Soc. Translations of Mathematical Monographs, vol. 41.

M.L.J. Hautus and M. Heymann [1978], "Linear feedback-an algebraic approach", *SIAM J. Control* 16, 83-105.

U. Helmke and P.A. Fuhrmann [1989], "Bezoutians", *Lin. Alg. Appl.*.

P. Khargonekar and E. Sontag [1982], "On the relation between stable matrix factorizations and regulable realizations of linear systems over rings", *IEEE Trans. on Auto. Contr.*,

27, 627-638.

D. McFarlane and K. Glover [1989], "Robust controller design using normalized coprime factor plant descriptions", *Lecture Notes in Control and Information Sciences*, vol. 10, Springer Verlag.

D. Meyer and G. Franklin [1987], "A connection between normalized coprime factorizations and linear quadratic regulator theory", *IEEE Trans. on Auto. Contr.* 32, 227-228.

N.K. Nikolskii [1985], *Treatise on the Shift Operator*, Springer.

R.J. Ober and J.A. Sefton [1991],"Stability of control systems and graphs of linear systems", *Systems and Control Letters*, 17, 265-280.

G. Picci and S. Pinzoni [1994]. "Acausal models of stationary processes", *Lin. Alg. Appl*, 205-206, 997-1043.

L.A. Sakhnovich [1976], "On the factorization of an operator valued transfer function", *Soviet Math. Dokl.*, 17, 203-207.

D. Sarason [1967], "Generalized interpolation in H^∞", *Trans. Amer. Math. Soc.* 127, 179-203.

J.M. Schumacher [1981], "Dynamic feedback in finite- and infinite-dimensional linear systems", Ph.D. thesis, Free University of Amsterdam.

B. Sz.-Nagy and C. Foias [1970], *Harmonic analysis of Operators on Hilbert Space*, North Holland, Amsterdam.

M. Vidyasagar [1988],"Normalized coprime factorizations for non strictly proper systems", *Automatica*, 85-94.

M. Vidyasagar [1985], *Control System Synthesis*, MIT Press.

J.C. Willems [1971], "Least squares stationary optimal control and the algebraic Ricatti equation", *Trans. Automat. Contr.*, 16, 621-634.

FAILURE DETECTION BY MEANS OF SINGULAR SYSTEMS

Ph. POIGNET and M. GUGLIELMI

Laboratoire d'Automatique de Nantes, U.R.A. C.N.R.S. 823
Ecole Centrale de Nantes/Université de Nantes
1 rue de la Noë, 44072 NANTES CEDEX, FRANCE
Tel:(33) 40 37 16 00/Fax:(33) 40 37 25 22
E-mail poignet@lan.ec-nantes.fr
Members of G.d.R. 134 "Traitement du Signal et des Images"

Abstract : This paper deals with the abrupt changes detection in the linear dynamical SISO systems. Our method is based on the singular state modeling which takes into account the parameter jumps in the model. The estimation is then achieved by an extended Kalman filter specifically developed for this class of models. Finally the designed detector is applied to the engine rate change detection of a water-steam engine.

Résumé : Le travail présenté porte sur la détection de ruptures dans les systèmes dynamiques linéaires monovariables. La méthode proposée repose sur une modélisation d'état singulier qui prend en compte structurellement les sauts paramétriques. L'estimation de ces sauts est alors réalisée grâce à un filtre de Kalman étendu développé spécifiquement. Le détecteur obtenu est ensuite appliqué à la détection de changement de régime de fonctionnement d'un générateur de vapeur.

Key Words : modeling, singular systems, identification, failure detection.
Mots clés : modélisation, systèmes singuliers, identification, détection de défauts.

1 Introduction

A usual technical way of detecting failure in dynamic systems is based on modeling input/output relations which generate residuals very useful for the decision. This paper investigates the detection of abrupt changes in dynamic characteristics of linear systems by means of a different modeling of an ARMAX model and a new algorithm of identification. The new modeling, exposed in the first section, takes into account the jumps parameters and the linear model is naturally transformed into an implicit system. The problem is to identify the unknown parameters of the descriptor model composed of static and dynamic equations. The main difficulty concerning these static equations is that it is impossible to define a predicted state when applying the Kalman filter. The second section presents the identification algorithm for this class of model. It is then easy to develop a detection algorithm based on the identification result. Finally we exhibit simulations about real process issued from identification water-steam cycle, modelized by an ARMAX model.

2 Modeling

In this section, we present the modeling of linear systems with stiff variation parameters which leads to a singular state system (Poignet, 1995). First, let's consider the classical system described by the ARMAX model :

$$y_k + a_1 y_{k-1} + \dots + a_n y_{k-n} = b_1 u_{k-1} + \dots$$
$$\dots + b_n u_{k-n} + v_k + c_1 v_{k-1} + \dots + c_n v_{k-n} \quad (1)$$

where : $(a_1,...,a_n,b_1,...,b_n,c_1,...,c_n)$ are parameters and v_k is a white centered sequence of variance σ^2.
In case of abrupt changes of parameters a_i, the linear system is modelized with a very similar equation (1) by introducing s_i the jump on the ith parameter :

$$y_k + (a_1+s_1)y_{k-1} + ... + (a_n+s_n)y_{k-n} = b_1u_{k-1} + ... + b_nu_{k-n} + v_k + c_1v_{k-1} + ... + c_nv_{k-n} \quad (2)$$

where s_i, the jump on the ith parameter, is expressed as: $s_i = \alpha_i H_{r_i}(k)$. α_i is the jump level, r_i the changing time and $H_{r_i}(k)$ the Heavyside function defined by:

$$\begin{aligned} H_{r_i}(k) &= 1, \; if \; k \geq r_i \\ &= 0, \; if \; k < r_i. \end{aligned}$$

From equation (2), it is possible, as in the classical case, to get a state model. Actually, the system (2) is equivalent to:

$$\begin{aligned} y_k + a_1y_{k-1} + ... + a_ny_{k-n} + s_1y_{k-1} + ... \\ ... + s_ny_{k-n} = \\ b_1u_{k-1} + ... + b_nu_{k-n} + v_k + c_1v_{k-1} + ... \\ ... + c_nv_{k-n} \quad (3) \end{aligned}$$

Let's define X_k^r and X_k^s such as:

$$\begin{aligned} X_k^r &= (x_1^r = y_{k-n+1}, ..., x_n^r = y_k)_k^T \quad (4) \\ X_k^s &= (x_1^s = s_ny_{k-n+1}, ..., x_n^s = s_1y_k)_k^T \quad (5) \end{aligned}$$

The system (3) leads to the following model :

$$X_{k+1}^r = \begin{bmatrix} 0 & & & & -c_n \\ 1 & \ddots & & & \vdots \\ & \ddots & \ddots & & \vdots \\ & & \ddots & 0 & \vdots \\ & & & 1 & -c_1 \end{bmatrix} X_k^r - X_k^s + ... \begin{bmatrix} b_n \\ \vdots \\ b_1 \end{bmatrix} u_k + \begin{bmatrix} c_n - a_n \\ \vdots \\ c_1 - a_1 \end{bmatrix} y_k \quad (6)$$

$$0 = X_k^s + \begin{bmatrix} -s_n \\ \vdots \\ -s_1 \end{bmatrix} y_k \quad (7)$$

$$y_k = [0 \dots 0 \; 1] X_k^r + v_k \quad (8)$$

Noting $X_k = X_k^r \doteq X_k^s$, it follows :

$$\left\{ \begin{aligned} \begin{bmatrix} I_n & 0 \\ 0 & 0 \end{bmatrix} X_{k+1} &= \begin{bmatrix} A_c & -I_n \\ 0 & I_n \end{bmatrix} X_k + ... \\ & B u_k + \begin{bmatrix} M_{c,a} \\ S \end{bmatrix} y_k \\ y_k &= [H \; 0] X_k + v_k \end{aligned} \right. \quad (9)$$

with :

$$A_c = \begin{bmatrix} 0 & & & & -c_n \\ 1 & \ddots & & & \vdots \\ & \ddots & \ddots & & \vdots \\ & & \ddots & 0 & \vdots \\ & & & 1 & -c_1 \end{bmatrix}, \; S = \begin{bmatrix} -s_n \\ \vdots \\ -s_1 \end{bmatrix} \quad (10)$$

and :

$$\begin{aligned} B &= \begin{bmatrix} b_n \\ \vdots \\ b_1 \end{bmatrix}, \; M_{c,a} = \begin{bmatrix} c_n - a_n \\ \vdots \\ c_1 - a_1 \end{bmatrix} \text{ and} \\ H &= [0 \dots 0 \; 1] \end{aligned} \quad (11)$$

Due to the fact of abrupt changes, the state model (9) is a singular one. The structural analysis of this class of models is widely studied in Dai (1989). In our case, the useful important properties are summarized in Poignet (1995).

3 Identification algorithm

Here we focus on the unknown parameters $\{a_i, b_i, c_i, s_i\}$ identification of the singular model (9). This will be achieved with a new algorithm based on the Extended Kalman Filtering (EKF) and especially developed for the singular systems. As the classical method, we define an extended state $z_k^T = (X_k^r, \theta_k, X_k^s)$ where :

$$\theta_k = (c_n, ..., c_1, c_n - a_n, ..., c_1 - a_1, b_n, ..., b_1, s_n, ..., s_1)_k^T \quad (12)$$

The dynamic equation of z_k is :

$$\begin{bmatrix} I_n & 0 & 0 & 0 & 0 & 0 \\ 0 & I_n & 0 & 0 & 0 & 0 \\ 0 & 0 & I_n & 0 & 0 & 0 \\ 0 & 0 & 0 & I_n & 0 & 0 \\ 0 & 0 & 0 & 0 & I_n & 0 \\ 0 & 0 & 0 & 0 & 0 & 0 \end{bmatrix} z_{k+1} = ... \begin{bmatrix} A_c & 0 & y_kI_n & u_kI_n & 0 & -I_n \\ 0 & I_n & 0 & 0 & 0 & 0 \\ 0 & 0 & I_n & 0 & 0 & 0 \\ 0 & 0 & 0 & I_n & 0 & 0 \\ 0 & 0 & 0 & 0 & I_n & 0 \\ 0 & 0 & 0 & 0 & -y_kI_n & I_n \end{bmatrix} z_k \quad (13)$$

and :

$$y_k = [H \; 0 \; 0 \; 0 \; 0 \; 0] z_k + v_k \quad (14)$$

In relation to the classical estimation problem, there is no prediction equation for X_k^s. To solve the problem of the identification, it needs a prediction of the state X_k^s. Since it is impossible to get it from the state space equation (13), we choose, a priori, the prediction equal to the previous estimation of X_k^s. Then the prediction of the extended state z_k is :

$$z_{k+1/k} = \cdots$$

$$\begin{bmatrix} A_{ck/k} & 0 & y_k I_n & u_k I_n & 0 & -I_n \\ 0 & I_n & 0 & 0 & 0 & 0 \\ 0 & 0 & I_n & 0 & 0 & 0 \\ 0 & 0 & 0 & I_n & 0 & 0 \\ 0 & 0 & 0 & 0 & I_n & 0 \\ 0 & 0 & 0 & 0 & 0 & I_n \end{bmatrix} z_{k/k} \quad (15)$$

The prediction error covariance matrix related to the singular state X_k^s is choosen a priori equal to ∞ (i.e. no information). The prediction error covariance matrix is :

$$P_{k+1/k} = \begin{bmatrix} P_{k+1/k}^c & 0 \\ 0 & \infty \end{bmatrix} \quad (16)$$

where $P_{k+1/k}^c$ is the usual prediction error covariance matrix concerning the prediction errors about the regular states (X_k^r) and the parameters (θ_k) (Poignet, 1995).
Then the method consists in considering the static equations of the model, i.e. the last one of the dynamic model of z_k, as an output measurement :

$$0 = [0\ 0\ 0\ 0\ -y_k I_n\ I_n] z_k \quad (17)$$

This one is associated with the output equation y_k. The new output vector is :

$$yp_k = \begin{bmatrix} 0 & 0 & 0 & 0 & -y_k I_n & I_n \\ H & 0 & 0 & 0 & 0 & 0 \end{bmatrix} z_k + \begin{bmatrix} 0 \\ 1 \end{bmatrix} v_k \quad (18)$$

The estimation equations for this model from the *a priori* conditions (prediction $z_{k+1/k}$ and covariance matrix $P_{k+1/k}$) are :

$$\begin{aligned} z_{k+1/k+1} &= z_{k+1/k} + \cdots \\ & P_{k+1/k}\mathcal{H}^T(\mathcal{H}P_{k+1/k}\mathcal{H}^T + \mathcal{R})^{-1} * \cdots \\ & (yp_{k+1} - \mathcal{H}z_{k+1/k}) \quad (19) \\ P_{k+1/k+1} &= P_{k+1/k} - P_{k+1/k}\mathcal{H}^T * \\ & (\mathcal{H}P_{k+1/k}\mathcal{H}^T + \mathcal{R})^{-1}\mathcal{H}P_{k+1/k} \quad (20) \end{aligned}$$

with :

$$\begin{aligned} \mathcal{H} &= \begin{bmatrix} 0 & 0 & 0 & 0 & -y_k I_n & I_n \\ H & 0 & 0 & 0 & 0 & 0 \end{bmatrix} \text{ and} \\ \mathcal{R} &= \begin{bmatrix} 0 & 0 \\ 0 & \sigma^2 \end{bmatrix} \quad (21) \end{aligned}$$

This new algorithm of identification for singular systems based on the Extended Kalman Filter is able to identify all the parameters of these systems (Poignet *et al.*, 1993).

4 Detection

Generally speaking, the changes detection problem is composed of two parts : First we have to identify the jump level and secondly we want to know the changing time. The new modelization only consider explicitly the jump which will be computed by the identification algorithm developed in §2. With regard to the changing time, it can be obtained by tracking the jump during the recursive loop of the identification algorithm.
Formally, observing the sample $(y_0, ..., y_n)$ up to time n, we have to decide between the two hypotheses :

$$\begin{aligned} For\ (1 \le k \le l)\ H_0 &: \theta = \theta_0 = (s_i) = (0) \quad (22) \\ H_1 &: \exists\, r_i\ (1 \le r_i \le l)\ / \quad (23) \\ For\ 1 \le k \le r_i - 1 &: \theta = \theta_0 \quad (24) \\ For\ r_i \le k \le l &: \theta = \theta_1,\ i.e.\ \exists s_i \ne 0 \quad (25) \end{aligned}$$

In practical way, the method is the following one :
In the first part, we identify the a_i parameters, all the jumps s_i are equal to zeros, given that the initial conditions for s_i are zeros. The user's parameters of E.K.F. are choosen such as s_i keep the zero value during the first step. Once the system is well identified, we track, in the second part, the estimation of each s_i, $\hat{s_i} = E(s_i/measures)$, estimated with the same filter. If a change occurs in the spectral characteristics, one or more s_i become different of zero. The decision function g_k^s (quadratic norm weigthed by the covariance matrix) is designed :

$$g_k^s = \hat{\theta}^T \Sigma^{-1} \hat{\theta} \quad (26)$$

where $\Sigma = E[\hat{\theta}\hat{\theta}^T]$.
If g_k^s becomes greater than a threshold λ (*a priori* fixed by the user), a change is decided.

5 Simulations

The previous algorithm is then implemented on a real process. This process, a water-steam engine, could be modelized by several linear models whose the parameters are function of the engine rate. More particularly, the water-steam engine models identified in Guglielmi (1976) are second order ARMAX models where the input (u_k) is the fuel delivery and the output (y_k) is the temperature, the whole process depending on the steam delivery (Q_s). We view the general diagram of such a process on figure (1) :

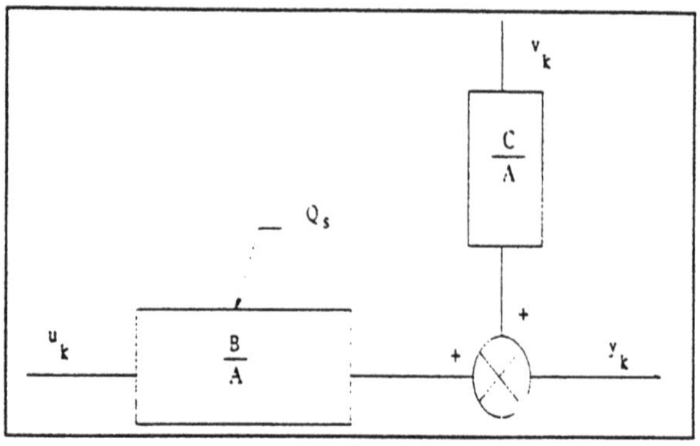

Figure 1 : ARMAX model

Issued from the models identified in Guglielmi (1976), we compute engine rate changes by modifying the parameters :

- before engine rate change (Q_s^1) :

$$A = 1 - 1.06z^{-1} + 0.17z^{-2} \quad (27)$$
$$B = 4.93z^{-1} - 4.74z^{-2} \quad (28)$$
$$C = 1 - 1.13z^{-1} + 0.48z^{-2} \quad (29)$$

- after engine rate change (Q_s^2) :

$$A = 1 + 0.173z^{-1} - 0.603z^{-2} \quad (30)$$
$$B = 4.93z^{-1} - 4.74z^{-2} \quad (31)$$
$$C = 1 - 1.13z^{-1} + 0.48z^{-2} \quad (32)$$

and v_k is a zero-mean white gaussian sequence. Our problem now is only to detect on-line water-steam engine model changes due to variation of Q_s that, we suppose, is unmeasured and not to know what happens (engine rate change or fault). This problem could be study further. From this set of parameters, two numerical studies were performed :

- The first one is computed with step input signal during the complete time of experiment. We achieve the identification and by the way expressed in the third section the detection.
- The second one considers only step input signal during the time required for the identification. When the process is correctly identified, we put the input equal to zero. Abrupt changes on parameters simulating a fault or engine rate change occur when the input is equal to zero. It is impossible to detect the changes. If the input signal is then non equal to zero for external reasons (engine rate change, process control,...) the jump parameters inform the user on the occurence of a change.

The following figures show inputs, outputs and parameters s_1 and s_2 which modelize the abrupt changes for the two cases.

First case :

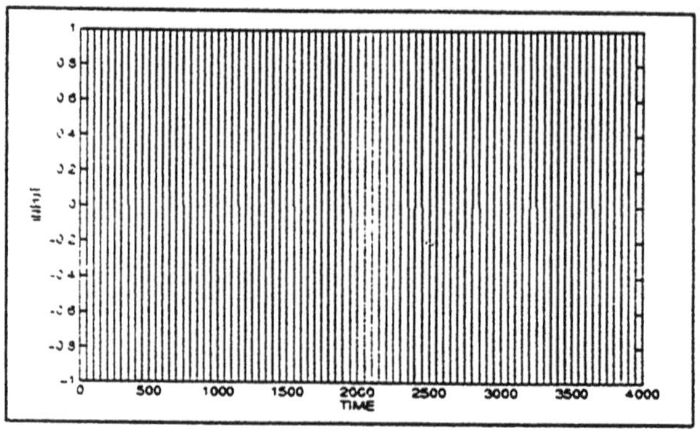

Figure 2 : Input

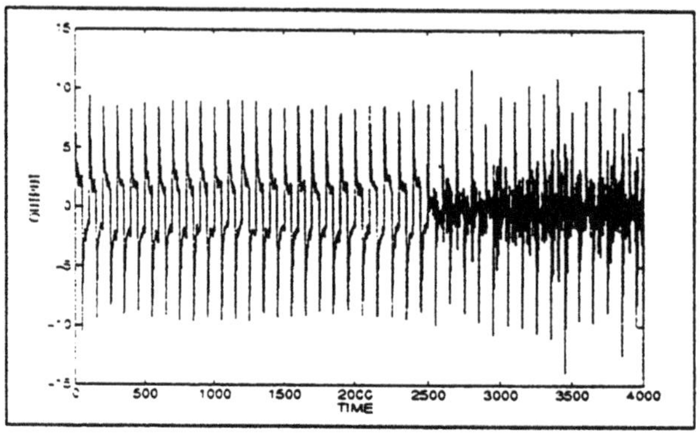

Figure 3 : Output

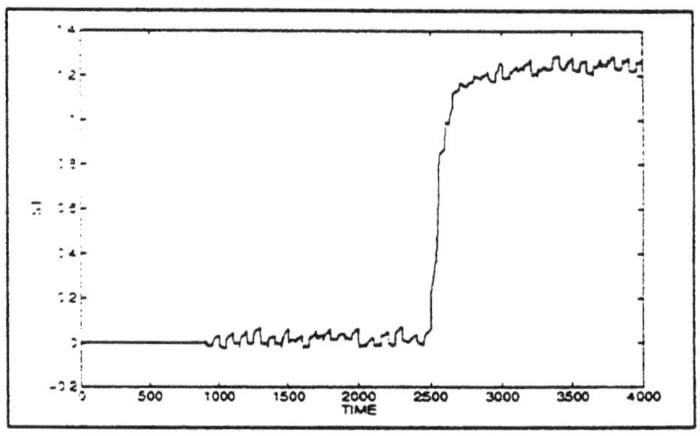

Figure 4 : Jump s_1

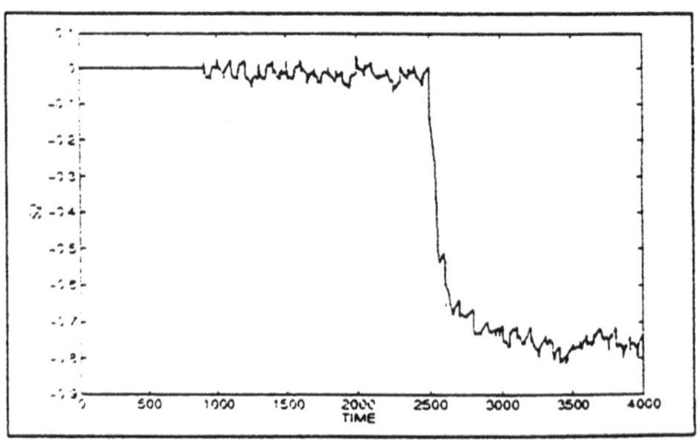

Figure 5 : Jump s_2

Second case :

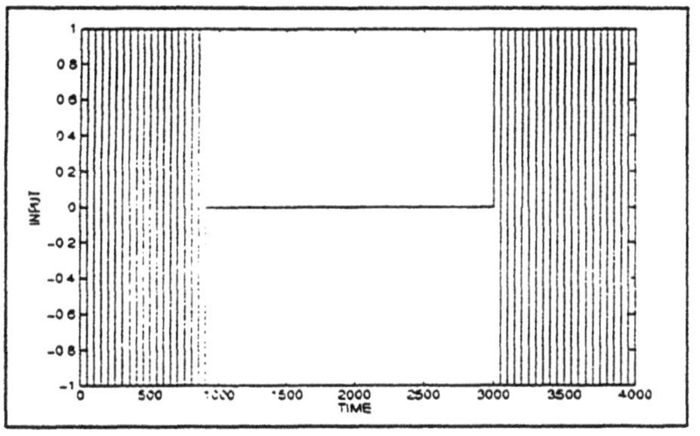

Figure 6 : Input

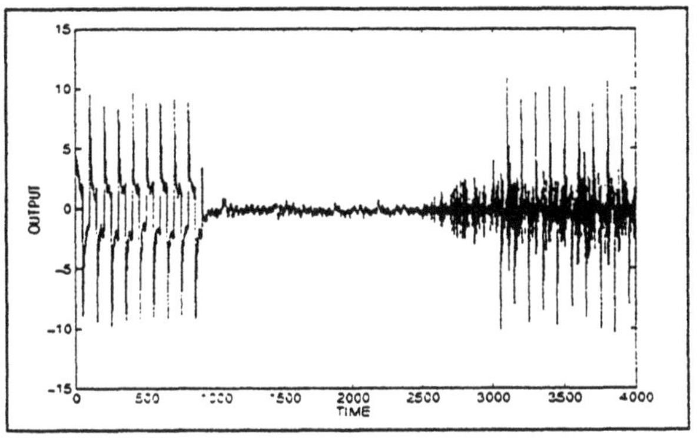

Figure 7 : Output

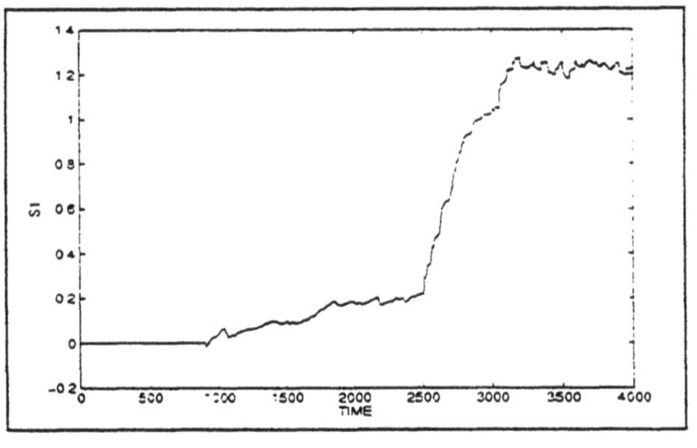

Figure 8 : Jump s_1

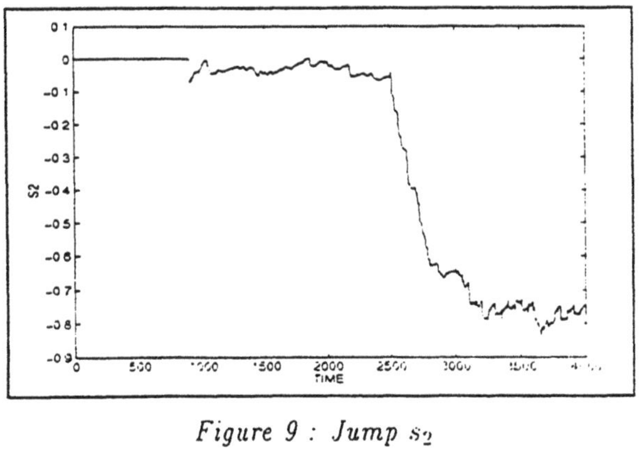

Figure 9 : Jump s_2

These results show the ability of the method to track abrupt changes.

6 Conclusions

This approach provides a new way to study failure detection. From the classical ARMAX model, we transform naturally the state system into a singular one. Applying specific algorithm for the two phases (identification and detection), we design a tool for detecting engine rate changes. Considering the system with an external point of view (i.e. only with the observation of the input and the output), the parameter change due to the steam delivery change could be the result of an engine rate change or process fault. The tracking of this model is not able to decide between the two hypotheses but is an element in the final diagnostic. The simulations show results with an easy task on a water-steam engine and allow us to hope further development of this method.

7 References

Dai, L. (1989). Singular control systems, *Lecture Notes in Control and Information Sciences*, Vol. 118, Springer Verlag.

Guglielmi, M. (1976). Contribution à la détermination et à la mise en oeuvre d'algorithmes de réalisation minimale : Application à la modélisation d'un générateur de vapeur d'un méthanier, Ph.D. Thesis, November.

Ljung, L. and T. Soderstrom (1983). Theory and pratice of recursive identification, MIT Press.

Poignet, Ph. and M. Guglielmi (1993). Identification of time-invariant descriptor systems by means of extended kalman filtering, I.E.E.E.-E.R.K., Portoroz.

Poignet, Ph. (1995). Détection de ruptures spectrales au moyen de l'analyse de systèmes singuliers, Ph.D. Thesis, January.

ON THE DEFINITION OF DECOUPLING FOR IMPLICIT AND GENERALIZED SYSTEMS†

E. DELALEAU* and P. S. PEREIRA da SILVA**

**Laboratoire des signaux et systèmes, CNRS - Supélec, Plateau de Moulon, F-91192 Gif-sur-Yvette, France. e-mail:* delaleau@lss.supelec.fr
***Dep. Eng. Eletrônica - Escola Politécnica - Universidade de São Paulo, Av. Prof. Luciano Gualberto, travessa 3, no. 158, Cidade Universitária, Sao Paulo - SP Cep 05508-900, Brazil. e-mail:* paulo@lac.poli.usp.br

Abstract. A definition of decoupling valid for implicit and generalized systems is presented and compared with many existing ones. It is shown that all these definitions are equivalent for *Kalman* dynamics, *i.e.*, dynamics with state equations not depending on the derivatives of the inputs.

Résumé. *Une définition de découplage applicable aux systèmes implicites ou généralisés est proposée et comparée la plupart des définitions existantes de découplage. Toutes ces définitons coïncident pour les dynamiques kalmaniennes, c'est-à-dire sans dérivées de l'entrée.*

Keywords. Disturbance decoupling, generalized systems, implicit systems, input-output decoupling, time-varying linear systems.

1 INTRODUCTION

The problems of input-output decoupling and disturbance decoupling have been studied in great detail in linear and nonlinear system theory. See Wonham (1985), Isidori (1989) or Nijmeijer and van der Schaft (1990) and the references therein for a survey.

This work deals with the definition of decoupling for implicit and generalized systems using the differential algebraic approach of Fliess (1989). Some antinomies with the standard notion of decoupling are exhibited via a simple example.

Consider a control system given by

$$\begin{aligned} \dot{x} &= f(x, v, \varpi), \\ y &= h(x), \end{aligned} \qquad (1)$$

where $x = (x_1, \ldots, x_n)$ is the state, $v = (v_1, \ldots, v_m)$ the control input, $\varpi = (\varpi_1, \ldots, \varpi_q)$ the disturbance input, and $y = (y_1, \ldots, y_p)$ the output. Essentially, the definition of a disturbance decoupled system (Wonham 1985, Isidori 1989, Nijmeijer and van der Schaft 1990) can be summarized as follows:

> *The output y is decoupled from the disturbance ϖ (or y is not affected by ϖ) if for each fixed initial condition $x(t_0) = x_0$, and each fixed control function $v(\cdot)$, the output function $y(\cdot)$ is the same for every disturbance function $\varpi(\cdot)$.*

The *disturbance decoupling problem* consists in finding a control law, *e.g.*, a dynamic state feedback in such a way that the output of the closed loop system is decoupled from the disturbance (see Respondek (1991), Cao and Zheng (1992), Huijberts *et al.* (1992), Delaleau and Fliess (1994), Perdon *et al.* (1993), Pereira da Silva (1993)).

This notion is perhaps too strong or not adapted for an implicit system or a system whose state variable representation depends on input derivatives (generalized systems) as can be seen on the following example.

†The work of P. S. Pereira da Silva has been financed by FAPESP, grants 92/4826-2 and 91/0508-3.

Example.[1] Consider the linear dynamics

$$\dot{x} = v - \dot{\varpi} \quad (2a)$$
$$y = x + \varpi \quad (2b)$$

where v is the control input, ϖ is the disturbance input and x is the state. Let $x_0 = 0, v(t) \equiv 0, \varpi_1(t) \equiv 0$ and $\varpi_2(t) \equiv 1$. The corresponding output functions will be $y_1(t) \equiv 0$ and $y_2(t) \equiv 1$. Hence, the output is not decoupled from the disturbance.

Computing the derivative of the output one gets $\dot{y} = v$. It is clear that, apart from an influence in the initial condition $y(t_0)$, the disturbance ϖ cannot affect the evolution of y. Taking the Laplace transform of $\dot{y} = v$ leads to $Y(s) = \frac{1}{s}V(s) + 0\,\Pi(s)$, where $Y(s), V(s)$ and $\Pi(s)$ denotes the Laplace transforms of y, v and ϖ. So, from an input-output point of view, the output of this dynamics is decoupled from the disturbance ϖ.

Notice that system (2a–2b) may appear in the context of implicit systems: Setting $\xi_1 = x$ and $\xi_2 = x - \varpi$ yields the implicit linear system $E\dot{\xi} = A\xi + Bv + H\varpi$, $y = C\xi + J\varpi$ where $\xi = (\xi_1, \xi_2)$, $E = \begin{pmatrix} 0 & 0 \\ 0 & 1 \end{pmatrix}$, $A = \begin{pmatrix} -1 & 1 \\ 0 & 0 \end{pmatrix}$, $B = \begin{pmatrix} 0 \\ 1 \end{pmatrix}$, $H = \begin{pmatrix} -1 \\ 0 \end{pmatrix}$, $C = \begin{pmatrix} 0 & 1 \end{pmatrix}$, and $J = \begin{pmatrix} 1 \end{pmatrix}$.

Generally speaking, a system (linear or not) can be described by a set of differential equations of the form

$$F_i(y, \ldots, y^{(\alpha_i)}, v, \ldots, v^{(\beta_i)}, \varpi, \ldots, \varpi^{(\gamma_i)}, z, \ldots, z^{(\delta_i)}) = 0$$

$i = 1, \ldots, s$, where y, v, and ϖ are the external variables, namely the output, the control input, and the disturbance input, where $z = (z_1, \ldots, z_\nu)$ are the latent variables (state variables, internal variables, ...), and where $\alpha_i, \beta_i, \gamma_i, \delta_i \in I\!N$, $i = 1, \ldots, s$.

From an input-output point of view, one can give the following definition[2], already considered in (Delaleau 1993, Delaleau and Fliess 1994, Delaleau and Pereira da Silva 1994):

> *The output y is decoupled of the disturbance ϖ if the system variables do not satisfy differential equations of the form*
>
> $$P(y, \ldots, y^{(\kappa)}, v, \ldots, v^{(\lambda)}, \varpi, \ldots, \varpi^{(\mu)}) = 0 \quad (3)$$
>
> *which depend explicitly on y and ϖ or their derivatives.*

[1]Delaleau (1993)

[2]This definition will be stated later in precise mathematical terms.

Based on the last statement, the output of system (2a–2b) is declared to be decoupled from the disturbance.

The paper establishes that the two notions of decoupling coincide for explicit systems without derivatives of the input like (1). However, the second notion is in general weaker than the first one, and is also valid for systems described by (algebro-)differential equations and depends neither on initial conditions nor on a particular state variable representation.

The paper is organized as follows: Section 2 gives a brief summary of the differential algebraic approach to linear (Fliess 1990) and nonlinear (Fliess 1989) systems. The notion of decoupling is presented in the Section 3 and comparisons with most of existing definitions are presented in the Section 4.

2 PRELIMINARIES

A. Linear systems (see Fliess (1990) for details)

2.1 A *linear system* is a finitely gererated $\mathbf{k}[\frac{d}{dt}]$-module, where $\mathbf{k}$ is the field of coefficients. An *output* of a linear system Λ is a finite subset $y = (y_1, \ldots, y_p)$ of Λ. An *input* of a linear system Λ is a set $e = (e_1, \ldots, e_s)$ of Λ such that the quotient $\mathbf{k}$-module $\Lambda/[e]$ is torsion. A *linear dynamics* is a linear system Λ where an input e is distinguished; it is denoted as $\Lambda/[e]$.

2.2 Any input can be decomposed into two parts: the *control input* (or *control* for short) $v = (v_1, \ldots, v_m)$ and the *disturbance input* (or *disturbance* for short) $\varpi = (\varpi_1, \ldots, \varpi_q)$. The disturbance represents the effects, generally undesirable, that influences the system. The control u represents the input variables that can be assigned freely and hence are supposed to be differentially independent. The control and disturbance inputs are supposed not to interact, *i.e.*, $[v] \cap [\varpi] = \{0\}$. For the sake of simplicity the disturbance is considered to be differentially independent[3].

2.3 A *state* of a linear dynamics $\Lambda/[v, \varpi]$ is a set $x = (x_1, \ldots, x_n)$ of Λ such that its canonical image in $\Lambda/[v, \varpi]$ is a basis of $\Lambda/[v, \varpi]$ considered as a $\mathbf{k}$-vector space. A state variables representation of the dynamics $\Lambda/[v, \varpi]$ reads

$$\dot{x} = Ax + \sum_{\text{finite}} B_i v^{(i)} + \sum_{\text{finite}} H_i \varpi^{(i)}$$

The state is said to be *Kalman* if $B_i = 0$ and $H_i = 0$ for $i > 0$ and *generalized* otherwise. For

[3]This hypothesis is implicit in most of the works.

an output y of Λ the output equation reads

$$y = C\,x + \sum_{\text{finite}} D_i\, v^{(i)} + \sum_{\text{finite}} J_i\, \varpi^{(i)}$$

B. Nonlinear systems (see Fliess (1989) for details)

2.4 A *nonlinear system* is a finitely generated differential field extension $\mathcal{K}/\mathbf{k}$, where $\mathbf{k}$ is the field of coefficients. An *output* of a linear system $\mathcal{K}$ is a finite set $y = (y_1, \ldots, y_p)$ of $\mathcal{K}$. An *input* e of a nonlinear system $\mathcal{K}$ is a set $e = (e_1, \ldots, e_s)$ such that the extension $\mathcal{K}/\mathbf{k}\langle e\rangle$ is differentially algebraic. A *nonlinear dynamics* is a nonlinear system $\mathcal{K}$ where an input e is distinguished. A nonlinear dynamics with input e is denoted by $\mathcal{K}/\mathbf{k}\langle e\rangle$.

2.5 As in the linear case any input can be decomposed into two parts: the *control input* $v = (v_1, \ldots, v_m)$ and the *disturbance input* $\varpi = (\varpi_1, \ldots, \varpi_q)$ and the same hypothesis are considered (see Paragraph 2.2).

2.6 A *state* of a nonlinear system $\mathcal{K}/\mathbf{k}\langle v, \varpi\rangle$ is a (non differential) transcendence basis $x = (x_1, \ldots, x_n)$ of the extension $\mathcal{K}/\mathbf{k}\langle v, \varpi\rangle$. A state variables represention of the dynamics $\mathcal{K}/\mathbf{k}\langle v, \varpi\rangle$ reads

$$F_i(\dot{x}_i, x, v, \ldots, v^{(\alpha_i)}, \varpi, \ldots, \varpi^{(\beta_i)}) = 0 \qquad (4)$$
$$i = 1, \ldots, n$$

where F_i, $i = 1, \ldots, n$ are polynomials over $\mathbf{k}$. The state x is said to be *Kalman* if $\alpha_i = \beta_i = 0$, $i = 1, \ldots, n$, and *generalized* otherwise. For an output y of $\mathcal{K}$ the outputs equations read

$$H_j(y_j, x, v, \ldots, v^{(\gamma_i)}, \varpi, \ldots, \varpi^{(\delta_i)}) = 0$$
$$j = 1, \ldots, p$$

3 DIST. DECOUPLING

This Section gives a precise mathematical statement to what is an output decoupled from a disturbance in the linear and nonlinear cases. Consequences and equivalent characterizations of these definitions are developped.

A. Linear systems — Recall the definition already considered in Delaleau and Fliess (1994).

Definition 3.1 *Let $\Lambda/[v, \varpi]$ be a linear dynamics with control v and disturbance ϖ. The output y of Λ is* decoupled from the disturbance *ϖ if $[y, v, \varpi] = [y, v] \oplus [\varpi]$*

3.2 This is an input-output definition completely independent of the representation of the system and possibly from the choice of the state and other latent variables. It is also valid for implicit or/and time-varying linear systems.

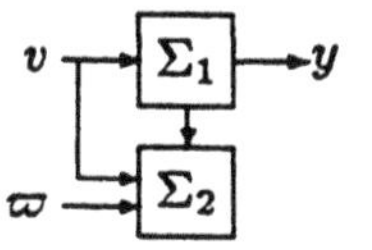

Fig. 1. Canonical form for decoupled system.

3.3 When the Condition of Definition 3.1 is satisfied, it is possible to exhibit a "canonical" state variables representation showing a decomposition for which the output is clearly seen to be decoupled from the disturbance. As $\dim \Lambda/[v, \varpi] = \dim \Lambda/[y, v, \varpi] + \dim\, [y, v, \varpi]/[v, \varpi]$, and because $[y, v, \varpi]/[v, \varpi] \cong [y, v]/[v]$, it is always possible to construct a state $z = (\bar{z}, \tilde{z})$ of $\Lambda/[v, \varpi]$ such that $\bar{z} = (\bar{z}_1, \ldots, \bar{z}_{n_1})$ is a state of $[y, v]/[v]$, and $\tilde{z} = (\tilde{z}_1, \ldots, \tilde{z}_{n_2})$ is a state of $\Lambda/[y, v, \varpi]$, with $n_1 + n_2 = \dim \Lambda/[v, \varpi]$. And in that case a state variables representation reads

$$\dot{\bar{z}} = A_1\,\bar{z} + \sum_{\text{finite}} B_{1l}\, u^{(l)} \qquad (5a)$$

$$\dot{\tilde{z}} = A_{21}\,\bar{z} + A_{22}\,\tilde{z} + \sum_{\text{finite}} B_{2l}\, u^{(l)} + \sum_{\text{finite}} H_l\, \varpi^{(l)} \qquad (5b)$$

$$y = C\,\bar{z} + \sum_{\text{finite}} D_l\, u^{(l)} \qquad (5c)$$

hence Λ is decomposed into two subsystems (see Figure 1), $\Sigma_1 = [y, v, \bar{z}]$, which is observable (Fliess 1990) with output y decoupled from ϖ, and $\Sigma_2 = [v, \varpi, \tilde{z}]$, which in influenced by the disturbance and is unobservable because $\tilde{z}_j \notin [y, v, \varpi]$, $j = 1, \ldots, n_2$ (Fliess 1990).

B. Nonlinear systems — The nonlinear counterpart of Definition 3.1 is

Definition 3.4 *Let $\mathcal{K}/\mathbf{k}\langle v, \varpi\rangle$ be a nonlinear dynamics with control v and disturbance ϖ. The output y of $\mathcal{K}/\mathbf{k}$ is* decoupled from the disturbance *ϖ if the fields $\mathbf{k}\langle\varpi\rangle$ and $\mathbf{k}\langle y, v\rangle$ are $\mathbf{k}$-algebraically disjoint.*

3.5 As in the linear case it is an *input-output* concept, completely independent from the choice of the state and other latent variables. It is equivalent to say that there is no polynomial differential equation of the form (3), with coefficients over $\mathbf{k}$, relating the output, the control, and the disturbance components.

3.6 For a dynamics $\mathcal{K}/\mathbf{k}\langle v, \varpi\rangle$ with output y, decoupled from the disturbance, the system $\mathcal{K}/\mathbf{k}$ can be decomposed into two subsystems (see Figure 1): the first one is decoupled from the disturbance and its output is y; the second one is influenced by the disturbance and is unobservable. Indeed, because $\mathbf{k}\langle\varpi\rangle$ and $\mathbf{k}\langle y, v\rangle$ are $\mathbf{k}$-algebraically disjoints and because tr.d° $\mathcal{K}/\mathbf{k}\langle v, \varpi\rangle = n$ is finite, tr.d° $\mathbf{k}\langle y, v\rangle/\mathbf{k}\langle v\rangle =$ tr.d° $\mathbf{k}\langle y, v, \varpi\rangle/\mathbf{k}\langle v, \varpi\rangle = n_1$ is finite and $n_1 \leq n$. Consequently it is always possible to construct a state $z = (\bar{z}, \tilde{z})$ of $\mathcal{K}/\mathbf{k}\langle v, \varpi\rangle$ where $\bar{z} = (\bar{z}_1, \ldots, \bar{z}_{n_1})$ is a transcendence basis of the extension $\mathbf{k}\langle y, v\rangle/\mathbf{k}\langle v\rangle$ and $\tilde{z} = (\tilde{z}_1, \ldots, \tilde{z}_{n_2})$ is a transcendence basis of the extension $\mathcal{K}/\mathbf{k}\langle y, v, \varpi\rangle$, with $n_1 + n_2 = n$. Notice that the components of $\tilde{z}$ are transcendent over $\mathbf{k}\langle y, v, \varpi\rangle$ and hence unobservable (Diop and Fliess 1991). With that choice, a state variables representation $\mathcal{K}/\mathbf{k}\langle v, \varpi\rangle$ with output y always reads

$$F_{1i}(\dot{\bar{z}}_i, \bar{z}, v, \ldots, v^{(\alpha)}) = 0 \qquad (6a)$$
$$i = 1, \ldots, n_1$$
$$F_{2j}(\dot{\tilde{z}}_i, \bar{z}, \tilde{z}, v, \ldots, v^{(\beta)}, \varpi, \ldots, \varpi^{(\gamma)}) = 0 \qquad (6b)$$
$$j = 1, \ldots, n_2$$
$$H_k(y_k, \bar{z}, v, \ldots, v^{(\beta)}) = 0 \qquad (6c)$$
$$k = 1, \ldots, p$$

where F_{1i}, F_{2j}, and H_k are polynomials over $\mathbf{k}$. Observe that it clear on those equations that y is decoupled from ϖ.

C. Comments and other characterization — This part applies for both linear and nonlinear systems, but for the sake of brevity it is written in the nonlinear language only.

3.7 In the decomposition of figure 1 the decoupling of the output from the disturbance appears undoubtly. If a system is already given under a state variables model it is would be necessary to perform generalized transformations, involving input (both control and disturbance input) and its derivatives, to obtain the state $z = (\bar{z}, \tilde{z})$ from the original state x. That is the source of the antinomy presented in the Introduction. For instance, in Example 1, $\bar{z} = x + \varpi$ and $\tilde{z} = \emptyset$, and the decomposition reads $\dot{\bar{z}} = v$, $y = \bar{z}$.

3.8 For dynamics given under state variable models, the dependence of the output on the disturbance can be checked on the state variables equations. Two cases arise whether the state is classical or generalized with respect to the disturbance input ($\beta_i = 0$ or $\beta_i \neq 0$ in (4)).

Proposition 3.9 *Let $\mathcal{K}/\mathbf{k}\langle v, \varpi\rangle$ be a nonlinear dynamics with state x such that the components of $\dot{x}$ are algebraic over $\mathbf{k}\langle v\rangle(x, \varpi)$. The output y is decoupled from the disturbance, according to Definition 3.4, if and only if $\mathbf{k}\langle y, v\rangle(x)$ is algebraic over $\mathbf{k}\langle v\rangle(x)$.*

Proof. Using Kähler differentials the statement "$\mathbf{k}\langle y, v\rangle$ and $\mathbf{k}\langle\varpi\rangle$ are $\mathbf{k}$-algebraically disjoint" is easily seen to be equivalent to $[dy, dv] \cap [d\varpi] = 0$ (see Johnson (1969)). In the same manner, the statement "$\mathbf{k}\langle y, v\rangle(x)$ is algebraic over $\mathbf{k}\langle v\rangle(x)$" is equivalent to $[dy] \subset \text{span}_{\mathcal{K}}\{dx, [dv]\}$.
Assume first that $[dy] \subset \text{span}_{\mathcal{K}}\{dx, [dv]\}$. As x is a state of the dynamics $\mathcal{K}/\mathbf{k}\langle v, \varpi\rangle$ and because v and ϖ are differentially independent, it follows easily that $[dy, dv] \cap [d\varpi] = 0$.
Assume now that $[dy, dv] \cap [d\varpi] = 0$. This implies that the $\mathcal{K}[\frac{d}{dt}]$-modules $[dy, dv, d\varpi]/[dv, d\varpi]$ and $[dy, dv]/[dv]$ are isomorphic and hence that $[dy, dv]/[dv]$ is torsion. Consider that $[dy]$ is not contained in $\text{span}_{\mathcal{K}}\{dx, [v]\}$. In particular there exists some $\eta \in [dy]$ of the form $\eta = \sum_{i=1}^{n} a_i dx_i + \sum_{i=1}^{m}\sum_{j=0}^{\alpha} b_{ij} dv_i^{(j)} + \sum_{i=1}^{q}\sum_{j=0}^{\beta} c_{ij} d\varpi_i^{(j)}$ where $a_i, b_{ij}, c_{ij} \in \mathcal{K}$ and $(c_{1\beta}, \ldots, c_{q\beta}) \neq 0$. Differentiating η and using $\text{span}_{\mathcal{K}}\{d\dot{x}\} \subset \text{span}_{\mathcal{K}}\{dx, [dv], d\varpi\}$, it is easy to show that any linear combination of the form $\sum_{i=0}^{\gamma} e_i \eta^{(i)}$, where $e_i \in \mathcal{K}$, will contain a nonzero term (of maximal order on ϖ) given by $\sum_{i=0}^{q} e_\gamma c_{i\beta} \varpi_i^{(\gamma+\beta)}$. So the canonical projection of η on $[dy, dv]/[dv]$ cannot be a torsion element and there is a contradiction. ∎

Proposition 3.10 *Let $\mathcal{K}/\mathbf{k}\langle v, \varpi\rangle$ be a nonlinear dynamics with state x such that the components of $\dot{x}$ are algebraic over $\mathbf{k}\langle v\rangle(x, \varpi, \ldots, \varpi^{(\alpha)})$. The output y is decoupled from the disturbance, according to Definition 3.4, if and only if $\mathbf{k}\langle y, v\rangle(x, \varpi, \ldots, \varpi^{(\alpha-1)})$ is algebraic over $\mathbf{k}\langle v\rangle(x, \varpi, \ldots, \varpi^{(\alpha-1)})$.*

Proof. It is quite analogous to the proof of the Proposition 3.9 and it is left to the reader. ∎

3.11 This last Proposition allows to explain the phenomenon of dependence on the initial conditions of the output components on the disturbance. Definition 3.4 implies that $\mathbf{k}\langle y, v\rangle/\mathbf{k}\langle v\rangle$ is differentially algebraic and hence each output component satisfies a differential equation of the form

$$Q_k(y_k, \ldots, y_k^{(\kappa_k)}, v, \ldots, v^{(\lambda_k)}) = 0 \qquad (7)$$

where Q_k, $k = 1, \ldots, p$, are polynomials over $\mathbf{k}$. This ensures that the evolution of the output is really decoupled from the disturbance, at least for nonsingular trajectories for which the implicit relation (7) can be made explicit with respect to $y_k^{(\kappa_k)}$.

However, if the system is given under the form of a generalized state variables representation, according to Proposition 3.10 the output components also satify

$$R_k(y_k^{(l)}, x, \varpi, \dots, \varpi^{(\alpha-1)}, v, \dots, v^{(\lambda_{kl})}) = 0 \quad (8)$$

where R_k, $k = 1, \dots, p$, are polynomials over $\mathbf{k}$. Setting $t = 0$ in (8) shows the dependence on the initial conditions of the output and its derivatives on a finite number (given by $\alpha - 1$) of derivatives of the components of the disturbance and on the state and control input and its derivatives. That arises only on state variable representation and that is the reason for which a state dependent definition cannot work for generalized of implicit systems.

4 COMPARISON

A. Linear case — In the linear case there exist two definitions of decoupling, one given in the transfer matrix approach, and the other on state equations.

Consider a linear time-invariant system with transfer equation

$$Y(s) = T_1 U(s) + T_2 \Pi(s) \quad (9)$$

where $Y(s)$, $U(s)$, and $\Pi(s)$, denote respectively the Laplace transforms of the output, the control input and the disturbance input.

Definition 4.1 [4] *The output is decoupled from the disturbance if* $T_2 \equiv 0$ *in* (9).

Consider a linear time-invariant system with state variable equations

$$\begin{aligned} \dot{x} &= Ax + Bv + H\varpi \\ y &= Cx \end{aligned} \quad (10)$$

Definition 4.2 [5] *The output of* (10) *is decoupled from the disturbance if*

$$C \int_{t_0}^{t} e^{[(t-\tau)A]} H \varpi(\tau) d\tau = 0 \quad (11)$$

Proposition 4.3 *The output of a linear system is decoupled from the disturbance according to Definition 4.1 if and only if it is decoupled according to Definition 3.1. The output of a proper linear time-invariant system is decoupled from the disturbance according to Definition 4.2 if and only if this output is decoupled from the disturbance according to Definition 3.1.*

Proof. As the disturbance input is differentially indepent, $[\varpi]$ is torsion-free and hence $[y, v, \varpi] = [y, v] \oplus [\varpi]$ is equivalent to $[\hat{y}, \hat{v}, \hat{\varpi}] = [\hat{y}, \hat{v}] \oplus [\hat{\varpi}]$, where $\hat{}: \Lambda \longrightarrow \hat{\Lambda}$ denotes the formal Laplace transform (see Fliess (1994)). As $[y, v]/[v]$ is torsion $\hat{y} = T\hat{u}$ and therefore Definition 3.1 is equivalent[6] to 4.1.
The second part of the Proposition follows from the fact that for proper time-invariant linear system described by (10), the matrix T_2 in (9) is the Laplace transform of $C \int_{t_0}^{t} e^{[(t-\tau)A]} H \varpi(\tau) d\tau$ in (11). ∎

B. Nonlinear case — In the nonlinear case there exists only state dependent definitions

Definition 4.4 [7] *The output y is decoupled from the disturbance ϖ if for every initial state x_0, for every control function $v(\cdot)$ and for all $t \leq 0$*

$$y(t, x_0, v, \varpi_1) = y(t, x_0, v, \varpi_2)$$

for every pair of disturbance functions $\varpi_1(\cdot), \varpi_2(\cdot)$.

In (Perdon *et al.* 1993, Pereira da Silva and Pinto Leite 1993), the notion of decoupling corresponds to the fact that the output y and its derivatives of any order can be determined as functions of the state and the control input v and its derivatives. This definition may be stated in the differential algebraic language as follows:

Definition 4.5 *Let $\mathcal{K}/\mathbf{k}\langle v, \varpi \rangle$ be a nonlinear dynamics with state x and output y that is classic. Then the output y is decoupled from the disturbance ϖ if $[dy] \subset \mathrm{span}_{\mathcal{K}} \{dx, [v]\}$.*

Proposition 4.6 *If the output y of a nonlinear system is decoupled from the disturbance ϖ according to Definition 4.4 then y is decoupled from ϖ according to Definition 3.4. For a system given under a Kalman form* (1)*, if y is decoupled from ϖ according to Definition 3.4 then y is decoupled ϖ according to Definition 4.4.*

[4] Bhattacharyya (1974)
[5] Wonham (1985)

[6] This proof show that Definition 4.1, originaly stated for time-invariant proper linear system extends to time-varying and/or non proper linear systems.
[7] Isidori (1989), Nijmeijer and van der Schaft (1990)

Proof. *(Sketch)* Assume that y is not decoupled from ϖ according to Definition 3.4. If x is a state of the dynamics $\mathcal{K}/\mathbf{k}\langle v, \varpi\rangle$, according to Propositions 3.9 and 3.10 the output components depend effectively on ϖ and hence y is not decoupled according to Definition 4.4

Consider a system, given under Kalman form (1). Assume now that y is decoupled from ϖ according Definition 3.4. Because of Proposition 3.9 a similar reasoning than the one of paragraph 3.11 can be used to write a differential equation of the form (7). Since $y_i^{(j)} = h_{ji}(x, v, \dot{v}, \ldots, v^{(j-1)})$, all the initial conditions of (7) do not depend on the disturbance and so y is decoupled from ϖ according to Definition 4.4. ∎

Proposition 4.7 *The notions of decoupling from the disturbance of Definitions 4.5 and 3.4 are equivalent for systems given under Kalman form* (1).

Proof. Apply Kälher differentials to the Condition of Proposition 3.9. ∎

5 CONCLUSION

All the ideas developed here can be adapted to deal with the input-output decoupling problem.

The definition of decoupling can be restated in the differential geometric approach of infinite jets and prolongations of Fliess *et al.* (1993), Fliess *et al.* (1994)). This allows to consider systems which are not only described by polynomial equations.

REFERENCES

Bhattacharyya, S. P. (1974). 'Disturbance rejection in linear systems'. *Internat. J. Control* 5, 633–637.

Cao, L. and Y.-F. Zheng (1992). 'Disturbance decoupling via dynamic feedback'. *Internat. J. Systems Sci.* **23**, 683–694.

Delaleau, E. (1993). *Sur les dérivées de l'entrée en représentation et commande des systèmes non linéaires.* Thèse de doctorat, Université Paris-XI. Orsay.

Delaleau, E. and M. Fliess (1994). Nonlinear disturbance rejection by quasi-static state feedback. In U. Helmke, R. Mennicken and J. Saurer (Eds.). 'Systems and Networks: Mathematical Theory and Applications (MTNS'93, Volume II, Invited and Contributed Papers)'. Akademie Verlag. Berlin. pp. 109–112.

Delaleau, E. and P. S. Pereira da Silva (1994). 'Conditions de rang pour le rejet dynamique de perturbations'. *C. R. Acad. Sci. Paris Sér. I Math.* **319**, 1121–1126.

Diop, S. and M. Fliess (1991). On nonlinear observability. In C. Commault, D. Normand-Cyrot, J.-M. Dion, L. Dugard, M. Fliess, A. Titli, G. Cohen, A. Benveniste and I. D. Landau (Eds.). 'Proc. of the 1st European Control Conference'. Hermès. Paris. pp. 152–157.

Fliess, M. (1989). 'Automatique et corps différentiels'. *Forum Math.* **1**, 227–238.

Fliess, M. (1990). 'Some basic structural properties of generalized linear systems'. *Systems Control Lett.* **15**, 391–396.

Fliess, M. (1994). 'Une intreprétation algébrique de la transformation de Laplace et des matrices de transfert'. *Linear Algebra Appl.* **203–204**, 429–442.

Fliess, M., J. Lévine, P. Martin and P. Rouchon (1993). 'Linéarisation par bouclage dynamique et transformations de Lie-Bäcklund'. *C. R. Acad. Sci. Paris Sér. I Math.* **t. 317**, 981–986.

Fliess, M., J. Lévine, P. Martin and P. Rouchon (1994). Nonlinear control and Lie-Bäcklund transformations: Towards a new differential geometric standpoint. In 'Proc. 33rd IEEE Conference on Decision and Control'. Lake Buena Vista, FL. pp. 981–986.

Huijberts, H. J. C., H. Nijmeijer and L. L. M. van der Wegen (1992). 'Dynamic disturbance decoupling for nonlinear systems'. *SIAM J. Control Optim.* **30**, 336–349.

Isidori, A. (1989). *Nonlinear Control Systems.* Springer-Verlag. Berlin.

Johnson, J. (1969). 'Kähler differentials and differential algebra'. *Ann. of. Math* **89**, 92–98.

Nijmeijer, H. and A. J. van der Schaft (1990). *Nonlinear Dynamical Control Systems.* Springer-Verlag. New York.

Perdon, A. M., Y.-F. Zheng, C. H. Moog and G. Conte (1993). 'Disturbance decoupling for nonlinear systems: a unified approach'. *Kybernetika (Prague)* **29**, 479–484.

Pereira da Silva, P. S. (1993). 'On the nonlinear dynamic disturbance decoupling problem'. *J. Math. Systems Estim. Control.* to appear.

Pereira da Silva, P. S. and V. M. Pinto Leite (1993). Disturbance decoupling by regular dynamic feedback for affine nonlinear systems: a linear algebraic approach. In 'Proc. IFAC-World Congress'. Vol. 6. Sidney. pp. 387–390.

Respondek, W. (1991). Disturbance decoupling via dynamic feedback. In B. Bonnard, B. Bride, J.-P. Gauthier and I. Kupka (Eds.). 'Controlled Dynamical Systems'. Vol. 8 of *Progr. Systems Control Theory.* Birkhäuser. pp. 347–357.

Wonham, W. M. (1985). *Linear Multivariable Control: a Geometric Approach.* 3rd edn. Springer-Verlag. New York.

SINGULAR SYSTEMS: THEIR ORIGINS, GENERAL FEATURES, AND NON NUMERICAL SOLUTION

XAVIER GRÀCIA, MIGUEL C. MUÑOZ-LECANDA, NARCISO ROMÁN-ROY

Departament de Matemàtica Aplicada i Telemàtica, Universitat Politècnica de Catalunya, Campus Nord edifici C3, 08071 Barcelona, Catalonia, Spain

Abstract. A brief review on singular systems of differential equations is presented. In particular, the problems they pose and their geometric features, including a geometric method for obtaining a non-numerical solution, are analized. The case of linear singular systems is especially studied, and a particular example is presented.

Key words. Singular differential equations, linear singular systems, stabilization algorithms.

1 INTRODUCTION

Roughly speaking, a singular system is a system of differential equations where the maximum order derivatives are not solved. Since any system can be made of first order by adjoining new variables, we can assume that our system is of first order. The general expression of such a system is $F(t, \mathbf{x}, \dot{\mathbf{x}}) = 0$, where t is the independent variable, usually understood as the *time*, and $\mathbf{x}$ is the *state vector* of the physical system. From now on we consider only autonomous systems, that is, systems where the variable t does not appear explicitly in the equations.

As it will be shown later, even in a local study we are led to work on submanifolds of the initial phase space. So we will suppose from the beginning that the phase space of the system is a differentiable manifold M whose dimension, m, is the number of free variables used to describe the system. Of course it may be assumed that M is embedded in an suitable euclidean space as a submanifold.

With these conventions, an autonomous differential equation is no more than a submanifold D of $\mathrm{T}(M)$, the tangent bundle to M, whose elements are the points $\mathbf{x}$ and their velocities $\dot{\mathbf{x}}$ with respect to t. This submanifold is locally defined by a relation $F(\mathbf{x}, \dot{\mathbf{x}}) = 0$.

A simple example of a singular system comes from a linear control system $\dot{\mathbf{x}} = A\mathbf{x} + B\mathbf{u}$, where $\mathbf{u}$ are the inputs. Suppose that a feedback of the type $\mathbf{u} = -C\dot{\mathbf{x}}$ is introduced. Then we obtain the system $E\dot{\mathbf{x}} = A\mathbf{x}$, where $E = I + BC$; this is a singular system if the matrix T is singular. In fact, this type of systems, those of the form $E\dot{\mathbf{x}} = A\mathbf{x}$ where E is a singular matrix, are often studied and used in several applications.

Besides control theory, there are many applications of singular differential equations to other fields like theoretical physics, cicuit theory, chemical engineering, industrial engineering, econometry ...

2 PROBLEMS ARISING WHEN SOLVING SINGULAR SYSTEMS

In these systems we have not a vector field representing the system and there appear some problems when trying to integrate it analytically or numerically. Some of these problems, which are different from the regular case, are the following:

Compatibility In general, the system of differential equations may not be compatible at all the points of the *initial phase space* M of the system. This means that not any set of values of the variables is admissible as initial conditions for the differential equation.

In order to solve the system, the first step of the procedure must be to identify the set of admissible initial conditions. This problem is not solved directly but in an algorithmic way. For instance the differential equation may imply some relations $\phi^a(\mathbf{x}) = 0$ between the variables $\mathbf{x}$; these relations, called *primary constraints* in the physical literature, define a subset $M_1 \subset M$, the *primary constraint submanifold.* Hence, the first question consists in finding these primary constraints.

Nevertheless, the problem does not finish at this point, since another consistency condition is required. In fact, the evolution of the variables is restricted to abride inside the primary constraint submanifold M_1, that is, for every value of the evolution parameter t, the functions $\mathbf{x}(t)$ must take values in that set. Hence, the problem is not solved until a set of admissible initial conditions verifying this consistency condition is found. This procedure is repeated until the *final constraint submanifold* M_f is obtained; on this submanifold there exist tangent vector fields representing the system. This is the key idea for constructing algorithms to solve the problem: the set of constraints is obtained by iteration of the tangency condition.

Uniqueness Once these problems have been solved, the initial singular system of differential equations is reduced to another one which is compatible and has consistent solution on the points of the final constraint submanifold. Then there are two possibilities:

1. If the singularity of the initial system of differential equations arises from an "inadequate" choice of the variables (that is, the initial state space is too large in order to describe the real degrees of freedom of the problem), then in the final constraint submanifold the system has a unique solution.

2. However in some physical problems, the singularity of the system is a consequence of the existence of a certain kind of internal symmetries (the *gauge symmetry* in theoretical physics terminology). Then the final system of equations may be *undetermined*; that is, the solution of the system may not be unique. This means that, for every point in the final constraint submanifold which is taken as an initial condition, the evolution of the system is not determined because a multiplicity of integral curves (of different vector fields solution) pass trough it. A *reduction procedure* can be used to remove this ambiguity: different points of M_f that can be reached through different solutions beginning at the same initial condition must be *identified*. This quotient is the *reduced phase space* of the system.

States The physical states of the systems are identified with the points of the reduced phase space. This coincides with the final constraint submanifold when the vector field solution of the system is unique.

Singular control systems Besides these problems, other questions are also of interest when dealing with singular control systems; namely: *reachability, controllability, stabilizability, observability, realizability* and *optimality*. All these concepts, which are well established for regular systems in control theory, must be reconsidered for singular systems and, as far as we know, they are not completely solved.

Other problems There are also other problems mainly concerning with the numerical solution of singular systems of differential equations. There are many articles concerning this problem, but, we are not going to deal with it —see for instance [Hairer *et al* (1989)] [Hairer and Wanner (1991)] for a detailed account.

3 GEOMETRIC ASPECTS OF SINGULAR DIFFERENTIAL EQUATIONS

As we have told above an autonomous first order differential equation is a relation $F(\mathbf{x}, \dot{\mathbf{x}})$ between several functions $\mathbf{x}$ in one variable t and their first derivatives $\dot{\mathbf{x}}$. In geometric terms, a path $\gamma: I \to M$ is a solution if

$$\dot{\gamma}(I) \subset D,$$

where D is a submanifold of $\mathrm{T}(M)$ decribing the differential equation.

Singular differential equations from this most general point of view have been seldom studied in the literature.

For instance, in [Menzio and Tulczijew (1978)] this general framework is applied to the hamiltonian dynamics of singular lagrangian systems. More recently the article [Marmo *et al* (1992)] studies symmetries and constants of motion for these singular equations, and the preprint [Mendella *et al* (1994)] obtains some conditions of integrability; in this paper "integrability" means the possibility of finding a solution through any point of D.

In the same way, but taking as M an euclidean space, [Rabier and Rheinboldt (1994)] studies the existence of solutions of the same problem, and in a certain sense it gives an algorithm to find these solutions under some particular conditions, namely, when D is what the authors call a "completely reducible π-submanifold" of $\mathrm{T}(\mathbf{R}^n)$; the aim of this article is to give a geometric treatment of the authors' previous works, especially [Rabier and Rheinboldt (1991)] —see also [Rheinboldt (1984)].

In the literature the most frequently considered singular differential equations are of a special type: they turn out to combine some restrictions on the base M and some linear transformations (usually singular) of the velocities in $\mathrm{T}(M)$. Moreover, the former can be absorbed in the latter, and so the local expression of these equations is

$$A(\mathbf{x})\dot{\mathbf{x}} = b(\mathbf{x}),$$

where $A(\mathbf{x})$ is a matrix, usually singular. We call them *linearly singular equations*.

The geometric study of these equations has been developed independently in the areas of Theoretical Physics and of Systems Theory.

In Theoretical Physics the initial problem [Dirac (1950)] was to obtain a hamiltonian description for singular lagrangian systems; these are lagrangian systems for which the hessian matrix of the lagrangian function with respect to the velocities is singular. Some progress was done in the geometrization of this problem —see for instance [Śniatycki (1974)] [Lichnerowicz (1975)], but the solution did not arrive until [Gotay *et al* (1978)]. In this article, an algorithm is presented to study the existence of solutions of a "presymplectic system"

$$i_X\omega = \alpha,$$

where the unknown is a vector field X in a manifold M, ω is a presymplectic form and α is a 1-form; for instance, one may have M a submanifold of $\mathrm{T}^*(Q)$, ω the pullback of the canonical symplectic form of Q to M and $\alpha = \mathrm{d}H$, where H is a hamiltonian function. Notice that, since the transformation $X \mapsto i_X\omega$ is not an isomorphism, there is no guarantee of neither existence nor uniqueness of the solutions. In fact, as the question of global existence has negative answer, the authors look for a submanifold of M where the equation is satisfied.

The same ideas are applied to the search of solutions in the lagrangian formalism for the singular case; then the presymplectic equation must be supplemented with a second-order condition [Gotay and Nester (1980)]. In this case a variation of the algorithm allows to include the second-order condition [Muñoz-Lecanda and Román-Roy (1992)]. Another way to treat this problem is to work in the manifold $M = \mathrm{T}^*(Q) \times_Q \mathrm{T}(Q)$ [Skinner (1983)] [Skinner and Rusk (1983)].

A more general framework is presented in [Gràcia and Pons (1991)] and with more details in [Gràcia and Pons (1992)]: a "linearly singular system" is an expression

$$A \circ \dot{\gamma} = \sigma \circ \gamma,$$

where the unknown is a path γ in M, σ is a section of a vector bundle $F \to M$, and A is a vector bundle morphism $\mathrm{T}(M) \to F$. This problem is also formulated in terms of vector fields. In these articles the algorithm for presymplectic systems is generalized to linearly singular systems —see section 5. This framework includes applications to other problems like singular lagrangian formalism and higher order lagrangians —see for instance [Gràcia *et al* (1991)].

A less general framework, but still including the presymplectic systems, is presented in [Muñoz-Lecanda and Román-Roy (1995)], which corresponds to a linearly singular system with $F = \mathrm{T}^*(M)$; besides extending the presymplectic algorithm, it includes a study of the symmetries of the equation.

On the other hand, some problems in circuit theory led several researchers to consider singular differential equations.

For instance, in [Takens (1976)] a "constrained differential equation" is a set of data including a linear restriction on the velocities as well as some constraints on the base manifold obtained from a certain potential function; the existence of discontinuous solutions of this problem is analized. The paper also includes a study of singularities of these constrained equations, and some applications to circuits and singular lagrangians. A related paper is [Shankar Sastry and Desoer (1981)].

The same geometric framework for constrained equations is presented in [Haggman and Bryant (1984)], with some simplifications. The problem is to find a curve $c: I \to E$ such that

$$\mathrm{T}(p) \circ \dot{c} = \chi \circ c \qquad c(I) \subset \Sigma,$$

where $p: E \to B$ is a smooth map, χ a vector field along p, and Σ is a submanifold of E with the same dimension as M. A particular class of these equations is studied, and the results are applied to circuit theory. Notice that this geometric framework can be understood as a linearly singular equation.

We finally mention the paper [Chua and Oka (1988)], which uses the concept of "generalized vector field", namely, a vector bundle endomorphism $A: \mathrm{T}(M) \to \mathrm{T}(M)$ and a vector field v in M. The equation of motion is then

$$A \circ \dot{\gamma} = v \circ \gamma.$$

Although the solution of this equation is not discussed, the first constraint submanifold is presented. The aim of the paper is to classify the normal forms for generalized vector fields.

4 THE CASE OF LINEAR SINGULAR SYSTEMS

One of the most studied kind of singular systems are the so called *linear singular systems*

$$E\dot{\mathbf{x}} = A\mathbf{x} + B\mathbf{u}, \qquad \mathbf{y} = C\mathbf{x} + D\mathbf{u},$$

where E, A, B, C and D are constant matrices, E singular. There are different approaches and problems about these systems.

For instance, in [Campbell (1980)] the system $A\dot{\mathbf{x}} = B\mathbf{x}$ with A and B constant matrices and A singular is studied. Different problems giving rise to this kind of systems are presented, problems coming from very different areas: optimal control, constrained control, electric circuits and so on. The approach therein is to convert the system into another one with A depending on a parameter ϵ such that $A(0)$ is singular but $A(\epsilon)$ is regular for $\epsilon \neq 0$. In fact these are the so called *perturbed problems* and they come from boundary layer problems.

One of the initial problems has been to identify the real state space for a linear singular system. Using operational methods, *inconsistent initial conditions* for the system $E\dot{\mathbf{x}} = A\mathbf{x} + B\mathbf{u}$, $\mathbf{y} = C\mathbf{x}$, where the matrix E is singular, are studied in [Verghese *et al.* (1981)]. Following the ideas of this work, the notions of observability and controllability are defined for such systems in [Cobb (1984)].

Another problem is studied in [Campbell (1989)]. Consider the regular system $\dot{\mathbf{x}} = A\mathbf{x} + B\mathbf{u}$, $\mathbf{y} = C\mathbf{x} + D\mathbf{u}$, and fix the output $\mathbf{y}$. Do there exist any inputs $\mathbf{u}$ giving this fixed output? There it is proved that the solution of this problem is equivalent to solve a singular system on the variables $(\mathbf{x}, \mathbf{u})$.

The methods in these articles are operational. Also wide use is made of matrix pencils. General properties of the set of solutions are not provided and these methods arrive to some inconsistent situations when applied to obtain numerical solutions of some systems.

One of the most complete references is [Kaczorek (1992, 1993)]. In these two volumes a complete study of singular linear systems is performed including the problems of pole-assignement and factor-assignement in closed-loop systems and the precise definitions of reachability, controllability, observability and constructibility for such systems.

5 GEOMETRIC SOLUTION OF SINGULAR EQUATIONS

Here we summarize the geometric algorithm for finding the solutions γ of a linearly singular equation $A \circ \dot{\gamma} = \sigma \circ \gamma$ as presented in [Gràcia and Pons (1991)] [Gràcia and Pons (1992)]:

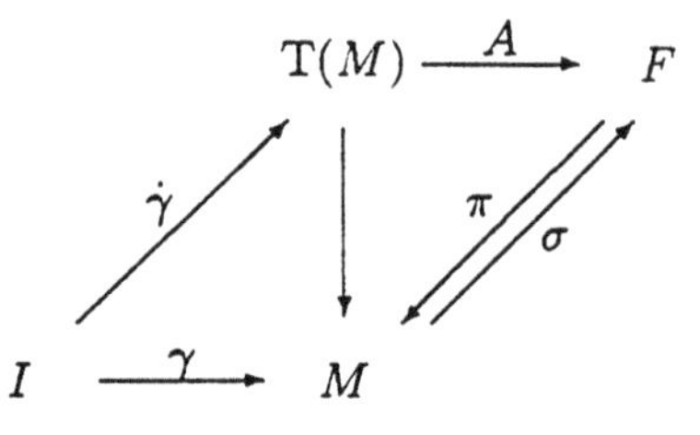

An alternative description of the dynamics can be given in terms of the equation of motion for a vector field: if $N \subset M$ is a submanifold contained in the motion set, and X is a vector field in M *tangent* to N, then the integral curves of X contained in N are solutions of the equation of motion if and only if X satisfies[1]

$$A \circ X \underset{N}{\simeq} \sigma. \tag{1}$$

In general this must be considered as an equation both for the submanifolds $N \subset M$ where the motion can take place and the vector fields X *tangent to* N.

To begin an algorithm to solve the equation of motion, it is assumed that $\operatorname{Ker} A \subseteq \mathrm{T}(M)$ and $\operatorname{Im} A \subseteq F$ are vector subbundles, which amounts to saying that A has locally constant rank.

[1] The notation $\underset{N}{\simeq}$ means equality at the points of N.

The equation $A_x \cdot X(x) = \sigma(x)$ for the unknown vector $X(x)$ can be solved only at the points $x \in M$ such that the compatibility condition

$$\sigma(x) \in \operatorname{Im} A_x \tag{2}$$

holds. This is equivalent to saying that $\sigma(x)$ is orthogonal (in the sense of duality) to the kernel of the transposed morphism,

$$\langle \operatorname{Ker} {}^t\!A_x, \sigma(x) \rangle = 0. \tag{3}$$

Let $M_1 = \{x \in M \mid \sigma(x) \in \operatorname{Im} A_x\}$. If $(s^a)_{1 \le a \le m}$ is a local frame for $\operatorname{Ker} {}^t\!A$, then $M_1 \subset M$ is locally described, using (3), by the vanishing of the m *primary constraint functions*

$$\phi^a = \langle s^a, \sigma \rangle. \tag{4}$$

As usual in the theory of constrained systems, one assumes that M_1 is a submanifold, the *primary constraint submanifold.*

The compatibility condition (2) has lead to consider solutions X which satisfy the equation of motion only on M_1, therefore one is interested only in the values of $X_1 = X|_{M_1}$. Since X (or X_1) must be *tangent to* M_1, the initial problem becomes the same problem for the subsystem defined by M_1:

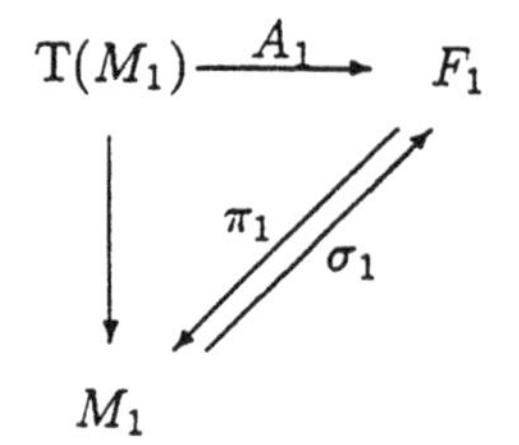

where $F_1 = F|_{M_1}$, and π_1, A_1 and σ_1 are the corresponding restrictions to M_1.

The procedure goes on similarly. The compatibility condition for this system yields a subset M_2, also assumed to be a submanifold... In general, let $M_0 = M$, $A_i = A|_{\mathrm{T}(M_i)}$ and $\sigma_i = \sigma|_{M_i}$, considered as mappings into $F_i = F|_{M_i}$, and define recursively

$$M_{i+1} := \{x \in M_i \mid \sigma_i(x) \in \operatorname{Im} A_{ix}\}. \tag{5}$$

This sequence ends in the ***final constraint submanifold*** M_f. Then $\sigma_f(x) \in \operatorname{Im} A_{fx}$ and therefore the equation $A_f \circ X_f = \sigma_f$ for a vector field X_f in M_f has solutions. Moreover, these solutions are not unique when $\operatorname{Ker} A_f \neq 0$: their multiplicity is described by the vector fields lying in $\operatorname{Ker} A_f$.

This algorithm may be presented explicitly in terms of the constraints: the constraints defining M_2 are constructed from those defining M_1, and so on.

6 AN EXAMPLE

In order to show the power of geometric methods in the study of singular systems, let us apply them to a particular linear singular system with constant coefficients. This constancy condition is not relevant to apply the algorithm but we assume it for simplicity.

Consider the system $E\dot{\mathbf{x}} = B\mathbf{x}$ with E and B square constant matrices, E singular. Following section 5, the data in this case are: $M = \mathbf{R}^m$ (and so $\mathrm{T}(M) = \mathbf{R}^m \times \mathbf{R}^m$), $F = \mathbf{R}^m \times \mathbf{R}^m$, $A(\mathbf{x}, \dot{\mathbf{x}}) = (\mathbf{x}, E\dot{\mathbf{x}})$ and $\sigma(\mathbf{x}) = (\mathbf{x}, B\mathbf{x})$.

Suppose that the defect of E is 2, and let $\mathrm{Ker}\,{}^tE = \mathrm{span}\{\Omega^1, \Omega^2\}$, $\mathrm{Ker}\, E = \mathrm{span}\{X_1, X_2\}$.

The application of the algorithm begins with the *compatibility conditions*, which are the linear forms $\phi^\alpha_{(1)}(\mathbf{x}) = \Omega^\alpha(B\mathbf{x})$, $\alpha = 1,2$; and the primary constraint manifold is

$$M_1 = \{\mathbf{x} \mid \phi^\alpha_{(1)}(\mathbf{x}) = 0,\ \alpha = 1,2\}.$$

On the submanifold M_1 the vector fields solution of the system are given by

$$X = X_0 + g^1 X_1 + g^2 X_2, \qquad (6)$$

where g^1 and g^2 are arbitrary functions.

Observe that both the primary constraints and the general vector field solution can be obtained by application of Gauss elimination method.

Now different situations have to be considered according to the rank of the primary constraints.

(1) *Both constraints $\phi^\alpha_{(1)}$ vanish.*

In this case the final constraint manifold is already $M_1 = \mathbf{R}^m$ and the system is undetermined. There are a family of vector fields solution depending on two arbitrary functions, g^1 and g^2. This family is given by equation (6).

(2) *The set of forms $\phi^\alpha_{(1)}$ has rank 1.*

Assume for instance that $\phi^1_{(1)} \neq 0$. The primary constraint manifold is $M_1 = \{\mathbf{x} \mid \phi^1_{(1)}(\mathbf{x}) = 0\}$, the solution is given by equation (6) and we must study the *tangency conditions.*

As the vector field solution must be tangent to M_1, we have $X(\phi^1_{(1)}) = 0$, that is,

$$X_0(\phi^1_{(1)}) + g^1 X_1(\phi^1_{(1)}) + g^2 X_2(\phi^1_{(1)}) = 0.$$

Notice that, since X_1 and X_2 are constant, the coefficients of g^1 and g^2 are indeed constant. Two possibilities are considered:

(2a) One of the numbers $X_1(\phi^\alpha_{(1)})$ is not zero.

If for instance $X_1(\phi^\alpha_{(1)})$ does not vanish on M_1 then we can solve the arbitrary function g^1 in terms of g^2; M_1 is the final constraint manifold and the family of vector fields solution of the system depends on one arbitrary function.

(2b) Both numbers $X_1(\phi^\alpha_{(1)})$ are zero.

Once again there are two different possibilities:

(2b$_1$) $X_0(\phi^1_{(1)}) \underset{M_1}{\simeq} 0$.

No arbitrary function can be determined, M_1 is the final constraint manifold and the vector field solution of the system depends on two arbitrary functions.

(2b$_2$) $X_0(\phi^1_{(1)}) \underset{M_1}{\not\simeq} 0$.

We have a new constraint given by $\phi^1_{(2)} = X_0(\phi^1_{(1)}) = 0$ and a new constraint submanifold M_2 defined by

$$M_2 = \{\mathbf{x} \in M_1 \mid \phi^1_{(2)}(\mathbf{x}) = 0\}.$$

Now the situation is the same as at the beginning of the point (2), and the procedure follows in the same way.

(3) *Both functions $\phi^\alpha_{(1)}$ are linearly independent.*

Then they define the submanifold M_1 where the system is compatible and its general solution is given by (6).

The *tangency condition* yields two equations:

$$\begin{pmatrix} X_1(\phi^1_{(1)}) & X_2(\phi^1_{(1)}) \\ X_1(\phi^2_{(1)}) & X_2(\phi^2_{(1)}) \end{pmatrix} \begin{pmatrix} g^1 \\ g^2 \end{pmatrix} = - \begin{pmatrix} X_0(\phi^1_{(1)}) \\ X_0(\phi^2_{(1)}) \end{pmatrix}$$

This is a linear system for g^1 and g^2. Let D be the matrix of this system. We have three different cases to study:

(3a) $\mathrm{rank}(D) = 2$.

Then both functions can be determined by solving the linear system. The subset M_1 is the final constraint manifold and the solution on M_1 is unique.

(3b) $\mathrm{rank}(D) = 1$.

In this case one of the arbitrary functions can be determined in terms of the other one, and in general there appears a new constraint that must vanish to ensure compatibility. If this constraint vanishes on M_1 then M_1 is the final constraint submanifold and the solution depends on an arbitrary function; otherwise a new constraint $\phi^1_{(2)}$ is obtained, this constraint defines a new submanifold M_2 and the tangency condition must be applied to this new constraint as in item (2).

(3c) $\mathrm{rank}(D) = 0$.

In this case both functions $X_0(\phi^\alpha_{(1)})$ are the compatibility conditions for the system. These functions are new constraints $\phi^\alpha_{(2)}$ defining a submanifold M_2; we are in the same situation as in the beginning of the algorithm, but notice that the rank of the new constraints is to be examined on the submanifold M_1.

As a final comment, observe that all the submanifolds obtained are linear submanifolds, and that all the constraints are linear. All the systems that have been solved are linear and, in fact, by iterated application of the Gauss reduction method to suitable systems the problem has been solved in an algorithmic way. If the defect of the matrix E is different from two then the different possibilities can be reduced to the above studied with minor modifications.

7 REFERENCES

S.L. Campbell (1980), *Singular systems of differential equations*, Research Notes in Mathematics, Pitman, San Francisco.

S.L. Campbell (1989) "Comments on functional reproducicility of time-variyng input-output systems", *IEEE Trans. Automat. Contr.* **34** 922–924.

L.O. Chua, H. Oka (1988), "Normal forms for constrained nonlinear differential equations. Part I: theory", *IEEE Trans. Circuits Syst.* **35** 881–901.

D. Cobb (1984), "Controllability, observability and dualty in singular systems", *IEEE Trans. Automat. Contr.* **AC-29** 1076–1082.

P.A.M. Dirac (1950), "Generalized hamiltonian dynamics", *Canad. J. Math.* **2** 129–148.

M.J. Gotay, J.M. Nester (1980), "Presymplectic lagrangian systems II: the second order equation problem", *Ann. Inst. H. Poincaré A* **32** 1–13.

M.J. Gotay, J.M. Nester, G. Hinds (1978), "Presymplectic manifolds and the Dirac-Bergmann theory of constraints", *J. Math. Phys.* **27** 2388–2399.

X. Gràcia, J.M. Pons (1991), "Constrained systems: a unified geometric approach", *Int. J. Theor. Phys.* **30** 511–516.

X. Gràcia, J.M. Pons (1992), "A generalized geometric framework for constrained systems", *Diff. Geom. Appl.* **2** 223–247.

X. Gràcia, J.M. Pons, N. Román-Roy (1991), "Higher order lagrangian systems: geometric structures, dynamics and constraints", *J. Math. Phys.* **32** 2744–2763.

B.C. Haggman, P.P. Bryant (1984), "Solutions of singular constrained differential equations: a generalization of circuits containing capacitor-only loops and inductor-only cutsets", *IEEE Trans. Circuits Syst.* **CAS-31**(12) 1015–1025.

E. Hairer, C. Lubich, M. Roche (1989), *The Numerical Solution of Differential-Algebraic Systems by Runge-Kutta Methods*, Lecture Notes in Mathematics **1409**, Springer, Berlin.

E. Hairer, G. Wanner (1991), *Solving Ordinary Differential Equations II*, Springer, Berlin.

T. Kaczorek (1992 and 1993), *Linear control systems*, volumes I and II, Research Studies Press, Taunton.

A. Lichnerowicz (1975), "Variété symplectique et dynamique associée à une sous-variété", *C.R. Acad. Sc. Paris* **280**(A) 523–527.

G. Marmo, G. Mendella, W.M. Tulczijew (1992), "Symmetries and constant of the motion for dynamics in implicit form", *Ann. Inst. H. Poincaré A* **57**(2) 147–166.

G. Mendella, G. Marmo, W.M. Tulczijew (1994), "Integrability of implicit differential equations", preprint.

M.R. Menzio, W.M. Tulczijew (1978), "Infinitesimal symplectic relations and generalized hamiltonian dynamics", *Ann. Inst. H. Poincaré A* **28** 349–367.

M.C. Muñoz-Lecanda, N. Román-Roy (1992), "Lagrangian theory for presymplectic systems", *Ann. Inst. H. Poincaré A* **57** 27–45.

M.C. Muñoz-Lecanda, N. Román-Roy (1995), "Linear singular differential systems: a geometrical approach", preprint DMAT-UPC.

P.J. Rabier, W.C. Rheinboldt (1991), "A general existence and uniqueness theory for implicit differential-algebraic equations", *Differential Integral Equations* **4** 563–582.

P.J. Rabier, W.C. Rheinboldt (1994), "A geometric treatment of implicit differential-algebraic equations", *J. Differential Equations* **109** 110–146.

W.C. Rheinboldt (1984), "Differential-algebraic systems as differential equations on manifolds", *Math. Comp.* **43**(168) 473–482.

S. Shankar Sastry, C.A. Desoer (1981), "Jump behavior of circuits and systems", *IEEE Trans. Circuits Syst.* **CAS-28**(12) 1109–1124.

R. Skinner (1983), "First order equations of motion for classical mechanics", *J. Math. Phys.* **24** 2581–2588.

R. Skinner, R. Rusk (1983), Generalized hamiltonian dynamics I: formulation on $T^*Q \otimes TQ$", *J. Math. Phys.* **24** 2589–2594.

J. Śniatycki (1974), "Dirac brackets in geometric dynamics", *Ann. Inst. H. Poincaré A* **20** 365–372.

F. Takens (1976), "Constrained equations: a study of implicit differential equations and their discontinuous solutions", in *Structural Stability, the Theory of Catastrophes and Application in the Sciences*, P. Hilton ed., Lectures Notes in Mathematics **525** Springer-Verlag, Berlin.

G.C. Verghese, B.C. Lévy, T. Kailath (1981), "A generalized state-space for singular systems", *IEEE Trans. Automat. Contr.* **AC-26** 811–831.

INDEX AND DECOMPOSITION OF NONLINEAR IMPLICIT DIFFERENTIAL EQUATIONS

Michel Fliess
Laboratoire des Signaux et Systèmes, C.N.R.S.-Supélec,
Plateau de Moulon, 91192 Gif-sur-Yvette, France. E-mail: fliess@lss.supelec.fr

Jean Lévine, Philippe Martin
Centre Automatique et Systèmes, E.N.S.M.P.,
35, rue Saint-Honoré, 77305 Fontainebleau, France. E-mail: {levine,martin}@cas.ensmp.fr

Pierre Rouchon
Centre Automatique et Systèmes, E.N.S.M.P.,
60, bld Saint-Michel, 75272 Paris cédex 06, France. E-mail: rouchon@cas.ensmp.fr

Abstract In engineering, as well as in physics, one often encounters *implicit* systems of ordinary differential equations of the form $F_i(\dot{y}_1, \ldots, \dot{y}_m, y_1, \ldots, y_m) = 0$, $i = 1, \ldots, m$, in the unknowns $y_1, \ldots, y_m$, where the Jacobian matrix $(\partial F_i/\partial \dot{y}_j)_{i,j}$ is identically singular. We first state a condition under which the above equation may be solved. Then, we give a decomposition, which provides as a byproduct a clear-cut definition of the *index*. Since the F_i's must be differentiated an arbitrary number of times, we employ tools stemming from the differential geometry of infinite jets and prolongations like Lie-Bäcklund morphisms.

Résumé. *Tant en ingéniérie qu'en physique se rencontrent des systèmes différentiels implicites ordinaires* $F_i(\dot{y}_1, \ldots, \dot{y}_m, y_1, \ldots, y_m) = 0$, $i = 1, \ldots, m$, *en les inconnues* $y_1, \ldots, y_m$, *où la matrice jacobienne* $(\partial F_i/\partial \dot{y}_j)_{i,j}$ *est identiquement singulière. Nous énonçons une condition de résolubilité et donnons une décomposition, qui fournit une définition claire de l'* index. *Les dérivations des* F_i *conduisent à la géométrie différentielle des jets et prolongations infinis, et aux transformations de Lie-Bäcklund.*

Keywords: nonlinear implicit differential equations, nonlinear DAE's, well-posed, index, decomposition, constraints, differential geometry, infinite jets, infinite prolongations, Lie-Bäcklund morphisms, exterior algebra, differential algebra.

Introduction

Consider a system of m *implicit* ordinary differential equations in m unknowns, which, for the sake of simplicity in this introduction, we assume to be first order,

$$F_i(\dot{y}_1, \ldots, \dot{y}_m, y_1, \ldots, y_m) = 0, i = 1, \ldots, m \qquad (1)$$

In many concrete situations, in engineering as well as in physics, which might result from algebro-differential systems, the Jacobian matrix $(\partial F_i/\partial \dot{y}_j)_{i,j}$ is identically singular. As a consequence, the implicit function theorem does not allow to write (1) in the usual *explicit* form

$$\dot{y}_i = \mathcal{F}_i(y_1, \ldots, y_m), i = 1, \ldots, m$$

Severe difficulties may then arise in the numerical integration of (1), which are studied in a huge literature, from which we extract the three following recent books [3, 26, 27].

An essential ingredient for understanding (1) is the *index* (see, e.g., [4, 5, 7]), a precise definition of which has been given for constant linear systems in [35] by means of matrix pencils, and, more recently, for time-varying linear systems in [23] (see, also, [9]) *via* noncommutative principal ideal rings (*cf.* [8]). Our main result is a precise derivation of the index for (1), which as a byproduct gives a decomposition of (1), as a vector-field on a manifold which expresses the *constraints*. Here is a more intuitive explanation:
We replace (1), as customary, by

$$F_i(\dot{y}_1, \ldots, \dot{y}_m, y_1, \ldots, y_m) = w_i, i = 1, \ldots, m \qquad (2)$$

where the w_i's are *perturbation*, or *input*, variables which serve, among other things, to mark the equations of (1)-(2). The index is the minimum number of derivations of the w_i's, i.e., of the F_i's, needed to write the explicit first order system and the manifold of constraints. In the following elementary time-invariant linear example

$$\begin{array}{rcl} y_1 & = & w_1 \\ \dot{y}_1 + y_2 & = & w_2 \\ & \cdots & \\ \dot{y}_{m-2} + y_{m-1} & = & w_{m-1} \\ \dot{y}_{m-1} + \dot{y}_m & = & w_m \end{array} \qquad (3)$$

the first order equation is $\dot{y}_m = w_m - \dot{w}_{m-1} + \ldots + (-1)^{m-1} w_1^{(m-1)}$, and the constraints are $y_1 = w_1$, $y_2 = w_2 - \dot{w}_1, \ldots, y_{m-1} = w_{m-1} - \dot{w}_{m-2} + \ldots + (-1)^{m-2} w_1^{(m-2)}$.

(3) confirms the necessity, already mentioned by several authors, to consider higher order derivatives of the system variables

and not to restrict oneself to first order derivatives, although we are working with first order differential equations. As we do not know *a priori* the bound of the order of derivations, we are working with *infinite jets* and *prolongations*, which are becoming nowadays a familiar tool in differential geometry (see, e.g., [1, 31, 33, 34, 36, 37, 38] and the references therein). A morphism between two such infinite-dimensional manifolds, which is compatible with the total derivation with respect to time, is said, according to Ibragimov (see [1] and, also, [38]), to be *Lie-Bäcklund*. This setting, which has already revealed itself as most useful in dynamic feedback linearization of nonlinear control systems (see [18, 19]), permits also here an elegant characterization of the *well-posedness*, i.e., the solvability, of (2): There exists a unique solution, with initial conditions satisfying the constraints, if the F_i's are *differentially functionally independent*, i.e., the F_i's and their derivatives of any order are functionally independent (compare with [6, 25]). This result derives from a rank theorem on Lie-Bäcklund morphisms, which may be viewed as an infinite-dimensional extension to Fréchet spaces of the classic rank theorem in finite-dimensional analysis (see, e.g., [10] and compare with [28]).

When the F_i's are analytic functions of their arguments, we prove that the well-posedness condition and the index of (2) are the same as the ones for the linear tangent differential systems. This establishes a close relationship of the general nonlinear situation and the linear time-varying case, already investigated in [23].

This communication is organized as follows. We first review the infinite-dimensional differential geometry in a quite elementary manner: this section is interesting in its own sake and is complementary to the differential-geometric approach to *flatness* [18, 19] [1]. We also demonstrate the equivalence between (2) and a Lie-Bäcklund morphism. This is applied to the solvability condition. The index and the decomposition are obtained *via* elementary exterior algebra. The final section studies the relationship with linear systems when the F_i's are analytic. The conclusion indicates some other directions of research. An appendix gives first hints on a related differential algebraic approach.

1 The mathematical framework

1.1 Infinite-dimensional differential geometry

1.1.1 Consider the dynamics

$$\begin{aligned}\dot{x}_1 &= f_1(x_1,\ldots,x_n,u_1,\ldots,u_m)\\ &\vdots\\ \dot{x}_n &= f_n(x_1,\ldots,x_n,u_1,\ldots,u_m)\end{aligned}\tag{4}$$

where $f_1,\ldots,f_n$ are smooth functions, i.e., C^∞-functions, of the input $u=(u_1,\ldots,u_m)\in\mathbb{R}^m$ and of the state $x=(x_1,\ldots,x_n)^t\in\mathbb{R}^n$. Take a "classic" output function $h(x)$, where $h:\mathbb{R}^n\to\mathbb{R}$ is C^∞. Then

$$\dot{y}=f_1\frac{\partial h}{\partial x_1}+\ldots+f_n\frac{\partial h}{\partial x_n}$$

[1] See [21] for more details.

In the differential geometric language, $\dot{y}$ is expressed as the *Lie derivative* of the function h with respect to the vector field $f_1\frac{\partial}{\partial x_1}+\ldots+f_n\frac{\partial}{\partial x_n}$. If $\overline{y}=\overline{h}(x,u)$, where $\overline{h}:\mathbb{R}^{n+m}\to\mathbb{R}$ is C^∞, then

$$\dot{\overline{y}}=f_1\frac{\partial\overline{h}}{\partial x_1}+\ldots+f_n\frac{\partial\overline{h}}{\partial x_n}+\dot{u}_1\frac{\partial\overline{h}}{\partial u_1}+\ldots+\dot{u}_m\frac{\partial\overline{h}}{\partial u_m}.$$

For an output $\tilde{y}:\mathbb{R}^{1+n+m(1+\alpha)}\to\mathbb{R}$

$$\tilde{y}=\tilde{h}(t,x_1,\ldots,x_n,u_1,\ldots,u_m,\ldots,u_1^{(\alpha)},\ldots,u_m^{(\alpha)}),$$

which depends on derivatives of arbitrary order of u and also on the time t, the derivative with respect to time reads

$$\begin{aligned}\dot{\tilde{y}} &= \frac{\partial\tilde{h}}{\partial t}+f_1\frac{\partial\tilde{h}}{\partial x_1}+\ldots+f_n\frac{\partial\tilde{h}}{\partial x_n}+\\ &\quad+\sum_{\nu=0}^{\alpha}(u_1^{(\nu+1)}\frac{\partial\tilde{h}}{\partial u_1^{(\nu)}}+\ldots+u_m^{(\nu+1)}\frac{\partial\tilde{h}}{\partial u_m^{(\nu)}})\end{aligned}$$

The Lie derivative may be thus interpreted as a derivative along the vector field

$$\begin{aligned}&\frac{\partial}{\partial t}+f_1\frac{\partial}{\partial x_1}+\ldots+f_n\frac{\partial}{\partial x_n}+\\ &+\sum_{\nu=0}^{\alpha}(u_1^{(\nu+1)}\frac{\partial}{\partial u_1^{(\nu)}}+\ldots+u_m^{(\nu+1)}\frac{\partial}{\partial u_m^{(\nu)}}).\end{aligned}$$

All the three above Lie derivative formulae may be unified when taking the derivatives with respect to time of h, $\overline{h}$ and $\tilde{h}$ as Lie derivative with respect to the <u>infinite</u> vector field

$$\begin{aligned}\frac{d}{dt} &= \frac{\partial}{\partial t}+f_1\frac{\partial}{\partial x_1}+\ldots+f_n\frac{\partial}{\partial x_n}+\\ &\quad+\sum_{j=0}^{m}\sum_{\nu_j\geq 0}u_j^{(\nu_j+1)}\frac{\partial}{\partial u_j^{(\nu_j)}}.\end{aligned}\tag{5}$$

It expresses the total derivative with respect to time of any function of a finite number of variables belonging to the infinite set $\{t,x_i,u_j^{(\nu_j)}\mid i=1,\ldots,n;\ j=1,\ldots,m;\ \nu_j\geq 0\}$. Three objects have been introduced:

- a space of infinite, but denumerable, dimension described by the coordinates $\{t,x_i,u_j^{(\nu_j)}\mid i=1,\ldots,n;\ j=1,\ldots,m;\ \nu_j\geq 0\}$;
- C^∞-functions of the above coordinates, which only depend on a finite number of these coordinates;
- infinite vector fields of the type $a\frac{\partial}{\partial t}+\sum_{i=1}^{i=n}b_i\frac{\partial}{\partial x_i}+\sum_{=1}^{m}\sum_{\nu_j\geq 0}c_j^{\nu_j}u_j^{(\nu_j+1)}\frac{\partial}{\partial u_j^{(\nu_j)}}$, where a, b_i and $c_j^{\nu_j}$ are smooth functions in the above sense. Among them, $\frac{d}{dt}$ in (5) plays an outstanding role.

Notice, furthermore, that any 1-form ω is a <u>finite</u> linear combination of $\{dt,dx_i,du_j^{(\nu_j)}\mid i=1,\ldots,n;\ j=1,\ldots,m;\ \nu_j\geq 0\}$, since $d\tilde{h}=\tilde{h}_t dt+\sum_{i=1}^{n}\tilde{h}_{x_i}dx_i+\sum_{j=1}^{m}\sum_{\nu_j=0}^{\alpha}\tilde{h}_{u_j^{(\nu_j)}}du_j^{(\nu_j)}$.

1.1.2 To obtain a more intrinsic description of the above considerations, i.e., a description which is coordinate-free, we need to introduce a notion of infinite-dimensional manifold (see, e.g., [36, 38]). Contrary to the usual case, where a k-dimensional manifold looks locally like $\mathbb{R}^k$, we allow here for an infinite, but denumerable, Cartesian product $\mathbb{R}^\infty$ of copies of $\mathbb{R}$. This means, in other words, that we construct a C^∞-manifold, i.e., a smooth manifold, which is locally defined by a denumerable set $\{\xi_\mu \mid \mu = 0, 1, 2, \ldots\}$ of coordinates. Charts and atlases are defined exactly in the same way as in the finite-dimensional case, once we recall that smooth functions locally depend only on a finite number of coordinates. A vector field is locally given by an expression $\sum a_\mu \frac{\partial}{\partial \xi_\mu}$, which, as in (5), may be infinite. On the other hand, a 1-form is always a finite expression $\sum b_\mu d\xi_\mu$.

1.1.3 REMARK. Most of the above features are close to standard finite-dimensional differential geometry (*cf.* [11]). The topology, of $\mathbb{R}^\infty$, however, is more complex than the usual Euclidean one of $\mathbb{R}^k$, and is called a *Fréchet topology*, with which are also endowed our infinite-dimensional manifolds. The implicit function theorem as well as the Frobenius theorem do not hold true in this context (see, e.g., [38]).

1.1.4 The control examples involve something more, namely a distinguished vector field $\frac{d}{dt}$, which represents the total derivation with respect to time. We are therefore led to the following definition:

Definition. *An* (ordinary) diffiety[2] *is a smooth manifold, which is equipped with a one-dimensional distribution, called its* Cartan distribution. *Any smooth section of the Cartan distribution is called a* Cartan field. *Any morphism between diffieties, which is compatible with the Cartan distribution, is said to be* Lie-Bäcklund[3]

1.1.5 EXAMPLE. Take again (4). The corresponding diffiety is described globally by the coordinates $\{t, x_i, u_j^{(\nu_j)} \mid i = 1, \ldots, n;\ j = 1, \ldots, m;\ \nu_j \geq 0\}$. Its Cartan distribution is spanned by the Cartan field $\frac{d}{dt}$ in (5).

1.1.6 EXAMPLE. Consider an uncontrolled dynamics

$$\begin{aligned} \dot{x}_1 &= f_1(t, x_1, \ldots, x_n) \\ &\vdots \\ \dot{x}_n &= f_n(t, x_1, \ldots, x_n) \end{aligned}$$

The associated diffiety is $\mathbb{R}^{n+1}$ with global coordinates $\{t, x_1, \ldots, x_n\}$ and Cartan distribution spanned by the Cartan field $\frac{\partial}{\partial t} + \sum_{i=1}^n f_i \frac{\partial}{\partial x_i}$. The diffiety here is finite-dimensional. The infinite-dimensional character of (4) is due to the input variables or, in other words, to the fact that a controlled system is an *underdetermined* system of differential equations: the number of variables is strictly greater than the number of equations.

1.1.7 EXAMPLE. Take the diffiety $\mathbb{R} \times \mathbb{R}_m^\infty$ which is globally defined by the coordinates $\{t, y_j^{(\nu_j)} \mid j = 1, \ldots, m;\ \nu_j \geq 0\}$ and the Cartan distribution spanned by the Cartan field

$$\frac{\partial}{\partial t} + \sum_{j=1}^m \sum_{\nu_j \geq 0} y_j^{(\nu_j+1)} \frac{\partial}{\partial y_j^{(\nu_j)}}$$

on $\mathbb{R} \times \mathbb{R}_m^\infty$, which plays an utmost important role, is called a *differential affine space*, since it appears to generalize the usual finite-dimensional affine space $\mathbb{R}^m$, or a *trivial diffiety*.

[2] The word *diffiety* is due to Vinogradov [37].

[3] This means that the associated tangent morphism maps the Cartan distribution of the first diffiety in the Cartan distribution of the second one.

1.2 Implicit equations as Lie-Bäcklund morphisms

1.2.1 Consider the general implicit system of m differential equations, with m unknowns $y_1, \ldots, y_m$ and m perturbation, or control, variables $w_1, \ldots, w_m$

$$F_i(t, y_1, \ldots, y_1^{(\alpha_1)}, \ldots, y_m, \ldots, y_m^{(\alpha_m)}) = w_i \qquad (6)$$
$$i = 1, \ldots, m$$

where the F_i's are (local) C^∞-functions of their arguments. Take the differential affine space A_1 (resp. A_2) with global coordinates $\{t, y_j^{(\nu_j)} \mid j = 1, \ldots, m;\ \nu_j \geq 0\}$ and Cartan field $\frac{d}{dt} = \frac{\partial}{\partial t} + \sum_{j=1,\ldots,m} \sum_{\nu_j \geq 0} y_j^{(\nu_j+1)} \frac{\partial}{\partial y_j^{(\nu_j)}}$ (resp. $\{t, w_j^{(\mu_j)} \mid j = 1, \ldots, m;\ \mu_j \geq 0\}$ and $\frac{d}{dt} = \frac{\partial}{\partial t} + \sum_{j=1,\ldots,m} \sum_{\mu_j \geq 0} w_j^{(\mu_j+1)} \frac{\partial}{\partial w_j^{(\mu_j)}}$). Associate to (6) the (local) Lie-Bäcklund morphism $\Phi : A_1 \to A_2$, which keeps t fixed and such that

$$w_i = F_i(t, y_1, \ldots, y_1^{(\alpha_1)}, \ldots, y_m, \ldots, y_m^{(\alpha_m)})$$

and

$$\frac{d^{\mu_i} w_i}{dt^{\mu_i}} = \frac{d^{\mu_i}}{dt^{\mu_i}} \left(F_i(t, y_1, \ldots, y_1^{(\alpha_1)}, \ldots, y_m, \ldots, y_m^{(\alpha_m)}) \right)$$

1.2.2 The above construction leads to the

Proposition. *There is an equivalence between implicit differential equations of type (6) and time-preserving Lie-Bäcklund morphisms between the differential affine spaces A_1 and A_2.*

2 Solvability

2.1 The functions F_i's in (6) are said to be (locally) *differentially functionally (in)dependent* if, and only if, the infinite set $\{\frac{d^{\mu_i}}{dt^{\mu_i}} F_i \mid i = 1, \ldots, m;\ \mu_i \geq 0\}$ of functions is (locally) functionally (in)dependent.

2.2 Denote by T^*A_ι the cotangent bundle of A_ι, $\iota = 1, 2$. Any (local) C^∞-section of T^*A_1 (resp. T^*A_2) is a (local) smooth 1-form of the type $a dt + \sum_{finite} b_{i,\nu_i} dy_i^{(\nu_i)}$ (resp. $a' + \sum_{finite} b'_{i,\mu_i} dw_i^{(\mu_i)}$). Let $\Phi^* : T^*A_1 \to T^*A_2$ be the cotangent morphism associated to Φ.

2.3 Φ is said to be a (local) *Lie-Bäcklund submersion* (*cf.* [38]) if, and only if, there exists in A_1 and A_2 local coordinates such that it is a projection.

2.4 THEOREM. *For the (local) Lie-Bäcklund morphism, the three following properties are equivalent:*

(i) Φ^* is a (local) injection;

(ii) Φ is a (local) Lie-Bäcklund submersion;

(iii) $F_1, \ldots, F_m$ are (locally) differentially functionally independent.

2.5 SKETCH OF THE PROOF. The equivalence between (i) and (iii) is clear. The equivalence between (ii) and (iii) relies on the impossibility of finding more than m (locally) differentially functionally independent functions $A_1 \to \mathbb{R}$.

2.6 COROLLARY. *If Φ is a (local) Lie-Bäcklund submersion, then any fiber of Φ is finite-dimensional.*

2.7 Assume that Φ is a (local) Lie-Bäcklund submersion. Let $x_1, \ldots, x_n$ be local coordinates of the fiber. Then

$$\begin{aligned} &\text{for } k = 1, \ldots, n: \\ &\frac{dx_k}{dt} = f_k(t, x_1, \ldots, x_n, w_1, \ldots, w_1^{(\gamma_1)}, \ldots, w_m, \ldots, w_m^{(\gamma_m)}); \\ &\text{for } i = 1, \ldots, m: \\ &y_i = g_i(t, x_1, \ldots, x_n, w_1, \ldots, w_1^{(\delta_1)}, \ldots, w_m, \ldots, w_m^{(\delta_m)}). \end{aligned} \tag{7}$$

(7) is the analogue of the generalized state-variable representation of [14, 15] obtained *via* differential algebra.

2.8 THEOREM AND DEFINITION. *The implicit equation (6) is said to be* well-posed *if, and only if, the F_i's are (locally) differentially functionally independent. Then, the y_i's may be obtained by the state-variable representation (7).*

2.9 REMARK. Initial conditions for (6) may be expressed in terms of the variables $x_1, \ldots, x_n$.

2.10 The "original" equation

$$F_i(t, y_1, \ldots, y_1^{(\alpha_1)}, \ldots, y_m, \ldots, y_m^{(\alpha_m)}) = 0, \quad i = 1, \ldots, m \tag{8}$$

where the w_s's have been set to zero, is said to be *regular* if, and only if, the two following conditions are verified:

- (6) is well-posed;
- the fiber of the (local) Lie-Bäcklund submersion Φ at the point $(t, 0, \ldots, 0, \ldots)$ in A_2 is the typical one.

Then, solutions of (8) are given by setting the w_i's and their derivatives to zero in (7), i.e.,

$$\begin{aligned} \frac{dx_k}{dt} &= f_k(t, x_1, \ldots, x_n, 0, \ldots, 0) \\ y_i &= g_i(t, x_1, \ldots, x_n, 0, \ldots, 0) \end{aligned}$$

The results of the next two sections are easily extended to (8) in the regular case.

3 Structure

3.1 Let us restrict to a well-posed first order implicit differential system

$$F_i(t, \dot{y}_1, \ldots, \dot{y}_m, y_1, \ldots, y_m) = w_i, \quad i = 1, \ldots, m \tag{9}$$

3.2 Denote by Ω the (local) exterior algebra (see, e.g., [11]) spanned by all the variables and their time derivatives, i.e., $\{t, y_i^{(\nu_i)}, w_i^{(\mu_i)} \mid i = 1, \ldots, m; \nu_i \geq 0; \mu_i \geq 0\}$. Let Ω_y be the (local) exterior algebra spanned by the variables $y_i = y_i^{(0)}$. We assume constant the ranks of the (local) codistributions used in the sequel.

3.3 Set $m_{-1} = 0$. Let m_0 be the dimension of the codistribution spanned by the 1-forms ω in Ω_y such that

$$\omega \wedge dw_1 \wedge \ldots \wedge dw_m \wedge dt = 0.$$

Clearly, $0 \leq m_0 \leq m$. There are two extreme possibilities:

1) $m_0 = 0$. This is equivalent to the usual (local) explicit form

$$\dot{y}_i = \psi_i(t, y_1, \ldots, y_m, w_1, \ldots, w_m), \quad i = 1, \ldots, m.$$

2) $m_0 = m$. This amounts to saying that the y_i's are (locally) functionally dependent on t and the w_i's:

$$y_i = f_i(t, w_1, \ldots, w_m), \quad i = 1, \ldots, m.$$

(9) degenerates then into a trivial, i.e., non-differential equation.

3.4 Let m_s be the dimension of the codistribution spanned by the 1-forms ω in Ω_y such that

$$\omega \wedge dw_1 \wedge \ldots \wedge dw_1^{(s)} \wedge \ldots \wedge dw_m \wedge \ldots \wedge dw_m^{(s)} \wedge dt = 0$$

Clearly, $0 = m_{-1} \leq m_0 \leq \ldots \leq m_s \leq \ldots \leq m$. There exists a least integer σ_0 such that, for any $s \geq \sigma_0$, $m_s = m_{\sigma_0}$. Then, $\sigma_0 + 1$ is called the *index* of (9). The following lemma gives a most useful upper bound for the index:

LEMMA. *For any $s \not\leq \sigma_0$, $m_s \not\leq m_{s+1}$.*

COROLLARY. $0 \leq \sigma_0 \leq m - 1$.

Notice that in the two extreme cases of section 3.3, the indices are respectively 0 and 1. Notice also that the upper bound $\sigma_0 = m - 1$ is reached, as shown in the following example:

$$\begin{aligned} y_1 &= w_1 \\ \dot{y}_1 + y_2 &= w_2 \\ &\cdots \\ \dot{y}_{m-2} + y_{m-1} &= w_{m-1} \\ y_{m-1} + y_m &= w_m \end{aligned}$$

which is equivalent to $y_1 = w_1, y_2 = w_2 - \dot{w}_1, \ldots, y_m = w_m - \dot{w}_{m-1} + \ldots + (-1)^{m-1} w_1^{(m-1)}$.

3.5 The set $\{y_1, \ldots, y_m\}$ may be divided in two disjoint parts $\{\xi, \eta\}$, where $\xi = \{\xi_1, \ldots, \xi_{m-m_{\sigma_0}}\}$ and $\eta = \{\eta_{m-m_{\sigma_0}+1}, \ldots, \eta_m\}$. Then, locally,

$$\begin{aligned} &\text{for } \alpha = 1, \ldots, m - m_{\sigma_0} \\ &\frac{d\xi_\alpha}{dt} = a_\alpha(t, \xi, w_1, \ldots, w_1^{(\sigma_0+1)}, \ldots, w_m, \ldots, w_m^{(\sigma_0+1)}) \\ &\text{for } \beta = m - m_{\sigma_0} + 1, \ldots, m \\ &\eta_\beta = g_\beta(t, \xi, w_1, \ldots, w_1^{(\sigma_0)}, \ldots, w_m, \ldots, w_m^{(\sigma_0)}). \end{aligned} \tag{10}$$

This yields the

THEOREM. *(9) may be (locally) written as (10), which gives a decomposition into an explicit first order system of differential equations and non-differential equations; (10) expresses the* constraints *on initial conditions.*

3.6 REMARK. (10) may be viewed as a nonlinear and time-varying generalization of the well-known Weierstrass canonical form for matrix pencils (*cf.* [24]).

4 Analytic coefficients and relationship with linear systems

4.1 We assume in this section that the F_i's in (6) and (9) are (local) analytic functions of their arguments.

4.2 Denote by $\hat{d}$ the *vertical differential* (see, e.g., [33]), i.e., the differential which "freezes" t, i.e., $\hat{d}t = 0$. Remember that $\hat{d}\frac{d}{dt} = \frac{d}{dt}\hat{d}$. Apply $\hat{d}$ to (6):

$$\begin{aligned} &\frac{\partial F_i}{\partial y_1}\hat{d}y_1 + \ldots + \frac{\partial F_i}{\partial y_1^{(\alpha_1)}}\hat{d}y_1^{(\alpha_1)} + \ldots \\ &+ \frac{\partial F_i}{\partial y_m}\hat{d}y_m + \ldots + \frac{\partial F_i}{\partial y_m^{(\alpha_m)}}\hat{d}y_m^{(\alpha_m)} = \hat{d}w_i, \\ &i = 1, \ldots, m \end{aligned} \tag{11}$$

4.3 Let K be the *differential field* (*cf.* [30]) generated by the $\frac{\partial F_i}{\partial y_i^{(\beta)}}$'s in (11) and their derivatives of any order with respect to $\frac{d}{dt}$. Then, (11) may be written as

$$(A_0 + A_1 \frac{d}{dt} + \ldots + A_\alpha \frac{d^\alpha}{dt^\alpha}) \begin{pmatrix} \hat{d}y_1 \\ \vdots \\ \hat{d}y_m \end{pmatrix} = \begin{pmatrix} \hat{d}w_1 \\ \vdots \\ \hat{d}w_m \end{pmatrix} \qquad (12)$$

where $\alpha = sup\ (\alpha_1, \ldots, \alpha_m)$ and the entries of the $m \times m$ matrices $A_0, \ldots, A_\alpha$ lie in K. (11)-(12) are called the *linearized equations* of (6). Denote by $K[d/dt]$ the noncommutative principal ideal ring (*cf.* [8]) of linear differential operators of the form $\sum_{finite} c_\gamma \frac{d^\gamma}{dt^\gamma}$; $K[d/dt]$ verifies the *Ore property* and, therefore, possesses a skew field of fractions $K(d/dt)$ [8]. The same arguments as in [23] yield the

THEOREM. *(6) is well-posed, i.e., the F_i's are differentially functionally independent, if, and only if, (11)-(12) is solvable, i.e., if $A_0 + A_1 \frac{d}{dt} + \ldots + A_\alpha \frac{d^\alpha}{dt^\alpha}$ is invertible as a matrix over $K(d/dt)$.*

4.4 Take, now, the linearized equation of (9). Applying, like in [23], techniques from [16] yields the

PROPOSITION. *The indices of (9) and of its linearized equation coincide.*

4.5 REMARK. The two above results permit the verification of the solvability and the calculation of the index *via* manipulations in the noncommutative Euclidean ring $K[d/dt]$ (*cf.* [8]), i.e., by methods of *computer algebra.*

Conclusion

A very similar treatment might be given for implicit control systems, i.e., systems of the form (6) where control variables appear in the F_i's. Let us, however, emphasize that most of the known examples in the literature, like the crane (*cf.* [22]), happen to be *differentially flat* (*cf.* [18, 19, 20]). Control synthesis problems are then easily solved by utilizing the *flat*, or *linearizing*, output (*cf.* [19, 20]).

The case where the F_i's are differentially functionally dependent will be developed elsewhere. It leads to the notion of *gauge degrees of freedom*, which plays a crucial role in physics (*cf.* [29]).

Similar results on implicit differential equations may be obtained *via* differential algebra (see the appendix below for first hints). Besides demonstrating the strong connections between differential algebra and infinite-dimensional differential geometry, this latter method might benefit from algorithmic procedures already utilized in nonlinear control (*cf.* [2, 12, 13, 17, 32]).

Appendix: Differential algebraic setting

Let $k\{y_1, \ldots, y_m, w_1, \ldots, w_m\}$ be the ring of differential polynomials, over the ordinary differential field k of characteristic zero, in the differential k-indeterminates $y_1, \ldots, y_m, w_1, \ldots, w_m$ (see [30] for the terminology and the notations). Let

$$\phi : k\{w_1, \ldots, w_m\} \to k\{y_1, \ldots, y_m\}$$

be the differential morphism defined by $w_i \mapsto F_i$, where $F_1, \ldots, F_m$ are differential polynomials in the indeterminates $y_1, \ldots, y_m$, i.e., elements of $k\{y_1, \ldots, y_m\}$. The differential field extension $k\langle y_1, \ldots, y_m\rangle / k\langle \phi w_1, \ldots, \phi w_m\rangle$ is *differentially algebraic* if, and only if, $\ker \phi = \{0\}$. The F_i's, in other words, must be *differentially algebraically independent* over k: We recover the well-posedness condition of 2.4.

The decomposition and the index are obtained by considering the transcendence degrees of the non-differential field extensions $k(y_1, \ldots, y_m)/k(\phi w_1, \ldots, \phi w_1^{(s)}, \ldots, \phi w_m, \ldots, \phi w_m^{(s)})$, for $s \geq 0$.

References

[1] R.L. Anderson and N.H. Ibragimov. *Lie-Bäcklund Transformations in Applications.* S.I.A.M., Philadelphia, 1979.

[2] F. Boulier. *Étude et implantation de quelques algorithmes en algèbre différentielle.* Thèse de doctorat, Université des Sciences et Technologies de Lille, Villeneuve-d'Ascq, 1994.

[3] K.E. Brenan, S.L. Campbell, and L.R. Petzold. *Numerical Solution of Initial-Value Problems in Differential-Algebraic Equations.* Elsevier, Amsterdam, 1989.

[4] S.L. Campbell. High index differential algebraic equations. *Mechanics Structures Machines*, 23:199–222, 1995.

[5] S.L. Campbell and C.W Gear. The index of general nonlinear DAE's. *Numerische Math.*

[6] S.L. Campbell and E. Griepentrog. Solvability of general differential algebraic equations. *SIAM J. Scientific Computing*, 16, 1995.

[7] S.L. Campbell and E. Moore. Constraint preserving integrations for general higher order DAEs. *Numerische Math.*, 69:383–399, 1995.

[8] P.M. Cohn. *Free Rings and their Relations.* Academic Press, London, second edition, 1985.

[9] E. Delaleau and J. Rudolph. An intrinsic characterization of properness for linear time-varying systems. *J. Math. Systems Estim. Control*, 5:123–126, 1995.

[10] J. Dieudonné. *Éléments d'analyse.* Gauthier-Villars, Paris, 1969. t. 1.

[11] J. Dieudonné. *Éléments d'analyse.* Gauthier-Villars, Paris, 1971. t. 4.

[12] S. Diop. Elimination in control theory. *Math. Control Signals Systems*, 4:17–32, 1991.

[13] S. Diop. Differential-algebraic decision methods and some applications to system theory. *Theoret. Comput. Sci.*, 98:137–161, 1992.

[14] M. Fliess. Automatique et corps différentiels. *Forum Math.*, 1:227–238, 1989.

[15] M. Fliess. Generalized controller canonical forms for linear and nonlinear dynamics. *IEEE Trans. Automat. Control*, 35:994–1001, 1990.

[16] M. Fliess. Some basic structural properties of generalized linear systems. *Systems Control Lett.*, 15:391–396, 1990.

[17] M. Fliess and S. T. Glad. An algebraic approach to linear and nonlinear control. In H. L. Trentelman and J. C. Willems, editors, *Essays on Control: Perspectives in the Theory and its Applications*, pages 223–267, Boston, 1993. Birkhäuser.

[18] M. Fliess, J. Lévine, P. Martin, and P. Rouchon. Linéarisation par bouclage dynamique et transformations de Lie-Bäcklund. *C.R. Acad. Sci. Paris*, I-317:981–986, 1993.

[19] M. Fliess, J. Lévine, P. Martin, and P. Rouchon. Nonlinear control and Lie-Bäcklund transformations: Towards a new differential geometric standpoint. In *Proc. IEEE Decision Control Conf.*, pages 981–986, Lake Buena Vista, FL, 1994.

[20] M. Fliess, J. Lévine, P. Martin, and P. Rouchon. Flatness and defect of nonlinear systems: Introductory theory and applications. *Internat. J. Control*, 1995.

[21] M. Fliess, J. Lévine, P. Martin, and P. Rouchon. Flatness and Lie-Bäcklund mappings: Elementary theory and examples. 1995.

[22] M. Fliess, J. Lévine, and P. Rouchon. A generalised state-variable description for a simplified crane description. *Internat. J. Control*, 58:277–283, 1993.

[23] M. Fliess, J. Lévine, and P. Rouchon. Index of time-varying linear differential equations: A noncommutative linear algebraic approach. *Linear Algebra Appl.*, 186:59–71, 1993.

[24] F.R. Gantmacher. *Théorie des matrices*. Dunod, Paris, 1966. t. 2.

[25] X. Gràcia and J.M. Pons. A generalized geometric framework for constrained systems. *Differential Geometry Appl.*, 2:223–247, 1992.

[26] E. Griepentrog and R. März. *Differential-Algebraic Equations and their Numerical Treatment*. Teubner, Leipzig, 1986.

[27] E. Hairer, C. Lubich, and M. Roche. *The Numerical Solution of Differential-Algebraic Systems by Runge-Kutta Methods*. Springer-Verlag, Berlin, 1989.

[28] R.S. Hamilton. The inverse function theorem of Nash and Moser. *Bull. Amer. Math. Soc.*, 7:65–222, 1982.

[29] M. Henneaux and C. Teitelboim. *Quantization of Gauge Systems*. Princeton University Press, Princeton, 1992.

[30] E.R. Kolchin. *Differential Algebra and Algebraic Groups*. Academic Press, New York, 1973.

[31] I.S. Krasil'shchik, V.V. Lychagin, and A.M. Vinogradov. *Geometry of Jet Spaces and Nonlinear Partial Differential Equations*. Gordon and Breach, New York, 1986.

[32] F. Ollivier. *Le problème de l'identifiabilité : approche théorique, méthodes effectives et bornes de complexité*. Thèse de doctorat, École Polytechnique, Palaiseau, 1990.

[33] P.J. Olver. *Applications of Lie Groups to Differential Equations*. Springer-Verlag, New York, second edition, 1993.

[34] M. Rahula. *New Problems in Differential Geometry*. World Scientific, Singapore, 1993.

[35] R.F. Sincovec, A.M. Erisman, E.L. Yip, and M.A. Epton. Analysis of descriptor systems using numerical algorithms. *IEEE Trans. Automat. Control*, 26:139–147, 1981.

[36] T. Tsujishita. Formal geometry of systems of differential equations. *Sugaku Expos*, 3:25–73, 1990.

[37] A.M. Vinogradov. Local symmetries and conservation laws. *Acta Appl. Math.*, 2:21–78, 1984.

[38] V.V. Zharinov. *Geometrical Aspects of Partial Differential Equations*. World Scientific, Singapore, 1992.

MINIMAX STATISTICAL ESTIMATION PROCEDURES IN INFINITE DIMENSIONAL SPACES

ANDREY V. BORISOV and ALEXEI R.PANKOV

Moscow State Aviation Institute, Department of Applied Mathematics
Volokolamskoje shosse 4, Moscow 125871, Russia CIS.

Résumé. La théorie d'évaluation statistique par la méthode du minimax des vecteurs aléatiores est étendue à un cas d'évaluation des éléments à valeurs dans des espaces hilbertiens séparables. Dans le rapport sont présentées les procédures numériques de calcul des évaluations basées sur les approximations fini- et infinidimensionnelles.

Abstract. The theory of minimax statistic estimation of random vectors is extended to the case of random elements with values in the separable Hilbert spaces. The numerical procedures of estimate calculation, based on both the finite and infinite dimensional approximations are suggested.

Keywords: estimation of random elements; minimax optimality criterion; finite and infinite dimensional approximating estimates.

1 Preliminaries.

In this paper, we shall use the following notations and concepts:

H_i is a separable Hilbert space with the inner product $< x, y >_{H_i}$ and norm

$$||x||_{H_i} = \sqrt{< x, x >_{H_i}};$$

$\mathcal{L}_2(H_i)$ is a Hilbert space of H_i*-valued random elements*(HRE) defined on the probability space $(\Omega, \mathcal{F}, \mathcal{P})$: $\theta \in \mathcal{L}_2(H_i)$ iff $\theta \in H_i$ P–a.s., and

$$||\theta||^2_{\mathcal{L}_2(H_i)} = E\{< \theta, \theta >_{H_i}\} =$$

$$\int_\Omega < \theta(\omega), \theta(\omega) >_{H_i} P(d\omega) < \infty;$$

$\mathcal{B}(H_i, H_j)$ is a space of bounded mappings from H_i into H_j;

$\mathcal{C}(H_i, H_j)$ is a space of continuous mappings from H_i into H_j;

$\mathcal{L}(H_i, H_j)$ is a space of linear bounded operators from H_i into H_j;

$\mathcal{L}_1^+(H_i)$ is a family of linear selfadjoint positive semi-definite nuclear operators from H_i into itself with the norm

$$||A||_{\mathcal{L}_1^+(H_i)} = tr\{A\} = \sum_{n=1}^{\infty} | < e_n, Ae_n >_{H_i} |,$$

where $\{e_n\}$ is an arbitrary basis of H_i;

the condition $A \leq B$ for $A, B \in \mathcal{L}_1^+(H)$ means that $(B - A) \in \mathcal{L}_1^+(H_i)$;

$\mathcal{L}_2(H_i, H_j)$ is a space of Hilbert-Schmidt operators from H_i into H_j;

A^+ is a pseudoinverse of $A \in \mathcal{L}_1^+(H_i)$ (Nashed, 1976):

$\mathcal{N}_{H_i}(m, R)$ is a Gaussian σ-additive measure with expectation m and covariance operator R, defined on the Borel σ-field of H_i;

$cl\{A\}$ is a bounded continuous extension of operator A, densely defined on H_i, to all of H_i;

$\mathcal{D}(A)$ and $\mathcal{R}(A)$ are the domain and the range of linear operator A, resectively.

2 Problem Statement.

Let us consider two HRE $\xi \in \mathcal{L}_2(H_1)$ and $\theta \in \mathcal{L}_2(H_2)$. The element θ should be estimated

This work is supported by INTAS grant No. 94-697

given the observation ξ. We suppose that the expectations $m_\xi = E\{\xi\}$ and $m_\theta = E\{\theta\}$ are known. Let us introduce the covariance operators $R_\xi = cov(\xi,\xi)$ and $R_\theta = cov(\theta,\theta)$ of ξ and θ, respectively, and the mutual one $R_{\theta\xi} = cov(\theta,\xi)$ (Curtain *et al.*, 1978). It is assumed that $R_\xi \in \mathcal{L}_1^+(H_1)$ and $R_\theta \in \mathcal{L}_1^+(H_2)$, hence $R_{\theta\xi} \in \mathcal{L}_2(H_1,H_2)$. Consider a block HRE $\eta = col(\theta,\xi)$, then the moment characteristics of η take the form

$$m_\eta = E\{\eta\} = col(m_\theta, m_\xi) \in H_3;$$

$$R_\eta = cov(\eta,\eta) = \begin{pmatrix} R_\theta & R_{\theta\xi} \\ R_{\theta\xi}^* & R_\xi \end{pmatrix} \in \mathcal{L}_1^+(H_3),$$

where $H_3 = H_2 \otimes H_1$.
The covariance operator R_η is unknown, but the following *a priori* information is available:

$$\underline{R} \le R_\eta \le \overline{R}, \qquad (1)$$

where $\underline{R}, \overline{R} \in \mathcal{L}^+(H_3)$ are given and have the form:

$$\underline{R} = \begin{pmatrix} \underline{R}_\theta & \underline{R}_{\theta\xi} \\ \underline{R}_{\theta\xi}^* & \underline{R}_\xi \end{pmatrix}, \; \overline{R} = \begin{pmatrix} \overline{R}_\theta & \overline{R}_{\theta\xi} \\ \overline{R}_{\theta\xi}^* & \overline{R}_\xi \end{pmatrix}. \quad (2)$$

From (1), (2) it follows, that

$$\underline{R}_\theta \le R_\theta \le \overline{R}_\theta, \quad \underline{R}_\xi \le R_\xi \le \overline{R}_\xi$$

and $\underline{R}_\xi, \overline{R}_\xi \in \mathcal{L}_1^+(H_1)$, $\underline{R}_\theta, \overline{R}_\theta \in \mathcal{L}_1^+(H_2)$, $\underline{R}_{\theta\xi}, \overline{R}_{\theta\xi} \in \mathcal{L}_2(H_1,H_2)$. Due to the condition $R_\eta \in \mathcal{L}_1^+(H_3)$ the HRE η induces a σ-additive measure (distribution) F_η on the Borel σ-field of H_3. Let Ξ_η be a family of all distributions F_η with given expectation $m_\eta \in H_3$ and covariance operator R_η, satisfying (1). It should be noted that $F_\eta = \mathcal{N}_{H_3}(m_\eta, R) \in \Xi_\eta$, if $\underline{R} \le R \le \overline{R}$. The corresponding sets of θ and ξ distributions are denoted by Ξ_θ and Ξ_ξ, respectively. Now we consider the class Θ of admissible estimates of HRE $\theta \in \Xi_\theta$ given the observations $\xi \in \Xi_\xi$.

Definition 1 *The estimate $\hat\theta$ belongs to Θ, if there exists a sequence $\{\psi_n(\cdot)\}_{n=1}^\infty$ of operators such that*

1. $\psi_n(\cdot) \in C(H_1,H_2)$, *and* $\psi_n(\xi) \in \mathcal{L}_2(H_2)$, $n = 1,2,\ldots$ *for any* $\xi \in \Xi_\xi$;
2. $\|\hat\theta - \psi_n(\xi)\|^2_{\mathcal{L}_2(H_2)} \to 0$, *as* $n \to \infty$.

According to Definition 1 we introduce the class Ψ of estimators:

$$\Psi = \{\psi(\cdot) : \; \psi(\xi) = \hat\theta \in \Theta, \; \forall \xi \in \Xi_\xi\}.$$

It should be noted that this class is quite broad and includes both the linear and nonlinear estimators. Let an estimator $\psi(\cdot) \in \Psi$ be fixed. Then we introduce the following optimality criterion for $\psi(\cdot)$:

$$J(F_\eta,\psi) = \|\theta - \psi(\xi)\|^2_{\mathcal{L}_2(H_2)}, \qquad (3)$$

where F_η is an exact distribution of $\eta = col(\theta,\xi)$. If $\underline{R} \ne \overline{R}$ in (1), then the exact value of criterion (3) under fixed $\psi(\cdot)$ is unknown, even if $\psi(\cdot)$ is linear, hence it is more natural to optimize $\psi(\cdot)$ using the minimax approach.

Definition 2 *The estimate $\tilde\theta = \tilde\psi(\xi)$, where $\tilde\psi(\cdot) \in \Psi$, is called the minimax one on the set (Ψ, Ξ_η) in the sense of criterion $J(F_\eta,\psi)$, if for any $\psi(\cdot) \in \Psi$ the inequality*

$$\sup_{F_\eta \in \Xi_\eta} \|\theta - \tilde\psi(\xi)\|^2_{\mathcal{L}_2(H_2)} \le \sup_{F_\eta \in \Xi_\eta} \|\theta - \psi(\xi)\|^2_{\mathcal{L}_2(H_2)} \qquad (4)$$

holds.

From Definition 2 it follows, that $\tilde\psi(\cdot)$ satisfies

$$\tilde\psi(\cdot) = \arg\min_{\psi(\cdot)\in\Psi} \sup_{F_\eta\in\Xi_\eta} J(F_\eta,\psi). \qquad (5)$$

The main difference of presented estimation problem from the traditional linear optimal estimation one (Curtain *et al.*, 1978) is the utilization of minimax optimality criterion and the extended class of admissible estimators. In (Pankov, 1993; Pankov *et al.* 1992, 1994) the corresponding finite dimensional minimax estimation problem has been investigated. In the next section we present the estimators, which satisfy (4) for the general infinite dimensional case.

3 Solution of Minimax Estimation Problem.

3.1 Infinite Dimensional Approximation of Minimax Estimator.

The pair $(\tilde F_\eta, \tilde\psi)$, where $\tilde F_\eta \in \Xi_\eta$, and $\tilde\psi(\cdot) \in \Psi$, is called a saddle point of $J(F_\eta,\psi)$ on (Ψ,Ξ_η), if the inequality

$$J(F_\eta, \tilde\psi) \le J(\tilde F_\eta, \tilde\psi) \le J(\tilde F_\eta, \psi) \qquad (6)$$

holds for all $F_\eta \in \Xi_\eta$ and $\psi(\cdot) \in \Psi$. If the saddle point $(\tilde F_\eta, \tilde\psi)$ exists then $\tilde\psi$ satisfies (5), and $\tilde F_\eta$ satisfies

$$\tilde F_\eta = \arg\max_{F_\eta\in\Xi_\eta} \inf_{\psi\in\Psi} J(F_\eta,\psi). \qquad (7)$$

Conditions (5) and (7) mean that $\tilde F_\eta$ is the least favourable distribution for estimation of θ given ξ, and $\tilde\psi$ is the best estimator, if the exact distribution of η coincides with $\tilde F_\eta$. The next Theorem characterizes a structure of minimax estimate $\tilde\theta = \tilde\psi(\xi)$ and describes a saddle point $(\tilde F_\eta, \tilde\psi)$ on (Ψ, Ξ_η).

Theorem 1 *Let $\eta = col(\theta, \xi)$ have a distribution $F_\eta \in \Xi_\eta$, and the sequence of estimators $\{\psi_n(\xi)\}_{n=1}^{\infty}$ be given by*

$$\psi_n(\xi) = m_\theta + \overline{R}_{\theta\xi}(\overline{R}_\xi + \gamma_n I)^{-1}(\xi - m_\xi), \quad (8)$$

where $\gamma_n \downarrow 0$ as $n \to \infty$. Then

1. *$\exists\, \tilde{\theta} \in \Theta$ such that*

$$\|\tilde{\theta} - \psi_n(\xi)\|^2_{\mathcal{L}_2(H_2)} \to 0$$

as $n \to \infty$ for any $F_\eta \in \Xi_\eta$, and $\tilde{\theta} = \tilde{\psi}(\xi)$, $\tilde{\psi}(\cdot) \in \Psi$;

2. *$\tilde{F}_\eta = \mathcal{N}_{H_3}(m, \overline{R})$, and the pair $(\tilde{F}_\eta, \tilde{\psi})$ satisfies (6) on (Ψ, Ξ_η);*

3. *for criterion $\tilde{J} = J(F_\eta, \tilde{\psi})$ the inequality*

$$\underline{J} \le \tilde{J} \le \overline{J}$$

holds for any F_η from Ξ_η, where

$$\underline{J} = \|\underline{R}_\theta - \overline{R}_{\theta\xi}(\overline{R}_\xi)^+(\underline{R}_{\theta\xi})^* - \underline{R}_{\theta\xi}(\overline{R}_\xi)^+(\overline{R}_{\theta\xi})^* +$$
$$\overline{R}_{\theta\xi}(\overline{R}_\xi)^+(\underline{R}_\xi)(\overline{R}_\xi)^+(\overline{R}_{\theta\xi})^*\|_{\mathcal{L}_1^+(H_2)}$$
$$\overline{J} = \|\overline{R}_\theta - \overline{R}_{\theta\xi}(\overline{R}_\xi)^+(\overline{R}_{\theta\xi})^*\|_{\mathcal{L}_1^+(H_2)}.$$

Proof of Theorem 1 is given in Appendix.
We note that $\underline{J}$ and $\overline{J}$ are the lower and upper bounds for the criterion $\tilde{J} = J(F_\eta, \tilde{\psi})$ with respect to $F_\eta \in \Xi_\eta$, respectively, hence $\overline{J}$ is the guaranteed accuracy of estimate $\tilde{\theta}$.
Remark 1 From Theorem 1 it follows that $\tilde{\theta}$ is the minimax estimate, and the corresponding estimator $\tilde{\psi}$ satisfies (5).
It should be mentioned, that the sequence $\{\gamma_n I\}_{n=1}^{\infty}$ of the regularized terms in (8) can be changed by an arbitrary sequence $\{C_n\}_{n=1}^{\infty}$ of nuclear operators which satisfies conditions $C_n\overline{R}_\xi = \overline{R}_\xi C_n$, $C_n > 0$ and $\|C_n\|_{\mathcal{L}_1^+(H_1)} \to 0$ as $n \to \infty$.
Remark 2 It can be easily shown that if $\overline{R}_{\theta\xi}(\overline{R}_\xi)^+ \in \mathcal{L}(H_1, H_2)$, then the estimate $\tilde{\theta}$ takes the following explicit form:

$$\tilde{\theta} = m_\theta + \overline{R}_{\theta\xi}(\overline{R}_\xi)^+(\xi - m_\xi). \quad (9)$$

The corresponding cases, which are often take place in practice, considered below.

Corollary 1 *If $\dim[\mathcal{R}(\overline{R}_\xi)] < \infty$, then the estimate $\tilde{\theta}$ is determined by the formula (9).*

Remark 3 This case corresponds to the problem of infinite dimensional HRE estimation given the finite dimensional observation vector.

Corollary 2 *If $\dim[\mathcal{R}(\overline{R}_\theta)] < \infty$, then*

$$\tilde{\theta} = m_\theta + G(\xi - m_\xi),$$

where $G = cl\{\overline{R}_{\theta\xi}(\overline{R}_\xi)^+\} \in \mathcal{L}(H_1, H_2)$.

Remark 4 This case corresponds to the problem of finite dimensional random vector estimation given the infinite dimensional observation. In particular, the filtering and smoothing problems in stochastic differential finite dimensional systems belong to this class (Liptser *et al.*, 1978).
Remark 5 In general, the estimate $\tilde{\theta}$ takes the form $\tilde{\theta} = K(\xi) + k$, where $K \in \mathcal{L}(H_1, H_2)$ and $k \in H_2$ is a known nonrandom element. From the assertion of Theorem 1 it follows that the extension of admissible estimator class by the nonlinear ones does not enhence the quality of estimate of θ given ξ.
Let us investigate the sensitivity of minimax estimate with respect to nonexact *prior* knowledge of the moment characteristics of esimated and observed HRE.
Let the expectation m_η and covariance operator R_η of partially observable HRE $\eta \in H_3$ be unknown, and there exists a sequence $\{\eta^n\}_{n=1}^{\infty}$ of the HRE η approximations, i.e. $\eta^n \in \mathcal{L}_2(H_3)$, and $l.i.m._{n\to\infty}\eta^n = \eta$. The moment characteristics of this sequence are known: $E\{\eta^n\} = m_\eta^n$, $cov(\eta^n, \eta^n) = R_\eta^n = \underline{R}_\eta^n = \overline{R}_\eta^n$.
The problem is to find a minimax estimate of HRE θ given ξ, using *a priori* information concerning the characteristics of sequence $\{\eta^n\}_{n=1}^{\infty}$ only, i.e. the admissible estimators $\{\psi_n\}_{n=1}^{\infty}$ can use only m_η^n and R_η^n as parameters, not m_η and R_η.

Theorem 2 *Let the sequence of positive real numbers $\{\gamma_n^*\}_{n=1}^{\infty}$ tends to zero as $n \to \infty$, and for all integer $n > 0$*

$$\sup_{m \ge n} \|\eta - \eta^m\|^2_{\mathcal{L}_2(H_3)} \le (\gamma_n^*)^2/4, \quad (10)$$

then the minimax estimate $\tilde{\theta}$ of HRE θ given the observation ξ, takes the form:

$$\tilde{\theta} = l.i.m._{n\to\infty}\left(m_\theta^n + R_{\theta\xi}^n(R_\xi^n + \gamma_n^* I)^{-1}(\xi - m_\xi^n)\right).$$

Proof of Theorem 2 is given in Appendix.

3.2 Finite Dimensional Approximations of Minimax Estimator.

Suppose, we have a sequence of *finite dimensional operators* $\{\psi_n(\cdot)\}$, i.e.

$$\dim[\mathcal{R}(\psi_n(\cdot))] < \infty$$

for each n, which produces a convergent sequence of estimates $\theta_n = \psi_n(\xi) : \|\tilde{\theta} - \theta_n\|_{\mathcal{L}_2(H_2)} \to 0$ as $n \to \infty$. It means that in this case we obtain a numerical algorithm of the finite dimensional approximation of the minimax estimation problem, and $\|\theta - \theta_n\|_{\mathcal{L}_2(H_2)} \to \|\theta - \tilde{\theta}\|_{\mathcal{L}_2(H_2)}$. In this subsection we present two types of finite dimensional approximations, obtained by either the projection of observation or estimated vector on the sequence

of finite dimensional subspaces.
To construct the first type of estimators, let us consider the basis $\{e_i\}_{i=1}^{\infty}$ of operator $\overline{R}_\xi$ eigenvectors. $\{P_n^1\}_{n=1}^{\infty}$ is an increasing sequence of projectors on the finite dimensional subspaces, induced by the subsets $\{e_i\}_{i=1}^{n}$. We construct the first sequence as follows:

$$\psi_n^1(\xi) = m_\theta + \overline{R}_{\theta\xi}(\overline{R}_\xi)^+ P_n^1(\xi - m_\xi). \qquad (11)$$

Note, that $\overline{R}_{\theta\xi}(\overline{R}_\xi)^+ P_n^1 \in \mathcal{L}(H_1, H_2)$ for each $n < \infty$.
Now consider the second type of estimators. Let us choose an arbitrary basis $\{g_i\}_{i=1}^{\infty}$ of H_2. Let $\{P_n^2\}_{n=1}^{\infty}$ be an increasing sequence of projectors on the finite dimensional subspaces, induced by the subsets $\{g_i\}_{i=1}^{n}$, $n = 1, 2, \ldots$. Now we can construct the second sequence of estimates

$$\begin{array}{c} \psi_n^2(\xi) = m_\theta + G_n(\xi - m_\xi), \\ G_n = cl\left\{P_n^2 \overline{R}_{\theta\xi}(\overline{R}_\xi)^+\right\}. \end{array} \qquad (12)$$

For each fixed $n < \infty$ the operator $G_n \in \mathcal{L}(H_1, H_2)$, since $P_n^2\overline{R}_{\theta\xi}(\overline{R}_\xi)^+$ is finite dimensional. Note, that the sequences $\{P_n^1\}$ and $\{P_n^2\}$ converge strongly to identity as $n \to \infty$.

Theorem 3 *Let $\eta = col(\theta, \xi)$ have some distribution $F_\eta \in \Xi_\eta$, and the sequence of estimates $\{\psi_n(\xi)\}_{n=1}^{\infty}$, where $\psi_n(\xi) = \psi_n^1(\xi)$ (11) or $\psi_n(\xi) = \psi_n^2(\xi)$ (12), be given. Then $l.i.m._{n\to\infty}\psi_n(\xi) = \tilde{\theta}$, where the estimate $\tilde{\theta}$ is defined in Theorem 1.*

Proof of Theorem 3 is given in Appendix.
Remark 6 From Theorem 3 it follows that the distribution $\tilde{F}_\eta$ satisfies (7), i.e. is the least favourable one within Ξ_η for estimation of θ in the sense of (3), and $\tilde{\theta} = E\{\theta|\mathcal{F}^\xi\}$ in the case $F_\eta = \tilde{F}_\eta$, where $\mathcal{F}^\xi$ is a σ-algebra, induced by the HRE ξ.
Remark 7 Consider the sequence of estimates $\{\psi_n^3(\xi)\}_{n=1}^{\infty}$:

$$\psi_n^3(\xi) = m_\theta + \overline{R}_{\theta\xi}(\overline{R}_\xi)^+ T_n(\xi - m_\xi), \qquad (13)$$

where $\{T_n\}_{n=1}^{\infty}$ is an arbitrary sequence of projectors, which strongly converges to the identity as $n \to \infty$ on H_1. It can be shown that

$$l.i.m._{n\to\infty}\psi_n^3(\xi) = \tilde{\theta},$$

if the sequence $\{\|\psi_n^3(\xi)\|_{H_2}^2\}_{n=1}^{\infty}$ is uniformly integrable.

4 Conclusion.

The obtained results in the field of minimax estimation of random elements in the separable Hilbert spaces provide also a posibility to solve various estimation and control problems for the generalized dynamic systems such as minimax filtering for the processes defined by the Ito-Volterra equations, filtering and generalized control for the distributed stochastic differential systems, conditionally-minimax nonlinear filtering and control.

5 Appendix.

5.1 Proof of Theorem 1.

1. First, we prove that $\overline{R}_{\theta\xi}\overline{R}_\xi^+\overline{R}_{\theta\xi}^* \in \mathcal{L}_1^+(H_2)$. Let $\{e_j^1\}_{j=1}^{\infty}$ - be the basis of eivenvectors of H_1, corresponding to the operator $\overline{R}_\xi$, and $\{e_i^2\}_{i=1}^{\infty}$ be an arbitrary basis of H_2. Note that $\forall v \in H_1$ and $\forall y \in \mathcal{D}(\overline{R}_\xi^+)$

$$\overline{R}_\xi v = \sum_{j=1}^{\infty} \lambda_j P_{e_j^1} v, \qquad \overline{R}_\xi^+ y = \sum_{j=1}^{\infty} \lambda_j^+ P_{e_j^1} y,$$

where $P_{e_j^1}$ are projectors to the one dimensional subspaces induced by e_j^1, and $\lambda_j \geq 0$ are the corresponding eigenvalues; $\lambda_j^+ = 1/\lambda_j$, if $\lambda_j > 0$, and $\lambda_j^+ = 0$ otherwise. Let $\{P_m^1\}_{m=1}^{\infty}$ be the sequence of projectors to the linear subspaces induced by the sets $\{e_j^1\}_{j=1}^{m}$. Obviously, this sequence converges strongly to the identity on H_1 as $n \to \infty$. Consider a sequence of operators $\{A_n\}_{n=1}^{\infty}$, $A_n \in \mathcal{L}(H_3, H_2)$ defined as follows: if $z = col(u, w) \in H_3$ is nonrandom element of space H_3, $u \in H_2$ and $w \in H_1$ then

$$A_n z = u - \overline{R}_{\theta\xi}\overline{R}_\xi^+ P_n^1 w.$$

Clearly,

$$A_n \overline{R}_\eta A_n^* = \overline{R}_\theta - \overline{R}_{\theta\xi}\overline{R}_\xi^+ P_n^1 \overline{R}_{\theta\xi}^* \in \mathcal{L}_1^+(H_2),$$

and

$$S_n = \|A_n\overline{R}_\eta A_n^*\|_{\mathcal{L}_1^+(H_2)} =$$

$$\sum_{i=1}^{\infty} < e_i^2, \overline{R}_\theta e_i^2 >_{H_2} - \sum_{j=1}^{n} \lambda_j^+ \|\overline{R}_{\theta\xi} e_j^1\|_{H_2}^2.$$

The sequence $\{S_n\}_{n=1}^{\infty}$ is nonincreasing and bounded from below, hence there exists a limit

$$S = \lim_{n\to\infty} S_n.$$

This means,

$$\sum_{j=1}^{\infty} \lambda_j^+ \|\overline{R}_{\theta\xi} e_j^1\|_{H_2}^2 = \qquad (14)$$

$$\|\overline{R}_{\theta\xi}\overline{R}_\xi^+\overline{R}_{\theta\xi}^*\|_{\mathcal{L}_1^+(H_2)} =$$

$$\|\overline{R}_\theta\|_{\mathcal{L}_1^+(H_2)} - S \leq \|\overline{R}_\theta\|_{\mathcal{L}_1^+(H_2)} < \infty.$$

Note that $\forall F_\eta \in \Xi_\eta$

$$\|\overline{R}_{\theta\xi}\overline{R}_\xi^+ cov(\xi, \xi)\overline{R}_\xi^+\overline{R}_{\theta\xi}^*\|_{\mathcal{L}_1^+(H_2)} \leq$$

$$\|\overline{R}_{\theta\xi}\overline{R}_\xi^+\overline{R}_{\theta\xi}^*\|_{\mathcal{L}_1^+(H_2)} \leq \|\overline{R}_\theta\|_{\mathcal{L}_1^+(H_2)} < \infty.$$

Further, consider the sequence of estimates

$$\bar{\psi}^m(\xi) = m_\theta + \overline{R}_{\theta\xi}\overline{R}_\xi^+ P_m^1(\xi - m_\xi) =$$

$$m_\theta + \overline{R}_{\theta\xi}(P_m^1 \overline{R}_\xi P_m^1)^+ P_m^1(\xi - m_\xi) = \quad (15)$$

$$m_\theta + \sum_{j=1}^{m} < \xi - m_\xi, e_j^1 >_{H_1} \lambda_j^+ \overline{R}_{\theta\xi} e_j^1,$$

and prove that $\{\bar{\psi}^m(\xi)\}_{m=1}^\infty$ is the Cauchy sequence. Let $n > m$, then

$$\|\bar{\psi}^n(\xi) - \bar{\psi}^m(\xi)\|^2_{\mathcal{L}_2(H_2)} =$$

$$\sum_{j=m+1}^{n} \| < \xi - m_\xi, e_j^1 >_{H_1} \lambda_j^+ \overline{R}_{\theta\xi} e_j^1 \|^2_{\mathcal{L}_2(H_2)} =$$

$$\|\overline{R}_{\theta\xi} \overline{R}_\xi^+ (P_n^1 - P_m^1) cov(\xi,\xi) \times$$

$$(P_n^1 - P_m^1) \overline{R}_\xi^+ \overline{R}_{\theta\xi}^* \|_{\mathcal{L}_1^+(H_2)} \leq$$

$$\|\overline{R}_{\theta\xi} \overline{R}_\xi^+ (P_n^1 - P_m^1) \overline{R}_{\theta\xi}^* \|_{\mathcal{L}_1^+(H_2)} =$$

$$\sum_{j=m+1}^{n} \lambda_j^+ \|\overline{R}_{\theta\xi} e_j^1\|^2_{H_2} \leq \sum_{j=m+1}^{\infty} \lambda_j^+ \|\overline{R}_{\theta\xi} e_j^1\|^2_{H_2} \to 0$$

as $m \to \infty$, since the serie (14) is convergent. Thus, there exists the limit

$$\tilde{\theta} = l.i.m._{m\to\infty} \bar{\psi}^m(\xi) =$$

$$m_\theta + l.i.m._{m\to\infty} \sum_{j=1}^{m} < \xi - m_\xi, e_j^1 >_{H_1} \lambda_j^+ \overline{R}_{\theta\xi} e_j^1.$$

Now we prove that the sequence $\{\psi_n^1(\xi)\}_{n=1}^\infty$ (8) converges to the same limit $\tilde{\theta}$. It should be noted that

$$\psi_n^1(\xi) = m_\theta + \overline{R}_{\theta\xi}(\overline{R}_\xi + \gamma_n I)^{-1}(\xi - m_\xi) =$$

$$m_\theta + l.i.m._{p\to\infty} \sum_{j=1}^{p} < \xi - m_\xi, e_j^1 >_{H_1} \times$$

$$(\lambda_j + \gamma_n)^{-1} \overline{R}_{\theta\xi} e_j^1.$$

Denote $B_n = \overline{R}_{\theta\xi}\left(\overline{R}_\xi^+ - (\overline{R}_\xi + \gamma_n I)^{-1}\right)$, then

$$\|\tilde{\theta} - \psi_n^1(\xi)\|^2_{\mathcal{L}_2(H_2)} =$$

$$l.i.m._{m\to\infty} \sum_{j=1}^{m} \| < \xi - m_\xi, e_j^1 >_{H_1} \times$$

$$\left(\lambda_j^+ - (\lambda_j + \gamma_n)^{-1}\right) \overline{R}_{\theta\xi} e_j^1 \|^2_{\mathcal{L}_2(H_2)} =$$

$$\sum_{i,j=1}^{\infty} [< \overline{R}_{\theta\xi} e_j^1, \overline{R}_{\theta\xi} e_i^1 >_{H_2} \times$$

$$(\lambda_j^+ - (\lambda_j + \gamma_n)^{-1})(\lambda_i^+ - (\lambda_i + \gamma_n)^{-1}) \times$$

$$cov(< \xi - m_\xi, e_j^1 >_{H_1}, < \xi - m_\xi, e_i^1 >_{H_1})] =$$

$$\|B_n cov(\xi,\xi) B_n^* \|_{\mathcal{L}_1^+(H_2)} \leq$$

$$\|B_n \overline{R}_\xi B_n^*\|_{\mathcal{L}_1^+(H_2)} = \sum_{j=1, \lambda_j \neq 0}^{\infty} \phi_j(\gamma_n) \lambda_j^+ \|\overline{R}_{\theta\xi} e_j^1\|^2_{H_2},$$

where $\phi_j(\gamma_n) = \frac{\gamma_n^2}{(\lambda_j + \gamma_n)^2}$.
Without loss of generality we assume that the eigenvalues λ_j are in descending order, i.e. $\lambda_j \downarrow 0$ as $j \to \infty$, and $\gamma_n \in [0,1]$ for all n. Further, for any $\gamma_n \in [0,1]$ the sequence $\{\phi_j(\gamma_n)\}_{j=1}^\infty$ is monotonic and uniformly bounded ($\phi_j(\gamma_n) \leq \phi_k(\gamma_n) \leq 1$ as $k \leq j$). Due to convergence of the serie (14) we have

$$\sum_{j=1}^{N} \phi_j(\gamma_n) \lambda_j^+ \|\overline{R}_{\theta\xi} e_j^1\|^2_{H_2} \to V(\gamma_n) < \infty$$

as $N \to \infty$, and this convergence is uniform with respect to $\gamma_n \in [0,1]$. So,

$$\|\tilde{\theta} - \psi_n^1(\xi)\|^2_{\mathcal{L}_2(H_2)} = \lim_{\gamma_n \downarrow 0} V(\gamma_n) =$$

$$\sum_{j=1}^{N} \lambda_j^+ \|\overline{R}_{\theta\xi} e_j^1\|^2_{H_2} \lim_{\gamma_n \downarrow 0} \phi_j(\gamma_n) = 0.$$

Finally, $l.i.m._{n\to\infty} \psi_n^1(\xi) = \tilde{\theta}$.
2. Now we prove that the pair $(\tilde{F}_\eta, \tilde{\psi})$ is a saddle point of criterion $J(F_\eta, \psi)$, i.e. the condition (6) is true.
Let the HRE η have the distribution $F_\eta = \tilde{F}_\eta = \mathcal{N}_{H_3}(m_\eta, \overline{R}_\eta)$. Now we remind the sequence of estimates $\{\bar{\psi}^m(\xi)\}_{m=1}^\infty$ (15). Denote $\xi^m = P_m^1 \xi$, then the following inclusions for the σ-algebras are true: $\mathcal{F}^{\xi^m} \subseteq \mathcal{F}^{\xi^p}$ as $m \leq p$, and $\mathcal{F}^\xi = \bigcup_{m=1}^\infty \mathcal{F}^{\xi^m}$. Note that for all $m < \infty$ the elements ξ^m are finite dimensional vectors, and the mutual distribution of θ and ξ^m is Gaussian. Then the optimal estimate of θ can by obtained by the theorem on normal correlation (Liptser *et al.*, 1978), and the conditional expectation coincides with $\bar{\psi}^m(\xi)$ P-a.s.:

$$\bar{\psi}^m(\xi) = E\{\theta | \mathcal{F}^{\xi^m}\}.$$

By the Levi theorem (Liptser *et al.*, 1978) the following equality is true P-a.s.:

$$\lim_{m\to\infty} \bar{\psi}^m(\xi) = \lim_{m\to\infty} E\{\theta|\mathcal{F}^{\xi^m}\} = E\{\theta|\mathcal{F}^\xi\}.$$

Consequently, the sequence $\{\bar{\psi}^m(\xi)\}_{m=1}^\infty$ converges both as surely and in mean square sense, hence both these limits should coincide P-a.s.:

$$\tilde{\theta} = E\{\theta|\mathcal{F}^\xi\}.$$

The right inequality (6) follows directly from the properties of conditional expectation. Obviously, the estimate $\tilde{\theta}$ is a mean square limit of *linear* estimates, hence the performance index $J(F_\eta, \tilde{\psi})$ depends only on the covariance operator $R_\eta = cov(\eta, \eta)$, and this dependence is linear one:

$$J(F_\eta, \tilde{\psi}) = tr\{DR_\eta D^*\} \leq tr\{D\overline{R}_\eta D^*\} = J(\tilde{F}_\eta, \tilde{\psi}), \quad (16)$$

where the densely defined operator $D : H_3 \to H_2$ is determined as follows: for any $u = col(w, v)$: $w \in H_2$, $v \in \mathcal{D}\{\overline{R}_\xi^+\}$

$$Du = w - \overline{R}_{\theta\xi} \overline{R}_\xi^+ v$$

Thus, the left inequality (6) is also valid, i.e. the pair $(\tilde{F}_\eta, \tilde{\psi})$ is a saddle point of criterion $J(F_\eta, \psi)$.
3. The bounds $\underline{J}$ and $\overline{J}$ can be directly verified using (16) .

5.2 Proof of Theorem 2.

Since $||\eta^n - \eta||^2_{\mathcal{L}_2(H_3)} \to 0$ as $n \to \infty$, for an arbitrary small $\epsilon > 0$ the inequality

$$||\eta^n - \eta||^2_{\mathcal{L}_2(H_3)} = ||\delta^n_\eta||^2_{\mathcal{L}_2(H_3)} < \epsilon$$

is valid, beginning from some integer $N = N(\epsilon)$, i.e. $\forall n \leq N(\epsilon)$

$$||m_{\eta^n} - m_\eta||^2_{H_3} = ||\delta^n_{m_\eta}||^2_{H_3} < \epsilon$$

and

$$||R_{\eta^n} - R_\eta||_{\mathcal{L}^+_1(H_2)} = ||\delta^n_{R_\eta}||_{\mathcal{L}^+_1(H_2)} < \epsilon.$$

Without loss of generality we suppose that $\epsilon < 1$. Clearly,

$$||\delta^n_{R_\xi}||_{\mathcal{L}(H_2,H_2)} \leq ||\delta^n_\eta||^2_{\mathcal{L}_2(H_3)} < \epsilon,$$
$$||\delta^n_{R_{\xi\theta}}||_{\mathcal{L}(H_1,H_2)} \leq ||\delta^n_\eta||^2_{\mathcal{L}_2(H_3)} < \epsilon,$$
$$||\delta^n_{m_\xi}||^2_{H_1} \leq ||\delta^n_{m_\eta}||^2_{H_3} < \epsilon,$$
$$||\delta^n_{m_\theta}||^2_{H_2} \leq ||\delta^n_{m_\eta}||^2_{H_3} < \epsilon.$$

Let the sequence $\{\gamma^*_n\}^\infty_{n=1}$ satisfy the condition (10), then

$$\tilde{\theta}^n = m^n_\theta + R^n_{\theta\xi}(R^n_\xi + \gamma^*_n I)^{-1}(\xi - m^n_\xi).$$

Using the Taylor expansion for the operator $(R^n_\xi + \gamma^*_n I)^{-1}$, we obtain

$$\tilde{\theta}^n = [m_\theta + R_{\theta\xi}(R_\xi + \gamma^*_n I)^{-1}\times (\xi - m_\xi)] + \Delta^n, \quad (17)$$

where

$$\Delta^n = \delta^n_{m_\theta} - R_{\theta\xi}(R_\xi + \gamma^*_n I)^{-1}\delta^n_{m_\xi} - R_{\theta\xi}\delta^n_{R_\xi}(R^n_\xi + \gamma^*_n I)^{-1}(\xi - m^n_\xi) + \delta^n_{R_{\theta\xi}}(R^n_\xi + \gamma^*_n I)^{-1}(\xi - m^n_\xi).$$

By Theorem 1 the term in the square brackets in (17) tends to the minimax estimate $\tilde{\theta}$ in the mean square sense. Further, $\lambda_{min}(R^n_\xi + \gamma^*_n I) \geq 2\sqrt{\epsilon} - \epsilon \geq \sqrt{\epsilon}$, and $\lambda_{min}(R_\xi + \gamma^*_n I) \geq 2\sqrt{\epsilon}$, i.e.

$$||(R^n_\xi + \gamma^*_n I)^{-1}||_{\mathcal{L}(H_1,H_1)} \leq 1/\sqrt{\epsilon},$$
$$||(R_\xi + \gamma^*_n I)^{-1}||_{\mathcal{L}(H_1,H_1)} \leq 1/(2\sqrt{\epsilon}),$$

and there exists a constant $0 < M < \infty$, such that $||\xi - m^n_\xi||^2_{\mathcal{L}_2(H_1)} \leq M$, $\forall n$. Then

$$||\Delta^n||^2_{\mathcal{L}_2(H_2)} \leq \left[\sqrt{\epsilon} + ||R_{\theta\xi}||_{\mathcal{L}(H_1,H_2)}(\sqrt{\epsilon}/2 + \sqrt{M\epsilon}) + \sqrt{M\epsilon}\right]^2$$

for all $n > N(\epsilon)$. Due to arbitrariness of ϵ we conclude that $||\Delta^n||^2_{\mathcal{L}_2(H_2)} \to 0$ as $n \to \infty$, i.e. $||\tilde{\theta}^n - \tilde{\theta}||^2_{\mathcal{L}_2(H_2)} \to 0$.
Theorem 2 is proved.

5.3 Proof of Theorem 3.

The assertion of Theorem 3 concerning the estimate $\theta^1 = l.i.m._{k\to\infty}\psi^1_k(\xi)$ follows from the proof of Theorem 1, because $\tilde{\psi}^k(\xi) = \psi^1_k(\xi)$.
Further,

$$\psi^2_n(\xi) = m_\theta + cl\{P^2_n\overline{R}_{\theta\xi}\overline{R^+_\xi}\}(\xi - m_\xi) =$$
$$m_\theta + l.i.m._{k\to\infty}P^2_n\overline{R}_{\theta\xi}\overline{R^+_\xi}P^1_k(\xi - m_\xi) =$$
$$(I - P^2_n)m_\theta + P^2_n l.i.m._{k\to\infty}\tilde{\psi}^k(\xi) =$$
$$(I - P^2_n)m_\theta + P^2_n\tilde{\theta}.$$

Thus,

$$l.i.m._{n\to\infty}\psi^3_n(\xi) = l.i.m._{n\to\infty}\left((I - P^2_n)m_\theta + P^2_n\tilde{\theta}\right) = \tilde{\theta}.$$

Theorem 3 is proved.

6 References

1. Curtain, R.F., and A.J. Pritchard (1978) *Infinite Dimensional Linear System Theory,* Springer-Verlag, New York.

2. Liptser, R.Sh., and A.N. Shiryayev (1978) *Statistics of Random Proccesses,* Springer-Verlag, New York.

3. Nashed Z., ed. (1976) *Generalized Inverses and Its Applications,* Academic Press, New York.

4. Pankov A.R. (1993) Conditionally - Minimax Nonlinear Filter for Differential System with Discrete Observations. *Advances in Modelling and Analysis,* **B, 28,** 31-39.

5. Pankov A.R., Bosov A.V., and A.V. Borisov (1992) Finite Dimensional Algorithms of Nonlinear System State Estimation, Preprint TR-92-13, University of Michigan, Ann Arbor, USA.

6. Pankov A.R., and A.V. Bosov (1994) Conditionally-Minimax Algorithm of Nonlinear System State Estimation. *IEEE Transactions on Automatic Control,* **AC-39,** 1617-1620.

W-STABILITY AND LOCAL SMALL GAIN THEOREM

H. BOURLÈS*, and F. COLLEDANI**

*Electricité de France, Direction des Etudes et Recherches, 1, Avenue du Général de Gaulle, 92141 Clamart, France
**Ecole Supérieure d'Electricité, Plateau du Moulon, 92192 Gif sur Yvette, France

Résumé. Une nouvelle sorte de stabilité entrée-sortie est définie sur la base d'un espace de Sobolev. Des versions locales du théorème des petits gains et du théorème de passivité sont établies dans ce formalisme.

Abstract. A new type of input-output stability is defined, based on the use of a Sobolev space. A local version of the Small Gain Theorem is established in this framework, as well as a local version of the Passivity Theorem.

Key Words. Local small gain; Local passivity: Sobolev space; Input-output stablity; Asymptotic stability

1. INTRODUCTION

As is well known, two different types of stability can be considered for a system (Vidyasagar, 1993). The first one is the "internal stability" (*e.g.*, asymptotic stability); the input of the system is then assumed to be zero, and, roughly speaking, one looks if the system state tends to zero from a non zero initial value; the basic tool to analyze this type of stability is the use of Lyapunov functions. This approach is well suited to obtain global or local stability results: the internal stability is global when (roughly speaking again) the state tends to zero from any initial value, and is local when this occurs for "sufficiently small" initial values of the state.

The second main type of stability is the "input-output stability". In this framework, the initial value of the state is assumed to be zero, so that the system output is a function of the input[1] ; and one looks if the output belongs to a specific function space (e.g. a Lebesgue space L_p, $1 \leq p \leq \infty$) when the input belongs to the same space; moreover, one looks if the norm of the output (in this function space) is smaller than some constant multiplied by the norm of the input. The greatest lower bound of the constants satisfying this property is called the gain of the system. The case p=2 is particularly interesting, because the L_2-gain of any linear time-invariant (LTI) system is the H_∞-norm of its transfer matrix [where H_∞ denotes the well known Hardy space (Francis, 1987)]; moreover, as the Fourier transform is an isomorphism of L_2, this space is well suited to lead to stability characterizations in the time domain as well as in the frequency domain. This approach was limited by the fact that only global stability results are available in the literature; the very aim of this paper is to establish local results in a (more or less) equivalent framework. Note that the internal and the input-output stability are closely related in the case of reachable and uniformly observable systems (Vidyasagar, 1993; Hill and Moylan, 1982; Vidyasagar and Vannelli, 1982).

A new input-output approach, leading to global as well as local stability results has recently been developed: this is the so-called "input-to-state stability" (Sontag, 1989, 1990; Pan *et al.*, 1993). In this framework, (though it seems promising), characterizations of the stability in the frequency domain cannot be obtained, because the function space which is considered is L_∞. Moreover, as L_∞ is not an inner product space, notions such as passivity cannot be defined in this context.

The function space used in this paper is the set of functions x such that x and its derivative belong to L_2. This is the well known Sobolev space $W_{1,2}$ (Trèves, 1967). This space is interesting for several reasons: on the one hand, it will be shown that all nice properties of L_2 are still satisfied by this space; in particular, the gain of any LTI system remains the H_∞-norm of its transfer matrix. On the other hand, if a function x belongs to $W_{1,2}$ and is smooth enough [more precisely, if it is absolutely continuous; see, e.g., (Alexeev *et al.*, 1982)], then x belongs to L_∞, and its norm in this latter space is upper bounded by its norm in $W_{1,2}$ if the initial value of x is zero. This is the key property which will enable us to obtain simple

[1] In what follows, this function is called the input-output operator associated to the system (Bourlès and Aïoun, 1994).

conditions for nonlinear input-output operators to be "locally stable" (in a sense precised below).

The paper is organized as follows: in Section 2, various signal spaces are defined, and their relations are clarified. In Section 3, "local W-stability" is defined, and then studied in various cases of systems. The notion of "local W-gain" is defined; the local-W-gain of a system G is denoted as $\gamma_{Wl}(G)$. The relationship between W-stability and asymptotic stability is studied in Section 4. In Section 5, a local version of the Small Gain Theorem is established. According to this theorem, if Systems G_1 and G_2 in Fig.1 are locally W-stable, and if $\gamma_{Wl}(G_1)\,\gamma_{Wl}(G_2) < 1$, then the closed loop system in Fig.1 is locally W-stable.

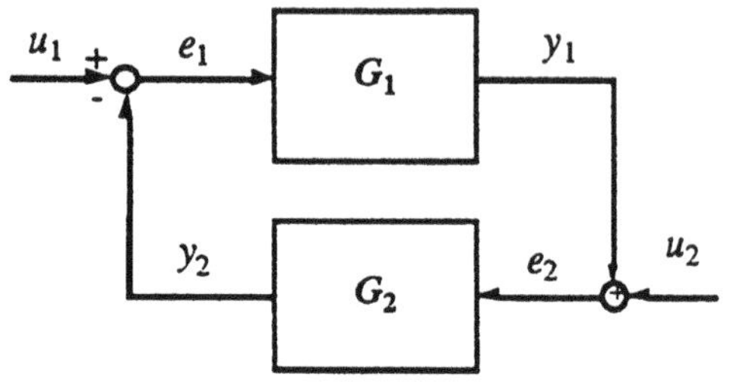

Fig. 1. Standard closed loop system

In Section 6, as a consequence of this theorem, a local version of the Passivity Theorem is obtained. Due to the lack of place, most of results are stated without proof. The complete proofs will be found in (Bourlès, 1995).

2. SIGNAL SPACES AND THEIR RELATIONS

2.1. Notation

The norm of any vector ξ in $\mathbb{R}^n$ or $\mathbb{C}^n$ is denoted as $\|\xi\|$. L_2^n denotes the usual Lebesgue-space of functions $x : \mathbb{R}^+ \to \mathbb{R}^n$, which are Lebesgue-measurable and square-integrable; $<x,y>_2$ denotes the inner product of two functions x and y in L_2^n, and we set $\|x\|_2 = [<x,x>_2]^{1/2}$. For any function $x : \mathbb{R}^+ \to \mathbb{R}^n$, x_T denotes the truncation of x to the interval [0,T], i.e. the function such that $x_T(t) = x(t)$ if $t \le T$, and $x_T(t) = 0$ if $t > T$; L_{2e}^n denotes the "extended L_2^n-space", i.e. the space of functions x such that for any T in $\mathbb{R}^+$, x_T belongs to L_2^n. For any functions x and y belonging to L_{2e}^n, we set $<x,y>_{2,T} = <x_T,y_T>_2$ and $\|x\|_{2,T} = \|x_T\|_2$. In the same manner, L_∞^n denotes the space of functions $x : \mathbb{R}^+ \to \mathbb{R}^n$, which are Lebesgue-measurable and essentially bounded; the norm in this space is defined as $\|x\|_\infty = \text{ess.sup}\ \{\|x(t)\|;\ t \in \mathbb{R}^+\}$. The "extended L_∞^n-space", $L_{\infty e}^n$, is the space of functions x such that for any T in $\mathbb{R}^+$, x_T belongs to L_∞^n and $\|x\|_{\infty,T}$ denotes the quantity $\|x_T\|_\infty$. $W_{1,2}^n$ denotes the Sobolev space of functions $x : \mathbb{R}^+ \to \mathbb{R}^n$ such that x and its distributional-derivative $\dot{x}$ belong to L_2^n. The inner product of two functions x and y in $W_{1,2}^n$ is defined as $<x,y>_W = <x,y>_2 + <\dot{x},\dot{y}>_2$ and the norm of x in $W_{1,2}^n$ is defined by $\|x\|_W = [<x,x>_W]^{1/2}$. Obviously, $W_{1,2}^n$ is included in L_2^n and for any function x in $W_{1,2}^n$ one has $\|x\|_2 \le \|x\|_W$. AC^n denotes the set of functions $x : \mathbb{R}^+ \to \mathbb{R}^n$ which are absolutely continuous.

2.2. Definitions and useful properties

By Lebesgue's theorem, if x belongs to AC^n, then x is almost everywhere differentiable (in the usual sense), and its derivative $\dot{x}$ is integrable on any bounded interval [a,b] included in $\mathbb{R}^+$; moreover, one has $\int_a^b \dot{x}(\tau)\,d\tau = x(b) - x(a)$. The following facts are well known (Alexeev *et al.*, 1982) : AC^n is a vector space; if x and y belong to AC^n, then the function $t \to x(t)^T y(t)$ belongs to AC^1. Let be a function $\Phi: \Omega \to \mathbb{R}^m$, where Ω denotes some open subset of $\mathbb{R}^n$; if Φ is Lipschitz-continuous and if x belongs to AC^n and takes its values in Ω, then the function $t \to \Phi(x(t))$ belongs to AC^m.

Definition 1. W^n is the space of functions x belonging to $W_{1,2}^n \cap AC^n$ and such that $x(0)=0$; W_e^n is the space of functions x belonging to AC^n and such that $x(0)=0$ and for any finite $T>0$, x_T and $(\dot{x})_T$ belong to L_2^n. For any functions x and y belonging to W_e^n, we set $<x,y>_{W,T} = <x,y>_{2,T} + <\dot{x}, \dot{y}>_{2,T}$ and $\|x\|_{W,T} = [<x,x>_{W,T}]^{1/2}$. Obviously, $\|.\|_W$ is a norm on W^n, whereas $(\|.\|_{W,T})_{T\ge 0}$ is a family of semi-norms on W_e^n. (These spaces can be replaced by their completions, and then become a Hilbert space and a Fréchet space, respectively).

Proposition 1. Let $x \in W_e^n$; then for any finite $T>0$ the following inequality holds :

$$\|x\|_{\infty,T} \le \|x\|_{W,T} \tag{1}$$

W^n is included in L_∞^n; for any $x \in W^n$, one has $\|x\|_\infty \le \|x\|_W$ and x(t) tends to 0 as t tends to ∞.

Let K_1 denote the causal LTI operator with transfer matrix $K_1(s) = (1+s)^{-1} I_n$, where I_n denotes the identity matrix of dimension n The following result gives a useful characterization of signals in W^n:

Proposition 2. $\mathbf{K_1}$ is one-to-one from L_{2e}^n onto W_e^n and $\mathbf{K_1}^{-1}$ is the operator $y \to y + \dot{y}$;[2] for any finite T>0, $\|y\|_{W,T} \le \|u\|_{2,T}$; $y = \mathbf{K_1}$ u belongs to W^n if and only if (iff) $u \in L_2^n$, and then $\|y\|_W = \|u\|_2$.

The notion of W-gain can be defined in an usual manner (Desoer and Vidyasagar, 1975): let $\mathbf{G} : W_e^n \to W_e^m$, and $K = \{k>0 : \|\mathbf{G}\ u\|_{W,T} \le k\ \|u\|_{W,T},\ \forall u \in W_e^n, \forall T>0\}$. If K is non empty, **G** is said to be W-stable (W-s), and $\gamma_W(\mathbf{G}) = \inf(K)$ is called the W-gain of **G** (note that, according to the terminology used in the reference above, this is the definition of the W-gain *with zero bias*); if K is empty, we set $\gamma_W(\mathbf{G})=\infty$. All operators considered in this paper are assumed to be causal, hence one has (Willems, 1971)

$$\gamma_W(\mathbf{G}) = \sup_{u \in W^n - \{0\}} \frac{\|\mathbf{G}\ u\|_W}{\|u\|_W} \qquad (2)$$

Let $\mathbf{G}$: $W_e^n \to W_e^m$; by Proposition 2, one has the following result, where γ_2 denotes the L_2-gain:

Proposition 3. **G** is W-stable iff $\mathbf{K_1}^{-1}\ \mathbf{G}\ \mathbf{K_1}$ is L_2-stable, and $\gamma_W(\mathbf{G}) = \gamma_2(\mathbf{K_1}^{-1}\ \mathbf{G}\ \mathbf{K_1})$.

Remark1. Assume that **G** is a LTI operator with transfer matrix G; as **G** and $\mathbf{K_1}$ commute, one has $\gamma_W(\mathbf{G}) = \gamma_2(\mathbf{G}) = \|G\|_\infty$ (note that with respect to transfer matrices, $\|.\|_\infty$ denotes the norm in H_∞, whereas with respect to time domain signals, this symbol denotes the norm in L_∞).

For any operator $\mathbf{G}$: $L_{2e}^n \to W_e^m$, one has by Proposition 2

$$\sup_{u \in L_2^n - \{0\}} \frac{\|\mathbf{G}\ u\|_W}{\|u\|_2} = \gamma_2(\mathbf{K_1}^{-1}\ \mathbf{G}) \qquad (3)$$

(this quantity being finite iff $\mathbf{K_1}^{-1}\ \mathbf{G}$ is L_2-stable). In particular, for $\tau>0$, let $\mathbf{K_\tau}$ be the LTI operator with transfer matrix $K_\tau(s) = (1+\tau s)^{-1}\ I_n$. Obviously, $\mathbf{K_1}^{-1}\ \mathbf{K_\tau}$ is L_2-stable, so that for any function $u \in L_2^n$, $\mathbf{K_\tau}\ u \in W^n$. Hence, W^n (rep. W_e^n) is the space of functions in L_2^n (resp. L_{2e}^n) filtered by first order low-pass filters. Note that by Proposition 3, one has $\min(\tau, \tau^{-1})\ \gamma_2(\mathbf{K_\tau}^{-1}\ \mathbf{G}\ \mathbf{K_\tau}) \le \gamma_W(\mathbf{G}) \le \max(\tau, \tau^{-1})\ \gamma_2(\mathbf{K_\tau}^{-1}\ \mathbf{G}\ \mathbf{K_\tau})$. The following result is a consequence of (3) and Proposition 1, and can be of practical interest:

Corollary 1. Assume that $\mathbf{G}$: $L_{2e}^n \to W_e^m$ is such that $\mathbf{K_1}^{-1}\ \mathbf{G}$ is L_2-stable, and let $y = \mathbf{G}\ u$; then, y belongs to L_∞^m if u belongs to L_2^n, and $\|y\|_\infty \le \gamma_2(\mathbf{K_1}^{-1}\ \mathbf{G})\ \|u\|_2$.

3. LOCAL W-STABILITY

Definition 2. Let $\mathbf{G} : W_e^n \to W_e^m$, and $K_l = \{k>0 : \exists\ \varepsilon >0, \forall u \in W_e^n, \forall T>0, \|u\|_{W,T} < \varepsilon \Rightarrow \|\mathbf{G}\ u\|_{W,T} \le k\ \|u\|_{W,T}\}$. If K_l is non empty, then we say that **G** is locally-W-stable (l-W-s), and $\gamma_{Wl}(\mathbf{G}) = \inf(K_l)$ is called the local-W-gain (l-W-g) of **G**. If K_l is empty, we set $\gamma_{Wl}(\mathbf{G}) = \infty$.[3]

Remark 2. This definition of the "local gain" is close, but not identical to the definition of the "small-signal-gain" in (Vidyasagar, 1993). This last definition corresponds to the case where the inequality $\|u\|_{W,T} < \varepsilon$ above is replaced by $\|u\|_{\infty,T} < \varepsilon$. By Proposition 1, any small-signal-W-stable system is l-W-s. Obviously, if **G** is linear, then **G** is l-W-s iff **G** is W-s, and $\gamma_{Wl}(\mathbf{G}) = \gamma_W(\mathbf{G})$.

Let us consider the case of a nonlinear memoryless operator **G**, defined as

$$(\mathbf{G}\ u)(t) = \Phi(t, u(t))\ \forall t \ge 0 \qquad (4)$$

where $\Phi : \mathbb{R}^+ \times \Omega \to \mathbb{R}^m$ (Ω denoting some open neighborhood of 0 in $\mathbb{R}^n$) is a C^1 function satisfying $\Phi(t,0) = 0$, $\forall t$, and where u is any function in W_e^n, taking its values in Ω. Let ∂_t and ∂_u respectively denote the partial derivative with respect to t and u. The following notion will be useful :

Definition 3. Let $u^* \in \Omega$; we say that Φ is differentiable with respect to u at point u* *uniformly with respect to t*, if there exists a function $\eta : \Omega \to \mathbb{R}^+$ such that $\eta(u^*+h)$ tends to 0 as h tends 0 and that $\forall t \ge 0$

$$\|\Phi(t,u^*+h) - \Phi(t,u^*) - \partial_u\Phi(t,u^*)\ h\| = \eta(u^*+h)\ \|h\|. \qquad (5)$$

The only difference, compared with the usual notion of differentiability with respect to u, is that η is independent of t.

Proposition 4. Assume that Φ is locally Lipschitz-continuous at point u*=0, is differentiable with respect to u at point 0 uniformly with respect to t, and satisfies the following condition:

[2] As $y \in W_e^n$, the derivative $\dot{y}$ is defined almost everywhere in the usual sense. Note that $\mathbf{K_1}^{-1}$ is causal.

[3] Note that $\gamma_{Wl}(\mathbf{G}) = \inf_{\varepsilon>0} \sup_{0<\|u\|_W<\varepsilon} \frac{\|\mathbf{G}\ u\|_W}{\|u\|_W}$, because **G** is causal: see (Willems, 1971).

$$\lim.\sup_{u \to 0} \frac{\sup_{t\geq 0} \|\partial_t \Phi(t,u)\|}{\|u\|} = \beta < \infty \qquad (6)$$

Then, G is l-W-s and

$$\gamma_{Wl}(G) \leq \beta + \sup_{t\geq 0} \bar{\sigma}\,(\partial_u \Phi(t,0)). \qquad (7)$$

For example, in the case n=1, m=1, assume that $\Phi(t,u) = \cos \omega t\,(e^u - 1)$, $\omega \geq 0$; one then has $\gamma_{Wl}(G) \leq \omega+1$.

Now, let us consider the case of an operator G associated to a nonlinear time-invariant system Σ described by state-space equations $\dot{x} = f(x,u)$, $y = g(x,u)$ (see footnote 1), where f and g are C^1 in a neihborhood of (0,0) and satisfy $f(0,0) = 0$, $g(0,0) = 0$. Set $\partial_x f(0,0) = A$, $\partial_u f(0,0) = B$, $\partial_x g(0,0) = C$, $\partial_u g(0,0) = D$, and let G(s) be the transfer matrix of the linear approximation Σ_l of Σ around (0,0). The theorem below is very useful:

Theorem 1. Assume that (C, A) is detectable, (A, B) is stabilizable and $G \in H_\infty$. Then, Σ is l-W-s and $\gamma_{Wl}(\Sigma) = \|G\|_\infty$.[4]

Sketch of the proof. The equilibrium 0 is exponentially stable for the unforced system Σ_l, and therefore for the unforced system Σ. Now, Σ can be considered as a perturbed version of Σ, with a perturbation whose local W-gain is equal to 0 (applying Proposition 4). The theorem is then a consequence of Remark 1.

4. RELATIONSHIP BETWEEN W-STABILITY AND ASYMPTOTIC STABILITY

4.1. Time-varying case

Proposition 5. Consider the system Σ, described by the state-space equations

$$\dot{x} = f(t,x,u) \qquad (8)$$

$$y = g(t,x,u) \qquad (9)$$

$x(t) \in \mathbb{R}^n$, $u(t) \in \mathbb{R}^m$, $y(t) \in \mathbb{R}^p$, where f is C^1, f and g are locally Lipschitz-continuous in a neighborhood of $z^*=0$, where $z = [x^T\ u^T]^T$, and $f(t,0,0) = 0$ and $g(t,0,0) = 0$, $\forall t$. Assume that g is differentiable with respect to z at point 0, uniformly with respect to t, and that

$$\lim.\sup_{z \to 0} \frac{\sup_{t\geq 0} \|\partial_t g(t,z)\|}{\|z\|} < \infty \qquad (10)$$

[4] For the sake of simplicity, the system Σ and the associated operator are denoted by the same symbol.

Then, if 0 is a locally exponentially stable equilibrium point of (8) (with u = 0), the operator G associated to Σ is l-W-s.

This result is closely related to (Vidyasagar, 1993, Sec. 6.3, Theorem 15); the main difference (apart from the fact that W-stability, instead of L_p-stability is obtained) is Condition (10). For instance, in the case of a linear output equation $y = C(t)\,x$, this condition reduces to $\sup_{t\geq 0} \bar{\sigma}\,(\dot{C}(t)) < \infty$.

4.2. Time-invariant case

We now assume that Σ is time-invariant, i.e. $f(t,x,u) = f(x,u)$, $g(t,x,u) = g(x,u)$, where f and g are C^1 in a neighborhood of $z^* = 0$ and satisfy $f(0,0) = 0$, $g(0,0) = 0$. Let us denote as $\phi(t, t_0, x_0, u)$ the state x(t) satisfying the initial condition $x(t_0) = x_0$.

Definition 4. Σ is locally W-reachable if there exists a function β of class K and a neighborhood U of 0 in $\mathbb{R}^n$ such that for any x in U, there exists a finite time τ and a control $u \in W^m$ such that $\phi(\tau, 0, 0, u) = x$, $u(t) = 0$ for $t \geq \tau$ and $\|u\|_W \leq \beta(\|x\|)$. If $U = \mathbb{R}^n$, Σ is said to be globally W-reachable.

W-reachability is a natural generalization of controllability in the case of linear systems possibily fed back by a time-invariant memoryless nonlinearity (see Proposition 6 and Remark 3):

Proposition 6. Assume that Σ is linear, i.e. $f(x,u) = A\,x + B\,u$, $g(x,u) = C\,x + D\,u$. Then, (A, B) is controllable iff Σ is globally W-reachable.

Remark 3. Consider the feedback system in Fig.1, where G_1 is the system Σ of Proposition 6, and where G_2 is a time-invariant memoryless nonlinearity defined by $G_2(u)(t) = \Phi(u(t))$, where Φ is differentiable in an open neighborhood U of 0 in $\mathbb{R}^m$ and satisfies $\Phi(0) = 0$. Assume that $u_2=0$ and consider the system Σ_f with input u_1. Then, Σ_f is W-reachable (locally is U is stricly included in $\mathbb{R}^m$, and globally if $U=\mathbb{R}^m$) if (A, B) is controllable.

Recall that the nonlinear time-invariant system Σ is said to be locally uniformly observable iff there exist a function α of class K and a neighborhood V of 0 in $\mathbb{R}^n$ such that $\forall x \in V$, $\|g(\phi(., 0, x, 0), 0)\|_2 \geq \alpha(\|x\|)$; if $V=\mathbb{R}^n$, then Σ is said to be globally uniformly observable (Vidyasagar, 1993). For LTI systems, uniform observability is equivalent to the standard notion of observability; moreover, the closed loop system Σ_f above is globally uniformly observable if (C, A) is observable. The theorem below can be considered as the reciprocal of Proposition 5 (with additional minimality assumptions on the state-space realization). Its proof is close to the one of (Vidyasagar, 1993, Sec. 6.3, Theorem 39), with appropriate modifications.

Theorem 2. Assume that the nonlinear time-invariant system Σ is locally (resp. globally) W-reachable, locally (resp. globally) uniformly observable and l-W-s (resp. W-s). Then, 0 is a locally (resp. globally) asymptotically stable equilibrium for the unforced system.

5. LOCAL SMALL GAIN THEOREM

Let us consider the standard closed loop system in Fig. 1, where $\mathbf{G_1} : W_e^n \to W_e^m$ and $\mathbf{G_2} : W_e^m \to W_e^n$ are l-W-s; set $u = [u_1^T \ u_2^T]^T$, $y = [y_1^T \ y_2^T]^T$ and $e = [e_1^T \ e_2^T]^T$. All signals are assumed to be zero at initial time 0. Assume that this closed loop system is well posed (Willems, 1971), i.e. there exist two operators $\mathbf{H_1}$ and $\mathbf{H_2} : W_e^{n+m} \to W_e^{n+m}$ such that $e = \mathbf{H_1}\, u$ and $y = \mathbf{H_2}\, u$. We will say that this closed loop system is l-W-s if $\mathbf{H_1}$ and $\mathbf{H_2}$ are l-W-s.

The following theorem is a local version of the well known Small Gain Theorem (Zames, 1966a). Its proof is very close to the one given in (Bourlès, 1994) for discrete-time systems.

Theorem 3. If $\gamma_{Wl}(G_1)\, \gamma_{Wl}(G_2) < 1$, then the closed loop system in Fig. 1 is l-W-s.

Remark 4. This theorem is still true if L_p-stability ($1 \le p < \infty$) is considered instead of W-stability. However, W-stability is more interesting because Proposition 4 and Theorem 1 do not hold in L_p framework. Consider for instance the case where one of the operators in the loop is memoryless and only *locally* bounded (in the sense where the associated function Φ satisfies the assumptions of Proposition 4); a signal x in L_2 can take very large values, even if $\|x\|_2$ is very small, and then nothing can be said about the signal $t \to \Phi(t,x(t))$; hence, local L_2-stability cannot be established (in an input-output approach). Small-signal-L_2-stability can be proved in this case by first proving internal stability (*via* Lyapunov theory) and then using (Vidyasagar, 1993, Sec. 9.3, Theorem 39). But to prove internal stability can be difficult when one of the systems in the loop is infinite dimentional (see all examples below).

Example 1. Assume that $\mathbf{G_1}$ is the LTI operator with transfer function $k\,\dfrac{e^{-\tau s}}{1 + Ts}$ ($\tau>0$, $T>0$) and that $\mathbf{G_2}$ is an operator of the form (4), where $\Phi(t,u) = \cos \omega t\,(e^u - 1)$, $\omega \ge 0$. By Theorem 3 and Proposition 4, the resulting closed loop system is l-W-s if $|k| < \dfrac{1}{1+\omega}$.

Theorem 3 gives a sufficient condition for l-W-stability; conversely, using Remark 1, a necessary condition can also be obtained:

Corollary 2. Let $\mathbf{G_1} : W_e^n \to W_e^m$ be a LTI operator. The closed-loop system in Fig. 1 is l-W-stable for any operator $\mathbf{G_2} : W_e^m \to W_e^n$ satisfying $\gamma_{Wl}(G_2) < 1$, iff $\gamma_W(G_1) \le 1$.

6. LOCAL PASSIVITY

As W^n is an inner product space, the notions of W-passivity and strict W-passivity can obviously be defined (Willems, 1971; Desoer and Vidyasagar, 1975). Local versions of these notions will now be defined. Let $\mathbf{G} : W_e^n \to W_e^n$ be a causal operator.

Definition 4. $\mathbf{G}$ is locally W-passive if there exists $\varepsilon>0$ such that $\langle u, \mathbf{G}\, u\rangle_{W,T} \ge 0$ whenever $u \in W_e^n$ and $T>0$ are such that $\|u\|_{W,T} < \varepsilon$ [5]; $\mathbf{G}$ is locally strictly W-passive if there exists $\eta>0$ such that $\mathbf{G} - \eta\, \mathbf{I}$ is locally W-passive.

Assume that $(I + G)^{-1} : W_e^n \to W_e^n$ is well defined, and let $\mathbf{H} = (G - I)(I + G)^{-1}$. It is well known (Desoer and Vidyasagar, 1975, p. 216) that $\gamma_W(\mathbf{H}) \le 1$ if $\mathbf{G}$ is W-passive, and $\gamma_W(\mathbf{H}) < 1$ if $\mathbf{G}$ is W-s and W-strictly passive. These properties still hold if local gains and local passivity are considered (instead of "global" ones). Therefore, by a standard scheme equivalence and by Theorem 3, one obtains the following result:

Theorem 4. Let us consider the standard closed loop system in Fig.1, where G_1 and $\mathbf{G_2} : W_e^n \to W_e^n$ are such that this closed loop system is well posed. If $\mathbf{G_1}$ is locally W-passive and if $\mathbf{G_2}$ is l-W-s and locally strictly W-passive, then this closed loop system is l-W-s.

Proposition 7. Let $\mathbf{G} : W_e^n \to W_e^n$ be a LTI operator; then, $\mathbf{G}$ is locally (strictly) W-passive iff $\mathbf{G}$ is (strictly) W-passive, and this condition is satisfied if $\mathbf{G}$ is (strictly) L_2-passive.

Therefore, W-passivity or strict W-passivity of a LTI operator can be analyzed in the frequency domain.

Now, let us consider the case of a nonlinear memoryless operator $\mathbf{G}$, defined by (4), where Φ is like in Proposition 4. Let $\mu(.)$ denote the usual matrix measure (Desoer and Vidyasagar, 1975), i.e. for any matrix $A \in \mathbb{C}^{n \times n}$, $\mu(A) = \lambda_{max}(A + A^*)/2$.

Proposition 8. $\mathbf{G}$ is locally strictly W-passive if i) $\inf_{t \ge 0} \{-\mu[-\partial_u\Phi(t,0)]\} = \delta > 0$ and ii) $\beta < 2\delta$, where β is defined by (6).

Example 2. Let $\mathbf{G_1}$ be the LTI operator with transfer function $\dfrac{b}{1 + a\,e^{-\tau s} + Ts}$, $|a| < 1$, $b>0$, $T>0$, and let

---

[5] In other words, $\mathbf{G}$ is locally W-passive iff $\inf_{\varepsilon>0} \ \inf_{0<\|u\|_W<\varepsilon} \langle u, \mathbf{G}\, u\rangle_W \ge 0$.

G_2 be the operator G defined by (4), with $\Phi(t,u) = (\delta + 1 + \cos \omega t) \sin u$; G_1 is W-passive and G_2 is l-W-s and locally strictly W-passive if $0 \leq \omega < 2\delta$ (by Proposition 8); hence, by Theorem 4, the closed loop system in Fig. 1 is l-W-s if the latter condition is satisfied.

Remark 5. Other standard loop transformations can be made, in order to obtain a local version of the Circle Criterion (Zames, 1966b; Safonov and Athans, 1981).

7. CONCLUSION

For several reasons, W-stability and local-W-stability are of practical interest. Roughly speaking, a system is l-W-s if, when excited by a signal with small energy and filtered by a first order low-pass filter K_τ, its output, "filtered" by the inverse filter, is of small energy (Section 2). Therefore, this type of stability is of signifiance in cases where disturbances, reference signals, etc., acting on the system can be considered as filtered signals (all signals are probably of this kind in practice). Moreover, W-stability and asymptotic stability are closely related in case of a minimal realization (Section 4). The local-W-gain of a time-invariant nonlinear system is nothing but the H_∞-norm of the transfer matrix of its linear approximation (Theorem 1), so this notion is clear. W-stability is well suited to formulate a useful Local Small Gain Theorem and a Local Passivity Theorem (Remark 4).

An interesting possible application of the local stability results stated in this paper is the robustness study of feedback systems perturbed by nonlinearities which are only locally bounded (Colledani *et al.*, 1993). Usual input-output theorems indeed cannot be applied in this case, and to prove internal stability can be difficult (especially in case of infinite dimentional systems).

Local input-output stability results can also be obtained for discrete-time systems (Bourlès, 1994).

The relationship between W-stability and asymptotic stability in the case of infinite-dimensional systems will be studied elsewhere.

7. REFERENCES

Alexeev, V. ,Tikhomirov, V., and Fomine, S. (1982). *Commande optimale*. Mir, Moscow.

Bourlès, H. (1994). A Local Small Gain Theorem for Discrete-Time Systems, Proc. 33rd CDC, December 14-16, Lake Buena Vista (Florida), 2137-2138.

Bourlès, H., and Aïoun, F. (1994). In: *La Robustesse, Analyse et Synthèse de Commandes Robustes*, (A. Oustaloup, Ed.), Chap. 3, pp. 163-235. Hermès, Paris.

Bourlès, H., and Colledani, F. (1995). W-Stability and Local Input-output Stability Results, *IEEE Trans. on Autom. Control* (to appear).

Colledani, F., Bourlès, H., and Vanhersecke, M. P. (1993). Robust Controller with Local Linearizing Feedback for Frequency/Power Control of Power Plants, Proc. 32nd CDC, December 13-17, San Antonio (Texas), 3740-3741.

Desoer, C. A., and Vidyasagar, M. (1975), *Feedback Systems: Input-Output Properties*. Academic Press, New-York .

Francis, B. A. (1987). *A Course in H_∞ Control Theory*. Springer Verlag, Berlin .

Hill, D. J., and Moylan, P. J. (1982). Connections Between Finite-Gain and Asymptotic Stability, *IEEE Trans. on Autom. Control*, **25**, 931-936.

Pan, D.-J., Han, Z.-Z., and Zhang, Z.-J. (1993). Bounded-input—bounded-output stabilization of nonlinear systems using state detectors, *Syst. Contr. Letters*, **21**, 189-198.

Safonov, M. G., and Athans, M. (1981). A Multiloop Generalization of the Circle Criterion for Stability Margin Analysis, *IEEE Trans. on Autom. Control*, **26**, 415-422.

Sontag, E. D. (1989). Smooth Stabilization Implies Coprime Factorization, *IEEE Trans. on Autom. Control*, **34**, 435-443.

Sontag, E. D. (1990). Further Facts About Input-To-State Stabilization, *IEEE Trans. on Autom. Control*, **35**, 473-476.

Trèves, F. (1967). *Topological Vector Spaces, Distributions and Kernels*. Academic Press, London.

Vidyasagar, M. (1993). *Nonlinear Systems Analysis* (second ed.). Prentice-Hall, Englewood Cliffs, N. J.

Vidyasagar, M., and Vannelli, A. (1982). New Relationships Between Input-Output and Lyapunov Stability, *IEEE Trans. on Autom. Control*, **27**, 481-483.

Willems, J. C. (1971). *The Analysis of Feedback Systems*. The MIT Press, Cambridge (MA) .

Zames, G. (1966a). On the Input-Output Stability of Time-Varying Nonlinear Feedback Systems. Part I: Conditions Derived Using Concepts of Loop Gain, Conicity, and Positivity", *IEEE Trans. on Autom. Control*, **11**, 228-238.

Zames, G. (1966b). On the Input-Output Stability of Time-Varying Nonlinear Feedback Systems. Part II: Conditions Involving Circles in the Frequency Plane and Sector Nonlinearities, *IEEE Trans. on Autom. Control*, **11**, 465-476.

A STOCHASTIC APPROACH FOR THE CONTROL OF A NONLINEAR SYSTEM

L. AUTRIQUE* and J.E. SOUZA DE CURSI**

* Laboratoire d'Automatique de Nantes, Ecole Centrale de Nantes, 1 rue de la noë, 44072 Nantes cedex 03, FRANCE, Tel: (33) 40.37.16.00, FAX: (33) 40.37.25.22, e-mail: autrique@lan.ec-nantes.fr
** Institut de Mécanique de Rouen, I.N.S.A de Rouen, Place Emile Blondel BP 8, 76131 Mont Saint Aignan cedex, FRANCE, Tel: (33) 35.52.83.70, FAX: (33) 35.52.84.21, e-mail: souza@insa-rouen.fr

Abstract. The industrial process of vulcanization is modelized by a system of two nonlinear equations depending on the temperature and on the reticulation rate. A control problem is proposed and written as a optimization one. Deterministic methods are presented to minimize a cost function. A stochastic approach is introduced in order to improve the numerical resolution. The numerical effect of the coupling between stochastic methods and a deterministic one is investigated. The interest of the introduction of random perturbations is shown.

Résumé. Le processus industriel de vulcanisation des élastomères est modélisé par un système de deux équations non-linéaires dépendant de la température et du degré de réticulation. Un problème de contrôle est décrit sous forme d'un problème de optimisation. Des methodes déterministes sont présentées afin de minimiser le critère. Une approche stochastique est proposée afin d'améliorer la résolution numérique. Les effets numériques du couplage entre les méthodes stochastiques et les méthodes purement déterministes sont étudiés. L'intérêt de l'adjonction de perturbations aléatoires est mis en évidence.

Keywords. Stochastic methods, Control problem, Nonlinear system, Distributed Parameter system.

1. INTRODUCTION

In this paper, control problems arising in the industrial process of vulcanization are presented. Since 1844, rubber materials are more and more used for high-technology components (cf Hou *et al.*, 1990). The thermal and chemical behaviours of the material during the industrial process leading to its elaboration are quite difficult to modelize (see Golay *et al.*, 1992). For the last thirty years, different developments have been carried out in order to propose a good model (cf Garnier *et al.*, 1990). Thus, different apects have to be taken into account. The vulcanization reaction is irreversible: that means that, the reticulation rate is a time growing function. Moreover, an exothermal phenomenon can occur during the industrial process and damage the material. In addition, thermal and mechanical properties are closely connected to the curing process. A sharp control of the reaction is essential to avoid the destruction of the material and to ensure good performances in practical utilisation (see Khouider *et al.*, 1988). The control problem of this reaction is a great challenge. In this paper, a two dimensional geometry is considered. The material behaviour is described by the evolution of the temperature and the curing degree. These parameters satisfy a set of non-lineary partial differential equations. So, when control or identification problems are considered, global optimization methods must be introduced due to the intrinsic non-convexity of the model. Simple descent methods cannot ensure the convergence to a global minimum. In this paper, we consider a stochastic perturbation of a classical method: the gradient method with a fixed step. When such a perturbation is suitably choosen, this can ensure, on the one hand, the convergence to a global minimum, and on the other hand can increase the convergence speed of the algorithm (cf Pogu *et al.*, 1994).

2. A MODEL FOR THE INDUSTRIAL PROCESS

2.1 The industrial process

In natural state, rubber is composed of long carbon molecules. The vulcanization reaction consists in linking these long chains between themselves, in order to obtain some desired mechanical behaviour. Elastomer materials are usually manufacturing by injection process. Initially, the upper chamber (also called the transfer pot) is filled by rubber samples. Then, rubber is transferred through injection holes. For some experimentations, the mold can also be pre-filled using sheets of rubber.

The cure process is performed by exposing the material to elevated (cure) temperatures and (cure) pressures for a given time (cf Bost, 1980). The cure temperature provides the heat required for initiating and maintaining the vulcanization reaction in the elastomers during the curing process. Then, the reticulation reaction start to occur. The heat comes from two differents phenomena:

- conduction from the mold (convection aspects are not taken into account),
- an exothermal term generated by the chemical reaction of vulcanization.

Rubber is injected under high pressure into the mold slabs at temperature between 80 and 160 Celsius degree. For tire manufacturing, cure temperatures are usually about 220 Celsius degree. It is important to notice that the material keep on curing even after removal from the mold (see Khouider *et al.*, 1988). Thus, partial states of cure are difficult to obtain. In the following paragraph, a modelization of such a process is proposed.

2.2 The model

The following model can be found, for example, in Delaunay *et al.* (1993) and Autrique *et al.* (1994). The real three-dimensionnal process is simplified in a 2-D geometry. Let us denote by Ω the domain: $\Omega = \Omega_1 \cup \Omega_2 \subset \mathbb{R}^2$, where Ω_1 is the part of the domain corresponding to the iron mold and Ω_2 is the domain containing the elastomer material. The contact between Ω_1 and Ω_2 is assumed to be perfect and thus, at the interface, continuity conditions have to be satisfied for the temperature and the heat flux. The boundary of the whole domain Ω is $\partial\Omega = \Gamma_D \cup \Gamma_N$, corresponding to Dirichlet data (given or controlled temperature) and Neumann data (given or controlled heat flux) (see Fig.1).

Let us denote by $X = (X_1; X_2)$ the space variable and $t \in [0; T]$ the time.

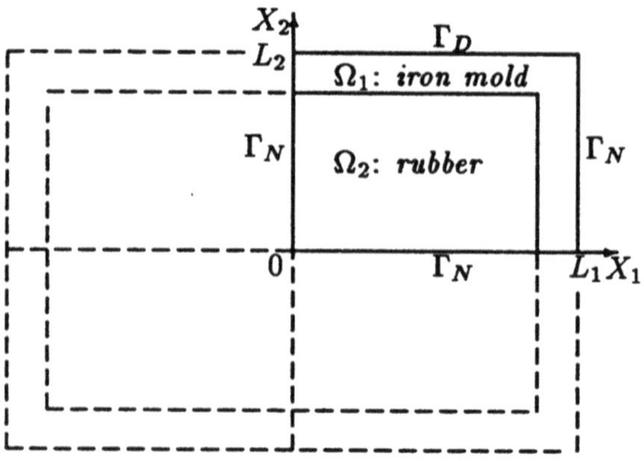

Fig.1 - *The domain Ω and the parts of its boundary corresponding to Dirichlet and Neumann data.*

The thermochemical behaviour of the elastomer is assumed to be described by the evolution of two different parameters: the temperature and the cure degree. This one, also called the reticulation rate, describes the chemical state of the material. Let us denote by $\Theta = \Theta(X;t)$ the temperature, and $\alpha = \alpha(X;t) \in [0;1]$ the cure degree, at point X and instant t. When α is equal to 0 the cross-linking reaction is not started in the rubber ; while for α equal to 1, the material is completely vulcanized and no more jonction between chains can be elaborated.

The evolution of the temperature Θ and the cure degree α is described by a set of equations:

- In the iron mold (domain Ω_1).
  The temperature evolution is described by a heat equation:

$$c_1.\frac{\partial\Theta}{\partial t} - div(\lambda_1.\nabla\Theta) = 0 \quad (1)$$

  where c_1 denotes the specific volumic heat (in $J.m^{-3}.K^{-1}$) and λ_1 is the thermal conductivity (in $W.m^{-1}.K^{-1}$) of the iron mold. ∎
- In the rubber (domain Ω_2).
  - The temperature in the domain satisfy a heat equation taking into account the exothermal reaction which can occur:

$$c_2.\frac{\partial\Theta}{\partial t} - div(\lambda_2.\nabla\Theta) = Q_\infty\frac{\partial\alpha}{\partial t} \quad (2)$$

    where c_2 denotes specific volumic heat (in $J.m^{-3}.K^{-1}$), λ_2 is the thermal conductivity (in $W.m^{-1}.K^{-1}$) of the rubber, and Q_∞ is the volumic enthalpy of the reaction (in $J.m^{-3}$).
  - The kinetic of the reaction is described by the following equation:

$$\frac{\partial\alpha}{\partial t} = \begin{cases} 0 & \text{if } t < t_{ind} \\ f(\Theta)\,g(\alpha) & \text{if } t > t_{ind} \end{cases} \quad (3)$$

where $g(\alpha)$ is called the reaction function such that $g(1) = 0$; thus when the cross-linking reaction is over, the speed of the reaction is equal to zero. The function $g(\alpha)$ is assumed to be continuous and we consider that:

$$\max_{\alpha \in (0;1)} g(\alpha) = 1 = g(\alpha^*)$$

The function f is given by equation 4 :

$$f(\Theta) = k_0 . \exp\left(\frac{-E_{act}}{R\Theta}\right) \quad (4)$$

where E_{act} is the energy of activation, k_0 a characteristic of the rate of the transformation and R the universal constant of ideal gases. The following notation (see equation 5) is proposed to obtain a dimensionless formulation:

$$f(\Theta) = k_{ref}.\exp\left(-A\left(\frac{\Theta_{ref}}{\Theta} - 1\right)\right) \quad (5)$$

where Θ_{ref} is a reference temperature in Kelvin. A physical interpretation of the parameter A is given by considering the variation of temperature $\delta\Theta$ around the reference Θ_{ref} for which the kinetic is twice faster, that is :

$$f(\Theta_{ref} + \delta\Theta) = 2.f(\Theta_{ref}) = 2.k_{ref}$$

Thus, the relationship between A, Θ_{ref} and $\delta\Theta$ is :

$$A = \ln 2 \frac{\Theta_{ref} + \delta\Theta}{\delta\Theta}$$

We observe with these notations that :

$$k_{ref} = f(\Theta_{ref}) = \left.\frac{\partial \alpha}{\partial t}\right|_{\alpha = \alpha^*}$$

■

In Ω_2, the evolution equations are coupled by their right members and involve a characteristic time t_{ind} called the *induction time*, before what a chemical reaction cannot occur. The induction time t_{ind} is the solution of the scalar equation:

$$h(t_{ind}, \Theta) = 0$$

where:

$$h(t, \Theta) = \int_0^t \exp\left(\frac{-E_{ind}}{R\Theta(X;\tau)}\right) d\tau - \delta_{ind}$$

E_{ind} is called the induction energy and δ_{ind} is a constant, characteristic of the material. Thus, t_{ind} depends on the *history* of the temperature. We observe that: $t \longmapsto h(t, \Theta)$ is increasing. Thus, we introduce:

$$W(\xi) = \begin{cases} 0 & \text{if } \xi < 0 \\ 1 & \text{if } \xi > 0 \end{cases}$$

and equation 3 can be rewritten as:

$$\frac{\partial \alpha}{\partial t} = W(h(t, \theta)) f(\Theta) g(\alpha) \quad (6)$$

In order to simplify equations 1 and 2, the following notations are introduced:

$$\chi(X) = \begin{cases} 0 & \text{if } X \in \Omega_1 \\ 1 & \text{if } X \in \Omega_2 \end{cases}$$

$$c(X) = \begin{cases} c_1 & \text{if } X \in \Omega_1 \\ c_2 & \text{if } X \in \Omega_2 \end{cases}$$

$$\lambda(X) = \begin{cases} \lambda_1 & \text{if } X \in \Omega_1 \\ \lambda_2 & \text{if } X \in \Omega_2 \end{cases}$$

Thus, equations 1 and 2 can be written as:

$$c(X).\frac{\partial \Theta}{\partial t} - \mathrm{div}(\lambda(X)\nabla\Theta) = \chi(X).Q_\infty.\frac{\partial \alpha}{\partial t} \quad (7)$$

The initial state in the mold and in the material are assumed to be known:

$$\Theta(X;0) = \Theta_0(X) \quad X \in \Omega \quad (8)$$
$$\alpha(X;0) = \alpha_0(X) \quad X \in \Omega_2 \quad (9)$$

On Γ_D, the part of the boundary corresponding to Dirichlet data, the controlled temperature is denoted by $u(t)$:

$$\Theta(X;t) = u(t) \quad (X;t) \in \Gamma_N \times [0;T] \quad (10)$$

On Γ_N, the part of the boundary corresponding to Neumann data, heat flux are given by equation 11:

$$\frac{\partial \Theta}{\partial \vec{n}} = \varphi(X;t) \quad (X;t) \in \Gamma_N \times [0;T] \quad (11)$$

Continuity of temperatures and heat flux at the interface Ω_1/Ω_2 is implicitely assumed in equation 7.

Then the temperature $\Theta(X;t)$ and the cure degree $\alpha(X;t)$ are solutions of the equations:

evolution equations:	equations 6,	7
initial states:	equations 8,	9
boundary conditions:	equations 10,	11

3. CONTROL PROBLEM CONNECTED TO THE INDUSTRIAL PROCESS

From the industrial standpoint, the goal is to obtain a desired cure degree at the end of the process. In fact, mechanical or chemical properties are connecteded to the reticulation rate. So, we consider as model situation the problem of determining the curing cycle $u(t)$ such that the final state of cure $\alpha(X,T)$ is a given $\tilde{\alpha}$. This formulation considers only a partial aspect of the industrial process. Other control problems can be introduced:

- the curing time may be minimized in order to reduce the manufacturing expenses,
- the optimal curing cycle may be calculated with constraints involved by the heat specifications of the mold,
- the exothermal reaction which can occur in the rubber, may lead to critical temperature and damages the material.

The control problem is written as an optimization one, where a functionnal J has to be minimized:

$$J = \int_{\Omega_2} (\alpha(X;t_f) - \tilde{\alpha}(X))^2 dX \qquad (12)$$

An optinal curing cycle $u(t)$ has to be determined. Thus, we consider $J = J(u)$, and the control problem is written:

> *Continuous control problem* $\mathcal{P}^{\star}$:
> Find $u^{\star}$ such that:
> $J(u^{\star})$ is a minimum of $u \longmapsto J(u)$
> under constraints of equations 6, 7, 8, 9, 10 and 11.

4. DISCRETIZATION OF THE PROBLEM

The optimization problem has been introduced in a continuous formulation. The numerical resolution is performed by considering the discretization of the problem $\mathcal{P}^{\star}$.

4.1 Discretization of the constraints

Equations 6, 7, 8, 9, 10 and 11 are considered as constraints for the control problem.

Discretization in time Let us denote by K_{max} the number of time interval and $\tau = \frac{T}{K_{max}}$. Equation 6 is discretized as follows: for $K = 0,\ldots,K_{max}-1$:

$$\begin{aligned}\alpha_{K+1} = \ & \alpha_K + \\ & \tau.f(\Theta_K).g(\alpha_K).W(h(t_K,\Theta_K))\end{aligned} \qquad (13)$$

Discretization of equation 7 is performed as follows for $K = 0,\ldots,K_{max}-1$:

$$\begin{aligned}& C_K(X).\Theta_{K+1}(X) - \tau.\lambda(X).\Delta\Theta_{K+1}(X) \\ & = \tilde{s_K}(X) + C_K(X).\Theta_K(X)\end{aligned} \qquad (14)$$

where:

$$\tilde{s_K}(X) = \tau.Q_{\infty}.f(\Theta_K(X)).g(\alpha_K(X)).W(h(t_K,\Theta_K(X)))$$

α_K and Θ_K are the approximated values of $\alpha(.;t_K)$ and $\Theta(.;t_K)$ respectively.

Discretization in space The 2-D domain is partitionned into triangular elements by introducing a grid made of N_1+1 vertical lines V_i and N_2+1 horizontal lines H_j. The intersection of V_i and H_j is the point P_{ij} of coordinates $((i-1)h_1,(j-1)h_2)$, where $h_i = \frac{L_i}{N_i}$. We denote by α_K^{ij} and Θ_K^{ij} the approximated values of $\alpha(P_{ij};t_K)$ and $\Theta(P_{ij};t_K)$ respectively. The discretized in time equations are discretized in space by a finite-element method (see Beauquet *et al.*, 1987 and Dautray *et al.*, 1988). This leads to a set of equations having the form:

$$E_K^{ij} = 0 \text{ and } F_K^{ij} = 0 \quad \text{for} \begin{cases} K = 0,\ldots,K_{max}-1 \\ i = 1,\ldots,N_1+1 \\ j = 2,\ldots,N_2 \end{cases} \qquad (15)$$

corresponding to the discretization of equations 14 and 13 respectively. By reasons of limitation of room, we do not give here the expressions of E_K^{ij} and F_K^{ij}.

4.2 Discretization of u

We consider the following discretization of u:

$$u(t) \simeq u_h(t) = \sum_{K=1}^{K_{max}} u_K S_K(t) \qquad (16)$$

where S_K are the continuous piecewise linear function such that:

for each $t_j = j\tau$, $j = 1,\ldots,Kmax$, S_K verify:

$$S_K(t_j) = \begin{cases} 1 & \text{if } K = j, \\ 0 & \text{if } K \neq j. \end{cases}$$

Thus, our unknowns are $(u_1,\ldots,u_{K_{max}})$.

4.3 Discretization of J

The continuous cost-functionnal is discretized as:

$$J(u) \simeq J_h(u) = h_1.h_2.\sum_{i=1}^{N_1}\sum_{j=1}^{N_2}\left(\alpha_{K_{max}}^{ij} - \tilde{\alpha}^{ij}\right)^2 \qquad (17)$$

4.4 The discretized control problem

The problem is formulated as follows:

Discretized problem $\mathcal{P}_h^{\star}$:
Find $\tilde{u}_h^{\star} = (u_1^{\star}, \ldots, u_{K_{max}}^{\star})$, such that $\tilde{u}_h$ minimizes $\tilde{u}_h \longmapsto J_h(\tilde{u}_h)$ under the constraints of system 15.

5. DESCENT METHODS AND STOCHASTIC PROCESSES

The numerical solution of the discretized problem is performed by an iterative descent method which reads as follows:

$$u_{m+1} = Q(u_m) \tag{18}$$

Here, lower subscripts refer to the iteration number. For example,

- the gradient method with a fixed parameter $\eta > 0$ involves:

$$Q(u) = u - \eta \frac{\partial J_h}{\partial u} \frac{1}{\left\| \frac{\partial J_h}{\partial u} \right\|} \tag{19}$$

- the newton method with a fixed parameter $\eta > 0$ involves:

$$Q(u) = u - \eta . \left[\frac{\partial^2 J_h}{\partial u^2} \right]^{-1} . \frac{\partial J_h}{\partial u} \tag{20}$$

Howewer, the convergence to the global minimum cannot be ensured without convexity assumptions which are not obvious to verify for the functionnal J given in equation 12. Recent developments (see Pogu *et al.*, 1994) have shown that the introduction of stochastic methods for such optimization problems can improve the numerical resolution.

Let us consider the random perturbations:

$$u_{m+1} = Q(u_m) + \lambda_m . Z_m \tag{21}$$

where λ_m is a real number and Z_m is a random vector. It has been shown (see Pogu *et al.*, 1994) that, for suitable λ_m and Z_m, these perturbations ensure, on the one hand, the convergence to a global minimum and, on the other hand, can improve the convergence speed.

6. NUMERICAL RESULTS

In order to investigate the interest of the coupling of stochastic methods to deterministic one, numerical results obtained by 18 and 21 are compared for the gradient method with a fixed parameter (see equation 19). The evaluation of the gradient $\frac{\partial J_h}{\partial u}$ is performed by two methods:

- (M1): Direct derivation method: differentiation of the discretized equations corresponding to the continuous equations 7 and 6. This leads to a linear system giving the values of $\frac{\partial \alpha_{K_{max}}^{ij}}{\partial u}$. These values are used in the differentiation of 17).
- (M2): Discretized Lagrangian method: penalty method for taking the discretized equations 15 into account.

The interest of the addition of random perturbations on the descent method is investigated and different results are presented. Physical parameters of the model are presented in table 1. Thickness of the mold slabs is taken equal to

$L_1 = 0,1m$	$L_2 = 0,06m$
$\lambda_1 = 50Wm^{-1}K^{-1}$	$c_1 = 3,6.10^6 Jm^{-3}K^{-1}$
$\lambda_2 = 0,27Wm^{-1}K^{-1}$	$c_2 = 1,5.10^6 Jm^{-3}K^{-1}$
$\tau = 300\ s$	$T_f = 10800\ s$
$Q_\infty = 2,18.10^7 J.m^{-3}$	$E_{ind} = 8,1.10^4 Jmol^{-1}$
$R = 8,3\ J.mol^{-1}.K^{-1}$	$\delta_{ind} = 2,52.10^{-8}\ s$
$A = 24,37$	$\Theta_{ref} = 400\ K$
$k_{ref} = 1,31.10^{-4}\ s^{-1}$	$g(\alpha) = 1 - \alpha$
$\Theta_0(X) = 353\ K$	$\alpha_0(X) = 0$
$\varphi(x,t)(X) = 0$	

TABLE 1 Physical parameters

$0,02m$. The desired state of cure $\tilde{\alpha}$ is given by solving a direct problem. The initial curing cycle is:

$$u_0(t) = \begin{cases} 353 + t/15 & \text{if } t \in [0; 1800] \\ 473 & \text{if } t \in [1800; 3600] \\ 593 - t/30 & \text{if } t \in [3600; 7200] \end{cases}$$

The minimization algorithm written as equations 18 and 19 is implemented with the following parameters:

- the fixed step of the gradient method is $\eta = 2$,
- the resolution algorithm is stopped when the functionnal J is lower than a given value (for example: $J < 0.5$) or for a maximum number of iteration (for example: $imax = 500$).

The stochastic process introduced in equation 21 is performed with the following characteristics:

- (λ_m) is a sequence of real numbers decreasing to zero: $\lambda_m = \tau/(2 * m)$,
- the random vector Z_m is choosen such that each coordinate verifies an uniform distribution,
- at each step of the minimization algorithm, results obtained with 50 stochastic perturbations are compared and the best curing cycle is choosen.

With the above parameters, the coupling of deterministic and stochastic methods seems to be effective, on the one hand in order to ensure the convergence to the global minimum, and on the other hand, to increase the speed of convergence of the descent algorithm presented in equations 18 and 19.

7. CONCLUSION

The interest of the coupling of deterministic and stochastic methods has been shown on an identification problem in a one-dimensional geometry (see Autrique *et al.*, 1994). The same methodology is presented in this paper for a control problem modelized by a system of two non-linear equations. The space domain is 2-D and a finite element method is proposed. For the control problem, written as a minimization one, the introduction of random perturbations on the gradient method with a fixed step can improve the numerical resolution. Different numerical experiments are actually implemented in order to illustrate that important result. Other possible developments are:

- the numerical effect of the coupling of stochastic approach with the newton method,
- the control of the real 3-D process,
- the introduction of genetic-like strategics which have shown to be effective to calculate in mechanical engineering problems.

REFERENCES

Autrique L., and E. Souza De Cursi (1994). Vulcanization process: a stochastic method for identification and control problem. *Proceedings of the 3rd I.E.E.E conference on Control applications, Glasgow, U.K.*, **1**, 533-535.

Beauquet G., and M. Pogu (1987). Programmation des éléments finis P1-2D. *Cepadues Editions, Toulouse, France.*

Bost J. (1980). Matières plastiques - Chimie - Applications. *Technique et Documentation, Ed. Lavoisier, Paris, France.*

Dautray R., and J.L. Lions (1988). Analyse Mathématique et calcul numérique pour les sciences et les techniques. *Ed. Masson, Paris, France.*

Delaunay D., B. Garnier, Y. Jarny, and J.S. Le Brizaut (1993). Implementation of an inverse method for identification of reticulation kinetics from temperature measurements on a thick sample. *International Journal Heat Mass Transfer*, **36-16**, 4039-4047.

Garnier B., F. Danes, and D. Delaunay (1990). Coupling between Heat Transfer and Rubber Vulcanization. *Proceedings of the IXth international Heat Transfer Conference*, Jerusalem, Israel.

Golay F., and O. Debordes (1992). Modélisation par éléments finis de la polymérisation de structures composites - Applications industrielles. *Revue Européenne des éléments finis*, **1-1**, 51-73.

Hou G.J.W., T.H. Hou, and J.S. Sheen (1990). Optimal Cure Cycle Design for the Fiber Reinforced Composite Lamination Processing. *International Polymer Processing*, **5-2**, 88-99.

Khouider A., and J.M. Vergnaud (1988). Increase in the state of cure in vulcanizates after removal from the mold. *Journal of Polymer engineering*, **8-1,2**.

Pogu M., and E. Souza De Cursi (1994). Global optimization by random perturbation of the gradient method with a fixed parameter, *Journal of Global optimization, Kluwer Academic Publishers*, **5**, 159-180.

STRUCTURAL ANALYSIS METHOD OF INVESTIGATION OF LINEAR COMPLEX CONSERVATIVE SYSTEMS

A.V. PESTEREV and G.A. TAVRIZOV

State Institute of Physics and Technology, 13/7, Prechistenka str. Moscow, 119034, Russia. E-mail: pesterev@dep11.ifftr.msk.su

Abstract. The structural analysis method developed at the State Institute of Physics and Technology is intended for investigation of complex mechanical systems. The method is well formalised and can be applied to solution of many problems in the field of vibrations of complex systems both passive and active. It allows one to investigate efficiently different complex systems consisting of distributed and finite-dimensional subsystems interacting at a finite number of points by using a unified formalism. In the paper the general formalism of the technique as applied to spectral problems for conservative systems is discussed. The method involves use of the subsystem Green's operators. Natural frequencies of the system are obtained by investigating some symmetric characteristic matrix whose order is equal to the number of linearly independent forces and moments of the interaction.

Abstract. La méthode de l'analyse structurale développée dans l'Institut d'Etat des Problèmes Physiques et Technologies est destinée pour l'investigation des systèmes méchaniques complexes. La méthode est bien formalisée et peut être utilisée pour la solution de bon nombre de problèmes dans le domaine des vibrations des systèmes actives comme passives. Elle permit d'utiliser une formalisme uniforme pour étudier efficacement les systèmes différents se composant des sous-systèmes aux paramètres distribués et concentrés. L'article expose le formalisme général de la méthode appliquée aux problèmes spectrals des systèmes conservatives. La méthode utilise l'opérateurs de Green des sous-systèmes. Les fréquences propres du système sont obtenues comme le résultat de l'investigation de quelque matrice symétriquee caractéristiquee dont l'ordre est égal au nombre des forces et des moments linéairement indépendants d'interaction parmi les sous-systèmes.

Key Words. Structural analysis method; distributed parameter systems; finite-dimensional interaction; Green's operators; characteristic matrix.

1. INTRODUCTION

The structural analysis method (or factorized perturbations method) is intended for investigation of complex physical (controllable) systems. The first results related to the method discussed were published in 1974 (Azarov *et al.*, 1974) with a more detailed discussion being found in the book by Azarov *et al.* (1975). Since then the method was developed successfully and applied to investigation of mathematical models of some real space structures.

The aim of this paper is to outline the current structural analysis technique as applied to solving spectral problems for conservative systems consisting of subsystems interacting at a finite number of points. The method is most convenient when the system is composed of different subsystems, e.g., if some components of the system are considered as distributed ones (rods, beams, plates and so on) whereas other components are approximated by finite-dimensional models. The "interaction space" is assumed to be "poor", i.e. distributed subsystems interact with each other via a finite number of points, the number of degrees of freedom (DOFs) of a finite-dimensional subsystem (FDS) taking part in the interaction between subsystems is small as compared to the total number of DOFs.

The most related to the method discussed are the Green's function method (Bergman and McFarland, 1988; McFarland and Bergman,1990) and the modal force technique (Yee and Tsuei, 1989), which reduce the problem of finding natural frequencies of a combined system to investigation of some characteristic matrix. The authors of the former method consider uniform distributed subsystems connected to discrete masses and/or supported at a finite number of points. To obtain the characteristic equation governing natural frequencies of the combined system they use explicit formulas for the Green's functions of the distributed subsystems. The latter method is formulated as applied to FDSs. The authors use the boundary coordinate FRF matrices of subsystems to obtain the modal force matrix. The order of both the characteristic equation and the modal force matrix is equal to the number of physical boundary coordinates between the subsystems. In both cases the procedure of obtaining the characteristic (modal force) equation is not formalized (at least in the cited papers) but only described by means of various examples.

In the context of the structural analysis method the procedure of calculating the characteristic matrix is well formalized. As soon as the system under investigation is divided into subsystems, such that the interaction between the subsystems is of finite rank, one can introduce notations described in the next section and obtain the characteristic matrix following a specified formal procedure. The characteristic matrix obtained for the case of conservative systems is always symmetric. This, as well as a minimal order of the matrix, is achieved by the use of appropriate factorization of the operator governing the interaction between the subsystems. Factorization of the interaction is a simple but very useful expedient, which, as far as the authors know, is not used in other related methods of investigating complex structures.

2. THEORY

2.1. Mathematical Formulation of the Problem

Let us consider a linear complex conservative system consisting of a finite number N interacting subsystems with a finite number of interaction points. As the equations governing vibrations of distributed and finite-dimensional subsystems are different it is reasonable to use operator terminology to unify the discussion. The hat ^ will mark operators, adjoint operators are denoted by the * superscript. Assuming stationary vibrations under action of external harmonic forces and eliminating the time dependent part of the motion one can write the equation governing vibrations of the jth isolated subsystem in the following operator form

$$\hat{A}_j(\lambda)Y_j = F_j \ , j = 1, \ldots, N \ , \tag{1}$$

where linear self-adjoint operator

$$\hat{A}_j(\lambda) = \hat{K}_j - \lambda \hat{M}_j \tag{2}$$

is the dynamic stiffness operator of the jth subsystem depending on spectral parameter $\lambda = \omega^2$, ω is a circular frequency, F_j is a vector-function of amplitudes of external harmonic generalized forces acting on the jth subsystem, Y_j is an amplitude response vector-function of the subsystem, $\hat{M}_j$ and $\hat{K}_j$ are mass and stiffness operators of the jth subsystem. In the case of a one-dimensional distributed subsystem $\hat{K}_j$ is a differential operator, $\hat{M}_j$ is an operator of multiplication by the mass distribution function $m_j(x)$. In the case of a FDS $\hat{K}_j$ and $\hat{M}_j$ are operators of multiplication by symmetric stiffness and mass matrices respectively.

It is evident that vibrations of the aggregate of the non-interacting subsystems satisfy the equation

$$\hat{A}(\lambda)Y = F \ , \tag{3}$$

where $\hat{A}(\lambda)$ is a diagonal operator matrix

$$\hat{A}(\lambda) \equiv \hat{K} - \lambda \hat{M} = \operatorname{diag}[\hat{A}_j(\lambda)]_{j=1}^{N} \ ,$$

Y and F are the N-component vector-functions of responses and external forces, jth components of which are vector-functions Y_j and F_j respectively.

Let now the subsystems interact elastically at a finite number of points (e.g., the subsystems are connected to each other by means of a finite number of conservative translational and rotational springs). The equation governing vibrations of the system can be obtained from (3) by adding forces and moments of the interaction between the subsystems to the right side of the equation. Clearly, the interaction forces and moments depend on only kinematic displacements of those points that take part in the interaction, i.e. the interaction forces and moments depend on the finite-dimensional vector Y^s rather than on the distributed response Y. Denote by $\hat{S}$ an operator transforming vector-function Y into Y^s : $Y^s = \hat{S}Y$. To illustrate this note that if a subsystem is a space beam the operator $\hat{S}$ acting on its response vector-function Y transforms it into an aggregate of Eulerian variables of those points of the beam at which springs are attached. Acting on a response vector of a FDS it "chooses" responses of those DOFs (or linear combinations of them) that take part in the interaction.

Let the interaction between the subsystems be given by a stiffness matrix K^s, i.e., an amplitude vector of interaction forces and moments F^s be given by the equation

$$F^s = -K^s Y^s \ .$$

Note that such a description of the interaction is the most convenient. On the one hand, if the interaction is described in terms of translational and/or torsional springs one can easily obtain the corresponding stiffness matrix. On the other hand, the description of the interaction in terms of the stiffness matrix allows one to formalize the mathematical treatment. To transform F^s into the right side of equation (3) the adjoint operator $\hat{S}^*$ is used. As a result one obtains

$$\hat{A}(\lambda)Y = F - \hat{S}^* K^s \hat{S} Y .$$

Assuming $F = 0$ one arrives at the following operator equation governing free vibrations of the system

$$\left(\hat{A}(\lambda) + \hat{S}^* K^s \hat{S}\right) Y = 0 . \qquad (4)$$

The discrete spectrum of the system is defined to be the set of numbers λ such that equation (4) has non-trivial solutions. An example of formulation of the spectral problem in form (4) for a space beam with oscillators is given in Pesterev and Tavrizov (1994a).

2.2. Factorization of the Interaction

As the interaction between the subsystems is conservative, matrix K^s is positive semidefinite and, hence, can be factorized in the form (Horn and Johnson, 1985)

$$K^s = B^T B , \qquad (5)$$

where B is the matrix of order $n \times \dim K^s$, B^T is its transpose. The number of rows n of matrix B is equal to the rank of K^s and is also equal to the number of linearly independent interaction forces. By using formula (5) equation (4) can be written in the explicit self-adjoint form

$$\left(\hat{A}(\lambda) + \hat{S}^* B^T B \hat{S}\right) Y = 0 . \qquad (6)$$

Factorization (5), which is sometimes referred to as the skeleton factorization, is not necessary but very useful and advantageous. As will be seen further to obtain the characteristic matrix one does not need the factorization. But with the factorization the characteristic matrix obtained will have two important properties. First, the characteristic matrix will be always symmetric. Second, it will have a minimal order. Factorization (5) can be obtained either analytically (for simple stiffness matrices) or numerically (for complicated matrices obtained both experimentally and by means of a spring model of the interaction).

2.3. Characteristic Matrix and the Green's Operators

According to the theorem on perturbation of an invertible operator (Azarov *et al.*, 1974, 1975), if $\hat{\Gamma}$, $\hat{\alpha}$ and $\hat{\beta}$ are linear operators (to save room the necessary conditions of the theorem are omitted here) and $\hat{\Gamma}$ is invertible then the perturbed operator $\hat{\hat{\Gamma}} = \hat{\Gamma} + \hat{\alpha}\hat{\beta}$ is invertible if, and only if, the characteristic operator

$$\hat{\chi} = \hat{I} + \hat{\beta}\hat{\Gamma}^{-1}\hat{\alpha} \qquad (7)$$

is invertible, where $\hat{I}$ is an identity operator. If $\hat{\hat{\Gamma}}$ is invertible then

$$\hat{\hat{\Gamma}}^{-1} = \hat{\Gamma}^{-1} - \hat{\Gamma}^{-1}\hat{\alpha}\hat{\chi}^{-1}\hat{\beta}\hat{\Gamma}^{-1} . \qquad (8)$$

Otherwise, dimensions of null spaces of $\hat{\hat{\Gamma}}$ and $\hat{\chi}$ are equal each other and there exists one-to-one relationship between vectors from the null spaces

$$\xi = \hat{\Gamma}^{-1}\hat{\alpha}\eta,\ \eta = -\hat{\beta}\xi,\ \xi \in \operatorname{Ker}\hat{\hat{\Gamma}},\ \eta \in \operatorname{Ker}\hat{\chi}. \quad (9)$$

Formula (8) is well-known in matrix analysis — it is used to calculate the inverse of an invertible matrix perturbed by a matrix of small rank (Horn and Johnson, 1985). If operator $\hat{\Gamma}$ is perturbed by an operator of finite rank, as in the problem discussed in the paper, the characteristic operator is a matrix.

Denote $\hat{W}(\lambda) \equiv \hat{A}^{-1}(\lambda)$. Applying formula (7) to the operator of the left side of equation (6) one arrives at the formula for the characteristic matrix

$$\chi(\lambda) = I + B\hat{S}\hat{W}(\lambda)\hat{S}^* B^T , \qquad (10)$$

where I is the identity matrix of order n.

Operator $\hat{W}(\lambda)$ is referred to as the Green's (dynamic flexibility, receptance) operator of the aggregate of the non-interacting subsystems. Clearly, it is (as well as $\hat{A}(\lambda)$) a diagonal matrix of operators: $\hat{W}(\lambda) = \operatorname{diag}[\hat{W}_k(\lambda)]_{k=1}^N$, where $\hat{W}_k(\lambda)$ is the Green's operator of the kth isolated subsystem. In the case of a one-dimensional distributed subsystem, such as a rod or a beam, $\hat{W}_k(\lambda)$ is an integral operator, the kernel of which is the Green's function of the subsystem $G^k(x, \xi; \lambda)$. For a FDS $\hat{W}_k(\lambda)$ is the operator of multiplication by the receptance matrix $G^k(\lambda) \equiv (K_k - \lambda M_k)^{-1}$. It is well known that the Green's operators can be calculated by means of the modal series

$$\hat{W}_k(\lambda) = \sum_i \frac{\varphi_{ki}\varphi_{ki}^*}{\lambda_{ki} - \lambda},\ \hat{W}(\lambda) = \sum_j \frac{\varphi_j\varphi_j^*}{\lambda_j - \lambda}, \quad (11)$$

where λ_{ki} and φ_{ki}, $i = 1, 2, \ldots$, are eigenvalues (natural frequencies squared) and M-orthonormal (normalized to unitary mass) eigenfunctions of the kth subsystem: $(\varphi_{ki}, \hat{M}_k\varphi_{kj}) = \delta_{ij}$. δ_{ij} is the Kronecker delta, the inner product in the formula depends on the subsystem under consideration and is defined in either functional or finite-dimensional space; the set of values λ_j is a union of eigenvalues λ_{ki}, eigenfunction φ_j is the N-component vector-function, such that if λ_j

is equal to λ_{ki} then the kth component of φ_j is vector-function φ_{ki} and the others are zeros. As in practice a limited number of eigenvalues and eigenfunctions is available to calculate $\hat{W}_k(\lambda)$ (or $\hat{W}(\lambda)$) and to diminish the truncation error one should use the series with accelerated convergence (Hirai *et al.*, 1973; Azarov *et al.*, 1974, 1975; Pesterev and Tavrizov, 1994a,c) instead of formula (11). To accelerate convergence (generalized) static Green's operators can be used (Azarov *et al.*, 1975; Pesterev and Tavrizov, 1994a–c).

2.4. Solving Spectral Problem in Intervals of Regularity

According to the aforementioned theorem, if $\hat{A}(\lambda)$ is invertible problem (6) has non-trivial solutions if and only if characteristic matrix (10) is singular. As $\hat{A}(\lambda)$ is invertible for all values of λ not belonging to the spectra of the isolated subsystems, "new" or "nondegenerate" (McFarland and Bergman, 1990) eigenvalues of the combined system (i.e., other than eigenvalues of the isolated subsystems) satisfy the determinant equation: $\det \chi(\lambda) = 0$. Note that in practice it is more convenient not to solve the determinant equation but use one of the two following ways.

As established by Pesterev (1992) eigenvalues of the characteristic matrix are monotonically non-decreasing continuous functions of λ in any interval $[\lambda_1, \lambda_2]$ not containing eigenfrequencies of the isolated subsystems. As a consequence, the number of non-negative eigenvalues of the characteristic matrix $N^+(\lambda)$ is monotonic non-decreasing function of λ in the interval and the number r of new eigenfrequencies in the interval can be calculated by the formula

$$r = N^+(\lambda_2) - N^+(\lambda_1)\,. \tag{12}$$

One way to calculate new eigenfrequencies is to solve the spectral problem for the characteristic matrix and to look for the values of λ at which the characteristic matrix eigenvalues (those that change sign in the interval) equal zero. Formula (12) and the aforementioned properties of monotonicity and continuity of the characteristic matrix eigenvalues make solving the problem not difficult and guarantee that all new eigenfrequencies of the combined system will be found.

Another way is to transform the matrix into a triangular form by means of some inertia-preserving (congruent) transformation, e.g., by Gaussian elimination without row or column interchanges, (Bergman and McFarland, 1988; McFarland and Bergman, 1990) and to look for those values of λ at which some diagonal elements change signs. Clearly, the number of non-negative diagonal elements of the triangular matrix at some value of λ is equal to $N^+(\lambda)$.

Let $\tilde{\lambda}_k$ be some new (nondegenerate) eigenvalue of the combined system of multiplicity $\tilde{r}_k$, which is to say that the equation $\chi(\tilde{\lambda}_k)h^k = 0$ has $\tilde{r}_k$ non-trivial linearly independent solutions $h_i^k \in \operatorname{Ker}\chi(\tilde{\lambda}_k)$, $i = 1, \ldots, \tilde{r}_k$. Then, as it follows from formulas (6) and (9), the corresponding linearly independent eigenfunctions of the system can be calculated by the formula

$$Y^i = \hat{W}(\tilde{\lambda}_k)\hat{S}^* B^T h_i^k\,,\ i = 1, \ldots, \tilde{r}_k\,. \tag{13}$$

2.5. Degenerate Eigenfrequencies and Corresponding Eigenfunctions

Some eigenfrequencies of the isolated subsystems may also belong to the spectrum of the combined system. As the Green's operator does not exist at these frequencies the theorem on perturbation of an invertible operator is not applicable to this case and such frequencies have to be analyzed separately.

Let λ_j be some eigenvalue of the aggregate of the noninteracting subsystems of multiplicity r_j. Rewrite formula (11) for $\hat{W}(\lambda)$ in a neighbourhood of λ_j separating those terms that have pole λ_j

$$\hat{W}(\lambda) = \hat{W}_{reg}(\lambda) + \frac{\varphi_j^0 (\varphi_j^0)^*}{\lambda_j - \lambda}\,, \tag{14}$$

where (functional) matrix φ_j^0 has r_j columns, and define matrix $\chi_{reg}(\lambda)$ by the formula

$$\chi_{reg}(\lambda) = I + B\hat{S}\hat{W}_{reg}(\lambda)\hat{S}^* B^T \equiv \chi(\lambda) - \frac{B\hat{S}\varphi_j^0 (B\hat{S}\varphi_j^0)^T}{\lambda_j - \lambda}\,. \tag{15}$$

Clearly, $\hat{W}_{reg}(\lambda)$ and $\chi_{reg}(\lambda)$ exist at λ_j.

It has been shown that to determine whether λ_j is also the eigenvalue of the combined system one should investigate the following (written in the partitioned form) matrix of order $n + r_j$ (Gould, 1966; Azarov *et al.*, 1974, 1975)

$$\chi_j = \begin{bmatrix} \chi_{reg}(\lambda_j) & B\hat{S}\varphi_j^0 \\ (\hat{S}\varphi_j^0)^T B^T & 0 \end{bmatrix}\,. \tag{16}$$

Eigenvalue λ_j belongs to the spectrum of the combined system if, and only if, χ_j is singular and its multiplicity $\tilde{r}_j$ is equal to the dimension of the null space of χ_j.

The different method, based upon investigation of limit behaviour of the characteristic matrix eigenvalues in a neighbourhood of λ_j, was proposed by Pesterev (1992). The multiplicity $\tilde{r}_j$ of λ_j in the spectrum of the combined system can be defined by calculating inertia of the characteristic matrix on the left and on the right of λ_j, and is given by the formula

$$\tilde{r}_j = N^+(\lambda_j + 0) - N^+(\lambda - 0) + r_j \ .$$

If $\tilde{r}_j = 0$, then λ_j does not belong to the combined system spectrum. Eigenfrequencies belonging both to the spectra of the isolated subsystems and to the spectrum of the combined system are called "degenerate" (McFarland and Bergman, 1990) or "conservative" (Gould, 1966) eigenfrequencies.

Let λ_j belong to the spectrum of the combined system and let its multiplicity be $\tilde{r}_j$, then there exist $\tilde{r}_j$ linearly independent vectors g_i^j satisfying the equation

$$\chi_j g_i^j = 0 \ , \ i = 1, \ldots, \tilde{r}_j \ . \qquad (17)$$

Write g_i^j in the partitioned form

$$g_i^j = \left\{ \begin{array}{c} h_i^j \\ z_i^j \end{array} \right\} , h_i^j \in R^n \ , z_i^j \in R^{r_j} \ . \qquad (18)$$

According to Pesterev (1992) $\tilde{r}_j$ linearly independent eigenfunctions of the combined system corresponding to λ_j are given by the formula

$$\tilde{\varphi}_i = \hat{W}_{reg}(\lambda_j)\hat{S}^* B^T h_i^j + \varphi_j^0 z_i^j \ . \qquad (19)$$

It can be seen from the formula that in the general case any system eigenfunction is the sum of some linear combination of the eigenfunctions of the resonating subsystems (the second term) and some "particular" solution. To illustrate formula (19) take, for simplicity, $r_j = 1$ (only one resonating subsystem at $\lambda = \lambda_j$) and consider, without going into details, two cases which can take place:
1) $h_i^j = 0$. The case takes place when the eigenfunction of the resonating subsystem corresponding to λ_j has nodes at the points of connections to the other subsystems. The corresponding eigenvibration of the combined system is as follows. The resonating subsystem vibrates as if it were isolated, the forces of interaction between the subsystems are equal to zero. Actually, acting on the both sides of (19) by operator $\hat{S}$ and premyltiplying by matrix K^s, using (5) and (16)-(18), one obtains

$$F^s \equiv K^s \hat{S} \tilde{\varphi}_i = B^T B \hat{S} \varphi_j^0 z_i^j = 0 \ , \qquad (20)$$

where (as $r_j = 1$) $\hat{S}\varphi_j^0$ is one-column matrix and z_i^j is a scalar. In terms of limit behaviour of eigenvalues of $\chi(\lambda)$ the case means that all the eigenvalues have removable singularities at λ_j and none of them equals zero (Pesterev, 1992).
2) $h_i^j \neq 0$. In this case the resonating subsystem exchanges forces with other subsystems. Analogously to (20) one can obtain $F^s = -B^T h_i^j$, the right side of the equation is non-zero as the rank of matrix B^T is equal to the number of its columns. It is not difficult to show that vector F^s is orthogonal to the corresponding eigenfunction of the resonating subsystem when isolated, and therefore the interaction forces do not cause resonance of the subsystem. In terms of limit behaviour of the characteristic matrix eigenvalues this case means that there exists an eigenvalue of $\chi(\lambda)$ such that its limit is equal to zero when λ tends to λ_j (Pesterev, 1992).

Eigenfunctions of the combined system corresponding to the first and to the second case are called by McFarland and Bergman (1990) "nondegenerate" and "degenerate" modes, respectively. It is possible for both degenerate and nondegenerate modes to exist at the same degenerate frequency (that is consistent with formula (19)), one relevant example of such a phenomenon is given by McFarland and Bergman (1990).

2.6. Orthogonalization and Normalization of Eigenfunctions

If some eigenvalue of the combined system is multiple then corresponding eigenfunctions in the general case may be not orthogonal. To orthogonalize (and normalize) them one has to calculate inner product in the space which the eigenfunctions belong to. That is, the orthogonality relations and the formulas for norms are given either in a functional space (in the case of distributed subsystems) or in R-dimensional space (in the case of finite-dimensional subsystems), where R is the number of DOFs of the combined system. It is established by Pesterev (1992) that in the context of the structural analysis method the problems of orthogonalizing eigenfunctions and calculating their norms can be solved in a finite-dimensional space even if the subsystems are distributed.

For each eigenvalue $\tilde{\lambda}_j$ of the combined system introduce inner product $< \cdot \ , \ \cdot >$ in a finite-dimensional space as follows. If λ_j is nondegenerate define the inner product in the n-dimensional space according to the formula

$$< h_i, h_k > = (h_i, \chi'(\tilde{\lambda}_j) h_k) \ . \qquad (21)$$

If λ_j is degenerate and r_j is its multiplicity in the spectra of the subsystems the inner product in $(n + r_j)$-dimensional space of vectors g of form (18) is given by the formula

$$< g_i, g_k > = (h_i, \chi'_{reg}(\tilde{\lambda}_j) h_k) + (z_i, z_k) \ . \qquad (22)$$

The round brackets in the right sides of formulas (21) and (22) denote usual (Euclidean) scalar product of vectors. Derivatives of the characteristic matrix $\chi'(\tilde{\lambda}_j) = B\hat{S}\hat{W}'(\tilde{\lambda}_j)\hat{S}^* B^T$ and its regular component $\chi'_{reg}(\tilde{\lambda}_j) = B\hat{S}\hat{W}'_{reg}(\tilde{\lambda}_j)\hat{S}^* B^T$ can be easily obtained by differentiating the series in (11). It is evident that matrices $\chi'(\tilde{\lambda}_j)$ and

$\chi'_{reg}(\tilde{\lambda}_j)$ are positive definite and, thus, formulas (21) and (22) define the inner products correctly.

To obtain orthonormal eigenfunctions of the combined system one should set the norms of vectors $h_i^j \in \operatorname{Ker}\chi(\tilde{\lambda}_j)$ (or $g_i^j \in \operatorname{Ker}\chi_j$) to unity and orthogonalize the vectors corresponding to the same frequency in the space with inner product (21) ((22)) and then take advantage of formulas (13) ((19)). It is shown that vectors h_i orthogonal in the space with the metrics defined by formula (21) are also orthogonal in the ordinary sense (Pesterev, 1992), and that orthogonality of vectors g_i in the space with the metrics defined by (22) results in Euclidean orthogonality of vectors h_i (Pesterev and Tavrizov, 1994c).

3. CONCLUSION

The general formalism of the structural analysis method as applied to spectral problems for complex conservative systems has been presented. The method is well suited to investigation of systems consisting of distributed and finite-dimensional subsystems interacting at a finite number of points. It reduces the problem of finding eigenfrequencies of the combined system to the problem of investigating some characteristic matrix. The order of the matrix is equal to the rank of the interaction (the number of linearly independent interaction forces and moments). The advanced formalism of the method allows one to obtain the characteristic matrix in a predetermined way. The use of factorization of the interaction results in a minimal order of the characteristic matrix and makes it symmetric. An efficient algorithm of calculating multiplicity of degenerate eigenfrequencies and established properties of the characteristic matrix eigenvalues guarantee finding all eigenfrequencies and orthonormal eigenfunctions of the combined system in a given frequency range. The problem of calculating both nondegenerate and degenerate modes corresponding to a degenerate eigenfrequency is solved successfully. It is shown how one can solve problems of orthogonalizing eigenfunctions and calculating their norms in the space, dimension of which is equal to the rank of the interaction.

4. REFERENCES

Azarov, V.L., L.N. Lupichev, and G.A. Tavrizov (1974). Factorized perturbations method in study of complex physical systems, *Proc. of VI IFAC Cong., Sec. III*, Yerevan, USSR, pp. 63–77.

Azarov, V.L., L.N. Lupichev, and G.A. Tavrizov (1975). *Mathematical Methods for Investigation of Complex Physical Systems.* Nauka, Moscow (in Russian).

Bergman, L.A., and D.M. McFarland (1988). On the vibration of a point supported linear distributed structure. *J. Vib., Acous., Stress and Reliab. Design*, **110**, 485–492.

Gould, S.H. (1966). *Variational Methods for Eigenvalue Problems.* Oxford Univ. Press, London.

Hirai, I., T. Yoshimura, and K. Takamura (1973). On a direct eigenvalue analysis for locally modified structures. *Int. J. Numer. Meth. Engin.*, **6**, 441–442.

Horn, R.A., and C.R. Johnson (1985). *Matrix Analysis.* Cambridge University Press, Cambridge.

McFarland, D.M., and L.A. Bergman (1990). Analysis of passive and active discrete-distributed linear dynamical systems using Green's function methods. *Technical Report AAE 90-4, UILU ENG 90-0504, University of Illinois*, Urbana, Illinois.

Pesterev, A.V. (1992). The method of guaranteed finding complex conservative systems discrete spectra. In: *Computational and Applied Mathematics I — Algorithms and Theory, Selected and revised papers from the 13th IMACS World Cong.* (C. Brezinski and U. Kulish, Eds), pp. 399–408. Elsevier.

Pesterev, A.V., and G.A. Tavrizov (1994a). Vibrations of beams with oscillators, I: Structural analysis method for solving the spectral problem. *J. of Sound and Vibr.*, **170**, 521–536.

Pesterev, A.V., and G.A. Tavrizov (1994b). Vibrations of beams with oscillators, II: Generalized static Green's operator. *J. of Sound and Vibr.*, **170**, 537–544.

Pesterev, A.V., and G.A. Tavrizov (1994c). On inversion of some meromorphic matrices. *Linear Algebra and its Applications*, **212/213**, 505–521.

Yee, E.K.L., and Y.G. Tsuei (1989). Direct component modal synthesis technique for system dynamic analysis. *AIAA J.*, **27**, 1083–1088.

A REALIZATION METHOD OF THE TRANSFER FUNCTIONS CONTAINING VARIABLE PARAMETER

ATSUSHI KAWAKAMI

Kanazawa Institute of Technology, Department of Electronics,
7-1, Ougigaoka, Nonoichi-machi, Ishikawa-ken, 921 Japan

Abstract. In this paper, we propose a method for realizing transfer functions containing variable parameter, by the state-space method. By using this method, variable transfer functions (VTF) can be often realized with a minimal dimension. In case that a minimal realization can not be obtained, the realization dimension can be fairly reduced.

Key Words. Variable parameter; variable transfer functions; statical feedback; minimal realization; dynamical feedback

1. INTRODUCTION

Circuits containing variable elements are the general name of the circuits containing elements whose values can be varied independently to time and frequency, by varing the values continuously, the designated input-output characteristics can be obtained continuously. Such variable circuits take an important place in the measurement-control system and the communication system. Especially, in the design of the variable equalizers, many researches have been done so far (Bode, 1938 et al).

Though, various methods to synthesize the circuits containing variable elements mentioned above exist, the synthesis method in terms of variable transfer functions (VTF) which consist of the complex angular frequency s and the variable parameter λ is thought to be an interesting method from the point of view of the unification of the theory.

Some methods for realizing VTF have been reported (Newcomb, 1968; Delansky, 1969) and they are very interesting. But, these methods have the difficulty that the realization dimension is very high. Hence, the circuit synthesis is very complicated.

Therefore, in this paper, in order to overcome the difficulty of the above-mentioned methods, we propose a method for realizing VTF by the state-space method. In this method, we perform the statical feedback on the state-space realization form of separable-denominator VTF, so that we realize the general VTF. By using this method, VTF can be often realized with a minimal dimension.

In case that minimal realization can not be obtained, we propose a method that we perform the dynamical feedback on it. By using this method, the realization dimension can be fairly reduced.

2. VOLTAGE TRANSFER FUNCTIONS CONTAINING VARIABLE PARAMETER

Circuit functions of circuit containing m gearing variable elements, can be expressed by

$$F(s,\lambda)=\sum_{i=0}^{m} g_i(s)\lambda^i / \sum_{i=0}^{m} f_i(s)\lambda^i \quad (1)$$

$$=\sum_{j=0}^{n} \hat{g}_j(\lambda)s^j / \sum_{j=0}^{n} \hat{f}_j(\lambda)s^j \quad (2)$$

(Kawakami et al, 1982a). We call this $F(s,\lambda)$ variable transfer functions (VTF).

3. STATE-SPACE REALIZATION FORM OF SEPARABLE-DENOMINATOR VTF

Separable-denominator VTF in the following, can be realized with a minimal dimension (Kawakami et al, 1982b).

$$F_S(s,\lambda)=g(s,\lambda)/\{a(s)\cdot b(\lambda)\} \quad (3)$$

A minimal dimension is a degree with respect to s, λ, of its denominator.

$F_S(s,\lambda)$ in (3) can be realized by the following two types of state-space form, by the method in (Kawakami et al, 1982b).

Controllable-Observable (C-O) state-space realization form:

$$\begin{cases} \begin{bmatrix} \dot{\mathbb{x}} \\ \mathbb{y}_i \end{bmatrix} = \begin{bmatrix} \mathbb{A}_{11} & \mathbb{A}_{12} \\ \mathbb{0} & \mathbb{A}_{22} \end{bmatrix} \begin{bmatrix} \mathbb{x} \\ \mathbb{u}_i \end{bmatrix} + \begin{bmatrix} \mathbb{b}_1 \\ \mathbb{b}_2 \end{bmatrix} u \\ y = [\mathbb{0} \;\; 1 \,\vdots\, \mathbb{C}_2][\mathbb{x}^T \vdots \mathbb{u}_i^T]^T + du \end{cases} \quad (4)$$

where

$$\mathbb{y}_i=[y_1 \cdots y_m]^T, \quad \mathbb{u}_i=[u_1 \cdots u_m]^T \quad (5)$$

$$y_i=\lambda u_i \qquad (i=1,2,\cdots,m) \quad (6)$$

$$\mathbb{b}_2=[\,\mathbb{0}\;\; 1]^T. \quad (7)$$

Observable-Controllable (O-C) state-space realization form:

$$\begin{cases} \begin{bmatrix} \dot{\mathbb{x}} \\ \mathbb{y}_i \end{bmatrix} = \begin{bmatrix} \mathbb{A}_{11}^T & \mathbb{0} \\ \mathbb{A}_{12}^T & \mathbb{A}_{22}^T \end{bmatrix} \begin{bmatrix} \mathbb{x} \\ \mathbb{u}_i \end{bmatrix} + \begin{bmatrix} \mathbb{b}_2 \\ \mathbb{C}_2^T \end{bmatrix} u \\ y = [\mathbb{b}_1^T \vdots \mathbb{0} \;\; 1][\mathbb{x}^T \vdots \mathbb{u}_i^T]^T + du. \end{cases} \quad (8)$$

4. MINIMAL REALIZATION OF VTF

In (1), it is assumed that $n \geq m$. Similarly, we can realize it in case that $n < m$.

4.1. Modification of VTF

We perform a bilinear transformation of λ in the following

$$\lambda=(c\lambda'+d)/(a\lambda'+b) \quad (9)$$

on $F(s,\lambda)$ in (1). And, we decide values of a, b, c, d so that $\hat{g}_m(s)$ of

$$\hat{F}(s,\lambda')=\sum_{i=0}^{m} \hat{g}_i(s)\lambda'^i / \sum_{i=0}^{m} \hat{f}_i(s)\lambda'^i \quad (10)$$

is (n-1)-degree polynomial and $\hat{f}_m(s)$ of (10) is n-degree polynomial.

Dividing the denominator and numerator of $\hat{F}(s,\lambda')$ in (10) by the coefficient of s^n of $\hat{f}_m(s)$, we obtain

$$\hat{F}(s,\lambda')=\sum_{i=0}^{m} \tilde{g}_i(s)\lambda'^i / \sum_{i=0}^{m} \tilde{f}_i(s)\lambda'^i \quad (11)$$

$$=\sum_{j=0}^{n} \bar{g}_j(\lambda')s^j / \sum_{j=0}^{n} \bar{f}_j(\lambda')s^j. \quad (12)$$

4.2. Realization of Separable-Denominator VTF Containing Unknown Quantities

Consider the case of

$$\hat{F}_S(s,\lambda',\mathbb{a}_i)=\sum_{i=0}^{m} \tilde{g}_i(s)\lambda'^i / \{(\sum_{i=0}^{n-1} a_i s^i + s^n)\cdot \bar{f}_n(\lambda')\} \quad (13)$$

where a_i $(i=0,1,\cdots,n-1)$ are unknown quantities.

We obtain the C-O state-space realization form of $\hat{F}_S(s,\lambda',a_i)$ in (13). It can be expressed as

$$\begin{cases}\begin{bmatrix}\dot{x}\\ \mathbb{y}_i\end{bmatrix}=\left[\begin{array}{c|c}\mathbb{A}_{11}(a_i) & \mathbb{A}_{12}(a_i)\\ \hline 0 & \mathbb{A}_{22}\end{array}\right]\begin{bmatrix}x\\ \mathbb{u}_i\end{bmatrix}+[\mathbb{b}_1^T \mid \mathbb{b}_2^T]^T u\\ y=[0 \mid \mathbb{C}_2][x^T \mid \mathbb{u}_i^T]^T.\end{cases} \tag{14}$$

4.3. Statical Feedback Containing Unknown Quantities

We perform the following statical feedback on (14).

$$u=\mathbb{F}(a_i)x+v \tag{15}$$

$\mathbb{F}(a_i)$ in (15) can be calculated by the method in (Bass *et al*, 1965).

Then, the following system can be obtained.

$$\begin{cases}\begin{bmatrix}\dot{x}\\ \mathbb{y}_i\end{bmatrix}=\left[\begin{array}{c|c}\hat{\mathbb{A}}_{11}(a_i) & \mathbb{A}_{12}(a_i)\\ \hline \hat{\mathbb{A}}_{21}(a_i) & \mathbb{A}_{22}\end{array}\right]\begin{bmatrix}x\\ \mathbb{u}_i\end{bmatrix}+[\mathbb{b}_1^T \mid \mathbb{b}_2^T]^T v\\ y=[0 \mid \mathbb{C}_2][x^T \mid \mathbb{u}_i^T]^T\end{cases} \tag{16}$$

Now,

$$\det\{s\mathbb{I}_n-\hat{\mathbb{A}}_{11}(a_i)\}=\tilde{f}_m(s) \tag{17}$$

$$\det(\lambda'\mathbb{I}_m-\mathbb{A}_{22})=\bar{f}_n(\lambda') \tag{18}$$

are satisfied.

4.4. Calculation of Unknown Quantities

We calculate the following characteristic polynomial of (16),

$$\det\left[\begin{array}{c|c}s\mathbb{I}_n-\hat{\mathbb{A}}_{11}(a_i) & -\mathbb{A}_{12}(a_i)\\ \hline -\hat{\mathbb{A}}_{21}(a_i) & \lambda'\mathbb{I}_m-\mathbb{A}_{22}\end{array}\right]$$

$$=\det\{s\mathbb{I}_n-\mathbb{A}_{11}(a_i)-\mathbb{b}_1\mathbb{F}(a_i)\}\cdot\det(\lambda'\mathbb{I}_m-\mathbb{A}_{22})+\hat{h}(s,\lambda',a_i). \tag{19}$$

From (17) and (18), equation (19) can be expressed as

$$\tilde{f}_m(s)\bar{f}_n(\lambda')+\hat{h}(s,\lambda',a_i). \tag{20}$$

Then, VTF of the system in (16) is

$$F(s,\lambda',a_i)=\sum_{i=0}^{m}\tilde{g}_i(s)\lambda'^i/\{\tilde{f}_m(s)\cdot\bar{f}_n(\lambda')+\hat{h}(s,\lambda',a_i)\}. \tag{21}$$

We express denominator polynomial of $\hat{F}(s,\lambda')$ in (11) as

$$\tilde{f}_m(s)\bar{f}_n(\lambda')+h(s,\lambda'). \tag{22}$$

Comparing (22) with (20), we calculate the values of a_i $(i=0,1,\cdots,n-1)$ so that

$$h(s,\lambda')=\hat{h}(s,\lambda',a_i). \tag{23}$$

In case that such a_i $(i=0,1,\cdots,n-1)$ don't exist, it is impossible to realize $F(s,\lambda)$ in (1) with a minimal dimension, by this method. Then, we apply the realization method which we will propose in 5.

Now, it is assumed that such a_i $(i=0,1,\cdots,n-1)$ (equation (23) is satisfied) exist. Substituting those a_i into (16), equation (16) is a minimal realization of $\hat{F}(s,\lambda')$ in (11).

In case that $n<m$ in $F(s,\lambda)$ in (1), we consider a separable-denominator VTF in the following.

$$\hat{F}_S(s,\lambda',a_i)=\sum_{i=0}^{m}\tilde{g}_i(s)\lambda'^i/\{\tilde{f}_m(s)\cdot(\sum_{i=0}^{m-1}a_i\lambda'^i+\lambda'^m)\} \tag{24}$$

We obtain O-C state-space realization form of $\hat{F}_S(s,\lambda',a_i)$ in (24), and after that, we calculate the values of a_i $(i=0,1,\cdots,m-1)$ through the similar steps.

5. REALIZATION METHOD BY DYNAMICAL FEEDBACK

5.1. Modification of VTF

We perform a bilinear

transformation of λ in the following

$$\lambda=(g\lambda''+h)/(e\lambda''+f) \qquad (25)$$

on $F(s,\lambda)$ in (1). And, we decide the values of e, f, g, h so that the following transformed VTF

$$\tilde{F}(s,\lambda'')=\sum_{i=0}^{m} q_i(s)\lambda''^i / \sum_{i=0}^{m} p_i(s)\cdot\lambda''^i \qquad (26)$$

satisfies the following two conditions

① $q_m(s)$ is (n-1)-degree polynomial.
② $p_m(s)$ is a Hurwitz polynomial.

Dividing the denominator and numerator of $\tilde{F}(s,\lambda'')$ in (26) by the coefficient of s^n in $p_m(s)$, we obtain

$$\tilde{F}(s,\lambda'')=\sum_{i=0}^{m} \hat{q}_i(s)\lambda''^i / \sum_{i=0}^{m} \hat{p}_i(s)\cdot\lambda''^i. \qquad (27)$$

Performing the steps in 4.2., 4.3. on $\tilde{F}(s,\lambda'')$ in (27), we obtain the system corresponding to (16).

5.2. Calculation of Dynamical Feedback

The system (corresponding to (16)) obtained in 5.1., can be expressed as

$$\begin{cases} \dot{\mathbb{x}}=\hat{\mathbb{A}}_{11}(a_i)\mathbb{x}+[\mathbb{A}_{12}(a_i) \mid \mathbb{b}_1]\cdot[\mathbb{u}_i^T \mid v]^T \\ \begin{bmatrix} \mathbb{y}_i \\ y \end{bmatrix}=\begin{bmatrix} \hat{\mathbb{A}}_{21}(a_i) \\ 0 \quad 1 \end{bmatrix}\mathbb{x}+\begin{bmatrix} \mathbb{A}_{22} & \mathbb{b}_2 \\ \mathbb{c}_2 & 0 \end{bmatrix}\begin{bmatrix} \mathbb{u}_i \\ v \end{bmatrix}. \end{cases} \qquad (28)$$

From (28), we calculate backward

$$\begin{bmatrix} \mathbb{y}_i \\ y \end{bmatrix}=\begin{bmatrix} \hat{\mathbb{A}}(s,a_i) & \hat{\mathbb{b}}(s,a_i) \\ \hat{\mathbb{c}}(s,a_i) & \hat{d}(s) \end{bmatrix}\cdot[\mathbb{u}_i^T \mid v]^T. \qquad (29)$$

From (29), we calculate

$$\hat{p}_m(s)\cdot\det\{\lambda''\mathbb{I}_m-\hat{\mathbb{A}}(s,\mathbb{a}_i)\} =\sum_{i=0}^{m-1}\tilde{h}_i(s,\mathbb{a}_i)\lambda''^i+\hat{p}_m(s)\lambda''^m. \qquad (30)$$

Then, VTF of the system (28) is

$$F(s,\lambda'',\mathbb{a}_i)=\sum_{i=0}^{m}\hat{q}_i(s)\lambda''^i / \{\sum_{i=0}^{m-1}\tilde{h}_i(s,\mathbb{a}_i)\lambda''^i+\hat{p}_m(s)\lambda''^m\}. \qquad (31)$$

By performing the following dynamical feedback on (28), we can let the denominator polynomial of VTF coincide with the denominator polynomial of (27),

$$v=\mathbb{f}(s,\mathbb{a}_i)\mathbb{u}_i+w. \qquad (32)$$

$\mathbb{f}(s,\mathbb{a}_i)$ in (32) can be calculated by the method in (Bass <u>et al</u>, 1965).

We decide the values of a_i (i=0,1, --,n-1) so that the following three conditions are satisfied.

③ Least common multiple of denominator polynomials of all entries of $\mathbb{f}(s,\mathbb{a}_i)$ is Hurwitz polynomial.
④ All entries of $\mathbb{f}(s,\mathbb{a}_i)$ are proper rational functions.
⑤ McMillan degree of $\mathbb{f}(s,\mathbb{a}_i)$ is as low as possible.

$\mathbb{f}(s,\mathbb{a}_i)$ designated in this way can be expressed as $\mathbb{f}(s)$.

5.3. Realization by Dynamical Feedback

The dynamical feedback becomes

$$v=\mathbb{f}(s)\mathbb{u}_i+w. \qquad (33)$$

We realize $\mathbb{f}(s)$ in (33) with a minimal dimension, and substitute the system into (33).

$$\begin{cases} \dot{x}_v=A_v x_v+[\mathbb{b}_{v1} \mid \mathbb{b}_{v2}]\,[\mathbb{u}_i^T \mid w]^T \\ v=c_v x_v+[\mathbb{d}_{v1} \mid d_{v2}]\,[\mathbb{u}_i^T \mid w]^T \end{cases} \qquad (34)$$

We substitute the values of a_i (i =0,1,--,n-1) obtained in 5.2 into the system corresponding to (16) in this section, and combine that system-(34).

6. CONCLUSION

In this paper, we proposed a method for realizing VTF, by the state-space method. By using this method, VTF can be realized with a very low dimension.

7. REFERENCES

Bode, H.W. (1938). Variable equalizers. Bell Syst. Tech. J., 17, 2, pp.229-244.

Newcomb, R.W. (1968). Active Integrated Circuit Synthesis. Englewood Cliffs, NJ: Prentice-Hall.

Delansky, J.F. (1969). Some synthesis methods for adjustable networks using multivariable techniques. IEEE Trans. Circuits Theory, Vol.CT-16, pp.435-443.

Kawakami, A., and S. Takahashi (1982a). Approximation of the characteristics of the networks containing variable elements. Trans. I.E.C.E., Japan, J65-A, 8, pp.857-863.

Kawakami, A., J. Tajima, and S. Takahashi (1982b). Approximation and construction of the network containing variable elements. Electron. Commun. Japan, Vol.64-A, No.4, pp.9-18.

Bass, R.W., and I. Gura (1965). Higher-order system design via state space considerations. JACC, pp.311-318.

THE GENERIC WIENER-HOPF INDICES OF A STRUCTURED POLYNOMIAL MATRIX

J.W. VAN DER WOUDE

Faculty of Technical Mathematics and Informatics, Delft University of Technology, Mekelweg 4, 2628 CD Delft, The Netherlands

Abstract. In this paper a graph theoretical method is presented for the computation of the generic Wiener-Hopf indices of a structured full row rank polynomial matrix. The method only requires the degree structure of the polynomial matrix.

Keywords: row proper matrices, matching, bipartite graphs, degree structure.

1. INTRODUCTION

This paper deals with polynomial matrices of which only the degree structure is given. Given this degree structure a method is proposed that for almost all possible coefficient combinations (= generically) results in the so-called right Wiener-Hopf indices.

This paper is inspired by the approach towards linear time-invariant systems initiated by J.C. Willems (1986). One of the features of this approach is that the systems are described by means of linear difference or differential equations with constant coefficients in the external variable $w(t)$ and are denoted by $R(\sigma)w(t) = 0$. Here $R(s)$ is a polynomial matrix and σ is either the shift operator or differentiation. The system with the polynomial matrix $R(s)$ is called an AR-system. The starting point in this paper is that only the degrees of the (nonzero) entries of $R(s)$ are known.

2. PRELIMINARIES

The set of polynomials with real coefficients in the indeterminate s is denoted by $\mathbb{R}[s]$. $\mathbb{R}^{g\times q}[s]$ stands for the set of $g \times q$ polynomial matrices with entries in $\mathbb{R}[s]$. The degree of a nonzero scalar polynomial $r(s) \in \mathbb{R}[s]$, or a polynomial row $r(s) \in \mathbb{R}^{1\times q}[s]$, is the exponent of the highest power of s that it is present in $r(s)$. This degree is denoted by $\deg(r(s))$ and it follows that $\deg(r(s)) \geq 0$. If $r(s) \equiv 0$ then $\deg(r(s)) = -\infty$.

3. WIENER-HOPF INDICES

3.1. Factorization

A polynomial matrix $R(s) \in \mathbb{R}^{g\times q}[s]$ with rank p can be factorized as follows (the so-called Wiener-Hopf factorization) (Clancey *et al.*, 1981):

$$R(s) = U(s)\begin{bmatrix} \mathrm{diag}(s^{t_1}, \ldots, s^{t_p}) & 0 \\ 0 & 0 \end{bmatrix} V(s),$$

where $U(s)$ is a unimodular polynomial matrix (square invertible with a polynomial inverse), $V(s)$ a biproper rational matrix (square invertible with a proper rational inverse) and $t_1, \ldots, t_p$ are integers such that $t_i \leq t_{i+1}$, $1 \leq i < p$. The integers $t_1, \ldots, t_p$ are called the right Wiener-Hopf indices of $R(s)$.

3.2. Row proper polynomial matrices

Let $R(s)$ be a polynomial matrix in $\mathbb{R}^{g\times q}[s]$ with rows $r_1(s), r_2(s), \ldots, r_g(s) \in \mathbb{R}^{1\times q}[s]$, i.e.

$$R(s) = \begin{bmatrix} r_1(s) \\ r_2(s) \\ \vdots \\ r_g(s) \end{bmatrix}.$$

Denote $\deg(r_i(s)) = n_i$ for $i = 1, 2, \ldots, g$, and write $r_i(s) = r_{i\,0}s^{n_i} + r_{i\,1}s^{n_i-1} + \ldots + r_{i\,n_i-1}s + r_{i\,n_i}$ with $r_{ij} \in \mathbb{R}^{1\times q}$ for $j = 0, 1, \ldots, n_i$, where $i = 1, 2, \ldots, g$. Then the constant $g \times q$-matrix

$$L = \begin{bmatrix} r_{1\,0} \\ r_{2\,0} \\ \vdots \\ r_{g\,0} \end{bmatrix}$$

is called the leading row coefficient matrix of $R(s)$. The polynomial matrix $R(s)$ is row proper if its leading row coefficient matrix has full row rank (Wolovich, 1974).

3.3. Full row rank polynomial matrices

Assume that $R(s) \in \mathbb{R}^{g\times q}[s]$ has full row rank p, i.e. $p = g \leq q$. Furthermore, assume that $R(s)$ has a Wiener-Hopf factorization as above. Then

$$U^{-1}(s)R(s) = \begin{bmatrix} \mathrm{diag}(s^{t_1}, \ldots, s^{t_p}) & 0 \end{bmatrix} V(s),$$

where 0 denotes a zero matrix of dimensions $p \times (q-p)$. Since $U(s)$ is a unimodular matrix, the left-hand side of the above is a polynomial matrix, and consequently, so is the right-hand side. Writing $V(s) = \begin{bmatrix} V_1(s) \\ V_2(s) \end{bmatrix}$ with $V_1(s)$ of dimensions $q \times p$, it follows easily from the fact that $V_1(\infty)$ has full row rank that $\mathrm{diag}(s^{t_1}, \ldots, s^{t_p})V_1(s)$ is a row proper polynomial matrix with a leading row coefficient matrix $V_1(\infty)$ and row degrees $t_1, t_2, \ldots, t_p$. Hence, for a full row rank polynomial matrix $R(s)$ the right Wiener-Hopf indices can be determined from the row degrees of a row proper form of $R(s)$.

4. ROW PROPER MATRICES

4.1. Algorithm of Wolovich

Assume that $R(s) \in \mathbb{R}^{g\times q}[s]$ has full row rank equal to p. If its leading row coefficient matrix L also has full rank, then $R(s)$ is already row proper and the right Wiener-Hopf indices of $R(s)$ follow directly from the row degrees. If $R(s)$ is not row proper, then it can be brought into a row proper form. To see how this can be done, assume for reasons of exposition that the row degrees n_i of $R(s)$ are ordered in a nondecreasing way, i.e. $n_i \leq n_{i+1}$, $1 \leq i < p$. Since $R(s)$ is not row proper, the rows in L are linearly dependent. Therefore, there exists an index $h < p$ such that the row $r_{h+1\,0}$ in L can be written as a linear combination of the rows $r_{1\,0}, r_{2\,0}, \ldots, r_{h\,0}$ in L. Consequently, there are coefficients α_i, $i = 1, 2, \ldots, h$ such that $r_{h+1\,0} + \sum_{i=1}^{h} \alpha_i r_{i\,0} = 0$. Given these coefficients, replace row $r_{h+1}(s)$ of $R(s)$ as follows :

$$r_{h+1}(s) := r_{h+1}(s) + \sum_{i=1}^{h} \alpha_i r_i(s) s^{d_i},$$

where $d_i = n_{h+1} - n_i$ for $i = 1, 2, \ldots, h$. It is easy to see that the row degree of row $r_{h+1}(s)$ of $R(s)$ becomes less than n_{h+1}. The other rows of $R(s)$ and their degrees remain unchanged. Hence, the *sum* of the row degrees of $R(s)$ has been decreased by at least one. The above transformation can be achieved by means of a multiplication from the left by a unimodular matrix $U(s)$. It can now checked whether or not $R(s)$ is row proper. If this is the case the right Wiener-Hopf indices of $R(s)$ can be obtained from its row degrees (for the original as well as for the new version of $R(s)$). If $R(s)$ is not row proper the above procedure must be repeated. Since in each step the sum of the row degrees decreases, but never can become less than zero, this process is guaranteed to end with a row proper matrix.

4.2. Example of the algorithm

Consider the polynomial matrix $R(s)$ with its leading row coefficient matrix L given by

$$R(s) = \begin{bmatrix} s^6 - 2s^2 & 3 \\ s^5 + s^4 + s & 2s^2 - 1 \end{bmatrix}, L = \begin{bmatrix} 1 & 0 \\ 1 & 0 \end{bmatrix},$$

where the row degrees of $R(s)$ are 6 and 5, respectively. Since L has rank 1 it follows that $R(s)$ is not row proper. To bring $R(s)$ into a row proper form subtract s times the second row from the first ($r_1(s) := r_1(s) - sr_2(s)$). The result is a new matrix $R(s)$ given by

$$R(s) = \begin{bmatrix} -s^5 - 3s^2 & -2s^3 + s + 3 \\ s^5 + s^4 + s & 2s^2 - 1 \end{bmatrix}.$$

The new matrix $R(s)$ has row degrees 5 and 5, respectively, and from its leading row coefficient matrix it follows that $R(s)$ is still not row proper. So, the above procedure must be repeated. It turns out that it has to be repeated twice. Then it follows that the right Wiener-Hopf indices of the original matrix $R(s)$ are 4 and 4. See also below.

5. GENERIC ASPECTS

In this section the above algorithm of Wolovich is reviewed, but now only the degrees of the nonzero entries of a polynomial matrix $R(s)$ are assumed to be available. Furthermore, it is assumed that when repeating the algorithm no unintended cancellations occur. Below the word 'generically' always means 'generically in the original coefficients'.

5.1. Example 1

First, consider the polynomial matrix

$$R(s) = \begin{bmatrix} a_6 s^6 + \ldots + a_0 & b_0 \\ c_5 s^5 + \ldots + c_0 & d_2 s^2 + d_1 s + d_0 \end{bmatrix}$$

where the coefficients $a_0, \ldots, a_6$, b_0, $c_0, \ldots, c_5$ and $d_0, \ldots, d_2$ can have any real value. Clearly, the generic row degrees of $R(s)$ are 6 and 5, respectively. The leading row coefficient matrix of $R(s)$ in generic cases is given by $\begin{bmatrix} a_6 & 0 \\ c_5 & 0 \end{bmatrix}$. Hence, it follows that $R(s)$ generically is not row proper. To bring $R(s)$ into a row proper form, subtract $s\frac{a_6}{c_5}$ times the second row from the first ($r_1(s) := r_1(s) - s\frac{a_6}{c_5}r_2(s)$). This results in a new matrix $R(s)$ that has the following form

$$R(s) = \begin{bmatrix} \tilde{a}_5 s^5 + \ldots + \tilde{a}_0 & \tilde{b}_3 s^3 + \ldots + \tilde{b}_0 \\ c_5 s^5 + \ldots + c_0 & d_2 s^2 + d_1 s + d_0 \end{bmatrix}.$$

Here for instance $\tilde{a}_5 = a_5 - \frac{a_6}{c_5}c_4$, $\tilde{a}_4 = a_4 - \frac{a_6}{c_5}c_3$, $\tilde{a}_3 = a_3 - \frac{a_6}{c_5}c_2$, $\tilde{b}_3 = -\frac{a_6}{c_5}d_2$ and $\tilde{b}_0 = b_0$.

The leading row coefficient matrix of the new matrix $R(s)$ generically is $\begin{bmatrix} \tilde{a}_5 & 0 \\ c_5 & 0 \end{bmatrix}$. Hence, it follows that the new matrix $R(s)$ still is generically not row proper. Therefore, continue with the algorithm and subtract $\frac{\tilde{a}_5}{c_5}$ times the second row from the first ($r_1(s) := r_1(s) - \frac{\tilde{a}_5}{c_5} r_2(s)$). This yields a new matrix $R(s)$ given by

$$R(s) = \begin{bmatrix} \hat{a}_4 s^4 + \ldots + \hat{a}_0 & \hat{b}_3 s^3 + \ldots. + \hat{b}_0 \\ c_5 s^5 + \ldots + c_0 & d_2 s^2 + d_1 s + d_0 \end{bmatrix},$$

where for instance $\hat{a}_4 = \tilde{a}_4 - \frac{\tilde{a}_5}{c_5} c_4 = a_4 - \frac{a_6}{c_5} c_3 - \frac{a_5}{c_5} c_4 + \frac{a_6}{c_5^2} c_4^2$ and $\hat{b}_3 = \tilde{b}_3 = -\frac{a_6}{c_5} c_3$. Now the matrix generically has row degrees 4 and 5, respectively, and still is generically not row proper. Therefore, in the matrix $R(s)$ subtract $\frac{c_5}{\hat{a}_4} s$ times the first row from the second ($r_2(s) := r_2(s) - \frac{c_5}{\hat{a}_4} s r_1(s)$). This results in

$$R(s) = \begin{bmatrix} \hat{a}_4 s^4 + \ldots + \hat{a}_0 & \hat{b}_3 s^3 + \ldots + \hat{b}_0 \\ \hat{c}_4 s^4 + \ldots + \hat{c}_0 & \hat{d}_4 s^4 + \ldots. + \hat{d}_0 \end{bmatrix}.$$

Here for instance $\hat{c}_4 = c_4 - \frac{c_5}{\hat{a}_4} \hat{a}_3$ and $\hat{d}_4 = -\frac{c_5}{\hat{a}_4} \hat{b}_3$. As before, it follows easily that the row degrees now are generically 4 and 4, respectively. The leading row coefficient matrix is then given by $\begin{bmatrix} \hat{a}_4 & 0 \\ \hat{c}_4 & \hat{d}_4 \end{bmatrix}$. The determinant of the leading row coefficient matrix equals $\hat{a}_4 \hat{d}_4 = a_6 d_2$, from which it is clear that the matrix $R(s)$ generically is row proper. It follows that the right Wiener-Hopf indices of the original $R(s)$ generically are 4 and 4.

The evolution of the degrees of the individual entries of the intermediate polynomial matrices yields the next sequence.

$$\begin{bmatrix} 6 & 0 \\ 5 & 2 \end{bmatrix} \begin{matrix} \uparrow \\ 1 \end{matrix} \longrightarrow \begin{bmatrix} 5 & 3 \\ 5 & 2 \end{bmatrix} \begin{matrix} \uparrow \\ 0 \end{matrix} \longrightarrow \begin{bmatrix} 4 & 3 \\ 5 & 2 \end{bmatrix} \begin{matrix} 1 \\ \downarrow \end{matrix}$$

$$\longrightarrow \begin{bmatrix} 4 & 3 \\ 4 & 4 \end{bmatrix}.$$

The entry 6 (the original degree) in the first matrix above goes over into an entry 5 (the new degree), while at the same time, the entry 0 (the original degree) in the matrix goes over into an entry 3 (the new degree), being the maximum of 0 (the original degree) and 2 + 1 (the degree of the term that is subtracted). In the same way the transformations from the second to the third, and from the third to the fourth matrix can explained.

5.2. Example 2

Next, consider the polynomial matrix $R(s) =$

$$\begin{bmatrix} a_1 s + a_0 & b_2 s^2 + b_1 s + b_0 & c_2 s^2 + c_1 s + c_0 \\ 0 & d_3 s^3 + \ldots. + d_0 & e_3 s^3 + \ldots. + e_0 \\ f_2 s^2 + f_1 s + f_0 & g_4 s^4 + \ldots.. + g_0 & 0 \end{bmatrix}$$

where the coefficients in $R(s)$ again can have any real value. The row degrees of $R(s)$ are generically 2,3 and 4. The leading row coefficient matrix in generic cases is given by

$$L = \begin{bmatrix} 0 & b_2 & c_2 \\ 0 & d_3 & e_3 \\ 0 & g_4 & 0 \end{bmatrix}$$

from which it is immediate that L generically does not have full row rank and consequently that $R(s)$ is generically not row proper. To bring $R(s)$ into row proper form, determine α_1 and α_2 such that $\alpha_1 [\, 0 \;\; b_2 \;\; c_2 \,] + \alpha_2 [\, 0 \;\; d_3 \;\; e_3 \,] + [\, 0 \;\; g_4 \;\; 0 \,] = 0$, and replace the third row as follows: $r_3(s) := r_3(s) + \alpha_1 s^2 r_1(s) + \alpha_2 s r_2(s)$. This yields $R(s) =$

$$\begin{bmatrix} a_1 s + a_0 & b_2 s^2 + b_1 s + b_0 & c_2 s^2 + c_1 s + c_0 \\ 0 & d_3 s^3 + \ldots. + d_0 & e_3 s^3 + \ldots. + e_0 \\ \tilde{f}_3 s^3 + \ldots. + \tilde{f}_0 & \tilde{g}_3 s^3 + \ldots. + \tilde{g}_0 & \tilde{h}_3 s^3 + \ldots. + \tilde{h}_0 \end{bmatrix}$$

with for instance $\tilde{f}_3 = \alpha_1 a_1$, $\tilde{g}_3 = g_3 + \alpha_1 b_1 + \alpha_2 d_2$ and $\tilde{h}_3 = \alpha_1 c_2 + \alpha_2 e_3$. Here

$$\alpha_1 = \frac{e_3 g_4}{b_2 e_3 - c_2 d_3} \text{ and } \alpha_2 = -\frac{c_2 g_4}{b_2 e_3 - c_2 d_3}.$$

From this it follows easily that $R(s)$ generically is row proper. Hence, the right Wiener-Hopf indices of $R(s)$ are generically 2,3 and 3. Finally, observe that the degrees evolve as follows.

$$\begin{bmatrix} 1 & 2 & 2 \\ -\infty & 3 & 3 \\ 2 & 4 & -\infty \end{bmatrix} \begin{matrix} 2 & \\ | & 1 \\ \downarrow & \downarrow \end{matrix} \longrightarrow \begin{bmatrix} 1 & 2 & 2 \\ -\infty & 3 & 3 \\ 3 & 3 & 3 \end{bmatrix}.$$

To conclude this example we note that if α_1 and α_2 in the above were chosen arbitrarily then the degree of the third row would have stayed 4. However, because of the specific choice for α_1 and α_2, the intended cancellation occurs and the degree of the third row becomes less than 4, and generically will become 3.

6. DEGREE BEHAVIOUR

6.1. Refinement of the algorithm of Wolovich

Consider the first step of the algorithm of Wolovich for the following polynomial matrix $R(s)$ with its leading row coefficient matrix L:

$$R(s) = \begin{bmatrix} 4s + 2 & 2 & 0 \\ s^2 - 3 & 1 & 2s \\ -s^2 & s + 3 & -3s + 2 \end{bmatrix},$$

$$L = \begin{bmatrix} 4 & 0 & 0 \\ 1 & 0 & 0 \\ -1 & 0 & 0 \end{bmatrix}.$$

Writing the third row of L as a linear combination of the first two rows can be done in many ways. In fact, only the first row can be used, but also only the second one. In the first case the coefficient α_2 will be zero, and in the second case the coefficient α_1 will be zero. This indicates that not every α_i in the algorithm needs to be nonzero and that the algorithm can be refined as described below.

Assume that $R(s)$ is not row proper. Then there is a 'smallest' set of rows of L that are linearly dependent. Here 'smallest' set means that the rows in the set are linearly independent after an arbitrary row has been deleted. Without loss of generality this 'smallest' set consists of the rows of L with row index $1, 2, \ldots, t, t+1$ and the corresponding row degrees in $R(s)$ are ordered in a nondecreasing way. By the definition of 'smallest' set, the row $t+1$ of L can be written as a linear combination of the other t rows of L, and all t rows are really required in this linear combination.

It is easy to see that the algorithm of Wolovich can be as before, but now with only the rows in the above 'smallest' set, i.e determine coefficients α_i, $i = 1, 2, \ldots, t$ such that $r_{t+1\,0} + \sum_{i=1}^{t} \alpha_i r_{i\,0} = 0$ and replace row $t+1$ of $R(s)$ as follows:

$$r_{t+1}(s) := r_{t+1}(s) + \sum_{i=1}^{t} \alpha_i r_i(s) s^{d_i}$$

where $d_i = n_{t+1} - n_i$ for $i = 1, 2, \ldots, t$. Note that every coefficient α_i, $i = 1, 2, \ldots, t$ is nonzero, since every row $r_i(s)$, $i = 1, 2, \ldots, t$ is really needed to reduce the degree of row $r_{t+1}(s)$. Hence, the row degree of row $r_{t+1}(s)$ becomes less than n_{t+1}. In fact, the elements in row $r_{t+1}(s)$ that had degree n_{t+1} will go over into elements that have degree less than n_{t+1}. In the generic situations in this paper, the degrees of these elements are assumed to be $n_{t+1} - 1$. The other elements of in row $r_{t+1}(s)$ that already had degree less than n_{t+1} and will go over into elements that generically also have degree less than n_{t+1}.

6.2. Generic behavior of degrees

To see how the degrees in the generic case behave, denote the (i, j) entry of $R(s)$ by $R_{ij}(s)$ and denote its degree by $\deg(R_{ij}(s)) = \delta_{ij}$ for $i = 1, 2, \ldots, p$ and $j = 1, 2, \ldots, q$. With this notation the next relation between the row degrees n_i of the rows of $R(s)$ and the degrees δ_{ij} of its individual entries is evident: $n_i = \max(\delta_{i\,1}, \ldots, \delta_{i\,q})$ for all $i = 1, 2, \ldots, p$. The following update scheme for row $r_{t+1}(s)$ follows in the context of the above refined version of the algorithm of Wolovich. Here it is assumed that there is no unintended cancellation or, in other words, that genericity holds.

$$\delta_{t+1\,j} := \min\{n_{t+1} - 1, \max(\delta_{t+1\,j}, \delta_{1\,j} + d_1, \ldots, \delta_{t\,j} + d_t)\},$$

for $j = 1, 2, \ldots, q$, where $d_i = n_{t+1} - n_i, i = 1, 2, \ldots, t$ are as previous. To understand the scheme first note that if the coefficients in the algorithm of Wolovich are chosen arbitrarily, then the degree of the entry $(t+1, j)$ generically will become the maximum of $\delta_{t+1\,j}$ and $\delta_{ij} + d_i$, where $i = 1, 2, \ldots, t$. However, since the coefficients are chosen in such a way that the coefficients corresponding $s^{n_{t+1}}$ are cancelled the degree of row $t+1$ becomes less than n_{t+1}. As mentioned before, in this paper *it is assumed that then this degree generically will be equal to* $n_{t+1} - 1$.

7. BIPARTITE GRAPHS

Consider the polynomial matrix $R(s) \in \mathbb{R}^{g \times q}[s]$ and assume that only its degree structure is known. This means that only the degree of each entry of $R(s)$ is given. Assume that the values of the coefficients present in the nonzero entries are unknown reals. Of course $R(s)$ may also contain entries that are identically equal to zero. These entries have degree $-\infty$. The previous means that each entry can be identified with its degree.

From a polynomial matrix $R(s)$ as above (or its degree matrix Δ), the zero-nonzero structure of its leading row coefficient matrix L can be determined.

The zero-nonzero structure of L can be represented by means of a bipartite graph G. This graph consists of the vertex sets $\mathcal{V}$ and $\mathcal{W}$ and the edge set $\mathcal{E}$, and is therefore also denoted as $G = (\mathcal{V}, \mathcal{W}, \mathcal{E})$. The edges in $\mathcal{E}$ connect vertices in $\mathcal{V}$ with vertices in $\mathcal{W}$. More specific, $\mathcal{V} = \{1, 2, \ldots, g\}$, $\mathcal{W} = \{1, 2, \ldots, q\}$ and $\mathcal{E} = \{(i, j) | \delta_{ij} = n_i\}$. Here (i, j) denotes the edge between vertex $i \in \mathcal{V}$ and vertex $j \in \mathcal{W}$ and $\delta_{ij} = n_i$ means that the degree of the (i, j) entry of $R(s)$ equals the degree of the i-th row of $R(s)$.

A matching in the graph $G = (\mathcal{V}, \mathcal{W}, \mathcal{E})$ is a subset $\mathcal{M}$ of $\mathcal{E}$ consisting of arcs that pairwise do not have any vertex in common. The number of arcs in a matching $\mathcal{M}$ is called the order of the matching.

The next result is proved in for instance Murota (1987).

Proposition 7.1 *The generic rank of L equals the order of a maximum matching in G.*

It is possible that L generically has full row rank equal to $|\mathcal{V}|$, with $|\mathcal{V}|$ the cardinality of $\mathcal{V}$. In that case there is a matching in G of which the edges cover all the vertices of $\mathcal{V}$. Moreover, then the matrix $R(s)$ is generically row proper and the generic Wiener-Hopf indices of $R(s)$ are its row degrees.

However, it is also possible that the generic rank of L is less than $|\mathcal{V}|$. Then L generically is not of full row rank and there is no matching in G of which the edges cover all the vertices of $\mathcal{V}$. From proposition 7.1 it follows now that the rows of L form a set of rows that generically are linearly dependent. In particular, there exists a 'smallest' subset of rows of L that generically are linearly dependent. As before, such a set is called a 'smallest' set of rows of L that generically are linearly dependent. The indices of the rows in such a set form a subset of $\mathcal{V}$ that is also known as a minimal structurally singular subset (Pantelides, 1988).

Let such a minimal structurally singular set be denoted by $\mathcal{I}$ and write $L_{\mathcal{I}}$ for the matrix consisting of the rows of L with index in $\mathcal{I}$. Further, denote $\mathcal{I}/\{i\}$ for the set $\mathcal{I}$ without element i and write $L_{\mathcal{I}/\{i\}}$ for the matrix consisting of the rows of L with index in $\mathcal{I}/\{i\}$. Then it follows that generically rank $L_{\mathcal{I}} < |\mathcal{I}|$ and rank $L_{\mathcal{I}/\{i\}} = |\mathcal{I}| - 1$ for all $i \in \mathcal{I}$.

For subset $\mathcal{I}'$ in $\mathcal{V}$ we denote $N(\mathcal{I}')$ for the set of vertices in $\mathcal{W}$ that are connected to vertices in $\mathcal{I}'$, i.e. $N(\mathcal{I}') = \{j \in \mathcal{W} | \ (i,j) \in \mathcal{E}$ for some $i \in \mathcal{I}'\}$. Then it can be shown that a minimal structurally singular subset $\mathcal{I}$ can be characterized by the property that $|\mathcal{I}| > |N(\mathcal{I})|$ (Pantelides, 1988).

8. MAIN RESULT

8.1. Graph theoretic algorithm

Given the polynomial matrix $R(s)$ (or its degree matrix Δ), construct the bipartite graph G associated to the zero-nonzero structure of its leading row coefficient matrix L. Let this graph be denoted $G = (\mathcal{V}, \mathcal{W}, \mathcal{E})$ with $\mathcal{V}$, $\mathcal{W}$ and $\mathcal{E}$ as before.

Next determine a 'smallest' set of row indices $\mathcal{I}$ in $\mathcal{V}$ such that for any $i \in \mathcal{I}$ there holds generically

$$\text{rank } L_{\mathcal{I}/\{i\}} = \nu \text{ where } \nu = \text{ rank } L_{\mathcal{I}} = |\mathcal{I}| - 1.$$

If L generically has full row rank, the set $\mathcal{I}$ does not exist. Then $R(s)$ is row proper and the generic right Wiener-Hopf indices of $R(s)$ follow immediately from the row degrees. If L generically does not have full row rank, $\mathcal{I}$ does exist. Renumber the elements of $\mathcal{V}$ in such a way that $\mathcal{I} = \{1, 2, \ldots, \nu, \nu+1\}$ and $n_1 \leq n_2 \leq \ldots \leq n_\nu \leq n_{\nu+1}$. Next denote $d_i = n_{\nu+1} - n_i$ for $i = 1, 2, \ldots, \nu$, and update the $\nu + 1$-th row of the degree matrix Δ. Hence, for $j = 1, 2, \ldots, q$, define

$$\delta_{\nu+1\,j} := \min \{n_{\nu+1} - 1, \max(\delta_{\nu+1\,j}, \delta_{1j} + d_1, \ldots, \delta_{\nu j} + d_\nu)\}$$

With the new degree matrix Δ a new polynomial matrix $R(s)$ can be associated and the whole procedure can be repeated until after a finite number of steps $R(s)$ has become row proper.

8.2. Minimal structurally singular subset

Note that the important step in the algorithm is the determination of a set $\mathcal{I}$, because $\mathcal{I}$ is a set of minimal size of indices of rows that are really needed in making one step in the algorithm of Wolovich. This can be done by looking for subsets $\mathcal{I}$ such that $|\mathcal{I}| = |N(\mathcal{I})| + 1$.

9. EXAMPLES

9.1. Example 2 using graphs

Consider the degree matrix Δ given by

$$\begin{bmatrix} 1 & 2 & 2 \\ -\infty & 3 & 3 \\ 2 & 4 & -\infty \end{bmatrix}.$$

Hence, generically $n_1 = \max(1,2,2) = 2$, $n_2 = \max(-\infty, 3, 3) = 3$ and $n_3 = \max(2, 4, -\infty) = 4$. From the degree matrix the form of the matrices $R(s)$ and L follow. Indeed, $R(s) =$

$$\begin{bmatrix} a_1 s + a_0 & b_2 s^2 + b_1 s + b_0 & c_2 s^2 + c_1 s + c_0 \\ 0 & d_3 s^3 + \ldots. + d_0 & e_3 s^3 + \ldots. + e_0 \\ f_2 s^2 + f_1 s + f_0 & g_4 s^4 + \ldots.. + g_0 & 0 \end{bmatrix}$$

and

$$L = \begin{bmatrix} 0 & b_2 & c_2 \\ 0 & d_3 & e_3 \\ 0 & g_4 & 0 \end{bmatrix}.$$

Having obtained L, the graph G can be constructed. Note that in G the row numbering is changed for reasons of convenience.

$\mathcal{W} = \{ \ 1 \ , \ 2 \ , \ 3 \ \}$

G:

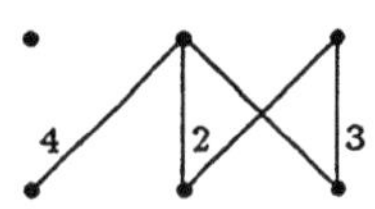

$\mathcal{V} = \{ \ 3 \ , \ 1 \ , \ 2 \ \}$

Now the set $\mathcal{I}$ in the algorithm of this paper equals $\{1, 2, 3\}$ with $N(\mathcal{I}) = \{2, 3\}$. Indeed, it follows easily that generically

$$\nu = \text{ rank } L_{\mathcal{I}} = \text{ rank } L_{\mathcal{I}/\{i\}} = 2$$

for any $i = 1, 2, 3$. Moreover, any other subset $\mathcal{I}'$ of $\mathcal{V}$ is such that $|\mathcal{I}'| \leq |N(\mathcal{I}')|$. From the row degrees in G it follows that the degree of the third row of $R(s)$ can be reduced by adding the first and second row multiplied by suitable factors of degree two and one, respectively. Then the degrees change as follows. For $j = 1, 2, 3$

$$\delta_{3j} := \min\{n_3 - 1, \max(\delta_{3j}, \delta_{1j} + d_1, \delta_{2j} + d_2)\},$$

where $d_1 = n_3 - n_1 = 2$ and $d_2 = n_3 - n_2 = 1$. Hence, $\delta_{31} := \min\{4-1, \max(2, 1+2, -\infty+1)\} = 3$, $\delta_{32} := \min\{4-1, \max(4, 2+2, 3+1)\} = 3$, $\delta_{33} := \min\{4-1, \max(-\infty, 2+2, 3+1)\} = 3$. The other degrees remain unchanged. So, the degree matrix Δ becomes

$$\begin{bmatrix} 1 & 2 & 2 \\ -\infty & 3 & 3 \\ 3 & 3 & 3 \end{bmatrix}.$$

The leading row coefficient matrix L corresponding to the new polynomial matrix $R(s)$ is given by

$$L = \begin{bmatrix} 0 & b_2 & c_2 \\ 0 & d_3 & e_3 \\ \tilde{f}_3 & \tilde{g}_3 & \tilde{h}_3 \end{bmatrix}$$

Again the bipartite graph associated to L can be easily constructed. Since this graph contains a matching of degree 3, it follows that L generically has full row rank, and consequently, that the matrix $R(s)$ generically is row proper. Hence, the algorithm can stop here and the generic right Wiener-Hopf indices of a polynomial matrix with the originally given degree structure are 2,3 and 3.

9.2. Example 3

To conclude this section a slightly modified example from van der Woude, (1992) is considered, given by $R(s) =$

$$\begin{bmatrix} 0 & \alpha(s) & 0 & 0 & 0 & \beta(s) & \gamma(s) \\ \phi(s) & \epsilon(s) & \zeta(s) & \eta(s) & 0 & \theta(s) & 0 \\ \kappa(s) & 0 & 0 & 0 & \lambda(s) & 0 & 0 \\ 0 & 0 & 0 & \mu(s) & 0 & 0 & 0 \\ \nu(s) & 0 & 0 & \xi(s) & \omega(s) & 0 & 0 \end{bmatrix},$$

with degree matrix $\Delta =$

$$\begin{bmatrix} -\infty & 1 & -\infty & -\infty & -\infty & 0 & 4 \\ 3 & 2 & 2 & 4 & -\infty & 3 & -\infty \\ 2 & -\infty & -\infty & -\infty & 0 & -\infty & -\infty \\ -\infty & -\infty & -\infty & 2 & -\infty & -\infty & -\infty \\ 4 & -\infty & -\infty & 1 & 1 & -\infty & -\infty \end{bmatrix}.$$

Hence, for instance $R_{24}(s) = \eta(s) = \eta_4 s^4 + \ldots + \eta_0$ and $R_{53}(s) = 0$ for all s. The row degrees of $R(s)$ are generically 4,4,2,2 and 4, respectively. The graph associated to the leading row coefficient matrix L of $R(s)$ can be depicted as follows.

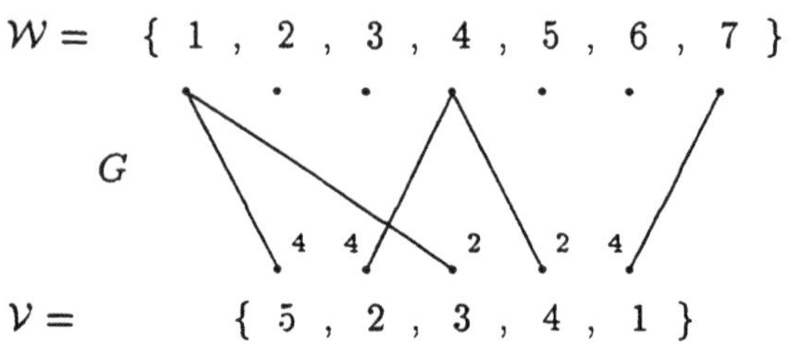

For reasons of convenience the rows in the graph G are permuted. The numbers in G stand for the row degrees. The set $\mathcal{I}$ from the algorithm can be taken as $\mathcal{I} = \{2, 4\}$ so that $N(\mathcal{I}) = \{4\}$. From the row degrees it follows that the degree of the second row can be reduced by adding the fourth row multiplied by a suitable factor of degree 2. The degree matrix Δ then generically becomes

$$\begin{bmatrix} -\infty & 1 & -\infty & -\infty & -\infty & 0 & 4 \\ 3 & 2 & 2 & 3 & -\infty & 3 & -\infty \\ 2 & -\infty & -\infty & -\infty & 0 & -\infty & -\infty \\ -\infty & -\infty & -\infty & 2 & -\infty & -\infty & -\infty \\ 4 & -\infty & -\infty & 1 & 1 & -\infty & -\infty \end{bmatrix}.$$

The generic row degrees are 4,3,2,2 and 4, respectively. The bipartite graph associated to the above degree matrix can be depicted as follows.

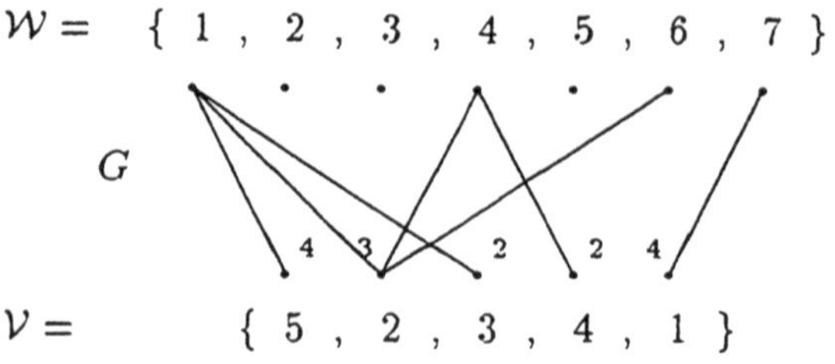

The graph does not contain a matching covering $\mathcal{V}$. Therefore, the previous step has to be repeated. It turns out that it has to repeated two times. It results in generic right Wiener-Hopf indices for the original matrix $R(s)$ equal to 4,3,2,2 and 2. Following the methods of van der Woude (1992) the generic minimal dimension of an underlying state space representation is computed to be 13. This result can also be obtained here by adding the obtained right Wiener-Hopf indices (Willems, 1986).

10. CONCLUDING REMARKS

In this paper a method is developed for the computation of the generic right Wiener-Hopf indices of a structured polynomial matrix. The method uses bipartite graphs and is based on elementary techniques from graph theory. Throughout this paper a polynomial matrix is called structured when it is only known which of its entries are not identically equal to zero and what the degree of each of these entries is. This means that if an entry has degree five, then in principle it may contain every power upto the fifth. Because of this, the method has to be applied with great care if more information on the form of the polynomial matrix is available. It should also be mentioned that the method is still not yet rigorously proved. A rigorous proof is under current investigation.

11. REFERENCES

Clancey, K. and I.C. Gohberg (1981). *Factorization of Matrix Functions and Singular Integral Operators*, Birkhäuser, Basel.

Murota, K. (1987). Systems Analysis by Graphs and Matroids - Structural Solvability and Controllability, *Algorithms and Combinatorics*, Vol. 3, Springer-Verlag, New York.

Pantelides, C.C. (1988). The consistent initialization of differential-algebraic systems, *SIAM J. Sci. Stat. Comput.*, **9**, pp. 213-231.

Willems, J.C. (1986). From time series to linear system - Part I : "Finite dimensional linear time invariant systems", *Automatica*, **22**, pp. 561-580.

Wolovich, W.A. (1974). *Linear Multivariable Systems*, Springer Verlag, New York.

van der Woude, J.W. (1992). The generic dimension of a minimal realization of an AR-system, *Faculty of Technical Mathematics and Informatics*, report 92-82 (to appear in MCSS).

DECOUPLING AND MINIMALITY OF STRUCTURED SYSTEMS IN A TRANSFER MATRIX FRAMEWORK

C.COMMAULT, J.M. DION, M. BENAHCENE

Laboratoire d'Automatique de Grenoble, ENSIEG - INPG - CNRS, BP 46, 38402 Saint-Martin d'Hères - France
E-mail : commault@lag.grenet.fr

Abstract. In this paper we consider linear systems described by transfer matrices. The prior knowledge on the system consists of the infinite zero order of each entry of the transfer matrix. We assume that the infinite frequency gains of the entries of the transfer matrix constitute a set of independent parameters. We are interested in studying the structural properties of such systems i.e. properties which are valid generically (true for almost any value of the parameters). These assumptions lead to the definition of a new class of structured systems within the transfer matrix approach. For such systems we look for diagonal precompensators which make the compensated system state feedback decouplable. We characterize such precompensators and give simple formulas for computing those of minimal order. We show that generically the minimal McMillan degrees achievable for the entries of the decoupled system are the initial system's essential orders. The results of this paper are expressed in terms of minimal weight of matchings in a bipartite graph associated with the structured system.

Keywords. Structured Systems; Graphs; Decoupling; Essential orders.

Résumé. Dans cet article on s'intéresse à l'étude des propriétés génériques (vraies pour presque toute valeur des paramètres) de systèmes décrits par des matrices de transfert. En particulier on caractérise les pré-compensateurs diagonaux qui permettent de rendre génériquement le système compensé découpable par retour d'état. Il est montré que génériquement le degré de McMillan minimal du système ainsi découplé est égal à la somme des ordres essentiels du système de départ. Les résultats donnés s'expriment en termes de couplages de poids minimaux dans un graphe biparti associé au système.

1. INTRODUCTION

The decoupling of linear time invariant multivariable systems is a problem of great theoretical and practical interest. Different ways are used for solving this problem namely the decoupling by dynamic output feedback, dynamic state feedback or static state feedback (see for example Commault *et al.*, 1986; Descusse and Dion, 1982; Dion and Commault, 1988; Falb and Wolovich, 1967; Hammer and Khargonekar, 1984; Hautus and Heymann, 1983; Willems, 1980; Wonham and Morse, 1970). The static state feedback is usually used in decoupling problems because it does not introduce additional states. When this is not possible one may decouple by pure precompensation. This is always feasible when the transfer matrix is square and nonsingular and it is shown in Dion and Commault (1988) that the minimal infinite structure achievable for the decoupled system coincides with the essential orders of the initial system. When the parameters are unknown or badly known it is of interest to find a parameter independent diagonal precompensator which makes the compensated system feedback decouplable. This is of particular interest in adaptive control since the analysis and synthesis of adaptive control laws are much simple for feedback decouplable systems (see Singh and Narendra, 1984). In this case the required prior knowledge to implement the adaptive control law is considerably reduced. Singh and Narendra (1984) first gave a characterization of such a precompensator under the assumption that the infinite zero order of each entry of the transfer matrix is known. Asano *et al.* (1990) have considered the same problem and have given a combinatorial algorithm which computes a precompensator of minimal order. Recently Gibbens et al. (1993) gave an interesting algorithm which determines if the system can be made decouplable by diagonal precompensation and compute such a precompensator.

In this paper we assume that the infinite frequency gains of the entries of the transfer matrix constitute a set of independent parameters. We are interested in studying the structural properties of such systems i.e. properties which are valid generically (true for almost any value of the parameters). These assumptions lead to the definition of a new class of structured systems within the transfer matrix approach. We associate to such a structured system a bipartite graph. On this graph the generic values of the classical structural invariants (infinite zero orders, essential orders) can be easily obtained. Notice that the hypotheses on our model are close to the ones used in Singh and Narendra (1984) or Asano *et al.* (1990) and fit with the concept of structured system introduced in Commault *et al.* (1987).

In this paper the design of a diagonal precompensator which makes the compensated system state feedback decouplable is revisited within our framework. With the given hypotheses such a precompensator always exists. Our contribution in this context is twofold :

- We give explicitely the minimal precompensator in terms of the system feedback invariants while in Singh and Narendra (1984) or in Asano *et al.* (1990) the precompensator is obtained as the result of a combinatorial algorithm.

- We prove that the minimal McMillan degree achievable generically for the decoupled system is the sum of the structural essential orders of the initial system. This last result is interesting since Ortega and Herrera (1994) have shown that for some particular systems the minimal McMillan degree achievable is strictly greater than the sum of the essential orders.

The results of this paper are expressed in terms of minimal weight of matchings in a bipartite graph associated with a structured system.
The layout of this paper is as follows. The problem statement and preliminaries are given in Section 2. Section 3 is devoted to structured systems and associated graphs in a transfer matrix framework. The infinite structure and essential orders of such systems are defined and characterized and graphical conditions for state feedback decouplability are given. The characterization of minimal diagonal decoupling precompensators is presented in Section 4. Concluding remarks end the paper. The proofs of all the results given in this paper can be found in the full version of this paper. See Commault *et al*, 1994.

2. PROBLEM FORMULATION AND PRELIMINARIES

We consider a linear time invariant system Σ defined by the triplet (A, B, C) whose transfer matrix $T(s) = C(sI - A)^{-1}B$ is a (pxp) square and nonsingular rational matrix.

2.1 *Definitions*

Let us recall first the infinite structure.
Let $R_p(s)$ be the ring of proper rational functions.
$t(s) = \dfrac{n(s)}{d(s)} \in R_p(s)$, $\beta = \deg(d(s)) - \deg(n(s))$, when deg $(n(s))$ denotes the polynomial degree of $n(s)$.

$\beta \geq 0$ is the order of the zero at infinity of $t(s)$.
Let $R_p^{pxm}(s)$ be the set of (pxm) proper rational transfer matrices. The units (invertible elements) $B(s)$ of the ring $R_p^{pxp}(s)$ are called bicausal matrices and are characterized by the following :

$$\det\left(\lim_{s\to\infty} B(s)\right) \neq 0$$

Definition 1
Let $P(s)$ be a (pxm) strictly proper nonsingular rational matrix then it can be factorized as follows :

$$P(s) = B_1(s)\Lambda(s)B_2(s)$$

$$\text{where } \Lambda(s) = \begin{bmatrix} \Delta(s) & 0 \\ 0 & 0 \end{bmatrix}$$

$\Delta(s) = \text{diag}\,(s^{-n_1}, \ldots, s^{-n_r})$, n_i, positive integers, $n_i \geq n_{i+1}$, $i = 1, \ldots, r-1$; $B_1(s)$ and $B_2(s)$ are bicausal matrices. Since the bicausal matrices are non singular at infinity, the infinite structure of $T(s)$ (i.e. the number of infinite zeros with multiplicities) is given by $\Lambda(s)$. From the definition the number of infinite zeros is the rank r of $T(s)$ over $R_p(s)$.

Introduce now the characterization of the essential orders denoted by n_{ie}, $i = 1, \ldots, p$ (see Commault *et al.* (1986) for details). Denote $T^i(s)$ the $((p-1)xp)$ transfer matrix obtained by removing the ith row from $T(s)$. Let n_k, $k = 1, \ldots, p$, be the infinite zero orders of $T(s)$ and n_k^i, $k = 1, \ldots, p-1$, be the infinite zero orders of $T^i(s)$.

Definition 2
The essential orders n_{ie}, $i = 1, \ldots, p$ of the transfer matrix $T(s)$ are defined as follows :

$$n_{ie} = \sum_{k=1}^{p} n_k - \sum_{k=1}^{p-1} n_k^i \qquad (1)$$

Let us denote by n_i', $i = 1, \ldots, p$ the infinite zero order of the ith row of $T(s)$ which is the minimal infinite zero order of the entries of this row.

2.2. *Decoupling*

We shall recall three ways of system decoupling given in the litterature namely decoupling by static state feedback, by pure precompensation or by diagonal precompensation with static state feedback on the compensated system.

Static state feedback
Notation : Let a_i $i = 1, \ldots, p$ be p integers. $\{a_i\}_{i=1,\ldots,p}$ denotes the non increasingly ordered list of these integers.

Theorem 1. (Descusse and Dion, 1982; Commault and Dion, 1993; Commault *et al.* 1986).
Let $T(s)$ be a square and nonsingular transfer matrix. There exist a control law $u = Fx+Gv$, G non singular such that $T_{F,G}(s) = C(sI-A-BF)^{-1}BG$ is diagonal and nonsingular if and only if one of the three equivalent conditions holds :

a) $\{n_i\}_{i=1,\ldots,p} = \{n_i'\}_{i=1,\ldots,p}$

b) $\sum_{k=1}^{p} n_k = \sum_{k=1}^{p} n_k'$

c) $n_{ie} = n_i'$, $i = 1, \ldots, p$

Pure precompensation (see Dion and Commault, 1988)
Let $T(s)$ be a square and nonsingular transfer matrix of the system Σ. There exists a proper precompensator $C(s)$ such that $T_c(s) = T(s)C(s)$ is diagonal and nonsingular. $T_c(s)$ is the transfer matrix of the decoupled system. The minimal McMillan degree achievable for the decoupled system is the sum of the essential orders n_{ie}.

Diagonal precompensation with static state feedback
Another way of decoupling is the combination of diagonal precompensation and static state feedback. Several authors solved the problem of diagonal precompensation which makes whenever possible the compensated system decouplable via static state feedback (see Singh and Narendra, 1984; Gibbens *et al.*, 1993). It has been assumed in Singh and Narendra (1984) that the parameters of the system Σ are unknown and only the infinite zero of each element of $T(s)$ is known in advance and the high frequency gain matrix is generically nonsingular. Asano *et al.* (1990) gave a combinatorial algorithm to compute such a precompensator with minimal order. Recently Ortega and Herrera (1993) showed that in general the minimal McMillan degree of the ith diagonal entry of the decoupled system may be greater than the ith essential order n_{ie} of the system Σ.

We introduce now the notions of structured system and genericity in a transfer matrix framework.

3. STRUCTURED SYSTEMS IN A TRANSFER MATRIX FRAMEWORK

3.1. *Basic definitions*

A structured matrix is a real matrix whose entries are either fixed zeros or independent parameters (Lin, 1974; Linneman, 1981).

Denote by Λ the vector composed of the k non-null parameters of a structured matrix, $\Lambda \in V$, where V is an open subset of R^k.

Let $T(s)$ be a strictly proper rational transfer matrix. The (i, j)th entry $t_{ij}(s)$ of $T(s)$ can be factorized as follows :

$t_{ij}(s) = \dfrac{n_{ij}(s)}{d_{ij}(s)} = \gamma_{ij}\, s^{-\beta_{ij}}\, b_{ij}(s)$ with $b_{ij}(s)$ rational function such that $\lim_{s\to\infty} b_{ij}(s) = 1$

where :

γ_{ij} are real numbers and β_{ij} are integers.
γ_{ij} is called the gain at infinity of $t_{ij}(s)$.
$\beta_{ij} = \deg(d_{ij}(s)) - \deg(n_{ij}(s))$ is the infinite zero order of $t_{ij}(s)$.

If $t_{ij}(s) = 0$ then $\beta_{ij} = \infty$ and $\gamma_{ij} = 0$.

Definition 3
Let $T(s)$ be a strictly proper rational transfer matrix with β_{ij} fixed and known. $T(s)$ is said to be a structured transfer matrix if the real matrix $\Gamma = (\gamma_{ij})$ is structured. In this case $T(s)$ is denoted by $T_\Lambda(s)$ where $\Lambda = (\lambda_1, \dots, \lambda_N)^T$ is a vector composed of the non-null parameters of the matrix Γ. The corresponding system is said to be a structured system and is denoted by Σ_Λ. The structure of the system Σ_Λ is the location of the non null entries $t_{ij}(s)$ of $T_\Lambda(s)$ associated with the corresponding infinite zero order β_{ij}.

For Σ_Λ a property which is true for any $\Lambda \in V$ except on the intersection of V with a proper algebraic variety of R^N is called a structural property.

Recall that a proper algebraic variety in R^N is defined as the set $\Lambda = (\lambda_1, \dots, \lambda_N)^T$ of common zeros of a finite number of nontrivial polynomials in λ_i with real coefficients. Then a structural property will be true for almost all systems with the required structure. Furthermore, if for a given system the property is not true it will be true for some arbitrarily close system. In other words the property is generic relative to the considered structure.

We will associate with the structured system Σ_Λ defined by the transfer matrix $T_\Lambda(s)$ a bipartite graph $B(\Sigma_\Lambda) = (U, Y, E)$ where the vertex set is $U \cup Y$ where :
$U = \{u_1, \dots, u_m\}$ denotes the inputs vertices of Σ_Λ
$Y = \{y_1, \dots, y_p\}$ denotes the outputs vertices of Σ_Λ and the edge set is defined as follows :

$$E = \{(u_j, y_i) \mid t_{ij}(s) \neq 0\}$$

The weight of the edge (u_j, y_i) in E is β_{ij}.

Note that $t_{ij}(s) \neq 0$ if and only if $\gamma_{ij} \neq 0$. Then there exists an edge (u_j, y_i) in $B(\Sigma_\Lambda)$ if and only if $\gamma_{ij} \neq 0$.

A matching is a set of edges in $B(\Sigma_\Lambda)$ such that no two of which have a common endvertex. The order or the size of a matching is the number of its edges.

The weight of a matching is the sum of its edge weights.

3.2. *Graph characterization of the infinite structure*

As we will see now the infinite structure of a structured system can be obtained from the associated weighted bipartite graph.

Theorem 2. Let Σ_Λ be a structured system of transfer matrix $T_\Lambda(s)$ and $B(\Sigma_\Lambda)$ the associated weighted bipartite graph. One has the following :

(i) The structural rank of Σ_Λ which is the number of infinite zeros of Σ_Λ is equal to the order r of a maximal matching in $B(\Sigma_\Lambda)$.
(ii) The structural infinite zero orders of Σ_Λ are characterized on $B(\Sigma_\Lambda)$ as follows :

$$n_1 = w_1$$
$$n_q = w_q - \sum_{j=1}^{q-1} n_j = w_q - w_{q-1}, \qquad q = 2, \dots, r \tag{2}$$

where w_q is the minimal weight of a matching of order q.

Proof. See Commault *et al*, 1994.

3.3. *Graph characterization of the essential orders*

In this subsection we will define the structural essential orders of Σ_Λ in a transfer matrix framework.

Since we are able to compute the structural infinite structure from the bipartite graph of the system using the equalities (2) it is then easy to determine the structural essential orders in the following way :

- Compute the structural infinite structure of the system without output y_i. This is obtained by application of the previous theorem on a subgraph obtained by deleting output vertex y_i and the edges incident to this vertex.

- The ith structural essential order is then the difference between the total structural infinite structure of the system and the total structural infinite structure of the system without output y_i.

These essential orders can be obtained as follows :

Proposition 1. Let Σ_Λ be a structured system with a $(p \times p)$ structured transfer matrix $T_\Lambda(s)$ of structural rank p and $B(\Sigma_\Lambda)$ be the associated bipartite graph.

Let Σ^i_Λ be the structured system obtained by removing the ith output and $B(\Sigma^i_\Lambda)$ the associated bipartite graph. The ith structural essential order is given by :

$$n_{ie} = \Pi - \Pi^i \tag{3}$$

where Π is the minimal weight of a maximal matching in $B(\Sigma_\Lambda)$, Π^i is the minimal weight of a maximal in $B(\Sigma^i_\Lambda)$.

Proof.

See Commault *et al*, 1994.

Example 1. Consider the structured system Σ_Λ described by the following structured transfer matrix with $\Lambda \in R^3$.

$$T_\Lambda(s) = \begin{bmatrix} \lambda_1 s^{-2} & 0 \\ \lambda_2 s^{-1} & \lambda_3 s^{-3} \end{bmatrix}$$

The associated weigthed bipartite graph is given in Figure 1.

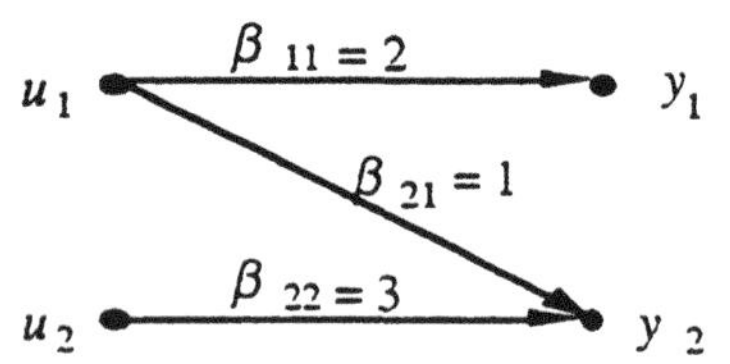

Fig. 1. Graph $B(\Sigma_\Lambda)$ of Example 2.

The minimal weigth of edges in $B(\Sigma_\Lambda)$ is $\beta_{21} = 1$, then $w_1 = 1$ and $n_1 = 1$. The minimal weigth of a matching of order 2 in $B(\Sigma_\Lambda)$ is $w_2 = \beta_{11} + \beta_{22} = 5$ then $n_2 = 4$. Then the structural infinite zero orders of Σ_Λ are 1 and 4.

The structural essential orders of Σ_Λ are $n_{1e} = 5 - 1 = 4$, $n_{2e} = 5 - 2 = 3$.

3.4. *State feedback decoupling of structured systems in a transfer matrix approach*

Proposition 2. Let Σ_Λ be a $(p \times p)$ structured system of structural rank p defined by a structured matrix $T_\Lambda(s)$ and $B(\Sigma_\Lambda)$ the weighted associated bipartite graph. Σ_Λ is structurally feedback decouplable by $u = Fx + Gv$, G non singular if and only if one of the following equivalent statements holds :

a') $\{w_k - w_{k-1}\}_{k=1,\dots,p} = \{w'_k\}_{k=1,\dots,p}$ with $w_0 = 0$

b') $\Pi = \sum_{k=1}^{p} w'_k$

c') $\Pi - \Pi^i = w'_i \qquad i = 1, \dots, p$

where $\Pi = w_p$ and $\Pi^i = w^i_{p-1}$ are the weights of a maximal matching in $B(\Sigma_\Lambda)$ and $B(\Sigma^i_\Lambda)$ respectively, w'_i is the minimal arc weight incident to the output vertex y_i and w_k is the minimal weight of a matching of order k. The statements a'), b') and c') are nothing but the graphical interpretation of statements a), b) and c) respectively given in theorem 1.

4. DECOUPLING WITH DIAGONAL PRECOMPENSATION AND STATIC STATE FEEDBACK FOR STRUCTURED SYSTEMS

Let $C(s) = \mathrm{diag}(s^{-\delta_i})$ be a diagonal precompensator in cascade with a structured system Σ_Λ whose transfer matrix is $T_\Lambda(s) = (\gamma_{ij}\, s^{-\beta_{ij}}\, b_{ij}(s))$. The precompensated system $\Sigma_{c\Lambda}$ is structured because its transfer matrix $T_{c\Lambda}(s) = (\gamma_{ij} s^{-(\beta_{ij}+\delta_j)} b_{ij}(s))$ is structured. $T_{c\Lambda}(s)$ has the same structure as $T_\Lambda(s)$ but it differs from it only by the fact that the infinite zero order of the entries of the jth column of $T_\Lambda(s)$ are increased by δ_j. As shown by Asano et *al*. (1990) there always exists a diagonal precompensator which makes the structured system Σ_Λ structurally decouplable via regular static state feedback.

It is known that in general it is possible to make the compensated system state feedback decouplable by diagonal precompensation. In this case the McMillan degree of the ith diagonal entry of the decoupled system is greater than or equal to the ith essential order n_{ie}. In particular this remains true generically.

Then we can state the following :

Lemma 1. Let Σ_Λ be a structured system defined by a nonsingular structured transfer matrix $T_\Lambda(s)$. There always exists a diagonal precompensator such that the compensated system is structurally state feedback decouplable. For any such a decoupling precompensator the McMillan degree achievable generically for the ith diagonal entry of the decoupled system is greater than or equal to the ith structural essential order n_{ie}.

Example 2. Consider the structured system Σ_Λ described by the following structured transfer matrix with $\Lambda \in R^5$.

$$T_\Lambda(s) = \begin{bmatrix} \lambda_1 s^{-1} & 0 & 0 \\ \lambda_2 s^{-1} & \lambda_3 s^{-1} & 0 \\ 0 & \lambda_4 s^{-1} & \lambda_5 s^{-2} \end{bmatrix}$$

The associated weigthed bipartite graph is given in Figure 2.

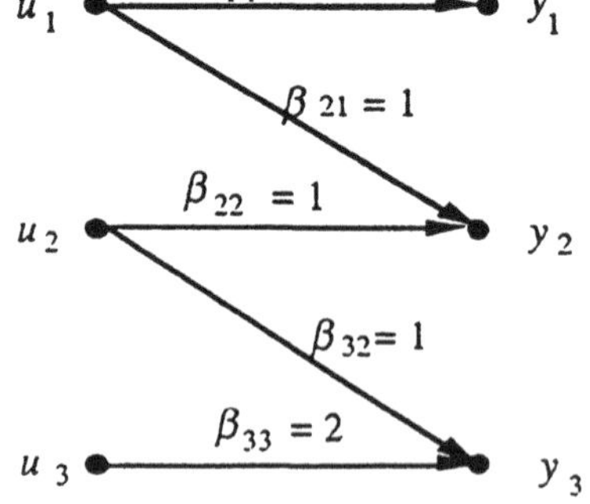

Fig.2. Graph $B(\Sigma_\Lambda)$ of Example 2.

The structural essential orders of Σ_Λ are : $n_{1e} = 4 - 2 = 2$, $n_{2e} = 4 - 2 = 2$, $n_{3e} = 4 - 2 = 2$.

We have $\Pi = w_3 = 4$, $\sum_{k=1}^{3} w'_k = 3$. The condition b') is not satisfied, then Σ_Λ is not structurally decouplable by static state feedback. Such a system can be made decouplable by the adjunction of a diagonal precompensator. Such a precompensator is computed via a combinatorial algorithm established in Asano *et al.* (1990) which gives one of minimal order.

As a result of this algorithm one gets the minimal McMillan degrees achievable for the entries of the diagonal decoupling precompensator : $\delta_1 = 1$, $\delta_2 = 1$, $\delta_3 = 0$. On the bipartite graph this precompensation is equivalent to add δ_i to the weight of the edges incident to the input vertex u_i in $B(\Sigma_\Lambda)$, we obtain the following associated bipartite graph $B(\Sigma_{c\Lambda})$ for the compensated system $\Sigma_{c\Lambda}$:

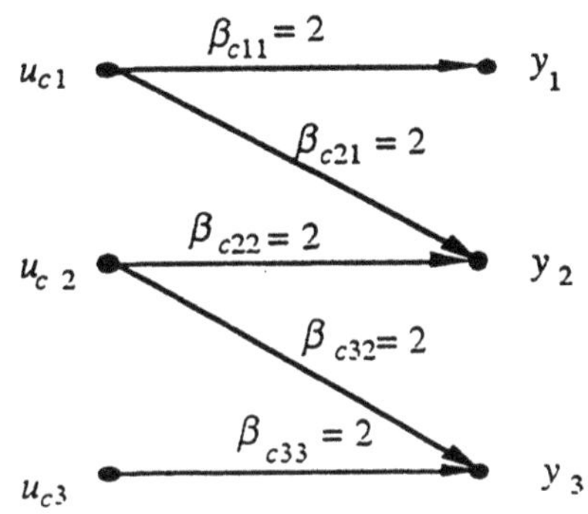

Fig.3. Graph $B(\Sigma_{c\Lambda})$ of Example 2.

One has $\Pi_c = w_{c3} = \sum_{k=1}^{3} w'_{ck} = 6$. The condition b') is now sastisfied, then $\Sigma_{c\Lambda}$ is structurally decouplable by static state feedback. Furthermore the McMillan degree of each entry of the decoupled system is equal to the corresponding essential order of Σ_Λ calculated above.

Let us now state the main result of this paper :

Theorem 3. Let Σ_Λ be a (pxp) structured system of structural rank p and $B(\Sigma_\Lambda)$ the associated bipartite graph. Reorder and rename the input vertices of $B(\Sigma_\Lambda)$ in such a way that $\{(u_1, y_1), \dots, (u_p, y_p)\}$ is a maximal matching of minimal weight.

i) The diagonal precompensator of minimal McMillan degree which makes the compensated system structurally state feedback decouplable is given by :

$$C(s) = \mathrm{diag}(s^{-\delta_1}, \dots, s^{-\delta_p})$$

where :

$$\delta_i = n_{ie} - \beta_{ii}, \; i = 1, \dots, p \qquad (4)$$

n_{ie} denotes the ith structural essential order of Σ_Λ, β_{ii} is the weight of the edge (u_i, y_i) in $B(\Sigma_\Lambda)$.
ii) The minimal McMillan degree achievable generically for the decoupled system is equal to the sum of the structural essential orders of the system.

Proof. See Commault *et al*, 1994.

To illustrate this result we give some examples.

Consider again example 2 we have obtained $n_{c1} = n_{c2} = n_{c3} = n_{1e} = n_{2e} = n_{3e} = 2$. $\{(u_1, y_1), (u_2, y_2), (u_3, y_3)\}$ is a maximal matching of minimal weight $\Pi = \beta_{11} + \beta_{22} + \beta_{33} = 4$.

The minimal McMillan degrees δ_i, $i = 1, 2, 3$ of the diagonal entries for the diagonal precompensator which makes $\Sigma_{c\Lambda}$ structurally decouplable by static state feedback are :

$$\delta_1 = n_{1e} - \beta_{11} = 2 - 1 = 1,$$
$$\delta_2 = n_{2e} - \beta_{22} = 2 - 1 = 1,$$
$$\delta_3 = n_{3e} - \beta_{33} = 2 - 2 = 0.$$

One can verify that the combinatorial algorithm of Asano *et al* (1990) gives the same values for the δ_i.

Example 3. Consider the structured system Σ_Λ described by the following transfer matrix with $\Lambda \in R^6$.

$$T_\Lambda(s) = \begin{bmatrix} \lambda_1 s^{-1} & \lambda_4 s^{-2} & 0 \\ \lambda_2 s^{-1} & 0 & 0 \\ \lambda_3 s^{-1} & \lambda_5 s^{-2} & \lambda_6 s^{-2} \end{bmatrix}$$

The associated weighted bipartite graph is given in Figure 4.

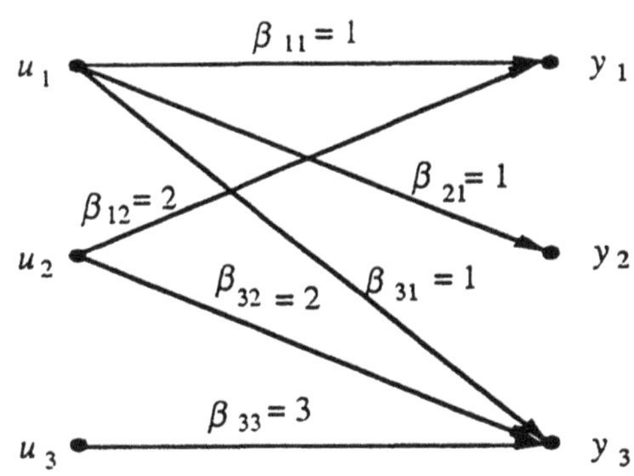

Fig. 4. Graph $B(\Sigma_\Lambda)$ of Example 3.

$\{(u_1, y_2), (u_2, y_1), (u_3, y_3)\}$ is a maximal matching of minimal weight $\Pi = \beta_{21} + \beta_{12} + \beta_{33} = 6$.

Permute and rename the input vertices u_1 and u_2 and weights of the arcs incident from those. We obtain the following bipartite graph where $\{(u_1, y_1), (u_2, y_2), (u_3, y_3)\}$ is a maximal matching of minimal weight :

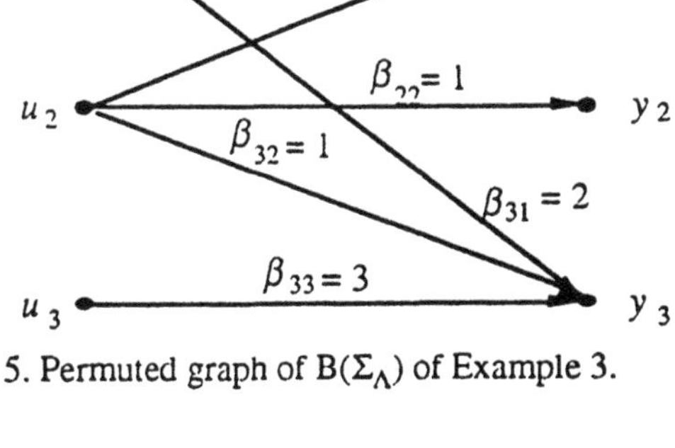

Fig. 5. Permuted graph of $B(\Sigma_\Lambda)$ of Example 3.

The essential orders of Σ_Λ are :

$$n_{1e} = \Pi - \Pi^1 = 6 - 3 = 3,$$
$$n_{2e} = \Pi - \Pi^2 = 6 - 3 = 3,$$
$$n_{3e} = \Pi - \Pi^3 = 6 - 3 = 3.$$

Using the formulas (6) to determine the McMillan degrees δ_i, $i = 1, 2, 3$ of diagonal entries for the diagonal precompensator of minimal order which makes $\Sigma_{c\Lambda}$ decouplable by static state feedback we obtain :

$$\delta_1 = n_{1e} - \beta_{11} = 1,\ \delta_2 = n_{2e} - \beta_{22} = 2,\ \delta_3 = n_{3e} - \beta_{33} = 0.$$

The permutation of the input vertices u_1 and u_2 in $B(\Sigma_\Lambda)$ is equivalent to the permutation of the first and second columns in $T_\Lambda(s)$.

Note that the transfer matrix given by Ortega and Herrera (1993), has the same structure as $T_\Lambda(s)$ but with fixed parameters $\lambda_1 = \lambda_4 = \lambda_6 = 1$, $\lambda_2 = \lambda_3 = \lambda_5 = -1$.

We have seen that generically the essential orders (structural essential orders) are equal to $n_{1e} = 3$, $n_{2e} = 3$, $n_{3e} = 3$ while for the specific values of the parameters λ_i chosen by Ortega and Herrera (1993) the essential orders are $n_{1e} = 3$, $n_{2e} = 2$, $n_{3e} = 3$.

For both cases (generically and particular example) the minimal McMillan degree achievable for the decoupled system is equal to 9 which is the sum of the structural essential orders.

5. CONCLUDING REMARKS

In this paper we have considered a new class of structured systems within the transfer matrix framework. We have characterized the minimal diagonal precompensators which makes the compensated system structurally feedback decouplable and computed those of minimal Mac Millan degree when only the infinite zero order of each element of the transfer matrix is known. We have also shown that the minimal McMillan degree achievable for the decoupled system is the sum of the structural essential orders of the initial system. Then instead of using the algorithm given in Asano *et al.* (1990) to determine the considered precompensator we have given a simple way to calculate it.

6. REFERENCES

Asano T., K.Yoshikawa, S. Suzuki (1990). The design of a preconpensator for multivariable adaptive control- A network-theoretic approach. *IEEE Trans. on Auto. Cont.*, **35**, N°6, pp. 706-710.

Berge C. (1973). *Graphs and Hypergraphs*. North-Holland/American Elsevier Amsterdam, New York.

Chen W.K (1971). *Applied Graph Theory*. North-Holland Publishing Company.

Commault C., J. Descusse, J.M. Dion, J.F. Lafay and M. Malabre (1986). New decoupling invariants : the essential orders. *Int. J. of Control*, **44**, pp. 689-700.

Commault C., J.M. Dion, A. Belmehdi (1987). Structured Systems within the Transfer Matrix Approach. *Syst. and Contr. Lett.*, **9**, pp. 335-339.

Commault C., J.M. Dion and M. Benahcène (1993). On dynamic decoupling of Structured Systems. Proceeding of 32th IEEE CDC Conference.

Commault, C., J.M. Dion, and M. Benahcène (1994). Decoupling and minimality of structured systems in a transfer matrix framework. *Internal Report LAG.*

Commault C., J.M. Dion, A. Perez (1991). Disturbance rejection for structured systems. *IEEE Trans. on Auto. Cont.*, **36**, pp. 884-887.

Descusse J., J.M. Dion (1982). On the structure at infinity of linear square decouplable systems. *IEEE Trans. on Auto. Contr.*, pp. 971-974.

Dion J.M., C. Commault (1988). The minimal delay decoupling problem, feedback implementation with stability. *SIAM J. on Cont.*, **26**, pp. 66-82.

Dion J.M., C. Commault (1993). Feedback decoupling of structured systems. *IEEE Trans. on Auto.Cont*, AC-38, pp. 1132-1135.

Falb P.L., W.A. Wolovich (1967). Decoupling in the design and synthesis of multivariable control systems. *IEEE Trans. on Auto. Cont.*, AC-12, pp. 651-659.

Gibbens P.W., C.A. Schwartz, M. Fu. (1993). Achieving diagonal interactor matrix for multivariable linear systems with uncertain parameters. *Automatica*, **29**, pp. 1547-1550.

Glover K., L.M. Silverman (1976). Characterization of Structural Controllability. *IEEE Trans. on Auto. Cont.*, **21**, pp. 534-537.

Gondran M., M. Minoux (1984). *Graphs and Algorithms*, Wiley, Chichester.

Hammer J., P. Khargonekar (1984). Decoupling of linear systems by dynamic output feedback. *Math. Systems Theory*, **17**, pp. 135-157.

Hautus M.L.J., M. Heymann (1983).Linear feedback decoupling -Transfer function analysis. *IEEE Trans. on Auto. Cont.*, AC-28, pp. 823-832.

Lin C.T. (1974). Structural Controllability. *IEEE Trans. on Auto. Contr.*, **19**, pp 201-208.

Linnemann A. (1981). Decoupling of Structured Systems. *Syst. and Contr. Lett.*, **2**, pp. 71-86.

Murota K. (1987). Systems analysis by graphs and matroids, structural solvability and controllability. *Algorithms and combinatorics*, **3**, Springer Verlag.

Ortega R., A. Herrera (1994). A solution to the continuous-time adaptive decoupling problem. *IEEE Trans. on Auto.. Contr.*, AC39, *8*, PP. 1639-1644.

Reinschke K.J. (1988). Multivariable control, A Graph-theoretic Approach. *Lectures Note in Control and Information Science*, Springer Verlag.

Rosenbrock H.H. (1970). *State Space and Multivariable Theory*. London, Nelson.

Singh R. P., K.S. Narendra (1984). Prior information in the design of multivariable adaptive controllers. *IEEE Trans. on Auto. Cont.*, AC-29, N° 12, pp. 1108-1111.

Van der Woude J.W. (1991). On the structure at infinity of a structured system. *Linear Algebra and its Applications*, **148**, pp. 145-169.

Willems J.C. (1980). Almost non interacting control design using dynamic state feedback. *Proc. of INRIA conference*, Springer Verlag, Berlin, pp. 555-561.

Wonham W.M., A.S. Morse (1970). Decoupling and pole assignment in linear multivariables systems : a geometric approach. *SIAM J. on Cont.*, **8**, pp. 1-18.

CONTROLLABILITY INDICES FOR BOND GRAPH MODELS

C. SUEUR and G. DAUPHIN-TANGUY

*LAIL URA CNRS D1440, Ecole Centrale de Lille, Cité scientifique, BP 48
59651 Villeneuve d'Ascq Cedex France, Tel: (33) 20 33 54 07 Fax: (33) 20 33 54 18*

Résumé. Une nouvelle caractérisation des indices de commandabilité des systèmes linéaires multivariables est proposée. Un nouveau symbolisme consiste à ne retenir que les numéros de la position des coefficients non nuls dans la représentation d'état. A l'aide de ce symbolisme et d'une représentation graphique, de type digraphe ou bond graph, nous caractérisons d'un point de vue purement structurel un ensemble de sous espaces commandables en relation directe avec l'ensemble des indices de commandabilité. Dans une première partie, nous calculons de manière structurelle les indices de commandabilité d'un modèle bond graph. Une nouvelle procédure de caractérisation de ces indices est ensuite proposée en relation avec les variables d'entrée.

Abstract. A new methodology is proposed for the characterization of the controllability indices of linear multivariable systems. Related to the state space representation, a new symbolism dealing only with numbers associated with the position of non null terms of matrices is proposed. This symbolism associated with the graphical representation which is either a digraph or a bond graph model allows to highlight, from a structural point of view, a list of dimensions of controllable subspaces corresponding one to one with the list of controllability indices. The purpose of this paper is two-fold. The first part is devoted to the characterization of the controllability indices of bond graph models from a structural point of view. In the second part the authors discuss possibilities to associate an input variable to a controllabilty indice.

Key words. Bond graph; linear systems; controllability indice; structural property.

1. INTRODUCTION

Controllability and observability indices for linear time-invariant systems have been widely studied in the past. They are introduced for the state space representation and they form a complete and minimal set of invariants with respect to the action of some groups.

Early works were done by Brunovsky (1970), Morse (1973), Heymann (1976) and Wohnam *et al* (1972).

The structural controllability study has been introduced for a graphical representation of linear systems by Lin (1974) and Shields *et al* (1976). Reinschke (1988) and Andrei (1985) investigated the structural properties in a general framework.

The aim of this article is to introduce the controllability and observability indices for linear bond graph models fom a structural point of view. In part two, we recall some properties of bond graph models investigated in other papers, Sueur *et al* (1991, 1992, 1993). In the third part, we establish a relation between the controllability indices and the dimension of some subspaces. A procedure for calculating the controllability indices is proposed.

Let us now recall some concepts. Consider a linear time-invariant dynamical system $\Sigma(A, B)$ described by equation (1) where $A \in \Re^{nxn}$, $B \in \Re^{nxm}$, B is supposed to be of full rank. We denote $\mathcal{B} = \mathrm{Im}B$.

$$\dot{x} = Ax + Bu \tag{1}$$

A linear system described by equation (1) is said to be state controllable if and only if the rank of the controllability matrix $\left[B\, AB\, A^2B \ldots\ldots\ldots A^{n-1}B\right]$ is n. When the rank is n, the object is to find n independent columns. This can be achieved by considering one at a time the columns A^kB_i of $\left[B\, AB\, A^2B \ldots\ldots\ldots A^{n-1}B\right]$, for $i = 1,\ldots,m$ and $k = 0,\ldots,n-1$ and by eliminating the columns linearly depending on the previous ones. The matrix obtained when reordering the columns can be written as $\left[B_1,\ldots,A^{\sigma_1-1}B_1,\ \ldots\ldots,\ B_m,\ldots,A^{\sigma_m-1}B_m\right]$.

The integers of the list $\{\sigma_1, \sigma_2, \ldots, \sigma_m\}$ are called the controllability indices. This list corresponds one to one to a list of dimensions of controllability subspaces.

Consider the space defined by

$$< A|B >_i = \mathcal{B} + A\mathcal{B} + \ldots + A^{i-1}\mathcal{B} \qquad i = 1, 2, \ldots \qquad (2)$$

The limit of equation (2), that is, when the dimension of the subspaces does no longer increase, is denoted by $< A|B >$ and is the controllable subspace of $\Sigma(A, B)$. Suppose now the list $\{\alpha_i\}$, defined by equation (3).

$$\begin{cases} \alpha_1 := m = \dim \mathcal{B} \\ \alpha_i := \dim\left(\dfrac{\langle A|B\rangle_i}{\langle A|B\rangle_{i-1}}\right) & i \geq 2 \end{cases} \qquad (3)$$

The two lists $\{\sigma_i\}$ and $\{\alpha_i\}$ are dual and are linked by the relations (4) and (5).

$$\sigma_i = \mathrm{card}\{j \mid \alpha_j \geq i\} \qquad i \geq 1 \qquad (4)$$

$$\alpha_i = \mathrm{card}\{j \mid \sigma_j \geq i\} \qquad i \geq 1 \qquad (5)$$

1. BOND GRAPH APPROACH

2.1. Causal properties

We consider bond graph models without information bonds for which a state space representation (1) can be derived formally by the simultaneous use of the causal relations in the bond graph model and of the Grassmann algebra (Sueur *et al* 1992). Some procedures will be recalled in the next section allowing to highlight some structural properties directly from the bond graph model.

Definition 1

Two dynamical elements I and C are said to be directly causally connected if there is a causal path between these two dynamical elements which does not meet any other dynamical element having an integral causality assignment.

The causal path gain is the product of the gains of the different elements belonging to the causal path. In the case of several causal paths linking the two same elements, the sum of the coefficients may cancel each other and in that case, the elements are not said causally connected.

We propose a definition for the causal path length between two elements of a bond graph model. It allows to identify these lengths with some integers found in the structural study. Then two properties are recalled which relate to the causal path lengths and the state space representation.

Definition 2

The causal path length between two elements of the set { I, C, Se, Det } is equal to the number of dynamical elements in integral causality met in the causal path between the two elements considered. If at least one of the two elements is a dynamical element with an integral causality assignment, add 1.

Property 1

The non-zero terms of the matrices A and B deduced from a bond graph model correspond to the direct causal paths without cancellation between the elements of the bond graph model.

Property 2 (Rahmani *et al* 1992)

If there is a causal path of length l between the kth dynamical element and the jth input source then $A^{l-1}B_j \neq 0$, where A is the state matrix and B_j is the input matrix associated with the input j, derived from the bond graph model.

As an example, consider the bond graph model of figure (1) and its corresponding state equation (6), with $x = [p_{I_1}, p_{I_2}, p_{I_3}, q_{C_4}, q_{C_5}, q_{C_6}]^t$ the state vector associated with the dynamical elements $I_1, I_2, I_3, C_4, C_5, C_6$ in integral causality and $u = [U_1, U_2]^t$.

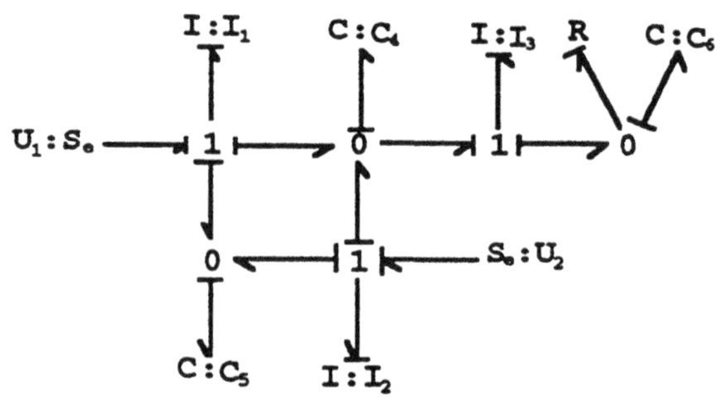

Figure 1. Bond graph model

$$\dot{x} = Ax + Bu \qquad (6)$$

with

$$A = \begin{bmatrix} 0 & 0 & 0 & -\frac{1}{C_4} & -\frac{1}{C_5} & 0 \\ 0 & 0 & 0 & -\frac{1}{C_4} & -\frac{1}{C_5} & 0 \\ 0 & 0 & 0 & \frac{1}{C_4} & 0 & -\frac{1}{C_6} \\ \frac{1}{I_1} & \frac{1}{I_2} & -\frac{1}{I_3} & 0 & 0 & 0 \\ \frac{1}{I_1} & \frac{1}{I_2} & 0 & 0 & 0 & 0 \\ 0 & 0 & \frac{1}{I_3} & 0 & 0 & -\frac{1}{RC_6} \end{bmatrix}$$

$$B^t = \begin{bmatrix} 1 & 0 & 0 & 0 & 0 & 0 \\ 0 & 1 & 0 & 0 & 0 & 0 \end{bmatrix}$$

2.2 . BG-rank.

In order to avoid numerical difficulties arising when computing the rank of characteristic matrices, the so-called structural study can be implemented. It supposes a graphical representation which is only concerned with the eventual relationships between variables, but not with the numerical values of parameters. Graph theoretic tools can be used in order to find the structural rank of these matrices, but they suppose that the parameters of the matrices are independent. A difficulty arises when a parameter corresponding to the same physical component exists in several relationships between the variables. The conclusion is therefore at times false.
A new concept of structural rank has been introduced for bond graph models, named the bond graph rank and denoted as BG-rank (Sueur *et al* 1991). It has been shown that the difficulty arising from the digraph approach is removed. The BG-rank is derived simply from causal manipulations on the bond graph model as recalled in property 3.

Let us assume a bond graph model leading to the state space representation (1). The integral causality can be assigned to n dynamical elements.

Property 3 (Sueur *et al* 1992).

The BG-rank of the state matrix A is $n-t$, where t is the number of dynamical elements which retain their integral causality when a derivative causality assignment is implemented.

Property 4 (Sueur *et al* 1992).

The BG-rank of $[A\vdots B]$ is equal to $n-t_r$ where t_r is the number of dynamical elements which retain their integral causality assignment on the bond graph model when a derivative causality assignment is performed and when the maximum number of input sources is dualized in order to eliminate the elements remaining in integral causality.

Theorem 1 (Sueur *et al* 1992).

The state space system (1) obtained from a bond graph model is structurally state controllable if and only if BG-rank $[A\vdots B]$=n

Consider the bond graph model of figure 2. A derivative causality assignment is implemented, figure 3. There is an algebraic causality loop with unity gain (non solvable junction structure). The BG-rank cannot be 6. In order to remove this loop, it is enough to assign the integral causality to the element I_1, for example. It is concluded that BG-rank$[A]$=5. In figure 4, the first input source is dualized and all the dynamical elements have a derivative causality assignment, then BG-rank$[A\vdots B]$=6. The system is structurally controllable, with only one input source.

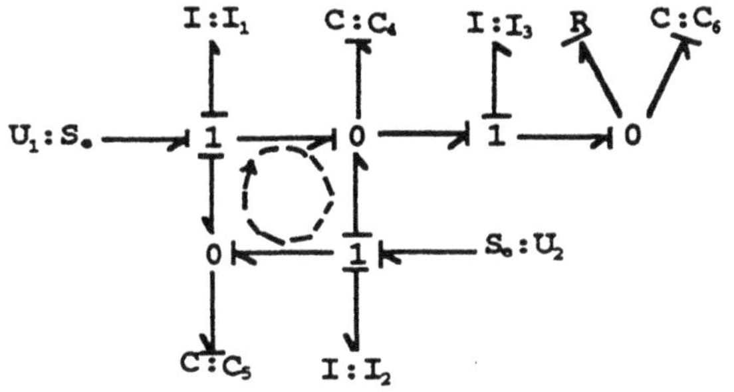

Figure 3 Derivative Causality

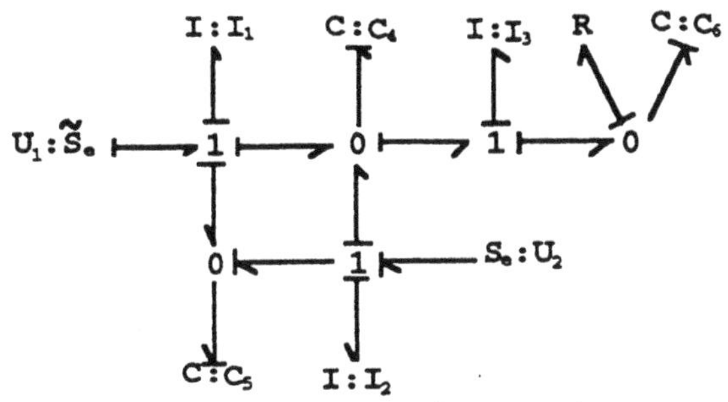

Figure 4 Derivative causality & dualization

3. CONTROLLABILITY INDICES

The problem is to calculate the dimension of the subspaces $\langle A|B\rangle_i$ and to select the essential (that is independent) columns of the associated matrix $[B, AB, \dots, A^{i-1}B]$. For large systems, the problem is too hard to be done formally. From a numerical point of view, the rank of the different matrices $[B, AB, \dots, A^{i-1}B]$ can be provided but is not right for non fixed values of parameters.

3.1 . New formulation

In order to simplify the formal calculations involved by the determination of the controllability matrix and of its rank, we propose to use a symbolism (See appendix A) dealing only with numbers associated with the position of the non null terms of the matrices. The property 1 on the bond graph model allows us to identify these positions. This symbolism is valid for the digraph representation.

3.2 . Example

Following the notations and the rules of appendix A, the state matrix and the input matrix of the bond graph model of figure (1) are respectively denoted as $\overline{A} = (4,5)\vdots(4,5)\vdots(4,6)\vdots(1,2,3)\vdots(1,2)\vdots(3,6)$ and $\overline{B} = (1)\vdots(2)$. The controllability matrix, $C_{A,B} = [B, AB, A^2B, A^3B, A^4B, A^5B]$ is denoted in a symbolic way as

$$\overline{C}_{A,B} = \begin{bmatrix} (1)\vdots(2), (4,5)\vdots(4,5), (1,2,3)\vdots(1,2,3), (4,5,6)\vdots \\ (4,5,6), (1,2,3,6)\vdots(1,2,3,6), (4,5,6)\vdots(4,5,6) \end{bmatrix}.$$

Note that this matrix can be written thanks to the property 2.

The dimension of the subspace $\langle A|B\rangle_i$ is directly calculated using the bond graph model thanks to the BG-rank concept. The steps are implemented on the bond graph model of figure (2). Tables 3 (a&b) describe a simplification procedure for the controllability matrix, that is by eliminating numbers which can not be a solution for the problem.

	B_1	B_2
$\mathcal{B}$	1	2
$A\mathcal{B}$	4,5	4,5
$A^2\mathcal{B}$	1,2,3	1,2,3
$A^3\mathcal{B}$	4,5,6	4,5,6
$A^4\mathcal{B}$	1,2,3,6	1,2,3,6

3.a Controllability matrix

	B_1	B_2
$\mathcal{B}$	1	2
$A\mathcal{B}$	4,5	4,5
$A^2\mathcal{B}$	1,2,3	1,2,3
$A^3\mathcal{B}$	4,5,6	4,5,6
$A^4\mathcal{B}$	1,2,3,6	1,2,3,6

3.b Simplified controllability matrix

Table 3 Controllability subspaces for the bond graph model

* Rank of $\mathcal{B}$

$\overline{B} = (1)\vdots(2)$. It is equivalent to state that the two dynamical elements I_1 and I_2 are directly causally connected to one of the input sources as shown in figure 5 and that $\text{BG}-\text{rank}[B] = 2$, because the two dynamical elements have a derivative causality assignment, when dualizing the effort sources (property 4). Then, $\mathcal{B}$ = Im(1,2), $\alpha_1 = 2$ and $\text{List}[U_1(1), U_2(2)]$.

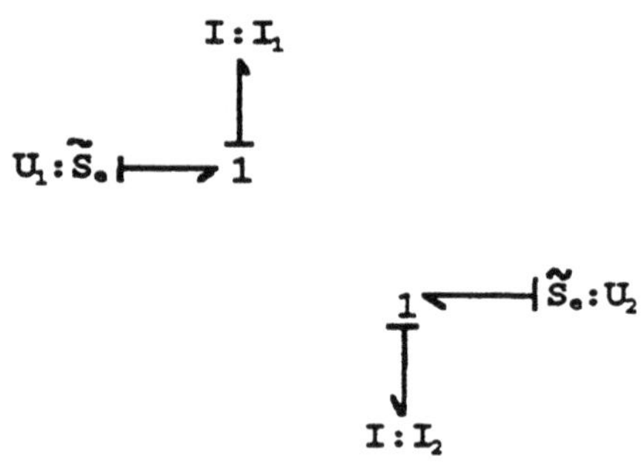

Figure 5. Subspace $\mathcal{B}$

* Rank of $\mathcal{B} + A\mathcal{B}$

The BG-rank of the controllable subspace associated with the four variables 1, 2, 4 and 5 cannot be 4 because there is a non solvable algebraic causality loop when all I-C elements have a derivative causality assignment, (even when dualizing the effort sources), (figure 6) . From figure 7, it can be concluded that the BG-rank of this subspace is 3. Then, $\mathcal{B}$+A$\mathcal{B}$ = Im(1,2,4) or $\mathcal{B}$+A$\mathcal{B}$ = Im(1,2,5) and $\alpha_2 = 1$, depending on the choice of the dualized input source. The different possible lists are $\text{List}[U_1(1,4), U_2(2)]$ or $\text{List}[U_1(1,5), U_2(2)]$ or $\text{List}[U_1(1), U_2(2,4)]$ or $\text{List}[U_1(1), U_2(2,5)]$.

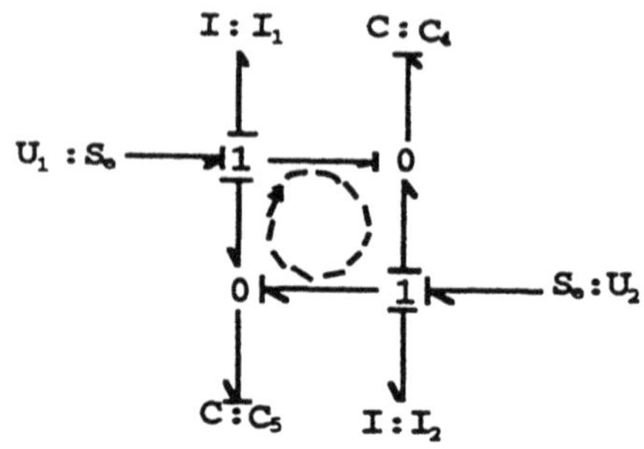

6. Algebraic causality loop

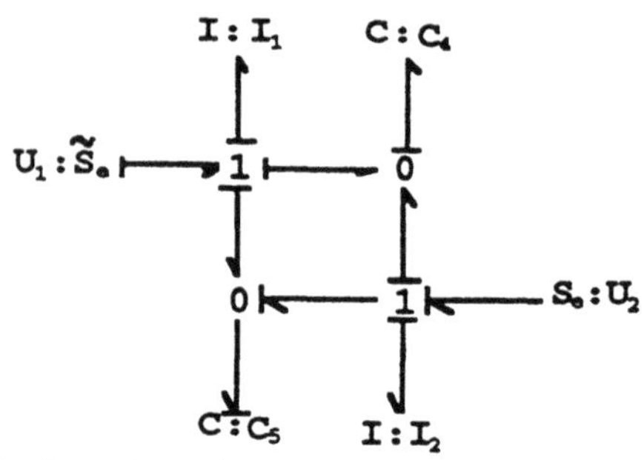

7. Dualization of one input source

* Rank of $\mathcal{B} + A\mathcal{B} + A^2\mathcal{B}$

Given that $\alpha_3 \leq \alpha_2$, the dimension of the subspace $\mathcal{B} + A\mathcal{B} + A^2\mathcal{B}$ is, at most, 4.
At most, the number of the new variable can be 3. The possible new lists are then $\text{List}[U_1(1,4,3), U_2(2)]$, $\text{List}[U_1(1,5,3), U_2(2)]$ or $\text{List}[U_1(1), U_2(2,4,3)]$ or $\text{List}[U_1(1), U_2(2,5,3)]$.
The bond graph model drawn in figure 8, corresponding to the first and third lists associated with numbers 1,2,3 and 4, is controllable.

The bond graph model corresponding to the two other lists is drawn in figure 9. The third dynamical element is alone, but the corresponding column is essential in A^2B. Therefore, we propose to add a fictitious input source in order to calculate the BG-rank of the matrix. The fictive source dissapears as soon as this part of the bond graph model is causally connected to one other input source, as in the digraph approach. The dualisation of this fictive input source allows to find a derivative causality for the dynamical element.
Then, $\mathcal{B} + A\mathcal{B} + A^2\mathcal{B} =$ Im(1,2,3,4) or $\mathcal{B} + A\mathcal{B} + A^2\mathcal{B} =$ Im(1,2,3,5) and $\alpha_3 = 1$. The four lists are retained.

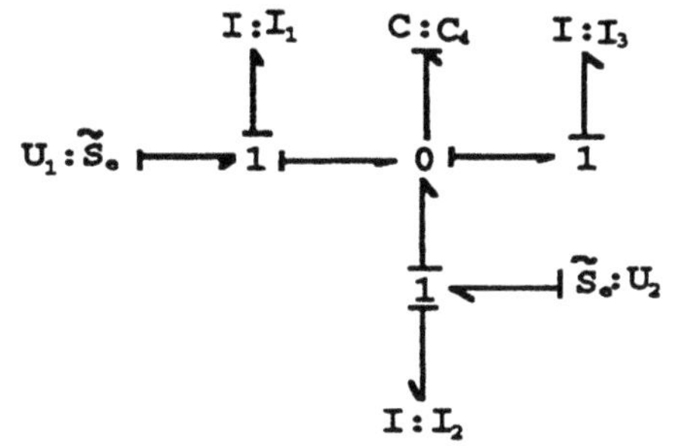

8. First choice for the controllable subspace

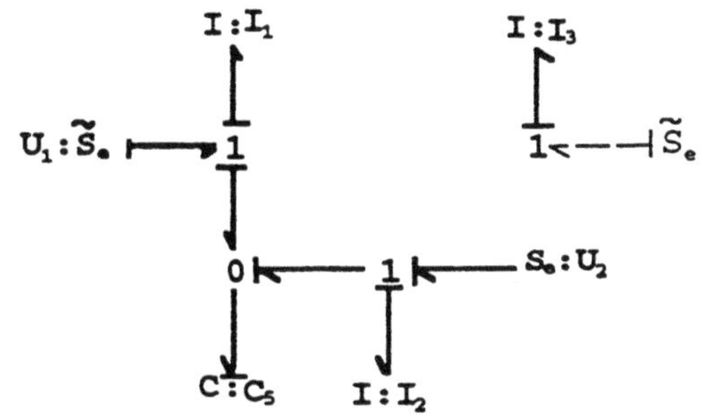

9. Fictive dualized input source

*<u>Rank of</u> $\mathcal{B}+A\mathcal{B}+A^2\mathcal{B}+A^3\mathcal{B}$

Given that the system is controllable, $\alpha_4 = 1$. There are eight possible lists, but number 6 must be chosen at the next step, because the three other numbers are 1,2 and 3. Thus, the lists are $\text{List}[U_1(1,4,3,5), U_2(2)]$, $\text{List}[U_1(1), U_2(2,4,3,5)]$, $\text{List}[U_1(1,5,3,4), U_2(2)]$, $\text{List}[U_1(1), U_2(2,5,3,4)]$

There are four lists in case of the controllable subspace $\mathcal{B}+A\mathcal{B}+A^2\mathcal{B}+A^3\mathcal{B}+A^4\mathcal{B}$. The results are gathered in table (4) in the case of $\text{List}[U_1(1,5,3,4,6), U_2(2)]$.

<u>Table 4. Controllability subspaces</u>

	B_1	B_2	Retained numbers		
$\mathcal{B}$	(1)	(2)	1	2	$\alpha_1 = 2$
$A\mathcal{B}$	4, (5)	4, 5	5		$\alpha_2 = 1$
$A^2\mathcal{B}$	1, 2, (3)	1, 2, 3	3		$\alpha_3 = 1$
$A^3\mathcal{B}$	(4), 5, 6	4, 5, 6	4		$\alpha_4 = 1$
$A^4\mathcal{B}$	1, 2, 3, (6)	1, 2, 3, 6	6		$\alpha_5 = 1$

<u>4.a Controllable subspaces</u>

$$\left[\begin{array}{cc:c:c:c:c}
\otimes & 0 & 0 & \times & 0 & \times \\
0 & \otimes & 0 & \times & 0 & \times \\
0 & 0 & 0 & \otimes & 0 & \times \\
0 & 0 & \times & 0 & \otimes & 0 \\
0 & 0 & \otimes & 0 & \times & 0 \\
0 & 0 & 0 & 0 & \times & \otimes \\
\hdashline
B_1 & B_2 & AB_1 & A^2B_1 & A^3B_1 & A^4B_1
\end{array}\right]$$

<u>4.b Essential columns with selected numbers</u>

The list $\{\alpha_i\}$ is $\{2,1,1,1,1\}$ and the dual list $\{\sigma_i\}$ is $\{5,1\}$. The controllability indices are then 1 and 5. There are two possible lists leading to two full-rank matrices $[B_1, AB_1, A^2B_1, A^3B_1, A^4B_1, B_2]$ and $[B_1, B_2, AB_2, A^2B_2, A^3B_2, A^4B_2]$ derived from the controllability matrice.This proves that even if the ordered list $\{\sigma_i\}$ is unique, the choice of the inputs is not unique. The controllability index is 5. It is not equal to the longest causal path length between the input sources and the dynamical variables which here is equal to 4.

3.3. Proposed new procedure

A procedure is now proposed which has been developed from a mathematical point of view with the symbolism of appendix A.

Procedure : Controllability indices

<u>Step 1:</u> Number the dynamical elements and verify the controllability property.

<u>Step 2:</u> Number the state matrix and the input matrix with the new formulation.

<u>Step 3:</u>Calculate the controllability table and simplify this table.

<u>Step 4:</u> Set $\text{List}[U_1(\text{numbers}), \ldots, U_m(\text{numbers})] = \text{List}[U_1(\varnothing), \ldots, U_m(\varnothing)]$. Set i=1.

<u>Step5:</u> Draw the graphical model with the dynamical elements whose numbers are in $\text{List}[U_1(\text{numbers}), \ldots, U_m(\text{numbers})]$ or in $\overline{A^{i-1}B}$, with the limiting constraint of at most one new number for each column of $[A^{i-1}B]$.

<u>Step 6:</u> Verify the controllability property and write all the possible lists $\text{List}[U_1(\text{numbers}), \ldots, U_m(\text{numbers})]$ according to the number of controllable variables. Calculate α_i.

<u>Step 7:</u> If i=n or if all the numbers are selected in $\text{List}[U_1(\text{numbers}), \ldots, U_m(\text{numbers})]$ then go to step 8, else i=i+1 and go to step 5.

<u>Step 8:</u> Derive the list $\{\sigma_i\}$ from the list $\{\alpha_i\}$ and write the possible full-rank matrices drawn from the controllability matrix $[B, AB, \ldots, A^{n-1}B]$.

During the implementation of the procedure, some conditions have to be verified. For example, the dimension of a controllable subspace cannot increase at a greater value than at the previous step, that is $\alpha_i \geq \alpha_j$ if $i \leq j$. If a new number of one column of $\overline{A^{i-1}B}$ cannot be retained, then the list of the corresponding input source cannot increase anymore. Additionally, at each step there may exist several lists List$[U_1(\text{numbers}),\ldots,U_m(\text{numbers})]$. Some of these lists can be eliminated at the following step of the procedure because they cannot be a solution of the problem.

Another procedure could be proposed from a bond graph point of view, using the causal path length concept and the BG-rank concept. For example, in step 4 of procedure 2, the numbers of each column of $\bar{A}^{i-1}\bar{B}$ are directly identified on the bond graph model by searching for the causal path of length i between the dynamical elements and the input sources. The symbolism proposed in appendix A allows to simplify the methodology and can be easily implemented on a computer.

4. CONCLUSION

In this paper, we have provided a new characterization of controllability indices for linear multivariable systems in a general framework and, more particularly, for bond graph models. It is shown that it may be relevant to employ a new symbolism for the matrices, for example, for the implementation of the algorithm on a computer. Nevertheless, the result can be achieved only from a graphical point of view.

Lastly, let us remark that the list of the observability indices can be obtained in exactly the same manner.

5. REFERENCES

Andrei, N. (1985). Sparse systems: Digraph approach of large scale linear systems theory. *ISR, Verlag TÜV Rheinland*, 1985.

Brunovsky, P. (1970). A classification of linear controllable systems. *Kybernetika*, **3**, 173-187.

Heymann, M. (1976). Controllability subspaces and feedback simulation. *SIAM J. of Control & Opt.*, Vol. 14, **4**, 769-789.

Lin, C.T. (1974). Structural controllability. *IEEE Trans. Auto. Cont.*, Vol AC-19, 201-208.

Morse, A.S. (1973). Structural invariants of linear multivariable systems. *SIAM J. of Control & Opt.*, Vol. 11, **3**, 446-465.

Rahmani, A., Sueur, C., Dauphin-Tanguy, G. (1992). Formal determination of controllability/ observability matrices for multivariables systems modelled by bond graph". *IMACS/SICE, Intern. Symp. on Robotics, Mechatronics and Manufacturing Systems'92 Kobe*, Japan, 573-580, sep. 16-20, 1992.

Reinschke, K.J. (1988). Multivariable control. A graph-theoretic approach. *Lecture Notes in Cont. and Inf. Sc.*, Vol. 108.

Shields, R.N., Pearson, J.B. (1976). Structural controllability of multi-input linear systems. *IEEE Trans. Auto. Cont.*, Vol AC-21, **2**, 203-212.

Sueur, C., Dauphin-Tanguy, G. (1992). Grassmann algebra and bond graph modelling. *J. Franklin Inst.*, Vol. 329, **6**, 1129-1146.

Sueur, C., Dauphin-Tanguy, G. (1991). Bond graph approach for structural analysis of MIMO linear systems. *J. Franklin Inst.*, Vol.328, **1**, 55-70.

Sueur, C., Dauphin-Tanguy, G. (1993). Bond Graph determination of controllability subspaces for pole assignment. *IEEE SMC, Inter. Conf. on Systems, Man and Cybernetics, Systems Engineering in the Service of Humans*, Le Touquet, France, 14-19, oct 1993.

Wonham, W.M., Morse, A.S. (1972). Feedback invariants of linear multivariable systems. *Automatica*, **8**, 93-100,.

Appendix A : Matrix symbolism

Suppose $M = \begin{bmatrix} a_{11} & a_{12} & a_{13} \\ a_{21} & a_{22} & a_{23} \\ 0 & a_{32} & 0 \end{bmatrix}$ and $N = \begin{bmatrix} b_{11} & 0 \\ 0 & b_{22} \\ b_{31} & b_{32} \end{bmatrix}$

A symbolism is proposed in order to deal only with the non-zero terms of matrices. The first column of M is $(a_{11}, a_{21}, O)'$. It is represented by (1,2), where the numbers correspond to the indices of the rows containing one non null variable.

In order to describe completely the matrix, we introduce a separator between each column of the matrix. It is denoted as $\vdots$. Then, the matrix M can be rewritten as $\overline{M} = (1,2)\vdots(1,2,3)\vdots(1,2)$. The upper bar means that the information regarding the value of the coefficients is lost.

The object is now to find the position of the non-zero elements of the product $M.N$. $\overline{M} = (1,2)\vdots(1,2,3)\vdots(1,2)$ and $\overline{N} = (1,3)\vdots(2,3)$. Suppose two matrices A and B, with $A = (a_{ij})$ and $B = (b_{ij})$. The product is $AB = (\sum_k a_{ik} b_{kj})$. The coefficients of the jth column of the matrix AB are then derived by summing the coefficients of each row of A multiplied by the non zero coefficient of the jth column of the matrix B.

In other words, the numbers of the jth column of the matrix $\overline{AB}$ are obtained by considering the numbers of the jth column of the matrix $\overline{B}$ and gathering all the numbers of the corresponding column of the matrix $\overline{A}$.

The matrix $M.N$ is a two column matrix. The non zero terms without numerical cancellation of the first column of $\overline{MN}$ are obtained by gathering the numbers in the first and third positions of matrix $\overline{M}$ because the first column of $\overline{N}$ is $(1,3)$. In the same way, the second column is obtained by gathering the numbers in the second and third positions of the matrix $\overline{M}$. Then $\overline{MN} = (1,2)\vdots(1,2,3)$.

ASYMPTOTIC STABILITY VIA ENERGY AND POWER.

Part II : Bond-Graph bridging for non linear systems.

G. DAUPHIN-TANGUY * and Lj. T. GRUJIĆ **.

* *LAIL - URA CNRS 1440, Ecole Centrale de Lille, BP 48, 59651 Villeneuve d'Ascq Cedex, France*
Tel : (33) 20 33 54 02, Fax : (33) 20 33 54 18, e-mail : gdt@laïls1.ec-lille.fr

** *ENIB, Espace Bartholdi, BP 525, 90016 Belfort Cedex, France*
Tel : (33) 84 58 23 48, Fax : (33) 84 58 23 42

Abstract. A systematic way, based on bond-graph modelling, is used to determine the total energy stored in the system and its time derivative, which is the power supplied to the dynamical elements. A car suspension non linear example shows how the two variables satisfy necessary and sufficient conditions for the system to be asymptotically stable, which are established in Part I.

Résumé. Une méthode systématique, fondée sur la modélisation par bond-graph (ou graphe à liens) est utilisée pour déterminer l'énergie totale stockée dans un système, et sa dérivée par rapport au temps, qui est la puissance fournie aux éléments dynamiques. Un exemple non linéaire de suspension automobile montre comment ces deux variables satisfont des conditions nécessaires et suffisantes permettant de conclure à la stabilité asymptotique pour le système, à partir de théorèmes établis en partie I.

Key Words. Bond Graphs, Energy, Power, Asymptotic Stability, Non Linear System.

1. INTRODUCTION

The bond-graph technique was created by H. Paynter to develop a general theory for engineering systems with the belief that energy and power alone are the fundamental dynamical variables, and that such variables are the ones allowing all physical interactions and transactions (Paynter, 1958). Since that time, a very large number of papers in the literature have shown the physical insight of this tool (Breedveld, 1991). Papers dealing with bond-graphs can be separated into two classes. The first one is composed of works presenting bond-graph as a very performing modelling and simulation tool. Such papers are more involved in applications. The second group of papers presents the bond-graph approach from a theoretical point of view, with the goal to show how the bond-graph model can be a helpful tool for analysis, control and monitoring the system to be studied (see for example Sueur, Dauphin-Tanguy, 1991a; Sueur, Dauphin-Tanguy, 1991b). Among them, only one paper (Breedveld, 1984) deals with stability study. It presents a bond-graph based algorithm to determine the steady (so called "equilibrium") state of a system.
Part I (Grujić, Dauphin-Tanguy, 1995) of this study established the necessary and sufficient conditions for asymptotic stability of a steady state, which are expressed in terms of the total energy stored in the system and the corresponding system power as the total time derivative of the energy. Inversely to the existing Lyapunov methodology that starts with (for example) the energy as a positive definite function and then tests for power negative (semi-) definiteness, the new methodology of Part I starts with the power test. If the power is negative semi-definite on a neighbourhood of the steady state relative to it, then it is asymptotically stable if and only if the system energy is positive definite.This transfers the fundamental problem of the Lyapunov stability theory - the construction of a system Lyapunov function, which has been an extremely complex mathematical problem, to an essentially physical problem of the determination of the system power and energy. This fundamental physical problem will be solved in what follows by using the bond-graph theory by Paynter (1958).

Firstly it will be shown how to determine the total energy and its time derivative directly by means of the variables defined in the bond-graph model. Secondly the results will be implemented on a non linear system, corresponding to an oilpneumatic car suspension.

2. POWER AND ENERGY IN A BOND- GRAPH

2.1. Definitions

Power and Energy are fundamental backgrounds in bond-graph modelling (Karnopp, Margolis, Rosenberg, 1990), in which power continuity and energy conservation properties are satisfied. If P_s

and P_d denote respectively the power supplied to the system by the power sources (active elements) and the power dissipated in the passive elements, then the property of power conservation in the bond-graph structure leads to the following relationship :

$$P_s - P_d = 0 \tag{1}$$

The dissipated power P_d is decomposed into three parts, according to the type of elements, I and C as energy storage elements and R as energy dissipative elements, which can be written

$$P_d = \sum_I P_I + \sum_C P_C + \sum_R P_R \tag{2}$$

The total system energy $E(t)$ is defined as the integral versus time of the power supplied to the energy storage elements.
It is expressed by

$$E(t) = \int_0^t \left(\sum_I P_I(\tau) + \sum_C P_C(\tau) \right) d\tau + E(0) \tag{3}$$

The bond-graph model of a system leads to the direct calculus of energy $E(t)$ expressed in terms of the state variables associated with the energy storage elements I and C. The state variables are called "energy variables" because they are involved in the determination of the different types of energies shown up in a physical system (electrical, potential, magnetic, kinetic).

2.2. Energy expression

Let us denote by $x = \left[x_I^T, x_C^T \right]^T$ the state vector. $x_I, x_I \in \mathcal{R}^{nI}$, associated with the I-elements is composed of the "generalized momenta" corresponding to inertia impulses in the mechanical domain or magnetic flux linkages in the electro-magnetic domain. $x_C, x_C \in \mathcal{R}^{nC}$, associated with the C-elements is composed of the "generalized displacements" corresponding to elongations of springs or electrical charges in capacitors.

x_I and x_C are defined as

$$x_I(t) = \int_0^t e_I(\tau) d\tau + x_I(0) \tag{4}$$

and

$$x_C(t) = \int_0^t f_C(\tau) d\tau + x_C(0) \tag{5}$$

where $e_I(\cdot)$ and $f_C(\cdot)$ are respectively the efforts (force, voltage,...) and the flows (velocity, current,...) applied to the I- and C-elements.

The power is obtained as the inner product of two complementary effort and flow vectors,
$P = e^T \cdot f$.

The vector Z complementary to x, defined by $Z = \left[Z_I^T, Z_C^T \right]^T$ is composed of $Z_I = \left[f_I \right]^T$ and $Z_C = \left[f_C \right]^T, Z_I \in \mathcal{R}^{nI}$ and $Z_C \in \mathcal{R}^{nC}$.

From those definitions it can be derived the power supplied to the I- and C-elements and transformed into stored energy, as

$$\left.\begin{aligned} \sum_I P_I &= Z_I^T \dot{x}_I \\ \sum_C P_C &= Z_C^T \dot{x}_C \end{aligned}\right\} \tag{6}$$

Indeed, if we change the independent variable t into x, the energy expression (3) becomes

$$\begin{aligned} \mathcal{E}\left(x_{(t)} \right) = & \int_{x_I(0)}^{x_I(t)} Z_I^T(x_I) d x_I \\ & + \int_{x_C(0)}^{x_C(t)} Z_C^T(x_C) d x_C + \mathcal{E}\left(x(0) \right) \end{aligned} \tag{7}$$

with $\mathcal{E}\left(x(0) \right) = E(0)$

The time derivative of total energy stored in I- and C-elements in a bond-graph model is, from (6),

$$\frac{d\mathcal{E}}{dt} = Z_I^T \dot{x}_I + Z_C^T \dot{x}_C \tag{8}$$

Z_I and Z_C are linked to x_I and x_C by the characteristic laws of the I- and C-elements as

$$\left.\begin{aligned} Z_I &= F_I(x_I) \\ Z_C &= F_C(x_C) \end{aligned}\right\} \tag{9}$$

If the state equation deduced from the bond-graph model is written as

$$\left\{\begin{aligned} \dot{x}_I &= f_1\left(x_I, x_C, U \right) \\ \dot{x}_C &= f_2\left(x_I, x_C, U \right) \end{aligned}\right. \tag{10}$$

with U : input vector, $U \in \mathcal{R}^m$,

then by combining (9) and (10), equation (8) becomes

$$\begin{aligned} \frac{d\mathcal{E}}{dt} = & F_I^T(x_I) f_1\left(x_I, x_C, U \right) \\ & + F_C^T(x_C) f_2\left(x_I, x_C, U \right) \end{aligned} \tag{11}$$

Remark. Equation (7) for total energy $\mathcal{E}(x)$ stored in the physical system and equation (11) for power defined as time derivative of the total energy $\mathcal{E}(x)$ are general and valid for all types of systems, whatever the physical domain (electrical, mechanical, hydraulic...).

It was shown in Part I of this paper (Grujić, Dauphin-Tanguy, 1995) that the total energy $\mathcal{E}(x)$ can be chosen as a Lyapunov function to study the asymptotic stability.
Let us recall Theorem 1 (Grujić, Dauphin-Tanguy, 1995) (for notations, see (Grujić, Dauphin-Tanguy, 1995)), after introduction of
$\mathcal{P}_0 = \{ x : x \in \mathcal{A}, \mathrm{P}(x) = \dot{\mathcal{E}}(x) = 0 \}, \quad \mathcal{A} \subseteq \mathcal{R}^n$.

Theorem 1.
Let consider a system modelled by $\dot{x} = \mathrm{f}(x)$, where $\mathrm{f}(\cdot) : \mathcal{R}^n \to \mathcal{R}^n$, $t \in \mathcal{R}$ and $x \in \mathcal{R}^n$. Let system power $\mathrm{P}(\cdot) = \dot{\mathcal{E}}(x)$ be negative semi-definite relative to x^* on $\mathcal{A} \subseteq \mathcal{R}^n$ and let singleton $\{x^*\}$ be the largest positive invariant set of the system in set $\mathcal{P}_0(\mathcal{A})$. In order for x^* to be asymptotically stable, it is both necessary and sufficient that there exists an open connected neighbourhood $\mathcal{S}$ of x^* such that
(i) $\mathcal{S} \subseteq [\mathcal{A} \cap \varepsilon \cap \mathcal{M}]$
(ii) $\mathrm{f}(x) = 0$ for $x \in \mathcal{S}$ if and only if $x = x^*$
(iii) energy $\mathcal{E}(\cdot)$ is positive definite relative to x^* on $\mathcal{S}$ and differentiable on $\mathcal{S}$.

The stability study problem is now summarized in the following :
- determine directly from the bond-graph model the total energy $\mathcal{E}(x)$ and its time derivative $\dot{\mathcal{E}}(x)$ equal to power P.
- determine their signs and the corresponding domains.
- conclude by applying the preceding theorem.

3. IMPLEMENTATION TO AN EXAMPLE

Let us consider the 1/4 car - oilpneumatic suspension system whose scheme is given in figure 1(a), and the associated bond-graph in figure 1(b).

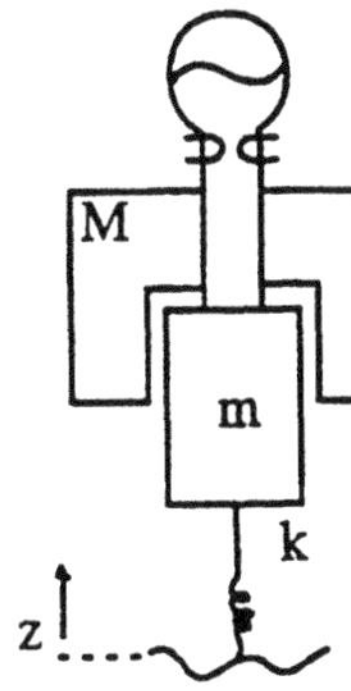

(a)

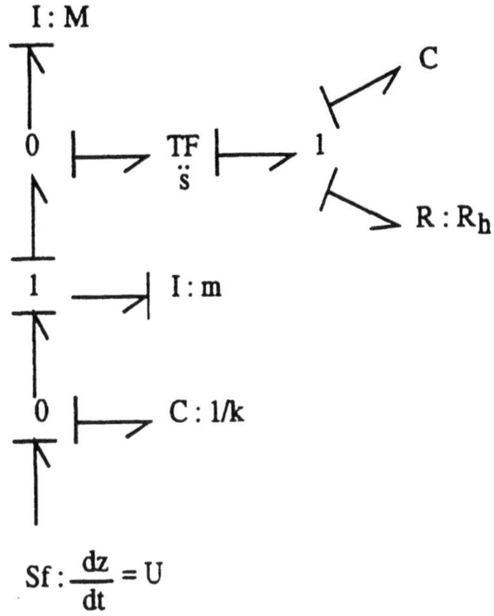

(b)

Fig. 1. Car suspension. Physical and bond-graph models.

M and m are respectively the 1/4 vehicle mass and the wheel mass. k is the stiffness of the spring modelling the tire. The suspension system is composed of an oilpneumatic sphere as spring (modelled by a C-element) and an hydraulic damper, modelled by the R-element.

3.1. State equation

The state vector associated with the bond-graph model is $x = [\mathrm{p}_m, \mathrm{p}_M, \Delta L, \Delta V]^T$ with $\Delta L = L - L_e$ and $\Delta V = V - V_e$, L_e being the spring lenght and V_e the sphere volume at equilibrium to compensate gravity effects. p_m and p_M are the impulses of masses m and M. The state equation deduced from the bond-graph model is thus

$$\dot{x}_I = \mathrm{f}_1(x_I, x_C, \mathrm{U})$$
$$\dot{x}_C = \mathrm{f}_2(x_I, x_C, \mathrm{U})$$

with $\quad x_I = [\mathrm{p}_m, \mathrm{p}_M]^T, \; x_C = [\Delta L, \Delta V]^T$

$$\mathrm{f}_1(\cdot) = \begin{bmatrix} -s^2 R_h \left(\frac{\mathrm{p}_m}{m} - \frac{\mathrm{p}_M}{M} \right) - s\,\psi_{sph}(\Delta V) \\ + k \Delta L \\ s^2 R_h \left(\frac{\mathrm{p}_m}{m} - \frac{\mathrm{p}_M}{M} \right) + s\,\psi_{sph}(\Delta V) \end{bmatrix} \tag{12}$$

$$\mathrm{f}_2(\cdot) = \begin{bmatrix} s \left(\frac{\mathrm{p}_m}{m} - \frac{\mathrm{p}_M}{M} \right) \\ -\frac{\mathrm{p}_m}{m} + \mathrm{U} \end{bmatrix}$$

s is the area of the piston composing the interface between the mechanical part and the hydraulic part. The input U is supposed null. The zero state $x^* = [p_m{}^*, p_M{}^*, \Delta L^*, \Delta V^*]^T = 0$ is the unique equilibrium state so that (ii) of Theorem 1 holds.
The characteristic law of the oilpneumatic sphere is

$$\Delta P = P_e\left[1-\left(\frac{V_e}{V_e+\Delta V}\right)^{\gamma}\right] = \psi_{sph}(\Delta V) \qquad (13)$$

with $\Delta P = P - P_e$, P pressure in the sphere, P_e pressure at equilibrium, γ gas isentropic constant.
R_h is the friction parameter of the hydraulic damper supposed here to be linear. From (13) follows that $\Delta P = \psi_{sph}(\Delta V)$ is well defined on $]-V_e,+\infty[$. Hence $\mathcal{M} = \{x : x \in \mathcal{R}^4, \Delta V \in]-V_e,+\infty[\}$. (Grujić, Dauphin-Tanguy, 1995).

3.2. Power $\dot{\mathcal{E}}(x)$

By defining $Z_I = [v_m, v_M]^T$ and $Z_C = [\Delta F, \Delta P]^T$, with v_m and v_M the two mass velocities, ΔF the force applied to the tire, equation (9) is expressed as

$$Z_I = F_I(x_I)$$
$$Z_C = F_C(x_C)$$

with $x_I = [p_m, p_M]^T$, $x_C = [\Delta L, \Delta V]^T$

$$F_I(\cdot) = \begin{bmatrix} 1/m & 0 \\ 0 & 1/m \end{bmatrix}\begin{pmatrix} p_m \\ p_M \end{pmatrix} \qquad (15)$$

$$F_C(\cdot) = \begin{bmatrix} k\Delta L \\ \psi_{sph}(\Delta V) \end{bmatrix}$$

Applying equation (11), it gives

$$P(x) = \dot{\mathcal{E}}(x) = -R_h s^2\left(\frac{p_m}{m} - \frac{p_M}{M}\right)^2 \qquad (16)$$

The power $P(x)$ is globally negative semi-definite so that it obeys theorem 1. Hence the set $\mathcal{A}$ (theorem 1) is found as $\mathcal{A} = \mathcal{R}^4$.

From (16), it follows that

$$\mathcal{P}_o = \{x : x \in \mathcal{S}, \dot{\mathcal{E}}(x) = 0\}$$
$$= \Big\{x : p_M = \frac{M}{m}p_m,$$
$$(p_m, p_M, \Delta L, \Delta V) \in \mathcal{R}\times\mathcal{R}\times\mathcal{R}\times\mathcal{R}\Big\} \qquad (17)$$

It is easy to verify that $\{x^*\}$ is the largest positive invariant set of the oilpneumatic suspension system in the set $\mathcal{P}_o$ (17).

3.3. Energy $\mathcal{E}(x)$

Following equation (7), it is found

$$\mathcal{E}(x) = \int_{p_{m_o}}^{p_m} \frac{p_m}{m} dp_m + \int_{p_{M_o}}^{p_M} \frac{p_M}{M} dp_M + \int_{\Delta L_o}^{\Delta L} k\Delta L\, d(\Delta L)$$
$$+ \int_{\Delta V_o}^{\Delta V} P_e\left[1-\left(\frac{V_e}{V_e+\Delta V}\right)^{\gamma}\right] d(\Delta V) + \mathcal{E}(x(0)) \qquad (18)$$

or, after calculation,

$$\mathcal{E}(x) = \frac{p_m^2}{2m} + \frac{p_M^2}{2M} + \frac{k}{2}(\Delta L)^2$$
$$+ P_e\left[\Delta V + \frac{V_e}{\gamma-1}\left(\frac{V_e}{V_e+\Delta V}\right)^{\gamma-1}\right] \qquad (19)$$

with

$$\mathcal{E}(x(0)) = \frac{p_{m_0}^2}{2m} + \frac{p_{M_0}^2}{2M} + \frac{k}{2}(\Delta L_0)^2$$
$$+ P_e\left[\Delta V_0 + \frac{V_e}{\gamma-1}\left(\frac{V_e}{V_e+\Delta V_0}\right)^{\gamma-1}\right]$$

It is obvious that the sign of $\mathcal{E}(x)$ depends on the sign of the energy stored in the oilpneumatic sphere, expressed by

$$I(\Delta V) = P_e\left[\Delta V + \frac{V_e}{\gamma-1}\left(\frac{V_e}{V_e+\Delta V}\right)^{\gamma-1}\right] \qquad (20)$$

Figure 2 shows the evolution of I in terms to ΔV.

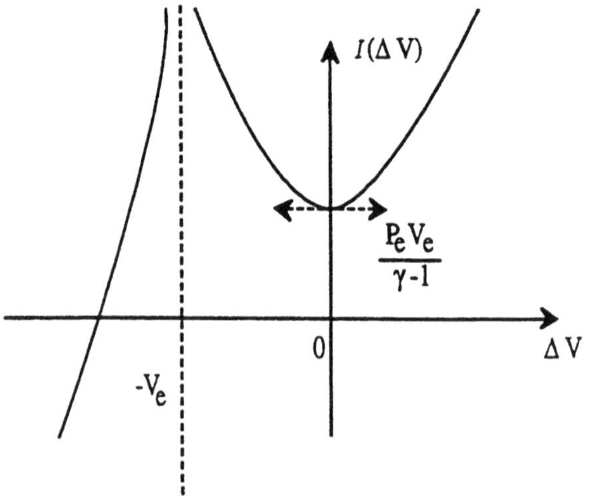

Figure 2. Energy stored in the oilpneumatic sphere.

It is immediate to conclude that

$$I(\Delta V)>0 \text{ for } \Delta V \in \mathcal{N}_v =]-V_e,\infty[\qquad (21)$$

which leads directly to $\varepsilon = \mathcal{M}$ (Grujic, Dauphin-Tanguy, 1995) and

$$\mathcal{E}(x) - \mathcal{E}(0) > 0 \text{ for } x \in \mathcal{S}$$

where $\mathcal{E}(0) = \frac{P_e V_e}{\gamma - 1}$, and $\mathcal{E}(x) \in C^{(1)}(\mathcal{S})$, and

$$\mathcal{S} = \{ x : (p_m, p_M, \Delta L) \in \mathcal{R}\times\mathcal{R}\times\mathcal{R}, \\ \Delta V \in]-V_e, +\infty[\} = \mathcal{M} \qquad (22)$$

because $\mathcal{S} \subseteq \mathcal{A} \cap \varepsilon \cap \mathcal{M} = \mathcal{M}$ (the condition (i) of theorem 1). The set $\mathcal{S}$ obeys all the conditions of theorem 1 having in mind that $\mathcal{E}(\cdot)$ is positive definite on $\mathcal{S}$ relative to $x^* = 0$ (Grujić, Dauphin-Tanguy, 1995).

Conclusion

From the system state equations (12) and $\mathcal{P}_o$ (17), it was deduced that {0} is the largest positive invariant subset of the system in $\mathcal{P}_o$. Hence all the conditions of Theorem 1 are satisfied. The equilibrium state $x = 0$ of the suspension system is monotone asymptotically stable with respect to the energy $\mathcal{E}(x)$ and its domain $\left| |\mathcal{E}(x) - \mathcal{E}(0)| \to \infty \right|$ is determined by $\mathcal{D}_m = \mathcal{S}$ (22) because $|\mathcal{E}(x) - \mathcal{E}(0)| \to \infty$ as $\|x\| \to \infty$ or as $x \to \delta\mathcal{S}$ the boundary of $\mathcal{S}$ (Grujić, Dauphin-Tanguy, 1995).

Simulation results (Tutsim)

$m =$ 75 kg, $M =$300 kg, k = 1.18 E5 N/m,
s = 2.69 E-3 m^2, P_e = 11.15 E5 Pa,
V_e =1.7 E-4 m^3
R_h = 2.6 E9 Pa.s/m^3, $\gamma = 1.4$

Result n°1

p_{m_0} = 75 kg.m/s, p_{M_0} = 300 kg.m/s which involves

$$\frac{p_{m_0}}{m} = \frac{p_{M_0}}{M} = 1 \text{m/s and } [\dot{\mathcal{E}}(x)]_{t=0} = 0$$

ΔL_0 = - 8.47 E-3 m,

$$\Delta V_0 = -\frac{V_e}{2} = -0.85 \text{ E-4 m}^3$$

so that

$$x(0) = [1, 1, -8.47\text{E}-3, -0.85\text{ E}-4]^T \in \mathcal{D}.$$

The curves given figure (3) show that

$$(\mathcal{E}[x(t)] - \mathcal{E}(0)) > 0, \forall t \in \mathcal{R}_+$$

with $\lim_{t\to\infty} \mathcal{E}[x(t)] = \mathcal{E}(0) = \frac{P_e V_e}{\gamma - 1}$,

and

$$P[x(t)] = \dot{\mathcal{E}}[x(t)] \leq 0 \text{ for all } t \in \mathcal{R}_+$$

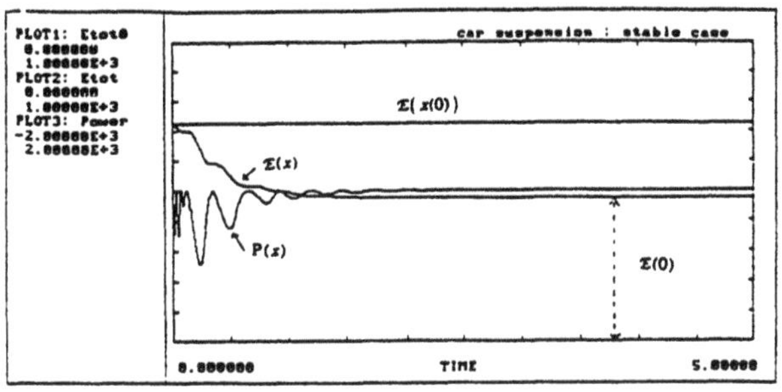

Fig. 3. Behaviour of power $P[x(t)]$ and energy $\mathcal{E}[x(t)]$ along system motions for initial state $x(0) \in \mathcal{D}$.

For the initial state $x(0) \in \mathcal{D}$ the motion obeys :

$$\lim_{t\to\infty} x(t) = 0 \text{ (figure 4)}$$

which illustrates asymptotic stability of $x^* = 0$.

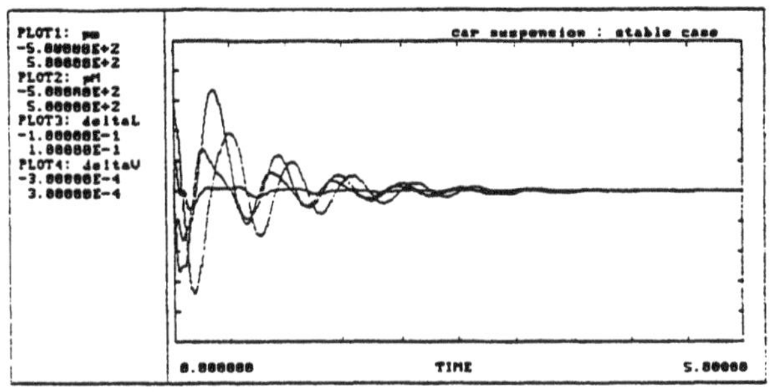

Fig. 4. System state variables for $x(0) \in \mathcal{D}$.

Result n°2

p_{m_0}, p_{M_0}, ΔL_0 unchanged

$\Delta V_0 = -1.088 V_e$= - 1.85 E-4 m^3
$\Delta V_0 \notin]-V_e, \infty[$

Hence $x(0) = [1, 1, -8.47\text{E}-3, -1.85\text{ E}-4] \notin \mathcal{D}$

The curves given in figure (5) show that

$$(\mathcal{E}[x(t)] - \mathcal{E}(0)) < 0 \text{ and } \lim_{t\to\infty} \mathcal{E}[x(t)] = -\infty.$$

for $\dot{\mathcal{E}}[x(t)] \leq 0$ for all $t \in \mathcal{R}_+$.

Hence the energy does not obey Theorem 1 and Theorem 2 of Grujić, Dauphin-Tanguy (1995) for $x(0) \notin \mathcal{D}$, which illustrates that $x(0) = [1, 1, -8.47\text{E}-3, -1.85\text{ E}-4]^T$ is out of the domain $\mathcal{D}$ of asymptotic stability of $x = 0$. This is also illustrated by

$$\lim_{t\to\infty} \Delta V_{(t)} = -\infty \text{ (figure 6).}$$

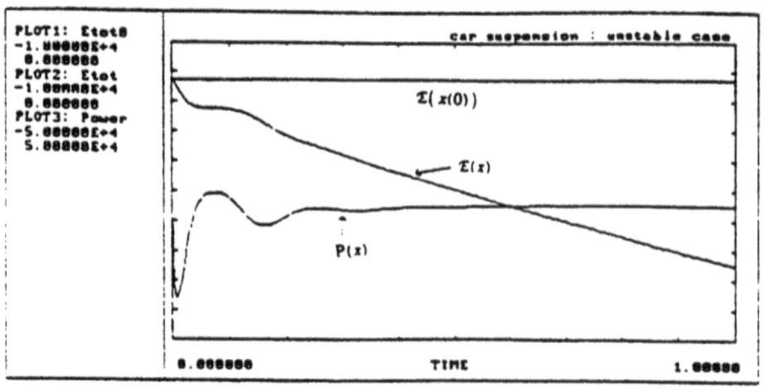

Fig. 5. Behaviour of power $P[\,x(t)\,]$ and energy $\mathcal{E}[\,x(t)\,]$ along system motions for initial state $x(0) \notin \mathcal{D}$.

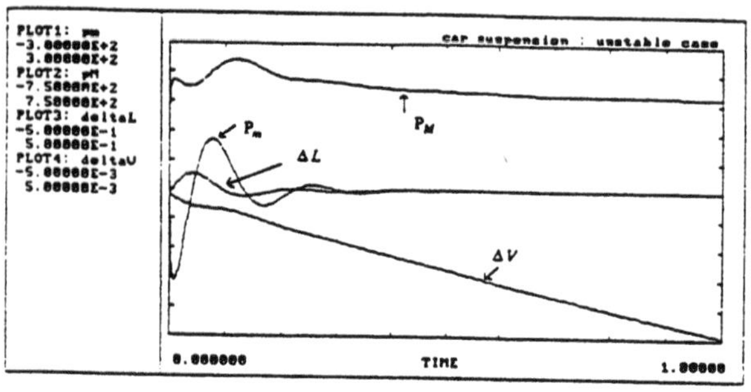

Fig. 6. System state variables for $x(0) \notin \mathcal{D}$.

4. CONCLUSION

In this paper, a systematic way was presented which allows the determination of a system Lyapunov functions with a physical insight for the asymptotic stability study. The obtained results are of fundamental importance, from a theoretical point of view but also from an industrial point of view. Directly, from the bond-graph model of a system, which is a very good knowledge model, the analyst is able to derive the power and energy, and to effectively test necessary and sufficient conditions for asymptotic stability of an equilibrium of a non linear system. The asymptotic stability domains have a physical meaning because the used variables are directly connected to physical phenomena. The proposed approach can also be very helpful in an integrated design phase, in which the model is not only used for simulation or analysis but also for control design.

5. REFERENCES

Breedveld, P.C. (1984). A bond-graph algorithm to determine the equilibrium state of a system. Journal of Franklin Instiute, vol. 318, no 2, pp. 71-75.

Breedveld, P.C., Rosenberg, R.C., and Zhou T. (1991). Bibliography of bond-graph theory and application. Journal of Franklin Institute, vol. 328, no 5/6, pp.1067-1119.

Grujič, Lj. T., and Dauphin-Tanguy G. (1995). Asymptotic stability via energy and power. Part I : new Lyapunov methodology for non-linear systems. IFAC Conference on System, Structure and Control, Nantes (France), July.

Karnopp, D.C., Margolis, D.L., and Rosenberg R.C. (1990). System dynamics. A unified approach. (second edition). John Wiley and sons, New-York.

Paynter, H.M. (1958). Generalizing the concepts of power transport and energy ports for system engineering. ASME paper 58-A-296, 1958 Annual Meeting, New-York.

Sueur, C., and Dauphin-Tanguy G. (1991a). Bond-graph approach for structural analysis of MIMO linear systems. Journal of Franklin Institute, vol. 328, no 1, pp. 55-70.

Sueur, C., and Dauphin-Tanguy G. (1991b). Bond-graph approach for multi-time scale systems analysis. Journal of Franklin Institute, vol. 328, no 5/6, pp.1005-1026.

Tutsim. User guide. Meerman Automatisering, Neede, NL.

H_∞ CONTROL OF A SATELLITE AXIS: LOOP-SHAPING, CONTROLLER REDUCTION AND μ-ANALYSIS

S. LE BALLOIS* and G. DUC**

** IUP Génie Électrique, Avenue du Parc, 8 le Campus, 95033 Cergy, France*

*** Supélec - Service Automatique, Plateau de Moulon, 91192 Gif-sur-Yvette, France*

Abstract. This paper consider the control of one axis of the satellite SPOT4, which exhibits lightly damped modes with large uncertainties in their natural frequencies. The proposed solution involves H_∞ synthesis with loop-shaping, modal reduction and μ-analysis of robustness. In both cases considered (the rate deviation is measured or not), low order controllers are obtained, which give the required performances while allowing large parameter deviations.

Résumé. Cet article considère la commande d'un axe du satellite SPOT4, qui présente des modes très peu amortis avec de larges incertitudes sur leurs fréquences. La solution proposée comprend une synthèse H_∞ avec modelage de boucle, une réduction modale et une μ-analyse de la robustesse. Dans les 2 cas considérés (la vitesse est mesurée, ou non), on obtient des correcteurs d'ordre faible, qui assurent les performances requises pour de larges déviations des paramètres.

Key Words. H_∞ Control, Satellite, Loop-Shaping, Reduction, μ-Analysis

1. INTRODUCTION

Space structures control is a challenging problem: such processes generally present lightly damped modes with large uncertainties, while performance requirements become higher. For that reason, the French Space Agency ("Centre National d'Études Spatiales") has been involved since 1990 in a program to assess potential benefits of new Guidance, Navigation and Control (GNC) concepts.

As a part of this program, a collaboration has been developed with the Automatic Control Department of Supélec ("École Supérieure d'Électricité"). The focus on the Earth observation satellite SPOT4 attitude control was retained because it includes interesting generic problems: a lot of flexible modes due to a large solar array, with badly known flexibility and low damping ratios, a lot of disturbances, a required high level of precision.

In order to handle the difficulties step by step, a SISO problem was first defined. The aim of this paper is to present the results and the knowledge obtained by such a way.

2. PROCESS AND DESIGN OBJECTIVES

The problem considered represents the control of one axis of the satellite SPOT4, for a given fixed position of the solar array. The 3 dominant flexible modes are retained in the model, with 5.10^{-4} damping ratios and uncertainties up to ±30% in the natural frequencies:

$$\begin{cases} \ddot{\theta}_x = -p^T\ddot{\eta} + \beta C_x \\ \ddot{\eta} = -M(I+\Delta)^2\eta - N(I+\Delta)\dot{\eta} - qC_x \end{cases} \quad (1)$$

with C_x: applied torque (Nm); θ_x: angular position (rd); η: state vector of flexible modes, and:

$$p^T = (-9.2673e-3 \;\; 4.6895e-3 \;\; 1.1338e-3);$$
$$q^T = (-1.1905e-2 \;\; 8.4060e-3 \;\; 1.6255e-3)$$

$$M = \begin{pmatrix} 7.4267e-1 & -1.9034e-1 & -7.3047e-1 \\ -2.0472e-2 & 9.3464e+0 & 1.5336e+0 \\ -1.4908e-2 & 2.9100e-1 & 2.8964e+1 \end{pmatrix}$$

$$N = \begin{pmatrix} 9.6885e-4 & -8.1434e-5 & -1.3613e-4 \\ -2.6707e-5 & 3.9987e-3 & 2.8581e-4 \\ -1.9448e-5 & 1.2450e-4 & 5.3979e-3 \end{pmatrix}$$

$\beta = 4.5216e-4$; I the 3×3 Identity matrix;

Δ a 3×3 diagonal matrix with $-0.3 \le \Delta_{ii} \le 0.3$.

A 0.165 s time delay is added to take into account the discretisation of the control law together with sensor and actuator delays. Fig. 1 shows the Bode diagram of different models, for deviations up to ±30% in the natural frequencies.

The control variable is a torque C_c generated by reaction wheel. The main disturbance C_d is a torque applied at the plant input (*i.e.* $C_x = C_c + C_d$), which occurs when the video recorder starts (see Fig. 2).

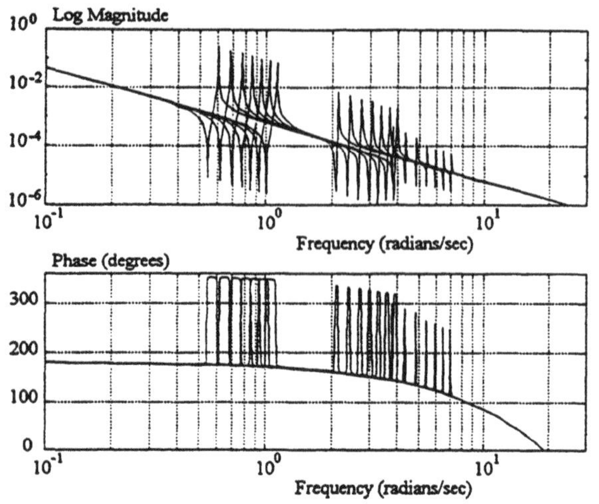

Fig. 1. Bode plots, for $-0.3 \le \Delta_{ii} \le 0.3$

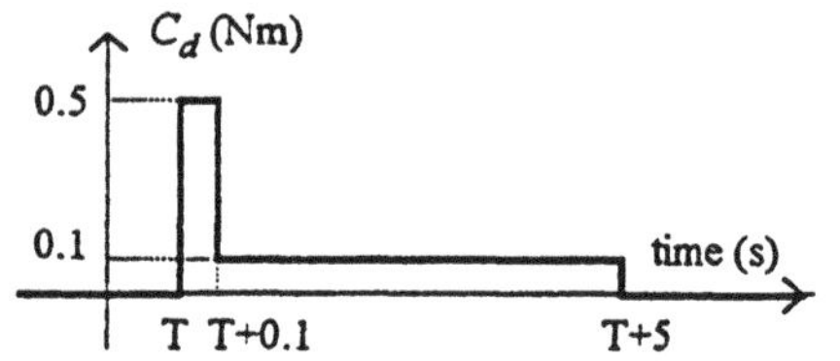

Fig. 2. Torque disturbance

The design objectives are the following: the *same* controller has to insure, for the largest possible deviations of the natural frequencies:

- closed-loop stability
- angle deviation θ_x less than $\pm 12.10^{-5}$ rd
- angular rate deviation $\dot{\theta}_x$ less than $\pm 5.10^{-5}$ rd/s
- control torque C_c less than ± 0.2 Nm
- settling time: from $T+11$, θ_x and $\dot{\theta}_x$ have to be less than 10% of the above maximal deviations.

Two case studies have been considered:

- **case 1**: only the angle θ_x is measured
- **case 2**: both θ_x and $\dot{\theta}_x$ are measured.

3. DESIGN OF THE CONTROL LAW

3.1 Loop-Shaping H_∞ Design

The method of McFarlane and Glover (1990, 1992) has been chosen as synthesis procedure. The reasons for this choice are the following :

- H_∞ methods have proved to be powerful for very lightly damped plants, see *e.g.* Safonov *et al.* (1991), Zhou *et al.* (1992)
- the method of McFarlane and Glover requires first to shape the plant model P_0, using compensators W_1, W_2 (see Fig. 3.*a*). This task is easily done by classical compensators (lead, lag, PI, ...), and following the usual rules of Automatic Control
- the following H_∞ problem is then considered:

$$\inf_{K \text{ stabilising}} \left\| \begin{pmatrix} K \\ I \end{pmatrix} (I - P_a K)^{-1} \begin{pmatrix} P_a & I \end{pmatrix} \right\|_\infty \qquad (2)$$

where the criterion includes the 4 main transfers of the closed-loop plant, from input and output disturbances to outputs of the controller and the plant respectively (see Fig. 3.*b*)

- the optimal value $\gamma_{\min}$ of the H_∞ criterion (2) can be computed in advance: no γ-iteration is required. Furthermore the obtained value for $\gamma_{\min}$ is a robustness and performance indicator (one has usually to try to obtain $\gamma_{\min}$ between 2 and 3).
- the actual controller is $W_1 K W_2$, where the computation of K involves 2 classical LQG-type Riccati equations; they are known to be better conditioned than *e.g.* the ones involved by the γ-iteration approach (Doyle *et al.*, 1989), especially for lightly damped plants. Numerical difficulties are therefore avoided.

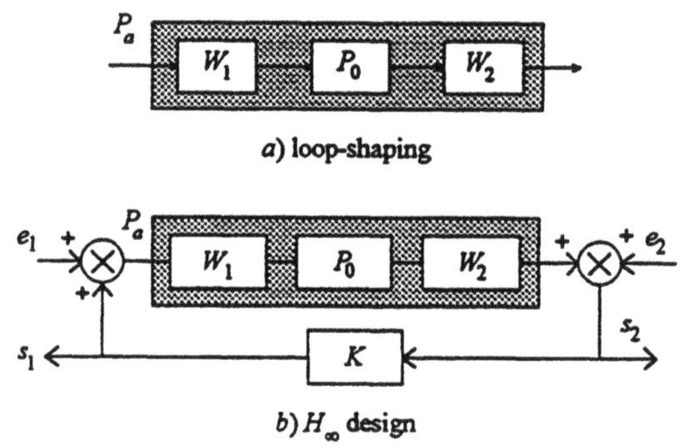

Fig. 3. Loop-shaping H_∞ design

3.2 Controller Reduction

The controller K so obtained is of the same order as P_a, which can therefore be high. It is thus desirable to apply a reduction procedure.

Instead of using the popular balanced truncation method (Anderson and Liu, 1989), a modal method is chosen (Siret *et al.*, 1979), which allows to select the most important poles of the controller, which will be exactly retained in the reduced order one.

Assuming that the system to be reduced has distinct eigenvalues $\lambda_1 \ldots \lambda_m$, $\lambda_{m+1} \ldots \lambda_n$, where $\lambda_1 \ldots \lambda_m$ are the eigenvalues to be retained, it can be written :

$$\begin{cases} \dot{x} = \Lambda x + Be = \begin{pmatrix} \Lambda_1 & 0 \\ 0 & \Lambda_2 \end{pmatrix} x + \begin{pmatrix} B_1 \\ B_2 \end{pmatrix} e \\ y = Cx + De = \begin{pmatrix} C_1 & C_2 \end{pmatrix} x + De \end{cases} \qquad (3)$$

with $\Lambda_1 = diag\{\lambda_1 \ldots \lambda_m\}$, $\Lambda_2 = diag\{\lambda_{m+1} \ldots \lambda_n\}$. Criteria are given by Siret *et al.*(1979) to choose conveniently the λ_i's to be retained.

The reduced order system is then :

$$\begin{cases} \dot{z} = \Lambda_1 z + B_1 e \\ \hat{y} = H x + E e \end{cases} \tag{4}$$

where matrices E and H are selected to obtain the same static gain for systems (3) and (4), and to optimise the integral of the square error between their step responses:

$$\begin{cases} H = C X L^T (L X L^T)^{-1} \\ E = D - C\Lambda^{-1} B + H \Lambda_1^{-1} B_1 \end{cases} \text{ with :} \tag{5}$$
$$L = (I_m \quad 0) \; ; \quad \Lambda X + X \Lambda^T + \Lambda^{-1} B B^T \Lambda^{-T} = 0$$

3.3 <u>μ-analysis</u>

μ-analysis (Doyle, 1982, Fan *et al.*, 1991) is then used to check the robustness: the closed-loop system is put on the standard form of Fig. 4, where all uncertainties are incorporated in $\Delta(s)$. For each $s_0 \in \mathbf{C}$, a general form is $\Delta(s_0) \in \underline{\Delta}$, where $\underline{\Delta}$ is the set defined as:

$$\underline{\Delta} = \left\{ diag\left\{ \Delta_1^C \ldots \Delta_C^C, \delta_1^c I_{c_1} \ldots \delta_c^c I_{c_c}, \delta_1^r I_{r_1} \ldots \delta_r^r I_{r_c} \right\}\right\},$$
$$\Delta_i^C \in \mathbf{C}^{n_i \times n_i} \quad \delta_i^c \in \mathbf{C} \quad \delta_i^r \in \mathbf{R} \tag{6}$$

which includes C full complex blocks, c complex repeated scalars and r real repeated scalars. Roughly speaking, the Δ_i^C's correspond to unmodeled dynamics, the δ_i^r's to parameter variations, whereas the δ_i^c's will be used in this paper to specify some stability margin (see Section 4).

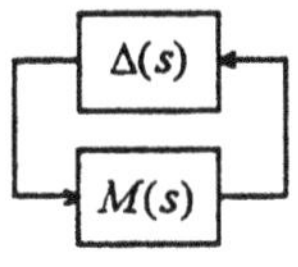

Fig. 4. μ-analysis

Define the structured singular value as:

$$\mu(M) = \left(\inf_{\Delta \in \underline{\Delta}} \{ \overline{\sigma}(\Delta); I - M\Delta \text{ singular} \} \right)^{-1} \tag{7}$$

and let $\overline{\mu} = \sup_{\omega \in \mathbf{R}} \mu(M(j\omega))$. The closed-loop system remains stable for all $\Delta(s)$ such that $\Delta(s_0) \in \underline{\Delta}$ for each $s_0 \in \mathbf{C}$, and such that $\|\Delta(s)\|_\infty < 1/\overline{\mu}$.

4. DESIGN 1: CONTROL FROM θ_x

The first step is to choose a precompensator W_1 to be applied to the nominal model P_0 of the plant. To this end, a lead filter was designed to obtain a −20 dB / decade roll-off around the crossover frequency, whose value has been chosen centred between the first and second flexible mode:

$$\omega_{0\,\text{dB}} = 1.5 \text{ rd/s}$$
$$W_1(s) = 3747 \frac{1 + s\sqrt{6}/1.5}{1 + s/1.5\sqrt{6}} \tag{8}$$

The resulting value for $\gamma_{\min}$ is 2.19. The H_∞ controller K is then computed for the shaped plant $P_a = P_0 W_1$ (for this computation, the time delay is replaced by a 1 order Padé approximant). The controller order is 10, which is equal to the order of P_a. The final controller obtained for the plant is $W_1 K$.

At this stage, the fact has to be emphasised that due to the presence of lightly damped poles and zeros, the controller has to be reduced with special care to obtain satisfying results. Consider first the Bode plot of controller $W_1 K$ (see Fig. 5). It clearly shows that it includes 3 pairs of lightly damped poles, each of them close to a pair of lightly damped zeros, whose natural frequencies are very close to those of the nominal model. This phenomenon is a common feature of central H_∞ controllers, even if exact cancellations do not occur (which is the case with the applied method).

For robustness purpose, this tendency can be very damaging. In fact it can be verified that +10% deviations of each of the flexible modes frequencies yield closed-loop instability. The reason is that such deviations modify the alternance of poles and zeros of the controller and the plant, altering significantly the open-loop frequency response.

This lack of robustness can be easily rectified by applying the reduction procedure presented in Section 3 to the H_∞ controller K. In fact each combination of lightly damped poles and zeros can be removed without significantly altering the controller frequency response (see Fig. 5). The corresponding Nichols chart is given in Fig. 6. The resulting controller order is then reduced to 4.

To perform μ-analysis, scaling factors on each uncertainty are first introduced in (1):

$$\Delta = S\Delta_r \quad \begin{matrix} S = diag\{s_1, s_2, s_3\} & s_i \in \mathbf{R} \\ \Delta_r = diag\{\delta_1, \delta_2, \delta_3\} & \delta_i \in \mathbf{R} \end{matrix} \tag{9}$$

such that if *e.g.* $\delta_i \in [-1\ +1]$, the frequency of mode i can take any value between $\omega_i(1-s_i)$ and $\omega_i(1+s_i)$, where ω_i is the nominal value.

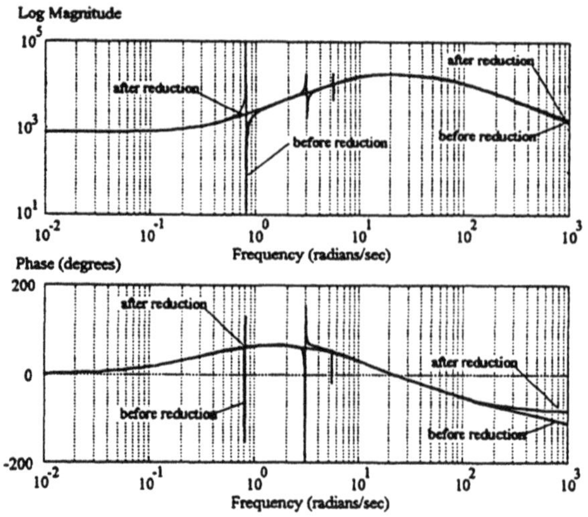

Fig. 5. Design 1: Bode plots of controller $W_1 K$

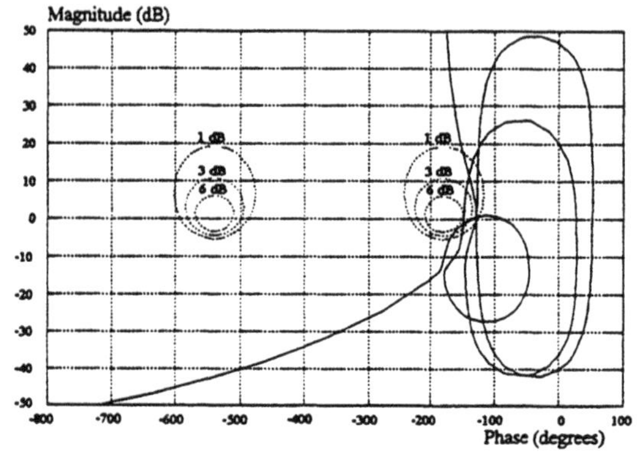

Fig. 6. Design 1: Nichols plot (reduced controller)

By linearising equations (1) using a first order approximation:

$$\begin{cases} \ddot{\theta}_x = -p^T \ddot{\eta} + \beta C_x \\ \ddot{\eta} = -M(I + 2S\Delta_r)\eta - N(I + S\Delta_r)\dot{\eta} - q C_x \end{cases} \quad (10)$$

extraction of matrix Δ_r can be obtained using the method of Morton (1985). Matrix *H*(*s*) (see Fig. 7) is then obtained by adding the Padé approximant representing the time-delay. On Fig. 7, *G*(*s*) is the model of the plant perturbed by real uncertainty Δ_r.

When all uncertainties are real, μ is known to be not necessarily a continuous function in ω; such a case is therefore to be avoided. Complex uncertainty δ_c is thus introduced (see Fig. 7), with $s_c \in \mathbf{R}$ as scaling factor: uncertainty $s_c\delta_c$ can be viewed as a gain/phase perturbation on the control applied to the plant. Furthermore, the transfer function from v_c to z_c is $s_c(1-W_1KG)^{-1}$: if *e.g.* $\mu(M(j\omega)) < 1$, the H_∞ norm $\left\|(1-W_1KG)^{-1}\right\|_\infty$ is guaranteed to be less than $1/s_c$, for all perturbed plants; or equivalently the modulus margin (*i.e.* the distance between the Nyquist plot and critical point −1, see Fig. 8) is greater than s_c for all perturbed plants.

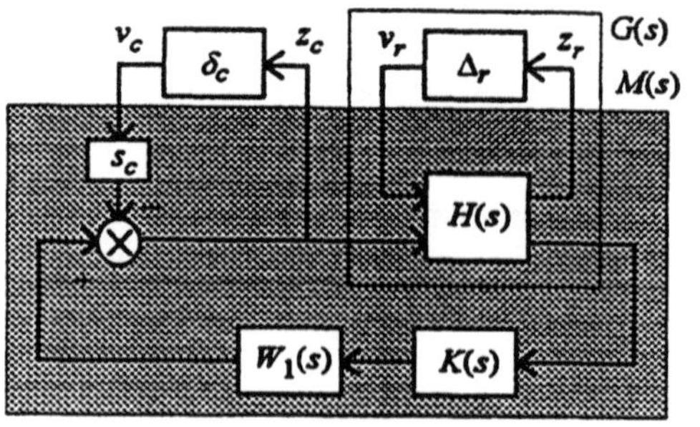

Fig. 7. μ-analysis: real & complex uncertainties

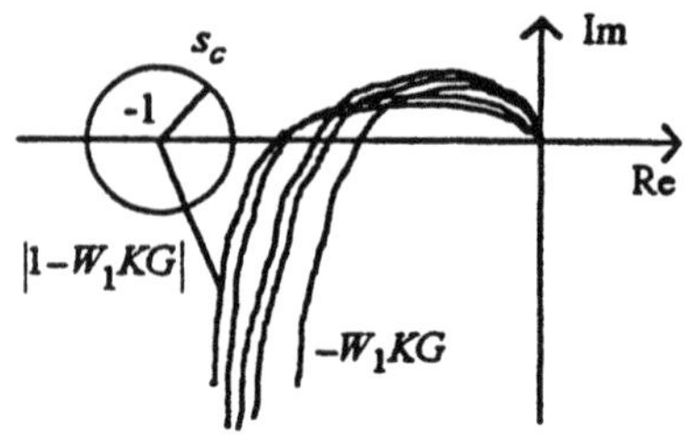

Fig. 8. Guaranteed modulus margin

Evaluation of $\mu(M(j\omega))$ has been performed using $s_1 = 0.3$, $s_2 = s_3 = 0.2$, $s_c = 0.1$. Fig. 9 gives the upper bound of μ obtained frequency by frequency (Balas *et al.*, 1993). According to the maximum of this plot, which is 0.92, closed-loop stability is guaranteed with a modulus margin at least equal to $0.1/0.92 = 0.109$, for frequency deviations up to:

- $\pm 0.3/0.92 = \pm 32.6\%$ for the mode 1
- $\pm 0.2/0.92 = \pm 21.7\%$ for modes 2 and 3

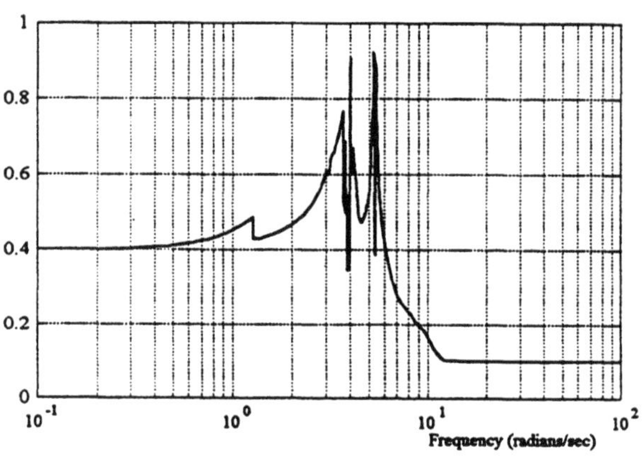

Fig. 9. Design 1: μ plot for 30%, 20%, 20%

Note that without reduction of the controller, the maximum of μ goes to 3.8, which allows only ±7.9%, ±5.3% and ±5.3% frequency deviations.

Finally, Fig. 10 gives the responses of θ_x, $\dot{\theta}_x$ and C_c to the disturbance defined by Fig. 2, for deviations of −30%, 0% and +30% on the frequency of each flexible mode. From these figures, it can be seen that the design objectives are almost satisfied, although the angular rate overtakes somewhat the allowed deviation after $T+11$.

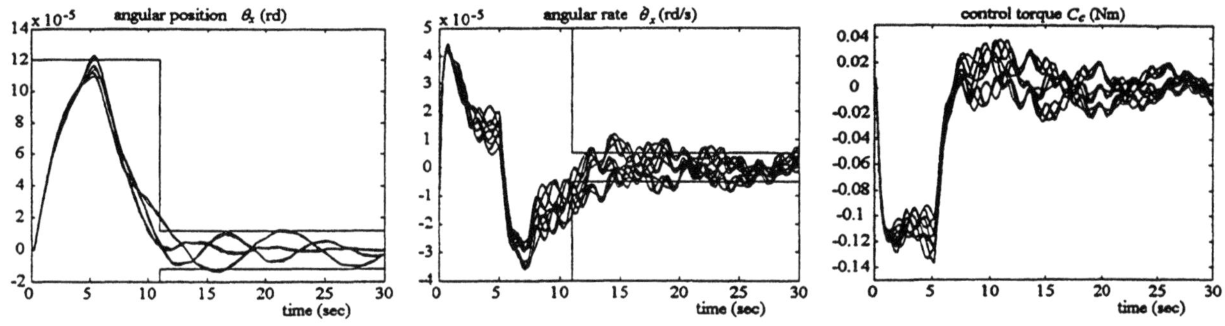

Fig. 10. Design 1: responses to torque disturbance

5. DESIGN 2 CONTROL FROM θ_x AND $\dot{\theta}_x$

Sensitivity to parameter variations can be reduced by using an inner loop. The control structure is given in Fig. 11: prior to the loop-shaping realised by precompensator W_1, a rate feedback is introduced by the gain K_i. Such a collocated feedback is a well-known technique to control flexible structures (Safonov *et al.*, 1991; Chiang *et al.*, 1993). Choosing K_i = 2250 gives a crossover between modes 2 and 3 for the inner loop.

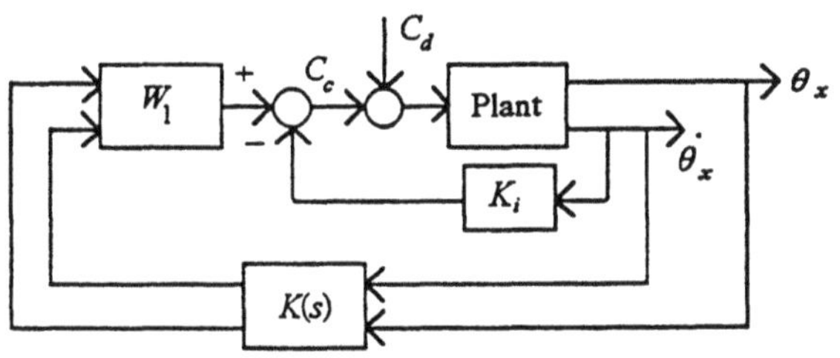

Fig. 11. Control structure for Design 2

This structure allows to remove the lead filter previously introduced in W_1: the precompensator includes only 2 gains K_p = 2000, K_r = 1, corresponding to the position and rate feedback respectively (with $K_r << K_i$ to slightly modify the inner loop). The resulting value for γ_{min} is only 1.74. A 9th order H_∞ controller K is then computed for the nominal model P_0 with inner loop gain K_i and precompensator $W_1 = \left[K_p \;\; K_r\right]$.

As for Design 1, this controller is reduced by removing all sets of closely related badly damped poles and zeros. A non-dominant real pole having also be removed, the reduced controller K has order 2. The quality of the reduction procedure is underlined in Fig. 12, which gives the Bode plots of the SISO transfer function from $\dot{\theta}_x$ to C_c, *i.e.* $T(s) = W_1 K(s)\left[1/s \;\; 1\right]^T - K_i$, before and after reduction. Clearly this frequency response is not altered by the reduction procedure.

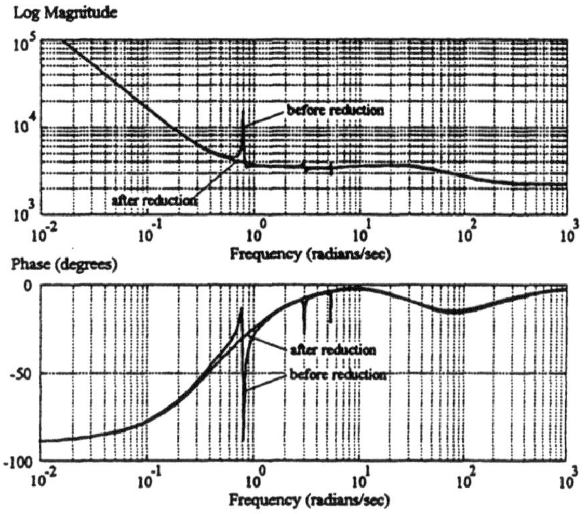

Fig. 12. Design 2: Bode plots of transfer $C_c/\dot{\theta}_x$

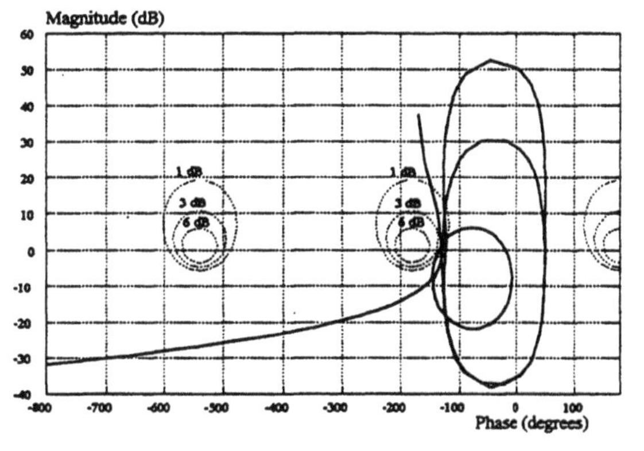

Fig. 13. Design 2: Nichols plot (reduced controller)

The corresponding Nichols chart is given in Fig. 13. It can be noted that the region of high closed-loop gain is better avoided than with Design 1.

A μ-analysis has been performed, by using s_1 = 0.3, $s_2 = s_3$ = 0.2, s_c = 0.3. Fig. 14 gives the upper bound obtained for μ, whose maximal value is 0.76. From this result closed-loop stability is guaranteed with at least $0.3/0.76 = 0.395$ modulus margin, for frequency deviations up to:

- $\pm 0.3/0.76 = \pm 39.5\%$ for mode 1
- $\pm 0.2/0.76 = \pm 24.1\%$ for modes 2 and 3.

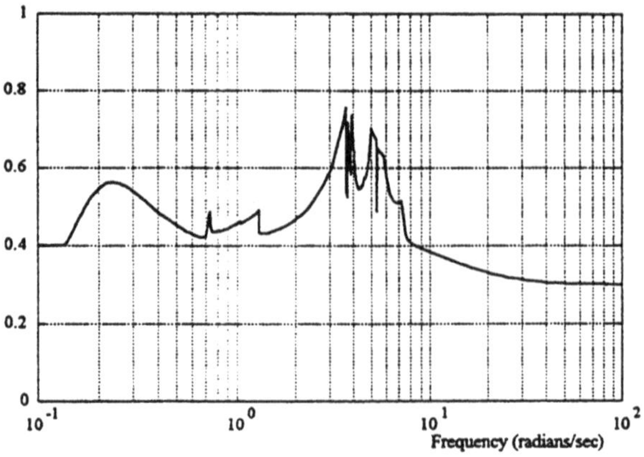

Fig. 14. Design 2: μ plot

Finally, Fig. 15 gives the responses of θ_x, $\dot{\theta}_x$ and C_c to the torque disturbance defined by Fig. 2, for the same deviations (–30%, 0%, +30%) than in Fig. 10. It can be seen that this second design reduces significantly the position and rate deviations, without increasing the control level.

6. CONCLUSION

Results obtained during a part of a collaboration between Supélec, CNES and MMS have been presented: the problem was to control one axis of the satellite SPOT4 despite lightly damped modes with large uncertainties in their natural frequencies.

The proposed solution involves 3 different tools. First a controller is determined by using the loop-shaping H_∞ design of McFarlane and Glover, which exhibits mathematical properties and is easy to use. Special care is then taken to reduce the controller, using a modal reduction method. Finally μ-analysis allows to guarantee simultaneously some stability margin and parameter deviations.

Following this procedure, low order controllers have been obtained, which give the required performances up to ±30% deviation in the frequencies of the modes.

7. REFERENCES

Anderson, B.D.O. and Y. Liu (1989). Controller reduction, concepts and approaches. *IEEE Trans. Autom. Contr.*, **34**, 802-812.

Balas, G.J., J.C. Doyle, K. Glover, A. Packard and R. Smith (1993). *μ-Analysis and Synthesis Toolbox*. The Math Works Inc.

Chiang, R.Y., M.G. Safonov, K. Haiges, K. Madden and J. Tekawy (1993). A fixed controller for a supermaneuverable fighter performing the herbst maneuver. *Automatica*, **29**, 111-127.

Doyle, J.C. (1982). Analysis of control systems with structured uncertainty. *IEE Proc. Part D*, **129**, 242-250.

Doyle, J.C., K. Glover, P.P. Khargonekar and B.A. Francis (1989). State-space solutions to standard H_2 and H_∞ control problems. *IEEE Trans. Autom. Contr.*, **34**, 831-846.

Fan, M.K.H., A. Tits and J.C. Doyle (1991). Robustness in the presence of mixed parametric uncertainty and unmodeled dynamics. *IEEE Trans. Autom. Contr.*, **36**, 25-38.

McFarlane, D. and K. Glover (1990). *Robust Controller Design Using Normalised Coprime Factor Plant Descriptions*. Lecture Notes in Control and Information Sciences, Springer Verlag.

McFarlane, D. and K. Glover (1992). A loop-shaping design procedure using H_∞ synthesis. *IEEE Trans. Autom. Contr.*, **37**, 759-769.

Morton, B.G. (1985). New applications of μ to real parameter variation problems. *24th IEEE Conf. on Decision and Control*, Ft Lauderdale, FL, 242-250.

Safonov, M.G., R.Y. Chiang and H. Flashner (1991). H_∞ robust control synthesis for a large space structure. *J. Guidance*, **14**, 513-520.

Siret, J.M., G. Michaïlesco and P. Bertrand (1979). On the use of aggregation techniques. In *Handbook of Large Scale Systems Engineering Applications*, 20-37, North-Holland.

Zhou, T., M. Fujita and F. Matsumura (1992). Robust control of a two-axis, magnetic suspension, flexible-beam system based on H_∞ optimization theory. *Int. J. of Robust and Nonlinear Control*, **2**, 165-182.

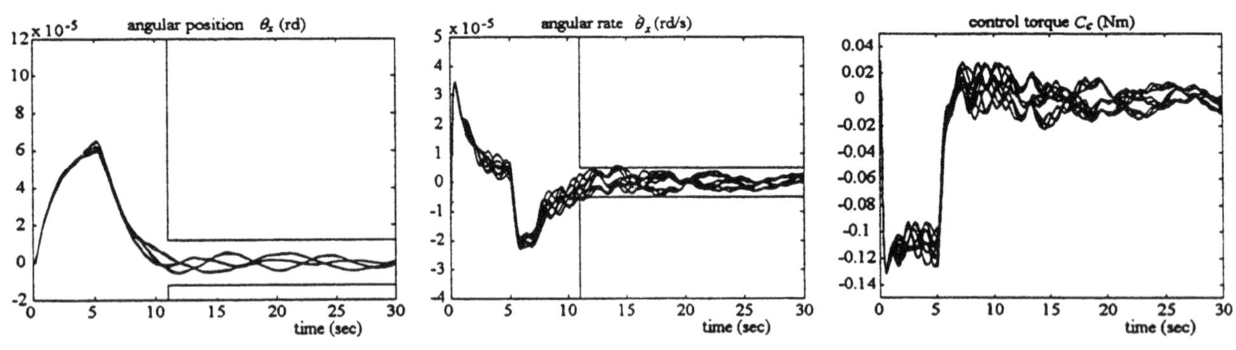

Fig. 15. Design 2: responses to torque disturbance

SLIDING MODE POSITION CONTROL OF AN ACTUATOR WITH BACKLASH AND COULOMB FRICTION.

J.C. CADIOU, N.K. M'SIRDI and P. BLAZEVIC

Laboratoire de Robotique de Paris. 10/12 Avenue de l'Europe. 78140 VELIZY. FRANCE

Résumé: Cet article présente une application expérimentale d'une commande à structure variable utilisant une approche par mode de glissement sur un actionneur électrique original présentant des jeux et des frottements secs. De bonnes performances et une grande robustesse face aux jeux ont été obtenues.

Abstract: This paper presents an experimental application of the variable structure control using the sliding mode technique to control an original actuator with backlash and Coulomb friction. Good robustness performances to reject the backlash effect, and quick responses are obtained.

Key words: Backlash, Coulomb friction, Original electric actuator, Robustness, Sliding mode control, Variable structure control.

1. INTRODUCTION

Few studies of servomechanism control on system with backlash and Coulomb friction have been presented so far. A descriptive study of non-linear phenomena (including both backlash and Coulomb friction) has been developed by Chan *et al.*(1973). A general representation of servomechanisms with the usual non-linearity has been proposed by Brandenburg *et al.*(1988). Different approaches have been presented. An adaptive approach is proposed by Tao et al. (1993). and a speed control using sliding mode control is presented by Razik et al. (1990). Other techniques has been studied as the dithering method by Desoer et al (1986) for example.

The purpose of this work is to study the performances of the sliding mode control for a non-linear system with large backlash and coulomb friction. In this paper a sliding mode position control using only the output position and velocity measurements is presented. At first the structure and a model of the actuator are presented. Then the control law is developed and experimental results are shown before concluding.

2. ACTUATOR DESCRIPTION

The actuator. developed in LRP laboratory (Blazevic 1991). has six DC motors connected in series which drive a two stage reducer (the first stage is a fixed one - the second is a planetary gear) with a gear ratio of about 60. Generally. the sensors are placed on the input shaft. In this case the gear ratio gives good precision and stability of the input part. But. if the reducer has imperfections like stiffness and backlash, the measured position cannot reflect the evolution of the output. In this case the position of the output shaft cannot be controlled with good performance.

The reason why the encoder is directly fixed on the output is to achieve good precision in the position of the load in spite of the presence of backlash and friction in the reducer affecting the stability of the system which becomes non-linear. The structure of the actuator permits easy measuring of the output position whereas in general that may not always be possible. For that purpose the encoder is directly fixed on the output shaft (see Fig.1) in the empty space between the six motors.

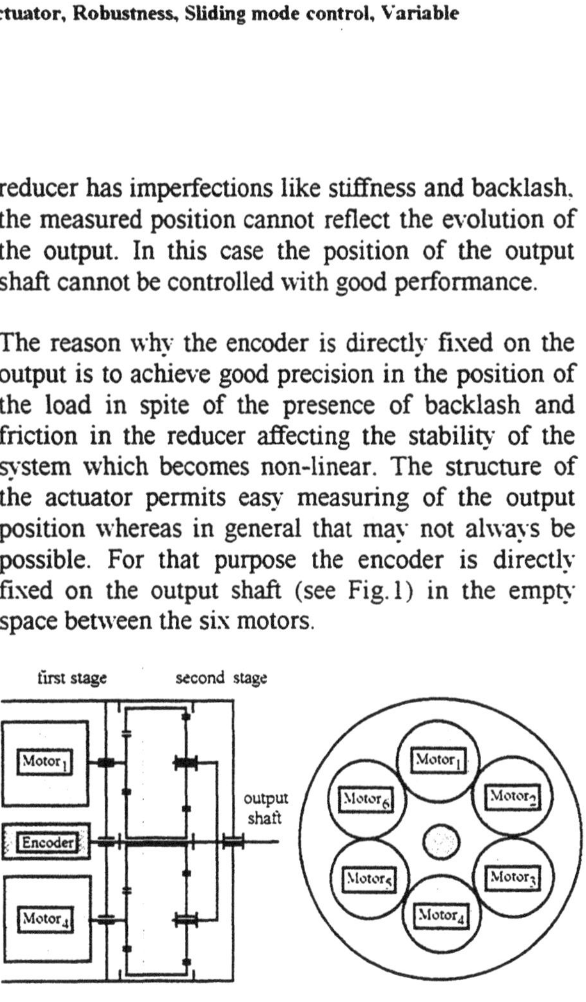

Fig.1. Architecture of the actuator.

With a classical position control loop the output is uncontrollable in the case where the system enters the deadzone. The system is then momentarily in an open loop (the exchanged torque becomes null). and a limit cycle may occur. These oscillations cause undesirable effects.

This special architecture have been studied in order to permit a complete integration of both electrotechnical and mechanical parts (where are included the sensor) in a compact electric machine. This actuator is the first step in the design of an integrated machine. In future, the actuator will be very compact. The six motors will be substituted by six rotors sharing the same magnetic flux created by permanent magnets. In the power supply there is a current loop with a time constant smaller than the actuator. This loop allow us to impose a torque reference to the system.

3. THE SYSTEM MODEL

Fig.1 shows that the only measurable states on the actuator are the output position and velocity. A non-linear model with backlash and Coulomb friction description (Cadiou 1994) can be developed, but in this case, it is easier to use the robustness of the sliding mode control. Only the parametric variation caused by the imperfections are take into account.

The system is modelled as a stiff axis with two parts connected by an elastic shaft. The first part corresponds to the equivalent motor formed by the set of the six motors and their connections. The second one contains the rest of the reducer. Due to the presence of the elastic joint the backlash can be represented by a classic function of an angle deviation. Fig.2 shows the general scheme of the retained model where the backlash and the Coulomb friction appear.

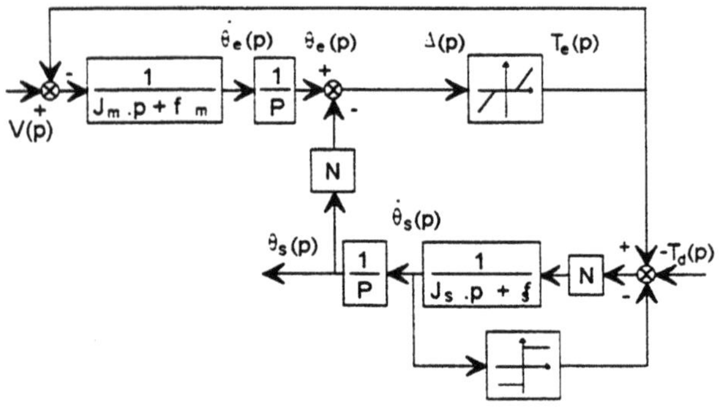

Fig.2. Structure scheme.

To obtain a model using only the two available states, the following assumptions must be assumed:
- the stiffness K is infinite therefore the system is consider as rigid;
- the time of the transfer between contact and non-contact zone (dead-zone) is supposed quasi-null.

The hardness of these two assertions is compensated by the robustness of the sliding mode control. Indeed it is well known that this method presents good performances towards parametrics variations, model imprecision and perturbations. The obtained model does not take into account the Coulomb friction, their effects are supposed to be compensated through the robustness of the control law. It involves only the backlash. With these assumptions the system representation has two equations. The first describes the dynamics of the actuator since the contact period and the second when the output part is in the dead-zone. These two equations may be reformulated in only one state space representation. In the respect of the matching condition (Leitmann 1979), the expression can be written using a nominal part plus a non-linear term which represents the uncertainties introduced by the presence of the backlash. So the system is:

$$\dot{x} = A.x + B.v + F \qquad y = C.x \qquad (1)$$

where the state vector $x = [\theta_s;\dot{\theta}_s]^T$, $v = K_m.I$ and the uncertain term $F = \Delta A(x) + \Delta B.v$. So it becomes:

$$\begin{bmatrix}\dot{\theta}_s\\ \ddot{\theta}_s\end{bmatrix} = \begin{bmatrix}0 & 1\\ 0 & -\dfrac{f_m.N^2+f_s}{(J_s+J_m.N^2)}\end{bmatrix}\cdot\begin{bmatrix}\theta_s\\ \dot{\theta}_s\end{bmatrix} + \begin{bmatrix}0\\ \dfrac{N}{(J_s+J_m.N^2)}\end{bmatrix}.v + \Delta A(X) + \Delta B.v \qquad (2)$$

$$y = [1\ 0].\begin{bmatrix}\theta_s\\ \dot{\theta}_s\end{bmatrix} \qquad \text{where}$$

$$\Delta A(x) = \begin{bmatrix}0\\ \left[\dfrac{f_m.N^2}{(J_s+J_m.N^2)} - \dfrac{f_s.J_m.N^2}{J_s.(J_s+J_m.N^2)}\right].\dot{\theta}_s.h\end{bmatrix} \qquad (3)$$

and

$$\Delta B = \begin{bmatrix}0\\ -\dfrac{N}{(J_s+J_m.N^2)}.h\end{bmatrix} \qquad (4)$$

with $h = 0$ if $|\theta_e - N.\theta_s| \le$ *dead-zone width* else $h = 1$

The variation of the incertitude is bounded. The parameter variations are also bounded so the system can be written as:

$$\begin{bmatrix}\dot{\theta}_s\\ \ddot{\theta}_s\end{bmatrix} = \begin{bmatrix}0 & 1\\ 0 & -a\end{bmatrix}\cdot\begin{bmatrix}\theta_s\\ \dot{\theta}_s\end{bmatrix} + \begin{bmatrix}0\\ b\end{bmatrix}.v, \quad y = [1\ 0].\begin{bmatrix}\theta_s\\ \dot{\theta}_s\end{bmatrix} \qquad (5)$$

with the following variations of the parameters:

$$2.14 \le a \le 2.87$$
$$0 \le b \le 0.426$$

The nominal estimation of the parameters may be chosen as the geometric mean for a and the arithmetic mean for b. But as it can easily seen, the previous expression must be divided by b to obtain the value of the input torque reference. But b may become null. To avoid this problem, we introduce a new expression w for the backlash disturbance. Let us note:

$$w = \begin{cases} -b.v & \text{if } |\theta_e - N.\theta_s| \leq \text{dead-zone width} \\ 0 & \text{otherwise} \end{cases}$$

The new state representation becomes:

$$\begin{bmatrix} \dot{\theta}_s \\ \ddot{\theta}_s \end{bmatrix} = \begin{bmatrix} 0 & 1 \\ 0 & -a \end{bmatrix} . \begin{bmatrix} \theta_s \\ \dot{\theta}_s \end{bmatrix} + \begin{bmatrix} 0 \\ b \end{bmatrix} . v + \begin{bmatrix} 0 \\ 1 \end{bmatrix} . w \qquad (6)$$

$$y = [1\ 0] . \begin{bmatrix} \theta_s \\ \dot{\theta}_s \end{bmatrix}$$

with the following variations of the parameter a:

$$2.14 \leq a \leq 2.87$$

The estimation of a may be considered as before, as the geometric mean, and the value of b as a constant equal to 0.425. The disturbance w is bounded, then it becomes possible to introduce this effect in the design of the Sliding mode control law. With this formulation it is preferable to consider w as a disturbance signal which can be compensated by the robustness of the control.

4. SLIDING MODE CONTROL

4.1 Sliding mode

A complete survey of the sliding mode theory can be found (Slotine *et al.* 1989; DeCarlo *et al.* 1988; Utkin 1987; Bühler 1986). In this case, the system is a single input-single output system. We consider it as:

$$\ddot{\theta}_s = - a.\dot{\theta}_s + b.v \qquad (7)$$

a: is the dynamic term not exactly known.
b: is the control gain supposed known and constant.
v: is the control input.

The method is based on a choice of a "sliding surface" defined in the phase space system passing by the equilibrium point. After that a control law is determined such that all state trajectories are directed towards it. And when the states of the system cross the surface the control is switched from one value v_{max} to one another v_{min} or inversely. When the state enters in sliding domain the trajectory move along the surface until the desired state. In the ideal case when the states stay on the surface, the system is in "sliding mode".

The system described above is of second order (n=2). In this case, the sliding surface equation $\sigma(\tilde{\theta})$ is defined by Slotine (1989):

$$\sigma(\tilde{\theta}) = \dot{\tilde{\theta}} + \lambda.\tilde{\theta} = 0 \qquad (8)$$

where $\tilde{\theta} = \theta_s - \theta_d$ $\dot{\tilde{\theta}} = \dot{\theta}_s - \dot{\theta}_d$ and $\ddot{\tilde{\theta}} = \ddot{\theta}_s - \ddot{\theta}_d$. σ is a weighted sum of the position and velocity errors. It represents the true distance between the state and the surface. The tracking trajectory problem is transformed into a σ stabilisation problem. In the ideal case the sliding mode occurs when $\sigma(\tilde{\theta})=0$, but in the real case the state stays only in the vicinity of the surface. This effect is due to the presence of switching delay or hysteresis which causes an oscillatory behaviour of the system around the surface. This phenomena is called "chattering". We will see later that the oscillations may be cancelled, by introducing a trade-off between the robustness and tracking precision.

It is possible to define an equivalent control needed to follow the surface. For this purpose $\dot{\sigma}$ must be written to introduce the dynamic of the system.

$$\dot{\sigma}(\tilde{\theta}) = \ddot{\tilde{\theta}} + \lambda.\dot{\tilde{\theta}} = 0 \qquad (9)$$

Substituting $\ddot{\theta}$ in (3) by (1) it becomes:

$$\dot{\sigma} = - a.\dot{\theta}_s + b.v - \ddot{\theta}_d + \lambda.\dot{\tilde{\theta}} = 0 \qquad (10)$$

from where $$v_{eq} = \frac{- a.\dot{\theta}_s + \ddot{\theta}_d - \lambda.\dot{\tilde{\theta}}}{b} \qquad (11)$$

If the input control is equal to u_{eq} then the system is in sliding mode. The dynamic becomes independent of the parameters of the system and is equal to:

$$\ddot{\tilde{\theta}} = \ddot{\theta}_d + \lambda.\dot{\tilde{\theta}}$$

In sliding mode the dynamic behaviour of the system is only dependent on the definition of the surface. the term a is uncertain, so an estimated equivalent control can be written:

$$v^*_{eq} = \frac{- \hat{a}.\dot{\theta}_s + \ddot{\theta}_d - \lambda.\dot{\tilde{\theta}}}{b} \qquad (12)$$

Where $\hat{a}$ represents the approximation of a. $\hat{a}$ and b define the nominal model. This value u^*_{eq} is the best approximation of the equivalent control to maintain the sliding mode and drive the system to the final state. With this input control the behaviour of the system is dependent on the definition surface but also on the bounds of the parameter variations. Then this control becomes insufficient to obtain the robustness. A discontinuous term must be added. This supplementary part allows to introduce a switching behaviour of control, and to lead the state toward the surface and compensate the effects of perturbations and parametric variations. The control law becomes:

$$v = v^*_{eq} - \Delta v \qquad (13)$$

The additional term with regard to the orientation of the state trajectories toward the surface, must follow the switching law defined as:

$$\Delta v = \begin{cases} \Delta v > 0 & \text{if} \quad \sigma > 0 \\ \Delta v < 0 & \text{if} \quad \sigma < 0 \end{cases} \tag{14}$$

4.2 Reachability and stability condition

The term Δv must be determined in order to make the surface attractive for any initial state. It is chose:

$$\Delta v = K(\theta).\text{sgn}(\sigma) \tag{15}$$

which correspond to (14) and where K is a gain and $\text{sgn} = -1$ if $\sigma < 0$ $\text{sgn} = 0$ if $\sigma = 0$ and $\text{sgn} = 1$ if $\sigma > 0$. The applied control law is now:

$$v = \dot{v}^*_{eq} - K(\theta).\text{sgn}(\sigma)$$

The gain K is determined by using the Lyapunov candidate function $V(\tilde{\theta}) = \frac{1}{2}\sigma^2(\tilde{\theta}.t)$:

$$\frac{1}{2}\frac{d\sigma^2(\tilde{\theta}:t)}{dt} = \sigma(\tilde{\theta},t).\dot{\sigma}(\tilde{\theta}:t) \le -\eta\, |\sigma(\tilde{\theta};t)| \tag{16}$$

where η is a positive constant corresponding to an acceleration (Slotine 1989). This constant gives response rate. This condition is called "reachability condition", it imposes a negative "speed" $\dot{\sigma}$ when the distance σ to the surface is positive and inversely a negative "speed" $\dot{\sigma}$ for positive distance. Thus, the control is computed such that the system state is always oriented toward the surface. This condition (16) allow also to assure the stability of the system. Indeed, with the Lyapunov function V, if the gain is choose K in respect to (16) $\dot{V} < 0$ (since η is strictly positive). Thus in conclusion the system is stable. The following control law is choose:

$$v = b^{-1}.(\dot{v}^*_{eq} - K(\theta).\text{sgn}(\sigma)) \tag{17}$$

with $\quad \dot{v}^*_{eq} = \hat{a}.\dot{\theta}_s + \ddot{\theta}_d - \lambda.\dot{\tilde{\theta}}$

In order to determine the gain K, (16) must be respected. Then it comes:

$$\sigma.\dot{\sigma} \le -\eta.|\sigma| \tag{18}$$

$$\sigma.\dot{\sigma} = \sigma.[(a - \hat{a}).\dot{\theta}_s - K.\text{sgn}(\sigma)] \tag{19}$$

and then $\sigma.[(a - \hat{a}).\dot{\theta}_s - K.\text{sgn}(\sigma)] \le -\eta.|\sigma|$ (20)

if (19) is majored, it comes:

$$|(a - \hat{a}).\dot{\theta}_s|.|\sigma| - K.|\sigma| \le -\eta.|\sigma| \tag{21}$$

from where proceeds the choice of K as:

$$K(\theta) = |(a - \hat{a}).\dot{\theta}_s| + \eta \tag{22}$$

posing that $|(a - \hat{a}).\dot{\theta}_s| \le F$, where F is the maximum variation of a. The control law becomes:

$$v = b^{-1}.[\dot{v}^*_{eq} - (|(a - \hat{a}).\dot{\theta}_s| + \eta).\text{sgn}(\sigma)] \tag{23}$$

4.3 Discontinuous input control smoothing

To avoid the Chattering phenomenon, the previous term Δv can be replaced by a new term which uses a saturation function (Slotine 1989).

$$\Delta v = K.\text{sat}(\sigma/\phi) \tag{24}$$

$$\text{sat}(\sigma/\phi) = \begin{cases} \sigma/\phi & \text{if} \quad |\sigma/\phi| \le 1 \\ \text{sgn}(\sigma/\phi) & \text{if} \quad |\sigma/\phi| > 1 \end{cases} \tag{25}$$

The saturation function introduce a boundary layer neighbouring the sliding surface. ϕ is the thickness of the boundary, ε is the width. This introduces a zone in which the control law can vary linearly then a smoothing of the control is obtained. ε is calculated from ϕ with the following relation (Slotine 1989):

$$\varepsilon = \phi/\lambda^{n-1} \tag{26}$$

Thus the tracking of the trajectory is guaranteed with a precision of order ε. So a trade-off between the robustness and the precision appears.

$$v = b^{-1}.[\dot{v}^*_{eq} - (|(a - \hat{a}).\dot{\theta}_s| + \eta).\text{sat}(\sigma/\phi)] \tag{27}$$

with $\quad \dot{v}^*_{eq} = \hat{a}.\dot{\theta}_s + \ddot{\theta}_d - \lambda.\dot{\tilde{\theta}}$

4.4 Variable boundary layer thickness

The smoothing of the control discontinuity is equivalent to a lowpass filter to the local dynamics of σ. This structure allows to tune the control law in order to obtain the best trade-off between the tracking precision and the robustness. Then a time-varying boundary layer thickness can be applied to exploit the available capacity at each time. Now it is not the surface which must be attractive but the boundary layer. In order to achieve this purpose (16) is modified and becomes:

$$\frac{1}{2}\frac{d\sigma^2(\tilde{\theta}:t)}{dt} \le (\dot{\phi} - \eta)\, |\sigma(\tilde{\theta}:t)| \tag{28}$$

the additional term $\dot{\phi}.|\sigma|$ reflects the fact that the boundary layer attraction is more stringent during layer contraction ($\dot{\phi} < 0$) and less stringent during layer expansion ($\dot{\phi} > 0$). In order to satisfy (27) $-\dot{\phi}$ is added to the control gain K. The new gain $\bar{K}$ becomes:

$$\bar{K}(\theta) = K(\theta) - \dot{\phi} \tag{29}$$

After different developments (Slotine 1989):

$$v = b^{-1}.[\dot{v}^*_{eq} - (|(a - \hat{a}).\dot{\theta}_s| + \eta - \dot{\phi}).\text{sat}(\sigma/\phi)] \tag{30}$$

with $\quad \dot{v}^*_{eq} = \hat{a}.\dot{\theta}_s + \ddot{\theta}_d - \lambda.\dot{\tilde{\theta}}$

and $\quad \dot{\phi} + \lambda.\phi = K(\theta_d)$ (31)

In this development different control laws has presented, where successively the chattering was cancelled and a trade-off between the robustness an the tracking precision applied. In the next paragraph the practical performances are presented.

5 EXPERIMENTAL RESULTS

This part presents experimental results of the three control laws in order to compare them and validate the application and the robustness of the sliding mode in our system in spite of the neglected perturbations. The parameters of the control law λ, η and ϕ have been chosen to obtain acceptable results in the three cases. In these experimental trial we have observed, as expected, that λ monitor the system behaviour in the sliding mode and that η impose the reaching rate. But the maximum speed allowed by the motor limits the effect of η. The sampling rate is 5 ms, and the reference trajectory imposes desired position, velocity and acceleration. There is a saturation on the control input current between -6 A and 6 A

5.1 Experimental part without load

The experimentation lead to the choice of control parameters:

$$\lambda = 120\ s^{-1},\ \eta = 1.10^{6}\ rad.s^{-2}.$$

In this first part of experimental show, the results are obtained without load, and with a final reference amplitude of 1 rad.

Sliding control with sign function. In this case the control law (17) is used with the above parameter values.

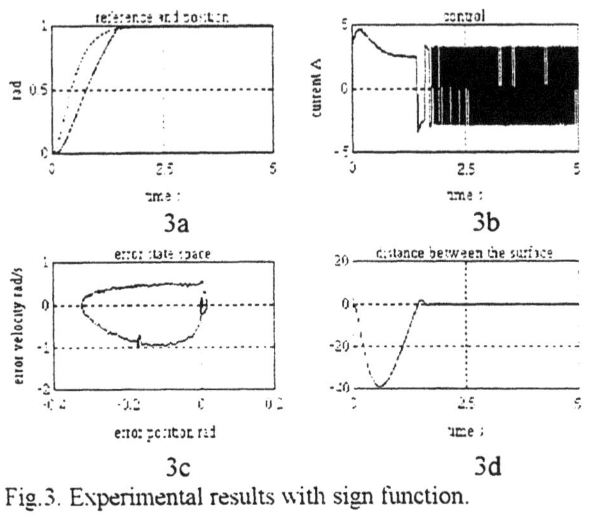

Fig.3. Experimental results with sign function.

Fig. 3a show the system response in continuous line (reference trajectory is in dash doted line), where we can see that it is slow. But it must be improved with other parameters choice. Fig. 3a and 3d show that the steady state error is very small. The main disadvantage is presented by Fig. 3b, where it can be observed that the control input is always in chattering, even under the saturation imposed by the power supply. This input oscillation must be cancelled by using a saturation function.

Sliding control with saturation function. In this case the control law (27) is used with the above parameter values. For this case a fixed boundary layer thickness is added.

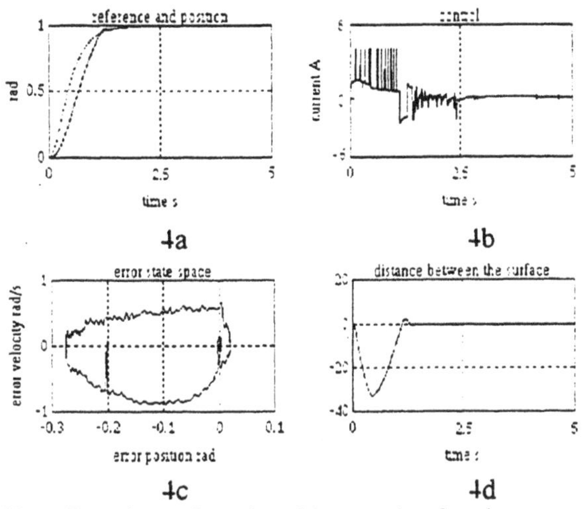

Fig.4. Experimental results with saturation function.

The response (in continuous line) on the Fig. 4a is slightly faster than the previous, but we can also improve them by changing the control parameters. The steady state error is also quasi null. Fig. 4b shows the main difference between the two control law. Indeed, in this case the chattering phenomenon disappear as expected, after the state reach the sliding surface. The parameter ϕ is small and the trade-off (26) enhance the precision despite the robustness, that is why the precision is better.

Sliding control with saturation function and variable boundary layer thickness. In this case the control law (30) is used with the above parameter values.

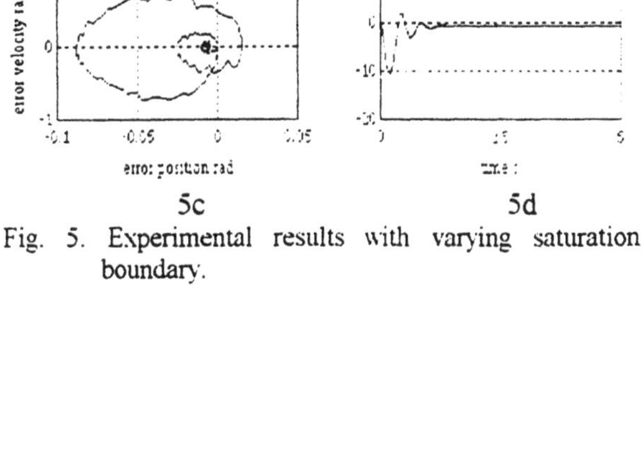

Fig. 5. Experimental results with varying saturation boundary.

In this case, we observe in Fig. 5a that the tracking performance are good. The system state follows the desired trajectory. The control law uses all available capacity to achieve the tracking. As we have seen in (26) we introduce with the boundary layer, as previous, a trade-off between the precision and the robustness. The system is robust because the parameter ϕ varies enough to lead good performances. This is not the case in the precedent law where ϕ is fixed. But as shown in Fig. 5d, the precision is worst than the previous control law. The state converge toward the boundary layer and not toward the surface.

The performance of the sliding approach applied on our system with backlash and Coulomb friction gives satisfactory results.

5.2 Experimental part with load

In this part, the results are obtained with the previous control law in which is used a varying boundary layer thickness. But now, there is an arm mounted on the output shaft as shown in Fig. 6. On this arm (m = 0.7 Kg) a load of about M = 3 Kg is fixed.

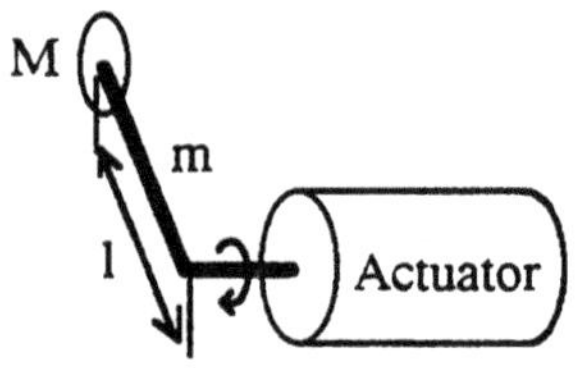

Fig.6. One degree of freedom arm.

For this experiment the results are obtained with a final reference amplitude of $2.\pi$ rad. The control parameters is the as previous. For the trial with the load, the initial and final position is such as the arm stand vertically. Fig. 7a Shows three curves. The dashed curve represents the reference trajectory, the dash doted curves show the position response of the actuator without load and the continuous curve indicates the position of the system under a load.

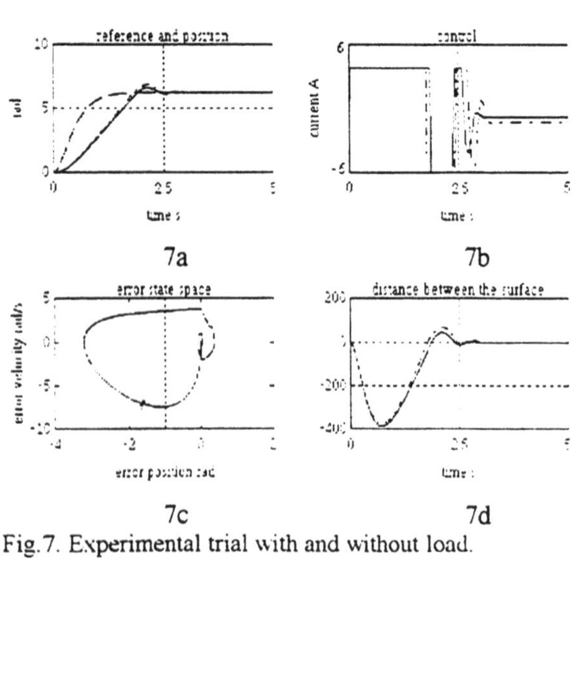

Fig.7. Experimental trial with and without load.

The response with or without load are similar, and for the two case the performance must be improve by a new tuning of the parameters. Fig. 7b shows that the control input is the same in the two cases. It confirm by both Fig. 7c and 7d. The control law is robust enough to cancelled the effect of the backlashes and the Coulomb friction effects.

6 CONCLUSION

The results proved the applicability of the sliding mode approach on our non-linear system type. This illustrate the interest of such control law in case of systems with backlash and Coulomb friction. The robustness of this method seems in this case more interesting than other control laws. A comparative study of several control approaches is under investigation. The control algorithm is easy to implement in a computer. It doesn't need a high performance processor because its computing time is short.

7. REFERENCES

Blazevic P. (1991). Contribution à l'étude d'un actionneur intégré. *Thèse de Doctorat de l'Univeristé Pierre et Marie Curie (Paris VI).*

Brandenburg G. and U. Schafer (1988). Influence and partial compensation of simultaneously acting backlash and coulomb friction in a speed and position controlled elastic two-mass system. *ICED '88, Proc. of Int. Conf. on electrical Drives*, Romania.

Bühler H. (1986). Réglage par mode de glissement. Presses Polytechniques Romandes.

Cadiou J.C. (1994). Etude et mise en oeuvre de commandes sur un actionneur électrique avec jeux et frottements secs. *Thèse de Doctorat de l'Univeristé Pierre et Marie Curie (Paris VI).*

Chan W.C. and H.T. Dorrah (1973). Describing function for servomechanism with backlash and multiple non-linearity. *International J. Control*, Vol. 17,n° 6.

DeCarlo R.A., S.H. Zak, Matthews G.P. (1988). Variable structure control of nonlinear multivariable systems: A tutorial. *Proc. of the IEEE*, Vol. 76, n°3, pp 212-232.

Desoer C.A. and S.M Shahruz (1986). Stability of dithered non linear systems with backlash or hysteresis. *Int. J. Control*, vol. 43, n°4, pp 1045-1060.

Leitmann G. (1979). Guaranteed asymptotic stability for some linear systems with bounded uncertainties. *ASME J. Dynamical Systems, Measurement and Control*, 101, pp 212-216.

Razik H. and C. Iung (1990). Speed control using sliding mode theory for an axis with constraints. *IECON 90, 16th Conf. on IEEE Industrial electronics society*, Pacific Grove, USA, Vol. 1, pp 52-56.

Slotine J.J.E. and W. Li (1989). Applied Nonlinear Control. *Prentice Hall.*

Tao G. and P.V. Kokotovic (1993). Adaptive control of systems with backslash. *Automatica*, Vol. 29, n° 2, pp 323-335.

Utkin V.I. (1987). Discontinuous control systems: State of art in theory and applications. *IFAC 10th triennial world congress*, Munich, Germany.

Modeling and Control of Eddy Current Brake

E. SIMEU, D. GEORGES
Laboratoire d'Automatique de Grenoble & Institut Supérieur d'Automatique et de Robotique de Valence

ADDRESS: ISAR-Rhône-Alpes ; 52-74, rue B. de Laffemas,
BP 1023- 26010 VALENCE, FRANCE.
Phone: (+33) 75 75 70 95; Fax: (+33) 75 43 56 42; E-mail: es@isar.fr & dg@isar.fr.

abstract
Recently a great deal of attention has been given to numerical algorithms for identification and control of nonlinear systems. The purpose of this paper is the application of some existing techniques for the modeling and regulation of an eddy current brake process. An approximate theoretical model is driven for the behavior of an eddy current disc brake in the low speeds zone. Input-output informations are used to obtain a polynomial state-affine behavior model for such a process. A dynamical and a static feedback compensator schemes are proposed for the process speed control in the presence of unknown braking resistant torque. The dragging torque value is estimated by an appropriate nonlinear observer. Experimental results are presented.
Key-words: Nonlinear systems, modeling, identification, nonlinear observer, nonlinear controler design, Eddy current brake.

Résumé
Depuis quelque temps, l'identification et la commande des procédés non linéaires par des algorithmes numériques connaît un essort important. Dans cet article, quelques techniques existantes sont appliquées pour modéliser et réguler un système de freinage à courant de Foucault. Un modèle de connaissance approché décrivant le comportement du frein dans la zone de faibles vitesses est proposé. Des séquences de données d'entrée-sortie correspondant à différents régimes de fonctionnement sont ensuite utilisées pour déterminer les paramètres d'un modèle polynomial à état affine du procédé. Un schéma de commande par retour dynamique et un compensateur statique sont proposés pour la régulation de la vitesse du dispositif en présence d'un couple d'entrainement inconnu. Une estimation de la valeur du couple résistant au freinage est fournie par un observateur non linéaire approprié. Les résultats obtenus sur la base d'un dispositif expérimental sont présentés.
Mots-clés: Systèmes non linéares, modelélisation, identification, observateur non linéaire, commande non linéaire, Frein à courants de Foucault.

1 Introduction

The problem of numerical control design for nonlinear processes which is under investigation here is a subject of growing interest. Although most industrial processes display nonlinear dynamics, in practice the linear control design methods are often used. When the nonlinearities of the process dynamics are not too severe, a linear approximation of its model in a neighborhood of the operating point is generally satisfactory. But in many cases, when large deviation from the operating point are possible or when the dynamical nonlinearities are too severe, the linear model becomes soon limited. In this case, the design of a nonlinear control scheme may be allowed by an accurate knowledge of a nonlinear model of the process under regulation.
One difficult problem in nonlinear control design is the modeling of nonlinear systems from a set of input-output data. Obviously this is often the only approach to modeling since, in most cases, only input-output data are available. In this paper, a theoretical polynomial-in-control and state-affine model is driven for an eddy current brake process. The model parameters are estimated by an " input upholding " technique.
The estimated parameters of the model are used to design a continuous time control law for speed regulation with the eddy current brake system. The corresponding numerical controller is implemented through a sampler and Zero-Order-Holder. The organization of this paper is the following: A theoretical parametric model of the process is proposed in section 2. In section 3, the coefficients of the previous model are estimated and section 4 deals with a dynamical and static continuous time compensator design for speed control and some experimental results.

2 Process description and theoretical model

An eddy current brake consists of a full copper ware or aluminum disc driven in rotative motion through the air gap of an electromagnet Fig. 1. When the disc is moved, an electrical field $E = v \times B$ is induced perpendicularly to both the tangential speed of movement $v = R\omega$ measured at the center of the pole and the magnetic induction under the pole B. When the air gap magnetic field is studied for different speeds, three different behavior regions are observed [1].

i) At low speeds region, the magnetic induction caused by the eddy current pattern is negligible compare with the original induction. The air gap magnetic induction B is then slightly less than one generated at zero speed B_0. In this case, the magnetic induction is almost perpendicular to the plane of the disc.

ii) The critical speed region is the speed zone at which the exerted brake force is maximal; the induction caused by the eddy current pattern is no longer negligible compared with B_0. The mean induction under the pole is already significantly less than B_0.

iii) At higher speeds region, the mean magnetic induction in the air gap tends to further decrease. Wouterse's experimental results in [1] show that for infinite speed, the original magnetic field will be completely cancelled by the induced eddy currents.

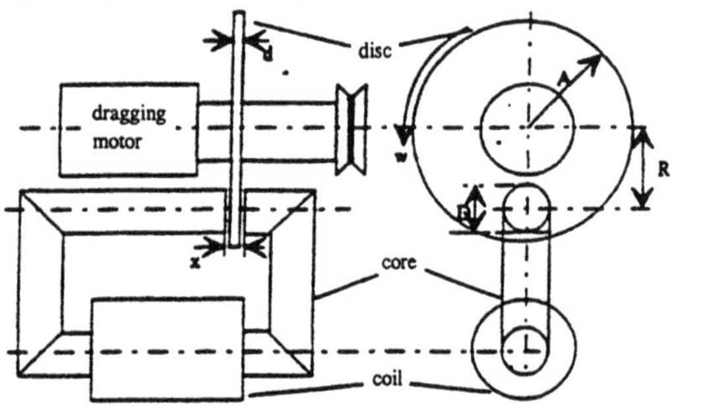

Fig.1. Eddy current brake sketch

The current density J under the pole in the 'ideal' case (super conducting ring around the pole slipping over the disc) a is given by the relation $J = 1/\rho(v \times B)$ ([1]) with ρ representing the specific resistance of disc material. The total power dissipation may be calculated simply by integrating ρJ^2 over the cylindrical volume $\frac{\pi D^2}{4}d$, where d and D denote respectively the disc thickness and the diameter of circle with same area as the pole face. So the total dissipation is

$$P_d = \frac{\pi}{4\rho}D^2 dB^2 v^2. \quad (1)$$

From equation (1) the brake force may be evaluated as follows:

$$f_d = \frac{P_d}{v} = \frac{\pi}{4\rho}D^2 dB^2 v. \quad (2)$$

This result means that for low speeds, the brake force is proportional to the speed.

In the real process, the return path of the induced currents is not of zero resistance. Several outstanding works dealing with the study of force in ferromagnetic eddy current brake exist in the literature ([1, 2, 3]); different results have been proposed. A factor C is generally introduced in equation (2), to take into account the resistance of the return path of induced eddy current, [1, 3].

$$f_d = \frac{P_d}{v} = \frac{\pi}{4\rho}D^2 dB^2 Cv. \quad (3)$$

Referring to **Fig. 1** for physical dimensions A, D and R, Wouterse provides the following value for coefficient C in [1]:

$$C = \frac{1}{2}\left[1 - \frac{1}{4}\frac{1}{\left(1+\frac{R}{A}\right)^2\left(\frac{A-R}{D}\right)^2}\right] \quad (4)$$

At low speed, an eddy current brake acts as a linear damper: the brake force exerted by a constant magnetic induction is proportional to the speed.

Assuming that the magnetic flux B of equation (3) is generated by a coil supplied by an exciting ampere turns $H = N.i_e$. We drive an approximative model giving the brake torque as a function of exciting current i_e and (angular) speed ω.

Although it is difficult to obtain an exact analytical solution $B = f(H)$ of the magnetization field curve of the material, the following approximation may be used:

$$B = \begin{cases} B_r + \epsilon\mu_0 H & for\ H > 0 \\ -B_r + \epsilon\mu_0 H & for\ H < 0 \end{cases} \quad (5)$$

in which B_r is the residual magnetic induction and the slope $\epsilon\mu_0$ of this function is the permeability of the material. Equation (5) has been used by Venkataratnam and Abdul Kadir in [2] and was shown to be a good qualitative and quantitative approximation of the nonlinear B/H curve.

Taking into account the exciting equation $H = N.i_e$ and the natural friction coefficient f_0 of the process, the total brake force expression may be deduced from equation (3).

$$f_b = (\alpha_0 + \alpha_1 i_e + \alpha_2 i_e^2)v. \quad (6)$$

in which

$\alpha_0 = f_0 + \frac{\pi}{4\rho}D^2 dB_r^2 C$;

$\alpha_1 = \pm\frac{\pi}{2\rho}D^2 dB_r CN\epsilon\mu_0$ and

$\alpha_2 = \frac{\pi}{4\rho}D^2 dCN^2(\epsilon\mu_0)^2$.

The following state model description, representing the input-output behavior may be deduced from expression (6).

$$\begin{cases} \dot{x}_1 = (a + bu + cu^2)x_1 + x_2 \\ \dot{x}_2 = 0 \\ y = x_1 \end{cases} \quad (7)$$

where input $u \in \Omega = [0\ \ i_{emax}]$ is the exciting current, $a = -\alpha_0/I$, $b = -\alpha_1/I$ and $c = -\alpha_2/I$ and I is the total inertia of the process. The output y to be controlled represents the process speed and the state x_2 is proportional to the dragging torque applied to the process. Notice that this model is states affine belongs to the following class of linearly parameterized nonlinear systems:

$$\dot{x} = F(\theta, u)x \quad and \quad y = H.x \tag{8}$$

With $x = \begin{pmatrix} x_1 \\ x_2 \end{pmatrix}$;

$$\begin{aligned} F(\theta, u) &= \begin{bmatrix} a + bu + cu^2 & 1 \\ 0 & 0 \end{bmatrix} \\ &= \begin{bmatrix} f(u) & 1 \\ 0 & 0 \end{bmatrix}; \end{aligned}$$

$H = [1 \quad 0]$ and $\theta = \begin{pmatrix} a \\ b \\ c \end{pmatrix}$ is the vector of the model parameters to be estimated.

3 Identification

In this section, we are concerned with the identification of model parameters θ. Most of the works that use an input-output approach to the modeling of nonlinear systems employ the Volterra kernels series representation [6]. The use of this method is however restricted by the difficulty of obtaining the Volterra kernels of practical systems. An algorithm using input-output informations to establish a polynomial state affine model for discrete time nonlinear systems is proposed by Diaz et al. in [4]. Our purpose of here is to make use of the polynomial structure of the process theoretical model in the identification process. Before dealing with problem of model parameters estimation it seem a good useless to explain the numerical integration scheme of continuous time process used in this application.

3.1 Choice of a sample period

The diagram of the retained digital control is the classical one depicted in Fig. 2.

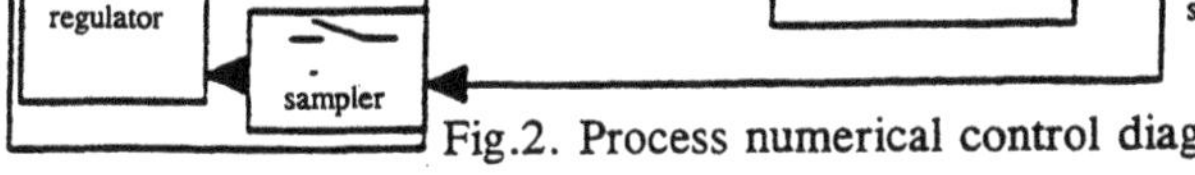

Fig.2. Process numerical control diagram

The state x(t) and the control u(t) are sampled at the same sampling period δ. The 'exact' sampled-data model [5], [7] corresponding to the continuous model 8 can be written as follows:

$$\begin{aligned} x_e(k+1) &= \left[1 + \sum_{j=1}^{\infty} \frac{(\delta F(\theta, u(k)))^j}{j!}\right] x_e(k) \\ &= e^{\delta . F(u(k))} . x_e(k) \quad and \\ y_e = H.x_e. \end{aligned} \tag{9}$$

At the sampling instants, the state $x_e(k)$ of the model discrete-time model exactly coincides with the state $x(t)$ ($t = k\delta$) of the continuous-time system of model (9).

Let suppose that the following natural assumptions hold:

Assumptions

a1) Any state x belongs to a compact set $C_s \subset \mathcal{R}^n$ $(= \mathcal{R}^2)$ containing $x = 0$ as interior point, i.e. $\exists k_1 \in \mathcal{R}^+$ such that $\|x\| \leq k_1$.

a2) The parameter θ belongs to a compact set C_θ of $\mathcal{R}^3$ containing θ_0 as interior point, i.e. $\exists k_2 \in \mathcal{R}^+$ such that $\|\theta_0 - \theta\| \leq k_2$.

Note that both the process speed (x_1) and the dragging torque (x_2)are naturally bounded in our application. Therefore, assumption **a1** is held. The control u is also bounded and belongs to a compact set $\Omega = [0 \;\; u_{max}] =\subset \mathcal{R}^+$.

Furthermore, since the braking force corresponding to any finite speed and control value has a finite value in practice, the coefficients a, b, and c of polynomial $f(u)$ are bounded and assumption **a2** is consequently held.

Since assumptions **a1** and **a2** are held in our application, it follows the next proposition:

Proposition 1 *For every $\epsilon \in \mathcal{R}^+$ there exists a small enough sampling period $\delta_\epsilon \in \mathcal{R}$ such that for all $\delta \leq \delta_\epsilon$ and for all $x \in C_s$:*
$\left\| \left[e^{\delta . F(\theta,u)} - (1 + \delta . F(\theta, u))\right] (x(k) \right\| \leq \epsilon$

Proof
$\left\| \left[e^{\delta . F(\theta,u)} - (1 + \delta . F(\theta, u))\right] (x(k)) \right\|$
$\leq \left\| e^{\delta . F(\theta,u)} - (1 + \delta . F(\theta, u)) \right\| k_1$

Since $F(\theta, u)$ is bounded in the compact set $C_\theta \times \Omega$, for small values of the sample period δ a first order expansion of $e^{\delta . F(\theta,u)}$ may be written:

$$\begin{aligned} &e^{\delta . F(\theta,u)} = 1 + \delta . F(\theta, u) + R(\delta); \\ &With \;\; \lim_{\delta \to 0} \|R(\delta)\| = 0 \end{aligned} \tag{10}$$

This result allows a choice of a classical Euler integration scheme, provided that the sample period is small enough.

For the eddy current process described above, the choice of the sample rate $f_e = 1/\delta$=40hertz gives very satisfactory results.

3.2 Parameter estimation

Our objective in this section is an evaluation of the parameters a, b, and c of the polynomial $f(u)$ defined above. Let ν be a constant input value of the process with model (8); from an initial state $x_i = \begin{pmatrix} x_{i1} \\ 0 \end{pmatrix}$ with $x_{i1} \neq 0$ the process output behavior may be described by the following model:
$\dot{y}(t) = f(\nu)y(t) = (a + b\nu + c\nu^2)y(t)$;
$with \;\; y(0) = x_{i1}$

Since ν is assumed to be constant, equation is linear and homogeneous with only one unknown parameter which is $f(\nu)$. For any fixed ν value, a measurement of the corresponding output data sequence allows an estimation of the unknown parameter $f(\nu)$

corresponding to the input value ν with the classical linear methods. The following identification process is repeated:

i) an exciting $i_e = \nu \in \bar{\Omega}$ is applied on the input (brake coil) of the process,
ii) a dragging torque is applied to bring the process speed to a wished initial value $y_{\nu 0} = y_\nu(0)$,
iii) the dragging torque is cancelled and a N size output sequence is recorded:
$S_\nu = [y_\nu(1), y_\nu(2), ..., y_\nu(N)]$
The input set $\bar{\Omega} = \{\nu_1, \nu_2, ...\nu_p\}$ is chosen such that matrix $\bar{M}^T\bar{M}$ is invertible. where $\bar{M}$ is the $p \times 3$ matrix defined by:

$$\bar{M}_{ij} = \nu_i^{j-1} \Bigg| \begin{array}{l} i = 1, 2, ...p \\ j = 1, 2, 3. \end{array} \qquad (11)$$

With the output sequence S_ν corresponding to any input value $\nu \in \bar{\Omega}$ let define the vectors X_ν and Y_ν as follows:

$$X_\nu = \frac{1}{2}\begin{pmatrix} y_\nu(0) + y_\nu(1) \\ y_\nu(1) + y_\nu(2) \\ \vdots \\ y_\nu(N-1) + y_\nu(N) \end{pmatrix}$$

and

$$Y_\nu = \begin{pmatrix} y_\nu(0) - y_\nu(1) \\ y_\nu(1) - y_\nu(2) \\ \vdots \\ y_\nu(N-1) - y_\nu(N) \end{pmatrix}$$

Proposition 2 *A least-squares estimation $\hat{\theta}$ of the parameter vector θ is given by the following expression:* $\hat{\theta} = \begin{pmatrix} \hat{a} \\ \hat{b} \\ \hat{c} \end{pmatrix} = \frac{1}{\delta}\left[\bar{M}^T\bar{M}\right]^{-1}\bar{M}^T\bar{Y}$ *Where $\bar{M}$ is the $p \times 3$ matrix defined above and $\bar{Y}$ is a $p \times 1$ column vector defined by:*
$\bar{Y}_i = \left[X_{\nu_i}^T X_{\nu_i}\right]^{-1} X_{\nu_i}^T Y_{\nu_i}\Big|_{i=1,2,...p}$

The estimation procedure described in this section was performed on the eddy current brake process of section 2. The identification algorithm computed the following θ parameter:

$$\hat{\theta} = \begin{pmatrix} \hat{a} \\ \hat{b} \\ \hat{c} \end{pmatrix} = \begin{pmatrix} -0.118 \\ 0.02 \\ -0.017 \end{pmatrix}$$

And the corresponding state space model is given by equation (8) with matrix $F(\theta, u)$ replaced by:

$$F(\hat{\theta}, u) = \begin{bmatrix} \hat{f}(u) & 1 \\ 0 & 0 \end{bmatrix} = \begin{bmatrix} \hat{a} + \hat{b}u + \hat{c}u^2 & 1 \\ 0 & 0 \end{bmatrix}$$

so we have

$$F(\hat{\theta}, u) = \begin{bmatrix} -0.118 + 0.02u - 0.0017u^2 & 1 \\ 0 & 0 \end{bmatrix}$$

We simulated the response of this nonlinear model to a random input sequence. The results are compared to the corresponding process output on Fig.3a. On fig.3b, polynomial $\hat{f}(u)$ is compared to the point-by-point measured values curve.

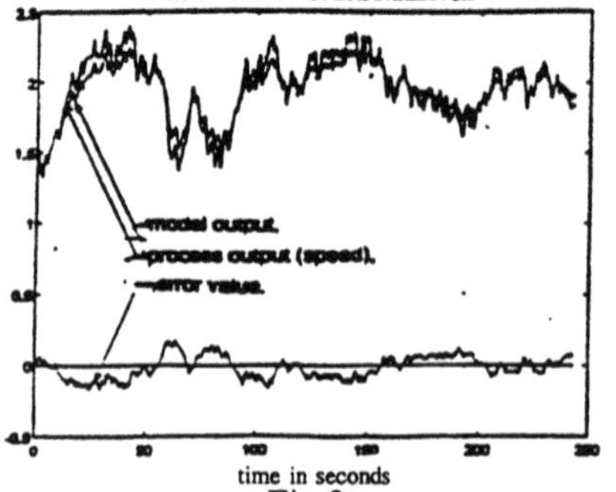

Fig.3a

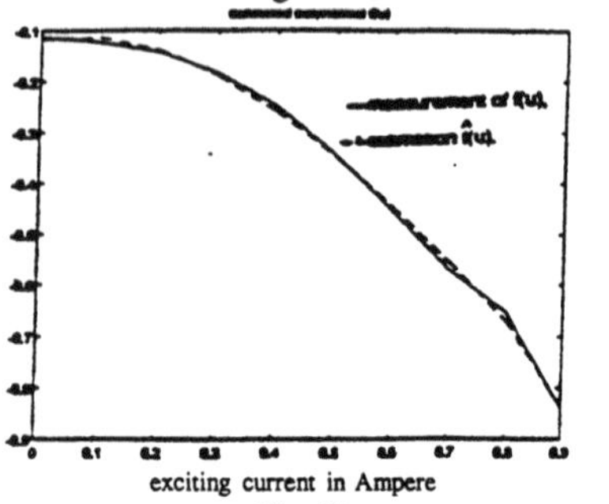

Fig.3b

4 Compensator design for speed control

In this section, we propose two observer-based nonlinear control schemes in order to regulate the speed around a reference speed value y_c, while estimating the torque to compensate. The first control scheme uses a dynamical feedback law in u and the second one, a static feedback law in u.
These control schemes are designed on the basis of the continuous-time model of the brake system given by equation (7):

$$\begin{cases} \dot{x}_1 = f(u)x_1 + x_2 \\ \dot{x}_2 = 0 \\ y = x_1 \end{cases} \qquad (12)$$

where x_1 is the system speed, u, the exciting current value and x_2, the torque, due to the exogenous system to be braked. In this model, x_2 is supposed to be constant or slowly time-varying, compared to the dynamics of the brake system. According to eddy brake current model, $f(u)$ is given by $f(u) = a + bu + cu^2$, where $a < 0$, $b > 0$ and $c < 0$. This dynamical model is linear in state and quadratic in the control input.

4.1 A dynamical feedback scheme

This control scheme is based on the idea of input-output linearization via feedback. As the system is not affine in the control input, we add an integrator in order to obtain an affine state-space model:

$$\begin{cases} \dot{x}_1 = f(u)x_1 + x_2 \\ \dot{u} = v \\ y = x_1 \end{cases} \qquad (13)$$

Then we apply linearization techniques by successive time-derivatives of the output y:

$$\begin{array}{l} y = x_1 \\ \dot{y} = f(u)x_1 + x_2 \\ \ddot{y} = v(b + 2cu)x_1 + f(u)^2 x_1 + f(u)x_2 \end{array} \qquad (14)$$

Consequently, a linearizing dynamical state-feedback v may be given by:

$$\dot{u} = v = \frac{-f(u)^2 x_1 - f(u)x_2 + w}{(b+2cu)x_1} \quad (15)$$

(15) leads to the following linear dynamics: $\ddot{y} = w$
Now if we choose $w = -K_1(y - y_c) - K_2\dot{y}$, where y_c is the reference speed to be obtained, the closed-loop system has the following dynamics:

$$\ddot{y} + K_2\dot{y} + K_1 y = K_1 y_c \quad (16)$$

Gains K_1 and K_2 are chosen such that the characteristic polynomial of (16) has asymptotically stable roots.
The associated change of coordinates is given by:

$$\begin{aligned} y &= x_1 \\ \dot{y} &= f(u)x_1 + x_2 \end{aligned} \quad (17)$$

This change of coordinates is only locally defined, because some states x_1^* are not reachable, i.e. there does not exist $u(t)$, $t \geq 0$, such that $x_1(x_1(0), u(t)) = x_1^*$.
In order to illustrate this problem, we consider the set E of the equilibrium points $x_1^e(u^e, x_2)$:

$$\begin{aligned} E &= \{x_1^e / f(u^e)x_1^e + x_2 = 0; x_2 \neq 0\} \\ &= \{x_1^e / \Delta = b^2 - 4c(a + \tfrac{x_2}{x_1^e}) \geq 0\} \end{aligned} \quad (18)$$

where Δ is the discriminant of second degree equation $cu^2 + bu + a + \frac{x_2}{x_1^e} = 0$. x_1^e is an equilibrium point, if and only if there exists a real u^e. This condition is met if and only if $\Delta \geq 0$. In this case, $u^e = \dfrac{-b \pm \sqrt{\Delta}}{2c}$.
E is a subset of the reachable set R, defined by: $R = \{x_1^* / \exists t \geq 0,\ \exists u(t) \in R;\ x_1(x_1(0), u(t)) = x_1^*\}$
Notice also that nonlinear feedback (15) is singular, when $y = 0$ or $f(u) = 0$, that represent a null measure set.
In practice, both change of coordinates (17) and feedback law (15) cannot be used directly because x_2 is unknown. We will suppose now that a estimated value $\hat{x}_2$ of x_2 is available, using an appropriate observer which will be defined later. We propose to consider the following modified change of coordinates:

$$\begin{aligned} y &= x_1 \\ \dot{y}' &= f(u)x_1 + \hat{x}_2 = \dot{y} - \bar{x}_2 \end{aligned} \quad (19)$$

where $\bar{x}_2 = x_2 - \hat{x}_2$. For the same reason, (15) is modified as follows:

$$\dot{u} = v = \frac{-f(u)^2 x_1 - f(u)\hat{x}_2 + w}{(b+2cu)x_1} \quad (20)$$

After some simple calculations and using

$$w = -K_1(y - y_c) - K_2\dot{y}'$$

the closed-loop system is given in the coordinates $(y, \dot{y})$ by:

$$\ddot{y} = -K_1(y - y_c) - K_2\dot{y} + (f(u) + K_2)\bar{x}_2 \quad (21)$$

We need now to define the torque-observer. We propose the following nonlinear observer:

$$\begin{aligned} \dot{\hat{x}}_1 &= f(u)x_1 + \hat{x}_2 + L_1(y - \hat{y}) \\ \dot{\hat{x}}_2 &= L_2(y - \hat{y}) \\ \hat{y} &= \hat{x}_1 \end{aligned} \quad (22)$$

where the gain matrix $L = (L_1, L_2)^t$ is chosen in order that the eigenvalues of $A_o - LC$ are stable, with $A_o = \begin{pmatrix} 0 & 1 \\ 0 & 0 \end{pmatrix}$ and $C = (1\ 0)$. Under these conditions, we can easily show that the estimation error dynamics is asymptotically stable around the origin. (13) and (22) may be rewritten as follows:

$$\begin{aligned} \dot{x} &= A_o x + Bf(u)x_1 \\ y &= Cx \end{aligned} \quad (23)$$

where $x = (x_1, x_2)^t$ and $B = (1\ 0)^t$. and:

$$\begin{aligned} \dot{\hat{x}} &= A_o\hat{x} + Bf(u)x_1 + L(y - \hat{y}) \\ \hat{y} &= C\hat{x} \end{aligned} \quad (24)$$

where $\hat{x} = (\hat{x}_1, \hat{x}_2)^t$. It follows that the error dynamics is linear and given by: $\dot{e} = (A_o - LC)e$ where $e = x - \hat{x}$. L is then chosen such that the spectrum of $A_o - LC$ is completely contained in the open left half-plane.
We sum up the structure of the here-proposed control scheme, in the original coordinates:

$$\begin{cases} u(t) = u(0) + \\ \displaystyle\int_0^t [\frac{-f(u(\tau))^2 x_1(\tau) - f(u(\tau))\hat{x}_2(\tau) + w(\tau)}{(b+2cu)x_1(\tau)}]d\tau \\ w = -K_1(y - y_c) - K_2(f(u)x_1 + \hat{x}_2) \\ \dot{\hat{x}}_1 = f(u)x_1 + \hat{x}_2 + L_1(y - \hat{y}) \\ \dot{\hat{x}}_2 = L_2(y - \hat{y}) \\ \hat{y} = \hat{x}_1 \end{cases} \quad (25)$$

Proposition 3 *The closed-loop system obtained from (25) and (22) is asymptotically stable in any invariant subset of the reachable set.*

Proof. Putting (21) and (22) together, we obtain the following state-space representation:

$$\begin{aligned} \dot{z} &= A_r z + D(u)e \\ \dot{e} &= (A_o - LC)e \end{aligned} \quad (26)$$

where $z = (y - y_c, \dot{y})^t$, $A = \begin{pmatrix} 0 & 1 \\ -K_1 & -K_2 \end{pmatrix}$ and $D = \begin{pmatrix} 0 & 0 \\ 0 & f(u) \end{pmatrix}$. As both A_r and $A_o - LC$ are Hurwitz, there exist some positive-definite matrices P and S such that $A^tP + PA = -q_r I_d$, $\forall q_r > 0$ and $(A_o - LC)^t S + S(A_o - LC) = -q_o I_d$, $\forall q_o > 0$, where I_d is the identity matrix.
Introducing $V = \frac{1}{2}z^t Pz + \frac{1}{2}e^t Se$ and using (26), we obtain:
$\dot{V} = -\frac{1}{2}q_r \parallel z \parallel^2 - \frac{1}{2}q_o \parallel e \parallel^2 + z^t PDe$ that may be rewritten as follows:

$$\dot{V} = -\frac{1}{2}(z^t e^t) M \begin{pmatrix} z \\ e \end{pmatrix}$$

Where $M = \begin{pmatrix} q_r I_d & -PD(u) \\ -D^t(u)P & q_o I_d \end{pmatrix}$.

Consider now a invariant subset W of the reachable set of (26), containing $z = 0$ and $e = 0$, defined by: $W = \{(z,e)/\|f(u)\| \leq \bar{f}\}$ We can show that there always exist q_r and q_o such that M is a positive-definite matrix. After simple calculations, we can state that the eigenvalues of M are positive if $q_r < \frac{q_o}{\alpha \bar{f}^2}$, where $\alpha = \frac{1}{4}\frac{(K_1+1)^2 + K_2^2}{K_1^2 K_2^2}$. Then it is always possible, for all $\bar{f}$ to find $q_r > 0$ and $q_o > 0$ such that M is positive-definite. Asymptotic stability follows from standard Lyapunov's stability theory.

4.2 A static feedback scheme

This control scheme is based on the idea of linearization of the term $f(u)x_1$.

Suppose that there exists $u \in R$ such that $f(u)x_1 = -K(y-y_c) - x_2$, with $K > 0$. In this case, dynamics (12) becomes linear and asymptotically stable:

$$\begin{cases} \dot{x}_1 = -K(x_1 - y_c) \\ y = x_1 \end{cases} \tag{27}$$

As previously, x_2 is unknown, so we need to introduce an observation $\hat{x}_2$ and to look for u, solution of $f(u)x_1 = -K(x_1 - y_c) - \hat{x}_2$.

Proposition 4 *The dynamics of $y(t) - y_c$ obtained from the following control scheme:*

$$\begin{cases} u(t) = \dfrac{-b \pm \sqrt{\Delta}}{2c} \\ \Delta = b^2 - 4c(a + \dfrac{K(y-y_c) + \hat{x}_2}{y})) \\ \dot{\hat{x}}_1 = f(u)x_1 + \hat{x}_2 + L_1(y - \hat{y}) \\ \dot{\hat{x}}_2 = L_2(y - \hat{y}) \\ \hat{y} = \hat{x}_1 \end{cases}$$

is asymptotically stable around the origin, providing that $y(t)$ and $\hat{x}_2$ belong to subset W, defined by:

$$W = \{(y, \hat{x}_2)/b^2 - 4c(a + \frac{K(y-y_c) + \hat{x}_2}{y}) \geq 0\}$$

Proof. Existence of some real-definite solutions of $f(u)x_1 = -K(y-y_c) - \hat{x}_2$, with $K > 0$, directly imposes trajectories $y(t)$ and $\hat{x}_2$ such that discriminant Δ of second degree equation $cu^2 + bu + a + \frac{K(y-y_c)+\hat{x}_2}{y} = 0$ is positive or null, i.e. $\Delta = b^2 - 4c(a + \frac{K(y-y_c)+\hat{x}_2}{y}) \geq 0$. Using observer (22) and separation principle, we can easily state that $y(t)$ converges asymptotically towards y_c.

For a constant reference speed value y_c and a time-varying dragging torque, the regulation curves resulting from the both dynamical and the static control scheme are represented on fig.4a and fig.4b respectively.

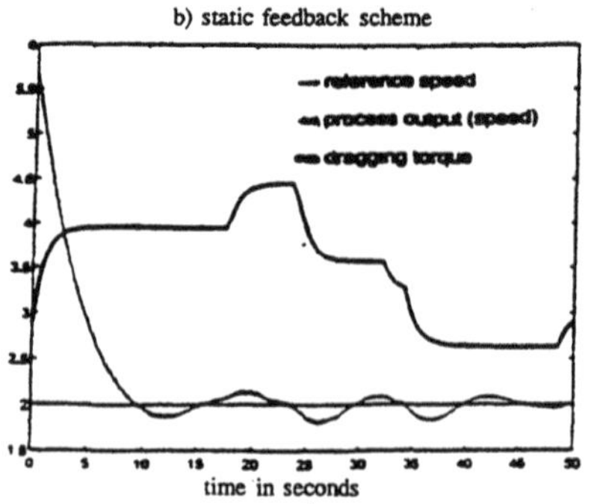

Fig.4b

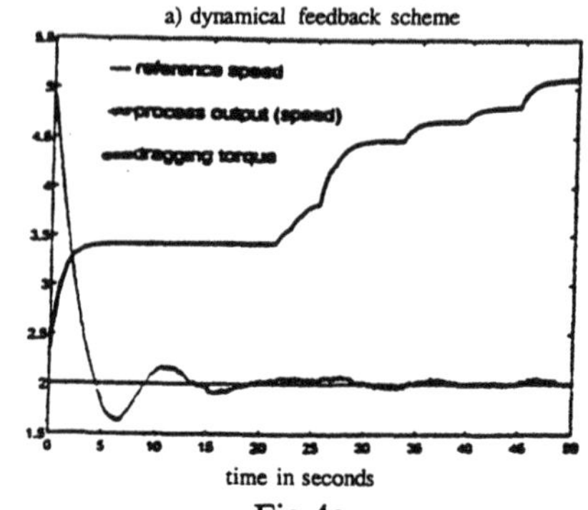

Fig.4a

Conclusion

The goal of this paper was to present a complete application of nonlinear control design procedure, to achievement a real process controller design. Modeling and identification problems have been considered as well as compensator design for regulation objectives. Application results obtained on an eddy current brake process are included to illustrate the modeling and the regulator design procedure.

References

[1] J.H. WOUTERSE; *Critical torque of eddy current brake with widely separated soft iron poles*; IEE Proceeding, Vol. 138, Pt. B, N° 4, July 1991.

[2] K.VENKATARATNAM, M.S. ABDUL KADIR; *Analysis and performance of eddy-current brake with loss drums: Part 1 & Part 2*; IEE Proceeding, Vol. 129, Pt. B, N° 3, May 1982.

[3] P.P. YANNOPOULOS - LASCARATOS, J.A. TEGOPOULOS; *Eddy-current distribution in cylindrical structures caused by rotating magnetic fields*; IEE Proceeding, Vol. 129, Pt. B, N° 2, March 1982.

[4] H. DIAZ, A.A. DESROCHERS; *Modeling of nonlinear discrete-time systems from input-output data*; Automatica, Vol.24, N° 5, pp. 629-641, 1988.

[5] S. MONACO, D. NORMAND - CYROT; *Quelques éléments sur la commande numérique non linéaire*, Systèmes non linéaires: *3. commande*; Masson 1993.

[6] Guy A. DUMONT, YE FU; *Nonlinear adaptative control via Laguerre expansion of Volterra kernels*; Inter. Journal of Adaptative Control and Signal Processing, Vol. 7, 367-382, 1993.

[7] A.M. GUILLAUME, G. BASTIN, G. CAMPION; *Discrete time adaptative control for a class of nonlinear continuous systems*; IFAC Nonlinear Control System Design, Capri, Italy 1989.

A Globally Stable Discrete-Time Controller for Current-fed Induction Motors

Romeo Ortega*

Université de Technologie de Compiègne
URA CNRS 817, BP 649, 60206, Compiègne
FRANCE
rortega@hds.univ-compiegne.fr

Keywords: Nonlinear control, control of electrical machines, discrete-time nonlinear systems.

Abstract

In this short paper we present a *discrete-time* field oriented controller (FOC) for current-fed induction motors which insures global asymptotic rotor position (or velocity) tracking. To the best of our knowledge, this is the first time such a result is rigourously established for a controller implemented in discrete-time. As expected in a discrete-time controller, an unavoidable condition relating the sampling rate with the rate of change of the references is imposed. This condition disappears as the sampling rate goes to zero. Interestingly enough, the condition also reflects the well known fact that to avoid magnetic saturation the flux norm must be increased when the machine is requested to deliver large torques. One important feature of our scheme is that it is obtained via a slight -but essential- modification to a first difference approximation of the classical indirect FOC, hence its additional computational burden is negligible.

1 Introduction

It is widely accepted that field oriented control (FOC), in one of its many forms, is the *de facto* industry standard for high dynamic performance induction motor drives [11], [10]. In particular, indirect FOC is a simple and highly reliable scheme which enjoys widespread popularity. The first rigorous stability proof for this controller was recently established in [14]. Namely, it was shown that in speed regulation tasks with constant load torque and current fed machines indirect FOC is *globally asymptotically stable* provided all motor parameters are exactly known. The fact that the rotor resistance is subject to significant changes during the machine operation, motivated our work in [3] where some remarkable robustness properties of this controller were found.

The computation of an FOC law, which is a nonlinear dynamic feedback, is invariably carried out in discrete-time (with microcomputers or digital signal processors). Since the machine model (and the controller) are nonlinear, the stability analysis of a discretized implementation of the controller is far from obvious. Furthermore, it is not even clear how such a discretization should be done. Some efforts in this direction may be found in [6], [7]. In this paper we present a *discrete-time* FOC for current-fed induction motors which insures global asymptotic rotor position (or velocity) tracking provided a condition relating the sampling rate with the rate of change of the references is satisfied. This condition, which is unavoidable due to the piece-wise nature of the control, disappears as the sampling rate goes to zero. Interestingly enough, the condition also reflects the well known fact that to avoid magnetic saturation the flux norm must be increased when the machine is requested to deliver large torques, i.e. the flux weakening policy. One fundamental feature of our scheme is that it is obtained via a slight -but essential for the analysis- modification to a first difference approximation of the classical indirect FOC, hence its additional computational burden is negligible. Establishing the stability of the discretized FOC, without this modification, remains a challenging open problem.

The remaining of the paper is organized as follows. In section 2 we briefly recall the classical continuous-time indirect FOC, this allows us to motivate the derivation of our discrete-time controller and to establish the connection with a discretized FOC. Our main result is then presented in section 3. We wrap up the

*This work was supported in part by the Comission of European Communities under contract ERB CHRX CT 93-0380.

paper with some concluding remarks in section 4. To facilitate the understanding of our result we have included two appendices explaining the motivation of indirect FOC and its stability analysis respectively.

Notation

For a given (constant) sampling rate $T \in \mathcal{R}_+$ we will use the notation $z_k = z(kT)$, $k \in \mathcal{Z}_+$, i.e., z_k denotes the samples of the continuous signal $z(t)$ at the time instants $t = kT$. Also, q will denote the advance operator, that is $q^i z_k = z_{k+i}$, $i \in \mathcal{Z}_+$, and $|\cdot|$ the Euclidean norm. To simplify the notation we define the (unitary rotation) matrices

$$\mathcal{J} := \begin{bmatrix} 0 & -1 \\ 1 & 0 \end{bmatrix}, \quad e^{-b\mathcal{J}} := \begin{bmatrix} \cos(b) & \sin(b) \\ -\sin(b) & \cos(b) \end{bmatrix}$$

Notice that $\mathcal{J}^2 = -I$, with I the 2×2 unity matrix. Also, we have that for all $z \in \mathcal{R}^2$, $z^T \mathcal{J} z = 0$. Further, $z^T \mathcal{J} e^{-b\mathcal{J}} z = |z|^2 \sin(b)$, for all $b \in \mathcal{R}$.

2 Continuous-time FOC

We consider in this paper current-fed induction machines described (in the fixed stator frame) by the well known equations[1]

$$\begin{aligned} \dot{\psi}_r &= -R_r \psi_r + \omega \mathcal{J} \psi_r + R_r i_s \\ \dot{\omega} &= \tau - \tau_L \\ \tau &= i_s^T \mathcal{J} \psi_r \end{aligned}$$

where $\psi_r \in \mathcal{R}^2$ is the rotor flux, $i_s \in \mathcal{R}^2$ is the stator current, which in current fed machines is the control input [10], ω is the rotor angular speed, τ is the generated torque, τ_L is the load torque, which we assume constant[2] though *unknown*, and $R_r > 0$ is the *known* rotor resistance.

To simplify the expressions below, and without loss of generality for the purposes of this study, all motor parameters have been set to unity except the rotor resistance and the load torque. Also, again for the sake of simplicity, we will consider only velocity control, the extension to the case of position control being straightforward.

We find convenient to make the (globally defined) change of coordinates

$$x = e^{-\mathcal{J}\theta} \psi_r, \quad u = e^{-\mathcal{J}\theta} i_s$$

[1] For further details and motivation of induction motors and FOC the reader is refered to [2], [10], in the electrical machines literature, and to [1], [14], [17] in the control journals.

[2] See discussion in section 4 for a relaxation of this assumption.

where $\dot{\theta} = \omega$. This yields the model

$$\begin{aligned} \dot{x} &= -R_r x + R_r u && (2.1) \\ \dot{\omega} &= \tau - \tau_L && (2.2) \\ \tau &= u^T \mathcal{J} x && (2.3) \end{aligned}$$

In *indirect* FOC the (rotated) stator currents are chosen as

$$u = e^{J\rho_d} \begin{bmatrix} \beta \\ \frac{\tau_d}{\beta} \end{bmatrix} \quad (2.4)$$

where $\beta > 0$ is the constant desired value of the rotor flux norm, ρ_d – which may be interpreted as the angle of the desired rotor flux (or an estimate of the actual rotor flux angle)– is given by

$$\dot{\rho}_d = \frac{R_r}{\beta^2} \tau_d \quad (2.5)$$

and τ_d is the desired torque which we define via a PI velocity loop as

$$\tau_d = \left(K_P + \frac{K_I}{p} \right) (\omega_d - \omega) \quad (2.6)$$

where $\omega_d(t)$ is the desired velocity, $p := \frac{d}{dt}$ is the derivative operator, and $K_P, K_I \geq 0$ are the PI tuning gains.

For the sake of completeness we have included in appendix A an explanation of the rationale of this scheme. Further, we give in appendix B the proof (established in [14]) of global stability of this controller for the case when ω_d is *constant*.

A simple discrete-time implementation of this controller is obtained from its *first order approximation* as

$$\begin{aligned} u_k &= e^{\mathcal{J}\rho_k^d} \begin{bmatrix} \beta \\ \frac{\tau_k^d}{\beta} \end{bmatrix} \\ \rho_{k+1}^d &= \rho_k^d + \frac{TR_r}{\beta^2} \tau_k^d \\ \tau_k^d &= K_P + \frac{K_I}{q-1} (\omega_k^d - \omega_k) && (2.7) \end{aligned}$$

The stability of the system (2.1)–(2.3) in closed loop with (2.7) remains to be established. On the other hand, we show below that adding a slight -but fundamental- modification to the scheme above yields a global tracking controller.

3 Main Result

Proposition 1

Consider the motor model (2.1)–(2.3) with control inputs the (rotated) stator currents u and measurable output the rotor speed ω. Let $\beta > 0$ denote the *constant* reference value for the rotor flux amplitude $|x|$, and $\omega_d(t)$ the desired rotor speed. Assume load torque τ_L *constant*, though unknown. Define the *discrete-time* controller as

$$\begin{aligned} u_k &= e^{\mathcal{J}\rho_k^d}\begin{bmatrix} \frac{\beta}{1-a}[\cos(b_k)-a] \\ \frac{\tau_k^d}{\beta} \end{bmatrix} && (3.1)\\ \rho_{k+1}^d &= \rho_k^d + b_k, \ \rho_0 = 0 && (3.2)\\ b_k &= \arcsin\left(\frac{1-a}{\beta^2}\tau_k^d\right) && (3.3)\\ \tau_k^d &= K_P + \frac{K_I}{q-1}(\omega_k^d - \omega_k) && (3.4)\end{aligned}$$

with sampling rate $T > 0$, $a := e^{-R_r T}$, $\omega_k^d = \omega_d(kT)$, $\omega_k = \omega(kT)$, and the PI gains chosen such that

$$2\left(\frac{1-a}{R_r}\right)^{-1} > K_P > K_I > 0 \qquad (3.5)$$

Under these conditions, for all motor initial conditions we have that all internal signals are bounded and

$$\lim_{t\to\infty}\omega = \omega_d, \ \lim_{t\to\infty}|x| = \beta$$

provided[3]

$$\frac{1-a}{\beta^2}|\tau_k^d| < 1, \qquad (3.6)$$

holds for all $k \in \mathcal{Z}_+$.

Proof

An (exact) discrete-time representation of (2.1)-(2.3) is readily obtained as

$$\begin{aligned} x_{k+1} &= ax_k + (1-a)u_k && (3.7)\\ \omega_{k+1} &= \omega_k - \left(\int_k^{k+1} x(s)ds\right)^T \mathcal{J}u_k - T\tau_L && (3.8)\\ \tau_k &= u_k^T\mathcal{J}x_k && (3.9)\end{aligned}$$

Now, using the fact that the difference equation

$$x_{k+1}^d = e^{b_k\mathcal{J}}x_k^d, \ x_0 = \begin{bmatrix}\beta\\0\end{bmatrix} \qquad (3.10)$$

has a unique solution

$$x_k^d = e^{\rho_k^d\mathcal{J}}\begin{bmatrix}\beta\\0\end{bmatrix} = \beta\begin{bmatrix}\cos(\rho_k^d)\\ \sin(\rho_k^d)\end{bmatrix}$$

and after some simple calculations, we can show that (3.1), (3.3) can be equivalently written as (3.10) and

$$u_k = \frac{1}{1-a}(x_{k+1}^d - ax_k^d) \qquad (3.11)$$

[3] Notice that the upper bounds on $|\tau_k^d|$ and K_P become arbitrarily large as $T \to 0$. See also section 4.

Notice that $|x_k^d| = \beta$ for all $k \in \mathcal{Z}_+$.

Replacing (3.11) in (3.7) we get $\tilde{x}_{k+1} = a\tilde{x}_k$ where $\tilde{x}_k := x_k - x_k^d$. Since $a < 1$ we have that $\tilde{x}_k \to 0$.

On the other hand, replacing (3.11) in (3.9) we get

$$\begin{aligned} \tau_k &= \frac{1}{1-a}(x_{k+1}^d)^T\mathcal{J}x_k^d + \epsilon_k\\ &= \frac{\beta^2}{1-a}\sin(b_k) + \epsilon_k\\ &= \tau_k^d + \epsilon_k && (3.12)\end{aligned}$$

where we have used (3.3) and assumption (3.6) to get the last identity, and ϵ_k is the exponentially decaying sequence

$$\epsilon_k := \frac{1}{1-a}(x_{k+1}^d - ax_k^d)^T\mathcal{J}\tilde{x}_k$$

We have thus shown that $\tau_k \to \tau_k^d$, we must now prove that this implies $\omega_k \to \omega_k^d$. To this end, we see from (2.1) that

$$\int_k^{k+1} x(s)ds = \frac{1-a}{R_r}x_k + (T - \frac{1-a}{R_r})u_k$$

consequently, (3.8) reduces to

$$\begin{aligned} \omega_{k+1} &= \omega_k + \frac{1-a}{R_r}u_k^T\mathcal{J}x_k - T\tau_L\\ &= \omega_k + \frac{1-a}{R_r}(\tau_k^d + \epsilon_k) - T\tau_L\end{aligned}$$

where we have used (3.9) and (3.12). This is a linear time invariant system with input τ_k^d, output ω_k and two unknown disturbances: τ_L, which is constant, and ϵ_k which exponentially goes to zero. The proof is completed closing the loop around this system with (3.4), recalling that a PI controller rejects the constant disturbance, and verifyng that the conditions on K_P, K_I insures stability of the characteristic polynomial.

□□□

4 Concluding Remarks

• The bound on the PI gains (3.5) reflects the natural restriction imposed by the sampling process on the *achievable closed-loop bandwidth*. A similar interpretation can be given to (3.6) if we notice that a bound on τ_k^d translates, via (3.4), on a bound on the speed of change of the desired speed. It is interesting to remark that the condition also captures the well known fact that to avoid magnetic saturation the flux norm must be increased when the machine is requested to deliver large torques. Thus condition (3.6) permits a systematic way for the practical implementation of a flux weakening policy.

• It is indeed remarkable that we can obtain a discrete-time implementation of the continuous-time indirect FOC with *guaranteed stability* properties by introducing only some "simple" modifications to its first order approximation. This, of course, stems from the fact that the first order approximation of the motor model provides an *exact* description of the motor behaviour. We believe this observation, together with our recent research on control of power electronic devices [16], will pave the way for the development of controllers at the "converter level", as advocated in [15] and [4].

• The *computational load* of our controller is esentially the same as the one required for classical discrete-time implementations of indirect FOC.

• One interesting feature of our analysis is that it formalizes the role of the two *tuning knobs*, K_P and K_I, which define the closed loop poles of the mechanical subsystem. This subsystem is perturbed by a disturbance that decays at an exponential rate determined by the rotor resistance and the sampling time. We should remark that this rate *decreases* as we sample faster. Further investigations are required to understand, and try to overcome, this counterintuitive behaviour.

• We have considered here only the problem of speed tracking. It is clear from the proof of the proposition that the extension to *position* control can be solved simply choosing a suitable compensator $C(q)$ in (3.4). Also, we can easily treat the case of *unknown non constant* load torque, e.g., a linearly parametrized function $\tau_L = \phi(\theta, \omega)^T P,\ P \in \mathcal{R}^q$, with a straightforward modification to (3.8) and classical adaptive control techniques. This extensions are left as an exercise to the reader.

• It is well known that to avoid converter saturation and/or maximize power efficiency it is convenient to have a *time-varying* desired rotor flux norm. In [5] the passivity-based controller of [13] was modified to handle this case. Current investigations are under way to find its discrete-time implementation.

• In spite of the remarkable *stability* robustness properties of indirect FOC [3] –naturally inherited by the present scheme– there is an obvious interest of considering adaptive controllers that estimate R_r to improve *performance*. Experimental evidence that substantiates this claim may be found in [9]. Providing a satisfactory solution to this fundamental problem is the main driving force of the field now. An important step towards its solution was given in [12] where a globally convergent adaptive observer is reported. In this author's opinion, in spite of its theoretical significance, there are two potential drawbacks to this result. First, since the problem is formulated in continuous-time, the resulting schemes (combining state and parameter estimation and control) are extremely complex to be of any practical use. In this respect we expect our discrete-time setting might alleviate this problem. Second, there is some sort of "inconsistency" since the approach taken in [12] seems to be applicable only to the full machine model. It would be interesting to see which modifications are needed to solve the problem for the simpler current-fed model considered here.

References

[1] Bodson, M., J. Chiasson and R. Novotnak, "High-performance induction motor control via input-output linearization", *IEEE Control Syst. magazine*, August 1994.

[2] Bose, B. K., **Power Electronics and AC Drives**, Prentice Hall, 1986.

[3] de Wit, P., R. Ortega and I. Mareels, "Indirect Field Oriented Control of Induction Motors is Robustly Globally Stable", *Submitted to Automatica*, 1995.

[4] Djemai, M., and J.P.Barbot, "Singularly perturbed methods for the control design of a synchronous motor with its PWM inverter", LSS Int. Report, 1995.

[5] Espinosa, G., P. Nicklasson and R. Ortega, "Control of induction motors in the field weakening region", *UNAM Int. Report*, Jan. 1995.

[6] Georgiou, G., "Sur des methodes de commande nonlineaire adaptative. Aspects echantillonnes et applications", *Ph. D. Thesis, U. Paris XI*, June 1992.

[7] Georgiou, G., Chelouah, A., S. Monaco and D. Normand-Cyrot, "Nonlinear Multirate Adaptive Control of a Synchronous Motor", *Proc. 31st IEEE CDC*, Dec. 1992.

[8] Khalil, H.,, **Nonlinear systems**, MacMillan Publ. House, N.J., 1992.

[9] Kim, K., R. Ortega, A. Charara and J.P. Vilain, "Experimental comparison of nonlinear controllers for induction motors", *UTC Int. Report*, Feb. 1995.

[10] Leonhard, W., **Control of electrical drives**, Springer-Verlag, Berlin, (1985).

[11] Lorenz, R., T. A. Lipo and D. W. Novotny, "Motion Control with Induction Motors", *Proc. of the IEEE*, (Special Issue on Power Electronics and Motion Control), Vol. 82, No. 8, pp. 1215-1240, 1994.

[12] Marino, R., S. Peresada and P. Tomei, "Adaptive observer-based control of induction motors with unknown rotor resistance", *Proc. IEEE CDC*, Orlando, FL, Dec. 1994.

[13] Ortega, R., Nicklasson, P.J. and G. Espinosa, "On Speed Control of Induction Motors", Proc. ACC, Seattle, WA.,June 21-23. 1995. (To be published in *Automatica*).

[14] Ortega, R., D. Taoutaou, R. Rabinovici and J. P. Vilain, "On Field Oriented and Passivity-based Control of Induction Motors: Downward Compatibility", *Proc. IFAC NOLCOS Conf. Tahoe City, CA, June 26-28, 1995.*

[15] Sabanovic, A., N. Sabanovic and K. Ohnishi, "Sliding modes in power converters and motion control systems", *Int. J. of Control*, Vol. 57, No. 5, 1237-1259, 1993.

[16] Sira-Ramírez, H., R. Ortega, M. Pérez-Moreno and M. García-Esteban, "Passivity-Based Controllers for the Stabilization of DC-to-DC Power Converters," *Automatica* (submitted for publication).

[17] Taylor, D., "Nonlinear control of electric machines: An overview", *IEEE Control Syst. Magazine*, Vol. 14, No. 6, pp. 41-51, 1994.

A From Direct to Indirect FOC

As discussed in [14] one way of explaining the rationale underlying indirect FOC is to compare it with direct[4] FOC where, assuming $|x| \neq 0$, we first define

$$i_{dq} := \begin{bmatrix} i_d \\ i_q \end{bmatrix} = e^{-\mathcal{J}\rho} u$$

with $\rho := \arctan \frac{x_2}{x_1}$ –the rotor flux angle. Some simple calculations with (2.1) and (2.3) show that

$$\begin{aligned} \tau &= i_q |x| \\ \tfrac{d}{dt}|x| &= -R_r|x| + R_r i_d \end{aligned}$$

which, together with the *time scale* separation between the mechanical and the magnetical dynamics, motivates a "decoupled" scheme where i_d regulates $|x|$ to β, while i_q is responsible for the torque control. This rationale leads to a direct FOC scheme with nested torque and velocity PI loops of the form

$$u = e^{\mathcal{J}\rho} i_{dq} = e^{\mathcal{J}\rho} \begin{bmatrix} \beta \\ -C(p)[\tau - \underbrace{C(p)(\omega_d - \omega)}_{\tau_d}] \end{bmatrix}$$

with $C(p) := k_P + k_I \frac{1}{p}$, $p := \frac{d}{dt}$. Notice that we have denoted the output of the second (outer) PI as τ_d to underscore its role as a "desired torque".

To obviate the measurement of rotor flux in *indirect* FOC first the PI loop around τ is removed. Further, noting that the rotor flux angle satisfies

$$\dot{\rho} = \frac{R_r}{|x|^2}\tau$$

we define a "desired" value for ρ as

$$\begin{aligned} \dot{\rho}_d &= \frac{R_r}{\beta^2}\tau_d \\ \tau_d &= C(p)(\omega_d - \omega) \end{aligned}$$

where we have replaced $|x|$ and τ by their desired value β and τ_d respectively. This estimate of the rotor flux is then used in the rotation matrix used for the calculation of u above to finally get

$$u = e^{\mathcal{J}\rho_d} \begin{bmatrix} \beta \\ \frac{\tau_d}{\beta} \end{bmatrix}$$

B Stability of Continuous-time Indirect FOC

Proposition B.1
Consider the system (2.1)-(2.3) in closed loop with (2.4)-(2.6). If ω_d and τ_L are constant, then for all initial conditions we have $\lim_{t\to\infty}\omega = \omega_d$ and $\lim_{t\to\infty}|x| = \beta$ with all signals uniformly bounded.

Proof [14]
First notice that (2.4), (2.5) can be alternatively described with the (nonminimal) representation

$$\dot{x}_d = \frac{R_r}{\beta^2}\tau_d \mathcal{J} x_d, \; x_d(0) = \begin{bmatrix} \beta \\ 0 \end{bmatrix} \quad \text{(B.1)}$$

$$u = (I + \frac{\tau_d}{\beta^2}\mathcal{J})x_d \quad \text{(B.2)}$$

Then, from (2.1), (B.1), (B.2) we see that $\dot{\tilde{x}} = -R_r\tilde{x}$, where we have defined $\tilde{x} := x - x_d$. From this equation it is clear that $\tilde{x}$ converges to zero (exponentially). Finally, defining $\tilde{\omega} := \omega - \omega_d$ as the speed tracking error,

[4] There is no common agreement in the literature as to how to distinguish direct and indirect FOC, here we adopt the convention that direct FOC assumes *measurement of rotor flux* while indirect FOC does not.

and recalling that ω_d is constant, we have from (2.2), (2.3) that

$$\begin{aligned}\dot{\tilde{\omega}} = \dot{\omega} &= u^T \mathcal{J}(\tilde{x} + x_d) - \tau_L \\ &= x_d^T \mathcal{J}\tilde{x} - \tau_L + (1 + \frac{1}{\beta^2} x_d^T \tilde{x})\tau_d \qquad \text{(B.3)} \\ &= -K_P(1 + \frac{1}{\beta^2} x_d^T \tilde{x})\tilde{\omega} - \\ &\quad -K_I(1 + \frac{1}{\beta^2} x_d^T \tilde{x})\frac{1}{p}\tilde{\omega} + x_d^T \mathcal{J}\tilde{x} - \tau \qquad \text{(B.4)}\end{aligned}$$

where we have used (3.1) and the skew-symmetry of $\mathcal{J}$ to obtain the second identity, and we have replaced (2.6) to get the third one. Noting that $|x_d(t)| = \beta$, we have that $\frac{1}{\beta^2} x_d^T \tilde{x}$ and $x_d^T \mathcal{J}\tilde{x}$ are exponentially decaying terms. Asymptotic stability of $\tilde{\omega} \to 0$ follows invoking standard arguments of linear time varying systems with exponentially convergent coefficients [8].

□□□

Remark
In [3] we proved, using a Lyapunov function, the stronger result that the equilibrium of the closed loop is globally asymptotically stable.

AN ADAPTIVE BILINEAR CONTROL METHOD IN THE DESIGN OF A POWER SYSTEM STABILISER

F. HE* and M.J. GIBBARD**

**The Flinders University of South Australia, School of Engineering, Bedford Park, South Australia 5042, AUSTRALIA*

***The University of Adelaide, Department of Electrical and Electronic Engineering, Adelaide, South Australia 5005, AUSTRALIA*

Abstract. This paper describes an *adaptive bilinear* control method for the design of a single-machine infinite-bus (SMIB) power system stabiliser (PSS). It is first shown that a bilinear model, which describes the nonlinear relationship between the electric power output of the generator and the excitation control input, *inherently* tracks the changes in the system operating point. This bilinear model uses a minimum set of measurable feedback signals, and is linear in its parameters. By incorporating the nonlinearities of the power system into the design of the control law, an adaptive bilinear PSS based on the bilinear model provides better damping performance than that of a corresponding adaptive linear device discussed in the literature.

Key Words. Adaptive control; nonlinear systems; feedback stabilisation; control applications; power system stabilisers

1. INTRODUCTION

It is well-known that the conventional automatic voltage regulator control of power systems through excitation modulation has the potential to introduce negative damping into the system dynamics. Power system stabilisers (PSSs) are designed to improve the system damping performance via utilisation of a supplementary damping signal through generator excitation. Conventional PSSs (C-PSSs), which are based on classical control theory, are *linear* and *fixed-parameter* devices. However, a power system is both *nonlinear* and *time-varying* in nature. The C-PSSs have difficulties in coping with the situations such as changes in the system configuration, random occurrences of faults, etc., because in these cases the *nonlinear* and/or *time-varying* natures of the power system are invoked. Adaptive *linear* control schemes have been proposed for the design of PSSs to overcome the shortcomings of the C-PSSs (Pierre, 1987; Chalam, 1987). However, because an adaptive linear PSS is based on an *assumed linear* regression model of the system, the inherent nonlinearities of the system are **not** included in the design of the control law. Consequently, its performance may not be as good as expected.

In order to handle the **nonlinear** and **time-varying** features of the power system, it is required that a control law be both **nonlinear** and **adaptive** in nature. The resulting controller (stabiliser) should include the inherent nonlinearities of the system in its control law while, at the same time, possess a mechanism to adapt to the system changes on-line. By doing so, it is anticipated that the deficiencies associated with the more conventional design methods would be overcome. In this paper, the design technique for such an **adaptive, nonlinear** PSS is described.

The power system is represented by a nonlinear SMIB model described in (He, 1991; He and Gibbard, 1992*a*); the governor and steam turbine are included. The electric power output of the generator is used as the stabilising signal. To include the inherent nonlinearities of the power system in the design of the control law, a *bilinear* model which describes the nonlinear relationship between the excitation control input and the electric power output of the generator is derived from the mathematical description of the SMIB power system in Section 3. An adaptive bilinear control law is developed for the bilinear model and is implemented in terms of a PSS; the desired adaptive bilinear PSS is thus formed in Section 4. Simulation studies are conducted in Section 5 to compare the performance of the adaptive bilinear PSS with that of a corresponding adaptive linear device briefly introduced in Section 2 (He, 1991; He and Gibbard, 1994).

2. AN ADAPTIVE LINEAR PSS

From system modelling point of view, the significance of utilising *adaptive linear* control strategies for the design of PSSs is that for small dynamics the *nonlinear time-varying* power system can be modelled **approximately** by a *time-varying linearised* model in which the on-line estimated parameters accommodate the nonlinearities of the system and/or the changes in the system operating condition. The structure of such a time-varying linearised model for electric power output of the generator can be derived from the nonlinear SMIB power system model described in (He, 1991; He and Gibbard, 1992*a*) through *linearisation*, discretisation, and simplifications about an arbitrary steady-state operating point. An input-output model of the nonlinear power system in a *linear regression form* is then derived as (He, 1991):

$$\bar{A}(q^{-1})\Delta y(k) = q^{-1}\bar{B}(q^{-1})\Delta u(k) + \epsilon(k) \quad (1)$$

where

$$\bar{A}(q^{-1}) = 1 + \bar{a}_1 q^{-1} + \bar{a}_2 q^{-2} + \bar{a}_3 q^{-3}, \quad (2)$$
$$\bar{B}(q^{-1}) = \bar{b}_0 + \bar{b}_1 q^{-1} + \bar{b}_2 q^{-2}. \quad (3)$$

The output signal $\Delta y(k)$ is the deviation in electric power of the generator from its steady-state level, and the control signal $\Delta u(k)$ is generated by the adaptive linear PSS to be designed. The term $\{\epsilon(k)\}$ is assumed to be a white noise which is unmeasurable. The coefficients of polynomials (2)-(3) are *time-varying* in nature, in accordance with the changes in the system operating point/condition. In this paper, model (1) is referred to as the *Linearised Power Model* (**LPM**) of the power system, from which the adaptive linear PSS is derived.

An optimal *Weighted Minimum Variance* (**WMV**) control scheme is chosen for the design of PSSs in this paper, based on the considerations: (i) by including a weighting coefficient λ, the control law can be used to stabilise a nonminimum phase system (which may result from the LPM for certain operating points/conditions); (ii) with a suitable choice of the value of λ, excessive control action can be penalised; (iii) due to its simplicity, the WMV control can be extended to a *nonlinear* case, which will facilitate the development of a corresponding nonlinear control law and the comparison of system performance of the *adaptive linear* and the *adaptive nonlinear* control strategies. The cost function is given by

$$J(k+1) = E\left\{[\Delta y(k+1)]^2 + \lambda[\Delta u(k)]^2\right\}. \quad (4)$$

The *linear optimal* WMV control law, which minimises (4), is derived from the LPM, i.e.,

$$\left[\bar{B}(q^{-1}) + \frac{\lambda}{\bar{b}_0}\right]\Delta u(k) + \left[1 - \bar{A}(q^{-1})\right] q\Delta y(k) = 0. \quad (5)$$

Note that in practice the control signal $\Delta u(k)$ in (5) is bounded due to limiting action.

Since the parameters of the LPM are *time-varying* and *unknown*, the recursive least squares algorithm with a time-varying forgetting factor, $\mu(k)$, is chosen to identify the model parameters on-line. The forgetting factor is used in the algorithm to improve the parameter tracking ability and to prevent the estimator from 'switching-off' in the steady state. Combining the linear optimal control law (5) with the selected parameter estimation algorithm, an algorithm for an *Adaptive Linear Power System Stabiliser* (**AL-PSS**) is derived.

Algorithm 1 [Adaptive Linear WMV PSS]

$$\hat{\Theta}(k) = \hat{\Theta}(k-1) + \mathbf{P}(k-1)\phi(k-1)e(k); \quad (6)$$

$$\mathbf{P}(k-1) = \frac{1}{\mu(k-1)}\left[\mathbf{P}(k-2) - \frac{\mathbf{P}(k-2)\phi(k-1)\phi(k-1)^T\mathbf{P}(k-2)}{\mu(k-1) + \phi(k-1)^T\mathbf{P}(k-2)\phi(k-1)}\right]; \quad (7)$$

$$\Delta\hat{y}(k) = \phi(k-1)^T\hat{\Theta}(k-1); \quad (8)$$

$$e(k) = \Delta y(k) - \Delta\hat{y}(k); \quad (9)$$

$$\mu_0(k) = 1 - \frac{1}{\Sigma_0} \times \frac{e(k)^2}{1 + \phi(k-1)^T\mathbf{P}(k-2)\phi(k-1)}, \quad (10)$$

$$\bar{\mu}(k) = \begin{cases} \mu_0(k) & \text{if } \mu_0(k) \geq \mu_{min} > 0 \\ \mu_{min} & \text{otherwise} \end{cases}, \quad (11)$$

$$\mu(k) = \begin{cases} \bar{\mu}(k) & \text{if } \dfrac{\text{trace}[\mathbf{P}(k-1)]}{\bar{\mu}(k)} < C \\ 1 & \text{otherwise} \end{cases}; \quad (12)$$

$$u^0(k) = \frac{\hat{\bar{b}}_0(k)}{\hat{\bar{b}}_0(k)^2 + \lambda} \times \left\{\left[\hat{\bar{a}}_1(k) + \hat{\bar{a}}_2(k)q^{-1} + \hat{\bar{a}}_3(k)q^{-2}\right]\Delta y(k) - \left[\hat{\bar{b}}_1(k)q^{-1} + \hat{\bar{b}}_2(k)q^{-2}\right]\Delta u(k)\right\},$$

$$\Delta u(k) = \begin{cases} u_{max} & \text{if } u^0(k) \geq u_{max} \\ u^0(k) & \text{if } u_{min} < u^0(k) < u_{max} \\ u_{min} & \text{if } u^0(k) \leq u_{min} \end{cases};$$

where $k \geq 1$, $\mathbf{P}(-1) = K_0\mathbf{I}$ $(0 < K_0 < C)$, and $\mu(0) = 1$. The parameters Σ_0, μ_{min}, C, λ, u_{max} and u_{min} are preselected positive constants. The vectors $\hat{\Theta}(k)$ and $\phi(k)$ are defined by

$$\hat{\Theta}(k)^T \triangleq \left[\ \hat{\bar{a}}_1(k) \quad \hat{\bar{a}}_2(k) \quad \hat{\bar{a}}_3(k) \quad \hat{\bar{b}}_0(k) \quad \hat{\bar{b}}_1(k) \quad \hat{\bar{b}}_2(k)\ \right],$$

$$\phi(k)^T \triangleq [\ -\Delta y(k) \quad -\Delta y(k-1) \quad -\Delta y(k-2) \quad \Delta u(k) \quad \Delta u(k-1) \quad \Delta u(k-2)\ ];$$

$\hat{\Theta}(0)$ is given. ▽ ▽ ▽

The control structure of the SMIB power system equipped with the AL-PSS is illustrated in Fig. 1. The stabilising signal $y(k)$ is the *measured* machine electric power $P_e(k)$ and the output trajectory $y^*(k)$ is set to be the reference power $P_{ref}(k)$.

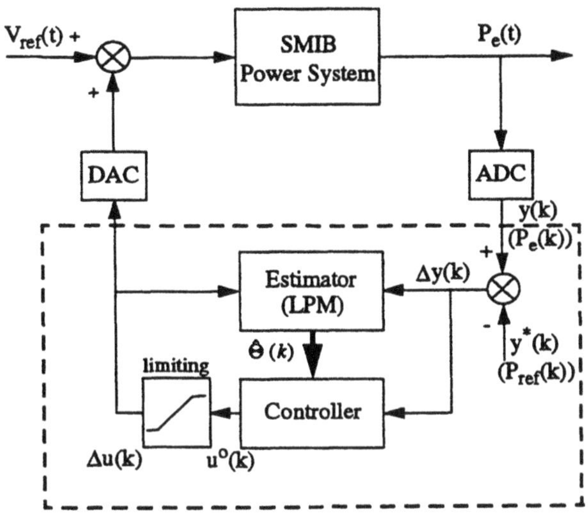

Fig. 1. Structure of the AL-PSS.

3. DESCRIPTION OF A SIMPLE BILINEAR POWER MODEL

The key feature in developing a nonlinear control scheme is that a nonlinear model for electric power output of the generator is derived **directly** from the mathematical description of the power system itself. Thus, *inherently*, the nonlinear power model will track any changes in electric power output due to the changes in the system operating point.

Assume that the relevant system variables are accessible. A *nonlinear input-output* regression equation which describes the nonlinear relationship between the control input, $u(k)$, and the actual electric power output, $y(k)$ $(P_e(k))$, is derived *directly* from the SMIB power system model (He, 1991; He and Gibbard, 1992*a*):

$$\begin{aligned}(1 + a_1q^{-1} + a_2q^{-2})\, y(k) &= (b_1q^{-1} \\ &+b_2q^{-2})\, z_1(k) + (c_1q^{-1} + c_2q^{-2})\, z_2(k) \\ &+d_1q^{-1}z_3(k) + e_1q^{-1}z_4(k)\, [V_{ref}(k) \\ &-V_t(k) + u(k)] + w(k), \end{aligned} \tag{13}$$

where $a_1, \ldots, e_1$ are functions of the power system parameters (X_q, X_e, etc.); $w(k)$ is a bounded noise signal due to discretisation, simplification, measurement errors, etc., and

$$\begin{aligned} z_1(k) &= \sin 2\delta(k), \\ z_2(k) &= \omega_s(k)E_q^{'}(k)\cos\delta(k), \\ z_3(k) &= \omega_s(k)E_{FD}(k)\cos\delta(k), \\ z_4(k) &= \sin\delta(k). \end{aligned}$$

It is seen that feedback of generator variables, namely rotor angle $\delta(k)$, speed deviation $\omega_s(k)$, field voltage $E_{FD}(k)$, transient voltage $E_q^{'}(k)$, and terminal voltage $V_t(k)$, is required by (13); some of these signals may be difficult to measure in practice. Hence, simplification in terms of eliminating or replacing some of the feedback signals in (13) is needed.

The first step in the simplification is to decompose the electric power output, $y(k)$, into the four components contributed by the signals $z_i(k)$, $(i = 1,2,3,4)$. From the results of a series of simulation studies, covering a wide range of operating conditions and disturbances, it is found that the contributions from the signals $z_1(k)$ and $z_3(k)$ are small and can be neglected. The second simplification is based on the assumption that, in $z_2(k)$, $E_q^{'}(k)$ remains more-or-less constant during fast transients and $\cos\delta(k)$ is bounded in region $[-1,1]$. The third simplification is to replace the feedback of rotor angle, $\delta(k)$, in the signal $z_4(k)$ by the electric power output, $y(k)$, through the relationship

$$z_4(k) = \tilde{K}(k)y(k) \tag{14}$$

where $\tilde{K}(k)$ is a function of the machine variable $E(k)$ and the infinite bus voltage V_∞, and is defined as

$$\tilde{K}(k) = \frac{X_e + X_q}{V_\infty E(k)}.$$

Equation (13) thus becomes

$$\begin{aligned} y(k) = &-(a_1q^{-1} + a_2q^{-2})\, y(k) + (\tilde{c}_1q^{-1} \\ &+\tilde{c}_2q^{-2})\, \omega_s(k) + e_{\tilde{K}}q^{-1}y(k)\, [V_{ref}(k) \\ &-V_t(k) + u(k)] + w(k), \end{aligned} \tag{15}$$

in which measured values of signals appear on the right-hand side, and

$$\begin{aligned} \tilde{c}_1 &= c_1E_q^{'}(k-1)\cos\delta(k-1), \\ \tilde{c}_2 &= c_2E_q^{'}(k-2)\cos\delta(k-2), \\ e_{\tilde{K}} &= e_1\tilde{K}(k-1). \end{aligned}$$

Equation (15) gives a simple *Bilinear Power Model* (**BPM**), from which an *adaptive bilinear* PSS is designed in the next section. It should be pointed out that the BPM retains the inherent nonlinearities of the system through the product term on the right-hand side of (15). The trigonometric nonlinearity of the output is taken into account in the relationship between $z_4(k)$ and $y(k)$, as shown in (14). The important feature associated with the BPM is that the model is *linear* in its parameters. Therefore, the parameter estimation and adaptive control algorithms developed for *linear* regression models can be extended to this *nonlinear* model.

4. DESIGN OF A PRACTICAL ADAPTIVE BILINEAR PSS

As a parallel study, the WMV control scheme is chosen for the development of the adaptive bilinear PSS. The *nonlinear optimal* WMV control, $u^*(k)$, which minimises the cost function

$$J(k+1) = E\left\{[y(k+1) - y^*(k+1)]^2 + \lambda u(k)^2\right\}$$

is derived from the BPM (He, 1991):

$$\begin{aligned}\left[e_{\bar{K}}^2 y(k)^2 + \lambda\right] u^*(k) &= e_{\bar{K}} y(k) \{y^*(k+1) \\ &+ \left(a_1 + a_2 q^{-1}\right) y(k) - \left(\tilde{c}_1 + \tilde{c}_2 q^{-1}\right) \omega_s(k) \\ &- e_{\bar{K}} y(k) \left[V_{ref}(k) - V_t(k)\right]\}, \qquad (16)\end{aligned}$$

where $y^*(k+1)$ is the desired output trajectory, $P_{ref}(k+1)$.

Since the parameters $\tilde{c}_1$, $\tilde{c}_2$, and $e_{\bar{K}}$ are generally time-varying and unknown, the parameter estimation algorithm introduced in Section 2 is used. In practice, the electric power, $y(k)$, may become negligible during certain system disturbances, such as three-phase faults and/or transmission line switching. The control law (16) is then modified by including a function, $f_y(k)$, to limit $y(k)$ to a minimum (non-zero) value y_o and to limit its rate of change during such disturbances. The resulting bounded *Adaptive Bilinear Power System Stabiliser*, namely the **AB-PSS**, is then described by (He and Gibbard, 1992*b*)

Algorithm 2 [Adaptive Bilinear WMV PSS]

$$\bar{f}_y(k) = \begin{cases} y(k) & \text{if } |y(k)| > y_o \\ 1 & \text{otherwise} \end{cases},$$

$$f_y(k) = \begin{cases} \bar{f}_y(k) & \text{if } [y(k-1) - y(k)] < C_y \\ 1 & \text{otherwise} \end{cases};$$

$$\begin{aligned}u^0(k) &= \frac{\hat{e}_{\bar{K}}(k) f_y(k)}{\hat{e}_{\bar{K}}(k)^2 f_y(k)^2 + \lambda} \times \{y^*(k+1) \\ &+ \hat{a}_1(k) y(k) + \hat{a}_2(k) y(k-1) \\ &- \hat{\tilde{c}}_1(k) \omega_s(k) - \hat{\tilde{c}}_2(k) \omega_s(k-1) \\ &- \hat{e}_{\bar{K}}(k) y(k) \left[V_{ref}(k) - V_t(k)\right]\}, \qquad (17)\end{aligned}$$

$$u(k) = \begin{cases} u_{max} & \text{if } u^0(k) \geq u_{max} \\ u^0(k) & \text{if } u_{min} < u^0(k) < u_{max} \\ u_{min} & \text{if } u^0(k) \leq u_{min} \end{cases};$$

where the parameters y_o, C_y, λ, u_{max}, and u_{min} are preselected positive constants. The model parameters $\hat{a}_1(k)$, ..., $\hat{e}_{\bar{K}}(k)$ in (17) are estimated by implementation of equations (6)-(12) in Algorithm 1, where the parameter vector $\hat{\Theta}(k)$ and the regression vector $\phi(k)$ are defined by

$$\begin{aligned}\hat{\Theta}(k)^T &\triangleq [\ \hat{a}_1(k) \quad \hat{a}_2(k) \quad \hat{\tilde{c}}_1(k) \\ &\quad \hat{\tilde{c}}_2(k) \quad \hat{e}_{\bar{K}}(k)\ ], \\ \phi(k)^T &\triangleq [\ -y(k) \quad -y(k-1) \\ &\quad \omega_s(k) \quad \omega_s(k-1) \\ &\quad y(k)\left(V_{ref}(k) - V_t(k) + u(k)\right)\ ],\end{aligned}$$

respectively, and $\hat{\Theta}(0)$ is given. ▽ ▽ ▽

The control structure of the SMIB power system equipped with the AB-PSS is illustrated in Fig. 2. The controller contains the inherent nonlinearities of the system while requires only a minimum set of additional feedback signals, all of which are measurable in practice. The number of the model parameters to be estimated is also reduced to five.

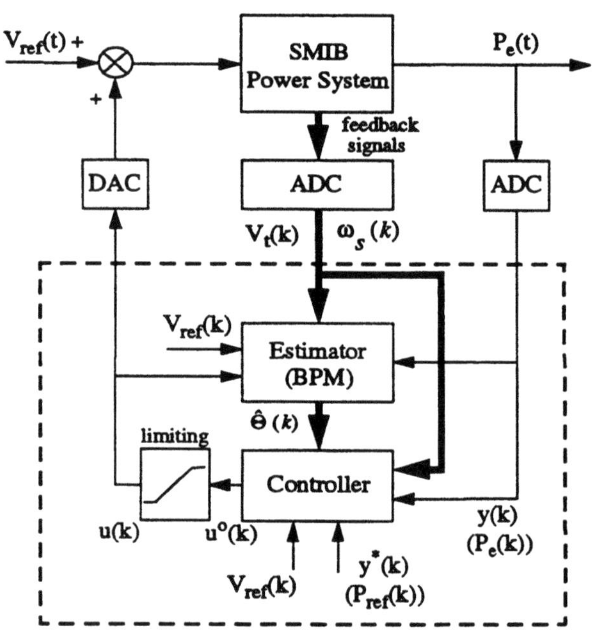

Fig. 2. Structure of the AB-PSS.

5. SIMULATION RESULTS

Simulation studies are conducted to examine: (a) on open-loop, the performance of the BPM in tracking the actual electric power output of the generator, with the estimates of parameters being continuously updated, (b) the closed-loop behaviour of the AB-PSS when the system is subjected to various disturbances, and (c) the robustness of the AB-PSS to unmodelled dynamics and modelling errors. In studies (b) and (c), the performance of the adaptive *bilinear* stabiliser (AB-PSS) is compared to that of the adaptive *linear* stabiliser (AL-PSS). The parameters and limits of the SMIB power system and the controllers are given in (He and Gibbard, 1992*a*). An identical value of the weighting coefficient λ is used in both the AB-PSS and the AL-PSS.

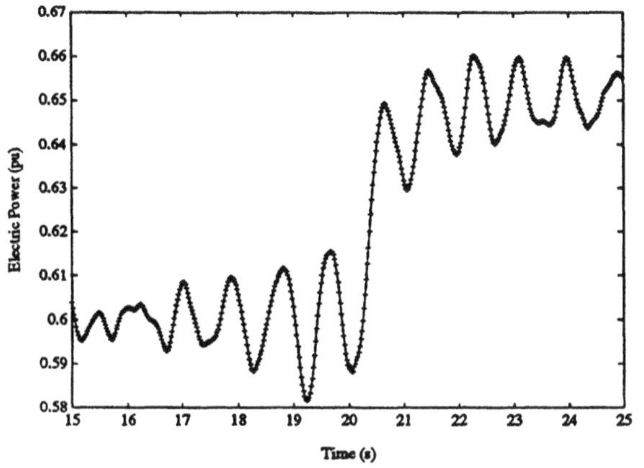

Fig. 3. Actual and estimated electric power responses ($P_t = 0.6$ pu, $Q_t = 0.3$ pu; 0.05 pu increase in reference power). $P_e(k)$ - solid line; $\hat{y}(k)$ - line marked by '+'.

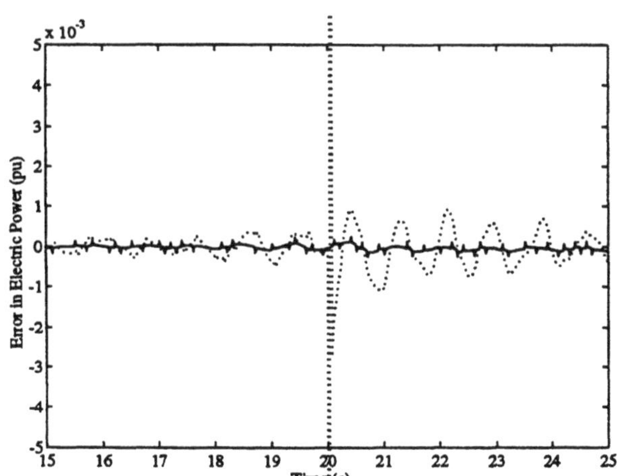

Fig. 4. Output prediction errors ($P_t = 0.6$ pu, $Q_t = 0.3$ pu; 0.05 pu increase in reference power). $e(k)$ of the LPM - dotted line; $e(k)$ of the BPM - solid line.

Four examples are chosen from a series of simulation studies covering a wide range of operating

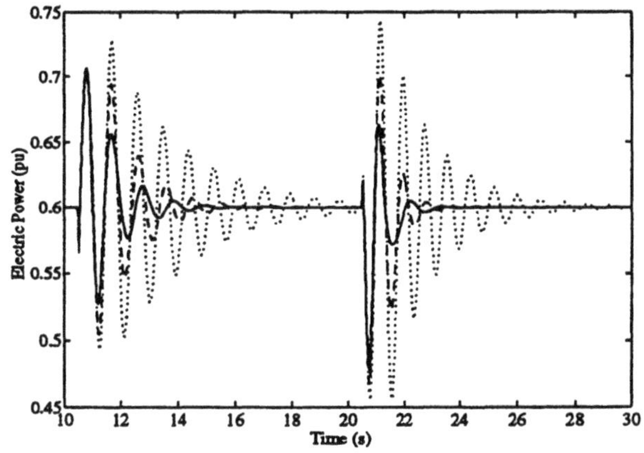

Fig. 5. Electric power responses ($P_t = 0.6$ pu, $Q_t = 0.3$ pu; periodic variations in the system operating point). No PSS - dotted line; with AL-PSS - dashed line; with AB-PSS - solid line.

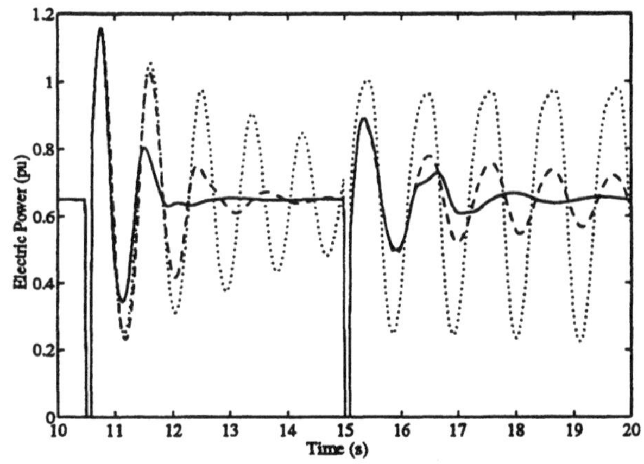

Fig. 6. Electric power responses ($P_t = 0.65$ pu, $Q_t = 0.3$ pu; a three-phase fault (0.1 s) at the receiving-end busbar with no line opening, followed by the same fault with line opening). No PSS - dotted line; with AL-PSS - dashed line; with AB-PSS - solid line.

conditions and types of disturbances. Fig. 3 shows the estimated electric power output of the BPM (on which the AB-PSS is based) when a 0.05 per unit (pu) step increase in reference power occurs at time $t = 20$ s. Since the BPM contains the nonlinearities associated with the electric power output of the generator, its estimated output *inherently* tracks the dynamics of the actual output even at $t = 20$ s when the system operating point changes. Further evidence of the output tracking ability of the BPM is shown in Fig. 4, in which the output prediction errors $e(k)$ associated with the BPM and the LPM (on which the AL-PSS is based) are produced for the same example. It is seen that while the LPM presents a *sudden* increase in its prediction error about $t = 20$ s when

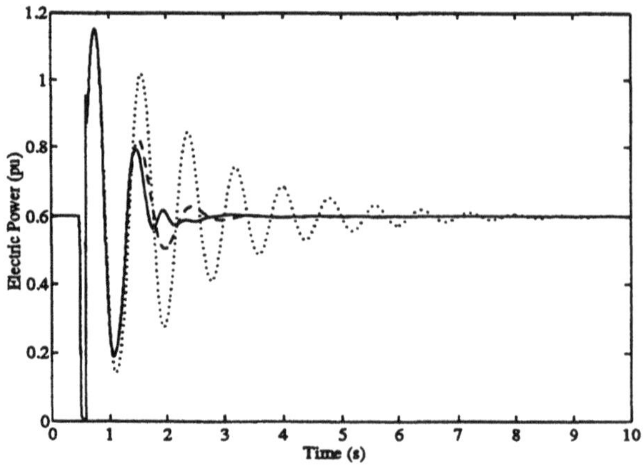

Fig. 7. Electric power responses ($P_t = 0.6$ pu, $Q_t = -0.1$ pu; a three-phase fault (0.1 s) at the machine terminal, no line opening). No PSS - dotted line; with AL-PSS - dashed line; with AB-PSS - solid line.

the system operating point has a step change calling upon the nonlinear nature of the system, the BPM shows a *consistent* small prediction error over the time horizon. Better output prediction of the BPM will result in better control effect of the adaptive *bilinear* stabiliser. This is demonstrated in Fig. 5, where the system operating point periodically changes between lagging and leading power factor operation. It is seen that the closed-loop damping performance of the AB-PSS is better than that of the AL-PSS; this is because the AB-PSS includes the inherent nonlinearities of the system in its control law. The closed-loop system responses to two successive three-phase faults with one of the parallel pair of the transmission lines opened after the second fault at time $t = 15$ s are shown in Fig. 6. In this case, both the time-varying and nonlinear features of the power system are invoked. It is shown that the AB-PSS together with the use of the function $f_y(k)$ in its control algorithm provides much stronger damping to the system oscillation than that of the AL-PSS. The robustness of the AB-PSS is tested in Fig. 7 in which a three-phase fault without line-opening is tested and a higher-order model of the SMIB power system (He, 1991) is employed — unmodelled dynamics and modelling errors of the system therefore being involved. The AB-PSS designed for the lower-order model of the SMIB power system can be used to control the higher-order model satisfactorily — the performance of the AB-PSS is shown consistently better than that of the AL-PSS.

6. CONCLUSION

From the simulation studies, it is concluded that the significance of the adaptive *bilinear* stabiliser (AB-PSS) lies in that (i) its nonlinear nature accounts for dynamic changes of the system due to its inherent nonlinearities and (ii) its adaptive feature accommodates changes in the system parameters and/or configuration. The overall performance of the AB-PSS is superior to that of the AL-PSS, which demonstrates the effect of including the inherent nonlinearities of the power system in the design of the control law. Due to its simplicity, the AB-PSS forms the basis for the development of a practical, adaptive nonlinear PSS.

7. REFERENCES

Chalam, V.V. (1987). *Adaptive control systems — techniques and applications.* Marcel Dekker Inc.. New York and Basel.

He, F. (1991). *Nonlinear adaptive control in the design of power system stabilisers.* Ph.D. Thesis, Dept. of Electrical & Electronic Engineering, The University of Adelaide. Australia.

He, F. and M.J. Gibbard (1992*a*). Adaptive nonlinear control in the design of a SMIB power system stabiliser. *Technical Report* **POW92/02**, Dept. of Electrical and Electronic Engineering, The University of Adelaide, Australia.

He, F. and M.J. Gibbard (1992*b*). On the design of a simple adaptive power system stabiliser. *Technical Report* **POW92/03**, Dept. of Electrical and Electronic Engineering, The University of Adelaide, Australia.

He, F. and M.J. Gibbard (1994). Adaptive nonlinear control in the design of a power system stabiliser. *AUPEC'94 Proceedings* **3**, 560–565.

Pierre, D.A. (1987). A perspective on adaptive control of power systems. *IEEE Trans. Power Systems* **PWRS-2(2)**, 387–396.

FIXED MODES IN QUASI-STATIC STATE FEEDBACK DECOUPLING

R. ANDIARTI*, C.H. MOOG* and H.J.C. HUIJBERTS**

**L.A.N. - CNRS URA 823, E.C.N. - Univ. de Nantes, 1 rue de la Noë, 44072 Nantes Cedex 03, France.*
***Dept. of Mathematics and Computing Science, Eindhoven University of Technology, P.O. Box 513, 5600 MB Eindhoven, The Netherlands.*

Abstract. We consider the problem of noninteracting control with stability for a square affine nonlinear system, whose its decoupling matrix is not necessarily invertible. The recent notion of quasi-static state feedback is used. The results of this paper show that the dimension of the fixed dynamics with respect to any quasi-static state feedback may be determined by using the so-called *controllability cospaces*.

Résumé. Le problème de découplage avec stabilité pour les systèmes non linéaires carrés et inversibles est considéré. On introduit la notion des *co-espaces de commandabilité* sous bouclage quasi-statique, et on montre que cette nouvelle notion permet de caractériser la dimension des modes fixes pour les systèmes dont la matrice de découplage n'est pas nécessairement inversible.

Key Words. Nonlinear systems, fixed modes, controllability cospaces

1. INTRODUCTION

The problem of noninteraction with stability of nonlinear systems by means of static state feedback has been considered by Isidori and Grizzle (1988). They have shown that there exists a fixed internal dynamics, called P^* *dynamics* whose stability is a necessary condition to solve the noninteracting control problem with stability. In the case where the P^* *dynamics* is unstable, Wagner (1991) has shown that there exists a well-defined dynamics, called Δ_{mix} *dynamics*, which cannot be eliminated by any regular dynamic feedback which renders the considered system noninteractive. The Δ_{mix} *dynamics* must then be asymptotically stable if noninteracting control and stability is achieved by means of dynamic state feedback. Glumineau *et al.* (1991) used dynamic compensator to remove a one-dimensional interconnection zero dynamics and showed that such compensator is able to cancel only the fixed dynamics which have a certain linearity property. A sufficient condition to solve the problem of noninteracting control with stability by means of dynamic state feedback was given by Battilotti (1991) and Battilotti and Dayawansa (1991). The problem of dynamic feedback noninteracting control with stability is solvable if Δ_{mix} *dynamics* is asymptotically stable and each decoupled subsystem is asymptotically stabilizable. Note that the class of dynamic feedbacks used in the above references are the one of exogenous dynamic feedbacks (in the sense of Martin (1992)).

All the above results are valid under the assumption that the decoupling matrix $A(x)$ is nonsingular. In the case where $A(x)$ is singular, and the system is square and invertible, Zhan *et al.* (1991) introduced the so-called Canonical Dynamic Decoupling Algorithm to construct a canonical dynamic extension. They have shown that the dynamically decoupled system is stable only if the Δ_{mix} *dynamics* of the canonical dynamic extension is stable, which is an intrinsic property of the given system.

In this paper, we investigate the case where the decoupling matrix is not necessarily invertible and study the fixed modes in the noninteracting control problem with stability by means of quasi-static state feedback. The goal is to get intrinsic geometric conditions generalizing the above ones. When considering the endogenous dynamic compensator (Martin, 1992) which decouples the system, it turns out that the dimension of the fixed dynamics of the decoupled system is equal to the dimension of the fixed dynamics with respect to any quasi-static state feedback which decouples the system.

In Section 2, we briefly review the definition of quasi-static state feedback introduced by Delaleau *et al.* (1992a;1992b), and in Section 3 a definition of controllability cospaces is given. Section 4 is devoted to characterize the fixed dynamics with respect to any quasi-static state feedback.

2. QUASI-STATIC STATE FEEDBACK

Consider a nonlinear control system of the form

$$\dot{x} = f(x) + g(x)u \tag{1}$$

where $x \in \mathbb{R}^n$, $u \in \mathbb{R}^m$ and where the entries of $f(x)$ and $g(x)$ are meromorphic functions of x. It is assumed that $\operatorname{rank} g(x) = m$ and that $n \geq 1$. We follow the notations and setting of Di Benedetto *et al.* (1989). Let $\mathcal{K}$ denote the field of meromorphic functions of $x, u, \dot{u}, \cdots, u^{(n-1)}$. $\mathcal{E}$ is the formal vector space spanned by $\{dx, du, d\dot{u}, \cdots, du^{(n-1)}\}$ over $\mathcal{K}$. The notation dx stands for $\{dx_1, \cdots, dx_n\}$ and $du^{(k)}$ for $\{du_1^{(k)}, \cdots, du_m^{(k)}\}$. Let $\mathcal{X} := \operatorname{span}_{\mathcal{K}}\{dx\}$ and $\mathcal{U} := \operatorname{span}_{\mathcal{K}}\{du, d\dot{u}, \cdots, du^{(n-1)}\}$.

A generalized static state feedback for (1) is a feedback of the form

$$u = \phi(x, v, \ldots, v^{(r)}) \tag{2}$$

where $v \in \mathbb{R}^m$ denotes the new controls. Such feedback transformations were previously considered by Perdon *et al.* (1989). Let $\mathcal{K}_v$ denote the field of meromorphic function of $\{x, \{v^{(k)} \mid k \geq 0\}\}$ and define the formal vector space $\mathcal{E}_v = \operatorname{span}_{\mathcal{K}_v}\{d\xi \mid \xi \in \mathcal{K}_v\}$. We define the following filtrations of $\mathcal{E}_v$:

$$\begin{aligned} \mathcal{V}_{-1} &= \operatorname{span}_{\mathcal{K}_v}\{dx\} \\ \mathcal{V}_k &= \operatorname{span}_{\mathcal{K}_v}\{dx, dv, \ldots, dv^{(k)}\} \ (k \geq 0) \end{aligned}$$

$$\begin{aligned} \mathcal{U}_{-1} &= \operatorname{span}_{\mathcal{K}_v}\{dx\} \\ \mathcal{U}_k &= \operatorname{span}_{\mathcal{K}_v}\{dx, d\phi, \ldots, d\phi^{(k)}\} \ (k \geq 0) \end{aligned}$$

The quasi-static state feedback which was formalized by Delaleau *et al.* (1992a;1992b) is defined as follows.

Definition 2.1 u given by (2) is said to be a ***quasi-static state feedback*** for (1) if the filtrations $\mathcal{U}_k$ and $\mathcal{V}_k$ have bounded difference, i.e., if there exists an integer $s \in \mathbb{N}$ such that for all $k \geq -1$, we have $\mathcal{U}_k \subset \mathcal{V}_{k+s}$, $\mathcal{V}_k \subset \mathcal{U}_{k+s}$.

This definition yields that there exists a function $\varphi(x, u, \ldots, u^{(s)})$ such that locally (2) is equivalent to

$$v = \varphi(x, u, \ldots, u^{(s)}) \tag{3}$$

3. CONTROLLABILITY COSPACES

Consider the control system (1) together with a quasi-static state feedback (2) and define $\mathcal{V} := \operatorname{span}_{\mathcal{K}_v}\{dv^{(k)} \mid k \geq 0\}$. In the rest of the paper, we denote $\Theta^{(k)}$ as the time derivative of order k of Θ along the trajectories of the system (1), and $\Theta^{[k]}$ as the time derivative of order k of Θ along the trajectories of the closed loop system (1) fed back with (2). We will write simply $\dot{\Theta}$ for $\Theta^{(1)}$.

Let us first review the definition of controlled invariance by means of quasi-static state feedback introduced by Huijberts *et al.* (1993;1994).

3.1. *Controlled invariance*

Definition 3.1 (Huijberts *et al.*, 1993;1994) A subspace $\Omega \subset \mathcal{X}$ is said to be controlled invariant for (1) if there exists a quasi-static state feedback (2) such that for (1),(2) one has

$$\Omega^{[1]} \subset \Omega + \mathcal{V} \tag{4}$$

The above definition is in accordance with the well known definition of a controlled invariant codistribution. Furthermore, it involves the non exact invariant subspaces. A special class of this subspace which plays an important role in practical applications is Ω_*, the smallest controlled invariant subspace containing the differential of the output. When considering the output $y = h(x)$, Theorem 3.9 of Huijberts *et al.* (1993;1994) gives

$$\Omega_* = \mathcal{X} \cap \operatorname{span}_{\mathcal{K}}\{dy, d\dot{y}, \cdots, dy^{(n-1)}\}$$

In this section, controllability cospaces are studied under quasi-static state feedbacks as a special class of controlled invariant subspaces previously defined. These controllability cospaces are related to the dual of dynamic controllability distributions (Respondek,1992).

3.2. *Controllability cospaces*

Controllability cospaces consist of vectors which are autonomous after applying a certain quasi-static state feedback $u = \phi(x, v, \cdots, v^{(r)})$ and zeroing certain input channels v_j, where $j \in \mathcal{J} \subset \{1, \cdots, m\}$. Such nonregular transformations are not defined for every element in $\mathcal{K}_v$. One possibility to circumvent this problem is to consider the module $\operatorname{span}_{\mathcal{A}}\{dx\}$ over the ring of analytic functions rather than the linear space over the field of meromorphic functions. Another way is chosen here; it consists in taking a particular basis of a given subspace of $\operatorname{span}_{\mathcal{K}}\{dx\}$ so that its time derivative is well defined when applying nonregular feedback. Such a basis always exists. More precisely, let $\Theta \subset \mathcal{X}$ be a subspace which admits a basis $\theta_1, \cdots, \theta_d$ with

$$\theta_i = \sum_{k=1}^{n} \frac{\alpha_{ik}(x, v, \cdots, v^{(k)})}{\beta_{ik}(x, v, \cdots, v^{(k)})} dx_i,$$

where α_{ik} and β_{ik} are in $\mathcal{A}$, the ring of all analytic functions of $\{x, \{v^{(k)} \mid k \geq 0\}\}$. Obviously, we can choose another basis for Θ, $\tilde{\theta}_1, \cdots, \tilde{\theta}_d$, in the module spanned by $\{dx\}$ over the ring $\mathcal{A}$ by taking

$$\tilde{\theta}_i = (\prod_{k=1}^{n} \beta_{ik})\theta_i$$

Definition 3.2 A subspace $\mathcal{C} \subset \mathcal{X}$ is said to be a controllability cospace for (1) if there exists a quasi-static state feedback (2) and a set of integers $\mathcal{J} \subset \{1, \cdots, m\}$ such that for (1),(2) one has

$$\mathcal{C} = \max\{\Theta \subset \mathcal{X} \mid \text{span}_{\mathcal{K}}\{\tilde{\theta}_i^{[1]} \mid_{v_j=0, j\in\mathcal{J}}\} \subset \Theta\}$$

where $\tilde{\theta}_i$ is defined as above.

This means that $\mathcal{C}$ is the largest autonomous subspace in $\mathcal{X}$ of the closed loop system. Moreover, by this definition, a controllability cospace is automatically controlled invariant.

3.3. *The smallest controllability cospace containing a given subspace*

As shown in Section 4, in practical applications, it is indispensable to know the smallest controllability cospace containing a given subspace. However, if an exact subspace $\Pi \subset \mathcal{X}$ is given, then there exists a smallest one containing Π. From Huijberts *et al.* (1993;1994), we know that the smallest controlled invariant subspace containing Π is $\Pi_* = \mathcal{X} \cap (\Pi + \cdots + \Pi^{(n-1)})$. The next theorem will relate Π_* to the smallest controllability cospace containing Π.

Theorem 3.3 Define the sequence $\mathcal{C}_\mu$ according to

$$\mathcal{C}_0 = \mathcal{X}$$

$$\mathcal{C}_\mu = \text{span}_{\mathcal{K}}\{\omega \in \mathcal{C}_{\mu-1} \mid \dot{\omega} \in \mathcal{C}_{\mu-1} + \dot{\Pi}_*\}$$

Then $\mathcal{C}_* = \lim_{\mu \to \infty} \mathcal{C}_\mu$ is the smallest controllability cospace containing Π.

For the output $y = h(x)$, the smallest controllability cospace containing the differential of the output is then given by the sequence $\mathcal{C}_\mu$ by replacing Π_* by $\Omega_* = \mathcal{X} \cap \text{span}_{\mathcal{K}}\{dy, \cdots, dy^{(n-1)}\}$.

Remark 3.4 When specialized to linear systems, the sequence $\mathcal{C}_\mu$ with $\Pi_* = \Omega_*$ turns out to be equal to the dual of the sequence $\mathcal{R}_\mu$ (the sequence computing the maximal controllability subspace in the kernel of the output mapping).

Now, we apply the notion of controllability cospace to solve the noninteracting control problem with stability.

4. FIXED MODES BY QUASI-STATIC STATE FEEDBACK

Let us consider a square invertible nonlinear affine system (Σ) defined by:

$$\Sigma : \begin{cases} \dot{x} = f(x) + \sum_{i=1}^{m} g_i(x)u_i \\ y_i = h_i(x), \; i = 1, \cdots, m, \end{cases} \tag{5}$$

where $x \in \mathbb{R}^n$, $u_i \in \mathbb{R}$ and $y_i \in \mathbb{R}$. Let $\{n_i'\}$ be the set of orders of zeros at infinity, where $n_1' > n_2' > \cdots > n_m'$. Permute if necessary y_i such that the corresponding order of zero at infinity is n_i'. Let $\mathcal{C}_{i*}$ be the smallest controllability cospace containing $\text{span}_{\mathcal{K}}\{dh_i(x)\}$. A first result is the following:

Lemma 4.1 Suppose that the system (Σ) is decouplable by a quasi-static state feedback $u = \psi(x, v, \cdots, v^{(s)})$. Then, there always exist coordinates $\xi = (\xi_0, \xi_1, \cdots, \xi_m, \hat{\xi})$ such that the system (Σ) is presented in the following form:

$$\begin{cases} \dot{\xi}_0 = f_0(\xi_0) \\ \dot{\xi}_1 = f_1(\xi_0, \xi_1, v_1) \\ \vdots \\ \dot{\xi}_m = f_m(\xi_0, \xi_m, v_m) \\ \dot{\hat{\xi}} = \hat{f}(\xi, v, \dot{v}, \cdots, v^{(s)}) \\ y_i = h_i(\xi_0, \xi_i) \end{cases} \tag{6}$$

Proof For a scalar output y_i, $\mathcal{C}_{i*}$ is an exact subspace. Thus, $\dot{\mathcal{C}}_{i*}$ is exact too as well as $\sum_{j=0}^{n_i'-1} \mathcal{C}_{i*}^{(j)}$. Let us define $\mathcal{C}_0$ as the dual of the strong accessibility distributions of the system (Σ). It is obvious that for each $i = 1, \cdots, m$

$$\mathcal{C}_0 = \sum_{j=0}^{n_i'-1} \mathcal{C}_{i*}^{(j)} \cap \sum_{k \neq i} \sum_{j=0}^{n_k'-1} \mathcal{C}_{k*}^{(j)}$$

Let $d\xi_0$ be a basis of $\mathcal{C}_0$, thus $\dot{\xi}_0 = f_0(\xi_0)$. For an invertible system, we can construct a quasi-static state feedback which decouples the system Σ by taking $v_i = y_i^{(n_i')}$. For $i = 1, \cdots, m$, then choose $d\xi_i$ such that $\{d\xi_0, d\xi_i\}$ is a basis of $\sum_{j=0}^{n_i'-1} \mathcal{C}_{i*}^{(j)}$.

Then one has

$$\dot{\xi}_i = f_i(\xi_0, \xi_i, v_i)$$

Complete now the new coordinates by choosing $\hat{\xi}$, such that $\{d\xi_0, d\xi_1, \cdots, d\xi_m, d\hat{\xi}\}$ is linearly independent. Thus, one has

$$\dot{\hat{\xi}} = \hat{f}(\xi, v, \dot{v}, \cdots, v^{(s)}),$$

and (6) is established. ■

Using this Lemma, the following corollary is obtained.

Corollary 4.2 The dimension of the fixed dynamics with respect to any quasi-static state feedback which decouples the system is equal to

$$\text{codim}(\sum_{i=1}^{m} \sum_{j=0}^{n_i'-1} \mathcal{C}_{i*}^{(j)}) \tag{7}$$

Remark 4.3 From the definition of the structure at infinity, one gets

$$\dim(\sum_{i=1}^{m} \sum_{j=0}^{n_i'-1} \mathcal{C}_{i*}^{(j)}) = \dim(\mathcal{X} \cap \sum_{i=1}^{m} \sum_{j \geq 0} \mathcal{C}_{i*}^{(j)}) \tag{8}$$

Remark 4.4 An invertible nonlinear system (Σ) can always be decoupled by an endogenous dynamic state feedback (Martin, 1992). Let Σ_e be a minimal dynamic extended decoupled system. One has necessarily

$$\text{codim}(\mathcal{X} \cap \sum_{i=1}^{m} \sum_{j \geq 0} \mathcal{C}_{i*}^{(j)}) = \dim(\mathcal{P}^*(\Sigma_e)) \tag{9}$$

where $\mathcal{P}^*(\Sigma_e)$ is as defined by Isidori *et el.* (1988) for the system Σ_e.

From Remark 4.3, Remark 4.4 and Corollary 4.2, the following theorem is derived.

Theorem 4.5 For a square invertible nonlinear system, the dimension of fixed dynamics with respect to any quasi-static state feedback (or endogenous dynamic feedback) is equal to

$$\text{codim}(\mathcal{X} \cap \sum_{i=1}^{m} \sum_{j \geq 0} \mathcal{C}_{i*}^{(j)})$$

The above theorem allows to characterize the fixed dynamics by computing only $\mathcal{C}_{i*}$. Under additional technical conditions as in (Isidori *et al.*, 1988), one describes the fixed dynamics. Suppose that the origin is an equilibrium point of f and the quasi-static state feedback rendering the system (Σ) noninteractive preserves this equilibrium point, then induced fixed dynamics are

$$\dot{\hat{\xi}} = \hat{f}(0, \cdots, 0, \hat{\xi}, 0, \cdots, 0)$$

And the asymptotic stability of these dynamics is a necessary condition for noninteracting control with internal stability. The next example illustrates Theorem 4.5.

Example 4.6 Let us consider a nonlinear system given by:

$$\left\{ \begin{array}{lcl} \dot{x}_1 & = & u_1 \\ \dot{x}_2 & = & x_4 + x_3 u_1 \\ \dot{x}_3 & = & x_3 + x_4 \\ \dot{x}_4 & = & u_2 \\ \dot{x}_5 & = & x_1 + x_2 \\ & & \\ y_1 & = & x_1 \\ y_2 & = & x_2 \end{array} \right. \tag{10}$$

We have $\{n_i'\} = \{2, 1\}$. Permute then y_i, and thus $\mathcal{C}_{1*} = \{dx_2\}$ and $\mathcal{C}_{2*} = \{dx_1\}$. The quasi-static feedback which decouples the system is $u_1 = v_1$ and $u_2 = v_2 - (x_3 + x_4)v_1 - x_3\dot{v}_1$, where $(v_1, v_2)^T$ is a new input vector. It is clear that $\mathcal{C}_0 = 0$. We choose $d\xi_1 = \{dx_2, d(x_4 + x_3 u_1)\}$ as a basis of $\{\mathcal{C}_{1*} + \dot{\mathcal{C}}_{1*}\}$, and thus

$$\dot{\xi}_1 = \begin{pmatrix} \dot{\xi}_{11} \\ \dot{\xi}_{12} \end{pmatrix} = \begin{pmatrix} \xi_{12} \\ v_2 \end{pmatrix} \tag{11}$$

Choose now $\{d\xi_2\} = \{dx_1\}$ as a basis of $\mathcal{C}_2^*$, and one has

$$\dot{\xi}_2 = v_1 \tag{12}$$

Complete the coordinates transformation by taking $\hat{\xi} = \begin{pmatrix} \hat{\xi}_1 \\ \hat{\xi}_2 \end{pmatrix} = \begin{pmatrix} x_4 \\ x_5 \end{pmatrix}$. So in the new coordinates $(\xi_1, \xi_2, \hat{\xi})$, the considered system becomes

$$\left\{ \begin{array}{lcl} \dot{\xi}_{11} & = & \xi_{12} \\ \dot{\xi}_{12} & = & v_2 \\ \dot{\xi}_2 & = & v_1 \\ \dot{\hat{\xi}}_1 & = & v_2 - (\xi_{12} - \hat{\xi}_1) - \hat{\xi}_1 v_1 - \\ & & \qquad (\xi_{12} - \hat{\xi}_1)\dot{v}_1 / v_1 \\ \dot{\hat{\xi}}_2 & = & \xi_2 + \xi_{11} \\ & & \\ y_1 & = & \xi_2 \\ y_2 & = & \xi_{11} \end{array} \right. \tag{13}$$

Clearly $\dim(\hat{\xi}) = 2$. This dimension equals $n - \dim(\mathcal{X} \cap (\sum_{i=1}^{m} \sum_{j \geq 0} \mathcal{C}_{i*}^{(j)})) = n - \dim(dx_1, dx_2, dx_4 + u_1 dx_3)$. Thus, the dimension of the fixed dynamics is two. Since the origin is an equilibrium point, the fixed dynamics are

$$\begin{cases} \dot{\hat{\xi}}_1 &= \hat{\xi}_1 \\ \dot{\hat{\xi}}_2 &= 0 \end{cases} \tag{14}$$

■

Similarly to Wagner and Battilotti's results, in the case where no quasi-static state feedback can render the system simultaneously noninteractive and stable, a suitable exogenous dynamic feedback may still solve the problem. This reduces to the results in (Zhan *et al.*, 1991).

Finally, we can then summarize the existing results related to the dimension of the fixed dynamics of a nonlinear square decoupled system in the following table:

Feedback	$A(x)$ invertible	$A(x)$ non-invertible
Static (Quasi-static or Endo. dynamic)	dim ($\mathcal{P}^*$) (Isidori *et al.*)	codim $\{\mathcal{X} \cap (\sum_{i=1}^{m} \sum_{j\geq 0} \mathcal{C}_{i*}^{(j)})\}$
Exogenous dynamic	dim (Δ_{mix}) (Wagner)	dim ($\Delta_{mix}(\Sigma_p)$) (Zhan *et al.*)

5. REFERENCES

Battilotti, S. (1991). A sufficient condition for nonlinear noninteracting control with stability via dynamic state feedback, *IEEE Trans. Automat. Control*, **36**, 1033-1045.

Battilotti, S. and W.P. Dayawansa (1991). Noninteracting control with stability for a class of nonlinear system, *Systems & Control Letters*, **17**, 327-338.

Delaleau, E., and M. Fliess (1992a). An algebraic interpretation of the structure algorithm with an application to feedback decoupling, *Proc. NOLCOS 1992*, Bordeaux, France, 489-494.

Delaleau, E., and M. Fliess (1992b). Algorithme de structure, filtrations et découplage, *C.R. Acad. Sci. Paris*, **t.315, Série I**, 101-106.

Di Benedetto, M.D., J.W. Grizzle and C.H. Moog (1989) Rank invariants of nonlinear systems, *SIAM J. Control Optimiz.*, **27**, 658-672.

Glumineau, A., C.H. Moog and T.J. Tarn (1990). Interconnected zero dynamics in nonlinear systems and their role in dynamic noninteracting control with stability, *Proc. Genoa Conf. New Trends in Systems Theory*, Birhäuser, pp. 316-323.

Huijberts, H.J.C. and C.H. Moog (1993). Controlled invariance of nonlinear systems: nonexact forms speak louder than exact forms, *in Systems and Networks: Mathematical Theory and Application*, Vol. II, U. Helmke, R. Mennicken and J. Saurer, eds., Akademie Verlag, Berlin, pp. 245-248.

Huijberts, H.J.C. and C.H. Moog (1994). Controlled invariance for nonlinear systems: generalized concepts, *Proceedings ACC*, Baltimore, USA.

Isidori, A. (1989). *Nonlinear control systems*, Springer Verlag, Berlin.

Isidori, A., A.J. Krener, C. Gori-Giorgi and S. Monaco (1981). Nonlinear decoupling via feedback: a differential geometric approach, *IEEE Trans. Automat. Control*, **26**, 331-345.

Isidori, A. and J.W. Grizzle (1988). Fixed modes and nonlinear noninteracting control with stability *IEEE Trans. Automat. Control*, **33**, 907-914.

Martin, P. (1992). *Contribution à l'étude des systèmes différentiellement plats*, Thèse de Doctorat, Ecole des Mines de Paris.

Nijmeijer, H. and A.J. van der Schaft (1990). *Nonlinear dynamical control systems*, Springer Verlag, New York.

Nijmeijer, H. (1982). Controllability distributions for nonlinear control systems, *Systems & Control letters*, **2**, 122-129.

Nijmeijer, H. and J.M. Schumacher (1986). The regular local noninteracting control problem for nonlinear control systems, *SIAM J. Control Optimiz.*, **24**, 1232-1245.

Perdon, A.M., G. Conte et C.H. Moog (1989). Some canonical properties of nonlinear systems, *Proceedings of the International Symposium MTNS-89*, pp. 89-96.

Respondek, W. (1992). Dynamic controllability distributions in nonlinear systems, *Proceding 2nd IFAC Workshop on System, Structure and Control*, Prague, pp. 256-258.

Wagner, K.G. (1991). Nonlinear noninteracting with stability by dynamic state feedback, *SIAM J. Control and Optimiz.*, **29**, 609-622.

Wonham, W.M. (1985). *Linear multivariable control: a geometric approach* (3rd Edition), Springer Verlag, Berlin.

Zhan, W., T.J. Tarn and A. Isidori (1991). A canonical dynamic extension for noninteraction with stability for affine nonlinear square systems, *Systems, & Control Letters*, **17**, 177-184.

Some results on singular feedback linearization *

Zhendong Sun
7th Research Div.,Beijing Univ. of Aero. & Astro., Beijing 100083,China
Xiaohua Xia
ISR,Univ. Stuttgart,Pfaffenwaldring 9,70550 Stuttgart,Germany

Résumé. Dans ce papier,on annonce la problème de transformer un affine système nonlinéaire à un retour d'état statique singulier et un difféomorphisme d'état statique.Mais, on établit une condition structurelle suffisante.

Mots-clés. Affine système nonlinéaire;linéarisation par un retour d'état régulier; linéarisation par un retour d'état singulier

Abstract. We address the problem of transforming an affine nonlinear system into a controllable one via singular static state feedback and state space diffeomorphism. A constructive sufficient condition is obtained.

Key Words. Affine nonlinear systems;regular feedback linearization;singular static feedback linearization

1 Introduction

In this paper,we address the problem of locally transforming an affine nonlinear system

$$\dot{x} = f(x) + \sum_{i=1}^{m} g_i(x)u_i = f(x) + G(x)u \quad (1)$$

with $x \in R^n, u \in R^m, f(0) = 0$ and $rankG(0) = m$ into a linear controllable system

$$\dot{z} = Az + Bv \quad z \in R^n, v \in R^{m'} \quad (2)$$

where $m' \leq m$.

Since Krener(1973) this problem has been studied using increasing more general transformations.State space diffeomorphisms

$$z = \phi(x) \quad \phi(0) = 0, x \in R^n, z \in R^n \quad (3)$$

were the first transformations to be studied.Brockett(1978) proposed to enlarge the class of transformation (3) by adding state feedback transformations

$$u = \alpha(x) + \beta v \quad v \in R^m, \alpha(0) = 0, det\beta \neq 0 \quad (4)$$

They were generalized in Jakubczyk and Respondek(1980) and Hunt *et al.*(1983) by

$$u = \alpha(x) + \beta(x)v \quad v \in R^m, det\beta(0) \neq 0 \quad (5)$$

where the nonsingular matrix β was allowed to depend the states as well.A system (1) is said to be locally regular static feedback linearizable if it can be transformed into (2) around the origin by (3) and (5).For those systems that are not regular static feedback linearizable,the problem of partial feedback linearization was posed by Krener *et al.* and was solved in Krener *et al.*(1983) and Marino(1986).

A more general class of state feedback transformations is given by dynamic state feedback transformations (Singh,1980):

$$\begin{cases} \dot{w} = a(x,w) + B(x,w)v & w \in R^q \\ u = \alpha(x,w) + \beta(x,w)v & v \in R^{m'} \end{cases} \quad (6)$$

with $a(0,0) = 0, \alpha(0,0) = 0$,and q being the order of the compensator. The extended system of (1) controlled by a dynamic compensator (6) can be written as

$$\dot{\bar{x}} = \begin{pmatrix} f + G\alpha \\ a \end{pmatrix}(\bar{x}) + \begin{pmatrix} G\beta \\ B \end{pmatrix}(\bar{x})v \quad (7)$$

with $\bar{x} = (x^T, w^T)^T$ being the extended state. The problem of transforming the system (1) into (2) via regular dynamic compensation (Di Benedetto,Grizzle and Moog,1989)

*This work has been supported in part by the Trans-Century Training Programme Foundation for the Talents by the State Education Commission (China),the second author has also been supported in part by the Alexander von Humboldt Foundation (Germany) and (extended) state space diffeomorphism is called (locally) regular dynamic feedback linearization problem.It was studied in Charlet,Levine and Marino(1989,1991):some sufficient conditions as well as necessary conditions were given.

The feedback transformations considered in the above cited papers all satisfy some 'regular' condition.The problem of transforming the system (1) into (2) via regular feedback transformation (5) or (6) and (may be extended) state space diffeomorphism (3),here will be called regular feedback linearization problem. Main results concerning this problem can be found in the survey paper Marino(1990).

In this paper,we investigate the problem of transforming an affine nonlinear multi-input system into a linear controllable one via singular static feedback transformation and state space diffeomorphism.Consider the following class of transformations

$$u = \alpha(x) + \beta(x)v, \quad \alpha(0) = 0, \beta(x) : m \times m', m' \leq m \tag{8}$$

The feedback transformations (8) are said singular,if they are not required to be regular.The importance of studying singular feedback linearization is obvious.It's a natural generalization of regular feedback linearization.There exist a fairly broad class of systems which are not regular feedback linearizable but are singular feedback linearizable,see Sun,Xia and Gao(1994).Even for those systems that are regular feedback linearizable,considering their singular feedback linearization is still meaningful in the engineering sense, for instance,it can reduce the number of input channels, and/or can avoid using dynamic compensators,etc..

Here we present some results about singular feedback linearization.To limit space,we either omit the proofs or just give the outline of the proofs.

2 Problem statement and basic results

We assume that all functions under consideration are defined and smooth (C^∞) in an open neighborhood of the origin.And our results will be stated in terms of an open neighborhood Γ of the origin in the respective Euclidean spaces,and we implicitly permit Γ to be made smaller to accommodate subsequent local argument.

The system (1) is said to be (locally) singular static feedback linearizable,if it can be transformed into (2) via some state feedback (8) and state space diffeomorphism (3).The problem of linearization via (8) and (3) is called singular static feedback linearization problem.

It is obvious that singular static feedback linearization is a natural generalization of the regular case.

The following proposition gives an equivalent statement of singular static feedback linearization problem.

Proposition 2.1. The system (1) is singular static feedback linearizable if, and only if,the $\beta(x)$ of (8) can be chosen to be a column vector (i.e. $m' = 1$).

Outline of the proof. We need only to prove the necessity.

Suppose under some transformations (8) and (3) the system (1) is changeable into (2). There must exists a linear feedback $v = Fz + Gv'$ such that system (2) will be changed into a linear controllable single-input system. Hence under state feedback

$$u = \alpha(x) + \beta(x)F\phi(x) + \beta(x)Gv'$$

and (3),the system (1) is transformable into a linear controllable single-input system. ◇

The following result states a Lie algebraic solution to the problem of singular static feedback linearization.It follows easily from Proposition 2.1.

Theorem 2.2. The system (1) is locally singular static feedback linearizable if, and only if, there exist a smooth distribution $\Delta(.)$ on R^n, $\Delta(x) \subseteq span\{g_1(x), \cdots, g_m(x)\}, x \in \Gamma$,and a smooth vector field f_0,$f_0(0) = 0$,and $f_0 - f \in span\{g_i, i = 1, \cdots, m\}$, such that the nested distributions

$$\Delta_0 = \Delta, \quad \Delta_i = \Delta_{i-1} + \mathrm{ad}_{f_0}^i \Delta, \quad i = 1, 2, \cdots$$

satisfy the following conditions in Γ:

(1)Δ_i is involutive and of constant rank for $0 \leq i \leq n-1$;

(2)$rank(\Delta_{n-1}) = n$.

Remark 2.1.In general,we don't know how to determine the desire distribution Δ and function f_0 if they exist.But under some further assumptions, explicit algorithm may be obtained.See,for example,Theorem 3.1 below.

Now we present an example which is singular static feedback linearizable but is not regular dynamic feedback linearible.

Example 2.3 Consider the system

$$\dot{x} = \begin{pmatrix} x_2 \\ x_3 \\ x_4 \\ x_5 \\ 0 \end{pmatrix} + \begin{pmatrix} 0 \\ 0 \\ 0 \\ 0 \\ 1 \end{pmatrix} u_1 + \begin{pmatrix} 1 \\ x_2^2 x_4 \\ 1 + x_2 + x_1 x_5^2 \\ 0 \\ 0 \end{pmatrix} u_2 \tag{9}$$

It can be easily verified that under the state feedback $u_1 = v, u_2 = 0$,the above system will be changed into a single–input Brunovsky canonical system.So it is singular static feedback linearizable. In the other hand,Sun,Xia and Gao (1994) has shown that (9) is not regular dynamic feedback linearizable.It follows that regular dynamic feedback linearizability does not imply singular static feedback linearizability,showing that the problem of singular static feedback linearization has independent theoretical interest of studying.

3 Main results

Theorem 2.2 gives a necessary and sufficient condition for singular static feedback linearization.But in general, it is difficult to compute the desire data. In this section,we will present sufficient conditions as well as certain methodologies to construct the related data.

Theorem 3.1. Suppose for a set of integers $\{\nu_1, \cdots, \nu_m\}$, the nested distributions

$$\begin{aligned} \Delta_0 &= span\{g_k : \nu_k = 0\} \\ \Delta_{i+1} &= \Delta_i + ad_f\Delta_i + span\{g_k : \nu_k = i+1\}, \\ & i \geq 0 \end{aligned}$$

are such that in Γ

(i) Δ_i is involutive and of constant rank for $0 \leq i \leq n-1$;

(ii) $rank(\Delta_{n-1}) = n$;

(iii) $[g_j, \Delta_i] \subseteq \Delta_{i+1}$ for all $j, 1 \leq j \leq m$,such that $1 \leq \nu_j \leq n-1$ and all $i, 0 \leq i \leq n-2$.

then the system (1) is locally singular static feedback linearizable.

Outline of the proof. We prove by induction.For convenience, given a smooth vector field $\underline{f}$,and smooth distributions $\underline{G}_0, \cdots, \underline{G}_s$,,we call the multiplet $(\underline{f}; \underline{G}_0, \cdots, \underline{G}_s)$ satisfying Condition($\aleph$),if the nested distributions

$$\underline{\Delta}_0 = \underline{G}_0, \underline{\Delta}_i = \underline{\Delta}_{i-1} + ad_{\underline{f}}\underline{\Delta}_{i-1} + \underline{G}_i, i = 1, \cdots$$

satisfy (i)–(iii) of Theorem 3.1 (using G_j instead of g_j in (iii)).

Denote $G_j = span\{g_k : \nu_k = j+1\}$ for $j = 0, 1, \cdots$. Suppose q is the integer such that $rank(\Delta_q) = n$ and $rank(\Delta_{q-1}) < n$.Define

$$\begin{aligned} l &= max\{i : i \leq q, rank(\Delta_{i-1} + ad_f\Delta_{i-1}) \\ &\quad < rank(\Delta_i)\}, \\ m_0 &= q - l \end{aligned}$$

It follows that either l is equal to or less than q.

(I)Suppose $l = q$.

Assume,without loss of generality,that $G_q = span\{g_1, \cdots, g_s\}$, and $rank(\Delta_{q-1} + ad_f\Delta_{q-1} + span\{g_s\}) > rank(\Delta_{q-1} + ad_f\Delta_{q-1})$. Thus there exists a real–value function λ,$\lambda(0) = 0$,such that

$$\lambda \perp \Delta_{q-1}, < d\lambda, g_s > (0) \neq 0$$

Let

$$\begin{aligned} g'_s &= \tfrac{1}{<d\lambda, g_s>} g_s, \\ f' &= f - (L_f\lambda)g'_s, \\ g'_i &= g_i - (L_{g_i}\lambda)g'_s, i = 1, \cdots, s-1. \end{aligned}$$

It is easy to verify that $(f'; G_1, \cdots, G_q)$ satisfies Condition($\aleph$) and still belongs to this case.

Let $\varphi, \varphi(0) = 0$ be such that $d\varphi \perp \Delta_{q-1}$ and $< d\varphi, ad_{f'}\Delta_{q-1} >$ has constant rank one. Set

$$\begin{aligned} \bar{f} &= f' + \varphi g'_s, \\ \bar{G}_q &= span\{g'_1, \cdots, g'_{s-1}\} \end{aligned}$$

One may check that the multiplet $(\bar{f}; G_1, \cdots, G_{q-1}, \bar{G}_q)$ still satisfies Condition($\aleph$).

(II)Suppose $l < q$.

We have

$$\begin{aligned} \Delta_l &= \Delta_{l-1} + ad_f\Delta_{l-1} + G_l \\ \Delta_{l+1} &= \Delta_{l-1} + ad_f\Delta_{l-1} + ad_f^2\Delta_{l-1} \\ &\quad + G_l + ad_fG_l \\ &\vdots \\ \Delta_q &= \Delta_{l-1} + \cdots + ad_f^{m_0+1}\Delta_{l-1} \\ &\quad + G_l + \cdots + ad_f^{m_0}G_l \end{aligned}$$

From the facts $\Delta_q = \Delta_{q-1} + ad_f^{m_0+1}\Delta_{l-1} + ad_f^{m_0}G_l$ and $rank(\Delta_q) > rank(\Delta_{q-1})$ it follows that at least one of the following two cases must hold.

Case(1):$rank(\Delta_{q-1} + ad_f^{m_0}G_l) > rank(\Delta_{q-1})$.

Case(2):$rank(\Delta_{q-1} + ad^{m_0+1}\Delta_{l-1}) = n$.

First we consider case(1). Without any loss of generality,we may assume that $G_l = span\{g_1,\cdots,g_s\}$ and $rank(\Delta_{q-1} + span\{ad_f^{m_0}g_s\}) = rank(\Delta_{q-1})+1$. Thus there exists a real-value function μ, $\mu(0) = 0$, such that

$$d\mu \perp \Delta_{q-1},\ < d\mu, ad_f^{m_0}g_s > (0) \neq 0$$

Define

$$g'_s = \frac{1}{<d\mu,ad_f^{m_0}g_s>}g_s,$$
$$f' = f - ((-1)^{m_0}L_f^{m_0+1}\mu_1)g'_s$$

It can be verified that $(f';G_1,\cdots,G_l)$ satisfies Condition($\aleph$),and still belongs to this case. For simplicity,here we denote f' by f.

Define

$$p_0 = min\{j - l : j \geq l, rank(\Delta_{l-1}+ \cdots + ad_f^{j-l}\Delta_{l-1} + G_l + \cdots + ad_f^{j-l}G_l) = rank(\Delta_j)\}$$

Set $min\phi = q - l$ if necessary.Thus $0 \leq p_0 \leq m_0$. There exists $g_0 \in \Delta_{l-1}$,such that

$$ad_f^{p_0}g_0\bar{\in}\Delta_{p_0-1} + \cdots + ad_f^{j-l}\Delta_{l-1} + G_l + \cdots + ad_f^{p_0-l}G_l$$

and such that there exists $\varphi,\varphi(0) = 0$,satisfying

$$d\varphi \perp \Delta_{l+p_0-2},\ < d\varphi, ad_f^{p_0}g_0 >= 1$$

Set

$$\bar{f} = f + \varphi g'_s,$$
$$g'_i = g_i - (L_{g_i}L_f^{m_0}\mu)g'_s,$$
$$\bar{G}_l = \{g'_1,\cdots,g'_{s-1}\}$$

One may check that $(\bar{f};G_0,\cdots,G_{l-1},\bar{G}_l)$ satisfies Condition($\aleph$).

Now we consider Case(2). Let $g \in \Delta_{l-1}$ be such that

$$rank(\Delta_{q-1} + ad_f^{m_0}G_l + span\{ad_f^{m_0+1}g\}) = rank(\Delta_{q-1} + ad_f^{m_0}G_l) + 1.$$

Define

$$p_0 = min\{j - l - 1 : j \geq l + 1, rank(\Delta_{l-1} + \cdots + ad_f^{j+l}\Delta_{l-1} + G_l + \cdots + ad_f^{j-l-1}G_l) = rank(\Delta_j)\}$$

Then $0 \leq p_0 \leq q - l - 1$.Without loss of generality,we assume

$$ad_f^{p_0}g_s\bar{\in}\Delta_{l-1} + \cdots + ad_f^{p_0+l}\Delta_{l-1} + G_l + \cdots + ad_f^{p_0-1}G_l$$

Hence there exists a real-value function λ,$\lambda(0) = 0$ such that

$$d\lambda \perp \Delta_{p_0+l-1},\ < d\lambda, ad_f^{p_0}g_s > (0) \neq 0$$

Set

$$g'_s = \frac{1}{<d\lambda,ad_f^{p_0}g_s>}g_s,$$
$$f' = f - (L_f^{p_0+1}\lambda)g'_s$$

one can verify that $(f';G_0,\cdots,G_l)$ satisfies Condition($\aleph$) and $L_{f'}^{p_0+1}\lambda = 0$.

Suppose $(f';G_0,\cdots,G_l)$ still belongs to this case. Let real-value function $\mu,\mu(0) = 0$ be such that

$$d\mu \perp \Delta_{q-1},\ < d\mu, ad_f^{q-l+1}g > (0) \neq 0.$$

Set

$$\bar{f} = f + \mu g'_s,$$
$$g'_i = g_i - (L_{g_i}L^{p_0}\lambda)g'_s, i = 1,\cdots,s-1,$$
$$\bar{G}_l = span\{g'_i : 1 \leq i \leq s-1\}$$

It can be verified that the multiplet $(\bar{f};G_1,\cdots,G_{l-1},\bar{G}_l)$ still satisfies Condition($\aleph$).

We have established that,if $(f;G_0,\cdots,G_s)$ satisfies Condition($\aleph$),then there exist $\bar{f} \in \{f\}+span\{G\}$, $\bar{f}(0) = 0$ and $\bar{G}_0,\cdots,\bar{G}_k (k \leq s)$ such that

$$\sum_{j=0}^{k} rank(\bar{G}_j) < \sum_{j=0}^{s} rank(G_j)$$

and $(\bar{f};\bar{G}_0,\cdots,\bar{G}_k)$ satisfies Condition($\aleph$). By induction,there exist $\bar{f}^* \in \{f\} + span\{G\}$ and $G_0^* \subseteq span\{G\}$, such that $(\bar{f}^*;G_0^*)$ satisfies Condition($\aleph$).By Theorem 2.2, system (1) is singular static feedback linearizable.◇

Remark 3.1. The conditions considered in Theorem 3.1 are similar to that of Charlet,Levine and Marino(1991) Theorem 4.2.But there exists one important difference between them:the later must take all g_i, $i = 1, \cdots, m$, into consideration,while the former may discard some of the input channels.Thus our conditions are milder(cf. Example 3.2 below).

Remark 3.2. The conditions stated in the above theorem are not necessary.For example,the following system

$$\dot{x} = \begin{pmatrix} x_2 + x_5 \\ x_3 + x_2^2 x_4 x_5 \\ x_4 + x_5 + x_2 x_5 + x_1 x_5^2 \\ x_5 \\ 0 \end{pmatrix} + \begin{pmatrix} 0 \\ 0 \\ 0 \\ 0 \\ 1 \end{pmatrix} u_1 + \begin{pmatrix} 1 \\ x_2^2 x_4 \\ 1 + x_2 + x_1 x_5^2 \\ 0 \\ 0 \end{pmatrix} u_2 \tag{10}$$

is singular static feedback linearizable (which follows from Theorem 2.2 by setting $f_0 = (x_2, x_3, x_4, x_5, 0)^T$ and $\Delta = span\{(0,0,0,0,1)^T\}$).But detailed computation exhibits that (10) does not satisfy the conditions of Theorem 3.1.

The proof of Theorem 3.1 is constructive.The computation of a desire feedback law follows from an inductive algorithm,as illustrated by the following example.

Example 3.2. Consider the following system

$$\dot{x} = \begin{pmatrix} x_4 + x_5^2 + (x_4 x_5 + x_2 x_6) u_4 \\ x_3 + x_3 u_3 + x_3^2 u_4 \\ x_6 + u_4 \\ u_3 \\ u_2 \\ u_1 \end{pmatrix} \tag{11}$$

Let $\nu_1 = 0, \nu_2 = 2, \nu_3 = 2, \nu_4 = 6$, one can easily verify that (i)–(iii) of Theorem 3.1 hold. Using the algorithm given by the proof of Theorem 3.1, we can compute a desire feedback law step by step.

Step 0:

$$f = \begin{pmatrix} x_4 + x_5^2 \\ x_3 \\ x_6 \\ 0 \\ 0 \\ 0 \end{pmatrix}, G_0 = span\{\begin{pmatrix} 0 \\ 0 \\ 0 \\ 0 \\ 0 \\ 1 \end{pmatrix}\},$$

$$G_1 = \emptyset, G_2 = span\{\begin{pmatrix} 0 \\ 0 \\ 0 \\ 0 \\ 1 \\ 0 \end{pmatrix}, \begin{pmatrix} 0 \\ x_3 \\ 0 \\ 1 \\ 0 \\ 0 \end{pmatrix}\}$$

Step 1:One may check that $q = 3, l = 2, m_0 = 1$. So it belongs to (II). Simple computation shows it is in Case(1). Let $\mu = x_1$, then $f' = f - (L_f^2 \mu) g_3 = f$. One can verify that $p_0 = 1$. Choose $g_0 = (0,0,1,0,0,0)^T$, then $\varphi = x_2$, hence $\bar{f} = f + \varphi g_3 = (x_4 + x_5^2, x_3 + x_2 x_3, x_6, x_2, 0, 0)^T$, $g_2' = g_2 - (L_{g_2} L_f \mu) g_3 = (0, -2x_3 x_5, 0, -2x_5, 1, 0)^T$. Denote

$$f^1 = \begin{pmatrix} x_4 + x_5^2 \\ x_3 + x_2 x_3 \\ x_6 \\ x_2 \\ 0 \\ 0 \end{pmatrix}, G_0^1 = span\{\begin{pmatrix} 0 \\ 0 \\ 0 \\ 0 \\ 0 \\ 1 \end{pmatrix}\},$$

$$G_1^1 = \emptyset, G_2^1 = span\{\begin{pmatrix} 0 \\ -2x_3 x_5 \\ 0 \\ -2x_5 \\ 1 \\ 0 \end{pmatrix}\}.$$

Step 2:It can easily checked that $(f^1; G_0^1, G_2^1)$ belongs to Case(2) of (II).Applying the algorithm presented in the proof of Theorem 3.1, one may compute

$$f^2 = \begin{pmatrix} x_4 + x_5^2 \\ x_3 + x_2 x_3 - 2x_1 x_3 x_5 \\ x_6 \\ x_2 - 2x_1 x_5 \\ x_1 \\ 0 \end{pmatrix},$$

$$G_0^2 = span\{\begin{pmatrix} 0 \\ 0 \\ 0 \\ 0 \\ 0 \\ 1 \end{pmatrix}\}.$$

Step 3:Now the system

$$\dot{x} = f^2 + g_1^2 v^1 \tag{12}$$

is (regular static) feedback linearizable,where $g_1^2 = (0,0,0,0,0,1)^T$.

Note that $f^2 = f^1 + x_1 g_2^1 = f + x_2 g_3 + x_1(g_2 - 2x_5 g_3) = f + x_1 g_2 + (x_2 - 2x_1 x_5) g_3, g_1^2 = g_1$. Hence under the feedback law

$$u = \begin{pmatrix} 0 \\ x_1 \\ x_2 - 2x_1x_5 \\ 0 \end{pmatrix} + \begin{pmatrix} 1 \\ 0 \\ 0 \\ 0 \end{pmatrix} v^1 \qquad (13)$$

the system (11) is changeable into (12).

Standard analysis exhibits that,under the feedback

$$v^1 = \frac{1}{1 + x_2 - 2x_1x_5}(v - \rho(x)) \qquad (14)$$

and state space change of coordinates

$$z = (x_5, x_1, x_4 + x_5^2, x_2, x_3 + x_2x_3 + 2x_1x_3x_5, \sigma(x))^T \qquad (15)$$

the system (12) will be changed into a canonical linear controllable system (here ρ and σ are tedious in expression but easily computed functions of x).

Therefore,under the feedback law

$$u = \begin{pmatrix} \frac{1}{1+x_2-2x_1x_5}\rho \\ x_1 \\ x_2 - 2x_1x_5 \\ 0 \end{pmatrix} + \begin{pmatrix} \frac{1}{1+x_2-2x_1x_5} \\ 0 \\ 0 \\ 0 \end{pmatrix} v$$

and state space diffeomorphism (15),the system (11) is transformable into a linear controllable system.◇

One can check that the system (11) does not satisfy the conditions of Theorem 4.2 of Charlet,Levine,Marino(1991),so the latter cannot apply to this system.In fact,further computation shows that (11) is not linearizable via prolongations.

4 Conclusions

In this paper,we presented some results about the problem of singular feedback linearization. Theorem 2.2 gives sufficient and necessary Lie algebraic conditions for singular static feedback linearization. Theorem 3.1 gives constructive sufficient conditions for singular static feedback linearization. The challenging problem of finding constructive solutions for singular feedback linearization problem is still open.We have solved the problem only when the number of controls is equal to the number of the states minus one.

5 References

Brockett,R.W.(1978). Feedback invariants for nonlinear systems.In:*Proc. VII IFAC Congress*,Helsinki,pp1115–1120.

Charlet,B.,J. Levine,R. Marino(1989). On dynamic feedback linearization. *Sys. Control Lett.* ,**13**,143–151.

—(1991). Sufficient conditions for dynamic state feedback linearization .*SIAM J. Control Optim.* ,**29**,38–57.

Di Benedetto,M.D.,J.W. Grizzle,C.H. Moog (1989).Rank invariants of nonlinear systems.*SIAM J. Contr. Optimiz.* ,**27**,658–672.

Hunt,L. R.,*et el.*(1983). Design for multi-input nonlinear systems.In: *Differential Geometric Control Theory*,R.Brockett,*et al.*(eds.),Birkhauser,basel ,pp286–298.

Isidori,A.(1989). *Nonlinear Control System.* Springer-Verlag,Berlin.

Jakubczyk,B.,W. Respondek(1980). On linearization of control systems. *Bull. Acad. Pol. Sci.,Ser. Sci. Math.* ,*28*,517–522.

Krener,A.J.(1973). On the equivalence of control systems and the linearization of nonlinear systems. *SIAM J. Control*,**11**,670–676.

Krener,A.J,*et al.*(1983). Partial and robust linearization by feedback.In:*Proc. 22ed IEEE CDC*,San Antonio,Texas,pp126–130.

Marino,R.(1986). On the largest feedback linearizable subsystem. *Sys. Control Lett.* ,**6**,345–351.

Marino,R.(1990). Static and dynamic feedback linearization of nonlinear systems. In: *Perspectives in Control Theory*,B. Jakabczyk,*et al.* (eds.),Birkhauser,Basel, pp249–260.

Singh,S.N.(1980). Decoupling of invertible nonlinear systems with state feedback and precompensation.*IEEE Trans. Automat. Control*,**25**,1237–1239.

Sun,Z.,X. Xia,W. Gao(1994). Some elementary results about singular feedback linearization. In:*Proc. Chinese Control Conf.*,Taiyuan,China,pp.308–313.

A GENERALIZED CONTROLLER FORM FOR LINEARIZATION BY QUASI-STATIC STATE FEEDBACK*

J. RUDOLPH, R. ROTHFUSS* and J.-B. POMET**

*Institut für Systemdynamik und Regelungstechnik, Universität Stuttgart, Postfach 80 11 40, D-70511 Stuttgart, GERMANY, E-mail: {rudolph,rothfuss}@isr.uni-stuttgart.de

**I.N.R.I.A., B.P. 93, F-06902 Sophia Antipolis Cedex, FRANCE, E-mail: pomet@sophia.inria.fr

Abstract. A generalized controller form for nonlinear systems is introduced which depends on the inputs and their time derivatives in a "block triangular" way. This form allows to linearize the system by a particular type of quasi-static state feedback. Sufficient involutivity type conditions are derived for the existence of this controller form which thus are sufficient also for dynamic state feedback linearization.

Résumé. Une forme de commande généralisée dépendant de manière "triangulaire par blocs" de dérivées temporelles des entrées est introduite pour les systèmes non linéaires. Cette forme permet de linéariser le système par un bouclage d'état quasi statique particulier. Des conditions "d'involutivité" suffissantes sont données pour l'existence de cette forme. Par conséquent, ces conditions sont aussi suffissantes pour la linéarisation par bouclages d'état dynamiques.

Key Words. Controller form, dynamic feedback linearization, flat systems, quasi-static state feedback.

1. INTRODUCTION

The problem of transforming nonlinear dynamics into linear controllable ones via state feedback and diffeomorphism in the state domain has been intensively studied during the last two decades and has led to a large variety of control methods (see (Charlet *et al.*, 1991; Isidori, 1989; Nijmeijer and van der Schaft, 1990) for surveys).

Recently, Fliess *et al.* (1992)–(1994) studied linearization by a particular kind of dynamic state feedback, called *endogenous*, and introduced the concept of *flatness*. This endogenous feedback is a generalization of the *quasi-static state feedback* as introduced by Delaleau and Fliess (1992). The advantage of the latter is that it preserves the state space. It is the type of feedback considered in the present paper.

In what follows, a generalized nonlinear controller form is defined which allows to construct a particular type of quasi-static state feedback linearizing the dynamics. By restricting this controller form to have a "block triangular" structure, the need of using concepts from infinite dimensional geometry (Fliess *et al.*, 1992; Fliess *et al.*, 1993) is avoided. If it exists, the transformation between the coordinates of a given nonlinear dynamics and the generalized controller form may depend on time derivatives of the inputs. Using finite prolongations of the system vector field, sufficient conditions of the involutivity type for the existence of the generalized controller form are derived. The latter thus form sufficient conditions for dynamic state feedback linearization. A simple example illustrates that the given conditions are not a consequence of those given by Charlet *et al.* (1991). For further references on dynamic state feedback linearization see e.g. (Charlet *et al.*, 1989; Jakubczyk, 1994; Pomet, 1995*a*; Pomet *et al.*, 1992; Rudolph, 1995; Delaleau and Rudolph, 1995; Shadwick, 1990).

The paper is organized as follows: section 2 is devoted to the introduction of some notations. In section 3 a generalized nonlinear controller form is defined and shown to be interesting because it can be linearized by quasi-static state feedback. Section 4 treats generalized state transformations used to transform a given dynamics into this form. After the definition of those transformations, the main result is given: geometric sufficient conditions for a system to be transformable via this kind of transformation into the generalized nonlinear controller form. This also provides sufficient conditions for dynamic feedback linearization. Section 5 is devoted to an example.

2. NOTATIONS

Consider a state representation of a multi-input nonlinear dynamics

$$\dot{x} = f(x, u). \tag{1}$$

The state x and the input u represent local coordinates in the neighborhood $\mathcal{M} \subseteq \mathcal{X} \times \mathcal{U}$ of the origin of the product of the state space $\mathcal{X} \subseteq I\!R^n$ and the input space $\mathcal{U} = I\!R^m$. The vector field $f : \mathcal{M} \to T\mathcal{X}$ associated with (1) is C^∞ with respect to all its arguments. For the sake of simplicity, one also assumes that $f(0,0) = 0$, rank $\frac{\partial f}{\partial u}(0) = m$, and restricts the

* This research has been supported by the Deutsche Forschungsgemeinschaft (DFG) (J. Rudolph as a postdoctoral fellow and R. Rothfuß as a research assistant with Prof. M. Zeitz). It has been completed while J. Rudolph was an invited fellow at the I.N.R.I.A, France.

considerations to a local neighborhood of the origin of $\mathcal{X} \times \mathcal{U}$.

For notational convenience, one introduces $\mathcal{U}_i^{r_i} \subseteq \mathbb{R}^{r_i+1}$ the coordinates on which will be $u_i^{(k)}$, $k = 0, \ldots, r_i$. Abusing notation, for a given m-tuple $r = (r_1, \ldots, r_m)$, one defines

$$\overset{\langle r_j \rangle}{u_j} = \begin{cases} (u_j, \ldots, u_j^{(r_j)}), & \text{for } r_j \geq 0 \\ () & \text{for } r_j < 0 \end{cases}, \quad (2)$$

$$\overset{\langle\langle r \rangle\rangle}{u} = \left(\overset{\langle r_1 \rangle}{u_1}, \ldots, \overset{\langle r_m \rangle}{u_m} \right), \quad (3)$$

and the manifolds

$$\overset{\langle\langle r \rangle\rangle}{\mathcal{U}} = \mathcal{U}_1^{r_1} \times \cdots \times \mathcal{U}_m^{r_m}. \quad (4)$$

As a convention, introduce $u_0 = ()$ and assume $\mathcal{U}_i^{r_i}$ to be zero for $r_i < 0$. The same conventions are used for the inputs $\overline{u}$ and v later on.

The prolonged vector field (*cf.* (Delaleau and Respondek, 1994; Olver, 1993))

$$\mathcal{F} = \sum_{j=1}^{n} f_j(x, u) \frac{\partial}{\partial x_j} + \sum_{i=1}^{m} \sum_{k=0}^{\tilde{N}-1} u_i^{(k+1)} \frac{\partial}{\partial u_i^{(k)}} \quad (5)$$

which is defined on the manifold[1] $\mathcal{X} \times \overset{\langle\langle N \rangle\rangle}{\mathcal{U}}$, where $N = (\tilde{N}, \ldots, \tilde{N})$ and $\tilde{N}$ is chosen large enough to ensure that $\mathcal{F}$ acts as time derivation on all functions considered, is associated with (1).

3. GENERALIZED NONLINEAR CONTROLLER FORM

3.1. *Definition of the generalized controller form*

The feedback design for a nonlinear system is often simpler if the representation in local coordinates has a particular structure. Therefore, for nonlinear systems, several generalizations of the linear controller form have been defined which consist of chains of integrators terminated by functions (Fliess, 1989; Fliess, 1990; Rudolph *et al.*, 1991; Sommer, 1980). These functions represent the "nonlinearity" of the dynamics and can be compensated by appropriate feedback since they are "directly accessible" by the control inputs.

In the sequel, a generalized controller form is defined that may depend on the inputs and their time derivatives. For this, let the integers σ_i and λ_i be such that for all i, $\sigma_i \leq i \leq \sigma_i + \lambda_i$, and, for all j such that $\sigma_i \leq j \leq \sigma_i + \lambda_i$, one has $\sigma_j = \sigma_i$ and $\lambda_i = \lambda_j$.

Definition 3.1 (Generalized nonlinear controller form) For given positive integers $\kappa_1, \ldots, \kappa_m$, with $\sum_{i=1}^{m} \kappa_i = n$, a (generalized) state representation consisting of m blocks of the form

$$\dot{z}^i = \begin{bmatrix} z_2^i \\ \vdots \\ z_{\kappa_i}^i \\ \varphi^i \left(z, \overset{\langle r_1 \rangle}{\overline{u}_1}, \ldots, \overset{\langle r_{\sigma_i - 1} \rangle}{\overline{u}_{\sigma_i - 1}}, \overline{u}_{\sigma_i}, \ldots, \overline{u}_{\sigma_i + \lambda_i} \right) \end{bmatrix} \quad (6)$$

with $i = 1, \ldots, m$, $z^i = [z_1^i, \ldots, z_{\kappa_i}^i]^T$, the r_i's any nonnegative integers, and the integers σ_i and λ_i with the property described above, is called a *generalized nonlinear controller form* with *generalized controllability indices* $(\kappa_1, \ldots, \kappa_m)$.

The *(generalized) state* z evolves on an n-dimensional manifold $\mathcal{X} \subseteq \mathbb{R}^n$ and the inputs $\overline{u}_i, i = 1, \ldots, m$ are the coordinates on $\mathcal{U}$. The functions $\varphi^i : \mathcal{X} \times \mathcal{U}_1^{r_1} \times \cdots \times \mathcal{U}_{\sigma_i - 1}^{r_{\sigma_i - 1}} \times \mathcal{U}_{\sigma_i} \times \cdots \times \mathcal{U}_{\sigma_i + \lambda_i} \to \mathbb{R}$, $i = \sigma_i, \ldots, \sigma_i + \lambda_i$ are C^∞ and are assumed to satisfy

$$\operatorname{rank} \left(\frac{\partial \varphi^i}{\partial \overline{u}_j}(0) \right)_{i,j=1,\ldots,m} = m. \quad (7)$$

Remark 3.2 Observe that in fact no derivative of $\overline{u}_m$ does appear in (6), *i.e.*, $r_m = 0$.

3.2. *Quasi-static state feedback*

The interest of the above form is the following: Introducing new input coordinates $v_i, i = 1, \ldots, m$ by

$$v_i = \varphi^i \left(z, \overset{\langle r_1 \rangle}{\overline{u}_1}, \ldots, \overset{\langle r_{\sigma_i - 1} \rangle}{\overline{u}_{\sigma_i - 1}}, \overline{u}_{\sigma_i}, \ldots, \overline{u}_{\sigma_i + \lambda_i} \right) \quad (8)$$

yields a state representation

$$\dot{z} = Az + Bv \quad (9)$$

with (A, B) a Brunovsky pair.

In the spirit of (Delaleau and Fliess, 1992), the equations (8) can be interpreted as defining a particular type of *quasi-static state feedback*. In order to see this, it is shown that $\overline{u}_i$, $i = 1, \ldots, m$ can be expressed as

$$\overline{u}_i = \tilde{\varphi}^i \left(z, \overset{\langle r_i \rangle}{v_1}, \ldots, \overset{\langle r_i \rangle}{v_m} \right), \quad (10)$$

i.e., as a function of z and a *finite* number of derivatives of v_j, $j = 1, \ldots, m$.

The functions φ^i, $i = 1, \ldots, m$ in (8) introduce new coordinates v_i, $i = 1, \ldots, m$ which may be used to replace $\overline{u}_i$, $i = 1, \ldots, m$ in an open neighborhood of the origin of $\mathcal{X} \times \overset{\langle\langle r \rangle\rangle}{\mathcal{U}}$. By prolongation this also introduces new coordinates $v_i^{(l)}$, $i = 1, \ldots, m$, $l > 0$ in such a way that the coordinates $(z, \overline{u}, \dot{\overline{u}}, \ddot{\overline{u}}, \overline{u}^{(3)}, \ldots)$ are replaced by $(z, v, \dot{v}, \ddot{v}, v^{(3)}, \ldots)$. In order to see this, observe that, using (7) and the implicit function theorem, for $i = 1, \ldots, m$, one gets

$$\overline{u}_i = \tilde{\varphi}^i \left(z, \overset{\langle r_1 \rangle}{\overline{u}_1}, \ldots, \overset{\langle r_{\sigma_i - 1} \rangle}{\overline{u}_{\sigma_i - 1}}, v_{\sigma_i}, \ldots, v_{\sigma_i + \lambda_i} \right). \quad (11)$$

[1] One might use an infinite prolongation here. However, in this case a restriction on $\tilde{N} = j^*$, with j^* the integer defined before equation (24), is sufficient.

For $i = 1$, equation (11) reads

$$\overline{u}_i = \bar{\varphi}^i(z, v_1, \ldots, v_{\lambda_1}), \quad i = 1, \ldots, \lambda_1 \tag{12}$$

which corresponds to (10). Suppose that (10) holds for $i = k$ and consider (11) with $i = k+1$. Then, after a short calculation, replacing the derivatives of $\overline{u}_i$, using (10) for $i = 1, \ldots, k$ and (9), yields

$$\overline{u}_{k+1} = \bar{\varphi}^{k+1}\left(z, \overset{\langle r_{k+1}\rangle}{v_1}, \ldots, \overset{\langle r_{k+1}\rangle}{v_m}\right). \tag{13}$$

Remark 3.3 The quasi-static state feedback (10) is invariant under an additional local diffeomorphism $\zeta : \mathcal{X} \to \mathcal{X}$ between two generalized states, as in the definition in (Delaleau and Fliess, 1992).

Remark 3.4 It is not difficult to see that with

$$z^i_{\kappa_i+l} = \overline{u}^{(l)}_i, \quad l = 0, \ldots, r_i - 1, \tag{14}$$

for $i = 1, \ldots, m$, new inputs $\bar{v}_i = u_i^{(r_i)}, i = 1, \ldots, m$, and (11) a dynamic state feedback compensator linearizing (6) can be defined (*cf.* (Charlet *et al.*, 1991)).

4. TRANSFORMATION INTO GENERALIZED CONTROLLER FORM

4.1. *Generalized State transformation*

Consider a *generalized coordinate transformation (local diffeomorphism)* $\tau : \mathcal{X} \times \overset{\langle\langle r-1\rangle\rangle}{\mathcal{U}} \to \mathcal{X} \times \overset{\langle\langle r-1\rangle\rangle}{\mathcal{U}}$. It is called a *generalized state transformation* if it preserves the inputs and their time derivatives, *i.e.*, if it is of the form[2]:

$$\tau(x, \overset{\langle\langle r-1\rangle\rangle}{u}) = (z, \overset{\langle\langle r-1\rangle\rangle}{\overline{u}})$$

with

$$z^i_k = \tau^i_k\left(x, \overset{\langle\langle r-1\rangle\rangle}{u}\right) \tag{15}$$

$$x_j = (\tau^{-1})_j\left(z, \overset{\langle\langle r-1\rangle\rangle}{\overline{u}}\right) \tag{16}$$

$$\overline{u}^{(l)}_i = u^{(l)}_i, \tag{17}$$

the latter for $i = 1, \ldots, m$ and $l = 0, \ldots, r_{i-1}$, and the first two ones for $i = 1, \ldots, m$, $k = 1, \ldots, \kappa_i$, and $j = 1, \ldots, n$. Upper and lower indices are used to denote the components of z with $z = (z^1_1, \ldots, z^1_{\kappa_1}, z^2_1, \ldots, z^m_{\kappa_m})$.

The tangent map of τ (at the origin) is denoted as $\tau_* : T_0(\mathcal{X} \times \overset{\langle\langle r-1\rangle\rangle}{\mathcal{U}}) \to T_0(\mathcal{X} \times \overset{\langle\langle r-1\rangle\rangle}{\mathcal{U}})$, its inverse as τ_*^{-1}.

Remark 4.1 If the components of z are the state coordinates of (6), then, by its integrator chain structure one must have:

$$\tau^i_k = L^{k-1}_{\mathcal{F}} \tau^i_1, \quad k = 1, \ldots, \kappa_i, \quad i = 1, \ldots, m. \tag{18}$$

[2] Of course, (17) naturally extends to $u^{(l)}_i, i = 1, \ldots, m, l > r_{i-1}$.

According to (Fliess *et al.*, 1992; Fliess *et al.*, 1993), the functions τ^i_1, $i = 1, \ldots, m$ are called ***flat outputs*** of the dynamics (1).

4.2. *Main result*

In order to derive conditions for the existence of generalized state coordinate transformations (15)-(17) into the generalized nonlinear controller form (6), define distributions as follows.

Definition 4.2 For given integers $r_1 \geq r_2 \geq \cdots \geq r_m = 0$, define

$$\Delta_{-1} = \operatorname{span}\left\{\frac{\partial}{\partial u_i^{(r_i+k)}}, i = 1, \ldots, m, k \geq 0\right\} \tag{19}$$

$$\Delta_j = \operatorname{span}\left\{[\mathcal{F}, Z] \,|Z \in \Delta_{j-1}\right\} \cup \Delta_{j-1}, \; j \geq 0. \tag{20}$$

In the sequel, all these distributions are assumed to be of constant dimension in the neighborhood of the origin.

By the successive application of $\mathcal{F}$ directional derivatives w.r.t. to time derivatives of input coordinates of lower order are introduced in the distributions Δ_j. Once $\frac{\partial}{\partial u_i}$ is attained, further application of $\mathcal{F}$ leads to the introduction of successive vector fields of the type $\mathrm{ad}^k_{\mathcal{F}} g_i$, where

$$g_i = \left[\frac{\partial}{\partial u_i}, f\right], \quad i = 1, \ldots, m. \tag{21}$$

One thus observes, that for $j = 1, \ldots, m$, $0 \leq l \leq k - r_i$

$$\mathrm{ad}^{k-r_j-l}_{\mathcal{F}} g_j \in \Delta_k, \quad k \geq r_j + l \tag{22}$$

and that for $j = 1, \ldots, m$, $0 \leq l < r_j$,

$$\frac{\partial}{\partial u_j^{(l)}} \in \Delta_k, \quad k \geq r_j - l - 1. \tag{23}$$

In both cases one may verify that the respective vector fields are not contained in Δ_k for k lower. It is thus clear that, for some j^* large enough, there exist integers $0 < \kappa_1, \ldots, \kappa_m$ such that for all $j \geq j^*$

$$\Delta_j \cap T_0\mathcal{X} = \operatorname{span}\left\{\mathrm{ad}^{k_i}_{\mathcal{F}} g_i, \begin{array}{l} i = 1, \ldots, m, \\ k_i = 0, \ldots, \kappa_i - 1 \end{array}\right\}. \tag{24}$$

Remark 4.3 It is in fact sufficient to consider time derivatives of the inputs up to some finite order in Δ_j. The order of those time derivatives required is upper bounded by the maximal order k of $\mathrm{ad}^k_{\mathcal{F}} g_j$ contained in Δ_j. For, the coefficients in the vector fields under consideration do not depend on time derivatives of the inputs of order higher than the maximal order k of $\mathrm{ad}^k_{\mathcal{F}} g_j$ involved. This is because the coefficients in f do not depend on derivatives of the inputs.

Now, the main result can be stated as follows.

Theorem 4.4 A nonlinear system (1) locally admits a generalized state coordinate transformation (15)-(17) into a generalized controller form (6) with positive generalized controllability indices $\kappa_1, \ldots, \kappa_m$ satisfying (24) if there exist nonnegative integers $r_1 \geq r_2 \geq \cdots \geq r_m = 0$, such that, up to renumbering the components of the input u, locally around the origin of $\mathcal{X} \times \overset{\langle\langle N \rangle\rangle}{\mathcal{U}}$

(i) for j large enough $\dim \Delta_j \cap T_0\mathcal{X} = n$,

(ii) the distributions $\Delta_j, j \geq 0$ are involutive and of constant dimension in the neighborhood of the origin.

The coordinates z_1^i, $i = 1, \ldots, m$ are determined by computing exact one forms $dz_1^i = d\tau_1^i$, $i = 1, \ldots, m$ satisfying

$$\langle d\tau_1^i, Z\rangle = 0, \quad Z \in \Delta_{r_i+\kappa_i-2}, \quad i = 1, \ldots, m, \tag{25}$$

and such that the $m \times m$ matrix

$$\left(\langle d\tau_1^i, \mathrm{ad}_{\mathcal{F}}^{\kappa_i-1} g_j\rangle\right)_{i,j=1,\ldots,m} \tag{26}$$

has full rank.

Proof: The functions $z_1^i = \tau_1^i, i = 1, \ldots, m$ are introduced according to the equations (25) and (26). This is possible by conditions (i), (ii) and the Frobenius Theorem. Now, since the r_i's are decreasing (see Definition 4.2), one may define the integers σ_i and λ_i by

$$r_j = r_i \Leftrightarrow \sigma_i \leq j \leq \sigma_i + \lambda_i, \tag{27}$$

so that they satisfy the property indicated before Definition 3.1.

The functions τ_j^i for $2 \leq j \leq \kappa_i$ are defined by differentiation of $\tau_1^i, i = 1, \ldots, m$, according to (18). Then define the functions φ^i, $i = 1, \ldots, m$ by

$$\varphi^i = \left(L_{\mathcal{F}}^{\kappa_i} \tau_1^i\right) \circ \tau^{-1}. \tag{28}$$

It is clear that defining the new coordinates according to (15) and (17) with the κ_i's defined above, one obtains a representation of the system as in (6), with φ^i given by (28). Hence, the only fact that remains to be proved is that the dependence of φ^i — as given by (28) — is the one prescribed by (6). In other words, it remains to prove that, as a function of $(z, \overline{u}, \dot{\overline{u}}, \ldots)$, φ^i does not depend on any $\overline{u}_j^{(l)}$ with $j \geq \sigma_i + \lambda_i + 1$ and $l \geq 0$ or with $\sigma_i \geq j \geq \sigma_i + \lambda_i$ and $l \geq 1$. This amounts to $L_{\frac{\partial}{\partial \overline{u}_j^{(l)}}} \varphi^i = 0$, or, by (28) and the fact that τ maps the coordinates $(x, u, \dot{u}, \ldots)$ to the coordinates $(z, \overline{u}, \dot{\overline{u}}, \ldots)$, to:

$$L_{\tau_*^{-1}\left(\frac{\partial}{\partial \overline{u}_j^{(l)}}\right)} L_{\mathcal{F}}^{\kappa_i} \tau_1^i = 0 \quad \text{for} \quad \begin{cases} j \geq \sigma_i + \lambda_i + 1 \text{ and} \\ l \geq 0 \\ \text{or} \\ \sigma_i \geq j \geq \sigma_i + \lambda_i \text{ and} \\ l \geq 1 \end{cases} \tag{29}$$

The sequel is devoted to the proof of (29).

Using (19) and (20), it follows

$$\langle d\tau_1^i, \mathrm{ad}_{\mathcal{F}}^k \frac{\partial}{\partial u_j^{(l)}}\rangle = 0, \tag{30}$$

for $i, j = 1, \ldots, m$, $0 \leq k \leq r_i + \kappa_i - 1$ and $l \geq r_j$ (and smaller than the highest order of derivation needed to write $\mathcal{F}$ in (5)). After an induction using the Leibniz rule (*i.e.*, $L_{\mathcal{F}}\langle dh, X\rangle = \langle d(L_{\mathcal{F}}h), X\rangle + \langle dh, [\mathcal{F}, X]\rangle$ for any function h and vector field X) exactly κ_i times, equation (30) yields

$$\langle d(L_{\mathcal{F}}^{\kappa_i} \tau_1^i), \mathrm{ad}_{\mathcal{F}}^{k-\kappa_i} \frac{\partial}{\partial u_j^{(l)}}\rangle = 0 \tag{31}$$

for $i, j = 1, \ldots, m$, $\kappa_i \leq k \leq r_i + \kappa_i - 1$ and $l \geq r_j$. In particular, using (27) and the fact that the r_i's are decreasing, and $\mathrm{ad}_{\mathcal{F}}^s \frac{\partial}{\partial u_j^{(l)}} = \frac{\partial}{\partial u_j^{(l-s)}}$ for $l \geq s \geq 0$, equation (31) implies

$$L_{\frac{\partial}{\partial u_j^{(l)}}} L_{\mathcal{F}}^{\kappa_i} \tau_1^i = 0 \quad \text{for} \quad \begin{cases} j \geq \sigma_i + \lambda_i + 1 \text{ and } l \geq 0 \\ \text{or} \\ \sigma_i \geq j \geq \sigma_i + \lambda_i \text{ and } l \geq 1 \end{cases} \tag{32}$$

Since the coordinates $u_j^{(l)}$ are preserved by τ, τ_* maps $T_0\mathcal{X}$ on itself, and, thus, the vector fields $\mathrm{ad}_{\mathcal{F}}^k g_j$ are restricted to be contained in $T_0\mathcal{X}$. This can easily be verified by checking the action of the directional derivatives on functions in local coordinates. Note that nonetheless the coefficients of $\mathrm{ad}_{\mathcal{F}}^k g_j$ may depend on the inputs and their time derivatives. Furthermore, as can be seen by a similar calculation, by preserving the input coordinates, one has

$$\tau_*\left(\frac{\partial}{\partial u_j^{(l)}}\right) = \frac{\partial}{\partial \overline{u}_j^{(l)}} + \sum_{i=1}^{m} \sum_{k=1}^{\kappa_i} \left(\frac{\partial \tau_k^i}{\partial u_j^{(l)}} \circ \tau^{-1}\right) \frac{\partial}{\partial z_k^i}, \tag{33}$$

for $j = 1, \ldots, m$, $l \geq 0$. But, for $i, j = 1, \ldots, m$, $1 \leq k \leq \kappa_i$, and $l \geq \max(0, r_j - r_i)$,

$$\begin{aligned} \frac{\partial \tau_k^i}{\partial u_j^{(l)}} &= \langle d(L_{\mathcal{F}}^{k-1} \tau_1^i), \frac{\partial}{\partial u_j^{(l)}}\rangle \\ &= \langle d\tau_1^i, \mathrm{ad}_{\mathcal{F}}^{k-1+r_i} \frac{\partial}{\partial u_j^{(l+r_i)}}\rangle = 0. \end{aligned} \tag{34}$$

Here the first equality follows from (18), the second one from applying the Leibniz rule $k-1$ times, together with the fact that $\frac{\partial}{\partial u_j^{(l)}} = \mathrm{ad}_{\mathcal{F}}^{r_i} \frac{\partial}{\partial u_j^{(l+r_i)}}$, for $r_i, l \geq 0$. The last equality follows from (30), replacing k in this equation by $k - 1 + r_i \leq \kappa_i + r_i - 1$, and l by $l + r_i$. Taking into account the definition of the integers σ_i and λ_i via the r_i's, it can be seen now that the equation (34) together with (32) and (33) implies (29).

Finally, the rank property (7) is a direct consequence of (26). □

Corollary 4.5 A nonlinear system (1) can be transformed by a generalized state coordinate transformation (15)–(17) and a quasi-static state feedback (10) into a state representation in Brunovsky form (9) with new input v and positive (generalized) controllability indices $\kappa_i, i = 1, \ldots, m$ if the conditions of Theorem 4.4 are satisfied.

The following lemma shows that the existence of regular static state feedback linearizing a "regular affine" system (Hunt *et al.*, 1983; Isidori, 1989; Jacubczyk and Respondek, 1980; Nijmeijer and van der Schaft, 1990) implies that the conditions given above are fulfilled, too.

Lemma 4.6 Every "affine" system

$$\dot{x} = g_0(x) + \sum_{i=1}^{m} g_i(x) u_i \tag{35}$$

with $\dim \operatorname{span}\{g_1, \ldots, g_m\} = m$, $g_i(0) = 0, i = 1, \ldots, m$ which is linearizable by a regular static state feedback (Hunt *et al.*, 1983; Isidori, 1989; Jacubczyk and Respondek, 1980; Nijmeijer and van der Schaft, 1990) locally around the origin of $\mathcal{X}$ admits a generalized controller form (6) locally around the origin of $\mathcal{X} \times \mathcal{U}$ via a local diffeomorphism (15)–(17).

Proof: A nonlinear system (35) is linearizable by a regular static state feedback if and only if the following conditions based on the distributions

$$G_i = \operatorname{span}\left\{ \operatorname{ad}_{g_0}^k g_j, \begin{array}{rcl} k & = & 0, \ldots, i, \\ j & = & 1, \ldots, m \end{array} \right\}, \quad i = 0, \ldots, n \tag{36}$$

are fulfilled (Hunt *et al.*, 1983; Isidori, 1989; Jacubczyk and Respondek, 1980; Nijmeijer and van der Schaft, 1990):

(i) G_{n-1} has dimension n at the origin of $\mathcal{X}$,
(ii) $G_i, i = 0, \ldots, n-2$ are constant dimensional and involutive in a neighborhood of the origin of $\mathcal{X}$.

Let the integers $r_i, i = 1, \ldots, m$ be equal to zero and let the integers $\kappa_1, \ldots, \kappa_m$ be the controllability indices of the Brunovsky canonical form for static state feedback equivalence corresponding to (35). Since (35) is affine, one has $[Z, \frac{\partial}{\partial u_j^{(l)}}] \in G_i$, for all $Z \in G_i$, $j = 1, \ldots, m, l \geq 0$. Moreover, by (ii), $\operatorname{ad}_{g_0}^k g_i = \operatorname{ad}_{\mathcal{F}}^k g_i \bmod G_{k-1}$, as can easily be verified by induction. It is now easy to observe that the involutivity of the G_j implies the involutivity of the Δ_j, $j \geq 0$. Finally, by the fact that $G_{n-1} \subseteq T_0\mathcal{X}$, it follows that the coordinates z can be introduced in such a way that the diffeomorphism (15)–(17) is independent of the inputs. □

Remark 4.7 The above corollary reflects the fact that the Theorem 4.4 is a generalization of the conditions from the static case. One observes that one might also first introduce new state coordinates (in an extended state space) according to Remark 3.4, and then apply the classical result from the static case cited above. This is a consequence of the restriction of τ to act as the identity on $\overset{\langle\langle r-1\rangle\rangle}{\mathcal{U}}$, which means that the extension consists of ("pure") integrators calculating the original inputs.

5. EXAMPLE

Consider the following bilinear system borrowed from (Charlet *et al.*, 1991) where it has been used to show that the sufficient conditions given there are not necessary[3]

$$\begin{aligned} \dot{x}_1 &= x_2 + x_3 \tilde{u}_2 \\ \dot{x}_2 &= x_3 + x_1 \tilde{u}_2 \\ \dot{x}_3 &= x_2 \tilde{u}_2 + \tilde{u}_1 \\ \dot{x}_4 &= \tilde{u}_2 \end{aligned}$$

Renumbering the inputs, $u_1 = \tilde{u}_2$, $u_2 = \tilde{u}_1$, and choosing $r_1 = 3$ and $r_2 = 0$, condition (i) from Theorem 4.4 can easily be verified. One obtains the generalized controllability indices $\kappa_1 = 1$ and $\kappa_2 = 3$. One thus has $\Delta_{\kappa_2+r_2-2} = \Delta_1$ and $\Delta_{\kappa_1+r_1-2} = \Delta_2$. The consecutive distributions are

$$\begin{aligned} \Delta_{-1} &= \operatorname{span}\left\{ \frac{\partial}{\partial u_2^{(k)}}, \frac{\partial}{\partial u_1^{(3+k)}}, k \geq 0 \right\} \\ \Delta_0 &= \operatorname{span}\left\{ g_2, \frac{\partial}{\partial u_2^{(k)}}, \frac{\partial}{\partial u_1^{(2+k)}}, k \geq 0 \right\} \\ \Delta_1 &= \operatorname{span}\left\{ \operatorname{ad}_{\mathcal{F}} g_2, g_2, \frac{\partial}{\partial u_2^{(k)}}, \frac{\partial}{\partial u_1^{(1+k)}}, k \geq 0 \right\} \\ \Delta_2 &= \operatorname{span}\left\{ \operatorname{ad}_{\mathcal{F}}^2 g_2, \operatorname{ad}_{\mathcal{F}} g_2, g_2, \frac{\partial}{\partial u_2^{(k)}}, \right. \\ &\qquad \left. \frac{\partial}{\partial u_1^{(k)}}, k \geq 0 \right\}. \end{aligned}$$

One easily verifies the involutivity of these distributions. For Δ_0 and Δ_1 this is trivial (for the latter because the system is affine in u.) and for the following ones it results

$$\begin{aligned} \Delta_1 &= \operatorname{span}\left\{ -u_1 \frac{\partial}{\partial x_1} - \frac{\partial}{\partial x_2}, \frac{\partial}{\partial x_3}, \frac{\partial}{\partial u_2^{(k)}}, \right. \\ &\qquad \left. \frac{\partial}{\partial u_1^{(1+k)}}, k \geq 0 \right\} \\ \Delta_2 &= \operatorname{span}\left\{ (1 - \dot{u}_1) \frac{\partial}{\partial x_1} + u_1^2 \frac{\partial}{\partial x_2} + u_1 \frac{\partial}{\partial x_3}, \right. \\ &\qquad \left. -u_1 \frac{\partial}{\partial x_1} - \frac{\partial}{\partial x_2}, \frac{\partial}{\partial x_3}, \frac{\partial}{\partial u_2^{(k)}}, \frac{\partial}{\partial u_1^{(k)}}, k \geq 0 \right\} \end{aligned}$$

The conditions of Thereom 4.4 are thus fulfilled for this example. This shows that they are not a consequence of those given by Charlet *et al.* (1991).

Remark 5.1 Observe that $\frac{\partial}{\partial u_1} \notin \Delta_1$ and else this distribution would not be involutive. This reflects the fact that $r_1 = 3$ is required.

Flat outputs $\tau_1^1 = z_1^1 = x_4$ and $\tau_1^2 = z_1^2 = x_1 - x_2 u_1$ can be calculated using (25) and (26). By differenti-

[3] See (Pomet, 1995*b*) for a complete solution of the affine case with four states, two inputs, and flat outputs depending on the state and the input.

ation of these flat outputs, the generalized controller form (6) can be calculated. One obtains the feedback laws from the relations $v_1 = \varphi^1(z, u_1) = u_1$ and $v_2 = \varphi^2(z, \overset{(3)}{u_1}, u_2)$.

Remark 5.2 Even in this simple example, the expression for $\varphi^2(z, \overset{(3)}{u_1}, u_2)$ is quite vast. In general, for the calculation of the controller form (6) and the feedback laws (11), as well as for checking the existence conditions, a computer algebra system is required. The recursive character of the conditions of Theorem 4.4 is particularly well-suited for this.

6. CONCLUDING REMARKS

The conditions given by Theorem 4.4 are sufficient conditions for dynamic linearizability, *i.e.*, flatness. They are far from necessary though, since they are not invariant by a static state feedback. This is obvious in (19) and (20) for instance, in which changing the inputs would change the distributions Δ_j.

It is reasonable to think that they are necessary for equivalence via a generalized state transformation to the generalized nonlinear controller form — *i.e.*, the converse of Theorem 4.4 is true — and in a sense this is linearizability by dynamic state feedback using an extension by "pure" integrators. This would also show that the conditions given in this paper are actually more general than those in Charlet *et al.* (1991), which could also be checked directly on the conditions themselves.

In order to derive more general sufficient conditions for the flatness, *i.e.*, for the feedback linearization based on the generalized controller form, one should consider more general transformations of the inputs, which corresponds to preliminary feedback.

7. REFERENCES

Charlet, B., J. Lévine and R. Marino (1989). On dynamic feedback linearization. *Systems Control Letters* **13**, 143–151.

Charlet, B., J. Lévine and R. Marino (1991). Sufficient conditions for dynamic state feedback linearization. *SIAM J. Control and Optim.* **29**, 38–57.

Delaleau, E. and J. Rudolph (1995). Decoupling and linearization by quasi-static feedback of generalized states. In: *Proceedings 3rd European Control Conference ECC '95, Roma/Italy.*

Delaleau, E. and M. Fliess (1992). Algorithme de structure, filtrations et découplage. *C. R. Acad. Sci. Paris* **Série I, 315**, 101–106.

Delaleau, E. and W. Respondek (1994). Lowering the orders of derivatives of controls in generalized state space systems. *Journal of Mathematical Systems, Estimation, and Control.* To appear.

Fliess, M. (1989). Généralisations non linéaires de la forme canonique de commande et linéarisation par bouclage. *C. R. Acad. Sci. Paris* **Série I, 308**, 377–379.

Fliess, M. (1990). Generalized controller canonical forms for linear and nonlinear dynamics. *IEEE Trans. Automat. Control* **35**, 994–1001.

Fliess, M., J. Lévine, P. Martin and P. Rouchon (1992). Sur les systèmes non linéaires différentiellement plats. *C. R. Acad. Sci. Paris* **Série I, 315**, 619–624.

Fliess, M., J. Lévine, P. Martin and P. Rouchon (1993). Linéarisation par bouclage dynamique et transformations de Lie-Bäcklund. *C. R. Acad. Sci. Paris* **Série I, 317**, 981–986.

Hunt, L. R., R. Su and G. Meyer (1983). Design for multi-input nonlinear systems. In: *Differential Geometric Control Theory* (R.W. Brockett, Ed.). Birkhäuser. pp. 258–298.

Isidori, A. (1989). *Nonlinear Control Systems.* 2nd ed. Springer–Verlag.

Jacubczyk, B. and W. Respondek (1980). On linearization of control systems. *Bull. Acad. Polonaise Sci. Ser. Sci. Math.* **28**, 517–522.

Jakubczyk, B. (1994). Dynamic feedback equivalence of nonlinear control systems. Preprint.

Nijmeijer, H. and A. J. van der Schaft (1990). *Nonlinear Dynamical Control Systems.* Springer–Verlag.

Olver, P. J. (1993). *Applications of Lie Groups to Differential Equations.* 2nd ed.. Springer–Verlag.

Pomet, J.-B. (1995*a*). A differential geometric setting for dynamic equivalence and dynamic linearization. In: *Banach Center Publications.* To appear.

Pomet, J.-B. (1995*b*). On dynamic feedback linearization of four-dimensional affine control systems with two inputs. Preprint. To appear.

Pomet, J.-B., C.H. Moog and A. Aranda (1992). A non-exact Brunovsky form and dynamic feedback linearization. In: *Proc. 31st CDC IEEE, Tucson.* pp. 2012–2017.

Rudolph, J. (1995). Well-formed dynamics under quasi-static state feedback. *Banach Center Publications, Warsaw.* To appear.

Rudolph, J., J. Birk and M. Zeitz (1991). Dependence on time derivatives of the input in the nonlinear controller canonical form. In: *New Trends in System Theory* (G. Conte, A.M. Perdon and B. Wyman, Eds.). Birkhäuser. pp. 636–643.

Shadwick, W. (1990). Absolute equivalence and dynamic feedback linearization. *Systems Control Letters* **15**, 35–39.

Sommer, R. (1980). Control design for multivariable non-linear time-varying systems. *Int. J. Control* **31**, 883–891.

NON-LINEAR FEEDBACK LAWS FOR OUTPUT REGULATION OF NON-HOLONOMIC SYSTEMS†

R. MAHONY*, I. MAREELS*, G. BASTIN** and G. CAMPION**

* *Department of Systems Engineering, R.S.I.S.E., A.N.U., A.C.T., 0200, Australia.*
** *CESAME, Batiment EULER, 4-6, Avenue G. Lemaitre, B1348 Louvain la Neuve, Belgium.*
† *The authors would like to acknowledge the funding of the Australian Commonwealth Government under the Cooperative Research Centres Program. The paper also presents research results of the Belgian Programme on Inter-university Poles of Attraction, initiated by the Belgian State, Prime minister's Office for Science Technology and Culture. The scientific responsibility rests with its authors.*

Résumé. On étudie dans cet article des lois de commande par rétroaction statique de l'état permettant de résoudre le problème de la stabilisation de la sortie de systèmes non linéaires non holonomes sans dérive. On montre que, sous la condition de l'existence d'une rétroaction dynamique découplant les entrées et sorties du système, il est possible de construire une rétroaction *statique* de l'état stabilisant les sorties du système. La procédure souligne les difficultés inhérentes à la conception de lois de commande assurant la stabilité interne d'un système non holonome. On présente une caractérisation du domaine de conditions intiales pour lesquelles un système commandé par une telle loi stabilisant les sorties présentes cette stabilité interne. La théorie est illsutrée par une application à la stabilisation des sorties d'un modèle simple de robot mobile.

Key Words: Non-holonomic systems, output regulation, Internal stability.

1 Introduction

When designing control laws for dynamical systems it is preferable to use the simplest most direct design methodology available. In the case of smooth finite dimensional nonlinear dynamical control systems, arguably the most direct form of feedback law is "static state feedback" where the control variable is computed as a time-invariant function of the state variable. If the system is smooth then one would naturally like the controller to also be smooth. Unfortunately, when one considers the problem of designing a controller to asymptotically stabilize the state of a non-linear dynamical control system there is a fundamental limitation to the efficacy of such a class of controllers. In particular, Brockett (1983) showed that for many systems where the number of inputs is strictly less than the number of states then no smooth static feedback law exists that asymptotically stabilizes the state. To overcome this difficulty two approaches are commonly employed, firstly the use of time-varying control laws, see for example (Pomet 1992), and secondly the use of discontinuous static control laws, see for example (Canudas de Wit & Sordalen 1991, Kolmanovsky, Reyhanoglu & McClamroch 1994). A connection between these methods is presented in (Coron & Rosier 1994). An important application for these control methodologies is the task of controlling mobile robots (d'Andréa-Novel, Bastin & Campion 1995).

In this paper we will be concerned with nonlinear dynamical control systems of the form

$$\begin{aligned} \frac{d}{dt}x &= f(x) + \sum_{i=1}^{m} g_i(x)u_i, \quad x(0) = x_0 \\ y &= h(x) \end{aligned} \tag{1.1}$$

where $x \in \mathbb{R}^n$, the system is square $y, u \in \mathbb{R}^m$, and the functions $f, g_i : \mathbb{R}^n \to \mathbb{R}^n$ and $h : \mathbb{R}^n \to \mathbb{R}^m$ are smooth. To simplify the technical details we further restrict our consideration to analytic systems. For typical models of mobile robots (d'Andréa-Novel et al. 1995) one has that the drift term $f(x) \equiv 0$ is equivalent to zero. It is convenient in the following development however, to derive the results for the general case. We investigate a related problem to the question of of full state stabilization, that of output stabilization. For example, in the case of a mobile robot one may wish to "park" the robot at the origin but not care about the orientation of the robot wheels. Thus, the output would be the Euclidean coordinates describing the position of the robot while the orientation of the wheels e.t.c. would be additional states.

Since the systems of interest to this paper are square systems in which one has as many control inputs as outputs, we tackle the controller design by utilising some of the input-output decoupling algorithms available in the literature. In particular, we rely heavily on the dynamic extension algorithm which is outlined in Nijmeijer and Van der Schaft (1990, pp. 263-264). In the case where the *input/output decoupling* matrix (or just *decoupling* matrix (Nijmeijer & Van der Schaft 1990, eq. 8.24)) of the system is full rank it is a trivial matter to assign linear stable dynamics to the output. Once a control law has been designed for the task of output stabilization, however, it is then necessary to consider the question of internal stability for the closed loop system. In a recent paper (Mahony, Mareels, Campion & Bastin 1993) we gave a result that characterises a set of initial conditions for which internal stability of the closed loop dynamics (given the trivial linearising control law) is guaranteed. In this paper we consider the case where the decoupling matrix is not full rank and yet it is possible to assign stable dynamics to outputs by utilising higher order system dynamics using the dynamic extension algorithm. The specific task considered in the design procedure results in important modifications to the basic algorithm and provides a way to study the difficulties associated with designing internally stable control laws for non-holonomic systems. Given the design procedure outlined a result characterising a set of initial conditions for which the system will remain internally stable, analogous to that in (Mahony et al. 1993), is given. Finally, the methods outlined in the paper are applied to a simple dynamical model of a mobile robot.

2 Static Output Stabilizing Feedback Laws for Square Systems

In this section we consider the problem of designing a locally smooth static control law $u := u(x)$ which exponentially stabilizes the output of a system for which the decoupling matrix is rank degenerate. The following algorithm is valid for those systems for which the dynamic extension algorithm (Nijmeijer & Van der Schaft 1990, Section 8.2) can be used to design a *dynamic* feedback law to input output decouple the system.

Consider the output equation $y = h(x)$, (1.1), and let $\rho_1^1, \ldots, \rho_m^1$ be the characteristic numbers of the outputs $y = (y_1, \ldots, y_m)$ (Nijmeijer & Van der Schaft 1990, pg. 246). Thus one can write

$$\begin{pmatrix} y_1^{\rho_1^1+1} \\ \vdots \\ y_m^{\rho_m^1+1} \end{pmatrix} = E^1(x) + F^1(x)u$$

where $E^1(x) = (L_f^{\rho_1^1+1}h(x), \ldots, L_f^{\rho_m^1+1}h(x))^T$ and $F^1(x)$ is the decoupling matrix

$$F^1(x) = \begin{pmatrix} L_{g_1}L_f^{\rho_1^1}h_1(x) & \cdots & L_{g_m}L_f^{\rho_1^1}h_1(x) \\ \vdots & & \vdots \\ L_{g_1}L_f^{\rho_m^1}h_m(x) & \cdots & L_{g_m}L_f^{\rho_m^1}h_m(x) \end{pmatrix}.$$

Since the system (1.1) is analytic then the rank of $F^1(x)$ is constant except on a set of measure zero in $\mathbb{R}^n$. Denote the generic rank of $F^1(x)$ (off the set of measure zero) by r_1. It is convenient to refer to the set of points at which the various rank conditions required for the construction of the output stabilizing feedback law, do not hold as the set of *singular points* in $\mathbb{R}^n$. Thus, the set of singular points includes all points at which rank $F^1(x) \neq r_1$ as well as other points defined in the sequel. The set of singular points will always be of zero measure in $\mathbb{R}^n$.

Given a point $x \in \mathbb{R}^n$ which is not a singular point, reorder and relabel the output functions $h_1, \ldots, h_m$ to ensure that the first r_1 rows of $F^1(x)$ are linearly independent. Reorder the inputs u (and hence the columns of F^1) to ensure that the first r_1 columns of F_{11}^1 are also linearly independent. Partition the output into two parts, $(h_1, \ldots, h_{r_1})$ the outputs associated with the first r_1 linearly independent rows of $F^1(x)$ and $(h_{r_1+1}, \ldots, h_m)$ the remaining outputs. Correspondingly, partition the vector $E^1(x)$ into its first r_1 entries $E_1^1(x)$ and its remaining entries $E_2^1(x)$ and the matrix F^1 into four submatrices where the upper left submatrix $F_{11}^1(x) \in \mathbb{R}^{r_1 \times r_1}$ is square. Thus one has

$$\begin{pmatrix} y_1^{\rho_1^1+1} \\ \vdots \\ y_m^{\rho_m^1+1} \end{pmatrix} = \begin{pmatrix} E_1^1(x) \\ E_2^1(x) \end{pmatrix} + \begin{pmatrix} F_{11}^1(x) & F_{12}^1(x) \\ F_{21}^1(x) & F_{22}^1(x) \end{pmatrix} u.$$

By construction $F_{11}^1(x)$ is a full rank invertible matrix and the last $m - r_1$ columns of $F^1(x)$ are are linearly dependent on the first r_1 columns of $F^1(x)$. The linear dependence can be computed from the dependence of $F_{12}^1(x)$ on $F_{11}^1(x)$. Consider the input transformation

$$u(x, u^1, \overline{u}^1) := \begin{pmatrix} -(F_{11}^1)^{-1}E_1^1 \\ 0 \end{pmatrix} + \begin{pmatrix} (F_{11}^1)^{-1} & -(F_{11}^1)^{-1}F_{12}^1 \\ 0 & I_{m-r_1} \end{pmatrix} \begin{pmatrix} u^1 \\ \overline{u}^1 \end{pmatrix}, \tag{2.1}$$

where the new inputs are denoted $(u^1, \overline{u}^1) \in \mathbb{R}^{r_1} \times \mathbb{R}^{m-r_1}$. Applying this feedback to (1.1) yields the following output dynamics

$$\begin{pmatrix} y_1^{\rho_1^1+1} \\ \vdots \\ y_m^{\rho_m^1+1} \end{pmatrix} = \begin{pmatrix} 0 \\ E_2^1 - F_{21}^1(F_{11}^1)^{-1}E_1^1 \end{pmatrix} + \begin{pmatrix} I_{r_1} & 0 \\ F_{21}^1(F_{11}^1)^{-1} & 0 \end{pmatrix}\begin{pmatrix} u^1 \\ \overline{u}^1 \end{pmatrix}.$$

Thus, after applying (2.1) the first r_1 outputs are input-output decoupled to the new inputs $u^1 \in \mathbb{R}^{r_1}$ and fully decoupled from the remaining inputs $\overline{u}^1$.

It is at this point that the present development differs significantly from the standard dynamic extension algorithm. Since our aim is simply to stabilize the output then it is possible to specify the first r_1 inputs u^1 to stabilize the first r_1 outputs explicitly. Define the inputs $u^1 = (u_1^1, \ldots, u_{r_1}^1)$ as follows

$$\begin{aligned} u_i^1(x) &= -\sum_{j=0}^{\rho_i^1} C_j^{\rho_i^1+1}\frac{d^j h_i}{dt^j}(x), \quad i = 1,\ldots,r_1, \\ &= -\sum_{j=0}^{\rho_i^1} C_j^{\rho_i^1+1} L_f^j h_i(x) \qquad (2.2) \end{aligned}$$

where $C_b^a = \frac{a!}{b!(a-b)!}$. Choosing $u_i^1(x)$ as given above ensures that $y_i(t)$ satisfies the linear homogeneous differential equation

$$\left(\frac{d}{dt}+1\right)^{\rho_i^1+1} y_i(t) = 0,$$

which has solution

$$y_i(t) = (c_i^0 + c_i^1 t + \cdots + c_i^{\rho_i^1} t^{\rho_i^1})e^{-t}$$

for constants $c_i^0 = y_i(0)$,

$$c_i^p = \frac{1}{p!}\sum_{j=0}^{p-1} C_j^p \frac{d^j y_i}{dt^j}(0), \quad \text{for } p = 1,\ldots,\rho_i^1. \quad (2.3)$$

Thus, one may define a new (partly closed-loop) system with inputs $u' \in \mathbb{R}^{m-r_1}$ and outputs $y' \in \mathbb{R}^{m-r_1}$

$$\dot{x} = f(x) + g(x)\left[\begin{pmatrix} -(F_{11}^1)^{-1}E_1^1 \\ 0 \end{pmatrix} + \begin{pmatrix} (F_{11}^1)^{-1} & -(F_{11}^1)^{-1}F_{12}^1 \\ 0 & I_{m-r_1} \end{pmatrix}\begin{pmatrix} u^1(x) \\ u' \end{pmatrix}\right], \quad (2.4)$$

$$\begin{pmatrix} y_1 \\ \vdots \\ y_{r_1} \end{pmatrix} = \begin{pmatrix} \sum_{j=0}^{\rho_1^1} c_1^j t^j e^{-t} \\ \vdots \\ \sum_{j=0}^{\rho_{r_1}^1} c_{r_1}^j t^j e^{-t} \end{pmatrix},$$

$$\begin{aligned} y' &= \begin{pmatrix} y_{r_1+1}^{\rho_{r_1+1}^1+1} \\ \vdots \\ y_m^{\rho_m^1+1} \end{pmatrix} \\ &= E_2^1(x) - F_{21}^1(x)(F_{11}^1(x))^{-1}E_1^1(x) + \\ &\quad F_{21}^1(x)(F_{11}^1(x))^{-1}u^1(x). \qquad (2.5) \end{aligned}$$

The new inputs $u' \in \mathbb{R}^{m-r_1}$ are just the inputs $\overline{u}^1$ that were not assigned in the first step of the algorithm while the new outputs y' are the highest order time derivatives of the unstabilized outputs that were obtained in the previous step. Observe that the new inputs u' do not enter directly into the output equation 2.5.

Remark 2.1 It is most convenient to continue subscripting both remaining inputs and outputs as though they are parts of the full input and output vectors. Thus the new input vector in $u' = (u_{r_1+1}, \ldots, u_m)$ and the output vector is analogous. Recall that these labels do not necessarily correspond to the original labels due to the relabelling and reordering procedures. □

To obtain a full control law it simply a question of applying the same step to the newly defined system (2.4) and (2.5). The new characteristic numbers are denoted ρ_i^2, $i = r_1+1,\ldots,m$ and the matrices E^2 and F^2 are analogous to those derived above. Certainly, the new system is analytic except on the set of singular points (in particular those where F^1 is rank deficient) and the generic rank of F^2 is denoted r_2. One includes all the points at which F^2 is rank deficient in the set of singular points and considers only points $x \in \mathbb{R}^n$ which do not lie in this set. An analogous input transformation to that used above will input-output decouple the first r_2 outputs of y', namely $(y_{r_1+1}, \ldots, y_{r_1+r_2})$ to the first r_2 inputs of u', denoted $u^2 = (u_{r_1+1}, \ldots, u_{r_1+r_2})$, as well as fully decoupling $(y_{r_1+1}, \ldots, y_{r_1+r_2})$ from the remaining inputs $\overline{u}^2 = (u_{r_1+r_2+1}, \ldots, u_m)$.

It is convenient at this stage to give the general form for the feedback used. Assume that one has assigned $k-1$ sets of inputs $u^1(x)$, $u^2(x), \ldots, u^{k-1}(x)$ (and consequently $k-1$ sets of outputs. Let

$$\Upsilon_i^k = \rho_i^1 + \rho_i^2 + \cdots + \rho_i^k,$$

for $i = (r_1 + \cdots + r_{k-1} + 1), \ldots, (r_1 + \cdots + r_k)$. Specify the k'th set of inputs as follows

$$u_i^k(x) = -\sum_{j=0}^{\Upsilon_i^k} C_j^{\Upsilon_i^k+1} k^{(\Upsilon_i^k+1-j)}\frac{d^j}{dt^j}h_i(x), \quad (2.6)$$

for $i = (r_1 + \cdots + r_{k-1} + 1), \ldots, (r_1 + \cdots + r_k)$. Choosing $u_i^k(x)$ as given above ensures that $y_i(t)$ satisfies the linear O.D.E.

$$\left(\frac{d}{dt}+k\right)^{\Upsilon_i^k+1} y(t) = 0,$$

which has solution

$$y_i(t) = e^{-kt} \sum_{j=0}^{\Upsilon_i^k} c_i^j t^j, \qquad (2.7)$$

for $i = (r_1 + \cdots + r_{k-1} + 1), \ldots, (r_1 + \cdots + r_k)$ and for constants c_i^j given by (2.3)

As a consequence of this choice the k'th set of outputs converge to zero strictly faster (exponentially with rate $-k$) than the previous sets of outputs (at best exponential with rate $-(k-1)$). This choice is made in the hope that the later outputs, most prone to numerical ill-conditioning (following from repeated singular input transformations), converge before any ill-conditioning in the algorithm occurs.

To fully determine the inputs one continues to apply the above procedure until all the inputs have been determined. Since the system has been assumed to be full rank (Nijmeijer & Van der Schaft 1990, Theorem 8.19) then it follows that each output can eventually be controlled by an input. Denote the total number of steps taken to assign every output as K. Then the evolution of the output of the system is given by

$$\begin{pmatrix} y_1(t) \\ \vdots \\ y_{r_1}(t) \\ y_{r_1+1}(t) \\ \vdots \\ y_m(t) \end{pmatrix} = \begin{pmatrix} \sum_{j=0}^{\rho_1^1} c_1^j t^j e^{-t} \\ \vdots \\ \sum_{j=0}^{\rho_{r_1}^1} c_{r_1}^j t^j e^{-t} \\ \sum_{j=0}^{\Upsilon_{r_1+1}^2} c_{\Upsilon_{r_1+1}^1}^j t^j e^{-2t} \\ \vdots \\ \sum_{j=0}^{\Upsilon_m^K} c_m^j t^j e^{-Kt} \end{pmatrix}$$

One of the consequences of focussing on the design of a practical control law using the algorithm outlined in this section is the structure of the set of singular points

$$\mathcal{S} = \{x \in \mathbb{R}^n \mid \det F_{11}^k(x) = 0, k \in \{1, 2, \ldots, K\}\}$$

in state space. On this set of points the above algorithm fails and one would expect that the output laws themselves would become unbounded in the vicinity of $\mathcal{S}$. This issue is not only relevant for the algorithm outlined above, but it is a fundamental limitation of any algorithm that relies on the decomposition of input-output dependence used by the dynamic extension algorithm. Observe, however, that in the method outlined above the new singularities introduced at each step, following from the equation $\det F_{11}^k(x) = 0$, are explicitly known before the control law is constructed. In this manner one can design the control to achieve the control objective as well as avoid the singularities which would cause the system to become unstable internally. The choice of control law given earlier appears to achieve this aim effectively in the case studies we have completed to date. A further observation is that the set of singular points is dependent on the choices made during the algorithm of which outputs to stabilize at each step and which inputs to use. This dependence is shown nicely by the example in Sections 3.

The algorithm outlined above results in a control that assigns exponentially stable dynamics to the output. It is of interest to obtain results guaranteeing the internal stability of a closed system where the controller was designed using this methodology. Though a global result appears difficult we have used the Bellman-Gronwall lemma to exploit the exponential convergence of the outputs to obtain a local stability result in the vicinity of the zero level set $\{x \mid h(x) = 0\}$. Unfortunately, space restrictions prevent us including the proof in the present publication.

Theorem 2.2 *Consider a square analytic system*

$$\begin{aligned} \frac{d}{dt}x &= \sum_{i=1}^{m} g_i(x)u_i, \quad x(0) = x_0 \\ y &= h(x) \end{aligned}$$

where $x \in \mathbb{R}^n$ and $y, u \in \mathbb{R}^m$. Assume that the system is full rank and use the algorithm outlined above to design an output stabilizing feedback law $u := u(x)$. Assume moreover, that there exists a neighbourhood $W \subseteq \mathbb{R}^n$ of the set $y \equiv 0$ which contains no singular points of the output feedback algorithm. Then there exists an open neighbourhood $\Omega \subset W$ of the set $y \equiv 0$ such that for any initial condition $x_0 \in \Omega$ the solution $x(t; x_0)$ to the closed loop system, with initial condition x_0, is well defined and bounded for all time and the output $h(x(t)) \to 0$ as $t \to \infty$.

3 The Unicycle

In this section a simple example is presented which indicates the manner in which the algorithm presented in Section 2 is applied. The system considered is a simple mobile robot, commonly known as the unicycle robot since it has the same dynamic behaviour as a mobile robot with a single orientable and driven wheel (accompanied, of course, by sufficient free support wheels to keep the robot upright). The robot is also often referred to as the box car robot.

Consider the unicycle robot shown in Figure 3.1. Denote its Euclidean position in $\mathbb{R}^2$ by (x, y) and its orientation (angle from x-axis to forward direction of the robot) by the angle α expressed as radians. One may write the system equations for the unicycle robot as follows (Canudas de Wit & Sordalen 1991, d'Andréa-Novel et al. 1995)

$$\frac{d}{dt}\begin{pmatrix} x \\ y \\ \alpha \end{pmatrix} = \begin{pmatrix} \cos(\alpha) \\ \sin(\alpha) \\ 0 \end{pmatrix} u_1 + \begin{pmatrix} 0 \\ 0 \\ 1 \end{pmatrix} u_2$$

The outputs in which one is interested are the Euclidean coordinates (x, y). The control task is to drive the output,

$$h(x, y, \alpha) = \begin{pmatrix} x \\ y \end{pmatrix} \rightarrow \begin{pmatrix} 0 \\ 0 \end{pmatrix},$$

to zero while the full state (x, y, α) remains bounded or at least well defined.

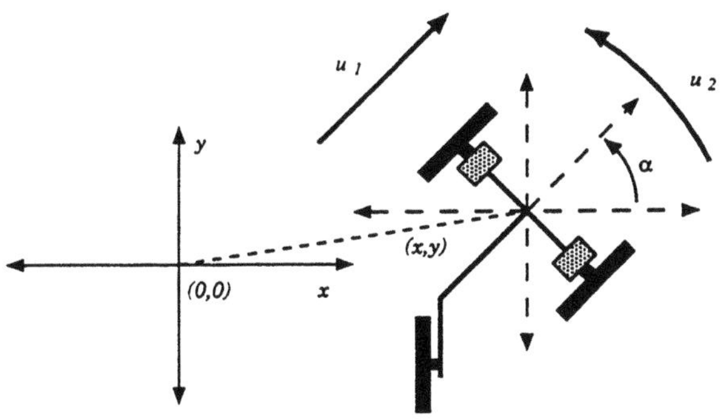

Figure 3.1: The Unicycle Robot, with forward velocity u_1 and angular velocity u_2.

Observe that the system cannot be input-output decoupled using static state feedback since

$$\begin{aligned} \begin{pmatrix} \dot{h}_1(x, y, \alpha) \\ \dot{h}_2(x, y, \alpha) \end{pmatrix} &= Dh(x, y, \alpha) g(x, y, \alpha) \\ &= \begin{pmatrix} \cos(\alpha) & 0 \\ \sin(\alpha) & 0 \end{pmatrix} \end{aligned}$$

is rank deficient. Following the algorithm presented in Section 2 one observes that as long as $\alpha \neq \frac{\pi}{2} + q\pi$ for any integer q (and hence $\cos(\alpha) \neq 0$) then no reordering of inputs nor an input transformation need be applied to input-output decouple the first output and the first input. According to (2.2) set

$$u_1(x, y, \alpha) = -\frac{x}{\cos(\alpha)}. \qquad (3.1)$$

The dynamics of the system become

$$\begin{aligned} \dot{x} &= -x \\ \dot{y} &= -x \tan(\alpha) \\ \dot{\alpha} &= u_2. \end{aligned}$$

The set of singular point contains all the points

$$\{(x, y, \alpha) \mid \alpha = \frac{\pi}{2} + q\pi, \text{ for } q \text{ any integer}\} \subset \mathcal{S}.$$

Define a new output function $h'(x, y, \alpha)$ to be

$$h'(x, y, \alpha) = -x \tan(\alpha) = \dot{y}.$$

Observe that

$$\begin{aligned} &Dh'(x, y, \alpha) g(x, y, \alpha) \\ &\quad = \begin{pmatrix} -\tan(\alpha)\cos(\alpha) & -x(1 + \tan(\alpha)^2) \end{pmatrix} \end{aligned}$$

and consequently, recalling that $u_1 = \frac{-x}{\cos(\alpha)}$, one obtains

$$\frac{d^2}{dt^2} y = x \tan(\alpha) - x(1 + \tan(\alpha)^2) u_2.$$

Set

$$u_2 = \frac{1}{x(1 + \tan(\alpha)^2)} (4y - 3x \tan(\alpha)), \qquad (3.2)$$

according to (2.6) and as long as $x(1+\tan(\alpha)^2) \neq 0$. Of course $(1 + \tan(\alpha)^2) > 0$ for all α and the final description of the singular points is

$$\mathcal{S} = \{(x, y, \alpha) \mid \alpha = \frac{\pi}{2} + q\pi, \text{ for } q \in \mathbb{Z}, \text{ or } x = 0\}.$$

Off the set of singular points the closed loop dynamics of the system are

$$\begin{aligned} \dot{x} &= -x \\ \ddot{y} &= -4\frac{d}{dt} y - 4y = 4x \tan(\alpha) - 4y \\ \dot{\alpha} &= \frac{4y \cos(\alpha)^2}{x} - 3 \sin(\alpha) \cos(\alpha). \end{aligned}$$

The evolution of the output of the system will be

$$\begin{aligned} x(t) &= x(0) e^{-t} \\ y(t) &= (y(0) + (2y(0) + t\dot{y}(0))) e^{-2t}, \\ \dot{\alpha} &= \frac{8y(0) + 4\dot{y}(0)}{x(0)} e^{-t} \cos(\alpha)^2 \qquad (3.3) \\ &\quad - \frac{3 \sin(2\alpha)}{2}. \end{aligned}$$

Observe that $x(t) \neq 0$ unless $x(0) = 0$. Moreover, the non-homogeneous term in the O.D.E. for α decays exponentially. It is easily verified that $\alpha = \frac{\pi}{2}(2q - 1)$, $q \in \mathbb{Z}$, are stationary points for the non-homogeneous O.D.E. (3.3). It follows that the solution $\alpha(t)$ of (3.3) will remain in some bounded interval $(\frac{\pi}{2}(2p - 1), \frac{\pi}{2}(2p + 1))$, where the integer p is determined by the initial condition α_0. Classical O.D.E. theory now ensures that the solution to the closed loop system exists for all time. The non-homogeneous term in (3.3) is exponentially decaying with time and the limiting dynamics will be given by the solution of the homogeneous O.D.E.

$$\dot{\alpha} = -\frac{3 \sin(2\alpha)}{2}.$$

Thus, $\alpha(t) \rightarrow p\pi$, (the only attractive equilibrium in the domain $(\frac{\pi}{2}(2p - 1), \frac{\pi}{2}(2p + 1))$,) as $t \rightarrow \infty$.

Remark 3.1 If $\cos(\alpha(0))$ is approximately zero then one may choose to stabilize the output y first and then stabilize x. The control laws generated by this approach are

$$\begin{aligned} u_1 &= \frac{-y}{\sin(\alpha)} \\ u_2 &= \frac{1}{y(1 + \cot(\alpha)^2)} (4x - 3y \cot(\alpha)) \end{aligned}$$

which are valid for $\alpha \neq \pi + q\pi$ for any integer q and $y \neq 0$. □

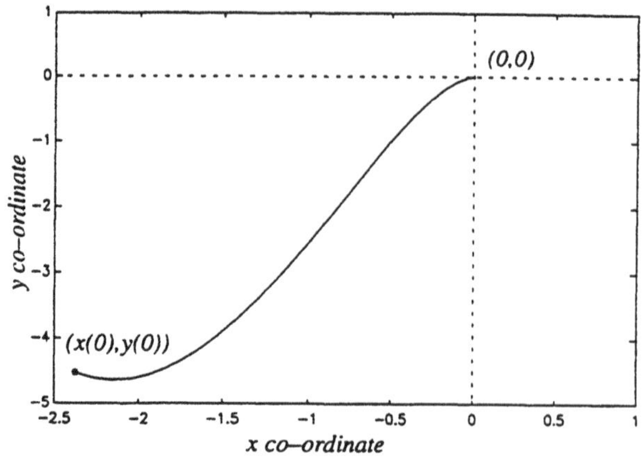

Figure 3.2: Plot of the evolution of the unicycle robot in Euclidean 2-space for Simulation 1. Initial condition $(-2.3755, -4.5254, 0.7515\pi)$.

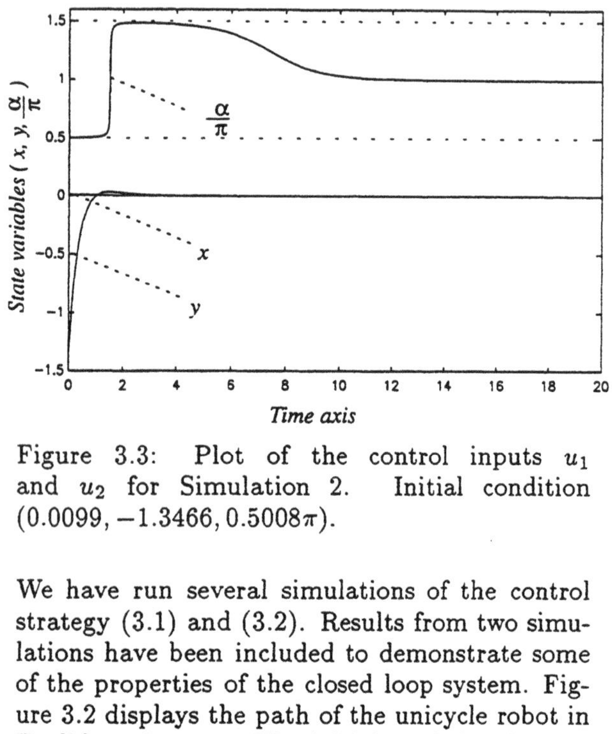

Figure 3.3: Plot of the control inputs u_1 and u_2 for Simulation 2. Initial condition $(0.0099, -1.3466, 0.5008\pi)$.

We have run several simulations of the control strategy (3.1) and (3.2). Results from two simulations have been included to demonstrate some of the properties of the closed loop system. Figure 3.2 displays the path of the unicycle robot in Euclidean 2-space. The initial condition for this example was

$$(x(0), y(0), \alpha(0)) = (-2.3755, -4.5254, 0.7515\pi).$$

Observe that the solution of the closed loop system does not pass through any singular points. Consequently the control scheme provides smooth bounded control laws. Indeed, even in the vicinity of singularities the closed behaviour is reasonable. In the second example, Figure 3.3, the initial condition was deliberately chosen to be nearly singular, (both $\cos(\alpha(0)) \approx 0$ and $x \approx 0$),

$$(x(0), y(0), \alpha(0)) = (0.0099, -1.3466, 0.5008\pi).$$

In particular, observe that despite the presence of numerical ill-conditioning (due to nearly singular input transformations) the behaviour of the system is smooth. Figure 3.3 also provides an excellent picture of the dynamics in the orientation α. Observe that $\alpha(t)$ remains in the interval $(\frac{\pi}{2}, \frac{3\pi}{2})$ for all time and that once the non-homogeneous terms of (3.3) have died away $t \geq 6$ then $\alpha(t) \longrightarrow \pi$ converges exponentially to π.

References

Brockett, R. W. (1983). Asymptotic stability and feedback stabilisation, *in* R. Brockett, R. Millman & H. Sussman (eds), *Differential Geometric Control Theory*, Vol. 27 of *Progress in Mathematics*, Birkhauser, Boston, USA.

Canudas de Wit, C. & Sordalen, O. J. (1991). Exponential stabilization of mobile robots with nonholonomic constraints, *Proceedings of the 30'th Conference on Decision and Control*, Brighton, England, pp. 692–697.

Coron, J.-M. & Rosier, L. (1994). A relation between continuous time-varying and discontinuous feedback stabilization, *Mathematics of Mathematical Systems, Estimation and Control* **4**(1): 67–84.

d'Andréa-Novel, B., Bastin, G. & Campion, G. (1995). Control of nonholonomic wheeled mobile robots by state feedback linearization, International Journal of Robotic Systems.

Kolmanovsky, I. V., Reyhanoglu, M. & McClamroch, H. (1994). Discontinuous feedback stailization of non-holonomic systems in power form, *Proceedings of the 33'rd Conference on Decision and Control.*

Mahony, R. E., Mareels, I. M., Campion, G. & Bastin, G. (1993). Output regulation for systems linear in the input, *Conference on Mathematical Theory of Networks and Systems*, Regensburg, Germany.

Nijmeijer, H. & Van der Schaft, A. (1990). *Nonlinear Dynamical Control Systems*, Springer Verlag, New York, USA.

Pomet, J.-B. (1992). Explicit design of time-varying stabilizing control laws for a class of controllable systems without drift, *Systems and Control Letters* **18**: 147–158.

DYNAMIC DISTURBANCE DECOUPLING FOR DISCRETE-TIME NONLINEAR SYSTEMS: A LINEAR ALGEBRAIC SOLUTION

E. ARANDA-BRICAIRE* and Ü. KOTTA**

* *CINVESTAV-IPN, Department of Electrical Engineering, Apartado Postal 14-740, 07000 Mexico, D.F., Mexico. E-mail: aranda@ctrl.cinvestav.mx*

** *Institute of Cybernetics, Akadeemia tee 21, EE0026, Estonia. E-mail: kotta@ioc.ee*

Abstract. The paper presents a linear algebraic solution of the dynamic disturbance decoupling problem for a submersive discrete-time nonlinear system. The solution is looked for in the class of regular (invertible) dynamic compensators. Necessary and sufficient conditions in terms of certain subspaces, defined by the system, are given both for the cases of measurable and unmeasurable input disturbances. The linear-algebraic solvability conditions are shown to be equivalent to the earlier conditions, formulated in terms of certain functions appearing in the inversion algorithm. The equivalence relies on the fact that the two versions of the inversion algorithm for systems with disturbances produce the basis for the above subspaces. Moreover, we establish that when the disturbance decoupling problem is solvable by a dynamic state feedback, then it is solvable by a quasi-static state feedback.

Résumé. Dans cet article nous présentons une solution au problème du rejet de perturbation par bouclage dynamique pour les systèmes non linéaires en temps discret. Nous considérons deux versions du problème : celle où la perturbation n'est pas mesurable et celle où la perturbation peut être mesurée et utilisée comme une entrée du compensateur dynamique. Nous étudions les systèmes génériquement submersifs et ne considérons que les compensateurs réguliers. La solution que nous proposons est donnée en termes de certains sous-espaces qui sont des invariants du système. Nous montrons que les conditions nécessaires et suffisantes que l'on obtient sont équivalentes aux conditions obtenues précedemment en utilissant l'algorithme d'inversion. Finalement, nous montrons que le problème du rejet de perturbation peut se résoudre par bouclage dynamique si et seulement si il peut se résoudre par bouclage quasi-statique.

Key Words. Discrete-time systems; nonlinear control systems; disturbance decoupling; linear algebraic methods; dynamic state feedback; quasi-static state feedback

1. INTRODUCTION

In this paper we shall consider the disturbance decoupling problem by the dynamic state feedback for discrete-time nonlinear generically submersive systems. The problem deals with the situation in which by means of dynamic state feedback compensator we want to achieve decoupling between the input disturbances entering the system and the outputs leaving the system. Clearly, the dynamic disturbance decoupling problem (DDDP) is essentially a nonlinear problem: for linear systems the DDDP is solvable if and only if the disturbance decoupling problem is solvable by static feedback (Huijberts *et al.*, 1992). The problem has been extensively studied for continuous-time systems (Huijberts *et al.*, 1991; Huijberts *et al.*, 1992; Huijberts, 1992; Respondek, 1991; Perdon *et al.*, 1993; Delaleau and Pereira da Silva, 1994; Pereira da Silva and Pinto Leite, 1993) as well as for discrete-time systems (Kotta and Nijmeijer, 1991; Kotta, 1992; Kotta, 1994a; Fliegner and Nijmeijer, 1994). The complete solution to the problem in the discrete-time case has been formulated either in terms of certain functions appearing in the inversion algorithm (Kotta, 1992) or in terms of the invertibility indices (Kotta, 1994a). Though the invertibility indices may be defined via the inversion algorithm, by the results of Grizzle (1993) they are actually system-intrinsic and algorithm-independent integer parameters of the system.

The purpose of this paper is to present a linear algebraic solution for the DDDP, i.e. a solution presented in terms of certain subspaces defined by the original system. The theory we develop extends the results of the recent paper (Perdon *et al.*, 1993) to the discrete-time case. Along with the case of unmeasurable disturbances we shall also consider the case when the disturbances can be measured. Note that the linear algebraic approach in the discrete-time case has been first applied in (Grizzle, 1993) for studing the invertibility problem; later this approach has been applied for studing the feedback linearization problem (Aranda-Bricaire *et al.*, 1994a; Aranda-Bricaire *et al.*, 1994b) and input-output decoupling problem (Kotta, 1993).

2. LINEAR ALGEBRAIC FRAMEWORK

Consider a discrete-time nonlinear system Σ, described by equations of the form

$$\begin{aligned} x(t+1) &= f(x(t), u(t), w(t)),\ x(0) = x_0, \\ y(t) &= h(x(t)), \end{aligned} \tag{1}$$

where the state $x(t) \in \mathbb{R}^n$, the control $u(t) \in \mathbb{R}^m$, the disturbance (either measurable or unmeasurable) $w(t) \in \mathbb{R}^r$ and the output $y(t) \in \mathbb{R}^p$. The mappings $f(\cdot)$ and $h(\cdot)$ are supposed to be real analytic. The linear algebraic framework, that we describe below, was extended by Grizzle (1993) for discrete-time nonlinear systems.

Let $\mathcal{R}$ denote the ring of analytic functions in a finite number of the variables $\{x(0), u(t), w(t), t \geq 0\}$ and let $\mathcal{K}$ be its associated quotient field, i.e. the field of meromorphic functions in a finite number of the variables $\{x(0), u(t), w(t), t \geq 0\}$.

Throughout the paper it is also assumed that the mapping $f(\cdot)$ generically defines a submersion, i.e. that

$$\text{rank}_{\mathcal{K}} \frac{\partial f}{\partial(x, u)} = n.$$

Over the field $\mathcal{K}$ one can define a difference vector space (Fliess, 1990), $\mathcal{E} := \text{span}_{\mathcal{K}}\{\text{d}\varphi, \varphi \in \mathcal{K}\}$. The space $\mathcal{E}$ can be decomposed into the direct sum of three subspaces, $\mathcal{E} = \mathcal{X} \oplus \mathcal{U} \oplus \mathcal{W}$, where

$$\begin{array}{rcl} \mathcal{X} & := & \text{span}_{\mathcal{K}}\{\text{d}x(0)\} \\ \mathcal{U} & := & \text{span}_{\mathcal{K}}\{\text{d}u(k),\ k \geq 0\} \\ \mathcal{W} & := & \text{span}_{\mathcal{K}}\{\text{d}w(k)\ , k \geq 0\}. \end{array} \tag{2}$$

Define the difference output space

$$\mathcal{Y} := \text{span}_{\mathcal{K}}\{\text{d}y(k),\ k \geq 0\}. \tag{3}$$

By (1), $\mathcal{Y} \subset \mathcal{E} = \mathcal{X} \oplus \mathcal{U} \oplus \mathcal{W}$.

The forward-shift operator $\delta : \mathcal{R} \to \mathcal{R}$ is defined by

$$\begin{array}{l} \delta\varphi(x(0), u(j), w(k)) = \\ \varphi(f(x(0), u(0), w(0)), u(j+1), w(k+1)). \end{array}$$

The operator δ induces a forward-shift operator $\Delta : \mathcal{E} \to \mathcal{E}$ by

$$\Delta(\sum_i a_i \text{d}\varphi_i) \mapsto \sum_i (\delta a_i)\text{d}(\delta\varphi_i), \quad a_i, \varphi_i \in \mathcal{K}.$$

3. PROBLEM FORMULATION

We shall consider separately the cases of unmeasurable and measurable disturbances. In the case of unmeasurable disturbances the compensator (dynamic state feedback) C used to control the system Σ is a discrete-time nonlinear system described by the equations of the form

$$\begin{array}{rcl} z(t+1) & = & \psi(z(t), x(t), v(t)), \ z(0) = z_0, \\ u(t) & = & \varphi(z(t), x(t), v(t)), \end{array} \tag{4}$$

with the state $z(t) \in \mathbb{R}^{\mu}$, with a new control $v(t) \in \mathbb{R}^m$ and with real analytic $\psi(\cdot)$ and $\varphi(\cdot)$.

We call the compensator C described by equations (4) regular, if the dynamical system

$$\begin{array}{rcl} x(t+1) & = & f(x(t), \varphi(z(t), x(t), v(t)), w(t)) \\ z(t+1) & = & \psi(z(t), x(t), v(t)) \\ u(t) & = & \varphi(z(t), x(t), v(t)), \end{array} \tag{5}$$

with inputs $v(t)$ and outputs $u(t)$ is invertible (see (Fliess, 1990; Grizzle, 1993; Kotta, 1990) for details about the notion of invertibility), or equivalently, that (5) defines a one-to-one (x, z, w)-dependent correspondence between the input variable $v(t)$ and output variable $u(t)$.

The closed-loop system (1),(4), initialized at (x_0, z_0) is denoted by $\Sigma \circ C$. For the closed-loop system $\Sigma \circ C$, define $\mathcal{V} = \text{span}_{\mathcal{K}}\{\text{d}v(k),\ k \geq 0\}$ and $\mathcal{Z} = \text{span}_{\mathcal{K}}\{\text{d}z(0)\}$.

Definition 1 (Regular dynamic disturbance decoupling problem in the case of unmeasurable disturbances (**RDDDPud**)). Find, if possible, a regular dynamic state feedback of the form (4) such that the difference output space $\mathcal{Y}^*$ of the closed-loop system $\Sigma \circ C$ satisfies $\mathcal{Y}^* \subset \mathcal{Z} \oplus \mathcal{X} \oplus \mathcal{V}$.

If the disturbances $w(t)$ are available for measurements, one can use the compensator (dynamic state feedback) C_w described by the equations

$$\begin{array}{rcl} z(t+1) & = & \psi(z(t), x(t), v(t), w(t)), \ z(0) = z_0, \\ u(t) & = & \varphi(z(t), x(t), v(t), w(t)), \end{array} \tag{6}$$

which besides to a feedback includes a feedforward on the disturbance $w(t)$. If this is the case, then decoupling the output from the disturbances is possible under conditions that are probably weaker than those for the case of unmeasurable disturbances.

The regularity of (6) is defined analogously to the regularity of (4). The closed-loop system (1),(6), initialized at (x_0, z_0) is denoted by $\Sigma \circ C_w$. With a slight abuse of notation, we also denote $\mathcal{V} = \text{span}_{\mathcal{K}}\{\text{d}v(k),\ k \geq 0\}$, $\mathcal{Z} = \text{span}_{\mathcal{K}}\{\text{d}z(0)\}$.

Definition 2 (Regular dynamic disturbance decoupling problem in the case of measurable disturbances (**RDDDPmd**)). Find, if possible, a regular dynamic state feedback of the form (6) such that the difference output space $\mathcal{Y}^*$ of the closed-loop system $\Sigma \circ C_w$ satisfies $\mathcal{Y}^* \subset \mathcal{Z} \oplus \mathcal{X} \oplus \mathcal{V}$.

4. THE INVERSION ALGORITHM FOR SYSTEMS WITH DISTURBANCES

The inversion algorithm for discrete-time nonlinear systems with disturbances, presented in (Kotta, 1994b), involves the solution of systems of nonlinear equations and therefore, in general, asks for the use of the Implicit Function Theorem. In this section we shall give another form of the algorithm, adapted from (Grizzle, 1993) where the systems without disturbances were considered. The basic idea of Grizzle (1993) was to work with differentials $\text{d}y(t)$ of the output $y(t)$ instead of working with the outputs themselves, thus linearizing the computations. Note that for the system with disturbances two versions of the algorithm can be considered (Moog *et al.*, 1991; Kotta, 1994b). The first version accomplishes inversion with respect to both types of inputs, the controls and the disturbances, whereas the other version considers disturbances as system parameters and accomplishes inversion with respect to the controls only. If we define $\nu = (u, w)$, then applying the first version of the inversion algorithm to system (1) means just applying the algorithm given in (Grizzle, 1993) with respect to ν. Here we present the second version of the inversion algorithm for system (1).

Denote

$$d\hat{y}_0(0) = dy(0) = \frac{\partial h}{\partial x(0)} dx(0) = \bar{a}_0(x(0))dx(0)$$

Step 1: Compute

$$\begin{aligned} d\hat{y}_0(1) &= \delta\bar{a}_0(x(0))dx(1) \\ &= \bar{a}_0(f(x(0),u(0),w(0))\left[\tfrac{\partial f}{\partial x(0)}dx(0) + \right. \\ &\quad \left. + \tfrac{\partial f}{\partial u(0)}du(0) + \tfrac{\partial f}{\partial w(0)}dw(0)\right] \\ &= a_1 dx(0) + b_1 du(0) + c_1 dw(0), \end{aligned} \tag{7}$$

where $a_1, b_1, c_1 \in \mathcal{K}$. Define $\rho_1 = \text{rank}_{\mathcal{K}} b_1$. Permute, if necessary, the components of $y(t)$ so that the first ρ_1 rows of b_1 are linearly independent. Decompose $y(1)$ so that

$$\begin{aligned} d\tilde{y}_1(1) &= \tilde{a}_1 dx(0) + \tilde{b}_1 du(0) + \tilde{c}_1 dw(0) \\ d\hat{y}_1(1) &= \hat{a}_1 dx(0) + \hat{b}_1 du(0) + \hat{c}_1 dw(0), \end{aligned}$$

where $\tilde{y}_1(1)$ has ρ_1 elements. Since the rows of $\hat{b}_1$ are $\mathcal{K}$-dependent on the rows of $\tilde{b}_1$, there exists a matrix $M_1(x(0), u(0), w(0))$ with entries in $\mathcal{K}$ such that $\hat{b}_1 = M_1\tilde{b}_1$ and thus

$$\begin{aligned} d\hat{y}_1(1) &= \hat{a}_1 dx(0) + M_1\left[d\tilde{y}_1(1) - \tilde{a}_1 dx(0) - \right. \\ &\quad \left. - \tilde{c}_1 dw(0)\right] + \hat{c}_1 dw(0) \\ &= \bar{a}_1 dx(0) + \bar{c}_1 dw(0) + \bar{e}_1^{11} d\tilde{y}_1(1). \end{aligned}$$

Step $k+1$ ($k \geq 0$): Suppose that in steps 1 through k, $\tilde{y}_1(1), \tilde{y}_2(2), \ldots, \tilde{y}_k(k), \hat{y}_k(k)$ have been defined so that

$$\begin{aligned} d\tilde{y}_1(1) &= \tilde{a}_1 dx(0) + \tilde{b}_1 du(0) + \tilde{c}_1 dw(0) \\ &\vdots \\ d\tilde{y}_k(k) &= \tilde{a}_k dx(0) + \tilde{b}_k du(0) + \textstyle\sum_{j=0}^{k-1} \tilde{c}_k^j dw(j) \\ &\quad + \textstyle\sum_{i=1}^{k-1}\sum_{j=i+1}^{k} \tilde{e}_k^{ij} d\tilde{y}_i(j) \\ d\hat{y}_k(k) &= \bar{a}_k dx(0) + \textstyle\sum_{j=0}^{k-1} \bar{c}_k^j dw(j) \\ &\quad + \textstyle\sum_{i=1}^{k}\sum_{j=i}^{k} \bar{e}_k^{ij} d\tilde{y}_i(j). \end{aligned}$$

Define $B_k = (\tilde{b}_1^T, \ldots, \tilde{b}_k^T)^T$ and let $\rho_k = \text{rank}_{\mathcal{K}} B_k$. Compute

$$\begin{aligned} d\hat{y}_k(k+1) &= (\delta\bar{a}_k)dx(1) + \textstyle\sum_{j=0}^{k-1} \delta\bar{c}_k^j dw(j+1) \\ &\quad + \textstyle\sum_{i=1}^{k}\sum_{j=i}^{k} \delta\bar{e}_k^{ij} d\tilde{y}_i(j+1) \\ &= a_{k+1}dx(0) + b_{k+1}du(0) \\ &\quad + \textstyle\sum_{j=0}^{k} c_{k+1}^j dw(j) + \\ &\quad \textstyle\sum_{i=1}^{k}\sum_{j=i}^{k} e_{k+1}^{ij} d\tilde{y}_i(j+1) \end{aligned}$$

and define $\rho_{k+1} = \text{rank}_{\mathcal{K}}[B_k^T, b_{k+1}^T]^T$. Permute, if necessary, the components of $\hat{y}_k(k+1)$ so that the first ρ_{k+1} rows of the matrix $[B_k^T, b_{k+1}^T]^T$ are linearly independent. Decompose $\hat{y}_k(k+1)$ so that

$$\begin{aligned} d\tilde{y}_{k+1}(k+1) &= \tilde{a}_{k+1}dx(0) + \tilde{b}_{k+1}du(0) \\ &\quad + \textstyle\sum_{j=0}^{k} \tilde{c}_{k+1}^j dw(j) \\ &\quad + \textstyle\sum_{i=1}^{k}\sum_{j=i+1}^{k+1} \tilde{e}_{k+1}^{ij} d\tilde{y}_i(j) \\ d\hat{y}_{k+1}(k+1) &= \hat{a}_{k+1}dx(0) + \hat{b}_{k+1}du(0) \\ &\quad + \textstyle\sum_{j=0}^{k} \hat{c}_{k+1}^j dw(j) \\ &\quad + \textstyle\sum_{i=1}^{k}\sum_{j=i+1}^{k+1} \hat{e}_{k+1}^{ij} d\tilde{y}_i(j), \end{aligned}$$

where $\tilde{y}_{k+1}(k+1)$ has $\rho_{k+1} - \rho_k$ elements. Denote $B_{k+1} = [B_k^T, \tilde{b}_{k+1}^T]^T$. Since the rows of $\hat{b}_{k+1}$ are $\mathcal{K}$-dependent on the rows of B_{k+1}, there exists a matrix M_{k+1} with entries in $\mathcal{K}$ such that $\hat{b}_{k+1} = M_{k+1}B_{k+1}$ and thus

$$\begin{aligned} d\hat{y}_{k+1}(k+1) &= \bar{a}_{k+1}dx(0) + \textstyle\sum_{j=0}^{k} \bar{c}_{k+1}^j dw(j) \\ &\quad + \textstyle\sum_{i=1}^{k+1}\sum_{j=i}^{k+1} \bar{e}_k^{ij} d\tilde{y}_i(j). \end{aligned}$$

End of step $k+1$.

Define the rank ρ^* of the system as $\rho^* = \max\{\rho_k, k \geq 1\}$ (see (Fliess, 1990; Grizzle, 1993) for details about the notion of the rank of the system) and let α be defined as the smallest $k \in N$ such that $\rho_k = \rho^*$. Recall that $\alpha \leq n$ (Kotta, 1994a).

Note that in general the components of $\tilde{y}_k(k)$ and $\hat{y}_k(k)$, $k \geq 0$, may be different for two versions of the inversion algorithm. In order to distinguish these vectors in case of different versions, from now on we shall use the upper indices u or uw respectively.

5. A LINEAR ALGEBRAIC SOLUTION

5.1. *The case of unmeasurable disturbances*

In (Kotta, 1992) necessary and sufficient conditions for local solvability of the RDDDPud have been presented in terms of certain functions appearing in the inversion algorithm. By a little modification of the proof in (Kotta, 1992), the result can be formulated as in the next Theorem.

Theorem 3 Assume that the nonlinear system Σ is generically submersive and apply to it the inversion algorithm with respect to the control. Then the RDDDPud for system Σ is almost everywhere solvable if and only if for $k \geq 0$

$$\begin{aligned} d\hat{y}_k^u(k+1) &= a_{k+1}dx(0) + b_{k+1}du(0) \\ &\quad + \textstyle\sum_{i=1}^{k}\sum_{j=i}^{k} e_{k+1}^{ij} d\tilde{y}_i^u(j+1). \end{aligned} \tag{8}$$

The condition (8) just amounts to saying that for every k, the function $\hat{y}_k^u(k+1)$ that appears in the $(k+1)$-th step of the inversion algorithm does not depend on $\{w(j),\ 0 \leq j \leq k\}$. Our purpose is to give an equivalent linear algebraic condition of (8).

Proposition 4 The following two conditions are equivalent

(i) For $k \geq 0$,

$$d\hat{y}_k^u(k+1) = a_{k+1}dx(0) + b_{k+1}du(0) + \sum_{i=1}^k \sum_{j=i}^k e_{k+1}^{ij} d\tilde{y}_i^u(j+1).$$

(ii) $\Delta(\mathcal{X} \cap \mathcal{Y}) \subset \mathcal{X} \oplus \mathcal{U}$.

The proof of Proposition 4 lies essentially on the fact that the inversion algorithm, applied to the system Σ with respect to the both inputs (u, w) produces a basis for a subspace $\mathcal{X} \cap \mathcal{Y}$.

Lemma 5

$$\begin{aligned}&\mathcal{X} \cap \mathcal{Y} = \\ &= \text{span}_{\mathcal{K}}\{d\hat{y}_k^{uw}(k) - \textstyle\sum_{i=1}^k \sum_{j=i}^k \bar{e}_k^{ij} d\tilde{y}_i^{uw}(j),\ k \geq 0\} \\ &= \text{span}_{\mathcal{K}}\{\bar{a}_k dx(0),\ k \geq 0\}.\end{aligned}$$

Proof of Proposition 4. First note that by Lemma 5 another way of stating (ii) is to say that

$$d\hat{y}_k^{uw}(k+1) - \textstyle\sum_{i=1}^k \sum_{j=i}^k \delta\bar{e}_k^{ij} d\tilde{y}_i^{uw}(j+1) \in \text{span}_{\mathcal{K}}\{dx(0), du(0)\}.$$

Next, by the two versions of the inversion algorithm, $\hat{y}_0^u = \hat{y}_0^{uw}$, and $\delta\bar{e}_k^{ij} = e_{k+1}^{ij}$.

The condition (i) says that the functions which we compute in the course of the inversion algorithm with respect to the control u, do not depend on w. Analogously, the condition (ii) says the same in regard to the other version of the inversion algorithm. This implies that either under (i) or under (ii),

$$\tilde{y}_k^u = \tilde{y}_k^{uw},\quad \hat{y}_k^u = \hat{y}_k^{uw},\quad k \geq 1,$$

and completes the proof.

We can now formulate our main result.

Theorem 6 Assume that the nonlinear system Σ is generically submersive. Then for system Σ with unmeasurable disturbances the regular DDDP is solvable if and only if

$$\Delta(\mathcal{X} \cap \mathcal{Y}) \subset \mathcal{X} \oplus \mathcal{U}. \tag{9}$$

5.2. *The case of measurable disturbances*

In (Kotta, 1994a) necessary and sufficient conditions for local solvability of the RDDDPmd have been presented in terms of certain functions appearing in the inversion algorithm. By a little modification of the proof in (Kotta, 1994a), the result can be formulated as in the next theorem.

Theorem 7 Assume that the nonlinear system Σ is generically submersive. Then the RDDDPmd for system Σ is almost everywhere solvable if and only if for $k \geq 0$

$$d\hat{y}_k^u(k) = \bar{a}_k dx(0) + \sum_{i=1}^k \sum_{j=i}^k \bar{e}_k^{ij} d\tilde{y}_i^u(j). \tag{10}$$

The condition (10) just amounts to saying that for every k, the function $\hat{y}_k(k)$ that appears in the k-th step of the inversion algorithm, does not depend on $\{w(j),\ 0 \leq j \leq k-1\}$. Again, our purpose is to give an equivalent linear algebraic condition of (10).

Proposition 8 The following two conditions are equivalent.

(i) For $k \geq 0$,

$$d\hat{y}_k^u(k) = \bar{a}_k dx(0) + \sum_{i=1}^k \sum_{j=i}^k \bar{e}_k^{ij} d\tilde{y}_i^u(j).$$

(ii) $(\mathcal{X} \oplus \mathcal{W}) \cap \mathcal{Y} \subset \mathcal{X}$.

The proof of the Proposition 8 lies essentially on the fact that the inversion algorithm, applied to the system Σ with respect to the controls u, produces a basis for the subspace $(\mathcal{X} \oplus \mathcal{W}) \cap \mathcal{Y}$.

Lemma 9

$$\begin{aligned}&(\mathcal{X} \oplus \mathcal{W}) \cap \mathcal{Y} = \\ &= \text{span}_{\mathcal{K}}\{d\hat{y}_k^u(k) - \textstyle\sum_{i=1}^k \sum_{j=i}^k \bar{e}_k^{ij} d\tilde{y}_i^u(j),\ k \geq 0\} \\ &= \text{span}_{\mathcal{K}}\{\bar{a}_k dx(0) + \textstyle\sum_{j=0}^{k-1} \bar{c}_k^j dw(j),\ k \geq 0\}.\end{aligned}$$

Proof of Proposition 8. By Lemma 9 it is immediate that another way of stating (ii) is to say that

$$d\hat{y}_k^u(k) - \sum_{i=1}^k \sum_{j=i}^k \bar{e}_k^{ij} d\tilde{y}_i^u(j) = \bar{a}_k dx(0)$$

which completes the proof.

Theorem 10 Assume that the nonlinear system Σ is generically submersive. Then for system Σ with measurable disturbances the regular DDDP is solvable if and only if

$$(\mathcal{X} \oplus \mathcal{W}) \cap \mathcal{Y} \subset \mathcal{X}. \tag{11}$$

6. DYNAMIC VERSUS QUASI-STATIC STATE FEEDBACK

The class of quasi-static feedback was introduced for the first time by (Perdon *et al.*, 1990) and was formalized by (Delaleau and Fliess, 1992; Delaleau and Fliess, 1993) for continuous-time systems. For discrete-time systems the quasi-static feedback is characterized by the fact that the control u is calculated from the state x and a finite number of forward shifts of the new control v, and vice versa:

$$u(t) = \varphi(x(t), v(t), v(t+1), \ldots, v(t+\beta)) \tag{12}$$

$$v(t) = \xi(x(t), u(t), u(t+1), \dots, u(t+\beta)). \quad (13)$$

Only equation (12) is used in the feedback synthesis, equation (13) reflects the properties of the feedback. Note that the control input u can be calculated without including additional dynamics in the feedback. For $\beta = 0$, the feedback is called static.

Analogously, we define a quasi-static state feedback with disturbance measurement

$$\begin{aligned} u(t) &= \varphi(x(t), v(t), v(t+1), \dots, v(t+\beta), w(t)) \\ v(t) &= \xi(x(t), u(t), u(t+1), \dots, u(t+\beta), w(t)). \end{aligned}$$

The purpose of this section is to obtain the solvability conditions of the disturbance decoupling problem by quasi-static state feedback both for the cases of unmeasurable and measurable disturbances, and compare the obtained conditions with the results of Section 5.

6.1. *The case of unmeasurable disturbances.*

Theorem 11 For generically submersive nonlinear system Σ disturbance decoupling is possible by quasi-static state feedback without disturbance measurement, i.e. by the feedback of the form (12)-(13), if and only if (9) holds.

Proof. Sufficiency. By Proposition 4, condition (9) guarantees that in applying the inversion algorithm to Σ with respect to u gives at the α-th (last) step of the algorithm

$$\begin{aligned} d\tilde{y}_1^u(1) &= \tilde{a}_1 dx(0) + \tilde{b}_1 du(0) \\ &\vdots \\ d\tilde{y}_\alpha^u(\alpha) &= \tilde{a}_\alpha dx(0) + \tilde{b}_\alpha du(0) \\ &\quad + \sum_{i=1}^{\alpha-1} \sum_{j=i+1}^{\alpha} \tilde{e}_\alpha^{ij} d\tilde{y}_i^u(j). \end{aligned} \quad (14)$$

which after a possible permutation of inputs can be uniquely solved with respect to $du^1 = (du_1, \dots, du_{\rho^*})^T$. Define $du^2 = (du_{\rho^*+1}, \dots, du_m)^T$.

The latter means that almost everywhere there exist the functions φ and A_α such that locally

$$u^1(t) = \varphi(x(t), \tilde{y}_i^u(t+j), u^2(t)), \quad (15)$$

where $1 \le i \le \alpha$, $i \le j \le \alpha$, and

$$\begin{bmatrix} \tilde{y}_1^u(t+1) \\ \vdots \\ \tilde{y}_\alpha^u(t+\alpha) \end{bmatrix} \equiv A_\alpha \left[x(t), \varphi(\cdot), u^2(t), \tilde{y}_k^u(t+\ell)\right] \quad (16)$$

where $1 \le k \le \alpha - 1$, $k+1 \le \ell \le \alpha$.

Now construct the quasi-static state feedback of the form (12) in the following way:

$$\begin{aligned} u^1(t) &= \varphi(x(t), \tilde{v}_i(t+r), v^2(t)) \\ u^2(t) &= v^2(t), \end{aligned} \quad (17)$$

where $1 \le i \le \alpha$, $0 \le r \le \alpha - i$.

By (16) it is easy to see that the compensator (17) applied to Σ yields for $k = 1, \dots, \alpha$

$$\tilde{y}_k^u(t+k) = \tilde{v}_k(t), t \ge 0.$$

Moreover, for $k \ge 0$,

$$d\hat{y}_k^u(k) - \sum_{i=1}^{k} \sum_{j=i}^{k} \tilde{e}_k^{ij} d\tilde{y}_i(j) \in \mathcal{X}$$

by assumption. Hence, the feedback (17) solves the disturbance decoupling problem.
Necessity. Assume that (9) does not hold, that is for some $k \ge 0$, $\hat{y}_k^u(k+1)$ depends on w. For simplicity assume $k = 0$, for the other values of k the proof is similar. So,

$$d\hat{y}_0(1) = a_1 dx(0) + b_1 du(0) + c_1 dw(0). \quad (18)$$

Since the feedback (12) solves the disturbance decoupling problem for (1), this w-dependence should disappear, if we substitute u in (18) by the right hand side of (12). Since (12) does not depend on w, this is not possible, except for the case if (12) is such that it imposes the constraint $c_1(x, u, w) = 0$. The latter will contradict (13).

Comparing the condition of Theorem 11 with the condition of Theorem 6 leads to

Proposition 12 For generically submersive nonlinear system Σ the disturbance decoupling problem without disturbance measurement is solvable by quasi-static state feedback if and only if it is solvable by regular dynamic state feedback.

6.2. *The case of measurable disturbances*

Theorem 13 For generically submersive nonlinear system Σ disturbance decoupling is possible by quasi-static state feedback with disturbance measurement if and only if (11) holds.

The proof is similar to the proof of Theorem 11 and will be omitted.

Comparing the condition of the above Theorem with the condition of Theorem 10 leads to

Proposition 14 For generically submersive nonlinear system Σ the disturbance decoupling problem with disturbance measurement is solvable by quasi-static state feedback if and only if it is solvable by regular dynamic state feedback.

7. CONCLUSIONS

Linear algebraic necessary and sufficient condition for the solvability of the DDDP have been given both for the cases of measurable and unmeasurable input disturbances. Moreover, we have shown that the solvability conditions of the disturbance decoupling problem by quasi-static state feedback are the same as the solvability conditions by dynamic state feedback. Our results are similar to those of (Perdon *et al.*, 1993), concerning the generalized disturbance decoupling problem where besides the feedback of the form (12)-(13) also a generalized state-transformation has been allowed.

Acknowledgements. The second author gratefully acknowledges that this work was partly supported by the Estonian Science Foundation under grant N 936.

8. REFERENCES

Aranda-Bricaire E., U. Kotta, and C.H. Moog (1994a). A linear algebraic framework for feedback linearization of discrete-time nonlinear systems. *Prep. 1st IFAC Workshop "New Trends in Design of Control Systems"*, Smolenice, Slovakia, 35–40.

Aranda-Bricaire, E., U. Kotta, and C.H. Moog (1994b). Accessibility and feedback linearization of discrete-time systems. *Proc. 33rd IEEE Conf. on Decision and Control*, **2**, 1627-1632.

Delaleau, E., and M. Fliess (1992). Algorithme de structure, filtrations et decouplage. *C.R. Acad. Sci. Paris*, Serie I, **315**, 101–106.

Delaleau, E., and M. Fliess (1993). An algebraic interpretation of the structure algorithm with an application to feedback decoupling. In: *Nonlinear Control Systems Design. Selected papers from the 2nd IFAC Symp.*, (Ed. M. Fliess) Pergamon Press, Oxford, 489–494.

Delaleau, E. and P.S. Pereira da Silva (1994). Rank conditions for the dynamic disturbance decoupling problem. *Proc. 33rd IEEE Conf. on Decision and Control*, Buena Vista, Florida, **2**, 1784-1789.

Fliegner, Th., and H. Nijmeijer (1994). Dynamic disturbance decoupling for nonlinear discrete-time systems. *Proc. 33rd IEEE Conf. on Decision and Control*,Buena Vista, Florida, **2**, 1790-1791.

Fliess, M.(1990). Automatique en temps discret et algebre aux differences. *Forum Mathematicum*, **2**, 213–232.

Grizzle, J.W. (1993). A linear algebraic framework for the analysis of discrete-time nonlinear systems. *SIAM J. Contr. and Optimiz.*, **31**, 1026–1044.

Huijberts, H.J.C. (1992). A nonregular solution of the nonlinear disturbance decoupling problem with an application to a complete solution of the nonlinear model matching problem. *SIAM J. Contr. and Optimiz.*, **30**, 350–366.

Huijberts, H.J.C., H. Nijmeijer, and L.L.M. van der Wegen (1991). Dynamic disturbance decoupling for nonlinear systems: the nonsquare and noninvertible case. In: *Controlled Dynamical Systems*, (B.Bonnard, B. Bride, J.P. Gauthier and I. Kupka, Eds.), Birkhäuser, Boston, 243–252.

Huijberts, H.J.C., H. Nijmeijer, and L.L.M. van der Wegen (1992). Dynamic disturbance decoupling for nonlinear systems. *SIAM J. Contr. and Optimiz.*, **30**, 336–349.

Kotta, U. (1990). On the right inverse of discrete-time nonlinear systems. *Int. J. Contr.*, **51**, 1–9.

Kotta, U. (1992). Dynamic disturbance decoupling for discrete-time nonlinear systems: the nonsquare and noninvertible case. *Proc. Estonian Academy of Sciences. Phys. Math.*, **41**, 14–22.

Kotta, U. (1993). Decoupling of discrete-time nonlinear systems by minimal order dynamic state feedback. *Proc. Estonian Academy of Sciences. Phys. Math.*, **42**, 308–319.

Kotta, U. (1994a). Dynamic disturbance decoupling for discrete-time nonlinear systems: a solution in terms of system invariants. *Proc. Estonian Academy of Sciences Phys. Math.*, **43**, 147–159.

Kotta, U. (1994b). Model matching of nonlinear discrete-time systems in the presence of disturbances. *Proc. Estonian Academy of Sciences. Phys. Math.*, **43**, 7–24.

Kotta, U., and H. Nijmeijer (1991). Dynamic disturbance decoupling for nonlinear discrete-time systems (In Russian). *Proc. Academy of Sciences of USSR. Technical Cybernetics*, 52–59.

Moog, C.H., A.M. Perdon, and G. Conte (1991). Model matching and factorization for nonlinear systems: a structural approach. *SIAM J. Control and Optimiz.*, **29**, 769–785.

Perdon, A. M., G. Conte, C. H . Moog (1990). Some canonical properties of nonlinear systems. In: *Realization and Modelling in System Theory*, (M. A. Kaashoek, J. H. van Schuppen, A. C. M. Ran, Eds.), Birkhäuser, Boston, 89–96.

Perdon, A.M., Y.F. Zheng, C.H. Moog, and G. Conte (1993). Disturbance decoupling for nonlinear systems: a unified approach. *Kybernetika*, **29**, 479–484.

Pereira da Silva, P.S., and V.M. Pinto Leite (1993). Disturbance decoupling by regular dynamic feedback for affine nonlinear systems: a linear algebraic approach. *Prepr. 12th IFAC World Congress*, Sydney, **6**, 387–390.

Respondek, W. (1991). Disturbance decoupling via dynamic feedback. In: *Controlled Dynamical Systems* (B. Bonnard. B. Bride, J.P. Gauthier and I. Kupka, eds.), Birkhauser, Boston, 347–357.

ALGEBRAIC CONDITIONS FOR LINEARIZATION BY INPUT-OUTPUT INJECTION

F. PLESTAN, A. GLUMINEAU and C.H. MOOG

Laboratoire d'Automatique de Nantes. U.R.A. C.N.R.S. 823. Ecole Centrale de Nantes-Université de Nantes. 1, rue de la Noë. 44072 Nantes cedex 03. FRANCE

Abstract. New insight is given on the linearization by input-output injection. Necessary and sufficient conditions of the existence of (generalized or not) state and output coordinates transformations are given. The approach is based on an algebraic analysis of the input-ouput differential equation. Exterior differentiation and product are extensively used throughout the paper.

Résumé. Une nouvelle approche est proposée pour la linéarisation par injection d'entrée-sortie. Des conditions nécessaires et suffisantes d'existence de changements de coordonnées d'état (classique ou généralisé) et de sortie sont données. L'approche est basée sur l'étude de l'équation différentielle entrée-sortie du système non linéaire et utilise la théorie des systèmes différentiels extérieurs.

Keywords. Exact linearization, input-output injection, nonlinear observers.

1. INTRODUCTION

The problem under interest in this paper is the (exact) linearization of a nonlinear system by state and output transformations and input-output injection. In the litterature, one can find two ways solving this problem.

The first way uses geometric tools: a solution for autonomous nonlinear systems has been given by Krener *et al.* (1983). Geometric necessary and sufficient conditions were given for the existence of a state diffeomorphism, transforming the nonlinear system into a linear one modulo an output injection. This approach has been extended to MIMO systems by Krener *et al.* (1985) and Xia *et al.* (1989): the linearization is made via a state transformation (Krener *et al.*, 1985; Xia *et al.*, 1989) and an output transformation (Krener *et al.*, 1985). In this latter reference, the existence of the output transformation is characterized by a necessary and sufficient condition which is not constructive, and the transformation is computed from a necessary condition. The same remark occurs for the state transformation. These techniques were not often used for practical applications: to our best knowledge, very few works used these approachs (Chiasson *et al.*, 1993; Chiasson, 1994).

A second way developed for the linearization by input-output injection uses algebraic tools. This approach has been stated in algebraic terms by Keller (1987), only for systems with state dimension equal to 2 or 3. Conditions on the dynamics of the system are given by Phelps (1991) and Proychev *et al.* (1993) to find linearizing state transformation: in Proychev *et al.* (1993), some time derivatives of the input appears in the input-output injection (it has been introduced by Williamson (1977)). For the algebraic way, the results given by these approachs can be applied only to precise dimension state system (Keller, 1987), or are not fully constructive. Note also that they never tackle the problem of output transformation.

In this paper, an algebraic characterization is given to solve the problem of linearization by input-output injection using state and output transformation: the main idea consists in verifying that the input-output differential equation belongs to a certain class of polynomials of $(y, \dot{y}, \cdots, y^{(k)}, u, \dot{u}, \cdots, u^{(l)})$, with $k \in I\!N$, $l \in I\!N$, y the system output and u the system input. It is shown that these polynomial conditions are easy to check and are constructive in practical cases (Plestan *et al.*, 1993). The problem of output transformation is also tackled.

Section 2 gives a necessary and sufficient condition for the existence of a linearizing change of state coordinates: it can be viewed as a dual result of Xia *et al.* (1989) for single output system, but seems more easy to apply, and fully constructive. Section 3 generalizes the results of Proychev *et al.* (1993) for single output systems. In this latter paper, the characterization of the existence of the state transformation is not fully constructive. The class of systems studied here is more general than the one considered by Proychev *et al.* (1993), where only the first time derivative of the input appears in the input-output injection. It is recasted in the framework of generalized state space representations (Fliess, 1990). Section 4 gives a necessary and sufficient condition of the existence of output and state linearizing transformations.

2. LINEARIZATION BY STATE TRANSFORMATION AND INPUT-OUTPUT INJECTION

2.1 Problem statement

Consider the nonlinear system:

$$\begin{array}{rcl} \dot{x} & = & f(x,u) \\ y & = & h(x) \end{array} \quad (1)$$

where $x \in \mathbb{R}^n$, $u \in \mathbb{R}^m$ and $y \in \mathbb{R}$. The components of $f(\cdot,\cdot)$ and $h(\cdot)$ are supposed to be meromorphic functions of the variables x and u. In the rest of the paper, let $y^{(k)}$ denote the k-th time derivative of y. The system is assumed to be observable.

Consider the system (1). This section gives a necessary and sufficient condition of the existence of a local diffeomorphism $\zeta = \phi(x)$, which transforms the system (1) into:

$$\begin{array}{rcl} \dot{\zeta} & = & A\cdot\zeta + \varphi(y,u) \\ y & = & C\cdot\zeta \end{array} \quad (2)$$

where:

$$A = \begin{pmatrix} 0 & 1 & 0 & 0 & \cdots & 0 \\ 0 & 0 & 1 & 0 & \cdots & 0 \\ \vdots & \vdots & \ddots & \ddots & \ddots & \vdots \\ 0 & 0 & \cdots & 0 & 1 & 0 \\ 0 & 0 & \cdots & 0 & 0 & 1 \\ 0 & 0 & \cdots & 0 & 0 & 0 \end{pmatrix},$$

$$\varphi(y) = \begin{pmatrix} \varphi_1(y,u) & \varphi_2(y,u) & \cdots & \varphi_n(y,u) \end{pmatrix}^T,$$

and

$$C = \begin{pmatrix} 1 & 0 & \cdots & 0 \end{pmatrix}.$$

An observer for this system is given by :

$$\dot{\hat{\zeta}} = A\cdot\hat{\zeta} + \varphi(y,u) + K\cdot C\cdot(\hat{\zeta}-\zeta) \quad (3)$$

where $\hat{\zeta}$ is an estimation of the state ζ. Let ϵ denote the error between the state ζ and the estimated state $\hat{\zeta}$. One gets the error dynamics :

$$\dot{\epsilon} = (A - K\cdot C)\cdot\epsilon$$

with $K = (k_1, k_2, \cdots, k_n)^T$ such that the eigenvalues $(A - K\cdot c)$ are in the half-left complex plan.

2.2 Result

Let the polynomial P denote, the differential input-output equation of the system (1):

$$y^{(n)} = P(y, \dot{y}, \cdots, y^{(n-1)}, u, \dot{u}, \cdots, u^{(n-1)})$$

The following algorithm computes the linearizing state transformation (if it exists) and input-output injections which solve the problem:

Input-Output Injection Algorithm (I.O.I.A.).

Set $P_0 := P$ et $\varphi_0(y,u) := 0$. For $k = 1$ to n, let:

$$P_k = P_{k-1} - [\varphi_{k-1}(y,u)]^{(n-k+1)} \quad (4)$$

The differential form ω_k (for $1 \le k \le n$) is defined as:

$$\omega_k = \frac{\partial P_k}{\partial y^{(n-k)}}\mathrm{d}y + \sum_{i=1}^{m} \frac{\partial P_k}{\partial u_i^{(n-k)}}\mathrm{d}u_i \quad (5)$$

- If $\mathrm{d}\omega_k \neq 0$, then the problem has no solution.
- If $\mathrm{d}\omega_k = 0$, then define a input-output injection $\varphi_k(y,u)$ as:

$$\begin{array}{rcl} \mathrm{d}\varphi_k(y,u) & = & \omega_k \text{ for } k = 1,\cdots,n-1 \\ \varphi_n(y,u) & = & P_n \end{array} \quad (6)$$

■

A necessary and sufficient condition of the existence of a state transformation, which transforms (1) into (2), is given next.

Theorem 1 *The nonlinear system (1) is locally equivalent to (2) under a state transformation $\zeta = \phi(x)$ if and only if:*

$$\mathrm{d}\omega_i = 0 \quad (7)$$

for $1 \le i \le n$, where ω_i is defined by (5).

From the comparison between (1) and (2), one can determine the linearizing state transformation, which writes as:

$$\begin{array}{rcl} \zeta_1 & = & h(x) = y \\ \zeta_2 & = & \dot{\zeta}_1 - \varphi_1(h(x),u) \\ & = & \dot{y} - \varphi_1(h(x),u) \\ & \cdots & \\ \zeta_n & = & \dot{\zeta}_{n-1} - \varphi_{n-1}(h(x),u) \\ & = & y^{(n-1)} - \varphi_{n-1}(y,u) - \dot{\varphi}_{n-2}(y,u) \\ & & - \cdots - \varphi_1^{(n-2)}(y,u) \end{array} \quad (8)$$

Remark 1 The condition on the existence of the state diffeomorphism is fully constructive once the input-output equation is obtained. It seems to be more easily applicable than (Krener *et al.*, 1983; Xia *et al.*, 1989). ■

Proof of necessity: Suppose that there exists a state transformation $\zeta = \phi(x)$ which transforms (1) into (2). Thus, P, which equals to $y^{(n)}$, can be written as $P = \varphi_1^{(n-1)} + \varphi_2^{(n-2)} + \cdots + \varphi_n = P_0$. Applying the first step of the I.O.I.A., one gets:

$$\begin{aligned} P_1 &= P_0 \\ &= \frac{\partial \varphi_1}{\partial y} \cdot y^{(n-1)} + \frac{\partial \varphi_1}{\partial u} \cdot u^{(n-1)} \\ &\quad + \chi_1(y, \dot{y}, \cdots, y^{(n-2)}, u, \dot{u}, \cdots, u^{(n-2)}) \end{aligned}$$

The differential form ω_1 is given by (5):

$$\begin{aligned} \omega_1 &= \frac{\partial P_1}{\partial y^{(n-1)}} \mathrm{d}y + \sum_{i=1}^{m} \frac{\partial P_1}{\partial u_i^{(n-1)}} \mathrm{d}u_i \\ &= \frac{\partial \varphi_1}{\partial y} \cdot \mathrm{d}y + \sum_{i=1}^{m} \frac{\partial \varphi_1}{\partial u_i} \cdot \mathrm{d}u_i \end{aligned}$$

The condition $\mathrm{d}\omega_1 = 0$ is satisfied. The proof for steps $2 \leq k \leq n$ follows the same lines than for $k = 1$. ■

Proof of sufficiency: If the condition of Theorem 1 is satisfied, then let $\varphi(y, u)$ be given by (6). The dynamics of the $n-1$ first state variables are derived from (8). Let us consider the last equation of (8), and compute its time derivative:

$$\dot{\zeta}_n = y^{(n)} - \dot{\varphi}_{n-1} - \cdots - \varphi_1^{(n-1)} \tag{9}$$

As the condition of Theorem (1) is satisfied, and from the definitions of the function P_i and $\varphi_i(y, u)$ $(1 \leq i \leq n)$, then the input-output differential equation $y^{(n)}$ can be written as:

$$y^{(n)} = \varphi_n + \varphi_{n-1}^{(1)} + \cdots + \varphi_1 \tag{10}$$

Then, from (9) and (10), one gets $\dot{\zeta}_n = \varphi_n(y, u)$. System (1) is then locally equivalent to (2) under a state diffeomorphism (8). It can be shown that the state transformation depends only on x (Glumineau *et al.*, 1994). ■

2.3 Example

Consider the following nonlinear system:

$$\begin{aligned} \dot{x}_1 &= x_1 \cdot \sin u_1 + u_2 + x_2 \\ \dot{x}_2 &= x_1 \cdot x_2 + u_1^2 \\ y &= x_1 \end{aligned} \tag{11}$$

The differential input-output equation is given by:

$$\begin{aligned} y^{(2)} &= \dot{y} \cdot (\sin u_1 + y) + y \cdot \cos u_1 \cdot \dot{u}_1 + \dot{u}_2 \\ &\quad - y \cdot (y \cdot \sin u_1 + u_2) + u_1^2 \\ &= P = P_0 \end{aligned}$$

Applying the I.O.I.A., then one gets:

$$\begin{aligned} P_1 &= P_0 \\ &= \dot{y} \cdot (\sin u_1 + y) + y \cdot \cos u_1 \cdot \dot{u}_1 + \dot{u}_2 \\ &\quad - y \cdot (y \cdot \sin u_1 + u_2) + u_1^2 \end{aligned}$$

The differential form ω_1 (from (5)):

$$\begin{aligned} \omega_1 &= \frac{\partial P_1}{\partial \dot{y}} \mathrm{d}y + \sum_{i=1}^{2} \frac{\partial P_1}{\partial \dot{u}_i} \mathrm{d}u_i \\ &= (\sin u_1 + y)\mathrm{d}y + (y \cdot \cos u_1)\mathrm{d}u_1 + \mathrm{d}u_2 \end{aligned}$$

satisfies the condition (7) pour $k = 1$, i.e. $d\omega_1 = 0$. The function $\varphi_1(y, u)$ can be derived from (6):

$$\varphi_1(y, u) = u_2 + \frac{1}{2} y^2 + \sin u_1 \cdot y$$

The second step of the algorithm yields the polynomial $P_2 = -y(y \cdot \sin u_1 + u_2) + u_1^2$. The differential form ω_2 is then computed from (5):

$$\begin{aligned} \omega_2 &= \frac{\partial P_2}{\partial y} \mathrm{d}y + \sum_{i=1}^{2} \frac{\partial P_2}{\partial u_i} \mathrm{d}u_i \\ &= (-2y \sin u_1 - u_2)\mathrm{d}y + (2u_1 - y^2 \cos u_1) \\ &\quad \mathrm{d}u_1 + (-y)\mathrm{d}u_2 \end{aligned}$$

Since $\mathrm{d}\omega_2 = 0$, the system is linearizable via a state diffeomorphism and an input-output injection. The input-output injection $\varphi_2(y, u)$ is derived from the definition (6). System (11) is then locally equivalent to:

$$\begin{aligned} \dot{\zeta}_1 &= \zeta_2 + u_2 + \frac{1}{2} y^2 + \sin u_1 \cdot y \\ \dot{\zeta}_2 &= -y u_2 - y^2 \sin u_1 + u_1^2 \\ y &= \zeta_1 \end{aligned}$$

The state transformation $\zeta = \phi(x)$ defined by (8) is $\zeta = (x_1, x_2 - \frac{1}{2} x_1^2)$.

3. LINEARIZATION BY GENERALIZED STATE TRANSFORMATION AND INPUT–OUTPUT INJECTION

3.1 Problem statement

This case is just a generalization of the previous one. The class of systems is enlarged with the use of generalized state transformation. The problem is to transform system (1) into:

$$\begin{aligned} \dot{\zeta} &= A \cdot \zeta + \varphi(y, u, \dot{u}, \cdots, u^{(s)}) \\ y &= c \cdot \zeta \end{aligned} \tag{12}$$

under a transformation of state coordinates $\zeta = \phi(x, u, \dot{u}, \cdots, u^{(s-1)})$. An observer for (12) can be computed:

$$\dot{\hat{\zeta}} = A \cdot \hat{\zeta} + \varphi(y, u, \dot{u}, \cdots, u^{(s)}) + K \cdot c \cdot (\zeta - \hat{\zeta})$$

where K is choosen as previously.

3.2 Result

Let the polynomial P denote:

$$y^{(n)} = P(y, \dot{y}, \cdots, y^{(n-1)}, u, \dot{u}, \cdots, u^{(n-1)})$$

The following algorithm is used to compute the input-output injection and the generalized state transformation which linearizes the nonlinear system (1).

Generalized Input-Output Injection Algorithm (G.I.O.I.A).

Set $P_0 := P$ and $\varphi_0(y, u, \dot{u}, \cdots, u^{(s_0)}) = 0$ with $s_0 \in I\!N$. For $k = 1$ to n, let:

$$P_k := P_{k-1} - [\varphi_{k-1}(y, u, \dot{u}, \cdots, u^{(s_{k-1})})]^{(n-k+1)}$$

The differential form ω_k is defined as:

$$\omega_k = \frac{\partial P_k}{\partial y^{(n-k)}} \mathrm{d}y \quad (13)$$

- If, for any integer s_k, one has $\mathrm{d}\omega_k \wedge \mathrm{d}u \wedge \mathrm{d}\dot{u} \cdots \wedge \mathrm{d}u^{(s_k)} \neq 0$, then the problem has no solution.
- If there exists an integer s_k such that $\mathrm{d}\omega_k \wedge \mathrm{d}u \wedge \mathrm{d}\dot{u} \cdots \wedge \mathrm{d}u^{(s_k)} = 0$, then let the function $\varphi_k(y, u, \dot{u}, \cdots, u^{(s_k)})$ $(1 \leq k \leq n-1)$ be defined such that:

$$\frac{\partial \varphi_k}{\partial y} = \frac{\partial P_k}{\partial y^{(n-k)}} \quad (14)$$

and, for $k = n$:

$$\varphi_n(y, u, \cdots, u^{(s_n)}) = P_n \quad (15)$$

■

Notation: $\wedge \mathrm{d}u^{(i)} \equiv \wedge \mathrm{d}u_1^{(i)} \wedge \mathrm{d}u_2^{(i)} \wedge \cdots \wedge \mathrm{d}u_p^{(i)}$, for $i \geq 0$.

Remark 2 s equals to $\mathrm{MAX}(s_1, s_2, \cdots, s_n)$. ■

A necessary and sufficient condition of the existence of a generalized state transformation ϕ, which changes system (1) into (12) is given in the next theorem:

Theorem 2 *The nonlinear system (1) is locally equivalent to the system (12) under a generalized state transformation* $\zeta = \phi(x, u, \dot{u}, \cdots, u^{(s-1)})$ *if and only if:*

$$\mathrm{d}\omega_i \wedge \mathrm{d}u \wedge \mathrm{d}\dot{u} \wedge \cdots \wedge \mathrm{d}u^{(s_i)} = 0 \quad (16)$$

for $1 \leq i \leq n$, *where* ω_i *is defined by (13).*

The diffeomorphism $\zeta = \phi(x, u, \cdots, u^{(s-1)})$ is obtained from:

$$\begin{aligned} \zeta_1 &= h(x) \\ \zeta_2 &= \dot{\zeta}_1 - \varphi_1(y, u, \cdots, u^{(s_1)}) \\ &= \dot{y} - \varphi_1(y, u, \cdots, u^{(s_1)}) \\ &\vdots \\ \zeta_n &= \dot{\zeta}_{n-1} - \varphi_{n-1}(y, u, \cdots, u^{(s_{n-1})}) \\ &= y^{(n-1)} - \varphi_{n-1}(y, u, \cdots, u^{(s_{n-1})}) \\ &\quad - \dot{\varphi}_{n-2}(y, u, \cdots, u^{(s_{n-2})}) - \cdots \\ &\quad - \varphi_1^{(n-2)}(y, u, \cdots, u^{(s_1)}) \end{aligned} \quad (17)$$

Remark 3 This result is a characterization of results which are partly contained in (Proychev *et al.*, 1993): it is more constructive (it needs only time differentiations of polynomials) and the functions φ_i are immediately derived:

$$\varphi_i(y, u, \cdots, u^{(s)}) = \int_{y_0}^{y} \frac{\partial P_i}{\partial y^{(n-i)}} \mathrm{d}y$$

for $1 \leq i \leq n-1$, and:

$$\varphi_n(y, u, \cdots, u^{(s)}) = P_n$$

The condition of Theorem 2 is independent of the choice of y_0. This result generalizes also the result proposed in (Proychev *et al.*, 1993) with the use of a generalized state transformation, and the use of higher derivative orders of the input u in the term φ_i. ■

Sketch of proof: It is based on the same remarks than in previous section: the input-output differential equation of the system (12) can be written as in the previous section, with the terms φ_i which can be written as:

$$\varphi_i^{(n-1)} = \frac{\partial \varphi_i}{\partial y} \cdot y^{(n-i)} + \chi_i(y, \dot{y}, \cdots, y^{(n-i+1)}, u, \dot{u}, \cdots, u^{(s)})$$

So, for each polynomial P_i, the coefficient multiplying $y^{(n-i)}$ has to be only a function of $(y, u, \dot{u}, \cdots, u^{(s)})$: Theorem 2 states this condition. A more precise proof of the necessity and sufficiency of Theorem 2 can be found in (Glumineau *et al.*, 1994). ■

3.3 Example

Let us consider the nonlinear system :

$$\begin{aligned} \dot{x}_1 &= x_2 \cdot u^2 \\ \dot{x}_2 &= 0 \\ y &= x_1 \end{aligned} \quad (18)$$

This system is observable for $u \neq 0$: the transformations are then given for $u \neq 0$. By definition, P is equal to the input-output differential equation $\ddot{y} = 2x_2u\dot{u} = 2\dot{y}\dot{u}/u = P = P_0$. Applying the G.I.O.I.A., one gets $P_1 := P_0 = 2\dot{y}\dot{u}/u$. The differential form ω_1 is derived from (13):

$$\omega_1 = \frac{\partial P_1}{\partial \dot{y}} \mathrm{d}y = \frac{2\dot{u}}{u} \mathrm{d}y$$

and verifies the condition $\mathrm{d}\omega_1 \wedge \mathrm{d}u \wedge \mathrm{d}\dot{u} = 0$, i.e. the integer s_1 is equal to 1. The function $\varphi_1(y, u, \dot{u})$ can be determined from (14) $\varphi_1(y, u, \dot{u}) = 2y\dot{u}/u$. P_2 can then be computed from the second step of the G.I.O.I.A. One gets $P_2 = 2(\dot{u}^2/u^2 - \ddot{u}/u) \cdot y$. The differential form ω_2 is determined:

$$\omega_2 = \frac{\partial P_2}{\partial y} \mathrm{d}y = 2(\frac{\dot{u}^2}{u^2} - \frac{\ddot{u}}{u})\mathrm{d}y$$

and verifies the condition of Theorem 2: then, the system (18) is linearizable. The function $\varphi_2(y, u, \dot{u}, \ddot{u})$ can then be derived from (14) $\varphi_2 = P_2 = 2(\dot{u}^2/u^2 - \ddot{u}/u) \cdot y$. The system (18) is then locally equivalent to the system:

$$\begin{aligned} \dot{\zeta}_1 &= \zeta_2 + 2y\dot{u}/u \\ \dot{\zeta}_2 &= 2(\dot{u}^2/u^2 - \ddot{u}/u)\cdot y \\ y &= \zeta_1 \end{aligned} \tag{19}$$

The change of state is given by the definition (17) $\zeta = (x_1, x_2u^2 - 2x_1 \cdot \dot{u}/u)$.

4. LINEARIZATION BY STATE AND OUTPUT TRANSFORMATION AND INPUT-OUTPUT INJECTION

4.1 Problem statement

In some cases, the necessary and sufficient conditions of the existence of linearizing state transformation and input-output injection are not satisfied with the original outputs, but the problem of linearization may have a solution if a change of output coordinates is made before the change of state. Consider the following example:

$$\begin{aligned} \dot{x}_1 &= x_2 \\ \dot{x}_2 &= x_2^2 + x_1^2 \cdot x_2 \\ y &= x_1 \end{aligned} \tag{20}$$

The output differential equation equals to $y^{(2)} = P = \dot{y}^2 + y^2 \cdot \dot{y}$. Applying the I.O.I.A., it is easy to check that $d\omega_1 \neq 0$: the condition of Theorem 1 fails: the system is not linearizable under only a state transformation and an input-output injection. However, considering a new output (which depends on y only) $\bar{y} = T(y) = e^{-y}$, the new output differential equation equals to $\bar{y}^{(2)} = \bar{P} = -e^{-y} \cdot y^2 \cdot \dot{y}$. Applying the I.O.I.A. to $\bar{P}$, the condition of Theorem 1 is fulfilled. Then, the nonlinear system (20) is locally equivalent to:

$$\begin{aligned} \dot{\zeta}_1 &= \zeta_2 + e^{-y}\cdot(y^2 + 2y + 2) \\ \dot{\zeta}_2 &= 0 \\ \bar{y} &= \zeta_1 \end{aligned} \tag{21}$$

under a change of state coordinates derived from (8) and an output transformation $\bar{y} = e^{-y}$.

This section gives a necessary and sufficient condition of the existence of a local change of output $\bar{y} = T(y)$ and a local change of state coordinates $\zeta = \phi(x)$, which transforms (1) into:

$$\begin{aligned} \dot{\zeta} &= A\cdot\zeta + \varphi(y,u) \\ \bar{y} &= c\cdot\zeta \end{aligned} \tag{22}$$

An observer as (3) can be built.

Remark 4 The problem can also be stated for linearization by generalized state and output transformations, and input-output injection. ∎

4.2 Result

Consider the nonlinear system (1). Let P denote the input-output differential equation, $y^{(n)} = P(y, \dot{y}, \cdots, y^{(n-1)}, u, \dot{u}, \cdots, u^{(n-1)})$.

Theorem 3 *If the nonlinear system (1) is locally equivalent to the system (22) under a state diffeomorphism $\zeta = \phi(x)$ and an output transformation $\bar{y} = T(y)$, then:*

$$d[\frac{\partial^2 P}{\partial\dot{y}\partial y^{(n-1)}}] \wedge dy = 0 \tag{23}$$

Moreover, if (23) is satisfied, then there exists a linearizing output transformation $\bar{y} = T(y)$ verifying:

$$\frac{d^2T}{dy^2} = -\frac{1}{n}\cdot\frac{\partial^2 P}{\partial\dot{y}\partial y^{(n-1)}}\cdot\frac{dT}{dy} \tag{24}$$

Remark 5 This condition is equivalent to a condition given by Krener *et al.*(1985), where a necessary and sufficient condition is given, but is not constructive. The construction of the output transformation is derived from a necessary condition.

Proof: The input-output differential equation of system (1) reads:

$$y^{(n)} = P(y, \dot{y}, \cdots, y^{(n)}, u, \dot{u}, \cdots, u^{(n-1)}) \tag{25}$$

The system (22) has an input-output differential equation equal to:

$$\begin{aligned} \bar{y}^{(n)} &= T^{(n)}(y) \\ &= \varphi_1^{(n-1)} + \varphi_2^{(n-2)} + \cdots + \varphi_n \end{aligned} \tag{26}$$

Assume that there exists an output transformation $\bar{y} = T(y)$ and a change of state transformation $\zeta = \phi(x)$ solving the problem. Then, one gets:

- For $n \geq 3$

$$\begin{aligned} T^{(n)} &= \frac{dT}{dy}\cdot y^{(n)} + n\cdot\frac{d^2T}{dy^2}\cdot\dot{y}\cdot y^{(n-1)} \\ &\quad + \Gamma(y, \dot{y}, \cdots, y^{(n-2)}) \\ &= \frac{dT}{dy}\cdot P + n\cdot\frac{d^2T}{dy^2}\cdot\dot{y}y^{(n-1)} \\ &\quad + \Gamma(y, \dot{y}, \cdots, y^{(n-2)}) \\ &= \varphi_1^{(n-1)} + \varphi_2^{(n-2)} + \cdots + \varphi_n \end{aligned}$$

- For $n = 2$

$$\begin{aligned} T^{(2)} &= \frac{dT}{dy}\cdot P + \frac{d^2T}{dy^2}\cdot\dot{y}^2 \\ &= \varphi_1^{(1)} + \varphi_2 \end{aligned}$$

For each $n \geq 2$, the second derivative with respect to $\dot{y}$ and $y^{(n-1)}$ of the two previous equations yields:

$$\frac{d^2T}{dy^2} = -\frac{1}{n}\cdot\frac{\partial^2 P}{\partial\dot{y}\partial y^{(n-1)}}\cdot\frac{dT}{dy} \tag{27}$$

Since dT/dy depends on y only, then necessarily:

$$d[\frac{\partial^2 P}{\partial\dot{y}\partial y^{(n-1)}}] \wedge dy = 0 \tag{28}$$

If this latter condition is satisfied, then the equation (27) has solutions. A solution for $T(y)$ is:

$$T(y) = \int \exp[-\frac{1}{n}\int \frac{\partial^2}{\partial \dot{y}\partial y^{(n-1)}}dy]dy \quad (29)$$

∎

Let $\bar{P}$ denote the input-output differential equation of system (1), with the output $\bar{y} = T(y)$, and $\bar{\omega}_k$ $(1 \leq k \leq n)$ the differential form derived from the I.O.I.A. applied to $\bar{P}$ (see (5)). A necessary and sufficient condition of the existence of the output transformation $\bar{y} = T(y)$ and the change of state coordinates $\zeta = \phi(x)$ is given next.

Theorem 4 *The nonlinear system (1) is locally equivalent to (22) under an output transformation $\bar{y} = T(y)$ and a state transformation $\zeta = \phi(x)$ if and only if:*

$$d\bar{\omega}_i = 0 \quad (30)$$

for $1 \leq i \leq n$.

Remark 6 A necessary and sufficient condition of the existence of a generalized state transformation and a change of output coordinates is found from Theorem 2 applied to $\bar{\omega}_k$ $(1 \leq k \leq n)$ derived from (13) of G.I.O.I.A. applied to $\bar{P}$. ∎

5. CONCLUSION

New necessary and sufficient conditions for the linearization by input-output injection, and state and output transformations are given and based on the theory of exterior differential systems. Moreover, the problem of linearization by input-output injection with time derivatives of u and generalized state and output transformations is tackled. An extension of these results to MIMO systems seems to be reachable.

REFERENCES

Chiasson, J.N., and R.T. Novotnak (1993). Nonlinear speed observer for the PM stepper motor. *IEEE Trans. on Autom. Contr.*, **38**, 1584-1588.

Chiasson, J.N. (1994). Nonlinear differential-geometric techniques for control of a series DC motor. *IEEE Trans. on Autom. Contr.*, **2**, 35-42.

Fliess, M. (1990). Generalized controller canonical forms for linear and nonlinear dynamics. *IEEE Trans. on Autom. Contr.*, **35**, 994-1001.

Glumineau, A., C.H. Moog, and F. Plestan (1994). New algebro-geometric conditions for the linearization by input-output injection. *Technical Report No 94-13.* Laboratoire d'Automatique de Nantes, Nantes, France.

Keller, H. (1987). Nonlinear observer design by transformation into a generalized observer canonical form. *Int. J. Control*, **46**, 1915-1930.

Krener, A.J., and A. Isidori (1983). Linearization by output injection and nonlinear observers. *Syst. and Contr. Letters*, **3**, 47-52.

Krener, A.J., and W. Respondek (1985). Nonlinear observers with linearizable error dynamics. *SIAM J. Contr. Optim.*, **23**, 197-216.

Phelps, A.R. (1991). On constructing nonlinear observers. *SIAM J. Contr. Optim.*, **29**, 516-534.

Plestan, F., and B. Cherki (1994). An observer for a one flexible joint robot by an algebraic method. *Proc. IFAC Workshop on New Trends in Design of Control Systems NTDCS'94*, 41-46, Smolenice, Slovakia.

Proychev, T. Ph., and R.L. Mishkov (1993). Transformation of nonlinear systems in observer canonical form with reduced dependency on derivatives of the input. *Automatica*, **29**, 495-498.

Williamson, D. (1977). Observation of bilinear systems with application to biological systems. *Automatica*, **13**, 243-254.

Xia, X.H., and W.B. Gao (1989). Nonlinear observers design by dynamic error linearization. *SIAM J. Contr. Optim.*, **1**, 199-216.

FEEDBACK INVARIANTS OF SUPLEMENTARY CONTROLLABLE PAIRS OF MATRICES

I. BARAGAÑA* and I. ZABALLA**

*Facultad de Informática, Universidad del País Vasco. Barrio de Ibaeta s/n. 20009. San Sebastián, Spain.
**Departamento de Matemática Aplicada y EIO, Universidad del País Vasco. Apdo. Correos 450. 01080 Vitoria-Gasteiz, Spain.

Abstract. Given a pair of matrices (A, B), with every (A, B)-invariant subspace and the correponding suplementary subspace we can associated pairs (A_1, B_1) and (A_2, B_2) which identify them up to feedback equivalence. A complete characterization of the controllability indices of (A, B) in terms of those of (A_1, B_1), (A_2, B_2) is provided unde the assumption of complete controllability.

Key Words. Invariant Subspaces; controllability indices; Brunovsky indices; majorization; invariant factors.

1. INTRODUCCION

Consider the following Linear System of Differential Equation with control:

$$\dot{x}(t) = A\,x(t) + B\,u(t) \qquad (1)$$

where $A \in \mathbb{F}^{n\times n}$, $B \in \mathbb{F}^{n\times m}$ and $\mathbb{F}$ is the field of the real or complex numbers. System (1) is usually identified with the matrix pair (A, B), and in turns this pair can be identified with the rectangular matrix $[A \;\; B]$. It has been shown in Gohberg *et al.*(1980) that a geometric interpretation for this type of matrices correponds to what the authors call a *Block*. This means that a matrix like $[A\, B]$ is the representation of a linear map $f : V \longrightarrow Im\, P$, where V is a vector space over $\mathbb{F}$ and P is a projection of V. A linear map like this is called a (P, I)-Block

The notion of (A, B)-invariant subspace is wellkown: A subspace $\mathcal{S}$ of $\mathbb{F}^n$ is said to be (A, B)-invariant if $A(\mathcal{S}) \subset \mathcal{S} + Im\, B$. As shown in Baragaña *et al.* (to appear) the corresponding geometric definition of a (P, I)-Block invariant subspace is as follows:

Definition 1.1 *Let $f : V \longrightarrow Im\, P$ a (P, I)-Block and let $\mathcal{S}$ be a subspace of $Im\, P$ (and then of V). $\mathcal{S}$ is said to be f-invariant if $f(\mathcal{S}+Ker\, P) \leq \mathcal{S} + f(Ker\, P)$.*

By using this definition, the notion of restriction of a matrix pair (A, B) to a (A, B)-invariant subspace makes sense. In fact the following result was proved in Baragaña *et al.* (to appear):

Proposition 1.2 *Let f be a (P, I)-block and let $\mathcal{S}$ be an f-invariant subspace of $Im\, P$. Then there exists a projection Q of V such that $Ker\, P = Ker\, Q$ and*

$$(i) \quad f^{-1}(\mathcal{S}) = (f^{-1}(\mathcal{S}) \cap Ker\, P) \oplus (f^{-1}(\mathcal{S}) \cap Im\, Q)$$

(ii) $Q(\mathcal{S})$ is isomorphic to $\mathcal{S}$

(iii) $f(Q(\mathcal{S})) \leq \mathcal{S}$.

A projection Q of V such that $Ker\, P = Ker\, Q$ and satisfying condition (i) of the above Proposition is called *adapted to the f-invariant subspace $\mathcal{S}$.*

Now we can give the following

Definition 1.3 *Let f be a (P, I)-block and let $\mathcal{S}$ be an f-invariant subspace. Let Q be a projection of V adapted to $\mathcal{S}$. The restriction of f to $\mathcal{S}$ is the linear map:*

$$f_{|\mathcal{S}} : \begin{array}{ccc} Q(\mathcal{S}) \oplus \mathcal{L} & \longrightarrow & Q(\mathcal{S}) \\ x & \rightsquigarrow & P^{-1}f(x) \end{array}$$

where $\mathcal{L} = (f^{-1}(\mathcal{S}) \cap Ker\, P)$.

We can see that the restriction of a (P, I)-block to an invariant subspace is a $(Q_{|\mathcal{S}}, I_{|\mathcal{S}})$-block, but unlike the restrictions of endomorphisms, is not unique. However all restrictions are Feedback Equivalent:

Definition 1.4 *Let $f_1 : V_1 \rightarrow Im\, P_1$ be a (P_1, I)-block and let $f_2 : V_2 \rightarrow Im\, P_2$ be a (P_2, I)-block.*

Then f_1 and f_2 are Feedback Equivalent if there is an isomorphism, φ, of V_1 into V_2, such that $\varphi(Ker\, P_1) = Ker\, P_2$ and $\bar{\varphi} \circ f_1 = f_2 \circ \varphi$, where $\bar{\varphi} = P_2 \circ \varphi_{|Im\, P_1}$.

This definition is, of course, the geometric interpretation in terms of (P, I)-blocks of the usual Feedback Equivalence definition of matrix pairs. Like for endomorphisms, associated with a restriction of a (P, I)-block to a invariant subspace, we can define the corresponding quotient block. In fact, if $f : V \longrightarrow Im\, P$ is a (P, I)-block, $\mathcal{S}$ is f-invariant and $\mathcal{L}$ is the subspace of Definition 3 then the linear map

$$\tilde{f} : \frac{V}{\mathcal{S} \oplus \mathcal{L}} \longrightarrow \frac{Im\, P \oplus \mathcal{L}}{\mathcal{S} \oplus \mathcal{L}}$$

is a $(\tilde{P}, I)$-block, where $\tilde{P}$ is the quotient projection:

$$\begin{array}{cccc} \tilde{P} : & \frac{V}{\mathcal{S}\oplus\mathcal{L}} & \longrightarrow & \frac{Im\, P\oplus\mathcal{L}}{\mathcal{S}\oplus\mathcal{L}} \\ & x + (\mathcal{S} \oplus \mathcal{L}) & \rightsquigarrow & Px + ((\mathcal{S} \oplus \mathcal{L}) \end{array}$$

Now if f is a (P, I)-block and $\mathcal{S}$ is f-invariant we can find bases of V and $Im\, P$ (see Baragaña *et al.* (to appear)) such that the matrix representation of f has the following form:

$$\left[\begin{array}{cc|cc} A_1 & A_2 & B_1 & B_2 \\ 0 & A_3 & 0 & B_3 \end{array}\right]$$

where $[A_1\ B_1]$ is the matrix of the restriction $f_{|\mathcal{S}}$ and $[A_3\ B_3]$ is the matrix of the quotient block $\tilde{f}$. As a conclusion we have that if (A, B) is a given pair of matrices and $\mathcal{S}$ is (A, B)-invariant then there is a nonsingular matrix T such that

$$(T^{-1}A\,T, T^{-1}B) = \left(\begin{bmatrix} A_1 & A_2 \\ 0 & A_3 \end{bmatrix}, \begin{bmatrix} B_1 & B_2 \\ 0 & B_3 \end{bmatrix} \right)$$

where (A_1, B_1) is the restriction of (A, B) to $\mathcal{S}$ and (A_3, B_3) is the quotient pair.

In Baragaña *et al.* (to appear) a complete characterization of the feedback invariants of a pair (A, B) was provided in terms of the quotient associated to the restriction pair of a given (A, B)-invariant subspace.

Theorem 1.5 *Let $(A_1, B_1) \in \mathbb{F}^{p\times p} \times \mathbb{F}^{p\times q}$ and let $k_1 \geq \ldots \geq k_r$ and $\gamma_1 \mid \ldots \mid \gamma_p$ be its controllability indices and invariant factors, respectively. Let n and m be positive integers such that $n \geq p$ and $m \geq q$. Assume that $l_1 \geq \ldots \geq l_s$ are positive integers and $\alpha_1 \mid \ldots \mid \alpha_n$ are monic polynomials satisfying that $l_1 + \ldots + l_s + d(\alpha_1) + \ldots + d(\alpha_n) = n$, where $d(.)$ stands for degree of. Then there exist matrices $A_2 \in \mathbb{F}^{(n-p)\times(n-p)}$, $A_3 \in \mathbb{F}^{(n-p)\times p}$, $B_2 \in \mathbb{F}^{(n-p)\times q}$ such that*

$$\left(\begin{bmatrix} A_1 & 0 \\ A_3 & A_2 \end{bmatrix}, \begin{bmatrix} B_1 & 0 \\ B_3 & B_2 \end{bmatrix} \right)$$

has $l_1, \ldots, l_s$ as controllability indices and $\alpha_1, \ldots, \alpha_n$ as invariant factors if and only if the following conditions hold:

$$\alpha_i \mid \gamma_i \mid \alpha_{i+n-p} \quad 1 \leq i \leq p$$
$$k_i \leq l_i \quad 1 \leq i \leq s$$

where $k_i := 0$ for $i > r$.

A more general and difficult problem is to characterize the feedback invariants of (A, B) when both the restriction and the quotient associated to an (A, B)-invariant subspace are known. The aim of this note is to provide a solution to this problem when both the restriction and the quotient are controllable pairs.

2. PRELIMINARY RESULTS

In this section we give some preliminary results that will be used in the next one. The aim of the first lemma is to show that we can replace the original pairs by anyother ones in the same class of feedback equivalece.

Lemma 2.6 .- *Let $(A_1, B_1), (\bar{A}_1, \bar{B}_1) \in \mathbb{F}^{n_1\times n_1} \times \mathbb{F}^{n_1\times m_1}$ and $(A_2, B_2), (\bar{A}_2, \bar{B}_2) \in \mathbb{F}^{n_2\times n_2} \times \mathbb{F}^{n_2\times m_2}$ Let $n = n_1 + n_2$ and $m = m_1 + m_2$. Let $(A, B) \in \mathbb{F}^{n\times n} \times \mathbb{F}^{n\times m}$ and assume that (A_i, B_i) is for $i = 1, 2$ feedback equivalent to $(\bar{A}_i, \bar{B}_i)$. Then there exist matrices $X \in \mathbb{F}^{n_1\times n_2}$ and $Y \in \mathbb{F}^{m_1\times m_2}$ such that (A, B) is feedback equivalent to*

$$\left(\begin{bmatrix} A_1 & X \\ 0 & A_2 \end{bmatrix}, \begin{bmatrix} B_1 & Y \\ 0 & B_2 \end{bmatrix} \right)$$

if and only if there exist matrices $\bar{X} \in \mathbb{F}^{n_1\times n_2}$ and $\bar{Y} \in \mathbb{F}^{m_1\times m_2}$ such that (A, B) is feedback equivalent to

$$\left(\begin{bmatrix} A_1 & \bar{X} \\ 0 & A_2 \end{bmatrix}, \begin{bmatrix} B_1 & \bar{Y} \\ 0 & B_2 \end{bmatrix} \right).$$

Proof.- Straightforward.

By means of this Lemma we can assume that (A_1, B_1) and (A_2, B_2) are in their corresponding Brunovsky canonical forms. Furthermore, since we can substitute the given pairs (A_1, B_1) and (A_2, B_2) by any pairs in their corresponding feedback equivalence class we could think that the solution of our problem will be given in terms of a complete system of invariants for the feedback equivalence. Such a system is described by the controllability indices of any pair in the class. It is known that if $(A, B) \in \mathbb{F}^{n\times n} \times \mathbb{F}^{n\times m}$ and $k_1 \geq \ldots \geq k_m$ are its controllability indices, then $(k_1, \ldots, k_m)$ is a partition of n, the size of A; i.e. $k_1 + \ldots + k_m = n$. The conjugate partition of $(k_1, \dots, k_m)$ is the sequence of nonnegative integers defined as follows:

$$r_i = \#\{j : k_j \geq i\}, \quad 1 \leq i \leq n,$$

where # stands for cardinality. We call these indices the Brunovsky indices of (A, B) (or the r-numbers as were originally called in Brunovsky (1970)) and, in general, we will use the notation $\overline{(n_1, \dots, n_m)}$ to mean the conjugate partition of $(n_1, \dots, n_m)$. The Brunovsky canonical form of (A, B) is associated to its controllability indices. Less known but very useful in our case is the following canonical form associated to the Brunovsky indices that can be found in Baragaña *et al.*(1990):

Lemma 2.7 .- *Let* $(A, B) \in \mathbb{F}^{n\times n} \times \mathbb{F}^{n\times m}$ *and let* $r_1 \geq \dots \geq r_k > 0$ *be its nonzero Brunovsky indices. Then* (A, B) *is feedback equivalent to a pair* (A_c, B_c) *with the following form:*

$$A_c = \begin{bmatrix} 0 & 0 & \dots 0 & 0 & \\ E_{r_2} & 0 & \dots & 0 & 0 \\ 0 & E_{r_3} & \dots & 0 & 0 \\ \vdots & \vdots & \ddots & \vdots & \vdots \\ 0 & 0 & \dots & E_{r_k} & 0 \end{bmatrix} \quad B_c = \begin{bmatrix} E_{r_1} \\ 0 \\ 0 \\ \vdots \\ 0 \end{bmatrix}$$

where $E_{r_i} = [I_{r_i} \quad 0] \in \mathbb{F}^{r_i \times r_{i-1}}$, $i = 1, \dots, k$ *and* $r_0 := m$.

Using Lemma 2.6 we can simplify the problem:

Lemma 2.8 .- *Let* $(A_1, B_1) \in \mathbb{F}^{n_1\times n_1} \times \mathbb{F}^{n_1\times m_1}$ *and* $(A_2, B_2) \in \mathbb{F}^{n_2\times n_2} \times \mathbb{F}^{n_2\times m_2}$ *be controllable matrix pairs. Let*

$$(A, B) = \left(\begin{bmatrix} A_1 & X \\ 0 & A_2 \end{bmatrix}, \begin{bmatrix} B_1 & Y \\ 0 & B_2 \end{bmatrix} \right).$$

Then there is a matrix $Z \in \mathbb{F}^{m_1\times m_2}$ *such that* (A, B) *is feedback equivalent to*

$$(A, B) = \left(\begin{bmatrix} A_1 & 0 \\ 0 & A_2 \end{bmatrix}, \begin{bmatrix} B_1 & Z \\ 0 & B_2 \end{bmatrix} \right)$$

Proof.- As said before we can assume that (A_1, B_1) is in the Brunovsky canonical form; i.e., if $c_1 \geq \dots \geq c_r$ are its controllability indices, then

$$A_1 = Diag(A_{11}, A_{12}, \dots, A_{1r}), \quad B_1 = \begin{bmatrix} B_{11} \\ B_{12} \\ \vdots \\ B_{1r} \end{bmatrix}$$

where

$$A_{1i} = \begin{bmatrix} 0 & I_{c_i-1} \\ 0 & 0 \end{bmatrix} \in \mathbb{F}^{c_i\times c_i}$$

$$B_i = \begin{bmatrix} 0 & \dots & 0 & \dots & 0 \\ \vdots & & \vdots & & \vdots \\ 0 & \dots & 1 & \dots & 0 \end{bmatrix} \in \mathbb{F}^{c_i\times m_1}$$

Put

$$X = \begin{bmatrix} X_1 \\ \vdots \\ X_r \end{bmatrix} \quad Y = \begin{bmatrix} Y_1 \\ \vdots \\ Y_r \end{bmatrix}$$

with $X_i \in \mathbb{F}^{c_i\times n_2}$ and $Y_i \in \mathbb{F}^{c_i\times m_2}$. By Lemma 2.10 of Zaballa (1987), for $i = 1, \dots, r$ there are matrices $T_i \in \mathbb{F}^{c_i\times n_2}$ such that

$$\begin{bmatrix} I_{c_i} & T_i \\ 0 & I_{n_2} \end{bmatrix} \begin{bmatrix} A_{1i} & X_i \\ 0 & A_2 \end{bmatrix} \begin{bmatrix} I_{c_i} & -T_i \\ 0 & I_{n_2} \end{bmatrix} = \begin{bmatrix} A_{1i} & D_i \\ 0 & A_2 \end{bmatrix}$$

with $D_i = \begin{bmatrix} 0 \\ d_i \end{bmatrix} \in \mathbb{F}^{c_i\times n_2}$ and d_i a row vector.

Put

$$P = \begin{bmatrix} I_{c_1} & \dots & 0 & T_1 \\ \vdots & \ddots & \vdots & \vdots \\ 0 & \dots & I_{c_r} & T_r \\ 0 & \dots & 0 & I_{n_2} \end{bmatrix}$$

then

$$PAP^{-1} = \left[\begin{array}{ccc|c} A_{11} & \dots & 0 & D_1 \\ \vdots & \ddots & \vdots & \vdots \\ 0 & \dots & A_{1r} & D_r \\ \hline 0 & \dots & 0 & A_2 \end{array}\right]$$

$$PB = \left[\begin{array}{c|c} B_1 & Z_1 \\ \vdots & \vdots \\ B_r & Z_r \\ \hline 0 & B_2 \end{array}\right]$$

with $Z_i = Y_i - T_i B_2$, $i = 1, \dots, r$. Bearing in mind that all the elements of B_i are zero except the one in position (c_i, i) it is easily seen that there is a feedbacck matrix $F \in \mathbb{F}^{m_1\times n}$ such that $PAP^{-1} + PBF = Diag(A_1, A_2)$. This proves the lemma. ■

Our next result is a basic tool for proving our main result. We will make the assumption that we are working with noncontrollable systems. For this kind of systems the controllability indices (or the Brunovsky indices) do not form a complete system of invariant for the feedback equivalence. Such a complete is system is given by the controllability (or the Brunovsky indices) and the invariant factors. By the invariant factors of (A, B) we mean the invariant factors of the matrix polynomial $[sI - A \quad -B]$ The proof of the following lemma can be found in Baragaña *et al.* (1990). Here and in the following section we will use the majorization in the sense of Hardy, Littlewood and Polya (Hardy *et al.* (1967)): Given to n-tuples of real numbers $a_1 \geq \dots \geq a_n$ and $b_1 \geq \dots \geq b_n$ we will say that $(b_1, \dots, b_n)$ majorizes $(a_1, \dots, a_n)$, and we write

$$(a_1, \dots, a_n) \prec (b_1, \dots, b_n)$$

if the two following conditions are satified:

$$\sum_{i=1}^{k} a_i \leq \sum_{i=1}^{k} b_i, \qquad 1 \leq k \leq n-1$$

$$\sum_{i=1}^{n} a_i = \sum_{i=1}^{n} b_i$$

Lemma 2.9 .- *Let $(A, B) \in \mathbb{F}^{n \times n} \times \mathbb{F}^{n \times m}$ and let $r_1 \geq \ldots \geq r_n$ and $\delta_1 \mid \ldots \mid \delta_n$ be its Brunovsky indices and invariant factors, respectively. Let $t_1 \geq \ldots \geq t_n$ be nonnegative integers, and put $s = t_1 - r_1$. Then there exists a matrix $C \in \mathbb{F}^{n \times s}$ such that $(A, [B \ \ C]) \in \mathbb{F}^{n \times n} \times \mathbb{F}^{n \times (m+s)}$ is controllable and $t_1, \ldots, t_n$ are its Brunovsky indices if and only if the following conditions hold:*

(i) $t_i \leq r_i + s \qquad 1 \leq i \leq n$

(ii) $\delta_i = 1 \qquad 1 \leq i \leq n-s$

(iii) $(r_1, \ldots, r_n) + \overline{(d(\delta_1), \ldots, d(\delta_{n-s+1})} \prec$

$$\prec (t_1, \ldots, t_n)$$

(iv) $\displaystyle\sum_{j=1}^{i}(t_j - r_j - p) \geq \sum_{j=1}^{i} z_j^p, \ 1 \leq i \leq h_p; ;$

$$1 \leq p \leq s$$

where $h_p := max\{i : t_i - r_i \geq p\}$, $1 \leq p \leq s$, $(z_1^p, z_2^p, \ldots) = \overline{(d(\delta_{n-p}, \ldots, d(\delta_{n-s+1})}$ and $d(.)$ stands for degree of.

3. MAIN RESULT

We now state the main result concerning the problem mentioned in Section 1. Our result will be given in terms of the Brunovsky indices.

Theorem 3.10 *Let $(A_1, B_1) \in \mathbb{F}^{n_1 \times n_1} \times \mathbb{F}^{n_1 \times m_1}$ and $(A_2, B_2) \in \mathbb{F}^{n_2 \times n_2} \times \mathbb{F}^{n_2 \times m_2}$ be two controllable matrix pairs, and let $r_1 \geq \ldots \geq r_{n_1}$ and $s_1 \geq \ldots \geq s_{n_2}$ (with $m_2 = s_1$) be their Brunovsky indices, respectively. Let $k_1 \geq \ldots \geq k_{s_1}$ be the controllability indices of (A_2, B_2) and put $n = n_1 + n_2$. Assume that $t_1 \geq \ldots \geq t_n$ are nonnegative integers. Then there exist matrices $X \in \mathbb{F}^{n_1 \times n_2}$ and $Y \in \mathbb{F}^{n_1 \times m_2}$ such that $t_1, \ldots, t_n$ are the Brunovsky indices of*

$$\left(\begin{bmatrix} A_1 & X \\ 0 & A_2 \end{bmatrix}, \begin{bmatrix} B_1 & Y \\ 0 & B_2 \end{bmatrix}\right)$$

if and only if the following conditions hold:

$$(r_1, \ldots, r_{n_1}) + (s_1, \ldots, s_{n_2}) \prec (t_1, \ldots, t_n) \qquad (2)$$

$$t_i \leq r_i + s_1 \quad 1 \leq i \leq n \qquad (3)$$

$$t_i \geq s_i, \quad 1 \leq i \leq n \qquad (4)$$

$$\sum_{j=1}^{i}(t_j - r_j - p) \geq \sum_{j=1}^{i} d_j^p, \ 1 \leq i \leq h_p, 1 \leq p \leq s_1 \qquad (5)$$

$$\sum_{j=1}^{i} r_j \leq (i-q)(t_q - s_q) + \sum_{j=1}^{q}(t_j - s_j)$$

$$1 \leq q \leq k_1, \ q+1 \leq i \leq n \qquad (6)$$

where $h_p = max\{i : t_i - r_i \geq p\}, 1 \leq p \leq s_1$, and $(d_1^p, \ldots) = \overline{(k_{1+p}, \ldots, k_{s_1})}, 1 \leq p \leq s_1$, i.e. the conjugate partition of $(k_{1+p}, \ldots, k_{s_1})$

Notice that we have made the assumption that $m_2 = s_1$. This means that B_2 has full rank.

The proof of this result is made by induction on s_1 and by using the following technical lemma:

Lemma 3.11 .- *Let $r_1 \geq \ldots \geq r_n \geq 0$, $s_1 \geq \ldots \geq s_k > 0 = s_{k+1} = \ldots = s_n$ and $t_1 \geq \ldots \geq t_n \geq 0$ be nonnegative integers and let $(k_1, \ldots, k_{s_1}) = \overline{(s_1, \ldots, s_k)}$. Assume that (2)-(6) hold and let*

$$t_i^1 = t_i - 1, \qquad 1 \leq i \leq u := max\{h_1, k\}$$

$$t_i^1 = min\{t_k - 1, r_i, t_i + \sum_{j=1}^{i-1}(t_j - t_j^1) - k\},$$

$$u+1 \leq i \leq n \qquad (7)$$

Then the following conditions are stisfied:

(i)

$$t_1^1 \geq \ldots \geq t_n^1 \geq 0 \qquad (8)$$

(i_A)

$$(r_1, \ldots, r_n) + (s_1 - 1, \ldots, s_k - 1) \prec (t_1^1, \ldots, t_n^1) \qquad (9)$$

(ii_A)

$$t_i^1 \leq r_i + s_1 - 1, \qquad 1 \leq i \leq n \qquad (10)$$

(iii_A)

$$\sum_{j=1}^{i}(t_j^1 - r_j - p) \geq \sum_{j=1}^{i} \tilde{d}_j^p,$$

$$1 \leq p \leq \tilde{h}_p; \ 1 \leq p \leq s_1 - 1 \qquad (11)$$

where

$$\tilde{h}_p = max\{i : t_i^1 - r_i \geq p\}, \quad 1 \leq p \leq s_1 - 1$$

$$(\tilde{d}_1^p, \ldots) = \overline{(\tilde{k}_{1+p}, \ldots, \tilde{k}_{s_1-1})}, \ 1 \leq p \leq s_1 - 1$$

and

$$(\tilde{k}_1, \ldots, \tilde{k}_{s_1-1}) = \overline{(s_1 - 1, \ldots, s_k - 1)}$$

(iv_A)

$$t_i^1 \geq s_i - 1, \; 1 \leq i \leq k \quad (12)$$

(v_A)

$$\sum_{j=1}^{i} r_j \leq \sum_{j=1}^{q}(t_j^1 - (s_j - 1)) + (i - q)(t_q^1 - (s_q - 1)), \quad 1 \leq q \leq k; \; q + 1 \leq i \leq n \quad (13)$$

(i_B)

$$(t_1^1, \ldots, t_n^1) + (\overline{k}) \prec (t_1, \ldots, t_n) \quad (14)$$

(ii_B)

$$t_i \leq t_i^1 + 1, \qquad 1 \leq i \leq n \quad (15)$$

(iii_B)

$$\sum_{j=1}^{i}(t_j - t_j^1 - 1) \geq 0, \qquad 1 \leq i \leq \bar{\bar{h}}_1 \quad (16)$$

where

$$\bar{\bar{h}}_1 = max\{i : t_i - t_i^1 \geq 1\}$$

Acknowledgement. Thanks are given to the DGICYT for partially supporting the work that has led to this manuscript.

French Summary: Donné un pair de matrices (A, B), avec chaque subspace (A, B)-invariant et le correspondant subspace supplementaire on peut associer des pairs (A_1, B_1) et (A_2, B_2), respectivement. Ces pairs identifient les subspaces àpart "feedback" equivalence. Dans cet travail on present une characterization complète des indices de commandabilité de (A, B) en terms des pairs (A_1, B_1) et (A_2, B_2) sous la condition d'être commandables.

4. REFERENCES

Brunovsky P. (1970) . Classification of Linear Controllable Systems. *Kybernetika. (Praga),* **3 (6)**, 173-188.

Hardy G. H., Littlewood J.E., Pólya G. (1967). *Inequalities.* Cambridge Univ. Press. Cambridge.

Gohberg I., Kaashoek M.A., van Schagen F. (1980). Similarity of Operator Blocks and Canonical Forms.I. General Results, Feedback Equivalence and Kronecker Indices. *Integral Equations and Operator Theory,* **3**, 350-396.

Baragaña I., Zaballa I. (1990). Column Completion of a Pair of Matrices. *Linear and Multilinear Algebra,* **27**, 243-273.

Baragaña I., Zaballa I. (To appear). Block Similarity Invariants of Restrictions to (A, B)-Invariant Subspaces. *Linear Algebra Appl.*.

Zaballa I. (1987). Matrices with Prescribed Rows and Invariant Factors. *Linear Algebra Appl.,* **87**, 113-146.

STRUCTURE ASSIGNMENT OF STATE-OUTPUT SYSTEMS BY CHOICE OF THE OUTPUT EQUATION.

S. Mondie* , **J.J. Loiseau****

**Sección de Control Automático Departamento de Ingeniería Eléctrica, Cinvestav IPN Av. IPN No. 2508, A.P. 14-740 , C.P. 07000, México D.F. MEXICO. E-mail:smondie@ctrl.cinvestav.mx*

***Laboratoire d'Automatique de Nantes, Ecole Centrale de Nantes/Université de Nantes, URA CNRS 823, 1 rue de la Noë, 44072 Nantes Cedex 03, FRANCE. E-mail:loiseau@lan-ec.nantes.fr*

Abstract. Necessary and sufficient conditions to assign a given finite and infinite zero structure to the state-output system $E\dot{x} = Ax + Bu$, $y = Cx + Dv$ by choice of C and D are given.

Résumé: Des conditions nécessaires et suffisantes pour le placement des zéros finis et infinis du système $E\dot{x} = Ax + Bu$, $y = Cx + Dv$ en choisissant C et D sont données.

Key Words. Linear systems, Invariant factors assignment, choice of output equations.

1. INTRODUCTION

We consider the linear system described by

$$E\dot{x} = Ax + Bu \quad (1)$$

$$y = Cx + Du \quad (2)$$

Where E, A are $q \times n$ matrices, B is a $q \times m$ matrix, C is a $p \times n$ matrix and D a $p \times m$ matrix over the field of real numbers.

We solve the problem of assigning the zero structure of the state-output system by choosing C and D . This problem is relevant in integrated system design as in Karcanias (1994), where it is suggested that good structural properties should be obtained at the system design stage, for instance through the choice of adequate outputs and instrumentation.

Rosenbrock first studied and solved this problem in the case of explicit ($E = I$) and controllable systems (see theorem 4.1, p. 186 in Rosenbrock, 1970). As noted in Rosenbrock (1970) and further in Syrmos and Lewis (1993) this problem is related to a pole placement problem. This fact and various other methods used to generalize Rosenbrock control structure theorem for explicit controllable systems to non controllable implicit systems (Zaballa, 1988), (Kučera and Zagalak, 1988), (Zagalak and Loiseau, 1992) and to implicit regularizable systems (Loiseau and Zagalak, 1991), are used in the solution presented here.

Notation. Throughout the paper, it is assumed that the invariant factors of polynomial matrices are non decreasingly ordered: $\alpha_1(s) \leq \alpha_2(s) \leq \ldots \leq \alpha_{k_\alpha}(s)$, where $\alpha_i(s) \leq \alpha_j(s)$ means that $\alpha_i(s)$ divides without remainder $\alpha_j(s)$. We sometimes assume, to simplify the notation, that $\alpha_i(s) = 1$ for $i \leq 0$ and $\alpha_i(s) = 0$ for $i > k_\alpha$. The infinite zeros degrees and minimal indices are non increasingly ordered.

2. PROBLEM FORMULATION

To handle both the finite and infinite zeros of polynomial matrices in a uniform way, we shall use the change of variable $w = \frac{1}{s-a}$. From Proposition A1 of the appendix, a pencil $(wJ - G)$ of dimension $q \times (n + m)$ is associated to $(sE - A, -B)$ and reciprocally, through

$$(wJ - G) = \mathcal{T}(sE - H)$$

where $\mathcal{T}$ is defined as in Proposition A1. If a is not a root of $\alpha_i(s)$, the pencil $(wJ - G)$ has no zero at infinity, and its Kronecker invariants are directly related to those of $(sE - A, -B)$. Let us respectively denote

- $\alpha_i(s), i = 1, 2, \ldots, \rho$, the invariant factors,
- $p_i, i = 1, 2, \ldots, k_p$, the infinite zeroes orders,
- $\varepsilon_i, i = 1, 2, \ldots, k_\varepsilon$, the column minimal indices,
- $\eta_i, i = 1, 2, \ldots, k_\eta$, the row minimal indices

of $(sE - A, B)$, ρ being the rank of $(sE - A, B)$. The Kronecker invariants of $(wJ - G)$ are then

- the column minimal indices $\varepsilon_i, i = 1, 2, \ldots, k_\varepsilon$,
- the row minimal indices $\eta_i, i = 1, 2, \ldots, k_\eta$,

- the invariant factors $\overline{\alpha_i}(w), i = 1, 2, ..., \rho$, where

$$\overline{\alpha_i}(w) = c_i w^{p_\rho - i+1} \alpha_i \left(\frac{1+aw}{w}\right) w^{\deg \alpha_i(s)}, \quad (3)$$

c_i being a normalizing constant so that $\alpha_i(w)$ be monic. Since

$$\begin{pmatrix} wJ - G \\ C \quad D \end{pmatrix} = \mathcal{T} \begin{pmatrix} sE - A & -B \\ C & D \end{pmatrix},$$

it appears that the pencil $\begin{pmatrix} wJ - G \\ C \quad D \end{pmatrix}$ whith invariant factors $\psi_i(s), i = 1, ..., \kappa$ and infinite zeroes orders $d_i, i = 1, ..., k_d$ is associated in the sense of Proposition A1 to the pencil $\begin{pmatrix} sE - A & -B \\ C & D \end{pmatrix}$. If a is not a root of $\psi_i(s)$, its invariant factors $\overline{\psi_i}(w)$ are

$$\overline{\psi_i}(w) = c_i' w^{\kappa - i+1} \psi_i \left(\frac{1+aw}{w}\right) w^{\deg \psi_i(s)} \quad (4)$$

The problem, which consists in choosing C and D to assign the zero structure of the system(1)-(2), that is nothing but the sets of $\psi_i(s)$s and of d_is, can now be stated as an eigenstructure assignment problem for the implicit non-controllable system $\left(\begin{array}{c|c} wJ - G & 0 \\ 0 & I_p \end{array}\right)$, by static feedback of the form $(C, D) = F$, of dimension $p \times (n+m)$.

We shall use the polynomial external description approach (Kučera et al., 1993) to give a polynomial formulation of the problem. To do this, the system must be amended so that it is controllable. This is done by adding artificial entries trough a matrix B' so that the pencil $(wJ - G, B')$ be irreducible ; i. e. the associated system be controllable. The extended system, which system matrix is $\left(\begin{array}{c|cc} wJ - G & B' & 0 \\ 0 & 0 & I_p \end{array}\right)$, is then also controllable. Let $N(w), D(w)$ be an external description for the system $(wJ - G, B')$; othersaid $\begin{pmatrix} N(w) \\ D(w) \end{pmatrix}$ is irreducible and forms a basis of the kernel of $(wJ - G, B')$.

Then $N(w)$ and $\begin{pmatrix} D(w) \\ 0 \end{pmatrix}$ form an external description for the extended system, and $N(w)$, $\begin{pmatrix} D(w) \\ -FN(w) \end{pmatrix}$ form an external description of the closed–loop system which system matrix is $\left(\begin{array}{c|cc} wJ - G & B' & 0 \\ F & 0 & I_p \end{array}\right)$. Thus (see Kučera and Zagalak, 1988) the non-unit invariant factors of $\begin{pmatrix} wJ - G \\ F \end{pmatrix}$ are exactly those of $\begin{pmatrix} D(w) \\ FN(w) \end{pmatrix}$.

With no loss of generality the pencil $wJ - G$ is assumed being in Kronecker canonical form (Gantmacher, 1966). The additional input matrix B' that makes the system controllable, and also the external description of the system, can be constructed for each block which appears in the diagonal of $wJ - G$.

- The invariant factors are associated to a matrix $wI - A_{\overline{\alpha}}$ in block diagonal Jordan form associated to the polynomials $\overline{\alpha_i}(w)$. The matrix $B_{\overline{\alpha}} := I$ is introduce so that the pair $(A_{\overline{\alpha}}, B_{\overline{\alpha}})$ be controllable.

$$\begin{aligned} &(wI - A_{\overline{\alpha}} | -B_{\overline{\alpha}}) \begin{pmatrix} N_{\overline{\alpha}}(w) \\ D_{\overline{\alpha}}(w) \end{pmatrix} \\ &= (wI - A_{\overline{\alpha}} | -I) \begin{pmatrix} I \\ \hline wI - A_{\overline{\alpha}} \end{pmatrix} = 0 \end{aligned}$$

- Each column minimal index corresponds to a controllable subsystem, which does not require any additional input.

$$\begin{aligned} &(wE_{\varepsilon_i} - A_{\varepsilon_i})(N_{\varepsilon_i}(w)) \\ &= \begin{pmatrix} w & -1 & & \\ & \ddots & \ddots & \\ & & w & -1 \end{pmatrix} \begin{pmatrix} 1 \\ w \\ \vdots \\ w^{\varepsilon_i} \end{pmatrix} = 0 \end{aligned}$$

- Each row minimal index requires the introduction of two additional inputs.

$$\begin{aligned} &(wE_{\eta_i} - A_{\eta_i} | -B_{\eta_i}) \begin{pmatrix} N_{\eta_i}(w) \\ D_{\eta_i}(w) \end{pmatrix} \\ &= \left(\begin{array}{cccc|cc} w & & & & 0 & -1 \\ -1 & \ddots & & & \vdots & 0 \\ & \ddots & w & & 0 & \vdots \\ & & -1 & & 1 & 0 \end{array}\right) \begin{pmatrix} w^{\eta_i - 1} \\ \vdots \\ w \\ 1 \\ \hline 1 \\ w^{\eta_i} \end{pmatrix} = 0 \end{aligned}$$

An external description for the overall controllable extended system is then :

$$\begin{pmatrix} N(w) \\ D(w) \\ 0 \end{pmatrix} = \begin{pmatrix} N_\varepsilon(w) & 0 & 0 \\ 0 & I & 0 \\ 0 & 0 & N_\eta(w) \\ 0 & D_{\overline{\alpha}}(w) & 0 \\ 0 & 0 & 0 \\ 0 & 0 & D_\eta(s) \\ 0 & 0 & 0 \end{pmatrix}$$

where $N_\varepsilon(w) = \text{diag}\{N_{\varepsilon_i}(w), i = 1, ..., k_\varepsilon\}$, $N_\eta(w) = \text{diag}\{N_{\eta_i}(w), i = 1, ..., k_\eta\}$ and $D_\eta(w) = \text{diag}\{D_{\eta_i}(w), i = 1, k_\eta\}$.
The invariant factors of the closed loop system are finally those of $\begin{pmatrix} D(w) \\ -FN(w) \end{pmatrix}$ that is equal to

$$\begin{pmatrix} 0 & \text{diag}\{\overline{\alpha_i}(w)\} & 0 \\ 0 & 0 & \text{diag}\{w^{\eta_i - 1}\} \\ 0 & 0 & I_{k_\eta} \\ D_1(w) & D_2(w) & D_4(w) \end{pmatrix}$$

and similar to

$$\begin{pmatrix} D_1(w) & D_2(w) & 0 \\ 0 & \mathrm{diag}\{\overline{\alpha_i}(w)\} & 0 \\ 0 & 0 & I_{k_\eta} \\ 0 & 0 & 0 \end{pmatrix}$$

where $\deg_{ci} D_1(w) \leq \epsilon_i$, $\deg_{ci} D_2(w) = 0$, $\deg_{ci} D_4(w) \leq \eta_i - 1$, $\deg_{ci}(.)$ denoting the degree of the ith column of the matrix (.).

The problem is now stated in terms of the existence of polynomial matrices in w, say $D_1(w), D_2(w)$ such that the overall matrix

$$\begin{pmatrix} D_1(w) & D_2(w) \\ 0 & \mathrm{diag}\{\overline{\alpha_i}(w)\} \end{pmatrix}$$

has invariant factors $\overline{\psi_i}(w)$ and such that $\deg_{ci} D_1(w) \leq \varepsilon_i$.

Conversely, if such matrices $D_1(w), D_2(w)$ exist, then a solution $(C, D) = F$ for $FN(w) = D_1(w)$ always exists, because of the particular form of $N(w)$.

This process can now be summarized.

Proposition 1 Given the system (1) and the monic polynomials $\psi_i(s), i = 1, ..., \kappa$ such that $\psi_1(s) \leq \psi_2(s) \leq ... \leq \psi_\kappa$ and the integers $d_i, i = 1, ..., k_d$ such that $d_1 \leq d_2 \leq ... \leq d_{k_d}$, and assuming that $(sE - A, -B)$ has Kronecker invariants

- $\alpha_i(s), i = 1, 2, ..., \rho$ invariant factors,
- $p_i, i = 1, 2, ..., k_p$ infinite zeroes,
- $\varepsilon_i, i = 1, 2, ..., k_\varepsilon$ column minimal indices,
- $\eta_i, i = 1, 2, ..., k_\eta$ row minimal indices,

there exist C, D such that the polynomials $\psi_i(s)$s and the integers d_is be respectively the invariant factors and the infinite zeroes orders of

$$\begin{pmatrix} sE - A & B \\ C & D \end{pmatrix}$$

if and only if there exist polynomial matrices $D_1(w)$ and $D_2(w)$ of dimension $p \times k_\varepsilon$, and $p \times \rho$ respectively, such that $\deg_{ci} D_1(w) \leq \varepsilon_i$, for $i = 1, ..., k_\varepsilon$ and such that

$$\begin{pmatrix} D_1(w) & D_2(w) \\ 0 & \mathrm{diag}\{\overline{\alpha_i}(w)\} \end{pmatrix}$$

has invariant factors $\overline{\psi_i}(w)$ given by (4).

3. MAIN RESULT

Theorem 2 There exist $p \times n$ and $p \times m$ matrices C and D as in the Proposition 1 if and only if the following conditions hold.

$$\psi_i(s) \leq \alpha_i(s) \leq \psi_{i+p}(s), i = 1, 2, ..., \rho, \quad (5)$$

$$d_{i+p} \leq p_i \leq d_i, \quad i = 1, 2, ..., k_p, \quad (6)$$

and, for $j = 1, ..., \kappa - \rho$,

$$\sum_{i=1}^{j} \varepsilon_{\kappa-\rho-i+1} + \sum_{i=1}^{\rho} (\deg \alpha_i(s) + p_i) \geq \sum_{i=1}^{\rho+j} (\deg \beta_i^j(s) + \max(p_{i-j}, d_i)), \quad (7)$$

where, for $j = 1, ..., p$ and $i = 1, ..., \rho + j$

$$\beta_i^j(s) = \mathrm{lcm}(\alpha_{i-j}(s), \psi_i(s)).$$

proof The necessity of conditions (5) and (6) follow from Proposition 1 and Corollary A4 of the appendix, which give, for $i = 1, ..., \rho$,

$$\overline{\psi_i}(w) \leq \overline{\alpha_i}(w) \leq \overline{\psi_{i+p}}(w).$$

Further it appears, from Proposition 1, Lemma A5 and Lemma A6, that

$$\sum_{i=1}^{j} \varepsilon_{\kappa-\rho-i+1} \geq \sum_{i=1}^{j} \deg \overline{\sigma_i}(w),$$

where, for $j = 1, ..., \kappa - \rho$ and $i = 1, ..., \rho + j$,

$$\overline{\sigma_i}(w) = \frac{\overline{\beta_1^i}(s)\overline{\beta_2^i}(s)...\overline{\beta_{\rho+i}^i}(s)}{\overline{\beta_1^{i-1}}(s)\overline{\beta_2^{i-1}}(s)...\overline{\beta_{\rho+i-1}^{i-1}}(s)},$$

and

$$\overline{\beta_i^j}(w) = \mathrm{lcm}(\overline{\alpha_{i-j}}(w), \overline{\psi_i}(w)).$$

(7) then follows since developping these expressions give, for $j = 1, ..., \kappa - \rho$,

$$\sum_{i=1}^{j} \deg \overline{\sigma_i}(w) = \sum_{i=1}^{\rho+j} \deg \overline{\beta_i^j}(w) - \sum_{i=1}^{\rho} \deg \overline{\alpha_i}(w).$$

The sufficiency of conditions (5), (6) and (7) comes remarking that (5), (6) together with Corollary A4 and Lemma A5 allow the construction of a matrix with invariant factors $\overline{\sigma_i}(w)$. Lemma A6 and condition (7) finally ensure the existence of a matrix $D_1(w)$ which satisfies the requirements enounced in Proposition 1. As it is explained in section 2, the matrices C and D are directly deduced from $D_1(w)$.

4. CONCLUSION

In this paper we solve the problem of assigning the zero structure of a State-output system by choosing matrices C and D . The result is obtained for general systems (E, A, B) and a construction procedure for C and D is given in the sufficiency proof. In addition, the results collected in the appendix are of independent interest.

In the particular case of explicit system the conditions of the main result slightly differ from those proposed by Rosenbrock (1970). This is due to the fact that only the case where $C = 0$, i.e. the system (A, B, C, D) is strictly proper,is considered by Rosenbrock.

Note also that, in the explicit case, we are not assigning the zeroes of the numerators of the Smith Mac Millan form of $C(sI - A)^{-1}B + D$, since they coincide with the zeroes of the system matrix if and only if the pair (C, A) is observable. As pointed out in Rosenbrock (1970), a sufficient, but not necessary, condition for observability is the coprimeness of $\alpha_\rho(s)$ and $\psi_\kappa(s)$.

5. APPENDIX

The change of variable $w = 1/(s - a)$ (note that it leads to $s = (1 + aw)/w$) is among our basic tools. The following result describes its effect in terms of Kronecker invariants.

Proposition A1 Given a $p \times m$ pencil $sE - H$ having the finite elementary divisors $(s - a_j)^{k_{ji}}, j = 1, ..., t$ and $i = 1, ..., k_j$, infinite zeros $p_i, i = 1, 2, ..., k_p$, column minimal indices $\varepsilon_i, i = 1, 2, ..., k\varepsilon$ and row minimal indices $\eta_i, i = 1, 2, ..., k_\eta$, let $w = \frac{1}{s-a}$, and let T be a square invertible matrix such that $TE = \begin{pmatrix} \widetilde{E} \\ 0 \end{pmatrix}$, where $\widetilde{E}$ is of full row rank, say r. Then the transformation $\mathcal{T}$ defined by

$$\mathcal{T}(sE - H) = \begin{pmatrix} wI_r & 0 \\ 0 & I \end{pmatrix} T \left(\frac{1 + aw}{w} E - H \right)$$

is biunivoque. In particular the Kronecker invariants of $\mathcal{T}(sE - H)$ (biunivoquely deduced from those of $sE - H$) are the finite elementary divisors $(w - 1/(a - a_j))^{k_{ji}}, j = 1, ..., t, a_j \neq a$ and $i = 1, ..., k_j$, and $w^{p_i}, i = 1, 2, ..., k_p$, the infinite zero orders k_{li}, if l is so that $a_l = a$, the column minimal indices $\epsilon_i, i = 1, 2, ..., k_\epsilon$, and the row minimal indices $\eta_i - 1, i = 1, 2, ..., k_\eta$.

proof. One can verify that the pencil $(sE - H)$ can be obtained from the pencil $(wJ - G) = \mathcal{T}(sE - H)$ as $(sE - H) = \begin{pmatrix} (s - a)I_{r'} & 0 \\ 0 & I \end{pmatrix} T'(\frac{1}{s-a}J - G)$ where $T'J = \begin{pmatrix} \widetilde{J} \\ 0 \end{pmatrix}$, $\widetilde{J}$ being of full row rank r'. $\mathcal{T}$ is thus invertible. We can further assume, without any loss of generality, that the pencil $sE - H$ is in Kronecker canonical form (Gantmacher, 1966). The calculus of $\mathcal{T}(sE - H)$ can thus be easily performed. Indeed each of the diagonal blocks in $sE - H$ is transformed through $\mathcal{T}$ in a diagonal block of the pencil $wJ - G$.

- Each finite elementary divisors $(s - a_j)^{k_{ji}}, j = 1, t$ and $i = 1, k_j$ corresponds in $sE - H$ to a diagonal block of the form

$$\begin{bmatrix} s - a_j & -1 & & \\ & \ddots & \ddots & \\ & & \ddots & -1 \\ & & & s - a_j \end{bmatrix}, k_{ji} \times k_{ji}$$

This block is transformed by T in a diagonal block of $wJ - G$ equivalent to

$$\begin{bmatrix} w - \frac{1}{a_i - a} & -1 & & \\ & \ddots & \ddots & \\ & & \ddots & -1 \\ & & & w - \frac{1}{a_i - a} \end{bmatrix}, \ k_{ji} \times k_{ji}$$

which corresponds to a finite elementary divisor of $wJ - G$ at $w = \frac{1}{a_i - a}$ of order k_{ji}, if $a \neq a_i$, and to an infinite zero of order k_{ji} if $a = a_j$.

- Each infinite zero order p_i corresponds to a diagonal block in $(sE - H)$, of the form

$$\begin{bmatrix} -1 & s & & \\ & \ddots & \ddots & \\ & & \ddots & s \\ & & & -1 \end{bmatrix}, (p_i + 1) \times (p_i + 1)$$

The transformed block is equivalent to

$$\begin{bmatrix} w & -1 & & \\ & \ddots & \ddots & \\ & & \ddots & -1 \\ & & & w \\ & & & & 1 \end{bmatrix}, \ (p_i + 1) \times (p_i + 1)$$

which corresponds to a finite elementary divisor of the form w_i^p for the pencil $wJ - G$.

- Similarly, each column minimal index ε_i corresponds to a block

$$\begin{bmatrix} s & 1 & & \\ & \ddots & \ddots & \\ & & s & 1 \end{bmatrix}, \varepsilon_i \times (\varepsilon_i + 1)$$

that $\mathcal{T}$ transforms in

$$\begin{bmatrix} w & -1 & & \\ & \ddots & \ddots & \\ & & w & -1 \end{bmatrix}, \ \varepsilon_i \times (\varepsilon_i + 1)$$

which correspond to a column minimal index of $wJ - G$ of order ε_i. The column minimal indices are actually not modified by $\mathcal{T}$.

- The row minimal index η_i is associated with the diagonal block of $sE - H$

$$\begin{bmatrix} s & & \\ 1 & \ddots & \\ & \ddots & s \\ & & 1 \end{bmatrix}, (\eta_i + 1) \times \eta_i$$

The transformed block is equivalent to

$$\begin{bmatrix} w & & & \\ 1 & \ddots & & \\ & \ddots & w & \\ & & 1 & \\ & & & 1 \end{bmatrix}, \ (\eta_i + 1) \times \eta_i$$

which corresponds to a row minimal index of $wJ - G$ of order $\eta_i - 1$.

The result clearly follows from this inspection.

Remark A2 The various lists of Kronecker invariants are generally assumed to be arranged in non-increasing order, that is $p_1 geq p_2 geq ...$, $k_{j1} geq k_{j2} geq ...$, and so on. If ρ denotes the rank of $sE - H$, the invariant factors of $sE - H$, arranged in non-decreasing order are the polynomials

$$\alpha_j(s) = \prod_{j=1}^{t} (s - a_j)^{k_{j,\rho-i+1}} \qquad (8)$$

where we take by convention $k_{ji} = 0$ for $i > k_j$.

The invariant factors of $wJ - G$, which rank is also ρ can similarly be expressed in terms of elementary divisors.

$$\overline{\alpha_i}(w) = \prod_{j | a_j \neq a} w^{p_{\rho-i+1}} \left(w - \frac{1}{a_j - a} \right)^{k_{ji}}. \qquad (9)$$

If there is no j such that $a_j = a$, a simpler expression is obtained,

$$\overline{\alpha_i}(w) = c_i w^{p_{\rho-i+1}} \alpha_i \left(\frac{1 + aw}{w} \right) w^{\deg \alpha_i(s)} \quad (10)$$

where c_i is a normalizing constant so that $\overline{\alpha_i}(w)$ be monic.

We shall now recall and develop some results concerning matrix completion, invariant factors and column degrees.

Theorem A3 (Sá, 1979, and Thompson, 1979). Given a $p \times m$ polynomial matrix A(s), having rank ρ and invariant polynomials $\alpha_1(s) \leq \alpha_2(s) \leq ... \leq \alpha_\rho(s)$, and a set of monic polynomials $\psi_1(s) \leq \psi_2(s) \leq ... \leq \psi_\kappa(s)$, there exists a $q \times m$ polynomial matrix $B(s)$ such that the composite matrix $\begin{pmatrix} B(s) \\ A(s) \end{pmatrix}$ have rank κ and invariant factors $\psi_1(s), \psi_2(s), ..., \psi_\kappa(s)$ if and only if the two following properties hold.

$$\psi_i(s) \leq \alpha_i(s) \leq \psi_{i+q}(s), \text{for } i = 1, ..., \rho$$

$$\kappa \leq min(m, q + \rho)$$

We shall use this famous result under the following form.

Corollary A4 Given a full rank $q \times q$ polynomial matrix $A(s)$ with invariant factors $\alpha_1(s) \leq \alpha_2(s) \leq ... \leq \alpha_q(s)$, and the monic polynomials $\psi_1(s) \leq \psi_2(s) \leq ... \leq \psi_\kappa(s)$, there exist polynomial matrices $D(s)$ and $Q(s)$ such that the $(p + q) \times (m + q)$ polynomial matrix

$$\begin{bmatrix} D(s) & Q(s) \\ 0 & A(s) \end{bmatrix}$$

has invariant polynomials $\psi_1(s), \psi_2(s), ..., \psi_\kappa$ if and only if $\kappa \leq q + min(p, m)$ and

$$\psi_i(s) \leq \alpha_i(s) \leq \psi_{i+p}(s), \text{for} i = 1, ..., q.$$

Lemma A5 Under the conditions of Corollary A4, the invariant factors of $D(s)$, say $\delta_1(s), \delta_2(s), ..., \delta_{\kappa-q}(s)$ are such that

$$\prod_{i=1}^{j} \delta_i(s) \geq \prod_{i=1}^{j} \sigma_i(s), j = 1, ..., \kappa - q$$

where

$$\sigma_j(s) = \frac{\beta_1^j(s)\beta_2^j(s)...\beta_{q+j}^j(s)}{\beta_1^{j-1}(s)\beta_2^{j-1}(s)...\beta_{q+j-1}^{j-1}(s)}, \quad j = 1, ..., \kappa - q$$

and, for $j = 0, ..., \kappa - q$ and $i = 1, ..., q + j$,

$$\beta_i^j(s) = \text{lcm}(\alpha_{i-j}(s), \psi_i(s)),$$

in addition, if $p = m$ then equality hold for $j = m$.

Proof Assume that the invariant factors of $D(s)$ are $\delta_1, \delta_2, ..., \delta_{\kappa-q}$. There then exist unimodular matrices $X(s)$ and $Y(s)$ such that

$$X(s)D(s)Y(s) = diag\{\delta_{\kappa-q}, ..., \delta_2, \delta_1\} = D_d(s)$$

Hence,

$$\begin{bmatrix} X(s) & 0 \\ 0 & I_q \end{bmatrix} \begin{bmatrix} D(s) & Q(s) \\ 0 & A(s) \end{bmatrix} \begin{bmatrix} Y(s) & 0 \\ 0 & I_q \end{bmatrix} = \begin{bmatrix} D_d(s) & Q_1(s) \\ 0 & A(s) \end{bmatrix} = G(s)$$

has invariant factors $\psi_1(s), \psi_2(s), ..., \psi_\kappa(s)$. Let $G_j(s)$ be the submatrix of $G(s)$ formed by its $m + q$ columns and by $j + q$ last rows. Let $\mu_1^j(s), \mu_2^j(s), ..., \mu_{q+j}^j(s)$ be the invariant factors of $G_j(s)$.

From the inbedment of $A(s)$ in $G_j(s)$ and of $G_j(s)$ in $G(s)$ and from Theorem A2, the following sets of interlacing inequalities hold for $j = 0, ..., p$ and $i = 1, ..., q + j$.

$$\psi_i(s) \leq \mu_i^j(s) \leq \psi_{i+p-j},$$

$$\mu_i^j(s) \leq \alpha_i(s) \leq \mu_{i+j}^j(s).$$

Define now, for $j = 0, ..., p$ and $i = 1, ..., q + j$, $\beta_i^j(s) = \text{lcm}(\alpha_{i-j}, \psi_i)$. Since $\psi_i(s) \leq \mu_i^j(s)$ and $\alpha_{i-j}(s) \leq \mu_i^j(s)$, it appears that, for $j = 0, ..., p$ and $i = 1, ..., q + j$, $\beta_i^j(s) \leq \mu_i^j(s)$.

Finally for $j = 1, 2, ..., p$, we obtain

$$\sigma_1\sigma_2...\sigma_j = \frac{\prod_{i=1}^{q+j}\beta_i^j}{\prod_{i=1}^{q}\alpha_i} \leq \frac{\prod_{i=1}^{q+j}\mu_i^j}{\prod_{i=1}^{q}\alpha_i} = \delta_1\delta_2...\delta_j,$$

which establishes the result. In addition, if $p = m$ and $j = m$ then $\prod_{i=1}^{q+m}\beta_i^m(s) = \prod_{i=1}^{q+m}\mu_i^m(s) = \prod_{i=1}^{q+m}\psi_i(s)$. Therefore equality holds for $j = m$.

Lemma A6 There exists a $p \times m$ polynomial matrix of rank r with invariant factors $\delta_1(s) \leq \delta_2(s) \leq ... \leq \delta_r(s)$ and which column degrees are smaller than or equal to $\varepsilon_1 \geq \varepsilon_2 \geq ... \geq \varepsilon_m$, if and only if

$$\sum_{i=1}^{j}\varepsilon_{r-i+1} \geq \sum_{i=1}^{j}\deg\delta_i(s), j = 1, ..., r$$

Proof The necessity of these conditions follows from the fact that, by definition of the invariant factors of a polynomial matrix, $\prod_{i=1}^{k}\delta_i(s)$ is the greatest common divisor of the minors of order k of the matrix.

To establish the sufficiency, let us remark that under these conditions there exist $k_1 \geq k_2 \geq ... \geq k_r$ such that $k_i \leq \varepsilon_i$ and $\sum_{i=1}^{j}k_{r-i+1} \geq \sum_{i=1}^{j}\deg\delta_i(s)$, for $j = 1, ..., r$ with equality for $j = r$. The result then follows applying the procedure used in Rosenbrock (1970) to solve the pole placement problem in the controllable case.

6. REFERENCES

Gantmacher F.R.(1966). Théorie des matrices, Dunod, Paris, 1966.

Karcanias N.(1994). The selection of input and output schemes for a system and the model projection problems, Kybernetica, **30**, 6, 585-596.

Karkanias N.(1994). Global process instrumentation–Issues and problems of a system and control theory framework, Measurement, **14**, 103-113.

Kučera V.and P. Zagalak (1988). Fundamental Theorem of state feedback for singular systems, Automatica, **24**, 5, 653-658.

Loiseau J.J., K. Özçaldiran, M. Malabre and N. Karcanias (1991). A feedback classification of singular systems, Kybernetika, **27**, 289-305.

Loiseau J.J.and P. Zagalak (1993). Eigenstructure assignment in linear and uncontrollable systems, proc. MTNS, Regensburg, Germany.

Marques de Sà E. (1979). Imbedding conditions for λmatrices, Lin. Alg. Appl., **24**, 33-50.

Rosenbrock H.H. (1970). State-space and Multivariable Theory, Wiley, New York.

Syrmos V. L. and F.L. Lewis (1993). Transmission zero assignment using semistate descriptions, IEEE Trans. Autom. Contr., **38**, 7, 1115-1120.

Thompson R.C. (1979). Interlacing inequalities for invariant factors, Lin. Alg. Applic., **24**, 1-31.

Zaballa I. (1987). Matrices with prescribed rows and invariant factors, Lin. Alg. Appl., **87**,113-146.

Zaballa I. (1988). Interlacing inequalities and control theory, Lin. Alg. Appl., **101**, 9-31.

Zagalak P.and J.J. loiseau (1992). Invariant factors assignment in linear systems, Proc. SINS'92, Ft.Worth, Texas, ARRI-UTA, 197-204.

Zagalak P., V. Kučera and J.J. Loiseau(1993). Polynomial methods in Optimal Control and Filtering, chapter 8, IEE Control Engineering Series 49, Peter Pelegrinus, London.

REDUCED SENSITIVITY SOLUTIONS TO THE GLOBAL LINEARISATION OF THE CONSTANT POLE ASSIGNMENT PROBLEM

J.LEVENTIDES and N.KARCANIAS

Control Eng. Dept., City University, London EC1V OHB

Résumé: Notre contribution s'inscrit dans le contexte de la linéarisation globale du placement de pôles par retour de sortie statique, qui a notamment conduit a une demonstration simple du résultat mp>n (condition nécessaire et suffisante de placement arbitraire des pôles pour un systéme générique a p entrées, m sortie et états) et a des méthodes permettant le calcul de lois de commande solutions. Un inconvénient de la méthode existante vient du fait que, pour approximer le polynôme désiré des pôles, il faut se rapprocher du gain dégénéré (singularité de la boucle) utilisé dans la linéarisation. Dans ce papier, nous améliorons cette méthode en utilisant des termes d'ordre supérieur qui assurent une certain distance par rapport a ce point dégénéré.

Abstract: The paper develops further the global linearisation framework of the pole assignment by constant output feedback, which has given a simple proof of the mp>n result (necessary and sufficient condition for arbitrary pole placement of a generic p-input, m-output and n-state system) and has introduced a computational framework for finding output feedbacks. Although the existing solution is in closed form it has the disadvantage that in order to approximate the desired pole polynomial we have to approach the degenerate gain (singularity of the feedback configuration) used for the linearisation. Here the original method is improved by using higher order terms in the original method which guarantees some distance from the degenerate point.

1. Introduction

The construction of constant output feedback compensators that place the poles of a MIMO system of p-inputs, m-outputs and n-states, to arbitrarily chosen positions was always a challenging open problem in Control theory. It is a multilinear problem in the gain parameters, and can be formulated as a problem of solving the algebraic equations $p(s)=\det\{D(s)+KN(s)\}$ (which is one form of the general determinantal assignment problem[3]). Where p(s) is the polynomial to be placed and $N(s)D(s)^{-1}$ is a coprime MFD of the transfer function. This equation has to be solved with respect to $K \in \mathbb{R}^{p\times m}$, and contrary to the state feedback case, it is not always solvable(assuming controllability). In fact, a necessary condition for its solvability is that $mp \geq n$ [12] where m,p,n are the number of outputs, inputs and states of the system. Although mp>n was proved to be a sufficient condition [10] for the generic solvability of the problem, the known constructive methods (by dyadic or full rank output feedback) work only in the restricted case when $m+p-1 \geq n$ [2,4,8]. Very recently a new constructive method was developed that treats the general case mp>n [5]. This method is based on a linearisation of the pole placement map [7,9,11] by considering special sequences of feedback compensators which, in the limit, converge to a so called degenerate compensator; The advantage of this approach is that it asymptotically reduces the overall pole placement map to a linear one and thus reduces the overall solvability of the problem to a linear set of equations (and this without simplifying the structure of the controllers). This type of global asymptotic linearization defined around singular solutions is different from the standard methodologies for linearizing the multilinear nature of the map (dyadic feedback, full rank linearization)[8]. The solutions worked out within the present framework are given in a closed form and they are of the type of a one parameter family of multivariable compensators. The essence of this new methodology is that we may rely on large gains in the case of strictly proper systems, but not necessarily large gains in the proper case and assignment is achieved not in an exact sense, but in small neighbourhoods (as small as we want) around a given set of poles. It should be stressed that the approach is sufficient, but quite general, since the condition mp>n is shown to guarantee the success of the method for a generic system. However a serious disadvantage of the method is that it is based on approaching a degenerate controller ie. a controller for which the closed loop system is not well defined; this has the effect that although the poles of closed loop system are very near to the desired ones, the sensitivity increases considerably and in fact tends to infinity as $\epsilon \to 0$ (ie. exact pole placement and sensitivity are conflicting).

The purpose of this paper is to improve the above linearisation method around a degenerate point, so that the desired poles can be approached whereas the feedback controller does not approach the degenerate point (and thus achieving pole placement which is less sensitive). This can be done if we consider feedback compensators of the form:

$L(\epsilon)=[A,K]+\epsilon[A_1,K_1]+\epsilon^2[A_2,K_2]+...+\epsilon^k[A_k,K_k]$

where $[A,K]$ is a degenerate point and $[A_i,K_i]$ are matrices to be calculated. Provided now that k is large enough, the roots of the closed loop polynomial $p_\epsilon(s)$ approach the roots of $p_1(s)$ without ϵ necessarily being close to zero ($\epsilon<1$ is only required), that is, without the controller, $L(\epsilon)$, to approach the degenerate one. Another important aspect of the method is that the coefficient matrix of the linear equations for every power of ϵ^i for all i is the same, the so called blow up matrix, which has mp rows and n columns and depends on the system as well as the specific degenerate point. A sufficient condition for arbitrary pole placement is that this matrix has to have full rank, in which case the above method works (in [7] it has been proven that if mp>n then this matrix has generically full rank). Next we briefly present the basic facts of the global linearisation method according to [7].

2. Global linearisation of the pole placement map.

The method of global asymptotic linearisation of the pole placement map was first intoduced in [5] and subsequently developed in [7,9,11] , was based on the remarkable property of the *degenerate gains* of a feedback configuration to "blow up" sequences of gains converging to them.

Consider the system described by the mxp transfer function matrix with McMillan degree n and represented by the right coprime MFD $G(s)=N(s)D(s)^{-1}$. The pole assignment by consatant output feedback involves the solution of:

$$p(s)=\det\left\{[I,\ K]\begin{bmatrix}D(s)\\N(s)\end{bmatrix}\right\} \qquad (1)$$

with respect to K. Degenerate gains were first defined in [1] in the framework of generalised gains ***rowspan***[A,K] instead of the ordinary gains K, as follows:

Definition (2.1) Let $G(s)=N(s)D(s)^{-1}$ is the transfer function of the open loop plant then, the generalised gain ***rowspan***[A,K] , is degenerate iff it satisfies equation:

$$\det\left\{[A,\ K]\begin{bmatrix}D(s)\\N(s)\end{bmatrix}\right\}=0 \quad \forall s \qquad (2)$$

(or in other words, iff the closed loop system for this gain is not well defined).

□

Although equation(2) is multilinear with respect to [A,K], degenerate gains can be constructed easily from the nullspaces of certain matrices [7]. Having constructed a degenerate gain [A,K], to achieve global linearisation, it is necessary to consider sequences of generalised gains (omitting ***rowspan*** for simplicity) : $L(\epsilon)=[A,K]+\epsilon\ [A_1,K_1]$ that converge to the degenerate gain [A,K] as ϵ tends to 0. After closing the loop using the gain matrix $(A+\epsilon A_1)^{-1}(K+\epsilon K_1)$, the closed loop polynomial has the same roots with the ones of the polynomial:

$$p_\epsilon(s)=\det\left\{L(\epsilon)\begin{bmatrix}D(s)\\N(s)\end{bmatrix}\right\}$$

Letting now $\epsilon \to 0$, $p_\epsilon(s)$ tends to a polynomial p(s), the so called *prime polynomial.* In the case of nondegenerate gains, when the left hand side of (2) is nonzero and equal to, say, $p_1(s)$, the prime polynomial does not depend on the direction $[A_1,K_1]$, and it is always equal to $p_1(s)$ (modulo dilations). This is a fact due to the continuity of the pole placement map at the nondegenerate gain (no matter how we approach the point on the domain we get the same limit on the range). On the contrary, when [A,K] is a degenerate gain, the prime polynomial p(s) is not unique and depends on the direction $[A_1,K_1]$. Furthermore, as the next theorem states the relation between the prime polynomial vector and the direction $[A_1,K_1]$ is linear.

Theorem(1)[7] Let **rowspan**[A,K] be a degenerate gain and $L(\epsilon)$ be a sequence of finite gains converging to it. Then the corresponding sequence of closed loop polynomial coefficient vectors $<\underline{p}_\epsilon>$ (modulo dilations) converges as $\epsilon \to 0$ to a vector $<\underline{p}> \in \mathbb{P}(\mathbb{R})^n$ which depends on $[A_1,K_1]$; furthermore the function τ which maps the direction $[A_1,K_1]$ to $<\underline{p}>$ is linear.

□

The above theorem is the basis for the global linearisation method for the pole placement problem. The matrix representation of the linear map τ can be deduced from the theorem below:

Theorem(2)[7] Let $\mathfrak{D}$=*rowspan*[A,K] be a degenerate point of a system defined by the coprime composite representation $M(s)=[D(s)^T,N(s)^T]^T$, then the prime polynomial of this system with respect to $\mathfrak{D}$ and the direction $[A_1,K_1] = (b_{ij})$ $1 \leq i \leq p$, $1 \leq j \leq p+m$, can be written as:

$$p(s)=\sum(b_{ij}\; p_{ij}(s)) \tag{3}$$

where $p_{ij}(s)$ is the determinant of the pxp polynomial matrix $D_{ij}(s)$ having the same rows with the matrix $AD(s) + KN(s)$ apart from the i-th, which is replaced by the j-th row of M(s).

□

Therefore the coefficient vector $\underline{p}^T$ of the prime polynomial can be written as:

$$\underline{p} = B\,\underline{x} \tag{4}$$

where $\underline{x}^T$ is the vector formed by all the columns of the unknown direction $[A_1,K_1]$ and B is the coefficient matrix of the polynomial vector $[p_{11}(s),\ p_{21}(s),\ \ldots,\ p_{p\ (p+m)}(s)]$ whose entries $p_{ij}(s)$ are given by theorem (2).

Thus the construction of an (approximate) solution for the arbitrary pole placement problem involves the steps: First we construct a degenerate gain [A,K] as in [7], then we find the matrix B for that gain as described previously and then solve the linear equations (4) with respect to $\underline{x}$ and thus get the direction $[A_1,K_1]$; finally the required one parameter family of feedback gains will be $K(\epsilon)=(A+\epsilon A_1)^{-1}(K+\epsilon K_1)$. After closing the loop with $K(\epsilon)$, the poles tend to the required ones as ϵ tends to 0.

Although the method works generically for the best possible case (ie mp>n) and as a constructive method allows considerably larger number of states in the open loop system compared to the already existing ones, it has a disadvantage which makes it difficult to implement in the form presented in [7]. This disadvantage lies on that the method is based on approaching a degenerate controller ie. a controller for which the closed loop system is not well defined. This has the effect that although the poles of the closed loop system can be placed very near to the desired ones, the norm of the sensitivity function $(I+K(\epsilon)G(s))^{-1}$ is very large as $\epsilon \to 0$, and thus, small perturbations in the parameters may result to very big pertrubations in the closed loop poles. Next we will examine how the global asymptotic linearisation method can be improved in order to overcome this obstacle.

3. Higher order approximations: the philosophy of the extended approach.

As we saw in the previous section the global asymptotic linearisation is based on calculating a certain matrix $[A_1,K_1]$ which serves the role of the direction via which the one parameter family $[A,K]+\epsilon\,[A_1,K_1]$ approaches the degenerate gain [A,K]. The appropriate calcualtion of this direction allows us to have a closed loop polynomial of the form: $p_\epsilon(s)=\epsilon p_1(s)+\epsilon^2 p_2(s)+\ldots$, where only the $p_1(s)$ is assigned (this is the prime polynomial). For ϵ large, the closed loop pole polynomial may be considerably far from the desired polynomial $p_1(s)$ and only when ϵ is small we have that $p_\epsilon(s) \approx \epsilon p_1(s)$ and therefore we get the desired poles. To be able to be close to the desired poles for ϵ large, we need also to appropriately assign the polynomials $p_2(s)$, $p_3(s)$, and so on. This can be done by introducing higher order perturbation series for the degenerate point [A,K] so that the additional degrees of freedom will contribute to the assignment of the additional polynomials. Therefore, we will try to improve the global linearisation method of [7] by using feedback compensators of the form:

$$L(\epsilon)=[A,K]+\epsilon[A_1,K_1]+\epsilon^2\,[A_2,K_2]+\ldots+\epsilon^k[A_k,K_k] \tag{5}$$

where [A,K] is a degenerate point and $[A_i,K_i]$ are matrices to be calculated. In this case the closed loop pole polynomial takes the form: $p_\epsilon(s)=\epsilon p_1(s)+\epsilon^2 p_2(s)+\epsilon^3 p_3(s)+\ldots$, and as it will be shown next, the polynomials $p_i(s)$ i=1,2,3,...,k depend linearly on $[A_i,K_i]$ i=1,2,...,k.

Theorem(3) Consider the feedback controller $L(\epsilon)$ to be as in (5) and let the closed loop pole polynomial be:

$$\det\left\{L(\epsilon)\begin{bmatrix} D(s) \\ N(s)\end{bmatrix}\right\}=\epsilon p_1(s)+\epsilon^2 p_2(s)+\epsilon^3 p_3(s)+\ldots$$

then the coefficient vectors $\underline{p}_1, \underline{p}_2, \ldots, \underline{p}_k$, of the polynomials $p_1(s)$, $p_2(s)$,...$p_k(s)$ are given by:

$$\underline{p}_1 = B\,\underline{x}_1 \tag{6a}$$

$$\underline{p}_2 = \underline{a}_2 + B\,\underline{x}_2 \tag{6b}$$

$$\vdots$$

$$\underline{p}_k = \underline{a}_k + B\,\underline{x}_k$$

where B is the so called *blow up matrix* defined as in eq.(4) , and if we denote by $\underline{x}_{ij}^T$ the j-th row of the matrix $[A_i,K_i]$, then

$$\underline{x}_i^T=[\underline{x}_{i1}^T,\ \underline{x}_{i2}^T,\ldots,\ \underline{x}_{ip}^T]$$

and

$$\underline{a}_i=\left(\sum \underline{x}_{i_1 1}^T \wedge \underline{x}_{i_2 2}^T \wedge \ldots \wedge \underline{x}_{i_p p}^T\right) P_S$$

where the sum is taken for all $0 \leq i_1, i_2, \ldots, i_p \leq i$ such that $i_k \neq i_f$ and $i_1+i_2+\ldots+i_p=i$.

□

The above theorem can be used as follows: Solve the first set of linear equations (6a) with respect to $[A_1,K_1]$ so that that $p_1(s)$ is the required polynomial to be assigned, then substitute this solution into $\underline{a}_2$ set also $p_2(s)=0$ and thus get the second linear equation (6b) which has to be solved with respect to $[A_2,K_2]$. Repeat this procedure gradually finding the $[A_i,K_i]$ $i=2,3,\ldots,k$ so that the $p_i(s)$ $i=2,3,\ldots,k$ are zero. This will produce a solution for which the closed loop pole polynomial will have the same roots with the polynomial

$$p_\epsilon(s)=p_1(s)+\epsilon^k p_k(s)+\ldots$$

and it is apparent that the more terms $[A_i,K_i]$ we use in the above algorithm the faster $p_\epsilon(s)$ converges to the required polynomial $p_1(s)$. This fast convergence will allow us to take a bigger ϵ and thus to stay away from the degenerate point which as we said is a singularity for the feedback configuration.

Remark(1) The methodology suggested by theorem(3) is clearly recursive but it involves only the computation of the left inverse of B (which exists since the Blow up matrix has full column rank). Note that the size of singular values of B depend on the properties of Plucker matrix[3] as well as the selection of the degenerate point. The optimal selection of degenerate points such that the condition number of B is minimised is still an open issue. □

For a measure of the sensitivity we will consider the norm of the Differential (or Jacobian) $\|D(\mathbf{F})_{K(\epsilon)}\|$ of the Pole Placement map $\mathbf{F}$ [6], evaluated at the compensator $K(\epsilon)=(A+\epsilon A_1\ldots)^{-1}(K+\epsilon K_1\ldots)$ found by the above method (although we can alternatively use the norm of the sensitivity function $(I+K(\epsilon)G(s))^{-1}$). On can prove that $\|D(\mathbf{F})_{K(\epsilon)}\| \to \infty$ as $\epsilon \to 0$ and therefore by taking the Laurent expansion at zero we have:

$$\|D(\mathbf{F})_{K(\epsilon)}\| \approx \frac{\alpha}{\epsilon^d}$$

for ϵ small. If we also consider the norm $\|\Delta p\|$ of the difference of the closed loop polynomial $p_\epsilon(s)$ (using $K(\epsilon)$ as compensator) and the required polynomial $p_1(s)$ we have that:

$$\|\Delta p\| \approx \beta \epsilon^k$$

for ϵ small. Therefore

$$\|D(\mathbf{F})_{K(\epsilon)}\| \approx c\|\Delta p\|^{-\frac{d}{k}}$$

for ϵ small. Thus we can have $\|\Delta p\|$ very close to zero (ie place the poles to the desired locations) and yet the sensitivity to be bounded (since the k-th root of $\|\Delta p\|$ tends to 1 as $k \to \infty$ no matter how small $\|\Delta p\|$ is). This indicates that we can reduce the sensitivity by introducing higher terms in the expansion (5)). As we will see in the next example the present modified global linearisation method indeed produces solutions with improved sensitivity.

Example(1) Consider the system defined by:

$$M(s)=\begin{bmatrix} s^3 & 0 \\ 1 & s^2 \\ s^2+1 & s+1 \\ s+3 & s \\ s+1 & 1 \end{bmatrix}=\begin{bmatrix} D(s) \\ N(s) \end{bmatrix}$$

A degenerate point for this system is defined by $\mathfrak{D}=$***rowspan***$[A,K]$ where:

$$[A,K]=\begin{bmatrix} 1 & 0 & 0 & 0 & 0 \\ 0 & 0 & 1 & -1 & -1 \end{bmatrix}$$

we used three types of perturbations
a) $[A,K]+\epsilon[A_1,K_1]$,
b) $[A,K]+\epsilon[A_1,K_1]+\epsilon^2[A_2,K_2]+\epsilon^3[A_3,K_3]+\epsilon^4[A_4,K_4]$
c) $[A,K]+\epsilon[A_1,K_1]+\epsilon^2[A_2,K_2]+\ldots+\epsilon^8[A_8,K_8]$

and the polynomial to be placed is $p(s)=s^5+5s^4+10s^3+13s^2+5.66667s+1$. Using the above described method we calculate the $[A_i,K_i]$ and we get three different one parameter families of controllers and for each of these we plot the maximum singular value of the Jacobian of the pole placement map [6] $\|D(\mathbf{F})_{K(\epsilon)}\|$, against the distance $\|\Delta p\|$ of the closed loop pole polynomial from the polynomial to be assigned. It is worth noting that both sensitivity and $\|\Delta p\|$ are functions of ϵ for all cases a),b) and c). Next diagram demonstrates the parametric plots of $\|D(\mathbf{F})_{K(\epsilon)}\|$ against $\|\Delta p\|$ for the three cases.

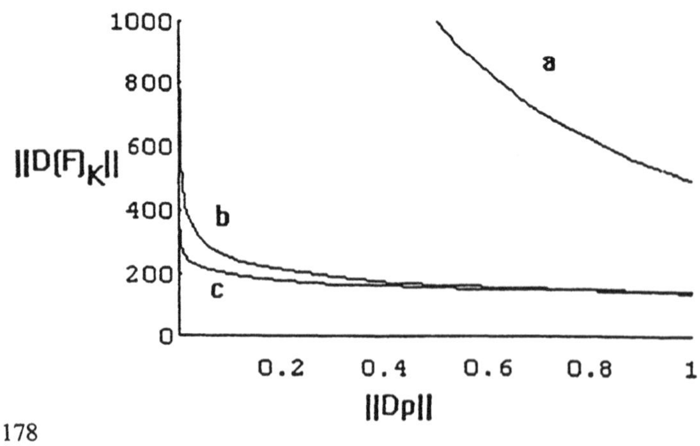

From these plots we can see that the 8th term series is the one we can achieve lower sensitivity with small ϵ. Therefore we can select an ϵ so that $\|\Delta p\| \leq 0.1$ without on the same time having very large $\|D(F)_{K(\epsilon)}\|$. For instance if we select $\epsilon=0.035$ we get the controller:

$$K=\begin{bmatrix} -5.21738 & 3.64006 & 4.75538 \\ 32.9542 & -22.7374 & -30.0227 \end{bmatrix}$$

and the closed loop polynomial for this controller is:
$s^5+4.99937s^4+10.0015s^3+12.9428s^2+5.6930s+0.960269$

which is very close to the required one p(s). It should be noted that to find this solution we solved eight sets of linear equations all of them having the same coefficient matrix B of linearisation. However since the sensitivity norm is close to 200 as the graph indicates, a possible error of ΔK in K may result to a perturbation $200\Delta K$ in the coefficients of the closed loop polynomial.

□

The further development of the method involves two different directions. The first is to obtain results linking the selection of degenerate point and condition number of the Blow up matrix and then relate them to the minimum sensitivity (or gain) requirements for a given distance between the required and the assigned pole polynomials. The second involves the characterisation of convergence of the infinite series around a selected degenerate compensator and the specification of degenerate compensators with best convergence properties.

4.Conclusions

An improved global linearisation method for pole assignment by constant output feedback has been given that involves high order terms in the original single parameter family. This leads to the definition of constant pole assignment schemes with improved sensitivity and reduced gains. The development of a systematic procedure for selection of the parameter ϵ with further improved sensitivity and low norm properties is under investigation at the moment.

REFERENCES

[1]Brockett R.W and Byrnes C.I.(1981). Multivariable Nyquist criteria, Root Loci and Pole Placement', I.E.E.E. Trans. on Aut. Control, Vol.26, 271.

[2]Davison E.J. and Wang S.H.(1975). On pole assignment in linear multivariable systems using output feedback. I.E.E.E. Trans. on Aut. Control, Vol.20,516.

[3]Karcanias N. and Giannacopoulos C.(1984). On Grassmann invariants and almost zeroes of linear systems and the determinantal zero, pole assignment problem. Int. J. Control, Vol. 40, 673.

[4]Kimura H.(1975). Pole assignment by gain output feedback, I.E.E.E. Trans. autom. Control , 20.

[5] Leventides J. (1993) Algebrogeometric and Topological methods in Control theory, Phd Thesis, City University London.

[6]Leventides J. and Karcanias N.(1993). The Pole Placement Map, Its properties and relationships to system invariants', IEEE Transactions AC, 38,1266.

[7]Leventides J. and Karcanias N.(sub. Jan 1994). Global symptotic linearisation of the pole placement map: a closed form solution for the output feedback problem (To appear in Automatica).

[8]Munro N.(1987). Pole assignment: A review of Methods', Systems and Control Encyclopedia, M.G.Singh, Pergamon Press, 3710.

[9] Rosenthal J., Schumacher J.M., Willems J.C.(Dec. 1994), Generic eigenvalue assignment by memoryless real output feedback, CWI report

[10]Wang X. (1992). Pole Placement by static output feedback. J. Math. Sys. Estim. Control, Vol.2, 205.

[11] Wang X. (July 1994) Grassmannian, Central Projection and Output Feedback Pole Assignment of Linear Systems, Preprint.

[12]Willems J.C. and Hesselink W.H.(1978). Generic properties of the pole placement problem. Proc. IFAC , Helsinki , Finland.

A REALIZATION THEORY FOR HOMOGENEOUS AR-SYSTEMS, AN ALGORITHMIC APPROACH

M. S. Ravi
Department of Mathematics, East Carolina University,
Greenville, NC 27858. USA. *e-mail:* maravi@ecuvax.cis.ecu.edu.

J. Rosenthal*
Department of Mathematics, University of Notre Dame,
Notre Dame. IN 46556, USA. *e-mail:* Joachim.Rosenthal@nd.edu.

J. M. Schumacher
CWI, P.O. Box 94079, 1090 GB Amsterdam, the Netherlands, and
Tilburg University, CentER and Dept. of Economics, P.O. Box 90153,
5000 LE Tilburg, the Netherlands. *e-mail:* Hans.Schumacher@cwi.nl.

Abstract. The set of m inputs, p outputs and n states homogeneous linear system has the structure of a smooth compact manifold. In this paper we associate to every homogeneous AR-system a generalized first order representation and we obtain a concrete interpretation as a space of discrete-time behaviors.

Résumé. La collection des systèmes linéaires avec m canaux d'entrée, p canaux de sorties et n états a la structure d'une variété différentiable compacte. Dans ce papier, nous associons chaque système homogène du type AR à une système du premier ordre et nous dérivons une interpretation concrete par une éspace des représentations digitales.

Keywords. Generalized state space system, homogeneous behavior, quot scheme.

1. Introduction

An analogy that has been instrumental in the development of the 'behavioral' theory of linear systems (Willems 1986, Willems 1989, Willems 1991) is that between linear behaviors and linear subspaces of $\mathbb{R}^n$. In fact a subspace of $\mathbb{R}^n$ can be looked at as a static behavior, and the theory of linear time-invariant behaviors may be viewed as a the most direct dynamic generalization of it. The analogy makes it natural for instance to look for 'kernel' and 'image' representations (corresponding to 'AR' and 'MA' representations in the terminology of J. C. Willems). It is also suggestive in defining a notion of convergence. Starting from the representation of a subspace of given dimension m as the column space of a matrix of size $n \times m$, which is determined up to right multiplication by nonsingular matrices, a natural notion of convergence for subspaces is obtained by the construction of the quotient topology. One may also use the kernel representation in the same way; fortunately, this leads to the same notion of convergence. In the same manner, one can define a topology for the the set of linear time-invariant systems of a given state space dimension n and a given number of inputs m. Starting for instance from a minimal first-order representation, which is unique up to similarity transformations, a notion of convergence is again provided by the quotient topology. The set of m-dimensional subspaces of $\mathbb{R}^n$, with the indicated topology, is usually denoted by $\mathrm{Grass}(m,n)$; we shall use the symbol $\mathrm{Sys}(m,p,n)$ to indicate the set of time-invariant linear systems with m inputs, p outputs, and an n-dimensional state space.

*This author was supported in part by NSF grant DMS-94-00965. The research for this paper was carried out in part while he was a visitor at CWI.

In spite of the analogy between 'subspaces' and 'systems', there is an important topological difference between Grass(m, n) and Sys(m, p, n). Whereas the set of subspaces is compact, so that every sequence must have a limit point, the set of systems is not compact. To give an example in the very simple case $n = 1$, $m = 0$, $p = 1$, the discrete-time behavior spanned by the sequence $(1, a^{-1}, a^{-2}, \cdots)$ does not converge to a limit if a tends to zero. (Compare this to the sequence of one-dimensional subspaces spanned by the vector $[1 \mid a^{-1} \mid \cdots \mid a^{-n}]$; as a tends to 0, this sequence does converge to a limit, namely the subspace spanned by $[0 \mid 0 \mid \cdots \mid 1]$.) This fact suggests that one somehow should add some 'generalized systems' which could serve as additional limit points. One may consider several possibilities to do this, but it would be in particular attractive to have a smooth compactification. It was recently discovered that it is possible to give such a compactification using either kernel representations by homogenous polynomial matrices (Ravi and Rosenthal 1994) or first-order representations under weakened minimality conditions (Geerts and Schumacher 1994*a*). Although a continuous-time interpretation in terms of impulsive-smooth behaviors was given in (Geerts and Schumacher 1994*b*), no discrete-time behaviors have so far been associated to the added points in the smooth compactification. We feel that it would be interesting to do so, since the availability of an interpretation that is as simple and as concrete as possible may contribute to the understanding of the compactified space of systems.

The purpose of the present paper is therefore to obtain a behavioral interpretation in discrete time of the systems given by homogeneous higher-order equations as in (Ravi and Rosenthal 1994) or by generalized first-order equations as in (Geerts and Schumacher 1994*b*). The resulting set of behaviors, which is slightly bigger than the set considered by J.C. Willems for instance in (Willems 1986), will be referred to as the set of 'homogeneous behaviors'.

The organization of this paper is as follows. The framework that we shall use will be introduced in section 2. Here we also introduce the various representations of homogeneous behaviors. The relations between these representations will be explored in some detail in section 3, which deals with realization theory. In section 4 we obtain the main result of the paper, establishing a one-to-one connection between (i) homogeneous polynomial matrices modulo homogeneous unimodular row transformations, (ii) generalized first-order representations modulo similarity transformations, and (iii) homogeneous behaviors. Conclusions follow in section 5.

2. Definitions and preliminaries

Denote by $\mathcal{L}$ the set of all functions from the integers $\mathbb{Z}$ to the reals $\mathbb{R}$ with support bounded on the left; that is, $\mathcal{L}$ is the set of two-sided infinite sequences of real numbers $(\cdots, a_{-1}, a_0, a_1, \cdots)$ for which there exists an N such that $a_i = 0$ for all $i < -N$. Of course, $\mathbb{Z}$ can be thought of as a time axis and $\mathcal{L}$ as a signal space. We shall write $\mathcal{L}^q$ for the analogous space of sequences with values in $\mathbb{R}^q$; in the terminology of (Willems 1986), this will be our *universum*.

We shall use two notations for the elements of $\mathcal{L}^q$. The first one is the sequence notation

$$w = (w_{-N}, \cdots, w_{-1}, w_0 \mid w_1, w_2, \cdots) \quad (1)$$

where it is understood that $w_i = 0$ for $i < -N$, and in which we use a vertical bar to indicate the position between the signal values at time points 0 and 1. The second notation is the representation as a formal Laurent series:

$$w(s) = \sum_{i=-N}^{\infty} w_i s^{-i}. \quad (2)$$

Such series may be thought of as consisting of a 'polynomial part' $w_+(s) := \sum_{i=-N}^{0} w_i s^{-i}$ and a 'strictly proper part' $w_-(s) := \sum_{i=1}^{\infty} w_i s^{-i}$, corresponding to the parts before and after the bar in the sequence notation. The Laurent series is written in powers of s^{-1} rather than in powers of s so that the polynomial part is a polynomial in s rather than in s^{-1}; of course this is just a matter of notation. We shall write $\mathcal{L}_+^q$ for the set of elements of $\mathcal{L}^q$ that appear as polynomials in the Laurent series notation, and the set of elements whose polynomial part is zero will be denoted by $\mathcal{L}_-^q$.

Elements of $\mathcal{L}$ can be multiplied in the way suggested by the Laurent series notation (2). With respect to this multiplication and the usual addition, the set $\mathcal{L}$ is a *field* and we shall make heavy use of this fact. Note that the field of rational functions $\mathbb{R}(s)$ can be viewed as a subfield of $\mathcal{L}$ by identifying $f(s) \in \mathbb{R}(s)$ with its Laurent expansion around infinity.

On $\mathcal{L}^q$ we shall consider two shift operators defined in the following manner:

$$\sigma_0: \quad (w_{-N},\cdots,w_{-1},w_0 \mid w_1,w_2,\cdots) \mapsto (w_{-N},\cdots,w_0,0 \mid w_2,w_3,\cdots)$$

$$\tau: \quad (w_{-N},\cdots,w_{-1},w_0 \mid w_1,w_2,\cdots) \mapsto (w_{-N},\cdots,w_{-2},w_{-1} \mid w_0,w_1,w_2,\cdots)$$

The same operators can be given in Laurent series notation by

$$\begin{aligned}(\sigma_0 w)(s) &= sw(s) - sw_-(s)|_{s=\infty} \\ (\tau w)(s) &= s^{-1}w(s).\end{aligned} \tag{3}$$

Consider now a homogenous polynomial row vector of degree ν, with q entries. Such a row vector may be written in the form

$$p(s,t) = \sum_{k=0}^{\nu} p_k s^k t^{\nu-k} \tag{4}$$

where the p_k are constant row vectors of length q. We want to associate a linear operator on $\mathcal{L}^q$ to this homogeneous vector, which in turn will determine a 'behavior' as the set of all elements in $\mathcal{L}^q$ that are mapped to zero by this operator. For this define:

$$p(\sigma_0,\tau) \stackrel{\text{def}}{=} \sigma_0^\nu \sum_{k=0}^{\nu} p_k \tau^{\nu-k}. \tag{5}$$

Note that in the scalar case $q = 1$, the dimension of the associated behavior on $\mathcal{L}$ is equal to the degree of the polynomial from which the operator is derived.

Remark 1 $p(\sigma_0,\tau)$ acts on one-sided sequences (interpreted as sequences in $\mathcal{L}^q$) in the same way as the standard operator associated to the dehomogenization $p(s) = p(s,1)$ of $p(s,t)$. In this context the effect of a factor t in $p(s,t)$ may be described as 'cancel one more element to the left'.

Remark 2 The association $p(s,t) \mapsto p(\sigma_0,\tau)$ is linear, but does not respect the multiplicative structure of homogeneous polynomials. For instance the operator associated to the homogeneous polynomial t^2 is not the square of the operator associated to t.

Remark 3 One might consider other sequence spaces than $\mathcal{L}^q$, which can be looked at as the space of bi-infinite sequences with support bounded to the left. In particular one might take the space of *all* bi-infinite sequences. It seems to be hard, though, to associate to homogeneous polynomials a linear operator on this space in such a way that the dimension of the kernel of this operator is equal to the degree of the homogeneous polynomial that one started with. If one associates to $p(s,t)$ the operator $p(\sigma_0,\tau)$ as defined above, then for instance the operator associated to $p(s,t) = s - t$, which is $\sigma_0(1-\tau)$, has a two-dimensional solution space associated to it (spanned by the sequences $(\cdots,1,1 \mid 0,0,\cdots)$ and $(\cdots,0,0 \mid 1,1,\cdots)$). On the other hand, if one would consider only sequences with finite support, then the same operator would have a zero-dimensional solution space. Finally, the use of sequences with support bounded to the right would give rise to a theory that is essentially the same as the one developed in this paper.

To a homogeneous polynomial matrix $P(s,t)$ we associate an operator $P(\sigma_0,\tau)$ by replacing each row $P_i(s,t)$ by the corresponding operator $P_i(\sigma_0,\tau)$. We now define:

Definition 4 The *homogeneous behavior* $\mathcal{B}(P)$ associated with a homogeneous polynomial matrix $P(s,t)$ is the set

$$\mathcal{B}(P) = \{w \in \mathcal{L}^q \mid P(\sigma,\tau)w = 0\}. \tag{6}$$

The dehomogenization $P(s,1)$ of a homogeneous polynomial matrix $P(s,t)$ will be written simply as $P(s)$. By identifying polynomials with elements of $\mathcal{L}_+$, we can look at the matrix of polynomials $P(s)$ as a matrix with entries in $\mathcal{L}$, and so (using the multiplicative structure of $\mathcal{L}$) as a mapping from $\mathcal{L}^q$ to $\mathcal{L}^p$ where p is the number of rows of $P(s,t)$. Immediately from the definitions we now have the following characterization of $\mathcal{B}(P)$.

Lemma 5 Let the row degrees of the homogeneous polynomial $P(s,t)$ be $\nu_1,\cdots,\nu_p$. An element $w(s) \in \mathcal{L}^q$ belongs to $\mathcal{B}(P)$ if and only if the entries of the p-vector $P(s)w(s)$ are polynomials of degrees at most $\nu_1 - 1,\cdots,\nu_p - 1$ respectively.

With a homogeneous polynomial matrix $P(s,t)$ of size $p \times q$ and of row degrees $\nu_1,\cdots,\nu_p$, we can associate the $\mathbb{R}$-linear space of p-vectors whose i-th entry is a homogeneous polynomial of degree $\nu_i - 1$, or is zero if $\nu_i = 0$. Obviously the dimension of this space is $n := \sum_{i=1}^{p} \nu_i$. We define the *canonical basis matrix* $X(s,t)$ as the matrix of size $p \times n$ given by

$$\begin{bmatrix} s^{\nu_1-1} & \cdots & t^{\nu_1-1} & 0 & 0 & \cdots & \cdots & 0 \\ 0 & & 0 & s^{\nu_2-1} & \cdots & & & \vdots \\ \vdots & & & & & & \ddots & 0 \\ 0 & \cdots & & & & & \cdots & t^{\nu_p-1} \end{bmatrix}.$$

The result stated in the lemma above can now be reformulated as

$$\mathcal{B}(P) = \{w(s) \in \mathcal{L}^q \mid P(s)w(s) \in \text{span}_{\mathbb{R}} X(s)\}.$$

It will be convenient to use polynomial representations that are of the above form but are not necessarily derived explicitly from a homogeneous polynomial matrix. For any pair of polynomial matrices $(R(s), V(s))$, where $R(s)$ and $V(s)$ have the same number of rows, we can define

$$\mathcal{B}(R, V) = \{w(s) \in \mathcal{L}^q \mid R(s)w(s) \in \text{span}_{\mathbb{R}} V(s)\}.$$

To make sure that such a behavior is also representable in the form $\mathcal{B}(P)$ for some homogeneous polynomial matrix $P(s,t)$, the pair $(R(s), V(s))$ will have to satisfy certain conditions; we shall come back to this later.

Finally, a third representation that we shall use is the first-order representation. Consider a triple of real matrices (F, G, H) where F and G both have size $n \times (n+m)$ and H has size $q \times (n+m)$. With this triple we associate a behavior as follows:

$$\mathcal{B}(F, G, H) = \\ \{w \in \mathcal{L}^q \mid w = Hz \text{ for some } z \in \mathcal{B}(sG - tF)\}.$$

Note that the operator associated to the homogeneous polynomial matrix $sG - tF$ is $\sigma_0(G - \tau F) = \sigma\tau_0(\sigma G - F)$. We may therefore also write the above definition in the form

$$\mathcal{B}(F, G, H) = \{w(s) \in \mathcal{L}^q \mid \exists z \in \mathcal{L}^{n+m}, x_0 \in \mathbb{R}^n \\ \text{s.t. } (sG - F)z(s) = x_0 \text{ and } w(s) = Hz(s)\}.$$

We now turn to the relation between homogeneous behaviors and behaviors defined on $\mathbb{Z}_+$ such as for instance in (Willems 1986). The latter are given in 'AR representation' as follows. Let $P(s)$ be a polynomial matrix and let S denote the standard left shift that takes $(w_1, w_2, \cdots)$ to $(w_2, w_3, \cdots)$; then define $\mathcal{B}_-(P) = \{w \mid P(S)w = 0\}$. Identifying the space of one-sided sequences with $\mathcal{L}^q_-$, we get the following simple embedding of the set of 'standard' behaviors in the space of homogeneous behaviors.

Proposition 6 Let $P(s,t)$ be a homogeneous polynomial matrix, and let $P(s)$ be its dehomogenization. We have:

$$\mathcal{B}_-(P) = \mathcal{B}(P) \cap \mathcal{L}^q_-. \tag{7}$$

To get a similar result for (R, V)-representations, we need to impose a condition involving the space X_R that is defined as follows (Fuhrmann 1981, Kuijper and Schumacher 1990, Geerts and Schumacher 1994*b*):

$$X_R = \{g(s) \in \mathcal{L}^p_+ \mid g(s) = R(s)w(s) \\ \text{for some } w(s) \in \mathcal{L}^q_-\}. \tag{8}$$

Using the same reasoning as above, we then obtain the following.

Proposition 7 Let $(R(s), V(s))$ be a pair of polynomial matrices having the same number of rows. If $X_R \subset \text{span}_{\mathbb{R}} V(s)$, then

$$\mathcal{B}_-(R) = \mathcal{B}(R, V) \cap \mathcal{L}^q_-. \tag{9}$$

A subspace of $\mathcal{L}^q$ may be given in image representation as $\text{im}\, M(s)$ or in kernel representation as $\ker N(s)$, where $M(s)$ and $N(s)$ are matrices over $\mathcal{L}$. If $M(s)$ and $N(s)$ are rational matrices, we may also look at $\text{im}\, M(s)$ and $\ker N(s)$ as curves, that is, as Grass(m, n)-valued mappings defined almost everywhere (i. e. everywhere except for a finite number of points) on $\mathbb{C}$. The following lemma, which will be needed below, essentially says that these two points of view are equivalent.

Lemma 8 Let $M(s)$ and $N(s)$ be rational matrices. In this case we have $\text{im}_{\mathbb{C}} M(s) = \ker_{\mathbb{C}} N(s)$ for almost all $s \in \mathbb{C}$ if and only if $\text{im}_{\mathcal{L}} M(s) = \ker_{\mathcal{L}} N(s)$.

3. Realization of homogeneous systems

We begin by noting the following relation between first-order representations and homogeneous polynomial representations.

Lemma 9 Let $P(s,t)$ be a homogeneous polynomial matrix, and let (F, G, H) a triple of real matrices in which F and G have the same size and H has the same number of columns as F and G. If the equality

$$\text{im} \begin{bmatrix} sG - tF \\ H \end{bmatrix} = \ker[-X(s,t) \mid P(s,t)] \tag{10}$$

holds for almost all $(s,t) \in \mathbb{C}^2 \setminus \{(0,0)\}$, then $\mathcal{B}(F, G, H)$ is equal to $\mathcal{B}(P)$.

Based on this Lemma we define:

Definition 10 We say G, F, H is a realization of $P(s,t)$ if there is a basis matrix $X(s,t)$ s.t.

$$P(s,t)H = X(s,t)(sG - tF). \tag{11}$$

We say the realization is minimal if G, F, H is admissible and observable.

The main theorem of this section is now as follows:

Theorem 11 Every homogeneous system $P(s,t)$ has a minimal realization G, F, H. Moreover if $\tilde{G}, \tilde{F}, \tilde{H}$ is a second minimal realization then there exist unique elements $S \in Gl_n$ and $T \in Gl_{m+n}$ s.t.

$$(\tilde{G}, \tilde{F}, \tilde{H}) = (SGT^{-1}, SFT^{-1}, HT^{-1}). \quad (12)$$

Finally $P(s_o, t_o)$ has rank p if, and only if $(s_oG - t_oF)$ has rank n, in particular $P(s,t)$ is controllable if, and only if $(sG - tF)$ is controllable.

The existence of a realization is shown via a realization algorithm.

Algorithm: First assume that $P(1,0)$ has full rank. In this case choose

$$G := (I_{n\times n}\ 0).$$

Choose a $(m+p) \times n$ matrix H_1 having the property that

$$P(1,0)H_1 = X(1,0)I_{n\times n}$$

and choose a $(m+p) \times m$ matrix H_2 having the property that

$$\text{colsp}H_2 = \ker(P(1,0))$$

Now define

$$H := (H_1 \quad H_2).$$

Note that by definition one has

$$P(1,0)H = X(1,0)G$$

(This is the condition at infinity.) Finally define F in a unique way through[1]

$$X(s,t)F := \frac{1}{t}(P(s,t)H - X(s,t)sG).$$

If $P(1,0)$ has not full rank choose a linear fractional transformation

$$(s,t) = (\bar{s},\bar{t})\begin{pmatrix} a & b \\ c & d \end{pmatrix}, \quad \begin{pmatrix} a & b \\ c & d \end{pmatrix} \in \mathbb{P}Gl_1$$

such that

$$\bar{P}(\bar{s},\bar{t}) := P(a\bar{s} + c\bar{t}, b\bar{s} + d\bar{t})$$

has the property that $\bar{P}(1,0)$ has full rank p. Define

$$\bar{X}(\bar{s},\bar{t}) := X(a\bar{s} + c\bar{t}, b\bar{s} + d\bar{t}).$$

[1] Note that F is unique since $X(s,t)$ is a basis matrix.

Proceeding in the same way as before one constructs a realization $\bar{G}, \bar{F}, \bar{H}$ s.t.

$$\bar{P}(\bar{s},\bar{t})\bar{H} = \bar{X}(\bar{s},\bar{t})(\bar{s}\bar{G} - \bar{t}\bar{F}).$$

But then G, F, H defined through

$$sG - tF = \bar{s}\bar{G} - \bar{t}\bar{F}, H = \bar{H}$$

is a realization of $P(s,t)$.

At this point we would like to remark that if $P(1,0)$ has not full rank then G has not full rank either, in particular it cannot be chosen in the form $G := (I_{n\times n}\ 0)$.

In the sequel we will show that the obtained realization is admissible, minimal and unique up to change of basis.

Admissibility and Minimality: Without loss of generality we can assume that $P(1,0)$ has rank p. (Otherwise we can work with the system $\bar{P}(\bar{s},\bar{t})$.) Since $G := (I_{n\times n}\ 0)$ admissability (i.e. $sG - tF$ has generically full row rank) is clear.

Consider now the dehomogenized $p \times (m+p+n)$ system

$$(X(s,1)\ P(s,1)),$$

which is a controllable AR-system of McMillan degree n. From the identity (11) and by construction it therefore follows that

$$\begin{bmatrix} sG - F \\ H \end{bmatrix}$$

is the "dual Hermann-Martin curve" ((Hermann and Martin 1977)). In particular $\left[\begin{smallmatrix} sG-F \\ H \end{smallmatrix}\right]$ has full column rank for all $s \in \mathbb{C} \cup \{\infty\}$. It therefore follows that G, F, H is a minimal representation.

Uniqueness: Without loss of generality we assume that $P(1,0)$ has rank p. Assume that the basis matrix $X(s,t)$ is fixed and consider two realizations (G, F, H) and $(\tilde{G}, \tilde{F}, \tilde{H})$. From the fact that $\left(\begin{smallmatrix} sG-F \\ H \end{smallmatrix}\right)$ and $\left(\begin{smallmatrix} s\tilde{G}-\tilde{F} \\ \tilde{H} \end{smallmatrix}\right)$ are both dual Hermann Martin curves of $(X(s,1)\ P(s,1))$ the existence of a unique transformation $T \in Gl_{m+n}$ s.t. $(\tilde{G}, \tilde{F}, \tilde{H}) = (GT^{-1}, FT^{-1}, HT^{-1})$ follows.

Finally we show that the realization is controllable whenever the original plant is controllable:

Controllability: First assume that $(s_oG - t_oF)$ has rank less than n. After a possible change of basis in the internal variable space Z we can assume that the last $m+1$ columns of $(s_oG - t_oF)$ are zero. By minimality it then follows that the last $m+1$ columns of H are linear independent. From the identity (11) it follows that $\ker P(s_o, t_o)$ has dimension at least $m+1$, i.e. $P(s_o, t_o)$ has rank at most $p-1$.

Conversely assume that $P(s_o, t_o)$ has rank at most $p-1$. Without loss of generality we will assume that all row indices ν_i of $P(s,t)$ are at least 1. (If this assumption is not fullfilled we delete the constant rows of $P(s,t)$ and the zero rows of $X(s,t)$ and proceed with a system where the assumption is fullfilled.) Consider again the identity

$$P(s,t)H = X(s,t)(sG - tF).$$

Since $X(s,t)$ has no zero rows it follows that $\text{rank}X(s_o, t_o) = p$. Since

$$\text{rank}(P(s_o, t_o)H) < p$$

it follows that

$$\text{rank}(X(s_o, t_o)(s_o G - t_o F)) < p \leq n$$

which completes the proof.

4. Homogeneous systems and their homogeneous behaviors

We shall need the following uniqueness theorem for polynomial representations (compare (Geerts and Schumacher 1994*a*; Thm. 3.10)).

Theorem 12 Let $(R_1(s), V_1(s))$ and $(R_2(s), V_2(s))$ be two pairs of polynomial matrices, and assume that for $i = 1, 2$ the following holds:

(i) $R_i(s)$ has full row rank,

(ii) $\text{span}_{\mathbb{R}} V_i(s) \supset X_{R_i}$,

(iii) $V_i(s)$ has full column rank over $\mathbb{R}$.

Under these conditions, we have $\mathcal{B}(R_1, V_1) = \mathcal{B}(R_2, V_2)$ if and only if there exists a unimodular polynomial matrix $U(s)$ and a nonsingular constant matrix S such that $R_2(s) = U(s)R_1(s)$ and $V_2(s) = U(s)V_1(s)S$.

The crucial property that we have been aiming for is stated in the theorem below.

Theorem 13 Two homogeneous polynomial matrices $P_1(s,t)$ and $P_2(s,t)$ are equivalent under left multiplication by homogeneous unimodular matrices if and only if $\mathcal{B}(P_1) = \mathcal{B}(P_2)$.

Example 14 The behavior associated to the homogeneous system $p(s,t) = as - t$ is given by the equation $\sigma_0(a - \tau)w = 0$ i.e. (in Laurent series notation) $(as - 1)w(s) = \text{constant}$. The set of solutions is the one-dimensional subspace spanned by $(as-1)^{-1}$ if $a \neq 0$, otherwise it is the one-dimensional subspace spanned by the constant function 1. This means in particular that we have

$$\lim_{a \to 0} (0 \mid 1, a^{-1}, a^{-2}, \cdots) = (1 \mid 0, 0, \cdots). \quad (13)$$

REFERENCES

Fuhrmann, P. A. (1981). *Linear Systems and Operators in Hilbert Space*. McGraw-Hill. New York, NY.

Geerts, A. H. W. and J. M. Schumacher (1994*a*). 'Impulsive-smooth behavior in multimode systems. Part II: Minimality and equivalence'. CWI Report BS-R9432.

Geerts, A. H. W. and J. M. Schumacher (1994*b*). 'Impulsive-smooth behavior in multimode systems. Part I: State-space and polynomial representations'. CWI Report BS-R9431.

Hermann, R. and C. F. Martin (1977). 'Applications of algebraic geometry to system theory part I'. *IEEE Trans. Automat. Control* **AC-22**, 19–25.

Kuijper, M. and J. M. Schumacher (1990). 'Realization of autoregressive equations in pencil and descriptor form'. *SIAM J. Control Optim.* **28**(5), 1162–1189.

Lomadze, V. G. (1990). 'Finite-dimensional time-invariant linear dynamical systems: Algebraic theory'. *Acta Appl. Math* **19**, 149–201.

Ravi, M. S. and J. Rosenthal (1994). 'A smooth compactification of the space of transfer functions with fixed McMillan degree'. *Acta Appl. Math* **34**, 329–352.

Ravi, M. S. and J. Rosenthal (1995). 'A general realization theory for higher order linear differential equations'. *Systems & Control Letters*. To appear.

Willems, J. C. (1986). 'From time series to linear system. Part I: Finite dimensional linear time invariant systems'. *Automatica* **22**, 561–580.

Willems, J. C. (1989). 'Models for dynamics'. *Dynamics Reported* **2**, 171–269.

Willems, J. C. (1991). 'Paradigms and puzzles in the theory of dynamical systems'. *IEEE Trans. Automat. Control* **AC-36**(3), 259–294.

A GEOMETRIC PROOF OF ROSENBROCK'S THEOREM ON POLE ASSIGNMENT

M.Bonilla E.*, J.J. Loiseau.**, R. Baquero S.*,**

*CINVESTAV-IPN, Sec. Control Automático. AP14-740. México 07000, MEXICO. Tel.: 747-7089. FAX: 747-7089, 747-7002; e-mail: MBONILLA@MVAX1.RED.CINVESTAV.MX

**LAN, URA CNRS 823, ECN, 1 rue de la Noë, 44072 NANTES, Cedex 03, FRANCE. Tl.: 40-37-1651. FAX: 40-74-7406; e-mail: LOISEAU@LAN.EC-NANTES.FR

***CINVESTAV-IPN. Sponsored by CoSNET-MEXICO.

Abstract. Résumé. Une nouvelle preuve du célèbre Théoreme de Rosenbrock sur le placement des poles par retour d'état statique est donnée. Cette preuve se sert seulement de resultats de base bien connus de l'approche géométrique, comme la forme canonique de Brunovsky d'un sistème commandable et la partition de l'espace d'état en sous–espaces cycliques liés aux facteurs invariants d'un opérateur.

Key Words. Geometric Approach, Linear Systems, Pole Assignment, Rosenbrock Theorem.

1. Introduction.

There already exist a lot of different proofs of Rosenbrock theorem. Most of them use polynomial arguments (Dickinson, 1974, Kučera, 1981, Rosenbrock, 1970). Actually all the generalizations of Rosenbrock theorem that was developed are also based on polynomial arguments (Zagalak and Kučera ,1991, Zagalak and Loiseau, 1992). For algorithmic conveniency it would be useful to develop a geometric proof. Indeed this was already done by Flamm (1980) and later by Ozcaldiran (1990). These two proofs use at key-points difficult results of linear algebra. Our aim here is to give a new proof based on results which are well-known by control theoretists, namely the canonical form of controllable systems which was described by Brunovsky (1970) and the splitting of the state-space accordingly to the invariant factors of a given map. Actually Rosenbrock's conditions are derived from considering the dimension of certain cyclic subspaces. Our proof of their sufficiency is constructive: it effectively permits to calculate a feedback which assigns to prespecified values the invariant factors of the closed–loop system. This construction is divided in two steps. We first construct a feedback which fixes the degrees of the invariant factors. A second feedback is designed which set up the coefficients of the invariant factors without changing their degrees. At the difference of the previous proofs of Rosenbrock Theorem (Flamm, 1980, Özçaldiran, 1990, Rosenbrock, 1970, and Zagalak and Loiseau, 1992) we emphasized the geometric aspects of that result

Rosenbrock theorem is recalled in section 2. Sections 3 and 4 are devoted to the proof of the necessity and the sufficiency, respectively, of the claim established by Rosenbrock. Finally we conclude.[1]

[1] **Notation.** Script capitals $\mathcal{V}, \mathcal{W}, \ldots$, denote linear spaces with elements $v, w, \ldots$; span $\{x_1, x_2, \ldots, x_m\}$ is the linear space spanned by the vectors $x_1, x_2 \ldots, x_m$. The dimension of a space $\mathcal{V}$ is denoted $\dim \mathcal{V}$. $A, B, \ldots$, is used to denote a given map or its matrix representation in a suitable base. The image of A is denoted Im A. $A \mid \mathcal{V}$ denotes the erstriction of A to $\mathcal{V}$. When $\mathcal{V} \subset \mathcal{W}$, $\frac{\mathcal{W}}{\mathcal{V}}$ or $\mathcal{W}/\mathcal{V}$ stands for the quotient space $\mathcal{W}$ module $\mathcal{V}$. The direct sum of independent spaces is gritten as $\oplus$. α and β being two polynomials, $\alpha \mid \beta$ means α divides β; $\deg \alpha$ denotes the degree of α. The $n \times n$ unity matrix diagonal$\{1, \ldots, 1\}$ is denoted I_n. J_n will stand for the $n \times n$ matrix $\begin{bmatrix} 0 & \mathrm{I}_{n-1} \\ 0 & 0 \end{bmatrix}$. $\{\alpha_i\} \setminus \{\beta_i\}$ denots the set $\{\gamma_i\}$ which contains the elements of the set $\{\alpha_i\}$ which are not in the set $\{\beta_i\}$. $\langle H \mid T \rangle$ denotes $\sum_{i=1}^{n} H^{i-1}T$ where $T \subset \mathcal{X}$, $H : \mathcal{X} \to \mathcal{X}$ and $n = \dim \mathcal{X}$.

2. Rosenbrock Theorem

Let us consider the state space description:

$$\dot{x}(t) = Ax(t) + Bu(t) \tag{2.1}$$

where $A : \mathcal{X} \to \mathcal{X}$ and $B : \mathcal{U} \to \mathcal{X}$ are real matrices, $\mathcal{X} \approx R^n$ and $\mathcal{U} \approx R^m$. We assume that (A, B) is controllable and that B is monic (Wohnam, 1985), $\langle A \mid \mathrm{Im}B \rangle = \mathcal{X}$ and $\ker B = \{0\}$.

One of the most classical problems of control theory is to find a *static state feedback* $F : \mathcal{X} \to \mathcal{U}$ which fix at will the spectrum of the map $(A + BF)$, i.e. the zeros of its characteristic polynomial $\pi(\lambda) = \det[\lambda \mathrm{I} - (A + BF)]$, under assumption of controllability of course.

A more general problem is to assign the full internal structure of the system, that is the set of the *invariant factors* (Rosenbrock, 1970 and Zagalak and Kučera, 1991) of $A + BF$.

Definition 2.1 **Gantmacher (1977).** Let $D_\mu(\lambda)$ $(\mu = 1, \ldots, n)$ the greatest common divisor of all the nonzero minors of order μ of the characteristic matrix

$\lambda I - A$. The *invariant factors* of A, $\psi_i(\lambda)$, are:

$$\psi_\mu(\lambda) = \frac{D_{n-\mu+1}(\lambda)}{D_{n-\mu}(\lambda)} \quad , \text{ for } \mu = 1, \ldots, n$$

where $D_0(\lambda) = 1$. □

Property 2.1 . The *invariant factors* satisfy:

(i) $\psi_2(\lambda) \mid \psi_1(\lambda), \psi_3(\lambda) \mid \psi_2(\lambda), \ldots, \psi_n(\lambda) \mid \psi_{n-1}(\lambda)$,

(ii) $\pi(\lambda) = \psi_1(\lambda)\psi_2(\lambda)\cdots\psi_n(\lambda)$,

(iii) The polynomials ψ_i form a complete set of invariants of A under change of basis on $\mathcal{X}$.

□

As it is well-known, $\mathcal{X}$ is decomposed into cyclic subspaces accordingly to the invariant factors of the map A. Wonham [11] restated this result as follows:

Theorem 2.1 **Wohnam (1985) : Ch 1.4).** Let (A, B) be controllable and let $\psi_1(\lambda), \ldots, \psi_r(\lambda)$ be those of the invariant factors of A which are not equal to 1 (r is called the cyclic index of A). Then $m \geq r$. There exist A-invariant subspaces $\mathcal{X}_i \subset \mathcal{X}$, and vectors $b_i \in \mathrm{Im}B$, for $i = 1, \ldots, n$, such that :

(i) $\mathcal{X} = \bigoplus_{j=1}^{r} \mathcal{X}_j$,

(ii) $A \mid \mathcal{X}_i$ is cyclic with minimal polynomial $\psi_i(\lambda)$, $i = 1, \ldots, r$, i.e. $\exists v_i \in \mathcal{X}_i =$ span $\{v_i, av_i, \ldots, A^{n_i-1}v_i\}$, $n_i := \deg \psi_i(\lambda)$

(iii) $\langle A \mid \text{span } \{b_1, \ldots, b_i\}\rangle = \bigoplus_{j=1}^{i} \mathcal{X}_j \quad ; i = 1, \ldots, r$

□

To each nonunit invariant factor $\psi_\mu(\lambda)$ is thus associated an isolated chain of integrators, spanning $\mathcal{X}_\mu$, having $\deg\{\psi_\mu(\lambda)\}$ for size and whose dynamic is given by $\psi_\mu(\lambda)$.

Let us recall the Brunovsky canonical form of a controllable pair (A, B) (Brunovsky, 1970, Wohnam, 1985).

Theorem 2.2 **Brunovsky (1970).** Let (A, B) be controllable and define :

$$k_i := \text{card}\{j \mid \bar{k}_j \geq i\} \quad ; i = 1, \ldots, m \tag{2.2}$$

where

$$\begin{cases} \bar{k}_i := \dim \left\{ \frac{\mathrm{Im}B - A\mathrm{Im}B + \cdots + A^{i-1}\mathrm{Im}B}{\mathrm{Im}B + \cdots + A^{i-2}\mathrm{Im}B} \right\} \\ \bar{k}_1 := \dim\{\mathrm{Im}B\} \end{cases} \tag{2.3}$$

with $i = 2, \ldots, n$. There then exists a feedback $F : \mathcal{X} \to \mathcal{U}$ and two isomorphisms $T : \mathcal{X} \to \mathcal{X}$, $G : \mathcal{U} \to \mathcal{U}$ such that (in the given basis) :

$$\begin{aligned} A_c &= T^{-1}(A + BF)T \\ &= \text{block diagonal matrix}\{A_1, \ldots, A_m\}_{n \times n} \\ B_c &= T^{-1}BG \\ &= \text{block diagonal matrix}\{b_1, \ldots, b_m\}_{n \times m} \end{aligned}$$

with $A_i = \mathrm{J}_{k_i}$, for $i = 1, \ldots, m$ and $b_i^T = [0, 1, 0, \ldots, 0]_{1 \times k_i}$ □

The lists of integers $\{k_1, \ldots, k_m\}$ and $\{\bar{k}_1, \ldots, \bar{k}_n\}$ are respectively called the *controllability indices* and the *Brunovsky indices* of the pair (A, B). Relation (2.2) defines the list $\{k_i\}$ as the *conjugate* of the list $\{\bar{k}_i\}$.

Property 2.2 **Brunovsky (1970), Brylawski (1973).** $\{k_i\}$ and $\{\bar{k}_i\}$ satisfy:

(i) $n \geq k_1 \geq k_2 \geq \cdots \geq k_m > 0$

(ii) $m = \bar{k}_1 \geq \bar{k}_2 \geq \cdots \geq \bar{k}_n \geq 0$

(iii) $\bar{k}_1 + \bar{k}_2 + \cdots + \bar{k}_n = n$

(iv) $\sum_{j=1}^{i} \bar{k}_j = \dim \left\{ \sum_{j=1}^{i} A^{j-1}\mathrm{Im}B \right\}$

(v) $\bar{k}_i = \text{card}\{j \mid k_j \geq i\} \quad ; i = 1, \ldots, m$

□

Let us recall another usefull result.

Property 2.3 **Loiseau (1988), Brylawski (1973).** Let $\{\alpha_i\}$ and $\{\beta_i\}$ be two finite lists of non negative integers ranged in non increasing order, and let $\{\bar{\alpha}_i\}$ and $\{\bar{\beta}_i\}$ their conjugated lists :

$$\begin{aligned} \bar{\alpha}_i &= \text{card}\{j \mid \alpha_j \geq i\} \quad , i \geq 1, \\ \bar{\beta}_i &= \text{card}\{j \mid \beta_j \geq i\} \quad . i \geq 1. \end{aligned}$$

The two following statements are equivalent.

$$\sum_{j=1}^{i} \alpha_j \geq \sum_{j=1}^{i} \beta_j \quad ; i \geq 1 \tag{2.4}$$

$$\sum_{j=i}^{\infty} \bar{\alpha}_j \geq \sum_{j=i}^{\infty} \bar{\beta}_j \quad ; i \geq 1 \tag{2.5}$$

□

We are now in position to formulate the RTPA.

Theorem 2.3 **RTPA, Rosenbrock (1970).** Let (2.1) be a controllable system with *controllability indices* $\{k_1, \ldots, k_m\}$ and $\{\psi_1(\lambda), \ldots, \psi_n(\lambda)\}$ be a set of monic polynomials satisfying the divisibility conditions

$$\psi_{i+1}(\lambda) \mid \psi_i(\lambda) \quad , i = 1, \ldots, n-1$$

There then exists a feedback $F : \mathcal{X} \to \mathcal{U}$ such that the *invariant factors* of $(A + BF)$ are precisely the above polynomials $\psi_i(\lambda)$ if and only if :

$$\sum_{j=i}^{n} n_j \geq \sum_{j=i}^{n} k_j \quad ; \; i = 2, \ldots, n \tag{2.6}$$

$$\sum_{j=1}^{n} n_j = \sum_{j=1}^{n} k_j \tag{2.7}$$

where $n_j = \deg \psi_j(\lambda)$ and $k_j = 0$ for $j \geq m+1$. □

3. Geometric Proof of the Necessity

3.1. *Preliminaries*

In this Section we are going to show that the *invariant factors* of A, $\psi_1(\lambda), \ldots, \psi_m(\lambda)$, satisfy (2.6)–(2.7).

We need the following Corollary of Theorem 2.1.

Corollary 3.1 For $i = 1, \ldots, r$:

$$\begin{aligned} &\langle A \mid \text{span}\,\{b_1, \ldots, b_i\}\rangle = \\ &\bigoplus_{\mu=1}^{i} \bigoplus_{j=1}^{n_\mu} A^{j-1}\text{span}\,\{b_\mu\} \end{aligned} \tag{3.1}$$

□

Proof of Corollary 3.1. (3.1) is true for $i = 1$ (see Theorem 2.1). Assume that (3.1) is true for $i-1$, i.e.

$$\begin{aligned} &\langle A \mid \text{span}\,\{b_1, \ldots, b_{i-1}\}\rangle = \\ &\bigoplus_{\mu=1}^{i-1} \bigoplus_{j=1}^{n_\mu} A^{j-1}\text{span}\,\{b_\mu\} \end{aligned} \tag{3.2}$$

Defining

$$Q_\mu : \mathcal{X} \to \mathcal{X} \text{ the projection on} \bigoplus_{j=\mu}^{r} \mathcal{X}_j \text{ along} \bigoplus_{j=1}^{\mu-1} \mathcal{X}_j$$

where the $\mathcal{X}_j$ are the subspaces given in Theorem 2.1, we obtain (see Theorem 2.1) :

$$\begin{aligned} \bigoplus_{j=1}^{i} \mathcal{X}_j &= \langle A \mid \text{span}\,\{b_1, \ldots, b_{i-1}\}\rangle \\ &\quad + \sum_{j=1}^{n} A^{j-1}\text{span}\,\{b_i\} \\ &\subset \langle A \mid \text{span}\,\{b_1, \ldots, b_{i-1}\}\rangle \\ &\quad + \sum_{j=1}^{n} A^{j-1}\text{span}\,\{(\mathrm{I} - Q_i)b_i\} \\ &\quad + \sum_{j=1}^{n} A^{j-1}\text{span}\,\{Q_i b_i\} \\ &= \langle A \mid \text{span}\,\{b_1, \ldots, b_{i-1}\}\rangle \\ &\quad + \bigoplus_{j=1}^{n_i} A^{j-1}\text{span}\,\{Q_i b_i\} \subset \bigoplus_{j=1}^{i} \mathcal{X}_j \end{aligned} \tag{3.3}$$

Last equality is due to: $\sum_{j=1}^{n} A^{j-1}\text{span}\,\{(\mathrm{I} - Q_i)b_i\} \subset \langle A \mid \text{span}\,\{b_1, \ldots, b_{i-1}\}\rangle$, $Q_i b_i \in \left(\bigoplus_{j=1}^{i} \mathcal{X}_j\right) \cap \left(\bigoplus_{j=i}^{r} \mathcal{X}_j\right) = \mathcal{X}_i$, and from the fact that the minimal polynomial of $\mathcal{X}_i$ has degree n_i. Thus the last inclusion follows since $Q_i b_i \in \mathcal{X}_i$ and $A\mathcal{X}_i \subset \mathcal{X}_i$.

From (3.3) and (3.2) we have :

$$\begin{aligned} \bigoplus_{j=1}^{i} \mathcal{X}_j &= \bigoplus_{\mu=1}^{i-1} \bigoplus_{j=1}^{n_\mu} A^{j-1}\text{span}\,\{b_\mu\} \\ &\quad + \bigoplus_{j=1}^{n_i} A^{j-1}\text{span}\,\{Q_i b_i\} \end{aligned} \tag{3.4}$$

$$= \bigoplus_{\mu=1}^{i-1} \bigoplus_{j=1}^{n_\mu} A^{j-1}\text{span}\,\{b_\mu\} + \bigoplus_{j=1}^{n_i} A^{j-1}\text{span}\,\{b_i\}$$

Indeed, let x be in $\bigoplus_{j=1}^{i} \mathcal{X}_j$ there then exists a vector z in $\langle A \mid \text{span}\,\{b_1, \ldots, b_{i-1}\}\rangle$, $\alpha_1, \ldots, \alpha_{n_i} \in R$, such that: $x = z + \sum_{j=1}^{n_i} A^{j-1} Q_i b_i \alpha_j = \left(z - \sum_{j=1}^{n_i} A^{j-1}(\mathrm{I} - Q_i) b_i \alpha_j\right) + \sum_{j=1}^{n_i} A^{j-1} b_i \alpha_j \in \langle A \mid \text{span}\,\{b_1, \ldots, b_{i-1}\}\rangle + \bigoplus_{j=1}^{n_i} A^{j-1}\text{span}\,\{b_i\}$. On the other hand $\langle A \mid \text{span}\,\{b_1, \ldots, b_i\}\rangle \supset \langle A \mid \text{span}\,\{b_1, \ldots, b_{i-1}\}\rangle + \bigoplus_{j=1}^{n_i} A^{j-1}\text{span}\,\{b_i\}$.

Now, since $\dim \left\{\bigoplus_{j=1}^{i} \mathcal{X}_j\right\} = \sum_{j=1}^{i} n_j$ (see Theorem 2.1) it appears that (3.4) implies (3.1). ■

3.2. *Proof of the necessity.*

Let us define :

$$\begin{cases} \lambda_\mu := \min\{i, n_\mu\} \\ \bar{n}_j := \text{card}\{\mu \mid n_\mu \geq j\} \quad ; \; j = 1, \ldots, i \end{cases} \tag{3.5}$$

We then have from (3.1), Theorem 2.1 and (3.5) :

$$\begin{aligned} \sum_{j=1}^{i} A^{j-1}\mathrm{Im}B &\supset \sum_{j=1}^{i} A^{j-1} \left(\bigoplus_{\mu=1}^{r} \text{span}\,\{b_\mu\}\right) \\ &= \sum_{\mu=1}^{r} \left(\sum_{j=1}^{i} A^{j-1}\text{span}\,\{b_\mu\}\right) \\ &\supset \bigoplus_{\mu=1}^{r} \left(\bigoplus_{j=1}^{\lambda_\mu} A^{j-1}\text{span}\,\{b_\mu\}\right) \\ &= \bigoplus_{j=1}^{i} A^{j-1} \left(\bigoplus_{\mu=1}^{\bar{n}_j} \text{span}\,\{b_\mu\}\right) \end{aligned} \tag{3.6}$$

Now, from claim (*iv*) of Property 2.2 and (3.6):

$$\sum_{j=1}^{i} \bar{k}_j \geq \sum_{j=1}^{i} \bar{n}_j \tag{3.7}$$

Hence, (3.7) and Property 2.3 imply (2.6).

On the other hand, from claim (*iii*) of Property 2.2 and Theorem 2.1 we have:

$$\sum_{j=1}^{n} \bar{k}_j = n = \sum_{j=1}^{n} \deg \psi_j(\lambda) \tag{3.8}$$

which, together with Property 2.3, imply (2.7) and ends the proof of necessity. ∎

4. Geometric Proof of the Sufficiency.

In this Section we are going to show that for a given set of polynomials , $\psi_1(\lambda), \ldots, \psi_n(\lambda)$, satisfying (2.6)–(2.7) and three claims of Property 2.1, there exists a feedback $F : \mathcal{X} \rightarrow \mathcal{U}$ such that the *invariant factors* of $(A + BF)$ be $\psi_1(\lambda), \ldots, \psi_n(\lambda)$. For this, we are first going to give some geometric interpretation of RTPA.

4.1. *Preliminaries.*

Theorem 2.2 expresses that $\mathcal{X}$ can be decomposed as follows (c.f. Theorem 5.10 of Wonham (1985)) :

$$\mathcal{X} = \bigoplus_{i=1}^{m} \mathcal{R}_i^* \tag{4.1}$$

where $\mathcal{R}_i^*$ satisfies for $i = 1, \ldots, m$:

$$\begin{array}{l} \mathcal{R}_i^* = \text{span } \{b_i, A_c b_i, \ldots, A_c^{k_i-1} b_i\} \approx R^{k_i}, \\ A_c \mathcal{R}_i^* \subset \mathcal{R}_i^*, \\ A_c^{k_i} \mathcal{R}_i^* = \{0\}. \end{array} \tag{4.2}$$

where $\{b_1, \ldots, b_m\}$ is some basis of Im B.

Let us establish the following result:

Lemma 4.1 Given a contrallable pair (A, B) and a list of integers $n_1, \ldots, n_\mu$ satysfying

$$\left\{ \begin{array}{l} n_1 \geq n_2 \geq \ldots \geq n_\mu \\ \sum_{i=1}^{j} n_i \geq \sum_{i=1}^{j} k_i \ , j = 1, \ldots, \min(\mu, m) \\ \sum_{i=1}^{\mu} n_i = \sum_{i=1}^{m} k_i = n \end{array} \right. \tag{4.3}$$

we can decompose $\mathcal{X}$ as follows

$$\mathcal{X} = \bigoplus_{i=1}^{\mu} \mathcal{I}_i \tag{4.4}$$

where for $i = 1, \ldots, \mu$:

$$\begin{array}{l} \mathcal{I}_i = \text{span } \{\bar{b}_i, \bar{A}\bar{b}_i, \ldots, \bar{A}^{n_i-1}\bar{b}_i\} \approx R^{n_i} \\ \bar{A}\mathcal{I}_i \subset \mathcal{I}_i \\ \bar{A}^{n_i}\mathcal{I}_i = \{0\} \end{array} \tag{4.5}$$

being the $\bar{b}_i$ some projections of the b_i on $\mathcal{I}_i$ and $\bar{A} = A_c + BF$ □

We shall now establish the result in the case $m = 2$.

Proof:

i) Let us first consider a pair (A, B) having only two controllability indices, namely k_1 and k_2 with $k_1 \geq k_2$. Under these conditions the state space $\mathcal{X}$ can be decomposed as in (4.1)-(4.2) with $m = 2$.

Now, j being an integer, $0 \leq j \leq k_2$, we define the feedback $F_j : \mathcal{X} \rightarrow \mathcal{U}$ and the map $\bar{A} : \mathcal{X} \rightarrow \mathcal{X}$ as follows:

$$\begin{array}{lcl} BF_j\mathcal{R}_1^* & = & \{0\} \\ BF_jA_c^{j-1}b_2 & = & b_1 \\ BF_jA_c^{i-1}b_2 & = & 0 \quad ; \ i \in \{1, \ldots, k_2\} \setminus \{j\} \end{array} \tag{4.6}$$

$$\bar{A} := A_c + BF_j, \tag{4.7}$$

we can construct the following subspace :

$$\mathcal{I}_1 := \text{span } \{b_2, \bar{A}b_2, \ldots, \bar{A}^{k_1-1+j}b_2\} \tag{4.8}$$

satisfying (see : (4.1), (4.2), (4.6)-(4.8)) :

$$\begin{array}{l} \mathcal{I}_1 = \text{span } \{b_2, A_cb_2, \ldots, A_c^{j-1}b_2, A_c^jb_2 + b_1, \ldots, \\ \quad A_c^{k_2-1}b_2 + A_c^{k_2-1-j}b_1, A_c^{k_2-j}b_1, \ldots, A_c^{k_1-1}b_1\} \\ \quad \approx R^{k_1+j}, \\ \\ \bar{A}\mathcal{I}_1 \subset \mathcal{I}_1, \\ \bar{A}^{k_1+j}\mathcal{I}_1 = \{0\}. \end{array} \tag{4.9}$$

Remark that $\mathcal{I}_1$ is an $\bar{A}$-invariant cyclic subspace generated by b_2 with dimension $k_1 + j$ and minimal polynomial λ^{k_1+j}.

Let us now construct another subspace, $\mathcal{I}_2$, as:

$$\mathcal{I}_2 := \text{span}\,\{A_c^j b_2, \ldots, A_c^{k_2-1} b_2\} \tag{4.10}$$

It is readily seen that

$$\left\{ \begin{array}{lcl} \mathcal{I}_2 & \approx & R^{k_2-j}, \\ \mathcal{R}_2^* & = & (\mathcal{R}_2^* \cap \mathcal{I}_1) \oplus \mathcal{I}_2 \\ \mathcal{X} & = & \mathcal{I}_1 \oplus \mathcal{I}_2 \end{array} \right. \tag{4.11}$$

(first and second claim follow from (4.10), (4.1) and (4.9), and for the third claim, we have from this two first claims and the first claim of (4.9): $\dim(\mathcal{I}_1 + \mathcal{R}_2^*) = \dim \mathcal{I}_1 + \dim \mathcal{I}_2 = (k_1 + j) + (k_2 - j) = n$).

Let us now define the following projection :

$$P_j : \mathcal{X} \to \mathcal{X} \quad \text{the projection on } \mathcal{I}_2 \text{ along } \mathcal{I}_1 \tag{4.12}$$

we then have from (4.8), (4.9) and (4.11) :

$$P_j b_1 = P_j \left(\bar{A}^j b_2 - A_c^j b_2 \right) = -A_c^j b_2 \tag{4.13}$$

and from (4.6) we have :

$$\bar{A} \left(A_c^{i-1} b_2 \right) = A_c^i b_2 \tag{4.14}$$

for $i = j+1, \ldots, k_2$, which implies :

$$A_c^{j+i} b_2 = \bar{A}^i \left(A_c^j b_2 \right) \quad ; \; i = 1, \ldots, k_2 - j \tag{4.15}$$

And thus, we have from (4.13), (4.15), (4.10) and (4.2.c) :

$$\begin{array}{l} \mathcal{I}_2 = \text{span}\,\{\bar{b}_1, \bar{A}\bar{b}_1, \ldots, \bar{A}^{k_2-j-1}\bar{b}_1\} \\ \quad \approx R^{k_2-j}, \\ \bar{A}\mathcal{I}_2 \subset \mathcal{I}_2, \\ \bar{A}^{k_2-j}\mathcal{I}_2 = \{0\}. \end{array} \tag{4.16}$$

with

$$\bar{b}_1 = P_j b_1 \tag{4.17}$$

Which implies that $\mathcal{I}_2$ is also an $\bar{A}$-invariant cyclic subspace generated by the projection of b_1 on $\mathcal{I}_2$ (see (4.12)) with dimension $k_2 - j$ and minimal polynomial λ^{k_2-j}. ■

For the case of more than two controllability indices it is enough to apply (i) successively.

4.2. *Proof of the Sufficiency.*

Without any loss of generality, we can assume that the pair (A, B) is already in Brunovsky canonical form (we can achieve that by means of a change of bases in $\mathcal{X}$ and $\mathcal{U}$ and with a first state feedback). We next apply a second state feedback as in (4.6)-(4.7) and express the state space in terms of the subspaces $\mathcal{I}_1, \mathcal{I}_2, \ldots$ (see (4.8)-(4.9) and (4.16)-(4.17)). We thus have obtained a matrix A which is block diagonal with minimal polynomials $\lambda^{n_1}, \lambda^{n_2}, \ldots, \lambda^{n_\mu}$ where the n_i satisfy (4.3), and each block is expressed in its rational canonical form (see Wohnam, 1985 and Gantmacher, 1977).

Let us now go back, without any loss of generality, to the case of two controllability indices.

After having applied the above mentioned procedure, we obtain from (4.8)-(4.9) and (4.16)-(4.17):

$$\begin{array}{l} [(\lambda \mathrm{I} - A_2) \mid B] \\ = \left[\begin{array}{cc|cc} \lambda \mathrm{I} - \mathrm{J}_{n_1} & 0 & \star & b_2 \\ 0 & \lambda \mathrm{I} - \mathrm{J}_{n_2} & \bar{b}_1 & 0 \end{array} \right] \end{array} \tag{4.18}$$

where $\star$ is a vector which value is not precised, $b_2^T = \left[\begin{array}{cccc} 0 & \cdots & 0 & 1 \end{array} \right]_{1 \times n_1}$, $\bar{b}_1^T = \left[\begin{array}{cccc} 0 & \cdots & 0 & 1 \end{array} \right]_{1 \times n_2}$ and $n_1 \geq n_2$.

Applying now the third state feedback, $F_3 = \left[\begin{array}{cccccccc} 0 & 0 & \cdots & 0 & a_1^2 & a_2^2 & \cdots & a_{n_2}^2 \\ a_1^1 & a_2^1 & \cdots & a_{n_1}^1 & 0 & 0 & \cdots & 0 \end{array} \right]$, in order to assign the coefficients corresponding to the desired invariant factors

$$\psi_i(\lambda) = \lambda^{n_i} - \left(a_1^i + a_2^i \lambda + \cdots + a_n^i \lambda^{n_i - 1} \right) \quad , \; i = 1, 2.$$

We have (with : $A_3 := A_2 + BF_3$):

$$\begin{array}{l} [(\lambda \mathrm{I} - A_3) \mid B] \\ = \left[\begin{array}{cc|cc} \lambda I - L_1 & X & \star & b_2 \\ 0 & \lambda I - L_2 & \bar{b}_1 & 0 \end{array} \right] \end{array} \tag{4.19}$$

where $L_i = \underline{\chi}_{n_i} \underline{a}^T + \mathrm{J}_{n_i}$, $\underline{a}^T = \left[\begin{array}{cccc} a_1^i & a_2^i & \cdots & a_{n_i}^i \end{array} \right]$, $\underline{\chi}_{n_i}^T = \left[\begin{array}{cccc} 0 & \cdots & 0 & 1 \end{array} \right]_{1 \times n_i}$, $j = 1, \ldots, n_2$, and where X is some $n_1 \times n_2$ real matrix, which entries will be denoted $x_{i,j}$ in the following.

Note that the matrices A_2, A_3 and B are expressed in the basis $\{\mathrm{e}_1, \ldots, \mathrm{e}_{n_1}\}$ of $\mathcal{X}_1$ and $\{\mathrm{e}_{n_1+1}, \ldots, \mathrm{e}_{n_1+n_2}\}$ of $\mathcal{X}_2$ (see Ch. 0.10 and 1.3 of Wohnam (1985)):

$$\left\{ \begin{array}{lll} \mathrm{e}_j & = \psi_1^{(j)}(L_1) b_2 & , \; j = 1, \ldots, n_1 \\ \mathrm{e}_{j+n_1} & = \psi_2^{(j)}(L_2) \bar{b}_1 & , \; j = 1, \ldots, n_2 \end{array} \right. \tag{4.20}$$

where for $i = 1, 2$: $\psi_i^{(0)}(\lambda) := \psi_i(\lambda)$ and $\psi_i^{(j)}(\lambda) := \frac{1}{\lambda}\left(\psi_i^{j-1}(\lambda) + a_j^i\right)$, with $j = 1, \ldots, n_i$.

The column vectors $\underline{x}_j := \left[\begin{array}{ccc} x_{1,j} & \cdots & x_{n_1,j} \end{array}\right]^T$ of matrix X are expressed in this basis as:

$$\underline{x}_j = \sum_{\kappa=1}^{n_1} x_{\kappa,j} e_\kappa = (L_1\Delta_j + \alpha_j)b_2 \tag{4.21}$$

with $j = 1, \ldots, n_2$, and where $\Delta_j = \sum_{\kappa=1}^{n_1-1} x_{\kappa,j}\psi_1^{\kappa+1}(L_1)$ and $\alpha_j = x_{n_1,j} - \sum_{\kappa=1}^{n_1-1} x_{\kappa,j}a_{\kappa+1}^1$.

If we now define the new basis :

$$\bar{e}_j = e_j \quad ; j = 1, \ldots, n_1 \tag{4.22}$$

$$\bar{e}_{j+n_1} = e_{j+n_1} - \Delta_j b_2 \quad , j = 1, \ldots, n_2$$

we then have :

$$\bar{e}_{n_1} = b_2 \tag{4.23}$$

$$A_3\bar{e}_j = \bar{e}_{j-1} + a_j^1 b_2 \quad , j = 1, \ldots, n_1$$

since $A_3 e_j = L_1 e_j$, for $j = 1, \ldots, n_1$. And

$$\left\{\begin{array}{ll} \bar{e}_{n_1+n_2} & = \bar{b}_1 - \Delta_{n_2} b_2 \\ A_3\bar{e}_{j+n1} & = \bar{e}_{j+n_1-1} + a_j^2\bar{b}_1 + \beta_j b_2 \ , \\ & \qquad j = 1, \ldots, n_2 \end{array}\right. \tag{4.24}$$

since $A_3 e_{j+n_1} = L_2 e_{j-n_1} + \underline{x}_j$, $\underline{x}_j = (L_1\Delta_j + \alpha_j)b_2$ and $A_3 b_2 = L_1 b_2$.

In this new basis (4.19) takes the form

$$[(\lambda I - A_3) \mid B] = \left[\begin{array}{cc|cc} \lambda I - L_1 & \chi_{n_1}\underline{\beta}^T & * & b_2 \\ 0 & \lambda I - L_2 & \bar{b}_1 & 0 \end{array}\right] \tag{4.25}$$

where $\underline{\beta} = \left[\begin{array}{ccc} \beta_1 & \cdots & \beta_{n_2} \end{array}\right]$.

Finally with the feedback $F_4 = \left[\begin{array}{cccccccc} 0 & 0 & \cdots & 0 & 0 & 0 & \cdots & 0 \\ 0 & 0 & \cdots & 0 & -\beta_1 & -\beta_2 & \cdots & -\beta_{n_2} \end{array}\right]$, we achieve the desired goal. ∎

5. Conclusion

We give a new proof of Rosenbrock Theorem on pole assignment. This proof is done within a very geometric framework; it is based on two well-known basic results of control theory (Theorems 2.1 and 2.2).

Our proof of the necessity provides a new interpretation of Rosenbrock inequalities. Actually invariant factors are associated with cyclic subspaces generated by projections of input vectors, the system being controllable, and the dimensions of subspaces of that kind necessarily satisfy Rosenbrock's inequalities.

The proof of the sufficiency, based on the construction of such cyclic subspaces, is constructive and leads to an efficient method for the design of a feedback which assigns the invariant factors of the system. This procedure can be sumarized as follows:

(i) Calculate a feedback F_1 which puts the system in Brunovsky cannonical form.
(ii) Calculate a feedback F_2, which as in (4.6), which leads to cyclic subspaces having the desired dimensions, that are the degrees of the invariant factors.
(iii) The feedback $F_3 + F_4$ calculated as in subsection 4.2 finally permits to adjust as specified the coefficients of the invariant factors.

6. References

Brunovsky (1970). A Classification of Linear Controllable Systems. Kybernetika, Vol. 6, No. 3, pp. 173-188.

Brylawski (1973). The Lattice of Integer Partitions. Discrete Mathematics Vol. 6, pp. 201-219.

Dickinson (1974). On the fundamental theorem of linear state variable feedback. IEEE TAC-19, pp. 577-579.

Flamm (1980). A New Proof of Rosenbrock's Theorem on Pole Assignment.IEEE-TAC, Vol. AC-25, No. 6, December

Gantmacher (1977). The Theory of Matrices, Vol. I.New York, Chelsea Publishing Co.

Kailath (1980). Linear Systems. Prentice-Hall, Inc. Englewood Cliffs, N.J.

Kučera (1981). Assigning the invariant factors by feedback. Kybernetika, 17, pp. 118-127.

Loiseau (1988). Sur la Modification de la Structure à l'Infini par un Retour d'Etat Statique. SIAM J. Control & Opt., Vol. 26 No.2, pp. 251-273.

Özçaldiran (1990). Fundamental Theorem of Linear State Feedback for Singular Systems. 29th. CDC, Honolulu, Hawaii, December.

Rosenbrock (1970). State Space and Multivariable Theory. New York, Wiley.

Wohnam (1985). Linear Multivariable Control, A Geometric Approach. New York, Springer-Verlag, 1985.

Zagalak and Loiseau (1992). On the Problem of Pole Assignment by Feedback in Linear Systems. Part 1: Constant Feedback. to be submitted, November 18.

Zagalak and Kučera (1991). Eigenstructure Assignment in Linear Descriptor Systems. European Control Conference ECC'91 Grenoble France.

SAMPLING AND STRUCTURAL PROPERTIES OF DISCRETISED LINEAR MODELS: SEGRE CHARACTERISTIC AND RELATED PROPERTIES

N. Karcanias and N. Tamvaclis
Control Engineering Centre, City University, London EC1V 0HB, U.K.

Résumé: Le choix de l'encantillonage est l'un des problèmes importants pour la conception et la réalisation de schémas de commande numériques et ce réglage est participalement déduit de la théorie du traitment du signal. Le but de ce papier est d'introduire quelques problèmes qui mettent en évidence les asoects structurels liés à ce choix et d'étudier certaines propriétés clés qui y sont associées. Nous examinons notamment la relation entre les caractéristiques de Segre des matrices d'état et les espaces invariants associés. Nous introduisons la notion d'échantillonage régulier et irrégulier et analysons les phénomènes de disparition de valeurs propres et de caractéristiques de Segre associées. Ces résultats permettent d'étudier les pertes de commandabilité ou d'observabilité sous l'action d'échantillonage irrégulier et conduisent à une classification dé phénomènes non commandables et non observables.

Abstract: The selection of sampling is one of the important issues in the design and implementation of digital control schemes and the rules for such selection come predominantly from the signal processing theory. The aim of this paper is to introduce a number of problems highlighting the structural theory aspects of selection of samples and examine certain associated key properties. In particular, we examine the relationship between the Segre characteristics of continuous and discretised state matrices and associated invariant spaces. The notion of regular and irregular sampling is introduced and phenomena referred to as collapsing of eigenvalues and associated Segre characteristics are examined. The results provide the means for studying the loss of controllability, observability under irregular sampling and introduces a classification of uncontrollability, unobservability phenomena.

1 Introduction

A key issue in the implementation of digital control schemes is the selection of sampling period T. Almost all the theory and practice on the selection of T comes from the signal processing area [1]. It is evident that sampling affects the nature of the discretised model and thus also may affect the properties and characteristics of what we understand as a system through the computer. The effect of sampling on the structural characteristics such as controllability, observability has been examined in [2] [3]; however, the analysis there, was restricted only in the derivation of sufficient conditions for preservation of controllability, observability rather than providing a detailed study of these properties under all possible values of T. Some more recent work [4] has considered the effect of sampling on the zeros of SISO discretised models as a function of sampling and some results on the asymptotic properties of zeros have been derived. The study of structural properties of discretised models as a function of the corresponding properties of the continuous model and the sampling rate T is however still in its early stages of development. A structural theory for selection of sampling rate is an important topic that may complement the existing signal aspects. This work aims at contributing to the development of such a structural theory.

In this paper we examine an important property of the mapping of continuous to discrete model structural characteristics which is referred to as "collapsing of eigenvalues". This phenomenon occurs for specific values of the sampling rate determined by the root range of the state matrix and defined as irregular sampling. The effect of irregular sampling on properties such as Segre characteristic, A-invariant spaces and controllability, observability are examined. It is shown that under irregular sampling there is merging of certain Segre characteristics, as well as associated invariant subspaces; this merging affects the controllability, observability of discretised models and notions of structural and parametric uncontrollability (unobservability) are introduced. These results clarify the role of irregular sampling and provide criteria for controllability, observability of discretised models for all values of T. The results in this paper are stated without proof and their proof may be found in [5]

2 Statement of the problem

We consider the standard computer control configuration [6] where a system is represented by a linear state space model

$$\mathcal{S}(A,B,C,D): \begin{array}{l} \dot{x} = Ax + Bu \\ y = Cx + Du \end{array} \tag{2.1}$$

where $A \in \Re^{n\times n}$, $B \in \Re^{n\times \ell}$, $C \in \Re^{m\times n}$, $D \in \Re^{m\times \ell}$ is to be controlled by a digital computer and the interface consists of a Digital to Analogue Converter (DAC) and an Analogue to Digital Converter (ADC). Both the ADC and DAC are assumed to be driven synchronously by clock signals uniformly spaced T seconds apart, where T is referred to as the sampling interval. For the digital computer the plant is seen as a device acting on the sequences $u(kT)$ (input to DAC), $y(kT)$ (output of ADC) and thus the plant is represented by an equivalent discretised model

$$\hat{\mathcal{S}}(\hat{A},\hat{B},\hat{C},\hat{D}): \begin{array}{l} x[(k+1)T] = \hat{A}x(kT) + \hat{B}u(kT) \\ y(kT) = \hat{C}x(kT) + \hat{D}u(kT) \end{array} \tag{2.2}$$

where $x(kT)$ is the system state at time $t = kT$. If the DAC is a zero order hold, it is known [6] that the discretised model parameters are given by

$$\hat{A} = e^{AT},\ \hat{C} = C,\ \hat{D} = D,\ \hat{B} = \{\int_0^T e^{A\tau}d\tau\}B \tag{2.3}$$

From the above, it is clear that the discretised model is a function of the sampling T and thus the structural invariants (continuous and discrete), system properties and the values of property indicators are also functions of T. The selection of T is one of the issues related to the implementation of digital schemes [7]–[9] and the nature of the discretised model is crucial for the design of digital control schemes. The long term objective of this work is the study of the structural properties and invariants of the discretised model S as a function of the corresponding properties and invariants of the continuous model S and the sampling T. This is referred to as Structure Mapping Problems (SMP) and some of the basic issues related to the effect of sampling on the eigenvalues, associated Segre Characteristic and eigenspaces, as well as effects on Controllability and observability are examined in this paper.

3 Spectral form of the discretised model

In this section we work out explicit forms for the discretised state parameters as functions of the sampling T and then we work out the Jordan canonical description of the discretised model. For the given matrix A we shall denote by $\Phi(A)$ the root range of A (set of distinct eigenvalues) and for every $\lambda \in \Phi(A)$ we shall denote by

$$\mathcal{S}_\lambda(A) = \{0 < d_{i_1} \leq \cdots \leq d_{i_\nu}\} \tag{3.1}$$

the λ-Segre characteristic (set of degrees of elementary divisors of A). The Jordan form of A may be expressed

$$A = UJV,\ J = \text{diag}\{\cdots; J(\lambda); \cdots\} \tag{3.2}$$

where U, V are the eigenframes, dual eigenframes, $\lambda \in \Phi(A)$, and if $J(\lambda, d_{i_k})$, $k \in \underline{\nu}$ is the standard $d_{ik} \times d_{ik}$, λ-Jordan block, then

$$J(\lambda) = diag\{J(\lambda, d_{i_1}); \cdots; J(\lambda, d_{i_\nu})\} \tag{3.3}$$

The computation of the $\hat{A}$, $\hat{B}$ discretised parameters in the spectral representation defined above is considered next.

<u>**Proposition (3.1):**</u> With the notation introduced above

$$\hat{A} = Ue^{JT}V = U\text{diag}\{\cdots; e^{J(\lambda)T}; \cdots\}V \tag{3.4}$$

where for every $\lambda \in \Phi(A)$ we have

$$e^{J(\lambda)T} = e^{\lambda T}\cdot \text{diag}\{A(d_{i_1},T); \cdots; A(d_{i_\nu},T)\} \tag{3.5}$$

$$A(d,T) = \begin{bmatrix} 1 & T & T^2/2! & \cdots & T^{d-1}/(d-1)! \\ 0 & 1 & T & \cdots & T^{d-2}/(d-2)! \\ \vdots & & \ddots & \ddots & \vdots \\ 0 & 0 & \cdots & \ddots & T \\ 0 & 0 & \cdots & \cdots & 1 \end{bmatrix} \tag{3.6}$$

□

<u>**Remark (3.1):**</u> For the $J(0,d)$ block we have

$$\int_0^T e^{J(0,d)\tau}d\tau = P_0(d,T) =$$

$$= \begin{bmatrix} T & T^2/2! & \cdots & T^d/d! \\ 0 & T & \cdots & T^{d-1}/(d-1)! \\ \vdots & \ddots & \ddots & \vdots \\ 0 & 0 & \cdots & T \end{bmatrix} \tag{3.7}$$

□

<u>**Lemma (3.1):**</u> For the $J(\lambda,d)$ block, $\lambda \neq 0$, we have

$$\int_0^T e^{J(\lambda,d)\tau}d\tau = J(d,\lambda)^{-1}(e^{J(d,\lambda)T} - I_d) =$$

$$= (e^{J(d,\lambda)T} - I_d)J(d,\lambda)^{-1} = P_\lambda(d,T) \quad (3.8)$$

Furthermore, $P_\lambda(d,T)$ has the following form

$$P_\lambda(d,T) = \begin{bmatrix} a_0 & a_1 & \cdots & \cdots & a_{d-1} \\ 0 & a_0 & a_1 & \cdots & a_{d-2} \\ \vdots & \ddots & \ddots & \ddots & \vdots \\ \vdots & & \ddots & \ddots & a_1 \\ 0 & \cdots & \cdots & 0 & a_0 \end{bmatrix} \quad (3.9)$$

and $a_0 = (e^{\lambda T} - 1)\lambda$ □

Proposition (3.2): With the notation previously defined assuming that $\Phi(A) = \{0, \cdots, \lambda, \cdots\}$, then

$$\hat{B} = U\text{diag}\{P_0(T); \cdots; P_\lambda(T); \cdots\}VB =$$

$$= UP(T)VB \quad (3.10)$$

and if $\mathcal{S}_0(A) = \{0 < d_{j_1} \leq \cdots \leq d_{j_\mu}\}$, then

$$P_0(T) = \text{diag}\{P_0(d_{j_1}, T); \cdots; P_0(d_{j_\mu}, T)\} \quad (3.11)$$

$$P_\lambda(T) = \text{diag}\{P_\lambda(d_{i_1}, T); \cdots; P_\lambda(d_{j_\nu}, T)\} \quad (3.12)$$

□

The above explicit formulae will be used for the study of system properties of the discretised model.

Lemma (3.2): For any $\lambda \in \mathcal{C}$ the matrix $exp\{J(\lambda, d)T\}$ has an eigenvalue $e^{\lambda T}$ with geometric multiplicity 1 and algebraic multiplicity d. □

By the above and Proposition (3.1) we have:

Proposition (3.3): If $\mathcal{S}_\lambda(A) = \{0 \leq d_{i_1} \leq \cdots d_{i_\nu}\}$ is the λ–Segre characteristic of A then there exist matrices $\tilde{U}$, $\tilde{V} = U^{-1}$ such that

$$\hat{A} = expAT =$$

$$\tilde{U}\text{diag}\{\cdots; J(e^{\lambda T}, d_{i_1}); \cdots; J(e^{\lambda T}, d_{i_\nu}); \cdots\}\tilde{V} \quad (3.13)$$

□

Although (3.13) defines the Jordan decomposition of A explicitly, the Segre characteristic $\mathcal{S}_z(\hat{A})$, $z = e^{\lambda T}$, is not automatically defined by (3.13) for reasons of eigenvalue collapsing, which will be examined in the following section.

4 Eigenvalue collapsing and merging of Segre characteristics

From the derivation of the discretised parameters it is evident that for any $\lambda = \sigma + j\omega \in \Phi(A)$, then $z = e^{\sigma T + j\omega T}$ is an eigenvalue of $\hat{A} = e^{AT}$, i.e., $z \in \Phi(\hat{A})$. However, for two distinct eigenvalues $\lambda_1, \lambda_2 \in \Phi(A)$ there may be values of T such that $e^{\lambda_1 T} = e^{\lambda_2 T} = z \in \Phi(\hat{A})$. This phenomenon, where distinct elements of $\Phi(A)$ are mapped to one element of $\Phi(\hat{A})$ is called collapsing of eigenvalues and any value of T for which such phenomenon occur will be referred to as irregular sampling. All values of T for which any $\lambda_1, \lambda_2 \in \Phi(A)$, $\lambda_1 \neq \lambda_2$, is mapped to $z_1 \neq z_2$ will be called regular sampling. The property of eigenvalue collapsing may occur not only on a pair of (λ_1, λ_2), but on a set $\mathcal{L}(A) = \{\lambda_i \in \Phi(A), \lambda_i \neq \lambda_j\}$, such sets for which there exists T such that $z_i = e^{\lambda_i T} = z \in \Phi(\hat{A})$, for all $\lambda_i \in \mathcal{L}(A)$, will be called collapsing sets and depending on whether $z \in \mathcal{C}$ or $\in \Re$ this collapsing will be called complex, or real. Clearly, this property depends on how we select T, as well as the nature of the set. The presence of collapsing sets in $\Phi(A)$, as well as the characterisation of values of associated irregular samplings, is of great importance in the development model based theory for sampling, since it affects the basic structure of $\hat{A}$, as well as all related properties and it is subject of this section. The structural significance of such phenomena is indicated by the following results:

Theorem (4.1): Let $\mathcal{L}$(A) $= \{\lambda_i \in \Phi(A),\ \lambda i \neq \lambda j, i \in \underline{r}\}$ be a collapsing set and let $\mathcal{S}_{\lambda i}$ (A), $\mathcal{U}_{\lambda i}(A)$ be the Segre characteristic and associated generalised eigenspace of λ_i. If T is a value of the irregular sampling for which collapsing of $\mathcal{L}(A)$ occurs to $z = e^{\lambda T} \in \Phi(\hat{A})$, then

$$\mathcal{U}_z(\hat{A}) = \mathcal{U}_{\lambda_1}(A) \oplus \cdots \oplus \mathcal{U}_{\lambda_r}(A) \quad (4.1)$$

$$\mathcal{S}_z(\hat{A}) = \mathcal{S}_{\lambda_1}(A) \cup \cdots \cup \mathcal{S}_{\lambda_r}(A) \quad (4.2)$$

Note that the above property expresses the merging of generalised eigenspaces and Segre characteristics (preservation of distinct values and multiplicities) and expresses the nature of the structure transformation that occurs. In the following we examine the properties of collapsing sets and associated values of T.

Remark (4.1): If A has all its eigenvalues real and distinct, the $\hat{A}$ also has all its eigenvalues real and distinct for all $T > 0$ and there is no collapsing set. □

The above indicates clearly that collapsing is associated with the presence of complex eigenvalues in $\Phi(A)$.

Remark (4.2): If $\lambda_1, \lambda_2 \in \Phi(A)$, then necessary condition for the existence of T such that collapsing occurs, i.e., that $Re(\lambda_1) = Re(\lambda_2)$. □

Definition (4.1): Let $\Phi(a) = \{\lambda \in C, \lambda = \sigma + jw\}$ and let $\Lambda(A) = \{\sigma_i, i \in \underline{\tau}, \sigma_i \in \Re\}$ be the set of distinct values of the real parts $\Phi(A)$, and $\Phi_\sigma(A) = \{\forall\lambda \in \Phi(A), Re(\lambda) = \sigma\}$. $\Lambda(A)$ will be called the real trace of $\Phi(A)$ and $\Phi_\sigma(A)$ its σ–root range. □

It is clear that $\Phi(A)$ may be partitioned as

$$\Phi(A) = \Phi_{\sigma_i}(A) \overset{r}{\underset{i=1}{\cup}}, \ \forall \sigma_i \in \Lambda(A) \qquad (4.3)$$

It is evident from Remark (4.2) that collapsing may occur only for subsets of $\Phi_\sigma(A)$ and this is why our attention is focused next on the subsets of $\Phi_\sigma(A)$.

Proposition (4.1): Let $\lambda_1 = \sigma + j\omega_1$, $\lambda_2 = \sigma + j\omega_2 \in \Phi_\sigma(A)$ and $z_i = e^{\lambda_i T}$, $i = 1, 2$. The following properties hold true:

(a) (i) For any $\lambda_1, \lambda_2 \in \Phi_\sigma(A)$, $z_1 = z_2 \in \mathcal{C}$, if and only if

$$T = 2k\pi/|\omega_1 - \omega_2|, \ k = 1, 2, 3 \cdots \qquad (4.4)$$

(ii) There exists T such that $z_1 = z_2 \in \Re$, if and only if $\omega_2/\omega_1 = \mu/\nu$, $\mu, \nu \in \mathcal{Z}_{\neq 0}$. Furthermore, if the latter condition is satisfied, then the corresponding T is

$$T = 2|\nu - \mu| \cdot \bar{k}\pi/|\omega_1 - \omega_2|, \ \bar{k} = 1, 2, 3, \cdots \qquad (4.5)$$

(b) If $\lambda_1 = \sigma + j\omega, \lambda_2 = \sigma - j\omega = \lambda_1^*$, then $z_1 = z_2$ if and only if

$$T = k\pi/\omega, \ k = 1, 2, 3, \cdots \qquad (4.6)$$

Furthermore, for all such T we have $z_1 = z_2 = e^{-\sigma T}$, if $k = 2, 4, 6, \cdots$ and $z_1 = z_2 = -e^{-\sigma T}$, if $k = 1, 3, 5, \cdots$.

(c) If $\lambda_1 = \sigma + j\omega_1$, $\lambda_1^* = \sigma - j\omega_1$, $\lambda_2 = \sigma + j\omega_2$, $\lambda_2^* = \sigma - j\omega_2 \in \Phi_\sigma(A)$ then for any T such that $z_1 = z_2$, then also $z_1^* = e^{\lambda_1^* T} = z_2^* = e^{\lambda_2^* T}$. Furthermore, if T i is selected as in (4.5), then $z_1 = z_2 = z_1^* = z_2^* \in \Re$. □

The above result clearly establishes the existence of irregular samplings for complex and real collapsing on simple subsets of a general $\Phi_\sigma(A)$ set. More specifically:

Corollary (4.1): Given any set $\Phi_\sigma(A)$ the following properties hold true for simple subsets of $\Phi_\sigma(A)$:

(i) For the set $\Phi_\sigma^\alpha(A) = \{\sigma, \sigma \pm j\omega\}$, there always exists T such that there is total real collapsing to $e^{\sigma T}$.

(ii) For the set $\Phi_\sigma^\beta(A) = \{\sigma \pm j\omega_1, \sigma \pm j\omega_2\}$, there is always a T such that we have complex collapsing to z, $z^* \in \mathcal{C}$. Furthermore, if ω_1/ω_2 is rational, there exists T such that we have total real collapsing to $e^{\sigma T}$.

(iii) For the set $\Phi_\sigma^\gamma(A) = \{\sigma, \sigma \pm j\omega_1, \sigma \pm j\omega_2\}$, there exists total real collapsing to $e^{\sigma T}$ for some T if and only if ω_1/ω_2 is rational. □

For sets with more elements the problem of total complex, or real collapsing depends on the structure of the set and this is established below.

Theorem (4.2): Let $\Phi_\sigma(A) = \{\lambda_i, \lambda_i^* = \sigma \pm j\omega_i, i \in \underset{\sim}{p}, 0 < \omega_1 < \omega_2 \cdots < \omega_p\}$. The following properties hold true:

(i) There exists T such that for all $z_i = e^{\lambda_i T}, z_i^* = e^{\lambda_i^* T}$ $z_1 = \cdots = z_p = z$, $z_1^* = \ldots = z_p^* = z^*$, z, $z^* \in \mathcal{C}$, if and only if

$$\omega_p - \omega_{p-1} = \cdots = \omega_2 - \omega_1 = \delta\omega \qquad (4.7)$$

If the above holds true, then the appropriate sampling is $T = 2k\pi/\delta\omega$, $k = 1, 2 \cdots$.

(ii) If condition (4.7) holds true and for some ω_i

$$\omega_i/\delta\omega = \mu/\nu, \ \mu, \nu \in \mathcal{Z}_{>0} \qquad (4.8)$$

then every $\omega_j/\delta\omega$ is rational and there exists T such that $z_i = z_i^* = e^{\lambda_i T} = e^{\lambda_i^* T} = e^{\sigma T}$ for all $i \in \underset{\sim}{p}$. □

A set $\Phi_\sigma(A)$ that satisfies conditions (4.7) will be called equally spaced, and if in addition condition (4.8) holds true, then it will be called rationally equally spaced. Equal spacing is equivalent to existence of T for total complex collapsing, whereas rational equal spacing implies the existence of T for total real collapsing.

Remark (4.3): If $\Phi_\sigma(A) = \{\sigma, \sigma \pm j\omega_i, \ i \in \underset{\sim}{p}\}$ and the subset $\{\sigma \pm j\omega_i, \ i \in \underset{\sim}{p}\}$ is rationally equally spaced, then there exists T such that there is total real collapsing of $\Phi_\sigma(A)$. □

Any symmetric subset of $\Phi_\sigma(A)$ that allows total complex, or total real collapsing for some appropriate value of T will be called a $\mathcal{C}-$, $\Re$-prime subset respectively and the set of all prime subsets of $\Phi_\sigma(A)$ will be denoted by $\langle\Phi_\sigma(A)\rangle$. Using the results so far and Theorem (4.1) we may now define

Definition (4.2): Let $A \in R^{n \times n}$, $\Phi(A)$ be its root range $\lambda \in \Phi(A)$ and let $\tau(\lambda)$ be the cardinality of $\mathcal{S}_\lambda(A)$. Then,

$$\gamma_A = \max\{\tau(\lambda), \ \forall\lambda \in \Phi(A)\} \qquad (4.9)$$

is called the Segre index of A. □

This definition may be extended to $\hat{A}$ as follows:

Definition (4.3): If $\Phi(A) = \overset{r}{\underset{i=1}{\cup}} \Phi_{\sigma_i}(A)$, then:

(a.i) If $\mathcal{P}_\sigma^{\mathcal{C}}(A) = \{\sigma \pm j\omega_i, \ \omega_i > 0, \ i \in \underset{\sim}{t}\}$ is a $\mathcal{C}-$ prime set and $\tau_\sigma(\omega_i)$ is the cardinality of the $\mathcal{S}_{\sigma+j\omega_i}(A)$ Segre characteristic, then we define as the character of the set

$$\pi(P_\sigma^{\mathcal{C}}) = \sum_{i=1}^{t} \tau_\sigma(\omega_i) \qquad (4.10)$$

(a.ii) If $\mathcal{P}_{\sigma}^{\Re}(A) = \{\sigma,\ \sigma \pm j\omega_i,\ \omega_i > 0,\ i \in \underline{t}\}$ is an $\Re$–prime set and $\tau_\sigma(\omega_i)$, $\tau_\sigma(0)$ are the cardinalities respectively of $\mathcal{S}_{\sigma+j\omega_i}(A)$, $\mathcal{S}_\sigma(A)$ Segre characteristics, then we define as the character of the set.

$$\pi(P_\sigma^{\Re}) = \tau(\sigma) + 2\sum_{i=1}^{t} \tau_\sigma(\omega_i) \qquad (4.11)$$

(b) We define as the σ–index of $\Phi_\sigma(A)$ the number

$$\theta_\sigma = \max\{\pi(P_\sigma),\ \forall P_\sigma \in \langle\Phi_\delta(A)\rangle\} \qquad (4.12)$$

where P_σ, is either $\mathcal{C}$–, or $\Re$–prime set. Finally, the number

$$\theta_{\hat{A}} = \max\{\theta_\sigma,\ \forall\sigma \in \Lambda(A)\} \qquad (4.13)$$

will be called the collapsing index of $\hat{A}$. □

Remark (4.4): The collapsing index of $\hat{A}$ expresses the maximal value of the Segre index of $\hat{A}$ that can be achieved under all possible values of T, for which collapsing of eigenvalues occur. By definition, for any value of T we have:

$$\gamma_A \leq \gamma_{\hat{A}} \leq \theta_{\hat{A}} \qquad (4.14)$$

□

The computation of $\theta_{\hat{A}}$ involves finding $\tau(\lambda) = \dim \mathcal{N}_r\ (\lambda I - A)$ for $\forall\lambda \in \Phi(A)$, as well as computing $\{\langle\Phi_\sigma(A)\rangle,\ \forall\sigma \in \Lambda(A)\}$. For certain cases we have:

Remark (4.5): If a matrix A is generic then $\gamma_A = 1$ and $\theta_{\hat{A}} = 2$. If a matrix A has all its eigenvalues real, then $\gamma_A = \theta_{\hat{A}}$. □

5 Controllability and observability of discretised models

In this section we examine the properties of controllability and observability of the discretised system $\hat{\mathcal{S}}(\hat{A}, \hat{B}, \hat{C})$ as functions of the sampling T and we shall examine in particular the effect of eigenvalue collapsing on these properties. We shall examine the properties of regular sampling and then phenomena associated with irregular sampling will be examined. Criteria for modal controllability, observability [3] are central to this study and are summarised below.

Following the notation of section (3), assume that (A, B, C) are expressed in the Jordan canonical representation $(J, \tilde{B}, \tilde{C})$ where $J = diag\{\cdots, J(\lambda); \ldots\}$, $\forall\lambda \in \Phi(A)$, $J(\lambda)$ determined by $\mathcal{S}_\lambda(A) = \{0 < d_{i_1} \leq \cdots \leq d_{i_\nu}\}$ and $\tilde{B}$, $\tilde{C}$ are partitioned according to the partitioning of J as $\tilde{B} = [\cdots; \tilde{B}_\lambda^t; \ldots]^t$, $\tilde{C} = [\ldots; \tilde{C}_\lambda; \cdots]$. Furthermore, for every $\lambda \in \Phi(A)$ we have the compatible partitioning

$$J(\lambda) = diag\{\cdots; J(\lambda, d_j); \cdots\}$$

$$\hat{B}(\lambda) = \begin{bmatrix} \vdots \\ \tilde{B}(\lambda, d_j) \\ \vdots \end{bmatrix},\ \tilde{C}(\lambda) = [\cdots; \tilde{C}(\lambda, d_j); \cdots] \qquad (5.1)$$

If $\underline{\beta}^t(d_j)$ is the last row of $\tilde{B}(\lambda, d_j)$ and $\underline{\gamma}(d_j)$ is the first column of $\tilde{C}(\lambda, d_j)$, $j \in \{i_1, \cdots, 1_\nu\}$ then we define the λ -input, λ -output - spectral matrices by

$$\Delta_\lambda = \begin{bmatrix} \underline{\beta}^t(d_{i_1}) \\ \vdots \\ \underline{\beta}^t(d_{i_\nu}) \end{bmatrix},\ \Gamma_\lambda = [\underline{\gamma}(d_{i_1}), \cdots, \underline{\gamma}(d_{i_\nu})] \qquad (5.2)$$

Lemma (5.1) [3]: For the system (A, B, C) the following properties hold true:

(i) $\lambda \in \Phi(A)$ is controllable, iff rank $\{\Delta_\lambda\} = \nu$

(ii) $\lambda \in \Phi(A)$ is observable, iff rank $\{\Gamma_\lambda\} = \nu$

□

Using the above result we may state:

Theorem (5.1): For all regular sampling T the following results hold true:

(i) The pair $(\hat{A}, \hat{B})$ is controllable, iff (A, B) is controllable

(ii) The pair $(\hat{A}, \hat{C})$ is observable, iff (A, C) is observable. □

The above result expresses sufficient conditions for preservation of controllability asnd observability and it is the essence of the existing results of effect of sampling on such properties [2], [3]; the case of irregular sampling has not been examined. For a given irregular sampling T, there is always a $\mathcal{C}$, or $\mathcal{R}$–maximal prime subset corresponding to a $\sigma \in \Lambda(A)$, characterised by $z_T = e^{\lambda T}, e^{\lambda^* T}$ ($\mathcal{C}$–case), or $z_T = e^{\sigma T}$ ($\Re$–case) eigenvalues and denoted by $\Psi(\sigma, T)$.$\Psi(\sigma, T)$ is called a σ–associate of T and the subsystem of the original system $\mathcal{S}(\sigma, T) = (J(\sigma, T), \tilde{B}(\sigma, T), \tilde{C}(\sigma, T))$ defined by all blocks $(J(\lambda), \tilde{B}(\lambda), \tilde{C}(\lambda))$ of the (5.1) description associated with all $\lambda \in \Psi(\sigma, T)$,

$$J(\sigma, T) = diag\{\cdots; J(\lambda); \cdots\}$$

$$\tilde{B}(\sigma, T) = \begin{bmatrix} \vdots \\ \tilde{B}(\lambda) \\ \vdots \end{bmatrix},\ \tilde{C}(\sigma, T) = [\cdots; \tilde{C}(\lambda); \cdots] \qquad (5.3)$$

is referred to as σ–associate subsystem of T. Clearly, for a given irregular sampling T, there may be more than one σ–associates $\Psi(\sigma, T)$ and thus subsystems $\mathcal{C}(\delta, T)$.

Proposition (5.1): For any irregular sampling T with σ–associate subsystem $\mathcal{S}(\sigma, T)$ the following hold true:

(i) $(\hat{A}, \hat{B})$ is controllable, iff every $(J(\sigma,T), \tilde{B}(\sigma,T))$ subsystem is controllable.

(ii) $(\hat{A}, \hat{C})$ is observable, iff every $(J(\sigma,T), \tilde{C}(\sigma,T))$ subsystem is observable □

The above result reduces the overall investigation to a set of simpler problems considered below:

Theorem (5.2): For any irregular sampling T for which there is an $\Re$-prime 0–associate $\Psi(0,T)$, the following properties hold true:

(a) The pair $(\hat{A}, \hat{B})$ is always uncontrollable

(b) Let π_0 be the character of $\Psi(0,T)$, then:

(i) If $\pi_0 > m$, $(\hat{A}, \hat{C})$ is always unobservable

(ii) if $\pi_0 \leq m$, then $(\hat{A}, \hat{C})$ is observable, iff

$$rank[\Gamma_\lambda,\ \forall\lambda \in \Psi(0,T)] = \pi_0 \qquad (5.4)$$

□

Theorem (5.3): For any irregular sampling T with a σ associate $\Psi(\sigma,T)$, $\sigma \neq 0$, and character π_σ, the following properties hold true:

(a) (i) The pair $(\hat{A}, \hat{B})$ is always uncontrollable, if $\pi_\sigma > \ell$.

(ii) If $\pi_\sigma \leq \ell$, then $(\hat{A}, \hat{B})$ is controllable, iff

$$rank[\Delta_\lambda^t,\ \forall\lambda \in \Psi(\sigma,T)]^t = \pi_\sigma \qquad (5.5)$$

(b) (i) The pair $(\hat{A}, \hat{C})$ is always unobservable, if $\pi_\sigma > m$

(ii) If $\pi_\delta \leq m$ then $(\hat{A}, \hat{C})$ is observable, iff

$$rank[\Gamma_\lambda,\ \forall\lambda \in \Psi(\sigma,T)] = \pi_\sigma \qquad (5.6)$$

□

The study of irregular samplings and the associated collapsing of eigenvalue phenomena has revealed the presence of two different types of uncontrollability and unobservability. The first type is of structural nature and depends on the value of the π character (see definition (4.3)), it is independent of the numerical values of the parameters and will be referred to as structual. On the other hand, if the character π is small or equal to the number of inputs, outputs, then uncontrollablity, unobservability become numerically depending properties (condition (5.4)–(5.6)) and will be referred to as parametric. It is also worth pointing out that the imaginary axis has special properties as far as uncontrollability, since any real collapsing implies loss of controllability, but not necessarily of observability.

6 Conclusions

The basic properties of a model based theory of sampling have been derived based on the characterisation of samplings into regular and irregular. Central to this approach is the property of collapsing of eigenvalues and the corresponding merging of Segre characteristics and eigenspaces. The merging of Segre characteristics is behind both the structural and the parametric loss of controllability and observability. The study of other system and characteristic properties such as minimal indices and zeros are under investigation at the moment.

7 References

[1] E. W. Kamen, 1990. "Introduction to Signals and Systems", MacMillan Publ. Company, New York, Second Edit.

[2] R. E. Kalman, Y.C. Ho and K. S. Narendra, 1963. "Controllability of linear Dynamical Systems" In contributions of Diff. Equat. 1, pp 189-213, New York, J. Wiley.

[3] C. T. Chen, 1984, "Linear Systems Theory and Design", Holt Rinehart and Winston, New York.

[4] K. J. Astrom, P. Hagander and J. Sternby, 1980. "Zeros and Sampled Systems". Proc. 19th IEEE CDC Conf, Albuquerque, pp 1077-81.

[5] N. Karcanias and N. Tamvaclis, 1995 . "Sampling and structural properties of discretised linear models: Segre characteristics, controllability, observability and related properties", Control Engineering Centre, City University, Research Report, March 1995.

[6] K. J. Astrom and B. Wittenmark, 1984. "Computer control systems: Theory and design", Prentice Hall, Englewood Cliffs, N.J.

[7] F. L. Lewis, 1992, "Applied optimal control and estimation: Digital design and implementation", Prentice Hall, Englewood Cliffs, NJ, Digital Signal Proc. Series.

[8] T. Chen and B. A. Francis, 1991, "H_2-optimal sampled data control", IEEE Trans. Aut. Control, Vol. 36, pp 387-397

[9] D. Williamson, 1991. "Digital Control and Implementation: Finite Word length Consideration", Prentice Hall, Brunswick, Australia.

A CONVEX CHARACTERIZATION OF SELF-SCHEDULED $\mathcal{H}_2$ CONTROLLERS

J.M. BIANNIC* and P. APKARIAN **

*CERT-DERA and ENSAE, Toulouse FRANCE
**CERT-DERA, Toulouse FRANCE

Abstract. Les trente dernières années ont vu éclore de nombreuses ramifications autour de la théorie de la commande optimale. Cependant, la prise en compte de grandes variations paramétriques reste un problème souvent délicat. Or, dans de nombreux cas pratiques, les paramètres sont mesurables en temps réel. Une stratégie de tabulation de gains peut alors se révéler efficace. Aussi proposons nous dans cet article d'étendre le cadre usuel de la synthèse $\mathcal{H}_2$ à la classe des systèmes linéaires à paramètres variants (LPV). Il s'agit plus précisément de rechercher un correcteur lui-même LPV - donc "auto-tabulé" le long des trajectoires paramétriques suivies par le système - et qui minimise une "extension LPV" d'une borne supérieure de la norme $\mathcal{H}_2$ que nous introduirons. Une caractérisation convexe des solutions sera ensuite proposée. Finalement, une application physique permettra de valider notre approche.

Key Words. Robust Control; LPV systems; H2 synthesis; LMI

1. INTRODUCTION

The Optimal Control Problem has been widely investigated since 1960. Most recent works on this topic focused on robustness aspects. At the beginning of the 1980's, the robustness of LQG regulators was studied both in linear and nonlinear cases (Doyle, 1978; Doyle and Stein, 1979; Safonov and Athans, 1978; Lehtomaki *et al.*, 1981). At the end of that decade, a modern formulation of the Optimal Control Problem was proposed to highlight similarities between $\mathcal{H}_2$ and $\mathcal{H}_\infty$ synthesis techniques (Doyle *et al.*, 1989*a*). The mixed $\mathcal{H}_2/\mathcal{H}_\infty$ control method immediately followed, which permits to formulate robustness specifications in the $\mathcal{H}_2$ design (Doyle *et al.*, 1989*b*; P. Khargonekar and M.A. Rotea, 1991). The problem of robust $\mathcal{H}_2$ performance in the face of dynamic and parametric uncertainty was considered in (Bernstein and Haddad, 1989*a*; Bernstein and Haddad, 1989*b*). More recently, bounds on the worst case $\mathcal{H}_2$ norm of a system subject to norm-bounded nonlinear and time-varying uncertainties were introduced (Peres *et al.*, 1992; Petersen and McFarlane, 1992; Petersen *et al.*, 1993; Stoorvogel, 1993). Exploiting further information about the perturbations, refinements, restricted to the analysis problem, are proposed in (D'Andrea *et al.*, 1993; Feron, 1994) to reduce the conservatism of available upper-bounds.

Yet, in many applications, the operating point of the plant is known or can be measured in real-time. This allows for more powerful control strategies which exploite the parameter measurements. Such approaches are well-know by engineers and are usually referred as gain scheduling control. This paper proposes a solution based on the concept of **LPV** controllers which was first introduced in (Packard, 1993) and allows to achieve high performance in the presence of large parameters variations. Theoretical, as well as practical results are already available in the framework of $\mathcal{H}_\infty$ synthesis (Apkarian and Gahinet, 1995; Biannic *et al.*, 1994). This last point is a strong encouragement in favor of LPV control structure and its extensions to the $\mathcal{H}_2$ case.

The remainder of the paper is organized as follows. Section 2 is devoted to a short presentation of LPV systems. Then, an $\mathcal{H}_2$-based performance measure, handling the class of LPV systems and its computation are presented. In section 4 the related control problem is stated and its solutions are characterized, under mild assumptions, via Linear Matrix Inequalities (**LMI**), which ensures numerical tractability (Gahinet *et al.*, 1994). Section 5 presents an application. It consists in submarine depth keeping in the presence of swell and over a large speed range.

2. A BRIEF DESCRIPTION OF LPV SYSTEMS

The purpose of this section is to develop general ideas and notions concerning the class of LPV systems which is central to our approach.

Definition 2.1 (LPV system)
A system is called Linear Parameter-Varying if and only if it assumes the following state-space description:

$$\dot{x} = A(\theta(t))\, x + B(\theta(t))\, u \quad (2.1)$$
$$y = C(\theta(t))\, x + D(\theta(t))\, u. \quad (2.2)$$

where $\theta(t)$ is a **measurable** time-varying parameter vector which evolves in $\Theta \in \mathbf{R}^r$.

■

It can be pointed out that, if the trajectory is known beforehand, the explicit parameter dependence may be suppressed and the system becomes Linear Time Varying (LTV).

The LPV system above will be denoted $\mathbf{G}(\theta)$, since it can be interpreted as an operator whose input signal is a trajectory $\theta(t) \in \Theta$. See (Apkarian *et al.*, 1995) for a more detailed discussion on LPV systems.

Property 2.2 (Θ-Stability)
$G(\theta)$ is **Θ-Stable** if and only if, for any trajectory in Θ, the "LTV-derived" system remains stable (see (D'Angelo, 1970) for the stability analysis of LTV systems).

■

From a practical viewpoint, it is interesting to derive sub-classes of LPV systems which offer more constructive properties. One of these, is the **polytopic characterization**. It is based on the following assumptions:

(i) the parameter dependence of $A(.), B(.), C(.), D(.)$ is affine.
(ii) Θ is a convex polytope of vertices $\theta_1,...,\theta_r$.

In that case, the LPV plant assumes the state-space description:

$$\dot{x} = \sum_{i=1}^{r} \alpha_i(t) A_i\, x + \sum_{i=1}^{r} \alpha_i(t) B_i\, u \quad (2.3)$$
$$y = \sum_{i=1}^{r} \alpha_i(t) C_i\, x + \sum_{i=1}^{r} \alpha_i(t) D_i\, u \quad (2.4)$$

where the α_i's are computed according to the convex decomposition:

$$\theta(t) = \sum_{i=1}^{r} \alpha_i(t)\theta_i \quad (2.5)$$

and the matrices A_i's, B_i's, C_i's, D_i's respectively correspond to $A(\theta_i)$, $B(\theta_i)$, $C(\theta_i)$, $D(\theta_i)$.

This characterization is more amenable to theoretical developments since the varying plant is defined as a convex polytope. Though not fully general, the polytopic representation encompasses many practical situations.

There exists another well-known sub-class of LPV systems which is based on the assumption that the parameter dependence is linear-fractional (see (Packard, 1993; Doyle *et al.*, 1991; Apkarian and Gahinet, 1995) for details).

3. CHARACTERIZATION OF THE PERFORMANCE MEASURE

For LTI systems, it is well-known that the $\mathcal{H}_2$ norm can be interpreted as the energy of the impulse response which reads:

Definition 3.1 (The usual $\mathcal{H}_2$ norm)
Consider a strictly proper and stable LTI system $G(s)$, then the $\mathcal{H}_2$ norm of G can be defined as follows:

$$\|G(s)\|_2^2 = \int_0^\infty tr(\Omega(t)^T \Omega(t))dt \quad (3.6)$$

where $\Omega(t)$ denotes the impulsive response matrix of G and tr the usual trace operator.

■

This definition can easily be extended to handle the class of LPV systems. Consider for example a paramter varying plant $G(\theta)$ and select a trajectory $\theta(t) \in \Theta$. This choice leads to an LTV system (see section 2) which assumes an impulsive response matrix (D'Angelo, 1970).

Definition 3.2 ("LPV-$\mathcal{H}_2$" norm)
Consider a strictly proper Θ-stable LPV system $G(\theta)$:

$$\dot{x} = A(\theta(t))\, x + B(\theta(t))\, w \quad (3.7)$$
$$z = C(\theta(t))\, x \quad (3.8)$$

and let $\Omega_\theta(t)$ denotes the impulsive response matrix of the "LTV-derived" system along a given parameter trajectory $\theta(t) \in \Theta$. Then, $G(\theta)$ assumes the following "extended $\mathcal{H}_2$" norm:

$$\gamma_{\mathcal{H}_2}(G(\theta)) = \sup_{\theta(t)\in\Theta} \int_0^\infty tr(\Omega_\theta(t)^T \Omega_\theta(t))dt \quad (3.9)$$

■

Note that the expression (3.6) assumes a well-known numerically-tractable expression:

$$\|G(s)\|_2^2 = trace(B^T X B) \quad (3.10)$$
$$X > 0 \;/\; A^T X + XA + C^T C = 0 \quad (3.11)$$

whereas (3.9) cannot be simplified. However a useful upper-bound for **Θ-quadratically stable**

LPV systems can be derived.

Property 3.3 (A bound for the "LPV-$\mathcal{H}_2$" norm)

Consider a strictly proper Θ-quadratically stable LPV system $G(\theta)$ and define the non-empty set $\mathcal{X}_G$ of positive definite matrices X such that, for all $\tau \in \Theta$:

$$A(\tau)^T X + XA(\tau) + C(\tau)^T C(\tau) < 0 \quad (3.12)$$

then $\gamma_{\mathcal{H}_2}(G(\theta))$ is bounded by:

$$\bar{\gamma}(G(\theta)) = \sup_{\theta(t)\in\Theta} \inf_{X\in\mathcal{X}_G} tr(B(\theta)^T X B(\theta)) \quad (3.13)$$

■

When $G(\theta)$ is a **polytopic** plant, this bound is easily computed as the solution of the following **convex program**:

$$\bar{\gamma}(G(\theta)) = \min_{P,X>0} tr(P) \quad (3.14)$$

$$\forall i = 1...r, \begin{pmatrix} P & B_i^T X \\ XB_i & X \end{pmatrix} \geq 0 \quad (3.15)$$

$$\forall i = 1...r, \begin{pmatrix} A_i^T X + XA_i & C_i^T \\ C_i & -I \end{pmatrix} < 0 \quad (3.16)$$

In other words, the computation of $\bar{\gamma}(G(\theta))$ is reduced to the minimization of a linear objective under LMI constraints which is a central fact in our approach.

4. THE LPV CONTROL PROBLEM

The synthesis problem addressed in this paper lies on the previously defined performance measure $\bar{\gamma}$, and can be stated as follows:

Problem Statement 4.1 Given an LPV "standard" plant $P(\theta)$:

$$\begin{aligned} \dot{x} &= A(\theta(t))x + B_1(\theta(t))w + B_2(\theta(t))u \\ z &= C_1(\theta(t))x + D_{11}(\theta(t))w + D_{12}(\theta(t))u \\ y &= C_2(\theta(t))x + D_{21}(\theta(t))w + D_{22}(\theta(t))u. \end{aligned}$$

such that the pairs $(A(\theta), B(\theta))$ and $(A(\theta), B(\theta))$ are quadratically stabilizable and detectable over Θ respectively, the problem is to find an LPV Controller which, for all trajectories $\theta(t) \in \Theta$:

(i) quadratically stabilizes the parameter dependent closed-loop system $\mathcal{T}_{cl}(\theta)$ over Θ.

(ii) minimizes $\bar{\gamma}(\mathcal{T}_{cl}(\theta))$

In the following, for numerical purposes (computation of the norm), it is supposed that the previous plant $P(\theta)$ is defined by the convex hull:

$$P(\theta) \in Co\left\{ \begin{pmatrix} A_i & B_{1i} & B_2 \\ C_{1i} & D_{11i} & D_{12i} \\ C_2 & D_{21i} & D_{22i} \end{pmatrix} \right\} \quad (4.17)$$

which, in addition to quadratic stabilizability and detectability, further satisfies:

(A1) $\forall i = 1...r,\ D_{11i} = 0\ ,\ D_{22i} = 0$
(A2) $\forall i = 1...r,\ D_{21i} = 0$

Note that assumption (A1) is the usual hypothesis of the $\mathcal{H}_2$ control problem (Doyle *et al.*, 1989*a*). Interestingly, (A2) which may appear as a restrictive condition can be achieved using strictly proper input weighting functions. Note that the use of such functions is quite common in many applications.

With these assumptions in mind, existence conditions of controllers are now derived:

Theorem 4.2 (existence conditions)
Consider the plant (4.17) and assume **(A1)-(A2)**. Let $\mathcal{N}_R$ and $\mathcal{N}_S$ denote bases of the null space of (B_2^T, D_{12}^T) and $(C_2, 0)$ respectively. Then, there exists a stabilizing LPV controller which ensures that $\bar{\gamma}(\mathcal{T}_{cl}(\theta)) = \gamma$ if and only if there exist symmetric matrices (R,S) and P such that $tr(P) = \gamma$ and the following 3r+1 LMI constraints are satisfied:

$$\mathcal{N}_R^T \begin{pmatrix} A_iR + RA_i^T & RC_{1i}^T \\ C_{1i}R & -I \end{pmatrix} \mathcal{N}_R < 0$$

$$\mathcal{N}_S^T \begin{pmatrix} A_i^T S + SA_i & C_{1i}^T \\ C_{1i} & -I \end{pmatrix} \mathcal{N}_S < 0$$

$$\begin{pmatrix} P & B_i^T S \\ SB_i & S \end{pmatrix} > 0$$

$$\begin{pmatrix} R & I \\ I & S \end{pmatrix} \geq 0$$

■

See (Apkarian and Gahinet, 1995) for a similar proof on the $\mathcal{H}_\infty$ control problem. LPV controllers solutions are characterized by the convex hull:

$$K(\theta) \in Co\left\{ \begin{pmatrix} A_{Ki} & B_{Ki} \\ C_{Ki} & D_{Ki} \end{pmatrix} i = 1...r, \right\} \quad (4.18)$$

where the vertices $K_i = \begin{pmatrix} A_{Ki} & B_{Ki} \\ C_{Ki} & D_{Ki} \end{pmatrix}$ are obtained from the pair (R,S), using standard algebraic manipulations or solving an additional LMI problem.

5. APPLICATION

In this section, the proposed technique is applied to achieve a depth-control autopilot design for a submarine which speed evolves from 6 to 15 knots.

5.1. *System Description*

Under usual assumptions (ie neglecting nonlinear terms in the set of differential equations), the submarine behavior in the vertical plane can be adequately described by the following 4th order speed-dependent state-space model:

$$\dot{x} = A(v, v^2)x + B_2(v^2)u \qquad (5.19)$$

$$y = C_2 x \qquad (5.20)$$

The state-space variables are the depth, the pitch (which are also the two outputs) and their first order derivatives. The control inputs are the bow and stern hydro-planes angles. A brief open-loop analysis shows that the vessel is unstable below 6.3 knots and badly damped at higher speeds. For more informations about the model, the reader is referred to (Le Pourhiet and Carton, 1993).

In addition to speed-dependence, the submarine behavior is greatly influenced by waves effects. The waves provide indeed large heave forces and pitch moments in a neighborhood of 0.7 Hz. Besides, smaller forces and moments also perturb the vessel in the low-frequency domain. Of course, these phenomena depend on the sea-state, the speed, the depth and the direction of the submarine. In the remainder of the paper, for a given sea-state (5) and by fixing the depth (-20m), it is assumed (using a worse case argument) that a unique model can acceptably describe the waves effects whatever the speed. Thus, denoting (A_F, B_F, C_F) a state-space description of the weighting functions, B_1 the input disturbance matrix and w_b a white noise input vector, the open-loop system including waves effects assumes the following characterization S(v):

$$\begin{aligned} \dot{x} &= A(v, v^2)x + B_2(v^2)u \\ \dot{x}_F &= A_F x_F + B_1 w_b \\ y &= C_2 x \end{aligned}$$

The weighting functions that were used in this application are shown on figure 1. Note that more details about this characterization can be found in (Carton *et al.*, 1993).

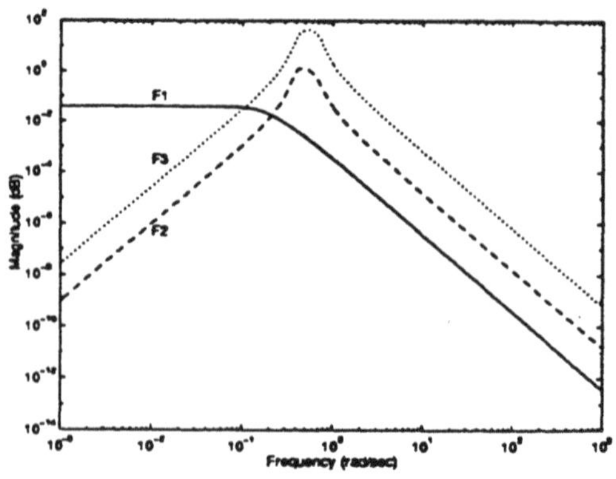

Fig. 1. Swell Characterization

5.2. *Problem Formulation*

With above description in mind, it is now desired to determine a speed-dependent control-law which stabilizes the submarine at a constant depth on its entire speed range and minimizes waves perturbations. Depth and pitch variations (induced by waves) must respectively remain below 1m and 1 degree. In addition, control signals (bow and stern hydro-planes angles) should not exceed 15 degrees.

The proposed method to tackle the problem consists in minimizing the "LPV-$\mathcal{H}_2$" norm of a speed-dependent closed-loop transfer from $\mathbf{w_b}$ to a regulated output z. It must be stressed that the extended $\mathcal{H}_2$ norm is appropriate in this case, because w_b is a white noise input vector.

Nevertheless, results of section 4 are not directly applicable since speed-dependence is not affine (see 5.19) and the control input matrix B_2 is also speed-dependent. A previous transformation is then required. The idea is quite simple and consists in considering $v(t)$ and $v^2(t)$ as separate variables and setting $\theta(t) = [v(t)\ ,\ v^2(t)]^T$. Thus, the parameter varying system $S(v)$ will further be denoted by $\mathbf{S}(\theta)$ and assumes a polytopic description with $2^{dim(\theta)} = 4$ vertices. Obviously, this characterization will be slightly conservative since it encompasses a larger class of systems than the one defined by $S(v)$. But, as θ is measurable, the quality of the results should not be over-much damaged. Note finally that the speed-dependence can be easily removed from B_2 and be integrated in the parameter varying controller.

Thus, a polytopic synthesis plant $P(\theta)$ of the form 4.17 can be derived (see figure 2). At this stage, it is important to note that for all realizable LTI weighting functions W_{add} and W_{track}, the plant $P(\theta)$ verifies assumptions **(A1)-(A2)**. This can be easily checked remarking that the waves perturbations model is strictly proper.

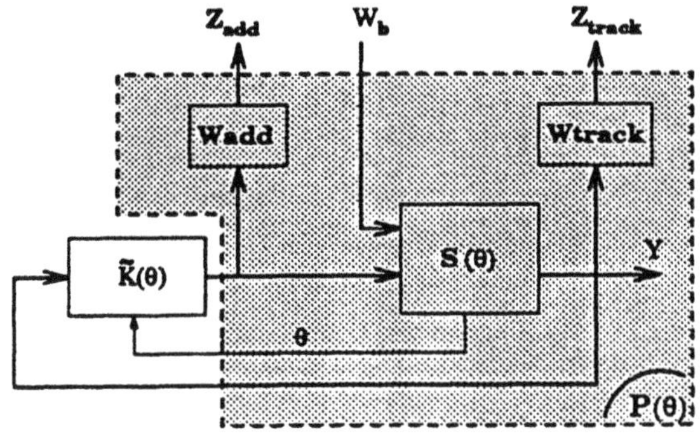

Fig. 2. Synthesis Model

The design method may thus be applied on $P(\theta)$ and provide as detailed in section 4, a controller $\tilde{K}(\theta)$ minimizing a bound of the "LPV-$\mathcal{H}_2$" norm of the transfer from w_b to $[z_{out}\ z_{add}]^T$. Finally, the controller of interest for the physical system $S(v)$ is readily obtained as follows: $K(v) = \tilde{K}([v\ ,\ v^2]^T)$.

5.3. *Results*

After a few trials, the following weighting functions presented on figure 3 have been retained and the optimization method used in LMI-Toolbox (Gahinet *et al.*, 1994) converged after 85 iterations. The bound ($\bar{\gamma} = 1.05$) obtained ensures that the parameter dependent energy of the output signals remains below 1.05. Observing figure 3, it is expected that the depth and the pitch (respectively weighted by W_{track1} and W_{track2}) variations should rarely exceed 1m and 1 degree (0.0175 rad). As for control signals (weighted by w_{add1} and w_{add2}), there should be no exceeding neither.

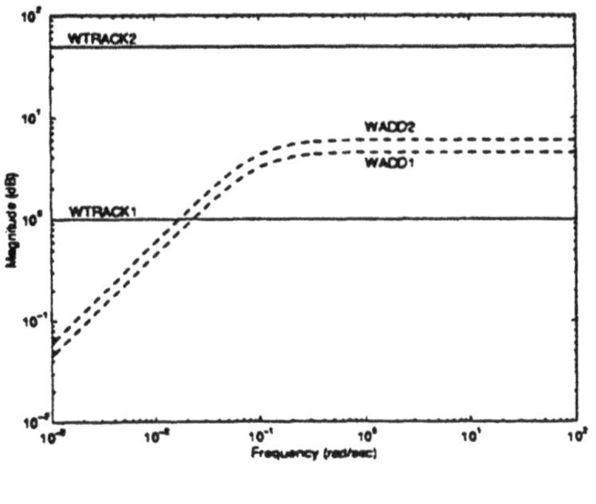

Fig. 3. Weighting Functions

To check the above comments it is now desirable to simulate the closed-loop system. The control law has then been implemented following the procedure of section 4. Every second, the speed is measured and θ coordinates are computed, which leads via elementary matrix manipulations to a new controller. A discretization is finally performed and outputs are computed until the next sampling time. For a sinusoidal speed trajectory (varying between 5 and 15 knots), results are presented on figure 4.

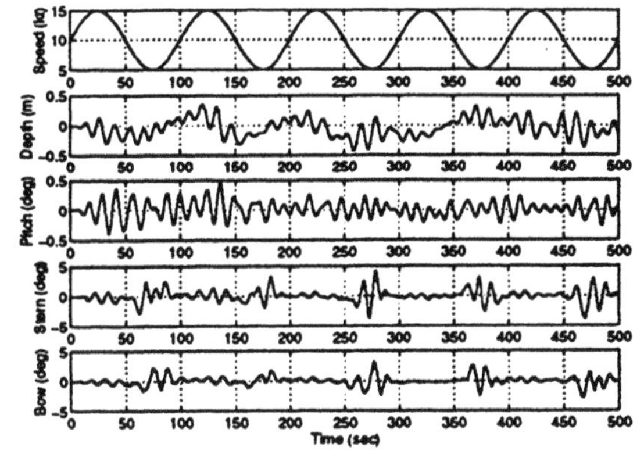

Fig. 4. Time Responses

As expected, the specifications are satisfied. Moreover, bow and stern hydro-planes angles variations do not exceed 4 deg/sec which is the actuators limit.

6. CONCLUSION

In this paper, a new synthesis method has been presented. It is based on the extension of the usual $\mathcal{H}_2$ norm to the class of linear-parameter-varying systems. The proposed technique provides then a useful answer to many control problems, especially for those implying LPV systems which external inputs are modelled as Gaussian white noises. It is interesting to remark that such signals are not square integrable and the "LPV-$\mathcal{H}_\infty$" (Apkarian and Gahinet, 1995; Biannic *et al.*, 1994) method fails to give a reliable solution in this case.

A real-world application has then been detailed. The method has been used to develop a speed-dependent depth control law, for a submarine which behavior is highly perturbed by swell effects. As explained in section 5, this type of problem falls into the scope of the technique which gave here quite promising results on a relatively large speed-range.

Further works on the "LPV-$\mathcal{H}_2$" control technique are thus strongly encouraged. In particular, it is currently studied how to relax the assumption (A2) which does not occur in "LPV-$\mathcal{H}_\infty$" synthesis. Extending the results of (Geromel and Debeche, 1994) a convex parametrization of LPV observer-based controllers can be derived. However, this approach leads to the usual orthogonality conditions and cannot be easily extended to handle parametric uncertainties. Subsequently, it would be of high interest to incorporate informations on the parameter rate of variations in the design. This could significantly reduce the conservatism of the method for slowly varying systems.

7. REFERENCES

Apkarian, P. and P.M. Gahinet (1995). A convex characterization of gain-scheduled $H\infty$ controllers. *To appear in IEEE Trans. Automat. Contr. Vol, 95.*

Apkarian, P., J.M. Biannic and P.M. Gahinet (1995). Self-scheduled $H\infty$ control of a missile via lmis. *To appear in AIAA Journal of Guid. Contr. and Dyn., 95.*

Bernstein, D.S. and W.M. Haddad (1989*a*). LQG control with a H^∞ performance bound: A riccati equation approach. *IEEE Trans Automat. Contr.* **AC-34**(3), 293–305.

Bernstein, D.S. and W.M. Haddad (1989*b*). Robust stability and performance analysis for linear dynamic systems. *IEEE Trans Automat. Contr.* **AC-34**(7), 751–758.

Biannic, J.M., P. Apkarian and L. Leletty (1994). A quadratic $H\infty$ performance approach to gain scheduling of a missile autopilot. In: *Proc. IEEE Workshop on Var. Struct. and Lyap. Tech.*. Benevento, Italy. pp. 140–147.

Carton, P., A. Le Pourhiet and C. Lambert (1993). Modélisastion spectrale des efforts de houle. *Rapport CERT/DERA 209/93 - Convention DCN Ingéniérie.*

D'Andrea, R., F. Paganini and J.C. Doyle (1993). Behavioral approach to robustness analysis.

D'Angelo, H. (1970). *LTV Systems: Analysis and Synthesis.* ALLYN and BACON Series in Electrical Engineeering.

Doyle, J.C. (1978). Guaranteed margins for lqg regulators. *IEEE Trans Automat. Contr.* **AC-23**, 756–757.

Doyle, J.C., A. Packard and K. Zhou (1991). Review of lft's, lmi's and μ. In: *Proc. IEEE Conf. Decision Contr.*. Brighton,England. pp. 1227–1232.

Doyle, J.C. and G. Stein (1979). Robustness with observers. *IEEE Trans Automat. Contr.*

Doyle, J.C., K. Glover, P. Khargonekar and B. A. Francis (1989*a*). State-space solutions to standard H_2 and H_∞ control problems. *IEEE Trans Automat. Contr.* **AC-34**(8), 831–847.

Doyle, J.C., K. Zhou and B. Bodenheimer (1989*b*). Optimal control with mixed H_2 / H_∞ performance objectives. In: *Proc. Amer. Contr. Conf.*. pp. 2065–2070.

Feron, E. (1994). Analysis of robust H_2 performance using multiplier theory. *Submitted to the SIAM Journal on Contr. and Optim.*

Gahinet, P.M., A. Nemirovskii, A.J. Laub and M. Chilali (1994). Lmi control toolbox, the math works inc.

Geromel, J.C. and M.S.J. Debeche (1994). Convex analysis and global optimization of output covariance constraint problems. In: *Proc. IEEE Conf. Decision Contr.*. Lake Buena Vista, FL. pp. 2603–2608.

Le Pourhiet, A. and P. Carton (1993). Pilotage vertical d'un sous-marin en immersion périscopique. *Rapport CERT/DERA 108/93 - Convention DCN Ingéniérie.*

Lehtomaki, N.A., N.R. Sandel and M. Athans (1981). Robustness results in linear quadratic gaussian based multivariable control design. *IEEE Trans Automat. Contr.* **AC-26(1)**, 75–92.

P. Khargonekar and M.A. Rotea (1991). Mixed H_2/H_∞ control: a convex optimization approach. *IEEE Trans Automat. Contr.* **AC-36**, 824–837.

Packard, A. (1993). Gain scheduling via linear fractional tranformations. *Syst. Contr. Lett.*

Peres, P.L.D., S.R. Souza and J.C. Geromel (1992). Optimal H_2 control for uncertain systems. In: *Proc. Amer. Contr. Conf.*. Chicago.

Petersen, I.R. and D.C. McFarlane (1992). Optimal guaranteed cost control for uncertain linear systems. In: *Proc. Amer. Contr. Conf.*. Chicago.

Petersen, I.R., D.C. McFarlane and M.A. Rotea (1993). Optimal guaranteed cost control of discrete uncertain systems. In: *In 12th IFAC World Congress.* Sydney, Australia. pp. 407–410.

Safonov, M.G. and M. Athans (1978). Robustness and computational aspects of nonlinear stochastic estimators and regulators. *IEEE Trans Automat. Contr.* **AC-23**, 717–725.

Stoorvogel, A.A. (1993). The robust H_2 problem: A worst case design. *IEEE Trans Automat. Contr.* **38**(9), 1358–1370.

NEW ALGORITHMS FOR THE QUADRATIC STABILIZATION OF UNCERTAIN LINEAR SYSTEMS

F. AMATO*, A. PIRONTI* and S. SCALA**

* *Università di Napoli Federico II, Dipartimento di Informatica e Sistemistica, via Claudio 21, 80125 Napoli, Italy*
** *Centro Italiano Ricerche Aereospaziali, via Maiorise, 81043 Capua, Italy*

Abstract. In this paper we consider a linear system depending in non-linear fashion on a vector of uncertain parameters. Assuming that the parameters range in a hyperrectangle, we provide a necessary and sufficient condition for the quadratic stabilizability of the system in terms of convex optimization procedures.

Résumé. Dans ce texte nous traitons un systéme linéaire qui dépend dans une maniére pas linéaire d'un vecteur de paramétres incertains. Supposé que les paramétres se trouvent dans un hyperectangle, nous donnons une condition nécessaire et suffisante pour la stabilisabilité quadratique du systéme en termes des procédures d'optimisation convexe.

Key Words. Linear systems; uncertain parameters; quadratic stabilization; linear matrix inequalities; convex programming

1. INTRODUCTION

The concept of quadratic stability dates back to the late Seventies with the germinal papers by Leitmann (1979) and Barmish (1983). An uncertain linear system is said to be quadratically stable if there exists a fixed (that is time-invariant) Lyapunov function whose derivative along the trajectories of the system is negative definite for all possible values of the uncertain parameters. In this case uniform asymptotic stability of the origin of the state space is guaranteed with respect to all possible temporal behaviour of the parameters within their bounding set.

In the same way, considered a forced linear plant with uncertain system and input matrices, it makes sense to consider the problem of finding a linear state feedback controller which quadratically stabilizes the overall closed loop system. A *sufficient* condition guaranteeing quadratic stabilizability via linear control is the so-called Matching Assumption (see Barmish, 1983). However, since this structural requirement is too restrictive, the effort of the researchers has been devoted to understand when it is possible to state conditions which are both *necessary* and sufficient. In Khargonekar *et al* (1990) it is shown that when the dependence on parameters is in the so-called norm bounded, one block form, the stabilizing controller exists if and only if a suitable algebraic Riccati equation has a positive definite solution.

A more general result can be found in Bernussou *et al* (1989), where a system depending affinely on parameters is proved to be quadratically stabilizable if and only if a certain convex optimization problem admits a solution.

In this paper we consider the case in which the system and input matrices depend non-affinely on parameters. First we show that if the dependence is multi-affine it is still possible to solve the quadratic stabilization problem via convex optimization algorithms. Then we deal with the case of non-affine dependence. In Section 3 we introduce a procedure presented in Amato *et al* (1994) to cover the image of a non-affine matrix valued function by the image of a function depending multi-affinely on parameters; via this procedure we can substitute the original non-affine system matrices with multi-affine ones and apply the previous algorithms. Since the covering introduces conservatism, the solvability of the original non-affine problem does not guarantee that of the problem with the new matrices; in Section 4 we show how to reduce this conservatism. Finally in Section 5 we provide an example of applicaton of our technique.

2. PRELIMINARIES

Consider an uncertain system in the form

$$\dot{x}(t) = A(p)x(t)\,, \quad t \in \mathbb{R}^+\,, \quad p \in \mathcal{R} \subset \mathbb{R}^{n_p}\,, \tag{1}$$

where $x(t) \in \mathbb{R}^n$, $p = (p_1 \ p_2 \ \dots \ p_{n_p})^T$ is the vector of uncertain parameters, $\mathcal{R}$ is a hyperrectangle, i. e.

$$\mathcal{R} \triangleq [\underline{p}_1, \overline{p}_1] \times [\underline{p}_2, \overline{p}_2] \times \dots \times [\underline{p}_{n_p}, \overline{p}_{n_p}], \quad (2)$$

and $A(\cdot)$ is continuous. We denote by $p_{(i)}$, $i = 1, \dots, 2^{n_p}$, the vertices of $\mathcal{R}$.

Definition 1: Quadratic stability.

The uncertain system (1) is said to be quadratically stable *if there exists a positive definite matrix* $P \in \mathbb{R}^{n \times n}$ *such that for all* $p \in \mathcal{R}$

$$L(p) \triangleq -(A^T(p)P + PA(p)) > 0. \quad (3)$$

Quadratic stability guarantees uniform asymptotic stability of the linear time-varying system

$$\dot{x}(t) = A(p(t))x(t), \qquad t \in \mathbb{R}^+ \quad (4)$$

for all Lebesgue measurable functions $p : \mathbb{R}^+ \mapsto \mathcal{R}$, $t \mapsto p(t)$.

In the same way, given a forced uncertain linear system in the form

$$\dot{x}(t) = A(p)x(t) + B(p)u(t), \quad (5)$$

with $u(t) \in \mathbb{R}^m$, and $A(\cdot)$ and $B(\cdot)$ continuous, we can state the following definition.

Definition 2: Quadratic Stabilizability.

The uncertain system (5) is said to be quadratically stabilizable via linear state feedback control *if there exists a matrix* $K \in \mathbb{R}^{m \times n}$ *such that the unforced system, obtained from (5) letting* $u = Kx$,

$$\dot{x}(t) = (A(p) + B(p)K)x(t)$$

is quadratically stable.

We recall that quadratic stabilizability via dynamic, time-varying state feedback linear control implies quadratic stabilizability via memoryless, time-invariant, linear state feedback control (see Petersen, 1988). Hence we do not lose any generality in Definition 2. On the contrary, as shown in Petersen (1985), quadratic stabilizability without any other specification does not imply quadratic stabilizability via linear control, hence this specification in Definition 2 is mandatory.

From Definition 2 follows that the uncertain system (5) is quadratically stabilizable via linear control if and only if there exist a positive definite $P \in \mathbb{R}^{n \times n}$ and a matrix $K \in \mathbb{R}^{m \times n}$ such that

$$L_K(p) \triangleq -((A(p) + B(p)K)^T P + P(A(p) + B(p)K)) > 0, \quad \forall p \in \mathcal{R}. \quad (6)$$

As shown in Bernussou *et al* (1989), letting

$$S = P^{-1}, \qquad L = KP^{-1} \quad (7)$$

it is readily seen that system (5) is quadratically stabilizable iff the following problem admits a solution.

Problem 1: *Find a symmetric matrix* $S \in \mathbb{R}^{n \times n}$ *and a matrix* $L \in \mathbb{R}^{m \times n}$ *such that*

i) $S > 0$;

ii)

$$L_S(p) \triangleq -(SA^T(p) + A(p)S + L^T B^T(p) + B(p)L) > 0 \quad \forall p \in \mathcal{R}.$$

When A and B depend multi-affinely on parameters we can state the next fundamental theorem. To this end we need the following well known result.

Fact 1: *A set of matrices is positive definite (i. e. each matrix in the set is positive definite) iff its convex hull is positive definite. When the convex hull is a polytope, the set of matrices is positive definite iff the vertices are.*

Theorem 1: *Assume that the dependence of* A *and* B *on parameters is multi-affine, i. e.*

$$A(p) = \sum_{i_1, i_2, \dots, i_{n_p}} A_{i_1, i_2, \dots, i_{n_p}} p_1^{i_1} p_2^{i_2} \cdots p_{n_p}^{i_{n_p}} \quad (8a)$$

$$B(p) = \sum_{i_1, i_2, \dots, i_{n_p}} B_{i_1, i_2, \dots, i_{n_p}} p_1^{i_1} p_2^{i_2} \cdots p_{n_p}^{i_{n_p}}, \quad (8b)$$

where $i_j \in \{0, 1\}$, $j = 1, \dots, n_p$, $A_{i_1, i_2, \dots, i_{n_p}} \in \mathbb{R}^{n \times n}$ *and* $B_{i_1, i_2, \dots, i_{n_p}} \in \mathbb{R}^{n \times m}$. *Then system (5) is quadratically stabilizable via linear control iff there exist* $S > 0$ *and a matrix* $L \in \mathbb{R}^{m \times n}$ *such that for all* $i = 1, \dots, 2^{n_p}$

$$L_S(p_{(i)}) > 0,$$

$p_{(i)}$ *being the i-th vertex of* $\mathcal{R}$.

Proof. Under the hypothesis of the theorem, the matrix function $L_S(p)$ defined in ii) of Problem 1 depends multi-affinely on parameters. The well known Mapping Theorem (Zadeh and Desoer, 1963) states that the convex hull of a multi-affine function, defined on a hyperrectangle, coincides with the convex hull of the images of the vertices of the hyperrectangle. The proof follows from Fact 1. ■

Remark 1: Note that Theorem 1 recovers as particular case Theorem 1 in Bernussou *et al* (1989), where the affine dependence on parameters was considered.

By virtue of Theorem 1, when A and B depend multi-affinely on parameters, Problem 1 is equivalent to the following.

Problem 2: *Find a symmetric matrix* $S \in \mathbb{R}^{n\times n}$ *and a matrix* $L \in \mathbb{R}^{m\times n}$ *such that*

i) $S > 0$;

ii) $L_S(p_{(i)}) > 0, \qquad i = 1, \dots, 2^{n_p}$.

Since the set of positive definite matrices is convex, and the constraints in ii) of Problem 2 contain linear matrix inequalities which are of a convex nature, Problem 2 can be solved via iterative algorithms based on *cutting plane* techniques. Recently Bernussou *et al* (1989) have proposed an algorithm which, in a finite number of iterations, establishes whether Problem 2 has a solution or not.

When the dependence on parameters of the system matrices in equation (5) is more complex than multi-affine, constraint ii) in Problem 1 is no longer reducible to a finite number of convex constraints as in ii) of Problem 2, with the obvious consequence that the previous algorithm cannot be utilized.

In the next section, making use of a procedure recently proposed in Amato *et al* (1994) (see Algorithm 1 below), we will show how to cover the image of a non-affine function by that of a multi-affine one; via this procedure we can substitute the original non-affine system matrices with multi-affine ones and apply the above-mentioned algorithm.

3. A PROCEDURE FOR NON-AFFINE FUNCTIONS

In this section we treat the case in which the system matrices in (5) depend non-affinely on parameters. The class of functions considered is specified in the next assumption.

Assumption 1: *The matrix valued functions* $A(\cdot)$ *and* $B(\cdot)$ *in (5) can be written in the following way*

$$A(p) = \sum_{i_1,i_2,\dots,i_\nu} A_{i_1,i_2,\dots,i_\nu} a_1^{i_1}(p) a_2^{i_2}(p) \cdots a_\nu^{i_\nu}(p) \quad (9a)$$

$$B(p) = \sum_{i_1,i_2,\dots,i_\mu} B_{i_1,i_2,\dots,i_\mu} b_1^{i_1}(p) b_2^{i_2}(p) \cdots b_\mu^{i_\mu}(p), \quad (9b)$$

where $i_j \in \{0,1\}$, $A_{i_1,i_2,\dots,i_\nu} \in \mathbb{R}^{n\times n}$, $B_{i_1,i_2,\dots,i_\mu} \in \mathbb{R}^{n\times m}$ *and the* a_j*'s and* b_j*'s are continuous scalar functions. For* $j = 1,\dots,\nu$, *if* a_j *is not multiaffine it is assumed to be convex or concave; the same holds for the* b_j*'s.*

Note that Assumption 1 recovers as a particular case the multi-affine dependence.

Remark 2: It is interesting to note that many functions are convex or concave only if the parameters on which they depend do not change sign. This fact does not cause any loss of generality in practical applications because in many situations the parameters have a physical significance and hence are inherently positive. In any case we can always split, prior to the application of the algorithm, the original hyperrectangle in smaller hyperrectangles, each contained only in one orthant of $\mathbb{R}^{n_p}$.

3.1. *An Algorithm to Cover the Image of a Non-Affine Function by That of a Multiaffine Function*

Consider a matrix valued function

$$F(p) = \sum_{i_1,i_2,\dots,i_\nu} F_{i_1,i_2,\dots,i_\nu} f_1^{i_1}(p) f_2^{i_2}(p) \cdots f_\nu^{i_\nu}(p) \quad (10)$$

satisfying Assumption 1. The next algorithm constructs a multi-affine function whose image covers that of $F(\cdot)$.

Algorithm 1:

Step 1
For $j = 1,\dots,\nu$ if $f_j(p)$ in (10) is not multi-affine, apply the procedure detailed below to construct a couple of *affine functions* $\underline{f}_j$ and $\overline{f}_j$ such that

$$\underline{f}_j(p) \le f_j(p) \le \overline{f}_j(p), \qquad \forall p \in \mathcal{R} \quad (11)$$

We show how to find these bounding functions in case of convex functions (the procedure is similar for the concave case).

Let $p^* \in \mathcal{R}$ and $\pi_j(p^*)$ the gradient of $f_j(p)$ in p^*. The following inequality holds (see Rockafellar, 1970, Theorem 25.1)

$$f_j(p) \ge f_j(p^*) + \pi_j(p^*)(p - p^*), \qquad p \in \mathcal{R}. \quad (12)$$

Then set

$$\underline{f}_j(p) = f_j(p^*) + \pi_j(p^*)(p - p^*). \quad (13)$$

Now consider the images of the vertices of $\mathcal{R}$ under f_j, namely $f_{j1} \triangleq f_j(p_{(1)}), \dots, f_{j2^{n_p}} \triangleq f_j(p_{(2^{n_p})})$, and order them with decreasing magnitude, say $f_{jh_1} \ge \dots \ge f_{jh_{2^{n_p}}}$. Simple considerations from linear algebra show that there exists an integer $r \in \{n_p + 1, \dots, 2^{n_p}\}$ such that the hyperplane $h_j(p)$ in the space $\mathbb{R}^{n_p+1}$, connecting the points $(p_{(h_1)}^T, f_{jh_1}), \dots, (p_{(h_r)}^T, f_{jh_r})$, is univocally determined. By virtue of the convexity of f_j we have that, for all $p \in \mathcal{R}$, $f_j(p) \le h_j(p)$; hence set

$$\overline{f}_j(p) = h_j(p). \quad (14)$$

Step 2
Construct the matrix function $\Psi(p, \delta)$ obtained from (10) substituting for $f_j(p)$, $j = 1,\dots,\nu$, the multi-affine function

$$f_j^m(p, \delta_j) \triangleq$$

$$\begin{cases} (1-\delta_j)\underline{f}_j(\boldsymbol{p})+\delta_j\overline{f}_j(\boldsymbol{p}) & \text{if } f_j \text{ is not multi-affine} \\ f_j(\boldsymbol{p}) & \text{if } f_j \text{ is multi-affine}, \end{cases} \tag{15}$$

where the $\underline{f}_j$'s and the $\overline{f}_j$'s are affine functions constructed according to Step 1 and

$$\delta_j \in I_j \triangleq \begin{cases} [0,1] & \text{if } f_j \text{ is not multi-affine} \\ \{0\} & \text{if } f_j \text{ is multi-affine}. \end{cases} \tag{16}$$

Let $\mathcal{D} \triangleq I_1 \times I_2 \times \cdots \times I_\nu$; hence $\boldsymbol{\Psi}(\boldsymbol{p},\boldsymbol{\delta})$ is defined over $\mathcal{R} \times \mathcal{D}$. It is simple to recognize that

$$\boldsymbol{\Psi}(\mathcal{R}\times\mathcal{D}) \supseteq \boldsymbol{F}(\mathcal{R}).$$

Observe that $\boldsymbol{\Psi}(\boldsymbol{p},\boldsymbol{\delta})$ is an affine function of the δ_i's and a polynomial function of the p_i's; moreover since in (10) there are at most ν products, the maximum exponent of each p_i cannot exceed ν.

Step 3
Replace each $p_i^{\alpha_i}$, $\alpha_i > 1$, with a multi-affine function as in (15); obviously in this case the bounding functions (which can be easily computed applying Step 1) will depend only on p_i. The number of replacements, say γ, may vary between zero (when $\boldsymbol{\Psi}$ is already multi-affine, that is $\boldsymbol{\Psi}$ does not contain powers of $p_1,\ldots,p_{n_p}$ greater than 1) and $n_p(\nu-1)$.

In this way we introduce a new vector of fictitious variables $\boldsymbol{\varepsilon}$ ranging in the hyperrectangle $\mathcal{E}$ defined in the same way as $\mathcal{D}$ and define the new function $\boldsymbol{\Phi}(\boldsymbol{p},\boldsymbol{\delta},\boldsymbol{\varepsilon})$; in Theorem 3 of Amato *et al* (1994) it is shown that $\boldsymbol{\Phi}$ is multi-affine, moreover

$$\begin{aligned} \boldsymbol{\Phi}(\mathcal{R}\times\mathcal{D}\times\mathcal{E}) &\supseteq \boldsymbol{\Psi}(\mathcal{R}\times\mathcal{D}) \\ &\supseteq \boldsymbol{F}(\mathcal{R}). \end{aligned}$$

Remark 3: Note that $\mathcal{R}\times\mathcal{D}\times\mathcal{E}$ is a hyperrectangle with $2^{n_p+\mu+\gamma}$ vertices, where μ is the number of non-multi-affine functions in the expression (10) of $\boldsymbol{F}$ and, as said, γ is the number of replacements in Step 3 of Algorithm 1.

3.2. *Solution of Problem 1 in the Non-Affine Case*

The solution of Problem 1 when $\boldsymbol{A}(\cdot)$ and $\boldsymbol{B}(\cdot)$ satisfy Assumption 1 is composed of two steps:

Step a)
Apply Algorithm 1 to construct two multi-affine functions

$$\begin{aligned} \boldsymbol{\Phi}_A(\boldsymbol{p},\boldsymbol{\delta}_A,\boldsymbol{\varepsilon}_A) &: \mathcal{R}\times\mathcal{D}_A\times\mathcal{E}_A \mapsto \mathrm{I\!R}^{n\times n} \\ \boldsymbol{\Phi}_B(\boldsymbol{p},\boldsymbol{\delta}_B,\boldsymbol{\varepsilon}_B) &: \mathcal{R}\times\mathcal{D}_B\times\mathcal{E}_B \mapsto \mathrm{I\!R}^{n\times m} \end{aligned}$$

such that

$$\boldsymbol{\Phi}_A(\mathcal{R},\mathcal{D}_A,\mathcal{E}_A) \supseteq \boldsymbol{A}(\mathcal{R}) \tag{17}$$
$$\boldsymbol{\Phi}_B(\mathcal{R},\mathcal{D}_B,\mathcal{E}_B) \supseteq \boldsymbol{B}(\mathcal{R}). \tag{18}$$

Let us define $\boldsymbol{\delta} \triangleq (\boldsymbol{\delta}_A^T \;\; \boldsymbol{\delta}_B^T)^T \in \mathrm{I\!R}^{\mu_A+\mu_B}$ and $\boldsymbol{\varepsilon} \triangleq (\boldsymbol{\varepsilon}_A^T \;\; \boldsymbol{\varepsilon}_B^T)^T \in \mathrm{I\!R}^{\gamma_A+\gamma_B}$; in the same way we define $\mathcal{D} \triangleq \mathcal{D}_A \times \mathcal{D}_B$ and $\mathcal{E} \triangleq \mathcal{E}_A \times \mathcal{E}_B$; note that $\mathcal{D}$ and $\mathcal{E}$ have $2^{\mu_A+\mu_B}$ and $2^{\gamma_A+\gamma_B}$ vertices respectively. Finally let $\Omega \triangleq \mathcal{R}\times\mathcal{D}\times\mathcal{E}$ and $2^{n_\Omega} = 2^{n_p+\mu_A+\mu_B+\gamma_A+\gamma_B}$ the number of its vertices. In order to simplify the notation, we will consider $\boldsymbol{\Phi}_A$ and $\boldsymbol{\Phi}_B$ functions of the whole vector $\boldsymbol{\omega} = (\boldsymbol{p}^T \;\; \boldsymbol{\delta}^T \;\; \boldsymbol{\varepsilon}^T)^T$.

Step b)
Define

$$\begin{aligned} \boldsymbol{\Phi}_{L_S}(\boldsymbol{\omega}) \triangleq\; & \boldsymbol{S}\boldsymbol{\Phi}_A^T(\boldsymbol{\omega}) + \boldsymbol{\Phi}_A(\boldsymbol{\omega})\boldsymbol{S} \\ & + \boldsymbol{L}^T\boldsymbol{\Phi}_B^T(\boldsymbol{\omega}) + \boldsymbol{\Phi}_B(\boldsymbol{\omega})\boldsymbol{L}. \end{aligned} \tag{19}$$

Then solve Problem 2 with $\boldsymbol{\Phi}_{L_S}$ replacing $\boldsymbol{L}_S$.

In this case, the existence of a solution to Problem 2 is no longer equivalent to the quadratic stabilizability of the system under consideration. This is obviously due to the fact that the immersion of the images of $\boldsymbol{A}$ and $\boldsymbol{B}$ introduces conservatism. In the next section we show how to reduce this conservatism.

4. A REFINEMENT PROCEDURE

Let us denote with "co$(\cdot)$" the operation of taking the convex hull of the argument; for a given $\boldsymbol{S}$ and $\boldsymbol{L}$, recalling that $\boldsymbol{\Phi}_{L_S}(\Omega) \supseteq \boldsymbol{L}_S(\mathcal{R})$ and that $\boldsymbol{\Phi}_{L_S}$ is multi-affine, application of Fact 1 enables to conclude that

$$\mathrm{co}\,(L_S(\mathcal{R})) \subseteq \mathrm{co}\left(\{\boldsymbol{\Phi}_{L_S}(\omega_{(i)}), i=1,\ldots,2^{n_\Omega}\}\right), \tag{20}$$

where, as usual, we denote the vertices of Ω by $\omega_{(i)}$.

In the next theorem we show that splitting the hyperrectangle $\mathcal{R}$ in smaller hyperrectangles and considering the union of the images of these sub-hyperrectangles, the right hand side in (20) converges to the left hand side when the splitting is made finer and finer.

We say that $\mathcal{T} = \{\mathcal{R}_1,\mathcal{R}_2,\ldots,\mathcal{R}_l\}$ is a rectangular covering of the hyperrectangle $\mathcal{R}$ if $\cup_{r=1}^{l}\mathcal{R}_r = \mathcal{R}$. We define $\boldsymbol{T}$ as the set of all rectangular coverings of $\mathcal{R}$.

Now let $\mathcal{C}(\mathrm{I\!R}^n)$ the set composed of all closed sets in $\mathrm{I\!R}^n$. We equip this set with the Hausdorff metric

$$d_H(S_1,S_2) \triangleq \max\left\{\max_{\mathbf{x}_1\in S_1} d(\boldsymbol{x}_1,S_2), \max_{\mathbf{x}_2\in S_2} d(S_1,\boldsymbol{x}_2)\right\} \tag{21}$$

with $S_1, S_2 \in \mathcal{C}(\mathrm{I\!R}^n)$ and where $d(\boldsymbol{x},S)$ denotes the usual Euclidean distance of the point $\boldsymbol{x}$ from the set S.

Let $\{\mathcal{T}_h\}_{h\in\mathrm{I\!N}}$, $\mathcal{T}_h \in \boldsymbol{T}$, a sequence of rectangular coverings, set up as follows. $\mathcal{T}_1 = \{\mathcal{R}\}$; $\mathcal{T}_2$ is constructed by splitting $\mathcal{R}$ into 2^{n_p} hyperrectangles by lines parallel to the coordinate axes; $\mathcal{T}_3$ is obtained applying the same procedure to each hyperrectangle in $\mathcal{T}_2$ and so on; note that $\mathcal{T}_h$ is composed of $2^{(h-1)n_p}$ hyperrectangles. We denote by $\mathcal{R}_{hr}$ the r-th hyperrectangle in $\mathcal{T}_h$.

The rigorous proof of the next theorem is long and technical and is omitted for brevity (the interested reader is referred to Amato, 1994).

Theorem 2: *Consider the sequence* $\{\mathcal{T}_h\}_{h\in\mathbb{N}}$, *then*

$$\lim_{h\to\infty} \bigcup_{r=1}^{2^{(h-1)n_p}} \mathrm{co}\left(\{\boldsymbol{\Phi}_{L_S}^{hr}(\omega_{(i)}^{hr}), i=1,\dots,2^{n_\Omega}\}\right) = \boldsymbol{L}_S(\mathcal{R})$$

in the metric space $(\mathcal{C}(\mathbb{R}^n), d_H)$, *where* $\boldsymbol{\Phi}_{L_S}^{hr}$ *is the function constructed on* $\mathcal{R}_{hr}$ *according to Algorithm 1 and* $\omega_{(i)}^{hr}$ *is the i-th vertex of* $\Omega_{hr} \triangleq \mathcal{R}_{hr} \times \mathcal{D} \times \mathcal{E}$.

Based on Theorem 2 we propose an algorithm for the solution of Problem 1 when the system matrices are non-affine and satisfy Assumption 1.

Given a set $\mathcal{S}$, $\mathcal{S}_j$ denotes the j-th element of $\mathcal{S}$ and card($\mathcal{S}$) the number of elements of $\mathcal{S}$. Given two sets $\mathcal{A}$ and $\mathcal{B}$, with $\mathcal{B} \subseteq \mathcal{A}$, $\mathcal{A} - \mathcal{B}$ denotes the set of the elements of $\mathcal{A}$ which do not belong to $\mathcal{B}$.

Algorithm 2:

Step 1 (Initialization)
Let $\mathcal{S} = \{\mathcal{R}\}$, $h = 0$.

Step 2
Let $h = h+1$; for $j = 1, \dots, \mathrm{card}(\mathcal{S}) \triangleq m_h$ repeat the following procedure: apply Algorithm 1 to $\boldsymbol{A}(\cdot)$ and $\boldsymbol{B}(\cdot)$ restricted to $\mathcal{S}_j$ and solve the following problem.

Problem 3: *Sufficient condition*

Find $\boldsymbol{S} > 0$ *and* $\boldsymbol{L} \in \mathbb{R}^{m\times n}$ *such that*

$$\boldsymbol{\Phi}_{L_S}(\omega_{j(i)}) > 0\,, i = 1,\dots,2^{n_\Omega}\,, j = 1,\dots,m_h\,,$$

where $\omega_{j(i)}$ *is the i-th vertex of* $\Omega_j \triangleq \mathcal{S}_j \times \mathcal{D} \times \mathcal{E}$.

If Problem 3 admits a solution then **stop**, the system is quadratically stabilizable and $\boldsymbol{K} = \boldsymbol{L}\boldsymbol{S}^{-1}$ is a compensator which quadratically stabilizes the system; otherwise test the necessary condition described below.

Problem 4: *Necessary condition*

Find $\boldsymbol{S} > 0$ *and* $\boldsymbol{L} \in \mathbb{R}^{m\times n}$ *such that*

$$\boldsymbol{L}_S(\boldsymbol{p}_{j(i)}) > 0\,, i = 1,\dots,2^{n_p}\,, j = 1,\dots,m_h\,,$$

where $\boldsymbol{p}_{j(i)}$ *is the i-th vertex of* $\mathcal{S}_j$.

If this problem does not admit solution then **stop**, the system is not quadratically stabilizable; otherwise goto Step 3.

Step 3
Split each element belonging to $\mathcal{S}$ in 2^{n_p} hyperrectangles, put them into $\mathcal{S}$ and go to Step 2.

The next theorem studies the convergence properties of Algorithm 2.

Theorem 3: *Algorithm 2 converges in a* finite number *of iterations.*

Proof. If system (5) is quadratically stabilizable there exist $\boldsymbol{S} > 0$ and $\boldsymbol{L} \in \mathbb{R}^{m\times n}$ such that $\boldsymbol{L}_S(\boldsymbol{p}) > 0$ for all $\boldsymbol{p} \in \mathcal{R}$. Following continuity arguments there exists an open set $\Theta \supset \mathcal{R}$ such that system (5) is quadratically stabilizable in Θ. Obviously $\boldsymbol{L}_S(\Theta) \supset \boldsymbol{L}_S(\mathcal{R})$; therefore by virtue of Theorem 2 there will exist an index s such that

$$\bigcup_{r=1}^{2^{(s-1)n_p}} \mathrm{co}\left(\{\boldsymbol{\Phi}_{L_S}^{sr}(\omega_{(i)}^{sr}), i=1,\dots,2^{n_\Omega}\}\right) \subset \boldsymbol{L}_S(\Theta)\,.$$

At iteration s Problem 3 admits a solution and therefore Algorithm 2 stops.

Conversely suppose that system (5) is not quadratically stabilizable. Since $\boldsymbol{L}_S(\cdot)$ is continuous, two situations can happen

i)
the set $\mathcal{R}$ equals the union of a finite number of subdomains Γ_i, $i = 1,\dots,r$, such that for each of these domains there exists a class of positive definite matrices $\mathcal{S}_i$ and a class of matrices $\mathcal{L}_i$ satisfying, for $i = 1,\dots,r$,

$$\boldsymbol{L}_S(\boldsymbol{p}) > 0\,, \quad \forall \boldsymbol{p} \in \Gamma_i \quad \forall \boldsymbol{S} \in \mathcal{S}_i \quad \forall \boldsymbol{L} \in \mathcal{L}_i\,, \tag{22}$$

and $\bigcap_{i=1}^r \mathcal{S}_i = \emptyset$ or $\bigcap_{i=1}^r \mathcal{L}_i = \emptyset$;

ii)
there is a subset $\mathcal{Q}$ of $\mathcal{R}$ such that the pair $(\boldsymbol{A}(\boldsymbol{p}), \boldsymbol{B}(\boldsymbol{p}))$ is not stabilizable for $\boldsymbol{p} \in \mathcal{Q}$.

For what concerns case i), suppose, for istance, that $r = 2$. It is obvious that in this case Problem 3 does not have solution. On the other hand, there certainly exists a finite iteration index s such that the points $\boldsymbol{L}_S(\boldsymbol{p}_{j(i)})$, $i = 1,\dots,2^{n_p}$, $j = 1,\dots,m_s$, belong in part to Γ_1 and in part to Γ_2. Therefore at iteration s Problem 4 does not have solution and the algorithm stops. The discussion of case ii) is similar. ∎

5. AN EXAMPLE

We present the following example to clarify how the algorithm works when the system depends non-affinely on parameters.

Example 1: Let

$$\boldsymbol{A}(\boldsymbol{p}) = \begin{pmatrix} a_{11}(\boldsymbol{p}) & 0 & 0 \\ 0 & 1 & -1 \\ 0 & a_{32}(\boldsymbol{p}) & 1 \end{pmatrix} \tag{23a}$$

$$\boldsymbol{B}(\boldsymbol{p}) = \begin{pmatrix} 1 & 1 & 2 \end{pmatrix}^T \tag{23b}$$

where

$$a_{11}(\boldsymbol{p}) = -1 + \exp(p_1)p_2 \quad (24a)$$
$$a_{32}(\boldsymbol{p}) = 2 + p_2 \sin(p_2) \quad (24b)$$

with $\boldsymbol{p} \in [0, .8]^2$. Note that $\boldsymbol{B}$ does not depend on parameters, so that $\boldsymbol{\Phi}_B(\omega) = \boldsymbol{B}$. Now we have to compute $\boldsymbol{\Phi}_A(\omega)$. The function $\boldsymbol{A}(\boldsymbol{p})$ can be written in the standard form (9) by letting

$$a_1(\boldsymbol{p}) = \exp(p_1)\,; \quad a_2(\boldsymbol{p}) = p_2\,; \quad a_3(\boldsymbol{p}) = \sin(p_2)\,. \quad (25)$$

Step 1
In this case the non-multi-affine functions are a_1, and a_3. Using Step 1 of Algorithm 1, a choice of the bounding functions is

$$\underline{a}_1(\boldsymbol{p}) = 0.82 + 1.64p_1\,; \quad \overline{a}_1(\boldsymbol{p}) = 1 + 1.72p_1$$
$$\underline{a}_3(\boldsymbol{p}) = 0.84p_2\,; \quad \overline{a}_3(\boldsymbol{p}) = 0.05 + 0.88p_2\,.$$

Step 2
Substituting a_j^m to a_j, $j = 1, 3$, we obtain

$$\boldsymbol{\Psi}_A(\boldsymbol{p}, \boldsymbol{\delta}) = \begin{pmatrix} \psi_{11}(\boldsymbol{p}, \boldsymbol{\delta}) & 0 & 0 \\ 0 & 1 & -1 \\ 0 & \psi_{32}(\boldsymbol{p}, \boldsymbol{\delta}) & 1 \end{pmatrix},$$

with

$$\psi_{11}(\boldsymbol{p}, \boldsymbol{\delta}) = 0.82 + 1.64p_1 + 0.18\delta_1 + 0.08p_1\delta_1 \quad (26a)$$

$$\psi_{32}(\boldsymbol{p}, \boldsymbol{\delta}) = 2 + 0.05p_2\delta_3 + 0.84p_2^2 + 0.04p_2^2\delta_3 \quad (26b)$$

Step 3
The only non-multi-affine function in (26) is $g(\boldsymbol{p}) = p_2^2$; as bounding functions we choose $\underline{g}(\boldsymbol{p}) = -0.25 + p_2$ and $\overline{g}(\boldsymbol{p}) = p_2$. Substituting $\underline{g}$ and $\overline{g}$ in (26b) we obtain that

$$\boldsymbol{\Phi}_A(\boldsymbol{p}, \boldsymbol{\delta}, \varepsilon) = \begin{pmatrix} \psi_{11}(\boldsymbol{p}) & 0 & 0 \\ 0 & 1 & -1 \\ 0 & \phi_{32}(\boldsymbol{p}, \boldsymbol{\delta}, \varepsilon) & 1 \end{pmatrix},$$

where

$$\phi_{32}(\boldsymbol{p}, \boldsymbol{\delta}, \varepsilon) = 1.79 + 0.84p_2 - 0.01\delta_3 + 0.21\varepsilon + 0.09p_2\delta_3 + 0.01\delta_3\varepsilon\,. \quad (27)$$

Note that $\boldsymbol{\Phi}_A$ is multi-affine, moreover $\boldsymbol{\Phi}_A : \Omega \to \mathbb{R}^{3\times 3}$ where

$$\Omega \triangleq \mathcal{R} \times \mathcal{D} \times \mathcal{E} = [0, 0.8]^2 \times [0, 1]^2 \times [0, 1]\,. \quad (28)$$

The algorithm solving Problem 2 converges after 55 iterations, and the resulting $\boldsymbol{K}$ is:

$$\boldsymbol{K} = (\,-24.8003 \quad 29.3263 \quad -7.2079\,)\,. \quad (29)$$

6. CONCLUSIONS

In this paper we have considered the problem of quadratically stabilizing a linear system whose system matrices depend non-affinely on parameters. Using an algorithm which covers the image of a non-affine function by that of a multi-affine function, we have reconducted our original problem to one which can be solved via standard convex optimization procedures. Moreover we have provided an algorithm which reduce at will the conservatism due to the covering procedure.

ACKNOWLEDGEMENTS

The work of the first author was supported by *Consiglio Nazionale delle Ricerche* under Grant n. 201.15.5.

7. REFERENCES

Amato, F. (1994). Stability analysis for linear systems depending on uncertain time-varying parameters. *PhD Thesis*, Dipartimento di Informatica e Sistemistica, Napoli.

Amato, F., F. Garofalo, L. Glielmo, and A. Pironti (1994). Robust and quadratic stability via Polytopic set covering. *Int. J. Robust and Nonlinear Control*, to appear.

Barmish, B. R. (1983). Stabilization of uncertain systems via linear control. *IEEE Trans. Aut. Contr.*, **AC-28**, 848–850.

Bernussou, J., P. L. D. Peres, and J. C. Geromel (1989). A linear programming oriented procedure for quadratic stabilization of uncertain systems. *Syst. Contr. Lett.*, **13**, 65–72.

Khargonekar, P. P., I. R. Petersen, and K. Zhou (1990). Robust stabilization of uncertain linear systems: quadratic stabilizability and H_∞ control theory. *IEEE Trans. Aut. Contr.*, **AC-35**, 356-361.

Leitmann, G. (1979). Guaranteed asymptotic stability for some linear systems with bounded uncertainties. *J. Dyn. Meas. Contr.,*, **101**, 212–216.

Petersen, I.R. (1985). Quadratic stabilizability of uncertain linear systems: existence of a nonlinear stabilizing control does not imply existence of a linear stabilizing control. *IEEE Trans. Aut. Contr.*, **AC-30**, 291–293.

Petersen, I. R. (1988). Quadratic stabilizability of uncertain linear systems containing both constant and time-varying uncertain parameters. *JOTA*, **57**, 439–461.

Rockafellar, R. T. (1970). *Convex Analysis.* Princeton University Press, Princeton.

Zadeh, L. A., C. A. Desoer (1963). *Linear System Theory.* McGraw-Hill, New York.

OPTIMAL GUARANTEED COST CONTROL OF DISCRETE-TIME UNCERTAIN NONLINEAR SYSTEMS

A. V. SAVKIN* and I. R. PETERSEN**

* Department of Electrical and Electronic Engineering and Cooperative Research Center for Sensor Signal and Information Processing, University of Melbourne, Parkville, Victoria 3052, Australia, email: savkin@ee.mu.oz.au.

** Department of Electrical Engineering, Australian Defence Force Academy, Campbell, 2600, Australia, email: irp@ee.adfa.oz.au.

Abstract. L'article donne une commande optimale pour une classe nouvelle de système incertain nonlinèaire. La solution inveloppe un problème de programmation dynamique avec paramètre dependence.

Keywords. Optimal Control, Uncertain Systems, Discrete-time Systems, Nonlinear Systems

1. INTRODUCTION

The dynamic programming solution to the state feedback optimal control problem is an important approach to many optimal control problems; e.g., see (Bryson and Ho 1975). With recent advances in robust control theory, it is of interest to extend this approach to the case of systems with uncertainty. One approach to the robust linear quadratic control problem is the guaranteed cost control approach of (Chang and Peng 1972, Bernstein and Haddad 1988, Petersen and McFarlane 1992, Savkin and Petersen 1995*a*). In these papers, a (continuous-time) linear quadratic optimal control problem is considered and a state feedback controller is designed to minimize a certain upper bound on the quadratic cost function.

In this paper, we consider a guaranteed cost control problem for nonlinear discrete-time uncertain systems defined over a finite time interval. The class of nonlinear uncertain discrete-time systems being considered, involves a new class of uncertainties satisfying a certain "Averaged Sum Quadratic Constraint". This notion extends the standard (continuous-time) Integral Quadratic Constraint which originated in the work of Yakubovich (Yakubovich 1988*b*, Yakubovich 1988*a*); see also, (Savkin and Petersen 1994*b*, Savkin and Petersen 1995*b*, Savkin and Petersen 1995*a*). The discrete-time version of this standard integral quadratic constraint could be referred to as the standard "Sum Quadratic Constraint". The standard sum quadratic constraint defines a class of uncertainties which is extremely rich and allows for nonlinear time-varying dynamic structured uncertainties. One interpretation of the averaged sum quadratic constraint is as follows: In the standard sum quadratic constraint, the uncertainty is required to satisfy a certain induced norm bound but with an allowed measure of mismatch. This mismatch may account for a non-zero initial condition on the uncertainty dynamics. In our averaged sum quadratic constraint, we bound only the average value of this measure of mismatch. This leads to an uncertainty description which combines some of the characteristics of a stochastic uncertainty description and the standard norm bounded uncertainty description. However, it does not lead to a mathematically intractable problem as is usual with stochastic uncertainty descriptions when applied to nonlinear systems.

The remainder of the paper proceeds as follows. In Section 2, we introduce the class of uncertain systems under consideration and define our corresponding notion of "Guaranteed Cost Control". In Section 3, we present certain S-procedure result which is used to prove our main result. Section 4 presents the main result of the paper which is a necessary and sufficient condition for the existence of a guaranteed cost controller. This result is given in terms of the existence of a positive-definite solution to a parameterized dynamic programming equation. If such a solution exists, then it is shown that it can be used to construct a corresponding non-dynamic state feedback guaranteed cost controller. The main result of this paper is proved by converting the guaranteed cost control problem into an equivalent nonlinear H^∞ control problem which can be solved using the methods of (Baras and James 1993).

2. PROBLEM STATEMENT

Consider the following discrete-time uncertain nonlinear system defined on the finite time interval $t = 0, 1, \ldots, N$:

$$\begin{aligned} x(t+1) &= b(x(t), u(t), \xi_1(t), \ldots, \xi_k(t)); \\ z_1(t) &= l_1(x(t), u(t)); \\ &\vdots \end{aligned}$$

$$z_k(t) = l_k(x(t), u(t)) \quad (1)$$

where $x(t) \in \mathbf{R}^n$ is the *state*, $u(t) \in \mathbf{R}^m$ is the *control input*, $z_1(t) \in \mathbf{R}^{h_1}$, $z_2(t) \in \mathbf{R}^{h_2}, \ldots, z_k(t) \in \mathbf{R}^{h_k}$ are the *uncertainty outputs*, $\xi_1(t) \in \mathbf{R}^{r_1}$, $\xi_2(t) \in \mathbf{R}^{r_2}, \ldots, \xi_k(t) \in \mathbf{R}^{r_k}$ are the *uncertainty inputs*, and $b(\cdot,\cdot,\cdot,\ldots,\cdot), l_1(\cdot,\cdot),\ldots,l_k(\cdot,\cdot)$ are continuous functions such that $b(0,0,0,\ldots,0) = 0, l_1(0,0) = 0,\ldots,l_k(0,0) = 0$.

Associated with the system (1) is the cost function

$$F_0\left(x(\cdot), u(\cdot), \xi_1(\cdot), \ldots, \xi_k(\cdot)\right) \triangleq \sum_{t=0}^{N} g(x(t), u(t)) \quad (2)$$

where $g(\cdot,\cdot) \geq 0$ is a given continuous weighting function such that $g(0,0) = 0$ and

$$\inf_{u \in \mathbf{R}^m} g(x,u) > 0$$

for all $x \neq 0$.

System Uncertainty The uncertainty in the above system is described by a set of equations of the form:

$$\begin{aligned} \xi_1(t) &= \phi_1(t, x(\cdot)) \\ &\vdots \\ \xi_k(t) &= \phi_k(t, x(\cdot)). \end{aligned} \quad (3)$$

Alternatively, the uncertainty inputs and outputs may be collected together into two vectors. That is, we define $\xi(t) \triangleq [\xi_1(t)'\ \xi_2(t)'\ \ldots\ \xi_k(t)'\ ]'$ and $z(t) \triangleq [z_1(t)'\ z_2(t)'\ \ldots\ z_k(t)'\ ]'$. Then (3) can be re-written in the more compact form

$$\xi(t) = \Phi(t, x(\cdot)). \quad (4)$$

The uncertainty described by the above equations is required to satisfy a certain "Averaged Sum Quadratic Constraint". That is, we consider finite sequences of uncertainty functions of the form (4) such that the following constraint is satisfied.

Definition 1 (Averaged Sum Quadratic Constraint) Let $d_1(x_0), d_2(x_0), \ldots, d_k(x_0)$ be given continuous functions associated with the system (1) such that $d_s(0) = 0$ and $d_s(x_0) > 0$ for all $x_0 \neq 0$ and $s = 1,\ldots,k$. We will consider sequences of uncertainty functions $\mathcal{S} = \{\Phi^1(\cdot), \Phi^2(\cdot), \ldots \Phi^q(\cdot)\}$ of arbitrary length q. A sequence of uncertainty functions of the form $\mathcal{S} = \{\Phi^1(\cdot), \Phi^2(\cdot), \ldots \Phi^q(\cdot)\}$ is an admissible *uncertainty sequence* for the system (1) if the following conditions hold: Given any $\Phi^i(\cdot) \in \mathcal{S}$, any control input $u^i(t)$ defined for $t = 0,1,\ldots,N$, and any corresponding solution $\{x^i(\cdot), \xi^i(\cdot)\}$ to equations (1), (4), then

$$\sum_{i=1}^{q} (\|\xi_s^i(\cdot)\|_2^2 - \|z_s^i(\cdot)\|_2^2) \leq \sum_{i=1}^{q} d_s(x^i(0)) \quad (5)$$

for all $s = 1,2,\ldots,k$. Here $\|\cdot\|$ denotes the standard Euclidean norm and $\|v(\cdot)\|_2$ denotes the $l_2[0,N]$ norm of $v(\cdot)$. That is, $\|v(\cdot)\|_2^2 \triangleq \sum_{t=0}^{N} \|v(t)\|^2$.

The class of all such admissible uncertainty sequences is denoted Ξ.

The class of controllers to be considered in this paper are nonlinear time-varying state feedback controllers of the form

$$u(t) = U(t, x(\cdot)|_0^t) \quad (6)$$

where $U(t,0) \equiv 0$.

Definition 2 Let $W(x_0) \geq$ be a given continuous weighting function such that $W(0) = 0$. A controller of the form (6) is said to be a *guaranteed cost controller* for the uncertain system (1), (3) with the cost function (2) and the weighting function $W(\cdot)$ if the following condition holds: Given any $\mathcal{S} \in \Xi$, let q be the number of elements in this uncertainty sequence and let $[x^i(\cdot), u^i(\cdot), \xi^i(\cdot)]$ be any corresponding solution to equations (1), (4), (6). Then,

$$\sup_{\mathcal{S} \in \Xi} \frac{\sum_{i=1}^{q} F_0\left(x^i(\cdot), u^i(\cdot), \xi^i(\cdot)\right)}{\sum_{i=1}^{q} W(x^i(0))} < 1. \quad (7)$$

3. NEW RESULTS ON THE S-PROCEDURE

In this section, we present a result which extends the so-called "S-procedure"; e.g., see (Yakubovich 1971). A proof of this result can be found in (Savkin and Petersen 1994*a*).

Theorem 1
Consider a set $P = \{x = [x_0\ x_1\ \cdots\ x_k]'\} \subset \mathbf{R}^{k+1}$ with the following property: Given any finite sequence $\{x^1, x^2, \ldots, x^q\} \subset P$ such that $\sum_{i=1}^{q} x_1^i \geq 0, \ldots, \sum_{i=1}^{q} x_k^i \geq 0$, then $\sum_{i=1}^{q} x_0^i \geq 0$. Then there exist constants $\tau_0 \geq 0, \tau_1 \geq 0, \ldots, \tau_k \geq 0$ such that $\sum_{s=0}^{k} \tau_s > 0$ and

$$\tau_0 x_0 \geq \tau_1 x_1 + \cdots + \tau_k x_k \quad (8)$$

for all $x = [x_0\ x_1\ \cdots\ x_k]' \in P$.

Observation 1 If assumption of the above theorem is satisfied and there exists an element $x = [x_0\ x_1\ \cdots\ x_k]' \in P$ such that $x_1 > 0, \ldots, x_k > 0$, then the constant τ_0 in (8) may be taken as $\tau_0 = 1$.

4. THE MAIN RESULT

In this section, we present the main result of this paper which establishes a necessary and sufficient condition for the existence of a nonlinear state feedback guaranteed cost control (6) for the uncertain time-varying system (1), (3) with associated cost function (2) and weighting function $W(\cdot)$. This condition is given in terms of the existence of a suitable solution to a parameter dependent dynamic programming equation. The equation under consideration is defined as follows: Let $\tau_1 \geq 0, \ldots, \tau_k \geq 0, \delta > 0$ be given constants and consider the equation

$$V_\delta(t,x) = \inf_u \sup_\xi \left\{ \begin{array}{l} V_\delta\left(t+1, b(x,u,\xi_1,\ldots,\xi_k)\right) \\ +G_\delta(x,u,\xi_1,\ldots,\xi_k) \end{array} \right\} \quad (9)$$

where

$$G_\delta(x,u,\xi_1,\ldots,\xi_k) \triangleq (1+\delta)g(x,u) + \sum_{s=1}^{k} \tau_s(\|z_s\|^2 - \|\xi_s\|^2). \quad (10)$$

Our solution to guaranteed cost control problem will involve constructing a static state feedback controller of the form

$$u(t) = u^*(t,x) \quad (11)$$

so that the following dynamic programming inequality is satisfied:

$$H_\delta(t,x) \geq \sup_\xi \left\{ \begin{array}{l} H_\delta\left(t+1, b(x,u^*(t,x),\xi)\right) \\ +G_\delta(x,u^*(t,x),\xi) \end{array} \right\} \quad (12)$$

for all $x \in \mathbf{R}^n$ and for $t = 0,1,\ldots,N$.

Our main result requires that the uncertain system (1), (5) satisfies the following additional assumption.

Assumption

4.1 There exist continuous functions $\xi_1^*(x),\ldots,\xi_k^*(x)$ such that $\xi_1^*(0) = 0,\ldots,\xi_k^*(0) = 0$ and $b(0,0,\xi_1^*(x),\ldots,\xi_k^*(x)) = x$ for all $x \in \mathbf{R}^n$.

Theorem 2 Consider the uncertain system (1), (3) with associated cost function (2) and weighting function $W(\cdot)$ and suppose that Assumption 4.1 is satisfied. Then the following statements are equivalent:

(i) There exists a nonlinear state feedback guaranteed cost control of the form (6).

(ii) There exist constants $\tau_1 \geq 0,\ldots,\tau_k \geq 0$ and $\delta^* > 0$ such that for all $\delta \in [0,\delta^*]$, the equation (9) with boundary condition $V_\delta(N+1,\cdot) \equiv 0$ has a solution $V_\delta(t,x) \geq 0$ such that

$$V_\delta(0,x_0) \leq W(x_0) - \sum_{s=1}^{k} \tau_s d_s(x_0) \quad (13)$$

for all $x_0 \in \mathbf{R}^n$, $V_\delta(t,0) = 0$, and the function $V_\delta(t,\cdot)$ is continuous at the origin for $t = 0,1,\ldots,N$. Furthermore, if $x = 0$, then the infimum in (9) is achieved at $u = 0$ for all t.

Moreover, if condition (ii) holds, then for any $\delta \in (0,\delta^*)$, there exists a static state feedback control of the form (11) such that $u^*(t,0) \equiv 0$ and $u^*(t,\cdot)$ is continuous at the origin. Also, there exists a function $H_\delta(t,x) \geq 0$ such that $H_\delta(t,0) = 0$, $H_\delta(t,\cdot)$ is continuous at the origin for $t = 0,1,\ldots,N$,

$$H_\delta(0,x_0) \leq W(x_0) - \sum_{s=1}^{k} \tau_s d_s(x_0) \quad (14)$$

and inequality (12) is satisfied. Furthermore, any such controller is a guaranteed cost controller for the uncertain system (1), (3) with associated cost function (2) and weighting function $W(\cdot)$.

Proof $(i) \Rightarrow (ii)$ Suppose the uncertain system (1), (3) is such that there exists a nonlinear state feedback guaranteed cost control of the form (6). It follows from (7) that there exists a $\delta^* > 0$ such that

$$\sum_{i=1}^{q}(1+\delta)F_0\left(x^i(\cdot),u^i(\cdot),\xi^i(\cdot)\right) \leq \sum_{i=1}^{q} W(x^i(0))$$

for all $\delta \in [0,\delta^*]$ and for all $\mathcal{S} \in \Xi$. Now let $\delta \in [0,\delta^*]$ be given. For the system closed loop system defined by (1) and (6) a corresponding set Ω is defined as follows:

$$\Omega \triangleq \left\{ \begin{array}{l} \lambda(\cdot) = [x(\cdot)\ u(\cdot)\ \xi(\cdot)] : \{x(\cdot),u(\cdot),\xi(\cdot)\} \\ \text{are related by (1), (6)} \end{array} \right\}$$

Also, we consider the following set of functionals mapping from Ω into $\mathbf{R}$:

$$\begin{aligned} f_0\left(x(\cdot),u(\cdot),\xi(\cdot)\right) &\triangleq -(1+\delta)F_0(x(\cdot),u(\cdot)) \\ &\quad +W\left(x(0)\right); \\ f_1\left(x(\cdot),u(\cdot),\xi(\cdot)\right) &\triangleq \|z_1(\cdot)\|_2^2 - \|\xi_1(\cdot)\|_2^2 \\ &\quad +d_1(x(0)); \\ &\vdots \\ f_k\left(x(\cdot),u(\cdot),\xi(\cdot)\right) &\triangleq \|z_k(\cdot)\|_2^2 - \|\xi_k(\cdot)\|_2^2 \\ &\quad +d_k(x(0)). \end{aligned}$$

Now consider a sequence of vector functions $\{x^i(\cdot),u^i(\cdot),\xi^i(\cdot)\}_{i=1}^q \subset \Omega$ such that

$$\begin{aligned} \sum_{i=1}^{q} f_1\left(x^i(\cdot),u^i(\cdot),\xi^i(\cdot)\right) &\geq 0, \\ &\vdots \\ \sum_{i=1}^{q} f_k\left(x^i(\cdot),u^i(\cdot),\xi^i(\cdot)\right) &\geq 0. \end{aligned}$$

It follows from Definition 1 that this sequence corresponds to an admissible uncertainty sequence $\mathcal{S} \in \Xi$. Hence, inequality (7) implies that

$$\sum_{i=1}^{q} f_0\left(x^i(\cdot),u^i(\cdot),\xi^i(\cdot)\right) \geq 0. \quad (15)$$

We now apply Theorem 1 and Observation 1 to the

set $P \subset \mathbf{R}^{k+1}$ defined as follows:

$$P \triangleq \left\{ \begin{array}{l} [f_0(x(\cdot),u(\cdot),\xi(\cdot)) \ldots f_k(x(\cdot),u(\cdot),\xi(\cdot))] : \\ \{x(\cdot),u(\cdot),\xi(\cdot)\} \in \Omega \end{array} \right\}$$

From this definition and condition (15), it follows that the condition required by Theorem 1 is satisfied. Furthermore using the assumption that $d_s(x_0) > 0$ for all $x_0 \neq 0$ and all $s = 1,\ldots,k$, it follows that if we choose a vector function $[x(\cdot),u(\cdot),0] \in \Omega$ corresponding to an uncertainty input $\xi(t) \equiv 0$ and any initial condition $x(0) \neq 0$, the corresponding values of the functionals $f_1(\cdot),\ldots,f_k(\cdot)$ will satisfy $f_1(x(\cdot),u(\cdot),0) > 0, f_2(x(\cdot),u(\cdot),0) > 0,\ldots, f_k(x(\cdot),u(\cdot),0) > 0$. Hence, there will exist a point $x \in P$ such that $x_1 > 0,\ldots x_k > 0$.

Using Theorem 1 and Observation 1, it now follows that there exist constants $\tau_1 \geq 0,\ldots,\tau_k \geq 0$ such that $x_0 \geq \tau_1 x_1 + \cdots + \tau_k x_k$ for all $x = [x_0\ x_1\ \cdots\ x_k]' \in P$. Hence,

$$f_0(x(\cdot),u(\cdot),\xi(\cdot)) \geq \sum_{s=1}^{k} \tau_s f_s(x(\cdot),u(\cdot),\xi(\cdot)) \quad (16)$$

for all $[x(\cdot),u(\cdot),\xi(\cdot)] \in \Omega$. The inequality (16) may be rewritten as

$$\sum_{t=0}^{N} G_\delta(x(t),u(t),\xi_1(t),\ldots,\xi_k(t)) \leq B(x(0)) \quad (17)$$

where $G_\delta(\cdot,\cdot,\cdot,\ldots,\cdot)$ is defined by (10) and

$$B(x_0) \triangleq W(x_0) - \sum_{s=1}^{k} \tau_s d_s(x_0). \quad (18)$$

Since $W(0) = 0$ and $d_s(0) = 0$ for $s = 1,\ldots,k$, we have $B(0) = 0$. Also, it follows from the continuity of the functions $W(\cdot)$ and $d_s(\cdot)$ that $B(\cdot)$ is a continuous function. Furthermore, we have from the condition $g(\cdot,\cdot) \geq 0$ that $G_\delta(x,u,0) \geq 0$. Hence, the inequality (17) implies that $B(x_0) \geq 0$ for all $x_0 \in \mathbf{R}^n$.

In order to establish the existence of a suitable solution to the equation (9), we consider the system

$$\begin{aligned} x(t) &= \hat{b}(x(t),u(t),w(t)); \\ \hat{z}(t) &= l(x(t),u(t)) \end{aligned} \quad (19)$$

where

$$\begin{aligned} \hat{b}(x,u,w) &\triangleq b(x,u,\sqrt{\tau_1}\xi_1,\ldots,\sqrt{\tau_k}\xi_k), \\ w &\triangleq [\ \sqrt{\tau_1}\xi_1' \ \cdots \ \sqrt{\tau_k}\xi_k'\ ]', \end{aligned}$$

$$\hat{z} \triangleq \begin{bmatrix} \sqrt{(1+\delta)g(x,u)} \\ \sqrt{\tau_1}z_1 \\ \vdots \\ \sqrt{\tau_k}z_k \end{bmatrix},$$

$$l(x,u) \triangleq \begin{bmatrix} \sqrt{(1+\delta)g(x,u)} \\ \sqrt{\tau_1}l_1(x,u) \\ \vdots \\ \sqrt{\tau_k}l_k(x,u) \end{bmatrix}.$$

It is straightforward to establish that with this notation, the state equations (19) are equivalent to the state equations (1). Furthermore, condition (17) with the functions (10), (18) may be rewritten as

$$\|\hat{z}(\cdot)\|_2^2 - \|w(\cdot)\|_2^2 \leq B(x(0)) \quad (20)$$

for all solutions to the closed loop system (19), (6) with any input $w(\cdot)$. Hence for all uncertainty inputs $w(\cdot)$, the controller (6) solves a nonlinear H^∞ control problem defined by the system (19) and the H^∞ norm bound condition (20), see (Baras and James 1993). Hence, Theorem 3.2 of (Baras and James 1993) implies that the equation (12) with bound condition $V_\delta(N+1,\cdot) \equiv 0$ has a solution $V_\delta(t,x) \geq 0$ for $t = 0,1,\ldots,N$ such that

$$V_\delta(0,x_0) \leq B(x_0) \quad (21)$$

for all $x_0 \in \mathbf{R}^n$. From this and (18), the inequality (13) follows immediately.

Now observe that it follows from (20) that

$$\|\hat{z}(\cdot)\|_2^2 - \|w(\cdot)\|_2^2 \leq 0 \quad (22)$$

for all solutions to the system (19), (6) with initial condition $x(0) = 0$ and with any input $w(\cdot)$. Now let $x_0 \in \mathbf{R}^n$ be any initial condition and let $w_0(t)$, $t = 1,2,\ldots,N$ be a given input. We consider a new input $w_1(t)$ defined so that $w_1(0)$ is such that $\hat{b}(0,0,w_1(0)) = x_0$ and $w_1(t) = w_0(t)$ for $t = 1,2,3,\ldots,N$. Assumption 4.1 implies that for all $x_0 \in \mathbf{R}^n$, the set $\{w : \hat{b}(0,0,w) = x_0\}$ is non-empty. Corresponding to the input $w_1(\cdot)$, let $[x_1(\cdot),\hat{z}_1(\cdot)]$ be the solution to the system (19), (6) with initial condition $x_1(0) = 0$. Using Assumption 4.1 and the fact that $U(t,0) \equiv 0$, we can conclude that $x_1(1) = x_0$ and $\hat{z}_1(0) = 0$. Hence,

$$\begin{aligned} &\sum_{t=1}^{N} \left(\|\hat{z}(\cdot)\|^2 - \|w_0(\cdot)\|^2\right) \\ &= \sum_{t=0}^{N} \left(\|\hat{z}_1(\cdot)\|^2 - \|w_1(\cdot)\|^2\right) - \|\hat{z}_1(0)\|^2 \\ &\quad + \|w_1(0)\|^2 \\ &= \sum_{t=0}^{N} \left(\|\hat{z}_1(\cdot)\|^2 - \|w_1(\cdot)\|^2\right) + \|w_1(0)\|^2. \end{aligned}$$

Furthermore, from condition (22) it follows that $\sum_{t=0}^{N} \left(\|\hat{z}_1(\cdot)\|^2 - \|w_1(\cdot)\|^2\right) \leq 0$. Hence, the inequality

$$\sum_{t=1}^{N} \left(\|\hat{z}_1(\cdot)\|^2 - \|w_0(\cdot)\|^2\right) \leq \bar{B}(x(1)) \quad (23)$$

holds for all solutions to the closed loop system (19),

(6) with any input $w_0(\cdot)$ where

$$\bar{B}(x) \triangleq \inf\{\|w\|^2 : \hat{b}(0,0,w) = x\}. \tag{24}$$

Hence, using a similar argument to that involved in deriving (20), it follows that the solution to (12) is such that

$$V(1, x(\bar{t})) \leq \bar{B}(x(1)) \tag{25}$$

for all $x(1) \in \mathbf{R}^n$.

It follows immediately from the definition of $\bar{B}(x)$ that $\bar{B}(x) \geq 0$ for all $x \in \mathbf{R}^n$. Furthermore, since $b(0,0,0,\ldots,0) = 0$, it follows that $\bar{B}(0) = 0$. Now let $\xi_1^*(x), \ldots, \xi_k^*(x)$ be vector functions defined as in Assumption 4.1. Given any $x \in \mathbf{R}^n$, it follows that

$$\begin{aligned} x &= b(0,0,\xi_1^*(x),\ldots,\xi_k^*(x)) \\ &= \hat{b}\left(0,0,\left[\sqrt{\tau_0}\xi_1^*(x)' \ldots \sqrt{\tau_k}\xi_k^*(x)'\right]'\right). \end{aligned}$$

Hence,

$$\bar{B}(x) \leq \sum_{s=0}^{k} \tau_s \|\xi_s^*(x)\|^2$$

for all $x \in \mathbf{R}^n$. However as in Assumption 4.1, the functions $\xi_1^*(x), \ldots, \xi_k^*(x)$ are continuous and satisfy $\xi_1^*(0) = 0, \ldots, \xi_k^*(0) = 0$. Hence, the function $\bar{B}(x)$ must be continuous at $x = 0$.

We now consider a delayed version of the controller (6) applied to the system (19) over the time interval $t = \bar{t}, \bar{t}+1, \ldots, N$ where $\bar{t} \in \{1, 2, \ldots, N-1\}$. To investigate this situation, we suppose that $w(t) = 0$ for $t = N+1, N+2, \ldots, N+\bar{t}$. Using the time invariance of the system (19), this is equivalent to applying the original controller (6) to the system (19) over the time interval $t = 0, 1, \ldots, N$ and letting $w(t) = 0$ for $t = N - \bar{t}+1, N-\bar{t}+2, \ldots, N$. In this situation, we can then apply inequality (23). Using (23) and the time invariance of the system, it follows that if the delayed version of controller (6) is applied to the system (19) over the time interval $t = \bar{t}, \bar{t}+1, \ldots, N$, then

$$\sum_{t=\bar{t}+1}^{N+\bar{t}} \|\hat{z}(t)\|^2 - \sum_{t=\bar{t}+1}^{N} \|w(t)\|^2 \leq \bar{B}(x(\bar{t}+1)).$$

Therefore

$$\sum_{t=\bar{t}+1}^{N} \|\hat{z}(t)\|^2 - \sum_{t=\bar{t}+1}^{N} \|w(t)\|^2 \leq \bar{B}(x(\bar{t}+1)).$$

Hence, using a similar argument to that involved in the derivation of (20), it follows that the solution to (12) is such that

$$V(t, x(t)) \leq \bar{B}(x(t)) \tag{26}$$

for all $x(t) \in \mathbf{R}^n$ and $t = 1, 2, \ldots, N$. Furthermore, we recall that $\bar{B}(0) = 0$ and $\bar{B}(\cdot)$ is continuous. Hence, $V_\delta(t, 0) = 0$ and $V_\delta(t, \cdot)$ is continuous at the origin, for $t = 1, 2, \ldots, N$. Also, it follows from (21) that $V_\delta(0,0) = 0$ and $V_\delta(0, \cdot)$ is continuous at the origin.

We will now show that for $x = 0$, the infimum in (9) is achieved at $u = 0$. Indeed, if we substitute $x = 0$ and $u = 0$ into the right hand side of (9) and use (10) and the properties of the functions $g(\cdot,\cdot)$ and $l_s(\cdot,\cdot)$, it follows that

$$\begin{aligned} &\sup_{\xi} \left\{ V_\delta(t+1, b(0,0,\xi)) + G_\delta(0,0,\xi) \right\} \\ &= \sup_{\xi} \left\{ \begin{array}{l} V_\delta(t+1, b(0,0,\xi)) \\ -\sum_{s=0}^{k} \tau_s \|\xi_s\|^2 \end{array} \right\}. \end{aligned} \tag{27}$$

However, it follows from (24) and inequality (26) that $V(t,x) \leq \|w\|^2$ for $t = 0, 1, \ldots, N$ and for all w such that $\hat{b}(0,0,w) = x$. Hence, $V_\delta(t+1, b(0,0,\xi_1,\ldots,\xi_k)) \leq \sum_{s=0}^{k} \tau_s \|\xi_s\|^2$ for $t = 0, 1, \ldots, N-1$ and for all $\xi_1 \in \mathbf{R}^{r_0}, \ldots, \xi_k \in \mathbf{R}^{r_k}$. Thus, it follows from (27) that

$$\sup_{\xi} \{V_\delta(t+1, b(0,0,\xi)) + G_\delta(0,0,\xi)\} \leq 0.$$

Moreover, we have already established that $V_\delta(t, 0) = 0$ for $t = 0, 1, \ldots, N$ and thus for $x = 0$, $u = 0$ must achieve the infimum in (9). This completes the proof of this part of the theorem.

$(ii) \Rightarrow (i)$ Suppose condition (ii) holds and consider the system defined by state equations (19). Also, let $\delta \in (0, \delta^*)$ be given. We will construct a corresponding control of the form (11) and a function $H_\delta(t,x)$ satisfying inequality (12) as follows: Choose any $\delta_1 \in (\delta, \delta^*]$ and let $H_\delta(t,x) \triangleq V_{\delta_1}(t,x)$ where $V_{\delta_1}(t,x)$ is the solution to equation (9) corresponding to $\delta_1 \in (0, \delta^*]$. Now, given any non-zero $x \in \mathbf{R}^n$ and any $t \in \{0, 1, \ldots, N\}$, choose $\varepsilon > 0$ such that

$$\varepsilon < (\delta_1 - \delta) \inf_{u \in \mathbf{R}^m} g(x,u).$$

It follows from (9) that $u^*(t,x)$ can be chosen so that

$$\begin{aligned} &V_{\delta_1}(t,x) \\ &\geq \sup_{\xi} \left\{ \begin{array}{l} V_{\delta_1}(t+1, b(x, u^*(t,x), \xi)) \\ +G_{\delta_1}(x, u^*(t,x), \xi) \end{array} \right\} - \varepsilon \end{aligned}$$

Hence,

$$\begin{aligned} &H_\delta(t,x) \\ &\geq \sup_{\xi} \left\{ \begin{array}{l} H_\delta(t+1, b(x, u^*(t,x), \xi)) \\ +G_\delta(x, u^*(t,x), \xi) \end{array} \right\} \\ &\qquad +(\delta_1 - \delta) g(x, u^*(t,x)) - \varepsilon \\ &> \sup_{\xi} \left\{ \begin{array}{l} H_\delta(t+1, b(x, u^*(t,x), \xi)) \\ +G_\delta(x, u^*(t,x), \xi) \end{array} \right\}. \end{aligned}$$

That is, with this definition of $u^*(t,x)$, $H_\delta(t,x)$ satisfies inequality (12) for all $x \neq 0$ and $t = 0, 1, \ldots, N$. Furthermore, it follows from the definition of $H_\delta(t,x)$ and the properties of $V_\delta(t,x)$ given in condition (ii) that $H_\delta(t,x) \geq 0$, $H_\delta(0,0) \equiv 0$, $H_\delta(0,\cdot)$ is continuous at the origin and inequality (14) is satisfied.

We now suppose $x = 0$. In this case, it follows from condition (ii) that $u = 0$ achieves the infimum in (9) for all t. Hence,

$$\begin{aligned} H_\delta(t,0) &= V_{\delta_1}(t,0) \\ &= \sup_\xi \left\{ \begin{array}{l} V_{\delta_1}(t+1, b(0,0,\xi)) \\ +G_{\delta_1}(0,0,\xi) \end{array} \right\} \\ &= \sup_\xi \left\{ \begin{array}{l} H_\delta(t+1, b(0,0,\xi)) \\ +G_\delta(0,0,\xi) \end{array} \right\}. \end{aligned}$$

That is for this case, if we let $u^*(t,0) \equiv 0$, $H_\delta(t,x)$ satisfies (12) with equality when $x = 0$.

Note, the above proof of the existence of a function $H_\delta(t,x)$ satisfying the dynamic programming inequality (12) amounts to a version of the Bounded Real Lemma for nonlinear systems.

We will now prove that the controller $u^*(t,x)$ constructed as above is a guaranteed cost controller for the uncertain system (1), (3). Indeed, using Theorem 3.3 of (Baras and James 1993), it follows that

$$\|\hat{z}(\cdot)\|_2^2 - \|w(\cdot)\|_2^2 \leq V(0, x(0)) \tag{28}$$

for all solutions of the closed loop system (19), (11). Furthermore, using the notation given above, inequality (28) may be rewritten as

$$\begin{aligned} &(1+\delta)F_0\left(x(\cdot), u(\cdot)\right) + \sum_{s=1}^{k} \tau_s(\|z_s(\cdot)\|_2^2 - \|\xi_s(\cdot)\|_2^2) \\ &\leq H_\delta(0, x(0)) \end{aligned} \tag{29}$$

for all solutions to the closed loop system (1), (11). Combining this inequality with inequalities (5) and (14) we obtain

$$\begin{aligned} &(1+\delta)\sum_{i=1}^{q} F_0\left(x^i(\cdot), u^i(\cdot), \xi^i(\cdot)\right) \\ &-\sum_{s=1}^{k} \tau_s \sum_{i=1}^{q} d_s(x^i(0)) \\ &\leq \sum_{i=1}^{q} H_\delta(0, x^i(0)) \\ &\leq \sum_{i=1}^{q} W(x^i(0)) - \sum_{s=1}^{k} \tau_s \sum_{i=1}^{q} d_s(x^i(0)). \end{aligned}$$

for all solutions to the closed loop uncertain system (1), (3), (11). Hence,

$$\frac{\sum_{i=1}^{q} F_0\left(x^i(\cdot), u^i(\cdot), \xi^i(\cdot)\right)}{\sum_{i=1}^{q} W(x^i(0))} \leq \frac{1}{1+\delta} < 1.$$

for all $S \in \Xi$. This leads directly to the satisfaction of condition (7). Thus, the controller (11) is a guaranteed cost controller. This completes the proof of the theorem. □

Acknowledgement This work was supported by the Australian Research Council.

REFERENCES

Baras, J. S. and M. R. James (1993). Robust output feedback control for discrete-time nonlinear systems: The finite time case. In 'Proceedings of the 32nd IEEE Conference on Decision and Control'. San Antonio, Tx. pp. 51–55.

Bernstein, D. S. and W. M. Haddad (1988). 'The optimal projection equations with Petersen-Hollot bounds: Robust stability and performance via fixed order dynamic compensation for systems with structured real valued parameter uncertainty'. *IEEE Transactions on Automatic Control* **33**(6), 578–582.

Bryson, A. E. and Y. C. Ho (1975). *Applied Optimal Control*. Hemisphere. New York.

Chang, S. S. L. and T. K. C. Peng (1972). 'Adaptive guaranteed cost control of systems with uncertain parameters'. *IEEE Transactions on Automatic Control* **AC-17**(4), 474–483.

Petersen, I. R. and D. C. McFarlane (1992). Optimal guaranteed cost control of uncertain linear systems. In 'Proceedings of the 1992 American Control Conference'. Chicago, Illinois. pp. 2929–2930.

Savkin, A. and I.R. Petersen (1994*a*). An uncertainty averaging approach to optimal guaranteed cost control of uncertain systems with structured uncertainty. In 'Proceedings of the Workshop on Robust Control via Variable Structure and Lyapunov Techniques'. Benevento, Italy.

Savkin, A. V. and I. R. Petersen (1994*b*). 'A connection between H^∞ control and the absolute stabilizability of uncertain systems'. *Systems and Control Letters* **23**(3), 197–203.

Savkin, A. V. and I. R. Petersen (1995*a*). Minimax optimal control of uncertain systems with structured uncertainty. Accepted for publication in the International Journal of Robust and Nonlinear Control.

Savkin, A. V. and I. R. Petersen (1995*b*). Nonlinear versus linear control in the absolute stabilizability of uncertain linear systems with structured uncertainty. Accepted for publications in the IEEE Transactions on Automatic Control.

Smith, R. S. and J. C. Doyle (1992). 'Model validation: A connection between robust control and identification'. *IEEE Transactions on Automatic Control* **37**(7), 942–952.

Yakubovich, V. A. (1971). 'S-procedure in nonlinear control theory'. *Vestnik Leningrad University, Series 1* **13**(1), 62–77.

Yakubovich, V. A. (1988*a*). 'Absolute stability of nonlinear systems with a periodically nonstationary linear part'. *Soviet Physics Doklady* **32**(1), 5–7.

Yakubovich, V. A. (1988*b*). 'Dichotomy and absolute stability of nonlinear systems with periodically nonstationary linear part'. *Systems and Control Letters* **11**(3), 221–228.

A CONNECTION BETWEEN $\mathcal{H}_\infty$ CONTROL AND THE ABSOLUTE STABILIZABILITY OF DISCRETE-TIME UNCERTAIN LINEAR SYSTEMS

S. O. Reza Moheimani*, Andrey. V. Savkin†, and Ian R. Petersen*

*Department of Electrical Engineering, Australian Defence Force Academy, Campbell, 2600, Australia, reza@ee.adfa.oz.au.
†Department of Electrical and Electronic Engineering and Cooperative Research Center for Sensor Signal and Information Processing, University of Melbourne, Parkville, Victoria 3052, Australia.

Abstract. L'article considère un problème de stabilisation absolu d'un système incertain. La solution invellope un problème de commande $\mathcal{H}_\infty$.

Keywords. Uncertain systems, robust control, $\mathcal{H}_\infty$ control, absolute stability

1. INTRODUCTION

One of the important problems in control theory and practice is the problem of stabilizing a system with uncertain parameters. So far, a number of methods have been proposed, which differ in the notion of stability and the form of uncertainty considered. One of these methods is based on the idea of quadratic stability. The uncertainty is assumed to be norm bounded and the idea is to find a common quadratic Lyapunov function for the closed loop uncertain system e.g., see (Khargonekar *et al.* 1990), (Petersen *et al.* 1993), (Packard and Doyle 1990).

In some more recent papers a broader class of uncertain systems were considered; e.g., see (Savkin and Petersen 1994), (Savkin and Petersen 1993). Here the uncertainty is assumed to satisfy a certain *Integral Quadratic Constraint.* This class of uncertain systems was first considered by V. A. Yakubovich and is studied extensively in the Russian literature; e.g. see (Krasovskii 1987), (Savkin 1991), (Yakubovich 1977), (Yakubovich 1988*b*), (Yakubovich 1988*a*) and also (Boyd *et al.* 1994). This class of uncertainties is richer than the class of uncertainties satisfying the corresponding norm bounded condition since it allows for nonlinear, time-varying, dynamic uncertainties.

The main result of (Savkin and Petersen 1994) shows that if an uncertain system can be absolutely stabilized via nonlinear output feedback control, then it can be absolutely stabilized via linear output feedback control. It also shows that the problem of absolutely stabilizing an uncertain system is equivalent to an output feedback $\mathcal{H}_\infty$ control problem, the solution of which is well known.

In practice, if one wants to implement these techniques via computer control, it is desirable to use the discrete-time version of these results. This motivates the main result of this paper which is a discrete-time version of the main result of (Savkin and Petersen 1994).

The remainder of the paper proceeds as follows. In Section 2 of the paper, we define the class of uncertain discrete-time systems under consideration. We require the uncertainty to satisfy a certain *Sum Quadratic Constraint* which is the discrete-time version of the Integral Quadratic Constraint. We also introduce a definition of *absolute stabilizability* which is the discrete-time version of the notion of absolute stability given in (Savkin and Petersen 1994). In Section 3, we present our main result, which shows that problem of output feedback absolute stabilizability of a discrete-time system is equivalent to a standard discrete-time $\mathcal{H}_\infty$ control problem. This result also shows that if the problem can be solved via nonlinear control then it can be solved via linear control.

2. ABSOLUTE STABILITY OF DISCRETE-TIME SYSTEMS

Consider the following discrete-time uncertain system

$$\begin{aligned} x(k+1) &= Ax(k) + Bu(k) + Dw(k); \\ z(k) &= Hx(k) + Gu(k); \\ y(k) &= Cx(k) + Ew(k). \end{aligned} \tag{1}$$

Here $x(k) \in \mathbf{R}^n$ is the state, $w(k) \in \mathbf{R}^p$ is the uncertainty input, $z(k) \in \mathbf{R}^q$ is the *uncertainty output* and $y(t) \in \mathbf{R}^l$ is the *measured output.*

The uncertainty in the above system is considered to be of the form

$$w(k) = \phi(k, x(\cdot)|_0^k, u(\cdot)|_0^k). \tag{2}$$

The uncertainty described by the above equation is required to satisfy a certain *Sum Quadratic Constraint,* described below. This is the discrete-time version of the integral quadratic constraint given in (Yakubovich 1977), (Yakubovich 1988*b*), (Yakubovich 1988*a*), (Savkin and Petersen 1994).

Definition 1 (Sum Quadratic Constraint; see also (Moheimani *et al.* 1994)) An uncertainty of the form (2) is an admissible uncertainty for the system (1) if given any control input $u(\cdot)$ and any corresponding solution to equation (1), (2), then there exists a sequence $\{k_i\}_{i=1}^{\infty}$ and a constant $d \geq 0$ such that $k_i \to \infty$, $k_i \geq 0$ and

$$\sum_{k=0}^{k_i} \|w(k)\|^2 \leq d + \sum_{k=0}^{k_i} \|z(k)\|^2 \qquad \forall i. \tag{3}$$

Here $\|\cdot\|$ denotes the Euclidean norm. It is also noted that k_i may be equal to infinity.

Remarks Given an admissible uncertainty of the form (2), then for any solution to the equations (1), (2), there will be a corresponding parameter $d \geq 0$ such that condition (3) is satisfied. Parameter d is a *measure of mismatch* between the resulting uncertainty input $w(\cdot)$ and the *Homogeneous Sum Constraint:* $\sum_{k=0}^{k_i} \|w(k)\|^2 \leq \sum_{k=0}^{k_i} \|z(k)\|^2$ for all i. Also, note that the description of uncertainty given above allows for the uncertainty input $w(k)$ to depend dynamically on $u(k)$, and $x(k)$. It can also be shown that the uncertain system (1), (2) includes uncertain systems in which the uncertainty satisfies a norm bounded condition:

$$\begin{aligned} x(k+1) &= [A + D\Delta(k)H]x(k) \\ &\quad +[B + D\Delta(k)G]u(k); \\ y(k) &= [C + E\Delta(k)H]x(k) \\ &\quad +E\Delta(k)Gu(k) \end{aligned} \tag{4}$$

where $\Delta(k)$ is the time-varying *matrix of uncertain parameters* satisfying $\|\Delta(k)\| \leq 1$; see (Savkin and Petersen 1994) for further details.

We are now able to introduce the definitions of absolute stability and stabilizability for the uncertain system (1), (2); see also (Savkin and Petersen 1994), (Yakubovich 1977), (Yakubovich 1988*b*), (Yakubovich 1988*a*).

Definition 2 Consider the uncertain system (1), (2) with control input $u(k) \equiv 0$. This uncertain system is said to be *absolutely stable* if there exists a constant $c > 0$ such that given any admissible uncertainty for the uncertain system (1), (2), then any solution to the equations (1), (2) satisfies $[x(\cdot), w(\cdot)] \in \mathbf{l}_{2+}$, and

$$\sum_{k=0}^{\infty}(\|x(k)\|^2 + \|w(k)\|^2) \leq c(\|x_0\|^2 + d). \tag{5}$$

Here $\mathbf{l}_{2+}$ denotes the Hilbert space of vector sequences $\{x(k)\}_{k=0}^{\infty}$ such that $\sum_{k=0}^{\infty} \|x(k)\|^2 < \infty$.

Remarks: Condition (5) of Definition 2 requires that the $\mathbf{l}_{2+}$ norm of any solution to (1) and (2) be bounded in terms of the initial condition and the parameter d.

In this paper, we consider the problem of absolute stabilization of the discrete-time uncertain system (1), (2) via a causal output feedback of the form

$$u(k) = U(k, y(\cdot)|_0^k) \tag{6}$$

or a linear output feedback controller of the form

$$\begin{aligned} \hat{x}(k+1) &= \hat{A}\hat{x}(k) + \hat{B}y(k); \\ u(k) &= \hat{C}\hat{x}(k); \\ \hat{x}(0) &= 0. \end{aligned} \tag{7}$$

Definition 3 The uncertain system (1), (2) is said to be *absolutely stabilizable via nonlinear control* if there exists an output feedback controller of the form (6) and a constant $c > 0$ such that for any given initial condition $x(0) = x_0$ and any admissible uncertainty $w(\cdot)$ for the uncertain system (1), (2), then any solution to the equations (1), (2), (6) satisfies $[x(\cdot), u(\cdot), w(\cdot)] \in \mathbf{l}_{2+}$ and

$$\begin{aligned} &\sum_{k=0}^{\infty}(\|x(k)\|^2 + \|u(k)\|^2 + \|w(k)\|^2) \\ &\leq \; c(\|x_0\|^2 + d). \end{aligned} \tag{8}$$

Definition 4 The uncertain system (1), (2) is said to be *absolutely stabilizable via linear control* if there exists a linear output feedback controller of the form (7) and a constant $c > 0$ such that given any admissible uncertainty $w(\cdot)$ for the uncertain system (1), (2), then any solution to the equations (1), (2), (7) satisfies $[x(\cdot), u(\cdot), w(\cdot)] \in \mathbf{l}_{2+}$ and condition (8) is satisfied.

Observation 1 If the discrete-time uncertain system (1), (2) is absolutely stabilizable, then the corresponding closed loop uncertain system will have the property that $[x(k), u(k), w(k)] \to 0$ as $k \to \infty$ for any admissible uncertainty $w(\cdot)$ since $[x(\cdot), u(\cdot), w(\cdot)] \in \mathbf{l}_{2+}$.

3. THE PROBLEM OF ABSOLUTE STABILIZATION AND THE $\mathcal{H}_\infty$ CONTROL PROBLEM

In this section we consider the problem of absolutely stabilizing the discrete-time uncertain system (1), (2). We will show that this problem is equivalent to a standard discrete-time $\mathcal{H}_\infty$ control problem. In this $\mathcal{H}_\infty$ control problem, the underlying linear system is described by the state-space equations (1). However, in this case, $w(k) \in \mathbf{R}^p$ is regarded as the *disturbance input* and $z(k) \in \mathbf{R}^q$ is regarded as the *error output.* The $\mathcal{H}_\infty$ controller is to be designed such that

$$J = \sup_{w(\cdot)\in \mathbf{l}_{2+}, x(0)=0} \frac{\|z(\cdot)\|_2^2}{\|w(\cdot)\|_2^2} < 1. \tag{9}$$

Here, $\|\cdot\|_2$ denotes the $\mathbf{l}_2$ norm.

A controller (6) is said to solve the $\mathcal{H}_\infty$ control problem defined by (1), (9) if the closed loop system is asymptotically stable with $w(k) \equiv 0$, and the inequality (9) is satisfied.

The following theorem is the main result of this paper. It shows that the problem of absolute stabilization of a discrete-time system of the form (1), (2) is equivalent to a discrete-time $\mathcal{H}_\infty$ control problem.

Theorem 1 Consider the uncertain system (1), (2) and assume that (A, D) is controllable, (A, H) is observable, $H'H = Q$, $E'E = N > 0$, $G'G = R > 0$, $H'G = P$, and $DE' = L$. Then the following statements are equivalent:

1. The uncertain system (1), (2) is absolutely stabilizable via a nonlinear control of the form (6).

2. The controller (6) solves the $\mathcal{H}_\infty$ control problem (1), (2).

3. Let $\bar{A} = A - BR^{-1}P$, and $\bar{Q} = Q - PR^{-1}P'$. There exist solutions $M > 0$ and $\Sigma > 0$ to the Riccati equations

$$\begin{aligned} M &= \bar{A}'(M^{-1} + BR^{-1}B' - DD')^{-1}\bar{A} \\ &\quad +\bar{Q}; \\ \Sigma &= \tilde{A}(\Sigma^{-1} + C'N^{-1}C - Q)^{-1}\tilde{A}' \\ &\quad +DD' - LN^{-1}L' \end{aligned} \qquad (10)$$

which satisfy the condition $\rho(\Sigma M) < 1$. Here $\rho(\cdot)$ denotes the spectral radius.

4. The linear controller (7) with

$$\begin{aligned} \hat{A} &= A - (B + \tilde{A}\Theta^{-1}H'G)R^{-1} \\ &\quad \times(B'\Gamma^{-1}\bar{A} + P')(I - \Sigma M)^{-1} \\ &\quad +\tilde{A}\Theta^{-1}(H'H - C'N^{-1}C) \\ &\quad -GN^{-1}C \\ \hat{B} &= (\tilde{A}\Theta^{-1}C' + G)N^{-1}; \\ \hat{C} &= -R^{-1}(B'\Gamma^{-1}\bar{A} + P') \\ &\quad \times(I - \Sigma M)^{-1} \end{aligned} \qquad (11)$$

where

$$\begin{aligned} \Gamma &:= M^{-1} + BR^{-1}B' - DD'; \\ \Theta &:= \Sigma^{-1} + C'N^{-1}C - Q; \\ \Pi &:= DD' - LN^{-1}L'; \\ \tilde{A} &= A - LN^{-1}C \end{aligned}$$

solves the $\mathcal{H}_\infty$ control problem (1), (9).

5. The uncertain system (1), (2) is absolutely stabilizable via the linear controller (7), (11).

Proof $(i) \Rightarrow (ii)$: We need the following claim for the proof.
Claim: If the controller (6) is absolutely stabilizing for the uncertain system (1), (2), then there exists a constant $\delta > 0$ such that

$$\|z(\cdot)\|_2^2 \leq (1 - \delta)\|w(\cdot)\|_2^2 \qquad (12)$$

for all solutions of the closed loop system (1), (2), (6) with $x(0) = 0$ and $w(\cdot) \in \mathbf{l}_{2+}$.

To prove the claim, let

$$d \triangleq \begin{cases} -(\|z(\cdot)\|_2^2 - \|w(\cdot)\|_2^2) & \text{if } \|z(\cdot)\|_2^2 - \|w(\cdot)\|_2^2 < 0; \\ 0 & \text{if } \|z(\cdot)\|_2^2 - \|w(\cdot)\|_2^2 \geq 0. \end{cases} \qquad (13)$$

With d defined as above, it can be seen that for all $w(\cdot) \in \mathbf{l}_{2+}$, the corresponding solution to the closed loop system (1), (2), (6) will satisfy condition (3) with $k_i = \infty$. Therefore, with $x(0) = 0$, condition (8) implies that there exists a constant $c > 0$ such that

$$\|x(\cdot)\|_2^2 + \|u(\cdot)\|_2^2 + \|w(\cdot)\|_2^2 \leq cd \qquad (14)$$

for all solutions of the closed loop system (1), (6) with $x(0) = 0$ and $w(\cdot) \in \mathbf{l}_{2+}$.

We now consider two cases:
Case 1: $\|z(\cdot)\|_2^2 - \|w(\cdot)\|_2^2 \geq 0$. In this case $d = 0$ and (13) imply that $x(k) \equiv 0$, $u(k) \equiv 0$, and $w(k) \equiv 0$. Hence inequality (12) will be satisfied.
Case 2: $\|z(\cdot)\|_2^2 - \|w(\cdot)\|_2^2 < 0$. In this case (13) and (14) imply that

$$\|w(\cdot)\|_2^2 \leq -c(\|z(\cdot)\|_2^2 - \|w(\cdot)\|_2^2)$$

which is equivalent to

$$\|z(\cdot)\|_2^2 - \|w(\cdot)\|_2^2 \leq -c^{-1}\|w(\cdot)\|_2^2.$$

Therefore, condition (12) is satisfied with $\delta = c^{-1}$. Thus the claim has been established in both cases.

Now since (6) is an absolutely stabilizing controller for the uncertain system (1), (2), inequality (12) implies that this controller solves the $\mathcal{H}_\infty$ control problem (1), (9).
$(ii) \Rightarrow (iii)$: This follows directly from Theorems 6.4 and 6.5 of (Basar and Bernhard 1991).
$(iii) \Rightarrow (iv)$: This is a standard result from $\mathcal{H}_\infty$ control theory; e.g., see Section 6.4.1 of (Basar and Bernhard 1991).
$(iv) \Rightarrow (v)$: Assume that the linear controller (7), (11) solves the discrete-time $\mathcal{H}_\infty$ control problem (1), (9), we will show that the same controller solves the absolute stabilization problem for the uncertain system (1), (2).

The closed loop uncertain system defined by (1), (2), (7), and (11) can be rewritten as

$$\xi(k+1) = U\xi(k) + Vw(k); \quad z(k) = T\xi(k) \qquad (15)$$

where

$$\begin{aligned} \xi &= \begin{bmatrix} x \\ \hat{x} \end{bmatrix}, \\ U &= \begin{bmatrix} A & B\hat{C} \\ \hat{B}C & \hat{A} \end{bmatrix}, \\ V &= \begin{bmatrix} D \\ \hat{B}E \end{bmatrix}, \\ T &= \begin{bmatrix} H & G\hat{C} \end{bmatrix}. \end{aligned}$$

Since the controller (7) solves the $\mathcal{H}_\infty$ problem (1),

(9), the matrix U is stable and

$$\|T(zI-U)^{-1}V\|_{\infty} < 1.$$

Here, $\|\cdot\|_{\infty}$ denotes the discrete-time $\mathcal{H}_{\infty}$ norm; i.e., $\|H(z)\|_{\infty} \triangleq \sup_{z\in C,|z|=1} \|H(z)\|$. Therefore, there exists a constant $\delta > 0$ such that

$$\left\| \begin{bmatrix} T \\ \delta I \\ 0 \end{bmatrix} (zI-U)^{-1}V + \begin{bmatrix} 0 \\ 0 \\ \delta I \end{bmatrix} \right\|_{\infty} \leq 1.$$

This inequality implies that

$$\begin{aligned} &\sum_{k=0}^{\infty}(\|z(k)\|^2 - \|w(k)\|^2) \\ &\leq -\delta^2 \sum_{k=0}^{\infty}(\|\xi(k)\|^2 + \|w(k)\|^2) \end{aligned}$$

for all $[\xi(\cdot), w(\cdot)] \in \mathbf{l}_{2+}$ connected by (15) with $\xi(0) = 0$. This last inequality, along with the stability of the matrix U and results of (Yakubovich 1977), which are applicable to both continuous and discrete-time systems, imply that the uncertain system (15) is absolutely stable. Now, since $\hat{x}(0) = 0$, it follows from Definition 2 that for any initial condition $x(0) = x_0$ and any admissible uncertainty $w(\cdot)$, then $[x(\cdot), \hat{x}(\cdot), w(\cdot)] \in \mathbf{l}_{2+}$ and $\sum_{k=0}^{\infty}(\|x(k)\|^2 + \|\hat{x}(k)\|^2 + \|w(k)\|^2) \leq c(\|x_0\|^2 + d)$. The above inequality implies that for any initial condition $x(0) = x_0$ and any admissible uncertainty $w(\cdot)$, $[x(\cdot), u(\cdot), w(\cdot)] \in \mathbf{l}_{2+}$ and $\sum_{k=0}^{\infty}(\|x(k)\|^2 + \|u(k)\|^2 + \|w(k)\|^2) \leq (\|\hat{C}\|^2 + 1)c(\|x_0\| + d)$ which guarantees that the system (1), (2) is absolutely stabilizable using the linear controller (7), (11).
$(v) \Rightarrow (i)$: This is obvious from the definitions.
◇◇◇

Corollary 1 The uncertain system with norm bounded uncertainty, (4) will be absolutely stabilizable via the linear controller (7), (11) if condition (iii) of Theorem 1 is satisfied.

REFERENCES

Basar, T. and P. Bernhard (1991). *$\mathcal{H}_{\infty}$-Optimal control theory and related minimax design problems: A dynamic game approach.* Birkhauser. Boston.

Boyd, S., L. El Ghaoui, E. Feron and V. Balakrishnan (1994). *Linear Matrix Inequalities in System and Control Theory.* SIAM. Philadelphia, PA.

Khargonekar, P. P., I. R. Petersen and K. Zhou (1990). 'Robust stabilization of uncertain systems and $\mathcal{H}_{\infty}$ optimal control'. *IEEE Transactions on Automatic Control* **AC-35**(3), 356–361.

Krasovskii, A. A. (1987). *Handbook of automatic control theory.* Nauka. Moscow.

Moheimani, S. O. R., A. V. Savkin and I. R. Petersen (1994). Minimax optimal control of discrete-time uncertain systems with structured uncertainty. Submitted for publication.

Packard, A. and J. C. Doyle (1990). 'Quadratic stability with real and complex perturbations'. *IEEE Transactions on Automatic Control* **AC-35**(2), 198–201.

Petersen, I. R., D. C. McFarlane and M. A. Rotea (1993). Optimal guaranteed cost control of discrete-time uncertain linear systems. In 'Proceedings of 12th IFAC world congress, Sydney, Australia'. pp. vol–1,407–410.

Savkin, A. V. (1991). 'Absolute stability of nonlinear control systems with nonstationary linear part'. *Automation and Remote Control* **3**(41), 362–367.

Savkin, A. V. and I. R. Petersen (1993). Nonlinear versus linear control in the absolute stabilizability of uncertain linear systems with structured uncertainty. In 'Proceedings of the 32nd IEEE Conference on Decision and Control'.

Savkin, A. V. and I. R. Petersen (1994). A connection between $\mathcal{H}_{\infty}$ control and the absolute stabilizability of uncertain systems. To appear in *Systems and Control Letters.*

Yakubovich, V. A. (1977). 'Abstract theory of absolute stability of nonlinear systems'. *Vestnik Leningrad University, Series 1* **13**(41), 99–118.

Yakubovich, V. A. (1988*a*). 'Absolute stability of nonlinear systems with a periodically nonstationary linear part'. *Soviet Physics Doklady* **1**(32), 5–7.

Yakubovich, V. A. (1988*b*). 'Dichotomy and absolute stability of nonlinear systems with periodically nonstationary linear part'. *Systems and Control Letters* **3**(11), 221–228.

QUADRATIC STABILIZATION OF UNCERTAIN LINEAR SYSTEMS BY UNKNOWN INPUT OBSERVER

M. HAYAR, M. ZASADZINSKI, M. DAROUACH
CRAN - ACS - CNRS URA 821 - Université Henri Poincaré-Nancy I - 186 rue de Lorraine, 54400 Cosnes-et-Romain, FRANCE

Abstract In this note, we present a feedback control law, based on unknown input observer, to stabilize a class of uncertain linear systems. Necessary and sufficient conditions for the existence of an unknown input observer-based controller are given. We show that the use of an unknown input observer-based controller permits to recover exactly the disturbance attenuation given by state feedback. This property is employed to the stabilization of uncertain linear systems. The use of a high gain and the reduction of the uncertainty bound are avoided.

Key Words Quadratic stabilization, Uncertain systems, Loop transfer recovery, Unknown input observer

Résumé Dans cet article, nous présentons une loi de commande, basée sur un observateur à entrées inconnues, afin de stabiliser une classe de systèmes incertains. Les conditions nécessaires et suffisantes pour l'existence de l'observateur à entrées inconnues sont données. Nous montrons que ce type d'observateur permet de restituer exactement le niveau d'atténuation de perturbations obtenu avec une loi de commande par retour d'état. Cette propriété est utilisée pour la stabilisation de systèmes incertains par retour de sortie, sans observateur à grand gain ni réductions des bornes des incertitudes.

Mots clés Stabilisation quadratique, Systèmes incertains, Restitution du transfert de boucle, Observateur à entrées inconnues

1. INTRODUCTION

The problem of designing stabilizing controller for uncertain systems has been received considerable attention in the last decade. Many design strategies based on algebraic Riccati equation to stabilize uncertain systems has been developed [3], [4], [5]. The notion of quadratic stabilization to deal with time-varying uncertainties has been used. Khargonekar et al [3] showed that quadratic stabilization problem is an H_∞ problem. Petersen et al [1] employed disturbance attenuation problem to stabilize uncertain systems via an output feedback controller.

In this note, we give a controller based on an unknown input observer to stabilize a class of uncertain systems. This paper is organized as follows. In section 2, we review the Riccati equation approach to the design of the stabilizing control law when the complete state vector is available and we give the necessary and sufficient conditions for the existence of unknown input observer-based controller. In section 3, we use the disturbance attenuation problem to stabilize uncertain systems. The advantage of this work is to avoid high gain and reduction of uncertainty magnitude. We can use full or reduced-order observer in the synthesis of an unknown input observer-based controller. We can achieve other performances, using the free location of the observer's poles, furthermore of the disturbance attenuation and stabilization of uncertain systems.

2. PROBLEM STATEMENT AND PRELIMINARIES

2.1. Quadratic stabilization by state feedback control

Consider the following uncertain system

$$\dot{x}(t) = (A + \Delta A(t))\, x(t) + B\, u(t) \qquad (1)$$
$$y(t) = C\, x(t)$$

where $x(t) \in \Re^n$, $u(t) \in \Re^q$, $y(t) \in \Re^p$ are respectively the state, the control input, the measured output. A, B and C are constant matrices with appropriate dimensions, they represent the nominal system. The uncertainty matrix $\Delta A(t)$ is defined as

$$\Delta A(t) = D\, \Delta(t)\, E_x$$

where D, E_x are known real matrices and $\Delta(t)$ is an unknown matrix satisfying $\Delta(t)^T \Delta(t) \leq I$.

In the sequel we adopt the following assumptions :

A1- (A,B) is stabilizable and (A,C) is detectable

A2- $\operatorname{rank} \begin{bmatrix} sI_n - A & D \\ C & 0 \end{bmatrix} = n + \operatorname{rank} D,\ \forall\, s \in \mathbb{C}\ \operatorname{Re}(s) \geq 0$

A3- rank CD = rank D

Assumption A1 is required for the existence of a stabilizing controller, assumptions A2 and A3 are necessary to design the unknown input observer.

Definition 1 [1] The uncertain system (1) (with u = 0) is said to be quadratically stable if there exists a positive definite symmetric matrix P and a positive constant a such that, for any admissible uncertainty $\Delta A(t)$, the derivative of the Lyapunov function $V(x) = x^T P x$ satisfies the bound

$$\dot{V}(x) = 2\, x^T P\, (A + \Delta A(t))x \leq -\alpha \|x\|^2$$
($\|.\|$ denotes the euclidean norm) ❐

Definition 2 [2] The uncertain linear system (1) is said to be quadratically stabilizable if there exists a feedback matrix $K \in \Re^{n \times q}$ such that the resulting closed-loop system is quadratically stable. ❐

The following lemma connects the notion of quadratic stability to H_∞ optimization problem.

Lemma 1 [3] The uncertain system (1) is quadratically stable if and only if it satisfies the following conditions

(i) A is a stability matrix
(ii) $\| E_x(sI_n-A)^{-1}D\|_\infty < 1$ ❑

Necessary and sufficient conditions for the quadratic stabilizability of system (1) via a state feedback are given by Petersen [4], [5] and Khargonekar et al. [3]. The following lemma provides an algorithm for designing a feedback control law to stablilize system (1).

Lemma 2 [5] Let $Q \in \Re^{n\times n}$, $R \in \Re^{q\times q}$ be given positive-definite symmetric weigthting matrices, the uncertain linear system (1) is quadratically stabilizable by a state feedback control if there exists a real ε>0 such that the following Riccati equation has a symmetric and positive definite solution

$$A^TP+PA+\varepsilon PDD^TP-PBR^{-1}B^TP+\frac{1}{\varepsilon}E_x^TE_x+Q=0 \quad (2)$$

The gain matrix is given by

$$K=R^{-1}B^TP \quad (3)❑$$

2.2. Unknown input observer

In this section, we review the results associated to the design of an unknown input observer. Sereval approaches are used to design an unknown input observer. Kudva et al. [6] have developed an algebraic approach. Battacharyya [7] has used a geometric approach to design an unknown input observer. Hautus [8] has investigated a frequential approach. Darouach et al. [10] present an easy method to design a full-order observer for systems subject to unknown inputs.

We use the frequential approach to give necessary and sufficient conditions for the existence of the unknown input observer.

In the sequel, we consider the uncertainty as unknown input w (i.e. w(t) = Δ(t) z(t)).

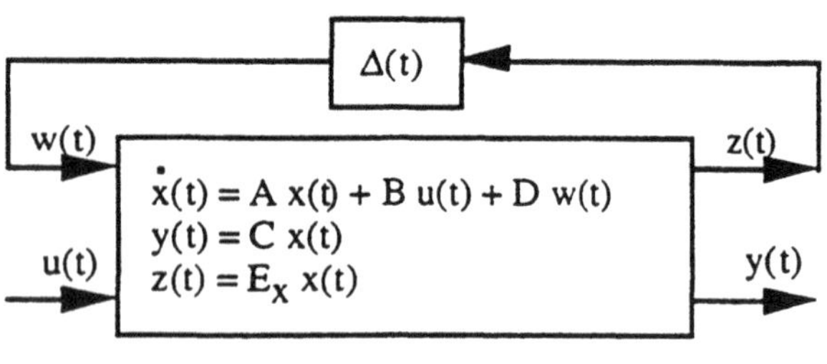

The linear uncertain system (1) can be written as

$$\dot{x}(t) = Ax(t) + B\,u(t) + D\,w(t) \quad (4.1)$$
$$y(t) = C\,x\,(t) \quad (4.2)$$
$$z(t) = E_x\,x(t) \quad (4.3)$$

Using Laplace representation, the two first relations of system (4) can be written as

$$\begin{bmatrix} sI_n-A & -D & -B \\ C & 0 & 0 \\ 0 & 0 & I_q \end{bmatrix}\begin{bmatrix} x \\ w \\ u \end{bmatrix}=\begin{bmatrix} x_0 \\ y \\ u \end{bmatrix} \quad (5)$$

where x_0 is the initial state vector.

Lemma 3 [8], [9] The unknown input observer converges exponentially and asymptotically if and only if there exists stable and proper matrices M(s), $N_y(s)$ and $N_u(s)$ such that

$$[M(s)\,N_y(s)\,N_u(s)]\begin{bmatrix} sI_n-A & -D & -B \\ C & 0 & 0 \\ 0 & 0 & I_q \end{bmatrix}=[I_n\;0\;0] \quad (6)❑$$

The estimator of x is given by the following form

$$x_e = N_y(s)\,y(s) + N_u(s)\,u(s) \quad (7)$$

with its associated observation error

$$e = x - x_e = M(s)\,x_0 \quad (8)$$

Since M(s) is stable and proper, then e will asymptotically converge to zero.

Lemma 4 [7], [9] The following statements are equivalent

(i) there exists a stable unknown input observer for system (1),
(ii) assumptions A2 and A3 are verified. ❑

The following theorem presents the separation principle for an unknown input observer-based controller.

Theorem 1 If M(s), $N_u(s)$, $N_y(s)$ and $(sI_n-A+BK)^{-1}$ are stable, then the closed-loop transfer function of system (4) is internally stable with the output feedback control

$$u = -(I_q + K\,N_u(s))^{-1}\,K\,N_y(s)\,y \quad (9)$$

where the state feedback gain matrix K, and the observer matrices $N_u(s)$ and $N_y(s)$ are given by relations (3) and (6) respectively. ❑

Proof. From (6), the state estimation error is given by (8) where the matrix M(s) is given by

$$M(s) = (I_n - N_y(s)\,C)\,(sI_n - A)^{-1} \quad (10)$$

which is strictly proper.

Let [Φ, G, H, 0] be a state-space realization of M(s) and ζ a state vector associated with the system M(s). By using the system equations (4), with $u = -Kx_e$, the closed-loop system has the following state equations

$$\begin{cases} \dot{x} = A\,x + D\,w + B\,u \\ \dot{\zeta} = \Phi\,\zeta + G\,x_0 \\ x_e = x - H\,\zeta \\ u = -K\,x_e \end{cases}$$

Then the closed-loop state matrix A_{cl}

$$A_{cl} = \begin{bmatrix} A - BK & BKH \\ 0 & \Phi \end{bmatrix}$$

is internally stable under assumptions of theorem 1. ■

The output feedback control given by relation (9) is an unknown input observer-based controller.

3. QUADRATIC STABILIZATION BY AN UNKNOWN INPUT OBSERVER

In this section, we use the unknown input observer given by (6) to propose a solution to the problem of quadratic stabilization by output feedback control.

Let $T_{Fzw}(s)$ be the closed-loop transfer functions from w to z given by a state feedback controller and $T_{zw}(s)$ the one given by an unknown input observer-based controller. First, we show that this kind of controller allows exact recovery of disturbance attenuation obtained by a state feedback control.

Theorem 2 Under assumptions A1, A2, and A3, the unknown input observer-based controller $u = F(s)\,y$, with $F(s) = -(I_q + K\,N_u(s))^{-1}\,K\,N_y(s)$, ensures that

$$T_{zw}(s) = T_{Fzw}(s)$$ ❑

Proof. From (6), we have

$$M(s) + N_y(s)\,C\,(sI_n - A)^{-1} = (sI_n - A)^{-1} \quad (11.1)$$

$$M(s)\,D = 0 \quad (11.2)$$

and $$M(s)\,B = N_u(s) \quad (11.3)$$

The transfer function from w to z is given by

$$T_{zw}(s) = E_x(sI_n - A + B(I_q + KN_u(s))^{-1}KN_y(s)C)^{-1}D$$

Using relation (11.3), we obtain

$$T_{zw}(s) = E_x(sI_n - A + B(I_q + KM(s)B)^{-1}KN_y(s)C)^{-1}D$$
$$= E_x((I_n + BKM(s))(sI_n - A) + BKN_y(s)C)^{-1}(I_n + BKM(s))D$$

and from (11.1) and (11.2), we have

$$T_{zw}(s) = E_x((I_n + BKM(s))(sI_n - A) + BKN_y(s)C)^{-1}D$$
$$= E_x(sI_n - A + BK)^{-1}D = T_{Fzw}(s)$$ ■

Theorem 3 Consider the uncertain system (1). Suppose that assumptions A1, A2 and A3 are verified. Assume that there exists a state feedback control $u = -Kx$ such that the resulting closed-loop system is quadratically stable. Then the closed-loop of system (1) with an unknown input observer-based controller (9) is quadratically stable. ❑

Proof. Using theorem 1 the eigenvalues of the closed-loop are those of state feedback and observer's eigenvalues, so the closed-loop of the nominal system is stable. Using theorem 2 and lemma 1, it follows that the closed-loop of system (1) is quadratically stable. ■

Remarks Petersen and Hollot [1] have established that a high gain observer-based controller ensures the quadratic stability of the closed-loop system if we replace the uncertainty bound by $1-\varepsilon$ ($\varepsilon>0$). The unknown input observer based-controller guarantees that the closed-loop uncertain system is quadratically stable without reducing the uncertainty bound. Other performances can be achieved using the free location of the observer's poles, and the use of high gain is avoided. ❑

4. CONCLUSION

We have presented a new design method for the stabilization of uncertain systems. This method is based on an unknown input observer. We have shown that the unknown input observer-based controller ensures the exact recovery of disturbance attenuation given by a state feedback and the stabilization of uncertain systems. This method can be easily extended to the case of feedthrough of controls in measurements.

5. REFERENCES

[1] I.R. Petersen, C.V. Hollot, "High-gain observers applied to problems in the stabilization of uncertain linear systems, disturbance attenuation and H∞ optimization", Int. J. Adapt. Cont. Sig. Proc., 2, 347-369, 1988.

[2] B.R Barmish, "Necessary and sufficient conditions for quadratic stabilizability of an uncertain linear systems", J. Opt. Th. Appl., 46, 399-408 1985.

[3] P.P Khargonekar, I.R Petersen, K Zhou, "Robust stabilization of uncertain systems : quadratic stabilization and H∞ control theory", IEEE Trans. Aut. Cont., 35, 356-361, 1990.

[4] I.R Petersen, "Stabilization of uncertain linear system in which uncertain parameters enter into the input matrix", SIAM J. Opt., 1988.

[5] I.R Petersen, "A stabilization algorithm for a class of uncertain linear systems", Syst. Cont. Lett., 8, 351-357, 1987.

[6] P. Kudva, N. Viswanadham, A. Ramakrishna, "Observers for linear systems with unknown inputs", IEEE Trans. Aut. Cont., 25, 113-115, 1980.

[7] S.P. Bhattacharyya, "Observer design for linear systems with unknown inputs", IEEE Trans. Aut. Cont., 23, 483-484, 1978.

[8] M.L.J. Hautus, "Strong detectability and observers", Lin. Alg. Appl., 50, 353-369, 1983.

[9] M. Zasadzinski, M. Darouach, M. Hayar, "Loop transfer recovery designs with an unknown input reduced-order observer-based controller", to be published in Int. J. Rob. Nonlin. Cont., 1995.

[10] M. Darouach, M. Zasadzinski, S.J. Xu, "Full-order observers for linear systems with unknown inputs", IEEE Trans. Aut. Cont., 39, 606-609, 1994.

A UNIFYING APPROACH TO SOME MODEL-BASED CONTROL STRUCTURES

R. GOREZ*, D. GALARDINI* and M. DE NEYER*

*Université de Louvain, Centre for Systems Engineering and Applied Mechanics, Bât. EULER, Avenue G. Lemaître, 4, B-1348 Louvain-la-Neuve, Belgium. E-mail: gorez@auto.ucl.ac.be

Abstract. Different control structures including an explicit model of the system to be controlled are particular cases of a generalized structure. This allows the derivation of common properties, especially a property of separation of the tracking and regulation dynamics which may be useful in the design of the control system. Problems related to internal stability and realization of various control structures are considered, and applications are quoted.

Résumé. Différentes structures de commande contenant un modèle explicite du système à commander sont des cas particuliers d'une structure généralisée. Il en résulte des propriétés communes, notamment une propriété de séparation des dynamiques de conduite et de régulation qui peut être utile dans la conception du système de commande. On examine les problèmes de stabilité interne et de réalisation de diverses structures de commande, et on mentionne certaines applications.

Key Words. Model-based control, Control system design, Robustness, Stability.

1. INTRODUCTION

An explicit model of the plant to be controlled can be embedded in the control system in view to disturbance rejection. This includes some well-known control structures such as *Internal Model Control* and *Model Reference Control*. Both have been introduced in the fifties (Moore, 1951; Smith, 1957; Giloi, 1959). Internal Model Control has gained an increasing interest over the two last decades, because of its inherent robustness (Brosilow and Joseph, 1978; Garcia and Morari, 1982; Morari and Zafiriou, 1989). Model Reference Control has been applied in adaptive control systems (Landau, 1979) and in control systems using disturbance compensators based on a global estimate of non measurable disturbances (Ostertag *et al.*, 1989; Gorez and Galardini, 1991). In both structures, comparing the outputs of the controlled plant and of a model yields an error signal which is fed back to the control system in order to correct the control variable. It turns out that these model-based control schemes differ mainly by the way the error signal is used inside the control system, hence they can be viewed as particular cases of a generalized control structure and an analysis of their characteristics and performances can be conducted in this general framework.

The paper is organized as follows. The next section introduces a generalized model-based control structure and derives some basic relations and general properties of such control systems. Section 3 compares four model-based control schemes combining Internal Model Control or Model Reference Control with an additional feedback compensator. In Section 4 some problems of realization and implementation of model-based control systems are considered and applications are quoted. The last section includes conclusions and comments.

2. GENERALIZED MODEL-BASED CONTROL STRUCTURE

2.1. *Control Structure*

Explicit model-based control systems differ by the location of the model, inside or outside the feedback loop providing a correction of the control variable applied to the process. In *Internal Model Control* systems the actual control variable is applied to the controlled process and to its model, the error signal resulting from the comparison of the process and model outputs being fed back to correct this common control input. In this case the model is embedded inside the correction feedback loop. On the contrary, in another class of model-based control systems, the model of the controlled process is outside the correction feedback loop, the error signal being used to correct the control variable applied to the process but not the model input. Such control systems are called *Model Reference Control* or *Model Following Control* systems, although, by comparison to the previous one, this control structure should be referred as *External Model Control*. As pointed in (Landau, 1979) and (Gorez *et al.*, 1992*a*), an External Model Control scheme is the dual of a Luenberger observer (Luenberger, 1971): in the latter the difference of the process and model outputs is used to force the state of the model to track that of the process; on the contrary, in External Model Control, it is used to force the state of the process to track that of the model.

Both previous control schemes, as well as Luenberger observers, can be viewed as particular cases:

$$\begin{aligned} &\textit{Internal Model Control}: & \boldsymbol{R}_p + \boldsymbol{R}_m = 0, \\ &\textit{External Model Control}: & \boldsymbol{R}_m = 0, \\ &\textit{Luenberger Observer}: & \boldsymbol{R}_p = 0, \end{aligned}$$

of the generalized model-based control scheme introduced in (Gorez *et al.*, 1992*a*) and represented in Fig. 1, where $\boldsymbol{P}$ and $\boldsymbol{M}$ denote the input-ouput transfer matrices of the process and of a model, $\boldsymbol{R}_p$ and $\boldsymbol{R}_m$ that of compensators which correct the control input vectors, $\boldsymbol{u}_p$ and $\boldsymbol{u}_m$, according to the difference of the process and model output vectors: $\boldsymbol{e} = \boldsymbol{y}_p - \boldsymbol{y}_m$; $\boldsymbol{u}_c$ is a common command input, $\boldsymbol{w}_p, \boldsymbol{w}_a, \boldsymbol{w}_m$ represent disturbances acting respectively on the outputs of the process, of the actuators and of the measurement lines (or of the model).

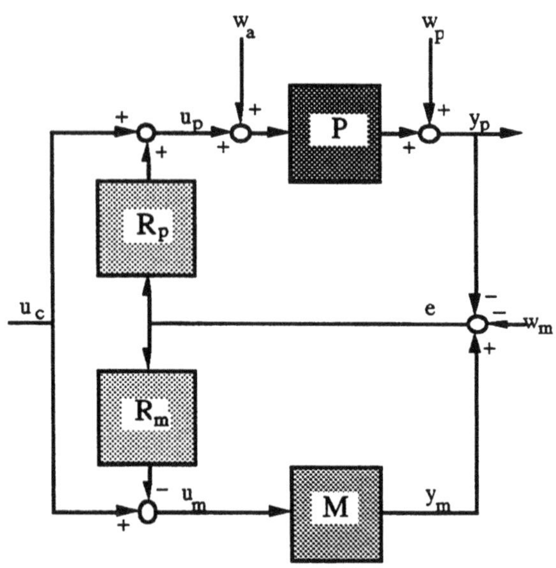

Fig. 1. Generalized Model-Based Control Structure

2.2. *Basic Relations*

Using the same symbols for denoting a variable and its Laplace transform or its z-transform, with $\boldsymbol{I}$ for unit matrices, yields the following relations, respectively for the error signal and the control

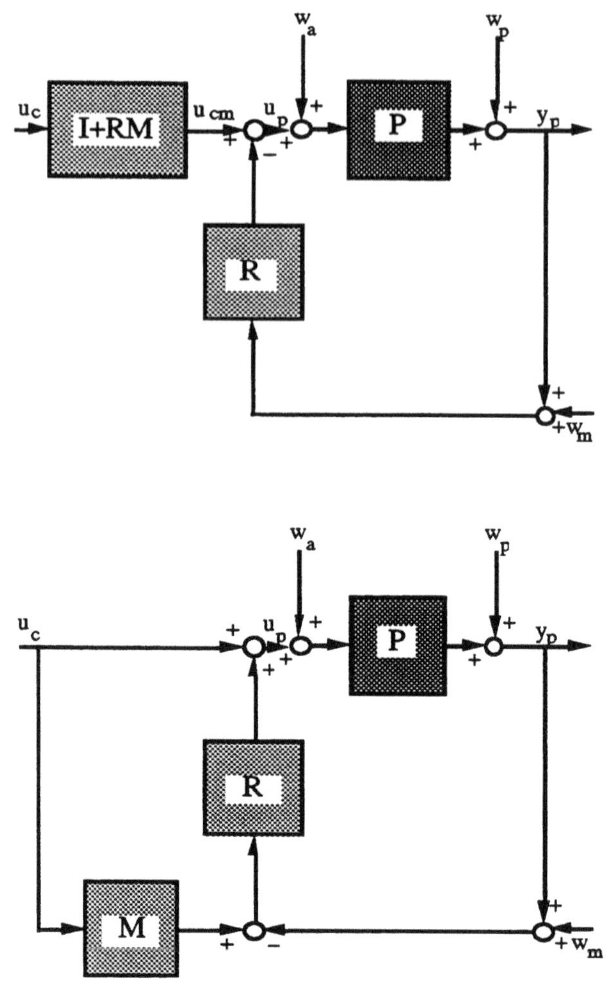

Fig. 2. Equivalent Schemes of the Generalized Model-Based Control Structure

applied to the plant:

$$\begin{aligned} \boldsymbol{e} &= (\boldsymbol{I} + \boldsymbol{M}\boldsymbol{R}_m)^{-1}[\boldsymbol{M}\boldsymbol{u}_c - (\boldsymbol{y}_p + \boldsymbol{w}_m)], \\ \boldsymbol{u}_p &= (\boldsymbol{I} + \boldsymbol{R}\boldsymbol{M})\boldsymbol{u}_c - \boldsymbol{R}(\boldsymbol{y}_p + \boldsymbol{w}_m), \end{aligned}$$

where $\boldsymbol{R} = \boldsymbol{R}_p(\boldsymbol{I} + \boldsymbol{M}\boldsymbol{R}_m)^{-1}$ and $\boldsymbol{I} + \boldsymbol{R}\boldsymbol{M}$ are the transfer matrices of feedback and feedforward compensators in the first equivalent scheme of Fig. 2. Therefore, all the model-based control systems which can be represented by the general scheme of Fig. 1 are equivalent (if $\boldsymbol{R}_p$, hence $\boldsymbol{R}$, $\neq \boldsymbol{0}$), and eventually reduce to a Model Reference Control scheme. Consequently, all properties on separation of tracking and regulation dynamics, robustness, ..., which have been proven for any one of such model-based control systems, hold for all of them.

Obviously, the properties of model-based control systems can also be established directly from Fig. 2. Denoting by $\boldsymbol{u}_{cm} = (\boldsymbol{I} + \boldsymbol{R}\boldsymbol{M})\boldsymbol{u}_c$ the output of the feedforward compensator, and using input-output relations given by (Maciejowski, 1989) and (Boyd and Barratt, 1991)), yields:

$$\begin{aligned} y_p &= S[P(u_{cm} + w_a - Rw_m) + w_p], \\ u_p &= \tilde{S}[u_{cm} - R(Pw_a + w_m + w_p)], \end{aligned}$$

or, through simple matrix transformations and identity relations:

$$\begin{aligned} y_p &= P(I + RP)^{-1}(I + RM)u_c \\ &\quad + S[P(w_a - Rw_m) + w_p], \\ u_p &= (I + RP)^{-1}(I + RM)u_c \\ &\quad + RS[Pw_a + (w_m + w_p)], \end{aligned}$$

where $S = (I + PR)^{-1}$ and $\tilde{S} = (I + RP)^{-1}$ are the *sensitivity* matrices of the feedback loop. In the nominal case ($P = M$), the above relations reduce to:

$$\begin{aligned} y_p &= Mu_c + SMw_a - Tw_m + Sw_p, \\ u_p &= u_c - RS[Mw_a + (w_m + w_p)], \end{aligned}$$

$S = (I + MR)^{-1}$ and $T = I - S = SMR$ being respectively the nominal *sensitivity* and *complementary sensitivity* matrices. Both can be calculated easily:

$$\begin{aligned} S &= (I + MR_m)[I + M(R_p + R_m)]^{-1}, \\ T &= MR_p[I + M(R_p + R_m)]^{-1}. \end{aligned}$$

2.3. *General properties of model-based control systems*

The previous results have a straightforward interpretation. Fig. 2 shows that the control system introduces a feedback loop characterized by sensitivity matrices S and $\tilde{S}$, and a feedforward compensator aiming at compensating $\tilde{S}$ in the transfer matrix relating the actual control variable u_p to the command variable u_c. Therefore, at least in the nominal case, model-based control systems have the same tracking transfer matrix $P = M$ as the controlled process. This is obvious from Fig. 2, as well as from the expression of the error signal:

$$\begin{aligned} e &= [I + MR_m]^{-1}S\{(M - P)u_c \\ &\quad -[Pw_a + (w_m + w_p)]\}. \end{aligned}$$

The latter expression has a straightforward intuitive interpretation: any deviation of the process output from the model output is due to external disturbances, noise and/or model mismatches (internal perturbations); otherwise, both outputs should be equal. From the above relations, it turns out that the sensitivity matrix S provides attenuation of the effects of noise, disturbances and model mismatches. Therefore, having determined the equivalent regulator R providing the desired sensitivity and having selected the control structure, which establishes a relation between R_p and R_m, the designer can obtain $R_p = R(I + MR_m)$. Consequently, any model-based control scheme allows the determination of an equivalent regulator resulting in the desired sensitivity matrix and regulation characteristics without changing the servo characteristics of the process.

Internal stability of a system (Morari and Zafiriou, 1989; Boyd and Barratt, 1991) implies that each of the transfer functions relating any one of the outputs (process output, controller output, error signal, ...) to any one of the inputs is bounded-input-bounded-output (B.I.B.O.) stable. For the general control structure of Fig. 2, this implies the stability of the sensitivity matrices and of all the transfer matrices relating the process, model and controller outputs and the error signals to the various inputs, including control variables and disturbances. Therefore, the model, the sensitivity matrices S and T, and other transfer matrices such as RS, SM, $(I + MR_m)^{-1}S$, $(I + MR_m)^{-1}SM$, must be stable. Besides, since the tracking transfer matrix of the plant should not be changed by model-based control structures, the latter can be used as such only if the controlled process has satisfactory tracking performances, which implies its stability. Otherwise, the tracking dynamics has to be corrected via an appropriate feedback loop.

2.4. *Changing the tracking performances*

If the tracking dynamics must be corrected through an additional feedback loop, the complete control system includes two feedback loops. The first one includes a model as shown in the previous figures, in the sequel it will be refered to as the *regulation loop*. The second loop is the *servo-loop* aiming at changing the tracking performances of the system. Obviously, the servo-loop can be inside or outside the regulation loop. In the first case (Fig. 3), the control applied to the process becomes: $u_p = u_{cm} - (R + C)(y_p + w_m)$. Therefore, one has only to replace R by $R + C$ in the sensitivity and transfer matrices. Using again simple matrix identities yields for the sensitivity matrices:

$$\begin{aligned} S &= (I + QR)^{-1}(I + PC)^{-1}, \\ T &= (I + QR)^{-1}Q(C + R), \end{aligned}$$

where $Q = (I + PC)^{-1}P = P(I + CP)^{-1}$ is the transfer matrix relating the process output y_p to the modified control variable u_{cm}, in other words the closed-loop transfer matrix of the servo-loop (inside the dotted frame in Fig. 3). Straightforward calculations lead to the following expression: $Q(I + RQ)^{-1}(I + RM)$ for the transfer matrix relating the process output to the command u_c; it reduces to Q if the model imbedded in the control structure is that of the process equipped with its servo-loop ($Q = M$). This could be expected from a simple comparison of the control schemes in Fig. 2 and 3. On the contrary, building the

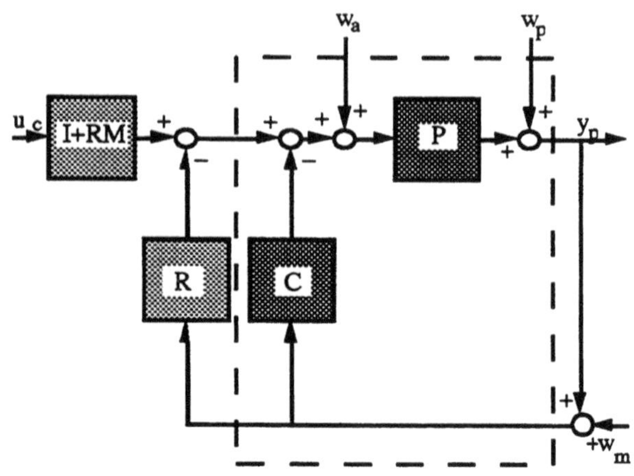

Fig. 3. Changing the tracking performances via an internal servo-loop

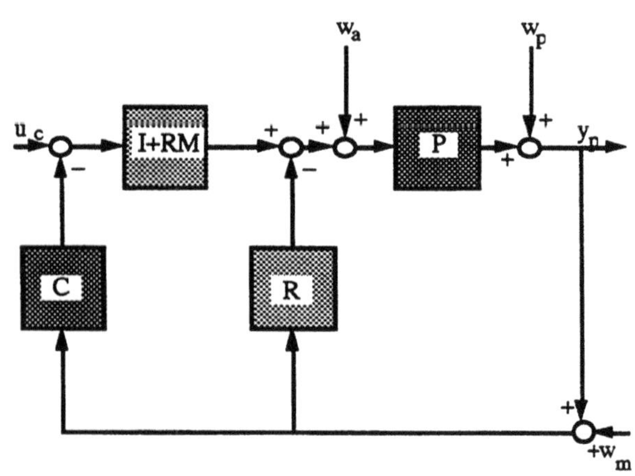

Fig. 4. Changing the tracking performances via an external servo-loop

servo loop around the regulation loop, hence outside the latter, leads to the scheme of Fig. 4. Now the model must be that of the controlled process itself: $M = P$. In the nominal case, this yields:

$$\begin{aligned} S &= (I + PR)^{-1}(I + PC)^{-1}, \\ y_p &= Qu_c + SPw_a - Tw_m + Sw_p, \end{aligned}$$

where Q again denotes the tracking transfer matrix of the process with its servo loop only. Here also this result can be explained by the fact that, in the nominal case, the regulation loop does not change the tracking transfer matrix; therefore only the outer feedback loop via the compensator C can change this matrix.

2.5. *Separation property*

In the nominal case, either with internal or external servo-loop, the tracking transfer matrix of the complete system has the same expression. The sensitivity matrix, which determines the regulation characteristics of the complete system, is given by: $S = (I+MR)^{-1}(I+PC)^{-1}$, where M is the model of the process, either alone ($M = P$) if the servo-loop is around the regulation loop, or equipped with its servo-loop ($M = Q$) if the latter is an internal loop inside the regulation loop. Consequently, the control system designer can adjust separately, on the one hand the tracking transfer matrix Q by means of the matrix C of the compensator in the servo loop, and on the other hand the sensitivity matrix S by means of the matrix R of the equivalent regulator. Defining $S_c = (I+PC)^{-1}$ and $S_r = (I+MR)^{-1}$, respectively as the ***tracking sensitivity*** and the ***additional regulation sensitivity*** matrix, one obtains the following relations for determining the tracking and regulation compensators, C and R_p, from the desired sensitivity matrices:

$$\begin{aligned} PC &= S_c^{-1} - I, \\ MR_p &= (S_r^{-1} - I)(I + MR_m), \end{aligned}$$

where P is the transfer matrix of the controlled process and either $M = P$ if the servo loop is outside the regulation loop or $M = Q = (I + PC)^{-1}P$ if the servo loop is inside the regulation loop. This ***separation property*** is the dual of the property separating the control and observation dynamics in state feedback control with estimation of the state via a Luenberger observer; it results from the duality of Reference Model Control and Luenberger observers.

3. COMPARATIVE ANALYSIS OF FOUR MODEL-BASED CONTROL STRUCTURES

If the designer wants to assign separately the tracking dynamics (Q) and the disturbance rejection characteristics (S_r), he can select one of the four following control schemes:

- Internal Model Control with Internal Servo-Loop: $QR_p = I - S_r$
- Model Reference Control with Internal Servo-Loop: $QR_p = S_r^{-1} - I = S_r^{-1}(I - S_r)$
- Internal Model Control with External Servo-Loop: $PR_p = I - S_r$
- Model Reference Control with External Servo-Loop: $QR_p = S_r^{-1} - I = S_r^{-1}(I - S_r)$.

For linear systems, these four control schemes seem to be equivalent. However, for Internal Model Control, the difference of the model and process outputs is given by $e = -[Mw_a + (w_m + w_p)]$, in the nominal case. Consequently, if the controlled process is not stable and there is no internal stabilizing servo-loop, direct implementations of Internal Model Control are not internally stable, since any disturbance $\boldsymbol{w}_a$ at the process input can give rise to an unbounded difference of the model and process outputs. Internal Model Control can be used as such only when the process to be controlled is stable or has been stabilized by an internal feedback loop. If not, the unstable model $\boldsymbol{M}$ cannot be implemented inside the control system (of course, one may think of selecting $\boldsymbol{R}_p$ in such a way that the unstable poles of $\boldsymbol{M}$ are cancelled by zeroes of $\boldsymbol{R}_p$ and by implementing the global transfer matrix $\boldsymbol{R}_p\boldsymbol{M}$). Such restrictions do not hold for Model Reference Control schemes; the difference of the model and process outputs being given by: $e = -(\boldsymbol{I} + \boldsymbol{M}\boldsymbol{R}_p)^{-1}[\boldsymbol{M}\boldsymbol{w}_a + (\boldsymbol{w}_m + \boldsymbol{w}_p)]$, it is sufficient that the regulation loop be stable.

However a control scheme using internal model regulation, and an outer stabilizing loop if needed, has a definitive advantage over all other model-based control schemes; it is the only scheme where it is possible to apply at the model input the true control variable which is applied to the process, simply by measuring the process control input and using this measured signal as the input of the model. In this way, nonlinearities in the control line such as saturation, hysteresis, dead-zone, ..., are automatically taken into account by the control system, and in particular, there is no problem of *reset wind-up*. In other control schemes such nonlinearities should be introduced in the model input line, which implies that they should be known and which would make model implementation more difficult.

4. IMPLEMENTATION OF THE CONTROL SYSTEM. - APPLICATIONS

A model-based control system may include two separate loops, one obtaining for the desired servo performances, the other one including a model for regulation. Therefore, there are in such a control system two or three different parts: model and compensators, which can be designed independently and implemented through various techniques: analog or digital, conventional arithmetics or qualitative reasoning (possibly using fuzzy logic concepts). Obviously, hybrid systems combining several of those realizations are also possible (Gorez *et al.*, 1992*b*). Furthermore, if some parts of this system have to process sampled-data, the sampling rates may be different, in particular one can implement a fast loop and a slow loop. The internal loop should normally be a fast "local" loop, implemented around a part of the system by using an intermediate output signal; this is the case for position control of mechanical systems (robotic manipulators, precision telescopes, ...) where the outputs to be controlled are positions or displacements and available intermediate outputs are velocities. In that case, the internal loop is a regulation loop aiming at rejecting rapidly local disturbances such, for example, friction, gravitational forces, coupling of various motions in a mechanical system with several degrees of freedom (Ostertag *et al.*, 1989; Gorez, 1989; Gorez and Galardini, 1991; De Neyer and Gorez, 1993). In this case, the model in the control system is that of the system or of the subsystem which must be controlled. On the contrary, if the designer is satisfied with a "slow" regulation, the internal loop if needed may be a stabilizing loop or a servo-loop giving the controlled system the desired tracking dynamics, while the outer loop will be a regulation loop for rejection of slow disturbances and progressive reset of the controlled output. In that case, the model implemented in the control system is a model of the process equipped with its servo-loop. Applications of Internal Model Control, with or without internal servo-loop, can be found in the process industry or in power plants, in the United-States and in some countries of the European Union, Germany, France and Belgium where it was sometimes called "functional regulation";see e.g. (Smith, 1957; Foureau, 1967; Brosilow and Joseph, 1978; Garcia and Morari, 1982; Stephanopoulos, 1985; Sebord *et al.*, 1989).

The model implemented in the regulation loop may be a detailed model, or a simplified model representing roughly the main features of the input-output response. Generally, model-based control systems have an intuitive interpretation which makes them comprehensive and attractive for engineers and technicians who have to control the process or to design an automatic control system. Therefore, it is relatively easy to introduce in such control systems some a-priori knowledge about the process to be controlled and its actuators, for example nonlinearities or variations of some parameters over time or with respect to some external conditions such as the load of the plant. Similarly, the control system designer can introduce in a rational way constraints and limitations aiming at improving the safety of the control system. All these features allow automatic control for all operating conditions of the plant, including automatic start-up and shut-down (Debelle, 1979). However, one can also use a simplified model either in an internal or in an outer regulation loop. For example, in the internal velocity regulation loop of a mechanical system, one can neglect friction, gravitational forces, coupling of motions along several degrees of freedom, flexibility effects and vibrations, ...; thereby, if the control variable is the driving force or torque acting upon the mechanical system, the model relating the velocity to the control variable reduces to a simple integrator. On the contrary, if the regulation loop is the outer loop, related data can be sampled at a slow rate for the purpose of a periodical update of the reference input of the internal servo-loop. Therefore, if the sampling interval is approximately equal to or longer than the response time of the servo-controlled system, or if it is between 1/3 and 2/3 of this response time, the servo-controlled system can be represented either by a static model or by a simplified first-order dynamical model (Gorez *et al.*, 1992*b*).

5. CONCLUSIONS

Various basic control structures based on a model of the process to be controlled have been considered. If the process and the control system are linear, all these control schemes are equivalent and have the same performances and the same robustness. Consequently, the choice of one among these control schemes should be guided by considerations on the implementation of the control system and on the sensitivity to noise and to nonlinearities; it could be influenced also by the relative weights of tracking and regulation performances and by the possibility of an intuitive interpretation of the control scheme. Various realizations and implementations of different parts of the control system are allowed. However, direct implementation of one of those control schemes may be forbidden by internal stability requirements if the process to be controlled is not stable. Analysis of the control schemes and relations for calculation of the components of the control system, given in Sections 3 and 4, bring some useful elements for the selection of the control scheme which will be the most appropriate in a given application.

Acknowledgement – This paper presents research results of the Belgian Programme on InterUniversity Poles of Attraction initiated by the Belgian State, Prime Minister's Office for Science, Technology and Culture. The scientific responsibility rests with its authors

6. REFERENCES

Boyd, S.P. and C.H. Barratt (1991). *Linear Control Design. Limits of Performances.* Prentice-Hall I.E. Englewood Cliffs, N.J.

Brosilow, C.B. and B. Joseph (1978). Inferential control of processes. *A. I. Ch. E. J.* **24**, 485–500.

Debelle, J. (1979). A control structure based upon process models. *Journal* **A 20**(2), 71–81.

De Neyer, M. and R. Gorez (1993). Fuzzy and quantitative model-based control systems for robotic manipulators. *International Journal of Systems Sciences* **24**(10), 1863–1883.

Foureau, A. (1967). Les régulations fonctionnelles de processus industriels ACEC-CHANCE. *Revue ACEC* (2), 31–44.

Garcia, C.E. and M. Morari (1982). Internal model control - a unifying review and some new results. *Ind. Eng. Chem. Process Des. Dev.* **21**, 308–323.

Giloi, W. (1959). Zur Theorie und Verwirklichung einer Regelung für Laufzeitstrecken nach den Prinzip der Ergänzenden Rückfürung. PhD thesis. Univ. Stuttgart. Germany.

Gorez, R. (1989). Hybrid control for d-c servomotors. In: *Proc. LCA'89: IFAC Symposium on Low Cost Automation.* Vol. 1. pp. 13–17.

Gorez, R. and D. Galardini (1991). Robot control with disturbance observers. In: *Proc. ICAR '91: Fifth IEEE International Conference on Advanced Robotics.* pp. 417–422.

Gorez, R., D. Galardini and K. Y. Zhu (1992*a*). Model-based control systems. In: *Mathematics of the Analysis and Design of Process Control* (P. Borne, S.G. Tzafestas and N.E. Radhy, Eds.). pp. 223–232. Elsevier Science Publishers. Amsterdam, The Netherlands.

Gorez, R., M. De Neyer, D. Galardini and J. Barreto (1992*b*). Model-based control systems: Fuzzy and qualitative realizations. In: *Proc. IFAC Symp. on Intelligent Components and Instruments for Control Applications.* Malaga. pp. 681–686.

Landau, Y. (1979). *Adaptive Control: The Model Reference Approach.* M. Dekker Inc. New York.

Luenberger, D.G. (1971). An introduction to observers. *IEEE Trans. on Automatic Control* **AC-16**(3), 596–602.

Maciejowski, J. M. (1989). *Multivariable Feedback Design.* Addison-Wesley. Reading, Ma, U.S.A.

Moore, J. H. (1951). Combination open-cycle closed-cycle systems. In: *Proc. IRE.* Vol. 39. pp. 1421–1432.

Morari, M. and E. Zafiriou (1989). *Robust process control.* Prentice-Hall. Englewood Cliffs, N.J.

Ostertag, E., N. Bakri and N. Becker (1989). Functional disturbance observer for simultaneous control and dry friction compensation. In: *Advanced Information Processing in Automatic Control* (R. Husson, Ed.). pp. 421–426.

Sebord, D. E., T. F. Edgar and D.L. Mellichamp (1989). *Process Dynamics and Control.* J. Wiley & Sons. New York.

Smith, O. J. M. (1957). Closer control of loops with dead-time. *Chemical Engng. Prog.* **53**(5), 217–219.

Stephanopoulos, G. (1985). *Chemical Process Control. An Introduction to Theory and Practice.* Prentice-Hall. Englewood Cliffs, N.J.

ON ROBUST POSITIVE INVARIANCE OF UNBOUNDED POLYHEDRA

C.E.T. DÓREA* and J.-C. HENNET

L.A.A.S. - C.N.R.S., 7 avenue du Colonel Roche, 31077 Toulouse FRANCE

Abstract. This study analyzes continuous-time linear systems with uncertain parameters on their state and control matrices (A and B), and subject to linear symmetrical constraints on their state (or output) vector. The presented work investigates the possibility to construct a state-feedback law that makes the constrained domain positively invariant with respect to the uncertain closed-loop system. In particular, conditions are given for robust positive invariance of unbounded symmetrical polyhedra, with stability, and a design technique is proposed.

Key Words. Linear systems, state feedback, positively invariant sets, constrained control, robust control.

SUR L'INVARIANCE POSITIVE ROBUSTE DES POLYEDRES NON BORNES

Résumé. Cette étude analyse les systèmes linéaires dont les matrices d'état et de commande (A et B) ont des paramètres incertains. Ces systèmes sont en outre soumis à des contraintes linéaires symétriques sur leur vecteur d'état (ou de sortie). Le travail présenté explore la possibilité de construire une loi de retour d'état rendant le domaine des contraintes positivement invariant en boucle fermée pour le système incertain. En particulier, cet article établit des conditions d'invariance positive robuste, avec stabilité, pour des polyèdres symétriques non bornés, et propose une technique de synthèse.

Mots Clés. Systèmes linéaires, retour d'état, domaines positivement invariants, commande contrainte, commande robuste.

1. INTRODUCTION

Many practical systems are modeled, for control purposes, by linear differential equations with several unperfectly known parameters. Such models are valid only in a limited range of the system evolution. Typically, control problems associated with such models are characterized by robustness requirements and constraints on state variables. Other constraints may arise, for linear as well as non-linear systems, from physical limitations on input or output variables. And it is often desirable not to reach too often the saturation levels.

It is only during the last decade that attempts have been made to combine constraints satisfaction and parametric uncertainties within an integrated design technique. Such a combination has been facilitated by the use of the *Positive Invariance* approach for solving constrained control problems (Bitsoris and Gravalou, 1993; Blanchini, 1991; Hennet and Castelan, 1992; Sznaier, 1993).

Some advantages of the *Positive Invariance* approach are :

-it provides a closed-form control solution of state-feedback type,
-it avoids frequent constraint saturations, because of the contractivity property of positive invariance.

It has also been shown that for compact positively invariant domains, strict contractivity of the invariance property implies robustness of this property with respect to small norm-bounded perturbations of the system parameters (Hennet and Castelan, 1992). On the contrary, positive invariance of unbounded domains is usually non-robust to unstructured perturbations, mainly because of the requirement of a subspace to be (A, B)-invariant (Castelan and Hennet, 1992). However, as shown in this paper, some significant results on robust positive invariance of unbounded domains can be obtained under structured uncertainties applying to certain parameters of the state and control matrices (A and B).

Some geometric conditions for robust invariance of a given subspace with respect to structured parametric variations have been given by Bhattacharyya (1983). In this study, similar structural conditions are obtained, through a change of vari-

*The research support for this author is provided by CNPq - Brazil.

ables, as a subset of the set of robust positive invariance conditions for an unbounded polyhedral domain. Moreover, through the construction of an appropriate basis of the state space, verification of the structural conditions becomes straightforward, and the problem of positive invariance with stability for the uncertain system is decomposed into two subproblems with a reduced size: a robust stabilizability problem and a problem of robust positive invariance for a bounded polyhedron. The first subproblem is classical, and the second one can be solved by standard Linear Programming.

2. PROBLEM PRESENTATION

The considered uncertain linear systems admit the following state representation :

$$\dot{x}(t) = A(\underline{\alpha})x(t) + B(\underline{\beta})u(t) \tag{1}$$

where $x \in \Re^n$ is the vector of state variables, $u \in \Re^m$ the vector of control variables. Uncertainties on matrices A and B are structured. They depend on 2 vectors of bounded uncertain parameters, denoted $\underline{\alpha} \in \Re^p$ and $\underline{\beta} \in \Re^q$, subject to the following componentwise interval constraints :

$$-\underline{\alpha}_M \leq \underline{\alpha} \leq \underline{\alpha}_M \tag{2}$$

$$-\underline{\beta}_M \leq \underline{\beta} \leq \underline{\beta}_M \tag{3}$$

Furthermore, all the parameters α_i, $i = 1, \cdots, p$ and β_j, $j = 1, \cdots, q$ are supposed independent. Such a representation of structured uncertainties is classical (Bhattacharyya, 1983).

The elements of $A(\underline{\alpha})$ and $B(\underline{\beta})$ are linear functions of these parameters, that is :

$$A(\underline{\alpha}) = A_0 + \alpha_1 A_1 + \cdots + \alpha_p A_p = A_0 + \Delta A(\underline{\alpha}) \tag{4}$$

$$B(\underline{\beta}) = B_0 + \beta_1 B_1 + \cdots + \beta_q B_q = B_0 + \Delta B(\underline{\beta}) \tag{5}$$

Conditions (2), (3) define lower and upper bounds for each uncertain component of $A(\underline{\alpha})$. The convex combination of these bounds define hypercubes $\mathcal{D}_A$ with $n_A = 2^p$ extreme points and $\mathcal{D}_B$ with $n_B = 2^q$ extreme points. For any possible pair of vectors ($\underline{\alpha}$, $\underline{\beta}$) satisfying (2), (3), $A(\underline{\alpha}) \in \mathcal{D}_A$ and $B(\underline{\beta}) \in \mathcal{D}_B$. Let A^i be the value of $A(\underline{\alpha})$ at extreme point i and B^j the value of $B(\underline{\beta})$ at extreme point j. All the considered matrices $A \in \mathcal{D}_A$ and $B \in \mathcal{D}_B$ can be written as a convex linear combination of matrices A^i and matrices B^j, respectively:

$$\mathcal{D}_A := \left\{ A : A = \sum_{i=1}^{n_A} \xi_i A^i ; \ \sum_{i=1}^{n_A} \xi_i = 1; \ \xi_i \geq 0 \right\} \tag{6}$$

$$\mathcal{D}_B := \left\{ B : B = \sum_{j=1}^{n_B} \mu_j B^j ; \ \sum_{j=1}^{n_B} \mu_j = 1; \ \mu_j \geq 0 \right\} \tag{7}$$

This convex representation of uncertainties has been used by Geromel *et al.* (1991) and Milani (1994).

The vector of state variables of system (1) is also subject to linear symmetrical constraints of the following type :

$$-\rho \leq Gx(t) \leq \rho \tag{8}$$

where $G \in \Re^{r \times n}$ has rank $r \leq n$ and the components of vector $\rho \in \Re^r$ are strictly positive. Such constraints define in the state space the following symmetrical convex polyhedron :

$$S[G, \rho] = \{x \in \Re^n \ / \ -\rho \leq Gx \leq \rho\} \tag{9}$$

Constraints satisfaction is guaranteed all along the system trajectory from any initial state $x(0) \in S[G, \rho]$ if the polyhedron $S[G, \rho]$ is positively invariant with respect to the closed-loop system.

By definition, *positive invariance* of a domain Ω w.r.t. (with respect to) a system evolution simply expresses the property that any trajectory starting at any point of Ω always belongs to Ω.

The objective of the LCRP (Linear Constrained Regulation Problem) is to find a static state feedback control law, $u(t) = Fx(t)$, under which the closed-loop system :

$$\dot{x}(t) = (A(\underline{\alpha}) + B(\underline{\beta})F)x(t) \tag{10}$$

is stable and the polyhedron (9) is positively invariant, for all the admissible values of uncertain vectors $\underline{\alpha}$ and $\underline{\beta}$.

Necessary and sufficient conditions for $S[G, \rho]$ to be positively invariant for the uncertain system described by (1)-(7) were established by Milani (1994):

Proposition 2.1 *The polyhedron $S[G, \rho]$ is positively invariant with respect to system (10) if and only if there exist matrices $F \in \Re^{m \times n}$ and $H^{ij}, K^{ij} \in \Re^{n \times n}$ such that, $\forall\, i = 1, \cdots, n_A$, $j = 1, \cdots, n_B$,*

$$(H^{ij} - K^{ij})G = G(A^i + B^j F) \tag{11}$$

$$(H^{ij} + K^{ij})\rho \leq 0 \tag{12}$$

with $H^{ij}_{kl} \geq 0$ for $k \neq l$ and $K^{ij}_{kl} \geq 0 \ \forall\, k, l$.

Note that in the case of precisely known systems ($n_A = n_B = 1$), equations (11), (12) correspond to the classical positive invariance relations (Bitsoris, 1991; Castelan and Hennet, 1993).

In the case of the uncertain systems analyzed in this study, it is possible to formulate a Linear Program with conditions (11), (12) as constraints (Milani, 1994). If such a problem is feasible, its solution provides a state feedback control law under which $S[G,\rho]$ becomes positively invariant w.r.t. system (10). Moreover, if rank$(G) = n$ and the inequalities (12) are strictly satisfied, the closed-loop uncertain system is asymptotically stable.

If rank $(G) = r < n$, positive invariance of $S[G,\rho]$ does not imply closed-loop stability. The problem of positive invariance with stability then has to be addressed for uncertain as well as for deterministic systems.

3. STRUCTURAL CONDITIONS FOR ROBUST POSITIVE INVARIANCE OF UNBOUNDED POLYHEDRA

Consider now the case rank $G = r < n$. In the deterministic case, when the pair (A, B) is supposed perfectly known, the existence of a pair of matrices $(H, K) \in \Re^{r\times r}$ satisfying relation (11) requires the closed-loop invariance of the subspace ker(G) (Castelan and Hennet, 1992).

In the uncertain case, consider the class of continuous linear systems described by (1)-(3). The state vector is subject to constraints (8), with the components of vector $\rho \in \Re^r$ strictly positive. Such constraints define an unbounded convex symmetrical polyhedron $S[G,\rho]$ (9).

The aim here is to find a single static state feedback, $u(t) = Fx(t)$, for which $S[G,\rho]$ is positively invariant and the closed-loop system (10) is asymptotically stable for all the $\underline{\alpha}$ and $\underline{\beta}$ satisfying (2), (3)). Consider first the structural conditions of existence of such a control. From relation (11), closed-loop positive invariance of $S[G,\rho]$ requires the existence of a pencil of matrices $M(\underline{\alpha},\underline{\beta}) = H(\underline{\alpha},\underline{\beta}) - K(\underline{\alpha},\underline{\beta})$ satisfying for all the admissible $\underline{\alpha}$ and $\underline{\beta}$:

$$M(\underline{\alpha},\underline{\beta})G = G(A(\underline{\alpha}) + B(\underline{\beta})F) \qquad (13)$$

Parameters $\underline{\alpha},\underline{\beta}$ being supposed independent, condition (13) can be re-written, using (4), (5):

$$(M_0 + \Delta M(\underline{\alpha}) + \Delta M(\underline{\beta}))G = G(A_0 + \Delta A(\underline{\alpha}) + (B_0 + \Delta B(\underline{\beta}))F) \qquad (14)$$

And the three following conditions should be simultaneously satisfied :

$$M_0 G = G(A_0 + B_0 F) \qquad (15)$$

$$\Delta M(\underline{\alpha})G = G\Delta A(\underline{\alpha}) \qquad (16)$$

$$\Delta M(\underline{\beta})G = G\Delta B(\underline{\beta})F \qquad (17)$$

Due to the independence of the components of $\underline{\alpha}$ and $\underline{\beta}$, it is not difficult to show (Castelan and Hennet, 1992) that these conditions are respectively equivalent to :

a) ker(G) is an (A_0, B_0)-invariant subspace.

b) ker(G) is an A_i-invariant subspace for $i = 1,\cdots,p$.

c) There exists a matrix F satisfying (15) and such that ker(G) is a B_jF-invariant subspace for $j = 1,\cdots,q$.

Condition (b) only depends on the matrix of constraints, G, and on matrices A_i, which characterize the structure of uncertainties on $A(\underline{\alpha})$. Therefore, if this condition is satisfied, the feedback gain matrix, F, should be selected so as to satisfy conditions (a) and (c). The conditions for existence of such a feedback gain matrix were established by Bhattacharyya (1983). Define

$$B_j^{-1}\ker(G) := \{v \in \Re^m \ / \ B_j v \in \ker(G)\}; \ j = 1,\cdots q \qquad (18)$$

$$\mathcal{R} := \cap_{i=1}^{q} B_j^{-1}\ker(G) \qquad (19)$$

where B_j^{-1} represents the "functional inverse" (Wonham, 1985) of map B_j. And let $P \in \Re^{m\times m}$ be the projection of $\Re^m$ onto $\mathcal{R}$.

Proposition 3.1 *(Bhattacharyya, 1983). There exists a matrix F such that* ker(G) *is $(A_0 + B_0F)$-invariant and B_jF-invariant, $j = 1,\cdots,q$, if and only if the subspace* ker(G) *is (A_0, B_0P)-invariant.*

The algebraic conditions of existence of a state feedback satisfying (13) can then be stated as follows :

Corollary 3.1 *There exist matrices F and $M(\underline{\alpha},\underline{\beta})$ such that $M(\underline{\alpha},\underline{\beta})G = G(A(\underline{\alpha}) + B(\underline{\beta})F)$ for all the admissible $\underline{\alpha}$ and $\underline{\beta}$ if and only if*

$$A_0\ker(G) \subset \mathrm{Im}(B_0P) + \ker(G)$$

$$A_i\ker(G) \subset \ker(G), \quad i = 1,\cdots,p$$

Such conditions are clearly difficult to satisfy in general. But they happen to be satisfied for some particular classes of problems of practical interest, as it will be shown in the next section.

4. CONDITIONS FOR A ROBUST SOLUTION OF LCRP

Based on the structural conditions of the preceding section, a systematic technique for testing the existence and constructing a solution to the original Robust Constrained Linear Regulation Problem will now be developed. This result will be obtained by extending to the uncertain case a method previously proposed by Dórea and Milani (1995) for deterministic systems. The basis of the method is a change of basis in the state space of the uncertain system (1)-(3):

$$x = Q\tilde{x} \qquad (20)$$

where $Q \in \Re^{n\times n}$ is an orthogonal matrix such that :

$$\tilde{G} = GQ = \begin{bmatrix} 0 & \tilde{G}_2 \end{bmatrix}; \ \tilde{G}_2 \in \Re^{r\times r} \qquad (21)$$

In the new basis, system (1)-(3) and the polyhedron $S[G,\rho]$ are represented as follows :

$$\dot{\tilde{x}}(t) = \tilde{A}(\underline{\alpha})\tilde{x}(t) + \tilde{B}(\underline{\beta})u(t) \qquad (22)$$

$$\tilde{A}(\underline{\alpha}) = \tilde{A}_0 + \alpha_1\tilde{A}_1 + \cdots + \alpha_p\tilde{A}_p = \tilde{A}_0 + \Delta\tilde{A}(\underline{\alpha}) \qquad (23)$$

$$\tilde{B}(\underline{\beta}) = \tilde{B}_0 + \beta_1\tilde{B}_1 + \cdots + \beta_q\tilde{B}_q = \tilde{B}_0 + \Delta\tilde{B}(\underline{\beta}) \qquad (24)$$

$$\tilde{A}_i = Q'A_iQ = \begin{bmatrix} \tilde{A}_{i11} & \tilde{A}_{i12} \\ \tilde{A}_{i21} & \tilde{A}_{i22} \end{bmatrix}; \; \tilde{A}_{i22} \in \Re^{r\times r} \qquad (25)$$

$$\tilde{B}_j = Q'B_j = \begin{bmatrix} \tilde{B}_{j1} \\ \tilde{B}_{j2} \end{bmatrix}; \; \tilde{B}_{j2} \in \Re^{r\times m} \qquad (26)$$

$$S[\tilde{G},\rho] = \{\tilde{x} \in \Re^n \; / \; -\rho \le \tilde{G}\tilde{x} \le \rho\} \qquad (27)$$

Proposition 4.1 *The polyhedron $S[G,\rho]$ is positively invariant with respect to system (10) if and only if*

1. *(a) There exists a matrix $\tilde{F}_1 \in \Re^{m\times(n-r)}$ such that*

$$\tilde{A}_{021} + \tilde{B}_{02}\tilde{F}_1 = 0 \qquad (28)$$

$$\tilde{B}_{j2}\tilde{F}_1 = 0; \; j = 1,\cdots,q$$

   (b) $\tilde{A}_{i21} = 0; \; i = 1,\cdots,p$

2. *There exist a matrix $\tilde{F}_2 \in \Re^{m\times r}$ and matrices $H(\underline{\alpha},\underline{\beta}), K(\underline{\alpha},\underline{\beta}) \in \Re^{r\times r}$, with $H_{kl}(\underline{\alpha},\underline{\beta}) \ge 0$ for $k \ne l$, $K_{kl}(\underline{\alpha},\underline{\beta}) \ge 0$ such that, $\forall\; \underline{\alpha},\underline{\beta}$:*

$$(\mathrm{H}(\underline{\alpha},\underline{\beta}) - \mathrm{K}(\underline{\alpha},\underline{\beta}))\tilde{\mathrm{G}}_2 = \tilde{\mathrm{G}}_2(\tilde{\mathrm{A}}_{22}(\underline{\alpha}) + \tilde{\mathrm{B}}_2(\underline{\beta})\tilde{\mathrm{F}}_2)$$

$$(\mathrm{H}(\underline{\alpha},\underline{\beta}) + \mathrm{K}(\underline{\alpha},\underline{\beta}))\rho \le 0$$

Proof : Consider the state feedback matrix for the transformed system (22):

$$\tilde{F} = FQ = \begin{bmatrix} \tilde{F}_1 & \tilde{F}_2 \end{bmatrix}; \; \tilde{F}_2 \in \Re^{m\times r} \qquad (29)$$

Substitution of matrices $\tilde{F}$, $\tilde{G}$ (21), $\tilde{A}$ (23) and $\tilde{B}$ (24) in equation (14) yields :

$$\begin{bmatrix} 0 & M_0 + \Delta M(\underline{\alpha}) + \Delta M(\underline{\beta}) \end{bmatrix}\tilde{G}_2 = \tilde{G}_2\begin{bmatrix} \tilde{A}_{F_{21}}(\underline{\alpha},\underline{\beta}) & \tilde{A}_{F_{22}}(\underline{\alpha},\underline{\beta}) \end{bmatrix} \qquad (30)$$

where: $\tilde{A}_{F_{21}}(\underline{\alpha},\underline{\beta}) = \tilde{A}_{021} + \Delta\tilde{A}_{21}(\underline{\alpha}) + (\tilde{B}_{02} + \Delta\tilde{B}_2(\underline{\beta}))\tilde{F}_1$; $\tilde{A}_{F_{22}}(\underline{\alpha},\underline{\beta}) = \tilde{A}_{022} + \Delta\tilde{A}_{22}(\underline{\alpha}) + (\tilde{B}_{02}+\Delta\tilde{B}_2(\underline{\beta}))\tilde{F}_2$. The columns of matrix $\tilde{G}_2$ being linearly independent, satisfaction of this equation is equivalent to :

$$\tilde{A}_{021} + \Delta\tilde{A}_{21}(\underline{\alpha}) + (\tilde{B}_{02} + \Delta\tilde{B}_2(\underline{\beta})\tilde{F}_1) = 0 \qquad (31)$$

and because of the assumed independence between $\underline{\alpha}$ and $\underline{\beta}$, relation (28) is obtained, and : $\Delta\tilde{A}_{21}(\underline{\alpha}) = 0$, $\Delta\tilde{B}_2(\underline{\beta})\tilde{F}_1 = 0$, $\forall\; \underline{\alpha},\underline{\beta}$. And that shows conditions 1. On the other hand, conditions 2 can be directly verified from equation (30) and relations (11), (12). □

Remarks :

• If conditions 1 in proposition 4.1 are verified, the original problem of robust positive invariance of the unbounded polyhedron $S[G,\rho]$ with respect to the pair $(A(\underline{\alpha}), B(\underline{\beta}))$ is reduced to a simpler problem of robust positive invariance of the bounded polyhedron $\tilde{S}[\tilde{G}_2,\rho]$, with respect to the pair $(\tilde{A}_{22}(\underline{\alpha}), \tilde{B}_2(\underline{\beta}))$. The conditions of existence of solutions to this problem have been presented in proposition 2.1.

• Proposition 4.1 also allows for an easier verification of structural conditions for robust positive invariance. Consider first the uncertainties on matrix A. It is clear that condition $\tilde{A}_{i21} = 0$; $i = 1,\cdots,p$, is not verified for whatever constraints or uncertainties. But it may be satisfied for classes of problems of practical interest. In particular, the example below presents a case where this condition applies, with the state matrix under a canonical companion form. Consider now the uncertainties on matrix B. The existence of a state feedback satisfying (13) is related to the existence of a solution to the $(n-r)$ systems of $r(1+q)$ linear equations with m unknown variables (see proposition 4.1):

$$\begin{bmatrix} \tilde{B}_{02} \\ \tilde{B}_{12} \\ \vdots \\ \tilde{B}_{q2} \end{bmatrix}\tilde{F}_1 = \begin{bmatrix} -\tilde{A}_{021} \\ 0 \\ \vdots \\ 0 \end{bmatrix} \qquad (32)$$

For these systems to admit solutions, it is generally necessary to have : $q \le (m/r)-1$ (the number of equations should not be bigger than the number of variables). This condition does not depend in general on the structure of matrices B_j, but on their number.

It is clear that as the number of uncertain parameters increases, the possibilities of existence of solutions decrease.

Example :

Consider the system of second order differential equations, with $y \in \Re^r$ as the output vector:

$$\frac{d^2y}{dt^2} + L_1\frac{dy}{dt} + L_0y = N_0u + N_1\frac{du}{dt}. \qquad (33)$$

The canonical observability companion form of the system is :

$$\dot{x} = \begin{bmatrix} 0 & -L_0 \\ I & -L_1 \end{bmatrix}x + \begin{bmatrix} N_0 \\ N_1 \end{bmatrix}u \qquad (34)$$

$$y = \begin{bmatrix} 0 & I \end{bmatrix}x \qquad (35)$$

A typical regulation specification could be to compute a state feedback gain matrix

$$F = \begin{bmatrix} F_1 & F_2 \end{bmatrix}; \; F_1 \in \Re^{m\times(n-r)}$$

so as to limit the outputs to : $-\rho \leq y \leq \rho$, or $-\rho \leq Gx \leq \rho$, with $G = \begin{bmatrix} 0 & I \end{bmatrix}$.

The system directly takes the form (21)-(27). And note that $\tilde{A}_{i21} = 0$ even if all the elements of L_0 and L_1 are uncertain. Moreover, if $r \leq m$ and N_1 has full rank, there always exists a matrix $\tilde{F}_1$ such that $\tilde{A}_{021} + \tilde{B}_{02}\tilde{F}_1 = I + N_1 F_1 = 0$. The structural conditions are thus verified independently from the uncertainties on matrices L_1 and L_2.

Also, using (34) it can be observed that only the uncertainties on N_1 play a role in the problem of existence of solutions to (13). Then, if the numbers of inputs and outputs are equal, no uncertainty on N_1 can be admitted. On the contrary, if the number of inputs is twice the number of outputs, the problem generally admits a solution if there is only one uncertain parameter, γ_1, in $N_1 = N_{10} + \sum_{j=1}^{q} \gamma_j N_{1j}$.

5. SYNTHESIS OF A ROBUST CONTROLLER

The objective is now to compute a state feedback law which renders $S[G,\rho]$ positively invariant and simultaneously stabilizes the uncertain system (10). To this end, the conditions of proposition 4.1 must be verified.

In order to find a solution to (32), it is always possible to compute orthogonal transformation matrices $Q_b \in \Re^{r(q+1)\times r(q+1)}$ and $Z \in \Re^{m\times m}$ such that:

$$\hat{B}_b = Q_b' \begin{bmatrix} \tilde{B}_{02} \\ \tilde{B}_{12} \\ \vdots \\ \tilde{B}_{q2} \end{bmatrix} Z = \begin{bmatrix} \hat{B}_{b1} & 0 \\ 0 & 0 \end{bmatrix} \tag{36}$$

$$\hat{A}_b = Q_b' \begin{bmatrix} -\tilde{A}_{021} \\ 0 \\ \vdots \\ 0 \end{bmatrix} = \begin{bmatrix} \hat{A}_{b1} \\ \hat{A}_{b2} \end{bmatrix} \tag{37}$$

where $\hat{B}_{b1} \in \Re^{b\times b}$ is non-singular. Singular Value Decomposition or QR techniques, found in most mathematical software packages, can be used to compute matrices Q_b and Z.

Solving (32) is clearly equivalent to solving :

$$\hat{B}_b \hat{F}_b = \hat{A}_b \tag{38}$$

where

$$\hat{F}_b = Z'\tilde{F}_1 = \begin{bmatrix} \hat{F}_{11} \\ \hat{F}_{21} \end{bmatrix}; \; \hat{F}_{11} \in \Re^{b\times(n-r)} \tag{39}$$

A solution to (39) exists if and only if $\hat{A}_{b2} = 0$, and is given by the unique solution of the systems of linear equations:

$$\hat{B}_{b1}\hat{F}_{11} = \hat{A}_{b1}. \tag{40}$$

Note that matrix $\hat{F}_{21} \in \Re^{(m-b)\times(n-r)}$ does not play any role in the robust positive invariance of $S[G,\rho]$ and may therefore be used to satisfy other design specifications.

Consider now the change of basis:

$$u = Z\hat{u} \tag{41}$$

such that:

$$\dot{\tilde{x}} = \tilde{A}(\underline{\alpha})\tilde{x} + \hat{B}(\underline{\beta})\hat{u} \tag{42}$$

$$\hat{B}_j = \tilde{B}_j Z = \begin{bmatrix} \hat{B}_{j11} & \hat{B}_{j12} \\ \hat{B}_{j21} & \hat{B}_{j22} \end{bmatrix}; \; \hat{B}_{j21} \in \Re^{r\times b} \tag{43}$$

$$\hat{F} = Z'\tilde{F} = \begin{bmatrix} \hat{F}_{11} & \hat{F}_{12} \\ \hat{F}_{21} & \hat{F}_{22} \end{bmatrix}; \; \hat{F}_{11} \in \Re^{b\times(n-r)} \tag{44}$$

Let **F** be the set of state feedback matrices F for which (13) is satisfied. The set $\hat{\mathbf{F}}$ is then uniquely defined by:

$$\hat{\mathbf{F}} \equiv \left\{ \hat{F} \;/\; \hat{B}_{b1}\hat{F}_{11} = \hat{A}_{b1} \right\} \tag{45}$$

For $\hat{F} \in \hat{\mathbf{F}}$, one has in closed-loop:

$$\tilde{A}_F(\underline{\alpha},\underline{\beta}) = \tilde{A}(\underline{\alpha}) + \hat{B}(\underline{\alpha})\hat{F} =$$

$$\begin{bmatrix} \tilde{A}_{F_{11}}(\underline{\alpha},\underline{\beta}) & \tilde{A}_{F_{12}}(\underline{\alpha},\underline{\beta}) \\ 0 & \tilde{A}_{F_{22}}(\underline{\alpha},\underline{\beta}) \end{bmatrix} \tag{46}$$

where: $\tilde{A}_{F_{11}}(\underline{\alpha},\underline{\beta}) = \tilde{A}_{11}(\underline{\alpha}) + \hat{B}_{11}(\underline{\beta})\hat{F}_{11} + \hat{B}_{12}(\underline{\beta})\hat{F}_{21}$; $\tilde{A}_{F_{12}}(\underline{\alpha},\underline{\beta}) = \tilde{A}_{12}(\underline{\alpha}) + \hat{B}_{11}(\underline{\beta})\hat{F}_{12} + \hat{B}_{12}(\underline{\beta})\hat{F}_{22}$; $\tilde{A}_{F_{22}}(\underline{\alpha},\underline{\beta}) = \tilde{A}_{22}(\underline{\alpha}) + \hat{B}_{21}(\underline{\beta})\hat{F}_{12} + \hat{B}_{22}(\underline{\beta})\hat{F}_{22}$. The eigenvalues of $\tilde{A}_F(\underline{\alpha},\underline{\beta})$ are then those of sub-matrices $\tilde{A}_{F_{11}}(\underline{\alpha},\underline{\beta})$ and $\tilde{A}_{F_{22}}(\underline{\alpha},\underline{\beta})$.

From the discussion of section 2, positive invariance of $\tilde{S}[\tilde{G}_2,\rho]$ implies stability of $\tilde{A}_{F_{22}}$. The original problem of robust positive invariance of $S[G,\rho]$ with stability may then be split into two distinct problems as follows:

Problem 1: Given $\hat{F}_{11}$ fixed by (45), find $\hat{F}_{21}$ such that, for all the admissible $\underline{\alpha},\underline{\beta}$, the eigenvalues of $\tilde{A}_{11}(\underline{\alpha}) + \hat{B}_{11}(\underline{\beta})\hat{F}_{11} + \hat{B}_{12}(\underline{\beta})\hat{F}_{21}$ have negative real parts.

Problem 2: Find $\hat{F}_{12}$ and $\hat{F}_{22}$ such that there exist $H(\underline{\alpha},\underline{\beta}), K(\underline{\alpha},\underline{\beta})$ that satisfy, for all the admissible $\underline{\alpha},\underline{\beta}$:

$$(H(\underline{\alpha},\underline{\beta}) - K(\underline{\alpha},\underline{\beta}))\tilde{G}_2 =$$

$$\tilde{G}_2(\tilde{A}_{22}(\underline{\alpha}) + \hat{B}_{21}(\underline{\beta})\hat{F}_{12} + \hat{B}_{22}(\underline{\beta})\hat{F}_{22}) \tag{47}$$

$$(H(\underline{\alpha},\underline{\beta}) + K(\underline{\alpha},\underline{\beta}))\rho < 0 \tag{48}$$

with $H_{kl}(\underline{\alpha},\underline{\beta}) \geq 0$ for $k \neq l$ and $K_{kl}(\underline{\alpha},\underline{\beta}) \geq 0 \; \forall \; k,l$.

Problem 1 is a classical problem of robust stabilization of the pair $(\tilde{A}_{11}(\underline{\alpha}) + \hat{B}_{11}(\underline{\beta})\hat{F}_{11}, \hat{B}_{12}(\underline{\beta}))$. Design techniques based on the concept of quadratic stabilizability (Barmish, 1985) have been lately the most applied ones. It should be noticed meanwhile that the uncertain parameters of $(\tilde{A}_{11}(\underline{\alpha}) + \hat{B}_{11}(\underline{\beta})\hat{F}_{11}$ and $\hat{B}_{12}(\underline{\beta})$ are not independents. In particular, the methodology proposed by Geromel *et al.* (1991) is able to treat efficiently this kind of problem.

For Problem 2, defining the convex domains:

$$\mathcal{D}_{\tilde{A}_{22}} := \left\{ \tilde{A}_{22} : \tilde{A}_{22} = \sum_{i=1}^{n_A} \xi_i \tilde{A}_{22}^i ; \; \sum_{i=1}^{n_A} \xi_i = 1; \; \xi_i \geq 0 \right\}$$

$$\mathcal{D}_{\hat{B}_2} := \left\{ \hat{B}_2 : \hat{B}_2 = \sum_{j=1}^{n_B} \mu_j \hat{B}_2^j ; \; \sum_{j=1}^{n_B} \mu_j = 1; \; \mu_j \geq 0 \right\}$$

where $\hat{B}_2 = [\hat{B}_{21} \;\; \hat{B}_{22}]$, the solution of Problem 2 can then be obtained from the following Linear Program (Dórea and Milani, 1995) :

$$\min_{\hat{F}_2, H^{ij}, K^{ij}, W} \sum_{k=1}^{m} \sum_{l=1}^{r} (\hat{F}_{2_{kl}} + 2W_{kl})$$

$$\text{subject to}: \; (H^{ij} - K^{ij})G = G(\tilde{A}_{22}^i + \hat{B}_2^j \hat{F}_2)$$

$$(H^{ij} + K^{ij})\rho \leq -\sigma\rho \qquad (49)$$

$$\hat{F}_2 + W \geq 0$$

$$H_{kl}^{ij} \geq 0 \text{ for } k \neq l; K_{kl}^{ij} \geq 0; W_{kl} \geq 0;$$

$$i = 1, \cdots, n_A; j = 1, \cdots, n_B$$

where $\hat{F}_2 = [\; \hat{F}_{12}' \;\; \hat{F}_{22}' \;]'$.

In view of the constraints involved, it can be easily verified that the minimization of the objective function yields $W_{ij} = 0$ for $\hat{F}_{2_{kl}} \geq 0$ and $W_{kl} = -\hat{F}_{2_{kl}}$ for $\hat{F}_{2_{kl}} \leq 0$. Therefore, at the optimum, $\hat{F}_{2_{kl}} + 2W_{kl} = |\hat{F}_{2_{kl}}|$. The sum of the absolute values of the elements of $\hat{F}_2$ is then minimized.

The positive number σ must be chosen so as to guarantee the strict satisfaction of inequalities (12) as well as to avoid an excessive proximity to the origin of the eigenvalues of $\tilde{A}_{22}(\underline{\alpha}) + \hat{B}_{21}(\underline{\beta})\hat{F}_{12} + \hat{B}_{22}(\underline{\beta})\hat{F}_{22}$.

6. CONCLUSION

The problem studied in this paper concerns the existence of a state-feedback law able to impose positive invariance of a given unbounded symmetrical polyhedron and stability for an uncertain linear system with convex structured uncertainties. The proposed method allows for a straightforward verification of the existence of a solution. And, if a solution exists, the proposed design technique solves the problem by decomposing it into two simpler problems : one of robust stabilization, for which several techniques are available, and one of positive invariance of a bounded polytope for a reduced system, which can be solved by Linear Programming.

7. REFERENCES

Barmish, B.R. (1985). Necessary and Sufficient Conditions for Quadratic Stabilizability of an Uncertain System. *JOTA*, **46**, 399-408.

Bhattacharyya, S.P. (1983). Generalized Controllability, (A, B)-invariant Subspaces and Parameter Invariant Control. *SIAM J. Alg. Disc. Meth.*, **4**, 529-533.

Bitsoris, G. (1991). Existence of Polyhedral Positively Invariant Sets for Continuous-time Systems. *Contr. - Theor. and Advanc. Tech.*, **7**, 407-427.

Bitsoris, G. and E. Gravalou (1992). Robust Linear Controller under State and Control Constraints. *Proc. of 31st CDC*, Tucson, USA, 2640-2642.

Blanchini, F. (1991). Constrained Control for Uncertain Linear System. *JOTA*, **71**, 465-484.

Castelan, E.B. and J.-C. Hennet. (1992). Eigenstructure Assignment for State Constrained Linear Continuous-time Systems. *Automatica*, **28**, 605-611.

Castelan, E.B. and J.-C. Hennet (1993). On Invariant Polyhedra of Continuous-time Linear Systems. *IEEE Trans. Automat. Contr.*, **38**, 1680-1685.

Dórea, C.E.T. and B.E.A. Milani (1995). Design of L-Q Regulators for State Constrained Continuous-time Systems. To appear in *IEEE Trans. Automat. Contr.*

Geromel, J.C., P.L.D. Peres and J. Bernussou (1991). On a Convex Parameter Space Method for Linear Control Design of Uncertain Systems. *SIAM J. Contr. Optimiz.*, **29**, 381-402.

Hennet, J.-C. and E.B. Castelan (1992). Robust Invariant Controllers for Constrained Linear Systems. *Proc. of 1992 ACC*, Chicago, USA, 993-997.

Milani, B.E.A (1994). Robust Linear Regulator Design for Continuous-time Systems under State and Control Constraints. *Proc. of 33st CDC*, Orlando, USA, 2067-2068.

Sznaier, M. (1993). A Set Induced Norm Approach to the Robust Control of Constrained Systems. *SIAM J. Contr. Optimiz.*, **31**, 733-746.

Wonham, W.M. (1985). *Linear Multivariable Control. A Geometric Approach.* Springer-Verlag, New York.

THE REGULATOR PROBLEM WITH ROBUST STABILITY

M. K. K. Cevik

Electrical and Electronics Engineering Faculty
Istanbul Technical University, 80626 Maslak, Istanbul, Turkey

J. M. Schumacher

CWI. P.O. Box 94079. 1090 GB Amsterdam, the Netherlands, and
Tilburg University. CentER and Department of Economics,
P.O. Box 90153, 5000 LE Tilburg, the Netherlands

Abstract. The design of a controller such that the closed-loop system will track reference signals or reject disturbance signals from a specified class is known as the 'servomechanism problem' or the 'regulator problem'. For the regulator problem to be solvable with robust closed-loop stability, the plant obviously needs to be such that the regulation problem and the robust stabilization problem are solvable separately. In this paper we determine the extra conditions that are necessary and sufficient for the two problems to be solved simultaneously. It turns out that these conditions can be given a simple geometric interpretation in terms of a multivariable version of the Nyquist curve of the plant.

Résumé. On donne des conditions nécessaires et suffisantes pour le problème de la stabilisation robuste d'une système linéaire sous une contrainte de régulation. Il est montré que les conditions obtenues admettent une interprétation géometrique.

Keywords. Regulation, robust stability, trade offs.

1. INTRODUCTION

Recent research in control theory has emphasized the need for a study of the trade-offs between various design objectives in order to work towards a unification of the classical and the modern theory. Various approaches have been suggested, including realizability constraints (Freudenberg and Looze 1988), optimization methods (Boyd and Barratt 1991), and loop shaping (McFarlane and Glover 1990, Doyle *et al.* 1992). In this paper we study the interaction between robust stability requirements and regulation requirements. It turns out that this particular interaction can be described in a remarkably simple way. The development will be based on the use of subspace-valued functions, as introduced in (Martin and Hermann 1978, Brockett and Byrnes 1981).

This paper is a condensed version of (Cevik and Schumacher 1995). We refer to that paper and the companion paper (Cevik and Schumacher 1994) for proofs.

2. PRELIMINARIES

The regulator problem may be formulated as follows, assuming that the observed outputs coincide with the regulated outputs (cf. for instance (Basile and Marro 1992; p. 317), where the problem below is referred to as the Autonomous Regulator Problem). Consider a finite-dimensional linear time-invariant system of the following form:

$$\dot{x}_1(t) = A_{11}x_1(t) + A_{12}x_2(t) + B_1u(t) \quad (2.1)$$
$$\dot{x}_2(t) = A_{22}x_2(t) \quad (2.2)$$
$$y(t) = C_1x_1(t) + C_2x_2(t). \quad (2.3)$$

The interpretation is as follows: x_1 denotes the state of the plant, whereas x_2 is the state of an 'exosystem' that generates signals which can be disturbances or references. The matrix A_{22} has its eigenvalues on the imaginary axis, allowing the reference/disturbance signals to be steps, ramps, sinusoids, etc. The variable $y(t)$ should converge to zero, irrespective of the presence of the signals generated by the exosystem. This is to be achieved by a linear time-invariant compensator of the form

$$\dot{z}(t) = Fz(t) + Gy(t) \quad (2.4)$$
$$u(t) = Hz(t) + Jy(t). \quad (2.5)$$

We shall consider the regulator problem under the following standing assumptions.

Assumptions The system (2.1–2.3) satisfies

(A1) the pair (A_{11}, B_1) is stabilizable;
(A2) the pair (C, A) given by

$$C = [C_1 \ C_2], \quad A = \begin{bmatrix} A_{11} & A_{12} \\ 0 & A_{22} \end{bmatrix} \quad (2.6)$$

is detectable;

(A3) all eigenvalues of A_{22} are on the imaginary axis;

(A4) for every eigenvalue λ of A_{22}, the matrix

$$\begin{bmatrix} \lambda I - A_{11} & -B_1 \\ C_1 & 0 \end{bmatrix}$$

has full column rank.

Assumption (A1) is necessary for the plant to be stabilizable by a feedback compensator, and so this is a natural assumption to make. Detectability of the pair (C_1, A_{11}) is necessary as well for closed-loop stability to be achieved by a compensator of the form (2.4–2.5); assumption (A2) requires a bit more, however. It can be argued that (A2) may be assumed without essential loss of generality in the regulator problem (cf. (Wonham 1979; §8.1)). Instead of (A3), the usual assumption is that the exosystem poles are in the closed right half plane (cf. for instance (Francis 1977)); although (A3) is of course stronger, it represents hardly a restriction from the applications point of view. Assumption (A4) is not quite so harmless, because it implies that the number of outputs is at least equal to the number of inputs, whereas it is well-known (Wonham 1979; Ch. 8) that the regulator problem can only be 'well-posed' if the number of outputs is at most equal to the number of inputs. One may therefore say that (A4) essentially limits one to the case in which the number of control inputs is equal to the number of regulated outputs. The assumption requires that the plant zeros do not coincide with the exosystem poles, which is a well-known condition in connection with the regulator problem (Wonham 1979; Thm. 8.3), (Basile and Marro 1992; Cor. 5.2-2).

An important role in our analysis will be played by certain subspace-valued functions associated to plant and controller. To define these, consider the following vector equations associated with the plant (2.1–2.3) and the controller (2.4–2.5):

$$\begin{bmatrix} sI - A_{11} & 0 & -B_1 \\ C_1 & -I & 0 \end{bmatrix} \begin{bmatrix} x \\ y \\ u \end{bmatrix} = 0 \tag{2.7}$$

$$\begin{bmatrix} sI - A_{11} & -A_{12} & 0 & -B_1 \\ 0 & sI - A_{22} & 0 & 0 \\ C_1 & C_2 & -I & 0 \end{bmatrix} \begin{bmatrix} x_1 \\ x_2 \\ y \\ u \end{bmatrix} = 0 \tag{2.8}$$

$$\begin{bmatrix} sI - F & -G & 0 \\ H & J & -I \end{bmatrix} \begin{bmatrix} z \\ y \\ u \end{bmatrix} = 0. \tag{2.9}$$

We now define

$$\mathcal{P}(s) = \{\begin{bmatrix} y \\ u \end{bmatrix} \mid \exists x \text{ s.t. (2.7) holds}\},$$
$$\mathcal{P}(\infty) = \text{im} \begin{bmatrix} 0 \\ I \end{bmatrix} \tag{2.10}$$

$$\mathcal{M}(s) = \{\begin{bmatrix} y \\ u \end{bmatrix} \mid \exists x_1, x_2 \text{ s.t. (2.8) holds}\},$$
$$\mathcal{M}(\infty) = \text{im} \begin{bmatrix} 0 \\ I \end{bmatrix} \tag{2.11}$$

$$\mathcal{C}(s) = \{\begin{bmatrix} y \\ u \end{bmatrix} \mid \exists z \text{ s.t. (2.9) holds}\},$$
$$\mathcal{C}(\infty) = \text{im} \begin{bmatrix} I \\ J \end{bmatrix}. \tag{2.12}$$

Note that all functions take values in the set of subspaces of the product space $\mathcal{Y} \times \mathcal{U}$, which is an $(m+p)$-dimensional space if m is the number of inputs and p is the number of outputs. The functions above may be considered as functions on the extended complex plane $\mathbb{C} \cup \{\infty\}$, but we shall only need their values on the closed right half plane

$$\mathbb{C}^+ \stackrel{\text{def}}{=} \{s \in \mathbb{C} \mid \text{Re}\, s \geq 0\} \cup \{\infty\}. \tag{2.13}$$

The *closed-loop system* takes the form

$$\frac{d}{dt} \begin{bmatrix} x_1 \\ z \\ x_2 \end{bmatrix} (t) = A_e \begin{bmatrix} x_1 \\ z \\ x_2 \end{bmatrix} (t) \tag{2.14}$$

$$y(t) = [C_1 \ \ 0 \ \ C_2] \begin{bmatrix} x_1 \\ z \\ x_2 \end{bmatrix} (t) \tag{2.15}$$

where

$$A_e = \begin{bmatrix} A_{11} + B_1 J C_1 & B_1 H & A_{12} + B_1 J C_2 \\ G C_1 & F & G C_2 \\ 0 & 0 & A_{22} \end{bmatrix}. \tag{2.16}$$

The compensator is said to satisfy the *internal stability requirement* if the closed-loop system is stable when $x_2(t) = 0$, that is, if the matrix

$$\begin{bmatrix} A_{11} + B_1 J C_1 & B_1 H \\ G C_1 & F \end{bmatrix}$$

is stable.

In order to define a requirement for *robust* stability, it is of interest to consider an equivalent formulation based on the subspace-valued functions (2.10) and (2.12), and on the notion of the minimal angle between subspaces. The minimal angle between two subspaces $\mathcal{Y}$ and $\mathcal{Z}$ of a unitary space $\mathcal{X}$ is defined as follows (see for instance (Gohberg and Krein 1969; p. 339)):

$$\sin \phi(\mathcal{Y}, \mathcal{Z}) = \min \{\|y - z\| \mid y \in \mathcal{Y},\ z \in \mathcal{Z},\ \|y\| = 1\}, \tag{2.17}$$

with $0 \leq \phi \leq \frac{1}{2}\pi$. Note that the minimal angle is nonzero if and only if the two subspaces intersect only in 0. If this condition holds, another formula for the minimal angle is given by (see again for instance (Gohberg and Krein 1969; p. 339))

$$\sin \phi(\mathcal{Y}, \mathcal{Z}) = \|\Pi_{\mathcal{Z}}^{\mathcal{Y}}\|^{-1} \tag{2.18}$$

where $\Pi_{\mathcal{Z}}^{\mathcal{Y}}$ denotes the skew projection along $\mathcal{Y}$ onto $\mathcal{Z}$, defined on $\mathcal{Y} + \mathcal{Z}$.

Lemma 2.1 The closed-loop system formed by the plant (A_{11}, B_1, C_1) and the compensator (2.4–2.5) is stable if and only if

$$\min_{s\in\mathbb{C}^+} \sin\phi(\mathcal{P}(s), \mathcal{C}(s)) > 0. \quad (2.19)$$

Another way to express the above result is that $\mathcal{P}(s)$ and $\mathcal{C}(s)$ should be complementary at each point $s \in \mathbb{C}^+$. It has been shown in (Schumacher 1992) that the minimal angle is the appropriate measure of the robustness of complementarity of two subspaces $\mathcal{Y}$ and $\mathcal{Z}$, in the sense that it gives exactly the distance (in the sense of the gap) of $\mathcal{Y}$ to the set of subspaces $\mathcal{Y}'$ that are not complementary to $\mathcal{Z}$. As a measure of robustness of stability, we shall therefore take

$$\sin\phi(\mathcal{P}, \mathcal{C}) \stackrel{\text{def}}{=} \min_{s\in\mathbb{C}^+} \sin\phi(\mathcal{P}(s), \mathcal{C}(s)). \quad (2.20)$$

The minimum is actually achieved on the imaginary axis or at infinity. The above measure can also be motivated in other ways and has been used for instance by Vidyasagar and Kimura (1986), Glover and McFarlane (1989), Georgiou and Smith (1990), and Vinnicombe (1993). As the notation suggests, the expression $\phi(\mathcal{P}, \mathcal{C})$ can be interpreted as an angle between linear spaces associated with plant and controller (Ober and Sefton 1991, Schumacher 1992).

Now consider the following problems.

PROBLEM 1 (regulator problem with internal stability: RPIS)
Given the plant and exosystem (2.1–2.3), find a compensator of the form (2.4–2.5) such that the closed-loop system (2.14–2.16) is internally stable and satisfies the *regulation requirement*

$$\mathcal{X}_+(A_e) \subset \ker [C_1 \;\; 0 \;\; C_2] \quad (2.21)$$

where $\mathcal{X}_+(A_e)$ denotes the unstable subspace of A_e.

PROBLEM 2 (robust stabilization problem with margin γ: RSP(γ))
Given the plant (2.1–2.2) and γ with $0 < \gamma < 1$, find a compensator of the form (2.4–2.5) that satisfies the *robust stability requirement*

$$\min_{s\in\mathbb{C}^+} \sin\phi(\mathcal{P}(s), \mathcal{C}(s)) > \gamma. \quad (2.22)$$

PROBLEM 3 (regulator problem with robust stability margin γ: RPRS(γ))
Given the plant and exosystem (2.1–2.3) and γ with $0 < \gamma < 1$, find a compensator of the form (2.4–2.5) such that both the regulation property (2.21) and the robust stability property (2.22) hold.

Necessary and sufficient conditions for RPIS and RSP(γ) to be solvable, along with synthesis procedures to obtain a suitable compensator, are well-known; for RPIS, see (Wonham 1979, Basile and Marro 1992) and the references therein, and for RSP(γ), see (Vidyasagar and Kimura 1986, Glover and McFarlane 1989, McFarlane and Glover 1990). One may refer in particular to (Wonham 1979; Thm. 8.1) for RPIS and (McFarlane and Glover 1990; Thm. 4.14) for RSP(γ). Our purpose in this paper is to get the same results for RPRS(γ).

To facilitate a geometric formulation, we introduce

$$\mathcal{K} = \{ \begin{bmatrix} y \\ u \end{bmatrix} \mid y = 0 \}. \quad (2.23)$$

A necessary condition for a controller (2.4–2.5) to satisfy the regulation requirement is that the eigenvectors corresponding to the eigenvalues of the closed-loop system coming from the exosystem poles should have zero output values associated to them. With the introduced notation, we van now formulate this necessary condition as

$$\mathcal{C}(\lambda) \cap \mathcal{M}(\lambda) \subset \mathcal{K} \quad (2.24)$$

for all λ in $\sigma(A_{22})$. (See (Cevik and Schumacher 1994) for conditions that are both necessary and sufficient.) The above condition leads to an upper bound on the achievable robustness of stability through the following analysis, which is purely based on finite-dimensional geometry.

Lemma 2.2 Consider the system (2.1–2.3), and assume that the pair (C_1, A_{11}) is detectable, and that all eigenvalues of A_{22} are in the closed right half plane. Under these conditions, which are in particular satisfied if assumptions (A2) and (A3) hold, assumption (A4) holds if and only if

$$\mathcal{P}(\lambda) \cap \mathcal{K} = \{0\}. \quad (2.25)$$

Lemma 2.3 Let $\mathcal{W}$ be a finite-dimensional vector space with given subspaces $\mathcal{M}$, $\mathcal{P}$, and $\mathcal{K}$. Assume that $\mathcal{P} \subset \mathcal{M}$ and $\mathcal{P} \cap \mathcal{K} = \{0\}$. Under these conditions, there exists a subspace $\mathcal{C}$ such that

$$\mathcal{C} \oplus \mathcal{P} = \mathcal{W} \text{ and } \mathcal{C} \cap \mathcal{M} \subset \mathcal{K} \quad (2.26)$$

if and only if $\mathcal{M} = \mathcal{P} + (\mathcal{K} \cap \mathcal{M})$.

Lemma 2.4 Let $\mathcal{W}$ be a unitary space with given subspaces $\mathcal{M}$, $\mathcal{P}$, and $\mathcal{K}$. Assume that $\mathcal{P} \subset \mathcal{M}$, $\mathcal{P} \cap \mathcal{K} = \{0\}$, and $\mathcal{M} = \mathcal{P} + (\mathcal{K} \cap \mathcal{M})$. Under these conditions, we have

$$\phi(\mathcal{P}, \mathcal{C}) \leq \phi(\mathcal{P}, \mathcal{K} \cap \mathcal{M}) \quad (2.27)$$

for all subspaces $\mathcal{C}$ satisfying (2.26), and equality is achieved for instance for $\mathcal{C} = (\mathcal{K} \cap \mathcal{M}) \oplus \mathcal{M}^\perp$.

To get a readily computable expression for the upper bound appearing in (2.27), take normalized kernel and image representations for the subspace $\mathcal{P}$, so

$$\mathcal{P} = \operatorname{im} P = \ker \tilde{P}, \quad P^*P = I, \quad \tilde{P}\tilde{P}^* = I. \quad (2.28)$$

Also take an image representation C for $\mathcal{C}$. Because $\mathcal{C}$ and $\mathcal{P}$ are complementary, the matrix $\tilde{P}C$ must be invertible, and since image representations are only determined up to right multiplication by nonsingular matrices, we may as well assume that

$$\tilde{P}C = I. \quad (2.29)$$

Under the assumptions we have made, the projection along $\mathcal{P}$ onto $\mathcal{K} \cap \mathcal{M} = \mathcal{C} \cap \mathcal{M}$ in $\mathcal{M}$ is given by $C\tilde{P}|_{\mathcal{M}}$, and so by (2.18) we have $\sin\phi(\mathcal{P}, \mathcal{K} \cap \mathcal{M}) = \|C\tilde{P}|_{\mathcal{M}}\|^{-1}$. The latter expression can be further evaluated as follows, using the fact that the matrix $[P \ \tilde{P}^*]$ is unitary:

$$\|C\tilde{P}|_{\mathcal{M}}\| = \|C|_{\tilde{P}\mathcal{M}}\| =$$

$$= \| \begin{bmatrix} P^* \\ \tilde{P} \end{bmatrix} C|_{\tilde{P}\mathcal{M}}\| = (1 + \|P^*C|_{\tilde{P}\mathcal{M}}\|^2)^{\frac{1}{2}}.$$

In all, we get (under the assumptions (2.28–2.29))

$$\sin\phi(\mathcal{P}, \mathcal{K} \cap \mathcal{M}) = (1 + \|P^*C|_{\tilde{P}\mathcal{M}}\|^2)^{-\frac{1}{2}}. \quad (2.30)$$

3. SOLUTION OF THE REGULATOR PROBLEM WITH ROBUST STABILITY

It will be convenient to always work with *normalized* image and kernel representations for the plant; that is, we shall require these representations to satisfy

$$P^*(s)P(s) = I, \qquad \tilde{P}(s)\tilde{P}^*(s) = I \quad (3.31)$$

where $M^*(s)$, for a real rational matrix $M(s)$, denotes $M^T(-s)$. For the controller we shall always work with image representations $C(s)$ normalized such that $\tilde{P}(s)C(s) = I$. Due to these normalizations, the stability margin achieved by a controller represented by $C(s)$ is

$$\sin\phi(\mathcal{P}, \mathcal{C}) = \|C\tilde{P}\|_\infty^{-1} = \|C\|_\infty^{-1}. \quad (3.32)$$

It is shown in (Cevik and Schumacher 1994) that the stabilizing compensators that also satisfy the regulation requirement can be parametrized in the form

$$C(s) = C_0(s) - P(s)\Psi(s)\tilde{H}(s) \quad (3.33)$$

where $C_0(s)$ is a particular solution, $\Psi(s)$ is any element of $RH_\infty^{m \times p}$, and $\tilde{H}(s)$ is a square and nonsingular RH_∞-matrix. The matrix $\tilde{H}(s)$ can be computed from the problem data: its zeros coincide with the exosystem poles.

In terms of this parametrization, we see from (3.32) that we should aim at minimizing $\|C_0 - P\Psi\tilde{H}\|$ over the RH_∞-matrices $\Psi(s)$. Applying the same trick as in the derivation of (2.30), we can also write

$$\sin\phi(\mathcal{P}, \mathcal{C}) = \|C_0 - P\Psi\tilde{H}\|_\infty^{-1} =$$

$$= (1 + \|P^*C_0 - \Psi\tilde{H}\|_\infty^2)^{-\frac{1}{2}}$$

and so we may as well minimize $\|P^*C_0 - \Psi\tilde{H}\|_\infty$.

There are two obvious lower bounds for this problem. First of all, since $\Psi(s)\tilde{H}(s)$ is an RH_∞-matrix, it follows from Nehari's theorem (cf. for instance (Francis 1987)) that a lower bound for the minimization problem is given by the norm of the Hankel operator associated with P^*C_0:

$$\|P^*C_0 - \Psi\tilde{H}\|_\infty \geq \|\Gamma_{P^*C_0}\|. \quad (3.34)$$

In the case in which we have no regulation constraints and so we have a pure robust stabilization problem, we can take $\tilde{H}(s) = I$ and then the Hankel norm is an exact lower bound (McFarlane and Glover 1990). At first sight the bound may seem to depend on the choice of the particular stabilizing compensator $C_0(s)$; however, another compensator $C(s) = C_0(s) - P(s)Q(s)$ would produce the symbol $P^*C = P^*C_0 - Q$ which differs only by an H_∞-matrix from P^*C_0 so that the Hankel norm would not be affected. Glover and McFarlane (1989) give an expression that is explicitly independent of the choice of the compensator: they show that

$$(1 + \|\Gamma_{P^*C_0}\|^2)^{-\frac{1}{2}} = (1 - \|\Gamma_{\tilde{P}^*}\|^2)^{\frac{1}{2}}. \quad (3.35)$$

The second lower bound that is immediately seen to hold is

$$\|P^*C_0 - \Psi\tilde{H}\|_\infty \geq \|P^*(\lambda)C_0(\lambda)|_{\ker \tilde{H}(\lambda)}\| \quad (3.36)$$

for all zeros $\lambda \in i\mathbb{R}$ of $\tilde{H}(s)$. From the analysis in the previous section, we see that this inequality is really of a geometric nature. Indeed, since $\tilde{P}(\lambda)$ is surjective for all $\lambda \in i\mathbb{R}$, we have $\tilde{P}(\lambda)\ker\tilde{M}(\lambda) = \tilde{P}(\lambda)\ker\tilde{H}(\lambda)\tilde{P}(\lambda) = \ker\tilde{H}(\lambda)$, so it follows from (2.30) that the above inequality may also be written as

$$\phi(\mathcal{P}, \mathcal{C}) \leq \phi(\mathcal{P}(\lambda), \mathcal{K} \cap \mathcal{M}(\lambda)). \quad (3.37)$$

The theorem below shows that the actual situation is good as one might hope on the basis of the above two inequalities: RPRS(γ) is solvable whenever RPIS is solvable and we have both

$$\gamma < (1 + \|\Gamma_{P^*C_0}\|^2)^{-\frac{1}{2}} \quad (3.38)$$

and

$$\gamma < \sin\phi(\mathcal{P}(\lambda), \mathcal{K} \cap \mathcal{M}(\lambda)) \quad \forall\lambda \in \sigma(A_{22}). \quad (3.39)$$

Theorem 3.1 Consider the problems RPIS, RSP(γ), and RPRS(γ), as described in section 2. Define subspace-valued functions $\mathcal{P}(s)$ and $\mathcal{M}(s)$ by (2.10) and (2.11) respectively, and define $\mathcal{K}$ by (2.23). Under assumptions (A1–A4), the problem RPRS(γ) is solvable if and only if the following conditions hold:

(i) RPIS is solvable;

(ii) RSP(γ) is solvable;

(iii) $\gamma < \sin\phi(\mathcal{P}(\lambda), \mathcal{K} \cap \mathcal{M}(\lambda))$ for all exosystem poles λ.

4. EXAMPLE

To illustrate the results of this paper in the simplest possible context, let us consider a first-order system that is to be regulated against a constant disturbance. After scaling, the equations can be written in the form

$$\dot{x}_1 = ax_1 + x_2 + u \quad (4.40)$$

$$\dot{x}_2 = 0 \quad (4.41)$$

$$y = x_1. \qquad (4.42)$$

The assumptions (A1–A4) are satisfied for all values of a. The subspace-valued function $\mathcal{M}(s)$ defined in (2.11) is given by

$$\mathcal{M}(s) = \ker [s(s-a) \quad -s]. \qquad (4.43)$$

This expression can be obtained symbolically by eliminating x_1 and x_2 from the equations $(s-a)x_1 = x_2+u$, $sx_2 = 0$, and $y = x_1$, considering s as a non-cancellable parameter. If we do allow cancellation, we get the expression for $\mathcal{P}(s)$ as in (2.10):

$$\mathcal{P}(s) = \ker [s-a \quad -1]. \qquad (4.44)$$

The upper bound on the achievable robustness of stability imposed by the regulation constraint is obtained from (3.37), noting that $\mathcal{M}(0) = \mathbb{C}^2 = \mathcal{Y} \times \mathcal{U}$ and $\mathcal{K} = \text{im} \begin{bmatrix} 0 \\ 1 \end{bmatrix}$:

$$\sin \phi(\mathcal{P}(0), \mathcal{K}) = \frac{1}{\sqrt{1+a^2}}. \qquad (4.45)$$

Normalized image and kernel representations for $\mathcal{P}(s)$ are given by

$$P(s) = \frac{1}{s+\sqrt{a^2+1}} \begin{bmatrix} 1 \\ s-a \end{bmatrix} \qquad (4.46)$$

and

$$\tilde{P}(s) = \frac{1}{s+\sqrt{a^2+1}} [s-a \quad -1]. \qquad (4.47)$$

An image representation $C_0(s)$ for a stabilizing controller, normalized such that $\tilde{P}(s)C_0(s) = I$, is found by solving the equation

$$(s-a)c_{01}(s) - c_{02}(s) = s + \sqrt{a^2+1} \qquad (4.48)$$

in the unknown RH_∞-functions $c_{01}(s)$ and $c_{02}(s)$. A simple solution is provided by

$$C_0(s) = \begin{bmatrix} 1 \\ -a-\sqrt{a^2+1} \end{bmatrix}. \qquad (4.49)$$

We get

$$(1+\|\Gamma_{P^* C_0}\|^2)^{-\frac{1}{2}} = (1-\|\Gamma_{\tilde{P}^*}\|^2)^{\frac{1}{2}} = = \tfrac{1}{2}\sqrt{2}\,(1-\frac{a}{\sqrt{a^2+1}})^{\frac{1}{2}}. \qquad (4.50)$$

If we re-parametrize by setting

$$a = \cot\theta, \qquad 0 < \theta < \pi \qquad (4.51)$$

we find for the two upper bounds (4.45) and (4.50):

$$\sin \phi(\mathcal{P}(0), \mathcal{K}) = \sin\theta \qquad (4.52)$$

$$(1+\|\Gamma_{P^* C_0}\|^2)^{-\frac{1}{2}} = \sin \tfrac{1}{2}\theta. \qquad (4.53)$$

Note that $\sin\theta > \sin\frac{1}{2}\theta$ for $0 < \theta < \frac{2}{3}\pi$, whereas the inequality is reversed for $\frac{2}{3}\pi < \theta < \pi$; the value $\theta = \frac{2}{3}\pi$ corresponds to $a = -\frac{1}{3}\sqrt{3}$. Taking into account that $-a^{-1}$ is the dc gain of the system (4.40) and that this quantity is not affected by the scaling we used to arrive at the form (4.40), we get the following conclusion:

> *for a first-order system regulated against a constant disturbance, the regulation requirement is restrictive with respect to the achievable robustness of stability if and only if the system is open-loop stable with a dc gain less than* $\sqrt{3}$.

Of course, similar rules may be derived for higher-order systems and for other types of regulation constraints.

The proof of Thm. 3.1 in (Cevik and Schumacher 1995) is constructive and so we can calculate a regulator that achieves a specified feasible robustness margin γ. For the above problem with $a = 0$, a solution for given $\gamma < \frac{1}{2}\sqrt{2}$ is given in (Cevik and Schumacher 1995) by the transfer function

$$c(s) = \frac{-2\alpha^2 s - \alpha^2 + 1}{(\alpha^2+1)s} \qquad (4.54)$$

where $\alpha = \frac{1}{\gamma}\sqrt{1-\gamma^2}$. This transfer function has a pole at 0, as it should have to satisfy the regulation requirement. The actual margin of robustness of stability achieved by the above controller is

$$\sin \phi(\mathcal{P}, \mathcal{C}) = (1+\left[\frac{2\alpha}{\alpha^2+1}\right]^2 \alpha^2)^{-\frac{1}{2}} \qquad (4.55)$$

and this is indeed better than the required margin $\gamma = (1+\alpha^2)^{-\frac{1}{2}}$. If we let γ tend to its optimal value $\frac{1}{2}\sqrt{2}$ then α tends to 1 and the controller tends to $c(s) = -1$; this controller optimizes the robustness margin, but it no longer satisfies the regulation constraint.

5. CONCLUSIONS

The problem of optimizing the robustness of stability with respect to coprime factor perturbations was posed by Vidyasagar and Kimura (1986) and was reduced in that paper to a certain H_∞ optimization problem. Later on it was shown by Glover and McFarlane (1989) that, if the perturbations are taken with respect to *normalized* coprime factors, an exact solution can be obtained in a relatively simple way. The fact that the optimal robust stabilization problem is such a nice one came as a surprise at the time. In the present paper we show that the problem even remains nice if we add regulation constraints to it; in view of the generally adverse behavior of optimization problems when side constraints are added, this outcome may be viewed as a new surprise.

Our techniques in this paper have relied in particular on the use of subspace-valued functions associated both with plant and controller. These can be seen as a multivariable generalization of the Nyquist curve, as may be argued in a mathematical sense using the identification of the extended complex plane with the Grassmannian $G^1(\mathbb{C}^2)$. The present paper however demonstrates more than that: it shows that the multivariable Nyquist curve continues to play the role of a mediator between various design objectives, as the scalar version does in classical control theory.

A noticeable difference between the scalar and the multivariable versions of the Nyquist curve is that geometry plays a much larger role in the latter than in the former. The reason for this is that the scalar version can be interpreted as a function whose values are one-dimensional subspaces of a two-dimensional space, and not much subspace geometry is possible in two dimensions; in particular there are no subspaces inclusions which are not trivial in one way or another. In the multivariable case, however, we do have a nontrivial setting, leading to formulations that would seem contrived in the scalar case. The role of finite-dimensional geometry is clear for instance in the characterization that we gave of the upper bound on the achievable robustness of stability due to regulation requirements.

A natural question to ask is whether it is possible to include robustness of regulation ((Wonham 1979; Ch. 8), (Vidyasagar 1985; §7.5)) along with robustness of stability. In forthcoming work it will be shown that robustness of regulation can indeed be incorporated by an appropriate sharpening of the interpolation conditions, and that the trade-off against robustness of stability can be assessed in much the same way as in the present paper.

Among our standing assumptions (A1–A4), the full column rank assumption (A4) is the most restrictive one. If this assumption does not hold, the set of compensators that achieve regulation with internal stability is no longer completely described by (3.33). An example of such a situation occurs in systems with hydraulic actuators, which ensure that the plant already has a zero at the origin of the complex plane so that tracking of step inputs is automatically guaranteed. Another extension of the present work that would be of interest is to consider the case in which the to-be-controlled outputs are not necessarily the same as the observed outputs.

REFERENCES

Basile, G. and G. Marro (1992). *Controlled and Conditioned Invariants in Linear System Theory.* Prentice Hall. Englewood Cliffs, NJ.

Boyd, S. P. and C. H. Barratt (1991). *Linear Controller Design.* Prentice Hall. Englewood Cliffs, NJ.

Brockett, R. W. and C. I. Byrnes (1981). 'Multivariable Nyquist criteria, root loci and pole placement: A geometric viewpoint'. *IEEE Trans. Automat. Control* **AC-26**, 271–284.

Cevik, M. K. K. and J. M. Schumacher (1994). 'Regulation as an interpolation problem'. Manuscript under review.

Cevik, M. K. K. and J. M. Schumacher (1995). 'The regulator problem with robust stability'. To appear in *Automatica.*

Doyle, J. C., B. A. Francis and A. R. Tannenbaum (1992). *Feedback Control Theory.* Macmillan. New York.

Francis, B. A. (1977). 'The linear multivariable regulator problem'. *SIAM J. Control* **15**, 486–505.

Francis, B. A. (1987). *A Course in H_∞ Control Theory.* Lect. Notes Contr. Inf. Sci. 88. Springer. Berlin.

Freudenberg, J. S. and D. P. Looze (1988). *Frequency Domain Properties of Scalar and Multivariable Feedback Systems.* Lect. Notes Contr. Inform. Sci. 104. Springer. New York.

Fuhrmann, P. A. (1981). *Linear Systems and Operators in Hilbert Space.* McGraw-Hill. New York, NY.

Georgiou, T. T. and M. C. Smith (1990). 'Optimal robustness in the gap metric'. *IEEE Trans. Automat. Control* **AC-35**, 673–685.

Glover, K. and D. C. McFarlane (1989). 'Robust stabilization of normalized coprime factor plant descriptions with H_∞ bounded uncertainty'. *IEEE Trans. Automat. Contr.* **AC-34**, 821–830.

Gohberg, I. C. and M. G. Krein (1969). *Introduction to the Theory of Linear Nonselfadjoint Operators.* Transl. Math. Monogr. 18. Amer. Math. Soc.. Providence, RI.

Hyde, R. A. and K. Glover (1993). 'The application of scheduled H_∞ controllers to a VSTOL aircraft'. *IEEE Trans. Automat. Contr.* **AC-38**, 1021–1039.

Martin, C. F. and R. Hermann (1978). 'Applications of algebraic geometry to system theory: The McMillan degree and Kronecker indices as topological and holomorphic invariants'. *SIAM J. Control Optim.* **16**, 743–755.

McFarlane, D. C. and K. Glover (1990). *Robust Controller Design Using Normalized Coprime Factor Plant Descriptions.* Lect. Notes Contr. Inform. Sci. 138. Springer. New York.

Ober, R. J. and J. A. Sefton (1991). 'Stability of control systems and graphs of linear systems'. *Syst. Contr. Lett.* **17**, 265–280.

Postlethwaite, I. and S. Skogestad (1993). Robust multivariable control using H^∞ methods: analysis, design and industrial applications. In H. Trentelman and J. C. Willems (Eds.). 'Essays on Control: Perspectives in the Theory and its Applications'. Birkhäuser. Boston. pp. 269–337.

Schumacher, J. M. (1992). 'A pointwise criterion for controller robustness'. *Systems & Control Letters* **18**(1), 1–8.

Vidyasagar, M. (1985). *Control System Synthesis: A Factorization Approach.* MIT Press. Cambridge, Mass.

Vidyasagar, M. and H. Kimura (1986). 'Robust controllers for uncertain linear multivariable systems'. *Automatica* **22**, 85–94.

Vinnicombe, G. (1993). 'Frequency domain uncertainty and the graph topology'. *IEEE Trans. Automat. Contr.* **AC-38**, 1371–1383.

Wonham, W. M. (1979). *Linear Multivariable Control: a Geometric Approach.* 2nd edn. Springer. New York.

STABILITY PROPERTIES OF CONE-PRESERVING DISCRETE-TIME SYSTEMS

Sophie TARBOURIECH and Christian BURGAT

L.A.A.S.-C.N.R.S., 7, Avenue du Colonel Roche, 31077 Toulouse Cedex, France.
Fax : (33) 61 55 35 77 - E-mail : tarbour@laas.fr

Abstract. The problem of the determination of stability conditions for a linear discrete-time system having the cone-preserving property is addressed. Indeed, necessary and/or sufficient conditions are proposed in geometrical form using cones in order to characterize the stability properties (i.e., asymptotic stability, stability, instability) of discrete-time linear systems described by $x_{k+1} = Ax_k$, where matrix A has the property of leaving positively invariant a proper cone. Such a characterization does not use the knowledge of the spectral radius of matrix A.

Keywords. Stability properties, linear discrete-time system, cone-preserving property.

Résumé. Des conditions nécessaires et/ou suffisantes sont proposées sous forme géométrique utilisant des cônes dans le but de caractériser les propriétés de stabilité (i.e., stabilité asymptotique, stabilité, instabilité) d'un système discret $x_{k+1} = Ax_k$ pour lequel la matrice A a la propriété de laisser positivement invariant un cône propre. Une telle caractérisation ne requiert pas la connaissance du rayon spectral de la matrice A

Mots-clefs. Propriétés de stabilité, système linéaire discret, propriété de cône invariant.

1 INTRODUCTION

Let us consider the discrete-time system described by :

$$x_{k+1} = Ax_k \qquad (1)$$

in which $k \in J_+$ the set of nonnegative integers and matrix $A \in \Re^{n \times n}$ possesses the property of leaving positively invariant a proper cone $K \subset \Re^n$, that is

$$AK \subset K \qquad (2)$$

In (Burgat and Benzaouia, 1990) and (Tarbouriech and Burgat, 1994b), some results allowing to characterize geometrically the stability properties, i.e., asymptotic stability, stability, instability, have been proposed when matrix A is assumed K-irreducible.

The objective of this note consists in proposing an extension of these results when no assumption on the K-irreducibility of matrix A is made.

This type of results is of interest in the field of control theory, since linear dynamical systems for which matrix A possesses the property of leaving positively invariant a proper cone are wide-spread, more especially when the considered proper cone is the positive orthant $\Re^n_+$, e.g., Rumchev and James (1989), Rouhani and Tse (1981). Recently, some papers dealing with characterization of their stability properties have been published, e.g., Yorke (1967), Stern (1982), Chéganças and Burgat (1986), Burgat and Benzaouia (1990), Tarbouriech and Burgat (1994b).

Another field of applications concerns dynamical constrained systems for which the use of such a class of matrices facilitates the construction of some positively invariant sets relative to their motions, e.g., Tarbouriech and Burgat (1994a), Burgat *et al.* (1990). These results become particularly important when the stability problem is considered, e.g., Gutman and Hagander (1985), Benzaouia and Burgat (1988), Vassilaki and Bitsoris (1989).

The paper is organized as follows. Section 2 gives some preliminaries. In Section 3, the extensions of the above preliminaries are addresses. Section 4 proposes concluding remarks. An appendix recalling the main useful definitions concerning stability and cones is then given.

2 PRELIMINARIES

This section recalls some preliminary results given in (Burgat and Benzaouia, 1990) and (Tarbouriech and Burgat, 1994b).

Lemma 2.1: *(Burgat and Benzaouia, 1990) Let system (1) with property (2). Then the following properties hold :*

(i) $\forall x \in K_+$, $-\Pi x \in K_- + x$,
(ii) $\forall x \in \partial K_+$, $-\Pi x \notin intK_+$ *(resp.* $\forall x \in \partial K_-$, $\Pi x \notin intK_+$*).*

Lemma 2.2: *(Burgat and Benzaouia, 1990) If* C_+ *(resp.* C_-*) is a nonempty cone, then* $AC_+ \subset C_+$ *(resp.* $AC_- \subset C_-$*).*

Lemma 2.3: *(Burgat and Benzaouia, 1990) If* $\rho(A) = 1$ *then :*

$$\nexists x \in K_+ \text{ such that } -\Pi x \in intK_+$$

or equivalently

$$\forall x \in K_+,\ -\Pi x \notin intK_+$$

Let us now add some useful necessary conditions in the case where $\rho(A) = 1$ or $\rho(A) > 1$.

Lemma 2.4: *If* $\rho(A) = 1$ *then* $(C_+ \cap C_-)\backslash\{0\} \cap K_+ \neq \emptyset$.

Proof : Let eigenvector x_0 of matrix A corresponding to $\rho(A) = 1$, x_0 is different from vector zero. From theorem 5.2, $x_0 \in K_+$ and since $\rho(A) = 1$ we get $-\Pi x_0 = \Pi x_0 = 0$, therefore x_0 belongs to C_+ and C_-. Thus $(C_+ \cap C_-)\backslash\{0\} \cap K_+ \neq \emptyset$.

Lemma 2.5: *If A is K-irreducible with* $\rho(A) = 1$ *then* $C_+ = C_-$ *and* $(C_+ \cap C_-)\backslash\{0\} \cap intK_+ \neq \emptyset$.

Proof : It readily follows from theorem 5.3 and lemma 2.4.

Lemma 2.6: *If A is K-irreducible with* $\rho(A) \geq 1$ *then* $\nexists x \in intK_+$ *such that* $-\Pi x \neq 0 \in K_+$.

Proof : Let x_0 the eigenvector of matrix A corresponding to $\rho(A)$, which can be either equal to 1 or > 1 ; $x_0 \in intK_+$. Assume also that there exists a vector x belonging to $intK_+$ such that $-\Pi x \neq 0 \in K_+$. We then construct the vector $y = \alpha x - x_0 \in \partial K_+$ with $\alpha > 0$; this scalar α always exists since x and x_0 belong to $intK_+$ (see Vandergraft, 1968). It follows $0 \overset{K}{\leq} Ay = \alpha Ax - \rho(A)x_0 \overset{K}{\leq} \alpha x - x_0 = y$ from the assumptions. This means that vector y belongs to ∂K_+, $Ay \in \partial K_+$, contradicting the K-irreducibility assumption.

Lemma 2.7: *If* $\rho(A) = 1$ *and system (1) is unstable, then there does not exist any* $y \in intK_+$ *such that* $y \in C_+$.

Proof : We slightly develop the proof given in (Burgat and Benzaouia, 1990). If we assume that $\rho(A) = 1$ in the unstable case, we can have without loss of generality :

$$Ax_1 = x_1$$
$$Ax_2 = x_2 + x_1$$

Hence, $x_1 \in (C_+ \cap C_-)\backslash\{0\} \cap K_+$ and $-\Pi x_2 \in (C_+ \cap C_-)\backslash\{0\} \cap K_-$ that implies $x_2 \in C_-$. If we also assume that $\exists y \in intK_+$ such that $-\Pi y \in K_+$ then we can construct an invariant set $\mathcal{D} = (K_+ - y) \cap (K_- + y)$, for which $\{0\} \subset int\mathcal{D}$ (lemma 5.1). There then always exists a scalar α such that $\alpha\mathcal{D}$ contains x_2. Hence, if we choose as initial condition x_2 by an iterative process it follows $A^k x_2 = kx_2 + x_1$ and $A^k x_2 \longrightarrow \infty$ as $k \longrightarrow \infty$: $A^k x_2 \notin \alpha\mathcal{D}$. We conclude that if system (1) is unstable, there cannot exist any $y \in (intK_+ \cap C_+)\backslash\{0\}$.

Remark 2.1: *Note that the result of lemma 2.7 remains valid in the case where* $\rho(A) > 1$*, that is, when system (1) is unstable.*

Lemma 2.8: *If A is K-irreducible with* $\rho(A) > 1$ *then there does not exist any* $x \in K_+\backslash\{0\}$ *such that* $-\Pi x \neq 0 \in K_+$.

Proof : With respect to vectors $x \in intK_+$, the result follows from the above lemma 2.6. Concerning vectors $x \in \partial K_+$, and since from lemma 2.1 we get $-\Pi x \notin intK_+$ the only possibility to obtain $-\Pi x \in K_+$ is therefore to have $-\Pi x \in \partial K_+$, that is, $0 \overset{K}{\leq} Ax \overset{K}{\leq} x$. Vector x belonging to ∂K_+, contradicts the assumption of K-irreducibility of A.

Lemma 2.9: *If* $\rho(A) > 1$ *then* $(C_- \cap K_+)\backslash\{0\} \neq \emptyset$.

Proof : Consider the eigenvector x_0 of matrix A corresponding to $\rho(A) > 1$. We have $x_0 \in K_+$ and thus $-\Pi x_0 = (1 - \rho(A))x_0 \in K_-$, therefore $x_0 \in C_-$ and lemma 2.9 follows.

Lemma 2.10: *(Burgat and Benzaouia, 1990) If* $\rho(A) < 1$ *then*

(i) C_+ *is a solid cone.*
(ii) $(C_+ \cap K_+)\backslash\{0\} \neq \emptyset$.

Lemma 2.11: *(Burgat and Benzaouia, 1990) If* $\rho(A) < 1$ *then there exists* $x \in intK_+$ *such that* $-\Pi x \in intK_+$.

Lemma 2.12: *(Burgat and Benzaouia, 1990) If* $\rho(A) < 1$ *then* $C_+ \subset K_+$.

3 STABILITY RESULTS

The first main result concerning the case $\rho(A) < 1$, that is, the asymptotic stability of system (1) is proposed in (Burgat and Benzaouia, 1990) and recalled as below.

Theorem 3.1: *(Burgat and Benzaouia, 1990) The system (1), with (2), is asymptotically stable if and only if :*

(i) C_+ is a solid cone,
(ii) $C_+ \subset K_+$.

This result remains valid whether matrix A be K-irreducible or not.
Consider now the characterization of the critical stability ($\rho(A) = 1$).

Lemma 3.1: *(Tarbouriech and Burgat, 1994b) If $(C_+ \cap C_-)\backslash\{0\} \cap K_+ \neq \emptyset$ then $\rho(A) \geq 1$.*

Theorem 3.2: *(Tarbouriech and Burgat, 1994b) The dynamical system (1), with (2) for which A is K-irreducible, is critically stable if and only if $(C_+ \cap C_-)\backslash\{0\} \cap intK_+ \neq \emptyset$.*

Remark 3.1: *Recall that in the case of a matrix K-irreducible $C_+ = C_- = Ker(\Pi)$.*

Theorem 3.3: *If matrix A is K-irreducible with $\rho(A) = 1$ then the critical case is stable in the sense of definition 5.1.*

If no assumption is made on the K-irreducibility property of matrix A, that is, in the general case, the following results lead to the characterization of the stability in the critical case, as finally stated by the following results.

Lemma 3.2: *If the following properties hold*

(i) $(C_+ \cap C_-)\backslash\{0\} \cap K_+ \neq \emptyset$
(ii) $K_+ \subset C_+$.

then $\rho(A) = 1$.

Proof : The first condition indicates the existence of a vector x such that $Ax = x$. The second condition implies that $\forall y \in K_+$, $-\Pi y \in K_+$. Let x_0 the eigenvector of A corresponding to $\rho(A)$: $x_0 \in K_+$, $-\Pi x_0 = (1-\rho(A))x_0 \in K_+$. This implies that $\rho(A) \leq 1$, but since there exists $\lambda_i(A) = 1$, it follows $\rho(A) = 1$.

Lemma 3.3: *If $C_+ = C_-$ then $\rho(A) = 1$.*

Proof : $\forall x \in C_+ = C_-$, $-\Pi x = \Pi x = 0$ because K_+ is pointed. Hence there exists $\lambda_i(A) = 1$. Let x_0 an eigenvector of A, $x_0 \in K_+$: $-\Pi x_0 = (1-\rho(A))x_0$; two solutions may arise : either $-\Pi x_0 \in K_+$ and $\rho(A) = 1$ or $-\Pi x_0 \in K_-$ and $\rho(A) > 1$. Nevertheless, since $C_+ = C_-$ and K_+ is pointed, the only possibility is $\rho(A) = 1$.

Theorem 3.4: *The equilibrium $x = 0$ of system (1) with (2) is stable ($\rho(A) = 1$) if and only if one of these following conditions is satisfied :*

(i) $C_+ = C_-$,
(ii) $(C_+ \cap C_-)\backslash\{0\} \cap intK_+ \neq \emptyset$,
(iii) $\begin{cases} (C_+ \cap C_-)\backslash\{0\} \cap K_+ \neq \emptyset \\ K_+ \subset C_+ \end{cases}$

Proof : (If). (i). If $C_+ = C_-$ then $\forall x \in C_+ = C_-$, $-\Pi x = \Pi x = 0$. Consider the eigenvector $x_0 \in K_+$ corresponding to $\rho(A) = 1$. Assume that the considered critical case is unstable, that is, without loss of generality : $Ax_0 = x_0$ and $Ax_1 = x_1 + x_0$. It follows that $-\Pi x_1 = -x_0 \in K_-$ hence $x_1 \in C_-$; since $C_+ = C_-$ then $x_1 \in C_+$ and thus $-\Pi x_1 = \Pi x_1 = 0$ from the pointedness of K_+. Vector x_1 thus corresponds to $\lambda = 1$ and from definition 5.1 the case where $\rho(A) = 1$ is stable.
(ii). This condition implies the existence of a vector $x \in intK_+$ such that $-\Pi x = \Pi x = 0 \Longleftrightarrow Ax = x$. Let us consider the eigenvector $x_0 \in K_+$ corresponding to $\rho(A)$: $-\Pi x_0 = (1-\rho(A))x_0$. Hence one may have either $\rho(A) > 1$ or $\rho(A) = 1$. Since $C_+ \cap intK_+ \neq \emptyset$, from lemma 2.7 and remark 2.1 we get $\rho(A) = 1$ and system (1) is stable (definition 5.1).
(iii). It readily follows from lemmas 3.2 and 2.7.
(Only if). In order to prove part (Only if), assume that $\rho(A) = 1$ and system (1) is stable, but no condition of theorem is true. Hence, from lemma 2.4 and the assumptions made, we get $(C_+ \cap C_-)\backslash\{0\} \cap \partial K_+ \neq \emptyset$ with $C_+ \cap intK_+ = \emptyset$ and $C_- \cap intK_+ \neq \emptyset$. Consider a vector $y_0 \in intK_+ \cap C_-$ such that $y_0 \overset{K}{\geq} x_0$, where $Ax_0 = x_0$, we get $-\Pi y_0 \in K_-$. It thus follows that $A^k y_0 \overset{K}{\geq} A^{k-1} y_0 \overset{K}{\geq} \dots \overset{K}{\geq} Ay_0 \overset{K}{\geq} y_0$. Hence, from our assumptions, we obtain $A^k y_0 \not\longrightarrow 0$ as $k \longrightarrow \infty$, $A^k y_0 \not\longrightarrow x_0$ as $k \longrightarrow \infty$ and $\nexists j$ such that $A^j y_0 = y_0$; the only possibility is therefore $A^k y_0 \longrightarrow \infty$ as $k \longrightarrow \infty$, that contradicting the stability assumption.

Remark 3.2: *Note that in the previous theorem 3.4, we cannot satisfy simultaneously either (i), (ii), (iii), or (ii), (iii), or still (i), (iii). On the other hand, when conditions (i) and (ii) hold simultaneously, we obtain the above theorem 3.2, that is, the case where A is K-irreducible.*

In the sequel, we first consider the unstable case, with $\rho(A) > 1$ (strictly instability), and next the unstable case where $\rho(A) = 1$ (critically instability).

Lemma 3.4: *(Tarbouriech and Burgat, 1994b) If $(C_+ \cap K_+)\backslash\{0\} = \emptyset$ then $\rho(A) > 1$.*

Theorem 3.5: *(Tarbouriech and Burgat, 1994b) If matrix A is K-irreducible, then system (1), with*

(2), is unstable ($\rho(A) > 1$) if and only if $(C_+ \cap K_+)\backslash\{0\} = \emptyset$.

The following results are required to show (in theorem 3.6) from what necessary and sufficient conditions the instability (critically or not) of system (1) is obtained.

Lemma 3.5: *If $K_+ \subset C_-$ then $\rho(A) \geq 1$.*

Proof : $\forall x \in K_+$, $-\Pi x \in K_-$. Consider the eigenvector x_0 of A such that $-\Pi x_0 = (1 - \rho(A))x_0$, then since by assumption $-\Pi x_0 \in K_-$, we must have $\rho(A) \geq 1$.

Corollary 3.1: *If the following conditions hold :*

(i) $K_+ \subset C_-$,
(ii) C_- is a solid cone.

then $\rho(A) > 1$.

Proof : The condition (ii) of this corollary permits us to conclude, in the proof of lemma 3.5, that the only solution is $\rho(A) > 1$.

Corollary 3.2: $\rho(A) \geq 1$ *if and only if* $(C_- \cap K_+)\backslash\{0\} \neq \emptyset$.

Proof : (If). Assume $(C_- \cap K_+)\backslash\{0\} \neq \emptyset$ and $\rho(A) < 1$. From theorem 3.1, this second assumption implies that C_+ is solid and $C_+ \subset K_+$. Hence, since $C_- = -C_+$ and K_+ is pointed, it is not possible to obtain simultaneously $y \in C_+ \subset K_+$ and $y \in C_- \cap K_+$, except for $y = 0$. This contradicts the assumption $(C_- \cap K_+)\backslash\{0\} \neq \emptyset$. It thus follows that $(C_- \cap K_+)\backslash\{0\} \neq \emptyset$ only if $\rho(A) \geq 1$, or equivalently, that if $(C_- \cap K_+)\backslash\{0\} \neq \emptyset$ then $\rho(A) \geq 1$.
(Only if). It follows from lemmas 2.4 and 2.9.

Corollary 3.3: $\rho(A) > 1$ *and* $\nexists \lambda_i = 1 \in \sigma(A)$ *if and only if the following conditions hold :*

(i) $(C_- \cap K_+)\backslash\{0\} \neq \emptyset$,
(ii) C_- is a solid cone.

Proof: (If). (i) implies that there exists a vector $x \neq 0 \in K_+$ such that $-\Pi x \in K_-$. Condition (ii) implies that there exists no vector $y \neq 0$ such that $-\Pi y = \Pi y = 0$, that is, there exists no eigenvalue $\lambda_i = 1$. Let $x_0 \in K_+$ such that $Ax_0 = \rho(A)x_0$, $-\Pi x_0 = (1 - \rho(A))x_0 \neq 0$. Two solutions are presented : either $-\Pi x_0 \in K_+$ or $-\Pi x_0 \in K_-$. Suppose that the first holds, i.e., $\rho(A) < 1$; however, $\rho(A) < 1$ implies that $C_+ \subset K_+$ and C_+ solid. But it is not possible to obtain simultaneously these conditions and conditions (i) and (ii) of corollary 3.3. Hence the only solution is $\rho(A) > 1$.
(Only if). Assume that $\rho(A) > 1$ with $\nexists \lambda_i = 1$, then, for the eigenvector $x_0 \in K_+$ corresponding to $\rho(A)$ we get $-\Pi x_0 \in K_-$ and, therefore, the condition (i). On the other hand, since $\nexists \lambda_i = 1$, there exists no vector $x \neq 0$ such that $-\Pi x = \Pi x = 0$ then condition (ii) follows.

Then, the following corollary 3.4 readily follows from lemmas 2.4 and 2.7.

Corollary 3.4: *If $\rho(A) = 1$ and system (1) is unstable then $(C_+ \cap C_-)\backslash\{0\} \cap intK_+ = \emptyset$, that is, $(C_+ \cap C_-)\backslash\{0\} \cap \partial K_+ \neq \emptyset$.*

Corollary 3.5: *If $\rho(A) = 1$ then $(C_+ \cap C_-)\backslash\{0\} \cap K_+ \neq \emptyset$, and if either*

(i) $K_+ \subset C_-$,

or

$$(ii) \begin{cases} intK_+ \cap C_- \neq \emptyset \\ intK_+ \cap C_+ = \emptyset \end{cases}$$

then system (1) is unstable (in the sense of (iv) of definition 5.1).

Proof : (i). Since $\rho(A) = 1$, we get from lemma 2.4 $(C_+ \cap C_-)\backslash\{0\} \cap K_+ \neq \emptyset$. Asume now that $K_+ \subset C_-$, which means that $\forall y_0 \in K_+$, $-\Pi y_0 \in K_-$. Suppose that $\rho(A) = 1$ and system (1) is stable, then for any $y_0 \in R^n$ we obtain either $A^k y_0 \longrightarrow 0$ or $A^k y_0 \longrightarrow x_0$ as $k \longrightarrow \infty$, where $Ax_0 = x_0$, or still $\exists j \in J_+$ such that $A^j y_0 = y_0$. Let us choose a vector $y_0 \neq 0 \in intK_+$ such that $y_0 \overset{K}{\geq} x_0$. Since $K_+ \subset C_-$, we obtain $-\Pi y_0 \in K_-$ and thus $A^k y_0 \overset{K}{\geq} A^{k-1} y_0 \overset{K}{\geq} ... \overset{K}{\geq} y_0$. Hence $A^k y_0 \not\longrightarrow 0$ as $k \longrightarrow \infty$, if not $y_0 \overset{K}{\leq} 0$, contradicting the assumption $y_0 \in intK_+$ (the case y_0 such that $Ay_0 = 0$ contradicts the assumption (i)). On the other hand, if $A^k y_0 \longrightarrow x_0$ as $k \longrightarrow \infty$ we obtain $x_0 \overset{K}{\geq} y_0$, which contradicts our assumption. Equivalently, if there exists an index j such that $A^j y_0 = y_0$ then the assumption $-\Pi y_0 \in K_-$ is not verified. Thus, the remaining possibility is $A^k y_0 \longrightarrow \infty$ as $k \longrightarrow \infty$, therefore $\rho(A) = 1$ is unstable.
(ii). This part, can be proved by selecting an initial vector $y_0 \in intK_+ \cap C_-$.

Theorem 3.6: *The system (1), with (2), is unstable (with $\rho(A) > 1$ or $\rho(A) = 1$) if and only if one of the following conditions is satisfied :*

(i) $(C_+ \cap K+)\backslash\{0\} = \emptyset$.
(ii) $K_+ \subset C_-$ (C_- solid or not).
(iii) $(C_- \cap K_+)\backslash\{0\} \neq \emptyset$ (C_- solid or not, and in the latter $C_+ \cap intK_+ = \emptyset$).

Proof : It follows from lemmas 2.4, 2.7, 2.9, theorem 3.4 and corollaries 3.2, 3.3 and 3.5.

4 CONCLUDING REMARKS

In this paper, the results of (Burgat and Benzaouia, 1990) and (Tarbouriech and Burgat, 1994b) have been extended to the case where matrix A is not assumed to be K-irreducible. These conditions of stability of system (1) have been express under a geometrical form from a structural property of the matrix A. In fact the main interest of this geometric approach appears when it is used to the determination of some local stability domains for linear constrained system controlled by state or output feedback. Hence this geometric approach relying cone C_+ and K_+ permits to have together the properties of stability and invariance ; thus an eigenstructure interpretation is possible : see for instance, Tarbouriech and Burgat (1994a), Tarbouriech and Castelan (1992).

5 APPENDIX

We first recall a precise definition on the stability properties of system (1) (see Lasalle, 1986).

Definition 5.1: *The equilibrium x=0 of dynamical system (1) is said to be :*

(i) asymptotically stable if all eigenvalues of matrix A, $\lambda_i(A)$ are such that $|\lambda_i(A)| < 1$.
(ii) strictly unstable if there exists at least one eigenvalue $\lambda_i(A)$ such that $|\lambda_i(A)| > 1$.

When there exist some eigenvalues $\lambda_i(A)$ such that $|\lambda_i(A)| = 1$, $|\lambda_j(A)| < 1$, $i \neq j$, a critical case, which can also be decomposed into two subcases, is obtained :

(iii) if the geometric multiplicity of each eigenvalue $\lambda_i(A)$ such that $|\lambda_i(A)| = 1$ is equal to its algebraic multiplicity, the critical stability property is obtained.
(iv) if not, the critical instability property is obtained.

We now give definitions on the proper cones in order to specify assumptions made in this paper.

Definition 5.2: *(Berman and Plemmons, 1979) Let $K \subset R^n$ and consider the two sets :*

(i) K^G, the set generated by K, which consists of all finite nonnegative linear combinations of elements of K,
(ii) K^*, the dual of K defined by :

$$K^* = \{y \mid y^t x \geq 0, \forall\ x \in K\} \qquad (3)$$

If $K^G = K$, K is said to be a cone.

Definition 5.3: *(Berman and Plemmons, 1979) A cone $K \subset R^n$ is said to be convex if :*

(i) $K + K \subset K$,
(ii) $\alpha K \subset K$, for all $\alpha \geq 0$.

Definition 5.4: *(Berman and Plemmons, 1979) A convex cone $K \subset R^n$ is said to be :*

(i) pointed if $K \cap (-K) = \{0\}$
(ii) solid if $intK \neq \emptyset$.
(iii) reproducing if $K - K = R^n$.

Theorem 5.1: *(Krasnoselskii, 1964) A closed convex cone in R^n is solid if and only if it is reproducing.*

Definition 5.5: *(Berman and Plemmons, 1979) A closed, solid, pointed and closed cone is called a proper cone.*

Throughout this paper, ∂K and $intK$ denote the boundary and the interior of K, respectively. Moreover, we denote $K_+ = K$ and $K_- = -K$. On the other hand, a proper cone induces a partial ordering in R^n, which is denoted $\overset{K}{\leq}$ and defined as : $x \overset{K}{\leq} y$ means $y - x \in K$ (see Vandergraft (1968), Berman and Plemmons (1979)). With respect to matrix A, $\rho(A)$ and $\sigma(A)$ denote the spectral radius and the spectrum of A, respectively.

In order to obtain a characterization of the positive invariance property for sets generated from the translation of a proper cone K_+, as Schneider and Vidyasagar (1970) we give :

$$K_+ + x = \{z \in R^n | z = y + x, y \in K_+, x \in R^n\}$$

To obtain the geometrical characterization of stability properties of system (1) in the sense of definitio 5.1, the following cone has to be defined :

$$C_+ = \{x \in R^n | -\Pi x \in K_+\},\ C_- = -C_+ \quad (4)$$

where $\Pi = A - I_n$, I_n being the identity matrix in R^n.

Lemma 5.1: *(Burgat and Benzaouia, 1990) Let system (1) with property (2). Then the set $K_- + x$ (resp. $K_+ - x$) has the properties :*

(i) $A(K_- + x) \subset K_- + x$ (resp. $A(K_+ - x) \subset K_+ - x$) if and only if $-\Pi x \in K_+$ (resp. $\Pi x \in K_+$) ;
(ii) $A(K_- + x) \subset int(K_- + x)$ (resp. $A(K_+ - x) \subset int(K_+ - x)$) if and only if $-\Pi x \in intK_+$ (resp. $\Pi x \in intK_+$).

Recall the results on the spectral properties of matrices leaving positively invariant a proper cone and the property of K-irreducibilty.

Theorem 5.2: *(Vandergraft, 1968) If K is a proper cone such that $AK \subset K$ then :*

(i) $\rho(A)$ is an eigenvalue of A,
(ii) the degree of $\rho(A)$ is not smaller than the degree of any other eigenvalue having the same modulus,
(iii) K contains an eigenvector of A corresponding to $\rho(A)$,

Remark 5.1: *In the above theorem 5.2, the degree of any eigenvalue $\lambda_i(A)$ of matrix A is the size of the largest diagonal block in the Jordan canonical form of A, which contains $\lambda_i(A)$ (the multiplicity of $\lambda_i(A)$ in the minimal polynomial of A).*

Definition 5.6: *(Berman and Plemmons, 1979) A matrix A such that $AK \subset K$ is K-irreducible if the only faces of K that it leaves invariant are $\{0\}$ or K itself. A is K-reducible if it is not K-irreducible.*

From definition 5.6 matrix A can be characterized by the following theorem 5.3.

Theorem 5.3: *(Berman and Plemmons, 1979) If A is K-irreducible then :*

(i) $\rho(A)$ is a simple eigenvalue and any other eigenvalue with the same modulus is also simple,
(ii) there is an eigenvector corresponding to $\rho(A)$ in int$K$, and no other eigenvector (up to scalar multiples) lies in K.

6 REFERENCES

Benzaouia, A., and C. Burgat (1988) : *Regulator problem for linear discrete-time systems with non-symmetric constrained control.* Int. J. of Control, Vol. 48, n°6, pp.2441-2451.

Berman, A., and R.J. Plemmons (1979) : *Nonnegative matrices in the mathematical sciences.* Academic Press, New York.

Burgat, C., and A. Benzaouia (1990) : *Stability properties of positively invariant linear discrete systems.* J. Math. Anal. Appl., vol 143, n°2,pp.587-596.

Burgat, C., A. Benzaouia and S. Tarbouriech (1990) : *Positively invariant sets of discrete-time systems with constrained inputs.* Int. J. of Syst. Science, vol.21, no.7, pp.1249-1271.

Chéganças, J., and C. Burgat (1986) : *Polyhedral cones associated to M-matrices and stability of time varying discrete-time systems.* J. Math. Anal. Appl., vol 118, n°1, pp.88-96.

Gutman, P.O., and P. Hagander (1985) : *A new design of constrained controllers for linear systems.* IEEE Trans. AC, vol AC-30,n°1, pp.22-33.

Krasnoselskii, M.A. (1964) : *Positive solutions of operator equations.* Noordhoff, The Netherlands.

Lasalle, J.P. (1986) : *The stability and control of discrete processes*, App. Math. Sci., 62, Springer Verlag.

Rouhani, R., and E. Tse (1981) : *Structural design for classes of positive linear systems.* IEEE Trans. on Systems, Man and Cybernetics, 11, pp.126-134.

Rumchev, V.G., and D.J.G. James (1989) : *Controllability of positive linear discrete-time systems.* Int. J. of Control, vol. 50, n°3, pp.845-857.

Schneider, H., and M. Vidyasagar (1970) : *Cross-positive matrices.* Siam J. Numer. Anal., 7, 4, pp.508-519.
Stern, R.J. (1982) : *A note on positively invariant cones.* Appl. Math. Syst. Theory, 3, pp.67-72.

Tarbouriech, S., and C. Burgat (1994a): *Positively invariant sets for constrained continous-time systems with cone properties,* IEEE T.A.C, vol.39, no.2, pp.401-405.

Tarbouriech, S., and C. Burgat (1994b) : *A note on stability properties of some discrete-time systems,* Journal Mathematical Analysis and Application, vol.187, n°2, pp.700-704.

Tarbouriech, S., and E.B. Castelan (1992) : *The linear constrained regulation problem for some linear continuous-time singular systems,* Proc. of 2nd IFAC Workshop on System Structure and Control, Prague (Czechoslovakia), pp.472-475.

Vandergraft, J.S. (1968) : *Spectral properties of matrices which have invariant cones.* Siam J. Appl. Math., 16, pp.1208-1222.

Vassilaki, M., and G. Bitsoris (1989) : *Constrained regulation of linear continuous-time dynamical systems.* Syst. and Control Letters, 13, pp.247-252.

Yorke, J.A. (1967) : *Invariance for ordinary differential equations.* Math. Syst. Theory, vol 1, n°4, pp.353-372.

The Sufficient Duplication Principle : an alternative issue to the Internal Model Principle

Philippe de LARMINAT*

*Laboratoire d'Automatique de Nantes, unité associée au C.N.R.S.,
Ecole Centrale de Nantes Université de Nantes, 1, Rue de la Noë,
44 072 Nantes Cedex 03, France.
(*also with the G.R. Automatique of the C.N.R.S.)*

Abstract - We consider the classical problem of robust regulation. As well known, this objective is equivalent to require that the controller contains an appropriate « internal model ». In this paper, it is shown that this internal model will appear, under an implicit form, if the controller is designed from a standard state reconstruction and feedback. The condition is that the design model duplicates the model of the disturbances as many times as the number of observations.

Résumé - On considère ici le problème classique de la régulation robuste. Comme on le sait, cet objectif équivaut à exiger que le régulateur incorpore un modèle interne approprié. Dans ce papier, on montre que ce modèle apparaîtra, sous une forme implicite, si le régulateur est conçu selon le principe de la reconstruction et du retour d'état. La condition est que le modèle de conception duplique le modèle des perturbations autant de fois qu'il y a d'observations.

Key words - Robust regulation; internal model principle; exogenous model; duplication.

1. Introduction

Before discussing the problem which is the focus of this paper, we briefly introduce the standard problem of robust regulation. This problem will be defined with respect to figure 1.

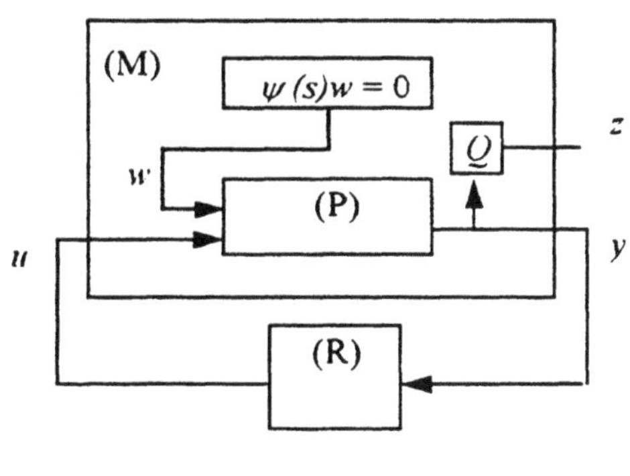

Fig. 1 : Setup for the robust regulation problem.

Here P is the plant and R is the controller to be designed. Both are finite dimensional linear time invariant systems, as are all systems throughout this paper. In the figure, w is the exogenous input, z is the controlled output, u is the control input, and y is the measurement.

It is assumed that $z = Qy$, where Q is an exactly known matrix, which could be equal to [I 0], without loss of generality. Then z is be a part of y, and the other component of y could be, either the set point measurement, either some measure of the disturbances, either some direct state observation. Thus in most cases, $n_y > n_z$, where n_y and n_z are the dimension of y and z.

In the figure 1, the whole block M represents the augmented process (plant + model of the disturbances). The disturbances are assumed to verify a differential equation $\psi(s)w = 0$. For example, a combination of polynomials and sinusoidal disturbances, leads to

$$\psi(s) = s^p(s^2-\omega_1^2)...(s^2-\omega_r^2).$$

Assumptions

- H1- The *observable part* of M admits a realization

$$\dot{x} = Ax + Bu \quad (1\text{-a})$$
$$y = Cx \quad (1\text{-b})$$
$$z = Dx \quad (1\text{-c})$$

where

$$x = \begin{bmatrix} x_1 \\ x_2 \end{bmatrix}, A = \begin{bmatrix} A_{11} & A_{12} \\ 0 & A_{22} \end{bmatrix}, B = \begin{bmatrix} B_1 \\ 0 \end{bmatrix} \quad (1\text{-d})$$
$$C = \begin{bmatrix} C_1 & C_2 \end{bmatrix}, D = QC = \begin{bmatrix} D_1 & D_2 \end{bmatrix} \quad (1\text{-e})$$

- H2- (A_{11}, B_1) is stabilizable.

- H3- A_{22} is assumed to be unstable. The unstabilizable part x_2 is due to the exogenous input w. Thus the minimal polynomial of A_{22} equals $\psi(s)$.

It is known that any controller R which internally stabilizes the nominal process P will, by virtue of local robustness, stabilize a non empty class of processes in the neighborhood of P (*robust stabilization*).
Now, the problem of *robust regulation* is to find R such that z reaches zero for any process P which is stabilized by R, only assuming that Q and $\psi(s)$ are exactly known. (This property is also known as the « structural robustness » [8])
It is well known that a controller solves the robust regulation problem if and only if it contains an appropriate « internal model ». See appendix A for a precise statement of the « internal model principle» (IMP).
On the other hand, *the regulator problem with internal stability* (RPIS) can be solved by a standard reconstruction and state feedback, as described in section 2. However, this solution of the RPIS does not necessarily solve the robust regulation problem.
The objective of this paper is to show that this standard reconstruction and state feedback actually solve the robust regulation problem if the design model (eq. 1) duplicates the disturbance model as many times as the number n_y of measurement. This condition is called the sufficient duplication principle (SDP)[3].
Because of some similarities, IMP and SDP must be carefully distinguished. The internal model principle concerns the *controller*, while the sufficient duplication principle concerns the *design model*, from which the controller will be deduced through a given procedure.
This standard design procedure for solving the RPIS is described and analyzed in section 2. The SPD in introduced in section 3. Illustrating examples are provided in section 4. The conclusion of section 5 ends the paper.

2. The standard design procedure for the RPIS.

It is assumed that the observable part of the model M admits the realization (1), and satisfies the assumptions H1 to H3.
Then one can find two matrices K and G_1 such that $(A\text{-}KC)$ and $(A_{11}\text{-}B_1G_1)$ be some stability matrices.
Let also G_a and T_a be the solution of the asymptotic rejection equations

$$A_{11}T_a - T_aA_{22} + B_1G_a = A_{12} \quad (2\text{-a})$$

$$D_1T_a = D_2 \quad (2\text{-b})$$

and define

$$G = [G_1\ G_2], \text{ where } G_2 = G_a + G_1T_a \quad (3)$$

The equations (2) are solvable when the problem is « well posed » [8].

Then the controller is defined as the following state reconstruction and feedback equations

$$\dot{\hat{x}} = A\hat{x} + Bu + K(y - C\hat{x}) \quad (4\text{-a})$$
$$u = -G\hat{x} \quad (4\text{-b})$$

Equations (2) and (3) are well known, but they are not widely used : they are not included in the standard tool boxes, although they are as important as the stabilization tools, when solving regulation problems. An other reason is that in the lack of the SDP, the only way for solving the robust regulation problem seems to be an explicit application of the IMP, in the context of which the asymptotic rejection equations are not essential.

Analysis

Define $\hat{X} = \begin{bmatrix} \hat{X}_1^T & \hat{X}_2^T \end{bmatrix}^T$,

$$\hat{X}_1 = \hat{x}_1 + T_a\hat{x}_1, \hat{X}_2 = \hat{x}_2, \hat{y} = C\hat{x} \quad (5)$$

Then

$$z = Qy = Q\hat{y} + Q(y - \hat{y})$$
$$z = QC\hat{x} + Q(y - \hat{y})$$
$$z = D_1\hat{x}_1 + D_2\hat{x}_2 + Q(y - \hat{y})$$
$$z = D_1(\hat{X}_1 - T_a\hat{X}_2) + D_2\hat{X}_2 + Q(y - \hat{y})$$

Then, using (2-b)

$$\boxed{z = D_1\hat{X}_1 - Q(y-\hat{y})} \quad (6\text{-a})$$

Similarly, manipulate

$$\dot{\hat{X}}_1 = \dot{\hat{x}}_1 + T_a\dot{\hat{x}}_2 = \cdots$$

Using (2-a), it follows easily (6-b)

$$\boxed{\dot{\hat{X}}_1 = (A_{11} - B_1G_1)\hat{X}_1 + (K_1 + K_2T_a)(y-\hat{y})}$$

The reconstruction equation (4-a) is asymptotically stable. It yields the (non robust) properties

$$\lim_{t\to\infty}(\hat{x} - x) = 0 \quad (7\text{-a})$$

and

$$\lim_{t\to\infty}(\hat{y} - y) = 0 \quad (7\text{-b})$$

and finally, from (6-b), where $(A_{11}\text{-}B_1G_1)$ is asymptotically stable, and from (6-a), it follows

$$\lim_{t\to\infty}\hat{X}_1 = 0$$

$$\boxed{\lim_{t\to\infty} z = 0} \quad (8)$$

Unfortunately, this last property (8) is not robust as long as (7-a) is not.

All the previous results are either well known, either obvious. Now, we can really enter in the scope of the paper.

3. The sufficient duplication principle

By itself, the purpose of robustifying (7-a) is meaningless, but that is possible for (7-b). Rewrite the controller realization on the base of (5). It follows

$$\dot{\hat{X}} = \begin{bmatrix} A_{11} - B_1G_1 & 0 \\ 0 & A_{22} \end{bmatrix}\hat{X} + \begin{bmatrix} K_1 + K_2T_a \\ K_2 \end{bmatrix}v \quad (9\text{-a})$$

$$\hat{y} = \begin{bmatrix} C_1 & C_2 - C_1T_a \end{bmatrix}\hat{X} \quad (9\text{-b})$$

where $v = y - \hat{y}$. (9-c)

The closed loop (P,R), combined with the filter equations (9-abc), can be arranged under the form of the figure 2

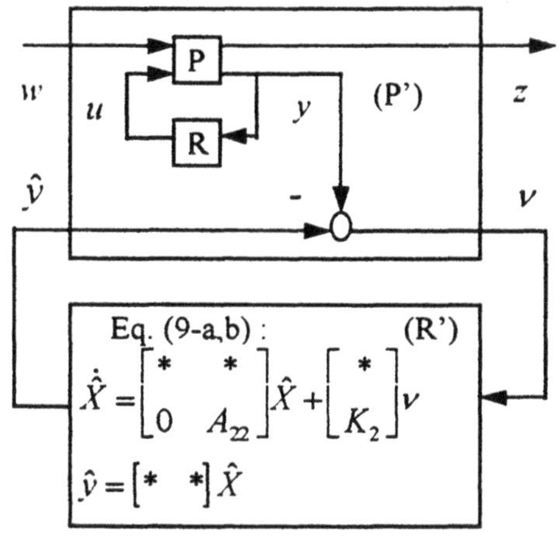

Fig 2. An equivalent loop.

On this figure, the « process » P' is asymptotically stable, since R internally stabilizes P. The rest of the loop is also stable, since it realizes the equations of an asymptotically stable filter. Now, the internal model principle (see appendix A) can be faced to the above loop.

- We have just set that the loop is internally stable.
- Actually, the « regulator » R' is under the required form (eq. A1 of the appendix A))
- The pair (A_{22}, K_2) is controllable : if not, the open loop unstable subsystem $\dot{\hat{X}}_2 = A_{22}\hat{X}_2 + K_2v$ never could lead to a stable loop, as it is done own to the return $v = y - C\hat{x}$.
- Finally, the only condition to be verified is the following :

> **A_{22} must be diagonalizable into n_y blocks, each of minimal polynomial $\psi(s)$.**

Under this condition, the IMP will be satisfied by the loop of figure 2, and implies that v robustly tends towards zero. Now, (7-b) is robust, and then the standard design of section 2 solves the problem of robust regulation.
As we shall see, the above condition is not necessarily satisfied by the model equations (1), but it will be possible to duplicate the exo-system in the design model, in order to involve a robust regulator. It is the reason why the above condition is called the *sufficient duplication principle*.

4. Examples

Example 1. A regulation problem is defined with respect to figure 3.

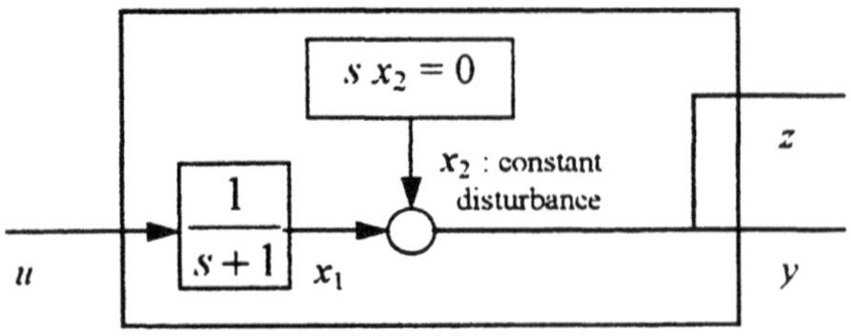

Fig. 3. A regulation problem.

The corresponding state space matrix are

$$A = \begin{bmatrix} -1 & 0 \\ 0 & 0 \end{bmatrix}, \quad B = \begin{bmatrix} 1 \\ 0 \end{bmatrix},$$

$$C = [1 \quad 1], \quad D = [1 \quad 1]$$

The SDP is satisfied, with $n_y = 1$, $\psi(s) = s$, $A_{22} = 0$.

We can take $G_1 = 0$, which stabilizes the pair $A_{11} = -1$, $B_1 = 0$.

Then, (3) leads to $G_a = -1$, and to $G = [\,0 \;\; 1\,]$.

We can take also $K = [\,0 \;\; 1\,]$, which leads to a stable filter.

From (4), it follows the controller equations

$$\begin{aligned} \dot{\hat{x}}_1 &= -\hat{x}_1 - \hat{x}_2 \\ \dot{\hat{x}}_2 &= -\hat{x}_1 - \hat{x}_2 + z \\ u &= -\hat{x}_2 \end{aligned}$$

The SDP was satisfied, thus we can conclude that a robust regulation is performed. However, this controller does not satisfy explicitly the IMP. In fact it satisfies it implicitly. Define

$$\begin{aligned} \zeta &= \hat{x}_2 - \hat{x}_1 \\ \xi &= \hat{x}_1 \end{aligned}$$

Then the controller becomes

$$\begin{aligned} \dot{\zeta} &= z \\ \dot{\xi} &= -2\xi - \zeta \\ u &= -\zeta - \xi \end{aligned}$$

which can be written under the form (A1)

$$\begin{bmatrix} \dot{\zeta} \\ \dot{\xi} \end{bmatrix} = \begin{bmatrix} 0 & 0 \\ * & * \end{bmatrix} \cdot \begin{bmatrix} \zeta \\ \xi \end{bmatrix} + \begin{bmatrix} 1 \\ * \end{bmatrix} z$$

$$u = [\,* \quad *\,] \begin{bmatrix} \zeta \\ \xi \end{bmatrix}$$

and which leads to the block diagram of fig.4, which reveals an integral controller.

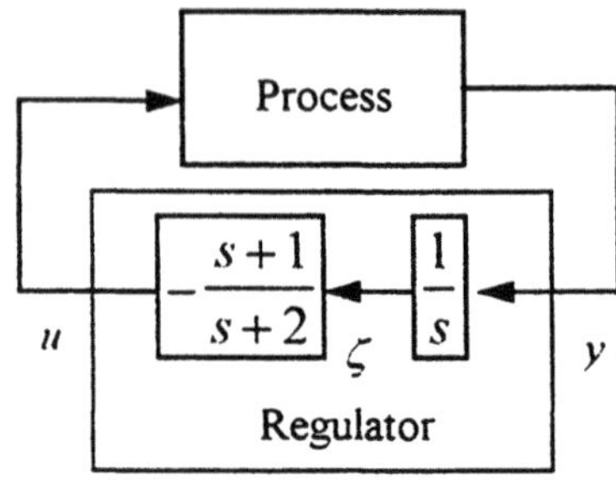

Fig. 4 : Explicit integral control

Example 2. Reconsider the previous problem, assuming now that the disturbance also is measured

$$\begin{aligned} y_1 &= z = x_1 + x_2 \\ y_2 &= x_2 \end{aligned}$$

Then, the most simple solution of the RPIS is now $u = y_2$, which is, in fact, an open loop (and non robust) feed-forward. The lack of robustness is in accordance with the fact that the SDP is not satisfied, since $n_y = 2$, $\psi(s) = s$, $A_{22} = 0$.

Example 3. If we want to keep the benefit of the feed-forward, without loosing the robustness, we can modelize a fictitious disturbance x_3, as shown in figure 5

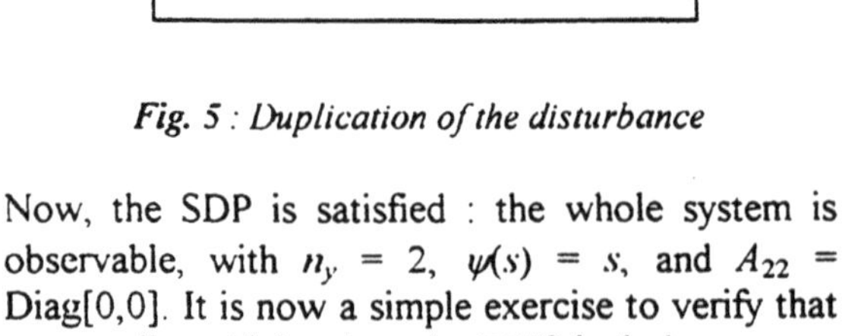

Fig. 5 : Duplication of the disturbance

Now, the SDP is satisfied : the whole system is observable, with $n_y = 2$, $\psi(s) = s$, and $A_{22} = \text{Diag}[0,0]$. It is now a simple exercise to verify that a controller which solves the RPIS includes now an integral term.

Example 4. Similarly, in the context of tracking problems, two measures are available : the set point and the controlled output. Then, an additional disturbance of the same type as the set point signal must be added into the initial model in order to satisfy the SDP. Then, it leads to a robust tracking controller.

5. Conclusion

Many well known approaches for solving regulation and tracking problems could be revisited in terms of RPIS, such as Predictive Control, Internal Model Control of Morari [7],(where *internal model* has not the same sense as for Wonham).
Typically these methods incorporate the reconstruction of a constant disturbance, and its compensation. It is well known that it leads to implicit integral controllers.
The aim of this paper was to provide a precise and general condition, under which RPIS actually solves the robust regulation problem. The observable part of the exo-system must duplicate the model of the disturbances as many times as the number of observations.
It often happens that this duplication is not necessary. In the other cases, it may be performed either on the control inputs, either on outputs (as on fig 5), either on any other point of the system. In any case, it must be verified that the observable part of the model increases , which means that the duplication is efficient, and does not make redundancy with the initial exo-system.

Appendix : The internal model principle

This principle is widely known, and has been investigated under various forms : the geometric approach [6,8], the algebraic approach [2], the frequency domain approach [4,5] , and the H_∞ approach [1].

Theorem.

A controller R solves the robust regulation problem, as stated on figure 1, if and only if

- R internally stabilizes P,
- R admits the realization

$$\dot{x}_c = \begin{bmatrix} \overline{A} & 0 \\ * & * \end{bmatrix} x_c + \begin{bmatrix} \overline{B}Q \\ * \end{bmatrix} y \qquad \text{(A1)}$$
$$u = \begin{bmatrix} * & * \end{bmatrix} x_c + [*] y$$

where * is for any matrix of appropriate dimension,

- $(\overline{A}, \overline{B})$ is controllable,
- $\overline{A}$ is diagonalizable into n_z blocks, each of minimal polynomial $\psi(s)$.

References

[1] J. ABEDOR, K. NAGPAL, P.P. KARGONEGAR, and K. POOLLA, « Robust regulation in the presence of norm-bounded uncertainty », *ACCC* 1994, pp. 773-777.
[2] E. J. DAVISON and A. GOLDENBERG, « Robust control of a general servomechanism problem : The servo compensator », *Automatica 11* (5), 1975, pp. 461-471.
[3] PH. DE LARMINAT, *Automatique - Commande des systèmes linéaires*, Hermès. Paris, 1993, pp. 280-286.
[4] C.A. DESOER and Y.T. WANG, « Linear time-invariant robust servomechanism problem: a self contained exposition », *Control and dynamic systems* C.T. Leondes ed., 1980.
[5] P.G. FERREIRA, « The servomechanism problem and the method of the state space in the frequency domain », *Int. J. of control 23* (2), pp. 245-255 (1976)
[6] B.A. FRANCIS, « The linear multivariable regulator problem », *SIAM J. control 15* (3), pp. 446-505 (1977)
[7] C.E. GARCIA and M. MORARI « Internal model control, 1 - A unifying review and some new results », *Ind. Eng. Chem. Process Des. Dev.*, vol. 21, pp. 308-323, 1982.
[8] W.M. WONHAM, *Linear multivariable control - a geometric approach*, Springer Verlag, New York,(1979)

ON FORCE MEASUREMENTS TIME-DELAYS IN CONTROL OF CONSTRAINED MANIPULATORS

Silviu-Iulian Niculescu*and Bernard Brogliato

Laboratoire d'Automatique de Grenoble, ENSIEG, BP 46, 38402 St.-Martin-d'Hères, France.
The authors are also with the GDR CNRS "Automatique".
E-mail: silviu@lag.grenet.fr; brogli@lag.grenet.fr

Abstract. In this note we consider stability issues of manipulators in contact with a rigid environment (i.e. submitted to holonomic kinematic constraints) when the interaction force feedback loop contains time-delays. Sufficient conditions (delay-dependent) are derived to guarantee that the robot's tip remains in contact with the surface. These conditions are found by taking into account the fact that due to unilateral nature of the constraints, the interaction force must have constant sign during the whole task. We analyze the cases of proportional and proportional-integral force-feedback.

Résumé. Ce travail concerne l'étude de la stabilité d'un robot à 1-ddl en contact avec un environnement rigide (costrainte holonome), lorsque la boucle de commande contient un retard de mesure. On prend explicitement en compte le caractère unilatéral de la contrainte en étudiant le signe de l'effort d'interaction.

Keywords. force-control, neutral functional differential equation, time-delay, stability, Lyapunov functional, holonomic constraints.

1 INTRODUCTION

Robot manipulators force control has motivated a lot of research studies during the last fifteen years. One of the first paper dealing with so-called hybrid control of manipulators subject to holonomic constraints was published in Raibert and Craig (1981). Since that time, many researchers in the field have extended, discussed and improved the pioneering work in Raibert and Craig (1981), see e.g. De Luca and Manes (1991), Duffy (1990), MacClamroch and Craig (1988), Fisher and Shahid Mujtaba (1992), Sinha and Goldberg (1993), Wen and Murphy (1991) to name only a few.

In this note, we shall consider the very simple example of a one degree of freedom (DOF) prismatic manipulator (i.e. a mass m) that contacts a one DOF environment. Notice that the study of such a simple model proves to be sufficient in higher dimensional case when the nonlinear robot dynamics are decoupled and compensated for Yoshikawa (1987), Mills and Lokhorst (1993); moreover it should be noted that problems on force control of manipulators are often studied and understood with one DOF models (masses, springs, dampers) Volpe and Khosla (1993), Eppinger and Seering (1992) (and the references therein). We shall treat only the rigid environment case.

The open-loop equation is an algebraic equation:

$$F + U = 0 \tag{1}$$

where U is the (force) control input, F is the interaction force between the mass and the environment. Assuming that $F = -kx$ is a valid relationship and that velocity $\dot{x}$ is available for feedback is equivalent to allow the designer to use the force derivative $\dot{F}(t) = -k\dot{x}(t)$ in the controller (provided k is known): This would however require to differentiate the measured force F_m in the rigid case and this is not a practically feasible procedure since F_m is often corrupted by noise Volpe and Khosla (1993). Thus, as we shall see later, closed-loop equations for the rigid case will typically contain F together with its succesive time integrals Yoshikawa *et al.* (1988). Since we are interested in the closed-loop equations when the measured quantities contain time-delays, these equations will typically become functional differential equations of neutral type (NFDE), which can be characterized by the fact that the rate of change of state depends not only on the past states, but also on the past rates of state (e.g. Kolmanovskii and Nosov, 1986; Hale,

*Also with the Department of Automatic Control, University "Politehnica" Bucharest, Romania.

1977).

It is important to note that in the particular case we are studying, stability of the closed-loop neutral functional differential equations is not sufficient for stability of the robotic task, in the sense that there can be loss of contact between the robot's tip and the environment's surface even if the closed-loop is stable.

This paper is organized as follows: In Section 2, we derive the closed-loop equations; Section 3 is devoted to study the stability of the associated functional differential equations; In Section 4, we examine the contact instability problem, i.e. conditions under which the interaction force has constant sign. Finally, some conclusions are given in Section 5.

Notations. $G_{[a,b)}(t)$ denotes the value of the function $G(t)$ in $[a,b)$. For $t \in [b,c)$, this value is given by $G_{[a,b)}(\alpha t+\beta)$, where $\alpha = \frac{b-a}{c-b}$, $\beta = \frac{ac-b^2}{c-b}$. For $a = k\tau$, $b=(k+1)\tau$, $c=(k+2)\tau$, we get $\alpha = 1$, $\beta = -\tau$.

2 CLOSED-LOOP EQUATIONS

Let us consider the following control input:

$$U = -F_d + \lambda_1 \tilde{F}_m(t) + \lambda_2 \int_\tau^t \tilde{F}_m(z)dz \tag{2}$$

where F_m is the measured force. We take $F_d < 0$ and constant. Note that integration is taken on $[\tau, t]$ because we need an initial condition given on $[0,\tau)$ to define the control law. It makes no sense to define U in (2) without defining an initial condition on the interaction force.

When there is no delay in the force measurements (i.e. $\tau = 0$) and U is applied from 0, then (1) is equal to

$$(1+\lambda_1)\tilde{F}(t) + \lambda_2 \int_0^t \tilde{F}(z)dz = 0 \tag{3}$$

which leads to the algebraic equation $F \equiv F_d$ if $U(0) = -F_d + \lambda_1\tilde{F}(t)$. Following Yoshikawa (1987) and MacClamroch and Wang (1988), the case when $\tau \neq 0$ is very different from the case $\tau = 0$. Indeed, if $\tau = 0$, the integral action is from a theoretical point of view useless since $F \equiv F_d$ for all $t \geq 0$ and it to be considered for practical purpose only. Conversely, when non-zero delay is present, the integral term plays a significant role in the closed-loop dynamics, as we shall see in this paper.

Now let us assume that the measured force $F_m(t) = F(t-\tau)$, where $\tau > 0$ represents a strictly positive time-delay. Then (3) becomes

$$\tilde{F}(t) + \lambda_1 \tilde{F}(t-\tau) + \lambda_2 \int_0^{t-\tau} \tilde{F}(z)dz = 0 \tag{4}$$

which can be rewritten defining $x(t) = \int_0^{t-\tau} \tilde{F}(z)dz$ as

$$\dot{x}(t) = -\lambda_1 \dot{x}(t-\tau) - \lambda_2 x(t-\tau) \tag{5}$$

Remark 2.1 We assume implicitely that the robot is in contact with the environment. In particular as long as the contact is maintained, then $x = \dot{x} = 0$. This is the reason why adding a damping term $-\lambda_2 \dot{x}$ in (2) is not necessary for the analysis. □

Remark 2.2 Instead of the control in (2), we could have also supposed that U is given by

$$U = -F(t) + \lambda_1 \tilde{F}(t) + \lambda_2 \int_\tau^t \tilde{F}(z)dz$$

which yields

$$\dot{x}(t) = (1-\lambda_1)\dot{x}(t-\tau) - \lambda_2 x(t-\tau) \tag{6}$$

Since both equations in (5) and (6) are quite similar, it is sufficient to study (5) □

Remark 2.3 Let us consider the following controller:

$$U = -F_d + \lambda_1 \tilde{F}_m(t) + \lambda_2 \dot{\tilde{F}}_m(t) \tag{7}$$

The closed-loop equation in the ideal case is

$$(1+\lambda_1)\tilde{F}(t) + \lambda_2 \dot{\tilde{F}}(t) = 0$$

which is a first order linear differential equation. Now assume $F_m(t) = F(t-\tau)$. Then we get

$$\tilde{F}(t) + \lambda_1 \tilde{F}(t-\tau) + \lambda_2 \dot{\tilde{F}}(t-\tau) = 0$$

which can be written with $x(t) = \tilde{F}(t-\tau)$ and $\lambda_2 > 0$ as

$$\dot{x}(t) = -\frac{\lambda_1}{\lambda_2}x(t) - \frac{1}{\lambda_2}x(t+\tau)$$

which is a non-causal differential equation.
The closed-loop system is ill-posed in this case, because the open-loop system in (1) is algebraic and the controller in (7) is a function of the output derivative. □

3 STABILITY OF THE FUNCTIONAL CLOSED-LOOP EQUATIONS

Using fundamental results of the stability theory for neutral functional differential equations (see Hale, 1977) we can state the following result:

Proposition 1 *Supposing $0 < \lambda_1 < 1$, $\lambda_2 > 0$ the trivial solution of the NFDE (5) is uniformly asymptotically stable for any constant delay τ satisfying*

$$\tau < \frac{1-\lambda_1}{\lambda_2} \tag{8}$$

The proof is given in the full version of the paper.

Remark 3.1 The condition $\lambda_1 < 1$ is an obvious necessary condition for stability analysis since that is equivalent to the stability of the operator $\mathcal{D}_1\phi = \phi(0) - \lambda_1\phi(-\tau)$. □

Remark 3.2 If $\lambda_1 = 0$ in (2), then stability of (3) remains unchanged. On the contrary, (5) becomes

$$\dot{x}(t) = -\lambda_2 x(t-\tau) \tag{9}$$

which is no longer a NFDE, but a RFDE (retarded functional differential equation, Kolmanovskii and Nosov, 1986). In this situation, the real condition of stability (Hale, 1977) is $\tau < 3/2\lambda_2$.

With $\lambda_1 = 0$ in (8) we obtain a 'relatively restrictive' condition $\tau < 1/\lambda_2$, which can be obtained using a Razumikhin type theorem (Hale, 1977). Furthermore, from (8) λ_2 is a "measure" for the admissible delay-size for a given λ_1. □

Remark 3.3 If $\lambda_2 = 0$ we obtain

$$\dot{x}(t) = -\lambda_1 \dot{x}(t-\tau) \tag{10}$$

which is a functional equation in $\dot{x}(t)$. A necessary and sufficient condition for the stability of (10) is $|\lambda_1| < 1$ (Hale, 1977; pp.291, Corollary 5.1). U in (2) is a simple proportional force feedback in this case. Wen and Murphy (1991) have already pointed out the necessity of having $|\lambda_1| < 1$ for stability of a proportional force feedback using a different argument. □

4 CONTACT INSTABILITY PHENOMENONS ANALYSIS

In the previous section we have given mathematical conditions that guarantee stability of closed-loop system. In this section we analyze conditions under which the robot may loose the contact, and we clearly identify time-delays to be a cause of possible bouncing of the robot's tip on the environment.

It is important to note that the stability conditions derived in the foregoing section do not guarantee a constant sign of the system's state, i.e. of the interaction force. Indeed the closed-loop system may be asymptotically stable but oscillating. In our case, a positive sign of F during a non-zero time-interval means that contact is lost, and the corresponding open-loop system is not the same (namely, the equation in (1) becomes a second order differential equation). Therefore mathematical stability conditions deduced in the previous section (which are sufficient conditions for the stability of the considered functional differential equation) are not sufficient for stability of the contact task.

4.1 Conditions for Interaction Force Constant Negative Sign

4.1.1 Proportional Force Feedback. Let us first analyze the case when $\lambda_2 = 0$, i.e. $U = -F_d + \lambda_1 \tilde{F}_m$. Since we assume that the robot is in contact with the environment before the first instant (taken here to be equal to τ) when U is applied, we must assume that on $[0,\tau]$ the contact force has a certain value, say $U_{[0,\tau]} = -F_0$, $F_0 < 0$. Note that if we do not suppose that contact is established before U is applied, then we must study the transition phase between unconstrained and constrained motion Brogliato *et al.* (1994).

Since we study the stability properties of the constrained-motion task, we must assume that contact has in fact always occured in the past, which is an implicit assumption in all stability analysis of hybrid force/position controllers (Yoshikawa, 1987; Raibert and Craig, 1981; MacClamroch and Wang, 1988). From a mathematical point of vue, the value of the interaction force F on $[0,\tau]$ is a necessary condition for the existence of a solution to the considered functional equation.

The problem can be formulated as follows: *For given F_0 and F_d, find conditions such that the sign of F remains constant for all $t \geq 0$.*

Since on $[0,\tau)$, we have $F_{[0,\tau)} = F_0 < 0$ and from (1) and (2) with $\lambda_2 = 0$ on $[\tau, 2\tau)$

$$F_{[\tau,2\tau)} = F_d - \lambda_1 F_{[0,\tau)} + \lambda_1 F_d \tag{11}$$

Thus we obtain:

$$|F_d| > \frac{\lambda_1}{1+\lambda_1} |F_0| \Leftrightarrow F_{[\tau,2\tau)} < 0$$

We have the following

Proposition 2

$$sgn F_{[k\tau,(k+1)\tau)}(t) = sgn F_{[0,\tau)} \Leftrightarrow |F_d| > \frac{\lambda_1}{1+\lambda_1} |F_0| \tag{12}$$

4.1.2 Proportional-Integral Force Feedback. We now examine the general case when $\lambda_1 > 0$, $\lambda_2 > 0$. Supposing that $U_{[0,\tau)} = -F_0 > 0$, we obtain

$$U_{[\tau,2\tau)}(t) = -F_d + \lambda_1(F_0 - F_d) + \lambda_2(t-\tau)(F_0 - F_d), \quad \forall t \in [\tau, 2\tau) \tag{13}$$

We also suppose that (12) is verified.

Assume first $F_0 - F_d < 0$. From (13) we get $U_{[\tau,2\tau)}(t) + F_d < 0$, for all $t \in [\tau, 2\tau)$. It follows that

$$|F_d| > \frac{\lambda_1 + \lambda_2\tau}{1+\lambda_1+\lambda_2\tau} |F_0| \Longrightarrow U_{[\tau,2\tau)}(t) > 0, \quad \forall t \in [\tau, 2\tau) \tag{14}$$

In this case for all $t \in [2\tau, 3\tau)$ we have

$$U_{[2\tau,3\tau)}(t) = -F_d + \lambda_1[F_{[\tau,2\tau)}(t-\tau) - F_d]$$
$$+\lambda_2 \int_0^\tau (F_0 - F_d)dz + \lambda_2 \int_\tau^{t-\tau} (F_{[\tau,2\tau)}(z) - F_d)dz \quad (15)$$

Since $F_0 - F_d < 0$ and $F_{[\tau,2\tau)}(t-\tau) - F_d > 0$ (from (13)), nothing can be straightforward concluded on the sign of $U_{[2\tau,3\tau)}(t)$.
However note that from (15) one sees that $U_{[2\tau,3\tau)}(t)$ is a polynomial of order 2 in t having coefficients depending on λ_1, λ_2, F_d, F_0.

Notice that we have for $t \in [k\tau, (k+1)\tau)$

$$\int_\tau^t F_m(z)dz = \sum_{j=0}^{k-2} \int_{j\tau}^{(j+1)\tau} F_{[j\tau,(j+1)\tau)}(z)dz$$
$$+ \int_{(k-1)\tau}^{t-\tau} F_{[(k-1)\tau,k\tau)}(z)dz \quad (16)$$

Therefore U in (1) is a polynomial of order k in t for $t \in [k\tau, (k+1)\tau)$ by simple induction from (15).

It is important to note that although $F_0 - F_d < 0$, from (13) and (14), $F_{[\tau,2\tau)}(t) - F_d > 0$ and $F_{[\tau,2\tau)}(t) < 0$. This means that there is a jump in the interaction force at $t = \tau$, but no loss of contact. One sees that it is difficult to draw such conclusions from (15). However note that since

$$U_{[k\tau,(k+1)\tau)}(t) = -F_d + \lambda_1(F_{[(k-1)\tau,k\tau)}(t-\tau) - F_d)$$
$$+\lambda_2 \int_\tau^t \tilde{F}_m(z)dz \quad (17)$$

we have the following

Proposition 3 *$U(t)$ is piecewise continuous with discontinuities at $t = j\tau$, for all integers $j \geq 1$. The jumps magnitudes are bounded and tend towards 0 as $t \to \infty$.*

Remark 4.1 From the proof of the Proposition 3, a necessary and sufficient condition to have no jumps in the closed-loop system's solution is that $U_{[\tau,2\tau)}(\tau) = -F_{[0,\tau)}$, which is equivalent to have $F_{[0,\tau)} = F_d$. In general such a condition is not verified, so the interaction force possesses discontinuities. Furthermore, the sign of the jump is alternatively positive and negative. □

Proposition 4 *There exist $\lambda_1{}^* > 0$, $\lambda_2{}^* > 0$ such that for any $0 < \lambda_1 < \lambda_1{}^* < 1$, $0 < \lambda_2 < \lambda_2{}^*$, we have $sgn(U(t)) = +1$ for all $t \geq 0$.*

Finally let us note that if $F_{[0,\tau)} - F_d > 0$, similar conclusions hold since $F_{[\tau,2\tau)}(t) - F_d < 0$ from (12), and so on.

4.2 Bouncing phase analysis

In this section we analyze the behaviour of the system when the "robot" looses contact with environment, which implies a phase of rebounds since we deal with the rigid case. The point is to determine if this phase is stable or not, i. e. will the mass be stuck on the environment again and will the rebounds stop?

4.2.1 Proportional Force Feedback. Let us consider $\lambda_2 = 0$. If (11) is not satisfied, then $F_{[\tau,2\tau)}(\tau) > 0$: the closed-loop equations are such that there is a jump in the interaction force at $t = \tau$, and the control force $U(\tau)$ is negative for a non-zero time-interval starting in τ. From (12) the control input on $t \in [\tau, 2\tau)$ is constant and given by

$$U_{[\tau,2\tau)}(t) = -(1+\lambda_1)F_d + \lambda_1 F_{[0,\tau)} < 0$$

In conclusion the contact is lost at $t = \tau$. For $t \geq 2\tau$, the measured interaction force is zero and the control input becomes

$$U_{[2\tau,t_f)}(t) = -(1+\lambda_1)F_d > 0$$

Note also that (1) becomes $m\ddot{x}(t) = U$ for $t \geq \tau$. We get on $[\tau, 2\tau)$

$$x(t) = \frac{U_{[\tau,2\tau)}}{2m}(t-\tau)^2 \quad (18)$$

and on $[2\tau, t_1)$

$$x(t) = \frac{2U_{[2\tau,t_f)}}{2m}(t-2\tau)^2 + 2\frac{U_{[\tau,2\tau)}}{m}\tau(t-\tau)$$
$$+\frac{2U_{[\tau,2\tau)}}{m}\tau^2 \quad (19)$$

where t_f is the final impact time and t_1 is the time of the first impact between the mass and the environment. Conditions under which $t_f < \infty$ are studied in Brogliato *et al.* (1994).

Let us simply note that since the input force applied to the mass is constant for $t \geq 2\tau$, the system is strictly identical to a bouncing ball submitted to gravity. Then a necessary and sufficient conditon to get $t_f < \infty$ is that the restitution coefficient $e > 0$ between the mass and the environment be < 1.

We suppose that $t_f < \infty$. Roughly, a relation similar to (19) is true between two impacts, the physical impact law providing initial conditions for $\dot{x}(t)$ after each impact (note that $x(t_k) = 0$ at impact times t_k).

On $t \in [t_f, t_f + \tau)$, we get

$$U_{[t_f,t_f+\tau)}(t) = -(1+\lambda_1)F_d > 0$$

and on $[t_f, t_f + 2\tau)$, we get

$$U_{[t_f+\tau,t_f+2\tau)}(t) = -(1-\lambda_1^2)F_d \quad (20)$$

which is positive since $0 < \lambda_1 < 1$.

By induction one can show that

$$U_{[t_f+k\tau,t_f+(k+1)\tau)}(t) = -(1+(-1)^{k+1}\lambda_1^{k+1})F_d > 0 \quad (21)$$

Thus there is no loss of contact any longer.

Remark 4.2 (11) is fullfilled if $F_0 = (1+\lambda_1)F_d$. We could have supposed that $U_{[0,\tau)} = -(1+\lambda_1)F_d$ and that the force feedback is switched on at $t = \tau$. Note however that the analysis done in the preceding subsection is true only if there is no force disturbance acting on the system. Assume that a disturbance $F_p = F_M\chi_\Delta$ acts on the system on $t \in \Delta$, where χ_Δ is the characteristic function of the interval $\Delta \subseteq [k\tau,(k+1)\tau)$ for some $k > 0$. The stability condition in (11) becomes

$$| F_d | > \frac{\lambda_1}{1+\lambda_1} \left| F_{[(k-1)\tau,k\tau)} - \frac{F_M}{\lambda_1} \right|$$

If $F_{[(k-1)\tau,k\tau)} - \frac{F_M}{\lambda_1} < 0$, and if $F_M < 0$, then $| F_M | < \lambda_1 | F_{[(k-1)\tau,k\tau)} |$: note that if this condition is not satisfied, then the robot will loose contact on $[k\tau,(k+1)\tau)$. Furthermore, the condition $| F_M | \leq | F_{[(k-1)\tau,k\tau)} |$ is necessary and sufficient for the robot not to loose the contact on $[(k-1)\tau,k\tau)$, but it is not sufficient for contact stability for $t \geq k\tau$. The stability analysis follows the same lines with F_0 replaced by $F_{[(k-1)\tau,k\tau)} - \frac{F_M}{\lambda_1}$. □

Remark 4.3 Following Remark 2.1, let us note that clearly the impact for transition phase should better behave if the input has a damping term. This however does not modify our analysis. □

4.2.2 Proportional-Integral Force Feedback. Let us analyze the case $\lambda_2 > 0$. For the sake of simplicity, we suppose that (14) is not verified, so that $U_{[\tau,2\tau)}(t) \leq 0$ for $t \in [t_l, 2\tau)$, i.e. there is loss of contact at $t = t_l$. Since $U_{[\tau,2\tau)}(t)$ is still given by (13), we get:

$$\begin{aligned} U_{[2\tau,t_l+\tau)}(t) &= -F_d + \lambda_1(F_{[\tau,t_l)}(t-\tau) - F_d) \\ &+\lambda_2 \int_0^\tau (F_{[0,\tau)} - F_d)dz \\ &+\lambda_2 \int_\tau^{t-\tau} (F_{[\tau,t_l)}(z) - F_{d[\tau,t_l)})dz \end{aligned} \quad (22)$$

We suppose that as soon as the measured interaction force $F_m(t)$ is zero, then we set $F_d \equiv 0$, i.e. on $[t_l+\tau, t_f+\tau)$, $F_m(t) \equiv 0$ and $F_{d[t_l+\tau,t_f+\tau)} \equiv 0$. For the mass to collide with the environment, it is necessary that $U_{[t_l+\tau,3\tau)}(t)$ be positive, that is:

$$\begin{aligned} U_{[t_l+\tau,3\tau)}(t) &= \lambda_2 \int_0^\tau (F_{[0,\tau)} - F_d)dz \\ &+\lambda_2 \int_\tau^{t_l} (F_{[\tau,t_l)}(z) - F_{d[\tau,t_l)})dz > 0 \end{aligned} \quad (23)$$

It should be noted that this control strategy is related to the fact that $\tau > 0$. If $\tau = 0$, there is in the ideal case no reason for the robot to take off than the surface. We do not analyze here the stability of a complete robotic task involving contact and non-contact phases: in this case the control strategy must be adapted suitably to guarantee stability. This is outside the scope of the present work.

From the fact that after $t_l+\tau$, the control input remains constant until $t_f+\tau$, we get $U_{[t_l+\tau,3\tau)} \equiv U_{[3\tau,t_f+\tau)}$. Thus we get

$$\begin{aligned} &-t_lF_d + \tau F_0 \\ &+ \int_\tau^{t_l} [(F_d - \lambda_1(F_0 - F_d) + \lambda_2(z-\tau)(F_0 - F_d)]dz > 0 \end{aligned}$$

that is

$$\begin{aligned} (F_0 - F_d)\left(\tau - \lambda_1(t_l - \tau) + \frac{\lambda_2}{2}(t_l-\tau)^2\right) > 0 \Leftrightarrow \\ \Leftrightarrow \tau - \lambda_1(t_l - \tau) + \frac{\lambda_2}{2}(t_l - \tau)^2 < 0 \end{aligned} \quad (24)$$

Note that from (13) we have

$$t_l = \frac{F_d - \lambda_1(F_0 - F_d)}{\lambda_2(F_0 - F_d)} + \tau \quad (25)$$

Therefore, after some algebraic manipulations, (24) is equivalent to

$$\frac{\lambda_1}{1+\lambda_1} < \frac{| F_d |}{| F_0 |} < \frac{3\lambda_1}{1+3\lambda_1}$$

then there exists $\lambda_2^* > 0$ such that for $0 \leq \lambda_2 \leq \lambda_2^*$, (24) is verified; λ_2^* is function of τ and of $\frac{|F_d|}{|F_0|}$.

Now we get

$$\begin{aligned} U_{[t_f+\tau,t_f+2\tau)} &= -F_d + \lambda_1(-U_{[3\tau,t_f)} - F_d) \\ &+\lambda_2 \int_0^\tau (F_0 - F_d)dz \\ &+\lambda_2 \int_\tau^{t_l} (F_{[\tau,t_l)}(z) - F_d)dz \\ &+\lambda_2 \int_{t_f+\tau}^{t} (-U_{[3\tau,t_f)}(z-\tau) - F_d)dz \end{aligned} \quad (26)$$

Note that $F_m = 0$ on $[t_f, t_f+\tau)$. Defining

$$K = \tau - \lambda_1(t_l - \tau) + \frac{\lambda_2}{2}(t_l - \tau)^2$$

from (26) we get:

$$U_{[t_f+\tau,t_f+2\tau)}(t) > 0 \Leftrightarrow \frac{| F_d |}{| F_0 |} > \frac{\lambda_2 | K |}{1 - \lambda_2 K} \quad (27)$$

From (25), (27) involves F_d, F_0, λ_1, λ_2 but explicit conditions on those parameters to guarantee that (27) is verified are difficult to be carried out.

The addition of an integral term makes the system significantly more difficult to analyze that when $\lambda_2 \equiv 0$.

The general tendancy is that λ_2 must be small enough to obtain a "stable" scheme. This was pointed out in Wen and Murphy (1991) and shown in Brogliato *et al.* (1994) for the stability of the impact phase.

Remark 4.5 We have analyzed each control law separately, i.e. as if only a proportional feedback or a proportional-integral feedback was applied all the time. We could also have supposed that both controllers are switched along a certain strategy, e.g. apply a proportional control during impact phases and integration during contact (Note that setting $F_d \equiv 0$ whenever $F_m \equiv 0$ is different from using proportional feedback since that i) both controllers need not necessarily have neither the same λ_1 gain nor desired force, ii) when a switch is applied, one may assume that the integrator is initialized). Usually the integral feedback is applied when contact is established. Thus the initial condition for the contact phase will depend on the proportional feedback value on the time instant when the switch is applied: the ideal switching time is t_f; in practice the controllers will be switched before or after t_f due to bad timing in the switching strategy: this reveals the complex behaviour of such systems and the role played both the low-level part (differential equations, delay in the force control loop) and the high-level part (the strategy that schedules the switches between several controllers, bad timing). These problems have already received some attention in Brogliato *et al.* (1994), Mills (1993). □

5 CONCLUSIONS

In this paper we have focused on the problem of stability of a robotic manipulator submitted to holonomic constraints. An important point is that the constraints are supposed to be unilateral, in the sense that if the interaction force has wrong sign, the contact is lost. Furthermore, we have assumed that the force-feedback loop contains time-delays. Sufficient delay-dependent conditions are derived to guarantee that the robot's tip remains in contact with the surface. These conditions are found by taking into account the fact that due to the unilateral nature of the constraints, the interaction force must have constant sign during the whole task. We analyze the cases of proportional and proportional-integral force-feedback. To the best of the authors' knowledge, this problem has been pointed out in Wen and Murphy (1991), Wilfinger *et al.* (1994), but had not been treated before in the robotics and control literature.

REFERENCES

Brogliato, B., P. Orhant and R. Lozano, (1994). *SYROCO '94, Capri, Italy, September 16-19, 1994.*

Chiaverini, S. and L. Sciavicco, (1993). *IEEE Trans. on Robotics and Automation*, **9**, 361-373.

De Luca, A. and C. Manes, (1991). 3rd *IFAC SYROCO, 16-18 September 1991, Wien.*

Duffy, J. (1990). *J. of Robotic Systems*, **7**, 139-144.

Eppinger, S. D. and W. P. Seering, (1992). *IEEE Trans. Robotics Automation*, **6**, 751-758.

Fisher, W. D. and M. Shahid Mujtaba, (1992). *Int. J. of Robotics Research*, **11**, 299-311.

Hale, J. (1977). *Theory of Functional Differential Equations*. Springer-Verlag, New York.

Kolmanovskii, V. B. and V. R. Nosov, (1986). *Stability of Functional Differential Equations*. Academic Press, New York.

MacClamroch, N. H. and D. Wang, (1988). *IEEE Trans. Automat. Contr.*, **AC-33**, 419-426.

Mills, J. K. (1993). *Automatica*, **26**, 861-874.

Mills, J. K. and D. M. Lokhorst, (1993). *Int. J. of Rob. Research*, **12**, 146-163.

Niculescu, S. -I. and B. Brogliato, (1994). *Internal Note L.A.G. 94-108* (the full version of the paper).

Raibert, M. H. and J. J. Craig, (1981). *Trans. ASME J. Dyn. Syst. Meas. Contr.*, **102**, 126-133.

Sinha, P. R. and A. A. Goldenberg, (1993). *IEEE Int. Conf. on Robotics and Automation*, 343-348.

Volpe, R. and P. Khosla, (1993). *Int. J. Robotics Res.*, **12**, 351-361.

Wen, J. T. and S. Murphy, (1991). *IEEE Trans. Automat. Contr.*, **36**, 365-371.

Wilfinger, L. S., T. J. Wen and S. H. Murphy, (1994). *IEEE Contr. Syst.*, **14**, 31-40.

Yoshikawa, T. (1987). *IEEE J. of Robotics and Automation*, **3**, 386-392.

Yoshikawa, T., T. Surgie and M. Tanaka, (1988). *IEEE J. of Robotics and Automation*, **4**, 699-705.

STABILIZATION CRITERIA FOR BILINEAR SYSTEMS WITH DELAYED STATE AND SATURATING ACTUATORS

Silviu-Iulian Niculescu,* Jean-Michel Dion and Luc Dugard

Laboratoire d'Automatique de Grenoble (URA CNRS 228),
ENSIEG, BP 46, 38402 St.-Martin-d'Hères, France.
The authors are also with the GDR CNRS "Automatique".

Abstract. In this paper, the problem of stabilization of a class of bilinear systems with delayed state and saturating actuators is considered. *Delay-independent* sufficient conditions for the stability of the closed-loop system via a memoryless static state feedback are developed. The stabilization conditions and the corresponding memoryless control laws are expressed in terms of symmetric positive definite solutions of appropriate finite dimensional algebraic Riccati equations.

Résumé: Dans cet article, on considère le problème de stabilisation d'une classe de systèmes bilinéaires à états retardés et à actionneurs saturés. Des conditions suffisantes, indépendantes de la taille du retard, sont développées pour la stabilité du système en boucle fermée via un retour d'état statique sans mémoire. Les conditions de stabilisation et les lois de commande correspondantes sont exprimées en termes de solutions définies positives d'équations algébriques de Riccati de dimension finie.

Keywords. Bilinear systems; saturating actuators; delay; memoryless stabilization; Lyapunov theory.

1 INTRODUCTION

In many practical situations, nonlinear actuators are encountered to follow the control signal computed from the design controllers. In the case when the control system was designed to be stable without considering the saturation, we can not guarantee, in general, the stability of the closed-loop system via the considered control law. An integral wind up or a limit cycle may occur in the system during the 'start-up', 'shift-down' or sudden changes (see, for example Glatfelder and Schaufelberger, 1983; Krikelis, 1980 and the references therein).

The stabilization of linear systems with saturating actuators has been investigated by many researchers in the past few years. A representative result is the work of Chen and Wang (1988) using a time-domain approach. This result was extended to the case of linear systems with delayed state and sufficient conditions for state feedback stabilization are given in Chen *et al.* (1988), Tissir and Hmamed (1992) (independent of the size of delay) or in Chen *et al.* (1988), Su *et al.* (1991) (dependent of the size of delay).

The great interest in the bilinear systems is due to their simplicity and capability to describe real processes (see Mohler, 1973). Conditions that guarantee the stabilization of a SISO bilinear system via constrained output feedback are given in Genesio and Tesi (1988); the constant feedback global asymptotic stabilization and practical stabilization are analyzed in Celikovsky (1993); structural stabilization conditions for bilinear systems of second order are given in Koditschek and Narendra (1983); the effect of quadratic state feedback on the global stability of the closed-loop system and on the transient response is given in Liu and Chen (1993) and the constrained quadratic state feedback case is considered in Quinn (1980), Ryan and Buckingham (1983).

In this paper, we consider a class of bilinear systems which contain saturating actuators and are described by delayed state equation. The delay is assumed to be constant, bounded, but unknown. The focal point of this paper is on investigating conditions for stabilization which are independent of the size of delay. The approach adopted here uses a Lyapunov-Krasovskii technique for analysing the uniform asymptotic stability of solutions of functional differential equations. Our main results establish *delay-independent* sufficient conditions for uni-

*The author is also with the Department of Automatic Control, University "Politehnica" Bucharest, Romania.

form asymptotic stability of the closed-loop system via memoryless static state feedback. These conditions are given in terms of solutions of appropriate finite dimensional algebraic Riccati equations.

The paper is organized as follows: Section 2 presents briefly the problem statement. In Section 3 a time-delay independent stabilization condition is given. In Section 4 bounds on the 'limits' of the saturating actuators to guarantee the stability of the closed-loop system are presented. Some final remarks are given in Section 5.

Notations. The following notations will be used throughout the paper. $\Re$ denotes the set of real numbers, $\Re^+$ is the set of non-negative real numbers, $\Re^n$ denotes the n dimensional Euclidean space, and $\Re^{n\times m}$ denotes the set of all $n \times m$ real matrices. The notation $X \geq Y$ (respectively, $X > Y$), where X and Y are symmetric matrices, means that the matrix $X - Y$ is positive semi-definite (respectively, positive definite). For a real matrix A, A^+ is the matrix obtained by replacing the entries of A with their absolute values. $\mathcal{C}_\tau = \mathcal{C}([-\tau, 0], \Re^n)$ denotes the Banach space of continuous vector functions mapping the interval $[-\tau, 0]$ into $\Re^n$ with the topology of uniform convergence. The following norms will be used: $\| \cdot \|$ refers to either the Euclidean vector norm or the induced matrix 2-norm; $\| \phi \|_c = \sup_{-\tau\leq t\leq 0} \| \phi(t) \|$ stands for the norm of a function $\phi \in \mathcal{C}_\tau$. Moreover, we denote by $\mathcal{C}_\tau^v$ the set defined by $\mathcal{C}_\tau^v = \{\phi \in \mathcal{C}_\tau \ : \| \phi \|_c < v\}$, where v is a positive real number.

2 PROBLEM STATEMENT

Consider the following bilinear system with delayed state described by:

$$\begin{aligned}\dot{x}(t) &= Ax(t) + A_d x(t-\tau) \\ &\quad + \sum_{i=1}^{m} sat\ (u_i(t)) N_i x(t) + Bsat\ (u(t)) \quad (1)\end{aligned}$$

with the initial condition

$$x(t_0 + \theta) = \phi(\theta),\ \forall\ \theta \in [-\tau, 0];\ (t_0, \phi) \in \Re^+ \times \mathcal{C}_\tau^v \quad (2)$$

where $x(t) \in \Re^n$ is the state, $u(t) \in \Re^m$ is the control input, τ is the time-delay of the system, A, A_d, $N_i (i = 1, \ldots, m)$ and B are known real constant matrices of appropriate dimensions.

$sat\ (u(t))$ is the saturating function which is defined as follows:

$$sat\ u(t) = [sat\ u_1(t)\ sat\ u_2(t)\ \ldots\ sat\ u_m(t)]^T \quad (3)$$

$$sat\ u_i(t) = \begin{cases} -U_i, & \text{if } u_i < -U_i < 0 \\ u_i, & \text{if } -U_i \leq u_i \leq U_i \\ U_i, & \text{if } u_i > U_i > 0 \end{cases} \quad (4)$$

Throughout this paper we shall use the following concept of stabilization for the bilinear systems of the form (1)-(4).

Definition 1 *The bilinear system with delayed state (1)-(4) is said to be stabilizable via memoryless state feedback control law if there exists a control law $u(t) = -Kx(t)$, $K \in \Re^{m\times n}$ such that the trivial solution $x(t) \equiv 0$ of the functional differential equation associated to the closed-loop system is uniformly asymptotically stable.*

The aim of this paper is to investigate conditions for stabilization via memoryless state feedback of the bilinear systems (1)-(4). The approach adopted here is based on the Lyapunov-Krasovskii Theorem (see Appendix A) for the uniform asymptotic stability of the trivial solution of the functional differential equation associated to the closed-loop system (see Hale, 1977; Kolmanovskii and Nosov, 1986).

3 STABILIZATION RESULTS

In this section we shall give sufficient conditions to guarantee the stabilization via a memoryless state feedback for the bilinear system with delayed-state (1)-(4). The corresponding control law is given in terms of a symmetric positive definite solution of a finite dimensional algebraic Riccati equation.

Theorem 1 *Consider the bilinear system described by the equations (1)-(4). If there exist two symmetric positive definite matrices Q and S such that the algebraic Riccati equation*

$$A^T P + PA + P(A_d S^{-1} A_d^T - BB^T)P + S + Q = 0 \quad (5)$$

has a symmetric positive definite solution P and if

$$\lambda_{min}(Q) - 2\sum_{i=1}^{m} U_i \| N_i P \| - \| B^T P \|^2 > 0 \quad (6)$$

then the bilinear system with delayed state and symmetric saturating actuators (1)-(4) is stabilizable via the memoryless state feedback $u(t) = -B^T Px(t)$.

Proof. With the state feedback control law $u(t) = -B^T Px(t)$, the closed-loop system (1) can be rewritten as

$$\begin{aligned}\dot{x}(t) &= (A - \frac{1}{2}BB^T P)x(t) + A_d x(t-\tau) \\ &\quad + \sum_{i=1}^{m} sat\ (u_i(t)) N_i x(t) \\ &\quad + B(sat\ (-B^T Px(t)) + \frac{1}{2}B^T Px(t)) \quad (7)\end{aligned}$$

Let introduce the following Lyapunov functional candidate:

$$V(x_t) = x(t)^T Px(t) + \int_{t-\tau}^{t} x(\theta)^T Sx(\theta) d\theta$$

where P and S satisfy the algebraic Riccati equation (5). It follows that

$$\beta_1 \parallel x(t) \parallel^2 \leq V(x_t) \leq \beta_2 \parallel x_t \parallel_c^2 \quad (8)$$

where $\beta_1 = \lambda_{min}(P)$ and $\beta_2 = \lambda_{max}(P) + \tau\lambda_{max}(S)$. The derivative of $V(x_t)$ along the trajectories of closed-loop system (7) is

$$\begin{aligned}\dot{V}(x_t) &= \dot{x}(t)^T Px(t) + x(t)^T P\dot{x}(t) \\ &+x(t)^T Sx(t) - x(t-\tau)^T Sx(t-\tau)\end{aligned}$$

Using the algebraic inequality

$$\parallel sat\ u(t) - \frac{1}{2}u(t) \parallel \leq \frac{1}{2} \parallel u(t) \parallel$$

after some algebraic manipulations, we get

$$\begin{aligned}\dot{V}(x_t) &\leq -(\lambda_{min}(Q) - 2\sum_{i=1}^{m} U_i \parallel N_i P \parallel \\ &- \parallel B^T P \parallel^2) \parallel x(t) \parallel^2\end{aligned}$$

From (6) it follows that there exists a $\beta_3 > 0$ such that

$$\dot{V}(x_t) \leq -\beta_3 \parallel x(t) \parallel^2 \quad (9)$$

From (8) and (9), using the Lyapunov-Krasovskii Theorem given in Appendix A (for a complete information see Hale, 1977 or Kolmanovskii and Nosov, 1986), it follows that the trivial solution of the functional differential equation associated to the closed-loop system is uniformly asymptotically stable. ∇∇∇

Remark 1 The condition (6) and the algebraic Riccati equation (5) do not contain any 'information' about the size of delay, and thus Theorem 1 is applicable for any time-delay $\tau \neq 0$. □

Remark 2 The result can be easily extended to the time-varying delay case: $0 \leq \tau(t) \leq \bar{\tau}$, $\dot{\tau}(t) \leq \beta < 1$. In this case, the condition (6) is not changed, but the algebraic Riccati equation (5) becomes:

$$\begin{aligned}&A^T P + PA + P(A_d S^{-1} A_d^T - BB^T)P \\ &+\frac{1}{1-\beta}S + Q = 0\end{aligned}$$

The proof follows the same ideas, with the following Lyapunov functional:

$$V(x_t) = x(t)^T Px(t) + \frac{1}{1-\beta}\int_{t-\tau(t)}^{t} x(\theta)^T Sx(\theta)d\theta$$

□

Remark 3 If the time-delay system (1) is linear and not bilinear ($N_i \equiv 0$), then in Theorem 1, condition (5) becomes:

$$\lambda_{min}(Q) - \parallel B^T P \parallel^2 > 0$$

which is independent of the saturating actuators' limits (U_i, $i = \overline{1,m}$). □

This result can be easily extended to the multiple delays case. Considering the following bilinear system

$$\begin{aligned}\dot{x}(t) &= Ax(t) + \sum_{i=1}^{r} A_{di}x(t-\tau_i) \\ &+ \sum_{i=1}^{m} sat\ (u_i(t))N_i x(t) + Bsat\ u(t) \quad (10)\end{aligned}$$

with the initial condition

$$\begin{aligned}&x(t_0+\theta) = \phi(\theta),\ \forall\ \theta \in [-\bar{\tau}, 0]; \\ &\bar{\tau} = \max_{i=\overline{1,r}} \tau_i\ (t_0, \phi) \in \Re^+ \times \mathcal{C}_\tau^v \quad (11)\end{aligned}$$

We have the following

Theorem 2 *Consider the bilinear system described by the equations (10)-(11), (3)-(4).*
If there exist $r+1$ symmetric positive definite matrices Q and S_i ($i = 1, \ldots, r$) such that the algebraic Riccati equation

$$\begin{aligned}&A^T P + PA + P(\sum_{i=1}^{r} A_{di} S_i^{-1} A_{di}^T - BB^T)P \\ &+\sum_{i=1}^{r} S_i + Q = 0\end{aligned}$$

has a symmetric positive definite solution P and if

$$\lambda_{min}(Q) - 2\sum_{i=1}^{m} U_i \parallel N_i P \parallel - \parallel B^T P \parallel^2 > 0$$

then the bilinear system with delayed state (10)-(11) and symmetric saturating actuators (3)-(4) is stabilizable via the memoryless state feedback $u(t) = -B^T Px(t)$. ∇∇∇

The proof follows the same steps as the proof of Theorem 1 and makes use of the following Lyapunov-Krasovskii functional candidate:

$$V(x_t) = x(t)^T Px(t) + \sum_{i=1}^{r}\int_{t-\tau_i}^{t} x(\theta)^T S_i x(\theta)d\theta$$

4 STABILITY BOUNDS ON $U_I (I = 1, \ldots, M)$

In this section we are interested in solving the following problem: give bounds on $U_i (i = 1, \ldots, m)$ that guarantee the uniform asymptotic stability of the closed-loop system.

We have the following:

Theorem 3 *Consider the bilinear system defined by equations (1)-(4). If there exist two symmetric positive definite matrices Q and S, such that the following algebraic Riccati equation*

$$A^T P + PA + P(A_d S^{-1} A_d^T - BB^T)P + S + Q = 0$$

has a symmetric positive definite solution P such that the matrix $Q_1 = Q - \| B^T P \|^2 I$ is a symmetric positive definite matrix and if

$$\sum_{i=1}^{m} U_i^2 < \frac{1}{\sum_{i=1}^{m} \| Q_1^{-\frac{1}{2}} F_i Q_1^{-\frac{1}{2}} \|^2}, \qquad (12)$$

(where $F_i = N_i^T P + P N_i$)
then the bilinear system (1)-(4) is stabilizable via the memoryless state feedback $u(t) = -B^T P x(t)$.

In the case when $U_i = U$, $i = 1, \ldots m$, (12) can be replaced by the following condition

$$U < \frac{1}{\lambda_{max}\left(\sum_{i=1}^{m} (Q_1^{-\frac{1}{2}} F_i Q_1^{-\frac{1}{2}})^+\right)}. \qquad (13)$$

▽▽▽

Note that conditions (12) or (13) provide "bounds" on $U_i (i = 1, \ldots, m)$ that guarantee closed-loop stability.

Proof. Using the same Lyapunov functional as in Theorem 1, we have for the derivative of $V(x_t)$:

$$\dot{V}(x_t) \le x(t)^T [-Q_1 + \sum_{i=1}^{m} sat\ (u_i(t)) F_i] x(t)$$

Since

$$Q_1 - \sum_{i=1}^{m} sat\ (u_i(t)) F_i =$$
$$= Q_1^{\frac{1}{2}} \left(I - \sum_{i=1}^{m} sat\ u_i(t) Q_1^{-\frac{1}{2}} F_i Q_1^{-\frac{1}{2}} \right) Q_1^{\frac{1}{2}}$$

we get

$$\dot{V}(x_t) \le$$
$$\le -x(t)^T Q_1^{\frac{1}{2}} \left(I - \sum_{i=1}^{m} sat\ (u_i(t)) Q_1^{-\frac{1}{2}} F_i Q_1^{-\frac{1}{2}} \right) Q_1^{\frac{1}{2}} x(t)$$

If (12) is satisfied, then using the Schwartz inequality, it follows

$$1 > \left(\sum_{i=1}^{m} \| sat\ (u_i(t)) Q_1^{-\frac{1}{2}} F_i Q_1^{-\frac{1}{2}} \| \right)^2$$
$$\ge \| \sum_{i=1}^{m} sat\ (u_i(t)) Q_1^{-\frac{1}{2}} F_i Q_1^{-\frac{1}{2}} \|^2, \quad \forall t$$

it follows that the matrix

$$I - \sum_{i=1}^{m} sat\ (u_i(t)) Q_1^{-\frac{1}{2}} F_i Q_1^{-\frac{1}{2}}$$

is positive definite for all values of t.

Thus, there exists a $\beta_3' > 0$ such that

$$\dot{V}(x_t) \le -\beta_3' \| x(t) \|^2 .$$

In conclusion (following the same steps as in the proof of Theorem 1) if (12) is satisfied, the trivial solution of the functional differential equation associated to the closed-loop system (7) is uniformly asymptotically stable.

If $U_i = U$, $i = 1, \ldots m$, then using the following property (Ortega, 1972):

For every $A, B \in \Re^{n \times n}$, if $A^+ [\le] B^+$, then $\lambda_{max}(A) \le \lambda_{max}(B^+)$, where '$\le$' is applied element by element to both matrices ($| a_{ij} | \le | b_{ij} |$).

we have

$$1 > \lambda_{max} \left(\sum_{i=1}^{m} (sat\ u_i(t) Q_1^{-\frac{1}{2}} F_i Q_1^{-\frac{1}{2}}) \right), \quad \forall t$$

and it follows that the matrix

$$I - \sum_{i=1}^{m} sat\ (u_i(t)) Q_1^{-\frac{1}{2}} F_i Q_1^{-\frac{1}{2}}$$

is positive definite for all values of t.

Thus, there exists a $\beta_3'' > 0$ such that

$$\dot{V}(x_t) \le -\beta_3'' \| x(t) \|^2 .$$

As previously, the uniform asymptotic stability of the trivial solution follows. ▽▽▽

Remark 4 The condition $Q_1 = Q_1^T > 0$ is less conservative than the condition (6) in the sense that we have a degree of freedom to 'choose' the saturating actuators' 'limits'. More precisely when the stability bounds U_i, $i = 1, \ldots, m$ do not satisfy inequality (12), they do not satisfy neither condition (6) . □

Remark 5 As in Remark 2, the result can be also easily extended to the time-varying delay case. □

Remark 6 To obtain 'better' bounds on $U_i (i = 1, \ldots, m)$, the same techniques as in Bien and Kim (1992), Luo *et al.* (1993) and the references therein can be used. This can be done by choosing an ad-hoc decomposition for Q_1. □

Remark 7 As in Remark 3, if the system (1) is linear and not bilinear ($N_i \equiv 0$), the existence of a symmetric positive definite solution for the algebraic Riccati equation (5) such that the matrix $Q - \| B^T P \|^2 I$ is positive definite is sufficient to guarantee the uniform asymptotic stability of the trivial solution for all finite time-delays and all finite upper bounds $U_i (i = 1, \ldots, m)$. □

5 CONCLUSION

In this paper we have derived conditions for the stabilization via a memoryless state feedback control law of a bilinear system with delayed state and saturating actuators. These conditions are independent of the size of delay and are expressed in terms of symmetric positive definite solutions of finite dimensional algebraic Riccati equations. The corresponding memoryless control laws are also given.

APPENDIX A: KRASOVSKII STABILITY THEOREM

Consider the functional differential equation of retarded type

$$\dot{x}(t) = f(t, x_t), \quad t \geq t_0 \tag{14}$$

$$x_{t_0}(\theta) = \phi(\theta), \quad \forall\, \theta \in [-\tau, 0] \tag{15}$$

where $x_t(\cdot)$, for a given $t \geq t_0$, denotes the restriction of $x(\cdot)$ to the interval $[t-\tau, t]$ translated to $[-\tau, 0]$, i.e.

$$x_t(\theta) = x(t+\theta), \quad \forall\, \theta \in [-\tau, 0].$$

It is assumed that $\phi \in C_\tau^v$ and the map $f(t, \phi) : \Re^+ \times C_\tau^v \to \Re^n$ is continuous and Lipschitzian in ϕ and $f(t, 0) = 0$.

Let us denote by $x(t_0, \phi)$ the solution of the functional differential equation (10) with the initial condition $(t_0, \phi) \in \Re^+ \times C_\tau^v$.

Definition A.1 *The trivial solution $x(t) \equiv 0$ of (13)-(14) is said to be 'uniformly asymptotically stable' if:*

(a) for every $\kappa > 0$ and for every $t_0 \geq 0$ there exist a $\delta = \delta(\kappa)$ independent of t_0 such that for any $\phi \in C_\tau^\delta$ the solution $x(t_0, \phi)$ of (13)-(14) satisfies $x_t(t_0, \phi) \in C_\tau^\kappa$ for all $t \geq t_0$;

(b) for every $\eta > 0$ and for every $t_0 \geq 0$ there exist a $T(\eta)$ independent of t_0 and an $v_0 > 0$ independent of η and t_0 such that for any $\phi \in C_\tau$, $\| \phi \|_c < v_0$ implies that $\| x_t(t_0, \phi) \|_c < \eta$, $\forall\, t \geq t_0 + T(\eta)$.

We recall that condition (a) implies uniform stability whereas (b) is required for the asymptotic stability property.

In the stability analysis of a system of the form (13)-(14), we shall make use of a Lyapunov functional that is continuous but not necessarily differentiable (in the usual sense of derivative). In this situation, the derivative used for a Lyapunov functional candidate, $W(t, x_t)$, is the largest Dini's time-derivative of $W(t, x_t)$, defined by

$$\dot{W}(t, x_t) = \lim_{\varepsilon \to 0^+} sup \frac{W(t+\varepsilon, x_{t+\varepsilon}) - W(t, x_t)}{\varepsilon}.$$

The above Dini's derivative is referred to as the Dini's time-derivative of $W(t, x_t)$.

Next, let $V_i(\cdot)$, $i = 1, 2, 3$ denote scalar, continuous and nondecreasing functions such that:

$$V_i(r) > 0, \forall\, r > 0; \qquad V_i(0) = 0.$$

Theorem A.1 Krasovskii Stability Theorem

Consider the functional differential equation (13)-(14). Let there exists a continuous functional $W(t, \phi)$ such that

(a) $V_1(\| \phi(0) \|) \leq W(t, \phi) \leq V_2(\| \phi(\theta) \|)$

(b) $\dot{W}(t, x_t) \leq -V_3(\| x(t) \|)$

where $\dot{W}(t, x_t)$ denotes the Dini's time-derivative of $W(t, x_t)$ along the solutions of (13)-(14). Then, the trivial solution of (13)-(14) is uniformly asymptotically stable. ∇∇∇

Note that condition (a) means that the functional $W(t, \phi)$ is *positive definite* and has an *infinitesimal upper limit*.

REFERENCES

Bien, Z. and J. H. Kim (1992). A robust stability bound of linear systems with structured uncertainty. *IEEE Trans. Automat. Contr.*, **37**, 1549-1551.

Chen, B. -S. and S. -S. Wang (1988). The stability of feedback control with nonlinear saturating actuator: Time Domain Approach. *IEEE Trans. Aut. Contr.*, **AC-33**, 483-487.

Chen, B. -S., S. -S. Wang and H. -C. Lu (1988). Stabilization of time-delay systems containing saturating actuators, *Int. J. Contr.*, **47**, 867-881.

Celikovsky S. (1993). On the stabilization of the homogeneous bilinear systems. *Syst. Contr. Lett.*, **21**, 503-510.

Genesio, R. and A. Tesi (1988). The output of SISO bilinear systems. *IEEE Trans. Automat. Contr.*, **33**, 950-952.

Glatffelder, A. H. and W. Schaufelberger (1983). Stability analysis of single loop control systems with saturation and antireset-windup circuits. *IEEE Trans. Automat. Contr.*, **AC-28**, 1074-1080.

Hale, J. K. (1977). *Theory of Functional Differential Equations*. Springer-Verlag, New York.

Koditschek, D. E. and K.S. Narendra (1983). Stabilizability of second-order bilinear systems. *IEEE Trans. Automat. Contr.*, **AC-28**, 987-989.

Kolmanovskii, V. B. and V. R. Nosov (1986). *Stability of Functional Differential Equations* (Academic Press, New York, 1986).

Krikelis, N. J. (1980). State feedback inegral control with intelligent integrator. *Int. J. Contr.*, **32**, 465-473.

Liu, J. S. and S. -L. Chen (1993). On global stability of quadratic state feedback controlled linear systems. *Syst. Contr. Lett.*, **21**, 371-379.

Luo, J. S., A. Johnson and P. P. J. van der Bosch (1993). On Lyapunov stability robustness bounds for linear continuous-time systems with structured uncertainty. *Proc. E.C.C. '93, Groningen (The Netherlands)*, **1**, 374-379.

Mohler, R. P. (1973). *Bilinear Control Processes*. Academic Press, New York.

Ortega, J. M. (1972). *Numerical Analysis*. Academic Press, New York.

Quinn, J. P. (1980). Stabilization of bilinear systems by quadratic feedback controls. *Journ. of Math. Anal. Appl.*, **75**, 66-80.

Ryan, E. P. and J. Buckingham (1983). On asymptotically stabilizing feedback control of bilinear systems. *IEEE Trans. Automat. Contr.*, **AC-28**, 863-864.

Su, T. J., P. -L. Liu and J. -T. Tsay (1991). Stabilization of delay-dependence for saturating actuator systems. *Proc. 1991 Conf. Dec. Contr.*, 2891-2892.

Tissir, E. and A. Hmamed (1992). Further results on the stabilization of time delay systems containing saturating actuators. *Int. J. Systems Sci.*, **23**, 615-622.

AN EXTENSION OF STABILITY CRITERIA FOR LINEAR AND NONLINEAR TIME DELAY SYSTEMS

A.GOUBET, M.DAMBRINE and J.P.RICHARD

Ecole Centrale de Lille, LAIL - URA CNRS D1440 BP 48, 59651 Villeneuve d'Ascq Cedex - FRANCE

Abstract. In this paper, two stability theorems for linear and nonlinear delay differential systems are presented; they can be considered as extensions of several criteria (Tokumaru *et al.*, 1975; Dambrine *et al.*, 1993-1994; Mori *et al.*, 1981).
Each of them gives a delay-dependent matrix criterion, respectively for linear and nonlinear systems. The originality of these criteria is that, unlike most, they do not consider the delayed terms as factors of destabilization. Actually, many control regulators are not memoriless and so it is sometimes essential to take advantage of the existing delayed terms in order to stabilize the system. As an example, unstable systems will be shown to be stabilizable by means of a feedback regulator with a delay.

Key words. time delay systems; stability; matrices; delay-dependence; vector Lyapunov functions

Résumé. Cet article présente deux critères originaux de stabilité (pour les systèmes linéaires et non linéaires à retards), qui peuvent être considérés comme des extensions de divers théorèmes (Tokumaru et al., 1975; Dambrine et al., 1993-1994; Mori et al., 1981).
Les critères proposés sont matriciels et dépendent du retard. Contrairement à ce qui peut être souvent lu dans la littérature, les termes retardés ne sont pas considérés a priori comme déstabilisants, ce qui permet d'envisager la stabilisation par un retour statique retardé de systèmes instables en boucle ouverte.
Enfin, trois exemples d'application sont proposés, dont un concerne la stabilisation d'un système instable en boucle ouverte à l'aide d'un retour proportionnel et dérivé retardé.

Mots clés. systèmes à retards; stabilité; critères dépendant du retard; fonctions de Lyapunov vectorielles

1. INTRODUCTION

Time lags often occur in real open-loop processes because a phenomenon rarely happens instantaneously after a stimulation. Some are also generated by the implementation of feedback control laws, because a finite time is required to sense information, then compute and send the control signal. Applications of time-delayed models are then numerous (see for example Driver (1977), Gopalsamy (1992), Kolmanovskii (1986)).

Small delays are often neglected during a stability and control study, or sometimes delayed terms are approximated by low-order rational functions (for instance $x(t-\tau) \simeq x(t) - \tau\,\dot{x}(t)$). But this way of dealing with such equations may lead to uncertain results, or even completely wrong ones. This is one of the reasons why many authors have studied stability properties of time-delay systems (see for example the references given by Dambrine (1994)).

The systems studied here are of the following form:

$$\frac{dx}{dt}(t) = A(.)\,x(t) + B(.)\,x(t-\tau(t)) \quad \text{for } t \geq 0 \text{ and } x \in \Re^n \qquad (E)$$

The following assumptions will be made:
<u>hypotheses H:</u>
-the delay $\tau(t)$, which depends on time only, has an upper limit τ;
-the matrices A(.) and B(.) may depend on time, on the state or on any disturbance parameter;
-$\tau(t)$, A(.) and B(.) are such that the solution of the system (E) is unique as soon as the initial function x(t) is known and continuous for t varying between $t_0 - \tau$ and t_0, where t_0 is the initial time.

Most of the available stability criteria applicable to these nonlinear systems state conditions that are independent of delays (Dambrine *et al.*, 1993-1994; Mori *et al.*, 1981; Lehman *et al.*, 1993). They involve the stability of the matrix A(.) and consider the delayed terms as perturbations that, if small, do not affect the stability of the considered solution. Only a few delay dependent criteria are stated in the literature (Su *et al.*, 1992; Wang *et al.*, 1987; Niculescu *et al.*, 1993).

This restriction can be overcome as shown in the sequel: the following paragraph (2) states conditions that prove the stability of some systems even if the influence of their delayed terms is not negligible. On the basis of the theorems developed by Tokumaru *et al.* (1975) and Mori *et al.* (1981),

one main theorem (Theorem 1) applicable to linear systems is stated, the proof of which involves a comparison principle. Corollaries are then given; one other theorem generalizes the former one to nonlinear systems.
At the end (section 3), three examples of application will be worked out : the results given here are compared to the ones given by Su and Huang (1992) and by Niculescu *et al.* (1994); afterwards, one second order nonlinear unstable system is proved to be stabilizable with a feedback regulator involving a delay.

Notations:
The entries of matrices and vectors used in this paper are real, and the elements of a matrix X or a vector x are respectively written X_{ij} and x_i. Moreover, $| X |_{ij} = | X_{ij} |$, $| x |_i = | x_i |$.

Let x(t) satisfy the equation (E). Then x(t) is called the instantaneous state of the system. The state of the system is the function x_t defined by $x_t(\theta) = x(t + \theta)$ where θ varies from $-\tau$ to 0. The initial state, which has to be known when trying to solve the equation, is x_{t_0}, where t_0 is the initial time.

With the matrix $A = (A_{ij})$ two matrices are associated: $\underset{d\in D}{\text{Sup}}\{A\} = (\underset{d\in D}{\text{Sup}}\{A_{ij}\})$ and $A^* = (A^*_{ij})$ which has the following entries: $A^*_{ii} = A_{ii}$; $A^*_{ij} = | A_{ij} |$ if $i \neq j$.
The inequality $A > B$ (respectively $A \geq B$) has to be understood as n^2 inequalities $A_{ij}>B_{ij}$ (resp $A_{ij}\geq B_{ij}$).

A^T denotes the transpose of the matrix A, det(A) its determinant.
I represents the identity matrix.

$\hat{p}(x(t))$ stands for
$[\underset{0\leq\lambda\leq 2\tau}{\text{Sup}} |x_1(t-\lambda)|, \ldots, \underset{0\leq\lambda\leq 2\tau}{\text{Sup}} |x_n(t-\lambda)|]^T$.

If χ is a real, sgn(χ) is the sign of χ: sgn(χ) = 1 iff $\chi \geq 0$; sgn(χ) = −1 iff $\chi < 0$.

2. DELAY-DEPENDENT STABILITY CRITERIA

Theorem 1:

Let us consider the following linear time-delay system (S1):

$$\frac{dx}{dt}(t) = A\,x(t) + B\,x(t - \tau(t))$$
$$\text{for } t \geq 0 \text{ and } x \in \Re^n$$
$$x_0(\lambda) = \varphi(\lambda) \text{ for } \lambda \in [-\tau ; 0] \qquad (S1)$$

On the hypotheses H, if the matrix
$$M = (A+B_1)^* + \tau(|B_1|.|A| + |B_1|.|B|) + |B_2|,$$
where B_1 and B_2 verify $B = B_1 + B_2$,
is the opposite of an M-matrix, then the equilibrium 0 of the system (S1) is asymptotically stable.

Proof:
If $s \geq \tau$ and x(t) is a solution of (S1), then
$$\frac{dx}{dt}(s) = (A + B_1)\,x(s) - B_1\,(x(s) - x(s - \tau(s))) + B_2\,x(s - \tau(s)),$$
and so
$$\frac{dx_i}{dt}(s) = \sum_{k=1}^{n}(A + B_1)_{ik}x_k(s) - \sum_{k=1}^{n}B_{1ik}\,(x_k(s) - x_k(s - \tau(s))) + \sum_{k=1}^{n}B_{2ik}x_k(s-\tau(s)) \qquad (1)$$
Let us consider an integer k between 1 and n.
By the mean value theorem of differential calculus, there exists a real $\alpha_k(s) \in\ ]0;\tau(s)[\ \subset\ ]0;\tau[$ such that
$$x_k(s) - x_k(s - \tau(s)) = \tau(s)\,\frac{dx_k}{dt}(s - \alpha_k(s)) \qquad (2)$$

As $s-\tau\geq 0$, then $s - \alpha_k \geq 0$, and (S1),for $t=s-\alpha_k(s)$, leads to:
$$\frac{dx_k}{dt}(s - \alpha_k(s)) = \sum_{j=1}^{n}A_{kj}\,x_j(s-\alpha_k(s)) + \sum_{j=1}^{n}(B_1+B_2)_{kj}x_j(s-\alpha_k(s)-\tau(s-\alpha_k(s))) \qquad (3)$$

Then, from (1), (2) and (3),
$$\frac{dx_i}{dt}(s) = \sum_{k=1}^{n}(A + B_1)_{ik}x_k(s) - \tau(s)\sum_{k=1}^{n}B_{1ik}\sum_{j=1}^{n}A_{kj}\,x_j(s-\alpha_k(s)) - \tau(s)\sum_{k=1}^{n}B_{1ik}\sum_{j=1}^{n}(B_1+B_2)_{kj}\,x_j(s-\alpha_k(s)-\tau(s-\alpha_k(s))) + \sum_{k=1}^{n}B_{2ik}\,x_k(s-\tau(s)).$$

The right-hand side derivative of the i-th component of $|x(s)|$ is:

$$\frac{d^+|x_i|}{dt}(s) = \sum_{k=1}^{n}(A+B_1)_{ik}\, x_k(s)\mathrm{sgn}(x_i(s))$$

$$- \tau(s)\sum_{j=1}^{n}\sum_{k=1}^{n} B_{1ik}A_{kj}\, x_j(s-\alpha_k(s))\mathrm{sgn}(x_i(s))$$

$$-\tau(s)\sum_{j=1}^{n}\sum_{k=1}^{n} B_{1ik}(B_1+B_2)_{kj}\; x_j(s-\alpha_k(s)-\tau(s-\alpha_k(s)))\mathrm{sgn}(x_i(s))$$

$$+ \sum_{k=1}^{n} B_{2ik}\, x_k\,(s-\tau(s))\,\mathrm{sgn}(x_i(s))$$

Using the same kind of overvaluation as in Dambrine *et al.* (1993, 1994), this equality yields:

$$\frac{d^+|x_i|}{dt}(s) \le \sum_{k=1}^{n}(A+B_1)^*_{ik}\,|\,x_k(s)\,|$$

$$+\,\tau\sum_{j=1}^{n}(|B_1|.|A|)_{ij}\,\hat{p}_j(x(s))$$

$$+\,\tau\sum_{j=1}^{n}(|B_1|.|B_1+B_2|)_{ij}\,\hat{p}_j(x(s))$$

$$+\sum_{k=1}^{n}|B_{2ik}|\,\hat{p}_k(x(s)).$$

So $\frac{d^+|x|}{dt}(s) \le (A + B_1)^*\,|x(s)|$

$$+\,(\tau\,(|\,B_1|.|\,A\,|+|\,B_1|.|\,B|)+|\,B_2|)\,\hat{p}(x(s)) \qquad (4)$$

As the off-diagonal elements of $(A + B_1)^*$ and all elements of

$$\tau\,(|\,B_1\,|.|\,A\,|+|\,B_1\,|.|\,B\,|)+|\,B_2\,|$$

are non-negative, Tokumaru *et al.* (1975) and Mori *et al.* (1981) allow us to conclude (see appendix).

Remark about the splitting up of the matrix B into B_1+B_2:

As can be seen in the previous theorem, the obtained maximum delay τ that maintains the system (S1) stable depends on the decomposition $B = B_1 + B_2$: the more negative the diagonal coefficients of B_1 are, the more numerous the chances to verify that $(A + B_1)^*$ is the opposite of an M-matrix will be. However, if the absolute values of the elements of B_1 are too big, the term $\tau\,(|\,B_1\,|.|\,A\,|+|\,B_1\,|.|\,B\,|)$ will prevent the matrix M from being the opposite of an M-matrix, unless τ is very small.

Considering the coefficients of B_1 as parameters, an optimization algorithm may be done to find the decomposition that leads to the domain of values of $\tau(t)$ that is the least constraining one.

The two following corollaries rewrite the former theorem for two particular decompositions: $B = B_1$ (efficient if B_1 is a diagonal-dominant matrix with negative elements), $B = B_2$ (efficient if B_2 mainly has non negative elements).

Corollary 1.1: case $B_2 = 0$

If the matrix

$$(A + B)^* + \tau\,(|\,B\,|.|\,A\,|+|\,B\,|^2)$$

is the opposite of an M_matrix, then the equilibrium 0 of the following time-delay system :

$$\frac{dx}{dt}(t)= A\,x(t) + B\,x(t-\tau(t))$$

is asymptotically stable.

This corollary is particularly useful if a stabilization with a delay forward or feedback regulator has to be realized: indeed, if the matrix A + B (of the closed loop system) is sufficiently stable,

$$\min_{\tau\ge 0}\,\{\tau\!:\ \det\,[(A + B)^*+\tau(|\,B\,|.|\,A\,|+|\,B\,|^2)] = 0\}$$

gives a value of the delay under which $\tau(t)$ can vary (see the example 3 of the paragraph 3).

Corollary 1.2: case $B_1 = 0$

If $A^* + |\,B\,|$ is the opposite of an M_matrix, then the equilibrium 0 of the time-delay system (S1) is asymptotically stable.

This result is already well known when dealing with a constant delay (Tokumaru *et al.*, 1975; Dambrine *et al.*, 1993-1994).

The theorem is also valid if several delays are noticed, as soon as these delays have τ as an upper limit:

Corollary 1.3:

Consider the following linear time-delay system (S2):

$$\frac{dx}{dt}(t) = A\,x(t) + B\,x(t-\tau_1(t)) + C\,x(t-\tau_2(t)) \qquad (S2)$$

If the matrix

$$M = (A + B_1 + C_1)^* + \tau\,(|\,B_1\,|.|\,A\,|+|\,B_1\,|.|\,B\,|+ |\,B_1\,|.|\,C\,|+|\,C_1\,|.|\,A\,|+$$

$|C_1|.|B| + |C_1|.|C|) + |B_2| + |C_2|$,
where B_1, B_2, C_1, C_2 are such that $B = B_1 + B_2$ and $C = C_1 + C_2$,
is the opposite of an M_matrix, then, on hypotheses H, the equilibrium 0 of the system (S2) is asymptotically stable.

The same kind of remark as the one written for the first theorem, about the decomposition of B, can be made here for the splittings up of B and C.

Theorem 2:

If the system considered is nonlinear but can be written in the following way:

$$\frac{dx}{dt}(t) = A(.)\, x(t) + B(.)\, x(t - \tau(t)) \quad \text{for } t \geq 0 \text{ and } x \in \Re^n \qquad (S3)$$
$$x_0(\lambda) = \varphi(\lambda) \text{ for } \lambda \in [-\tau; 0]$$

where $A(.) = A(t, x_t,.)$, $B(.) = B(t, x_t,.)$ depend on t, on the state and eventually on other parameters.
If the hypotheses H are verified, and if

$$\text{Sup}\,\{(A(.) + B_1(.))^*\} + \tau\, \text{Sup}\{|B_1(.)|.\text{SUP}\{|A(.)| + |B(.)|\} + |B_2(.)|\},$$

is the opposite of an M_matrix,
where $B_1(.)$ and $B_2(.)$ verify $B(.) = B_1(.) + B_2(.)$, where $\text{SUP}\{|A(.)| + |B(.)|\}$ denotes the upper value of $|A(t-\lambda, x_{t-\lambda},.)| + |B(t-\lambda, x_{t-\lambda},.)|$ when $\lambda \in [0; \tau]$, then the equilibrium 0 of the system (S3) is asymptotically stable.

Proof: The proof of this assertion is almost the same as that of the first theorem, using the same tools as Dambrine *et al.* (1993-1994) (i.e. majoration of some matrices, say E, by $\text{Sup}_{t \geq 0, x \in \Re^n} \{E\}$).

Remark: Instead of using the norm $|x|$ in the proofs of the theorems, the system could have aggregated in order to obtain a more easily verifiable criterion (see Borne *et al.* (1974) and Grujic *et al.* (1976) for the use of p-dimensional vector norms). However this criterion would have been less precise. This method will not be developed in this article.

3 APPLICATIONS

Example 1: Comparison with Su and Huang ' criterion:

Su and Huang (1992) developed a sufficient delay-dependent condition for the stability of linear uncertain time-delay systems. They presented an example:

$$\frac{dx}{dt}(t) = (A + \Delta A)\, x(t) + (B + \Delta B)\, x(t - \tau),$$
$$x(\theta) = [2 \;\; 2]^T \text{ if } \theta \in [-\tau; 0],$$

where $A = \begin{bmatrix} -2 & 0 \\ 0 & -1 \end{bmatrix}$, $\Delta A = \begin{bmatrix} 0.3\cos t & 0 \\ 0 & 0.2\sin t \end{bmatrix}$,

$B = \begin{bmatrix} -1 & 0 \\ -1 & -1 \end{bmatrix}$, $\Delta B = \begin{bmatrix} 0.2\cos t & 0 \\ 0 & 0.3\sin t \end{bmatrix}$.

Using their theorem, they affirmed that the system was asymptotically stable for each value of τ satisfying $\tau < 0.1614$.
A better result can be obtained with the method developed in this paper. In fact, applying the theorem 2 with $B_2 = 0$, the following result is stated: the system is asymptotically stable if $\tau < 0.4615$.

Example 2: Comparison with Niculescu ' s criterion:

Niculescu *et al.* have presented some results concerning the robust stability analysis of a class of uncertain time-delay systems.
They considered the following example:

$$\frac{dx}{dt}(t) = A\, x(t) + B\, x(t - \tau),$$

$$A = \begin{bmatrix} -2+\alpha_1\cos t & 0 \\ 0 & -1+\alpha_2\sin t \end{bmatrix},$$

$$B = \begin{bmatrix} -1+\beta_1\cos t & 0 \\ -1 & -1+\beta_2\sin t \end{bmatrix},$$

and where α_1, α_2, β_1, β_2 are unknown parameters satisfying $|\alpha_1| \leq 1.6$; $|\alpha_2| \leq 0.05$; $|\beta_1| \leq 0.1$; $|\beta_2| \leq 0.3$.
In the refered paper the zero solution was proved to be uniformly asymptotically stable for any constant time-delay $\tau < 0.1036$.
Using our criteria, it can easily be proved to be stable for any varying delay $\tau(t) < 0.2514$.

Example 3: Stabilization of an unstable open-loop process:

Let us study the nonlinear unstable system described by:
$\ddot{y}(t) - a_1(.)\dot{y}(t) = u(t)$
that is to be stabilized using a feedback regulator with a delay (see figure 1).

The parameter $a_1(.)$ varies between 1.5 and 2. It may depend on t, on the state or eventually on a disturbance parameter. The output y(t) is measured at the instant t, but its derivative is computed only after an unknown or varying time lag $\tau(.)$, whose upper bound is τ.

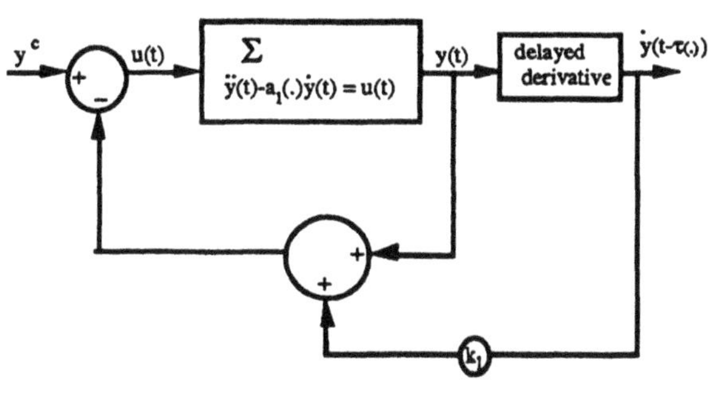

Figure 1

The problem is to obtain a relation between k_1 and τ that assures the stability of the closed-loop system.

If the system is asymptotically stable, then $\lim_{t \to +\infty} y(t) = y^C$ and y^C is set to 0 during the stability study.

A state equation of the closed-loop system is:

$$\dot{z}(t) = M_Z z(t) + N_Z z(t - \tau)$$

with $M_Z = \begin{bmatrix} 0 & 1 \\ -1 & a_1(.) \end{bmatrix}$ and $N_Z = \begin{bmatrix} 0 & 0 \\ 0 & -k_1 \end{bmatrix}$

$$y(t) = [1\ 0]\ z(t)$$

The Tokumaru *et al.*. criterion (this is theorem 2 with $B_1(.) = 0$) cannot be successfully applied in this case.

Let us define the following change of state variables: $x = Pz$, with $P = \begin{bmatrix} 1 & \lambda \\ 0 & 1 \end{bmatrix}$.

The condition of theorem 2,

" $\mathrm{Sup}((PM_Z P^{-1} + PN_Z P^{-1})^*)$

$+ \mathrm{Sup}(|PM_Z P^{-1}|.|PN_Z P^{-1}| + |PN_Z P^{-1}|^2)$

is the opposite of an M-matrix ",
is satisfied if and only if

(a) $k_1 > 2$

(b) $0 \le \lambda < 2$

(c) $k_1 > \dfrac{2\lambda^2+4\lambda+1}{2\lambda}$

(d) $\tau < \dfrac{k_1-2}{2\lambda k_1+k_1^2+2k_1}$

(e) $\tau < \dfrac{-2\lambda^2+\lambda(2k_1-4)-1}{k_1(2k_1\lambda+2\lambda^2+4\lambda+1)}$

(f) $\tau < \dfrac{1-0{,}5\lambda}{k_1(4k_1\lambda+0{,}5\lambda-1)}$

The graph (figure 2) gives the domain of convergence that is found with our method.

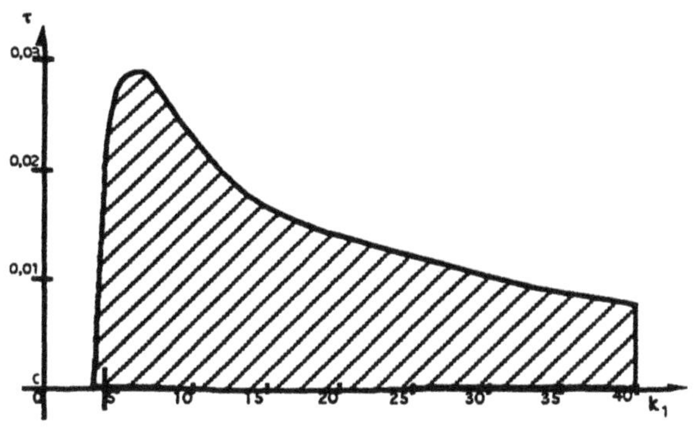

Figure 2

CONCLUSION

The different criteria given in this paper can be useful for the stability and control study of different delay systems; they can be applied to nonlinear, nonstationary, linear with a varying or unknown delay, as well as to classical linear systems. The method is the same for all these systems, because an overestimation of the nonlinear terms is made in order to simplify the study and deal with linear methods (refering to the concept of vector norms). The results are all the more precise and easy to use than the matrices have dominant principal negative diagonals. They are applicable to systems with important delay terms and give an estimation of the maximum delay that guarantees the stability of the system.

APPENDIX

The theorems developed in Tokumaru *et al.*(1975) and Mori *et al.*(1981) that are used in the proof of the theorem 1 are recalled in this appendix.

Consider a linear differential difference system

$$\dot{x}_i(t) = -\alpha_i\, x_i(t) + \sum_{j=1}^{m} \{\beta_{ij}\, x_j(t) + \gamma_{ij}\, x_j\,(t - \tau_{ij})\},$$

$i = 1, 2, \ldots, m,$

where α_i, β_{ij}, γ_{ij} are non-negative constants.

Theorem: (Mori *et al.*, 1981)

If the matrix $A - D(0)$ defined by

$A = \{a_{ij}\}$, $a_{ij} = \alpha_i - \beta_{ii}$, $i = j$

$\quad - \beta_{ij}$, $i \ne j$

$D(s) = \{d_{ij}(s)\} = \{\gamma_{ij} \exp(-\tau_{ij}\, s)\}$

is an M-matrix, then the system is asymptotically stable.

The following definition and theorem may be found in (Tokumaru *et al.*, 1975). They deal with a comparison method that enabled us to study the stability of the solutions of the differential inequality (4), in the proof of the first theorem. Consider the differential inequality of the form:

$$\dot{x}(t) \leq g(x(t), \bar{x}(t), t), \qquad (5)$$

where $x(t) = (x_1(t), ..., x_m(t))^T$,

and $\bar{x}(t) = (\sup_{t-\tau\leq\theta\leq t} x_1(\theta), ..., \sup_{t-\tau\leq\theta\leq t} x_m(\theta))^T$

Definition: (Tokumaru *et al.*, 1975)

Let $g: \Re^m \times \Re^m \times \Re^+ \rightarrow \Re^m$ be a continuous vector function defined on $\Re^m \times \Re^m \times \Re^+$. Then g is said to belong to the class H_m if the following conditions are satisfied:

(i) $g(x, y^1, t) \leq g(x, y^2, t)$, for any vectors $x, y^1, y^2 \in \Re^m$ such that $y^1 \leq y^2$,

(ii) if $x^1 \in \Re^m$ and $x^2 \in \Re^m$ are such vectors that $x^1 \leq x^2$ and $x^1_i = x^2_i$, then the i-th component of g satisfies the relation: $g_i(x, y^1, t) \leq g_i(x, y^2, t)$ for all $y \in \Re^m$ and $t \geq 0$.

Theorem: (Tokumaru *et al.*, 1975)

Let $x(t)$ $(t \geq 0)$ be a solution of the differential inequality (5) and $z(t)$ be a function satisfying the differential equation for $t \geq 0$:

$$\dot{z}(t) = g(z(t), \bar{z}(t), t).$$

Assume that the following conditions are satisfied:

(i) the function $g(x, y, t)$ belongs to the class H_m,

(ii) the solution of the differential equation $\dot{y}(t) = g(y(t), \bar{y}(t), t) + \varepsilon$ uniquely exists for any continuous initial function $y(\theta)$, $(-\tau \leq \theta \leq 0)$ and for any sufficiently small $\varepsilon \geq 0$.

Then $z(t)$ and $x(t)$ satisfy for all $t \geq 0$ the inequality $x(t) \leq z(t)$ if $x(\theta) \leq z(\theta)$ for $-\tau \leq \theta \leq 0$.

REFERENCES

Borne, P. and J.C. Gentina (1974). Absolute stability of non-linear non-stationary large scale systems. *Proc. Joint Automatic Control Conf.*, 317-326.

Dambrine, M. and J.P. Richard (1993). Stability Analysis of Time-Delay Systems. *Dynamic Systems and Applications 2*, 405-414.

Dambrine, M. and J.P. Richard (1994). Stability and Stability Domains Analysis for Nonlinear Differential-Difference Equations. *Dynamic Systems and Applications* 3, 369-378.

Dambrine, M. (1994). Contribution à l'étude de la stabilité des systèmes à retards. *Thesis*, University of Lille I, France, n° 1386.

Driver, R.D. (1977). Ordinary and Delay Differential Equations. *Springer Verlag*. New York.

Gopalsamy, K. (1992). Stability and Oscillations in Delay Differential Equations of Population Dynamics. *Mathematics and Its Applications.* Kluwer Academic Publishers.

Grujić, Lj.T., J.C. Gentina and P. Borne (1976). General aggregation of large-scale systems by vector Lyapunov functions and vector norms. *Int.J.Control* 24 4, 529-550.

Kolmanovskii, V.B. and V.R. Nosov (1986). Stability of Functional Differential Equations. *Academic Press*. INC.

Lehman, B. and K. Shujaee (1993). Robust Stability of Time Varying Functional Differential Equations. *Proceedings of the American Control Conference*. San Francisco. California.

Mori,T., N. Fukuma, M. Kuwahara (1981). Simple stability criteria for single and composite linear systems with time delays. *Int.J.Control* 34 6, 1185-1194.

Niculescu, S.I., C.E. de Souza, J.M. Dion and L. Dugard (1993). Robust Stability and Stabilization of Uncertain Linear Systems with State Delay: Single Delay Case. *Internal Report*. Laboratoire d'Automatique de Grenoble, 93-91.

Su, T-J. and C.G. Huang (1992). Robust Stability of Delay Dependence for Linear Uncertain Systems. IEEE *Transactions on Automatic Control*. Vol.37, No.10.

Tokumaru, H., N. Adachi and T. Amemyia (1975). Macroscopic stability of interconnected systems. 6-th IFAC *Congress*, ID44.4.

Wang, S.S., B.S. Chen and T.P. Lin (1987). Robust stability of uncertain time-delay systems. *Int.J.Control*. Vol.46. No.3, 963-976.

On the characterisation of 2-D least order polynomial realisations

A.C. Pugh[†], M.S. Boudellioua[†] and G.E. Hayton[‡]

[†] Department of Mathematical Sciences
Loughborough University
Loughborough, Leics.
LE11 3TU, U.K..

[‡] Faculty of Information & Engineering Systems
Leeds Metropolitan University
Leeds, W. Yorks.
LS1 3HE, U.K..

Abstract

The generalisation of Rosenbrock's characterisation of all least order polynomial realisations of a transfer function matrix is considered for the case of two dimensional (2-D) systems. The definition of what constitutes the least order is shown to be crucial.

La Généralisation de la caractérisation de Rosenbrock de toutes réalisations polynomiales du moindre ordre d'une matrice d'une fonction de transfer est considérée dans le cas des systèmes à deux dimensions (2-D). Il est démontré que la définition de ce qui constitue le moindre ordre d'une matrice d'un système polynomial à deux dimensions est tout à fait essentielle.

Keywords: 2-D systems; Polynomial system matrix; Least order; Strict system equivalence; Zero coprimeness; Minor coprimeness.

1. Introduction

In systems theory, an important concept is that of equivalence of systems. An important definition due to Rosenbrock [1] is that of strict system equivalence denoted s.s.e. while a more general and complete definition of this notion was given by Fuhrmann [3]. For 1-D systems, Rosenbrock [1] presented necessary and sufficient conditions under which two polynomial system matrices which give rise to the same transfer function matrix are s.s.e.. From another point of view these conditions can be regarded as describing the connection between all least order polynomial realisations of a given transfer function matrix. In this work the concept of relative primeness between polynomial matrices over $R[s]$ (the ring of polynomials in the indeterminate s with coefficients in the field of real numbers R) is crucial. It is the condition of relative primeness which determines that a polynomial realisation has least order.

In the 2-D case, the systems are represented by system matrices over the polynomial ring $R[z_1, z_2]$ and the main difficulty encountered in the extension of results from the 1-D case to the 2-D one is due to the fact that $R[z_1, z_2]$ is not a Euclidean division ring. Furthermore there are two different and non equivalent concepts of relative primeness, namely *minor coprimeness* and *zero coprimeness*, written respectively MC and ZC for short. In this paper, it is intended to illustrate the difficulties encountered in the extension of Rosenbrock's result to the 2-D case, some of these difficulties were originally noted in [4]. It is shown, however, that given a certain new definition of least order and using the 2-D version of the definition of s.s.e. in the sense of Fuhrmann, then a characterisation of all such least order realisations of a given transfer function matrix can be obtained.

2. Preliminaries

Consider the 2-D system matrix in the general form:

$$P(z_1, z_2) = \begin{pmatrix} T(z_1, z_2) & U(z_1, z_2) \\ -V(z_1, z_2) & W(z_1, z_2) \end{pmatrix} \tag{2.1}$$

where $T(z_1, z_2), U(z_1, z_2), V(z_1, z_2)$ and $W(z_1, z_2)$ are respectively $r \times r, r \times m, p \times r$ and $p \times m$ polynomial matrices, with $T(z_1, z_2)$ invertible. An obvious extension to 2-D systems of Rosenbrock's definition of s.s.e. is

Definition 2.1: Two polynomial system matrices $P_i(z_1, z_2), i = 1, 2$ of the form (2.1) and of the same dimensions are said to be *s.s.e. over $R[z_1, z_2]$ in the sense of Rosenbrock* if they are related by the following transformation:

$$\begin{pmatrix} M(z_1, z_2) & 0 \\ X(z_1, z_2) & I_p \end{pmatrix} P_1(z_1, z_2) =$$

$$P_2(z_1, z_2) \begin{pmatrix} N(z_1, z_2) & Y(z_1, z_2) \\ 0 & I_m \end{pmatrix} \tag{2.2}$$

where $M(z_1, z_2), N(z_1, z_2)$ are $r \times r$ unimodular matrices over $R[z_1, z_2]$ and $X(z_1, z_2), Y(z_1, z_2)$ are respectively $p \times r, r \times m$ polynomial matrices over $R[z_1, z_2]$. □

In the following, we shall generally omit (z_1, z_2) from the notation of matrices and therefore all matrices will be from now on assumed over $R[z_1, z_2]$ unless stated otherwise. The two distinct definitions of coprimeness for the case of 2-D polynomial matrices are then as follows [5].

Definition 2.2: The matrices T, U in (2.1) are said to be *minor left coprime* (MLC) in case the $r \times r$ minors of the compound matrix $(T \quad U)$ have no non trivial common factors in $R[z_1, z_2]$. Similarly, the matrices T, V in (2.1) are said to be *minor right coprime* (MRC) in case the $r \times r$ minors of the matrix $(T^T \quad -V^T)^T$ have no non trivial common factors in $R[z_1, z_2]$. □

It follows from the above definition that the following statements are equivalent.

- T, U are MLC

- any polynomial factorisation $(T \quad U) = A(T^* \quad U^*)$ implies that the matrix A is unimodular over $R[z_1, z_2]$.

- the matrix $(T \quad U)$ has Smith form $(I_r \quad 0)$.

Definition 2.3: The matrices T, U in (2.1) are said to be *zero left coprime* (ZLC) in case the compound matrix $(T \quad U)$ has rank r for all values of the indeterminate pair (z_1, z_2). Similarly, *zero right coprimeness* (ZRC) can be defined for the pair T, V occurring in (2.1). □

It follows from this definition, that the pair T, U are *ZLC* if and only if the matrix $(T \quad U)$ is equivalent by unimodular multiplication to the Smith form $(I_r \quad 0)$.

It is evident, from the above definitions, that *zero coprimeness* implies *minor coprimeness*. It is zero coprimeness which forms the basis for the extension of Fuhrmann's definition of s.s.e. to the 2-D setting.

Definition 2.4: $P_i(z_1, z_2), i = 1, 2$ are said to be s.s.e. over $R[z_1, z_2]$ in the sense of Fuhrmann if they are related by (2.2) where $M(z_1, z_2), T_2(z_1, z_2)$ are zero left coprime and $T_1(z_1, z_2), N(z_1, z_2)$ are zero right coprime. □

In this form s.s.e. is an equivalence relation in the formal sense of being reflexive, transitive and symmetric. It should be noted that if minor coprimeness is used in place of zero coprimeness in the above defintion then a relation results which is only reflexive and transitive [6], [7].

3. Least Order in 2-D Systems Theory

A general polynomial system matrix of the form (2.1) is obtained from the realisation

$$G = VT^{-1}U + W \tag{3.1}$$

of a $p \times m$ transfer function matrix $G(z_1, z_2)$.

A first question concerns the relationship which exists between any two realisations in the form (2.1) of a given $G(z_1, z_2)$. In fact an exact characterisation of this relationship can be given, as the following result shows.

Theorem 3.1: The system matrices in the form (2.1) given by:

$$P_i(z_1, z_2) = \begin{pmatrix} T_i(z_1, z_2) & U_i(z_1, z_2) \\ -V_i(z_1, z_2) & W_i(z_1, z_2) \end{pmatrix}, i = 1, 2$$

are realisations of the same transfer function matrix $G(z_1, z_2)$ if and only if there exist rational matrices $Q_1(z_1, z_2)$, $Q_2(z_1, z_2)$, $R_1(z_1, z_2)$, $R_2(z_1, z_2)$ such that

$$\begin{pmatrix} Q_1 & 0 \\ R_1 & I \end{pmatrix} P_2 = P_1 \begin{pmatrix} Q_2 & R_2 \\ 0 & I \end{pmatrix} \tag{3.2}$$

Proof: See [7]. □

It should be noted that the transformation described by (3.2) is based on rational transforming matrices and describes the basic connection between two polynomial system matrices which give rise to the same transfer function matrix. It is not required that the two system matrices be of the same dimensions nor that the transforming matrices Q_1, Q_2 be invertible or even square. The further conditions which should be imposed on two such system matrices to ensure that the transformation (3.2) is based on polynomial matrices are not known, and in fact are not even known in the case of polynomial matrices in a single indeterminate. Whatever conditions are necessary to ensure a polynomial connection of the form (3.2), are likely to be much more restrictive than their counterparts in the single indeterminate polynomial matrix theory.

The main purpose of this paper is to describe such conditions in a specific case. Recal that in conventional multivariable systems theory a principal result is that two least order polynomial realisations of a transfer function matrix are related by strict system equivalence in the sense of Fuhrmann [3]. If the two system matrices are of the same dimensions then the nature of their s.s.e. connection is reduced to that of Rosenbrock [1]. The corresponding result in 2-D systems theory differs from this classical result in at least two respects, as will be noted from the following examples.

The first way in which the 2-D result differs from its 1-D counterpart arises from the possible choice of the coprimeness used in the definition of least order. In two dimensional systems theory the term "least order " might most reasonably be attributed to the system matrix

$$P(z_1, z_2) = \begin{pmatrix} T(z_1, z_2) & U(z_1, z_2) \\ -V(z_1, z_2) & W(z_1, z_2) \end{pmatrix}$$

in case $T(z_1, z_2)$ and $U(z_1, z_2)$ are minor left coprime and $T(z_1, z_2)$ and $V(z_1, z_2)$ are minor right coprime. This is because minor coprimeness can always be guaranteed (by common matrix factor cancellation) whereas zero coprimeness cannot. Consider then the following example.

Example 3.1: The system matrices

$$P_1 = \begin{pmatrix} z_2^2 & z_1 \\ -(z_1 - 1) & 0 \end{pmatrix} \; P_2 = \begin{pmatrix} z_2^2 & (z_1 - 1) \\ -z_1 & 0 \end{pmatrix} \tag{3.3}$$

are both least order realisations corresponding to the transfer function $G(z_1, z_2) = z_1(z_1 - 1)/z_2^2$. By theorem 2.1, $P_1(z_1, z_2)$ and $P_2(z_1, z_2)$ will be related as in (3.2) through rational transforming matrices. Thus

$$\begin{pmatrix} q_1 & 0 \\ r_1 & 1 \end{pmatrix} P_1 = P_2 \begin{pmatrix} q_2 & r_2 \\ 0 & 1 \end{pmatrix} \tag{3.4}$$

for the particular choice of rational functions $q_1 = q_2 = 1$ and $r_1 = -r_2 = -1/z_2^2$. Now equation (1,2) of (3.4) is

$$q_1(z_1, z_2)z_1 - r_2(z_1, z_2)z_2^2 = z_1 - 1 \tag{3.5}$$

It follows from Hilbert Nullstellensatz that (3.5) has no polynomial solution for $q_1(z_1, z_2)$ and $r_2(z_1, z_2)$, since the polynomial $z_1 - 1$ has no zeros in common with z_1, ${z_2}^2$. Thus there will be no polynomial connection between $P_1(z_1, z_2)$ and $P_2(z_1, z_2)$ of the form (3.5).

If in (3.4) the roles of $P_1(z_1,z_2)$ and $P_2(z_1,z_2)$ are interchanged then it can be seen in a similar way (this time from the (2,1) equation) that no polynomial connection is possible. Thus although P_1 and P_2 are both least order realisations of the same transfer function matrix, there is no polynomial based transformation connecting them. □

It follows from the above example that the classical multivariable result that such realisations are strictly system equivalent will not generalise to two dimensional systems if least order in the minor coprime sense is used. Thus a redesignation of the term "least order" is necessary.

The second way in which the 2-D characterisation of least order realisations of the transfer function matrix will differ from the 1-D case is that s.s.e. in the sense of definition (2.1) will not suffice even if the two system matrices are of the same dimension. It is shown by the following example that if the s.s.e. is taken in the sense of definition (2.1), it is not possible to establish a polynomial connection of the type (3.2) even if least order is defined in the strictest possible terms, of the zero coprimeness of both T, U and T, V.

Example 3.2: The system matrices

$$P_1 = \begin{pmatrix} z_1z_2 - 1 & 1 \\ -z_1 & 0 \end{pmatrix}, \; P_2 = \begin{pmatrix} z_1z_2 - 1 & z_1 \\ -1 & 0 \end{pmatrix} \quad (3.6)$$

are clearly least order in the strongest sense ie. both T_i, U_i are zero left coprime for i=1,2 and T_i, V_i are zero right coprime for i=1,2. Further both system matrices are realisations of the same transfer function $G(z_1,z_2) = z_1/(z_1z_2 - 1)$, and yet P_1 and P_2 are not s.s.e. in the sense of Rosenbrock (definition (2.1). For suppose they are s.s.e. and linked via the transformation

$$\begin{pmatrix} m & 0 \\ x(z_1,z_2) & 1 \end{pmatrix} \begin{pmatrix} z_1z_2 - 1 & 1 \\ -z_1 & 0 \end{pmatrix} =$$

$$\begin{pmatrix} z_1z_2 - 1 & z_1 \\ -1 & 0 \end{pmatrix} \begin{pmatrix} n & y(z_1,z_2) \\ 0 & 1 \end{pmatrix} \quad (3.7)$$

where because of the requirement of unimodularity, m and n are non zero constants. From the (2,1) equation of (3.7), we have

$$x(z_1,z_2)(z_1z_2 - 1) - z_1 = -n$$

where n is a non zero constant. It thus follows that

$$x(z_1,z_2) = \frac{z_1 - n}{z_1z_2 - 1}$$

cannot be polynomial under the constraint that n is a constant. Therefore P_1 and P_2 are not s.s.e. in the sense of Definition (2.1). □

Clearly from what has been said in Example (3.1), it will not be appropriate to use the term "least order" for two dimensional realisations of the form (2.1) in which $T(z_1,z_2)$ and $U(z_1,z_2)$ are minor left coprime and $T(z_1,z_2)$ and $V(z_1,z_2)$ are minor right coprime. Further the term "least order" cannot be ascribed to the situation in which coprimeness occurs in the zero coprime sense, for such realisations can only give rise to limited forms of transfer function matrices which possess no nonessential singularities of the second kind. One way out of this difficulty is to adopt the following terminology.

Definition 3.1: A 2-D polynomial system matrix $P(z_1,z_2)$ of the form (2.1) will be said to be *least order* in case $T(z_1,z_2)$ and $U(z_1,z_2)$ are zero left coprime and $T(z_1,z_2)$ and $V(z_1,z_2)$ are minor right coprime. □

The definition expresses an intention to realise the nonessential singularities of the second kind in the transfer function matrix so that they always occur in the $(T(z_1,z_2)^t \; -V(z_1,z_2)^t)^t$ part of the system matrix. Note that $G(z_1,z_2)$ always possesses least order realisations of this type since any right (necessarily minor) coprime matrix fraction description (MFD) gives rise to a system matrix which is least order in the above sense. There is of course no particular necessity to handle the nonessential singularities of the second kind in this way and they could equally be taken to be realised in the $(T(z_1,z_2) \;\; U(z_1,z_2))$ part of the system matrix. The important point at issue, seems to be that is necessary to agree at the outset of the investigation exactly how these singularities of $G(z_1,z_2)$ are to be realised within the system matrix and the definition 2.4 takes one such view.

4. A 2-D Version of Rosenbrock's Theorem

We present two results which together establish the required generalisation. The first theorem establishes a standard form for least order realisations of the given transfer function matrix.

Theorem 4.1 (Standard form under s.s.e.): Given a system matrix P in the form (2.1) and having least order, then P can be reduced by s.s.e. to the system matrix in the standard form $\bar{P}$ given by:

$$\bar{P}(z_1,z_2) = \begin{pmatrix} \bar{T}(z_1,z_2) & I_m \\ -\bar{V}(z_1,z_2) & 0 \end{pmatrix}, \quad (4.1)$$

where $\bar{T}, \bar{V}$ are respectively $m \times m$ and $p \times m$ MRC polynomial matrices.

Proof: Since T, U are ZLC, there exist ZLC polynomial matrices N, Y such that the matrix:

$$K = \begin{pmatrix} T & U \\ -N & -Y \end{pmatrix} \quad (4.2)$$

is unimodular. It follows that there also exist ZLC polynomial matrices $M, \bar{T}$ and ZLC polynomial matrices C, D such that:

$$\begin{pmatrix} C & D \\ M & \bar{T} \end{pmatrix} \begin{pmatrix} T & U \\ -N & -Y \end{pmatrix} = \begin{pmatrix} I_r & 0 \\ 0 & I_m \end{pmatrix} \quad (4.3)$$

Hence we have,

$$(M \;\; \bar{T}) \begin{pmatrix} T & U \\ -N & -Y \end{pmatrix} = (0 \;\; I_m) \quad (4.4)$$

which yields:

$$MT = \bar{T}N \quad (4.5)$$

$$MU = \bar{T}Y + I_m \quad (4.6)$$

Now given V, W and that the matrix K is unimodular, it is always possible to find polynomial matrices $X, \bar{V}$

such that:

$$\begin{pmatrix} X & -\bar{V} \end{pmatrix} \begin{pmatrix} T & U \\ -N & -Y \end{pmatrix} = \begin{pmatrix} V & -W \end{pmatrix} \qquad (4.7)$$

which gives:

$$XT - V = -\bar{V}N \qquad (4.8)$$

$$XU + W = -\bar{V}Y \qquad (4.9)$$

From the relationships (4.5), (4.6), (4.8) and (4.9), it is deduced that P and $\bar{P}$ are linked via the following polynomial transformation:

$$\begin{pmatrix} M & 0 \\ X & I_p \end{pmatrix} \begin{pmatrix} T & U \\ -V & W \end{pmatrix} = \begin{pmatrix} \bar{T} & I_m \\ -\bar{V} & 0 \end{pmatrix} \begin{pmatrix} N & Y \\ 0 & I_m \end{pmatrix} \qquad (4.10)$$

where $M, \bar{T}$ are ZLC and T, N are ZRC (since the matrix K is unimodular). The matrices $\bar{T}, \bar{V}$ are MRC, because s.s.e. preserves the nature of coprimeness of the pairs T, U and T, V in the system matrix P. □

Now we present the main result of this paper which can be regarded as the extension of Rosenbrock's result to the 2-D case.

Theorem 3.2 (Rosenbrock Least Order Characterisation): Two least order system matrices $P_i(z_1, z_2)$, $i = 1, 2$ in the form (2.1), not necessarily of the same size, are strictly system equivalent in the sense of Definition (2.4) if and only if they give rise to the same transfer function matrix.

Proof: Necessity follows immediately from Theorem (3.1) and it remains therefore to prove the sufficiency. Assuming that $P_i, i = 1, 2$ have least order, then by Theorem 4.1, they can be reduced by s.s.e. in the sense of Definition (2.4) to the respective standard forms

$$\bar{P}_i = \begin{pmatrix} \bar{T}_i & I_m \\ -\bar{V}_i & 0 \end{pmatrix}, i = 1, 2 \qquad (4.11)$$

where the pairs $\bar{T}_i, \bar{V}_i, i = 1, 2$ are both MRC. Now since both P_1 and P_2 give rise to the same G, then from (4.11), G can be decomposed into the right minor coprime matrix fraction descriptions:

$$G = \bar{V}_1 \bar{T}_1^{-1} = \bar{V}_2 \bar{T}_2^{-1} \qquad (4.12)$$

Now all coprime matrix fraction descriptions of the form (4.12) of a 2-D rational matrix differ only by a unimodular factor [2]. Hence we have,

$$\begin{pmatrix} \bar{T}_1 \\ -\bar{V}_1 \end{pmatrix} = \begin{pmatrix} \bar{T}_2 \\ -\bar{V}_2 \end{pmatrix} Q \qquad (4.13)$$

where Q is a unimodular matrix over $R[z_1, z_2]$. It follows therefore that the system matrices in (4.11) are linked via the the following s.s.e. transformation (in the strict sense of Definition (2.1)):

$$\begin{pmatrix} \bar{T}_1 & I_m \\ -\bar{V}_1 & 0 \end{pmatrix} = \begin{pmatrix} \bar{T}_2 & I_m \\ -\bar{V}_2 & 0 \end{pmatrix} \begin{pmatrix} Q & 0 \\ 0 & I_m \end{pmatrix} \qquad (4.14)$$

Finally, since the system matrices in (4.11) were obtained by s.s.e. transformations of the form of Definition (2.4) from P_1 and P_2 respectively, it follows by the symmetry and transitivity of that relationship that P_1 and P_2 are also s.s.e. □

5. Conclusions

It has been shown in this paper that Rosenbrock's result concerning the characterisation of all least order realisations of a given transfer function matrix cannot be extended to the 2-D case using Rosenbrock's precise definition of s.s.e., based on unimodular matrices, irrespective of which concept of coprimeness (minor or zero) is used. However, by using Fuhrmann's definition of s.s.e. and introducing a specific definition of least order, it has been shown that it is possible to characterise all least order realisations of a given 2-D transfer function matrix via s.s.e.. The condition imposed on the pair of matrices $T(z_1, z_2), U(z_1, z_2)$ in (2.1) to be zero left coprime is not as restrictive as it may seem, since the class of such realisations contain all system matrices which correspond to right (minor) coprime MFD's of the given transfer function matrix. In fact such system matrices emerge as standard forms for least order 2-D polynomial realisations under the notion of strict system equivalence given in Definition (2.4).

Acknowledgments

The authors of this paper wish to express their thanks to the British Council (Algiers) for their support in carrying out this research work.

References

[1] Rosenbrock, H.H., 1970, *State space and multivariable theory*, Nelson-Wiley.

[2] Morf, M., Levy, B.C. and Kung, S.Y., 1977, "New Results in 2-D Systems Theory: Part I", Proc. of the IEEE, 65, pp. 861-872.

[3] Fuhrmann, P.A. 1977, " On Strict System Equivalence and Similarity", Int. J. Control, 25, pp. 5-10.

[4] Pugh, A.C., Johnson, D.S. and Hayton, G.E., 1992, "2-D System Structure and Applications", Proc. of the 2nd IFAC Workshop on System Structure and Control, Pergamon Press, pp.432-436.

[5] Youla, D.C. and Gnavi, G., 1979, "Notes on n-dimensional Systems", IEEE Trans. on Circuits and Systems,CAS-26, pp.105-111.

[6] Levy, B.C., 1981, "2-D Polynomial and Rational Matrices and Their Applications for the Modelling of 2-D Dynamical Systems", Ph.D. Thesis, Stanford University, U.S.A..

[7] Johnson, D.S., 1993, "Coprimeness in Multidimensional System Theory and Symbolic Computation", Ph.D. Thesis, Loughborough University of Technology, U.K..

ROBUST OUTPUT FEEDBACK CONTROL FOR UNCERTAIN DYNAMICAL SYSTEMS WITH UNKNOWN TIME DELAYS

Youyi Wang, Lihua Xie, Yeng Chai Soh
School of Electrical and Electronic Engineering, Nanyang Technological University, Singapore 2263

Abstract. This paper investigates robust output feedback controller design for a class of uncertain dynamical systems with unknown time delays. The uncertainties considered in this paper contain parametric uncertainties, unknown nonlinear functions, unknown time delays and unknown nonlinear functions. The problems of robust stabilization and robust H_∞ performance are addressed. It is shown that the solutions to the above problems can be obtained from the solutions to algebraic Riccati inequalities. A numerical example indicates that the proposed robust controller can ensure the stability of the overall system.

Résumé Ce Papier présente un contrôleur pour contrôler des systèmes avec des dynamiques incertaines. Ces incertains peuvent être causés par des paramètres incertains, fonctions non-linéaires inconnues et retards inconnus. Des problèmes de stabilisation robuste et H_∞-performance robuste sont discusés. En même temps, des solutions pour les problèmes sont obtenues en calculant une inégalité algébrique de Riccati. Un exemple de la simulation démontre que le contrôleur robuste proposé dans ce papier permet de stabiliser les systèmes.

Keywords. Time delay systems, uncertain systems, nonlinear systems, robust control, Riccati equation approach

1. INTRODUCTION

In the past two decades, a great deal of interest has been focused on the problem of robust control of uncertain systems (see Barmish, 1985, Doyle et. al., 1989, Khargonekar, et. al. 1990, Petersen, 1987, 1988, Xie and de Souza, 1992, Xie et. al. 1992, and the references therein). This includes the studies in Cheres et. al. (1989), Hasanul Basher et. al. (1986), and Shen et. al. (1991), etc. where various controller design approaches for uncertain dynamical delay systems were considered. In this paper, the robust output feedback stabilization and robust H_∞ performance problems for the uncertain dynamical delay system are considered. The uncertainties under consideration contain time-varying parametric uncertainties, unknown time delays and unknown nonlinear functions. The interests are focused on the design of robust linear dynamic output feedback control laws. The Riccati equation approach (see e.g. Khargonekar et. al. 1990, Xie et. al. 1992, Petersen 1987, Xie and de Souza 1992) will be applied to design output dynamical feedback controllers. The main results in this paper are that both the robust stabilization and the robust H_∞ performance problems for uncertain dynamical delay systems can be solved in terms of two algebraic Riccati inequalities (ARIs). One of these ARIs is for constructing the state feedback laws and another for the estimation of the state feedback control laws. In particular, when the state variables of the uncertain dynamical delay system are available for feedback, only one ARI is needed.

2. PROBLEM STATEMENT

Consider the following class of uncertain dynamical systems with unknown time delays

$$\dot{x}(t) = (A + \Delta A(t))x(t) + (A_d + \Delta A_d(t))x_h$$
$$+(B + \Delta B(t))u(t) + (G + \Delta G(t))g(x)$$
$$+ (G_d + \Delta G_d(t))g_1(x_\tau) + B_1 w(t) \quad (1)$$
$$z(t) = C_1 x(t) + D_1 u(t) \quad (2)$$
$$y(t) = (C + \Delta C(t))x(t)$$
$$+ (D + \Delta D(t))u(t) + D_2 w(t) \quad (3)$$

where $x(t) \in \mathbf{R}^n$ is the state; $x_h = x(t-h)$ and $x_\tau = x(t-\tau)$ with h and τ representing the unknown state delays in the system, $u(t) \in \mathbf{R}^m$ is the input; $w(t) \in R^q$ is the disturbance input; $y(t) \in \mathbf{R}^r$ is the measured output; $z(t) \in R^p$ is the controlled output; A, A_d, B, C, C_1, D, D_1, D_2, G and G_d are known real constant matrices with appropriate dimensions, $\Delta A(t)$, $\Delta A_d(t)$, $\Delta B(t)$, $\Delta C(t)$, $\Delta D(t)$, $\Delta G(t)$ and $\Delta G_d(t)$ are real matrices which represent time-varying uncertainties of the system; and $g(x)$ and $g_1(x_\tau)$ are unknown nonlinear vector functions.
The parametric uncertainty matrices $\Delta A(t)$, $\Delta A_d(t)$, $\Delta B(t)$, $\Delta C(t)$, $\Delta D(t)$, $\Delta G(t)$ and $\Delta G_d(t)$ are of the following forms:

$$\begin{bmatrix} \Delta A(t) & \Delta B(t) \\ \Delta C(t) & \Delta D(t) \end{bmatrix} = \begin{bmatrix} L_1 \\ L_2 \end{bmatrix} F(t)[E_1, \ E_2] \quad (4)$$

$$\Delta A_d(t) = L_{A_d} F_{A_d}(t) E_{A_d} \quad (5)$$
$$\Delta G(t) = L_G F_G(t) E_G \quad (6)$$
$$\Delta G_d(t) = L_{G_d} F_{G_d}(t) E_{G_d} \quad (7)$$

where E_1, E_2, E_{A_d}, E_G, E_{G_d}, L_1, L_2, L_{A_d}, L_G and L_{G_d} are known real constant matrices, and

$F(t)$, $F_{A_d}(t)$, $F_G(t)$ and $F_{G_d}(t)$ are unknown matrix functions satisfying

$$F^T(t)F(t) \le I; \quad F_{A_d}(t)F_{A_d}^T(t) \le I \quad (8)$$
$$F_G(t)F_G^T(t) \le I; \quad F_{G_d}(t)F_{G_d}^T(t) \le I \quad (9)$$

with their elements being Lebesgue measurable.

We also make the following assumptions concerning the unknown nonlinear functions and structural matrices E_2 and L_2:

Assumption 2.1:

(a) There exist known constant matrices W_1 and W_2 such that for all $x \in \mathbf{R}^n$

$$\|g(x)\| \le \|W_1 x(t)\|$$
$$\|g_1(x(t-\tau))\| \le \|W_2 x(t-\tau)\|$$

(b) $R_1 = E_2^T E_2 > 0$ and $R_2 = L_2 L_2^T > 0$.

Note that Assumption 2.1(b) is made only for the simplification of presentation. If Assumption 2.1(b) does not hold, a similar techniuqe employed in Khargonekar, *et. al.* (1990), can be used.

3. ROBUST STABILIZATION

In this paper, we will address the following two problems:

1. *Robust stabilization problem:*
Design a linear dynamic output feedback control law $u = G_c(s)y$ for the uncertain dynamical delay system (1) and (3) such that the resulting closed-loop system is asymptotically stable for all admissible uncertainties. In this case the system (1) and (3) is said to be robustly stabilizable via dynamic output feedback controller $G_c(s)$ and the closed loop system is said to be *robustly stable.*

2. *Robust H_∞ performance problem:*
Given a scalar $\gamma > 0$, design a linear dynamic output feedback control law $u = G_c(s)y$ for the uncertain dynamical delay system (1)-(3) such that the closed loop system is robustly stable and guarantees that under zero initial conditions $\|z\|_2 < \gamma\|w\|_2$ for all nonzero $w \in L_2[0, \infty)$ and for all admissible uncertainties given in (4)-(11). In this case, we say that the uncertain delay system (1)-(3) is *robustly stabilizable with disturbance attenuation γ via dynamic output feedback control* and the closed loop system is *robustly stable with disturbance attenuation γ.*

In this section we will present our main results on robust stabilization of the uncertain delay system (1) and (3). Consider the system (1) and (3) with $w(t) \equiv 0$. Motivated by the results in Xie *et. al.* (1992) where the robust stabilization problem for systems without time-delays and unknown nonlinear functions are considered, the linear time-invariant output feedback controller for (1) and (3) is expressed in the form of

$$\dot{\hat{x}}(t) = [A + \hat{A}]\hat{x}(t) + L[y(t) - \hat{y}(t)] + Bu(t) \quad (10)$$
$$\hat{y}(t) = [C + \hat{C}]\hat{x}(t) + Du(t) \quad (11)$$
$$u(t) = -K\hat{x}(t) \quad (12)$$

where $\hat{A}$, $\hat{C}$, L and K are to be chosen with $\hat{A}$ and $\hat{C}$ reflecting the effects of parametric uncertainties $\Delta A(t)$ and $\Delta C(t)$ on the controller structure, and K and L are respectively the appropriate controller and observer gains. When $\Delta A(t)$ and $\Delta C(t)$ of (1) and (3) disappear, $\hat{A}$ and $\hat{C}$ in (10)-(11) will naturally be set to be zero.

For the robust stabilization of the system (1) and (3), we define the following two algebraic Riccati inequalities

$$A^T P + PA + P\bar{B}\bar{B}^T P - \nu^{-2} B_P^T R_1^{-1} B_P + \nu^2 E_1^T E_1 + W_1^T W_1 + W_2^T W_2 < 0 \quad (13)$$

and

$$(A + \bar{B}\bar{B}^T P)^T Y + Y(A + \bar{B}\bar{B}^T P) + Y\bar{B}\bar{B}^T Y - \nu^2 \bar{C}^T R_2^{-1} \bar{C} + \nu^{-2} B_P^T R_1^{-1} B_P < 0 \quad (14)$$

where ν, λ_1, λ_2 and λ_3 are positive scaling paramters,

$$B_P = B^T P + \nu^2 E_2^T E_1$$

$$\bar{B}\bar{B}^T = \nu^{-2} L_1 L_1^T + [A_d(I - \lambda_1^2 E_{A_d}^T E_{A_d})^{-1} A_d^T + \lambda_1^{-2} L_{A_d} L_{A_d}^T] + [G(I - \lambda_2^2 E_G^T E_G)^{-1} G^T + \lambda_2^{-2} L_G L_G^T] + [G_d(I - \lambda_3^2 E_{G_d}^T E_{G_d})^{-1} G_d^T + \lambda_3^{-2} L_{G_d} L_{G_d}^T] \quad (15)$$

and

$$\bar{C} = C + \nu^{-2} L_2 L_1^T P + \nu^{-2} L_2 L_2^T Y \quad (16)$$

Our main result for robust stabilization of the system (1) and (3) is stated as follows:

Theorem 3.1: Consider the uncertain delay system (1) and (3) with $w(t) \equiv 0$ satisfying (4)-(9) and Assumption 2.1. Then, the system is robustly stabilizable via a linear dynamic output feedback controller if there exist positive scaling parameters ν, λ_1, λ_2 and λ_3 such that

(i) $\lambda_1^2 E_{A_d}^T E_{A_d} < I$, $\lambda_2^2 E_G^T E_G < I$ and $\lambda_3^2 E_{G_d}^T E_{G_d} < I$; and

(ii) There exist positive definite solutions P and Y to (13)-(14) respectively.

Moreover, if conditions (i) and (ii) hold, a suitable stabilizing controller is given by (10)-(12) with

$$\hat{A} = \bar{B}\bar{B}^T P \quad (17)$$
$$\hat{C} = \nu^{-2} L_2 L_1^T P \quad (18)$$
$$K = \nu^{-2} R_1^{-1}(B^T P + \nu^2 E_2^T E_1) \quad (19)$$
$$L = \nu^2 Y^{-1} \bar{C}^T R_2^{-1} \quad (20)$$

Proof: Define that $e(t) = \hat{x}(t) - x(t)$. Then from (1), (3) and (10)-(12) we have

$$\dot{e} = (A - LC + \hat{A} - L\hat{C})e + (L_1 - LL_2)FE_2Ke + (\hat{A} - L\hat{C})x - (L_1 - LL_2)F(E_1 - E_2K)x - \hat{A}_d x_h - \hat{G}g(x) - \hat{G}_d g_1(x_\tau)$$

where $\hat{A}_d = A_d + L_d F_d E_d$, $\hat{G} = G + L_G F_G E_G$ and $\hat{G}_d = G_d + L_{G_d} F_{G_d} E_{G_d}$.

Combining the above equation with (1) gives the closed loop system of the form

$$\dot{\xi} = \bar{A}\xi + \bar{L}F\bar{E}\xi + (\bar{A}_d + \bar{L}_{A_d} F_{A_d} E_{A_d}) x_h$$

$$+(\bar{G}+\bar{L}_G F_G E_G)g(x)+(\bar{G}_d+\bar{L}_{G_d}F_{G_d}E_{G_d})g_1(x_\tau)$$
$$= \tilde{A}\xi + \tilde{A}_d x_h + \tilde{G}g(x) + \tilde{G}_d g_1(x_\tau) \qquad (21)$$

where $\xi = [x^T \quad e^T]^T$,

$$\bar{A} = \begin{bmatrix} A - BK & -BK \\ \hat{A} - L\hat{C} & A - LC + \hat{A} - L\hat{C} \end{bmatrix}$$

$$\bar{L} = \begin{bmatrix} L_1 \\ -(L_1 - LL_2) \end{bmatrix}, \quad \bar{E} = [E_1 - E_2K, \; -E_2K]$$

$$\bar{A}_d = \begin{bmatrix} A_d \\ -A_d \end{bmatrix}, \quad \bar{L}_{A_d} = \begin{bmatrix} L_{A_d} \\ -L_{A_d} \end{bmatrix}$$

$$\bar{G} = \begin{bmatrix} G \\ -G \end{bmatrix}, \quad \bar{L}_G = \begin{bmatrix} L_G \\ -L_G \end{bmatrix}$$

$$\bar{G}_d = \begin{bmatrix} G_d \\ -G_d \end{bmatrix}, \quad \bar{L}_{G_d} = \begin{bmatrix} L_{G_d} \\ -L_{G_d} \end{bmatrix}$$

and

$$\tilde{A} = \bar{A} + \bar{L}F\bar{E}, \qquad \tilde{A}_d = \bar{A}_d + \bar{L}_{A_d}F_{A_d}E_{A_d}$$

$$\tilde{G} = \bar{G} + \bar{L}_G F_G E_G, \qquad \tilde{G}_d = \bar{G}_d + \bar{L}_{G_d}F_{G_d}E_{G_d}$$

Next, define

$$U = \begin{bmatrix} U_{11} & U_{12} \\ U_{21} & U_{22} \end{bmatrix}$$
$$= \bar{A}^T X + X\bar{A} + X\bar{B}\bar{B}^T X + \bar{W}^T\bar{W} + \nu^2\bar{E}^T\bar{E}$$

where X is a positive definite matrix, and

$$\bar{W}^T\bar{W} = \begin{bmatrix} W_1^T W_1 + W_2^T W_2 & 0 \\ 0 & 0 \end{bmatrix}$$

$$\bar{B}\bar{B}^T = \nu^{-2}\begin{bmatrix} L_1 \\ -(L_1 - LL_2) \end{bmatrix}[L_1^T \; -(L_1 - LL_2)^T]$$
$$+ \begin{bmatrix} A_d \\ -A_d \end{bmatrix}(I - \lambda_1^2 E_{A_d}^T E_{A_d})^{-1}\begin{bmatrix} A_d \\ -A_d \end{bmatrix}^T$$
$$+\lambda_1^{-2}\begin{bmatrix} L_{A_d} \\ -L_{A_d} \end{bmatrix}[L_{A_d}^T \quad -L_{A_d}^T]$$
$$+ \begin{bmatrix} G \\ -G \end{bmatrix}(I - \lambda_2^2 E_G^T E_G)^{-1}\begin{bmatrix} G \\ -G \end{bmatrix}^T$$
$$+\lambda_2^{-2}\begin{bmatrix} L_G \\ -L_G \end{bmatrix}[L_G^T \quad -L_G^T]$$
$$+ \begin{bmatrix} G_d \\ -G_d \end{bmatrix}(I - \lambda_3^2 E_{G_d}^T E_{G_d})^{-1}\begin{bmatrix} G_d \\ -G_d \end{bmatrix}^T$$
$$+\lambda_3^{-2}\begin{bmatrix} L_{G_d} \\ -L_{G_d} \end{bmatrix}[L_{G_d}^T \quad -L_{G_d}^T]$$

By setting

$$X = \begin{bmatrix} P & 0 \\ 0 & Y \end{bmatrix} \qquad (22)$$

where P and Y are the positive definite solutions to (13)-(14) respectively. Then, using straightforward but tedious manipulations gives that

$$U_{11} = A^T P + PA + P\bar{B}\bar{B}^T P - \nu^{-2}B_P^T R_1^{-1} B_P^T$$
$$+\nu^2 E_1^T E_1 + W_1^T W_1 + W_2^T W_2 < 0$$
$$U_{21}^T = U_{12} = 0$$

and

$$U_{22} = (A+\bar{B}\bar{B}^T P)^T Y + Y(A+\bar{B}\bar{B}^T P) + Y\bar{B}\bar{B}^T Y$$
$$-\nu^2\bar{C}^T R_2^{-1}\bar{C} + \nu^{-2}B_P^T R_1^{-1} B_P^T < 0$$

Therefore, $U < 0$, i.e.

$$\bar{A}^T X + X\bar{A} + X\bar{B}\bar{B}^T X + \bar{W}^T\bar{W} + \nu^2\bar{E}^T\bar{E} < 0 \qquad (23)$$

By applying Lemma 2.2 in Wang, *et. al.* (1992) to (23), we conclude that

$$\tilde{A}^T X + X\tilde{A} + \bar{W}^T\bar{W}$$
$$+ X(\tilde{A}_d\tilde{A}_d^T + \tilde{G}\tilde{G}^T + \tilde{G}_d\tilde{G}_d^T)X < 0 \qquad (24)$$

Now, in order to prove the closed loop system (21) is asymptotically stable, we choose the following Lyapunov function candidate

$$V(\xi) = \xi^T X\xi + \int_{t-h}^{t} x^T x ds + \int_0^t x^T W_1^T W_1 x ds$$
$$- \int_0^t g^T g ds + \int_0^{t-\tau} x^T W_2^T W_2 x^T ds - \int_0^{t-\tau} g_1^T g_1 ds$$

where X is given in (22).
Note that $V(\xi) > 0$ whenever $\xi \neq 0$. Then we have, in the view of (21),

$$\frac{d}{dt}V(\xi) = \xi_g^T A_S \xi_g$$

where $\xi_g = [\xi^T \quad x_h^T \quad g^T \quad g_1^T(x_\tau)]^T$ and A_S denotes the left hand side of (24).
By taking into account (24), it follows immediately that

$$\frac{d}{dt}V(\xi) < 0$$

whenever $\xi \neq 0$. Therefore, $V(\xi)$ is a Lyapunov function of the closed loop system (21) and thus the closed loop system (21) is asymptotically stable for all admissible uncertainties. Hence, the uncertain dynamical system with unknown time delay (1) and (3) is robustly stabilizable via the dynamic output feedback controller (10)-(12).
The result of Theorem 3.1 can be easily specialized to the case of state feedback as follows:

Corollary 3.1: Consider the uncertain delay system (1) with $w(t) \equiv 0$ satisfying (4)-(9) and Assumption 2.1. Suppose that the state variables in the system (1) are available for feedback. Then, the uncertain delay system (1) is robustly stabilizable via a linear state feedback control law $u(t) = -Kx(t)$ if there exist positive scaling parameters ν, λ_1, λ_2 and λ_3 such that
(i) $\lambda_1^2 E_{A_d}^T E_{A_d} < I$, $\lambda_2^2 E_G^T E_G < I$ and $\lambda_3^2 E_{G_d}^T E_{G_d} < I$, and
(ii) There exists a positive definite solution P to (13).
Moreover, if conditions (i) and (ii) hold, a suitable stabilizing controller gain is given by $K = \nu^{-2}R_1^{-1}(B^T P + \nu^2 E_2^T E_1)$.

4. ROBUST H_∞ PERFORMANCE RESULTS

In this section the results on robust stabilization will be extended for solving the problem of robust H_∞ performance for the uncertain delay system (1)-(3). Similar to the case of robust stabilization, we introduce dynamic output feedback controller of the form:

$$\begin{aligned}\dot{\hat{x}}(t) &= [A+\hat{A}]\hat{x}(t)+B_1\hat{w}(t)\\ &+ L[y(t)-\hat{y}(t)]+Bu(t) \quad (25)\\ \hat{y}(t) &= [C+\hat{C}]\hat{x}(t)+D_2\hat{w}(t)+Du(t) \quad (26)\\ u(t) &= -K\hat{x}(t) \quad (27)\end{aligned}$$

where $\hat{A}$, $\hat{C}$, L and K are to be chosen with $\hat{A}$ and $\hat{C}$ reflecting the effects of parametric uncertainties $\Delta A(t)$ and $\Delta C(t)$ on the controller structure, K and L are respectively the controller and observer gains, $\hat{w}(t)$ can be interpreted as the estimated worst-case disturbance input (see Petersen, 1988). In connection with the robust H_∞ performance problem for system (1)-(3), we define the following two algebraic Riccati inequalities:

$$\begin{aligned}&A^TP+PA+P(\bar{B}\bar{B}^T+\gamma^{-2}B_1B_1^T)P\\ &+C_1^TC_1+\nu^2E_1^TE_1+W_1^TW_1+W_2^TW_2\\ &-B_D^T(D_1^TD_1+\nu^{-2}R_1)^{-1}B_D<0 \quad (28)\end{aligned}$$

and

$$\begin{aligned}&(A+\bar{B}\bar{B}^TP+\gamma^{-2}B_1B_1^TP)^TY\\ &+Y(A+\bar{B}\bar{B}^TP+\gamma^{-2}B_1B_1^TP)+Y\bar{B}\bar{B}^TY\\ &-\bar{C}_P^T[\gamma^{-2}D_2D_2^T+\nu^{-2}R_2]^{-1}\bar{C}_P\\ &+B_D^T(D_1^TD_1+\nu^{-2}R_1)^{-1}B_D<0 \quad (29)\end{aligned}$$

where ν, λ_1, λ_2 and λ_3 are positive scaling paramters, and

$$B_D=B_P+D_1^TC_1$$

$$\bar{C}_P=\bar{C}+\gamma^{-2}D_2B_1^TP+\gamma^{-2}B_1D_2^TY$$

$\bar{B}\bar{B}^T$ and $\bar{C}$ are as in (15)-(16) .

Then we have the following result:

Theorem 4.1: Consider the uncertain time delay system of (1)-(3) satisfying (4)-(9) and Assumption 2.1 and let $\gamma>0$ be the prescribed level of disturbance attenuation. Then, the uncertain delay system (1)-(3) is robustly stabilizable with disturbance attenuation γ via dynamic output feedback controller (25)-(27) if there exist positive scaling parameters ν, λ_1, λ_2 and λ_3 such that

(i) $\lambda_1^2E_{A_d}^TE_{A_d}<I$, $\lambda_2^2E_G^TE_G<I$ and $\lambda_3^2E_{G_d}^TE_{G_d}<I$;

(ii) There exist positive definite solutions P and Y to (28) and (29) respectively.

Moreover, if the conditions (i) and (ii) hold, a suitable stabilizing controller is given by (25)-(27) with

$$\begin{aligned}\hat{A} &= \bar{B}\bar{B}^TP, \quad \hat{C}=\nu^{-2}L_2L_1^TP\\ K &= R^{-1}(B^TP+\nu^2E_2^TE_1+D_1^TC_1)\\ R &= D_1^TD_1+\nu^{-2}R_1>0\\ L &= Y^{-1}\bar{C}^T(\gamma^{-2}D_2D_2^T+\nu^2R_2)^{-1}\\ \hat{w}(t) &= \gamma^{-2}B_1^TP\hat{x}(t)\end{aligned}$$

Proof: Define that $e(t)=\hat{x}(t)-x(t)$ and consider that $\hat{w}(t)=\gamma^{-2}B^TPx(t)$. Then from (1) and (25)-(27) we have

$$\begin{aligned}\dot{e}&=(A-LC+\hat{A}-L\hat{C}+\gamma^{-2}(B_1-LD_2)B^TP)e\\ &+(L_1-LL_2)FE_2Ke\\ &+(\hat{A}-L\hat{C}+\gamma^{-2}(B_1-LD_2)B^TP)x\\ &-(L_1-LL_2)F(E_1-E_2K)x\\ &-\hat{A}_dx_h-\hat{G}g(x)-\hat{G}_dg_1(x_\tau)+(LD_2-B_1)w\end{aligned}$$

Combining the above equation with (1) gives the closed loop system equations

$$\begin{aligned}\dot{\xi}&=\bar{A}\xi+\bar{L}F\bar{E}\xi+(\bar{A}_d+\bar{L}_{A_d}F_{A_d}E_{A_d})x_h\\ &+(\bar{G}+\bar{L}_GF_GE_G)g(x)\\ &+(\bar{G}_d+\bar{L}_{G_d}F_{G_d}E_{G_d})g_1(x_\tau)+\bar{B}_1w\\ &=\tilde{A}\xi+\tilde{A}_dx_h+\tilde{G}g(x)+\tilde{G}_dg_1(x_\tau)+\bar{B}_1w \quad (30)\end{aligned}$$

where

$$\bar{A}=\begin{bmatrix}A-BK & -BK\\ T_1 & T_2\end{bmatrix},\quad \bar{B}_1=\begin{bmatrix}B_1\\ LD_2-B_1\end{bmatrix}$$

$T_1=\hat{A}-L\hat{C}+\gamma^{-2}(B_1-LD_2)B^TP$ and $T_2=A-LC+\hat{A}-L\hat{C}+\gamma^{-2}(B_1-LD_2)B^TP$. The matrices $\bar{L}$, $\bar{E}$, $\bar{A}_d$, $\bar{L}_{A_d}$, $\bar{G}$, $\bar{L}_G$, $\bar{G}_d$, $\bar{L}_{G_d}$, $\tilde{A}$, $\tilde{A}_d$, $\tilde{G}$ and $\tilde{G}_d$ are as in (21).

From (2) we have

$$z=(C_1-D_1K)x-D_1Ke$$

$$=\begin{bmatrix}C_1-D_1K\\ -D_1K\end{bmatrix}^T\begin{bmatrix}x\\ e\end{bmatrix}=\bar{C}_1\xi \quad (31)$$

We shall prove that the closed loop system (30) is asymptotically stable. To this end, let

$$X=\begin{bmatrix}P & 0\\ 0 & Y\end{bmatrix} \quad (32)$$

where P and Y are respectively positive definite solutions to (28)-(29).

A similar proof as in that of Theorem 3.1 leads to

$$\bar{A}^TX+X\bar{A}+X\bar{B}_1\bar{B}_1^TX+\bar{W}^T\bar{W}+\nu^2\bar{E}^T\bar{E}+\bar{C}_1^T\bar{C}_1<0$$

where

$$\bar{B}_1\bar{B}_1^T=\bar{B}\bar{B}^T+\gamma^{-2}\begin{bmatrix}B_1\\ LD_2-B_1\end{bmatrix}[B_1^T \ D_2^TL^T-B_1^T]$$

The above inequality, by Lemma 2.2 in Wang, *et. al.* (1992), implies that

$$\begin{aligned}&X\tilde{A}+\tilde{A}^TX+\bar{C}_1^T\bar{C}_1+\bar{W}^T\bar{W}+\gamma^{-2}X\bar{B}_1\bar{B}_1^TX\\ &+X(\tilde{A}_d\tilde{A}_d^T+\tilde{G}\tilde{G}^T+\tilde{G}_d\tilde{G}_d^T)X<0 \quad (33)\end{aligned}$$

Hence, the asymptotic stability of the closed loop system (30) follows from the proof of Theorem 3.1.

To establish the robust H_∞ performance index, $\|z\|_2 < \gamma\|w\|_2$ whenever $\|w\|_2 \neq 0$, let us introduce

$$J = \int_0^\infty (z^T z - \gamma^2 w^T w)dt$$

where $\gamma > 0$ is the prescribed levels of disturbance attenuation.
Note that the boundedness of J is guaranteed by the asymptotic stability of (30) and the fact that $w \in L_2[0, \infty)$.
As in the proof of Theorem 3.1, let

$$V(\xi) = \xi^T X \xi + \int_{t-h}^t x^T x ds + \int_0^t x^T W_1^T W_1 x ds$$

$$-\int_0^t g^T g ds + \int_0^{t-\tau} x^T W_2^T W_2 x^T ds - \int_0^{t-\tau} g_1^T g_1 ds$$

where X is given in (32).
Then with zero initial conditions in (30)-(31), the index J can be rewritten as

$$J = \int_0^\infty \{[z^T z - \gamma^2 w^T w] + \frac{d}{dt}V(t)\}dt - V(\infty)$$

It follows that

$$J = \int_0^\infty \{\xi^T \bar{C}_1^T \bar{C}_1 \xi - \gamma^2 w^T w + \frac{d}{dt}V(t)\}dt - V(\infty)$$

Define that

$$\mu(t) \triangleq \xi^T \bar{C}_1^T \bar{C}_1 \xi - \gamma^2 w^T w + \frac{d}{dt}V(t)$$

$$= \xi^T \bar{C}_1^T \bar{C}_1 \xi - \gamma^2 w^T w + \xi^T(\bar{A}^T X + X\bar{A})\xi$$

$$+ x_h^T \bar{A}_d^T X \xi + \xi^T X \bar{A}_d x_h + g^T \bar{G}^T X \xi + \xi^T X \bar{G} g$$

$$+ g_1^T \bar{G}_d^T X \xi + \xi^T X \bar{G}_d^T g_1$$

$$+ \xi^T X \bar{B}_1 w + w^T \bar{B}_1^T X \xi + [x^T x - x_h^T x_h]$$

$$+ x^T W_1^T W_1 x - g^T g + x_\tau^T W_2^T W_2 x_\tau - g_1^T g_1$$

Then we have

$$\mu(t) = \xi_T^T A_T \xi_T$$

where $\xi_T = [\xi^T \quad x_h^T \quad g^T \quad g_1^T(x_\tau) \quad w^T]^T$ and A_T stands for the left hand side of (33).
It then follows from (33) that

$$\mu(t) < 0$$

whenever $\|w\|_2 \neq 0$. Therefore, by noting the fact that $V(\infty)$ is bounded and nonnegative, we conclude that

$$J < 0$$

i.e.

$$\|z\|_2 < \gamma\|w\|_2$$

whenever $\|w\|_2 \neq 0$ and for all admissible uncertainties.
The result of Theorem 4.1 can be easily specialized to the case of state feedback as follows:
Corollary 4.1: Consider the system (1)-(3) satisfying (4)-(9) and Assumption 2.1, and assume that all state variables are available for feedback and let $\gamma > 0$ be the prescribed level of disturbance attenuation. Then, this system is robustly stablizable with disturbance attenuation γ via a linear state feedback control law$u(t) = -Kx(t)$ if there exist positive scaling parameters ν, λ_1, λ_2 and λ_3 such that
(i) $\lambda_1^2 E_{A_d}^T E_{A_d} < I$, $\lambda_2^2 E_G^T E_G < I$ and $\lambda_3^2 E_{G_d}^T E_{G_d} < I$; and
(ii) There exists a positive definite solution P to (28).
Moreover, if the conditions (i) and (ii) hold, a suitable controller gain is given by $K = R^{-1}(B^T P + \nu^2 E_2^T E_1 + D_1^T C_1)$ where $R = D_1^T D_1 + \nu^{-2} R_1$.

5. AN EXAMPLE

Consider the problem of robust stabilization of an uncertain system with unknown time delays:

$$\begin{aligned} \dot{x}_1 &= x_2 + c x_1(t-\tau)\sin(x_2(t-\tau)) \\ \dot{x}_2 &= (-3+a)x_1 + x_2 \\ &\quad + b\sqrt{x_1^2 + x_2^2}\sin(dx_1) + (a+2)u \end{aligned}$$

where a, b, c and d are unknown but bounded with $a \in [-1, 1]$, $b \in [-1, 1]$ and $c \in [-1, 1]$, and τ is the unknown time delay. The state is available for feedback.
This system is of the form of (1) with

$$A = \begin{bmatrix} 0 & 1 \\ -3 & 1 \end{bmatrix}, \quad B = \begin{bmatrix} 0 \\ 2 \end{bmatrix}, \quad G = G_d = \begin{bmatrix} 0 \\ 0 \end{bmatrix}$$

$$L_1 = \begin{bmatrix} 0 \\ 1 \end{bmatrix}, \quad E_1 = [1 \quad 0], \quad E_2 = 1$$

$$L_G = L_{G_d} = \begin{bmatrix} 0 \\ 1 \end{bmatrix}, \quad E_G = E_{G_d} = 1$$

$$g(x) = \sqrt{x_1^2 + x_2^2}\sin(dx_1)$$

$$g_1(x(t-\tau)) = x_1(t-\tau)\sin(x_2(t-\tau))$$

$g(x)$ and $g_1(x(t-\tau))$ satisfy Assumption 2.1(a) with

$$W_1 = \begin{bmatrix} 1 & 0 \\ 0 & 1 \end{bmatrix}, \quad W_2 = [1 \quad 0]$$

and $R_1 = E_2^T E_2 = 1$.
Equation (13) becomes

$$A^T P + PA + P\bar{B}\bar{B}^T P - \nu^{-2} B_P^T B_P$$

$$+ \nu^2 E_1^T E_1 + W_1^T W_1 + W_2^T W_2 < 0$$

where

$$\bar{B}\bar{B}^T = \nu^{-2} L_1 L_1^T + \lambda_2^{-1} L_G L_G^T + \lambda_3^{-1} L_{G_d} L_{G_d}^T$$

Choose $\nu = 1$, $\lambda_2 = 0.99$ and $\lambda_3 = 0.99$. Note that the conditions, $\lambda_2^2 E_G^T E_G < I$ and $\lambda_3 E_{E_d}^T E_{G_d} < I$ gives that $\lambda_2 < 1$ and $\lambda_3 < 1$. To minimize $\bar{B}\bar{B}^T$, we choose $\lambda_2 = 0.99$ and $\lambda_3 = 0.99$.
Solving the above inequality gives $K = [1.3926 \quad 5.2709]$
The simulation results with $x_1(0) = x_2(0) = 0.5$ are shown in Fig. 1 to Fig. 4 respectively, which indicates the effectiveness of the proposed robust

stabilization technique.

ACKNOWLEDGEMENT This work was supported by the Nanyang Technological University Research Found.

6. CONCLUSION

Robust stabilization and robust H_∞ performance problems of uncertain systems with unknown time delays are considered in this paper. The uncertainties we considered in this paper comprise real time-varying parametric uncertainties, unknown nonlinear functions and unknown time delays. Linear dynamic output feedback robust control techniques for such a class of systems have been developed. We have shown that both the problems of robust control and the robust H_∞ performance can be solved in term of Riccati inequalities. Simulation study on an illustrative example indicates that the proposed robust controller can ensure the stability of the overall system.

REFERENCES:

Barmish, B. R., (1985). Necessary and sufficient conditions for quadratic stabilizability of an uncertain system, *J. Optimaz. Theory Appli.*, **46**, 399-408.

Cheres, E., S. Gutman and Z. J. Palmor, (1989). Stabilization of uncertain dynamic systems including state delay, *IEEE Trans. Auto. Control*, **AC-34**, 1199-1203.

Doyle, J. C., K. Glover, P. P. Khargonekar and B. A. Francis, (1989). State-space solutions to the standard H_2 and H_∞ control problems, *IEEE Trans. Auto. Control*, **AC-34**, 831-847.

Hasanul Basher, A. M., R. Mukundan and D. A. O'Connor, (1986). Memoryless feedback control in uncertain dynamic delay systems, *Int. J. Syst. Sci.*, **17**, 409-415.

Khargonekar, P. P., I. R. Persen and K. Zhou, (1990). Robust stabilization of uncertain systems: quadratic stabilizability and H^∞ control theory, *IEEE Trnas. Auto. Control*, **AC-35**, 356-361.

Petersen, I. R., (1987). Disturbance attenuation and H_∞ optimization: a design method based on the algebraic Riccati equation, *IEEE Trans. Auto. Control*, **AC-32**, 397-411.

Petersen, I. R., (1988). Stabilization of an uncertain linear system in which uncertain parameters enter into the input matrix, *SIAM J. Contr. Optimaz.*, **26**, 1257-1264.

Shen, J., B. Chen and F. Kung, (1991). Memoryless stabilization of uncertain dynamic delay systems: Riccati equation approach, *IEEE Trans. Auto. Control*, **AC-36**, 638-640.

Wang, Y., L. Xie and C. E. de Souza, (1992). Robust control of a class of uncertain nonlinear systems, *Syst. & Control Lett.*, **19**, 139-149.

Xie, L., M. Fu and C. E. de Souza, (1992). H^∞ control and quadratic stabilization of systems with parameter uncertainty via output feedback, *IEEE Trans. Auto. Control*, **AC-37**, 1253-1256.

Xie, L. and C. E. de Souza, (1992). Robust H_∞ control for linear systems with norm-bounded time-varying uncertainty, *IEEE Trans. Auto. Control*, **AC-37**, 1188-1191.

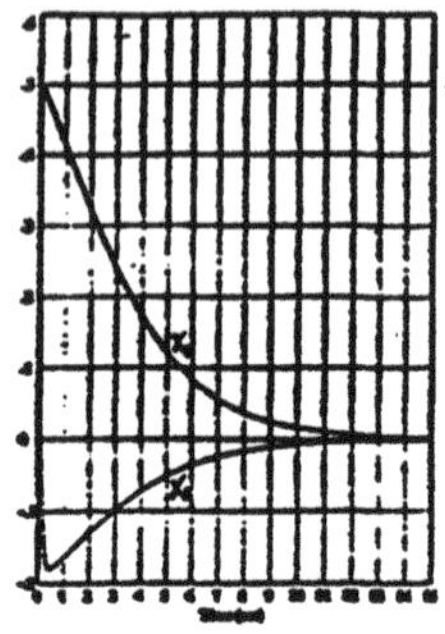

Fig. 1 $a = 1$ and $b = c = d = 1$

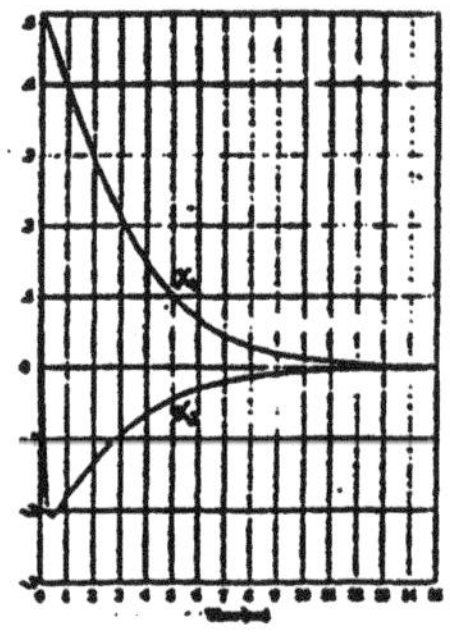

Fig. 2 $a = -1$ and $b = c = d = 1$

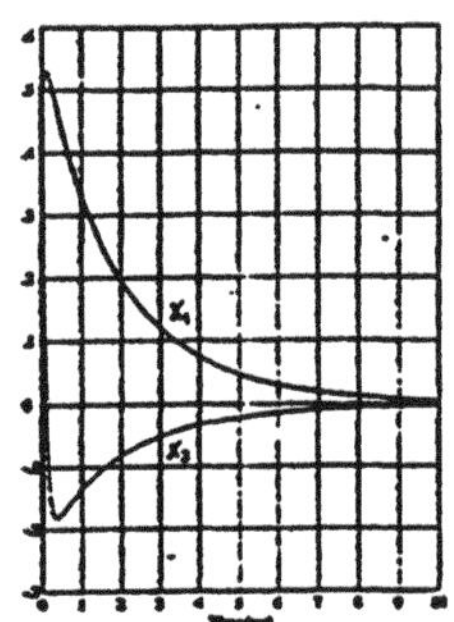

Fig. 3 $a = 1, b = -1$ and $c = d = 1$

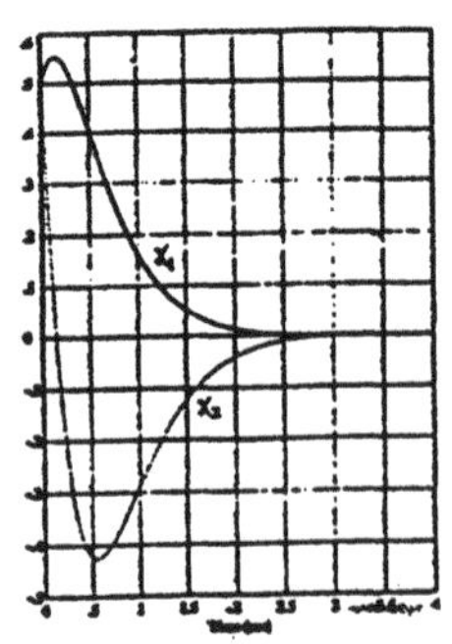

Fig. 4 $a = 1, b = c = -1$ and $d = 1$

STABILITY OF SYSTEMS WITH DISTRIBUTED DELAYS

E. I. VERRIEST *

*Georgia Tech Lorraine, Technopôle de Metz 2000, F57070 METZ, France.

Abstract. For linear systems with distributed delays, sufficient conditions are given for the stability and asymptotic stability independent of the matrix valued weight function on the delayed state. All these sufficient conditions, obtained via the Lyapunov-Krasovskii theory, are obtained in the form of the existence of some positive definite matrix functions satisfying certain Riccati-type differential equations. Connections are made with the theory of robust control, and its frequency domain criteria. New results on stabilizabilty with distributed feedback are derived.

Pour les systèmes avec retards distribués, des conditions suffisantes sont dérivées pour la stabilité asymptotique indépendante de la fonction matricielle multipliant l'état retardé. Celles-ci sont obtenues par la théorie de Lyapunov-Krasovskii, et apparaissent sous la forme d'existance des matrices positive semi-définies, satisfaisants une équation différentielle de Riccati. Ces résultats sont reliés à la théorie de contrôle robuste, et ses critères dans le domaine fréquentiel. Des nouveaux résultats sur la stabilisation des systèmes avec retards distribués sont démontrés.

Key Words. Differential-delay system; distributed delay; robust stability; stabilizability.

1. Introduction

In earlier work (Florchinger *et al.* 1993, Verriest *et al.* 1993, Verriest 1994a 1994b, Verriest *et al.* 1991, 1994, Verriest *et al.* 1995) the theory of Lyapunov functionals was used to obtain sufficient conditions for the stability of time-invariant and time-varying differential delay systems, independent of the delay time. The resulting conditions were 'nice', since they involved the existence of a triple of positive definite matrices satisfying a certain Riccati equation. This Riccati equation has a positive sign in its quadratic term, and is thus of the type encountered in robust control theory (Hewer 1993). In fact, connections between the stability of the delay differential equation, and the theory of robust control were obtained in Verriest *et al.* (1993). This paper extends this work to distributed delays. We do not assume knowledge of the exact matrix weight function of the delayed state. This problem is significant, as many continuum systems (infinite dimensional) are approximated by delay systems e.g., propagation delays in vibrating plates and beams. In viscoelastic structures the stress-strain states of the materials are modeled by such equations: e.g., the Piola-Kirchoff stress function in the equation of motion of a one-dimensional viscoelestic body which moves longitudinally has a distributed delay integral of the history of the displacement. Many other applications in man-machine systems, process control, remote control and robotics, involve delays due to transportation lags, and conduction or communication times which may be distributed due to reflection at boundaries whose distance smoothly varies with the angle (line of sight). In economy, distributed delay systems have been used to model the dynamics of commodity price fluctuations, and distributed models have been proposed in which advertising expenditures are regarded as an investment in a goodwill stock. Here and in applications to population dynamics, robustness with respect to the shape information is important, as the latter is seldom precisely known (Kolmanovskii *et al.* 1992).

In order to fix the ideas, we analyze in this paper the distributed delay system:

$$\dot{x}(t) = Ax(t) + \int_0^T B(\tau)x(t-\tau)\,d\tau. \qquad (1)$$

It is first assumed that the $\mathbb{R}^{n\times n}$-valued matrix weight function $B(\cdot)$ and the effective interference time T are known. As will be apparent, the sufficient conditions derived only require partial information, and hence can be exploited again to obtain *robust* stability conditions. It is also assumed that the initial data $\phi \in C([-T,0],\mathbb{R}^n)$ is given. $C([-T,0],\mathbb{R}^n)$ is the Banach space of continuous functions $[-T,0] \to \mathbb{R}^n$, with the norm $\|\phi\| = \sup_t |\phi(t)|$. The Euclidean norm of $\phi(t) \in \mathbb{R}^n$ is denoted by $|\phi(t)|$.

In Section 2, some background material on *retarded functional differential equations* and the Lyapunov-Krasovskii theory is introduced. Section 3 then proceeds with the derivation of sufficient conditions for stability, in terms of Riccati-like equations. In Section 4 the results are reinterpreted as robust stability conditions, and frequency domain criteria are derived via the positive real lemma. Finally, some comments on the feedback stabilization of differential delay systems are given in Section 5.

2. Retarded Functional Differential Equations

Consider the equation (1). It is common usage to consider a delay differential equation as a functional differential equation (Hale *et al.* 1984). For $t \geq t_0 - T$, let $x(t;t_0,\phi)$ denote its solution at time t with the initial data ϕ, specified at time t_0, i.e., $x(t_0+\theta,t_0,\phi) = \phi(\theta)$ for $\theta \in [-T,0]$. Because of the time-invariance of the dynamical system, $x(t;t_0,\phi) = x(t-t_0,0,\phi)$ for all $t > t_0$. As customary, $x(t+\theta)$ for $\theta \in [-T,0]$ is denoted by $x_t(\theta)$, which is the state (infinite dimensional) of the delay system. Equation (1) is then of the general form

$$\dot{x}(t) = F(x_t) \quad , \quad x_{t_0} = \phi \tag{2}$$

with $F : \mathcal{D}_T = C([-T,0],\mathbb{R}^n) \to \mathbb{R}^n$, linear in the argument. It is well known that this retarded functional Cauchy problem has a unique solution, defined for all $t \in [t_0 - T, \infty)$, and which depends continuously on F and ϕ (Hale *et al.* 1984, Kolmanovskii *et al.*, 1992). The equilibrium solution is $x_t \equiv 0$. The following is standard, for $Q_\delta = \{\phi \in C[-T,0] \mid \|\phi\| < \delta\}$.

Definition 1: The equilibrium solution, $x_t \equiv 0$, of the delay-differential equation (1) is said to be

1. *(uniformly) stable* if for any $\epsilon > 0$, there is a $\delta = \delta(\epsilon) > 0$, such that $|x(t;0,\phi)| \leq \epsilon$ for any initial function $\phi \in Q_\delta$, and $\forall t > 0$. In the opposite case, it is called *unstable*.
2. *(uniformly) asymptotically stable* if it is (uniformly) stable, and there is a $K > 0$ such that for any $\gamma > 0$ there is a $T(\gamma,K) > 0$ such that $|x(t;0,\phi)| \leq \gamma$, $\forall t \geq T(\gamma,K)$ and $\phi \in Q_K$.

Note that because of the time-invariance of the dynamical equation, the above definitions for stability and asymptotic stability are in fact automatically *uniform*. We introduce now the following definition:

Definition 2: Let $\mathcal{B} \in \mathbb{R}^{n\times n}$ be symmetric and positive definite. The distributed-delay differential system (1) is $\mathcal{B}$*-robustly stable* (respectively *robustly asymptotically stable*) if the equilibrium solution is stable (respectively asymptotically stable) for *all* bounded delay intervals $[0,T]$, and weight matrix functions $B(\cdot) : [0,T] \to \mathbb{R}^{n\times n}$ satisfying

$$\min_{\beta(\cdot)>0} \int_0^T \frac{B(t)B(t)'}{\beta(t)}\,dt \int_0^T \beta(t)\,dt \leq \mathcal{B}. \tag{3}$$

The main results of this paper hinge on the following facts, repeated here to make the paper self contained: Let Ω be the class of scalar nondecreasing functions $\alpha \in C([0,\infty),\mathbb{R})$ such that $\alpha(r) > 0$ for $r > 0$, and $\alpha(0) = 0$.

Definition 3: Let $V : Q_K \to \mathbb{R}$ be a continuous functional such that $V(0) \equiv 0$. The functional $(\psi) \to V(\psi)$ is called ***positive definite*** (*negative definite*) if there is a function $\alpha \in \Omega$ such that $V(\psi) \geq \alpha(|\psi(0)|)$ (respectively, $V(\psi) \leq -\alpha(|\psi(0)|)$) for all $\psi \in Q_K$. It is said to have *infinitesimal upper bound* if $|V(\psi)| \leq \alpha(\|\psi\|)$, for all $\psi \in Q_K$.

Fact 1 (Kolmanovskii *et al.* 1992, p. 103): *If there is a continuous positive definite functional $V : Q_K \to \mathbb{R}$ with derivative $\dot{V} \leq 0$, then the trivial solution of (2) is (uniformly) stable.*

Fact 2 (Kolmanovskii *et al.* 1992, p. 103): *Assume that for some $T > 0$, there exists a positive definite continuous functional $(\psi \to V(\psi)) : Q_K \to \mathbb{R}$ which has infinitesimal upper bound and whose derivative $\dot{V}$ is a negative definite functional on Q_K. Then the trivial solution of (2) is (uniformly) asymptotically stable.*

We have *global* stability in its various guises, if the the set Q_K may be replaced by *all* of $C([-T,0],\mathbb{R}^n)$, i.e., the bound K may be arbitrary large.

3. Riccati-type Equations as Sufficient Conditions

In order to facilitate the derivation of the sufficient stability conditions, we first prove the

Lemma 1: *Given a differentiable matrix valued function $Q(t) = Q(t)'$, satisfying $Q(T) = 0$ and $\frac{dQ(t)}{dt} = -C(t)'C(t)$, and a symmetric matrix P in $\mathbb{R}^{n\times n}$. The derivative of the Lyapunov-Krasovskii functional $V : C([-T,0],\mathbb{R}^n) \to \mathbb{R}$:*

$$V(\psi) = \psi(0)'P\psi(0) + \int_{-T}^{0} \psi(\theta)'Q(-\theta)\psi(\theta)\,d\theta, \tag{4}$$

along the solutions of the system (1), satisfies the

inequality

$$\dot{V}(t) \leq -x(t)'Rx(t) \tag{5}$$

where

$$-[R + A'P + PA + Q(0)] = \tag{6}$$
$$= P \int_0^T B(\tau)C(\tau)^{-1}C(\tau)^{-T}B(\tau)'\,d\tau\,P.$$

Proof: Consider the *Lyapunov-Krasovskii functional* (4). Along trajectories of (1) we have

$$\begin{aligned}\dot{V}(t) &= \dot{x}(t)'Px(t) + x(t)'P\dot{x}(t) + x(t)'Q(0)x(t) \\ &-x(t-T)'Q(T)x(t-T) + \\ &+ \int_{t-T}^{t} x(\tau)'\frac{\partial Q(t-\tau)}{\partial t}x(\tau)\,dt.\end{aligned}$$

Substituting (1) gives:

$$\begin{aligned}\dot{V}(t) &= x(t)'\,[A'P + PA + Q(0)]\,x(t) + \\ &+ \int_0^T x(t-\tau)'B(\tau)'Px(t)\,d\tau + \\ &+ \int_0^T x(t)'PB(\tau)x(t-\tau)\,d\tau + \\ &-x(t-T)'Q(T)x(t-T) + \\ &\int_0^T x(t-\tau)'\dot{Q}(\tau)x(t-\tau)\,d\tau.\end{aligned}$$

With $\dot{Q} = -C'C$, and 'completing the squares', the above expression is rewritten as

$$\begin{aligned}\dot{V} &= -x(t)'R(t)x(t) - \int_0^T [q(t)][q(t)]'\,dt \\ &-x(t-\tau)'Q(T)x(t-\tau),\end{aligned} \tag{7}$$

where $q(t) = x(t-\tau)'C(\tau)' - x(t)'PB(\tau)C(\tau)^{-1}$. Since by assumption $Q(T) = 0$, and the second term is negative semi-definite, eqn. (5) follows. □

The following sufficient conditions are now readily established.

Theorem 1: *Consider the system (1). If there exists a pair of positive definite (symmetric) matrices P, and R, and a matrix function* $C(\cdot)$: $[0,T] \to \mathbb{R}^{n\times n}$ *such that* $C(t)$ *is nonsingular for all* $t \in [0,T]$, *and*

$$A'P + PA + R + \int_0^T Q(\tau)\,d\tau = 0, \tag{8}$$

where
$Q(\tau) = C(\tau)C(\tau)' + PB(\tau)C(\tau)^{-1}C(\tau)^{-T}B(\tau)'P$, *then the system is uniformly stable.*

Proof: Let $Q(t) = \int_t^T C(\tau)'C(\tau)\,d\tau$. From condition (8) we have by lemma 1: $-\dot{V}(x_t) \geq 0$. Thus, under the given conditions the trivial solution is stable by fact 1. □

Theorem 2: *Consider the system (1). If there exists a pair of positive definite (symmetric) matrices P and R, and a nonsingular matix function* $C(\cdot) : [0,T] \to \mathbb{R}^{n\times n}$, *satisfying*

$$\| \int_0^T C(t)C'\,dt \| < \infty, \tag{9}$$

such that, for Q as in Theorem 1,

$$A'P + PA + R + \int_0^T Q(\tau)\,d\tau = 0 \tag{10}$$

then the system (1) is uniformly asymptotically stable.

Proof: We already established in Theorem 1 the positive definiteness of the functional (4). Bounding the other way:

$$\begin{aligned}V(\psi) &\leq \|\psi\|^2\,[\lambda_{\max}(P) + \int_0^T \lambda_{\max}(Q(\theta))\,d\theta\,] \\ &\leq C\|\psi\|^2\end{aligned}$$

shows that under the given conditions V has infinitessimal upper bound. Finally, by Lemma 1,

$$\begin{aligned}-\dot{V}(x_t) &\geq x_t(0)'Rx_t(0) \\ &\geq |x_t(0)|^2[\lambda_{\min}(R)].\end{aligned}$$

Thus, under the given conditions the trivial solution is asymptotically stable by fact 2. □

Condition (9) is implied by the spectral norm ccondition $\int_0^T \|C(t)\|^2\,dt < \infty$. Some simple (but more conservative) sufficient conditions for stability may be obtained by suitable selection of some of the matrices in the Riccati condition.

Corollary 1: *The delay system (1) is robustly stable, if there exists a positive scalar function* $\beta(t)$ *such that* (A_S denotes the symmetric part of A):

$$2A_S(t) < -\left[\int_0^T \beta(t)\,dt\,I + \int_0^T \frac{B(t)B'(t)}{\beta(t)}\,dt\right]. \tag{11}$$

Proof: Letting $C(t)'C(t) = \beta(t)I$ and $P = I$, the condition (11) implies the existence of a triple of positive definite matrices P, Q and R for which (8) and the conditions of Theorem 1 are satisfied. Theorem 1 then implies stability. □

Note that for $\beta(t)^* = \int_0^T \|B(t)\|\,dt$, the quantity $\int_0^T \beta(t)\,dt\,I + \int_0^T \frac{B(t)B(t)'}{\beta(t)}\,dt$ has a *minimal* upper

bound $2\int_0^T \|B(t)\|\,dt\,I$.
Alternatively, if the matrix $B(t)$ is nonsingular for all $t \in [0,T]$, then one can chose

$$\mathcal{C}(t)'\mathcal{C}(t) = \frac{B(t)'PB(t)}{\zeta(t)}, \tag{12}$$

for some positive function ζ. The Riccati equation is then *linear* in P, i.e., it becomes a (generalized) Lyapunov equation

$$A'P + PA + R + \int_0^T \zeta(t)\,dt\,P + \\ + \int_0^T \frac{B(t)'PB(t)}{\zeta(t)}\,dt = 0. \tag{13}$$

This *algebraic* Riccati equation can be iteratively solved,
with positive definite solution, if $A + \frac{1}{2}\int_0^T \zeta(t)\,dt\,I$ is stable and the mapping $P \rightarrow \mathcal{R}(P;\zeta) = \int_0^\infty e^{A't}[\int_0^T \frac{B(\tau)'PB(\tau)}{\zeta(\tau)}\,d\tau]e^{At}e^{\int_0^T \zeta(\tau)\,d\tau}\,dt$ is contractive (Wonham 1968). Note that the larger $\int_0^T \zeta(t)\,dt$, the more restrictive the condition on A, and the more relaxed the one on B becomes. We conclude thus by stating the following result:

Corollary 2: *The distributed delay system $\dot{x}(t) = Ax(t) + \int_0^T B(\tau)x(t-\tau)\,d\tau$ is robustly stable if there exists a positive scalar function $\zeta(\cdot)$, such that* $\mathcal{R}(P;\zeta)$ defined above is contractive, and $\mathrm{Re}\,\lambda(A) < -\int_0^T \zeta(t)\,dt$.

Example 1: Consider the scalar system

$$\dot{x}(t) = ax(t) + \int_0^T b(\tau)x(t-\tau)\,d\tau,$$

The associated Riccati equation (8) is

$$2ap + \int_0^T \beta(t)\,dt + p^2 \int_0^T \frac{b(t)^2}{\beta(t)}\,dt + r = 0,$$

Letting $\pi = \frac{p_0}{\int_0^T \beta(t)\,dt}$, and $\rho = \frac{r}{\int_0^T \beta(T)\,dt}$, the normalized Riccati equation is

$$2a\pi + 1 + \rho + \pi^2 \int_0^T \frac{b(t)^2}{\beta(t)}\,dt \int_0^T \beta(t)\,dt = 0,$$

which always has a positive definite solution for π if $-a > \sqrt{\int_0^T \frac{b(t)^2}{\beta(t)}\,dt \int_0^T \beta(t)\,dt}$. The right hand side is minimal $\beta(t) = |b(t)|$, hence the scalar distributed system is robustly asymptotically stable for all weights $b(t)$ satisfying $-a > \int_0^\infty |b(t)|\,dt$.

Example 2: Consider the n-th order system

$$\dot{x}(t) = Ax(t) + B\int_0^T \mu(\tau)x(t-\tau)\,d\tau,$$

where A and B are constant matrices, and $\mu(\cdot)$ is a *scalar* weigth-function. By Theorem 2, this system is asymptotically stable if one can find a matrix P and and a positive scalar function $\beta(\cdot)$, such that

$$0 > A'P + PA + I + \\ + PBB'P \int_0^T \frac{\mu(t)^2}{\beta(t)}\,dt \int_0^T \beta(t)\,dt$$

The integral is minimal for $\beta(\cdot) = |\mu(\cdot)|$. Its minimal value is $(\int_0^T |\mu(t)|\,dt)^2$.

4. Robust Stability and Frequency Domain Criteria

The basic conditions obtained in the previous section involve an algebraic Riccati equation of the form $A'P + PA + Q_1 + R + PQ_2P = 0$. The difference with the Riccati equations in the LQG theory is that the quadratic term appears with the plus sign. Such Riccati equations are known to appear in robust control theory. In fact, the specific *time dependence* of the weight matrix $B(\cdot)$ and the exact support $[0,T]$ of the distributed delay are immaterial in the sufficient conditions that were derived. This makes room for *robust* stability conditions. All theorems and corollaries stated in section 2 can thus be rephrased as robust (asymptotic) stability theorems. In particular we state:

Theorem 3: *The distributed delay system (1) is $\mathcal{B}$-robustly asymptotically stable, if there exists a positive definite matrix P such that*

$$A'P + PA + I + P\mathcal{B}P < 0. \tag{14}$$

Proof: Follows directly from Theorem 2 with $\mathcal{C}(\tau) = c(\tau)I$, where $c(t)$ is a nonzero scalar function such that

$$\int_0^T c(t)^2\,dt = 1. \tag{15}$$

Indeed, then $c(t)^2 = \beta(t)$ in definition 2. □

Corollary 3: *The distributed delay system (1) is $\mathcal{B}$-robustly asymptotically stable if A is stable and*

$$\sup_\omega \|[j\omega I - A]^{-1}\mathcal{B}^{1/2}\| < 1. \tag{16}$$

Proof: This is a direct consequence of the strict bounded real lemma (Hewer, 1993; Willems, 1971), and provides a variant to the frequency domain robust stability criterion for delay systems, developed in Verriest *et al.* (1993) □

Again, at the expense of obtaining more conservative results, these conditions can be simplified. The larger the positive definite matrix $\mathcal{B}$, the smaller the chance of finding a positive definite solution P to the Riccati equation. Hence it is important to derive a (tight) upper bound for the quantity

$$\min_{\beta(\cdot)>0} \int_0^T \frac{B(t)B(t)'}{\beta(t)}\,dt \int_0^T \beta(t)\,dt. \tag{17}$$

Note that it suffices to find the minimum over the set $\{\beta(\cdot) \mid \int_0^T \beta(t)\,dt = 1\}$ since (17) is scale invariant. Obviously we have for all $x \in \mathbb{R}^n$ with $\|x\| = 1$:

$$\begin{aligned} x' \int_0^T \frac{B(t)B(t)'}{\beta(t)}\,dt \int_0^T \beta(t)\,dt\, x \\ \leq \int_0^T \frac{\|B(t)\|^2}{\beta(t)}\,dt \int_0^T \beta(t)\,dt \\ \leq \left(\int_0^T \|B(t)\|\,dt\right)^2 \end{aligned} \tag{18}$$

The last step follows from a simple variational argument, the minimizer being $\beta^*(t) = \sqrt{\|B(t)\|}$. We have then the following:

Corollary 4 : *The distributed delay system (1) is asymptotically stable for all $B(\cdot)$, such that*

$$\int_0^T \|B(t)\|\,dt \leq b_0 \tag{19}$$

if there exists a positive definite matrix P such that either

$$A'P + PA + I + b_0^2 P^2 < 0. \tag{20}$$

or

$$\sup_{\omega \geq 0} \|(j\omega I - A)^{-1}\| < 1/b_0. \tag{21}$$

Proof: By virtue of Corollary 3, both conditions are equivalent. The proof follows from the upper bound (18). □

Example 2 (continued): By Corollary 4, the equivalent condition for asymptotic stability is:

$$\sup_{\omega} \|(j\omega I - A)^{-1}B\| < \frac{1}{\int_0^T |\mu(t)|\,dt}. \tag{22}$$

Consequently, for fixed matrices A and B, asymptotic stabilty independent of the precise delay distribution holds, as long as

$$\int_0^\infty |\mu(t)|\,dt < \frac{1}{\sup_\omega \|(j\omega I - A)^{-1}B\|}. \tag{23}$$

If n=2, and $A = B$, then as derived in Verriest *et al.* (1993), robust asymptotic stability holds if
- in case Spec(A) $\subset \mathbb{R}$:

$$\int_0^\infty |\mu(t)|\,dt < 1$$

- in case Spec(A) $= \{\sigma \pm j\omega\}$, $\omega \neq 0$:

$$\int_0^\infty |\mu(t)|\,dt < \frac{|\sigma|}{\sqrt{\sigma^2 + \omega^2}}.$$

5. Stabilization with Delayed Feedback

In this section the stabilizability by feedback of the state is analyzed, in case the state is perturbed by delays (e.g., due to multipath). To fix the ideas, assume that a finite dimensional, single input *unstable* system

$$\dot{x} = Ax + bu \tag{24}$$

is controlled by feeding back the *available state information* $\bar{x}(t)$ over a gain k. Let this available state information be a *delay perturbed* version of the instantaneous state, i.e.,

$$\bar{x}(t) = \lambda x(t) + \int_0^T \mu(\tau)x(t-\tau)\,d\tau, \tag{25}$$

where $\mu \in L_2[0,T]$ has "total weight"

$$\int_0^T |\mu(\tau)|\,d\tau = 1 - \lambda, \text{ for } \lambda \in [0,1]. \tag{26}$$

The closed loop dynamics are given by

$$\dot{x} = (A - bk\lambda)x - bk \int_0^T \mu(\tau)x(t-\tau)\,d\tau, \tag{27}$$

which is of the form (1) with $B = -bk$. Applying Corollary 3, one finds that the delay perturbed system is asymptotically stable if the matrix $A - \lambda bk$ is Hurwitz i.e., stable, and, using (22) and (26),

$$\sup_{\omega} \|(j\omega I - A + \lambda kb)^{-1}bk\| < \frac{1}{1-\lambda}. \tag{28}$$

Noting that, using Woodbury's lemma:

$$\begin{aligned} &\|(sI - A + \lambda bk)^{-1}bk\| \\ &= \|(k(sI - A + \lambda bk)^{-1}b)\| \\ &= \left|\frac{k(s)}{1 + \lambda k(s)}\right|, \end{aligned} \tag{29}$$

where $k(s) := k(sI - A)^{-1}b$, the stability condition can be given a geometric interpretation. Recall that the loci in the $k(s)$-complex plane where the closed loop gain $|\frac{k(s)}{1+k(s)}|$ is constant are the so-called M-circles: They constitute a circle bundle with centers at 0 (closed loop gain 0) and at -1 (closed loop gain infinite). Let $\mathcal{M}_\lambda$ be the domain containing the origin, bounded by the M-circle for $M = \frac{\lambda}{1-\lambda}$. The above robust stabilizability condition may be restated as $\left|\frac{\lambda k(j\omega)}{1+\lambda k(j\omega)}\right| < \frac{\lambda}{1-\lambda}$, or equivalently, $k(j\omega) \in \mathcal{L}_\lambda \overset{\text{def}}{=} \frac{1}{\lambda}\mathcal{M}_\lambda$ for all ω. The right hand side, being a conformal transformation of the M-circles, constitutes another circle bundle of 'L-circles' (L for lambda). The boundary of $\mathcal{L}_\lambda$ is a circle with center $(\frac{\lambda}{1-2\lambda}, 0)$ and radius $R_\lambda = \frac{1-\lambda}{|1-2\lambda|}$. This bundle is tangent to the vertical through -1. We conclude:

Theorem 4: *If the system $\dot{x} = Ax + bu$ is stabilized by a delay free state feedback over gain k, and the Nyquist plot of its associated system (the total open loop gain), $k(s) = k(sI-A)^{-1}b$, lies entirely in the domain $\mathcal{L}_{\lambda_0}$, then the system is stabilized for* all *delay perturbations of the state knowledge (25) with $\lambda \geq \lambda_0$.*

The figure shows the L-circles and the $k(s)$-polar plots for the double integrator $\dot{y} = v, \dot{v} = u$, with nominal feedback $k_1 v + y$ for $k_1 = 1$, 1.5 and 2. Note that the larger k_1, the lower the bound λ_0, and the more robust with respect to delay-perturbations. In this example, $\lambda = \frac{1}{2}$ is a limit to the guaranteed performance. As our conditions are only sufficient, the actual robustness region may still be greater.

6. Conclusions

We have studied the stability properties of a linear system with distributed delay. Our main results are sufficient condition for stability and asymptotic stability in terms of solvability of certain Riccati equations. Via the positive real lemma, frequency domain criteria were derived as well. The conditions are such that the precise form of the weight matrix function and the effective extend of the delay are immaterial. These results led then to conditions for *robust* stability of distributed delay systems. Some reflections were also given to the problem of stabilizability of systems with delayed feedback. A graphical test results, which is useful in optimizing the feedback gain with respect to delay uncertainty. Finally, it can be shown that if the weight matrix function $B(\cdot)$ is *impulsive*, the conditions of robust stability for systems with multiple delays are retrieved (Verriest *et al.*, 1993, 1994) .

7. REFERENCES

Florchinger,P., and E.I. Verriest (1993). Stabilization of nonlinear stochastic systems with delay feedback. In: *Proceedings of the 32-nd IEEE Conf. Dec. Control*, San Antonio, TX, 859-860.

Hale, J.K., L.T. Magelhães and W.M. Oliva (1984) *An Introduction to Infinite Dimensional Dynamical Systems - Geometric Theory*, Springer-Verlag, New York.

Hewer, G. (1993). Existence Theorems for Positive Semidefinite and Sign Indefinite Stabilizing Solutions of H_∞ Riccati Equations. *SIAM J. Control and Optimization*, **31**, 1, 16-29.

Kolmanovskii, V. and A. Myshkis (1992). *Applied Theory of Functional Differential Equations*, Mathematics and Its Applications, Kluwer Academic Publishers, Dordrecht.

Verriest, E.I. (1994a). Robust Stability of Time-Varying Systems with Unknown Bounded Delays. In: *Proceedings of the 34th IEEE Conference on Decision and Control*, Orlando, FL, 417-422.

Verriest, E.I. (1994b). Stabilization of Deterministic and Stochastic Systems with Uncertain Time Delays. In: *Proceedings of the 34th IEEE Conference on Decision and Control*, Orlando, FL, 3829-3834.

Verriest, E.I., M.K.H. Fan and J. Kullstam (1993). Frequency domain robust stability criteria for linear delay systems. In: *Proceedings of the 32nd Conference on Decision and Control*, San Antonio, TX, 3473-3478.

Verriest, E.I. and P. Florchinger (1995). Stability of Stochastic Systems with Uncertain Time Delays. *Systems & Control Letters*, **23**, **6**, 41-47.

Verriest, E.I., and A. F. Ivanov (1991). Robust stabilization of systems with delayed feedback. In: *Proceedings of the 2-nd International Symposium on Implicit and Robust Systems*, Warsau, Poland, 190-193.

Verriest, E.I., and A. F. Ivanov (1994). Robust stability of systems with delayed feedback. *Circuits, Systems and Signal Processing*, **13**(2)/**13**(3), 1994 .

Willems, J.C. (1971). Least Squares Stationary Optimal Control and the Algebraic Riccati Equation. *IEEE Trans. Automatic Control*, **16**, 621-634.

Wonham, W.M., (1968). On a matrix Riccati equation of stochastic control. *SIAM J. Control*, **4**, 4, 681.

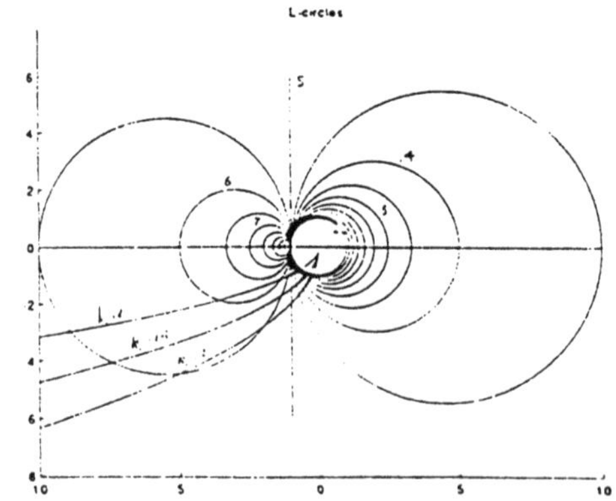

IMPLICIT CONTROL FOR LINEAR TIME VARYING SISO SYSTEMS (WITHOUT FINITE ZEROS).

Moisés Bonilla Estrada*, René Galindo Orozco *, Sergio Salazar Cruz**

* *CINVESTAV-IPN. Sección de ControlAutomático. A.P. 14-740 México 07000. MEXICO Tel.:747-7089, FAX:747-7089,747-7002. e-mail:"MBONILLA@MVAX1.RED.CINVESTAV.MX"*

** *CINVESTAV-IPN. Spounsored by CONACyT-MEXICO*

Abstract. Résumé. Bonilla et Malabra (1991b) ont montré qu'il est possible de modeler et commander les systèmes à structure variable (SSV) en utilisant les systèmes implicites. On a concideré deux types de systèmes, a savoir: (*i*) les systèmes dont sa structure interene change entre deux modèles differents, et (*ii*) les systèmes lineaires avec parametres variables dans le temps sans zeros finits. Bonilla et al (1991a) ont donné des conditions géométriques pour commander les SSV avec retour d'état proportionel–dérivé. Quand aux systèmes varaible dans le temps, Bonilla et al (1993) ont montré une façon de faire la mise au point dans le cas d'un système de premier ordre. Dans ce papier nous etudions le cas d'un système lineaire d'ordre n avec parametres variable dans le temps. On modéle d'abord le système par une description implicite rectangulaire (invariante dans le temps), on commande ensuite le système avec une "loi de commande idéale", et on montre finalement comment réalizer cette loi au moyen de séquences de lois de commande pratiques (propes) en garantissant la stabilitée interne.

Key Words. Implicit systems, Variable time linear systems, Poleplacement, Lyapunov Stability.

1. Introduction

Let us consider the system:[1]

$$y^{(n)}(t) + \sum_{i=1}^{n} a_i(t) y^{(n-i)}(t) = u(t); \quad t \geq 0 \qquad (1)$$

H1 $-\infty < a_{1,i} \leq a_i(t) \leq a_{2,i} < \infty \quad \forall t \geq 0$,
H2 $(a_{1,i}, a_{2,i})$ are known,
H3 $a_i(t)$ are unknown,
H4 $a_i(t)$ are at least one time derivable,
H5 $|a_i(t_2) - a_i(t_1)| \leq L\,|t_2 - t_1|$; $L < \infty, \forall t_1, t_2 > 0$.

($i = 1, \ldots, n$) with the following state description:

$$\dot{\xi}(t) = A(t)\xi(t) + \underline{\chi}_n u(t) \quad ; \quad y(t) = \underline{\chi}_1^T \xi(t) \qquad (2)$$

$$\left[A \mid \underline{\chi}_n \mid \underline{\chi}_1\right] = \qquad (3)$$

$$\left[\begin{array}{ccccc|c|c} 0 & 1 & 0 & \cdots & 0 & 0 & 1 \\ \cdot & \cdot & \cdot & \cdots & \cdot & \cdot & 0 \\ 0 & \cdot & \cdot & \cdot\cdot 0 & 1 & 0 & \cdot \\ -a_n(t) & -a_{n-1}(t) & \cdot & \cdots & -a_1(t) & 1 & 0 \end{array}\right]$$

Normalizing the parameters $a_i(t)$ as follows:

$$a_i(t) = a_{1,i} + \theta_i \alpha_i(t) \quad ; \quad i = 1, \ldots, n \qquad (4)$$

$$\theta_i = \begin{cases} a_{2,i} - a_{1,i} & if \quad a_{2,i} \neq a_{1,i} \\ 1 & if \quad a_{2,i} = a_{1,i} \end{cases} \qquad (5)$$

The state description (2) can be also modeled by the following flat implicit description:

$$\left.\begin{array}{r} [\ I \quad I\ ]\dot{x}(t) = [\ A_1 \quad A_2\ ]x(t) + \underline{\chi}_n u(t) \\ y(t) = [\ \underline{\chi}_1^T \quad \underline{\chi}_1^T\ ]x(t) \end{array}\right\} \qquad (6)$$

$$\begin{array}{l} x^T(t) = [\ \hat{x}^T(t) \quad \tilde{x}^T(t)\ ] \\ \hat{x}(t) = (I - \Lambda(t))\xi(t) \quad ; \quad \tilde{x}(t) = \Lambda(t)\xi(t) \end{array} \qquad (7)$$

$\Lambda(t) =$ block diagonal matrix $\{\alpha_n(t), ..., \alpha_1(t)\}$

$$A_i = \left[\begin{array}{ccccc} 0 & 1 & 0 & \cdots & 0 \\ \cdot & \cdot & \cdot & \cdots & \cdot \\ 0 & \cdot & \cdot & \cdot\cdot 0 & 1 \\ -a_{i,n} & -a_{i,n-1} & \cdot & \cdots & -a_{i,1} \end{array}\right]$$

with $i = 1, 2$. Note that (6) is externally equivalent to (2) when $x(t)$ satisfy the following internal restriction:

$$[\ \Lambda(t) \quad -(I - \Lambda(t))\ ]x(t) = 0 \qquad (8)$$

[1] We use the following notation: $\lambda(X)$ is the eigenvalues of matrix X, $r(X)$ is the spectral radius of X, $\lambda_{\max}(X)$ is the maximum eigenvalue of Hermitian matrix X, $\lambda_{\min}(X)$ is the minimum eigenvalue of Hermitian matrix X, $\|x\|_2$ is the euclidean vector norm, $\|X\|_{i_\infty}$ is the induced $\infty-$matrix norm, $\|X\|_{i_2}$ is the induced 2−matrix norm.

Let us consider the following ideal control law:

$$u^*(t) = [\ (\underline{a}_o^T - \underline{a}_1^T) \quad (\underline{a}_o^T - \underline{a}_2^T)\ ]x(t) + r(t) \quad (9)$$

(10)

$$\left.\begin{array}{l} \underline{a}_i^T = [-a_{i,n} \ldots - a_{i,1}], \quad i = 0,1,2 \\ \lambda^n + \sum_{i=1}^{n} a_{0,i}\lambda^{(n-i)} \text{ is a Hurwitz polynomial} \end{array}\right\}$$

Appliying (9) to (6) we have:

$$\left.\begin{array}{l} [\ I \quad I\ ]\dot{x}(t) = [\ A_o \quad A_o\ ]x(t) + \underline{\chi}_n r(t) \\ y(t) = [\ \underline{\chi}_1^T \quad \underline{\chi}_1^T\ ]x(t) \end{array}\right\} \quad (11)$$

$$A_o = \begin{bmatrix} 0 & 1 & 0 & \cdots & 0 \\ \cdot & \cdot & \cdot & \cdots & \cdot \\ 0 & \cdot & \cdot & \cdot\cdot 0 & 1 \\ -a_{o,n} & -a_{o,n-1} & \cdot & \cdots & -a_{o,1} \end{bmatrix} \quad (12)$$

Namely: $\dot{\xi}(t) = A_o\xi(t) + \underline{\chi}_n r(t)$, $\bar{x}(t) = \Lambda(t)\xi(t)$, and $y(t) = \underline{\chi}_1^T \xi(t)$. Thus $\xi(t)$ is exponentially stable and $\bar{x}(t)$ is stable (A_o is Hurwitz and $0 \le \alpha_i(t) \le 1$).

But (9) is equivalent to (see (7)):

$$u^*(t) = (\underline{a}_o^T - \underline{a}_1^T)\xi(t) - \underline{\theta}^T \bar{x}(t) + r(t) \quad (13)$$

$$\underline{\theta}^T = [-\theta_n \ldots - \theta_1] \quad (14)$$

$\bar{x}(t)$ is not available because of assumption H3 (see (7) and (4)), and we have thus to estimate it.

In next section we will give one practical solution of (13). We will show that the estimated control law tends to the ideal control law (13) when a certain parameter ε tends to zero. In section 3 we will establish the internal stability of the practical control law. And in last sections we conclude.

2. Practical Control Law

Noting that (see (3), (4), (7) and (14)): $A(t) = A_1 + \underline{\chi}_n \underline{\theta}^T \Lambda(t)$, so (2) can be rewritten as:

$$\left.\begin{array}{l} \dot{\xi}(t) = A_1\xi(t) + \underline{\chi}_n(\underline{\theta}^T \bar{x}(t) + u(t)) \\ y(t) = \underline{\chi}_1^T \xi(t) \end{array}\right\} \quad (2.1)$$

Let us propose the following practical control law:

$$\bar{u}(t) = (\underline{a}_o^T + a_{1,1}\underline{\chi}_n^T)\xi(t) + \theta_1\varphi(t) + r(t) \quad (2.2)$$

where $\varphi(t)$ is the input of the auxiliary system:

$$\dot{\zeta}(t) + a_{1,1}\zeta(t) = u(t) - \theta_1\varphi(t) \quad (2.3)$$

where $\zeta(t)$ is the output, $\varphi(t)$ the input and $u(t)$ is a measurable disturbance (in fact is the input of (2.1)). If $\zeta(t) \equiv \underline{\chi}_n^T \xi(t)$ we would have from (2.3) and (2.1):

$$\theta_1\varphi(t) + \underline{\theta}^T \bar{x}(t) = -(\underline{a}_1^T + a_{1,1}\underline{\chi}_n^T)\xi(t) \quad (2.4)$$

(just substitute $\zeta(t)$ by $\underline{\chi}_n^T \xi(t)$ in (2.3), apply $\underline{\chi}_n^T$ to (2.1a) and take into account (7e), (10) and (3)). Which implies when substituting it in (2.2):

$$\bar{u}(t) = (\underline{a}_o^T - \underline{a}_1^T)\xi(t) - \underline{\theta}^T \bar{x}(t) + r(t) \equiv u^*(t) \quad (2.5)$$

In order to achieve $\zeta(t) = \underline{\chi}_n^T \xi(t)$, let us define the following filter (where $p := d/dt$ and β_o, $\varepsilon > 0$):

$$\left.\begin{array}{rl} (p+1)\varphi(t) = & -\frac{\beta_o}{\theta_1 \varepsilon}\left(\frac{\varepsilon}{\beta_o}p + 1\right)\psi(t) \\ \varepsilon\psi(t) = & e(t) \\ e(t) = & \underline{\chi}_n^T \xi(t) - \zeta(t) \end{array}\right\} \quad (2.6)$$

we have from (2.1)–(2.3), (2.6) and (13):

$$\left.\begin{array}{l} \varepsilon(p+1)\varphi(t) = -\frac{\beta_o}{\theta_1 \varepsilon}\left(\frac{\varepsilon}{\beta_o}p + 1\right)e(t) \\ \dot{e}(t) + a_{1,1}e(t) = \bar{u}(t) - u^*(t) \end{array}\right\} \quad (2.7)$$

We then get the following lemma:

Lemma 1 If $h(t) := \dot{\varphi}(t) + \varphi(t) \in L_\infty$ we have:

$$\left.\begin{array}{l} \lim_{\varepsilon \to 0} \sup_{t > t_o > 0} |e(t)| = 0 \\ \lim_{\varepsilon \to 0} \sup_{t > t_o > 0} |\bar{u}(t) - u^*(t)| = 0 \end{array}\right\} \quad (2.8)$$

□

Proof. Since $h(t) \in L_\infty$ there then exists $0 < K < \infty$ such that $|h(t)| \le K \ \forall t \ge 0$. From (2.7a) we have:

$$\left.\begin{array}{l} |e(t)| \leq |e(0)|\, e^{-\frac{\beta_o}{\varepsilon}t} + \varepsilon^2 \left(\frac{\theta_1 K}{\beta_o}\right) \\ |\dot{e}(t)| \leq |e(0)|\, \frac{\beta_o}{\varepsilon} e^{-\frac{\beta_o}{\varepsilon}t} + \varepsilon \left(2\theta_1 K\right) \end{array}\right\} \quad (2.9)$$

Therefore from (2.9) and (2.7b) we have:

$$\begin{array}{l} \lim_{\varepsilon\to 0} \sup_{t>t_o>0} |e(t)| = \\ \qquad = \lim_{\varepsilon\to 0} \left(|e(0)|\, e^{-\frac{\beta_o}{\varepsilon}t_o} + \varepsilon^2 \left(\frac{\theta_1 K}{\beta_o}\right)\right) = 0 \\ \lim_{\varepsilon\to 0} \sup_{t>t_o>0} |\bar{u}(t) - u^*(t)| = \\ = \lim_{\varepsilon\to 0} \left(|e(0)| \left(\frac{1}{t_o}\left(\frac{\beta_o t_o}{\varepsilon}\right) e^{-\frac{\beta_o}{\varepsilon}t_o} + |a_{1,1}|\, e^{-\frac{\beta_o}{\varepsilon}t_o}\right) \right. \\ \qquad \left. + \varepsilon\theta_1 K \left(2 + \varepsilon\frac{|a_{1,1}|}{\beta_o}\right)\right) = 0 \end{array}$$

■

We have proved in this way that the practical control law tends to the ideal one when $h(t) \in L_\infty$. In next section we prove the internal stability of the closed loop system and thus $h(t) \in L_\infty$.

3. Internal Stability

In this section we are going to establish the internal stability of system (2) when it is controlled by (2.2), (2.3) and (2.6). We will show also that for a sufficiently small ε we will be very near from the ideal situation. For this we will meet the behaviours of $\xi(t)$, $e(t)$ and $\varphi(t)$ in a state description form and then we will conclude stability using a Lyapunov function.

3.1. *Closed Loop Internal Behaviour*

When we apply (2.2) to (2.1a) we have (see also (7c) and note that: $A_1 + \underline{\chi}_n \underline{a}_o^T = A_o + \underline{\chi}_n \underline{a}_1^T$):

$$\dot{\xi}(t) = \left(A_o + \underline{\chi}_n \underline{f}^T(t)\right)\xi(t) + \underline{\chi}_n \theta_1 \varphi(t) + \underline{\chi}_n r(t) \quad (3.1)$$

$$\underline{f}^T(t) = \underline{a}_1^T + a_{1,1}\underline{\chi}_n^T + \underline{\theta}^T \Lambda(t) \quad (3.2)$$

Next, from (2.7b), (13), (2.2), (7c), and (3.2) we have:

$$\dot{e}(t) = -a_{1,1}e(t) + \underline{f}^T(t)\xi(t) + \theta_1\varphi(t) \quad (3.3)$$

Finally, from (2.7a) and (3.3) we have:

$$\begin{array}{l} (\bar{\varepsilon}p + 1)\,\varphi(t) = -\frac{(1-\bar{\varepsilon})}{\theta_1}\underline{f}^T(t)\xi(t) \\ \quad -\frac{(1-\bar{\varepsilon})}{\theta_1}\left(\frac{(1-\bar{\varepsilon})}{\bar{\varepsilon}}\beta_o - a_{1,1}\right) e(t) \end{array} \quad (3.4)$$

where: $\bar{\varepsilon} = \varepsilon/(1+\varepsilon)$. And thus the internal behaviour of the closed loop system is given by (3.1)–(3.4).

3.2. *Stability Analysis.*

In order to study the internal stability of (3.1)–(3.4), let us carry then into the following state description (with $r(t) = \dot{r}(t) = 0$):

$$\dot{z}(t) = \frac{1}{\bar{\varepsilon}}\left(G_o + \bar{\varepsilon}G(t)\right) z(t) \quad ; \quad w(t) = H_o z(t) \quad (3.5)$$

$$(3.6)$$

$$\begin{array}{l} z^T(t) = \left[\omega^T(t) \quad \dot{\omega}^T(t)\right] \\ \omega^T(t) = \left[\ \xi^T(t) \quad e(t)\ \right] \ ; \ H_o = \left[\ I \quad 0\ \right] \\ \left[G_o \mid G(t)\right] = \left[\begin{array}{cc|cc} 0 & \bar{\varepsilon}I & 0 & 0 \\ F_o & -I & F_2(t) & F_1(t) \end{array}\right] \\ \left[F_o \mid F_1(t)\right] = \\ \left[\begin{array}{cc|cc} A_o & \underline{\chi}_n\left(\frac{\beta_o}{\bar{\varepsilon}} - a_{1,1}\right) & (A_o + \underline{\chi}_n \underline{f}^T(t) & 0 \\ 0 & -\left(\frac{1-2\bar{\varepsilon}}{\bar{\varepsilon}}\right)\beta_o & -\underline{f}^T(t) & -a_{1,1} \end{array}\right] \\ F_2(t) = \left[\begin{array}{cc} \underline{\chi}_n(\dot{\underline{f}}^T(t) + \underline{f}^T(t)) & \underline{\chi}_n(\beta_o + a_{1,1}) \\ (\dot{\underline{f}}^T(t) + \underline{f}^T(t)) & (\beta_o + a_{1,1}) \end{array}\right] \end{array}$$

The following two lemmas are proved in the appendix:

Lemma 2 The matrix G_o is exponentially stable if $\bar{\varepsilon}$ and β_o are related with a ρ_0 in the following way:

$$\bar{\varepsilon} < 1/2 \quad (3.7)$$

$$\left.\begin{array}{l} 0 < \rho_0 < \frac{1}{2} \quad ; \quad r(I + \bar{\varepsilon}4A_o) < 4\rho_o^2 \\ \frac{1-4\rho_o^2}{4(1-2\bar{\varepsilon})} < \beta_o < \frac{1+4\rho_o^2}{4(1-2\bar{\varepsilon})} \end{array}\right\} \quad (3.8)$$

Furtheremore:

$$-\frac{1}{2}(1 + 2\rho_o) < Re\lambda(G_o) < -\frac{1}{2}(1 - 2\rho_o) \quad (3.9)$$

□

Lemma 3 The state variable $z(t)$ and $\dot{z}(t)$ of system (3.5) are exponentially stable if in adition to conditions (3.7) and (3.8) there holds:

$$\bar{\varepsilon}v^* < \frac{(1 - 2\rho_o)^{(2n^*-1)}}{(2)^{2n^*}n(n^*)^2\,(2n^* - 1)^{2(n^*-1)}} \quad (3.10)$$

where:

$$\begin{array}{l} v^* := \max_{t\geq 0}\left\{\left\|T^{-1}G(t)T\right\|_{i_2}\right\} \\ n^* := \max\{n_1, ..., n_m\} \end{array} \quad (3.11)$$

$n_1, ..., n_m$ are the number of repetitions of the eigenvalues $\lambda_1, ..., \lambda_m$ (with: $\lambda_i \neq \lambda_j$ if $i \neq j$) of the Hurwitz matrix G_o. And T is the transformation matrix which takes G_o to its Jordan normal form J_o. □

From lemma 3 we have the following two corollaries:

Corollary 4 If conditions (3.7), (3.8) and (3.10) are satisfied $\xi(t)$, $e(t)$, $\dot{\xi}(t)$, $\dot{e}(t)$, $\ddot{\xi}(t)$, $\ddot{e}(t)$, $\zeta(t)$, and $\dot{\zeta}(t)$ are exponentially stable. Moreover $\bar{x}(t)$, $u(t)$ and $\varphi(t)$ are bounded. □

Corollary 5 If conditions (3.7), (3.8) and (3.10) are satisfied and if $r(t)$ is at least one time derivable and $r(t)$, $\dot{r}(t) \in L_\infty$ we have:

$$\left.\begin{array}{r} \lim_{\bar{\varepsilon}\to 0} \sup_{t>t_o>0} |e(t)| = 0 \\ \lim_{\bar{\varepsilon}\to 0} \sup_{t>t_o>0} |\dot{e}(t)| = 0 \\ \lim_{\bar{\varepsilon}\to 0} \sup_{t>t_o>0} |\bar{u}(t) - u^*(t)| = 0 \\ \lim_{\bar{\varepsilon}\to 0} \sup_{t>t_o>0} \|\xi(t) - \xi^*(t)\| = \\ = \left\|e^{A_o t_o}\right\| \|\xi(0) - \xi^*(0)\| \end{array}\right\} \quad (3.12)$$

where $\xi^*(t)$ is the solution of:

$$\dot{\xi}^*(t) = A_o\xi^*(t) + \underline{\chi}_n r(t) \quad (3.13)$$

□

Proof of Corollary 4 From Lemma 3 we have that $\xi(t)$, $e(t)$, $\dot{\xi}(t)$, $\dot{e}(t)$, $\ddot{\xi}(t)$, and $\ddot{e}(t)$ are exponentially stable (see: (3.6)). Which imply that $\zeta(t)$ and $\dot{\zeta}(t)$ are exponentially stable (see (2.6.c)). Now, since $\xi(t)$ is exponentially stable and $\Lambda(t)$ is bounded we have that $\bar{x}(t)$ is bounded (see: (7c), (7d), (4) and H1) and then from (2.1a) we have that $u(t)$ is also bounded. Finally from (2.2) we have that $\varphi(t)$ is bounded. ■

Proof of Corollary 5 From (3.5)–(3.6) we have:

$$\dot{e}(t) = -\frac{1-2\bar{\varepsilon}}{\bar{\varepsilon}}\beta_o e(t) + \bar{\varepsilon}g(t) \quad (3.14)$$

where $g(t)$ is a function of $\ddot{e}(t)$, $\dot{e}(t)$, $e(t)$, $\dot{\xi}(t)$, $\xi(t)$, $\underline{f}(t)$ and $\dot{\underline{f}}(t)$. In view of Corollary 1 and assumptions H1, H4 and H5 there exist $0 < K_g < \infty$ such that $|g(t)| < K_g \quad \forall t > t_o > 0$. From (3.14):

$$\begin{array}{l} \lim_{\bar{\varepsilon}\to 0} \sup_{t>t_o>0} |e(t)| \leq \\ \lim_{\bar{\varepsilon}\to 0} \left(|e(0)|\, e^{-\frac{1-2\bar{\varepsilon}}{\bar{\varepsilon}}\beta_o t_o} + \frac{\bar{\varepsilon}^2}{1-2\bar{\varepsilon}}\right) = 0 \end{array} \quad (3.15)$$

$$\lim_{\bar{\varepsilon}\to 0} \sup_{t>t_o>0} |\dot{e}(t)| \leq \quad (3.16)$$

$$\lim_{\bar{\varepsilon}\to 0} \left(\frac{|e(0)|}{t_o}\left(\frac{1-2\bar{\varepsilon}}{\bar{\varepsilon}}\beta_o t_o\right) e^{-\frac{1-2\bar{\varepsilon}}{\bar{\varepsilon}}\beta_o t_o} + \bar{\varepsilon}\,(1 + K_g)\right) = 0$$

From (3.15), (3.16) and (2.7b) we have (3.12c). Finally defining: $\tilde{\xi}(t) = \xi(t) - \xi^*(t)$, we have from (3.1) and (3.13):

$$\dot{\tilde{\xi}}(t) = A_o\tilde{\xi}(t) + \underline{\chi}_n(\bar{u}(t) - u^*(t)) \quad (3.17)$$

(since: $\bar{u}(t) - u^*(t) = \underline{f}^T(t)\xi(t) + \theta_1\varphi(t)$; see: (3.2), (2.2), (13) and (7c)), which implies (3.12d). ■

4. Conclusion

In this paper we have extended the result of Bonilla et al (1993) to the case of a linear time varying parameter SISO system of order n without finite zeros. In the case that there exist finite zeros but Hurwitz, $y^{(n)}(t) + \sum_{i=1}^{n} a_i(t)y^{(n-i)}(t) = \sum_{j=0}^{m} b_j v^{(m-j)}(t)$, we can define the prefilter $\sum_{j=0}^{m} b_j v^{(m-j)}(t) = u(t)$ and we will have (1).

In this paper we have proved stability without neglect small terms as in Bonilla et al (1993), we think that this gives a more solid proof. From (3.1), (3.3) and (3.4) we can arrive to the same goal using singular perturbations theory as in Bonilla et al (1993). Another improvement with respect to Bonilla et al (1993), it is that we have given a more accurate sense to the "convergence of systems" as Bonilla and Malabre (1994) have done.

Future directions on this work are to extend this result to the MIMO case and to take into account variable parameters in the zero polynomial. In fact in this same conference we are considering the case of a variable time input gain.

5. References

Bonilla and Malabre (1991a). Variable Structure Systems via Implicit Descriptions. 1st European Control Conference, Grenoble France.

Bonilla, Lebret and Malabre (1991b). Some Robust Control via Implicit Descriptions. 2nd International Symposium on Implicit and Robust Systems ISIRS'91 Warsaw, Poland.

Bonillla, Malabre and Velázquez (1993). Controlling a Linear First Order Time Varying Parameter System Using a Linear Constant Parameters Control Law. 32-th IEEE-CDC, San Antonio, Texas.

Bonilla and Malabre (1994). Convergence in Behavior to non Proper Systems. 33rd IEEE CDC

Gantmacher (1977). The Theory of Matrices. Vols. I and II, Chelsea Publishing Co., N.Y. U.S.A.

Stewart (1973). Introduction to Matrix Computations. Academic Press.

Vidyasagar (1993). Nonlinear Systems Analysis. Prentice Hall, N.J. U.S.A.

APPENDIX

A. Proof of lemma 2

Let us note that: $\det(\lambda I - G_o) = \bar{\varepsilon}^{n+1}\det(\mu I - F_o)$, where $\lambda^2 + \lambda - \bar{\varepsilon}\mu = 0$, namely $\lambda = -\frac{1}{2} \pm \frac{1}{2}\sqrt{1+\bar{\varepsilon}4\mu}$. We have from (3.6e) and (3.8): $|1+\bar{\varepsilon}4\mu| \leq 4\rho_o^2 < 1$, and then: $\left|\lambda(G_o) + \frac{1}{2}\right| < \rho_o < \frac{1}{2}$ ■

B. Proof of Lemma 3

Before proving Lemma 3, let us first express system (3.5) in a Jordan normal form as follows:

$$\dot{\bar{z}}(t) = \frac{1}{\bar{\varepsilon}}\left(J_o + \bar{\varepsilon}\bar{G}(t)\right)\bar{z}(t) \tag{B.1}$$

$$\left(\bar{z}(t), J_o, \bar{G}(t)\right) = \left(T^{-1}z(t), T^{-1}G_oT, T^{-1}G(t)T\right) \tag{B.2}$$

$$J_o = \text{ block diagonal matrix } \{J_1, ..., J_m\} \tag{B.3}$$

$$J_k = \begin{bmatrix} \lambda_k & 1 & 0 & \cdots & 0 \\ 0 & \lambda_k & 1 & 0\cdots & 0 \\ . & . & . & \cdots & . \\ 0 & . & . & \cdots 0 & \lambda_k \end{bmatrix} \quad ; \quad k = 1, \ldots, m$$

We need the following results:

Result B.1 Under the same conditions as Lemma 2 there exists a positive definite matrix, P, such that: $J_o^T P + PJ_o = -I$, furthermore: $P = \int_o^\infty e^{J_o^* t}e^{J_o t}dt$. □

Result B.2 $-\|X\|_i \leq Re\lambda_i(X) \leq \|X\|_i$. □

Result B.3 Let β be a positive real number, then:

$$\left(\frac{t}{\beta}\right)^k e^{-\frac{t}{\beta}} \leq \left(1 + (2k)^k e^{-k}\right)e^{-\frac{t}{2\beta}} \tag{B.4}$$

□

Result B.4

$$\lambda_{\max}(P) \leq \frac{(n^*)^2}{(2n^*-1)}\left[\frac{2(2n^*-1)}{1-2\rho_o}\right]^{(2n^*-1)} \tag{B.5}$$

□

Result B.5 Let us define:

$$R(t) := \bar{G}^T(t)P + P\bar{G}(t) \tag{B.6}$$

we then have:

$$\left|z^T R(t) z\right| \leq 2n\lambda_{\max}(P)\left\|\bar{G}(t)\right\|_{i_2} z^T z \tag{B.7}$$

□

Prof of Result B1. Directly from Lyapunov Theorem (eg Gantmacher, 1977, and Vidyasagar, 1993). ■

Prof of Result B2. This result follows from Theorem 2.2.16 of Vidyasagar (1993). ■

Prof of Result B3. Let us consider the function: $h_k(t) = K_k e^{-\frac{t}{2\beta}} - \left(\frac{t}{\beta}\right)^k e^{-\frac{t}{\beta}}$, which is positive if and only if the fuction $\bar{h}_k(t) = K_k - \left(\frac{t}{\beta}\right)^k e^{-\frac{t}{2\beta}}$ is also positive. And it is positive if and only if: $K_k > (2k)^k e^{-k}$. And thus for $K_k = 1 + (2k)^k e^{-k}$ we have (B.4). ■

Prof of Result B4. This result is proved in 4 steps:

i) Given that P is a positive definite matrix we then have that its eigenvalues are reals and positives. And then from Result B2 we have:

$$\lambda_{\max}(P) \leq \|P\|_i \tag{B.8}$$

ii) Since the J_k are Jordan's blocks, we have that:

$$e^{J_k t} = e^{\lambda_k t}\Pi(t) \quad ; \quad e^{J_k^* t} = e^{\bar{\lambda}_k t}\Pi^T(t)$$
$$\Pi(t) = \begin{bmatrix} 1 & t & t^2/2! & \cdots & t^{(n_k-1)}/(n_k-1)! \\ 0 & 1 & t & \cdots & t^{(n_k-2)}/(n_k-2)! \\ . & . & . & \cdots & . \\ 0 & . & . & \cdots 0 & 1 \end{bmatrix} \tag{B.9}$$

which implies: $\left\|e^{J_k t}\right\|_{i\infty} = \left\|e^{J_k^* t}\right\|_{i\infty} = \sum_{k=0}^{n_k-1} \frac{1}{k!}t^k e^{Re\lambda_k t}$. We have from (B.3), (3.11b) and (3.9):

$$\left\|e^{J_o t}\right\|_{i\infty} = \left\|e^{J_o^* t}\right\|_{i\infty} \leq \sum_{k=0}^{n^*-1} \frac{1}{k!}t^k e^{-\frac{1}{2}(1-2\rho_o)t} \tag{B.10}$$

iii) From (B.4) we have that: $\frac{1}{k!}t^k e^{-\frac{1}{2}(1-2\rho_o)t} \leq m_k e^{-\frac{1}{4}(1-2\rho_o)t}$, where: $m_k = \left(1 + (2k)^k e^{-k}\right)\frac{2^k}{(1-2\rho_o)^k k!}$, for $k = 0, 1, ..., n^* - 1$. Thus: $\frac{1}{k!}t^k e^{-\frac{1}{2}(1-2\rho_o)t} \leq \left(\frac{2(2n^*-1)}{1-2\rho_o}\right)^{(n^*-1)} e^{-\frac{1}{4}(1-2\rho_o)t}$. And from (B.10) we then have:

$$\left\|e^{J_o t}\right\|_{i\infty} = \left\|e^{J_o^* t}\right\|_{i\infty} \leq n^*\left(\frac{2(2n^*-1)}{1-2\rho_o}\right)^{(n^*-1)} e^{-\frac{1}{4}(1-2\rho_o)t} \tag{B.11}$$

iv) From Result B1, (B.8) and (B.11):

$$\lambda_{\max}(P) \leq \int_0^{\infty} (n^*)^2 \left(\frac{2(2n^*-1)}{1-2\rho_0}\right)^{2(n^*-1)} e^{-\frac{1}{2}(1-2\rho_0)t} dt = \frac{2(n^*)^2}{(1-2\rho_0)} \left(\frac{2(2n^*-1)}{1-2\rho_0}\right)^{2(n^*-1)} \quad \text{(B.12)}$$

and we then have (B.5). ■

Proof of Result B5. This result is proved in 3 steps:

i) Since P is a positive definite matrix, its eigenvectors are linearly independents. Taking as a basis the n normalized eigenvectors of P: $\mathcal{B} := \{v_1, ..., v_n : Pv_i = \lambda_i(P)v_i \ \& \ \|v_i\|_2 = 1\}$, we have:

$$\begin{aligned}
\left|v_i^T R(t) v_i\right| &= \left|\lambda_j(P) v_i^T \bar{G}^T(t) v_j + \lambda_i(P) v_i^T \bar{G}(t) v_j\right| \\
&\leq \lambda_j(P) \left|v_j^T \bar{G}^T(t) v_i\right| + \lambda_i(P) \left|v_i^T \bar{G}^T(t) v_j\right| \\
&\leq 2\lambda_{\max}(P) \max_{\|x\|_2=\|y\|_2=1} \left|x^T \bar{G}(t) y\right| \\
&\leq 2\lambda_{\max}(P) \left\|\bar{G}(t)\right\|_{i_2}
\end{aligned} \quad \text{(B.13)}$$

Last row follows from Theorem 2.10 of Stewart (1973).

ii) Let z be any vector expressed in the basis $\mathcal{B}$, namely: $z = \sum_{i=1}^n \alpha_i v_i$, we then have from (B.13):

$$\begin{aligned}
\left|z^T R(t) z\right| &= \left|\sum_{i=1}^n \alpha_i v_i^T R(t) \sum_{j=1}^n \alpha_j v_j\right| \\
&\leq \left(\sum_{i=1}^n \alpha_i\right) \left(\sum_{j=1}^n \alpha_j\right) 2\lambda_{\max}(P) \left\|\bar{G}(t)\right\|_{i_2} \\
&= \|z\|_1^2 \, 2\lambda_{\max}(P) \left\|\bar{G}(t)\right\|_{i_2} \\
&\leq 2n\lambda_{\max}(P) \left\|\bar{G}(t)\right\|_{i_2} \|z\|_2^2
\end{aligned}$$

(last row follows from example 4.1.10 of Stewart, 1973). ■

We are now in a position to prove lemma 3:

Let us consider the following Lyapunov function:

$$V(t) = \bar{z}^T(t) P \bar{z}(t) > 0 \quad \text{(B.14)}$$

We have from (B.14), (B.1) and Results B1, B4, B5:

$$\begin{aligned}
\dot{V}(t) &= \dot{\bar{z}}^T(t) P \bar{z}(t) + \bar{z}^T(t) P \dot{\bar{z}}(t) \\
&= \tfrac{1}{\bar{\varepsilon}} \bar{z}^T(t) \left(\left(J_0 + \bar{\varepsilon}\bar{G}(t)\right)^T P + P\left(J_0 + \bar{\varepsilon}\bar{G}(t)\right)\right) \bar{z}(t) \\
&= \tfrac{1}{\bar{\varepsilon}} \bar{z}^T(t) \left(-I + \bar{\varepsilon} R(t)\right) \bar{z}(t) \\
&\leq -\tfrac{1}{\bar{\varepsilon}} \left(1 - 2n\lambda_{\max}(P) \left\|\bar{G}(t)\right\|_{i_2} \bar{\varepsilon}\right) \bar{z}^T(t) \bar{z}(t) \\
&\leq -\tfrac{1}{\bar{\varepsilon}} \left(1 - 2n \frac{(n^*)^2}{(2n^*-1)} \left(\frac{2(2n^*-1)}{1-2\rho_0}\right)^{(2n^*-1)} \left\|\bar{G}(t)\right\|_{i_2} \bar{\varepsilon}\right) \bar{z}^T(t) \bar{z}(t)
\end{aligned} \quad \text{(B.15)}$$

Now in view of H1, H4 and H5 we have that $G(t)$ is a matrix with bounded elements for all $t \geq 0$ (see: (3.6),(3.2), (7d) and (4). Then:

$$\nu^* := \max\left\{\left\|\bar{G}(t)\right\|_{i_2} : \forall t \geq 0\right\} < \infty \quad \text{(B.16)}$$

We have from (B.15) and (B.16):

$$\dot{V}(t) \leq -\frac{1}{\bar{\varepsilon}} \lambda^* \bar{z}^T(t) \bar{z}(t) \quad \text{(B.17)}$$

where: $\lambda^* = 1 - 2n \frac{(n^*)^2}{(2n^*-1)} \left(\frac{2(2n^*-1)}{1-2\rho_0}\right)^{(2n^*-1)} \bar{\varepsilon}\nu^*$.
Then, from (3.10): $\lambda^* > 0$, and then:

$$\begin{aligned}
\dot{V}(t) &\leq -\tfrac{1}{\bar{\varepsilon}} \lambda^* z^T(t) z(t) \\
&= -\tfrac{1}{\bar{\varepsilon}} \lambda^* z^T(t) P^{\frac{1}{2}} P^{-1} P^{\frac{1}{2}} z(t) \\
&\leq -\tfrac{1}{\bar{\varepsilon}} \lambda^* \frac{1}{\lambda_{\max}(P)} V(t) < 0
\end{aligned} \quad \text{(B.18)}$$

Finally from (B.18) and (B.14) we have:

$$\begin{aligned}
\|z(t)\|_2^2 &\leq \frac{1}{\lambda_{\min}(P)} V(t) \\
&\leq \frac{1}{\lambda_{\min}(P)} V(0) e^{-\frac{\lambda^*}{\bar{\varepsilon}\lambda_{\max}(P)} t}
\end{aligned} \quad \text{(B.19)}$$

Therefore $z(t)$ is exponcntially stable. Exponencial stability of $\dot{z}(t)$ comes finally from (3.5a) and the fact that $G(t)$ is bounded. ■

C. Illustrative Example

Let us consider the following system:

$$\dddot{y}(t) + a_1(t)\ddot{y}(t) + a_2(t)\dot{y}(t) + a_3(t)y(t) = u(t)$$

where: $a_1(t) = \sum_{j=1}^5 \frac{1}{2j-1} \sin(2j-1)t$, $a_2(t) = \sum_{j=1}^5 \frac{(-1)^{j-1}}{j} \sin j\frac{t}{2}$ and $a_3(t) = \sum_{j=1}^5 \frac{j}{(2j-1)(2j-1)} \sin 4jt$. The implicit control squeme to have as characteristic polynomial $(\lambda+1)^3$ is:

$$\begin{aligned}
&\dot{\varsigma}(t) = -a_{1,1}\varsigma(t) + u(t) - \theta_1\varphi(t) \\
&\dot{\varphi}(t) + \varphi(t) = -\tfrac{\beta_o}{\theta_1 \varepsilon}\left(\tfrac{\varepsilon}{\beta_o}\dot{\psi}(t) + \psi(t)\right) \\
&\psi(t) = \tfrac{1}{\varepsilon}\left(\hat{\xi}_3(t) - \varsigma(t)\right) \\
&u(t) = \left(-y(t) - 3\hat{\xi}_2(t) + (a_{1,1} - 3)\hat{\xi}_3(t)\right) + \theta_1\varphi(t)
\end{aligned}$$

Where $\hat{\xi}(t)$ is obtained from the reconstructor:

$$\begin{bmatrix} 0 & 0 & 0 \\ 1 & 0 & 0 \\ 0 & 1 & 0 \end{bmatrix} \dot{\hat{\xi}} = \begin{bmatrix} 1 & 0 & \varepsilon_o^2 \\ 0 & 1 & 2\varepsilon_o \\ 0 & 0 & 1 \end{bmatrix} \hat{\xi} + \begin{bmatrix} -1 \\ 0 \\ 0 \end{bmatrix} y$$

with: $a_{1,1} = -1$ and $\theta_1 = 2$, $\varepsilon_o = \varepsilon = 0.01$ and $\beta_o = 0.1$. ■

IMPLICIT CONTROL FOR LINEAR SISO SYSTEMS WITH TIME VARYING INPUT GAIN

Galindo R.* , Bonilla M.**

* CINVESTAV-IPN. spounsored by CONACyT-MEXICO

**CINVESTAV-IPN. Secciónde Control Automático. A.P. 14-740 México 07000. MEXICO Tel.:747-7089, FAX:747-7089,747-7002. e-mail:"MBONILLA@MVAX1.RED.CINVESTAV.MX"

Abstract. Résumé. Bonilla et Malabra (1991b) ont montré qu'il est possible de modeler et commander les systèmes à structure variable (SSV) en utilisant les systèmes implicites. On a concideré deux types de systèmes, a savoir: (*i*) les systèmes dont sa structure interene change entre deux modèles differents, et (*ii*) les systèmes lineaires avec parametres variables dans le temps sans zeros finits. Bonilla et al (1991a) ont donné des conditions géométriques pour commander les SSV avec retour d'état proportionel–dérivé. Quand aux systèmes varaible dans le temps, Bonilla et al (1993) ont montré une façon de faire la mise au point dans le cas d'un système de premier ordre. Dans ce papier nous etudions le cas d'un système lineaire d'ordre n avec gain d'entrée variable dans le temps. On modéle d'abord le système par une description implicite rectangulaire (invariante dans le temps), on commande ensuite le système avec une "loi de commande idéale", et on montre finalement comment réalizer cette loi au moyen de séquences de lois de commande pratiques (propes) en garantissant la stabilitée interne.

Key Words. Implicit systems, variable time linear systems, singular perturbations, poleplacement.

1. Introduction

Let us consider the system:

$$y^{(n)}(t) + \sum_{i=1}^{n} a_i y^{(n-i)}(t) = b(t)u(t); \quad \forall t \geq 0 \qquad (1)$$

H1 $0 < b_1 \leq b(t) \leq b_2 < \infty \quad \forall t \geq 0$,
H2 b_1, b_2 are bounded, known and different,
H3 $b(t)$ is unknown,
H4 $|b(t_2) - b(t_1)| \leq L\,|t_2 - t_1|$; $L < \infty$, $\forall t_1, t_2 \in \mathcal{R}^+$,
H5 $b(t)$ is at least one time derivable.

with the following state description:

$$\dot{\xi}(t) = A\xi(t) + \underline{\chi}_n b(t)u(t) \quad ; \quad y(t) = \underline{\chi}_1^T \xi(t) \qquad (2)$$

$$[A \mid \chi_n \mid \chi_1] = \qquad (3)$$

$$= \left[\begin{array}{ccccc|c|c} 0 & 1 & 0 & \cdots & 0 & 0 & 1 \\ \cdot & \cdot & \cdot & \cdots & \cdot & \cdot & 0 \\ 0 & \cdot & \cdot & \cdot\cdot 0 & 1 & 0 & \cdot \\ -a_n & -a_{n-1} & \cdot & \cdots & -a_1 & 1 & 0 \end{array}\right]$$

Normalizing $b(t)$ and defining $\widetilde{x}(t)$ as follows:

$$\left.\begin{array}{c} b(t) = \bar{b} + \widetilde{b}\beta(t) \\ \bar{b} = \frac{(b_2+b_1)}{2} \quad ; \quad \widetilde{b} = \frac{(b_2-b_1)}{2} \end{array}\right\} \qquad (4)$$

$$\widetilde{x}(t) = \beta(t)u(t) \qquad (5)$$

system (2) is also well described by:

$$\left.\begin{array}{l} \dot{\xi}(t) = A\xi(t) + \underline{\chi}_n \widetilde{b}\widetilde{x}(t) + \underline{\chi}_n \bar{b}u(t) \\ y(t) = \underline{\chi}_1^T \xi(t) \end{array}\right\} \qquad (6)$$

which is externally equivalent to (2) as long as the auxiliary variable, $\widetilde{x}(t)$, is moving in the surface:

$$S \hat{=} \{\widetilde{x}(t) \mid \widetilde{x}(t) - \beta(t)u(t) = 0\} \qquad (7)$$

Note that (see: H1 and H2):

$$-1 \leq \beta(t) \leq 1, \quad \bar{b} \neq 0 \text{ and } \widetilde{b} \neq 0 \qquad (8)$$

The state description (6) can be also modeled by the following flat implicit description:

$$\left.\begin{array}{r} \left[\begin{array}{cc} I & 0 \end{array}\right] \dot{x}(t) = \left[\begin{array}{cc} A & \underline{\chi}_n \widetilde{b} \end{array}\right] x(t) + \underline{\chi}_n \bar{b}u(t) \\ y(t) = \left[\begin{array}{cc} \underline{\chi}_1^T & 0 \end{array}\right] x(t) \end{array}\right\} \qquad (9)$$

where: $x^T(t) = \left[\begin{array}{cc} \xi^T(t) & \widetilde{x}(t) \end{array}\right]$, and restricted to:

$$0 = \begin{bmatrix} 0 & 1 \end{bmatrix} x(t) + [-\beta(t)]\, u(t) \tag{10}$$

Let us consider the following ideal control law:

$$u^*(t) = \frac{1}{\bar{b}}\left(\begin{bmatrix} \left(\underline{a}_o^T - \underline{a}^T\right) & -\widetilde{b} \end{bmatrix} x(t) + r(t) \right) \tag{11}$$

$$\left.\begin{array}{l} \underline{a}^T \hat{=} \begin{bmatrix} -a_n & -a_{n-1} & \cdots & -a_1 \end{bmatrix} \\ \underline{a}_o^T \hat{=} \begin{bmatrix} -a_{o,n} & -a_{o,n-1} & \cdots & -a_{o,1} \end{bmatrix} \\ \lambda^n + \sum_{i=1}^{n} a_{0,i}\lambda^{(n-i)} \text{ is a Hurwitz polynomial} \end{array}\right\} \tag{12}$$

Applying (11) to (9) we have:

$$\left.\begin{array}{r} \begin{bmatrix} I & 0 \end{bmatrix} \dot{x}(t) = \begin{bmatrix} A_o & 0 \end{bmatrix} x(t) + \underline{\chi}_n r(t) \\ y(t) = \begin{bmatrix} \underline{\chi}_1^T & 0 \end{bmatrix} x(t) \end{array}\right\} \tag{13}$$

$$A_o = \begin{bmatrix} 0 & 1 & 0 & \cdots & 0 \\ \cdot & \cdot & \cdot & \cdots & \cdot \\ 0 & \cdot & \cdot & \cdots 0 & 1 \\ -a_{0,n} & -a_{0,n-1} & \cdot & \cdots & -a_{0,1} \end{bmatrix} \tag{14}$$

and restricted to (see: (10), (11), H1 and (4)):

$$0 = \begin{bmatrix} -\frac{\beta(t)}{b(t)}\left(\underline{a}_o^T - \underline{a}^T\right) & 1 \end{bmatrix} x(t) - \frac{\beta(t)}{b(t)} r(t) \tag{15}$$

We have obtained in this way the input-output behavior of a linear time invariant system of order n, where the state $\widetilde{x}(t)$, which we have done inobservable, absorbs the structure variation.

From (13) we can see that $\xi(t)$ is exponentially stable. And from (15), $\widetilde{x}(t)$ is bounded, since H1, H2 and (8) (assuming a bounded reference $r(t)$, of course).

But we have to estimate $\widetilde{x}(t)$ since it is unkown (H3). This could be done with the non proper filter, (see: (6)):

$$\widetilde{x}(t) = \frac{1}{\widetilde{b}}\left(y^{(n)}(t) + \sum_{i=1}^{n} a_i y^{(n-i)}(t) - \bar{b}u(t) \right) \tag{16}$$

In this paper we show how to implement the overall nonproper control strategy (16), (11) with a sequence of proper control laws (parametrized by an $\varepsilon > 0$) tending to the ideal one and guaranteeing internal stability. In section 2 we propose a practical proper control law. In section 3 we realize the internal stability analysis. And in last section we give an illustrative simulation.

2. Practical Control Law

Let us propose the following practical control law:

$$\bar{u}(t) = \frac{1}{\bar{b}}\left(\left(\underline{a}_o^T + a_1\underline{\chi}_n^T\right)\xi(t) - \widetilde{b}\varphi(t) + r(t) \right) \tag{2.1}$$

where $\varphi(t)$ is the input of the auxiliary system:

$$\dot{\varsigma}(t) + a_1\varsigma(t) = \widetilde{b}\varphi(t) + \bar{b}u(t) \tag{2.2}$$

where $\varsigma(t)$ is the output, $\varphi(t)$ is the input, and $u(t)$ is a measurable disturbance (in fact is the input of (6)). If $\varsigma(t) \equiv \underline{\chi}_n^T\xi(t)$ we would have from (2.2) and (6):

$$\widetilde{b}\varphi(t) = \left(\underline{a}^T + a_1\underline{\chi}_n^T\right)\xi(t) + \widetilde{b}\widetilde{x}(t) \tag{2.3}$$

(just do $\varsigma(t) = \underline{\chi}_n^T\xi(t)$ in (2.2), apply $\underline{\chi}_n^T$ to (6) and remember (3)). From (2.3) and (2.1) we have:

$$\begin{aligned} \bar{u}(t) &= \tfrac{1}{\bar{b}}\left(\left(\underline{a}_o^T - \underline{a}^T\right)\xi(t) - \widetilde{b}\widetilde{x}(t) + r(t) \right) \\ &\equiv u^*(t) \end{aligned} \tag{2.4}$$

In order to achieve $\varsigma(t) = \underline{\chi}_n^T\xi(t)$, let us define the following filter (where $p := d/dt$ and $\beta, \varepsilon > 0$):

$$\left.\begin{array}{rl} (p+1t)\varphi(t) = & \frac{\beta}{\widetilde{b}}\left(\frac{1}{\beta}p + 1\right)\psi(t) \\ \varepsilon\psi(t) = & e(y) \\ e(t) = & \underline{\chi}_n^T\xi(t) - \varsigma(t) \end{array}\right\} \tag{2.5}$$

we have from (6), (2.2), (2.5), (2.1) and (11):

$$\left.\begin{array}{l} \varepsilon(p+1)\varphi(t) = \frac{\beta}{\widetilde{b}}\left(\frac{1}{\beta}p + 1\right)e(t) \\ \dot{e}(t) + a_1 e(t) = \widetilde{b}\left(\bar{u}(t) - u^*(t)\right) \end{array}\right\} \tag{2.6}$$

We then get the following lemma:

Lemma 1 If $h(t) \hat{=} \dot{\varphi}(t) + \varphi(t) \in L_\infty$ we have:

$$\lim_{\varepsilon\to 0} |e(t)| \le |e(0)|\, e^{-\beta t} \tag{2.7}$$

$\lim_{\varepsilon\to 0} |\bar{u}(t) - u^*(t)| \le \frac{1}{\bar{b}}(\beta + |a_1|)\,|e(0)|\, e^{-\beta t}$ □

Proof. Since $h(t) \in L_\infty$ there then exists $0 < K < \infty$ such that $|h(t)| \le K \ \forall t \ge 0$. From (2.6a) we have:

$$\left.\begin{array}{l} |e(t)| \le |e(0)|\, e^{-\beta t} + \varepsilon \left(\frac{\widetilde{b}K}{\beta}\right) \\ |\dot{e}(t)| \le |e(0)|\, \beta e^{-\beta t} + \varepsilon \left(2\widetilde{b}K\right) \end{array}\right\} \tag{2.8}$$

Therefore from (2.8) and (2.6b) we have (2.7) ∎

We have proved in this way that the practical control law tends to the ideal one when $h(t) \in L_\infty$. In next section we prove the internal stability of the closed loop system and thus $h(t) \in L_\infty$.

3. Internal Stability

We are going to study now internal stability of system (2) when it is controlled by (2.5), (2.2) and (2.1). For this, let us consider the singular perturbation model (Kokotović, 1986) of (9) when we apply it the control law (2.1) (see also: (2.6), (2.1), (2.2), (2.5), and (10)):

$$\begin{array}{l} \begin{bmatrix} I & 0 & 0 \\ 0 & 0 & 0 \\ 0 & 0 & 1 \end{bmatrix} \dot{\bar{x}} = \begin{bmatrix} \left(A_o + \underline{\chi}_n \underline{\alpha}^T\right) & \underline{\chi}_n \widetilde{b} & 0 \\ \beta(t)\underline{\alpha}_o^T & -\widetilde{b} & 0 \\ \underline{\alpha}^T & \widetilde{b} & -a_1 \end{bmatrix} \bar{x} + \begin{bmatrix} -\underline{\chi}_n \widetilde{b} \\ -\widetilde{b}\beta(t) \\ -\widetilde{b} \end{bmatrix} \varphi + \begin{bmatrix} \underline{\chi}_n \\ \beta(t) \\ 0 \end{bmatrix} r \\ \dot{\varphi} = \begin{bmatrix} \frac{1}{\widetilde{b}}\underline{\alpha}^T & 1 & -\frac{1}{\widetilde{b}}\bar{a}_1 \end{bmatrix} \bar{x} - \frac{1}{\lambda}\varphi \end{array} \tag{3.1}$$

$$\left.\begin{array}{c} \underline{\alpha}^T \hat{=} \underline{a}^T + a_1 \underline{\chi}_n^T, \quad \underline{\alpha}_o^T \hat{=} \underline{a}_o^T + a_1 \underline{\chi}_n^T, \\ \lambda \hat{=} 1/(1+\varepsilon), \quad \bar{a}_1 \hat{=} a_1 - \beta \\ \bar{x}^T(t) \hat{=} \begin{bmatrix} \xi^T(t) & \widetilde{x}(t) & e(t) \end{bmatrix} \end{array}\right\} \tag{3.2}$$

3.1. *Stability Analysis*

The stability proof is inspired from Tikhonov's one (see: Kokotović, 1986 and O'Malley, 1991).

In order to analyze the internal stability we make $r(t) = 0$, $\dot{r}(t) = 0$, and we will first decouple the fast variable, $\varphi(t)$, from de slow one, $\bar{x}(t)$, and we will then conclude with the help of a Lyapunov function.

3.1.1. *Fast and Slow Modes.*

From (3.1b) we have:

$$\varphi(t) = e^{-\frac{t}{\varepsilon\lambda}}\varphi(0) + \frac{1}{\varepsilon}\int_0^t e^{-\frac{(t-\tau)}{\varepsilon\lambda}} f(\tau)d\tau \tag{3.3}$$

$$f(t) \hat{=} \begin{bmatrix} \frac{1}{\widetilde{b}}\underline{\alpha}^T & 1 & -\frac{1}{\widetilde{b}}\bar{a}_1 \end{bmatrix} \bar{x}(t) \tag{3.4}$$

Integrating (3.3) by parts we have (under H4 and H5):

$$\varphi(t) = \lambda f(t) + v(t) - \lambda \int_0^t e^{-\frac{(t-\tau)}{\varepsilon\lambda}} \dot{f}(\tau)d\tau \tag{3.5}$$

$$\varepsilon\lambda\dot{v}(t) + v(t) = 0 \quad ; \quad v(0) = \varphi(0) - \lambda f(0) \tag{3.6}$$

Let us note from (3.1) that for $\varepsilon = 0 : e(t) = e(0)e^{-\beta t}$, $\xi(t) = e^{A_o t}\xi(0) + \int_0^t e^{A_o(t-\tau)}\underline{\chi}_n \bar{a}_1 e(\tau)d\tau$ and $\widetilde{x}(t) = \frac{\beta(t)}{b(t)}\left(\bar{a}_1 e(0)e^{-\beta t} + \left(\underline{\alpha}_o^T - \underline{\alpha}^T\right)\right)\xi(t)$ for all $t > 0$ and that $\left|p\left(\frac{\beta(t)}{b(t)}\right)\right| \le \frac{\left(b_2+\widetilde{b}\right)L}{\widetilde{b}b_1^2}$ (see: H4 and H1). We then have that there exists a sufficiently small ε such that: $e^{-\frac{(t-\tau)}{\varepsilon\lambda}}\dot{f}(\tau) = e^{-\frac{(t-\tau)}{\varepsilon\lambda}}\left(\dot{f}(t) + O(\varepsilon)\right), \forall\tau \le t$. And thus, (3.5) becomes:

$$\varphi(t) = (1-\varepsilon) f(t) + v(t) - \varepsilon\rho_1(t)\dot{f}(t) + O(\varepsilon^2) \tag{3.7}$$

$$\rho_1(t) \hat{=} 1 - e^{-\frac{t}{\varepsilon\lambda}} \tag{3.8}$$

(since: $\lambda = 1 - \varepsilon\lambda = 1 - \varepsilon + \varepsilon^2\lambda$ and $\lambda^2 = 1 - 2\varepsilon\lambda + \varepsilon^2\lambda^2$, see: (3.2)). On the other hand, from (3.4) and (3.1a):

$$\begin{array}{rl} \dot{f}(t) & = \frac{1}{\widetilde{b}}\underline{\alpha}^T\dot{\xi}(t) + \dot{\widetilde{x}}(t) - \frac{1}{\widetilde{b}}\bar{a}_1\dot{e}(t) \\ & = \frac{1}{\widetilde{b}}\left(\underline{\alpha}^T A_o - \bar{a}_1\underline{\alpha}^T\right)\xi(t) - \bar{a}_1\widetilde{x}(t) \\ & + \frac{a_1}{\widetilde{b}}\bar{a}_1 e(t) + \bar{a}_1\varphi(t) + \dot{\widetilde{x}}(t) \end{array} \tag{3.9}$$

since: $\underline{\alpha}^T\underline{\chi}_n = \underline{a}^T\underline{\chi}_n + a_1 = 0$, (see: (3.2) and (12)), and from (3.1) and (3.4) we have (see also: (3.2)):

$$\begin{array}{rl} \widetilde{b}\dot{\widetilde{x}} & = \dot{\beta}(t)\left(\underline{\alpha}_o^T\xi - \widetilde{b}\varphi\right) + \beta(t)\left(\underline{\alpha}_o^T\dot{\xi} - \widetilde{b}\dot{\varphi}\right) \\ & = -\frac{1}{\varepsilon}\widetilde{b}\beta(t)\left(f(t) - \varphi(t)\right) \\ & + \left(\dot{\beta}(t)\underline{\alpha}_o^T + \beta(t)\left(\underline{\alpha}_o^T A_o + \widetilde{a}_1\underline{\alpha}^T\right)\right)\xi(t) \\ & + \widetilde{a}_1\widetilde{b}\beta(t)\bar{x}(t) - \widetilde{b}\left(\dot{\beta}(t) + \beta(t)\left(\widetilde{a}_1 - 1\right)\right)\varphi(t) \end{array} \tag{3.10}$$

$$\widetilde{a}_1 \hat{=} \underline{\alpha}_o^T\underline{\chi}_n = a_1 - a_{o,1} \tag{3.11}$$

substituting now (3.10) in (3.9):

$$\begin{array}{rl} \dot{f}(t) & = -\frac{1}{\varepsilon}\frac{\widetilde{b}}{b}\beta(t)\left(f(t) - \varphi(t)\right) \\ & + \underline{g}^T(t)\bar{x}(t) + \bar{a}(t)\varphi(t) \end{array} \tag{3.12}$$

$$
\left.\begin{aligned}
&\underline{g}^T(t) = \left[\ \underline{g}_1^T(t) \quad g_2(t) \quad g_3\ \right] \\
&\underline{g}_1^T(t) = \tfrac{\dot{\beta}(t)}{b}\underline{\alpha}_o^T + \left(\tfrac{\underline{\alpha}^T}{\widetilde{b}} + \tfrac{\beta(t)\underline{\alpha}_o^T}{b}\right) A_o \\
&\qquad + \left(\tfrac{a_1}{b}\beta(t) - \tfrac{\bar{a}_1}{\widetilde{b}}\right)\underline{\alpha}^T \\
&g_2(t) = \tfrac{\widetilde{a_1}\widetilde{b}}{b}\beta(t) - \bar{a}_1 \quad ; \quad g_3 = a_1\bar{a}_1/\widetilde{b}
\end{aligned}\right\} \tag{3.13}
$$

$$
\bar{a}(t) \hat{=} -\frac{\widetilde{b}}{\bar{b}}\dot{\beta}(t) + \bar{a}_1 + \frac{\widetilde{b}}{\bar{b}}\beta(t)(1 - \widetilde{a}_1) \tag{3.14}
$$

We then have from (3.4), (3.12) and (3.7):

$$
\begin{aligned}
&\left(\tfrac{\hat{b}(t)}{\bar{b}} + \varepsilon\rho_1(t)\bar{a}(t)\right)\varphi(t) = \tfrac{\hat{b}(t)}{\bar{b}} f(t) \\
&-\varepsilon\left(f(t) + \rho_1(t)\underline{g}^T(t)\bar{x}(t)\right) + v(t) + O(\varepsilon^2)
\end{aligned} \tag{3.15}
$$

$$
\hat{b}(t) \hat{=} \bar{b} + \rho_1(t)\widetilde{b}\beta(t) \tag{3.16}
$$

Note that (see: H1, (4), (3.8) and (3.16)): $0 < b_1 \leq \hat{b}(t) \leq b_2$, and that $\bar{a}(t)$ is also bounded:

$$
|\rho_1(t)\bar{a}(t)| < a_M \tag{3.17}
$$

$$
a_M \hat{=} \frac{L}{\bar{b}} + |\bar{a}_1| + \frac{\widetilde{b}}{\bar{b}}|1 - \widetilde{a}_1| \tag{3.18}
$$

(see: (3.14), H4, (8), and (4)). Taking now an ε sufficiently small such that: $\varepsilon\frac{\bar{b}}{b_1}a_M < 1$, and defining:

$$
\bar{\varepsilon}(t) \hat{=} \frac{\varepsilon}{1 + \varepsilon\frac{\bar{b}}{\hat{b}(t)}\rho_1(t)\bar{a}(t)} \tag{3.19}
$$

we have from (3.15) and (3.19):

$$
\begin{aligned}
\varphi(t) = f(t) - \bar{\varepsilon}(t)\left[\ \underline{\alpha}_1^T(t) \quad \alpha_2(t) \quad \alpha_3(t)\ \right]\bar{x}(t) \\
+ \alpha_o(t)v(t) + O(\varepsilon^2)
\end{aligned} \tag{3.20}
$$

$$
\left.\begin{aligned}
&\underline{\alpha}_1^T(t) \hat{=} \tfrac{\bar{b}}{\hat{b}(t)}\left(\tfrac{(1+\rho_1(t)\bar{a}(t))}{b}\underline{\alpha}^T + \rho_1(t)\underline{g}_1^T(t)\right) \\
&\alpha_2(t) = \tfrac{\bar{b}}{\hat{b}(t)}\left(1 + \rho_1(t)\bar{a}(t) + \rho_1(t)g_2(t)\right) \\
&\alpha_3(t) = \tfrac{\bar{b}}{\hat{b}(t)}\left(-\left(1 + \rho_1(t)\bar{a}(t)\right)\tfrac{\bar{a}_1}{b} + \rho_1(t)g_3\right) \\
&\alpha_o(t) = \tfrac{\bar{b}}{\hat{b}(t)}\left(1 - \bar{\varepsilon}(t)\tfrac{\bar{b}}{\hat{b}(t)}\rho_1(t)\bar{a}(t)\right)
\end{aligned}\right\} \tag{3.21}
$$

We have from (4), (3.2), (3.20), (3.6), (3.4) and (3.1):

$$
\begin{aligned}
&\begin{bmatrix} I & 0 & 0 \\ 0 & 0 & 0 \\ 0 & 0 & 1 \end{bmatrix} \dot{\bar{x}}(t) = \\
&= \begin{bmatrix} A_o & 0 & \underset{\sim}{\chi}_n\bar{a}_1 \\ \beta(t)\left(\underline{\alpha}_o^T - \underline{\alpha}^T\right) & -b(t) & \bar{a}_1\beta(t) \\ 0 & 0 & -\beta \end{bmatrix}\bar{x}(t) \\
&+\bar{\varepsilon}(t)\begin{bmatrix} \underset{\sim}{\chi}_n\widetilde{b}\underline{\alpha}_1^T(t) & \underset{\sim}{\chi}_n\widetilde{b}\alpha_2(t) & \underset{\sim}{\chi}_n\widetilde{b}\alpha_3(t) \\ \widetilde{b}\beta(t)\underline{\alpha}_1^T(t) & \widetilde{b}\beta(t)\alpha_2(t) & \widetilde{b}\beta(t)\alpha_3(t) \\ \widetilde{b}\underline{\alpha}_1^T(t) & \widetilde{b}\alpha_2(t) & \widetilde{b}\alpha_3(t) \end{bmatrix}\bar{x}(t) \\
&\qquad - \begin{bmatrix} \underset{\sim}{\chi}_n \\ \beta(t) \\ 1 \end{bmatrix}\widetilde{b}\alpha_o(t)v(t) + O(\varepsilon^2) \\
&\qquad\qquad \varepsilon\dot{v}(t) = -\tfrac{1}{\lambda}v(t)
\end{aligned} \tag{3.22}
$$

Now, since $\bar{\varepsilon}(t) = \varepsilon - \varepsilon^2\frac{\bar{b}}{\hat{b}(t)}\rho_1(t)\bar{a}(t) / \left(1 + \varepsilon\frac{\bar{b}}{\hat{b}(t)}\rho_1(t)\bar{a}(t)\right) = \varepsilon + O(\varepsilon^2)$ there then exists a sufficiently small ε such that $O(\varepsilon^2) \approx 0$ and thus $\bar{\varepsilon}(t) \approx \varepsilon$. Additionaly from (3.6) and (3.8) $v(t) \approx 0$ and $\rho_1(t) \approx 1$ for all $t > t_1 \approx 5\varepsilon$ which implies (see also: (3.21), (3.16) and (4)):

$$
\left.\begin{aligned}
&\begin{bmatrix} I & 0 & 0 \\ 0 & 0 & 0 \\ 0 & 0 & 1 \end{bmatrix} \dot{\bar{x}}(t) = \\
&= \begin{bmatrix} A_o & 0 & \underset{\sim}{\chi}_n\bar{a}_1 \\ \beta(t)\left(\underline{\alpha}_o^T - \underline{\alpha}^T\right) & -b(t) & \bar{a}_1\beta(t) \\ 0 & 0 & -\beta \end{bmatrix}\bar{x}(t) \\
&+\varepsilon\begin{bmatrix} \underset{\sim}{\chi}_n\widetilde{b}\bar{\underline{\alpha}}_1^T(t) & \underset{\sim}{\chi}_n\widetilde{b}\bar{\alpha}_2(t) & \underset{\sim}{\chi}_n\widetilde{b}\bar{\alpha}_3(t) \\ \widetilde{b}\beta(t)\bar{\underline{\alpha}}_1^T(t) & \widetilde{b}\beta(t)\bar{\alpha}_2(t) & \widetilde{b}\beta(t)\bar{\alpha}_3(t) \\ \widetilde{b}\bar{\underline{\alpha}}_1^T(t) & \widetilde{b}\bar{\alpha}_2(t) & \widetilde{b}\bar{\alpha}_3(t) \end{bmatrix}\bar{x}(t)
\end{aligned}\right\} \tag{3.23}
$$

$$
\left.\begin{aligned}
&\bar{\underline{\alpha}}_1^T(t) \hat{=} \tfrac{\bar{b}}{b(t)}\left(\tfrac{(1+\bar{a}(t))}{b}\underline{\alpha}^T + \underline{g}_1^T(t)\right) \\
&\bar{\alpha}_2(t) = \tfrac{\bar{b}}{b(t)}\left(1 + \bar{a}(t) + g_2(t)\right) \\
&\bar{\alpha}_3(t) = \tfrac{\bar{b}}{b(t)}\left(-\left(1 + \bar{a}(t)\right)\tfrac{\bar{a}_1}{b} + g_3\right)
\end{aligned}\right\} \tag{3.24}
$$

Let us note that $\beta(t)\bar{\alpha}_2(t)$ is bounded: $|\beta(t)\bar{\alpha}_2(t)| \leq \alpha_M$, where (see: (4), (8), (3.24), (3.17) and (3.13)): $\alpha_M \hat{=} \frac{\bar{b}}{b_1}\left(1 + a_M + |\widetilde{a}_1|\frac{\widetilde{b}}{\bar{b}} + |\bar{a}_1|\right)$. Let us define: $\gamma_o(t) \hat{=} b(t) - \varepsilon\widetilde{b}\beta(t)\bar{\alpha}_2(t)$, which satisfies:

$$
b_1 - \varepsilon\widetilde{b}\alpha_M \leq \gamma_o(t) \leq b_2 + \varepsilon\widetilde{b}\alpha_M \tag{3.25}
$$

Let us assume ε sufficiently small such that:

$$
\varepsilon < \min\left\{\frac{b_1}{\widetilde{b}\alpha_M}, \frac{b_1}{\bar{b}a_M}\right\} \tag{3.26}
$$

We then have from (3.23):

$$\widetilde{x}(t) = \underline{\gamma}_1^T(t)\xi(t) + \gamma_2(t)e(t) \tag{3.27}$$

$$\left.\begin{aligned} \underline{\gamma}_1^T(t) &\hat{=} \tfrac{\beta(t)}{\gamma_o(t)} \left(\underline{\alpha}_o^T - \underline{\alpha}^T + \varepsilon\widetilde{b}\underline{\bar{\alpha}}_1^T(t)\right) \\ \gamma_2(t) &\hat{=} \tfrac{\beta(t)}{\gamma_o(t)} \left(\bar{a}_1 + \varepsilon\widetilde{b}\bar{\alpha}_3(t)\right) \end{aligned}\right\} \tag{3.28}$$

And thus from (3.23) and (3.27) we have:

$$\dot{\hat{x}}(t) \approx (F_o + \varepsilon F_1(t))\,\hat{x}(t) \quad ; \quad \forall t > t_1 \tag{3.29}$$

$$\begin{aligned} F_o &\hat{=} \begin{bmatrix} A_o & \underline{\chi}_n \bar{a}_1 \\ 0 & -\beta \end{bmatrix} \quad ; \quad \hat{x}(t) \hat{=} \begin{bmatrix} \xi(t) \\ e(t) \end{bmatrix} \\ F_1(t) &\hat{=} \begin{bmatrix} \underline{\chi}_n \widetilde{b}\left(\underline{\bar{\alpha}}_1^T(t) + \bar{\alpha}_2(t)\underline{\gamma}_1^T(t)\right) & \underline{\chi}_n \widetilde{b}\left(\bar{\alpha}_3(t) + \bar{\alpha}_2(t)\gamma_2(t)\right) \\ \widetilde{b}\left(\underline{\bar{\alpha}}_1^T(t) + \bar{\alpha}_2(t)\underline{\gamma}_1^T(t)\right) & \widetilde{b}\left(\bar{\alpha}_3(t) + \bar{\alpha}_2(t)\gamma_2(t)\right) \end{bmatrix} . \end{aligned} \tag{3.30}$$

Note that F_o is Hurwitz, indeed:

$$\sigma(F_o) = \sigma(A_o) \cup \{-\beta\} \tag{3.31}$$

We prove in the appendix:

Lemma 2 Let us consider the following system:

$$\dot{x}(t) = (F_o + \varepsilon F_1(t))\, x(t) \quad ; \quad \forall t \geq 0 \tag{3.32}$$

H1*: $\varepsilon > 0$,

H2*: F_o is Hurwitz, and

H3*: all the elements of $F_1(t)$ are bounded $\forall t \geq 0$.

Then:

i) $\exists P^T = P, \quad P > 0$, such that:

$$F_o^T P + P F_o = -I \tag{3.33}$$

ii) System (3.32) is exponentially stable if:

$$\varepsilon \mu^* < 1 \tag{3.34}$$

where $\mu^* \hat{=} \max\{|\lambda(R(t))| \ , \forall t \geq 0\}$, and $R(t) \hat{=} (PF_1(t))^T + PF_1(t)$ □

Then from lemma 2 and from (3.31), (3.26), (3.25), (3.17), (3.18), (3.13), H4, H5 and (8) we have that $\xi(t)$ and $e(t)$ are exponentially stable which implies that $\zeta(t)$ and $\varphi(t)$ are stable (see: (3.20)). Therefore we conclude the internal stability of the closed loop system. ■

4. Illustrative Simulation

Let us consider the system:

$$\begin{aligned} \dot{\xi}(t) &= \begin{bmatrix} 0 & 1 \\ 0 & -a_1 \end{bmatrix} \xi(t) + \begin{bmatrix} 0 \\ b(t) \end{bmatrix} u(t) \\ y(t) &= \begin{bmatrix} 1 & 0 \end{bmatrix} \xi(t) \end{aligned} \tag{4.1}$$

where $a_1 = -1$, $\bar{b} = 0.5$, $\widetilde{b} = 0.4$,, and $\beta(t) = 0.568126(\sin(t) + \sin(2t))$. With the control law:

$$\left.\begin{aligned} & u(t) = \tfrac{1}{\bar{b}}\Big(-\widetilde{b}\varphi(t) \\ & + \begin{bmatrix} -\tfrac{1}{\tau_o^2} & \left(a_1 - \tfrac{2}{\tau_o}\right) \end{bmatrix} \xi(t) + \tfrac{1}{\tau_o^2} r(t)\Big) \\ & \dot{\zeta}(t) = -a_1\zeta(t) + \widetilde{b}\varphi(t) + \bar{b}u(t) \\ & \widetilde{b}\,(p+1)\,\varphi(t) = \beta\left(\tfrac{1}{\beta}p + 1\right)\psi(t) \\ & \psi(t) = \tfrac{1}{\varepsilon}\left(\xi_2(t) - \zeta(t)\ \right) \end{aligned}\right\} \tag{4.2}$$

the system behaves as a second order time invariant linear system with both poles in $-1/\tau_o$ when $\varepsilon \to 0$.

In figure 1 we show a PC 's simulation with $\varepsilon = 0.01$, $\tau_o = 0.5$, $\beta = 10$. We have used the following state reconstructor:

$$\begin{bmatrix} 0 & 0 \\ 1 & 0 \end{bmatrix} \dot{\bar{\xi}}(t) = \begin{bmatrix} 1 & 2\varepsilon_o \\ 0 & 1 \end{bmatrix} \bar{\xi}(t) + \begin{bmatrix} -1 \\ 0 \end{bmatrix} \dot{y}(t)$$

with $\varepsilon_o = 0.01$. Note that, when $\varepsilon_o = 0$ we have an ideal state reconstructor (see Bonilla et al, 1992). ■

5. Conclusion

In this paper we have shown that it is possible to control a linear system of order n with a time varying input gain by means of a proper linear time invariant control law. We only have assumed that we know the positive bounds of the variable time input gain and that it is Lipschitz.

Let us note, that the principal role of ε is to guarantee stability and to increase the time gap between small and fast modes and the parameters $\beta(t)$ and $\dot{\beta}(t)$ appear as some "constats" for the fast subsystem and in the slow subsystem they are "bounded" by ε.

6. References

Bonilla, Lebret and Malabre (1991a). Some Robust Control via Implicit Descriptions. 2nd

International Symposium on Implicit and Robust Systems, ISIRS'91. Warsaw, Poland.

Bonilla and Malabre (1991b). Variable Structure Systems via Implicit Descriptions. 1st. European Control Conference. Grenoble, France. 2-5 July.

Bonilla, Fonseca and Malabre (1992). Implementing non Proper Control Laws for Proper Systems. International Symposium of Implicit and Nonlinear Systems, SINS'92, Ft. Worth, U.S.A.
32th IEEE-CDC San Antonio Texas U. S. A.

Bonilla, Malabre and Velázquez (1993b). Controlling a linear first order time varying parameter system using a linear constant parameters control law. 32th IEEE-CDC San Antonio Texas U. S. A.

Kokotović, Khalil and O'Relly (1986). Singular perturbations Methods in Control. Ac. Press Inc. London LTD.

O'Malley (1991). Singular Perturbations Methods for Ordinary Differential Equations. Springer-Verlag, N.Y. Inc., V. 89.

APPENDIX

A. Proof of lemma 2

i)The first statement follows from the Lyapunov's Theorem (under H2*).

ii)Let us consider the Lyapunov's function: $V(t) = x^T(t)Px(t)$, note that $V(t) > 0$. We then have that: $\dot{V}(t) = \dot{x}^T(t)Px(t) + x^T(t)P\dot{x}(t)$. We then have from (3.32): $\dot{V}(t) = x^T(t)\left(F_o + \varepsilon F_1(t)\right)^T Px(t) + x^T(t)P\left(F_o + \varepsilon F_1(t)\right)x(t)$, and from (3.33): $\dot{V}(t) = x^T(t)\left(-I + \varepsilon R(t)\right)x(t)$.

From the fact that $R(t)$ is an Hermitian matrix, we have that: $\|x(t)\|_2^2 \lambda_{\min}(R(t)) \leq R(t) \leq \|x(t)\|_2^2 \lambda_{\max}(R(t))$, we then have:

$$\dot{V}(t) \leq -\left(1 - \varepsilon\mu^*\right) x^T(t) P^{\frac{1}{2}} P^{-1} P^{\frac{1}{2}} x(t) \qquad \text{(A.1)}$$

and since (3.34): $\dot{V}(t) \leq -\frac{(1-\varepsilon\mu^*)}{\lambda_{\max}(P)} V(t) < 0$, i.e.:

$$\|x(t)\|_2^2 \leq \frac{1}{\lambda_{\min}(P)} V(t) \leq \frac{V(0)}{\lambda_{\min}(P)} e^{-\frac{(1-\varepsilon\mu^*)}{\lambda_{\max}(P)} t} \qquad \blacksquare$$

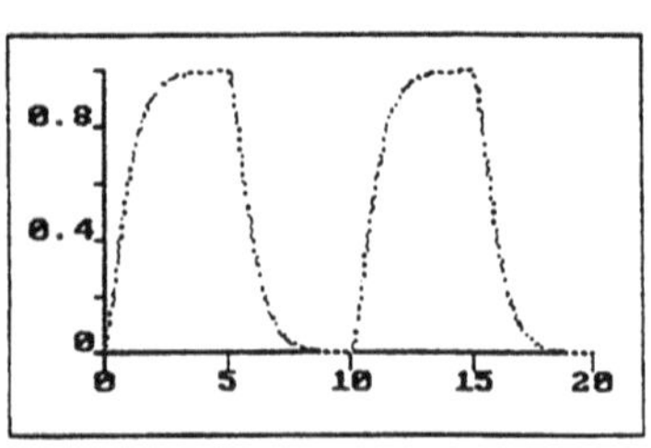

$y(t)$

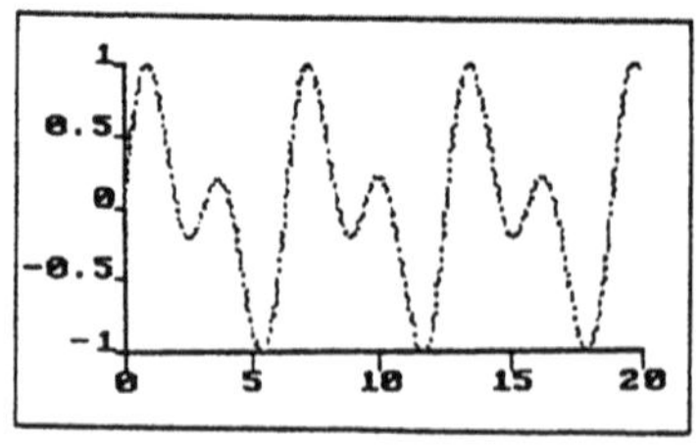

$\beta(t)$

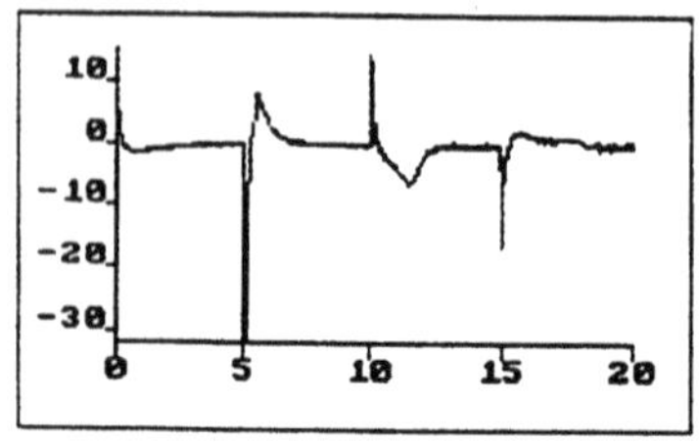

$u(t)$

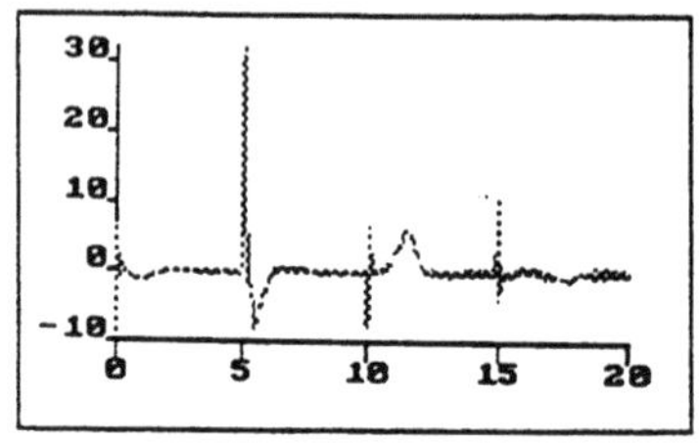

$\varphi(t)$

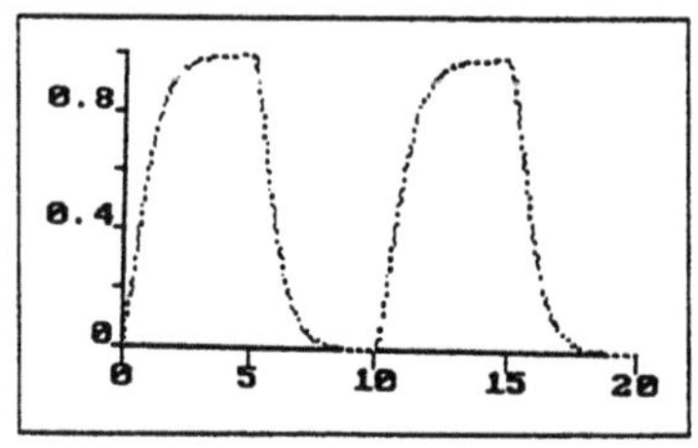

$\xi_1(t)$

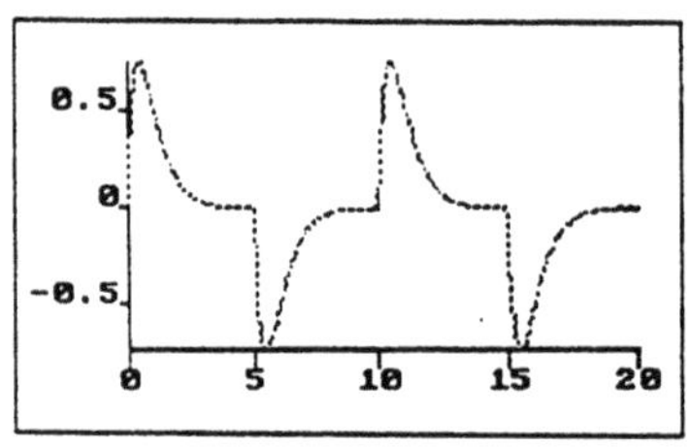

$\xi_2(t)$

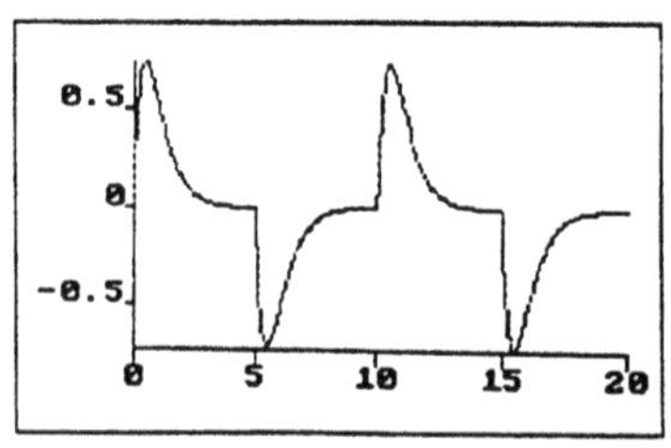

$\zeta(t)$

Figure 1.

A CONTROLLABILITY TEST FOR BEHAVIOR SYSTEMS

U. Helmke*

Department of Mathematics, University of Regensburg
D-93040 Regensburg, Germany

J. Rosenthal†

Department of Mathematics, University of Notre Dame
Notre Dame, IN 46556, USA

J. M. Schumacher

CWI
P.O. Box 94079, 1090 GB Amsterdam, the Netherlands, and
Tilburg University, CentER and Dept. of Economics,
P.O. Box 90153, 5000 LE Tilburg, the Netherlands

Abstract. In this paper we derive a new controllability rank test for general first-order representations. The criterion generalizes the well-known controllability rank test for linear input-state systems as well as a controllability rank test by Mertzios *et al.* for descriptor systems.

Résumé. Dans ce papier, nous dérivons un nouveau test de contrôle du rang d'une représentation générale du premier ordre. Le critère généralise à la fois le test de contrôle bien connu du rang des systèmes linéaires du type état d'entrée et le test de Mertzios pour les systèmes du type descriptif.

Keywords. Controllability, reachability space, generalized state space system, behavior.

1. INTRODUCTION

As is well-known (see for instance (Aplevich 1991, Kuijper 1994, Willems 1991)), a general form for linear time-invariant dynamical systems is the following:

$$K\sigma x + Lx + Mw = 0 \qquad (1)$$

where K, L, and M are matrices of sizes $(n+p)\times n$, $(n+p)\times n$, and $(n+p)\times(m+p)$ respectively. (We follow the notation of (Kuijper 1994) and therefore denote the parameter matrices for this representation by (K, L, M) rather than (E, F, G).) Specifically, Prop. VII.3 in (Willems 1991) states that a system with latent variables x and manifest variables w, over the time axis $\mathbb{Z}$, is a linear time-invariant complete state-space dynamical system if and only if it can be represented in the form (1) with σ denoting shift. In the continuous-time case one should interpret σ as differentiation. By various transformation algorithms it has been shown that all of the behaviors that are represented by any of the forms used in linear system theory (including matrix fraction descriptions, implicit systems, etc.) admit also a representation of

*Research partially supported by GIF grant I 184-078.6/91.

†This author was supported in part by NSF grant DMS-94-00965. The research for this paper was carried out in part while he was a visitor at CWI.

the form (1), with appropriate identification of variables; for instance the external variable w usually denotes a vector consisting of inputs and outputs.

Properties such as observability and controllability can of course be expressed in terms of the matrices K, L, and M. In particular it is stated in (Willems 1991; Prop. VII.11(v)) that (1) is a minimal representation of a controllable external behavior if and only if the following two conditions hold:

(i) $\lambda K + \mu L$ has full column rank for all $(\lambda, \mu) \in \mathbb{C}^2 \setminus \{(0,0)\}$;
(ii) $[\lambda K + \mu L \mid M]$ has full row rank for all $(\lambda, \mu) \in \mathbb{C}^2 \setminus \{(0,0)\}$.

Condition (i) is the observability condition whereas (ii) is the controllability condition; one readily verifies that these conditions do indeed reduce to the usual ones for the standard state space case which is obtained by taking

$$K = \begin{bmatrix} I \\ 0 \end{bmatrix}, \quad L = \begin{bmatrix} -A \\ -C \end{bmatrix},$$
$$M = \begin{bmatrix} 0 & -B \\ I & -D \end{bmatrix}. \tag{2}$$

It is the purpose of the present paper to develop a *rank test* for condition (ii) that is analogous to the well-known Kalman controllability test for standard state space systems. Actually the test that we shall present is a direct generalization of Kalman's rank test in that it calls for checking that a certain matrix with n rows has full column rank. In analogy to the standard state space situation it is possible to characterize the column space of this matrix as a reachability space (see Section 3 for details) and for this reason we will call this matrix a *reachability matrix*. Finally the test we are presenting is also a generalization of a test due to Mertzios *et al.* (Mertzios *et al.* 1988) (see also (Helmke 1993)), which applies to systems of the form (1) with $p = 0$.

Of course if one only is interested in an algebraic test which is capable of deciding if a particular generalized state space system of the form (1) is controllable then there are also other methods available. One possibility is the computation of all $(n+p) \times (n+p)$ full size minors of the pencil $[\lambda K + \mu L \mid M]$ and applying a classical "multi-resultant"-test due to Macaulay (Macaulay 1903). A controllability rank test of a somewhat different nature for systems of the general form (1) is also due to Lomadze (Lomadze 1990) (see also (Ravi and Rosenthal 1995)); this test involves a matrix of size $n(n+p) \times n(n-1+m+p)$.

By duality, the proposed controllability test can also be interpreted as an observability test. As such it applies to systems of the form

$$\begin{aligned} G\sigma z &= Fz \\ w &= Hz \end{aligned} \tag{3}$$

where F, G, and H are matrices of sizes $n \times (n+m)$, $n \times (n+m)$, and $q \times (n+m)$ respectively. We emphasize that systems of the form (3) (sometimes called the 'pencil form'), have the same descriptive power for smooth behaviors as the representation (1). The pencil form has been used recently in an investigation of 'impulsive-smooth' behaviors (Geerts and Schumacher 1994*a*, Geerts and Schumacher 1994*b*), which allow solutions in a space of generalized functions. Because solutions are allowed in a larger space than usual, the resulting minimality conditions are weaker than the standard ones. In fact, the following conditions for minimality are given in (Geerts and Schumacher 1994*b*; Thm. 4.4):

(i) $\lambda G + \mu F$ has full row rank for some $(\lambda, \mu) \in \mathbb{C}^2 \setminus \{(0,0)\}$;

(ii) $\begin{bmatrix} \lambda G + \mu F \\ H \end{bmatrix}$ has full column rank for all $(\lambda, \mu) \in \mathbb{C}^2 \setminus \{(0,0)\}$.

Condition (ii) is the observability condition, whereas condition (i) might be called an *admissibility condition*. The set of triples (F, G, H) satisfying conditions (i) and (ii) above, considered modulo similarity equivalence, has the interesting property of being a smooth and compact projective variety (Strømme 1987, Lomadze 1990, Helmke 1993, Ravi and Rosenthal 1995). Obviously the observability condition (ii) for systems in pencil form is related by duality to the controllability condition for systems in the form (1), and so after simple transposition the controllability test that will be derived below can also be used to test for observability of a triple (F, G, H) in the representation (3).

For complete proofs of the results we refer to the full paper (Helmke *et al.* 1995).

2. THE CONTROLLABILITY TEST

Consider behaviors represented by

$$K\dot{x} + Lx + Mw = 0 \qquad (4)$$

where $K, L \in \mathbb{R}^{(n+p)\times n}$ and $M \in \mathbb{R}^{(n+p)\times(m+p)}$ $(n > 0, m > 0, p \geq 0)$. The system will be called 'admissible' if the rank condition

$$\operatorname{rank}(\lambda K + \mu L) = n \qquad (5)$$

holds for some $(\lambda, \mu) \in \mathbb{C}^2$. This condition is implied by various forms of observability (Willems 1991; p. 270). Recall (Willems 1991; Prop. VII.11) that a minimal (in the sense of (Willems 1991)) system of the form (4) is *controllable* if and only if the rank condition

$$\operatorname{rank}[\lambda K + \mu L \mid M] = n + p \qquad (6)$$

holds for all $(\lambda, \mu) \in \mathbb{C}^2 \setminus \{(0,0)\}$. Motivated by the application to develop an observability test for pencil form descriptions of impulsive-smooth behaviors we shall assume only admissibility, rather than minimality.

For $p = 0$ a Kalman-type controllability matrix for the system (4) was introduced by Mertzios *et al.* (Mertzios *et al.* 1988), see also Helmke (Helmke 1993). It has been shown that the system is controllable if and only if the associated controllability matrix has full rank. Here we seek to extend that construction to the general case, where p is arbitrary.

Recall, given an $n \times n$ matrix A, that the *adjoint* of A (see for instance (Wedderburn 1934; p. 7)) is the $n \times n$ matrix defined as

$$\operatorname{adj} A := \left((-1)^{i+j} \det A_{ji}\right)_{i,j=1}^{n}. \qquad (7)$$

Here $\det A_{k\ell}$ denotes the $(n-1) \times (n-1)$ minor of A defined by omitting the k-th row and ℓ-th column from A. The adjoint of a 1×1 matrix is always 1.

For any multi-index $\alpha = (\alpha_1, \ldots, \alpha_p) \in \mathcal{I}(p, m+p)$ let M_α denote the $(n+p) \times p$ submatrix of M formed by selecting the α_1-th, α_2-th, $\ldots, \alpha_p$-th columns of M. Let $\alpha' := \{1, \ldots, m+p\} \setminus \alpha$ denote the complementary index and let $M_{\alpha'}$ denote the associated $(n+p) \times m$ submatrix of M. Given any $(n+p) \times p$ submatrix M_α of M, write

$$\operatorname{adj}[\lambda K + \mu L \mid M_\alpha] = \begin{bmatrix} R_\alpha(\lambda,\mu) \\ S_\alpha(\lambda,\mu) \end{bmatrix} \qquad (8)$$

where $R_\alpha(\lambda,\mu)$ and $S_\alpha(\lambda,\mu)$ are formed by the first n and last p rows of $\operatorname{adj}(\lambda K + \mu L \mid M_\alpha)$ respectively; thus $R_\alpha(\lambda,\mu)$ has size $n \times (n+p)$ and $S_\alpha(\lambda,\mu)$ has size $p \times (n+p)$. From the identity

$$\operatorname{adj}[t\lambda K + t\mu L \mid M_\alpha] =$$

$$\begin{bmatrix} t^{n-1}I & 0 \\ 0 & t^n I \end{bmatrix} \operatorname{adj}[\lambda K + \mu L \mid M_\alpha]$$

we obtain that $R_\alpha(\lambda,\mu)$ and $S_\alpha(\lambda,\mu)$ are matrices of homogeneous polynomials in (λ,μ) of degrees $n-1$ and n respectively. So in particular

$$R_\alpha(\lambda,\mu) M_{\alpha'} = \sum_{i=0}^{n-1} \Gamma_{i\alpha} \lambda^i \mu^{n-1-i}. \qquad (9)$$

for $(n \times m)$-matrices $\Gamma_{i\alpha}$, $i = 0, \ldots, n-1$. The *reachability matrix* of (K, L, M) is defined as the matrix of size $n \times nm\binom{m+p}{p}$ obtained by putting all matrices $\Gamma_{i\alpha}$ $(i = 0, \cdots, n-1;\ \alpha \in \mathcal{I}(p, m+p))$ next to each other:

$$\mathcal{R}(K, L, M) :=$$

$$[(\Gamma_{0\alpha} \mid \cdots \mid \Gamma_{n-1,\alpha}) \mid \alpha \in \mathcal{I}(p, m+p)]. \quad (10)$$

We can now state our main result.

Theorem 1 An admissible system (K, L, M) is controllable if and only if

$$\operatorname{rank} \mathcal{R}(K, L, M) = n.$$

A complete proof is provided in (Helmke *et al.* 1995).

3. THE REACHABLE SPACE

In this section we provide a dynamic interpretation of the matrix $\mathcal{R}(K, L, M)$ which also will justify the name reachability matrix. For simplicity we will restrict ourselves to discrete time systems, i.e. we will restrict our considerations to systems of the form

$$Kx_{t+1} + Lx_t + Mw_t = 0. \qquad (11)$$

Definition 2 A state vector $\hat{x}$ is said to be a reachable state if there exists a sequence of states

$$\Sigma := \{x_i \in \mathbb{R}^n \mid i \in \mathcal{Z}\}$$

having the property that

1. At most finitely many vectors $x_i \in \Sigma$ are nonzero.
2. There is a set of external variables such that (11) is satisfied for all $t \in \mathcal{Z}$.
3. $\hat{x} \in \Sigma$.

The set of all reachable states is denoted by $\hat{\mathcal{R}}(K, L, M)$.

The set $\hat{\mathcal{R}}(K, L, M)$ can also be characterized in the following way: $\hat{x} \in \hat{\mathcal{R}}(K, L, M)$ if and only if there is a vector polynomial

$$x(\lambda) = \sum_{i=0}^{k-1} x_i \lambda^i \in \mathbb{R}^n[\lambda]$$

having $\hat{x}$ as component and a vector polynomial

$$w(\lambda) = \sum_{i=0}^{k} w_i \lambda^i \in \mathbb{R}^{m+p}[\lambda]$$

such that

$$[\lambda K + L \mid M] \begin{bmatrix} x(\lambda) \\ w(\lambda) \end{bmatrix} = 0. \qquad (12)$$

If we introduce

$$\mathcal{K} := \begin{bmatrix} K & 0 & \dots & 0 \\ L & K & \ddots & \vdots \\ 0 & L & \ddots & 0 \\ \vdots & \ddots & \ddots & K \\ 0 & \dots & 0 & L \end{bmatrix}$$

and

$$\mathcal{M} := \begin{bmatrix} M & 0 & \dots & 0 \\ 0 & M & \ddots & \vdots \\ \vdots & \ddots & \ddots & 0 \\ 0 & \dots & 0 & M \end{bmatrix},$$

this last equation can also be written componentwise in the form

$$[\mathcal{K} \mid \mathcal{M}] \begin{bmatrix} x_0 \\ \vdots \\ x_{k-1} \\ w_0 \\ \vdots \\ w_k \end{bmatrix} = 0. \qquad (13)$$

The following lemma is now a simple consequence of the description (12).

Lemma 3 $\hat{\mathcal{R}}(K, L, M) \subset \mathbb{R}^n$ is a linear subspace.

From the description (12) it is also clear that $\hat{\mathcal{R}}(K, L, M)$ is invariant under transformation of the system equation and under change of basis in the external variables, i.e. we have for $T \in Gl_{n+p}$ and $U \in Gl_{m+p}$ that

$$\hat{\mathcal{R}}(TK, TL, TMU^{-1}) = \hat{\mathcal{R}}(K, L, M). \qquad (14)$$

Moreover, it is easy to show that $\hat{\mathcal{R}}(K, L, M)$ is also invariant under scaling transformations; i.e. $\hat{\mathcal{R}}(aK + bL, cK + dL, M) = \hat{\mathcal{R}}(K, L, M)$ holds for all invertible matrices

$$\begin{bmatrix} a & b \\ c & d \end{bmatrix}.$$

We are now in a position to establish the connection between the subspace $\hat{\mathcal{R}}(K, L, M)$ of reachable states and the reachability matrix $\mathcal{R}(K, L, M)$ as introduced in (10).

Theorem 4 The vector space $\hat{\mathcal{R}}(K, L, M)$ of reachable states is equal to the column space of the reachability matrix, i.e.

$$\hat{\mathcal{R}}(K, L, M) = \operatorname{colsp}(\mathcal{R}(K, L, M)). \qquad (15)$$

4. AN APPLICATION FROM CODING THEORY

In this section we present an application for our results which comes from coding theory. For this reason we first extend the main results obtained so far in this paper to discrete time linear systems defined over an arbitrary commutative base field $\mathbb{F}$ with algebraic closure $\bar{\mathbb{F}}$.

Definition 5 Let K, L, and M be matrices of sizes $(n+p)\times n$, $(n+p)\times n$, and $(n+p)\times(m+p)$ respectively having entries from $\mathbb{F}$. Then the discrete time linear system

$$Kx_{t+1} + Lx_t + Mw_t = 0 \quad (16)$$

is called observable if

(i) $\lambda K + \mu L$ has full column rank for all $(\lambda, \mu) \in \bar{\mathbb{F}}^2 \setminus \{(0,0)\}$.

It is called controllable if

(ii) $[\lambda K + \mu L \mid M]$ has full row rank for all $(\lambda, \mu) \in \bar{\mathbb{F}}^2 \setminus \{(0,0)\}$.

Conditions (i) and (ii) extend naturally the definitions given in the introduction and they were also used in (Ravi and Rosenthal 1995). By following the same reasoning as in the proof of Theorem 1 and Theorem 4 we establish the result:

Theorem 6 An admissible system of the form (16) is controllable if and only if the reachability matrix $\mathcal{R}(K, L, M)$ has full rank. Moreover the column space of $\mathcal{R}(K, L, M)$ is equal to the set $\hat{\mathcal{R}}(K, L, M)$ of reachable states.

In the remainder of this section we will show how the results of this paper are relevant to the design of so called convolutional codes and convolutional encoders. Since convolutional codes are easily implemented and have good error correcting properties, they are among the most widely used codes in data transmission. In the sequel we explain the most basic properties of convolutional codes from a systems theory point of view. For a detailed treatment of this active research area we refer to the excellent monograph of Piret (Piret 1988).

From a systems theory point of view convolutional codes are essentially the behavior of a discrete time linear system defined over a finite field. To be precise let $\mathbb{F} = \mathbb{F}_q$ be the Galois field with q elements and consider a $(m+p) \times m$ matrix $G(D)$ defined over the polynomial ring $\mathbb{F}[D]$ in a single variable D. The image of $G(D)$ generates a convolutional code as:

$$C := \{w(D) \mid w(D) = G(D)\ell(D)\}. \quad (17)$$

In this so called MA-representation, the matrix $G(D)$ is called the encoder of the convolutional code C and the set of $(m+p)$-vectors $w(D)$ describe the code words, i.e. the behavior of the system. One says the convolutional encoder is catastrophic (see (Piret 1988)) if the system (17) is not observable, i.e. if there is an element $f \in \bar{\mathbb{F}}$ having the property that $\text{rank}G(f) < m$. This is also equivalent to the condition that the full size minors of $G(D)$ have a nontrivial greatest common divisor. As the name indicates it it is most important to have non-catastrophic convolutional encoders.

Instead of studying convolutional encoders through higher order representations of the form (17) it is equally well possible to study them in terms of first order representations of the form (16). For this note that the realization theory as presented in (Ravi and Rosenthal 1995) is valid over any commutative base field. The design of non-catastrophic encoders is therefore equivalent with the design of triples (K, L, M) which are simultaneously observable and controllable.

Although first order representations of the exact form (16) have not been used so far in the coding literature we would like to remark that (16) exactly describes the dynamics of the so called *state diagram.* (See (Piret 1988) for details). The vector $x(t)$ describes in this representation the state of the state diagram and reachability of all possible states is certainly an important design issue.

As already mentioned earlier it is typically not so difficult to design observable triples (K, L, M). Indeed it is enough to search for matrices K, L of the form

$$[\lambda K + \mu L] = \begin{bmatrix} \lambda I - \mu A \\ \mu C \end{bmatrix}$$

where A, C form an observable pair in the traditional sense. Controllability is however an issue and in a concluding example we illustrate how Theorem 1 and Theorem 4 can successfully be applied.

Example 7 Consider the (binary) base field $\mathbb{F} = \mathbb{F}_2 = \{0, 1\}$ and consider the encoder

$$[\lambda K + \mu L \mid M] := \left[\begin{array}{ccc|cc} \lambda & \mu & 0 & 0 & 1 \\ \lambda+\mu & \lambda & 0 & 1 & 1 \\ \mu & 0 & \lambda & 0 & 0 \\ \lambda & \mu & \mu & 1 & 1 \end{array}\right].$$

A straightforward calculation shows that the 3×3 minors of $[\lambda K + \mu L]$ are coprime, i.e. we deal with a non-catastrophic (observable) encoder of McMillan degree $n = 3$. In order to verify controllability we use Theorem 1 and calculate the reachability matrix:

$$\mathcal{R}(K, L, M) = \begin{bmatrix} 1 & 1 & 0 & 1 & 1 & 0 \\ 0 & 1 & 1 & 0 & 1 & 1 \\ 0 & 1 & 1 & 0 & 1 & 1 \end{bmatrix}.$$

Obviously this matrix has not full rank and the reachable space $\hat{\mathcal{R}}(K, L, M)$ is according to Theorem 4 spanned by the vectors

$$\left\{ \begin{bmatrix} 1 \\ 0 \\ 0 \end{bmatrix}, \begin{bmatrix} 0 \\ 1 \\ 1 \end{bmatrix} \right\}.$$

5. CONCLUSIONS

We have presented a rank test for controllability of behaviors described by equations of the form $K\sigma x + Lx + Mw = 0$; similarly the dual form leads to an observability test for systems in pencil form. The test is in the spirit of Kalman's classical controllability condition; it requires checking that a certain matrix with n rows has full row rank. Moreover, the column span of this matrix has the interpretation of a reachable space. We extended the results to discrete time systems defined over an arbitrary base field and we illustrated the results with an application from coding theory.

REFERENCES

Aplevich, J. D. (1991). *Implicit Linear Systems.* Lect. Notes Contr. Inform. Sci. 152. Springer. Berlin.

Geerts, A. H. W. and J. M. Schumacher (1994*a*). 'Impulsive-smooth behavior in multimode systems. Part I: State-space and polynomial representations'. CWI Report BS-R9431.

Geerts, A. H. W. and J. M. Schumacher (1994*b*). 'Impulsive-smooth behavior in multimode systems. Part II: Minimality and equivalence'. CWI Report BS-R9432.

Helmke, U. (1993). 'The cohomology of moduli spaces of linear dynamical systems'. Regensburger Mathematische Schriften 24, Fak. Math. Univ. Regensburg.

Helmke, U., J. Rosenthal and J. M. Schumacher (1995). 'A controllability test for general first-order representations'. CWI Report BS-R95xx. Submitted to *Automatica.*

Kuijper, M. (1994). *First-Order Representations of Linear Systems.* Birkhäuser. Boston.

Lomadze, V. G. (1990). 'Finite-dimensional time-invariant linear dynamical systems: Algebraic theory'. *Acta Appl. Math* **19**, 149–201.

Macaulay, F. S. (1903). 'Some formulae in elimination'. *Proc. London Math. Soc.* **XXXV**, 3–27.

Mertzios, B. G., M. A. Christodoulou, B. L. Syrmos and F. L. Lewis (1988). 'Direct controllability and observability time domain conditions of singular systems'. *IEEE Trans. Automat. Contr.* **AC-33**, 788–791.

Piret, P. (1988). *Convolutional Codes, an Algebraic Approach.* MIT Press. Cambridge, MA.

Ravi, M. S. and J. Rosenthal (1995). 'A general realization theory for higher order linear differential equations'. *Systems & Control Letters.* To appear.

Strømme, S. A. (1987). On parametrized rational curves in Grassmann varieties. In F. Ghione, C. Peskine and E. Sernesi (Eds.). 'Space Curves'. Lecture Notes in Mathematics # 1266. Springer Verlag. pp. 251–272.

Wedderburn, J. (1934). *Lectures on Matrices.* Coll. Publ. XVII. Amer. Math. Soc.. Providence, R.I.

Willems, J. C. (1991). 'Paradigms and puzzles in the theory of dynamical systems'. *IEEE Trans. Automat. Control* **AC-36**(3), 259–294.

CANONICAL FORMS FOR SINGULAR SYSTEMS WITH OUTPUTS UNDER RESTRICTED SYSTEM EQUIVALENCE

D. VAFIADIS and **N. KARCANIAS**
Control Engineering Centre, City University, London EC1V 0HB, U.K.

Résumé: Nous considérons une forme canonique du type Popov pour des systèmes singuliers avec sorties, sous l' action de l' équivalence restreinte des systèmes. Nous décrivons les transformations qui conduisent à cette forme canonique et nous montrons que celle-ci est directement liée à la matrice composée, issue d' une factorisation de la matrice du transfert du système. Elle met en évidence, de manière tout à fait transparente, les indices d' atteignabilité du système et permet de classifier, par simple inspection, les indices de commandabilité propres et nonpropres.

Abstract: The problem of Popov type canonical forms for singular systems under restricted system equivalence is considered. The transformations leading to the canonical form are described and it is shown that the canonical form is directly related to the echelon form of the composite matrix of an MFD of the transfer function of the system. The proposed canonical form displays transparently the reachability indices of the system and allows the classification of the controllability indices into proper and nonproper by inspection.

Key Words: Singular systems; canonical forms; restricted system equivalence; echelon form; minimal bases

1 Introduction

The aim of the present paper is the construction of a canonical form for singular systems with outputs under restricted system equivalence (r.s.e.) transformations (Rosenbrock, 1974). This problem may be considered as the extension of canonical forms with outputs for state–space systems under similarity transformations.

Canonical forms of state–space models are related to MFD descriptions of strictly proper transfer functions via the echelon canonical forms for polynomial matrices under unimodular transformations (Popov, 1969; Forney,1975). This relation shows that the input–output and internal variable (state) descriptions are related in a straightforward manner and that some results of state–space theory may be derived directly via the transfer function approach.

In the present paper the relation between canonical forms of singular systems and MFD descriptions of nonproper transfer functions is established. First, a systematic way for the construction of the canonical form is given. The r.s.e. transformations leading to that form are described and finally it is shown that the canonical form may be obtained by inspection of the echelon canonical form (Forney, 1975) of the composite matrix $[N^T(s), D^T(s)]^T$ of a MFD of the transfer function $G(s) = N(s)D^{-1}(s)$ of the system (Vafiadis and Karcanias, 1995a). The proposed canonical form is of Popov type, in the sense that the discrete invariants (reachability and controllability indices) are obtained by the dimensions of the blocks of E, A, and the continuous invariants appear in the matrices E, A in a way analogous to that of Popov form for state–space systems. Furthermore the classification of the controllability indices into proper and nonproper (Karcanias and Heliopoulou, 1990; Malabre *et al.*, 1990), is obtained by simple inspection of E. It is mentioned that the problem of canonical forms for descriptor type systems with outputs has been considered in (Lebret and Loiseau, 1994). In that paper the transformation groups used are richer than the r.s.e. transformation group, since they allow output and state feedback, output injection, change of basis in the input and output spaces.

In the present paper, most of the results are given without proof. The proofs may be found in (Vafiadis and Karcanias, 1995b).

2 Statement of the problem

Consider the reachable singular system $\mathcal{S}(E, A, B, C)$ described by the equations

$$\mathcal{S} : E\dot{x} = Ax + Bu, \quad y = Cx \tag{2.1}$$

where $x \in \mathcal{X} \approx \Re^n$, $u \in \mathcal{U} \approx \Re^\ell$, $y \in \mathcal{Y} \approx \Re^m$, $(E, A, B, C) \in \Re^{n\times n} \times \Re^{n\times n} \times \Re^{n\times \ell} \times \Re^{m\times n}$, rank $\{B\} = \ell$ and rank $\{C\} = m$. System (2.1) is assumed to be minimal, i.e. $[sE - A, -B]$, $[sE^T - A^T, C^T]^T$ do not have finite Smith zeros and rank$[E, B]$=rank$[E^T, C^T]^T = n$ and regular i.e. $\det\{sE - A\} \not\equiv 0$. Let $P(s)$ be the system matrix of (2.1) (Rosenbrock, 1974).

$$P(s) = \begin{bmatrix} sE - A & -B \\ C & 0 \end{bmatrix} \tag{2.2}$$

The action of r.s.e. transformations on the system $\mathcal{S}(E, A, B, C)$ is defined as follows

$$(P, Q) \circ (E, A, B, C) = (PEQ, PAQ, PB, CQ) \tag{2.3}$$

where $(P, Q) \in \Re^{n\times n} \times \Re^{n\times n}$ and $\det\{P\} \neq 0$, $\det\{Q\} \neq 0$. The result of the above transformation is the equivalent system

$$\mathcal{S}' : PEQ\dot{x}' + PBu \quad y = CQx'$$

The action of the transformations (2.3) on the system matrix $P(s)$ results to $P'(s)$ which is as follows:

$$P'(s) = \begin{bmatrix} sPEQ - PAQ & -PB \\ CQ & 0 \end{bmatrix} \tag{2.4}$$

It is well known that r.s.e. transformations define an equivalence relation which partitions the set of all quadruples $(E, A, B, C) \in \Re^{n\times n} \times \Re^{n\times n} \times \Re^{n\times \ell} \times \Re^{m\times n}$ to equivalence classes or orbits. Our aim is to find the canonical element of each orbit and the invariants characterising the orbit. Clearly, this problem is equivalent to the problem of canonical form of the pencil $P(s)$ under the transformation shown in (2.4). Let $N \in \Re^{(n-\ell)\times n}$ be a maximal left annihilator of B and $B^\dagger \in \Re^{\ell \times n}$ a left inverse of B, i.e $NB = 0$ and $B^\dagger B = I_\ell$. Then the matrix $\begin{bmatrix} N \\ B^\dagger \end{bmatrix}$ is invertible. Consider the pencil

$$\begin{bmatrix} N \\ B^\dagger \end{bmatrix} [sE - A, -B] = \begin{bmatrix} sNE - NA, & 0 \\ sB^\dagger E - B^\dagger A, & -I \end{bmatrix} \tag{2.5}$$

which is clearly strictly equivalent (s.e.) to $T(s)$. The pencil $sNE - NA$ in (2.5) has only column minimal indices (c.m.i.) which coincide with the reachability indices of the system $\mathcal{S}$ (Karcanias, 1990). Thus, the pencil (2.5) is s.e. to a pencil of the form:

$$\begin{bmatrix} L(s) & 0 \\ sK - \Lambda & -I \end{bmatrix} \tag{2.6}$$

where

$$L(s) = \text{block} - \text{diag}\{\ldots, L_{\varepsilon_i}(s), \ldots\} \tag{2.7}$$
$$\varepsilon_1 \leq \varepsilon_2 \leq \cdots \leq \varepsilon_\ell$$

$$L_{\varepsilon_i}(s) = \begin{bmatrix} s & -1 & 0 & \cdots & 0 \\ & \ddots & \ddots & & \\ 0 & \cdots & 0 & s & -1 \end{bmatrix}_{\varepsilon_i \times (\varepsilon_i + 1)} \tag{2.8}$$

and $sK - \Lambda$ is unspecified. Similarly, the pencil $P(s)$ may be transformed to the form

$$P(s) \sim \left[\begin{array}{c|c} L(s) & 0 \\ sK - \Lambda & -I \\ \hline C & 0 \end{array}\right] \tag{2.9}$$

From the above discussion it is clear that the problem of canonical form of system (2.1) may be reduced to the problem of canonical form of (2.9) under the transformations defined by (2.4).

3 The stabilizer of $L(s)$

In this section we consider the matrix pencil $L(s)$ (see (2.7)) and we find its stabilizer. The stabilizer plays an important role on the determination of the canonical form of $T(s)$ since it leads to the transformations that bring $T(s)$ to the canonical form.

Definition 3.1 *(MacLane and Birkhoff, 1967) Let $\mathcal{X}$ be a set and G be a transformation group acting on $\mathcal{X}$ with the action $g : G \times \mathcal{X} \to \mathcal{X}$, $gx \mapsto y$, $g \in G$ $x, y \in \mathcal{X}$. Consider a fixed $x_0 \in \mathcal{X}$. Then the set $G_{x_0} \subset G$ with the property $gx_0 \mapsto x_0$, $g \in G_{x_0}$ is defined as the stabilizer of x_0. The set G_{x_0} is a subgroup of G and it is denoted by $Stab(x_0)$.* □

The above definition means that the stabilizer is the subgroup of the transformation group for which the action on x_0, leaves x_0 unchanged.

Definition 3.2 *The stabilizer of $L(s)$ is defined as the set of all pairs (P, Q) of invertible matrices such that $PL(s)Q = L(s)$.* □

The following results are related to the derivation of the stabilizer of $L(s)$.

Lemma 3.1 *Let P_{ii}, Q_{ii} be such that $P_{ii}L_{\varepsilon_i}(s) = L_{\varepsilon_i}(s)Q_{ii}$ where $L_{\varepsilon_i}(s)$ is the standard c.m.i. block. Then the forms of P_{ii}, Q_{ii} are the following:*

$$P_{ii} = \lambda_i I_{\varepsilon_i}, \; Q_{ii} = \lambda_i I_{\varepsilon_i + 1}, \tag{3.1}$$

where λ_i is a non zero constant. □

Lemma 3.2 *Let P_{ij}, Q_{ij} be such that $P_{ij}L_{\varepsilon_i}(s) = L_{\varepsilon_j}(s)Q_{ij}$ where $\varepsilon_i > \varepsilon_j$. Then P_{ij}, Q_{ij} have the following forms*

$$P_{ij} = \begin{bmatrix} \lambda_1^{ij} \cdots \lambda_{\varepsilon_i - \varepsilon_j + 1}^{ij} & & \\ \ddots & & \ddots \\ & \lambda_1^{ij} & \cdots \lambda_{\varepsilon_i - \varepsilon_j + 1}^{ij} \end{bmatrix}_{\varepsilon_j \times \varepsilon_i} \tag{3.2}$$

$$Q_{ij} = \begin{bmatrix} \lambda_1^{ij} \cdots \lambda_{\varepsilon_i-\varepsilon_j+1}^{ij} & & \\ \ddots & \ddots & \\ & \lambda_1^{ij} & \cdots \lambda_{\varepsilon_i-\varepsilon_j+1}^{ij} \end{bmatrix}_{(\varepsilon_j+1)\times(\varepsilon_i+1)} \tag{3.3}$$

□

Lemma 3.3 *Let P_{ij}, Q_{ij} be such that $P_{ij}L_{\varepsilon_i}(s) = L_{\varepsilon_j}(s)Q_{ij}$ where $\varepsilon_i < \varepsilon_j$. Then $P_{ij} = 0$ and $Q_{ij} = 0$.* □

Taking into account that a block matrix block$\{Q_{ij}\}$ with blocks as in the above lemmas has an inverse with the same type and structure of blocks, we have the following result.

Theorem 3.1 *The stabilizer of the canonical element $L(s)$ has the form: P and Q are block matrices with blocks P_{ij} and Q_{ij} respectively, where P_{ii} and Q_{ii} are as in (3.1) and P_{ij} and Q_{ij} are zero matrices if $\varepsilon_i < \varepsilon_j$ or as in (3.2), (3.3) if $\varepsilon_i \geq \varepsilon_j$* □

4 A canonical form for constant matrices

In this section a canonical form under a special transformation group for constant matrices is discussed. This canonical form is important for the derivation of the canonical form. Let A be a constant matrix and consider the following elementary operations on the columns of A

(i) Multiplication of a column by a non zero constant

(ii) Addition of a multiple of the i-th column of A to the j-th column, where $j > i$

The matrix that corresponds to the above transformations is an upper triangular invertible matrix. Consider now the following **reduction procedure** on A:

1. Multiply the first nonzero column by an appropriate constant such that the upper nonzero element of this column is 1.
2. By elementary column transformations eliminate all the entries to the right of the first (upper) nonzero element of the first column.
3. Repeat the above procedure for all the other columns

The resulting matrix has the following form

$$\begin{bmatrix} 0 & 0 & 0 & 0 & 0 & \cdots & \cdots & 0 \\ \vdots & \vdots & 0 & \boxed{1} & 0 & \cdots & \cdots & 0 \\ \vdots & \vdots & \boxed{1} & 0 & \cdots & \cdots & \cdots & 0 \\ \vdots & \vdots & \times & 0 & 0 & \cdots & \cdots & \times \\ \vdots & \vdots & \times & \times & 0 & \cdots & \cdots & 0 \\ \vdots & \vdots & \vdots & \times & \vdots & \boxed{1} & 0 & \cdots \\ 0 & 0 & \times & \times & \boxed{1} & 0 & \cdots & \cdots \end{bmatrix} \tag{4.1}$$

Observe that all the entries to the right of the 1's marked by boxes are zero and all the entries above the 1's are zero as well. The 1's in the boxes will be referred to as *pivot elements.*

Theorem 4.1 *Consider the set of matrices A_i of the same dimensions, which are related by the transformations described above. Then, application of the elimination procedure to any of A_i leads to a unique matrix A_c, which is in canonical form. This canonical form will be referred to as C-canonical form.* □

The above canonical form combined with the transformations corresponding to the stabilizer of $L(s)$ are the main tools for the development of the canonical forms in the following section.

5 Derivation of the canonical form

In this section the construction of the canonical form of the quadruple (E, A, B, C), or equivalently of the pencil $P(s)$, is developed. It is shown that the canonical form is directly related to the echelon form of the composite matrix of a coprime and column reduced MFD of the transfer function $G(s) = C(sE-A)^{-1}B$. Without loss of generality we may always assume that $P(s)$ is in form (2.9). This pencil is not unique since $sK - \Lambda$, C are not uniquely defined. Thus the problem of finding the canonical form of $P(s)$ under r.s.e. transformations is equivalent to the problem of finding a canonical form for the pencil $sK - \Lambda$ and the matrix C in (2.9) under the transformation group preserving the block structure of (2.9), i.e. leaves the blocks (1,1), (1,2), (2,2), (3,3) of $P(s)$ invariant.

Proposition 5.1 *Let $P(s)$ and $P'(s)$ be two pencils of the form* (2.9) *in the same orbit. Then they are related by s.e. transformations of the following form: If $PT\hat{Q} = T'$ then*

$$P = \left[\begin{array}{cc|c} P_1 & 0 & \\ P_3 & I & \\ \hline & & I \end{array}\right], \quad \hat{Q} = \begin{bmatrix} Q & 0 \\ 0 & I \end{bmatrix} \tag{5.1}$$

where $(P_1, Q) \in Stab(L(s))$. □

In order to proceed with the transformations leading to the canonical form it is convenient to consider a partitioning of $sK - \Lambda$ and C conformable to the block-partitioning of $L(s)$. Let

$$sK - \Lambda = [sK_1 - \Lambda_1, sK_2 - \Lambda_2, \ldots, sK_\ell - \Lambda_\ell] \tag{5.2}$$

$$C = [C_1, \ldots, C_\ell] \tag{5.3}$$

where

$$K_i \in \Re^{\ell\times r_i} \text{ and } \Lambda_i \in \Re^{\ell\times r_i},\ i = 1, \ldots, \ell,\ r_i = \varepsilon_i + 1 \tag{5.4}$$

$$C_i \in \Re^{m\times r_i} \tag{5.5}$$

and denote by $\underline{k}_j^i$, k_{qj}^i, $\underline{\lambda}_j^i$, λ_{qj}^i and $\underline{c}_j^i$, c_{qj}^i the columns and the entries of K_i, Λ_i and C_i respectively. Then $P(s)$ may be written as follows

$$P(s) = \left[\begin{array}{c|c|c|c} L_{\varepsilon_1}(s) & & & \\ & \ddots & & \\ & & L_{\varepsilon_\ell}(s) & \\ \hline sK_1 - \Lambda_1 & \cdots & sK_\ell - \Lambda_\ell & -I_\ell \\ \hline C_1 & \cdots & C_\ell & \end{array}\right] \quad (5.6)$$

Proposition 5.2 *The transformations of the type (5.1) can always be expressed as combinations of the following elementary transformations:*

$\boldsymbol{s(\alpha, i)}$: *This is a transformation of the type (5.1) with:*

$$P_1 = bl\text{-}diag\{I_{\varepsilon_1}, \cdots, I_{\varepsilon_{i-1}}, \alpha I_{\varepsilon_i}, I_{\varepsilon_{i+1}}, \cdots, I_{\varepsilon_\ell}\}$$

$$Q = bl\text{-}diag\{I_{\varepsilon_1+1}, \cdots, I_{\varepsilon_{i-1}+1}, 1/\alpha I_{\varepsilon_i+1}, I_{\varepsilon_{i+1}+1}, \cdots, I_{\varepsilon_\ell+1}\}$$

$$P_3 = 0_{\ell\times n}$$

$\boldsymbol{c(\alpha, i, j, k)}$: *This is a transformation of the type (5.1) with:*

$$P_1 = block\{P_{\mu\nu}\}, P_{\mu\mu} = I_{\varepsilon_\mu}, \; P_3 = 0_{\ell\times n}$$

$$P_{\mu\nu} = 0 \text{ if } \mu \neq \nu, \; (\mu, \nu) \neq (i, j)$$

$$P_{ij} = [0_{\varepsilon_i\times(\varepsilon_j-\varepsilon_i-k)}, -\alpha I_{\varepsilon_i}, 0_{\varepsilon_i\times k}]$$

$$Q = block\{Q_{\mu\nu}\}, Q_{\mu\mu} = I_{\varepsilon_\mu+1},$$

$$Q_{\mu\nu} = 0 \text{ if } \mu \neq \nu, \; (\mu, \nu) \neq (i, j)$$

$$Q_{ij} = [0_{(\varepsilon_i+1)\times(\varepsilon_j-\varepsilon_i-k)}, \alpha I_{\varepsilon_i+1}, 0_{(\varepsilon_i+1)\times k}]$$

This transformation is defined only in the case where $r_j \geq r_i$

$\boldsymbol{p(i, j)}$: *This is a transformation of the type (5.1) with:*

$$P_3 = 0_{\ell\times n}, \; P_1 = block\{P_{\mu\nu}\}, \; P_{ii} = P_{jj} = 0_{\varepsilon_i\times\varepsilon_i}$$

$$P_{ij} = P_{ji} = I_{\varepsilon_i}, \; P_{\mu\mu} = I_{\varepsilon_\mu} \text{ if } \mu \neq i, j$$

$$P_{\mu\nu} = 0 \text{ if } \mu \neq \nu, \; (\mu, \nu) \neq (i, j)$$

$$Q = block\{Q_{\mu\nu}\}, \; Q_{ii} = Q_{jj} = 0_{(\varepsilon_i+1)\times(\varepsilon_i+1)},$$

$$Q_{ij} = Q_{ji} = I_{\varepsilon_i+1}, \; Q_{\mu\mu} = I_{\varepsilon_\mu+1} \text{ if } \mu \neq i, j$$

$$Q_{\mu\nu} = 0 \text{ if } \mu \neq \nu, \; (\mu, \nu) \neq (i, j)$$

The above transformation is defined only if $r_i = r_j$.

$\boldsymbol{r(\alpha, i, j, k)}$: *This is a transformation of the type (5.1) with:*

$$P_1 = I_{n-\ell}$$

$$P_3 = [0_{\ell\times\varepsilon_1}, \cdots, 0_{\ell\times\varepsilon_{i-1}}, \cdots, \tilde{P}_{ijk}, 0_{\ell\times\varepsilon_{i+1}}, \cdots, 0_{\ell\times\varepsilon_\ell}]$$

where $\tilde{P}_{ijk}$ is an $\ell \times \varepsilon_i$ matrix with the kj-th entry equal to α and all the other entries zero. Note that in all of the above block{ } denotes block matrix and α is a real constant. □

Remark 5.1 *The above transformations do not affect the top $n-\ell$ rows of the matrix $P(s)$. They affect only the block $[sK^T - \Lambda^T, C^T]^T$ as follows:*

(i) $\boldsymbol{s(\alpha, j)}$: *multiplication of the columns of the $[sK_i^T - \Lambda_i^T, C^T]^T$ block by the scalar α.*

(ii) $\boldsymbol{c(\alpha, i, j, k)}$: *addition of α times the columns r_i, $r_i - 1, \ldots, 1$ of the block $[sK_i^T - \Lambda_i^T, C^T]^T$ to the columns $r_j - k$, $r_j - k - 1, \ldots, r_j - k - r_i + 1$, $[sK_j^T - \Lambda_j^T, C^T]^T$ $k \leq r_j - r_i$, $r_j \geq r_i$. This transformation is defined only in the case where $r_j \geq r_i$.*

(iii) $\boldsymbol{p(i, j)}$: *permutation of the blocks $[sK_i^T - \Lambda_i^T, C^T]^T$, $[sK_j^T - \Lambda_j^T, C^T]^T$ where $r_j = r_i$.*

(iv) $\boldsymbol{r(\alpha, i, j, k)}$: *addition of α times the j-th row of block $L_{\varepsilon_i}(s)$ to the k-th row of $sK_i - \Lambda_i$.* □

The first step of the construction of the canonical form is the elimination of the $1, 2, \ldots, r_i - 1$ columns of the matrices K_i using transformations of the type $r(\alpha, i, j, k)$. The resulting pencil $P(s)$ has the block form (5.6). where

$$K_i = [\; 0_{\ell\times(r_i-1)}, \;\; \underline{k}_{r_i}^i \;] \quad (5.7)$$

Next, by transformations of the form $s(\alpha, j)$ and $c(\alpha, i, j, k)$ we take an equivalent pencil $P(s)$ of the form (5.6) and (5.7) such that the matrix

$$K^* = [\underline{k}_{r_1}^1, \underline{k}_{r_2}^2, \ldots, \underline{k}_{r_\ell}^\ell] \quad (5.8)$$

is in $\mathcal{C}$-canonical form. Let now $r_i = r_{i+1} = r_{i+2} = \ldots = r_i + f_i$ for some $i \in \{1, \ldots, \ell\}$. By transformations of the form $c(\alpha, j)$, $r(\alpha, j, i, 0)$, $p(i, j)$ on $P(s)$ we take a pencil $P_1(s)$ of the form (5.6), (5.7) such that the submatrices of K^* consisting of the columns $i, i+1, \ldots, i+f_i$ $i \in \{1, \ldots, \ell\}$ are in the usual echelon form for constant matrices. Note that all other columns of K^* remain unaltered by the latter transformations. The matrix obtained after the above transformations for all sets of equal r_i, is canonical and will be referred to as *$\mathcal{K}$-canonical form* of K^*.

Proposition 5.3 *Let $P(s)$ be as in (2.9). There exist transformations of the type $s(\alpha, j)$, $c(\alpha, i, j, k)$, $p(i, j)$, $r(\alpha, i, j, k)$ such that when they are applied to $P(s)$ the result is a pencil with the same block structure and $sK - \Lambda$, C are as follows:*

(i) The matrix K^* is in the $\mathcal{K}$-canonical form

(ii) The matrix formed by the columns $\underline{c}_{r_i}^i$, where $\underline{k}_{r_i}^i = 0$ is in the C-canonical form.

(iii) If φ_i is the smallest row index of $\underline{c}_{r_i}^i$, ($\underline{k}_{r_i}^i = 0$), such that $c_{\varphi_i, r_i}^i \neq 0$ then $c_{\varphi_i, q}^j = 0$, $j > i$, $q \geq r_i$.

(iv) If ρ_i is the row index of the first nonzero entry of $\underline{k}_{r_i}^i, \underline{k}_{r_i}^i \neq 0$, then the entries $\lambda_{\rho_i, \nu}^i = 0, \nu > r_i$ □

Let the pencil of the above proposition be denoted by

$$P_c(s) = \left[\begin{array}{c|c} L(s) & 0 \\ sK_c - \Lambda_c & -I \\ \hline C_c & 0 \end{array}\right] \qquad (5.9)$$

Then, we have the following:

Proposition 5.4 *The transfer function of the system (2.1) can be expressed in MFD form $G(s) = N_c(s)D_c^{-1}(s)$ by inspection of the pencil $P_c(s)$ as follows (Vafiadis and Karcanias, 1995a):*

$$N_c(s) = C_c S(s) \qquad (5.10)$$

$$S(s) = bl\text{-}diag\{[1, s, \cdots, s^{r_i-1}]^T\},\ i = 1, \cdots, \ell \qquad (5.11)$$

$$D_c(s) = [\underline{d}_1(s), \cdots, \underline{d}_\ell(s)] \qquad (5.12)$$

$$\underline{d}_i(s) = \underline{k}^i_{r_\ell} s^{r_i} + \underline{\lambda}^i_{r_i} s^{r_i-1} + \cdots + \underline{\lambda}^i_1,\ i = 1, \cdots, \ell \qquad (5.13)$$

where $\underline{k}^i_j$, $\underline{\lambda}^i_j$ are derived from $sK_c - \Lambda_c$. Furthermore the matrix $[N^T(s), D^T(s)]^T$ is in echelon canonical form with pivot indices p_i (Forney, 1975):

$$p_i = \begin{cases} r_1 + \ldots + r_i & \text{if } \underline{k}^i_{r_i} = 0 \\ n + q_i & \text{if } \underline{k}^i_{r_i} \neq 0 \end{cases} \qquad (5.14)$$

where q_i are the pivot indices of the matrix K^* which is in $\mathcal{K}$-canonical form. □

Theorem 5.1 *The canonical form of the system (2.1) under r.s.e. transformations is the system corresponding to the pencil $P_c(s)$*

Proof: In order to prove that the system corresponding to $P_c(s)$ is in canonical form we have to show that it is uniquely defined and that every system $\mathcal{S}(E', A', B', C')$ related by r.s.e. transformations to the original system $\mathcal{S}(E, A, B, C)$ may be transformed to the form defined by $P_c(s)$.

Since $\mathcal{S}(E, A, B, C)$ and $\mathcal{S}(E', A', B', C')$ are related by strict equivalence transformations they have the same transfer function. Thus, the echelon form of the composite matrix $[N^T(s), D^T(s)]^T$ of the transfer function is common to both systems. Since the echelon form is canonical (Forney, 1975), and therefore unique, and $P_c(s)$ is uniquely defined from this echelon composite matrix it follows that $P_c(s)$ is in canonical form. □

The invariants of the system are the controllability indices σ_i of (E, A, B), the reachability indices ε_i of (E, A, B) and the entries of K_c, Λ_c and C_c. The reachability indices r_i are the column minimal indices of the pencil $L(s)$ plus one, and thus, they may be identified immediately since $L(s)$ is in canonical form. On the other hand, the continuous invariants i.e. the entries of K_c, Λ_c, C_c are uniquely defined from the coefficients of the polynomial entries of $[N_c^T(s), D_c^T(s)]$. The canonical form $P_c(s)$ allows the classification of the controllability indices into proper and nonproper (Karcanias and Heliopoulou, 1990; Malabre *et al.*, 1990) by inspection, as follows.

Proposition 5.5 *Let σ_i, $i = 1, \cdots, \ell$ denote the controllability indices of the system (2.1). Then*

$$\sigma_i = \begin{cases} r_i & \text{if } \underline{k}^i_{r_i} \neq 0 \\ r_i - 1 & \text{if } \underline{k}^i_{r_i} = 0 \end{cases} \qquad (5.15)$$

where $\underline{k}^i_{r_i}$ are determined from $P_c(s)$. Furthermore, if $\sigma_i = r_i$, σ_i is a proper index and if $\sigma_i = r_i - 1$ then σ_i is nonproper. □

Remark 5.2 *In order to find the canonical form of $\mathcal{S}(E, A, B, C)$ we may find any MFD of the transfer function $G(s) = C(sE - A)^{-1}B$, form the composite matrix $[N^T(s), D^T(s)]^T$ and find a minimal basis of the col-span of this matrix in echelon form. Then, the canonical form is obtained by inspection.* □

The following example clarifies the construction the canonical form.

Example 5.1 Consider the singular system with

$$P(s) = \left[\begin{array}{ccc|ccccc|cc} s & -1 & 0 & & & & & & & \\ 0 & s & -1 & & & & & & & \\ \hline & & & s & -1 & 0 & 0 & 0 & & \\ & & & 0 & s & -1 & 0 & 0 & & \\ & & & 0 & 0 & s & -1 & 0 & & \\ & & & 0 & 0 & 0 & s & -1 & & \\ \hline 1 & s & 0 & 3 & 0 & 1 & -1 & -s & -1 & 0 \\ 2 & 1 & 0 & 3 & 0 & 4 & -2 & 0 & 0 & -1 \\ \hline 0 & 0 & 1 & 3 & 15 & 1 & 8 & 0 & & \\ 0 & 5 & 1 & 3 & 0 & 0 & 0 & -5 & & \end{array}\right]$$

which is in the form (2.9). Applying the transformation $r(-1, 1, 2, 1)$, we get

$$P^{(1)}(s) = \left[\begin{array}{ccc|ccccc|cc} s & -1 & 0 & & & & & & & \\ 0 & s & -1 & & & & & & & \\ \hline & & & s & -1 & 0 & 0 & 0 & & \\ & & & 0 & s & -1 & 0 & 0 & & \\ & & & 0 & 0 & s & -1 & 0 & & \\ & & & 0 & 0 & 0 & s & -1 & & \\ \hline 1 & 0 & 1 & 3 & 0 & 1 & -1 & -s & -1 & 0 \\ 2 & 1 & 0 & 3 & 0 & 4 & -2 & 0 & 0 & -1 \\ \hline 0 & 0 & 1 & 3 & 15 & 1 & 8 & 0 & & \\ 0 & 5 & 1 & 3 & 0 & 0 & 0 & -5 & & \end{array}\right]$$

The matrix $[\underline{k}^1_{r_1},\ \underline{k}^2_{r_2}] = \begin{bmatrix} 0 & -1 \\ 0 & 0 \end{bmatrix}$ is not in $\mathcal{K}$-canonical form. Apply the transformation $s(-1, 2)$ and take

$$P^{(2)}(s) = \left[\begin{array}{ccc|ccccc|cc} s & -1 & 0 & & & & & & & \\ 0 & s & -1 & & & & & & & \\ \hline & & & s & -1 & 0 & 0 & 0 & & \\ & & & 0 & s & -1 & 0 & 0 & & \\ & & & 0 & 0 & s & -1 & 0 & & \\ & & & 0 & 0 & 0 & s & -1 & & \\ \hline 1 & 0 & 1 & -3 & 0 & -1 & 1 & s & -1 & 0 \\ 2 & 1 & 0 & -3 & 0 & -4 & 2 & 0 & 0 & -1 \\ \hline 0 & 0 & 1 & -3 & -15 & -1 & -8 & 0 & & \\ 0 & 5 & 1 & -3 & 0 & 0 & 0 & 5 & & \end{array}\right]$$

Now $[\underline{k}_{r_1}^1, \ \underline{k}_{r_2}^2] = \begin{bmatrix} 0 & 1 \\ 0 & 0 \end{bmatrix}$ is in $\mathcal{K}$–canonical form

In order to eliminate the entry $c_{14}^2 = -8$, we apply $c(8, 1, 2, 1)$. The resulting pencil is

$$P^{(3)}(s) = \left[\begin{array}{ccc|ccccc|cc} s & -1 & 0 & & & & & & & \\ 0 & s & -1 & & & & & & & \\ \hline & & & s & -1 & 0 & 0 & 0 & & \\ & & & 0 & s & -1 & 0 & 0 & & \\ & & & 0 & 0 & s & -1 & 0 & & \\ & & & 0 & 0 & 0 & s & -1 & & \\ \hline 1 & 0 & 1 & -3 & 8 & -1 & 9 & s & -1 & 0 \\ 2 & 1 & 0 & -3 & 16 & 4 & 2 & 0 & 0 & -1 \\ \hline 0 & 0 & 1 & -3 & -15 & -1 & 0 & 0 & & \\ 0 & 5 & 1 & -3 & 0 & 40 & 8 & 5 & & \end{array}\right]$$

Next, we apply $c(1, 1, 2, 2)$ in order to eliminate $c_{13}^2 = -1$ and we take

$$P^{(4)}(s) = \left[\begin{array}{ccc|ccccc|cc} s & -1 & 0 & & & & & & & \\ 0 & s & -1 & & & & & & & \\ \hline & & & s & -1 & 0 & 0 & 0 & & \\ & & & 0 & s & -1 & 0 & 0 & & \\ & & & 0 & 0 & s & -1 & 0 & & \\ & & & 0 & 0 & 0 & s & -1 & & \\ \hline 1 & 0 & 1 & -2 & 8 & 0 & 9 & s & -1 & 0 \\ 2 & 1 & 0 & -1 & 17 & 4 & 2 & 0 & 0 & -1 \\ \hline 0 & 0 & 1 & -3 & -15 & 0 & 0 & 0 & & \\ 0 & 5 & 1 & -3 & 5 & 41 & 8 & 5 & & \end{array}\right] =$$

$$\left[\begin{array}{cc|c} L_2(s) & & \\ & L_4(s) & \\ \hline sK_1 - \Lambda_1 & sK_2 - \Lambda_2 & I_2 \\ \hline C_1 & C_2 & \end{array}\right]$$

The above pencil is in canonical form. The echelon form of the composite matrix of the transfer function of the system is

$$\left[\frac{N(s)}{D(s)}\right] = \left[\begin{array}{cc} s^2 & -15s - 3 \\ s^2 + 5s & 5s^4 + 8s^3 + 41s^2 + 5s - 3 \\ \hline s^2 + 1 & s^5 + 9s^3 + 8s - 2 \\ s + 2 & 2s^3 + 4s^2 + 17s - 1 \end{array}\right]$$

We see that the entries of $[C_1, C_2]$ may be found from the numerator matrix $N(s)$ and the entries of $[sK_1 - \Lambda_1, sK_2 - \Lambda_2]$ from the denominator $D(s)$ by inspection. The controllability index σ_1 is nonproper since $\underline{k}_3^1 = 0_{2\times 1}$ while σ_2 is proper since $\underline{k}_5^2 = [1, 0]^T$ and the reachability indices are $r_1 = 3$ and $r_2 = 5$.

6 Conclusions

The problem of canonical forms under restricted system equivalence for minimal singular systems with outputs has been considered. The obtained canonical form is an extension of the Popov canonical form for state–space systems. The relation of the canonical form and the echelon canonical form of the composite matrix of any MFD of the transfer function of the system has been established and thus a generalisation of the results of Forney for the case of singular systems has been obtained. The canonical form allows the determination of reachability and controllability indices and their classification into proper and nonproper, by inspection.

7 References

Forney G. D. (1975). "Minimal Bases of Rational Vector Spaces, with Applications to Multivariable Linear Systems", *SIAM J. Control*, vol. 13, pp. 493–520, 1975.

Karcanias N. (1990). "Proper Invariant Realisations of Singular Systems Problems", *IEEE Trans. Automatic. Control*, vol. AC–35, pp. 230–233.

Karcanias N. and **H. Heliopoulou (1990).** "A Classification of Minimal Bases for Singular Systems", *Proc. of Int. Symp. MTNS–89*, vol. I, Birkhauser.

Lebret G. and **J. J. Loiseau 1994).** "Proportional and Proportional–Derivative Canonical Forms for Descriptor Systems with Outputs", *Automatica*, vol. 30, no. 5, pp. 847–864.

MacLane S. and **G. Birkhoff (1967).** *"Algebra"*, Chelsea.

Malabre M., V. Kučera and **P. Zagalak (1990).** "Reachability and Controllability Indices for Linear Descriptor Systems", *Syst. Contr. Letters*, vol. 1, pp. 119–123.

Popov V. M. (1969). "Some properties of the control systems with irreducible matrix transfer functions", in *Lecture Notes in Mathematics*, Springer, Vol. 144, pp. 250–261.

Rosenbrock H.H. (1974). "Structural Properties of Linear Dynamical Systems", *Int. J. Contr.*, vol. 20, no. 2, pp. 191–202.

Vafiadis D. and **N. Karcanias (1995a)** "Generalized State-Space Realizations from Matrix Fraction Descriptions", *IEEE Trans. Automat. Control*, to appear.

Vafiadis D. and **N. Karcanias (1995b)** "Canonical Forms and Invariants for Singular Systems", *City Univ. Res. Rep.*, No. CEC DV/NK–136.

INTERPRETATION AND COMPARISON OF VARIOUS TYPES OF DELAY SYSTEM CONTROLLABILITIES*

Michel Fliess, Hugues Mounier

Laboratoire des Signaux et Systèmes, C.N.R.S.–Supélec
plateau de Moulon, 91192 Gif-sur-Yvette, France
e-mail: fliess@lss.supelec.fr, mounier@lss.supelec.fr

Abstract. The controllability of linear delay differential systems is attacked via module theory. Many of the notions which have been proposed in the literature are interpreted and compared via a change of the ground ring. New definitions are also suggested which, as demonstrated by several case studies, turn out to be often encountered in practice.

Résumé. *On étudie la commandabilité des systèmes linéaires à retards grâce à la théorie des modules. Le changement de l'anneau de base permet d'interpréter et de comparer bien des notions introduites dans la littérature. De nouvelles définitions sont proposées dont la pertinence pratique est démontrée par plusieurs exemples.*

Keywords: Linear delay systems, controllability, reachability, weak controllability, spectral controllability, module theory, change of ring.

Introduction

Many concurrent definitions have been proposed for the controllability of linear systems with delays, which, roughly speaking, stem either from a functional analytic approach (*see, e.g.*, [1], [2], [3], [19], [29]) or from a more algebraic viewpoint (*see, e.g.*, [4], [6], [14], [27]). We here suggest a unified framework which permits to recover most of the past definitions and to compare them. It also leads to many other natural notions of controllability, which do not seem to have been considered in the literature.

We generalize to the delay case the module theoretic approach of [9], [10], which contrary to the classic Kalman setting [13], encompass all system variables (and not only the state-space). A linear delay system is then a finitely generated module Λ over the ring $k[\frac{d}{dt}, \boldsymbol{\delta}]$ of polynomials where k is the ground field, $\mathbb{R}$ for instance, and $\boldsymbol{\delta} = (\delta_1, \ldots, \delta_r)$ are the delay operators.

The classic reachability property [4] is nothing else than the freeness of Λ, as shown by the resolution due to Quillen [25] and Suslin [28] of Serre's famous conjecture [26]. For dealing with weak controllability, we need to consider the tensor product $k(\boldsymbol{\delta})[\frac{d}{dt}] \otimes_{k[\frac{d}{dt},\boldsymbol{\delta}]} \Lambda$ where $k(\boldsymbol{\delta})$ is the usual field of fractions of the ring of polynomials $k[\boldsymbol{\delta}]$: $k(\boldsymbol{\delta})[\frac{d}{dt}] \otimes_{k[\frac{d}{dt},\boldsymbol{\delta}]} \Lambda$ is a module over $k(\boldsymbol{\delta})[\frac{d}{dt}]$. Weak controllability [20] turns out to be equivalent to the freeness of this latter module. Varying the ring in this way is our key method. It can be shown that under mild conditions any system may be defined to be controllable in the following sense: There exists a polynomial $\pi \in k[\boldsymbol{\delta}]$ such that $k[\frac{d}{dt}, \boldsymbol{\delta}, \pi^{-1}] \otimes_{k[\frac{d}{dt},\boldsymbol{\delta}]} \Lambda$ is free, where $k[\frac{d}{dt}, \boldsymbol{\delta}, \pi^{-1}]$ is obtained from $k[\frac{d}{dt}, \boldsymbol{\delta}]$ by a classical algebraic localization operation. We will then call this system π-free.

An analogous treatment may be proposed in the functional analytic setting, by replacing, as customary, a delay of amplitude h by e^{-hs} (where s denotes $\frac{d}{dt}$). Spectral controllability [3] may then be viewed as the system module being torsion free over a suitable subring of the ring of entire functions, containing (in addition to s and e^{-hs}) certain types of distributed delays as operators. Approximate controllability [19] is interpreted along the same lines.

We report here detailed calculations on various case studies, namely the stabilization of a ship, crude oil mixing for octane number control, a wind tunnel model for control of the Mach number, and a two reflector antenna. They are seen to be spectrally controllable, weakly controllable, and, except for the first one, not reachable. We demonstrate that they are δ-free: this property, which conveys deep physical properties, is shared by a number of practical examples.

The paper is organized as follows. After some preliminaries, we give controllability definitions in our setting and interpretations of the above mentionned notions. A comparision diagram illustrates the preceeding results. Finally the examples are examined.

Full details as well as many other examples may be found in [22]. See also [11] for a first announcement and [21], [23], [12] for other aspects.

*Work partially supported by the G.D.R. "Automatique" of the C.N.R.S.

1 Preliminary notions

1.1. Let k be a commutative field with algebraic closure $\bar{k}$. We will consider the commutative polynomial ring $k[\frac{d}{dt},\boldsymbol{\delta}]$ in $r+1$ indeterminates $\frac{d}{dt}$, $\boldsymbol{\delta}=(\delta_1,\ldots,\delta_r)$, where $\delta_1,\ldots,\delta_r$ play the role of localized delay operators with constant amplitude.
Remark. – Let $k=\mathbb{R}$ and, for every map $f:\mathbb{R}\to\mathbb{R}$, $(\delta_i.f)(t)=f(t-h_i)$, $h_i\in\mathbb{R}$, $h_i>0$, $i=1,\ldots,r$. The h_i's are supposed to be non-commensurate, *i.e.*, the $\mathbb{Q}$-vector space generated by $h_1,\ldots,h_r$ is r-dimensional.

1.2. Except otherwise stated, every module will be a finite type $k[\frac{d}{dt},\boldsymbol{\delta}]$-module. $[w]$ will denote the module generated by a set w.

1.3. A straightforward generalization of [9] leads to the following. A *linear (differential) system with delays* is a module. A *linear dynamics with delays* is a module Λ where one distinguishes a finite set $\boldsymbol{u}=(u_1,\ldots,u_m)$, *the input*, such that the quotient module $\Lambda/[\boldsymbol{u}]$ is torsion. The input will be called *independent* if $[\boldsymbol{u}]$ is free of rank m. An *output* $\boldsymbol{y}=(y_1,\ldots,y_p)$ is a finite set of elements in Λ.
Remark. – The analogue of infinite dimension in our context reads as follows: a module such as $\Lambda/[\boldsymbol{u}]$, will not, in general, be a finite dimensional k-vector space.

1.4. *Remark.* – We refer to [8] for ring and module notions we are utilizing.

2 Controllability

2.1 First definitions and freeness

2.1.1. It appears from [9] that usual controllability corresponds to the lack of torsion elements. Let A be a $k[\frac{d}{dt},\boldsymbol{\delta}]$-algebra. We will thus call a system Λ *A-torsion free controllable* if $A\otimes_{k[\frac{d}{dt},\boldsymbol{\delta}]}\Lambda$ is torsion free. It will be said *A-projective controllable* (resp. *A-free controllable*) if $A\otimes_{k[\frac{d}{dt},\boldsymbol{\delta}]}\Lambda$ is projective (resp. free).

2.1.2. Let Λ be a system given by generators and relations. More precisely $\Lambda=[\boldsymbol{w}]$ with $\boldsymbol{w}=(w_1,\ldots,w_s)$ and

$$P(\tfrac{d}{dt},\boldsymbol{\delta})\boldsymbol{w}=0 \tag{1}$$

where P is a matrix with coefficients in $k[\frac{d}{dt},\boldsymbol{\delta}]$, the rank of which equals l.
Some considerations involving the ideal generated by the $l\times l$ minors of P (*see* [5]) as well as the affirmative answer to Serre's conjecture imply the following (*see* [11], [22]):

Theorem. – *(1) is $k[\frac{d}{dt},\boldsymbol{\delta}]$-free controllable if, and only if,* $\forall(s,z_1,\ldots,z_r)\in\bar{k}^{r+1}$, $\mathrm{rk}_{\bar{k}}P(s,z_1,\ldots,z_r)=l$.

2.2 π-freeness

Consider a $k[\frac{d}{dt},\boldsymbol{\delta}]$-torsion free controllable system Λ. One can wonder wether it is possible to have Λ A-free controllable over a suitable ring A containing $k[\frac{d}{dt},\boldsymbol{\delta}]$. The answer is given by a usual localization procedure (*see* [12], [22])

Theorem. – *Let Λ be a $k[\frac{d}{dt},\boldsymbol{\delta}]$-torsion free controllable delay system. Then, there exists $\pi\in k[\boldsymbol{\delta}]$ such that Λ is $k[\frac{d}{dt},\boldsymbol{\delta},\pi^{-1}]$-free controllable.*

The preceeding system will then be called *π-free*.

2.3 Interpretations of algebraic notions

2.3.1. Consider the usual dynamics

$$\dot{\boldsymbol{x}}=A(\boldsymbol{\delta})\boldsymbol{x}+B(\boldsymbol{\delta})\boldsymbol{u} \tag{2}$$

where $\boldsymbol{x}=(x_1,\ldots,x_n)$, $\boldsymbol{u}=(u_1,\ldots,u_m)$, $A(\boldsymbol{\delta})\in k[\boldsymbol{\delta}]^{n\times n}$ and $B(\boldsymbol{\delta})\in k[\boldsymbol{\delta}]^{n\times m}$. Denote by $\langle B(\boldsymbol{\delta})\mid A(\boldsymbol{\delta})\rangle$ the $k[\boldsymbol{\delta}]$-module generated by the columns of the $[B(\boldsymbol{\delta}),A(\boldsymbol{\delta})B(\boldsymbol{\delta}),\ldots,A^{n-1}(\boldsymbol{\delta})B(\boldsymbol{\delta})]$ matrix. The algebraic *reachability* notion (*see* [4], [27]) is expressed by any one of the two equivalent conditions:
$\langle B(\boldsymbol{\delta})\mid A(\boldsymbol{\delta})\rangle=k[\boldsymbol{\delta}]^n$,
$\forall(s,z_1,\ \ldots,z_r)\in\bar{k}^{r+1}$,
$\mathrm{rk}_{\bar{k}}[sI-A(z_1,\ldots,z_r)\mid B(z_1,\ldots,z_r)]=n$.
Thus (*see* [11], [22]):

Proposition. – *(2) verifies the preceeding conditions if, and only if, it is $k[\frac{d}{dt},\boldsymbol{\delta}]$-free controllable.*

2.3.2. *Examples.* – Set $r=1$, $h_1=1$ and $m=1$. The system $\dot{y}(t)-y(t-1)=u(t)$ is strongly controllable. Indeed, the corresponding module is free, with basis y. And the system $\dot{y}(t)=u(t-1)$ is controllable but not strongly controllable. We effectively have $\mathrm{rk}_{\bar{k}}[s\mid z]=0$ for $s=z=0$.

2.3.3. Let $k(\boldsymbol{\delta})$ be the field of fractions of $k[\boldsymbol{\delta}]$. For every module Λ, the canonical map $\Lambda\to k(\boldsymbol{\delta})\otimes_{k[\boldsymbol{\delta}]}\Lambda$ is injective if, and only if, Λ, considered as a $k[\boldsymbol{\delta}]$-module, is torsion free. This latter hypothesis is natural for delay systems. We will assume it to hold in the next proposition.

2.3.4. The *weak controllability* notion (*see* [20]) is derived for (2) from Kalman's criterion: $\mathrm{rk}_{k(\boldsymbol{\delta})}[B,AB,\ldots,A^{n-1}B]=n$. Then (*see* [11], [22]):

Proposition. – *(2) satisfies Kalman's criterion if, and only if, it is $k(\boldsymbol{\delta})[\frac{d}{dt}]$-free controllable.*

2.4 Interpretations of functional analytic notions

2.4.1. Assume $k=\mathbb{R}$ in (2). Replace, as usual, $\frac{d}{dt}$ by s and δ_i by e^{-h_is}, $i=1,\ldots,r$. The *spectral controllability* (*see* [3]) of (2) can be seen as a generalization of Hautus's criterion. It is equivalent to the fact that, for all $s \in \mathbb{C}$,

$$\mathrm{rk}_{\mathbb{C}}[sI - A(e^{-h_1 s}, \ldots, e^{-h_r s}) \mid B(e^{-h_1 s}, \ldots, e^{-h_r s})] = n.$$

Introduce the following two entire complex functions rings:

$$E_{\mathbb{F}} = \left\{ \sum_{i=0}^{N} p_i(s) e^{-\sum_{j=1}^{r} \alpha_{ij} h_j s} \mid p_i \in \mathbb{C}[s], \alpha_{ij} \in \mathbb{F} \right\},$$

$$S_{\mathbb{F}} = \left\{ \frac{f(s)}{s - s_0} \mid f \in E_{\mathbb{F}}, s_0 \in \mathbb{C}, f(s_0) = 0 \right\}$$

where $N \in \mathbb{N}$ and $\mathbb{F}$ is an additive submonoid of $\mathbb{R}$, such as $\mathbb{Q}$ or $\mathbb{Q}_+$. The divisibility properties of $E_{\mathbb{Q}_+}$, (*see* [24]), and the Skolem-Mahler-Lech theorem on the zeros of exponential polynomials (*see* [17]) yield (*see* [11], [23]):

Theorem. - *(2) is spectrally controllable if, and only if, it is $S_{\mathbb{Q}_+}$-torsion free controllable.*

Note that the elements of $S_{\mathbb{Q}_+}$ can be interpreted as differential operators with distributed delays (*see* [15]).

2.4.2. *Approximate controllability* corresponds to the density of the reachable space in a given function space. For a dynamics of the form:

$$\begin{aligned} \dot{\boldsymbol{x}}(t) = A_{-2}\dot{\boldsymbol{x}}(t-h_2) + A_{-1}\dot{\boldsymbol{x}}(t-h_1) \\ +A_0\boldsymbol{x}(t)A_1\boldsymbol{x}(t-h_1) + A_2\boldsymbol{x}(t-h_2) \\ +B_0\boldsymbol{u}(t) + B_1\boldsymbol{u}(t-a_1) \end{aligned} \quad (3)$$

($k = \mathbb{R}$, $\boldsymbol{x} \in \mathbb{R}^n$, $\boldsymbol{u} \in \mathbb{R}^m$, $a_i \in \mathbb{R}$, $h_i \in \mathbb{R}, 0 < a_1$, $0 < h_1 < h_2$, $a_1 \leqslant h_2$), a matrix criterion (*see* [1], [29]) is given by:
$\mathrm{rk}_{\mathbb{C}}[sI - A_{-1}se^{-h_1 s} - A_{-2}se^{-h_2 s} - A_0 - A_1 e^{-h_1 s} - A_2 e^{-h_2 s} \mid B_0 + B_1 e^{-a_1 s}] = n$, for all $s \in \mathbb{C}$, and $\mathrm{rk}_{\mathbb{C}}(sA_{-2} + A_2 \mid B_1) = n$, for almost all $s \in \mathbb{C}$. Consider the dynamics $\widetilde{\Lambda} = E_{\mathbb{Q}} \otimes_{\mathbb{R}[s, e^{-h_i s}]} \Lambda$ where Λ stands for (3). Let δ_{h_i} and δ_{a_i} be the delays of respective amplitude h_i and a_i. Let also $\boldsymbol{v}$ be a new input for $\widetilde{\Lambda}$ such that $\boldsymbol{u} = \delta_{h_2 - a_1}\boldsymbol{v}$. Then:

Proposition. - *(3) is approximately controllable if, and only if, the two following conditions are satisfied:*
(i) it is spectrally controllable,
(ii) the $E_{\mathbb{Q}}$-module $\delta_{h_2}^{-1}\widetilde{\Lambda}/\delta_{h_1}^{-1}\widetilde{\Lambda}$ is torsion free.

3 A comparison diagram

Assume a single delay, *i.e.*, $r = 1$, in (2) and set $\delta = e^{-h\frac{d}{dt}}$; then

$$\frac{d}{dt}\boldsymbol{x} = A(e^{-h\frac{d}{dt}})\boldsymbol{x} + B(e^{-h\frac{d}{dt}})\boldsymbol{u}$$

where A and B are matrices of appropriate sizes, with coefficients in $\mathbb{R}[\frac{d}{dt}, e^{-h\frac{d}{dt}}]$, and $A(e^{-h\frac{d}{dt}}) = \sum_{i=0}^{N} A_i e^{-ih\frac{d}{dt}}$. Setting

$$\begin{aligned} \langle A(\delta), B(\delta) \rangle &= [B(\delta), \ldots, A(\delta)^{n-1}B(\delta)] \\ \Delta(s, z) &= [sI - A(z) \mid B(z)] \\ \Delta(s, e^{-hs}) &= [sI - A(e^{-hs}) \mid B(e^{-hs})], \end{aligned}$$

yields the implications and equivalences diagram depicted in figure 1 on the next page.

4 Case studies

4.1 Ship stabilization

A ship is rolling in the waves. A servomechanism is utilized to reduce the angle of tilt θ and its velocity to zero, by ballast displacement. An effect

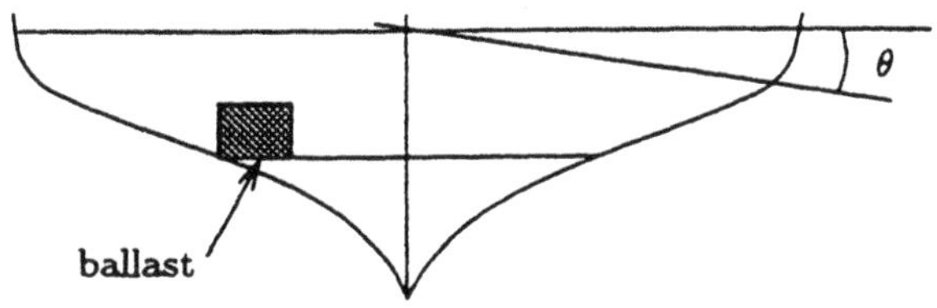

Figure 2: Ship stabilization

of this contrivance is to introduce a term proportional to $\dot{\theta}$ at an earlier instant $t - h$. It also introduces a control u, yielding the following equations (*see* [7]):

$$\begin{aligned} \dot{x}_1(t) &= x_2(t) \\ \dot{x}_2(t) &= -\frac{b}{m}x_2(t) - \frac{q}{m}x_2(t-h) - \frac{k}{m}x_1(t) + u(t) \end{aligned}$$

where m stands for the mass of the boat, b is a damping factor, q and k are constants. The associated matrices are

$$A(\delta) = \begin{bmatrix} 0 & 1 \\ -\frac{k}{m} & -\frac{b}{m} - \delta\frac{q}{m} \end{bmatrix}, \quad B(\delta) = \begin{bmatrix} 0 \\ 1 \end{bmatrix}.$$

This dynamics is $\mathbb{R}[\frac{d}{dt}, \delta]$-free controllable (*i.e.*, reachable):

$$\forall (s, z) \in \mathbb{C}^2, \quad \mathrm{rk}_{\mathbb{C}} \begin{bmatrix} s & -1 & 0 \\ \frac{k}{m} & \frac{b}{m} + \frac{q}{m}z & 1 \end{bmatrix} = 2.$$

Indeed, the previous condition is equivalent to the minors $s(b/m + (q/m)z) - q/m$, $-s$ and 1 of $[sI - A(z) \mid B(z)]$ having no common zeros in $\mathbb{C}^2$. Consequently, this system is also spectrally and weakly controllable.

4.2 Crude oil mixing

The following example is due to P. Rouchon (personal communication). Take two tanks of crude

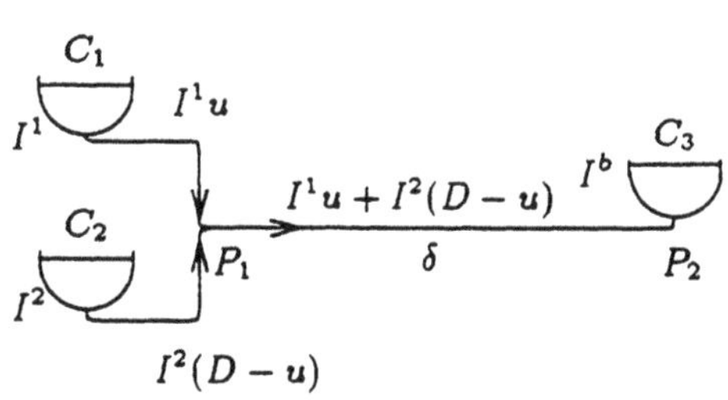

Figure 3: Crude oil mixing

oil C_1 and C_2 with respective octane numbers I^1

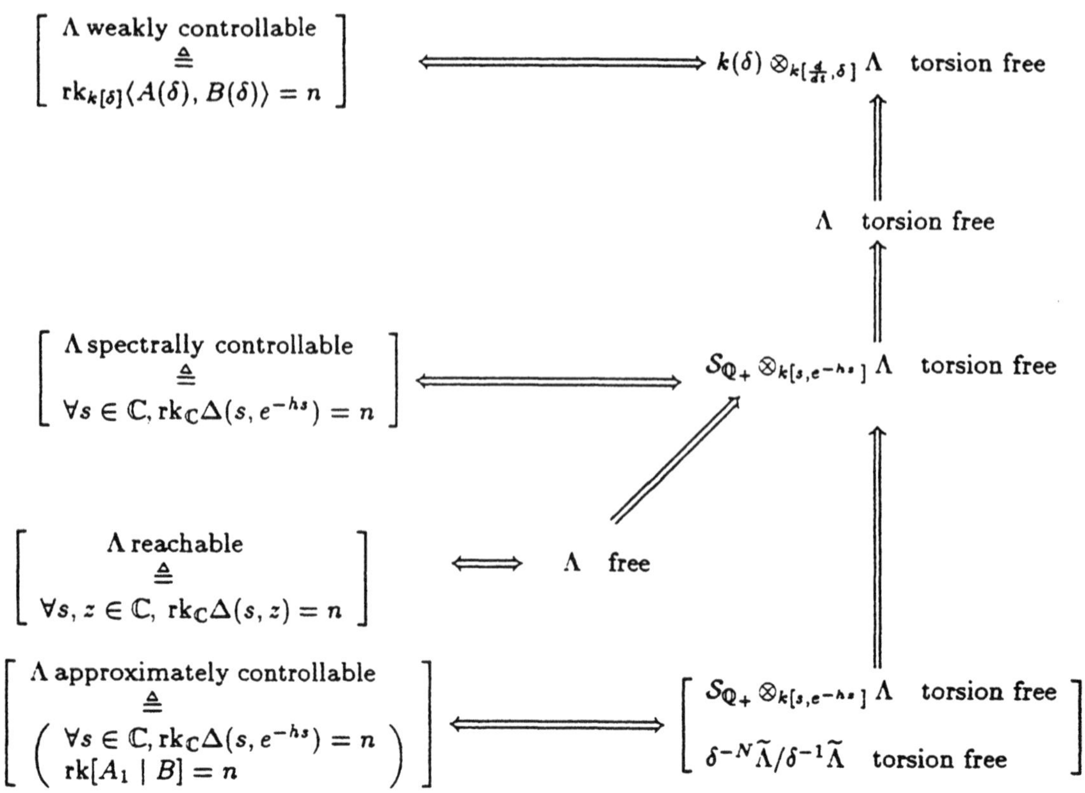

Figure 1: Controllability implications and equivalences diagram

and I^2. The output of C_1 (resp. C_2) is u (resp. $D-u$) where D is a constant. The two outputs mix in a pipe and are transported to a tank C_3. The transportation lag corresponds to a delay δ, and the mixture in C_3 has the octane number I^b. Supposing that, at time t, the mixture's volume is V_t and the octane number in C_3 is I_t^b, one has, from t to $t+dt$:

$$I_t^b V_t + \delta(I^1 u + I^2(D-u)) = V_{t+dt} I_{t+dt}^b.$$

Thus:

$$\frac{d}{dt}(I^b V)(t) = I^2 D + (I^1 - I^2)u(t-h) \qquad (4)$$

where h designates the amplitude of the delay δ. Set $y = I^b V$, and $v = u + I^2 D/(I^1 - I^2)$:

$$\dot{y} = (I^1 - I^2)v(t-h). \qquad (5)$$

This dynamics is weakly controllable, since

$$\mathrm{rk}_{\mathbb{R}(\delta)}[(I^1 - I^2)\delta] = 1.$$

It is not reachable, as

$$\mathrm{rk}_{\mathbb{C}}[s \mid (I^1 - I^2)z] = 0$$

for $s = z = 0$. It is spectrally controllable, the full rank condition

$$\forall s\in\mathbb{C}, \quad \mathrm{rk}_{\mathbb{C}}[s \mid (I^1 - I^2)e^{-hs}] = 1$$

being verified. And it is δ-free, with basis y:

$$v = \frac{1}{(I^1 - I^2)}\delta^{-1}\dot{y}.$$

4.3 Wind tunnel model

We consider the following linearized model of Mach number dynamics (*see* [18]):

$$\begin{aligned}\dot{x}_1 &= -ax_1 + kax_2(t-h)\\ \dot{x}_2 &= x_3\\ \dot{x}_3 &= -\omega^2 x_2 - 2\zeta\omega x_3 + \omega^2 u.\end{aligned}$$

where ω and ζ are constants. Call δ the delay operator with amplitude h. We have, as dynamics and control matrices

$$A(\delta) = \begin{bmatrix} -a & ka\delta & 0\\ 0 & 0 & 1\\ 0 & -\omega^2 & -2\zeta\omega\end{bmatrix}, \quad B(\delta) = \begin{bmatrix}0\\0\\\omega^2\end{bmatrix}.$$

This system is weakly controllable, since $[B(\delta), A(\delta)B(\delta), A^2(\delta)B(\delta)]$ has full rank:

$$\mathrm{rk}_{\mathbb{R}(\delta)}\begin{bmatrix} 0 & 0 & ka\omega^2\delta\\ 0 & \omega^2 & -2\zeta\omega^3\\ \omega^2 & -2\zeta\omega^3 & (4\zeta^2-1)\omega^4\end{bmatrix} = 3.$$

It is also spectrally controllable, *i.e.*, $\forall s\in\mathbb{C}$

$$\mathrm{rk}_{\mathbb{C}}\begin{bmatrix} s+a & -ake^{-hs} & 0 & 0\\ 0 & s & -1 & 0\\ 0 & \omega^2 & s+2\zeta\omega & \omega^2\end{bmatrix} = 3.$$

But it is not reachable (the precceding matrix with $e^{-hs} = z$ exhibits a loss of rank for $s = -a$,

$z = 0$).

Finally, this dynamics is δ-free, with basis x_1:

$$\begin{aligned} x_2 &= \frac{1}{ak}\delta^{-1}(\dot{x}_1 + ax_1) \\ x_3 &= \frac{1}{ak}\delta^{-1}(\ddot{x}_1 + a\dot{x}_1) \\ u &= \frac{1}{ak\omega^2}\delta^{-1}(x_1^{(3)} + (2\zeta\omega + a)\ddot{x}_1 \\ &\quad +(2a\zeta + \omega)\omega\dot{x}_1 + a\omega^2 x_1). \end{aligned}$$

4.4 A two reflector antenna

We consider a two reflector antenna with a big immovable reflector and a little movable one, furnished with a control system. This little reflector has a radiator situated in its focus, wich is placed on a massive platform having a diameter of 7m. Three screws can move three points of the platform in the vertical direction and the position errors in the screw servomechanisms are measured by hydrostatic pickups having a great time lag, until 2 seconds.

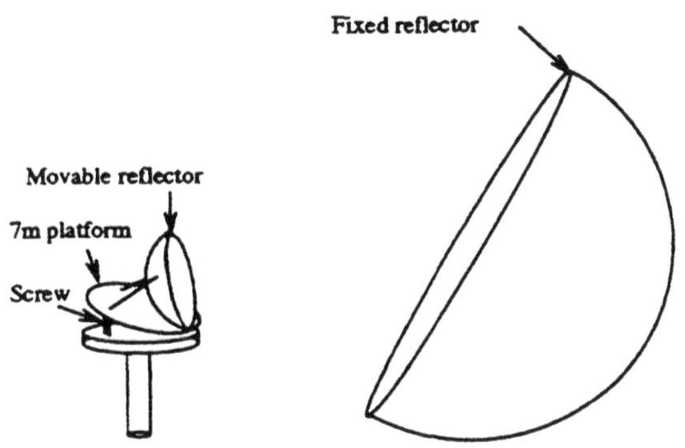

Figure 1 : Two reflector antenna.

This dynamics is 6-dimensional (*see* [16]):

$$\dot{\boldsymbol{x}} = A\boldsymbol{x} + B\delta\boldsymbol{u}$$

where $\boldsymbol{x} = (x_1, \ldots, x_6)^T$, $\boldsymbol{u} = (u_1, u_2, u_3)^T$,

$$A = \begin{bmatrix} 0 & K_1 & 0 & 0 & 0 & 0 \\ 0 & -T_e^{-1}K_2 & 0 & 0 & 0 & 0 \\ 0 & 0 & 0 & K_1 & 0 & 0 \\ 0 & 0 & 0 & -T_e^{-1}K_2 & 0 & 0 \\ 0 & 0 & 0 & 0 & 0 & K_1 \\ 0 & 0 & 0 & 0 & 0 & -T_e^{-1}K_2 \end{bmatrix}$$

$$B = T_e^{-1}\begin{bmatrix} 0 & 0 & 0 \\ K_p & K_c & K_c \\ 0 & 0 & 0 \\ K_c & K_p & K_c \\ 0 & 0 & 0 \\ K_c & K_c & K_p \end{bmatrix}$$

The parameters K_1 and K_2 are given by $K_1 = K_v K_{p_0} K_e K_r$ and $K_2 = 1 + K_{p_0} K_e K_f$, with K_v a voltage gain, K_{p_0} a power gain, K_e for the engine, K_r a reduction coefficient, and K_f a local feedback gain. The K_c and K_p are respectively an interconnection gain and the pickups gain. The system is not $\mathbb{R}[\frac{d}{dt}, \delta]$-free controllable, since the matrix

$$\begin{bmatrix} s & -K_1 & 0 & 0 & 0 & 0 & 0 & 0 & 0 \\ 0 & s+\frac{K_2}{T_e} & 0 & 0 & 0 & 0 & \frac{K_p}{T_e}z & \frac{K_c}{T_e}z & \frac{K_c}{T_e}z \\ 0 & 0 & s & -K_1 & 0 & 0 & 0 & 0 & 0 \\ 0 & 0 & 0 & s+\frac{K_2}{T_e} & 0 & 0 & \frac{K_c}{T_e}z & \frac{K_p}{T_e}z & \frac{K_c}{T_e}z \\ 0 & 0 & 0 & 0 & s & -K_1 & 0 & 0 & 0 \\ 0 & 0 & 0 & 0 & 0 & s+\frac{K_2}{T_e} & \frac{K_c}{T_e}z & \frac{K_c}{T_e}z & \frac{K_p}{T_e}z \end{bmatrix}$$

has losses of rank for $s = z = 0$ and $s = -T_e^{-1}K_2$, $z = 0$. But it is spectrally controllable, as the preceeding matrix with $z = e^{-hs}$ has no loss of rank. It is again δ-free, with basis (x_1, x_3, x_5):

$$\begin{bmatrix} u_1 \\ u_2 \\ u_3 \end{bmatrix} = \begin{bmatrix} K_p & K_c & K_c \\ K_c & K_p & K_c \\ K_c & K_c & K_p \end{bmatrix}^{-1} p(\frac{d}{dt})I_3\delta^{-1}\begin{bmatrix} x_1 \\ x_3 \\ x_5 \end{bmatrix}$$

$$\begin{bmatrix} x_2 \\ x_4 \\ x_6 \end{bmatrix} = K_1^{-1}I_3\frac{d}{dt}\begin{bmatrix} x_1 \\ x_3 \\ x_5 \end{bmatrix}$$

where $p(\frac{d}{dt}) = K_1^{-1}\frac{d}{dt}(T_e\frac{d}{dt} + K_2)$ and I_3 is the 3×3 identity matrix.

Conclusion

Our examples and other ones studied elsewhere [22] indicate that δ-free and, more generally, π-free delay systems might play a fundamental role. This type of controllability, wich seems to be new, is most easily expressed in our formalism. Other contributions [23], [22], [12] will demonstrate that several natural control problems, such as path tracking, may then be easily solved.

References

[1] Z. Bartosiewicz. Approximate controllability of neutral systems with delays in control. *J. Diff. Eq.*, 51:295–325, 1984.

[2] A. Bensoussan, G. Da Prato, M.C. Delfour, and S.K. Mitter. *Representation and Control of Infinite Dimensional Systems.* vol. 1 & 2, Birkhäuser, Boston, 1992-1993.

[3] K.P.M Bhat and H.N. Koivo. Modal characterizations of controllability and observability for time-delay systems. *IEEE Trans. Automat. Contr.*, 21:292–293, 1976.

[4] J.W. Brewer, J.W. Bunce, and F.S. Van Vleck. *Linear Systems over Commutative Rings.* Marcel Dekker, New York, 1986.

[5] D.A. Buchsbaum and D. Eisenbud. What makes a complex exact ? *J. Alg.*, 25:259–268, 1973.

[6] C.I. Byrnes. On the control of certain deterministic, infinite-dimensional systems by algebro-geometric techniques. *Amer. J. Math.*, 100:1333–1381, 1978.

[7] E.N. Chukwu. *Stability and Time-Optimal Control of Hereditary Systems.* Academic Press, Boston, 1989.

[8] P.M. Cohn. *Algebra.* 2nd ed., vol. 1 to 3, Wiley, Chichester, 1982–1989–1991.

[9] M. Fliess. Some basic structural properties of generalized linear systems. *Systems Control Lett.*, 15:391–396, 1990.

[10] M. Fliess. Une interprétation algébrique de la transformation de Laplace et des matrices de transfert. *Linear Alg. Appl.*, 203–204:429–442, 1994.

[11] M. Fliess and H. Mounier. Quelques propriétés structurelles des systèmes linéaires à retards constants. *C. R. Acad. Sci. Paris*, I–319:289–294, 1994.

[12] M. Fliess, H. Mounier, and P. Rouchon. Controllability and motion planning for linear systems with delays. In preparation.

[13] R.E. Kalman, P.L. Falb, and M.A. Arbib. *Topics in Mathematical Systems Theory.* McGraw-Hill, New York, 1969.

[14] E.W. Kamen. On an algebraic theory of systems defined by convolution operators. *Math. Syst. Theory*, 9:57–74, 1975.

[15] E.W. Kamen, P.P. Khargonekar, and A. Tannenbaum. Proper stable bezout factorization and feedback control of linear time-delay systems. *Internat. J. Control*, 43:837–857, 1986.

[16] V.B. Kolmanovskii and V.R. Nosov. *Stability of Functional Differential Equations.* Academic Press, London, 1986.

[17] K. Mahler. On the Taylor coefficients of rational functions. *Proc. Camb. Phil. Soc.*, 52:51–60, 1956.

[18] A. Manitius. Feedback controllers for a wind tunnel model involving a delay: Analytical design and numerical simulations. *IEEE Trans. Automat. Contr.*, 29:1058–1068, 1984.

[19] A. Manitius and R. Triggiani. Function space controllability of retarded systems : a derivation from abstract operator conditions. *SIAM J. Contr. Opt.*, 16:599–645, 1978.

[20] A.S. Morse. Ring models for delay-differential systems. *Automatica*, 12:529–531, 1976.

[21] H. Mounier. An algebraic interpretation of spectral controllability of linear delay systems. *Proc. 3rd European Control Conference*, Rome, 1995.

[22] H. Mounier. *Propriétés structurelles des systèmes linéaires à retards : aspects théoriques et pratiques.* Thèse, Université Paris-Sud, 1995.

[23] H. Mounier and J. Rudolph. A flexible rod as a linear delay system. *Proc. 3rd European Control Conference*, Rome, 1995.

[24] A.J. van der Poorten and R. Tijdeman. On common zeros of exponential polynomials. *L'enseignement mathématique*, 21:57–67, 1975.

[25] D. Quillen. Projective modules over polynomial rings. *Inv. Math.*, 36:167–171, 1976.

[26] J.P. Serre. Faisceaux algébriques cohérents. *Annals. Math.*, 61:197–278, 1955.

[27] E. D. Sontag. Linear systems over commutative rings: a survey. *Richerche di Automatica*, 7:1–34, 1976.

[28] A. A. Suslin. Projective modules over a polynomial ring are free (in Russian). *Dokl. Akad. Nauk. S.S.S.R.*, 229:1063–1066, 1976. (English translation: *Soviet Math. Dokl.*, 17:1160–1164).

[29] Y. Yamamoto. Reachability of a class of infinite-dimensional linear systems: an external approach with applications to general neutral systems. *SIAM J. Contr. Opt.*, 27:217–234, 1989.

MODEL MATCHING PROBLEM FOR SYSTEMS OVER A RING AND APPLICATIONS TO DELAY-DIFFERENTIAL SYSTEMS

G. CONTE*, A.M. PERDON**

** Dip. Elettronica Automatica, Università di Ancona, via Brecce Bianche, 60131 Ancona, ITALY*
*** Dip.Matematica "V. Volterra", Università di Ancona, via Brecce Bianche, 60131 Ancona, ITALY*

Abstract. Le résultat plus important contenu dans ce papier est une condition géometrique necessaire et suffisante pour l'existence de solutions du Problème de la Poursuite de Modèle dans le cas de systèmes sur un anneau. Le résultat est obtenu en considérant un Problème de Rejet de Pertourbations associé L'applications au cas de systèmes à rétard est illustré par un example.

Abstract. The main result of the paper is a geometric necessary and sufficient condition for the existence of solution of the Model matching Problem for systems over a ring. The result is obtained by considering an associated Disturbance Decoupling problem. An application to the case of delay-differntial systems is illustrated by an example.

Key Words. Systems over rings; geometric methods; model matching

1. INTRODUCTION

Many interesting situations, involving delay-differential systems, families of systems depending on one parameter, time-varying systems and so on, provide motivations for studying systems with coefficients in a ring, or systems over a ring (see Kamen (1976), Sontag (1976), Sontag (1981)). It is therefore useful to extend to this class of systems tools and techniques previously developed for systems with coefficients in a field. In particular, some authors (Conte et al. (1982), Conte et al. (1995), Hautus (1982), Hautus (1984), Inaba et al. (1992)) have considered the possibility of extending to systems over a ring the so called geometric approach (Wohnam (1985)).

In this paper we will present some recent contributions to the geometric theory of systems over a ring that are applicable, in particular, to delay-differential systems.

The first results concerning a geometric theory for systems over a ring were obtained by M. Hautus and by the authors in 1982 (see Hautus (1982)). The main difference encountered in dealing with coefficients in a ring instead of coefficients in a field concerns the fundamental geometric notion of (A,B)-invariant subspace of the state space, or controlled invariant subspace (Wohnam (1985)). In the case of coefficients in a field, any (A,B)-invariant subspace V can be made invariant with respect to a closed loop dynamics by chosing a suitable state feedback. This property is lost in the case of coefficients in a ring and, as a consequence, many results obtained by means of geometric techniques for systems with coefficients in a field can be generalized only partially to systems over a ring. An illustration of this, with regard to noninteracting control problems, is provided in Conte et al. (1995) and Hautus (1982).

Here, we will consider from a geometric point of view the Model Matching Problem for systems over a ring. We show that, similarly to what happens in the case of systems over a field, this problem is equivalent to a suitable Disturbance Decoupling

Problem with Measurable Disturbances. The basic difference with respect to the field case is the fact that the equivalence holds only if dynamic feedback solutions for the Disturbance Decoupling Problem are allowed. Thank to the results of Conte et al. (1995), it is then possible to state, in geometric terms, necessary and sufficient conditions for the solvability of the Model Matching Problem.

The above result apply in particular to the case of delay-differential systems with commensurable delays. Systems of that kind can in fact be modeled as systems with coefficients in a PID and the solutions found over the ring can be naturally interpreted in the delay-differential context. Examples concerning delay-differential systems will be presented and described in the paper.

2. PRELIMINARIES AND STATEMENT OF THE PROBLEM

Let R denote a commutative ring and let Σ be the system over R described by the set of equations

$$(2.1)\qquad \begin{cases} x(t+1) = Ax(t) + Bu(t) \\ y(t) = Cx(t) \end{cases}$$

where x, u, y belong respectively to the finitely generated, free R-modules $X = R^n$, $U = R^m$, $Y = R^p$ and A, B, C are matrices of suitable dimensions with entries in R.

Let us recall (Hautus(1982)) that a submodule V of the state module X is called
- (A,B)-*invariant* , or *controlled invariant* , if and only if $AV \subset V + \mathrm{Im}\, B$;
- (A,B)-*invariant of feedback type* if and only if there exists an R-linear map $F : X \rightarrow U$ such that $(A + BF)V \subset V$.

Such a feedback F, if any exists, is called a *friend* of V.

It is important to remark that, for systems over rings, controlled invariance and feedback type invariance are not equivalent properties, as it happens for systems with coefficients in a field. In particular, the maximum (A,B)-invariant submodule in Ker C, denoted usually by V^*, is not in general of feedback type.

Various characterization of (A,B)-invariant submodules of feedback type are possible (see Conte et al. (1995), Hautus(1982)). Here we are interested in pointing out that (A,B)-invariant submodules that are direct summand of X (i.e. that possess a basis that can be completed to a basis of X) are of feedback type.

The synthesis problem we consider in this paper is stated as follows.

2.1 STATEMENT OF THE PROBLEM Given a systems Σ of the form (2.1) and a system Σ_M, having the same number of outputs as Σ, defined over the ring R by the equations

$$(2.2)\qquad \begin{cases} x_M(t+1) = A_M x_M(t) + B_M u_M(t) \\ y(t) = C_M x_M(t) \end{cases}$$

the *Model Matching Problem* (MMP) for the *plant* Σ and the *model* Σ_M consists in finding a compensator Σ' of the form

$$(2.3)\qquad \begin{cases} z(t+1) = F_1 z(t) + F_2 x(t) + F_M x_M(t) + G_M u_M(t) \\ u(t) = K_1 z(t) + K_2 x(t) + K_M x_M(t) + L_M u_M \end{cases},$$

where z belongs to the a finitely generated, free R-module Z, and $F_1, F_2, F_M, G_M, K_1, K_2, K_M, L_M$ are matrices of suitable dimensions with entries in R, such that the transfer function of the compensated system $\Sigma_C = (A_C, B_C, C_C)$, described by the following equations

$$(2.4)\qquad \begin{cases} \begin{pmatrix} x(t+1) \\ x_M(t+1) \\ z(t+1) \end{pmatrix} = \begin{pmatrix} A+BK_2 & BK_M & BK_1 \\ 0 & A_M & 0 \\ F_2 & F_M & F_1 \end{pmatrix} \begin{pmatrix} x(t) \\ x_M(t) \\ z(t) \end{pmatrix} + \begin{pmatrix} BL_M \\ B_M \\ 0 \end{pmatrix} u_M \\ y(t) = \begin{pmatrix} C & -C_M & 0 \end{pmatrix} \begin{pmatrix} x(t) \\ x_M(t) \\ z(t) \end{pmatrix} \end{cases},$$

is zero. This is to say, in other words, that the input/output behavior of the plant Σ compensated by Σ' coincides with that of the model Σ_M.

The way we follow in approaching the study of the MMP consists in establishing a correspondence between the MMP itself and an associated Disturbance Decoupling Problem. To this aim, let us start by considering (following Malabre (1982)) the extended system Σ_E

$$(2.5)\quad \begin{cases} x_E(t+1) = A_E x_E(t) + B_E u(t) + D u_M(t) \\ y(t) = C_E x_E(t) \end{cases}$$

whose state module is the finitely generated free R-module $X_E = X \oplus X_M$, where

$$(2.6)\quad x_E = \begin{pmatrix} x(t) \\ x_M(t) \end{pmatrix}, A_E = \begin{pmatrix} A & 0 \\ 0 & A_M \end{pmatrix}, B_E = \begin{pmatrix} B \\ 0 \end{pmatrix},$$
$$D = \begin{pmatrix} 0 \\ B_M \end{pmatrix}, C_E = (C \;\; -C_M),$$

and where u_M is seen as a disturbance. A compensator Σ' of the form (2.3) that solves the MMP can be equivalently viewed as a dynamic feedback compensator for Σ_E that decouples the output from the disturbance. It is therefore clear that the MMP concerning Σ and Σ_M is solvable if and only if the Disturbance Decoupling Problem with Measurable Disturbance (DDPMD) concerning Σ_E is solvable by a dynamic state feedback (see Conte et al. (1995) for a precise statement of the Disturbance Decoupling Problem with Measurable Disturbance). Remark that for systems with coefficients in a field the correspondence between MMP and DDPMD does not requires to consider dynamic feedback solutions of the latter. For systems over a ring, however, a DDPMD may admit dynamic feedback solutions without admitting static feedback solutions (see Conte et al. (1995) Example 7.3).

3. SOLUTION OF THE MODEL MATCHING PROBLEM

The equivalence established in the previous Section between the MMP and an associated DDPMD can now be used for stating geometric necessary and sufficient condition for the existence of solutions to the MMP.

<u>4.1 THEOREM</u> The MMP concerning the plant Σ and the model Σ_M is solvable if and only if, denoting by V_E^* the maximum (A_E,B_E)-invariant submodule of the state module of Σ_E contained in Ker C_E, one has Im $D \subset V_E^* + \text{Im } B_E$.

<u>4.2 REMARK</u> The proof follows that of Conte et al. (1995) Theorem 6.4, that, although stated for systems over a principal ideal domain, holds for systems over any commutative ring. The first part of the proof, that is given here in details, provides a construction of the compensator that solves a given MMP.

Proof of 4.1. Let V be a suitable matrix whose, say, k columns form a basis of V_E^*. Then, there exist a suitable $k \times k$ matrix A_{22} and a suitable $m \times k$ matrix H such that the following matrix equality holds

$$(4.1)\qquad A_E V = V A_{22} + B_E H.$$

Now, by

$$(4.2)\quad \begin{bmatrix} A_E & 0 \\ 0 & A_{22} \end{bmatrix}\begin{bmatrix} V \\ I_k \end{bmatrix} = \begin{bmatrix} V \\ I_k \end{bmatrix} A_{22} + \begin{bmatrix} B_E \\ 0_{k\times m} \end{bmatrix} H,$$

one has that the submodule V', spanned in $X' = X_E \oplus R^k$ by the columns of $\begin{bmatrix} V \\ I_k \end{bmatrix}$, is $\left(\begin{bmatrix} A_E & 0 \\ 0 & A_{22} \end{bmatrix}, \begin{bmatrix} B_E \\ 0_{k\times m} \end{bmatrix}\right)$-invariant. Clearly, V' is contained in Ker $C' = [C_E \; 0_{p\times k}]$ and, since its basis can be completed to a basis of X', it is also of feedback type. Let $[K_E \; K_1] = [[K_2 \; K_M] \; K_1]$ be a friend of V', that is a matrix of suitable dimensions such that V' is $\left(\begin{bmatrix} A_E & 0 \\ 0 & A_{22} \end{bmatrix} + \begin{bmatrix} B_E \\ 0_{k\times m} \end{bmatrix}[K_E \; K_1]\right)$- invariant, and, assuming that q denotes the dimension of the vector $u_M(t)$, let G_M and L_M be matrices of dimensions respectively $k \times q$ and $m \times q$ such that $D = V G_M + B_E L_M$. This implies that Im $\begin{bmatrix} D \\ G_M \end{bmatrix}$ is contained in V', as shown by the equality $\begin{bmatrix} D \\ G_M \end{bmatrix} = \begin{bmatrix} V \\ I_k \end{bmatrix} G_M + \begin{bmatrix} B_E \\ 0_{k\times m} \end{bmatrix} L_M$, and then the compensator Σ' defined by

$$(4.3)\quad \begin{cases} z(t+1) = A_{22}z(t) + G_M u_M(t) \\ u(t) = K_1 z(t) + K_2 x(t) + K_M x_M(t) - L_M u_M(t) \end{cases}$$

decouples the output of Σ_E from $u_M(t)$. It follows from the discussion at the end of Section 2 that Σ' solves the MMP.

Conversely, if the geometric condition Im $D \subset V_E^* + \text{Im } B_E$ holds, Theorem 6.4 of Conte et al. (1995) assure the existence of a compensator of the form (2.3) that decouples the output of Σ_E from $u_M(t)$, and hence solves the MMP.

<u>4.3 REMARK</u> It is clear that the dimension of the compensator Σ' one obtains by applying the procedure described in the above proof equals the dimension of V_E^*. Smaller compensators can obviously be find, e.g. when V_E^* is of feedback type. Therefore dim V_E^* represents an upper bound for the complexity of the smaller compensator that solves the MMP. In case the plant Σ is injective (that is different inputs produce different outputs), the extended system Σ_E turns out to be injective with respect to the input u(t). If R is a principal ideal domain, this implies that there exists a minimum element, denoted by $V_*(\text{Im } D)$, in the family of all (A_E,B_E)-invariant submodules V contained in Ker C_E and such that Im $D \subset V + \text{Im } B_E$. Applying the procedure described in the above proof to $V_*(\text{Im } D)$

one gets a compensator of dimension equals to that of V*(Im D) itself and this becomes an upper bound for the complexity of the smaller compensator that solves the MMP.

4. EXAMPLE

Let us describe an application of the previous results to the case of delay-differential systems. To this aim, let us consider the delay-differential plant Σ

$$(5.1)\quad \begin{cases} \dot{x}(t) = x(t-\partial) + u(t-2\partial) - u(t-\partial) \\ y(t) = x(t-\partial) \end{cases}$$

and the model Σ_M

$$(5.2)\quad \begin{cases} \dot{x}_M(t) = x_M(t) + u_M(t-3\partial) - u_M(t-2\partial) \\ y(t) = x_M(t) \end{cases}$$

For studying the existence of a compensator Σ' such that Σ compensated by Σ' has the same input/output behavior of the model Σ_M, it is convenient to state the problem in terms of systems over a principal ideal domain. By introducing the delay operator Δ, defined for any time function f by $\Delta(f)(t) = f(t-\partial)$, this can be done by associating formally with Σ and Σ_M the following systems over $R[\Delta]$:

$$(5.3)\quad \begin{cases} x(t+1) = \Delta x(t) + (\Delta^2 - \Delta)u(t) \\ y(t) = \Delta x(t) \end{cases}$$

$$(5.4)\quad \begin{cases} x_M(t+1) = x_M(t) + (\Delta^3 - \Delta^2)u_M(t) \\ y(t) = x_M(t) \end{cases}.$$

In the extended system Σ_E

$$\begin{cases} x_E(t+1) = \begin{pmatrix} \Delta & 0 \\ 0 & 1 \end{pmatrix} x_E(t) + \begin{pmatrix} \Delta^2-\Delta \\ 0 \end{pmatrix} u(t) + \begin{pmatrix} 0 \\ \Delta^3-\Delta^2 \end{pmatrix} u_M(t) \\ y(t) = (\Delta \ \ -1) x_E(t) \end{cases}$$

one has that $V_E^* = \text{span}\left\{ \begin{pmatrix} \Delta \\ \Delta^2 \end{pmatrix} \right\}$ is the maximum (A_E,B_E)-invariant submodule contained in Ker C_E.

Remark that V_E^* is not of feedback type and, since no smaller (A_E,B_E)-invariant submodules of feedback type exists, no static compensator can solve the MMP concerning (5.3) and (5.4). On the other hand, since the condition

$$\text{Im}\begin{pmatrix} 0 \\ \Delta^3-\Delta^2 \end{pmatrix} \subset V_E^* + \text{Im}\begin{pmatrix} \Delta^2-\Delta \\ 0 \end{pmatrix}$$

holds, Theorem 4.1 assures the existence of a solution, given by a dynamic compensator which decouples the output of Σ_E from the disturbance u_M.

Following the proof of Theorem 4.1, we take $V' = \text{span}\left\{ \begin{pmatrix} \Delta \\ \Delta^2 \\ 1 \end{pmatrix} \right\}$ and we chose two matrices, $A_{22} = (1)$ and $K = (K_2\ K_M\ K_1) = (-\Delta\ \ 1\ \ -1)$, such that V' is $\left(\begin{pmatrix} A_E & 0 \\ 0 & A_{22} \end{pmatrix} + \begin{pmatrix} B_E \\ 0 \end{pmatrix} K \right)$ - invariant.

Then, we take $G_M = (\Delta\ \ -1)$ and $L_M = (-1)$ and we get the compensator

$$\begin{cases} z(t+1) = z(t) + (\Delta - 1)u_M(t) \\ u(t) = -z(t) - (\Delta)x(t) + x_M(t) + u_M \end{cases},$$

or, in the original delay-differential context, the compensator

$$\Sigma' = \begin{cases} \dot{z}(t) = z(t) + u_M(t-\partial) - u_M(t) \\ u(t) = -z(t) - x(t-\partial) + x_M(t) + u_M(t) \end{cases}.$$

It is now possible to check that the input $u_M(t)$ does not appears in the time derivative of any order of the output of the compensated system $\Sigma_C = (A_C,B_C,C_C)$

$$\Sigma_C = \begin{cases} \dot{x}(t) = x(t-\partial) - z(t-2\partial) - x(t-3\partial) + x_M(t-2\partial) + u_M(t-2\partial) - z(t-\partial) + x(t-2\partial) - x_M(t-\partial) - u_M(t-\partial) \\ \dot{x}_M(t) = x_M(t) + u_M(t-3\partial) - u_M(t-2\partial) \\ \dot{z}(t) = z(t) + u_M(t-\partial) - u_M(t) \\ y(t) = x(t-\partial) - x_M(t) \end{cases}$$

This means that y(t) is independent from $u_M(t)$ and, as a consequence, the MMP concerning Σ and Σ_M is solved , in the original delay-differential framework, by the compensator Σ'.

5. REFERENCES

Conte G. and A.M. Perdon (1982). Systems over Principal Ideal Domains. A Polinomial Model Approach, *SIAM J. Control Opt.* 20

Conte G. and A.M. Perdon (1991). The disturbance decoupling problem for systems over a principal ideal domain, New Trends in Systems Theory, G. Conte, A.M. Perdon and B. Wyman Eds., Birkhauser

Conte G. and A.M. Perdon (1995). The Disturbance Decoupling Problem for Systems Over a Principal Ideal Domain, to appear on *SIAM J. Control Opt.* , vol. 33

Hautus M.L.J.(1982). Controlled invariance in sytems over rings, Springer Lecture Notes in Control and Info. Sci. 39

Hautus M.L.J. (1984). Disturbance rejection for systems over rings, Springer Lecture Notes in Control and Info. Sci. 58

Inaba H. and N. Ito (1992). Disturbance rejection of linear systems over rings, *Dynamical System Theory*, 15

Kamen E. (1976). Lectures on Algebraic System Theory. Linear Systems over Rings, NASA Contractor Report 3016

Malabre M. (1982). Structure all'infini des triplets invariants. Application a la pursuite parfaite de modele, Prod. INRIA Conf., Versailles, Lect. Notes Control Info. Sci., Springer-Verlag

Sontag E. (1976). Linear Systems over Rings : a Survey, *Ricerche di Automatica* 7 (1976)

Sontag E. (1981). Linear Systems over Rings : a (Partial) Update Survey, Proc. IFAC/81, Kyoto

Wohnam M. (1985). Linear multivariable control : a geometric approach, 3rd Ed. Springer Verlag

Structural Properties and Controllabillity for Delay Systems of Neutral Type

Rabah RABAH *
Laboratoire d'Automatique de Nantes, URA C.N.R.S. 823
Ecole Centrale-Université-Ecole des Mines, Nantes
1, rue de la Noë, 44072 Nantes Cedex 3
E-mail: rabah@emn.fr

Abstract

Linear systems with delays of neutral type are considered. Structure at infinity and application for several control problems are given. Functional controllability to zero is studied. The main results are based on the representaion of systems of neutral type using subsystems without delays and with special initial conditions.

Résumé

Pour des systèmes à retards de type neutre on considère la structure à l'infini et ses applications. On étudie également le problème de la commandabilité fonctionnelle en zéro. Les résultats essentiels sont basés sur la représentation du système par des sous-systèmes sans retards avec des conditios initiales adéquates.

Keywords: Delay systems, Neutral type, Structure at infinity, Decoupling, Null-controllability.

1 Introduction

The present paper is concerned with delay systems of neutral type, i.e. systems described by equations

$$\left\{\begin{array}{rcl} \dot{x}(t) & = & A_0x(t)+A_1x(t-h) \\ & & +A_{-1}\dot{x}(t-h)+Bu(t) \\ y(t) & = & Cx(t) \end{array}\right. , \qquad (1)$$

where $x(t) \in \mathbb{R}^n$, $u(t) \in \mathbb{R}^p$ and $y(t) \in \mathbb{R}^m$; h stands for the delay. The matrices A_0, A_1, A_{-1}, B and C are of suitable dimensions. In this paper we assume that matrices B and C are of full rank.

There is more general forms of delay systems of neutral type: with several delays, with distributed delays... Our considerations may be formulated for systems with several commensurate delays in the state, input or output. For other approaches concerning systems with delay of neutral type see for example [2, 5].

The second section is devoted to the description of solutions of the system and transfer matrix function in several ways. The new approach is the generalization of the description given by Olbrot [5] and Zmood [12] for systems with simple delays in state (see [5, 10] for more detals on different approaches).

The third part describe structure at infinity of the system (1) and geometric tool which allows to characterize this structure. Applications to control problem are considered. The last section is devoted to the problem of controllability to zero function, i.e. to the characterization of the existence of a control function $u(t)$ such that the solution vanishes for $t \geq T$. We give here a generalization of the results given by Olbrot [7] for simple delay systems.

*Supported in part by ESPRIT Basic Research Project No 8924 (SESDIP).

2 Preliminaries

The system (1) defines one and only one solution for the following initial conditions:

$$x(0)=x_0, \quad x(t)=\varphi(t) \quad \text{for } t\in[-h,0[. \qquad (2)$$

The function $\varphi(t)$ is assumed to be absolutely continuous. The solution is given then by

$$\begin{array}{rcl} x(t) & = & \Phi(t)x_0+\int_0^h \Phi(t-\tau)A_1\varphi(\tau)d\tau \\ & & +\int_0^h \Phi(t-\tau)A_{-1}\dot{\varphi}(\tau)d\tau \\ & & +\int_0^t \Phi(t-\tau)Bu(\tau)d\tau, \end{array}$$

where the matrix $\Phi(t)$, called the fundamental matrix of solutions, satisfyies the differential equation in (1) with the initials conditions:

$$\Phi(0) = I, \quad \Phi(t) = 0 \quad \text{for} \quad t < 0.$$

Let us give the expression of this matrix. Let Q_j^i be the matrices defined by the relations [3, 4]:

$$Q_j^i = A_0 Q_j^{i-1} + A_1 Q_{j-1}^{i-1} + A_{-1} Q_{j-1}^i$$
$$Q_0^0 = I, \quad Q_j^i = 0 \quad \text{if} \quad ij < 0.$$

Then the matrix $\Phi(t)$ may be written as (see [10, 5])

$$\Phi(t) = \sum_{i=0}^{\infty} \sum_{j=0}^{k} Q_j^i \frac{(t-j)^i}{i!} \quad \text{for} \quad t \in [kh, (k+1)h[.$$

This relation may be verified by induction. Another expression of the fundamental matrix may be obtained via the solutions of the system

$$\left\{ \begin{array}{rcl} E_k \dot{z}_k(t) & = & F_k z_k(t) + G_k v_k(t) \\ w_k(t) & = & H_k z(t) \end{array} \right. \tag{3}$$

where the matrices E_k, F_k, G_k and $H_k, k \in \mathbb{N}$ are composed by $(k+1)^2$ blocks as follows.

$$E_k = \begin{bmatrix} I & 0 & 0 & \dots & 0 \\ -A_{-1} & I & 0 & \dots & 0 \\ 0 & -A_{-1} & I & \dots & 0 \\ \vdots & \vdots & \vdots & \ddots & \vdots \\ 0 & 0 & 0 & \dots & I \end{bmatrix},$$

$$F_k = \begin{bmatrix} A_0 & 0 & 0 & \dots & 0 \\ A_1 & A_0 & 0 & \dots & 0 \\ 0 & A_1 & A_0 & \dots & 0 \\ \vdots & \vdots & \vdots & \ddots & \vdots \\ 0 & 0 & 0 & \dots & A_0 \end{bmatrix},$$

$$G_k = \begin{bmatrix} B & 0 & 0 & \dots & 0 \\ 0 & B & 0 & \dots & 0 \\ 0 & 0 & B & \dots & 0 \\ \vdots & \vdots & \vdots & \ddots & \vdots \\ 0 & 0 & 0 & \dots & B \end{bmatrix},$$

$$H_k = \begin{bmatrix} C & 0 & 0 & \dots & 0 \\ 0 & C & 0 & \dots & 0 \\ 0 & 0 & C & \dots & 0 \\ \vdots & \vdots & \vdots & \ddots & \vdots \\ 0 & 0 & 0 & \dots & C \end{bmatrix}.$$

As E_k is invertible, the system (3) is regular for all integers k. This system allows to give another expression for the fundamental solution.

Proposition 2.1 *The fundamental matrix of solutions may be written as*

$$\Phi(t) = [0 \quad \dots \quad 0 \quad I] e^{E_k^{-1} F_k (t-kh)} Z_k$$

where the matrices Z_k are given by the relations

$$Z_0 = I, \quad Z_k = \begin{bmatrix} I \\ e^{E_k^{-1} F_{k-1} h} Z_{k-1} \end{bmatrix}.$$

Proof. Let $z_k(t)$ be the solution of the system (3) with the control $v_k(t) = 0$ and the initial condition

$$z_k(0) = \begin{bmatrix} x_0 \\ z_{k-1}(h) \end{bmatrix}.$$

Let $x(t)$ be the function with value in $\mathbb{R}^n$ defined by

$$x(t) = [0 \quad 0 \quad \dots \quad 0 \quad I] z_k(t - kh)$$

for $t \in [kh, (k+1)h[, k \in \mathbb{N}$. Let us denote by $x_i \in \mathbb{N}, i = 0, 1, \dots, k$ the vector components of z_k. Then we have from (3)

$$\begin{array}{rcl} \dot{x}_0 & = & A_0 x_0, \\ -A_1 \dot{x}_{i-1} + \dot{x}_i & = & A_0 x_i + A_1 x_{i-1}, \quad i > 0. \end{array}$$

Moreover, $x_i(0) = x_{i-1}(h)$. This means that for all $t \neq (i-1)h$ the function $x(t)$ is absolutely continuous and verifies the equation

$$\dot{x}(t) = A_0 x(t) + A_1 x(t-h) + A_{-1} \dot{x}(t-h)$$

almost every where (in fact for all $t \neq (i-1)h$) with the initial conditions $x(0) = x_0$ and $x(t) = 0$ for $t < 0$. Hence

$$x(t) = \Phi(t) x_0$$

and then

$$\Phi(t) = [0 \quad \dots \quad 0 \quad I] e^{E_k^{-1} F_k (t-kh)} Z_k.$$

Consider now the transfer matrix function of the system (1). It is given by the expression

$$T(s) = C(Is - A_0 - A_1 e^{-hs} - A_{-1} s e^{-hs})^{-1} B.$$

and may be decomposed as follows

$$T(s) = \sum_{j=0}^{\infty} C R_0(s) R_1^j(s) B e^{-jhs},$$

where $R_0(s) = (Is - A_0)^{-1}$ and $R_1(s) = (A_1 + A_{-1}s)(Is - A_0)^{-1}$. Let $T_j(s) = CR_0(s)R_1(s)^j B$ and $\Theta_k(s)$ be the transfer matrix function of the system (3). As in the case of a linear system with delay [6] , one has the following result.

Proposition 2.2 *The transfer matrix function of (3) verifies*

$$\Theta_k(s) = \begin{bmatrix} T_0(s) & 0 & \dots & 0 \\ T_1(s) & T_0(s) & \dots & 0 \\ \vdots & \vdots & \ddots & \vdots \\ T_k(s) & T_{k-1}(s) & \dots & T_0(s) \end{bmatrix}.$$

Proof. It follows from the equality

$$(sE_k - F_k)^{-1} =$$

$$\begin{bmatrix} Is - A_0 & 0 & \dots & 0 \\ -A_1 - A_{-1}s & Is - A_0 & \dots & 0 \\ 0 & -A_1 - A_{-1}s & \dots & 0 \\ \vdots & \vdots & \ddots & \vdots \\ 0 & 0 & \dots & Is - A_0 \end{bmatrix}^{-1}$$

$$= \begin{bmatrix} R_0(s) & 0 & \dots & 0 \\ R_0(s)R_1(s) & R_0(s) & \dots & 0 \\ R_0(s)R_1^2(s) & R_0R_1(s) & \dots & 0 \\ \vdots & \vdots & \ddots & \vdots \\ R_0(s)R_1^k(s) & R_0(s)R_1^{k-1}(s) & \dots & R_0(s) \end{bmatrix}.$$

The calculation of $\Theta(s) = H_k(sE_k - F_k)^{-1}G_k$ gives the desired result. ∎

As for the fundamental solution one can write the transfer function trough the matrices Q_j^i.

Proposition 2.3 *The rational matrices $T_j(s)$ verify*

$$T_j(s) = \sum_{i=0}^{\infty} Q_j^i s^{-(i+1)}.$$

Proof. Let us remark first that

$$\Theta_k(s) = \sum_{i=0}^{\infty} H_k(E_k^{-1}F_k)^i s^{-(i+1)}. \qquad (4)$$

In the other hand by induction and using the definition of the matrices Q_j^i one may show the following relation.

$$(E_k^{-1}F_k)^i = \begin{bmatrix} Q_0^i & 0 & \dots & 0 \\ Q_1^i & Q_0^i & \dots & 0 \\ \vdots & \vdots & \ddots & \vdots \\ Q_k^i & Q_{k-1}^i & \dots & Q_0^i \end{bmatrix}. \qquad (5)$$

Combining (4) and (5) one obtains the desired result. ∎

This proposition gives a relation between the approach of Gabasov and Kirillova based on matrices Q_j^i [3, 4] and thus of Olbrot, Zmood and other authors [7, 10, 12]

3 Structure at infinity and applications

For systems of neutral type the structure at infinity is introduced like the same one for systems with delays. As in [6, 8] we say that a function $f(s)$ defined and analytical for Re $s > s_0$ is proper if

$$\lim_{s \to +\infty} sf(s) = f_0,$$

where f_0 is a constant. $f(s)$ is strictly proper if $f_0 = 0$. A matrix $P(s)$ is said biproper if it is proper, invertible and $P^{-1}(s)$ is also proper.

One can now define the so called canonical form at infinity for the system (1).

Proposition 3.1 *There exist two biproper matrices $B_1(s)$ and $B_2(s)$ such that $B_1T(s)B_2(s) =$*

$$\begin{bmatrix} \Lambda_0(s) & 0 & \dots & 0 & 0 \\ 0 & \Lambda_1(s)e^{-hs} & \dots & 0 & 0 \\ \vdots & \vdots & \ddots & \vdots & \vdots \\ 0 & 0 & \dots & \Lambda_p(s)e^{-phs} & 0 \\ 0 & 0 & \dots & 0 & 0 \end{bmatrix},$$

where $\Lambda_j(s) = \text{diag}\,[s^{n_{1j}}, s^{n_{2j}}, \dots, s^{n_{r_j j}}]$. The integer n_{ij} may be nonnegative if $i > 0$.

The proof is the same as in [6].

The obtained form is unique if we assume that the integers n_{ij} are increasing with i. This form is called structure at infinity of the system (1) because that it has the same behavior at infinity as $T(s)$.

The structure at infinity permits to characterize several control problems: disturbance rejection, model matching, row-by-row decoupling. The results given in [6, 8] for simple delay systems may be formulated in a same way for systems with delay of neutral type. In order to illustrate that we consider the problem of row-by-row decoupling for square system (1), i.e. when $p = m$.

Definition 3.2 *The problem of row-by-row decoupling is solvable by precompensation if there exists a biproper compensator $C(s)$ such that*

$$T(s)C(s) = \text{diag}\,[t_1(s), \dots, t_m(s)].$$

Let us note by $\Sigma_\infty S$ the set of the integers defined by the structure at infinity of the system (1) and by $\Sigma_\infty S_i$ the set of integers given by the structure at infinity of the system with the same input and state as the system (1) but with the output $y_i(t) = c_i x(t)$, where c_i are the corresponding rows of the matrix C.

Proposition 3.3 *The problem of row-by-row decoupling is solvable if and only if*

$$\Sigma_\infty S = \dot{\cup}\{\Sigma_\infty S_i, i = 1, \ldots, m\}.$$

Proof. The proof is in fact the same as in [8] for systems with simple delay. Suppose that there exists a biproper matrix $K(s)$ such that

$$T(s)C(s) = \text{diag}[h_1(s), \ldots, h_p(s)].$$

Then the structure at infinity of $T(s)$ is the same as the structure at infinity of the matrix $\text{diag}[h_1(s), \ldots, h_p(s)]$. On the other hand the structure at infinity of the diagonal matrix is the union of the structures at infinity of the rows of the given matrix.

Conversely, assume that the condition of the proposition holds. Then $T(s)$ may be factorized in the following way:

$$T(s) =$$

$$\text{diag}\left[s^{-n_1}e^{-k_1 s}, \ldots, s^{-n_p}e^{-k_p s}\right](D_0 + W(s)),$$

with a non singular matrix D_0. Let us put $C(s) = (D_0 + W(s))^{-1}$. Then

$$T(s)C(s) = \text{diag}\left[s^{-n_1}e^{-k_1 s}, \ldots, s^{-n_p}e^{-k_p s}\right],$$

which means that the system is row-by-row decouplable by precompensation. ∎

Let us remark that some partial problems may be considered using the representation of the system (1) through the systems (3). By partial problems we mean solvability of the same problems (model matching, disturbance rejection or row-by-row decoupling) on $[0, kh]$ for some integer k. The approaches developped in [6, 9] can be generalized for systems of neutral type.

A geometric tool

A geometric tool which correspond to the given structure at infinity is represented by the subspace $\mathcal{V}_\Sigma(\mathcal{K})$ which elements are $x \in \mathbb{R}^n$ such that

$$x = (sI - A_0 - A_1 e^{-hs} - A_{-1} s e^{-hs})\xi(s) - B\omega(s),$$

for $s > s_0$, where s_0 is a real number and ξ, ω are strictly proper functions, such that $\xi(s) \in \mathcal{K}$.

If $\mathcal{K} = \text{Ker } C$ then the rank r of the system (1) is given by

$$r = \sum_{j=0}^{p} r_j = \dim \text{Im } B - \dim\ (\text{Im } B \cap \mathcal{V}_\Sigma(\mathcal{K})).$$

The proof is the same as in [6], where the canonical form is used. The row-by-row decoupling problem and other control problems mentioned is this paper may be formulated using the subspace $\mathcal{V}_\Sigma(\mathcal{K})$ with an adequate choice of the subspace $\mathcal{K}$ (see [8]). The considerations are the same and are omitted here.

4 Null-controllability

The conditions of functional controllability for delay systems are usually difficult to verify. However, for controllability to zero function Olbrot [7] gave conditions which are more easy to calculate. The initial function was assumed to be zero. Later, in [5] it aws pointed out that this assumption is not restrictive. Our purpose is to generalize the given results for systems of neutral type. First we give the corresponding definitions and then, using the representation of the system 1 by systems 3 we give the main result.

Definition 4.1 *The system (1) is called IP null-controllable at time T if for all initial conditions 2 with $\varphi = 0$ there exists a control $u(t)$ such that $x(t) = 0, t \geq T$.*

This a particular case of the following concept.

Definition 4.2 *The system (1) is called null-controllable at time T if for all initial conditions 2 there exists a control $u(t)$ such that $x(t) = 0, t \geq T$.*

It is clear that null-controllability implies IP null-controllability at the same time T. The converse is true in the following sense.

Lemma 4.3 *If the system (1) is IP null-controllable at time T, then it is null-controllable at time $T + h$.*

Proof. Let the system (1) be IP null-controllable at time T. Then for a basis $\{e_1, e_2, \ldots, e_n\}$ of $\mathbb{R}^n$ there exist control functions $u_1, u_2, \ldots, u_n$ such that

$$\Phi(t)e_i + \int_0^t \Phi(t-\tau)Bu_i(\tau)d\tau = 0 \qquad (6)$$

for $t \geq T$.

Consider the functions $v_i(t)$ defined by

$$\begin{aligned} v_i(t) &= Bu_i(t), \quad t \in [0, T] \\ v_i(t) &= 0, \quad t > T. \end{aligned}$$

Let $V(t)$ be the matrix $V(t) = [u_1 \quad u_2 \quad \dots \quad u_n$, then (6) gives

$$\Phi(t) + \int_0^t \Phi(t-\tau)BV(\tau)d\tau = 0 \qquad (7)$$

for $t \geq T$.

Let (x_0, φ) be arbitrary initial conditions. Then, for $t \geq h$ the solution is

$$\begin{aligned} x(t) &= \Phi(t)x_0 + \int_0^h \Phi(t-\tau)f(\tau)d\tau \\ &+ \int_0^t \Phi(t-\tau)Bu(\tau)d\tau \end{aligned}$$

where

$$f(\tau) = A_1\varphi(\tau) + A_{-1}\dot{\varphi}(\tau).$$

Denote

$$W(t) = V(t) + \int_0^h V(t-\tau)f(\tau)d\tau.$$

$W(t)$ take values in Im B and then there exists a function $u(t)$ such that $W(t) = Bu(t)$. Using this value of $Bu(t)$ in the expression of $x(t)$ we obtain

$$\begin{aligned} x(t) &= \Phi(t)x_0 + \int_0^t \Phi(t-\tau)V(\tau)x_0 d\tau \\ &+ \int_0^t \Phi(t-\tau) \int_0^h V(\tau-\theta)f(\theta)d\theta d\tau \\ &+ \int_0^h \Phi(t-\tau)f(\tau)d\tau \end{aligned}$$

for $t \geq h$. And then, by (7), we have

$$\begin{aligned} x(t) &= \int_0^t \Phi(t-\tau) \int_0^h V(\tau-\theta)f(\theta)d\theta d\tau \\ &+ \int_0^h \Phi(t-\tau)f(\tau)d\tau. \end{aligned}$$

And then, by changing the order of integration, we get

$$\begin{aligned} x(t) &= \int_0^h \int_0^t \Phi(t-\tau)V(\tau-\theta)f(\theta)d\tau d\theta \\ &+ \int_0^h \Phi(t-\tau)f(\tau)d\tau. \end{aligned}$$

Changing variables in the first integral we have

$$\begin{aligned} x(t) &= \int_0^h \int_{-\theta}^{t-\theta} \Phi(t-\tau-\theta)V(\tau)f(\theta)d\tau d\theta \\ &+ \int_0^h \Phi(t-\tau)f(\tau)d\tau. \end{aligned}$$

The second part of this equality may be written as

$$\int_0^h \left[\int_0^{t-\theta} \Phi(t-\tau-\theta)V(\tau)d\tau + \Phi(t-\theta) \right] f(\theta)d\theta.$$

The term in brackets vanishes for $t - \theta \geq T$, with $\theta \in [0, h]$. Hence, for $t \geq T+h$ we obtain $x(t) = 0$. This ends the proof. ∎

In order to formulate the main result of this section, let us introduce the following notations. By $\mathcal{M}_k^*$ we denote the supremal $(E_k^{-1}F_k, E_k^{-1}G_k)$-invariant subspace contained in the subspace $\mathcal{N}_k$ for which the k-th vectorial components is 0, i.e. each elements of $\mathcal{N}_k$ is of the form $[x_0 \quad x_1 \quad \dots \quad x_{k-1} \quad 0]^*$, where $(^*)$ denote the transpose. The subspace $\mathcal{L}_k^*$ is the $(E_k^{-1}F_k, E_k^{-1}G_k)$-controllability subspace contained in $\mathcal{N}_k$. Those subspaces may be computed using the wellknown algorithms [1, 11]. It is wellknown that there exist a feedback R_k such that $\mathcal{M}_k^*$ is $(E_k^{-1}F_k + E_k^{-1}G_kR_k)$-invariant. Let $\overline{F_k}$ be the matrix $(E_k^{-1}F_k + E_k^{-1}G_kR_k)$. Note by $\mathcal{N}_k^0$ the subspace of the elements $[x_0 \quad 0 \quad \dots \quad 0 \quad 0]^*$.

The solutions of the system (3) gives the solutions of the system (1) with $\varphi = 0$ as initial function in (2) if and only if the initial condition of the system (3) is given by

$$z_k(0) = \begin{bmatrix} x_0 \\ z_{k-1}(h) \end{bmatrix} = \begin{bmatrix} x_0 \\ 0 \end{bmatrix} + I_k^0 z_k(h), \qquad (8)$$

where I_k is the matrix

$$I_k^0 = \begin{bmatrix} 0 & 0 & 0 & \dots & 0 & 0 \\ I & 0 & 0 & \dots & 0 & 0 \\ 0 & I & 0 & \dots & 0 & 0 \\ \vdots & \vdots & \vdots & \ddots & \vdots & \vdots \\ 0 & 0 & 0 & \dots & I & 0 \end{bmatrix}.$$

In the other hand the value $z_k(h)$ may be written as

$$z_k(h) = e^{\overline{F_k}h} z_k(0) + \bar{z}_k(h),$$

with

$$\bar{z}_k(h) = \int_0^h e^{\overline{F_k}(h-\tau)} E_k^{-1} G_k v_k(\tau) d\tau \in \mathcal{L}_k.$$

Then the relation (8) is verified if and only if

$$(I_k - I_k^0 e^{E_k^{-1}F_k h}) z_k(0) = \begin{bmatrix} x_0 \\ 0 \end{bmatrix} + I_k^0 \bar{z}_k(h). \quad (9)$$

Proposition 4.4 *The system (1) is IP null-controllable at time kh if and only if*

$$\mathcal{N}_k^0 \subset (I_k - I_k^0 e^{E_k^{-1}F_k h})\mathcal{M}_k + I_k^0 \mathcal{L}_k. \qquad (10)$$

Proof. If the system (1) is IP null-controllable at time kh then for all $x_0 \in \mathbb{R}^n$ the corresponding condition $z_k(0)$ given by (8) is in $\mathcal{M}_k^*$ and $\bar{z}_k(h) \in \mathcal{L}_k^*$. Then from (9) we have the needed relation.

Conversely, let the condition (10) of the proposition is satisfied. Then, for all $x_0 \in \mathbb{R}^n$, there exist z_k and $\bar{z}_k$, respectively in $\mathcal{M}_k^*$ and $\mathcal{L}_k^*$ such that

$$(I_k - I_k^0 e^{E_k^{-1}F_k h})z_k - I_k^0 \bar{z}_k = \begin{bmatrix} x_0 \\ 0 \end{bmatrix}.$$

Hence for some control function $v_k(t)$ one has

$$\bar{z}_k = \int_0^h e^{\overline{F_k}(h-\tau)} E_k^{-1} G_k v_k(\tau) d\tau$$

and

$$z_k - I_k^0 z_k(h) = \begin{bmatrix} x_0 \\ 0 \end{bmatrix},$$

where $z_k(h) = e^{\overline{F_k}h} z_k + \bar{z}_k$. In other words, the relation (8) is satisfied for $z_k(0) = z_k \in \mathcal{M}_k^*$ and $z_k(h)$ remains in $\mathcal{N}_k$. Hence $z_k(h) \in \mathcal{N}_k$ and the corresponding initial condition verifies (8). The solution $x(t)$ with initial condition x_0, initial function $\varphi = 0$ and the corresponding control function calculated from v_k the feedback R_k verifies

$$x(t) = 0, \quad t \in [kh, (k+1)h],$$

i.e. the system is IP null-controllable at time $T = kh$. ∎

Using the Lemma we can have the condition of null-controllability.

Let us remark that the subspaces $\mathcal{L}_k^*$ and $\mathcal{M}_k^*$ may be computed in a finite number of steps. The feedback R_k may be computed in such a way that it include only "delay" blocks (see [6]).

5 Conclusion

For linear systems with delay of neutral type some techniques used for simple delay systems may be used. This allows to consider for delay systems the possibility to use feedback with delayed derivative and then, remaining in fact in the same class of systems, to study all the possibility which may be given by systems of neutral type. For example, with additional conditions, the null-controllability is equivalent to functional controllablity. For simple delay system, as it is well known, it is not the case: only smooth functions may be reachable.

References

[1] G. Basile, G. Marro, *Controlled and conditionned invariants in linear system theory.* Prentice Hall, Englewood Cliffs, 1991.

[2] A. Bensoussan, G. Da Prato, M.C. Delfour, S.J. Mitter, *Representation and control of infinite dimensional systems.* Birkhäuser, Boston, 1992.

[3] R. G. Gabasov, F. M. Kirillova, *Qualitative theory of optimal processes,* (in Russian). Nauka, Moscow, 1971.

[4] F. Kirillova, S. Churakova, *On controllability problem for linear systems with aftereffect.* Diff. Uravnenya, **3**, No 3, 1967.

[5] H. Górecki, S. Fuksa, P. Grabowski, A Korytowski, *Analysis and synthesis of time delay systems,* J. Wiley Ed., 1989.

[6] M. Malabre, R. Rabah, *Structure at infinity, model matching and disturbance rejection for linear systems with delays.* Kybernetika, **29**, No 5, pp. 485–498, 1993.

[7] A. W. Olbrot, *Algebraic criteria of controllability to zero function for linear constant time-lag systems.* Control and Cybernetics, **2**, No 1/2, 1973, pp. 60-76.

[8] R. Rabah, M. Malabre, *Structure at infinity for infinite dimensional systems.* Rapport Interne, Laboratoire d'Automatique de Nantes, No 94.6, mars 1994.

[9] O. Sename, R. Rabah, J.F. Lafay, *Decoupling without prediction of linear systems with delays: a structural approach.* Systems and Control Letters, to appear.

[10] A. C. Tsoi, *Recent advances in the algebraic system theory of delay differential equations,* in the book: Recent theoretical developments in control, M.J. Gregson Ed. Academic Press, 1978, pp. 67-127.

[11] W. M. Wonham, *Linear multivariable control: a geometric approach.* Springer-Verlag, New York, 1979.

[12] R. B. Zmood, *The euclidean space controllability of control systems with delays.* SIAM J. of Control, vol. 12. No 4, 1974, pp. 609-623.

ROBUST ADAPTIVE CONTROL OF ROBOT MANIPULATORS

Abdelhamid Laib and Françoise Lamnabhi - Lagarrigue

Laboratoire des Signaux et Systèmes, C.N.R.S., SUPELEC, 91192 Gif-sur-Yvette, Cedex, France.

Abstract. This paper proposes an algorithm for a robust adaptive control of robot manipulators. It is based on a computed torque adaptive method avoiding the acceleration measurement. Robustness of the algorithm to unmodelled dynamics is also investigated

Resumé.
Dans cet article un algorithme de commande adaptative robuste pour des robots manipulateurs est proposé. Cette méthode basée sur la méthode du couple calculé, n'exige pas la mesure d'accélération pour son implementation. De plus, la robustesse à des dynamiques non modélisées a été étudiée.

Keywords. Adaptive Control, Update Law, Robot Manipulators, Passivity

1 INTRODUCTION

Several algorithms for the adaptive control of robot manipulators have been proposed in the literature, see for instance *Ortega and Spong (1989)* for an overview. Some of these algorithms *Craig (1988)*, *Hsu* et al. *(1987)*, *Middleton and Goodwin (1987)*, *Spong and Ortega (1988)* require the acceleration measurements and/or the computation of the inverse of the moment inertia matrix containing estimated parameters. Later, new methods based on the passivity have been developed (see for instance *Slotine and Weiping (1991)*, *Johansson (1989)*, *Whitcomb* et al. *(1993)*, *Berghuis* et al. *(1993)*) which avoid these drawbacks, but these algorithms present a weak robustness to unmodelled dynamics . The problem that we address here is to design a controller which avoids acceleration measurements and which is at the same time robust with respect to unmodelled dynamics.

2 PROBLEM STATEMENT

Let us consider the following n-DOF robot manipulators *Craig (1988)*,

$$M(q,\theta)\ddot{q} + C(q,\dot{q},\theta)\dot{q} + G(q,\theta) = \tau \tag{1}$$

where $q \in \mathbf{R}^{n\times 1}$, $M \in \mathbf{R}^{n\times n}$ is the inertia matrix, $C \in \mathbf{R}^{n\times n}$ represents the Coriolis, centripetal and frictional forces, $G \in \mathbf{R}^{n\times 1}$ denotes the gravitational force vector. Moreover, it is assumed that the matrices M, C, and G have a known structure and contain constant but unknown parameters $\theta_1, \theta_2, \ldots, \theta_p$, $p < n$ where the vector $\theta = (\theta_1, \theta_2, \ldots, \theta_p)^T \in \Omega \subset \mathbf{R}^p, p < n$ belongs to a given compact set Ω. The components of the matrices M, C, and G are linear in the unknown parameters. In the following, we will also assume that the given trajectory $q_d(t)$ which is tracked by the manipulator is twice differentiable.

Let us define the acceleration, velocity and position errors as follows,

$$\begin{cases} \ddot{\tilde{q}}(t) = \ddot{q}(t) - \ddot{q}_d(t) \\ \dot{\tilde{q}}(t) = \dot{q}(t) - \dot{q}_d(t) \\ \tilde{q}(t) = q(t) - q_d(t). \end{cases}$$

The control objective is to track the given reference trajectory $q_d(t)$ such that $\tilde{q}(t) \to 0$ and $\dot{\tilde{q}}(t) \to 0$ when $t \to \infty$. We also define the parameter estimation error vector as

$$\tilde{\theta}(t) = \hat{\theta}(t) - \theta.$$

Another objective is

$$\tilde{\theta}(t) \to 0 \quad \text{when} \quad t \to \infty.$$

Let us first recall some of the main steps of the computed torque method developed by *Craig (1988)*.

3 COMPUTED TORQUE ADAPTIVE CONTROL

The control law is obtained by substituting H, C and G by their parameter estimates. This leads to

$$M(q,\hat{\theta})(\ddot{q_d} - K_v\dot{q} - K_p q) + C(q,\dot{q},\hat{\theta})\dot{q} + G(q,\hat{\theta}) = \tau \quad (2)$$

where the controller gains K_v and K_p are symmetric, positive definite matrices.

If this control is applied to the input of system (1), we obtain

$$\ddot{e} + K_v\dot{e} + K_p e = M(q,\hat{\theta})^{-1}Y(q,\dot{q},\ddot{q})\tilde{\theta} \quad (3)$$

where $e = q - q_d$ represents the tracking error and

$$Y(q,\dot{q},\ddot{q})\tilde{\theta} = (\hat{M} - M)\ddot{q} + (\hat{C} - C)\dot{q} + (\hat{G} - G)$$

or

$$\dot{X} = AX + BU$$

with

$$X^T = (e,\dot{e}), \quad B^T = (0, I)$$

$$\text{and} \quad A = \begin{bmatrix} 0 & I \\ -K_v & -K_p \end{bmatrix}.$$

By choosing the following update law

$$\dot{\tilde{\theta}} = -\Gamma^{-1}Y(q,\dot{q},\ddot{q})^T M(q,\hat{\theta})^{-1}B^T PX \quad (4)$$

where Γ is a symmetric definite positive matrix and P is a definite positive matrix solution of the equation

$$A^T P + PA = -Q,$$

it is easy to see that the error of tracking tends to zero $(e,\dot{e}) \rightarrow (0,0)$ when Q is definite positive.

4 CONTROLLER DESIGN

The main constraints of the previous controller are that it needs the computation of the inverse of the matrix $M(q,\hat{\theta})$ and acceleration measurements. This second constraint may be avoided by writing the error dynamics of system (3) as follows

$$\begin{aligned} \ddot{e} + {} & K_v\dot{e} + K_p e = \\ & M(q,\hat{\theta})^{-1}(C(q,\dot{q},\hat{\theta})\dot{q} + G(q,\hat{\theta})) \\ & -M(q,\theta)^{-1}(C(q,\dot{q},\theta)\dot{q} + G(q,\theta)) \\ & -(M(q,\hat{\theta})^{-1} - M(q,\theta)^{-1}))\times \\ & \tau(q,\dot{q},q_d,\dot{q_d},\ddot{q_d}) \end{aligned} \quad (5)$$

Now the parametrization is nonlinear. In order to go further, we consider only the first order linear approximation,

$$\ddot{e} + K_v\dot{e} + K_p e \simeq M(q,\hat{\theta})^{-1}\frac{\partial\tau}{\partial\hat{\theta}}(q,q_d,q_d,\dot{q},\ddot{q_d})\tilde{\theta} \quad (6)$$

Let us consider the following update law,

$$\dot{\hat{\theta}} = -\Gamma^{-1}\frac{\partial\tau}{\partial\hat{\theta}}(q,q_d,\dot{q_d},\dot{q},\ddot{q_d})M(q,\hat{\theta})^{-1}B^T PX + \sigma\Gamma^{-1}|e|\hat{\theta}$$

with

$$\begin{cases} \sigma & = \sigma_0 \quad \text{if } ||\hat{\theta}|| > C_0 = \sup||\theta|| \\ & = 0 \quad \text{elsewhere} \end{cases}$$

Clearly, this choice leads to the local stabilization of the system with robustness to noise measurement. Let us now study the robustness with respect to unmodelled dynamics. In order to do this, let us write the system (1) in the state space,

$$\begin{aligned} \dot{q_1} &= q_2 \\ \dot{q_2} &= M(q,\theta)^{-1}(-C(q,\dot{q},\theta)\dot{q} - G(q,\theta) + \tau). \end{aligned}$$

With the previous definition of e this system may be written

$$\begin{aligned} \dot{e_1} &= e_2 \\ \dot{e_2} &= M(q,\theta)^{-1}(-C(q,\dot{q},\theta)\dot{q} - G(q,\theta) - M(q,\theta)\ddot{q_d} + \tau) \end{aligned}$$

As in some recent works of *Taylor* et al. *(1989)* the unmodelled dynamics is assumed to be fast and stable. The analysis starts by perturbing the system in the following way,

$$\begin{aligned} \dot{e_1} &= e_2 \\ \dot{e_2} &= M(q,\theta)^{-1}(-C(q,\dot{q},\theta)\dot{q} - G(q,\theta) \\ & \quad -M(q,\theta)\ddot{q_d} + \zeta) \\ \mu\dot{\zeta} &= -\zeta + \tau. \end{aligned} \quad (7)$$

The change of variable $\zeta = \tau + \eta$ transforms the perturbed system (7) into,

$$\begin{aligned} \dot{e_1} &= e_2 \\ \dot{e_2} &= M(q,\theta)^{-1}(-C(q,\dot{q},\theta)\dot{q} - G(q,\theta) \\ & \quad -M(q,\theta)\ddot{q_d} + \tau + \eta) \\ \mu\dot{\eta} &= -\eta - \mu\dot{\tau} \end{aligned} \quad (8)$$

If the control (2) is applied to the system (8) we obtain,

$$\begin{aligned} \dot{e_1} &= e_2 \\ \dot{e_2} &= -K_v e_2 - K_p e_1 + M(q,\hat{\theta})^{-1}\tfrac{\partial\tau}{\partial\hat{\theta}}(\hat{\theta} - \theta) \\ & \quad +M(q,\theta)^{-1}\eta \\ \mu\dot{\eta} &= -\eta - \mu\dot{\tau} \end{aligned} \quad (9)$$

The stability analysis of the system is now performed using the Lyapunov function

$$V = C_1(X^T P_1 X + \tilde{\theta}^T\Gamma\tilde{\theta}) + C_2\eta^T P_2\eta \quad (10)$$

with $X^T = (e,\dot{e})$ and where P_1 and P_2 are definite positive matrix solutions of

$$P_1 A + A^T P_1 = -I_1$$

and

$$P_2 = \frac{1}{2} I_2$$

with I_1 and I_2 definite positive matrix. Therefore,

$$\dot{V} = C_1(-X^T X + 2X^T P_1 B[M(q,\hat{\theta})^{-1}\frac{\partial\tau}{\partial\theta}(\hat{\theta}-\theta)$$
$$+M(q,\theta)^{-1}\eta] + (\hat{\theta}-\theta)^T \Gamma\dot{\hat{\theta}})$$
$$+C_2(-\frac{1}{\mu}\eta^T\eta - 2\eta^T P_2\dot{\tau})$$

Now, the function $\dot{\tau}$ is such that

$$||\dot{\tau}|| \leq d_1||X|| + d_2||\eta||$$

provided that for all $X \in B_X$ and every pair $\theta, \hat{\theta} \in B_\theta$, the following inequalities hold,

$$2||P_1 BM(q,\theta)^{-1}|| \leq C_2$$
$$2||P_2||d_2 \leq C_3$$
$$2||P_2||d_1 \leq C_1.$$

Finally, if we choose the update law as follows,

$$\dot{\hat{\theta}} = -\Gamma^{-1}\frac{\partial\tau}{\partial\hat{\theta}}(q, q_d, \dot{q_d}, \dot{q}, \ddot{q_d}) \quad M(q,\hat{\theta})^{-1}B^T P_1 X +$$
$$\sigma\Gamma^{-1}|e|\hat{\theta} \qquad (11)$$

with

$$\begin{cases} \sigma & = \sigma_0 \quad \text{if } ||\hat{\theta}|| > C_0 = sup||\theta|| \\ & = 0 \quad \text{elsewhere} \end{cases}$$

we obtain,

$$\dot{V} = -C_1||X||^2 + C_1C_2||X||||\eta|| - \frac{C_2}{\mu}||\eta||^2$$
$$+C_1C_2||X||||\eta|| + C_2C_3||\eta||^2 - \sigma(\hat{\theta}-\theta)|e|\hat{\theta}.$$

The following result can then be proved,

Theorem : The equilibrium point $X^*, \hat{\theta} = \theta, \eta = 0$ of the dynamical system (9) and (11) is locally asymptotically stable with respect to the manipulator position and velocity for every

$$\mu \in (0, \frac{1}{C_1C_2 + C_3}).$$

5 SIMULATIONS

In order to illustrate the algorithm, numerical simulations have been carried out for a planar two-linked robot

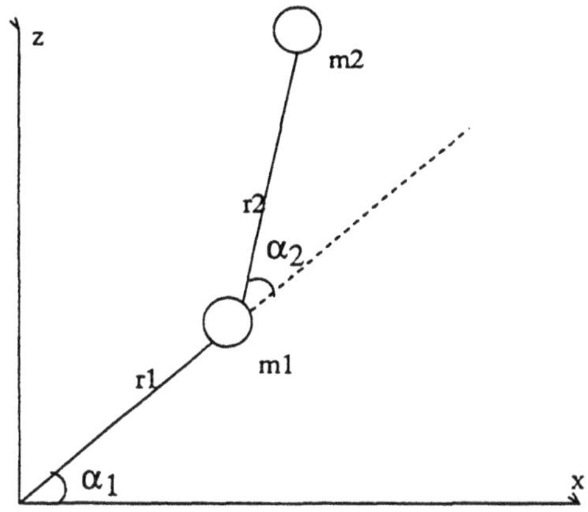

FIG. 1- Two-Link Robot

The model of the robot manipulator was taken from (*Craig* 1986) where the matrix C is written as in *Slotine and Weiping (1991)*

$$M(q,\theta)\ddot{q} + C(q,\dot{q},\theta)\dot{q} + G(q,\theta) = \tau$$

where

$$q = [\alpha_1, \alpha_2]^T$$

$$M = \begin{bmatrix} \theta(r_1^2 + r_2^2 + 2r_1r_2cos\alpha_2) + m_1r_1^2 & \theta(r_2^2 + r_1r_2cos\alpha_2) \\ \theta(r_2^2 + r_1r_2cos\alpha_2) & \theta r_2^2 \end{bmatrix}$$

$$C = \begin{bmatrix} -\theta r_1r_2\dot{\alpha_2}sin\alpha_2 & -\theta r_1r_2\dot{\alpha_1}sin\alpha_2 - \theta r_1r_2\dot{\alpha_2}sin\alpha_2 \\ \theta r_1r_2\dot{\alpha_1}sin\alpha_2 & 0 \end{bmatrix}$$

and

$$G = \begin{bmatrix} \theta r_2gcos(\alpha_2+\alpha_1) + (m_1+\theta)r_1gcos\alpha_1 \\ \theta r_2gcos(\alpha_2+\alpha_1) \end{bmatrix}$$

The values of parameters used in similation are :

$$m_1 = 0.5, \theta = 1, r_1 = 1, r_2 = 0.8, K_v = 18, K_p = 80$$

and the desired trajectories :

$$\begin{cases} \alpha_{1r} & = 1 \\ \alpha_{2r} & = 0.5\sin(4t) \end{cases}$$

$$\text{error}_1 = \alpha_1 - \alpha_{1r} \quad \text{error}_2 = \alpha_2 - \alpha_{2r} \quad U1 = \tau(1)$$

The results of the simulations are shown in Figure 2 where the continuous line corresponds to $\mu = 0.01$ and the dashed line to $\mu = 0.001$.

6 CONCLUSION

In this paper we have proposed a new stable direct adaptive control law for a robot manipulator based on the algorithm described by *Craig (1988)*. The controller has the advantage of being robust with respect to unmodelled dynamics and to noise measurement.

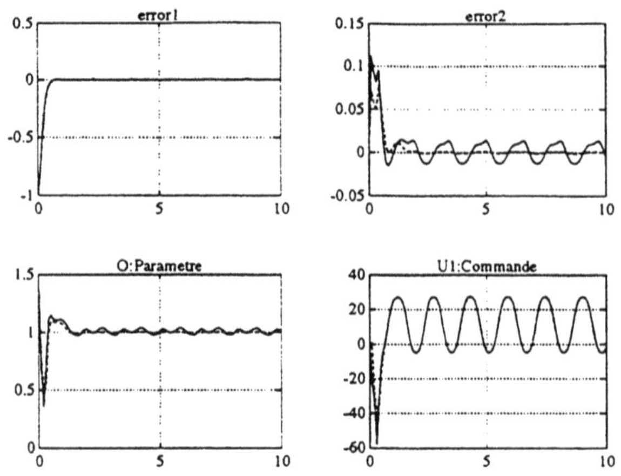

FIG. 2- Simulation of the robot. All graphs versus time[s]

REFERENCES

Berghuis, H., R. Ortega and H. Nijmeijer (1993). 'A robust adaptive robot controller'. *IEEE Robotics and Automation* **9**, 825–830.

Bridges, M., D.M. Dawson and X. Gao (1993). 'Adaptive control of rigid link electrically driven robots'. *Proc. 32nd. IEEE CDC* **9**, 159–165.

Craig, J. J. (1988). *Adaptive Control of Mechanical Manipulators*. Addisson-Wesley Publishing Company.

Garcia Benitez, E., J. Watkins and S. Yurkovich (1993). 'Non-linear control with acceleration feedback for a two link flexible robot'. *Control Eng. Practice* **1**, 989–997.

Hsu, P., Marc. Bodson, S. Sastry and B. Paden (1987). 'Adaptive identification and control for manipulators without using joint accelerations'. *IEEE Int Conf.Robotics and Automation* pp. 1210–1215.

Johansson, R. (1989). 'Lyapunov design for adaptive control of robot'. *NOLCOS Symposium on Nonlinear Control* pp. 209–214.

Luh, J., M. Walker and R. Paul (1985). 'Resolved-acceleration control of mechanical manipulators'. *Proc. 25nd IEEE CDC* pp. 469–474.

Middleton, R. and G. Goodwin (1987). 'Adaptive computed torque control for rigid link manipulators'. *Systems & Control Letters* **6**, 49–59.

Ortega, R. and M. Spong (1989). 'Adaptive motion control of rigid robots : a tutorial'. *Automatica* **25**, 877–888.

Slotine, J. and Li. Weiping (1087). 'On the adaptive control of robot manipulators'. *Int. J. of Robotics Research* **6**, 49–59.

Slotine, J. and Li. Weiping (1991). *Applied Nonlinear Control*. Prentice Hall.

Spong, M. and R. Ortega (1988). 'On adaptive inverse dynamics control of rigid robots'. *IEEE Int Conf.Robotics and Automation*.

Tang, Y. and M. A. Arteaga (1993). 'Adaptive control of robot manipulators based on passivity'. *Proc. 32nd IEEE CDC* pp. 148–152.

Taylor, D., P Kokotovic and R. Marino (1989). 'Adaptive regulation of nonlinear systems with unmodeled dynamics'. *IEEE Trans. Auto. Control* **34**, 405–412.

Whitcomb, L., A. A. Rizzi and D.E. Koditschek (1993). 'Comparative experiments with a new adaptive controller for robot arms'. *IEEE Robotics and Automation* **9**, 59–69.

ANALYSIS AND SYNTHESIS OF AN OUTPUT FEEDBACK FOR AN UNCERTAIN ROBOT MODEL WITH FLEXIBLE JOINTS*

A. AILON
Department of Electrical and Computer Engineering, Ben Gurion University of the Negev, Beer Sheva 84105, Israel

Abstract. This study considers the set-point regulation problem of a flexible-joint robot with uncertain parameters and unknown gravity forces when only position measurements are available for the controller. An essential tool in the present approach is the contraction mapping theorem. An analytical solution to the underlying control problem is presented. Some practical issues associated with the implementation of the controller-observer are considered.

Résumé. Nous considérons le problème du contrôle point-à-point d'un robot à jointures flexibles avec paramètres incertains et forces de gravité inconnues, quand le contrôleur n'a que des données de position. L'outil essentiel dans notre approche est le théorème des applications contractantes. Nous présentons une solution analytique à notre problème de contrôle. Nous considérons aussi des questions pratiques associées à la réalisation du contrôleur-observateur.

Key Words- Flexible-joint robot; controller-observer; point-to-point control; contraction mapping theorem.

1. INTRODUCTION

In the majority of current robot applications, the basic set-point regulation problem has been solved using a proportional plus derivative (PD) controller with gravity compensation. For example: Takegaki and Arimoto (1982) and Tomei (1991) consider stabilizing controllers for a rigid robot model, while Tomei (1991) presents controllers for stabilizing a robot with elastic joints. The schemes presented in this reference is based on the exact knowledge of the gravity vector. In contrast, De Luca and Panzieri (1992) demonstrates an iterative scheme for generating gravity compensation at the desired set-point. However, the resulting control law is based on the action of a PD controller, and requires noisy measurements of joint velocity.

The solution to the set-point regulation problem when only the output signals are available for the controller, has been solved by Berghuis and Nijmeijer (1993) for the rigid robot case and by Ailon and Ortega (1993) and Kelly *et al.* (1994) for flexible-joint robots.

We consider here the set-point regulation problem of a flexible-joint robot whose model contains uncertain parameters and unknown gravity forces, and only a set of *possible* values of these terms is known. The objective is to design a controller which implements only position measurements on the *motor side*, and yet ensures either *globally ultimate boundedness* such that every system response converges to an *arbitrarily small neighborhood* of the *operating point* or, under mild physical conditions, *global asymptotic stability*. An analytical approach associated with the underlying control problem is presented. Some practical issues concerning the application of the controller are considered.

*This work was supported in part by the joint French-Israeli collaboration project "Arc en ciel" 1994 under contract FN 49035.

2. PRELIMINARY RESULTS

From Spong and Vidyasagar (1989) the model of a flexible-joint robot is given by:

$$\begin{aligned} D(q_1)\ddot{q}_1 + C(q_1,\dot{q}_1)\dot{q}_1 + g(q_1) &= K(q_2 - q_1), \\ J\ddot{q}_2 + K(q_2 - q_1) &= u, \end{aligned} \tag{1}$$

where $q_1, q_2 \in \Re^n$ represent the link and the motor angles, $D(q_1) > 0$ is the robot's inertia matrix, $J > 0$ is the actuator inertias matrix, $C(q_1, \dot{q}_1)\dot{q}_1$ represents the Coriolis and centrifugal forces, $g(q_1)$ is the gravitational force vector, $K > 0$ is the joint stiffness coefficients diagonal matrix, and $u \in \Re^n$ is the applied torques. The *Euclidean* norm of x is denoted by $||x||$ and $||A||_i$ is the induced norm of the matrix A with $||A||_i^2 \doteq \lambda_{\max}(A^T A)$, where $\lambda_{\max}(A^T A)$ is the largest eigenvalue of $(A^T A)$. The state space representation of (1) is given by

$$\begin{aligned} \dot{x}_1 &= x_2 \\ \dot{x}_2 &= D(x_1)^{-1}(-C(x_1,x_2)x_2 - g(x_1) - K(x_1 - x_3)) \\ \dot{x}_3 &= x_4 \\ \dot{x}_4 &= J^{-1}(-K(x_3 - x_1) + u), \end{aligned} \tag{2}$$

where $x_1 \doteq q_1$, $x_2 \doteq \dot{q}_1$, $x_3 \doteq q_2$, $x_4 \doteq \dot{q}_2$. The output map is given by

$$y = x_3. \tag{3}$$

The suggested *n-dimensional linear controller-observer* is given by

$$\dot{z} = -S(z - x_3) - Rz + v, \quad (4)$$

$$u = -S(x_3 - z), \quad (5)$$

where the constant matrices $S = S^T, R = R^T > 0$ and the vector $v \in \Re^n$, are to be determined.

Remarks. The controller-observer (4) is different in its structure from the ones given in Ailon and Ortega (1993) and Kelly *et al.* (1994). Note that the vector v (which appears in the first reference but not in the second) is essential for the approach below and the application of the contraction mapping theorem.

We shall establishe the main results first for the case of a rigid robot, and then extend it to the flexible-joint robot model. For a rigid robot the model (2)-(3) reduces to

$$\begin{aligned} \dot{x}_1 &= x_2 \\ \dot{x}_2 &= D(x_1)^{-1}(-C(x_1, x_2)x_2 - g(x_1) + u), \end{aligned} \quad (6)$$

$$y = x_1, \quad (7)$$

and the controller (4)-(5) becomes

$$\dot{z} = -S(z - x_1) - Rz + v, \quad (8)$$

$$u = -S(x_1 - z). \quad (9)$$

Consider (6), (8)-(9) and define a scalar function H_r

$$\begin{aligned} H_r(x_1, x_2, z) = \tfrac{1}{2}[x_2^T D(x_1)x_2 + (x_1 - z)^T S(x_1 - z) \\ +(z - R^{-1}v)^T R(z - R^{-1}v)] + U_g(x_1) \end{aligned} \quad (10)$$

Evaluation of $\partial H_r/\partial x_r \mid_{x_2=z_2=0}; x_r \doteq [x_1^T, z^T]^T$ yields

$$\begin{aligned} S(x_1 - z) + g(x_1) = 0 \\ -S(x_1 - z) + R(z - R^{-1}v) = 0. \end{aligned} \quad (11)$$

From (10) and (11) we have

$$\frac{\partial^2 H_r}{\partial x_r^2} \mid_{x_2=0} = \begin{bmatrix} S + \frac{\partial g(x_1)}{\partial x_1} & -S \\ -S & S + R \end{bmatrix}, \quad (12)$$

where it is assumed that there exists $\beta \geq 0$ with

$$\beta \doteq \sup_{x_1 \in \Re^n} \|\frac{\partial g(x_1)}{\partial x_1}\| > 0. \quad (13)$$

Following *Sylvester's law of inertia*, $\frac{\partial^2 H_r}{\partial x_r^2} \mid_{x_2=0} > 0$ iff

$$\begin{aligned} &\begin{bmatrix} I_n & 0 \\ I_n & I_n \end{bmatrix} \left\{ \begin{bmatrix} S + \frac{\partial g(x_1)}{\partial x_1} & -S \\ -S & S + R \end{bmatrix} \right\} \begin{bmatrix} I_n & I_n \\ 0 & I_n \end{bmatrix} \\ &= \begin{bmatrix} S & 0 \\ 0 & R \end{bmatrix} + \begin{bmatrix} I_n & I_n \\ I_n & I_n \end{bmatrix} diag. \left\{ \frac{\partial g(x_1)}{\partial x_1} \right\} > 0 \end{aligned} \quad (14)$$

for all x_r. But from (13), (14) is satisfied if

$$\begin{bmatrix} S & 0 \\ 0 & R \end{bmatrix} > \beta \begin{bmatrix} I_n & I_n \\ I_n & I_n \end{bmatrix}.$$

Since there exists a permutation matrix P such that

$$P \begin{bmatrix} I_n & I_n \\ I_n & I_n \end{bmatrix} P^T = block - diag.\{U_r\};$$
$$\text{where } U_r \doteq \begin{bmatrix} 1 & 1 \\ 1 & 1 \end{bmatrix},$$

where $\left| \sqrt{\lambda_{\max}(U_r U_r^T)} \right| = 2$, $\frac{\partial^2 H_r}{\partial x_r^2} \mid_{x_2=0} > 0$ if

$$\min\{\lambda_{\min}(S), \lambda_{\min}(R)\} > 2\beta. \quad (15)$$

If $\{S, R\}$ satisfies (15) then, for a given vector v the scalar function $H_r(x_1, 0, z)$ has a unique *global minimum point* $\bar{x}_r = [\bar{x}_1^T, \bar{z}^T]^T$. Let the minimal value of H_r be denoted by $\bar{H}_r$ and define a *Lyapunov candidate function*

$$V_r(x_1, x_2, z) = H_r(x_1, x_2, z) - \bar{H}_r. \quad (16)$$

Along the system's trajectory we have

$$\begin{aligned} \dot{V}_r = -(S(z - x_1) + Rz - v)^T (S(z - x_1) + Rz - v) \\ \leq 0. \end{aligned} \quad (17)$$

Invoking *LaSalle's invariance theorem*, we conclude that the system is *globally asymptotically stable*.

For the flexible-joint robot (2) with the controller (4)-(5) we define the following scalar function

$$\begin{aligned} H_f(x_1, x_2, x_3, x_4, z) = \tfrac{1}{2}[x_2^T D(x_1)x_2 + x_4^T J x_4 \\ +(x_1 - x_3)^T K(x_1 - x_3) + (x_3 - z)^T S(x_3 - z) \\ +(z - R^{-1}v)^T R(z - R^{-1}v)] + U_g(x_1). \end{aligned} \quad (18)$$

For $x_f \doteq [x_1^T, x_3^T, z^T]^T$, $\partial H_f/\partial x_f \mid_{x_2=x_4=0} = 0$ implies

$$\begin{aligned} K(x_1 - x_3) + g(x_1) = 0 \\ -K(x_1 - x_3) + S(x_3 - z) = 0 \\ -S(x_3 - z) + R(z - R^{-1}v) = 0. \end{aligned} \quad (19)$$

From (18)-(19) we have

$$\frac{\partial^2 H_f}{\partial x_f^2} \mid_{x_2=x_4=0} = \begin{bmatrix} K + \frac{\partial g(x_1)}{\partial x_1} & -K & 0 \\ -K & K + S & -S \\ 0 & -S & S + R \end{bmatrix}. \quad (20)$$

Now, using previous arguments one concludes that the *Hermitian matrix* $\frac{\partial^2 H_f}{\partial x_f} \mid_{x_2=x_4=0}$ is positive definite for all x_f iff the matrix

$$\begin{aligned} &\begin{bmatrix} I_n & 0 & 0 \\ I_n & I_n & 0 \\ I_n & I_n & I_n \end{bmatrix} \left[\frac{\partial^2 H_f}{\partial x_f^2} \mid_{x_2=x_4=0} \right] \begin{bmatrix} I_n & I_n & I_n \\ 0 & I_n & I_n \\ 0 & 0 & I_n \end{bmatrix} \\ &= \begin{bmatrix} K & 0 & 0 \\ 0 & S & 0 \\ 0 & 0 & R \end{bmatrix} + \begin{bmatrix} I_n & I_n & I_n \\ I_n & I_n & I_n \\ I_n & I_n & I_n \end{bmatrix} diag. \left\{ \frac{\partial g(x_1)}{\partial x_1} \right\}, \end{aligned}$$

is positive definite. But again it is easy to see that there exists a permutation matrix Q such that

$$Q\begin{bmatrix} I_n & I_n & I_n \\ I_n & I_n & I_n \\ I_n & I_n & I_n \end{bmatrix} Q^T = block - diag.\{U_f\};$$

$$where\ U_f \doteq \begin{bmatrix} 1 & 1 & 1 \\ 1 & 1 & 1 \\ 1 & 1 & 1 \end{bmatrix},$$

where in the present case $\left|\sqrt{\lambda_{\max}(U_f U_f^T)}\right| = 3$. Hence, $\frac{\partial^2 H_f}{\partial x_f}|_{x_2=x_4=0} > 0$ if

$$\min\{\lambda_{\min}(K), \lambda_{\min}(S), \lambda_{\min}(R)\} > 3\beta. \quad (21)$$

Suppose the triplet $\{K, S, R\}$ satisfies (21). Then, for a given v the function $H_f(x_1, 0, x_3, 0, z)$ has a unique *global minimum*. Let $\bar{H}_f$ be the minimal value of H_f and define a *Lyapunov candidate function*

$$V_f(x_1, x_2, x_3, x_4, z) = H_f(x_1, x_2, x_3, x_4, z) - \bar{H}_f\ . \quad (22)$$

The derivative of $V_f(x_1, x_2, x_3, x_4, z)$ along the trajectories is now given by

$$\dot{V}_f = -\left(S(z - x_3) + Rz - v\right)^T \left(S(z - x_3) + Rz - v\right) \leq 0. \quad (23)$$

Invoking LaSalle's invariance theorem, we conclude that the system is *globally asymptotically stable*.

Finally, the following Theorem (Vidyasagar (1993)), is an essential tool along this study:

Theorem 2.1. (Global Contraction Mapping). Let $(X, \|.\|)$ be a Banach space, and let $T : X \rightarrow X$. Suppose there exists a fixed constant $\eta < 1$ such that

$$\| Tx - Ty \| \leq \eta \| x - y \|, \forall x, y \in X.$$

Under these conditions, there exists exactly one $x^* \in X$ such that $Tx^* = x^*$. For each $x_0 \in X$, the sequence $\{x_n\}$ in X defined by $x_{n+1} = Tx_n$ converges to x^*. Moreover,

$$\|x^* - x_n\| \leq \frac{\eta^n}{1-\eta} \|Tx_0 - x_0\|.$$

3. Basic Analytical Results

3.1. Case a: Rigid Robot.

Consider (6), and assume the pair $\{S, R\}$ satisfies (15). Then, from the first equation of (11) we have

$$Sz = Sx_1 + g(x_1) \Rightarrow z = x_1 + S^{-1}g(x_1).$$

Hence, for every $x_1 \in \Re^n$ the unique vector v that satisfies (11) is given by

$$v = L_r(x_1) \doteq (I_n + RS^{-1})g(x_1) + Rx_1. \quad (24)$$

From (24) and the fact that for each $v \in \Re^n$ (19) has (by recalling (20)), a unique solution the map $L_r : \Re^n \rightarrow \Re^n$ is *bijective*. In particular we may write

$$x_1 = G_r(v) \doteq L_r^{-1}(v). \quad (25)$$

Let x_{1d} be a given constant vector and define a map $T_r : \Re^n \rightarrow \Re^n$ by

$$T_r(v) = v - R(G_r(v) - x_{1d}). \quad (26)$$

Using Theorem 2.1 we have the following result.

Lemma 3.1. Consider the map $T_r(v)$ defined by (24)-(26), and assume that $g(x)$ satisfies (13). Then a pair of matrices $\{S, R\}$ can be determined such that T_r is a global contraction with a unique fixed point v^*, i.e.,

$$T_r(v^*) = v^*, \quad (27)$$

which implies

$$x_1^* \doteq G_r(v^*) = x_{1d}. \quad (28)$$

Proof. Let the pair $\{S, R\}$ be selected according to the following rule:

$$R \doteq rI_n;\ r > 2\beta,\ S \doteq R^2. \quad (29)$$

Since L_r in (24) is bijective, we may write using (25) $G_r(v_1) = x_1^1$, and $G_r(v_2) = x_1^2$. Let $\Gamma(x_1^1, x_1^2) \doteq \{w \mid w = \theta x_1^1 + (1-\theta)x_1^2,\ 0 < \theta < 1\}$ be a line segment connecting x_1^1 to x_1^2. By the *mean value theorem*

$$g(x_1^1) - g(x_1^2) = \Pi(x_1^1 - x_1^2);$$
$$\Pi \doteq \int_0^1 \frac{\partial g}{\partial x_1}\left(\theta x_1^1 + (1-\theta)x_1^2\right) d\theta. \quad (30)$$

From (24)-(26), and (29) we have

$$T_r(v_1) - T_r(v_2) = (I_n + R^{-1})g(x_1^1) + Rx_1^1 - R(x_1^1 - x_{1d}) - ((I_n + R^{-1})g(x_1^2) + Rx_1^2 - R(x_1^2 - x_{1d})) = (I_n + R^{-1})(g(x_1^1) - g(x_1^2)), \quad (31)$$

where, by definition $x_1^i = G_r(v_i)$. Hence, by (30)

$$T_r(v_1) - T_r(v_2) = (I_n + R^{-1})\Pi(x_1^1 - x_1^2). \quad (32)$$

Recalling (24) one obtains

$$v_1 - v_2 = (I_n + R^{-1})g(x_1^1) + Rx_1^1 - ((I_n + R^{-1})g(x_1^2) + Rx_1^2) = (I_n + R^{-1})(g(x_1^1) - g(x_1^2)) + R(x_1^1 - x_1^2)$$

and using (30) we arrive at

$$\|v_1 - v_2\| = \|[(I_n + R^{-1})\Pi + R](x_1^1 - x_1^2)\|. \quad (33)$$

By *Schwarz inequality* and its direct consequence

$$\|[(I_n + R^{-1})\Pi + R](x_1^1 - x_1^2)\| \geq \\ | \, \|R(x_1^1 - x_1^2)\| - \|(I_n + R^{-1})\Pi(x_1^1 - x_1^2)\| \, | \, . \tag{34}$$

For every $\rho > 1$ one can select $r_1 > 0$ such that

$$0 < \frac{r+1}{r} < \rho; \; \forall r \geq r_1. \tag{35}$$

Select arbitrarily $\rho > 1$ and $r > \max\{r_1, 2\rho\beta\}$. Then, by (13) and (29) the right hand side of (34) satisfies

$$| \, \|R(x_1^1 - x_1^2)\| - \|(I_n + R^{-1})\Pi(x_1^1 - x_1^2)\| \, | \\ = \|R(x_1^1 - x_1^2)\| - \|(I_n + R^{-1})\Pi(x_1^1 - x_1^2)\| \\ \geq r\|x_1^1 - x_1^2\| - \rho\beta\|x_1^1 - x_1^2\|,$$

and as a result of (33)-(34)

$$(r - \rho\beta) \parallel x_1^1 - x_1^2 \parallel \leq \parallel v_1 - v_2 \parallel . \tag{36}$$

From (32) and (30) one also concludes that

$$\|T_r(v_1) - T_r(v_2)\| \leq \rho\beta\|x_1^1 - x_1^2\|. \tag{37}$$

Therefore, since $G_r(v)$ is bijective (i.e. $\|x_1^1 - x_1^2\| \neq 0 \Leftrightarrow \|v_1 - v_2\| \neq 0$), for every $r > \max\{r_1, 2\rho\beta\}$ the following holds

$$\parallel T_r(v_1) - T_r(v_2) \parallel \leq \frac{\rho\beta}{r - \rho\beta} \parallel v_1 - v_2 \parallel, \tag{38}$$

and we clearly have

$$\eta_r \doteq \frac{\rho\beta}{r - \rho\beta} < 1. \tag{39}$$

Since the space $\Re^n$ with the *Euclidean* norm is a *Banach* space, following Theorem 2.1, T_r in (26) is a global contraction, and the sequence $\{v_n\}$ defined by

$$v_{n+1} = T_r(v_n), \; n = 0, 1, \ldots \tag{40}$$

converges to a unique fixed point v^* which satisfies (27)-(28). The rate of convergence of the sequence $\{v_n\}$ to v^* is determined by the parameter r, and there exists:

$$\|v^* - v_n\| \leq \frac{\eta_r^n}{1 - \eta_r}\|T_r(v_0) - v_0\|, \tag{41}$$

where $v_0 \in \Re^n$ is an arbitrary initial vector. ◇◇

Let $\{v_i\}$ be the sequence determined by (40). For any given initial position and for each v_i the closed-loop system trajectory $\varphi_{v_i}(\cdot)$ converges asymptotically to a unique equilibrium point $\bar{\varphi}_{v_i}$. The projection of $\bar{\varphi}_{v_i}$ into the subspace x_1 is given by x_1^i and the limit of the sequence $\{x_1^i\}$ is x_{1d}.

For every $v = v_i$ in (4), the system converges asymptotically to a unique equilibrium point, $\bar{\varphi}_{v_i} = \left[\bar{x}_1^T, 0, \bar{z}^T\right]^T$ with $\bar{x}_1 = x_1^i$. In the following algorithms the system's motion toward its equilibrium point will be represented by $\left(\overset{motion}{\Longrightarrow}\right)$. The vector x_1^i determines v_{i+1} through $v_{i+1} \doteq T_r(v_i) = v_i - R(x_1^i - x_{1d})$. The computation of v_i is based on measurable variables and will be represented below by $\left(\overset{comp.}{\Longrightarrow}\right)$. Using these notations we state the following Algorithm.

Algorithm 3.1. Select constants $\rho > 1$ and $r > \max\{r_1, 2\rho\beta\}$ where r_1 is any constant that satisfies (35), and determine according to (29) a pair $\{S, R\}$. Then, an arbitrary selection of a constant vector $v_0 \in \Re^n$ initiates the following chain:

$$v_0 \overset{motion}{\Longrightarrow} x_1^0 \overset{comp.}{\Longrightarrow} v_1 = T_r(v_0) \overset{motion}{\Longrightarrow} x_1^1 \overset{comp.}{\Longrightarrow} \\ \cdots \overset{comp.}{\Longrightarrow} v_n = T_r(v_{n-1}) \overset{motion}{\Longrightarrow} x_1^n \cdots .$$

In addition, (36), (39) and (41) ensure that:

$$\|x_{1d} - x_1^n\| \leq \tfrac{1}{(r-\rho\beta)}\|v^* - v_n\| \\ \leq \tfrac{\eta_r^n}{(1-\eta_r)(r-\rho\beta)}\|T_r(v_0) - v_0\|,$$

where v^* is the limit of the sequence $\{v_i\}$ and respectively x_{1d} is the limit of $\{x_1^i\}$. ◇◇

3.2. Case b: Flexible-Joint Robot.

Now consider (2), and assume that the triplet $\{K, S, R\}$ satisfies (21). From (19) we obtain

$$v \doteq L_f(x_1) = (I_n + RS^{-1} + RK^{-1})g(x_1) + Rx_1 \tag{42}$$

and by previous considerations the map $L_f : \Re^n \to \Re^n$ is *bijective*. In particular we may write

$$x_1 = G_f(v) \doteq L_f^{-1}(v). \tag{43}$$

Let x_{1d} be a given constant vector and define a map $T_f : \Re^n \to \Re^n$ as follows:

$$T_f(v) = v - R(G_f(v) - x_{1d}). \tag{44}$$

Lemma 3.2. Consider the map $T_f(v)$ defined by (42)-(44) and assume that $g(x)$ satisfies (13) and $\lambda_{\min}(K) > 3\beta$. Then a pair of matrices $\{S, R\}$ can be determined such that T_f is a global contraction with a unique fixed point v^*, i.e.,

$$T_f(v^*) = v^* \tag{45}$$

which implies

$$x_1^* \doteq G_f(v^*) = x_{1d}. \tag{46}$$

Proof. Let the pair $\{S, R\}$ be selected according to the following rule

$$R = rI_n; r > 3\beta, S = R^2. \tag{47}$$

From (21) L_f in (42) is bijective. Let $G_f(v_i) = x_1^i$. From (42)-(44), and (29) we have

$$\begin{aligned} &T_f(v_1) - T_f(v_2) \\ &= (I_n + R^{-1} + RK^{-1})g(x_1^1) + Rx_1^1 \\ &-R(x_1^1 - x_{1d}) - ((I_n + R^{-1} + RK^{-1})g(x_1^2) \\ &+Rx_1^2 - R(x_1^2 - x_{1d})) \\ &= (I_n + R^{-1} + RK^{-1})(g(x_1^1) - g(x_1^2)), \end{aligned} \tag{48}$$

and using (30)

$$T_f(v_1) - T_f(v_2) = (I_n + R^{-1} + RK^{-1})\Pi(x_1^1 - x_1^2). \tag{49}$$

Recalling (42) one obtains

$$\begin{aligned} &v_1 - v_2 \\ &= (I_n + R^{-1} + RK^{-1})g(x_1^1) + Rx_1^1 \\ &-((I_n + R^{-1} + RK^{-1})g(x_1^2) + Rx_1^2) = \\ &(I_n + R^{-1} + RK^{-1})(g(x_1^1) - g(x_1^2)) + R(x_1^1 - x_1^2). \end{aligned}$$

Again, using (30) we have

$$v_1 - v_2 = (I_n + R^{-1} + RK^{-1})\Pi(x_1^1 - x_1^2) + R(x_1^1 - x_1^2),$$

and hence

$$\|v_1 - v_2\| = \|[(I_n + R^{-1} + RK^{-1})\Pi + R](x_1^1 - x_1^2)\|. \tag{50}$$

By *Schwarz inequality* one gets

$$\begin{aligned} &\|[(I_n + R^{-1} + RK^{-1})\Pi + R](x_1^1 - x_1^2)\| \geq \\ &|\, \|R(x_1^1 - x_1^2)\| - \|(I_n + R^{-1} + RK^{-1})\Pi(x_1^1 - x_1^2)\| \,|. \end{aligned} \tag{51}$$

Arbitrarily select $\rho > 1$, and define r_1 such that (35) holds, and let $r > \max\{r_1, 6\rho\beta, \frac{1}{2}\rho\lambda_{\min}(K)\}$. Clearly r satisfies (47). Next we claim that

$$r > \|I_n + R^{-1} + RK^{-1}\|\beta. \tag{52}$$

Indeed, since $\lambda_{\min}(K) > 3\beta$ and $r > \frac{1}{2}\rho\lambda_{\min}(K)$, one obtains by (35):

$$\begin{aligned} &2r^2\beta > (r+1)\lambda_{\min}(K)\beta \Rightarrow \\ &r^2(\lambda_{\min}(K) - \beta) > (r+1)\lambda_{\min}(K)\beta, \end{aligned}$$

and the last inequality yields

$$r > \frac{r^2 + r\lambda_{\min}(K) + \lambda_{\min}(K)}{r\lambda_{\min}(K)}\beta. \tag{53}$$

The last inequality together with the relation (recall that by definition K and R are diagonal matrices)

$$\begin{aligned} \|I_n + R^{-1} + RK^{-1}\| &\leq \xi \doteq 1 + \tfrac{1}{r} + \tfrac{r}{\lambda_{\min}(K)} \\ &= \tfrac{r^2 + r\lambda_{\min}(K) + \lambda_{\min}(K)}{r\lambda_{\min}(K)}, \end{aligned} \tag{54}$$

assert the claim.

Therefore, if r is sufficiently large we have

$$\begin{aligned} &|\, \|R(x_1^1 - x_1^2)\| - \|(I_n + R^{-1} + RK^{-1})\Pi(x_1^1 - x_1^2)\| \,| \\ &= \|R(x_1^1 - x_1^2)\| - \|(I_n + R^{-1} + RK^{-1})\Pi(x_1^1 - x_1^2)\| \\ &\geq r\|x_1^1 - x_1^2\| - \xi\beta\|x_1^1 - x_1^2\|, \end{aligned}$$

which implies, using (50)-(51):

$$(r - \xi\beta)\, \| x_1^1 - x_1^2 \| \leq \| v_1 - v_2 \|\,. \tag{55}$$

From (49) and (54) we have

$$\|T_f(v_1) - T_f(v_2)\| \leq \xi\beta\|x_1^1 - x_1^2\|. \tag{56}$$

Hence, by observing the last two equations and recalling that $G_f(v)$ is bijective we arrive at:

$$\|T_f(v_1) - T_f(v_2)\| \leq \frac{\xi\beta}{r - \xi\beta}\|v_1 - v_2\|. \tag{57}$$

It remains to show that

$$\eta_f \doteq \frac{\xi\beta}{r - \xi\beta} < 1. \tag{58}$$

Clearly (58) holds if $r > 2\beta\xi$ or (see (54)):

$$r - \frac{2}{\lambda_{\min}(K)}\beta r > \frac{2r+2}{r}\beta. \tag{59}$$

But recalling (35) and using the fact that $\lambda_{\min}(K) > 3\beta$, (59) is satisfied if $r(1 - \frac{2}{3}) > 2\rho\beta$ or $r > 6\rho\beta$.

With this result one concludes by Theorem 2.1 that T_f in (44) is a global contraction, and the sequence $\{v_n\}$ in $\Re^n$ defined by

$$v_{n+1} = T_f(v_n), \;\; n = 0, 1, \ldots \tag{60}$$

converges to a unique fixed point v^* which satisfies (45). Hence by (44), (46) holds and there exists:

$$\|v^* - v_n\| \leq \frac{\eta_f^n}{1 - \eta_f}\|T_f(v_0) - v_0\|, \tag{61}$$

where $v_0 \in \Re^n$ is an arbitrary initial vector. ◇◇

Following Lemma 3.2 and using previous notations we have the following algorithm for the present case.

Algorithm 3.2. Select constants $\rho > 1$ and $r > \max\{r_1, 6\rho\beta, \frac{1}{2}\rho\lambda_{\min}(K)\}$ where r_1 is any number that satisfies (35). Then determine according to (47) a pair $\{S, R\}$. Then, an arbitrary selection of a constant vector $v_0 \in \Re^n$ initiates the following chain:

$$\begin{aligned} &v_0 \overset{motion}{\Longrightarrow} x_1^0 \overset{comp.}{\Longrightarrow} v_1 = T_f(v_0) \overset{motion}{\Longrightarrow} x_1^1 \overset{comp.}{\Longrightarrow} \\ &\cdots \overset{comp.}{\Longrightarrow} v_n = T_f(v_{n-1}) \overset{motion}{\Longrightarrow} x_1^n \cdots. \end{aligned}$$

In addition, (54), (55), (58) and (61) ensure that

$$\begin{aligned} \|x_{1d} - x_n\| &\leq \tfrac{1}{(r - \xi\beta)}\|v^* - v_n\| \\ &\leq \tfrac{\eta_f^n}{(1-\eta_f)(r-\xi\beta)}\|T_f(v_0) - v_0\|, \end{aligned}$$

where v^* is the limit of the sequence $\{v_i\}$ and respectively x_{1d} is the limit of $\{x_1^i\}$. ◇◇

4. Final Remarks Concerning the Practical Solution to the Control Problem

Some remarks with regard to the applications of the previous analytical results to the framework established by synthesis are in order.

(i)- In the case of a rigid robot the rate of convergence is determined by the parameter η_r which, as observed in (39), depends on r. If, for a given $\rho > 0$, one takes a relatively large r, η_r becomes as small as desired and the robot converges rapidly to x_{1d}.

(ii)- While the rate of convergence of the rigid robot to the final destination can be tuned arbitrarily, this is not the case for a flexible-joint robot. In fact, the key parameter with respect to the rate of convergence in this case is η_f in (58). Since ξ depends also on r (see (54)) there is a lower bound to η_f. By a direct calculation it can be shown that

$$\eta_f = \frac{(r^2 + r\lambda_{\min}(K) + \lambda_{\min}(K))\beta}{r^2\lambda_{\min}(K) - (r^2 + r\lambda_{\min}(K) + \lambda_{\min}(K))\beta}.$$

Following the assumption that $\lambda_{\min}(K) > 3\beta$ it can be seen that if for example $\lambda_{\min}(K) \cong 3\beta$, then for a large r, $\eta_f \cong \frac{1}{2}$. If $\lambda_{\min}(K)$ is relatively a large number, that lower bound becomes respectively small.

(iii)- Since asymptotic convergence is associated with *infinite-time process*, the above algorithms cannot be implemented in a straight-forward manner in *practical applications*. However, this problem can be overcomed using the following approach. (Due to limitation in space we present here just the outlines of the relevant approach.)

Let $\varepsilon > 0$ be a (small as desired) constant. Since from (17) and (23) the functions $\dot{V}_r\ (\cdot)$ and $\dot{V}_f\ (\cdot)$ depend just on measurable signals, assuming some upper bound on the system initial 'energy' is known, it can be shown formally that, based on available accumulated data it is possible to determine when the system's state vector belongs to an $\varepsilon -$ *neighborhood* of an (unknown) equilibrium point. Now, whatever the initial condition is, it takes finite time for the system's trajectory to enter that neighborhood. Following this observation, the algorithm can be synthesized such that in each step of the process a vector $\bar{v}_{n+1}$, rather than $v_{n+1} = T(v_n)$, is determined. Moreover, it can be shown that if in each step $\|\ \bar{v}_{n+1} - v_{n+1}\|$ is sufficiently small then, finally the system will converge to an *arbitrarily small neighborhood* of the *desired equilibrium point*.

Furthermore, if $\varepsilon > 0$ was selected sufficiently small, then, as soon as the system's trajectory enters into an $\varepsilon -$ *neighborhood* of the desired equilibrium point, the control strategy can be switched to a new mode, and a different controller-observer which is based on the principle of stability in the first approximation (Vidyasagar (1993)) can be initiated by, say the *decision maker*. This action ensures asymptotic convergence to the desired target.

References

[1] Takegaki, M. and S. Arimoto (1981). A new feedback method for dynamic control of manipulators, *Trans. ASME J. on Dynamic Systems, Measurement, and Control* **102**, 119-125.

[2] Tomei P. (1991). Adaptive PD controller for robot manipulators, *IEEE Trans. on Robotics and Automation*, **7**, 565-570.

[3] Tomei P. (1991). A simple PD controller for robots with elastic joints, *IEEE Trans. on Automatic Control,* **36**, 1208-1213.

[4] De Luca A. and S. Panzieri (1992). An iterative scheme for learning gravity compensation in flexible robot arms, *Cancun Workshop on Nonlinear Systems and Robotics,* Cancun, Mexico.

[5] Berghuis H. and H. Nijmeijer (1993). Global regulation of robots using only position measurements, *Systems and Control Letters*, **21**, 289-293.

[6] Ailon A. and R. Ortega (1993). An observer-based set-point controller for robot manipulators with flexible joints, *Systems and Control Letters*, **21**, 329-335.

[7] Kelly R., R. Ortega, A. Ailon, and A. Loria (1994). Global regulation of Flexible joints robots using approximate differentiation, *IEEE Trans. on Automatic Control,* **39**, 1222-1224.

[8] Spong M. and M. Vidyasagar (1989). *Robot Dynamics and Control*, John Wiley & Sons, New York.

[9] Vidyasagr M. (1993). *Nonlinear Systems Analysis*, Prentice-Hall, New York.

REDUCED-ORDER OBSERVER-BASED POINT-TO-POINT AND TRAJECTORY CONTROLLERS FOR ROBOT MANIPULATORS

M.F. KHELFI†, M. ZASADZINSKI†, M. DAROUACH†, H. RAFARALAHY† and E. RICHARD†‡
† CRAN - ACS - CNRS URA 821 - Université Henri Poincaré-Nancy I - 186 rue de Lorraine, 54400 Cosnes-et-Romain, FRANCE
‡ INRIA Lorraine - Université de Metz - 4 rue Marconi, 57070 Metz, FRANCE

Abstract. This note presents a design procedure of a reduced-order observer-based controller for a n-joint robot manipulators. We assume that only the joint angular positions are measured. The joint angular velocities are estimated via an exponential reduced-order observer. Two types of control laws based on this observer are studied: a point-to-point control with gravity compensation and a trajectory control. Sufficient conditions to ensure the closed-loop stability are given. Performances of the reduced-order observer used with these two control laws are illustrated in a simulation study of a two degrees-of-freedom robot manipulator.

Key Words. Rigid robot manipulator; exponential reduced-order observer; controller design; asymptotic stability.

Résumé. Cet article présente une procédure de synthèse d'un observateur exponentiel d'ordre réduit pour une classe de systèmes non linéaires incluant les modèles de robots manipulateur à n degrés de liberté. Seules les positions articulaires du robot manipulateur étant mesurées, les vitesses articulaires sont estimées via un observateur d'ordre réduit à convergence exponentielle. Le comportement de l'observateur dans une boucle de régulation est étudié avec deux correcteurs: une loi de commande point à point avec compensation de l'effet de la gravité et une loi de commande avec suivi de trajectoire. Les conditions suffisantes de stabilité asymptotique du système en boucle fermée sont énoncées. Les performances de l'observateur d'ordre réduit en boucle fermée avec ces deux lois de commande sont montrées en simulation sur un exemple de robot manipulateur à deux degrés de liberté.

Mots clés. Robot manipulateur rigide; observateur exponentiel d'ordre réduit; synthèse de correcteur; stabilité asymptotique.

1. INTRODUCTION

In most cases, the robot controllers presented in the literature are based on complete state feedback. But in many practical situations, the joint angular velocities are obtained either by differentiating the measured joint angular positions or through tachometers. In both cases, the velocity signal is contaminated by noise. This can degrade the closed-loop robot system performances. To overcome this problem, state estimators are used to reconstruct the joint angular velocities [1], [2], [7].

In this paper, we propose a new exponential reduced-order observer for a class of nonlinear systems, including robot manipulator dynamics. The design of this observer is straightforward, without needing a nonlinear change of coordinates.

This note is organized as follows. In section 2, we recall the dynamic model of a rigid robot manipulator. Section 3 is devoted to the reduced-order observer design. Sufficient conditions are given for the exponential convergence of the observer error. In section 4, this observer is used with two classical state feedback controllers: point-to-point control with gravity compensation and trajectory control [6], [9], [10]. For these two reduced-order observer-based controllers, sufficient conditions for the closed-loop asymptotic stability are given. In section 5, a simulation with a two degrees-of-freedom robot manipulator [1] is presented. The efficiency of the proposed reduced-order observer-based control schemes is shown by a comparison with the same state feedback controllers using the full-order state estimator given in [3] and [4].

2. SYSTEM DESCRIPTION AND STRUCTURAL PROPERTIES OF THE ROBOT MANIPULATOR

The dynamics of a n-joint rigid robot manipulator can be described by a second-order nonlinear vector differential equation [9]

$$\phi(q)\ddot{q} + C(q,\dot{q})\dot{q} + F_v\,\dot{q} + G(q) = \Gamma \qquad (2.1)$$

where $q \in \mathbb{R}^n$ is the vector of joint angular positions, $\Gamma \in \mathbb{R}^n$ is the vector of applied joint torques, $\phi(q) \in \mathbb{R}^{n\times n}$ is the symmetric positive definite inertia matrix, $C(q,\dot{q})\dot{q} \in \mathbb{R}^n$ is the Coriolis and centrifugal torque vector, $G(q) \in \mathbb{R}^n$ is the gravity loading vector, and $F_v \in \mathbb{R}^{n\times n}$ is the diagonal matrix of the viscous friction coefficients.

Let us examine the structural properties of each of the terms in the robot dynamics equation (2.1) [5]. These properties will offer a great deal of insight which will be used in deriving robot control schemes.

P1: $\varphi_1 I_n \le \phi(q) \le \varphi_2 I_n$ for some strictly positive constants φ_1 and φ_2. (2.2)

P2: $\dot{\phi}(q) - 2\,C(q,\dot{q})$ is a skew symmetric matrix. (2.3)

P3: $C(q,x_1)x_2 = C(q,x_2)x_1$ for all $x_1, x_2 \in \mathbb{R}^n$. (2.4)

P4: $\|C(q,x)\| \le \mu\,\|x\|$ for all $x \in \mathbb{R}^n$ (for a revolute joint robot) (2.5)

where μ is a positive real constant and $\|\,.\,\|$ denotes the Euclidean vector norm. Applied to a matrix, $\|\,.\,\|$ corresponds to the p-2 induced norm.

3. REDUCED-ORDER OBSERVER DESIGN

3.1- State-space representation

A first step in the design of an output feedback controller for the robot system (2.1) is the reconstruction of the joint angular velocities using the joint angular position measurements. The proposed observer is of reduced-order and based on a state-space observable canonical form of (2.1) given by

$$\dot{x} = A\,x + f(x) + h(x)\,u \qquad (3.1.1)$$

Since the matrix h(x) is composed only of the elements of the inverse inertia matrix $\phi^{-1}(q)$ which depends exclusively on the joint angular positions, then the state vector x can be replaced by the output measured vector y, thus equation (3.1.1) can be rewritten as

$$\dot{x} = A\,x + f(x) + h(y)\,u \qquad (3.1.2)$$
$$y = H\,x \qquad (3.1.3)$$

with, for i = 1 ... n,

$$A = \mathrm{diag}(A_i) \text{ and } A_i = \begin{bmatrix} 0 & 1 \\ 0 & 0 \end{bmatrix} \qquad (3.2)$$
$$H = \mathrm{diag}(H_i) \text{ and } H_i = [1\ \ 0] \qquad (3.3)$$
$$f(x) = [....f_i(x)^T....]^T \text{ and } f_i(x) = [0\ \ \alpha_i(x)]^T \qquad (3.4)$$
$$h(y) = [....h_i(y)^T....]^T \text{ and } h_i(y) = \begin{bmatrix} 0 \\ b_i(y) \end{bmatrix} \qquad (3.5)$$

where $x^T = [...x_i^T...]$ with $x_i = [q_i\ \dot{q}_i]^T$, $\alpha_i(x)$ is the ith component of the vector $-\phi^{-1}(q)[C(q,\dot{q})\dot{q} + F_v\,\dot{q} + G(q)]$, and $b_i(y)$ is the ith row of the inverse inertia matrix $\phi^{-1}(q)$. $x_i \in \mathbb{R}^2$, $x \in \mathbb{R}^{2n}$, $y \in \mathbb{R}^n$, $\alpha_i(x) \in \mathbb{R}$, $f_i(x) \in \mathbb{R}^2$, $b_i^T(y) \in \mathbb{R}^n$, and $h_i(y) \in \mathbb{R}^{2\times n}$. One can see that the pair (A,H) is observable.

Since $\alpha_i(x)$ is a combination of cosine and sine functions (robot with revolute joints), f(x) is Lipschitz in all domain $\mathcal{D}_0$ where $\dot{q}$ is bounded, i.e. f(x) satisfies the following property

P5: $\|f(x_1) - f(x_2)\| \le K\,\|x_1 - x_2\|$ for all $x_1, x_2 \in \mathcal{D}_0$ (3.6)

where $\mathcal{D}_0 = \{x \in \mathbb{R}^{2n} \,/\, \|\dot{q}\| < V_M\}$, with $V_M \le \sup_t \|\dot{q}(t)\|$ and K is the Lipschitz real positive constant.

3.2- Observer design

To reconstruct the joint angular velocities of a robot manipulator, the proposed reduced-order observer, which ensures an exponential convergence of the state observation error $(\hat{x} - x)$, has the following state-space model

$$\dot{z} = N\,z + L\,y + g(z,y) + j(y)\,u \qquad (3.7)$$
$$\hat{x} = M\,z + E\,y \qquad (3.8)$$
$$z = T\,\hat{x} \qquad (3.9)$$

where $N = \mathrm{diag}(N_i)$, $L = \mathrm{diag}(L_i)$, $M = \mathrm{diag}(M_i)$, $E = \mathrm{diag}(E_i)$, $T = \mathrm{diag}(T_i)$, $g(z,y) = [....g_i(z,y)....]^T$, and $j(y) = [....j_i(y)^T....]^T$. We have, $z \in \mathbb{R}^n$, $u \in \mathbb{R}^n$, $\hat{x} \in \mathbb{R}^{2n}$, $N_i \in \mathbb{R}$, $L_i \in \mathbb{R}$, $M_i \in \mathbb{R}^2$, $E_i \in \mathbb{R}^2$, $T_i^T \in \mathbb{R}^2$, $g_i(z,y) \in \mathbb{R}$, and $j_i^T(y) \in \mathbb{R}^n$. The exponential convergence of the observation error is established by the following theorem.

***Theorem* 1.** Assuming that the state of the robot remains into $\mathcal{D}_0$, the system (3.7)-(3.9) is an exponential observer for the robot system (3.1.2)-(3.1.3) if the following conditions are satisfied (for i=1...n)

$$N_i\,T_i + L_i\,H_i = T_i\,A_i \qquad (3.10)$$
$$g_i(z,y) = T_i\,f_i(\hat{x}) \qquad (3.11)$$
$$j_i(y) = T_i\,h_i(y) \qquad (3.12)$$
$$\begin{bmatrix} T_i \\ H_i \end{bmatrix} [\,M_i\ \ E_i\,] = \begin{bmatrix} 1 & 0 \\ 0 & 1 \end{bmatrix} \qquad (3.13)$$
$$\max(N_i) < -K \qquad (3.14)$$

where K is the Lipschitz constant (see (3.6)).

Then, we can deduce that

$$\|\hat{x} - x\| \le \eta_1 \exp(\eta_2 t) \qquad (3.15.1)$$

where

$$\eta_1 = \|M\|\sqrt{\frac{\min(N_i)}{\max(N_i)}}\,\|z(0) - T\,x(0)\| \qquad (3.15.2)$$
$$\eta_2 = \max(N_i) + K \qquad (\eta_2 < 0) \qquad (3.15.3)\square$$

Proof. Consider the observer reconstruction error

$$e = z - T\,x \qquad (3.16)$$

Then, the observer error dynamic is given by

$$\dot{e} = N\,e + (N\,T + L\,H - T\,A)\,x + (g(z,y) - T\,f(x)) + (j(y) - T\,h(y))\,u \qquad (3.17)$$

From (3.10)-(3.12), equation (3.17) becomes

$$\dot{e} = N\,e + T\,(f(\hat{x}) - f(x)) \qquad (3.18)$$

Let V(e) > 0 be a Lyapunov function candidate

$$V(e) = e^T S\,e \qquad (3.19)$$

where $S = S^T > 0$ is the solution of the Lyapunov equation

$$N^T S + S\,N = -2\,Q \qquad (3.20)$$

with N a stability matrix (according to (3.14)) and Q some diagonal positive definite matrix

$$Q = \mathrm{diag}(Q_i) \qquad (Q_i > 0) \qquad (3.21)$$

The time derivative of the Lyapunov function V along the error dynamics (3.18) is

$$\dot{V}(e) = -2\,e^T Q\,e + 2\,e^T S\,T\,(f(\hat{x}) - f(x)) \qquad (3.22)$$

The sign of $\dot{V}$ depends on matrices $N = \mathrm{diag}(N_i)$ and $T = \mathrm{diag}(T_i)$. Introducing R_i such that $\begin{bmatrix} R_i \\ H_i \end{bmatrix} \in \mathbb{R}^{2\times 2}$ is a regular matrix, there exists a scalar k_i satisfying

$$\begin{bmatrix} T_i \\ H_i \end{bmatrix} = \begin{bmatrix} 1 & -k_i \\ 0 & 1 \end{bmatrix} \begin{bmatrix} R_i \\ H_i \end{bmatrix} \qquad (3.23)$$

Then, we have

$$T_i = R_i - k_i\,H_i \qquad (3.24)$$

By the use of (3.13), equation (3.10) becomes

$$N_i - T_i\,A_i M_i = 0 \qquad (3.25.1)$$
$$L_i - T_i\,A_i\,E_i = 0 \qquad (3.25.2)$$

The substitution of equation (3.24) into (3.25.1) gives

$$N_i = \psi_i - k_i\,\Omega_i \qquad (3.26.1)$$

where

$$\psi_i = R_i A_i M_i \qquad (3.26.2)$$
$$\Omega_i = H_i A_i M_i \qquad (3.26.3)$$

The vector R_i can be written as

$$R_i = [\,R_{i1}\ \ R_{i2}\,] \qquad (3.27)$$

By making use of equations (3.3) and (3.27), equation (3.24) becomes

$$T_i = [\,R_{i1} - k_i\ \ R_{i2}\,] \qquad (3.28)$$

From (3.3), (3.13) and (3.27), M_i is given by

$$M_i = \begin{bmatrix} T_i \\ H_i \end{bmatrix}^{-1} \begin{bmatrix} 1 \\ 0 \end{bmatrix} = \begin{bmatrix} 0 \\ R_{i2}^{-1} \end{bmatrix} \qquad (3.29)$$

Hence, N_i can be obtained from (3.2), (3.25.1), (3.28) and (3.29)

$$N_i = \frac{R_{i1} - k_i}{R_{i2}} \qquad (3.30)$$

The combination of equations (3.4) and (3.28) entails a simplification in the nonlinear part of equation (3.22), which allows to bound more easily $\dot{V}$,

$$T\,(f(\hat{x}) - f(x)) = \underline{R}\,\,(\alpha(\hat{x}) - \alpha(x)) \qquad (3.31.1)$$

where

$$\underline{R} = \mathrm{diag}(R_{i2}) \qquad (i = 1...n) \qquad (3.31.2)$$
$$\alpha(x) = [....\alpha_i(x)....]^T \qquad (3.31.3)$$

As $x \in \mathcal{D}_0$, the property **P5** can be rewritten as

$$\|\alpha(\hat{x}) - \alpha(x)\| \le K\,\|\hat{x} - x\| \qquad (3.32)$$

By the use of equation (3.31.1), relation (3.22) is equivalent to

$$\dot{V}(e) = -2\, e^T Q e + 2 e^T S\, \overline{R}\, (\alpha(\hat{x}) - \alpha(x)) \qquad (3.33)$$

A relation between the state-space error $\hat{x} - x$ and the reconstruction error $e = z - T x$ is obtained from (3.8), (3.13) and (3.16)

$$\hat{x} - x = M e \qquad (3.34)$$

By using property **P5** and equation (3.34), $\dot{V}$ can be bounded as

$$\dot{V}(e) \le -2\, \lambda_m(Q)\, \|e\|^2 + 2\, K\, \lambda_M(S)\, \|\overline{R}\|\, \|M\|\, \|e\|^2 \qquad (3.35)$$

where $\lambda_M(\,.\,)$ $(\lambda_m(\,.\,))$ denotes the largest (smallest) eigenvalue of the corresponding matrix. From (3.28) and (3.29) we have

$$\|\overline{R}\| = R_{2M} \qquad (3.36.1)$$

$$\|M\| = \frac{1}{R_{2m}} \qquad (3.36.2)$$

where $R_{2M} = \max(R_{i2})$ and $R_{2m} = \min(R_{i2})$, with R_{i2} chosen positive for $i = 1....n$.

Taking the same positive value for all R_{i2} so that $\|\overline{R}\|\, \|M\| = 1$, simplifies equation (3.35)

$$\dot{V}(e) \le -2\, (\lambda_m(Q) - K\, \lambda_M(S))\, \|e\|^2 \qquad (3.37)$$

Then $\dot{V}$ is a semi-negative definite function if

$$\lambda_m(Q) > K\, \lambda_M(S) \qquad (3.38)$$

In order to relax the conservatism in (3.38), we maximize $\frac{\lambda_m(Q)}{\lambda_M(S)}$, this can be done by taking $Q = I_n$ [8]. To this fact, the elements Q_i of matrix Q are given by

$$Q_i = 1 \qquad (3.39)$$

Then $\lambda_M(S)$ is given by

$$\lambda_M(S) = \frac{1}{\min|N_i|} = -\frac{1}{\max(N_i)} \qquad (N_i<0) \qquad (3.40)$$

It is easy to see that the inequality (3.38) holds if the condition (3.14) is satisfied. Using the Rayleigh principle, $\frac{e^T S e}{\lambda_M(S)} \le \|e\|^2$, and the definition of V (3.19), equation (3.37) becomes

$$\dot{V}(e) \le -2\, \left(\frac{1}{\lambda_M(S)} - K\right) V(e) \qquad (3.41)$$

This last equation proves the exponential convergence of the reduced-order observer, and its integration gives

$$V(t) \le \gamma_1 \exp(2\alpha t) \qquad (3.42)$$

where

$$\alpha = -\left(\frac{1}{\lambda_M(S)} - K\right) = \max(N_i) + K < 0 \qquad (3.43.1)$$

$$\gamma_1 = (z(0)-Tx(0))^T(-\max(N_i))^{-1}(z(0)-Tx(0)) \qquad (3.43.2)$$

Since V(t) is a quadratic function of the observer reconstruction error e(t), the norm of e(t) can be deduced from relations (3.42) and (3.43.2)

$$\|e(t)\| \le \gamma_2 \exp(\alpha t) \qquad (3.44.1)$$

$$\text{where } \gamma_2 = \sqrt{\frac{\min(N_i)}{\max(N_i)}}\, \|z(0) - T\, x(0)\| \qquad (3.44.2)$$

Exponential decrease of the state observation error $(\hat{x} - x)$ results from equations (3.34) and (3.44.1)

$$\|\hat{x} - x\| \le \eta_1 \exp(\eta_2 t) \qquad (3.45.1)$$

$$\text{where } \eta_1 = \|M\|\, \gamma_2 \qquad (3.45.2)$$

$$\eta_2 = \alpha = \max(N_i) + K \qquad (\eta_2 < 0) \qquad (3.45.3)$$

This completes the proof of the theorem. ■

Observer design procedure:

- Choose R_i such that $\begin{bmatrix} R_i \\ H_i \end{bmatrix}$ is non singular, R_{i2} have the same positive value (for $i = 1....n$).
- M_i is given by (3.29).
- Choose k_i such that N_i (3.30) satisfies (3.14).
- T_i is given by equation (3.28).
- The combination of (3.3), (3.13), (3.24) and (3.30) gives

$$E_i = \begin{bmatrix} T_i \\ H_i \end{bmatrix}^{-1} \begin{bmatrix} 0 \\ 1 \end{bmatrix} = \begin{bmatrix} 1 \\ -N_i \end{bmatrix} \qquad (3.46)$$

- L_i is obtained from equation (3.25.2)

$$L_i = [\, R_{i1}-k_i \quad R_{i2}\,] \begin{bmatrix} 0 & 1 \\ 0 & 0 \end{bmatrix} \begin{bmatrix} 1 \\ -N_i \end{bmatrix} = \frac{(R_{i1}-k_i)^2}{R_{i2}} \qquad (3.47)$$

- $g_i(z,y)$ is given by (3.11), $g_i(z,y) = R_{i2}\, \alpha_i(\hat{x})$
- $j_i(y)$ is obtained from (3.12), $j_i(y) = R_{i2}\, b_i(y)$ □

***Remark* 1.** For a fixed R_{i2}, the choice of N_i can be made arbitrarily. It depends only on scalars R_{i1} and k_i. □

***Remark* 2.** The stability condition (3.14) shows explicitly that the choice of the observer gains is made locally for each axis of the robot system, independently of the others axis. The unique global constraint is the Lipschitz constant K. □

***Remark* 3.** This observer design can be extended to the case where matrices A_i and H_i are not given by (3.2) and (3.3), since the pairs (ψ_i,Ω_i) and (A_i,H_i) have the same unobservable poles. Indeed, by defining the following non-singular matrices

$$W_i = \begin{bmatrix} R_i & -sk_i+R_iA_iE_i \\ H_i & -sI_{p_i}+H_iA_iE_i \\ 0 & I_{p_i} \end{bmatrix} \text{ and } U_i = [\, M_i \quad E_i \,],$$

we have

$$\text{rank}\begin{bmatrix} sI_{n_i}-A_i \\ H_i \end{bmatrix} = \text{rank}\, W_i \begin{bmatrix} sI_{n_i}-A_i \\ H_i \end{bmatrix} U_i = \text{rank}\begin{bmatrix} sI_{n_i}-\psi_i \\ \Omega_i \end{bmatrix} + p_i$$

where $A_i \in \mathbb{R}^{n_i \times n_i}$, $H_i \in \mathbb{R}^{p_i \times n_i}$, n_i and p_i are the dimensions of x_i and y_i respectively. Hence, the detectability of the pair (A_i,H_i) is necessary and sufficient to stabilize the linear part of this observer (see (3.20)). In this case, a sufficient condition for the exponential convergence of the reduced-order observer is that the unobservables poles of the pair (A_i,H_i), i.e. the fixed modes of N_i, must satisfy condition (3.14). □

4. CONTROLLER DESIGN VIA THE REDUCED-ORDER OBSERVER

This section is devoted to the stability analysis of state feedback controllers, when the joint angular velocities are estimated. Two types of control laws are proposed: the so-called point-to-point control with gravity compensation and the trajectory control [1], [6], [7], [9], [10].

4.1- Point-to-point control with gravity compensation

The controller consists of the combination of a gravitation compensation and a linear proportional-derivative static state feedback. This control law is given by

$$\Gamma = G(q) - K_p\, \tilde{q} - K_v\, \dot{q} \qquad (4.1)$$

where K_p and K_v are diagonal positive definite matrices for proportional and derivative gains respectively, $\tilde{q} = q - q_d$ represents the position tracking error, with q the (measured) joint angular positions, and q_d the fixed constant desired joint angular positions. Then we have $\dot{\tilde{q}} = \dot{q}$.

Now, let us replace the state feedback controller (4.1) by the output feedback controller given by the use of the reduced-order observer (3.7)-(3.9)

$$\Gamma = G(q) - K_p \tilde{q} - K_v \hat{\dot{q}} \qquad (4.2)$$

where $\hat{\dot{q}} \in \mathbb{R}^n$ is the estimated joint angular velocities. From equations (3.29), (3.31.2) and (3.34), we have

$$\hat{\dot{q}} = \overline{R}^{-1} e + \dot{q} \qquad (4.3)$$

Thus, the new expression of the control law (4.2) can be written as

$$\Gamma = G(q) - K_p \tilde{q} - K_v \overline{R}^{-1} e - K_v \dot{q} \qquad (4.4)$$

Substituting this control law into the robot dynamics equation (2.1), we obtain the closed-loop system

$$\phi(q)\ddot{q} + C(q,\dot{q})\dot{q} + (F_v + K_v)\dot{q} + K_p \tilde{q} + K_v \overline{R}^{-1} e = 0 \qquad (4.5)$$

Now, we can state and prove the following theorem.

***Theorem* 2.** Under conditions (3.10)-(3.13) and

$$1 + \frac{K}{\max(N_i)} > \frac{\sigma_1}{R_{2m}^2} \qquad (4.6.1)$$

$$\text{with} \quad \sigma_1 = \frac{\|K_v\|^2}{4\,\lambda_m(F_v + K_v)} \qquad (4.6.2)$$

the controller (4.2) stabilizes asymptotically the robot system (2.1) at the equilibrium point $(\dot{q},\tilde{q},e) \equiv 0$ into the domain $\mathcal{D}_1$ given by

$$\mathcal{D}_1 = \{(\dot{q}(0),\tilde{q}(0),e(0)) \,/\, V_1(\dot{q}(0),\tilde{q}(0),e(0)) < \frac{\varphi_1 V_M^2}{2}\}$$

where φ_1 is defined in property **P1** and V_1 is the Lyapunov function given in the proof. ❑

Proof. Consider the Lyapunov function candidate

$$V_1(\dot{q},\tilde{q},e) = \frac{1}{2}\dot{q}^T \phi(q)\dot{q} + \frac{1}{2}\tilde{q}^T K_p \tilde{q} + \frac{1}{2} e^T S e \qquad (4.7)$$

Using property **P2**, equations (3.33) and (4.5), the time derivative of V_1 is equivalent to

$$\dot{V}_1(\dot{q},\tilde{q},e) = -\dot{q}^T (F_v + K_v)\dot{q} - e^T K_v \overline{R}^{-1} \dot{q} - e^T Q e + e^T S \overline{R} \,(\alpha(\hat{x}) - \alpha(x)) \qquad (4.8)$$

Using relations (3.32), (3.34), (3.39) and (3.40), with the facts that $\|\overline{R}\|\,\|M\| = 1$ and $(\dot{q},\tilde{q},e) \in \mathcal{D}_1$, $\dot{V}_1$ can be bounded as

$$\dot{V}_1(\dot{q},\tilde{q},e) \le -\lambda_m(F_v + K_v)\|\dot{q}\|^2 + \|K_v\|\,\|\overline{R}^{-1}\|\,\|e\|\,\|\dot{q}\| - (1 + \frac{K}{\max(N_i)})\|e\|^2 \qquad (4.9)$$

which is equivalent to

$$\dot{V}_1(\dot{q},\tilde{q},e) \le - \begin{bmatrix} \|\dot{q}\| \\ \|e\| \end{bmatrix}^T \begin{bmatrix} \lambda_m(F_v+K_v) & -\dfrac{\|K_v\|\,\|\overline{R}^{-1}\|}{2} \\ -\dfrac{\|K_v\|\,\|\overline{R}^{-1}\|}{2} & \dfrac{\max(N_i)+K}{\max(N_i)} \end{bmatrix} \begin{bmatrix} \|\dot{q}\| \\ \|e\| \end{bmatrix} \qquad (4.10)$$

With (3.31.2) and (4.10), $\dot{V}_1$ is semi-negative definite if

$$1 + \frac{K}{\max(N_i)} > \frac{\|K_v\|^2}{4\,\lambda_m(F_v + K_v)\, R_{2m}^2} \qquad (4.11)$$

By virtue of hypothesis (4.6.1)-(4.6.2), and by applying Lasalle's theorem [11], the theorem is proved. ■

4.2- Trajectory control

Suppose that the robot manipulator tracks a given trajectory $q_d(t)$, assigned in the joint space. Assuming that the time derivative $\dot{q}_d(t)$ and $\ddot{q}_d(t)$ are known, a state feedback control law is given as [7]

$$\Gamma = \phi(q)\ddot{q}_d + C(q,\dot{q})\dot{q}_d + F_v \dot{q}_d + G(q) - K_p \tilde{q} - K_v \dot{\tilde{q}} \qquad (4.12)$$

where K_p, K_v and $\tilde{q}$ have been defined above, and $\dot{\tilde{q}} = \dot{q} - \dot{q}_d$ is the velocity tracking error.

If we use, in place of the state feedback controller (4.12), an output feedback control law, we introduce the estimated joint angular velocities, given by the reduced-order observer, in this control law. By using equations (2.1) and (4.3), the resulting closed-loop system is

$$\phi(q)\ddot{\tilde{q}} + C(q,\dot{q})\dot{q} - C(q,\hat{\dot{q}})\dot{q}_d + (F_v+K_v)\dot{\tilde{q}} + K_p\tilde{q} + K_v \overline{R}^{-1} e = 0 \qquad (4.13)$$

where $\ddot{\tilde{q}} = \ddot{q} - \ddot{q}_d$ is the acceleration tracking error.

Owing to property **P3** and equation (4.3), we have

$$C(q,\dot{q})\dot{q} - C(q,\hat{\dot{q}})\dot{q}_d = C(q,\dot{q})\dot{q} - C(q,\dot{q})\dot{q}_d + C(q,\dot{q})\dot{q}_d - C(q,\dot{q}_d)\hat{\dot{q}} = C(q,\dot{q})\dot{\tilde{q}} - C(q,\dot{q}_d)\overline{R}^{-1} e \qquad (4.14)$$

that substituted in (4.13) yields

$$\phi(q)\ddot{\tilde{q}} + C(q,\dot{q})\dot{\tilde{q}} - C(q,\dot{q}_d)\overline{R}^{-1} e + (F_v + K_v)\dot{\tilde{q}} + K_p \tilde{q} + K_v \overline{R}^{-1} e = 0 \qquad (4.15)$$

The asymptotic stability of the closed-loop system (4.15) is proved by the following theorem.

***Theorem* 3.** Under the conditions $\|\dot{q}_d\| < \frac{V_M}{2}$, (3.10)-(3.13), and

$$1 + \frac{K}{\max(N_i)} > \frac{\sigma_2}{R_{2m}^2} \qquad (4.16.1)$$

$$\text{with} \quad \sigma_2 = \frac{(\mu \frac{V_M}{2} + \|K_v\|)^2}{4\,\lambda_m(F_v + K_v)} \qquad (4.16.2)$$

the system (4.15) is asymptotically stable at the equilibrium point $(\dot{\tilde{q}},\tilde{q},e) \equiv 0$ into the domain $\mathcal{D}_2$ given by

$$\mathcal{D}_2 = \{(\dot{\tilde{q}}(0),\tilde{q}(0),e(0)) / V_2(\dot{\tilde{q}}(0),\tilde{q}(0),e(0)) < \frac{\varphi_1 V_M^2}{8}\}$$

where φ_1 is defined in property **P1** and V_2 is the Lyapunov function given in the proof. ❑

Proof. Let $V_2(\dot{\tilde{q}},\tilde{q},e)$ be a Lyapunov function candidate

$$V_2(\dot{\tilde{q}},\tilde{q},e) = \frac{1}{2}\dot{\tilde{q}}^T\phi(q)\dot{\tilde{q}} + \frac{1}{2}\tilde{q}^T K_p \tilde{q} + \frac{1}{2} e^T S e \qquad (4.17)$$

By using property **P2**, equations (3.33) and (4.15), the time derivative of the Lyapunov function is

$$\dot{V}_2(\dot{\tilde{q}},\tilde{q},e) = -\dot{\tilde{q}}^T(F_v + K_v)\dot{\tilde{q}} - \dot{\tilde{q}}^T(C(q,\dot{q}_d) - K_v)\overline{R}^{-1} e - e^T Q e + e^T S \overline{R}\,(\alpha(\hat{x}) - \alpha(x)) \qquad (4.18)$$

Using the relations (3.32), (3.34), (3.39) and (3.40), with the facts that $\|\overline{R}\|\,\|M\| = 1$ and $(\dot{\tilde{q}},\tilde{q},e) \in \mathcal{D}_2$, $\dot{V}_2$ can be bounded as

$$\dot{V}_2(\tilde{q}, \dot{\tilde{q}}, e) \le -\lambda_m(F_v + K_v)\|\dot{\tilde{q}}\|^2 - \left(1 + \frac{K}{\max(N_i)}\right)\|e\|^2$$

$$+ \|\overline{R}^{-1}\| \|C(q,\dot{q}_d) - K_v\| \|e\| \|\dot{\tilde{q}}\| \quad (4.19)$$

that can be rewritten as

$$\dot{V}_2(\tilde{q},\dot{\tilde{q}},e) \le$$

$$-\begin{bmatrix} \|\dot{\tilde{q}}\| \\ \|e\| \end{bmatrix}^T \begin{bmatrix} \lambda_m(F_v+K_v) & -\dfrac{\|\overline{R}^{-1}\|\|C(q,\dot{q}_d)-K_v\|}{2} \\ -\dfrac{\|\overline{R}^{-1}\|\|C(q,\dot{q}_d)-K_v\|}{2} & \dfrac{\max(N_i)+K}{\max(N_i)} \end{bmatrix} \begin{bmatrix} \|\dot{\tilde{q}}\| \\ \|e\| \end{bmatrix} \quad (4.20)$$

$C(q,\dot{q}_d)$ is bounded according to properties **P4** then, with (3.31.2) and (4.20), function $\dot{V}_2$ is semi-negative definite if

$$1 + \frac{K}{\max(N_i)} > \frac{(\mu \frac{V_M}{2} + \|K_v\|)^2}{4\,\lambda_m(F_v + K_v)\,R^2_{2m}} \quad (4.21)$$

which holds by virtue of hypothesis (4.16.1)-(4.16.2). The proof can be concluded by using the same arguments as those of the preceding theorem. ■

5. SIMULATION RESULTS

Both observer based-controllers of sections 4.1 and 4.2 have been tested by simulation referred to a two degrees-of-freedom rigid robot manipulator described in the appendix.

Figures 1 to 4 and figures 5 to 8 correspond to the output feedback control laws of sections 4.1 and 4.2 respectively. In these figures the solid line is related to controllers using the proposed reduced-order observer, and the dotted and dashed lines are related to the controllers using the full-order observer given in [3] and [4]. In order to compare these controllers, the poles of the linear part of the reduced-order observer are assigned to -30, while those of the full-order observer are assigned to -30 (dashed line) and -60 (dotted line), (which can be obtained by choosing the design parameter θ equal to 30 and 60 [3], [4]).

Figures 1 and 2 show the angular velocity observer error for links 1 and 2. It can be noted that the reduced-order observer presents a dynamic behaviour better than the full-order observer one for a same poles assignment. In order to obtain similar performances, the poles of the full-order observer must be chosen twice as large as those of the reduced-order observer.

Figures 3 and 4 show the responses obtained for the two links with point-to-point output feedback control. Although, the responses in all cases seem similar, a change of the curve scale shows that the obtained performances are in keeping with those of figures 1 and 2.

Figures 5 and 6 represent the position tracking error for both links, figures 7 and 8 the velocity tracking error with trajectory output feedback control. For these figures, the tracking error is greater with the full-order observer for the same poles assignment than in the case of the reduced-order observer.

Through these simulations, it can be seen that the proposed observer provides better closed loop performances for both controllers for the considered robot manipulator than the full order observer [3] and [4].

6. CONCLUSION

In this paper, we have proposed a new exponential reduced-order observer for a class of nonlinear systems and given the sufficient conditions for its exponential convergence. This observer has been used with two types of state feedback controllers for robot manipulators: point-to-point control with gravity compensation and trajectory control. Sufficient conditions for closed-loop asymptotic stability are given. A simulation example shows the efficiency of the reduced-order observer-based controller designs.

APPENDIX

The considered robot [1] moving in the vertical plane is represented in figure a-1 and characterized by

$$\phi(q) = \begin{bmatrix} 8.77+1.02\cos(q_2) & 0.76+0.51\cos(q_2) \\ 0.76+0.51\cos(q_2) & 0.62 \end{bmatrix}$$

$$C(q,\dot{q}) = \begin{bmatrix} -0.51\sin(q_2)\dot{q}_2 & -0.51\sin(q_2)(\dot{q}_1+\dot{q}_2) \\ 0.51\sin(q_2)\dot{q}_1 & 0 \end{bmatrix}$$

$$G(q) = 10\begin{bmatrix} 7.6\sin(q_1)+0.63\sin(q_1+q_2) \\ 0.63\sin(q_1+q_2) \end{bmatrix}$$

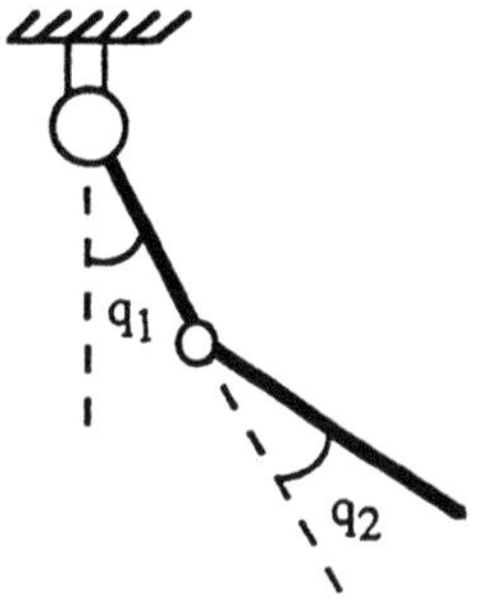

Figure a-1 : Two degrees-of-freedom robot system

REFERENCES

[1] H. Berghuis and H. Nijmeijer, "Global regulation of robots using only position measurements", Syst. Cont. Lett., 21, 289-293, 1993.

[2] C. Canudas de Wit, N. Fixot, K.J. Åström, "Trajectory tracking in robot manipulators via nonlinear estimated state feedback", IEEE Trans. Rob. Autom., 8, 138-144, 1992.

[3] F. Deza and J.P. Gauthier, "A simple and robust nonlinear estimator", IEEE Conf. Dec. Cont., Brighton, 453-454, 1991.

[4] F. Deza, D. Bossanne, E. Busvelle, J.P. Gauthier, and D. Rakotopara, "Exponential observers for nonlinear systems", IEEE Trans. Aut. Cont., 38, 482-484, 1993.

[5] D.E. Koditschek, "Natural motion of robot arms", IEEE Conf. Dec. Cont., Las Vegas, 1984.

[6] F.L. Lewis, C.T. Abdallah and D.M. Dawson, "Control of robot manipulators", Macmillan, 1993.

[7] S. Nicosia and P. Tomei, "Robot control by using only joint position measurements", IEEE Trans. Aut. Cont., 35, 1058-1061, 1990.

[8] R.V. Patel and M. Toda, "Quantitative measures of robustness for multivariable systems", Joint Aut. Cont. Conf., San Francisco, 1980.

[9] M.W. Spong and M. Vidyasagar, "Robot dynamics and control", John Willey & Sons, 1989.

[10] M. Takegaki and S. Arimoto, "A new feedback method for dynamic control of manipulators", ASME J. Dyn. Syst. Meas. Cont., 102, 119-125, 1981.

[11] M. Vidyasagar, "Nonlinear systems analysis- Second Edition", Prentice-Hall International, 1993.

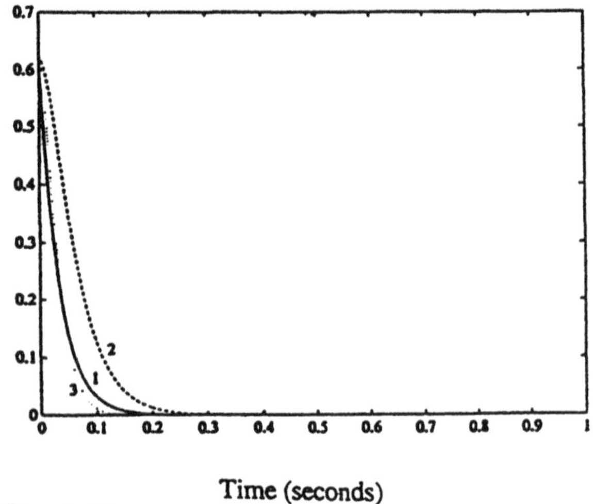

Fig. 1. Observer convergence for the first link.

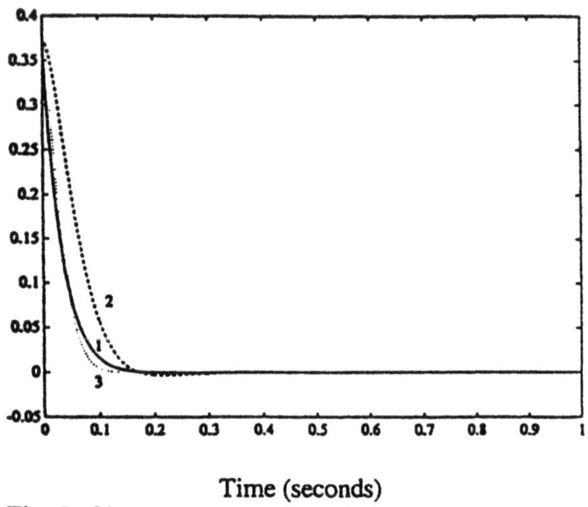

Fig. 2. Observer convergence for the second link.

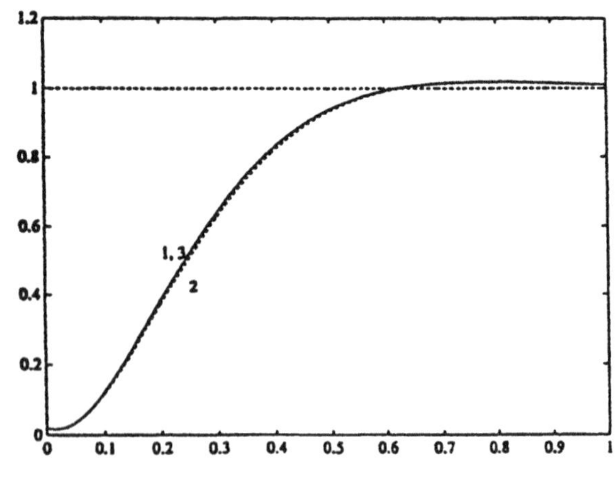

Fig. 3. Responses (first link) with observer-based point-to-point control law with gravity compensation.

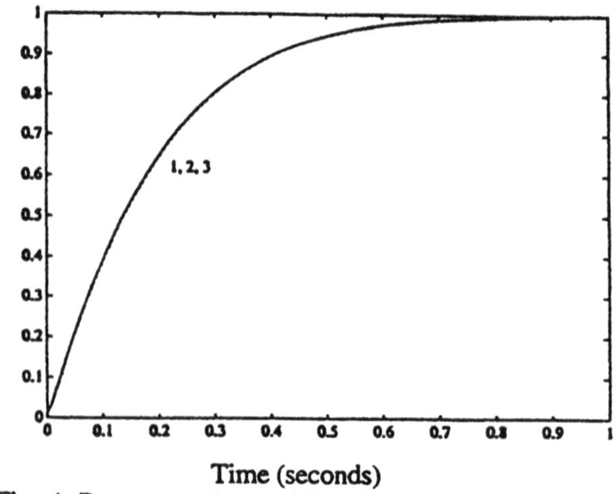

Fig. 4. Responses (second link) with observer-based point-to-point control law with gravity compensation.

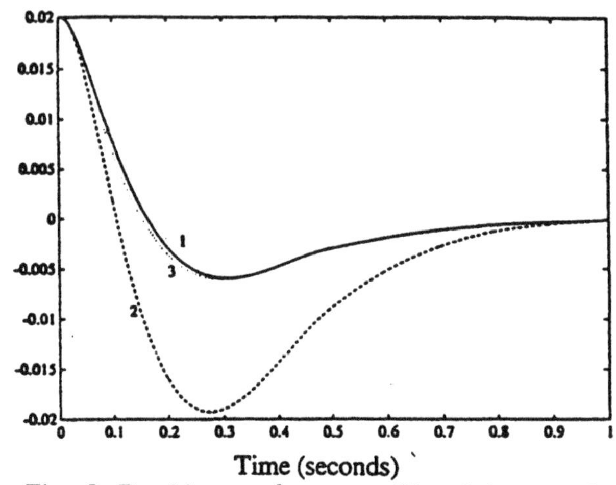

Fig. 5. Tracking performance (first joint angular position) with observer-based trajectory control law.

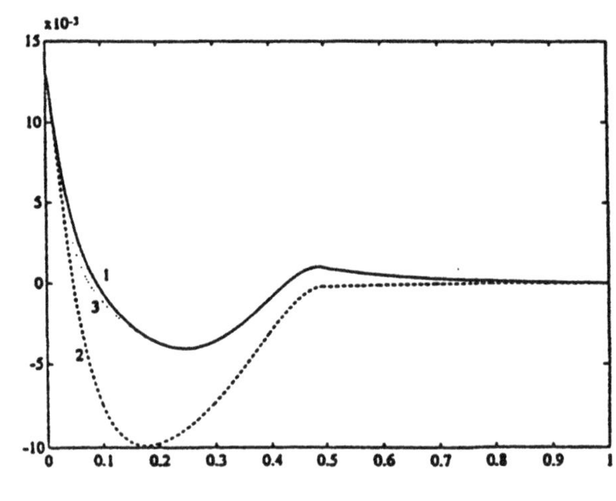

Fig. 6. Tracking performance (second joint angular position) with observer-based trajectory control law.

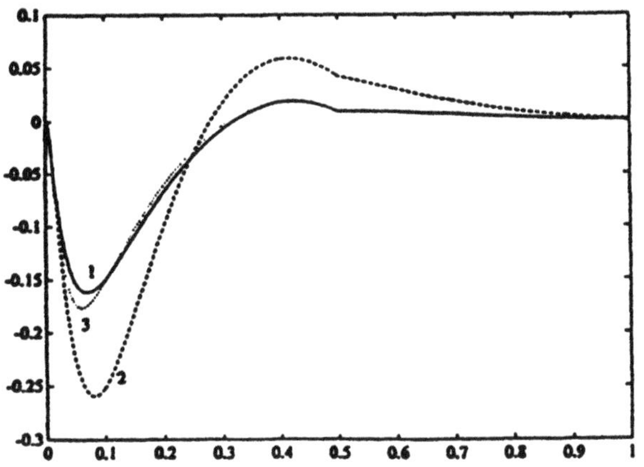

Time (seconds)

Fig. 7. Tracking performance (first joint angular velocity) with observer-based trajectory control law.

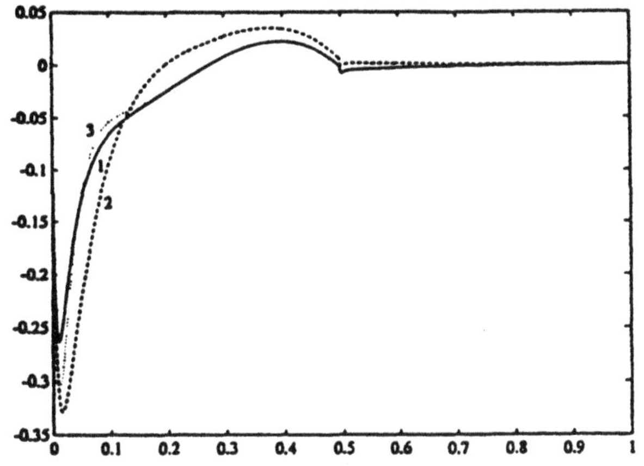

Time (seconds)

Fig. 8. Tracking performance (second joint angular velocity) with observer-based trajectory control law.

ROBUST TIME-OPTIMAL POSITIONAL STABILIZATION CONTAINING A SPEED LIMITATION TASK

Piotr KULCZYCKI

Cracow University of Technology, Faculty of Electrical Engineering,
Warszawska 24, PL-31-155 Cracow, POLAND

Abstract. The present paper deals with the time-optimal positional control for objects whose dynamics are described by a random differential inclusion with discontinuous right-hand side. Considerations include the speed limitation problem. The existence and characteristics of the almost certain time-optimal control, proved in recent works, are commented upon. In compliance with the results obtained, the switching curve, well-known from the (deterministic) classical case, has been "blurred" by the random factor introduced here to the switching area. Empirical examinations confirm numerous advantages of the control systems designed, especially in the area of robustness.

Keywords. Mathematical system theory; minimum time control; position control; discontinuous differential inclusion; stochastic control; random processes; control system design; suboptimal control; robustness

Résumé. Le présent article étudie le contrôle de position optimal en temps des objets dont la dynamique est décrite par l'inclusion différentielle aléatoire avec le côté droit discontinu. Ces considérations incluent la question de la limitation de la rapidité. L'existence et la caractéristique du contrôle optimal en temps presque sûr, démontrés dans des travaux récents, sont commentées. En accord avec les résultats obtenus la courbe de commutation, bien connue, du cas classique (déterministe), a été "délavée" par l'introduction du facteur aléatoire à l'aire des commutations. Les expériences empiriques ont confirmés plusieurs avantages du système de contrôle projeté, spécialement dans le cadre de la robustesse.

1. INTRODUCTION

The dynamics of a broad class of time-optimal controlled objects are described by the following differential inclusion, given below in the operator form:

$$\ddot{y} \in H + u \ , \qquad (1)$$

where u is a bounded control function, y denotes the position of the object, and the function H represents a model of motion resistances. If one omits this factor, i.e. when $H \equiv 0$, formula (1) naturally expresses the second law of Newtonian mechanics for straight-line or rotary motion. In many practical problems the speed $\dot{y}$ has to be limited.

The simplest example of those objects consists of various industrial automata and robots, reversing mills as well as many other plants operating mainly through a change of position in particular mechanisms. Such devices are called positional.

The time-optimal stabilization of system (1), or the task of reaching the equilibrium state $\dot{y}(t) = y(t) = 0$ in a minimal and finite time, will be considered in this paper.

The essential element of model (1) is the multivalued (set-valued) function H describing motion resistances. For the majority of practically appearing types of these resistances, that function can be expressed in the following form:

$$H(\dot{y}(t), y(t), t) = v(\dot{y}(t), y(t), t)\ F(\dot{y}(t)) \ , \qquad (2)$$

where v denotes a real bounded continuous function, and F is a real bounded piecewise continuous function, additionally multivalued in the points of discontinuities. (For example, suppose for the sake of illustration the simplest form of this function

$$F(\dot{y}(t)) = SGN(\dot{y}(t)) = \\ = \begin{cases} 1 & if \;\; \dot{y}(t) > 0 \\ [-s,s] & if \;\; \dot{y}(t) = 0 \\ -1 & if \;\; \dot{y}(t) < 0 \end{cases} \quad , \tag{3}$$

where the parameter $s > 1$ is connected with static friction.)

In many technical problems, the form of the applied model of motion resistances has great influence on the complexity or even the feasibility of a successful analysis. In the model adopted here, it is assumed that the function v introduced in equation (2) is the realization of a given stochastic process V with almost all the realizations being continuous and jointly bounded. Therefore, the dependence of the function v on $\dot{y}(t)$, $y(t)$ and t, is replaced by the dependence on a random factor. Moreover, such a model also regards as probabilistic uncertainty the dependence of motion resistances on a number of other factors, not only $\dot{y}(t)$, $y(t)$ and t, but also those which are usually omitted in the deterministic approach, due to the necessity to simplify the model. The probabilistic concept proposed here also considers perturbations and noise occurring in the system.

The present paper constitutes a continuation of the work described in (Kulczycki, 1992, 1993) additionally supplemented with the task of the speed limitation (Kulczycki, 1995).

2. MAIN RESULTS

First, some notions used in the following will be made more precise.

Let T be an interval with nonempty interior, $t_0 \in T$, and $\Re$ denote the real space.

Consider the deterministic differential inclusion

$$\dot{x}(t) \in g(x(t),t) \tag{4}$$

with the initial condition

$$x(t_0) = x_0 \quad , \tag{5}$$

where $g : \Re^n \times T \to \mathcal{P}(\Re^n)$, $\mathcal{P}(A)$ denotes the set of subsets of A, and $x_0 \in \Re^n$. The function $x : T \to \Re^n$ absolutely continuous on every compact subinterval of the set T, is a solution of differential inclusion (4):

- in the Caratheodory sense (C-solution), if it fulfills inclusion (4) almost everywhere in T,
- in the Filippov sense (F-solution), if

$$\dot{x}(t) \in F[g](x(t),t) \quad \text{almost everywhere in } T \quad , \tag{6}$$

where the operator F is defined as

$$F[g](x(t),t) = \bigcap_{\substack{Z \subset \Re^n: \\ m(Z)=0}} conv\big[g((x(t)+eB)\setminus Z,t)\big] \quad , \tag{7}$$

B denotes the open unit ball in the space $\Re^n$, m is the Lebesgue measure, and $conv[D]$ means the convex closed hull of the set D. The C- or F-solutions of deterministic differential inclusion (4) with initial condition (5) are unique, if all C- or F-solutions, respectively, are identically equal functions.

Let (Ω, Σ, P) be a probability space.

Suppose the random differential inclusion

$$\dot{X}(\omega,t) \in G(\omega, X(\omega,t),t) \quad , \tag{8}$$

with the initial condition

$$X(\omega,t_0) = X_0(\omega) \quad \text{for almost all } \omega \in \Omega \quad , \tag{9}$$

where $G : \Omega \times \Re^n \times T \to \mathcal{P}(\Re^n)$ and X_0 is an n-dimensional random variable defined on (Ω, Σ, P). An n-dimensional stochastic process X defined on (Ω, Σ, P) and T, is an almost certain C- or F-solution of random differential inclusion (8), if almost all its realizations are C- or F-solutions, respectively, of the related deterministic differential inclusions received at the fixed $\omega \in \Omega$. Almost certain C- or F-solutions of random differential inclusion (8) with initial condition (9) are unique, if all almost certain C- or F-solutions, respectively, are equivalent stochastic processes (i.e. $P(\{\ \omega \in \Omega : \tilde{X}(\omega,t) = \tilde{\tilde{X}}(\omega,t)\ \}) = 1$ for every $t \in T$).

Finally, let $U_a \subset \{u : T \to \mathcal{R}^m\}$ denote a set of admissible controls. Consider the function $G : \Omega \times \mathcal{R}^n \times \mathcal{R}^m \times T \to \mathcal{P}(\mathcal{R}^n)$ which defines the dynamics of the random system submitted to the control U

$$\dot{X}(\omega,t) \in G(\omega, X(\omega,t), U(\omega,t), t) \ , \tag{10}$$

and initial condition (9). Then, the m-dimensional stochastic process U^o defined on (Ω, Σ, P) and T, will be called an almost certain time-optimal control, if almost all its realizations are time-optimal controls for proper deterministic systems obtained from dependencies (9)-(10) at the fixed $\omega \in \Omega$.

The theorem proved by Kulczycki (1995), whose thesis constitutes the solution of the time-optimal positional stabilization problem considered here, will be presented now.

Theorem

Let:

(a) $t_0 \in \mathcal{R}$, $T = [t_0, \infty)$, $w > 0$, $x_0 \in \mathcal{R} \times [-w, w]$, and (Ω, Σ, P) be a complete probability space;

(b) origin of coordinates constitute a target set;

(c) $U_a = \{u : T \to [-1,1]\}$ represent a set of admissible controls, and $X_a = \mathcal{R} \times [-w, w]$ a set of admissible states;

(d) V mean a real stochastic process defined on (Ω, Σ, P) and T, with almost all realizations being continuous and fulfilling the boundary condition $V(\omega,t) \in [v_-, v_+]$ for $t \in T$, where $\varnothing \neq [v_-, v_+] \subset (-1,1)$;

(e) $f : \mathcal{R} \to [-1,1]$ denote a piecewise continuous function fulfilling locally a Lipschitz condition except at discontinuity points and $z \cdot f(z) \geq 0$ for every $z \in \mathcal{R}$, and also $F : \mathcal{R} \to \mathcal{P}([-1,1])$ be such that

$$F(z) = \begin{cases} f(z) & if \quad z \neq z_i \\ F_i & if \quad z = z_i \ , \end{cases} \tag{11}$$

where F_i means a subset of the interval $[-1,1]$ and z_i is any real number, for $i = 1, 2, ..., k$;

(f) the random differential inclusion describing the dynamics of the system submitted to the control U

$$\dot{X}_1(\omega,t) = X_2(\omega,t) \tag{12}$$

$$\dot{X}_2(\omega,t) \in U(\omega,t) - V(\omega,t)\, F(X_2(\omega,t)) \tag{13}$$

with the initial condition

$$\begin{bmatrix} X_1(\omega,t_0) \\ X_2(\omega,t_0) \end{bmatrix} = x_0 \quad \text{for almost all } \omega \in \Omega \ , \tag{14}$$

be given.

Then:

(A) there exists an almost certain time-optimal control U^o;

(B) realizations of this control take on the following forms:

$$U^o(\omega,t) = \begin{cases} -1 & for \ t \in [t_0, t_1(\omega)) \\ V(\omega,t)\, F(-w) & for \ t \in [t_1(\omega), t_2(\omega)) \\ +1 & for \ t \in [t_2,(\omega), \infty) \end{cases} \tag{15}$$

or

$$U^o(\omega,t) = \begin{cases} +1 & for \ t \in [t_0, t_1(\omega)) \\ V(\omega,t)\, F(w) & for \ t \in [t_1(\omega), t_2(\omega)) \\ -1 & for \ t \in [t_2,(\omega), \infty) \ , \end{cases} \tag{16}$$

where $t_0 \leq t_1(\omega) \leq t_2(\omega) < \infty$ for every $\omega \in \Omega$;

(C) the above control generates a unique almost certain C-solution;

(D) the above solution is also a unique almost certain F-solution;

(E) the values of almost all realizations of these solutions belong to the set X_a for every $t \in T$. ∎

In the proof of the above Theorem (Kulczycki, 1995), the set of admissible states X_a has been divided into the following disjoint subsets: Q_+, Q_-, R_+, R_-, and the origin being a target (Fig. 1). First, let K_{+-}, K_{++}, denote sets of all states from X_a which can be brought to the origin by the control $u \equiv +1$, if $v \equiv v_-$ or $v \equiv v_+$, respectively; analogically K_{--} and K_{-+} for $u \equiv -1$, if $v \equiv v_-$ or $v \equiv v_+$, respectively. Next, let the following sets be given:

$$Q_+ = \{[x_1, x_2]^T \in X_a \text{ such that there exist } [x_1', x_2]^T \in K_{++} \text{ and } [x_1'', x_2]^T \in K_{+-} \text{ with } x_1' \le x_1 \le x_1''\} \tag{17}$$

$$Q_- = \{[x_1, x_2]^T \in X_a \text{ such that there exist } [x_1', x_2]^T \in K_{--} \text{ and } [x_1'', x_2]^T \in K_{-+} \text{ with } x_1' \le x_1 \le x_1''\} \tag{18}$$

$$R_+ = \{[x_1, x_2]^T \in X_a \backslash Q \text{ such that there exists } [x_1', x_2]^T \in Q \text{ with } x_1 < x_1'\} \tag{19}$$

$$R_- = \{[x_1, x_2]^T \in X_a \backslash Q \text{ such that there exists } [x_1', x_2]^T \in Q \text{ with } x_1' < x_1\} \tag{20}$$

where $Q = Q_+ \cup \{[0,0]^T\} \cup Q_-$.

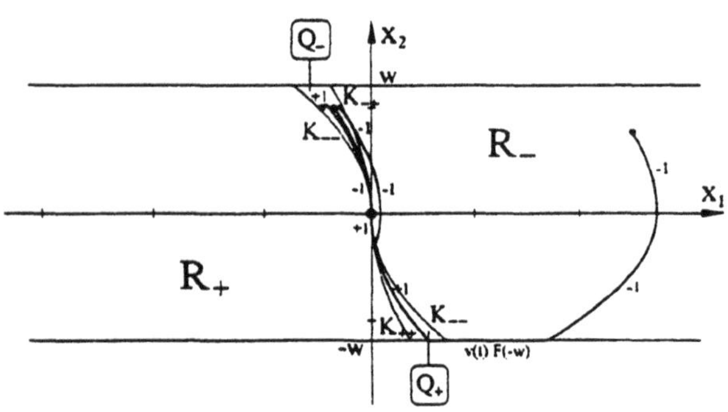

Fig. 1. Almost certain time-optimal control

Let now $\omega \in \Omega$, or thereby $V(\omega, \cdot) \equiv v$, be fixed. If $x_0 \in R_-$ (Fig. 1), then there exist t_1 and t_2 such that the solution generated by the control

$$u^\circ(t) = \begin{cases} -1 & \text{for } t \in [t_0, t_1) \\ v(t)\, F(-w) & \text{for } t \in [t_1, t_2) \\ +1 & \text{for } t \in [t_2, \infty) \end{cases} \tag{21}$$

reaches the target in a minimum time t_f, with $t_0 < t_1 \le t_2 < t_f < \infty$, $x(t_2) \in Q_+$, and if $t_1 \ne t_2$ then also $x(t) \in \Re \times \{-w\}$ for $t \in [t_1, t_2]$. Analogically, if $x_0 \in R_+$, then there exist t_1 and t_2 such that the solution generated by the control

$$u^\circ(t) = \begin{cases} +1 & \text{for } t \in [t_0, t_1) \\ v(t)\, F(w) & \text{for } t \in [t_1, t_2) \\ -1 & \text{for } t \in [t_2, \infty) \end{cases} \tag{22}$$

reaches the target in a minimum time t_f, with $t_0 < t_1 \le t_2 < t_f < \infty$, $x(t_2) \in Q_-$, and if $t_1 \ne t_2$ then also $x(t) \in \Re \times \{w\}$ for $t \in [t_1, t_2]$. Let $x_0 \in Q_-$ (Fig. 1); for particular $\omega \in \Omega$ the time-optimal control can be of the form (21) or (22), with additional conditions as above, or the following:

$$u^\circ(t) = -1 \quad \text{for } t \in [t_0, \infty) \ . \tag{23}$$

The case $x_0 \in Q_+$ is similar; the counterpart of control (23) is

$$u^\circ(t) = +1 \quad \text{for } t \in [t_0, \infty) \ . \tag{24}$$

The proof of the time-optimality of the above described control functions u° is based on the theory of differential inequalities.

The function $U^\circ : \Omega \times T \to [-1, +1]$ defined by the formula $U^\circ(\omega, \cdot) \equiv u^\circ$, where u° was assigned above at fixed $\omega \in \Omega$, is a stochastic process and the required almost certain time-optimal control fulfilling the conditions formulated in the thesis of the Theorem presented.

In assumption (d) of this Theorem there appears the stochastic process V with almost all realizations being continuous and jointly bounded. The condition of boundedness can be examined by one-dimensional distributions of that stochastic process; however, the Kolmogorov theorem formulates the sufficient condition for the continuity of almost all realizations of a stochastic process on the basis of properties of two-dimensional distributions. Finally, the requirements formulated above with respect to the stochastic process V are thus mutually independent and identifiable on the basis of finite dimensional distributions (Wong, 1971; Chapter II).

The relation $[v_-, v_+] \subset (-1,1)$ ensures the controllability of the system. The assumption that the function F fulfills Lipschitz condition has been introduced to guarantee the uniqueness and equality of C- and F-solutions.

The condition $z \cdot F(z) \geq 0$ included in assumption (e) has been formulated only for the sake of clarity of notation. At any rate, this inequality is physically justified, because with positive values of the stochastic process V it is consistent with the property of the energy dissipation.

3. CONCLUSIONS AND APPLICATIONS

The switching curve γ, well-known from the classical case of time-optimal transfer of a mass (Athans and Falb, 1966; Section 7.2), has been generalized in the above Theorem to a switching area Q ($\gamma = Q$ if $v_- = v_+ = 0$ and $w = \infty$). Namely, the function H, defined in the Introduction and representing the model of motion resistances, has been decomposed into two factors: $F(\dot{y}(t))$ and $V(\omega,t)$. The former, deterministic one, having only insignificant influence on the complexity of the theoretical analysis, made it possible to incorporate the properties of discontinuity and multivalence of friction phenomena. The latter one, thanks to its probabilistic nature, includes among other things approximations and identification errors (of the first factor too), motion resistances dependence on position, time and temperature, as well as perturbations and noise naturally occurring in real systems. The switching curve (even more general than γ, i.e. only with the condition $v_- = v_+ \in (-1,1)$) which is implied by the first, deterministic factor, has been "blurred" by the second, random one to the switching area.

An almost certain time-optimal control ensures the realization of the minimum of the expected value of the time for reaching a target set; however, it is additionally dependent on the random factor, in practice unknown *a priori*. The result of this dependence is that the above control is difficult to apply directly, but it constitutes a useful basis for the creation of technical constructions of suboptimal structures in which the direct dependency of the control function on the random factor is eliminated.

Namely, according to the results of the Theorem presented, the suboptimal feedback controller function, independent of a random factor, can be defined by the following formula:

$$U'(X(\omega,t)) = \begin{cases} 1 & if\ X(\omega,t) \in R_+ \cup \mathfrak{R} \times (-\infty, -w) \\ d_1 & if\ X(\omega,t) \in R_+ \cap \mathfrak{R} \times [w-e, w] \\ d_2 & if\ X(\omega,t) \in Q_+ \\ 0 & if\ X(\omega,t) \in \{[0,0]^T\} \\ -d_2 & if\ X(\omega,t) \in Q_- \\ -d_1 & if\ X(\omega,t) \in R_- \cap \mathfrak{R} \times [-w, -w+e] \\ -1 & if\ X(\omega,t) \in R_- \cup \mathfrak{R} \times (w, \infty) \end{cases} ,$$

(25)

where $-1 < d_1 < 1$, $0 < d_2 \leq 1$, and $0 < e < w$ (Fig. 2). In practice these values can be obtained heuristically. Particularly, the parameter d_1 would be approximately equal to the mean of motion resistances; however, the parameter d_2 should be close to 1. Only almost certain F-solutions occur in the system thus obtained; C-solutions do not exist in general case. These F-solutions are unique.

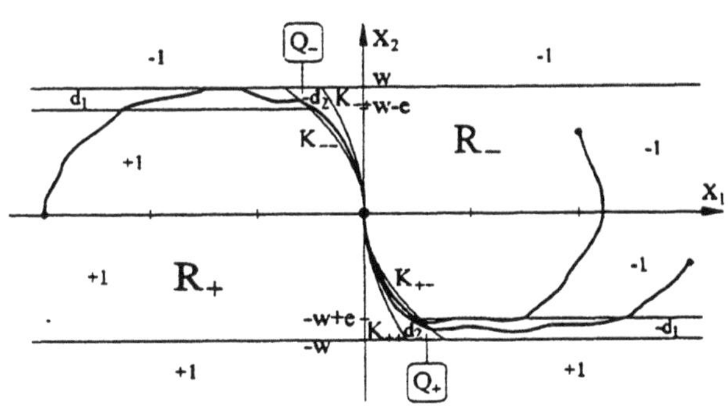

Fig. 2. Suboptimal control with constant parameters d_1 and d_2

The parameters d_1 and d_2 can be constant as in the above concept, but they can also vary, increasing continuously from the value $1 - v_+ + v_-$ on the set K_{+-} up to the value 1 on the set K_{++}, from -1 on K_{-+} to $-1 + v_+ - v_-$ on K_{--}, and from -1 on the sets

$R_+ \cap \Re \times \{w\}$, $R_- \cap \Re \times \{-w+e\}$ to the value 1 on $R_+ \cap \Re \times \{w-e\}$, $R_- \cap \Re \times \{-w\}$, respectively (Fig. 3). This makes it possible to achieve a result similar to the bicycle-racing track or bob-sleigh track, which are horizontal on the interior part, and become more vertical the farther they go to the outside. The value of the parameter d_2 should be equal to 1 even in the neighborhood of the sets K_{++} and K_{-+}, with the aim of neutralizing the most unfavorable realizations of the stochastic process V. Almost certain F-solutions occur in a system thus designed. Moreover, the rules for variations of the parameters d_1 and d_2 values postulated above ensure also the existence of almost certain C-solutions, equal to those F-solutions. Both types of the above solutions are unique.

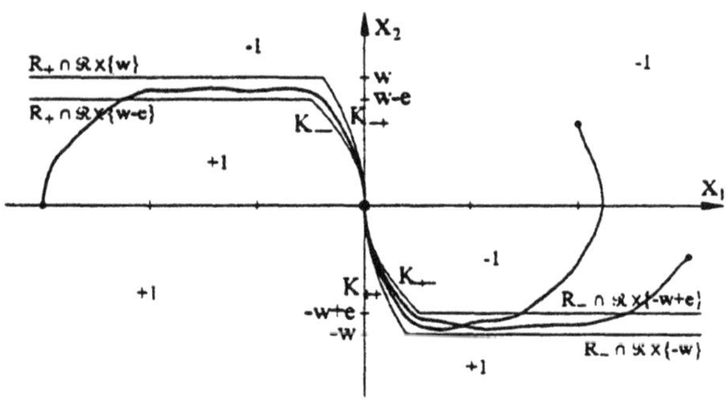

Fig. 3. Suboptimal control with inconstant parameters d_1 and d_2

If the constraints of an actuator limit the control to the extreme values of the admissible set, the results of the Theorem may be modified according to the physical observation that the influence of motion resistances in both periods of time - before and after a switching - can be averaged. Thus, after performing a detailed analysis of the sensitivity of the control system to the values of motion resistances, one can use elements of statistical decision theory, where a loss function is connected with extending the time for reaching the target if the control switching has been too late or too early. A detailed description of such a control structure will be the subject of a separate publication. In the general case there are no C-solutions in the system thus obtained; almost certain F-solutions exist, and they are unique.

The probabilistic concept of the control systems designed in the present paper have been successfully empirically verified (Kulczycki, 1995). It should be underlined that the control system constructed turned out to be only slightly sensitive to the inaccuracy resulting from identification and perturbations - robustness is a very valuable property of random control systems.

4. REFERENCES

Athans, M., and P.L. Falb (1966). *Optimal Control.* McGraw-Hill, New York.

Kulczycki, P. (1992). *Time-Optimal Stochastic Control of Discontinuous Dynamic Object.* Wydawnictwa Politechniki Krakowskiej, Cracow, (in Polish).

Kulczycki, P. (1993). Time-Optimal Stochastic Positional Control. In: *Proc. IFAC 12th World Congress*, Sydney, **VII**, 443-448.

Kulczycki, P. (1995). Robust Time-Optimal Positional Control. *In press.*

Wong, E. (1971). *Stochastic Processes in Information and Dynamical Systems.* New York, McGraw-Hill.

AN EXPERIMENTAL VALIDATION OF A NONLINEAR VELOCITIES OBSERVER

B CHERKI, Y AOUSTIN and PH LEMOINE

Laboratoire d'automatique de Nantes-URA CNRS 823, Ecole Centrale de Nantes, Université de Nantes
1 rue de la Noë 44072 Nantes Cedex 03 - Nantes, Email : Aoustin@lan10.ec-nantes.fr
Fax : 40 74 74 06, Phone Number : 40 37 25 12

résumé : La mesure des positions angulaires pour la commande des robots est disponible avec précision, cependant la mesure de vitesse est souvent bruitée ou non disponible. Pour remédier à ce problème des observateurs de vitesse sont utilisés. Dans cet article on présente une validation expérimentale d'un observateur de vitesse. Les vitesses estimées sont comparées aux vitesses déduites par dérivation numérique des positions filtrées à l'aide d'un filtre de type F.I.R. L'observateur utilisé est un observateur non linéaire à grands gains, la loi de commande est de type couple calculé. les manipulations sont réalisées sur un robot SCARA construit au laboratoire.

Abstract : High precision measurements of the joint displacement are available to control the robot manipulators. However the signal delivered by sensors velocities are often contaminated by noise and not so useful in contol laws. To overcome this problem velocities observers are used. In this paper we demonstrate experimentaly the applicability of the method. The estimated velocities by the observer are compared with the estimated velocities obtained by numerical derivation of the filtered positions with a F.I.R.The observer used is a nonlinear high gain one based on the notion of uniform observability.
As control law we use the linearizing and decoupling computed torque law. The experiencies are conducted on a SCARA robot manipulator built at the laboratory.

Key words : nonlinear velocity, high gain observer, computed torque, finite impulse response filter, trajectory tracking.

1.INTRODUCTION

The theory of nonlinear observability and the nonlinear observer problem design attracted the attention of many reseachers. One of the first important work in this field was done by Hermann and Krener, (1977) who gave the very known rank condition to test the observability of a system. More recently Diop and Fliess, (1991) have given an important contribution to the nonlinear observability.
Krener and Isidori, (1983) proposed a nonlinear observer for systems without input. This work was extended by Krener and Respondek, (1985), Xia-Hua-Xia and Wei-Bing-Gao, (1989) to systems with inputs. The technique used to build the observer is based on the linearization of a nonlinear system, via a state diffeomorphism and output injection. An application on flexible joint robots has been proposed by Plestan and Cherki, (1993). The observer proposed is based on the concept of uniform observability for any input introduced earlier by Gauthier and Bornard (1991). Different other observers were proposed in the litterature for different classes of nonlinear systems Ciccarella *et al,* (1993), Jankovic, (1993), and Lewis, (1992). They rely on change of coordinates or approximation or extended Kalman filter. Recently an observer based on a minimization technique was proposed by Zimmer, (1993). More specialized observers to robotics were proposed to estimate velocities for rigid robots or the velocities and the unmeasured link position for elastic joint robots. These observers are high gain, Tornambe, (1992), or sliding modes Canudas De Wit and Slotine, (1991) or based on the dynamic model of the robot Tomei, (1990), or passivity approach Berghuis H., Nijmeijer H. (1993). We use a MIMO nonlinear observer developed by Bornard et al (1991) to estimate the velocities of a rigid robot, and a computed torque as control law. The applicability of this method on a two rigid arms robot model has been demonstrated in simulation, Aoustin and Cherki (1993). We have implemented this observer on a SCARA robot built at the laboratory. One of the

main advantages with this control in our case, is that the velocity observer equations become linear . They can be discretized with classical linear techniques to minimize the computing time on line. We show that the experimental observer results are good compared to the classical numerical differentiation of the filtered signal joint positions.

2.THEORITICAL PRELIMINARIES

The observer we will use has been proposed in the general context of non linear systems by Bornard *et al* (1991). It is based on high gain output injection and the notion of uniform observability introduced earlier by Gauthier *et al* (1981). we recall here some important features related to this observer. The robot model is given by :

$$A(q)\ddot{q}+H(q,\dot{q})= \Gamma \quad (1)$$

$A(q)$ is the nxn positive definite inertial matrix.

$H(q,\dot{q})$ is the vector of centrifugal, Coriolis, gravity and friction forces.

Γ : the vector of torques applied to the robot.

In a state representation the model reads as:

$$\begin{cases} \dot{x}_1=x_2 \\ \dot{x}_2=\phi(x,\Gamma) \end{cases} \quad (2)$$

with $x=(x_1,x_2)^T=(q,\dot{q})^T$

$\phi(x,\Gamma)=A^{-1}(x_1)[\Gamma-H(x_1,x_2)]$

The nonlinear function $\phi(x,\Gamma)$ contains square terms which are only localy lipschitz, but the saturation of the velocities make them globaly lipschitz.

This form meets the requirement of the uniform observability as reported in Bornard *et al* (theorem 1), to work the observer proposed necessite a global Lipschitz assumption for the non linear function $\phi(x,\Gamma)$, this assumption is meeted as soon as we suppose the velocities of the robot bounded. The velocities observer is given by:

$$\begin{cases} \dot{\hat{x}}_1=\hat{x}_2+\frac{K_1}{T}(x_1-\hat{x}_1) \\ \dot{\hat{x}}_2=\bar{\phi}(\hat{x},\Gamma)+\frac{K_2}{T^2}(x_1-\hat{x}_1) \end{cases} \quad (3)$$

where $\bar{\phi}(x,\Gamma)=\begin{cases} \phi(x,\Gamma) & x_2 \in \Omega \\ \phi_1(x_1,\Gamma) & \text{otherwise} \end{cases}$

$\hat{x}=[x_1,\hat{x}_2]$

$\phi_1(x_1,\Gamma)$ is obtained from $\phi(x,\Gamma)$ by replacing x_2 by its maximum value on Ω.

It was proved that this observer work well, the states are estimated very fast if the gains K_1, K_2, and the parameter T are adequately chosen. The gains K_1, K_2 are determinated such as the eigenvalues of the linear part of the states equation of the observer lie in the left half plane, see Bornard *et al* (1991).

3.COMPUTED TORQUE WITH ESTIMATED VELOCITIES

The control law we will use is the decoupling and linearizing one

$$\Gamma=A(x_1)[\dot{x}_{2d}-K_v(x_2-x_{2d})-K_p(x_1-x_{1d})]+H(x_1,x_2)$$

where we replace x_2 by $\hat{x}_2$ since x_2 is not measurable. This control leads to a linear observer

$$\begin{cases} \dot{\hat{x}}_1=\hat{x}_2+\frac{k_1}{T}(x_1-\hat{x}_1) \\ \dot{\hat{x}}_2=w+\frac{k_2}{T^2}(x_1-\hat{x}_1) \end{cases} \quad (4)$$

-where $w=\dot{x}_{2d}-K_v(\hat{x}_2-x_{2d})-K_p(x_1-x_{1d})$

The equations of the robot in closed loop are:

$$\begin{cases} \dot{x}_1=x_2 \\ \dot{x}_2=w+A^{-1}(x_1)[H(x_1,\hat{x}_2)-H(x_1,x_2)] \end{cases} \quad (5)$$

To assert the convergence of the closed loop we will use a theorem of Teel and Praly (1992). We must precise some importants facts:

- In the case of initial errors on the positions, the observer begins by giving estimates which can be far from the true values, it is the well known peaking phenomenon.
- To avoid the peaking we use an idea of and F. Esfandiari *et al* (1992) reported also in A. Teel *et al* (1992, 1994)which consists of saturating the estimated variables to the maximum known values of the true variables. It is equivalent to say that we disregard the estimated values which are unrealistic.

Let $\tilde{x}=x-x_d$, $e_1=\frac{1}{T}(x_1-\hat{x}_1)$, and $e_2=x_2-\hat{x}_2$ the equations of the tracking and estimation errors are then:

$$\begin{cases} \dot{\tilde{x}}_1=\tilde{x}_2 \\ \dot{\tilde{x}}_2=-K_p\tilde{x}_1-K_v\tilde{x}_2+K_ve_2 \\ \qquad +A^{-1}(x_1)[H(x_1,\hat{x}_2)-H(x_1,x_2)] \end{cases} \quad (6)$$

$$\begin{cases} \dot{e}_1=\frac{-K_1}{T}e_1+\frac{1}{T}e_2 \\ \dot{e}_2=\frac{-K_2}{T}e_1+A^{-1}(x_1)[H(x_1,\hat{x}_2)-H(x_1,x_2)] \end{cases} \quad (7)$$

(6) and (7) are of the form :

$$\begin{cases} \dot{\tilde{x}}=h(\tilde{x},e,t) \\ \dot{e}=LAe+g(\tilde{x},e,t) \end{cases} \quad (8)$$

with $L=\frac{1}{T}$ and $A=\begin{pmatrix} -k_1 & 1 \\ -k_2 & 0 \end{pmatrix}$ Now we recall the theorem of Teel and Praly for the convenience of the reader. One hypothesis is made H: The system $\dot{z}=h(z,0,d(t))$ is globally asymptotically stabilizable, and a Lyapunov function $v(z)$ proving the result is known. This hypothesis is true for the robot since the computed torque control law is a globally asymptotically stabilizing one. Since x_1 is measurable, it is reasonable to initialize the system such that $x_1(0)=\hat{x}_1(0)$, so the initial conditions can be bounded independently of T.

Theorem Teel *et al*. (1992, 1994)
Consider the non linear system

$$\begin{cases}\dot{z}=h(z,e,t)\\ \dot{e}=LAe+g(z,e,t)\end{cases}$$

h and g are C^1 vectors fields. We suppose the hypothesis H verified and the matrix A Hurwitz, we define $\Gamma=\{z: v(z)\leq c+1\}$, $c\geq 1$, let the function:

$$W(z,e)=c\frac{v(z)}{c+1-v(z)}+\mu\frac{e^TPe}{\mu+1-e^TPe},\ \mu\geq 1$$ and P is solution of the matrix equation $A^TP+PA=-I$.

Let $A1=\{z: v(z)<c+1\}\times\{e: e^TPe<\mu+1\}$ then for each $L>0$ $W(z,e): A1\rightarrow R^+$ is definite positive on $A1\backslash\{0\}$ and proper on A1. Moreover for each $\rho>0$ there exists $L_*>0$ such that for each $L>L_*$ $\dot{W}\leq-\Phi(z,e)$, where $\Phi(z,e)$ is definite positive on $\{(z,e): \rho\leq W\leq c^2+\mu^2+1\}$.

The theorem proves that the trajectories of the system (8) will enter in a small compact around the origin. To prove the semiglobal asymptotic stability, we must prove that (8) is locally exponentially stable.

Due to the form of $h(\tilde{x},e,d(t))$ and $g(\tilde{x},e,d(t))$ the linearized system around the origin will be of the form $\dot{X}=B(t)X(t)$ with $X=[\tilde{x},e]$,

$$B(t)=\begin{bmatrix}0 & 1 & 0 & 0\\ -kp & -kv & 0 & f(t)\\ 0 & 0 & \frac{-k1}{T} & \frac{1}{T}\\ 0 & 0 & \frac{-k2}{T} & f(t)\end{bmatrix}=\begin{bmatrix}B1 & B3\\ 0 & B2\end{bmatrix}$$

$$f(t)=-A^{-1}(x_1)\left.\frac{\partial H(x_1,x_2)}{\partial x_2}\right|_{x=x_d}$$

$\sigma(B(t))=\sigma(B1)\cup\sigma(B2)$, $\sigma(B)$ is the spectrum of B. f(t) is bounded because $x_d(t)$ is bounded, so the eigenvalues of A(t) and $A^T(t)$ can always be made negative with the gains K_p, K_v, K_1, K_2 and T. This proves that the closed loop is semi globally asymptotically stable.

4. EXPERIMENTAL CASE STUDY

The experimental study was conducted on a two degrees of freedom robot manipulator (SCARA) moving in the horizontal plane, see figure 1. This robot was manufactured at the laboratory, There is an encoder to measure each link position, but no tachymetric dynamo for the velocities.

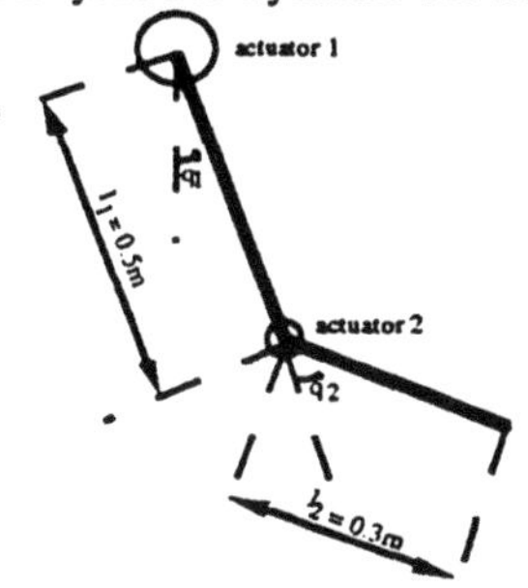

The two degrees of freedom scara robot.

The maximum allowed torques are : $\Gamma_{1max}=17$ N.m, $\Gamma_{2max}=8$ N.m. The minimal inertial parameters of the dynamic model robot are, zzr1=3.7 $Kg.m^2$, zzr2=0.08$Kg.m^2$, mxr2=0.272 Kg.m. For each link the Coulomb and viscous terms are identified respectively to: Fs1=0.62 N.m, Fs2=0.17 N.m and Fv1=0.07.N.m/rd/s, Fv2=0.013 N.m/rd/s. Then the components of the dynamic robot model are:

$a_{11}(q)=zzr1+zzr2+2mxr2L_1\cos(q_2)$,
$a_{12}(q)=zzr2+mxr2L_1\cos(q_2)$,
$a_{21}(q)=zzr2+mxr2L_1\cos(q_2)$, $a_{22}(q)=zzr2$.

$h_1(q,\dot{q})=-mxr2L_1[2\sin(q_2)\dot{q}_1\dot{q}_2+\sin(q_2)\dot{q}_2^2]+Fv1\dot{q}_1+Fs1sign(\dot{q}_1)$

$h_2(q,\dot{q})=mxr2l_1\sin(q_2)\dot{q}_1^2+Fv2\dot{q}_2+Fs2sign(\dot{q}_2)$.

The actual sampling frequency was set at 100hz for the controller and 4kHz for the joint position measurements. The implementation has been done with a fast digital signal processor from the DSPACE society and a 486PC.

The linear equation of the observer (4) is discretized by using the zoh method under the MATLAB logiciel to minimise the calculation time. The gains are chosen as $K_{p1}=25$, $K_{v1}=10$, for the first arm $K_{p2}=625$, $K_{v2}=50$, for the second arm, $K_1=2$, $K_2=1$ and T=0.01 for the two arms, these values seems to be very adequate. We obtain finally for each link,

$\hat{x}_{k+1}=B\hat{x}_k+Du_k$ with $u_k=[x_{1k},w_k]$, B and D are obtained from the discretization of (4). The equations of the observer are linear and decoupled.

$$w_k=\ddot{q}_{dk}-Kv(\hat{\dot{q}}_k-\dot{q}_{dk})-Kp(q_k-q_{dk})$$

and $x_{1k}=[q_{1k},q_{2k}]$. The indices d and k mean desired and value of the variable at $t=t_k$.

The finite impulse response filters have been computed with the help of the Matlab logiciel on forty points of joint link measurement, $y_k=\sum_{i=1}^{40}b_iq_{ki}$. The cut-off frequency of filters is chosen equal to 10Hz. Then it is easier to compute the numerical derivative: $\frac{y_k-y_{k-1}}{\delta}$, δ is the sampling period of the controller.

5. EXPERIMENTAL RESULTS

The experiments are always done with initial errors equal to zero which is quite natural since the robot start from position zero with a null velocity. So no peaking is expected and we need not to saturate the estimated variables.

The trials are done with a desired trajectory of a trapezoidal velocity profile, the desired joint positions are represented on figures 1 and 5.

In a first trial the loop is closed with the F.I.R and secondly the loop is closed with the observer, the control sampling period is 10 ms and the measurement sampling period is 250 µs in the two cases. We see that the error position on joint one is slightly smaller when the observer is in the closed loop, but it is bigger on joint two (fig 3, 7) than the case where the filter is in the closed loop (fig 2, 6). We think the reason of the value of the error position on joint two is error modeling in particular error in friction coefficient. The third trial is done with a sampling period equal to 1 ms (fig 4, 8) the errors on the two joints are less than the errors when the F.I.R is used in the closed loop despite the errors modeling which are necessary present.

The F.I.R on forty points can not be used with a sampling period equal to 1 ms because we need a period of 25 µs for joint measurement which is less than the necessary time to compute the control.

6. CONCLUSION

We have proved the experimental applicability of a nonlinear high gain observer to obtain the joint velocities of a two degrees of freedom SCARA robot. The calculation of the estimated state variables is easy due to the linearity of the equations. The time calculation is very short so we can use short sampling period if needed. The estimation is as good as the one obtained by the numerical differentiation of the filtered positions which need more time calculation.

7.ACKNOWLEDGEMENT

We warmly thank professors W.Khalil and M.Gautier for helpful discussions.

8.REFERENCES

Aoustin. Y, Cherki. B, (1994). Computed torque of robots with estimated velocities *SYROCO'94* Capri, Italy, september 94

Berghuis. H, Nijmeijer. H, (1993). A passivity approach to controller-observer design for robots. *IEEE Transactions on Automatic Control*, Vol., 09, N°6 pp.740-754.

Bornard. G, Hammouri. H, (1991). A High Gain Observer for a Class of Uniformly Observable Systems. *Proc. 30th Conf. Dec. Control.*, Brighton, England, pp 1494-1495.

Canudas De Wit. C, Slotine. J.J.E, (1991). Sliding Observers For Robot Manipulators. *Automatica* vol.27, N°5, pp 859-864.

Canudas De Wit. C. Astrom, K. J. Fixot, N. (1990) Computed torque control via non-linear observer. *Int. J. of Adaptive control and signal processing*, vol. 4, pp 443-452

Canudas De Wit, C. N Fixot. Astrom, K. J. (1992) Trajectory tracking in robots manipulators via nonlinear estimated state feedback. *IEEE Transactions on robotics and automation* vol.8, N°1, pp138-144

Ciccarella. G, Dalla Mora. M, and Germani. A, (1993). A Luenberger-like Observer For Nonlinear System. *Int. J. Control*, Vol. 57, N°3, pp 537-556.

Diop. S, Fliess. M, (1991). Nonlinear Observability, Identifiability, and Persistent Trajectories. *Proc. 30th Conf. Dec. Control.*, Brighton, England, pp 714-719.

Dombre. E,Khalil.W, (1988). Modelisation et commande des robots. Editons Hermès, Paris ISBN 2-86601-1482.

Esfandiari. F and Khalil. H, (1992). Output feedback stabilization of fully linearizable systems. Int. J. Control, vol 56, No. 5, pp 1007-1037.

Gauthier J.P, and Bornard G. (1981). Observability for any *u(t)* of a class of bilinear systems. *IEEE Transactions on Automatic Control*, Vol., 26, pp.922-926

Gauthier J.P, Hammouri H., and Othman S., (1992). A simple observer for non linear systems applications to bioreactors. *IEEE Transactions on Automatic Control*, vol. 37, N° 6, pp. 875-880.

Hermann.R and Krener.A.J, (1977). Non linear controllability and observability.*IEEE transactions on automatic control, 22*, pp 722-740.

Jankovic. M, (1993). A new Observer for a class of Nonlinear Systems. *Journal of Mathematical Systems, Estimation, and Control*, Vol.3, N°2, pp. 225-246.

Krener. A.J., and Isidori. A., (1983). Linearisation by output injection and nonlinear observers. *Systems and Control Letters*, Vol 3, pp. 47-52.

Krener A.J. and Respondek W., (1985). Nonlinear observers with linearizable error dymamics. *SIAM J. Control optim.* 23, pp 197-216.

Lewis F.L, (1992). Applied optimal control and estimation. *Prentice Hall edition*, N° ISBN 0-13-040361-X

Plestan. F and Cherki. B, (1993). An observer for a one flexible joint robot by an algebraic method. *1st IFAC workshop, New trends in Design of Control Systems NTDCS'94*, Sept. 8-10, 1994, Smolenice, Slovaquie.

Teel. A and Praly. L, (1992) Tools for semi global stabilization by partial state and output feedback. *Submitted for publication in S.I.A.M journal of control and optimization.*

Teel. A and Praly. L,(1994) Global stabilizability and observability imply semi -global stabilizability by output feedback. *Systems and control letters*, vol 22, pp 313-325

Tornambe. A, (1992). High-gain observer for non-linear systems. *Int.J.Systems SCI., Vol.23, N°9, pp* 1475-1489.

Tomei. P, (1990). An Observer for Flexible Robots. *IEEE Transactions on Automatic Control,, Vol.35, N°6, pp* 739-743.

Xiao-Hua Xia, Wei-Bin Gao, (1989). Non linear observer design by observer error linearisation. *SIAM J. control and optimization, vol 27, 1, pp. 199-216.*

Zimmer.G, (1994). State observation by on line minimization. *Int. J. Control, vol 60, N° 4, pp* 595-600

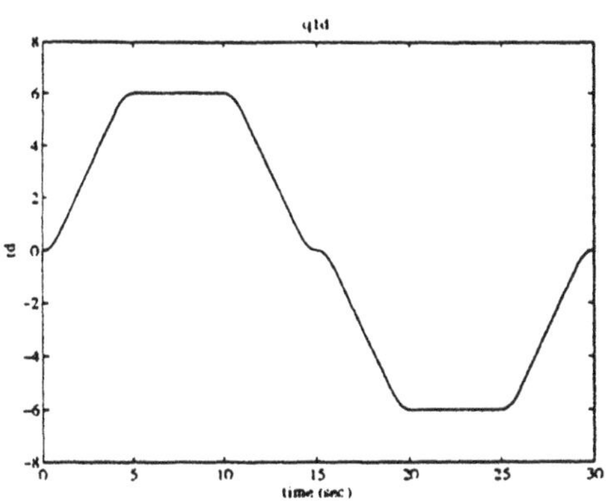

fig 1

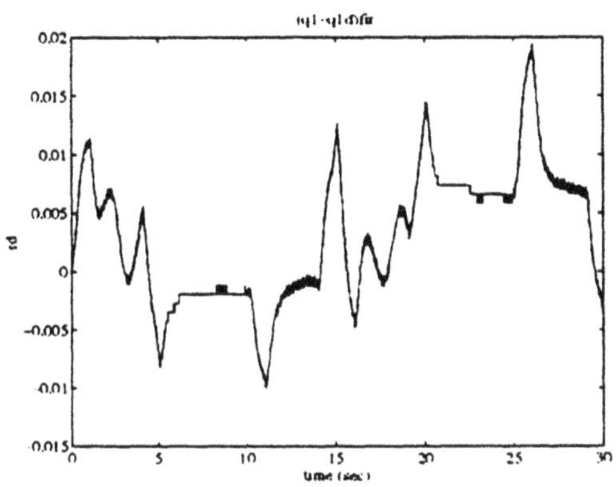

fig 2

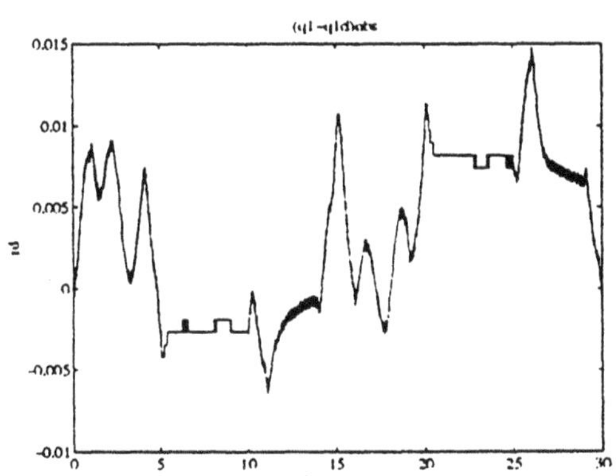

fig 3

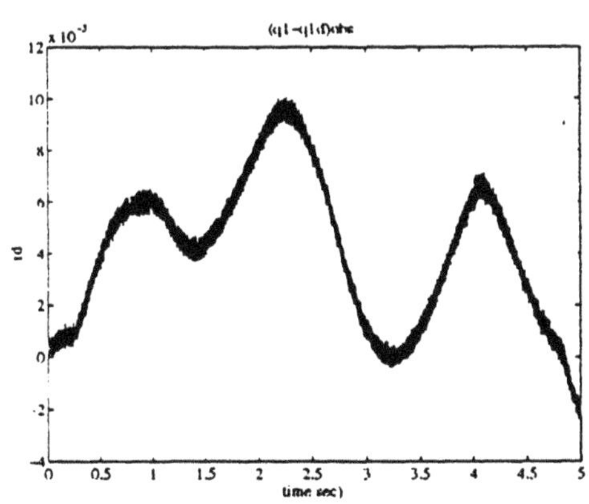

fig 4

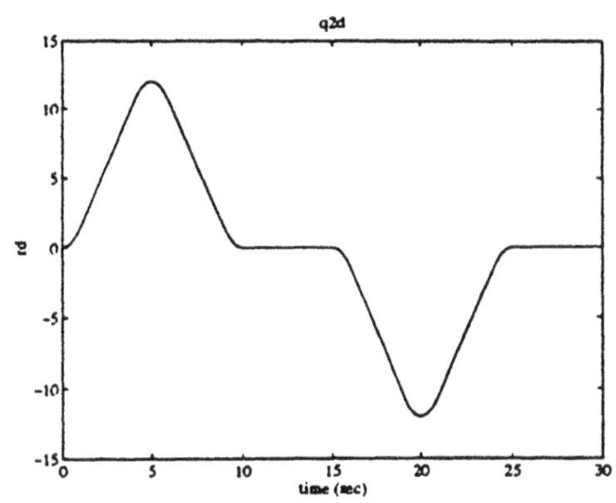

fig 5

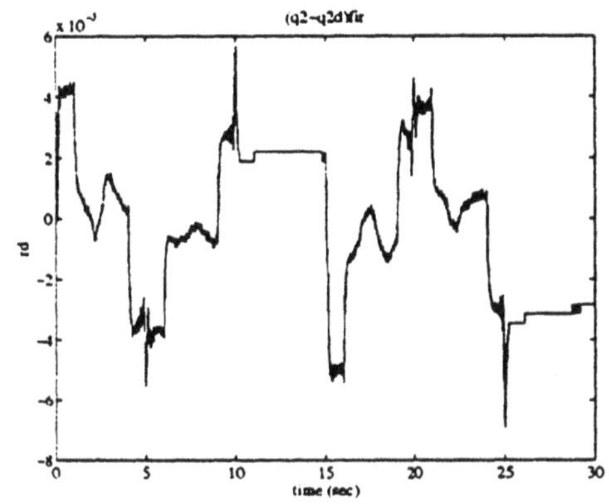

fig 6

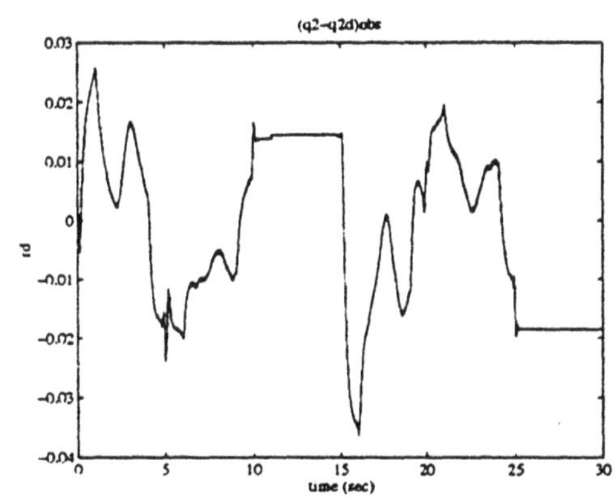

fig 7

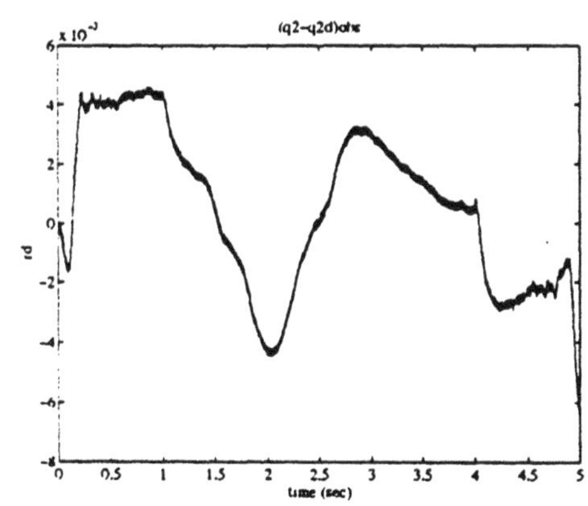

fig 8

NEURAL NETWORKS FOR OPTIMAL CONTROL

OLE SØRENSEN

Aalborg University
Inst. of Electronic Systems, Dept. of Process Engineering
Fredrik Bajers vej 7, DK-9220 Aalborg Ø, Denmark
phone +45 98 15 85 22, telefax +45 98 15 17 39, E-mail os@control.auc.dk

Abstract. Two neural networks are trained to act as an observer and a controller, respectively, to control a non-linear, multi-variable process. The process model is the well-known Innovation State Space model. Firstly, the observer network is trained with a Recursive Prediction Error Method using a Gauss-Newton search direction to minimize the the prediction error. Next, the trained observer network is applied in a closed-loop simulation to train another neural network, the controller. During this training an optimal control cost function is minimized using a recursive, off-line, backward training method, similar to the Back Propagation Through Time (BPTT) method. Finally, a practical, non-linear, noisy and multi-variable example confirms, that the model and the training methods are a promising technique to control non-linear processes, which are difficult to model.

Resume. 2 reseaux neuraux sont developpes pour agir respectivement comme un observateur et un controlleur, afin de controller un processus non lineaire et multi-variable. Le modele de ce processus est le bien connu modele des espaces d'etats (Innovation State Space Model). Premierement le reseau observateur est mis en place avec une methode recursive d'erreur de prediction (Recursive Prediction Error Method) utilisant une recherche de Gauss-Newton pour minimiser l'erreur de prediction. Ensuite ce reseau est applique dans une simulation a boucle fermee pour former un autre reseau neural, le controlleur. Pendant cette sequence de formation une fonction de cout de controle optimal est minimisee, utilisant une methode recursive, a priori, similaire a la propagation arriere a travers le temps (Back Propagation Through Time Method). Finallement un exemple pratique, non lineaire, bruite, et multi-variable confirme que le modele et les methodes de mise en place de ce modele sont une technique promettante pour controler des processus non lineaires, qui sont difficiles a modeliser.

Key Words. Neural network, innovation model, extended Kalmann filter, recursive prediction error method, non-linear control, optimal control.

1. INTRODUCTION

During the last couple of years much research has been published concerning the application of neural networks for non-linear control. For processes, which are difficult or perhaps impossible to model, the application of neural networks seems very attractive, since a trained neural network is considered to form a non-parametric model of the process.

(Narendra *et al.*, 1990) minimizes the error between a desired model reference and the output from the process. This method is known from linear control theory as MRAS, Model Reference Adaptive System.

(Psaltis *et al.*, 1988) minimizes the error between the reference and the output from the process. In this way the neural network is trained to act as the inverse process, and a *minimum variance* or *dead beat* controller is obtained.

(Hunt *et al.*, 1992) minimizes a cost function, in which, besides the control error, also the increment of the control signal is 'punished'. This is a generalization to the non-linear case of the LQ-design, known from linear control theory.

This paper belongs to the last-mentioned category. The process model is the Innovation State Space model, which possesses great generality and is well-known from linear system theory.

Based on a training data set, collected from measurements on the process, a neural network observer is trained to minimize the prediction error. The training method is the Recursive Prediction Error Method using a Gauss-Newton search direction, also well-known from linear system identification theory.

After training, this observer model is applied in a closed-loop simulation to train another neural network, the controller. The purpose of this training is to minimize not only the control error, but also the increment of the control signal. This is a generalization of the cost function from linear LQ control. A new off-line method, where errors successively are Backward Propagated Through Time (BPTT) for the training data set, is applied.

In the sections to come the neural network structure, the process model and the training methods are described. An illustrative practical example shows the ability of the trained neural networks to control a non-linear process.

2. THE MULTI LAYER PERCEPTRON

A Multi Layer Perceptron (MLP) with only one hidden layer has the capability to act as a universal approximator, (Hornik *et al.*, 1989). A MLP with this structure is in a short matrix notation shown in fig.1.

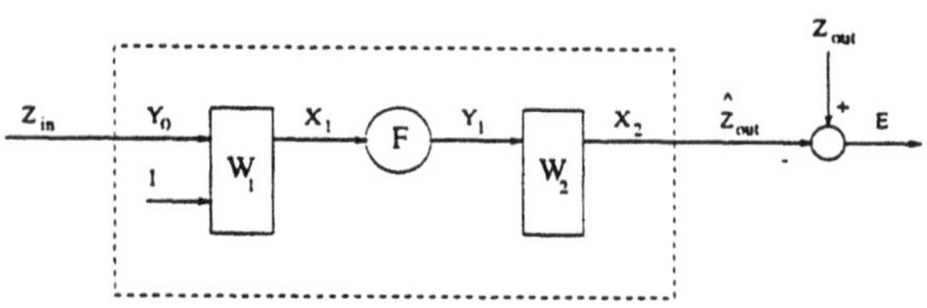

Fig. 1. *A one hidden layer MLP in matrix notation*

The matrix W_1 represents the weights between the input and hidden layers, the '1' shown in fig.1 together with the last column in W_1 giving the necessary offset in the net. The matrix W_2 represents the weights between the hidden and output layers. F represents a vector neural function with all elements chosen to the tanh-function. The input pattern is represented by the vector Z_{in} and the net output is represented by the vector $\hat{Z}_{out}$. The mismatch between the desired output Z_{out} and $\hat{Z}_{out}$ is the error E. Using this notation the output of the net is:

$$\hat{Z}_{out} = W_2\, F\left(W_1 \left\{ \begin{array}{c} Z_{in} \\ 1 \end{array} \right\}\right) \qquad (1)$$

A training set comprises one epoch consisting of N samples of matched pairs of input Z_{in} and desired output Z_{out}.

From a trained MLP it is possible to extract a so-called gain matrix G, the derivative of the output vector with respect to the input vector of the net,

$$G = \frac{d\hat{Z}_{out}}{dZ_{in}^T} = W_2\, \frac{dY_1}{dX_1^T}\, W_1(excl.\ last\ column) \qquad (2)$$

Though training of a MLP is considered to form a non-parametric and non-linear curve fitting, the above mentioned extraction of the gain matrix G allows an actual parameter identification. Note, that G is not a constant matrix, but depends on the actual value of X_1 and consequently on the actual value of the input vector Z_{in}.

3. THE PROCESS MODEL

For a *linear* process with m inputs, m_d disturbances, n states and p outputs the following model of great generality is often used.

$$\begin{aligned} \hat{x}(k+1) &= \Phi\hat{x}(k) + \Gamma u(k) + \Gamma_d d(k) + K\epsilon(k) \\ \hat{y}(k) &= H\hat{x}(k) \\ y(k) &= \hat{y}(k) + \epsilon(k) \end{aligned} \qquad (3)$$

In (3) $\hat{x}$ is the state vector (order n), u is the input vector (order m), d is the disturbance vector (order m_d), y is the output vector (order p) and ϵ is the prediction error vector (order p).

For a *linear* process Φ, Γ, Γ_d, K and H are constant matrices of dimension $n \times n$, $n \times m$, $n \times m_d$, $n \times p$ and $p \times n$ respectively.

It is assumed, that $\hat{x}$ is not measurable (Incomplete State Information), and the model is named the Innovation State Space model.

For the sake of simplicity and without loss of generality the matrix H here is chosen to $H = \{I_{p,p}O_{p,n-p}\}$, where $I_{p,p}$ is a $p \times p$ unity matrix and $O_{p,n-p}$ is a $p \times (n-p)$ zero matrix. This means, that the first p elements of the state vector $\hat{x}$ is filled with $\hat{y}$.

Given only matched pairs of input measurements $u(k)$, disturbance measurements $d(k)$ and desired output measurement $y(k)$, $k = 1 \cdots N$, learning this model is equivalent to solving the Extended Kalmann Filter problem. Even for linear models this is a non-linear problem without any assurance of convergence. In spite of this fact an attempt is made to apply these principles from linear models to be extended to non-linear models.

Generally a multivariable Non-linear Innovation State Space (NISS) model is formulated as

$$\begin{aligned} \hat{X}(k+1) &= \mathcal{F}\left(\hat{X}(k), U(k), D(k), E(k), \theta(k)\right) \\ \hat{Y}(k) &= H\hat{X}(k), \quad H = \{I_{p,p}\ O_{p,n-p}\} \\ Y(k) &= \hat{Y}(k) + E(k) \end{aligned} \qquad (4)$$

where $\mathcal{F}$ is a non-linear vector function and $\theta(k)$ is a vector containing all weights organized in some structure. In order to train a MLP to learn this model the configuration shown in fig.2 is applied.

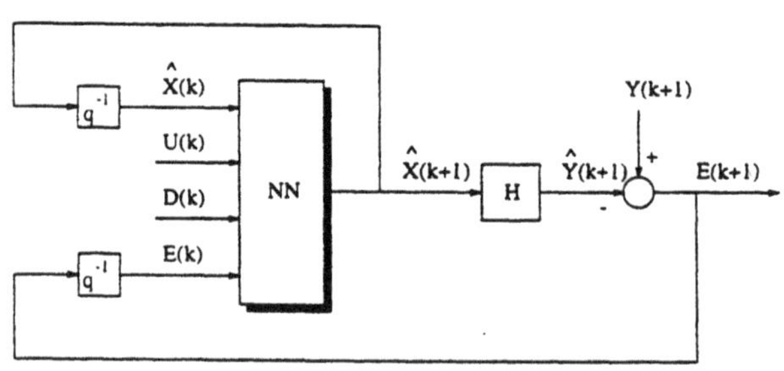

Fig. 2. *The Innovation State Space Model*

The output vector is $\hat{Y}(k+1) = H\hat{X}(k+1)$, the desired output vector is $Y(k+1)$ and the prediction error is $E(k+1) = Y(k+1) - H\hat{X}(k+1)$. During and after training the actual (on-line) gain matrix $G(k)$ can be extracted from the MLP using

(2) and partitioned by

$$
\begin{aligned}
G(k) &= \frac{d\hat{X}(k+1)}{dZ_{in}^T(k)} \\
&= \left\{ \frac{\partial \hat{X}(k+1)}{\partial \hat{X}^T(k)} \; \frac{\partial \hat{X}(k+1)}{\partial U^T(k)} \right. \\
&\qquad \left. \frac{\partial \hat{X}(k+1)}{\partial D^T(k)} \; \frac{\partial \hat{X}(k+1)}{\partial E^T(k)} \right\} \\
&= \left\{ \hat{\Phi}(k) \; \hat{\Gamma}(k) \; \hat{\Gamma}_d(k) \; \hat{K}(k) \right\} \qquad (5)
\end{aligned}
$$

For a *non-linear* process the parameters Φ, Γ, Γ_d and K are not constant matrices, but depend on the actual values of the input, disturbance, state and output vectors.

3.1. *Training method*

Since fig.2 comprises feedback loops around the MLP, it is a Recurrent Network, and training a Recurrent Network is rather more complicated than training a normal 'static' MLP by the Back Propagation algorithm, (Hecht-Nielsen, 1989), (Narendra *et al.*, 1990), (Billings *et al.*, 1991).

A second order Recursive Prediction Error Method (RPEM) using a Gauss-Newton search direction has been applied, since this method is dominating in linear system identification.

Although the training set only comprises one epoch containing N samples, a recursive method is preferred, performing recursive training on the same epoch over and over again, until the cost value is sufficiently low.

The method is rather involved and it is explained in several textbooks (Ljung, 1987) and papers (Chen *et al.*, 1990), (Billings *et al.*, 1991). In this algorithm the model specific derivative $\psi(k) = d\hat{X}^T(k)/d\theta$, where θ is a long column vector containing all weights, plays an important role.

Defining

$$
\psi(k) = \frac{d\hat{X}^T(k)}{d\theta} \quad and \quad \phi(k) = \frac{\partial \hat{X}^T(k)}{\partial \theta}, \qquad (6)
$$

a calculation of $\psi(k+1)$ yields

$$
\begin{aligned}
\psi(k+1) &= \phi(k+1) \\
&\quad + \psi(k)\left\{ \hat{\Phi}^T(k) - H^T \hat{K}^T(k) \right\} \qquad (7)
\end{aligned}
$$

It is strait forward to transform the organizing of the weights in fig.1 from the two matrices W_1 and W_2 to the long column vector θ in order to calculate $\phi(k)$, and for simplicity this is not shown here.

Below the Recursive Prediction Error Method using a Gauss-Newton search direction is summarized, and for simplicity all the arguments 'θ' are neglected. λ is the forgetting factor.

$$
\left.
\begin{array}{l}
1)\ E(k) = Y(k) - \hat{Y}(k) \\
2)\ R(k) = \lambda\, R(k-1) + \psi(k)\, \psi^T(k) \\
3)\ \theta(k) = \theta(k-1) + R^{-1}(k)\, \psi(k)\, E(k) \\
4)\ predict\ \hat{X}(k+1)\ from\ (4) \\
5)\ predict\ \psi(k+1)\ from\ (6)\ and\ (7)
\end{array}
\right\} \qquad (8)
$$

Note, that 4)-5) depend on the chosen specific prediction model, whereas 1)-3) are independent of the model. More practical versions of the algorithm is mentioned in (Ljung, 1987) and (Billings *et al.*, 1991).

4. NON-LINEAR OPTIMAL CONTROL

The principle of linear optimal control, when applying state space description, is illustrated in fig.3.

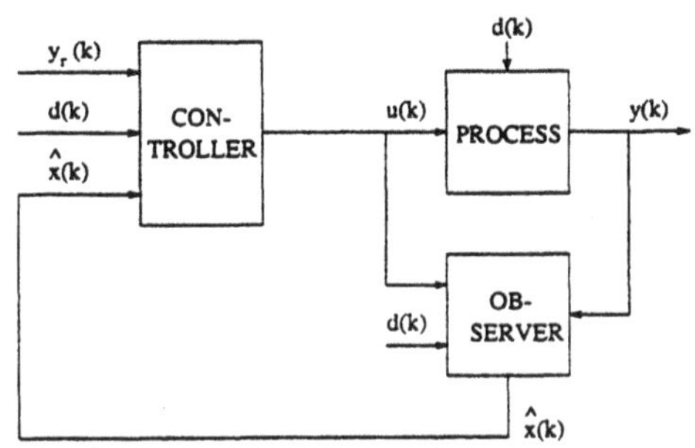

Fig. 3. *The principle of optimal control*

The observer error is $e(k) = y(k) - \hat{y}(k)$, the control error is $e_c(k) = y_r(k) - y(k)$ and the control signal increment is $\Delta u(k) = u(k) - u(k-1)$.

Details about the design method can be found in many textbooks, e.g. (Åström and Wittenmark, 1990).

4.1. *Training method for non-linear optimal control*

It will now be attempted to generalize these principles from linear control theory to the non-linear case.

The observer problem has already been solved in the previous section, since the process model (4) was trained just to minimize the variance of the prediction error, thus acting as an optimal prediction observer.

And now the same trained process model is applied in a closed-loop simulation in order to train a neural network controller. The connections between the controller NN2 and the process model NN1 is shown in fig.4, which shows the k'th section of an unfolding in time. Each section contains the same two neural networks, the controller and the process model, but at different samples, and the total unfolding is considered to form a large 'sandwich' containing N sections.

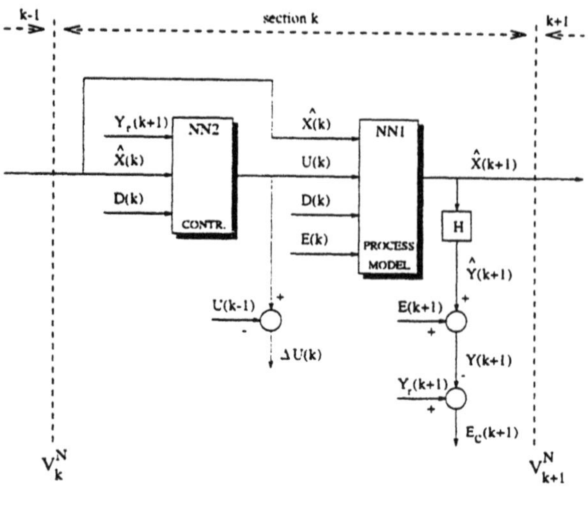

Fig. 4. *One section unfolded in time*

The process model also included stochastic modelling. The prediction error $E(k), k = 1 \ldots N$, obtained after training the process model, should be almost white noise. This prediction error is now reused, see the noise inputs $E(k+1)$ and $E(k)$ in fig.4, in order to provide the process model with a relevant noise, thus allowing the training of the controller NN2 to take place in a realistic, noisy environment.

The outcome, when reusing the prediction error as described, is a more robust controller, compared to a situation, where the controller is trained by a noise-free deterministic model, $E(k) = 0, k = 1 \ldots N$.

The rest of this section describes, see fig.4, how to train the neural network controller NN2,

$$U(k) = \mathcal{G}\left(Y_r(k+1), \hat{X}(k), D(k)\right) \tag{9}$$

based on a trained neural network process model NN1, described by (4), in accordance with the optimal control cost function

$$\begin{aligned} V_1^N &= \sum_{k=1}^{N} \frac{1}{2} E_c^T(k+1) E_c(k+1) \\ &\quad + \rho \sum_{k=1}^{N} \frac{1}{2} \Delta U^T(k) \Delta U(k) \end{aligned} \tag{10}$$

where

$$\begin{aligned} E_c(k+1) &= Y_r(k+1) - Y(k+1) \\ \Delta U(k) &= U(k) - U(k-1) \end{aligned} \tag{11}$$

For simplicity, only a scalar 'punishment' ρ is introduced in (10), since all involved signals, as usual, are scaled to the same dimension-less order of magnitude. This represents no loss of generality, since individual matrix 'punishment' can easily be applied, if desired.

The minimizing of (10) represents a new problem, since it contains a sum of two coupled contributions. This means, that the minimizing can not be based on a 'sample-based' updating of the weights in the controller network, but rather has to be based on an 'epoch-based' updating, and for that purpose a recursive expression for (10) is found.

Letting k, as usual, denote the actual sample number, the last part of the cost function, the sum from k to N, is written as

$$\begin{aligned} V_k^N &= \sum_{i=k}^{N} \frac{1}{2} E_c^T(i+1) E_c(i+1) \\ &\quad + \rho \sum_{i=k}^{N} \frac{1}{2} \Delta U^T(i) \Delta U(i) \\ &= \frac{1}{2} E_c^T(k+1) E_c(k+1) \\ &\quad + \rho \frac{1}{2} \Delta U^T(k) \Delta U(k) \; + \; V_{k+1}^N \end{aligned} \tag{12}$$

For the 'multi-multi layer perceptron' in fig.4 this means, that firstly a *forward* pass through all N sections is performed, saving all signals, and next a *backward* pass including a sectional updating of the controller weights is performed, while keeping the weights of the simulator fixed.

During this long *backward* pass the recursive expression (12) for V_k^N is used, and the weights of the controller are currently updated based on the gradient of V_k^N with respect to $U(k)$.

Since $V_k^N = V_k^N\left(\hat{X}(k+1), U(k)\right)$,

$$\begin{aligned} \frac{\partial V_k^N}{\partial \hat{X}(k)} &= \frac{\partial \hat{X}^T(k+1)}{\partial \hat{X}(k)} \frac{\partial V_k^N}{\partial \hat{X}(k+1)} \\ &\quad + \frac{\partial U^T(k)}{\partial \hat{X}(k)} \frac{\partial V_k^N}{\partial U(k)} \end{aligned} \tag{13}$$

where, from (12),

$$\frac{\partial V_k^N}{\partial \hat{X}(k+1)} = - H^T E_c(k+1) + \frac{\partial V_{k+1}^N}{\partial \hat{X}(k+1)} \tag{14}$$

$$\frac{\partial V_k^N}{\partial U(k)} = \rho \, \Delta U(k) + \frac{\partial \hat{X}^T(k+1)}{\partial U(k)} \frac{\partial V_k^N}{\partial \hat{X}(k+1)} \tag{15}$$

These partial gradients explain, how the errors recursively are *backward* propagated from section to section, and they are easily understood by comparing with fig.4.

At each iteration (one epoch) the following have to be done

1. A forward pass through all sections, $k = 1 \ldots N$, is calculated, saving all signals.
2. A backward pass through all sections, $k = N \ldots 1$, is performed, and for each section the following have to be done
   - $\frac{\partial V_k^N}{\partial \hat{X}(k+1)}$ is calculated by (14)
   - $\frac{\partial V_k^N}{\partial U(k)}$ is calculated by (15)

- The weights in the controller network NN2 are updated by BPEA with a gradient factor η by applying $\frac{\partial V_k^N}{\partial U(k)}$
- $\frac{\partial V_k^N}{\partial \hat{X}(k)}$ is calculated by (13)

The gradient factor η is gradually decreased from 0.005 to 0.0001 for this method. The value is considerably lower than for the normal Back Propagation algorithm.

The described recursive, backward, off-line training method has many similarities to the way in which the Riccati equations from linear optimal control are solved by Dynamic Programming, see (Åström and Wittenmark, 1990).

The method has also many similarities to a neural network training algorithm named Back Propagation Through Time (BPTT), (Hunt *et al.*, 1992), (Zbikowski and Gawthrop, 1992), although these two references consider recurrent neural networks, in which the time delays are placed *inside* the network around each neuron, whereas in this paper they are placed *around* the whole network.

4.2. *Implementation of non-linear optimal control*

The implementation of the total control structure is shown in fig.5. All signals in the figure are indicated by capital letters to emphasize that it concerns large-signals. Signals which are measured in physical units are supplied with an index 'm', contrary to signals which are scaled to dimension-less quantities.

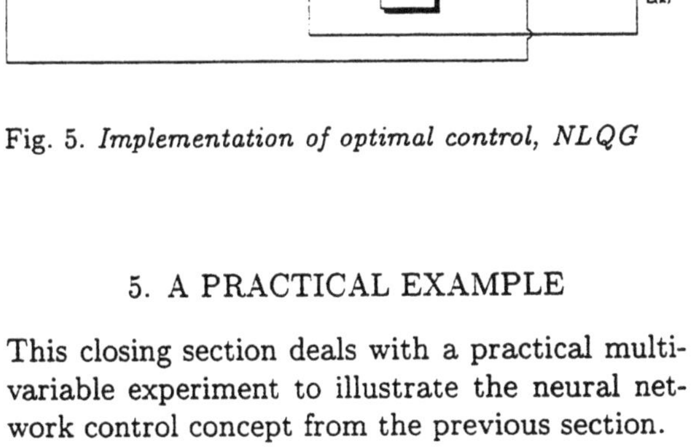

Fig. 5. *Implementation of optimal control, NLQG*

5. A PRACTICAL EXAMPLE

This closing section deals with a practical multivariable experiment to illustrate the neural network control concept from the previous section.

A sketch of the multi-variable process is shown in fig.6.

The process considered is a laboratory setup, in which cold and hot water are mixed in a cylindrical tank containing an outlet in the bottom.

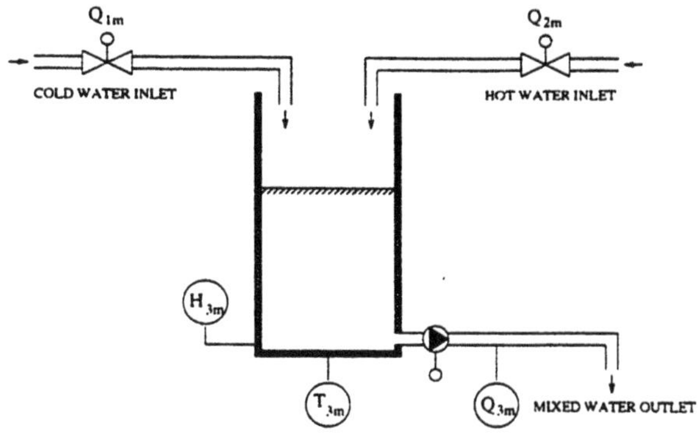

Fig. 6. *The multivariable process*

The control purpose is to maintain the water level and the temperature of the mixed water in spite of several disturbances, mainly the water outlet. The process is equipped with industrial actuators and sensors, but unfortunately they are contaminated with considerably electrical noise. However, the process is accepted as it is. No efforts have been made to establish inner loops around the valves, no filtering of the measured signals are performed, and no knowledge of the process model is assumed. The difficulties are considered to be extra challenges to a neural network controller.

- The controlling inputs are the voltages Q_{1m} and Q_{2m} to the valves, supplying the cold and hot water, respectively.
- The disturbing input is the outlet flow Q_{3m} of the mixed water. In fact, also the temperatures of the water inlets are disturbances, but they are rather constant and therefore dismissed.
- The measurable outputs are the water level H_{3m} and the temperature T_{3m}. The sensor for T_{3m} is rather slow, it contains a time constant of approximately ten seconds.
- The state is chosen to fourth order, the first two states being the measurable outputs H_{3m} and T_{3m}.
- The references are chosen to 0.25 m for H_{3m} and 25 C for T_{3m}.

As mentioned, the electrical noise is considerable. The measurement noise alone is white with a standard deviation of 0.23 cm for H_{3m} and 0.28 C for T_{3m}. The noise is considered to be an extra challenge to a neural network controller.

The sampling interval is 2 $sec.$, and without any knowledge of the process model, two traditional PI-controllers are tuned 'by hand' to give the 'best' performance, letting one PI-controller control H_{3m} by Q_{1m}, and the other PI-controller control T_{3m} by Q_{2m}.

The result, which is used as training data for the neural network controller, is shown in fig.7.

The unknown hidden coupling between the two

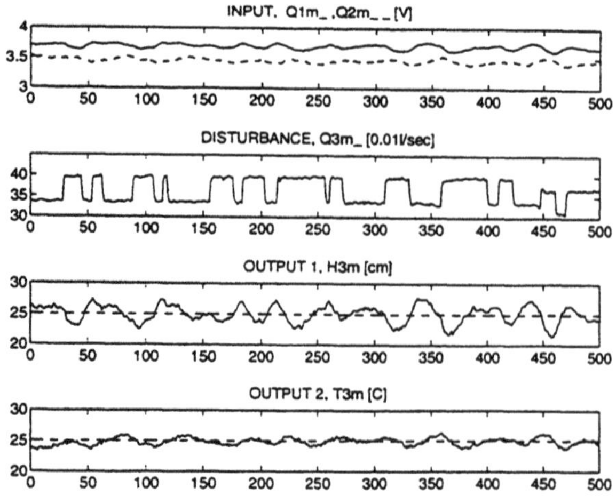

Fig. 7. *Two individually tuned PI-controllers. The data are used for training a neural network controller*

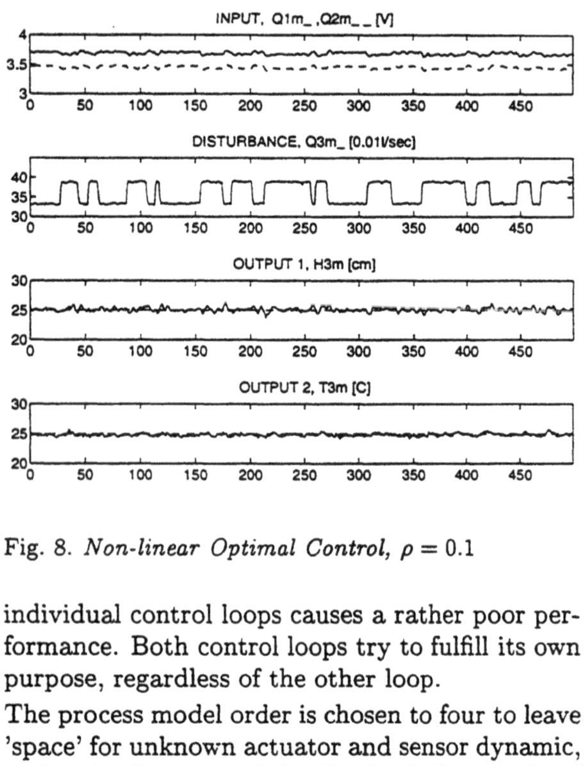

Fig. 8. *Non-linear Optimal Control,* $\rho = 0.1$

individual control loops causes a rather poor performance. Both control loops try to fulfill its own purpose, regardless of the other loop.

The process model order is chosen to four to leave 'space' for unknown actuator and sensor dynamic, and ρ is chosen to 0.1. In fig.8 the result of the neural network non-linear optimal control is shown.

Taking the considerable disturbance and noise into account, a nice performance is obtained. The control signals are cooperating to obtain the common goal. Without any explicit modelling, the neural networks were trained to learn the hidden coupling between the water level and the temperature. Further, the effect of the considerably disturbing outlet is nicely suppressed.

6. CONCLUSION

In this paper the neural network controller was trained, based on a trained neural network process model, to behave as a more or less 'detuned' inverse of the process, which was accomplished by choosing a cost function, known from linear optimal control, to be minimized during training of the neural network controller. In the cost function, besides the control error, the increment of the control signal is also 'punished'.

Minimizing this cost function was more complicated, and an off-line method, similar to Back Propagation Through Time (BPTT) was explained.

The method suffers from the fact, that re-tuning of the controller demands a quite new training session, and consequently a future research work will focus on a simple variant of the optimal control cost function, allowing tuning of the neural network controller *after* training.

7. REFERENCES

Åström, K. J. and Björn Wittenmark (1990). *Computer Controlled Systems: Theory and Design.* second edn. Prentice-Hall.

Billings, S. A. et al. (1991). 'A comparison of the backpropagation and recursive prediction error algorithms for training neural networks'. *Mechanical Systems and Signal Processing* **3**, 233–255.

Chen, S. et al. (1990). 'Non-linear system identification using neural networks'. *Int. J. Control* **6**, 1191–1214.

Hecht-Nielsen, R. (1989). *Neurocomputing.* Addison Wesley.

Hornik, K. et al. (1989). 'Multilayer feedforward networks are universal approximators'. *Neural Networks* **2**, 359–366.

Hunt, K. J. et al. (1992). 'Neural networks for control systems - a survey'. *Automatica* **6**, 1083–1120.

Ljung, L. (1987). *System identification, theory for the user.* first edn. Prentice-Hall.

Narendra, K. S. et al. (1990). 'Identification and control of dynamical systems using neural networks'. *IEEE Transaction on Neural Networks* **1**(1), 4–27.

Psaltis, D. et al. (1988). 'A multilayererd neural network controller'. *IEEE Control System Magazine* **8**(3), 17–21.

Rumelhart, D. E. et al. (1988). *Parallel Distributed Processing.* Vol. 1,2. MIT Press. Cambridge, London.

Zbikowski, R. and P.J. Gawthrop (1992). A survey of neural networks for control. In 'Neural networks for control and systems'. Peter Peregrinus Ltd.,London, United Kingdom. pp. 31–50.

AGGREGATIVE POLE PLACEMENT OF COMPLEX CONJUGATE POLES WITHIN CONTINUOUS SYSTEM OPTIMAL CONTROL

M. FRANCESCHI, G. ENEA, J. DUPLAIX

MS/LAIAT, Université de Toulon et du Var, B.P. 132, 83957 La Garde, France

Abstract. We present a new method to place complex conjugate poles within the framework of continuous systems optimal control. Our demonstration is based on a well known aggregative method. The pole-placement is determined by the means of the reduced system and it simultaneously optimises a quadratic criterion. Desired poles can be complex or real. We precisely explicit the whole area of shifting poles. We illustrate the method by the use of a program which gives the shifted poles area to the designer.

Key Words. Optimal pole placement, quadratic criterion, aggregation.

PLACEMENT PAR AGREGATION DE PÔLES COMPLEXES CONJUGUES EN COMMANDE OPTIMALE SUR UN SYSTEME CONTINU.

Résumé. Nous proposons une méthode systématique pour déplacer des pôles complexes conjugués dans le cadre d'une commande optimale sur un système continu. Après avoir rappelé une méthode générale de placement de pôle par agrégation, nous montrons que nous pouvons placer les pôles sur le système réduit en vérifiant la minimisation d'un critère quadratique. Les pôles cibles peuvent être complexes ou réels. Nous explicitons la totalité du domaine ou les pôles peuvent être déplacés. Nous illustrons la méthode par l'utilisation d'un programme qui sur un exemple donne au concepteur le domaîne des pôles cibles.

Mots Clé. Placement de pôle optimal, critère quadratique, agrégation.

1. INTRODUCTION

Many authors have proposed different pole placement methods for optimal control. Some methods have allowed to choose the poles of the full system (Solheim, 1972), afterwards other methods have authorised to move only selected poles by aggregative method (Alexandridis, 1987). These methods do not allow to determine the whole area of shifting poles and they require to resolve Riccati equation (Saif, 1989-1). Satisfying solutions have been given for real pole assignment, but proposed solutions for complex poles are very restrictive (Amin, 1985) or the determination of shifted poles must be obtained by successive choices to get an optimal control (Saif, 1989-2).
In this paper a systematic method which allows complex poles shifting is presented. An aggregative approach which does not need Riccati equation resolution is used. The region of closed loop poles, which can be real or complex, is obtained by solving an unequation set.
The linear system is described by the state space formulation :

$$\dot{X} = A\,X + B\,U \qquad (1)$$

X : state vector U : input vector

dim(X) = (n,1) dim(U)=(m,1)
dim(A) = (n,n) dim(B)=(n,m)

Let us assume a completely controllable system .The optimal modal control problem consists to obtain the desired poles and to minimize the quadratic criterion :

$$J = \int_{t=0}^{\infty} \left(X^T Q X + U^T R U\right) dt \qquad (2)$$

with R a fixed symmetric positive-definite matrix, and Q a symmetric matrix which must be determined to satisfy the pole assignment.

The form of the solution is :

$$U = -L\,X = -R^{-1}B^{T}K\,X \qquad (3)$$

With K the stable solution of Algebraic Matrix Riccati Equation (AMRE) which is associated to this problem :

$$K\left(A - BR^{-1}B^{T}K\right) + A^{T}K + Q = 0 \qquad (4)$$

laying $V = BR^{-1}B^{T}$ ($V \geq 0$) The closed loop gives:

$$Af = A - BL = A - VK \qquad (5)$$

The aim of this problem consists to determine
- a symmetric positive-definite matrix K,
- a symmetric matrix Q

which both,
- verify a Riccati Equation (4) for a fixed R,
- realise a closed loop with prescribed eigenvalues.

<u>Remark</u>
In the present work, only the shifting of two complex conjugate poles is considered.
In the following it is assumed that two complex conjugate eigenvalues λ and $\bar{\lambda}$ must be modified in two real or complex conjugate eigenvalues μ_1 and μ_2 .
If a real pole must be shifted, the method which has been presented in (Duplaix, 1994) will be used.

2. AGGREGATIVE METHOD

The method consists to aggregate the original system in a reduced system, in order to shift only the selected poles, without altering the rest of the eigenspectrum (Alexandridis, 1987; Saif,1989).

For shifting a pair of conjugate poles, the original system will be reduced to second-order system.

In the original system, let us consider the complex conjugate eigenvalues to shift :

$$\begin{cases} \lambda_k = \lambda = \lambda_r + j\lambda_i \\ \lambda_{k+1} = \bar{\lambda} = \lambda_r - j\lambda_i \end{cases} \quad \text{with} \quad \begin{matrix} \lambda_i, \lambda_r \in \Re \\ \lambda_i > 0 \end{matrix}$$

2.1. <u>Aggregation matrix</u>

The aggregation matrix is given by :

$$Ca = T_k W^{-1} = \begin{bmatrix} u_k^T \\ \bar{u}_k^T \end{bmatrix} \quad \text{and} \quad Ca^T = \begin{bmatrix} u_k & \bar{u}_k \end{bmatrix}$$

with :

$$T_k = \begin{bmatrix} 0 & I_{2,k} & 0 \end{bmatrix} = \begin{bmatrix} 0 & ..0 & 1 & 0 & 0.. & 0 \\ 0 & ..0 & 0 & 1 & 0.. & 0 \end{bmatrix}$$

$$\qquad\qquad 1 \;... \qquad k \quad k+1 \;..... \; n$$

and W is the eigenvector matrix of A.
Let us u_k^T and u_{k+1}^T the k and k+1 lines of W^{-1} then we have :

$$W^{-1} = \begin{bmatrix} \vdots \\ u_k^T \\ u_{k+1}^T \\ \vdots \end{bmatrix} = \begin{bmatrix} \vdots \\ u_k^T \\ \bar{u}_k^T \\ \vdots \end{bmatrix} \qquad (\text{with } \lambda_{k+1} = \bar{\lambda}_k)$$

2.2. <u>Reduced system</u> :

It is obtained by using (5) and Ca :

$$\begin{aligned} Ca \cdot Af \cdot Ca^{\oplus} &= Ca \cdot A \cdot Ca^{\oplus} - Ca \cdot BL \cdot Ca^{\oplus} \\ &= Ca \cdot A \cdot Ca^{\oplus} - Ca \cdot VK \cdot Ca^{\oplus} \end{aligned}$$

Where $Ca^{\oplus}$, the pseudo-inverse of Ca, is given by: $Ca^{\oplus} = Ca^{T}(Ca\,Ca^{T})^{-1}$

The reduced system is defined by the following transformations (Duplaix,1994)

$$\left.\begin{aligned} &\hat{A} = Ca\,A\,Ca^{\oplus} \qquad \hat{B} = Ca\,B \\ &\hat{A}f = Ca\,Af\,Ca^{\oplus} \\ &L = \hat{L}\,Ca \qquad K = Ca^{T}\hat{K}Ca \\ &\hat{V} = Ca\,V\,Ca^{T} = \hat{B}R^{-1}\hat{B}^{T} \end{aligned}\right\} \qquad (6)$$

hence the following relation is obtained :

$$\hat{A}f = \hat{A} - \hat{V}\hat{K} = \hat{A} - \hat{B}\hat{L} \qquad (7)$$

Let us express J criterion using $\hat{X}$ the state vector of reduced system. J may be written as

$$J = \int_{t=0}^{\infty} \left(\hat{X}^{T}\hat{Q}\,\hat{X} + U^{T}RU\right)dt$$

with $Q = Ca^{T}\hat{Q}\,Ca$ and $\hat{X} = Ca\,X$.

Hence Riccati Equation of the reduced system is given by :

$$\hat{K}(\hat{A} - \hat{B}R^{-1}\hat{B}^{T}\hat{K}) + \hat{A}^{T}\hat{K} + \hat{Q} = 0 \qquad (8)$$

The feedback L of the full system may be directly obtained from $\hat{K}$ by $L = R^{-1}\hat{B}^{T}\hat{K}\,Ca$.

Now the problem is reduced to a second order system. The new aim is to get $\hat{K}$ and $\hat{Q}$

matrices, which are solution of (8) and satisfy the pole assignment, with :

$\dim(\hat{K})=(2,2) \qquad \dim(\hat{Q})=(2,2)$.

An equivalent problem consists to determine the form of the $\hat{A}f$ matrix which verifies the structure given by (7) and which has the assigned eigenvalues. Afterwards $\hat{K}$ matrix will be deduced from (7), and at last $\hat{Q}$ matrix will be obtained from (8).

3. DETERMINATION OF $\hat{A}f$

In the following, to determine the form of $\hat{A}f$, we assume that assigned poles are fixed in the reachable domain.

Form of $\hat{V}$ and $\hat{K}$ matrices :

V is a symmetric positive semi-definite matrix. Hence $\hat{V}$ obtained from (6) has the following structure :

$$\hat{V}=\begin{bmatrix} v_1 & v_2 \\ \bar{v}_2 & v_1 \end{bmatrix} \quad \text{with} \begin{cases} v_1 \text{ real} \\ v_2 = |v_2|\cdot e^{j\psi} \end{cases} \quad \text{and}$$

$V \geq 0 \;\Rightarrow\; \hat{V} \geq 0$.

The semi-positive definite matrix $\hat{V}$ is **hermitian** with **equal diagonal terms** . The notation "h.e.d.t." will be used.

Conversely to get a K symmetric positive semi-definite matrix from relation (6), $\hat{K}$ must be of the form "h.e.d.t."

if $\hat{K}=\begin{bmatrix} k_1 & k_2 \\ \bar{k}_2 & k_1 \end{bmatrix}$ with k_1 positive real,

and if $\hat{K}>0 \;\Rightarrow\; K \geq 0$.

3.1. Structure of $\hat{A}f$

$\hat{A}f$ will be the result of an optimal control law determined from the minimisation of (2). Therefore the structure of $\hat{A}f$ can be deduced from (7) and "h.e.d.t." form of $\hat{V}$ and $\hat{K}$.

Development of (7) yields on the one hand:

$$\hat{V}\hat{K}=\begin{bmatrix} v_1k_1+v_2\bar{k}_2 & v_1k_2+v_2k_1 \\ \bar{v}_2k_1+v_1\bar{k}_2 & \bar{v}_2k_2+v_1k_1 \end{bmatrix}=\begin{bmatrix} a & b \\ \bar{b} & \bar{a} \end{bmatrix} \qquad (9)$$

and on the other hand :

$$\hat{A}-\hat{A}f=\begin{bmatrix} \lambda & 0 \\ 0 & \bar{\lambda} \end{bmatrix}-\begin{bmatrix} a_{11} & a_{12} \\ a_{21} & a_{22} \end{bmatrix}$$

It follows that : $\begin{matrix} a_{11}=\bar{a}_{22}=\lambda-a=a_1 \\ a_{12}=\bar{a}_{21}=-b=a_2 \end{matrix}$

then $\hat{A}f$ takes the form $\boxed{\hat{A}f=\begin{bmatrix} a_1 & a_2 \\ \bar{a}_2 & \bar{a}_1 \end{bmatrix}}$ (10)

with

$$\begin{cases} a_1=\rho_1e^{j\varphi_1}=\rho_1(\cos\varphi_1+j\sin\varphi_1)=a_{1r}+ja_{1i} \\ a_2=\rho_2e^{j\varphi_2}=\rho_2(\cos\varphi_2+j\sin\varphi_2)=a_{2r}+ja_{2i} \end{cases}$$

So the problem of $\hat{A}f$ determination is reduced to a_1 and a_2 determination.

3.2. Constraints on $\hat{A}f$

To get a stable closed loop system , the eigenvalue targets μ_1, μ_2 must be assigned with negative real part. Then :

for complex targets

$$\begin{cases} \mu_1=\mu e^{j\theta}=\mu_r+j\mu_i \\ \mu_2=\mu e^{-j\theta}=\mu_r-j\mu_i \end{cases} \quad \text{with } \mu_r<0 \;\; \mu_i>0$$

for real targets

$$\begin{cases} \mu_1<0 \\ \mu_2<0 \end{cases}$$

This choice of eigenvalues leads to verify the following conditions :

$\mathrm{trace}(\hat{A}f)=\mu_1+\mu_2=a_1+\bar{a}_1=a_{1r}<0$ (11)

$\det(\hat{A}f)=\mu_1\mu_2=a_1\bar{a}_1-a_2\bar{a}_2>0$ (12)

These two conditions yield to verify :

(11) => $\boxed{a_{1r}=\dfrac{\mu_1+\mu_2}{2}<0}$ (13)

(12) => $\boxed{\rho_2{}^2=a_{1i}^2+\left(\dfrac{\mu_1-\mu_2}{2}\right)^2>0}$ (14)

3.3. Case $\det(\hat{V})\neq 0$

Moreover the eigenvalue targets must be chosen in a domain which implies $\hat{K}$>0.

$$\hat{V}>0 \Leftrightarrow \begin{cases} \mathrm{trace}(\hat{V})>0 \\ \text{et } \det(\hat{V})>0 \end{cases}$$

Positive definiteness of $\hat{K}$ requires relations on $\hat{A}f$. To determine these relations consider,

$$\hat{K}>0 \Leftrightarrow \begin{cases} (1) \;\; \det(\hat{K})>0 \\ (2) \;\; \mathrm{trace}(\hat{K})>0 \end{cases}$$

3.3.1. Relation to have $\det(\hat{V})\neq 0$

The equation (7) allows to write,

$\det(\hat{K})\det(\hat{V}) = \det(\hat{A} - \hat{A}f)$

This last relation and $\hat{V} > 0$ imply :

$\det(\hat{K}) > 0 \iff \det(\hat{A} - \hat{A}f) > 0$

therefore (10) gives
$\det(\hat{A} - \hat{A}f) = \left((\lambda - a_1)(\bar{\lambda} - \bar{a}_1) - a_2\bar{a}_2\right) > 0$

(12) and (13) yields : $\boxed{a_{1i} < a_{1i\,sup}}$ (15)

$$\text{with } a_{1i\,sup} = \left(\frac{\lambda\bar{\lambda} + \mu_1\mu_2 - \lambda_r(\mu_1 + \mu_2)}{2\lambda_i}\right)$$

3.3.2. Relation to have $\mathrm{trace}(\hat{K}) > 0$

Since $\det(\hat{V}) > 0$ is assumed, it follows that

$$\hat{K} = \hat{V}^{-1}\left(\hat{A} - \hat{A}f\right)$$
$$\begin{bmatrix} k_1 & k_2 \\ \bar{k}_2 & k_1 \end{bmatrix} = \frac{1}{\det(\hat{V})}\begin{bmatrix} v_1 & -v_2 \\ -\bar{v}_2 & v_1 \end{bmatrix}\begin{bmatrix} \lambda - a_1 & -a_2 \\ -\bar{a}_2 & \bar{\lambda} - \bar{a}_1 \end{bmatrix}$$

then $\mathrm{trace}(\hat{K}) > 0$

$$\iff k_1 = \frac{1}{\det(\hat{V})}\left(v_1(\lambda - a_1) + v_2\bar{a}_2\right) > 0$$

let us $r = |v_2|/v_1$ ($r<1$), we have :

$$\iff \begin{cases} \lambda_r - a_{1r} + r\cdot\rho_2\cdot\cos(\psi - \varphi_2) > 0 \\ \qquad \iff \mathrm{Re}(k_1) > 0 \\ \lambda_i - a_{1i} + r\cdot\rho_2\cdot\sin(\psi - \varphi_2) = 0 \\ \qquad \iff \mathrm{Im}(k_1) = 0 \end{cases}$$

Conditions to have $\mathrm{Im}(k_1) = 0$

$$\mathrm{Im}(k_1) = 0 \iff \frac{\lambda_i - a_{1i}}{r\cdot\rho_2} = \sin(\varphi_2 - \psi) \qquad (16)$$

$\sin(\psi - \varphi_2)$ existence condition into (16) requires :

$$-1 < \frac{(\lambda_i - a_{1i})}{r\cdot\rho_2} < 1 \iff \frac{(\lambda_i - a_{1i})^2}{r^2} < \rho_2^{\,2}$$

This relation is more restrictive for ρ_2 than (13) inequality. Hence we have :

$$\left(1 - r^2\right)a_{1i}^{\,2} - 2\lambda_i\, a_{1i} + \left(\lambda_i^{\,2} - r^2\left(\frac{\mu_1 - \mu_2}{2}\right)^2\right) < 0 \qquad (17)$$

the polynomial study in a_{1i} leads to note :

- for real assignment this relation is always verified
- for complex assignment :
  - μ_i must be chosen to verify :
    $\lambda_i^{\,2}/1 - r^2 > \mu_i^{\,2}$ therefore

$$\boxed{|\mu_i| < \mu_{i\max} \text{ avec } \mu_{i\max} = \lambda_i/\sqrt{1 - r^2}} \qquad (18)$$

  - a_{1i} must be chosen to verify :

$$\boxed{a_{1im} < a_{1i} < a_{1iM}} \qquad (19)$$

$$\text{with } \begin{cases} a_{1im} = \dfrac{\lambda_i - \sqrt{\delta'}}{1 - r^2}, \quad a_{1iM} = \dfrac{\lambda_i + \sqrt{\delta'}}{1 - r^2} \\ \delta' = r^2\left(\lambda_i^{\,2} + (1 - r^2)\left(\dfrac{\mu_1 - \mu_2}{2}\right)^2\right) \end{cases}$$

then φ_2 can be computed by :

$$\boxed{\begin{aligned} &\varphi_2 = \arcsin\left(\frac{(\lambda_i - a_{1i})}{r\cdot\rho_2}\right) + \psi \\ &\text{or } \quad \varphi_2 = \pi - \arcsin\left(\frac{(\lambda_i - a_{1i})}{r\cdot\rho_2}\right) + \psi \end{aligned}} \qquad (20)$$

Condition to have $\mathrm{Re}(k_1) > 0$

is verified without further condition.

3.4. Case $\det(\hat{V}) = 0$

Conditions on $\hat{A}f$ to have $\hat{K} > 0$.

3.4.1. Constraint on $\hat{A}f$

$\det(\hat{V}) = 0$ case leads to have
$v_1^{\,2} = v_2\bar{v}_2 \Rightarrow v_2 = v_1 e^{j\psi}$ and $r = 1$

$$\hat{V}\hat{K} = \left(\hat{A} - \hat{A}f\right) = \begin{bmatrix} \lambda - a_1 & -a_2 \\ -\bar{a}_2 & \bar{\lambda} - \bar{a}_1 \end{bmatrix}$$
$$= \begin{bmatrix} v_1 & v_1 e^{j\psi} \\ v_1 e^{-j\psi} & v_1 \end{bmatrix}\begin{bmatrix} k_1 & k_2 \\ \bar{k}_2 & k_1 \end{bmatrix}$$

$$\Rightarrow \begin{cases} \bar{\lambda} - \bar{a}_1 = v_1(k_1 + k_2 e^{-j\psi}) \\ -a_2 e^{-j\psi} = v_1(k_1 + k_2 e^{-j\psi}) \end{cases} \qquad (21)$$

k_1 and k_2 are linearly dependent.
There is an infinite number of solutions.
furthermore : $\bar{\lambda} - \bar{a}_1 + a_2 e^{-j\psi} = 0$

$$\boxed{\Rightarrow a_2 = (\bar{a}_1 - \bar{\lambda})e^{j\psi}} \qquad (22)$$

Necessary conditions on $\hat{A}f$:

$$\left.\begin{array}{l}\det(\hat{A}-\hat{A}f)=\det(\hat{K})\det(\hat{V})\\ \det(\hat{V})=0\end{array}\right\}$$

$$\Rightarrow \det(\hat{A}-\hat{A}f)=0$$

developments are identical to the ones of paragraph 3.3.1, then :

$$\boxed{a_{1i}=\lambda\bar{\lambda}+\mu_1\mu_2-\lambda_r(\mu_1+\mu_2)/2\lambda_i} \quad (23)$$

3.4.2. Condition on $\hat{A}f$ to have $\hat{K}>0$

Choice of k1 positive real is necessary and sufficient to have $\mathrm{trace}(\hat{K})>0$

Relation $\det(\hat{K})>0 \Rightarrow k_1^2-k_2\bar{k}_2>0$ must be verified

From (21) it is

$$k_2=-\frac{v_1k_1e^{j\psi}+a_2}{v_1}=-\frac{k_1v_2+a_2}{v_1}$$

Finally the choice must be done with :

$$\boxed{k_1>\frac{-a_2\bar{a}_2}{a_2\bar{v}_2+\bar{a}_2v_2}>0} \quad (24)$$

4. DOMAIN OF PRESCRIBED POLES

Closed loop poles require to be prescribed in a stable domain. Furthermore in the present study they must verify simultaneously (15) and (19) inequalities. In case of complex prescribed poles (18) must also be verified. This relation set yields determination of prescribed pole domain.

4.1. Relations between μ_1 and μ_2

(15) and (19) relations must be simultaneously verified. There are solutions if : $a_{1\inf}<a_{1i\sup}$, which yields into explicit form :

$$\boxed{\begin{array}{l}2\lambda_i^2-(1-r^2).\left(\lambda_i^2+(\mu_1-\lambda_r)(\mu_2-\lambda_r)\right)\\ <2\lambda_i\sqrt{r^2\cdot\left(\lambda_i^2+\left(\frac{\mu_1-\mu_2}{2}\right)^2(1-r^2)\right)}\end{array}}$$

4.2. Complex prescribed poles

Development of calculation leads to 2 unequations which describe the possible pole domain in the plane of μ_i and μ_r

$$\left(\lambda_r-\mu_r-\frac{\lambda_i r}{\sqrt{(r^2-1)}}\right)^2+\mu_i^2-\lambda_i^2\left(1-\frac{r^2}{(r^2-1)}\right)$$

$$<0 \qquad \text{(with C1 boundary circle)}$$

$$\left(\lambda_r-\mu_r+\frac{\lambda_i r}{\sqrt{(r^2-1)}}\right)^2+\mu_i^2-\lambda_i^2\left(1-\frac{r^2}{(r^2-1)}\right)$$

$$>0 \qquad \text{(with C1 boundary circle)}$$

Let D the point set which verifies these two unequations (intersection of disc bounded by C1 and of disc complementary bounded by C2)

It then turns out that E the set of the prescribed pole domain is formed by points verifying :

$\boxed{\mu_i<\lambda_i/\sqrt{1-r^2}}$ and $\boxed{\mu_r<\inf(0,\lambda_r)}$

and which belong :

- to D set

or

- to D complementary set which verifies the unequation :

$$\lambda_i^2(1+r^2)-(1-r^2)\cdot\left((\lambda_r-\mu_r)^2+\mu_i^2\right)<0$$

(with C3 boundary circle)

An illustration of such a domain is given in figure 1 , which is corresponding to the numerical example developed in § 5. In complex plane (μ_i,μ_r), for two cases of pole placement, domain bounds have been drawn from C1 and C2.

5. APPLICATION

The method is applied to an example used by Saif in (Saif, 1989). With his method Saif places effectively desired closed-loop poles, but the K Riccati matrix with Q final matrix is not positive definite. The poles have been chosen in unreachable domain.

With the presented method the possible pole placement domain will be made explicit before assigning the closed loop poles.

Given:

$$A=\begin{bmatrix}0&1&0&1&0\\0&0&1&0&4\\-2&-1&4&2&-2\\1&0&0&0&1\\10&-1&0&4&-2\end{bmatrix}\quad B=\begin{bmatrix}0&0\\0&-1\\1&1\\0&0\\1&0\end{bmatrix}\quad R=\begin{bmatrix}1&0\\0&1\end{bmatrix}$$

Initial eigenvalues of the system are :
$\Lambda(A)=\{-2.5405\pm j3.0409,\ 3.6423\pm j0.7028,\ -0.2037\}$

The purpose is to determine an optimal control from a unit weighting matrix

R and with pole assignment to $\Lambda(Af)=\{-4\pm j2,-2\pm j0.5,-0.2037\}$.

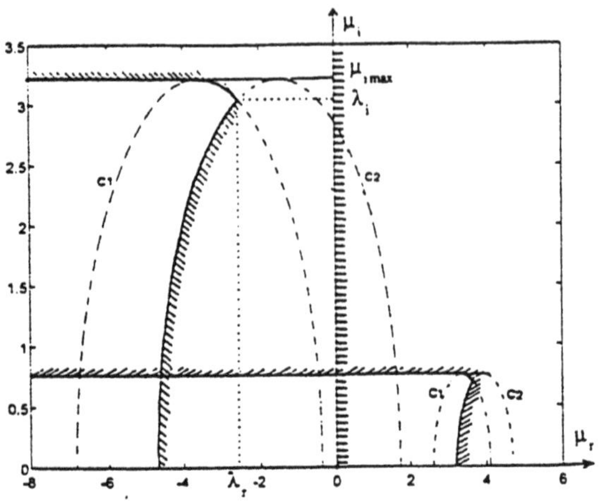

Fig. 1. Prescribed pole domain

a) Start by placing the unstable pair to $-2\pm j0.5$. Domain of possible prescribed poles is given on figure 1. The only constraint to have an optimal control with unit matrix R is to choose $\mu_i < \mu_{imax1} = 0.7614$, the wanted placement is possible.

It remains to the designer to choose a_{1i}, $0.5855 < a_{1i} < 1.0640$; let us choose $a_{1i} = 0.7$. Further to that first placement, K1 matrix is effectively positive definite. The eigenvalues of the resulting matrix $A1 = A-BL1$ are $\Lambda(A1)=\{-2.5405\pm j3.0409, -2\pm j0.5, -0.2037\}$.

b) Next move the pair $-2.5405\pm j3.0409$. The reachable domain is bounded by $\mu_i < \mu_{imax2} = 3.2186$ and by circle arcs (figure 1). The placement to the fixed pole $-4\pm j2$ is not possible. Let us prescribe reachable poles near the initial aim, for example: $-5\pm j2$. To have the requirement $2.532 < a_{1i} < 3.1727$, choose $a_{1i}=2.8$. Further K2 matrix is positive definite. The eigenvalues of the resulting matrix $Af=A1-BL2$ are $\Lambda(Af)=\{-5\pm j2, -2\pm j0.5, -0.2037\}$. Finally feed back matrix $Lf=L1+L2$ and Riccati matrix $Kf=K1+K2$ gives numerically :

$$Lf = \begin{bmatrix} 8.4652 & 2.0028 & 3.8380 & 7.8953 & 3.0317 \\ -16.1912 & -3.8259 & 5.5081 & -5.4778 & -3.6729 \end{bmatrix}$$

$$Kf = \begin{bmatrix} 53.1120 & 11.8785 & -4.3127 & 28.0323 & 12.7779 \\ 11.8785 & 2.8158 & -1.0101 & 6.3404 & 3.0129 \\ -4.3127 & -1.0101 & 4.4980 & 0.8626 & -0.6600 \\ 28.0323 & 6.3404 & 0.8626 & 17.2658 & 7.0326 \\ 12.7779 & 3.0129 & -0.6600 & 7.0326 & 3.6917 \end{bmatrix}$$

Kf is effectively positive semi-definite.

6 CONCLUSION

The presented method allows for a continuous system to determine an optimal control with prescribed poles. This method has the following qualities :

- chosen poles are shifted without modifying the other poles
- for each pole the full placement domain is made explicit.
- computation of Riccati equation solution is not required.

The designer cannot chose initially the weighting matrix Q, but a_{1i} parameter allows a relative degree of freedom to fix this matrix. The final choice must be done according to robust performance.

The choice of the amplitudes of eigenvectors is another freedom degree. These magnitudes must be chosen to lead a good numerical conditioning into W matrix inversion.

7. REFERENCES

Solheim, O. A. (1972). Design of optimal control systems with prescribed eigenvalues. *int. J. Control* , Vol. 15, n°1, pp. 143-160.

Alexandridis, A.T., G.D. Galanos (1987). Optimal pole placement for linear multi-input controllable systems. *IEE Trans* , CAS 34, pp. 1602-1604.

Saif, M. (1989-1). Optimal regulator pole-placement by weight selection,. *Int. J. Control,* vol. 50, n° 1, pp. 399-414.

Saif, M. (1989-2). Optimal modal controller design by entire eigenstructure assignment. *IEE Proc.,* Vol. 136, Pt. D, n° 6, pp341-344

Amin , M. H. (1985). Optimal pole shifting for continuous multivariable linear systems, *Int.. J. Control,* 1985, Vol. 40, n°4, pp. 783-794.

Duplaix , J., G. Enéa, M. Franceschi (1994). Commande optimale sous contrainte modale, *APII-RAIRO*,Vol. 28, n°3, pp. 247-262.

ROBUST CONTROL WITH A TERMINAL STATE CONSTRAINT

A. V. SAVKIN* and I. R. PETERSEN**

* Department of Electrical and Electronic Engineering and Cooperative Research Center for Sensor Signal and Information Processing, University of Melbourne, Parkville, Victoria 3052, Australia, email: savkin@ee.mu.oz.au.
**Department of Electrical Engineering, Australian Defence Force Academy, Campbell, 2600, Australia, email: irp@ee.adfa.oz.au.

Abstract. L'article donne une solution d'un problème de commande robuste avec une constrainte sur l'etat dernier. L'article donne une condition nècessaire et suffisant pour la solution. Cette condition invellope une èquation différentielle Riccati.

Keywords. H^∞ Control, Uncertain Systems, State Constraints

1. INTRODUCTION

This paper considers a new robust control problem referred to as "robust control with a terminal state constraint." Given an uncertain linear system, it is desired to construct an output feedback controller which steers a given initial state to the vicinity of the origin in a specified time. Since the underlying system is uncertain and only output feedback is available, it may be unreasonable to expect a controller to steer a given initial state exactly to the origin in a finite time. For example, in (Petersen *et al.* 1992), it was shown that even in the state feedback case, an uncertain system must satisfy a very strong assumption for this to be possible. Thus, we consider the problem of steering a given initial state into an ellipsoidal bounding set. Furthermore, the size of this bounding set will depend on the size of the initial condition.

The robust control problem with terminal state constraints described above may arise in a number of practical situations. For example, it may be desired to steer a robotic arm from a given initial state to the equilibrium point with a specified degree of accuracy and within a specified time; e.g., see (Canon and Schmitz 1984). Furthermore, in such a situation, the mathematical model describing the robotic arm may be uncertain due to variations in load mass and the truncation of high frequency flexible modes. Thus, it is desirable to find a robust controller to steer the initial state into the terminal bounding set.

The class of uncertain systems considered in this paper are uncertain time-varying linear systems in which the uncertainty is described by an *Integral Quadratic Constraint*; e.g., see (Yakubovich 1973, Savkin and Petersen 1995, Savkin and Petersen 1994, Megretski 1993, Boyd *et al.* 1994). This class of uncertain systems originated in the work of Yakubovich and is a particularly rich uncertainty class allowing for nonlinear, time-varying, dynamic uncertainties. For this class of uncertain systems, we define the robust control problem with a terminal state constraint described above. For a given uncertain system, if this problem has a solution, the system is said to be *robustly controllable* with the given terminal state constraint. The main result of the paper gives a necessary and sufficient condition for an uncertain linear system to be robustly controllable with a given terminal state constraint. This condition is given in terms of the existence of solutions to a pair of parameter dependent Riccati differential equations of the game type. This result is established by converting the problem into a finite time H^∞ control problem which can be solved using the results of (Khargonekar *et al.* 1991). A key step in converting the robust control problem into a corresponding H^∞ control problem is the use of a certain "S-procedure Theorem" for two quadratic forms (also referred to as Finsler's Theorem); see, (Yakubovich 1971, Uhlig 1979).

The remainder of the paper proceeds as follows. In Section 2, we introduce the class of uncertain systems under consideration. Also, in this section we define our notion of robust control with a terminal state constraint. In Section 3 we present the main result of the paper which is a necessary and sufficient condition for robust controllability with respect to a given state constraint.

2. PROBLEM STATEMENT

Consider an uncertain system defined on the finite-time interval $[0,T]$ as follows:

$$\begin{aligned} \dot{x}(t) &= A(t)x(t) + B_1(t)w(t) + B_2(t)u(t); \\ z(t) &= K(t)x(t) + G(t)u(t); \\ y(t) &= C(t)x(t) + v(t) \end{aligned} \quad (1)$$

where $x(t) \in \mathbf{R}^n$ is the *state*, $w(t) \in \mathbf{R}^p$ and $v(t) \in \mathbf{R}^k$ are the *uncertainty inputs*, $u(t) \in \mathbf{R}^m$ is the *control input*, $z(t) \in \mathbf{R}^q$ is the *uncertainty output*, $y(t) \in \mathbf{R}^k$ is the *measured output*, $A(\cdot), B_1(\cdot), B_2(\cdot), K(\cdot), G(\cdot)$ and

$C(\cdot)$ are bounded piecewise continuous matrix functions.

System Uncertainty The uncertainty in the above system is described by the equation

$$\begin{bmatrix} w(t) \\ v(t) \end{bmatrix} = \phi(t, x(\cdot)|_0^t, u(\cdot)|_0^t) \quad (2)$$

where the following Integral Quadratic Constraint is satisfied:

Definition 1 (Integral Quadratic Constraint) Let $D = D' > 0$ be a given positive definite matrix, $Q(\cdot) = Q(\cdot)'$ and $R(\cdot) = R(\cdot)'$ be given bounded piecewise continuous weighting matrix functions associated with the system (1), (2). An uncertainty of the form (2) is an admissible uncertainty for the system (1) if the following conditions hold: Given any control input $u(\cdot) \in \mathbf{L}_2[0,T]$, then any corresponding solution to equations (1), (2) is defined on $[0,T]$ and

$$\begin{aligned} &\int_0^T (w(t)'Q(t)w(t) + v(t)'R(t)v(t))dt \\ &\leq \int_0^T \|z(t)\|^2 dt + x(0)'Dx(0). \end{aligned} \quad (3)$$

Here $\|\cdot\|$ denotes the standard Euclidean norm. The class of all such admissible uncertainty inputs is denoted Ξ.

Note that the uncertainty description given above allows for $w(t)$ and $v(t)$ to depend dynamically on $x(t)$ and $u(t)$. Also, it is clear that the uncertain system (1), (2) allows for uncertainty satisfying a norm bound condition. That is, the uncertain system would be described as follows:

$$\begin{aligned} \dot{x}(t) &= [A(t) + B_1(t)\Delta(t)K_1(t)]x(t) \\ &\quad +[B_2(t) + B_1(t)\Delta(t)G_1(t)]u(t); \\ y(t) &= [C(t) + \Delta(t)K_2(t)]x(t) \\ &\quad +\Delta(t)G_2(t)u(t); \quad \|\Delta(t)\| \leq 1 \end{aligned}$$

where $\Delta(t)$ is a time-varying uncertainty matrix and $\|\cdot\|$ denotes the standard induced matrix norm; e.g., see (Khargonekar *et al.* 1990). To verify that such uncertainty is admissible for the uncertain system (2.1), (2.2), let $w(t) = \Delta(t)[K_1(t)x(t) + G_1(t)u(t)]$, $v(t) = \Delta(t)[K_2(t)x(t) + G_2(t)u(t)]$, $K(t)' = \begin{bmatrix} K_1(t)' & K_2(t)' \end{bmatrix}'$, $G(t)' = \begin{bmatrix} G_1(t)' & G_2(t)' \end{bmatrix}'$ where $\|\Delta(t)\| \leq 1$ for all $t \geq 0$. Then $w(\cdot)$ and $v(\cdot)$ satisfy condition (3) with $Q(\cdot) \equiv I$ and $R(\cdot) \equiv I$ and with any $D > 0$.

We now consider a problem of robust control with a terminal state constraint for the uncertain system (1), (2). The class of controllers to be considered are finite dimensional time-varying linear output feedback controllers of the form

$$\begin{aligned} \dot{x}_c(t) &= A_c(t)x_c(t) + B_c(t)y(t), \quad x_c(0) = 0, \\ u(t) &= C_c(t)x_c(t) + D_c(t)y(t) \end{aligned} \quad (4)$$

where $A_c(\cdot), B_c(\cdot), C_c(\cdot)$ and $D_c(\cdot)$ are bounded piecewise continuous matrix functions defined on $[0,T]$. Note that the dimension of the controller state x_c may be arbitrary.

Definition 2 Let $P_0 = P_0' > 0$ and $P_T = P_T' > 0$ be given positive definite matrices associated with the system (1), (2). The controller (4) is said to be a *robust controller with respect to the terminal state constraint defined by the matrices P_0 and P_T*, if the following conditions hold:

(i) There exists a constant $c > 0$ such that

$$\int_0^T (\|x(t)\|^2 + \|u(t)\|^2)dt \leq c\|x(0)\|^2 \quad (5)$$

for all solutions to the closed loop uncertain system (1), (2), (4) with admissible uncertainty inputs $[w(\cdot), v(\cdot)] \in \Xi$.

(ii) The closed loop uncertain system (1), (2), (4) satisfies the state constraint

$$\begin{aligned} &\sup_{x(0)\in\mathbf{R}^n\setminus\{0\},[w(\cdot),v(\cdot)]\in\Xi} \frac{x(T)'P_Tx(T)}{x(0)'P_0x(0)} \\ &< 1. \end{aligned} \quad (6)$$

An uncertain system (1), (2) which admits a robust controller (4) with a terminal state constraint defined by the matrices P_0 and P_T is said to be *robustly controllable* with respect to this terminal state constraint.

3. THE MAIN RESULT

In this section, we present the main result of this paper. Our main result requires that the system (1) satisfy a number of additional assumptions.

Assumptions

3.1 The matrix functions $K(\cdot)$ and $G(\cdot)$ satisfy the condition $K(\cdot)'G(\cdot) \equiv 0$.

3.2 The matrix function $E(\cdot) \triangleq G(\cdot)'G(\cdot)$ satisfies the following condition: There exists a constant $\delta > 0$ such that $E(t) \geq \delta I$ for all $t \in [0,T]$.

3.3 There exists a constant $\delta > 0$ such that $Q(t) \geq \delta I$ and $R(t) \geq \delta I$ for all $t \in [0,T]$.

3.4 The matrix function $B_1(\cdot)$ satisfies the condition $\int_0^T \|B_1(t)\|dt > 0$.

Remarks Assumptions 3.1-3.4 are technical assumptions required to ensure that the underlying H^∞ control problem is "non-singular"; e.g., see (Khargonekar *et al.* 1991).

We will consider the following pair of Riccati differential equations

$$\begin{aligned} -\dot{X} &= A'X + XA + K'K \\ &\quad +X\left(B_1Q^{-1}B_1' - B_2E^{-1}B_2'\right)X, \quad (7) \\ \dot{Y} &= AY + YA' + B_1Q^{-1}B_1' \\ &\quad +Y\left(K'K - C'RC\right)Y. \quad (8) \end{aligned}$$

Theorem 1 Consider the uncertain system (1), (2) and suppose that Assumptions 3.1-3.4 are satisfied. Also, let $P_0 = P_0' > 0$ and $P_T = P_T' > 0$ be given matrices. Then the uncertain system (1), (2) is robustly controllable with respect to the terminal state constraint defined by the matrices P_0 and P_T if and only if there exists a constant $\tau > 0$ such that the following conditions hold:

(i) There exists a solution $X(\cdot)$ to Riccati differential equation (7) defined on the interval $[0,T]$ with $X(T) = \tau P_T$. Furthermore, this solution satisfies $\tau P_0 - D > X(0) > 0$.

(ii) There exists a solution $Y(\cdot)$ to Riccati differential equation (8) defined on the interval $[0,T]$ with $Y(0) = (\tau P_0 - D)^{-1}$. Furthermore, this solution satisfies $Y(t) > 0$ for all $t \in [0,T]$.

(iii) The spectral radius, $\rho(Y(t)X(t))$, satisfies $\rho(Y(t)X(t)) < 1$ for all $t \in [0,T]$.

If conditions (i)-(iii) hold then a controller of the form (4) with coefficient matrices

$$\begin{aligned} A_c(t) &= A(t) + B_2(t)C_c(t) - B_c(t)C(t) \\ &\quad + (B_1(t) - B_c(t))\, Q(t)^{-1}B_1(t)'X(t); \\ B_c(t) &= (I - Y(t)X(t))^{-1}\, Y(t)C(t)'R(t); \\ C_c(t) &= -E(t)^{-1}B_2(t)'X(t); \\ D_c(t) &= 0 \qquad (9) \end{aligned}$$

is a robust controller with respect to the above terminal state constraint.

Before proving this theorem, we first present a lemma referred to as the S-procedure Theorem for two quadratic forms or Finsler's Theorem; e.g., see (Yakubovich 1971) and (Uhlig 1979).

Lemma 1 Let $\mathcal{F}_0(l)$ and $\mathcal{F}_1(l)$ be symmetric continuous quadratic forms defined on a Hilbert space $\mathcal{L}$. Suppose there exists a vector $l_0 \in \mathcal{L}$ such that $\mathcal{F}_1(l_0) > 0$ and also suppose that $\mathcal{F}_0(l) \leq 0$ for all $l \in \mathcal{L}$ such that $\mathcal{F}_1 \geq 0$. Then there exists a constant $\mu \geq 0$ such that

$$\mathcal{F}_0(l) + \mu\mathcal{F}_1(l) \leq 0 \quad \forall l \in \mathcal{L}. \qquad (10)$$

Proof of Theorem 3.1 *(Necessity)* Suppose that there exists a controller of the form (4) such that conditions (5) and (6) hold.. The condition (6) implies that there exists a constant $\delta > 0$ such that

$$(1+\delta)x(T)'P_Tx(T) \leq (1-\delta)x(0)'P_0x(0) \qquad (11)$$

for all solutions to the system (1), (2), (4). Furthermore, inequality (11) together with condition (5) and the positive-definiteness of P_0 implies that there exists a constant $\epsilon > 0$ such that

$$\begin{aligned} &(1+\delta)x(T)'P_Tx(T) + \epsilon\int_0^T (\|x(t)\|^2 + \|u(t)\|^2)dt \\ &\leq x(0)'P_0x(0) \qquad (12) \end{aligned}$$

for all solutions to the closed loop uncertain system (1), (2), (4). Now let $\mathcal{L}$ be the Hilbert space of all vector valued functions $l(\cdot) \triangleq [x(\cdot), u(\cdot), w(\cdot), v(\cdot)] \in \mathbf{L}_2[0,T]$ connected by the equations (1), (4). We also introduce continuous quadratic functionals $\mathcal{F}_0(\cdot)$ and $\mathcal{F}_1(\cdot)$ defined on $\mathcal{L}$ by the equations

$$\begin{aligned} &\mathcal{F}_0(l(\cdot)) \\ &\triangleq (1+\delta)x(T)'P_Tx(T) \\ &\quad +\epsilon\int_0^T (\|x\|^2 + \|u\|^2)dt - x(0)'P_0x(0); \\ &\mathcal{F}_1(l(\cdot)) \\ &\triangleq \int_0^T (\|z\|^2 - w'Qw - v'Rv)dt \\ &\quad +x(0)'Dx(0). \qquad (13) \end{aligned}$$

Since condition (12) holds for all solutions to the system (1), (4) with inputs $w(\cdot)$ and $v(\cdot)$ satisfying condition (3), then we have $\mathcal{F}_0(l(\cdot)) \leq 0$ for all $l(\cdot) \in \mathcal{L}$ such that $\mathcal{F}_1(l(\cdot)) \geq 0$. Since $D > 0$, if we let $w(t) \equiv 0$, $v(t) \equiv 0$ and $x(0) \neq 0$, the corresponding vector function $l(\cdot) \in \mathcal{L}$ is such that $\mathcal{F}_1(l(\cdot)) > 0$. Hence, using Lemma 3.1, it follows that there exists a constant $\mu \geq 0$ such that inequality (10) is satisfied for all $l(\cdot) \in \mathcal{L}$. Also, Assumption 3.4 implies that we can choose an input $w(\cdot)$ such that $B_1(\cdot)w(\cdot) \neq 0$. Then, for the corresponding solution to the system (1), (2) with $x(0) = 0$, we have $\int_0^T (\|x(t)\|^2 + \|u(t)\|^2)dt > 0$. Therefore, the corresponding vector function $l(\cdot) \in \mathcal{L}$ is such that $\mathcal{F}_0(l(\cdot)) > 0$. Referring to inequality (10), this implies that $\mu \neq 0$. However since, $\mu \geq 0$ we can now conclude that $\mu > 0$. Hence, if we define $\tau \triangleq 1/\mu > 0$, it follows from (10) that

$$\mathcal{F}_1(l) + \tau\mathcal{F}_0(l) \leq 0 \quad \forall l \in \mathcal{L}. \qquad (14)$$

Now with $\mathcal{F}_0(\cdot)$ and $\mathcal{F}_1(\cdot)$ defined as in (13), inequality (14) may be rewritten as

$$\begin{aligned} &x(0)'[\tau P_T - D]x(0) \\ &\geq \tau(1+\delta)x(T)'P_Tx(T) \\ &\quad + \int_0^T \left[\begin{array}{c} \tau\epsilon(\|x\|^2 + \|u\|^2) + \|z\|^2 \\ -w'Qw - v'Rv \end{array} \right] dt. \end{aligned}$$

If we let $w(t) \equiv 0$ and $v(t) \equiv 0$ it follows from this inequality that $\tau P_T - D > 0$. Furthermore, this inequality also implies that there exists a constant $\delta_0 > 0$ such that

$$\begin{aligned} &x(0)'[\tau P_T - D]x(0) \\ &\geq (\tau+\delta_0)x(T)'P_Tx(T) \\ &\quad + \int_0^T \left[\begin{array}{c} \delta_0(\|x\|^2 + \|u\|^2) + \|z\|^2 \\ -w'Qw - v'Rv \end{array} \right] dt. \qquad (15) \end{aligned}$$

Now consider the system

$$\dot{x}(t) = A(t)x(t) + \hat{B}_1(t)\hat{w}(t) + B_2(t)u(t);$$

$$\hat{z}(t) = K(t)x(t) + G(t)u(t);$$
$$y(t) = C(t)x(t) + D(t)\hat{w}(t) \qquad (16)$$

where

$$\hat{w}(t) \triangleq \begin{bmatrix} Q(t)^{\frac{1}{2}}w(t) \\ R(t)^{\frac{1}{2}}v(t) \end{bmatrix},$$
$$\hat{B}_1(t) \triangleq \begin{bmatrix} B_1(t)Q(t)^{-\frac{1}{2}} \\ 0 \end{bmatrix},$$
$$D(t) \triangleq \begin{bmatrix} 0 \\ R(t)^{-\frac{1}{2}} \end{bmatrix}.$$

For this system, we let $\hat{w}(\cdot)$ be the disturbance input and let $z(\cdot)$ be the controlled output. Using these definitions, it follows that inequality (15) may be rewritten as

$$\begin{aligned} &x(0)'[\tau P_0 - D]x(0) \\ &\geq (\tau + \delta_0)x(T)'P_Tx(T) \\ &\quad + \int_0^T [\delta_0(\|x\|^2 + \|u\|^2) + \|z\|^2 - \|\hat{w}\|^2]dt. \end{aligned} \qquad (17)$$

This implies that the following condition holds

$$\begin{aligned} &J \triangleq \\ &\sup\left\{\frac{x(T)'P_Tx(T) + \int_0^\infty \|z\|^2dt}{x(0)'[\tau P_0 - D]x(0) + \int_0^\infty \|\hat{w}\|^2dt}\right\} \\ &< 1 \end{aligned} \qquad (18)$$

where the supremum is taken over all $x(0) \in \mathbf{R}^n$ and $\hat{w}(\cdot) \in \mathbf{L}_2[0,\infty)$ such that

$$\|x(0)\|^2 + \int_0^T \|\hat{w}(t)\|^2dt > 0.$$

That is, the controller (4) solves a H^∞ control problem defined by the system (16) and the H^∞ norm bound condition (18), see (Khargonekar *et al.* 1991). Therefore, Theorem 2.3 of (Khargonekar *et al.* 1991) implies that conditions (i)-(iii) are satisfied. This completes the proof of this part of the theorem.

(Sufficiency) Suppose there exists a constant $\tau > 0$ such that conditions (i)-(iii) hold. Then Theorem 2.3 of (Khargonekar *et al.* 1991) implies that the controller (4) with the coefficients (9) solves the H^∞ control problem (18) for the uncertain system (16). It follows from condition (18) that there exists a constant $\delta_0 > 0$ such that the inequality (17) holds for all $[x(\cdot), u(\cdot), \hat{w}(\cdot)] \in \mathbf{L}_2[0,T]$ connected by (16), (4), (9). This fact implies that inequality (15) is satisfied for the closed loop system (1), (4), (9) with any uncertainty inputs $[w(\cdot), v(\cdot)] \in \mathbf{L}_2[0,T]$. Combining inequality (15) with condition (3), we conclude that for any admissible uncertainty inputs $[w(\cdot), v(\cdot)] \in \Xi$, conditions (5) and (6) hold. □

Acknowledgement This work was supported by the Australian Research Council.

REFERENCES

Boyd, S., L. El Ghaoui, E. Feron and V. Balakrishnan (1994). *Linear Matrix Inequalities in System and Control Theory*. SIAM. Philadelphia, PA.

Canon, R. and E. Schmitz (1984). 'Initial experiments on the endpoint control of a flexible one-link robot'. *International Journal of Robotics* **3**(3), 62–67.

Khargonekar, P. P., I. R. Petersen and K. Zhou (1990). 'Robust stabilization of uncertain systems and H^∞ optimal control'. *IEEE Transactions on Automatic Control* **AC-35**(3), 356–361.

Khargonekar, P. P., K. M. Nagpal and K. R. Poolla (1991). 'H_∞ control with transients'. *SIAM Journal on Control and Optimization* **29**(6), 1373–1393.

Megretski, A. (1993). Power distribution approach in robust control. In 'Proceedings of the 12th IFAC World Congress'. Sydney, Australia. pp. 399–402.

Petersen, I. R., M. Corless and E. P. Ryan (1992). A necessary and sufficient condition for quadratic finite time feedback controllability. In M. Mansour, S. Balemi and W. Truöl (Eds.). 'Robustness of Dynamic Systems with Parameter Uncertainties: Proceedings of the International Workshop on Robust Control'. Birkhäuser. Ascona, Switzerland. pp. 165–173.

Savkin, A. and I.R. Petersen (1994). Time-varying problems of H^∞ control and absolute stabilizability of uncertain systems. In 'Proceedings of the 1994 American Control Conference'. Baltimore, MD.

Savkin, A. V. and I. R. Petersen (1995). Nonlinear versus linear control in the absolute stabilizability of uncertain linear systems with structured uncertainty. Accepted for publications in the IEEE Transactions on Automatic Control.

Uhlig, F. (1979). 'A recurring theorem about pairs of quadratic forms and extensions: A survey'. *Linear Algebra and its Applications* **25**, 219–237.

Yakubovich, V. A. (1971). 'S-procedure in nonlinear control theory'. *Vestnik Leningrad University, Series 1* **13**(1), 62–77.

Yakubovich, V. A. (1973). 'Minimization of quadratic functionals under the quadratic constraints and the necessity of a frequency condition in the quadratic criterion for absolute stability of nonlinear control systems'. *Soviet Mathematics Doklady* **14**, 593–597.

MULTIVARIABLE TWO DEGREE OF FREEDOM CONTROL METHODOLOGY: THE R2M2 APPROACH

E. PREMPAIN* and B. BERGEON*

**Laboratoire d'Automatique et de Productique- Université Bordeaux I. 351, Cours de la Libération - 33405 Talence Cedex - FRANCE. Fax: (33).56.84.66.44, E-mail: prempain@lap.u-bordeaux.fr*

Abstract. We propose a methodology for designing robust two-degree-of-freedom-controllers. Our method is based on the Robust Multiple Refrence Model scheme which consists in the separation of nominal tracking and regulation objectives. Such a separation leads to two reference models: a nominal model which sets the desired nominal closed loop and a robust reference model which sets the robust performances specifications. Robust tracking is our major concern in this paper. In order to design the nominal reference model we propose an H_∞/H_2 matching approach, while we deal with the robust performance problem by using μ-synthesis techniques.

Key Words. Two-degree-of-freedom-controller; robust performance; robust stability; H_2,H_∞ optimization; structured singular values; μ -synthesis; model matching.

1. INTRODUCTION

During this last decade, usefull robust synthesis techniques appeared, among the most developed of which we can mention H_∞ synthesis which allows to compute a feedback controller ensuring nominal performances under unstructured robust stability constraint (e.g: mixed sensitivity problem); and the μ-synthesis which deals with the robust structured stability and also with performances problems (Doyle 1982a; Safonov 1982). Nonetheless, these new techniques were only applied to the feedback control design and little research work were led to extend them to Two Degree of Freedom controllers (T.D.F). Recently, Limebeer (1993) has proposed to extend H_∞ synthesis to T.D.F configuration. His methods are based on the McFarlane and Glover (1992) loop shaping design procedure, where uncertainties are modelled as H_∞-bounded perturbations in the normalized coprime factors of the plant. However, it seems difficult to extend them to other kind of uncertainties. More, it is well known that H_∞ formalism is not always suitable to deal with robust performance criterion. (Doyle, 82b).

The Robust Multiple Models Reference (R2M2) allows us to take into account a tracking criterion from the feedback synthesis phase. (Bergeon *et al.*, 1990; Bergeon, 1994). In fact, for the tracking point of view, R2M2 synthesis is equivalent to a robust Model Matching which takes into account nominal specifications. We focus our attention on the pure robust tracking problem (we consider only H_∞-bounded perturbations of the nominal plant.). That is, designing robust TDF controller confering to the perturbed closed loop behavior to not move off to a reasonable dynamic reference chosen by the designer (e.g. the nominal reference model).

The R2M2 methodology can be summarized as follows: first, we propose a model matching procedure to design a nominal controller which sets the desired nominal behavior. In a second step the problem is posed as a robust performance problem with an output weigthed error function as the performance criterion. The error function is the difference between the reasonable input/output transfer designed in the first step and the corresponding perturbed input/output transfer. We propose to design the feedback controller for stability and performances robustness by μ-synthesis.

The paper is organized as follows: we present in section 2 the robust tracking problem; our R2M2 scheme is presented in section 3. The design methodology is described in section 4., and finally, section 5 handles a numerical example in order to show the efficiency of the proposed method.

Notation

For all transfer functions the s Laplace variable is omitted.

$\mathbf{T}_{y,r}$ denotes the closed loop transfer matrix between the input r and the output y.

Let $\mathbf{P}$ be a partionned transfer matrix:

$$\mathbf{P}=\left(\begin{array}{c|c}\mathbf{P}_{11} & \mathbf{P}_{12}\\ \hline \mathbf{P}_{21} & \mathbf{P}_{22}\end{array}\right).$$

The lower Linear Fractionnal Transformation (L.F.T) of $\mathbf{P}$ and $\mathbf{K}$ is defined as:

$F_l(\mathbf{P},\mathbf{K}) := \mathbf{P}_{11}+\mathbf{P}_{12}\mathbf{K}(\mathbf{I}-\mathbf{P}_{22}\mathbf{K})^{-1}\mathbf{P}_{21}$,

H_∞: denotes the set of real rationnal matrices analytic in the R. H. P.

$\|\ \|_\infty$, (resp $\|\ \|_2$) is the H_∞, (respH_2) norm of transfer matrix in H_∞.

Π: is the set of all possible models of the unknown plant $\mathbf{G}$, and $\mathbf{G}_0$ denotes the nominal plant.

The structured singular value of a complex matrix $\mathbf{M}$ for a given block-diagonal structure $\underline{\Delta}=blocdiag(\Delta_1,\ldots,\Delta_n)$ is defined as:

$$\mu_{\underline{\Delta}}(\mathbf{M})=\frac{1}{\min\limits_{\Delta\in\underline{\Delta}}(\overline{\sigma}(\Delta)/\ \det(\mathbf{I}+\mathbf{M}\Delta)=0)},$$

$\mu_{\underline{\Delta}}(\mathbf{M})=0$ if no Δ in $\underline{\Delta}$ makes $\mathbf{I}+\mathbf{M}\Delta$ singular.

2. PROBLEM STATEMENT

2.1. *The pure robust tracking problem*

The pure robust tracking problem is considered. Interest is focused on the design of a robust controller $\mathbf{C}(s)$=[$\mathbf{F}(s)$,-$\mathbf{K}(s)$] (see Fig.1) so that for each plant $\mathbf{G}$ belonging to Π:

the closed loop is internally stable, (1)

and for some given positive function $\alpha(\omega)$, $\beta(\omega)$ and a transfer matrix $\mathbf{M}$ in H_∞,

$$\left\|\alpha((\mathbf{I}+\mathbf{GK})^{-1}\mathbf{GF}-\mathbf{M})\right\|_\infty \le 1, \tag{2}$$

$$(\mathbf{I}+\mathbf{G}_0\mathbf{K})^{-1}\mathbf{G}_0\mathbf{F}=\mathbf{M}, \tag{3}$$

$$\left\|\beta(\mathbf{I}+\mathbf{KG}_0)^{-1}\mathbf{F}\right\|_{\infty,2}\le 1. \tag{4}$$

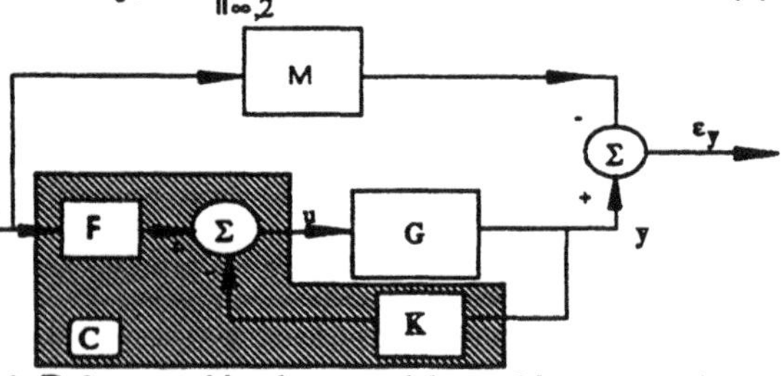

Fig. 1. Robust tracking by a model matching approach

The transfer matrix $\mathbf{M}$ stands for a target model which has good time responses. $\alpha(\omega)$ is a weight on the error transfer $\mathbf{T}_{\varepsilon_y,ref} := (\mathbf{I}+\mathbf{GK})^{-1}\mathbf{GF}-\mathbf{M}$.

(3) and (4) are nominal constraints:

The constraint (3) prescribes that the nominal closed loop transfer $\mathbf{T}_{y,ref}$ to be equal to the reference model $\mathbf{M}$.

The wheight function $\beta(\omega)$ in (4) is used to penalize the nominal control signal.

2.2. *Some difficulties to pose the pure tracking problem on classical T.D.F configuration*

The choice of a realistic reference model $\mathbf{M}$ is not a priori really easy. In fact, if $\mathbf{M}$ is chosen too optimistic, it leads to take $\alpha(\omega)$ too large at frequencies of interest (in the bandwidth). In such case a model matching approach loses all interest. Such incompatibilities may come from several reasons such as:

-the different structure of $\mathbf{G}$ and $\mathbf{M}$ (Dion, 1983) (for example $\mathbf{G}$ is a non minimum phase plant whereas $\mathbf{M}$ is minimum phase target);

- but above all, the admissibility of the control input.

Moreover, it is impossible to take into account the constraints (3) and (4) directly in a synthesis based on an augmented plant deduced from the Fig. 1.

The aims of this paper is to show how to solve the pure robust tracking problem stated in 2.1., and how to find a consistent reference model $\mathbf{M}$ for a given problem.

3. THE R2M2 SCHEME

3.1. *Preliminary*

We propose to characterize all stabilizing controllers of $\mathbf{G}_0$ [$\mathbf{F}$,$\mathbf{K}$] which satisfy (3) and (4).

Let

$$\mathbf{G}_0=\mathbf{A}_0^{-1}\mathbf{B}_0=\tilde{\mathbf{B}}_0\tilde{\mathbf{A}}_0^{-1}, \tag{5}$$

any left and right fractional coprime representations of $\mathbf{G}_0$.

and the corresponding Bezout identities are:

$$\begin{bmatrix}\tilde{\mathbf{V}}_0 & \tilde{\mathbf{U}}_0\\ -\mathbf{B}_0 & \mathbf{A}_0\end{bmatrix}\begin{bmatrix}\tilde{\mathbf{A}}_0 & -\mathbf{U}_0\\ \tilde{\mathbf{B}}_0 & \mathbf{V}_0\end{bmatrix}=\begin{pmatrix}\mathbf{I} & \mathbf{0}\\ \mathbf{0} & \mathbf{I}\end{pmatrix}. \tag{6}$$

Then all stabilizing controllers of $\mathbf{G}_0$ are given by:

$$\mathbf{K}=(\mathbf{U}_0+\tilde{\mathbf{A}}_0\mathbf{Q}_\mathbf{K})(\mathbf{V}_0-\tilde{\mathbf{B}}_0\mathbf{Q}_\mathbf{K})^{-1}, \tag{7}$$

$$\mathbf{F}=-(\tilde{\mathbf{A}}_0+\mathbf{K}\tilde{\mathbf{B}}_0)\mathbf{Q}_\mathbf{F}, \tag{8}$$

where

$\mathbf{Q}_\mathbf{K}\in H_\infty$ $\mathbf{Q}_\mathbf{F}\in H_\infty$, (Youla, 1985; Limebeer 1993).

Substituting (7) and (8) on the left side of (3) comes

$$(\mathbf{I}+\mathbf{G}_0\mathbf{K})^{-1}\mathbf{G}_0\mathbf{F}=-\tilde{\mathbf{B}}_0\mathbf{Q}_\mathbf{F}, \tag{9}$$

and in (4) it comes

$$(\mathbf{I}+\mathbf{KG}_0)^{-1}\mathbf{F}=-\tilde{\mathbf{A}}_0\mathbf{Q}_\mathbf{F}. \tag{10}$$

Now, let us suppose there exists $[\mathbf{F}_0,\mathbf{K}_0]$ so that (3) and (4) are satisfied, then there exists by identifying (9) and (3) a stable parameter $\mathbf{Q}_{\mathbf{F}_\bullet}$ so that

$$-\tilde{\mathbf{B}}_0\mathbf{Q}_{\mathbf{F}_\bullet}=\mathbf{M}, \tag{11}$$

and

$$\left\|\beta\tilde{A}_0 Q_{F_0}\right\|_{\infty,2} \le 1. \quad (12)$$

Therefore, if a solution Q_{F_0} exists for (11) and (12) then all sabilizing controllers of G_0 such that (3) and (4) are satisfied are given by:

$$F = -\tilde{A}_0 Q_{F_0} + KM, \quad (13)$$

and

$$K = (U_0 + \tilde{A}_0 Q_K)(V_0 - \tilde{B}_0 Q_K)^{-1}, \quad (14)$$

where

$$Q_K \in H_\infty, \quad (15)$$

(see Fig. 2).

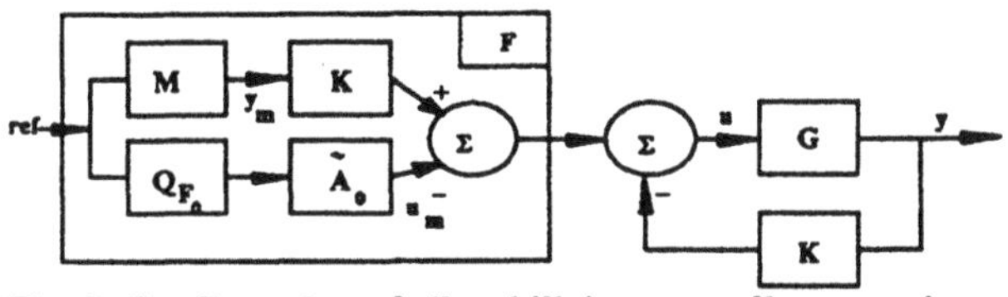

Fig. 2. Configuration of all stabilizing controllers ensuring (3) and (4).

Remarks:

If **K** stabilizes every **G** in Π then a controller [**F**,**K**] with **F** given by (13) statisfies also (3) and (4). Therefore, the original T.D.F problem exposed by (1), (2), (3), and (4) is reduced to an one degree of freedom problem, since the constraints (3) and (4) fix the Q_F parameter.

3.2. *The R2M2 scheme*

We can't make use for the configuration illustrated in Fig. 2 for doing a synthesis of **K**, because it appears both in feedback and feedforward parts. Though, an altnernate representation of this scheme is given in Fig. 3.

This scheme (R2M2) was introduced first by (Bourles, 1982) in a L.Q.G context and has been well studied by Bergeon *et al.* (1990), Bergeon, (1994) and Ygorra, (1991) in the SISO case.

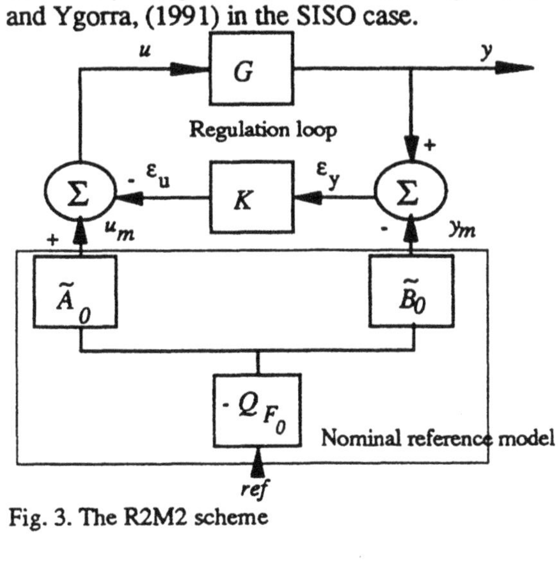

Fig. 3. The R2M2 scheme

From Fig. 3, the closed loop transfer to the reference input (*ref*) from the output error ε_y is given by:

$$T_{\varepsilon_y,ref} = S(G_0 - G)\tilde{A}_0 Q_{F_0}, \quad (16)$$

where

$$S = (I + GK)^{-1}, \quad (17)$$

is the output sensibility matrix.

Proposition 1

If there exists Q_F solution of (11) and (12) and a regulator **K** so that for all **G** belonging to Π :

the closed loop is internally stable (18)

and,

$$\left\|T_{\varepsilon_y,ref}\,\alpha\right\|_\infty \le 1, \quad (19)$$

then the controller [**F**,**K**] with **F** given by (13) is a solution to the pure robust tracking problem defined by (1), (2), (3).and (4).

Proof

If there exists Q_F so that (11) and (13) are satisfied then (3) and (4) are also satisfied for all [**F**, **K**] with **F** given by (13).

Now it is sufficient to show the equality of the closed loop transfer functions in the scheme (1) and (3).

From Fig. 3 we have

$$T_{\varepsilon_y,ref} = S(G_0 - G)\tilde{A}_0 Q_{F_0}, \quad (20)$$

$$= -SG\tilde{A}_0 Q_{F_0} + S\tilde{B}_0 Q_{F_0}, \quad (21)$$

$$= -SG\tilde{A}_0 Q_{F_0} + (I - T)\tilde{B}_0 Q_{F_0}, \quad (22)$$

by using (11)

$$T_{\varepsilon_y,ref} = SG(-\tilde{A}_0 - K\tilde{B}_0)Q_{F_0} - M, \quad (23)$$

and by substituting (13) into (23) it comes

$$T_{\varepsilon_y,ref} = SGF - M. \quad (24)$$

4. CONTROLLER SYNTHESIS

4.1. *Synthesis of the nominal reference model*

The goal of this section is to compute a stable Q_{F_0} parameter so that the nominal transfers, whoses corresponding to the nominal tracking dynamics:

$$T_{y_m,ref} = -\tilde{B}_0 Q_{F_0}, \quad (25)$$

and

$$T_{u_m,ref} = -\tilde{A}_0 Q_{F_0}, \quad (26)$$

give reasonable time responses (see Fig. 3).

As in Limebeer (1993) this problem can be viewed as a model matching one in the following way:

For a fixed M_0, compute a stable parameter Q_{F_0} so that:

$$-\tilde{B}_0 Q_{F0} \cong M_0, \quad (27)$$

under the nominal control admissibility constraint:

$$\left\|\tilde{A}_0 Q_{F_0} \beta\right\|_{\infty,2} < 1, \tag{28}$$

and, where M_0 is a target model chosen by the designer in order to introduce time domain specifications.

To handle both (27) and (28), a model matching approach (H_∞ or H_2) can be performed on the following augmented plant:

$$P = \left[\begin{array}{c|c} M_0 W_1 & \tilde{B}_0 W_1 \\ 0 & \tilde{A}_0 \beta \\ \hline I & 0 \end{array}\right], \tag{29}$$

The computation of Q_{F_0} may be done by the Glover-Doyle algorithms (Doyle, *et al.* 1988; Glover, *et al.* 1989). When W_1 and β are constant weights then

$$\mathrm{order}(Q_{F_0}) < \mathrm{order}(M_0) + \mathrm{order}(G_0). \tag{30}$$

The synthesis can be performed on the following iterative procedure:

a) Select the target model M_0.

b) Compute the time responses of (25) and (26). If there are acceptable then go to step (c), else ajust W_1 or β, or choose an another target model and retry step (a).

(c) Let $M = -\tilde{B}_0 Q_{F_0}$ and stop.

Remarks:

Q_{F_0} is always stable since Glover Doyle's algorithms ensures internal stability (our problem is like an open loop synthesis one).
The reference model M considered is $M = -\tilde{B}_0 Q_{F_0}$ and not M_0.
We can apply directly Glover-Doyle's algorithm if the weights are stable and have no zero on the $j\omega$ axis.

4.2. *Synthesis of the robust regulator* **K**

Among all the controllers ensuring (3) and (4), the goal of this section is to find a special regulator **K** so that the performance objective (31) is satisfied for every **G** in Π.

$$\left\|T_{\varepsilon_y, ref}\,\alpha\right\|_\infty \le 1, \tag{31}$$

We deal with this kind of problem by μ-synthesis techniques (Doyle, 1985).
Suppose whithout loss of generality that

$$\Pi = \{G \mid G = G_0(I + lm.\Delta_G), \|\Delta_G\|_\infty \le 1\} \tag{32}$$

which corresponds to multplicative input uncertainty.

The weight lm is chosen such that $\|\Delta_G\|_\infty \le 1$ generates the familly of possible plants to be considered.
Now, consider the following augmented plant P_1 (see Fig. 4) so that:

$$\begin{bmatrix} y_G \\ \alpha\varepsilon_y \\ \varepsilon_y \end{bmatrix} = P_1 \begin{bmatrix} u_G \\ ref \\ \varepsilon_u \end{bmatrix}, \tag{33}$$

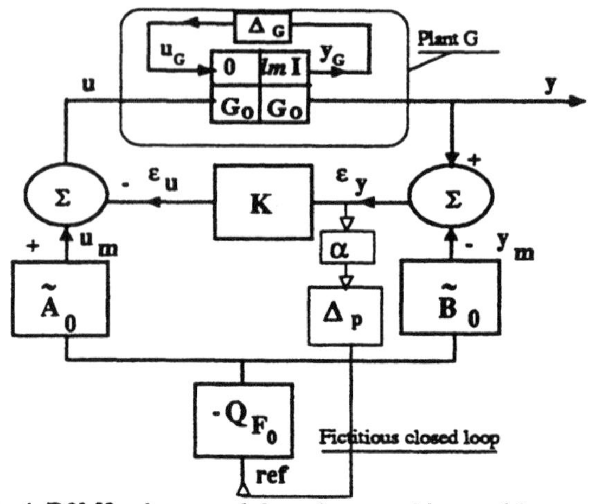

Fig.4. R2M2 scheme and the robust tracking problem

Where Δ_P is a fictitious uncertainty block introduced in Fig.4 in order to incorporate the performance objective (31).

From the Fig. 4. it is easy to show that:

$$P_1 = \left[\begin{array}{cc|c} 0 & -lm\tilde{A}_0 Q_{F_0} & -lmI \\ \alpha G_0 & 0 & -\alpha G_0 \\ \hline G_0 & 0 & -G_0 \end{array}\right]. \tag{34}$$

Proposition 2

If $F_l(P_1, K)$ is internally stable for all of **G** in Π and if $\mu_{\underline{\Delta}}(F_l(P_1,K)(j\omega)) < 1 \quad \forall\omega$, for the following structure $\underline{\Delta} = blocdiag(\underline{\Delta}_G, \underline{\Delta}_P)$ then,

$$\left\|\alpha T_{\varepsilon_y, ref}\right\|_\infty \le 1,$$

and the controller [**F**,**K**] whith **F** defined by (13) is a solution to the robust tracking problem defined by (1), (2), (3), and (4).

Proof

By the main loop theorem (Packard, 1993; Doyle, 1982b) and from proposition 1.

The synthesis of such a **K** may be performed by the Doyle, (1985) *D-K* iteration procedure, implemented in Matlab (Balas, *et al.* 1993), on the augmented system P_1.

Remarks:
It follows from Fig. 4 that the scaled closed loop system is:

$$DF_l(\mathbf{P}_1,\mathbf{K})D^{-1} = \begin{bmatrix} -lm\mathbf{K}\mathbf{S}_0\mathbf{G}_0 & -lm\tilde{\mathbf{A}}_0\mathbf{Q}_{\mathbf{F}0}/d(s) \\ -d(s)\alpha^{-1}\mathbf{S}_0\mathbf{G}_0 & \mathbf{0} \end{bmatrix}, \quad (35)$$

where $\mathbf{S}_0 = (\mathbf{I}+\mathbf{G}_0\mathbf{K})^{-1}$, (36)
denotes the nominal output sensibilty matrix.
Note that the (1,2) block of (35) is independant of **K**. This fact is of great importance because, it allows us to make an a priori choice of the D-scale function. Clearly, the initial D-scale may be chosen to be:

$$|d(\omega)| := \bar{\sigma}(lm\tilde{\mathbf{A}}_0\mathbf{Q}_{\mathbf{F}_0}(j\omega)) \quad \forall\omega. \quad (37)$$

Such a D-scale reduces the number of iterations substancially. $d(s)$ can be approximated by strandard curve fitting algorithms in magnitude, which produce stable minimum phase transfer function.

5. EXAMPLE

The following example has been well studied by several authors. (Morari, 1989; Stogestad, 1989; Limebeer, 1993). The plant is the distillation column whose nominal model :

$$\mathbf{G}_0 = \frac{1}{75s+1}\begin{bmatrix} 0.878 & -0.864 \\ 1.082 & -1.096 \end{bmatrix}, \; s = j\omega, \; \omega \text{ in } rad/\min \quad (38)$$

Unstructured multiplicative uncertainty is supposed as in (32).
In this problem we have:

$$lm(s) = 1.2\frac{1.2s+0.2}{0.5s+1}, \quad (39)$$

which captures unmodelled dynamics as time delays (up to 1 minute) and relative gains (up to 0.2) uncertainty at the plant input (see Morari, 1989 for more details).

5.1 *The nominal model reference synthesis*

The aim of this section is to obtain a reasonable nominal reference model **M**.
The target model selected is (Limebeer, 1993.)

$$\mathbf{M}_0 = \frac{0.12}{s+0.12}\mathbf{I}_2, \quad (40)$$

since it has idealized time responses for our problem.
The weighting functions were chosen to be:

$$W_1 = \frac{35}{s}, \; \beta = \frac{2}{3}. \quad (41)$$

Integral action is used to force the nominal input/output transfer to be close to $\mathbf{M}_0$ at low frequencies, leading to a perfect matching at steady states.
We have decided a H_2 model matching synthesis on the augmented plant given in (29), which leads to $\mathbf{Q}_{\mathbf{F}_0}$ being a 6th order parameter.
Fig. 5 represents the time responses of the nominal control input (u_m), the corresponding output (y_m), and the target model $\mathbf{M}_0$ for the unit step demand equal to $\upsilon(t) = \begin{bmatrix} 1 \\ 0 \end{bmatrix}\Gamma(t)$, where $\Gamma(t)$ is the Heaviside function. Note that $v(t)$ corresponds to the worst (the smallest) gain direction for the nominal plant.

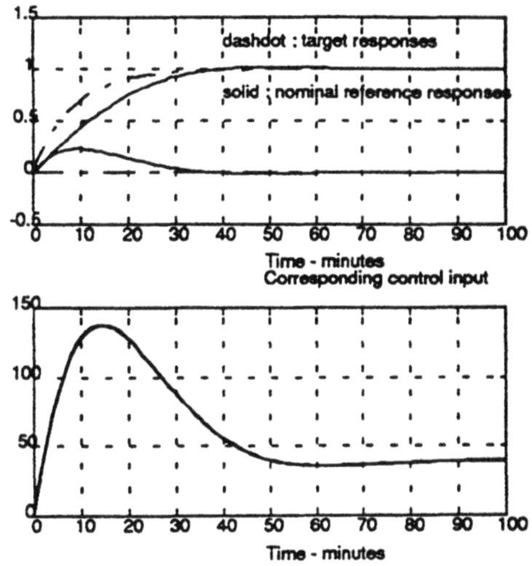

Fig. 5. Time responses: the nominal reference model (solid line), the target model (dashdotted line)

5.2 *Robust regulator synthesis*

The weight of the robust matching problem was decided to be:

$$\alpha\;(s) = 1/4s. \quad (42)$$

According to (37) the pre-scaling computed is:

$$d(s) = 1.3\frac{0.0116s^3 \; +86.7s^2 + \; 23.9s \; + \;\; 0.28}{s^3 + 2.31s^2 \;\; + \; 0.36s \;\;\; 0.0225}. \quad (43)$$

Only one DK is necessary for the pre-scaled plant (the Hinfini step) to obtained a **K** so that (31) is satisfied. Nevertheless, it leads to a very high order (24th order) regulator. So, by standard model reduction procedure (Glover, 1984) it was possible to reduce it greatly (10th order).
We have displayed on Fig. 6 the highest singlular value across the frequency of $\mathbf{T}_{\varepsilon_y,ref}$ for the worst case uncertainty. (see Balas (1993) for more details about the computation of such a perturbation).
The corresponding time responses for the set point tracking input *ref*=v(t) are given in Fig. 7.

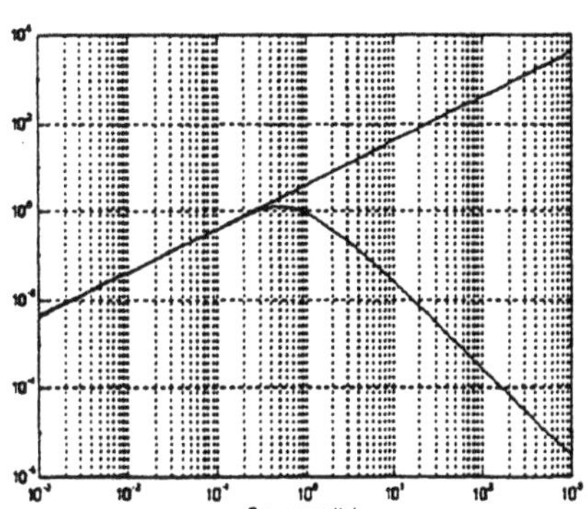

Fig. 6. Frequency responses of: $1/\alpha$ and $\mathbf{T}_{\varepsilon_y,ref}$ in the worst case

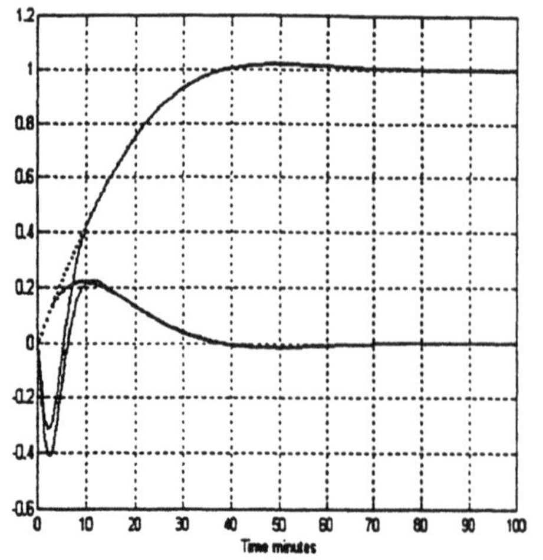

Fig. 7. Time responses for $ref = v(t)$ -worst case perturbed closed loop (solid line), -nominal reference model (dotted line)

6. CONCLUSION

This paper deals with the synthesis of two degree of freedom controllers. It focuses on the pure robust tracking problem, pointing out the difficulty to solve this problem on a standard scheme (involving a mixture of nominal constraints and a robust performance criterion). The Youla parametrization allowed us to justify the R2M2 scheme. The main advantage of our procedure is to reduce the T.D.F problem synthesis formulated in section 2 as an equivalent two single degree of freedom synthesis one of which is a simple precompensation problem.
Moreover, for the robust tracking problem, the special structure of the closed loop permits an initial pre-scaling. The proposed pre-scaling (37) reduces substantially the number of DK iterations in the μ–synthesis procedure. More, this scaling usually avoids the DK-synthesis step as shown in the example.
The resulting controller has nice properties:
- for the nominal plant, the closed loop matches exactly the nominal reference model;
- for all plant models considered the controller ensures $\left\|\alpha((I+GK)^{-1}GF-M)\right\|_{\infty} \leq 1$.

Finally, our simulation results are similar to those obtained by other authors.

7. REFERENCES

Balas, G.J. *et al.* (1993). μ-Analysis and Synthesis Toolbox. *The Math Works Inc.*

Bergeon, B., M. Monsion, S.Ygorra (1990). Frequential synthesis of Robust Multiple Reference Model control. *Proceeding of A.C.C.* vol 3 pp 3041-3043.

Bergeon, B. (1994). Commande monovariable robuste H_∞ et QFT. *A.P.I.I vol* 28. 2.

Bourles, H. (1982). Quantification et amélioration de la robustesse des régulateurs multivariables, avec application au pilotage des avions. *Thèse de doctorat. INPG*, Grenoble, France.

Dion, J.M. (1983). Sur la structure à l'infini des systèmes linéaires. *Thèse d'Etat*, INPG, Grenoble, France.

Doyle, J.C. (1982a). Analysis of feedback system with structured uncertainties *IEEE Proc. Part. D.* Vol 129.

Doyle, J.C., J.E. Wall and G. Stein (1982b). Performance and Robustness Analysis for Structured Uncertainty, " *Proc. IEEE conf. on Decission and Control*, Orlando, FL.

Doyle, J.C. (1985). Structured uncertainty in control system design. *Proc. CDC, Ft. Lauderdale* PP 260-265.

Doyle, J.C., Glover K., Khargonekar P., Francis B.A. (1989). State space solutions to standard H_2 and H_∞ control problems. *IEEE Trans. Aut. Control. Vol* 34.

Glover, K. (1984). All optimal Hankel-norm approximations of Linear Multivariable systems and their H_∞ error bounds. *Intenational Journal of Control*, vol 39 pp 1115-1195.

Glover, K., Doyle J.C. (1988). State Space Formulae for all Stabilizing Controllers that satisfy an H_∞-norm bound, and relations to Risk Sensitivity. *Systems and Control Letters*, 11. North-Holland.

Limebeer, D.J.N., E.M. Kasenally, Perkins J.D. (1993). On the Design of Robust Two Degree of Freedom Controllers. *Automatica*, vol 29, pp 157-168.

McFarlane, D.C. and K. Glover (1992). A loop Shaping Design Procedure Using H_∞ Synthesis. *IEEE Trans. Auto. Control*, Vol. 37, no 6, pp. 759-769.

Morari, M., Zafiriou E. (1989). Robust Process Control. *Prentice-hall.*

Packard, A. and J.C. Doyle (1993). The complex structured singular value, *Automatica*, pp. 71-109, vol. 29.

Skogestad, S., M. Morari and J.C. Doyle (1989). Robust control of ill-conditionned plants: high-purity distillation. *IEEE Trans. on Auto. Control*, 33, 1092-1105.

Ygorra, S., Bergeon and M. Monsion (1991). Design Methodology for Multiple Robust Reference Model. *Proc. ECC'91.*

Youla, D. C., H. A. Jabr and J. J. Bongiorno (1985). A feedback theory of two degree of freedom optimal Wiener-Hopf design. *IEEE Trans. Aut. Control*, 30, 662, 665.

POLES AND ZEROS OF LINEAR SYSTEMS: AN INVARIANT APPROACH

H. BOURLÈS*, and M. FLIESS**

*Electricité de France, Direction des Etudes et Recherches, 1, Avenue du Général de Gaulle, 92141 Clamart, France
**Laboratoire des Signaux et Systèmes, CNRS/Supélec, Plateau de Moulon, 92192 Gif-sur-Yvette, France

Résumé. Les poles et les zéros sont définis, et les relations qu'ils vérifient sont étudiées, dans une "approche invariante". De plus, une nouvelle relation est établie.

Abstract. Poles and zeros are defined, and their relationships are studied, in an "invariant approach". In addition, a new relationship is derived.

Key Words. Linear systems; Poles; Zeros; Hidden modes; Decoupling zeros; Module.

1. INTRODUCTION

Many kinds of poles and zeros were defined in the literature. The most important among them are the transmission poles, the system poles, the blocking zeros, the transmission zeros, the decoupling zeros, the invariant zeros and the system zeros. In the case of linear time-invariant (LTI) systems considered in this paper, the location of the system poles and of the transmission poles respectively govern the internal stability and the input-output stability with zero initial conditions[1]. The location of the blocking zeros plays a crucial role in the "strong stabilization" problem (Youla et al., 1974). The well known "Internal Model Principle" (Wonham, 1984) requires a condition on the transmission zeros. The existence of hidden modes, i.e. of input- or output-decoupling zeros, is related with a lack of controllability or observability. The invariant zeros can be viewed as the system poles of the inverse system (when it exists) and are also important; for example, one of the basic assumptions of the well known "Glover-Doyle algorithm", in H_∞ theory, is that two subsystems have no invariant zeros on the imaginary axis (Glover and Doyle, 1988); see (Kailath, 1980), (Shrader and Sain, 1989), (Wonham, 1984) for other uses of invariant zeros. Finally, the set of system zeros is, roughly speaking, the set of all zeros.

Most of these notions were defined, in the general framework of "polynomial matrix descriptions" (PMD) of linear systems, by Rosenbrock (1970, 1974), apart the notions of invariant zero and of blocking zero, respectively due to McFarlane and Karcanias (1976) and to Ferreira and Bhattacharyya (1977). However, the topic of poles and zeros has been somewhat confused during many years, since certain authors used the same terminology to denote different notions. The situation was clarified by MacFarlane and Karkanias (1976), (Kailath, 1980), and Schrader and Sain (1989).

More recently, some authors defined poles and zeros as beeing modules (instead of numbers): see, e.g., (Conte and Perdon, 1985), (Wyman and Sain, 1987), and related references. In this more sophisticated approach, of course poles and zeros contain more information, and this is useful in exploring, for instance, the question of multivariable pole-zero cancellations (Wyman and Sain, 1981), (Conte and Perdon, 1986).

In this paper, poles and zeros are considered as beeing numbers, like in Rosenbrock's point of view. However, they are defined and studied in an *invariant approach*, i.e. without reference to a peculiar choice of equations. This approach was developed, in the time-varying case, by Fliess (1990, 1991, 1992); see also (Fliess and Glad, 1993), where the nonlinear case is considered as well. For a system defined by a PMD, an important question is whether a notion is unchanged when another equivalent description is chosen[2]. All the notions which are considered here are invariant by nature. Infinite poles and zeros will be considered elsewhere.

[1] If the initial conditions are not zero, the "stability" of all transmission poles, i.e. their location in the open left half plane, is not sufficient to ensure the input-output stability of the system: see (Bourlès, 1994) and § 3.1.4.

[2] We are refering here to Fuhrmann's or Rosenbrock's system equivalence, which were proved to be equivalent (Kailath, 1980).

2. LINEAR SYSTEMS AND DYNAMICS

2.1 A brief overview of module theory

2.1.1. In what follows, $R = \mathbb{R}[\frac{d}{dt}]$ denotes the ring of linear differential polynomials with real coefficients. An element $\pi \in R$ is nothing but a polynomial $\pi(s) \in \mathbb{R}[s]$, where the derivation $\frac{d}{dt}$ has been substituted to the indeterminate s. As is well known, R is a principal ideal domain.

2.1.2. Let $w_1, ..., w_q$ be elements of a R-module M. The column matrix $[w_1, ..., w_q]^T$ and the module spanned by $w_1, ..., w_q$ are respectively written w and [w]. If $v_1, ..., v_r$ are other elements of M, [w, v] denotes the module spanned by $w_1, ..., w_q, v_1, ..., v_r$. All modules considered here are finitely generated R-modules. The properties of modules recalled below are well known (McLane and Birkhoff, 1968).

2.1.3. An element m of a module M is *torsion*, if, and only if (iff) there exists $\pi \in R$, $\pi \neq 0$, such that $\pi\, m = 0$. A *torsion module* only contains torsion elements. The set of all torsion elements of a module is a submodule, called the *torsion submodule*. A module is *free* iff its torsion submodule is trivial, i.e., equal to $\{0\}$[3].

A module M can be written as a direct sum

$$M = T \oplus \Phi \tag{1}$$

where T is its torsion submodule and Φ a free module, which is unique up to isomorphism.

The *rank* of a module M, which is written rk(M), is the rank of the free submodule Φ: it is equal to the cardinality of any basis of Φ. A module is torsion iff its rank is zero.

2.1.4. For every module M, there exists an exact short sequence

$$0 \longrightarrow E \xrightarrow{f} F \xrightarrow{\phi} M \longrightarrow 0 \tag{2}$$

where E and F are free modules. This means that f is a monomorphism, that Im f = Ker ϕ and that ϕ is an epimorphism. As a result, $M = \phi(F)$ is isomorphic to F/Im f = coker f (and we write $M \cong$ coker f). The R-morphism f is called a *presentation* of M. In the presentation (2), let n = rk(E), k = rk(F), and let α_i, $1 \le i \le r$, be the non zero invariant factors of f. In the decomposition (1), one then has $k \ge n$ and

$$T \cong \bigoplus_{i=1}^{r} R/(\alpha_i), \qquad \Phi \cong R^{k-n} \tag{3}$$

where (α_i) denotes the R-ideal generated by α_i. Note that every α_i is a differential polynomial $\alpha_i(\frac{d}{dt})$ and that α_i divides α_{i+1}.

2.1.5 Let $\{\varepsilon_1, ..., \varepsilon_n\}$ and $\{\underline{w}_1, ..., \underline{w}_k\}$ be bases of E and F, respectively. In these bases, f is represented by a matrix $S(\frac{d}{dt})$; S is called a *presentation matrix* of M. The Smith form of S is of dimension k×n and can be written $\Sigma = \begin{bmatrix} \Delta \\ 0 \end{bmatrix}$ where $\Delta = \mathrm{diag}(\alpha_1, ... \alpha_r)$ (with r=n). Set $e_i = f(\varepsilon_i)$, $1 \le i \le n$, so that $e = S(\frac{d}{dt})^T \underline{w}$, and let $w_i = \phi(\underline{w}_i)$, $1 \le i \le k$; then, $M = [w] = [\underline{w}]/[e]$, and one has

$$S(\frac{d}{dt})^T\, w = 0 \tag{4}$$

Definition[4]: Equation (4) is the *equation of the module* M in the chosen bases.

2.2 Linear systems

2.2.1. A linear system is a module (Fliess, 1990)[5]. It may be useful to recall in this context the following algebraic approach to Laplace transform and transfer matrices (Fliess, 1994):

The *transfer vector space* of a system Λ is the tensor product $\hat{\Lambda} = \mathbb{R}(\frac{d}{dt}) \otimes \Lambda$, which is a $\mathbb{R}(\frac{d}{dt})$-vector space, where $\mathbb{R}(\frac{d}{dt})$ is the quotient field of $\mathbb{R}[\frac{d}{dt}]$. The Laplace transform of $\lambda \in \Lambda$ is $\hat{\lambda} = 1 \otimes \lambda$.

2.2.2. Let Λ be a system, and consider its equation (4). According to (Willems 1983, 1991), the solutions of this equation on $\mathbb{R}$ (or on an open interval of $\mathbb{R}$) define the *behavior* of Λ; see also (Blomberg and Ylinen, 1983), (Fliess, 1992).

2.3 Dynamics

2.3.1. A *(linear) dynamics* $\mathcal{D}$ is a system where we distinguish a finite set $\{u_1, ..., u_m\}$ of input variables such that the quotient module $\mathcal{D}/[u]$ is

[3]This characterization of free modules is valid for finitely generated modules over principal domains, where any torsion free module is free.

[4]New definitions (in the framework of our "invariant approach) are pointed out in this manner.

[5]We only consider in this paper finite-dimensional linear systems. The case of linear systems with delays is discussed in (Fliess and Mounier, 1994).

torsion (Fliess, 1990, 1992). The torsion of $\mathcal{D}/[u]$ means that any element in $\mathcal{D}$ can be calculated from u by a linear differential equation (see § 2.3.2 for a concrete interpretation). The input u is *independent* iff the module [u] is free of rank m. We also distinguish a finite set $\{y_1, ..., y_p\}$, called the output. See (Fliess, 1990) for the connection with state-variable representation.

2.3.2 As $\mathcal{D}/[u]$ is torsion, $rk(\mathcal{D})=m$. Assume that $\mathcal{D}$ is the module M in § 2.1.4; by (3), k-n = m, so that there exist n elements $\xi_1,..., \xi_n$ of $\mathcal{D}$ such that $[\xi, u] = \mathcal{D}$; ξ is called a *partial state* of $\mathcal{D}$; ξ is not unique. Consider the equation(4) of $\mathcal{D}$ in appropriate bases. It can be written

$$D(\frac{d}{dt})\ \xi = N(\frac{d}{dt})\ u \tag{5}$$

where $D(\frac{d}{dt}) \in R^{nxn}$ and $N(\frac{d}{dt}) \in R^{nxm}$.

The equation of $\mathcal{D}/[u]$ is $D(\frac{d}{dt})\ \bar{\xi} = 0$, where $\bar{\xi}$ is the column matrix whose components are the images of the elements ξ_i by the canonical epimorphism $\mathcal{D} \to \mathcal{D}/[u]$. As $\mathcal{D}/[u]$ is torsion, D has no zero invariant factors, i.e. its rank over $\mathbb{R}(\frac{d}{dt})$ is n.

As $y_i \in \mathcal{D}$, $1 \le i \le p$, there exist polynomial matrices $Q(\frac{d}{dt}) \in R^{pxn}$ and $W(\frac{d}{dt}) \in R^{pxm}$ such that

$$y = Q(\frac{d}{dt})\ \xi + W(\frac{d}{dt})\ u \tag{6}$$

The equations (5) and (6) constitute a *polynomial matrix description* (PMD) of $\mathcal{D}$.

2.3.3. As $\mathcal{D}/[u]$ is torsion, $\hat{\mathcal{D}} = [\hat{u}]$. Therefore, there exists a rational matrix $G \in \mathbb{R}(\frac{d}{dt})^{pxm}$ such that $\hat{y} = G\ \hat{u}$. In everything that follows, u is assumed to be independant (§ 2.3.1); as a consequence, the matrix G is unique, because $\{\hat{u}_1, ..., \hat{u}_m\}$ is a basis of $\hat{\mathcal{D}}$, and G is called the *transfer matrix* of $\mathcal{D}$ (Fliess, 1994). In the case where $\mathcal{D}$ is described by (5), (6), one obtains

$$G = Q\ D^{-1}\ N + W \tag{7}$$

2.3.4. A system Λ or a dynamics $\mathcal{D}$ are said to be *controllable* iff they are free R-modules (Fliess, 1990). This definition of the controllability was proved in (Fliess, 1992) to be equivalent to the controllability "à la Willems" (Willems, 1991).

Definition: The free module $\mathcal{D}/T \cong \Phi$ is called the *controllable dynamics* of $\mathcal{D}$.

Proposition 1: Assume that the dynamics $\mathcal{D}$ is given by the PMD (5), (6). Then, $\mathcal{D}$ is controllable iff D and N are left-coprime.

Proof : Let L be a greatest common left divisor (GCLD) of D and N, i.e. $D = L\ D^\circ$, $N = L\ N^\circ$, where D° and N° are left-coprime. Equation (5) is equivalent to

$$D^\circ(\frac{d}{dt})\ \xi - N^\circ(\frac{d}{dt})\ u = v$$
$$L(\frac{d}{dt})\ v = 0 \tag{8}$$

Equation (8) is the equation of the torsion submodule T of $\mathcal{D}$. Therefore, T is equal to {0} iff L is unimodular, i.e. iff D and N are left-coprime.

Remark: Let $\bar{\bar{\xi}}$ and $\bar{\bar{u}}$ denote the column matrices whose elements are the images of the elements of ξ and u by the canonical epimorphism $\mathcal{D} \to \mathcal{D}/T$. The equation of $\mathcal{D}/T$ is

$$D^\circ(\frac{d}{dt})\ \bar{\bar{\xi}} = N^\circ(\frac{d}{dt})\ \bar{\bar{u}} \tag{9}$$

Now, $\mathcal{D}/T \cong \Phi$, and $[u] \subset \Phi$, so that $\bar{\bar{u}}$ can be identified with u. Note that $\bar{\bar{u}}$ (or u) and $\bar{\bar{y}}$ are the natural inputs and outputs of $\mathcal{D}/T$; hence, this system is a dynamics, and the above definition is consistent.

2.3 5. The dynamics $\mathcal{D}$ is said to be *observable* iff $\mathcal{D} = [y, u]$ (Fliess, 1990).

Definition: The dynamics [y, u] is called the *observable dynamics* of $\mathcal{D}$.

The proof of the following proposition is similar to the one of Proposition 1:

Proposition 2: Assume that the dynamics $\mathcal{D}$ is given by the PMD (5), (6). Then, $\mathcal{D}$ is observable iff D and Q are right-coprime.

2.3.6. The module $\Phi \cap [y, u]$ (which is unique up to isomorphism) is the observable dynamics of the controllable dynamics of $\mathcal{D}$. This module is also the free submodule of [y, u], i.e. the controllable dynamics of the observable dynamics of $\mathcal{D}$. Therefore, the following definition makes sense:

Definition: $\Phi \cap [y, u]$ is called the *controllable-and-observable dynamics* of $\mathcal{D}$.

The output y clearly is a partial state of [y, u], so that the controllable-and-observable dynamics can be described by an equation

$$D_0(\frac{d}{dt})\ \bar{\bar{y}} = N_0(\frac{d}{dt})\ u \tag{10}$$

(with the notation from § 2.3.4). The matrices D_0 and N_0 are left-coprime, and the transfer matrix of $\mathcal{D}$ then is equal to the irreductible fraction

$$G = D_0^{-1} N_0. \qquad (11)$$

3. POLES AND ZEROS

The following definition will be useful: let M be a module, and $\alpha_1, ..., \alpha_n$ be its non zero invariant factors (i.e. the invariant factors of its torsion submodule). Set $\alpha = \alpha_1 ... \alpha_n$.

Definition: The *invariant roots* of M are the roots of $\alpha(s)$ over $\mathbb{C}$, including multiplicities. Now, consider a matrix $S = S(\frac{d}{dt})$, let $\alpha_1, ..., \alpha_n$ be its non zero invariant factors, and set $\alpha = \alpha_1, ..., \alpha_n$. The *invariant roots* of S are the roots of $\alpha(s)$ over $\mathbb{C}$, including multiplicities. Obviously, the invariant roots of a module M are the invariant roots of any presentation matrix S of M, or of its transposate S^T.

3.1. Poles

3.1.1. Definition: The *system poles* of a dynamics $\mathcal{D}$ are the invariant roots of $\mathcal{D}/[u]$.

If $\mathcal{D}$ is given by the PMD (5), (6) its system poles are the invariant roots of D.

3.1.2. Definition: $\mathcal{D}$ is *internally stable* iff all its system poles belong to the open left half plane (OLHP).

3.1.3. Definition: The *controllable poles* (resp. the *observable poles*) of $\mathcal{D}$ are the system poles of its controllable dynamics (resp. of its observable dynamics).

3.1.4. Definition: $\mathcal{D}$ is *input-output stable* iff all its observable poles belong to the OLHP.

3.1.5. Definition: The *transmission poles* of $\mathcal{D}$ are the system poles of its controllable-and-observable dynamics. They are equal to the invariant roots of the matrix D_0 defined in (10).

Let

$$G^\bullet = \begin{bmatrix} \mathrm{diag}\{\varepsilon_i/\psi_i\} & 0 \\ 0 & 0 \end{bmatrix} \qquad (12)$$

be the Smith-McMillan form of the tranfer matrix G. Let ρ be the rank of G over $\mathbb{R}(\frac{d}{dt})$. By (11), the ε_i, $1 \leq i \leq \rho$, are the invariant factors of N_0, whereas the ψ_i, $1 \leq i \leq \rho$, are the ρ last invariant factors of D_0 in reverse order; the other invariant factors of D_0 are equal to 1. As a result, the transmission poles of $\mathcal{D}$ are the roots over $\mathbb{C}$ of the "transmission pole polynomial" $\psi(s) = \psi_1(s) ... \psi_\rho(s)$.

3.2. Decoupling zeros (or hidden modes)

3.2.1. The *input-decoupling-zeros* (i.d.z.) of $\mathcal{D}$ are the invariant roots of $\mathcal{D}$ (Fliess, 1991). Obviously, $\mathcal{D}$ is controllable iff it has no i.d.z. If $\mathcal{D}$ is defined by the PMD (5), (6), its i.d.z. are the invariant roots of L (see the proof of Proposition 1). They are also the invariant roots of [D N].

3.2.2. Definition: $\mathcal{D}$ is said to be *stabilizable* iff all its i.d.z. belong to the OLHP.

3.2.3. The *output-decoupling-zeros* (o.d.z.) of $\mathcal{D}$ are the invariant roots of $\mathcal{D}/[y, u]$ (Fliess, 1991). The dynamics $\mathcal{D}$ is observable iff it has no o.d.z. If $\mathcal{D}$ is given by the PMD (5), (6), its o.d.z. are the invariant roots of $[D^T \; Q^T]^T$.

3.2.4. Definition: $\mathcal{D}$ is said to be *detectable* iff all its o.d.z. belong to the OLHP.

3.2.5 Definition: The *input-output-decoupling zeros* (i.o.d.z.) are the invariant roots of $T/(T \cap [y, u])$.

3.2.6. The set of *decoupling zeros* (or *hidden modes*) is defined by

$$\{\text{hidden modes}\} = \{\text{i.d.z.}\} + \{\text{o.d.z.}\} - \{\text{i.o.d.z}\}^6 \qquad (13)$$

3.3. Invariant zeros, transmission zeros and blocking zeros

3.3.1 Definition: The *invariant zeros* of $\mathcal{D}$ are the invariant roots of $\mathcal{D}/[y]$.

Assume that $\mathcal{D}$ is given by the PMD (5), (6), and let $\check{\xi}$, $\check{u}$ denote the column matrices whose components are the canonical images of the elements of ξ and u by the canonical epimorphism $\mathcal{D} \to \mathcal{D}/[y]$. The equation of $\mathcal{D}/[y]$ is

$$P(\frac{d}{dt}) \begin{bmatrix} \check{\xi} \\ \check{u} \end{bmatrix} = 0$$

where P is the "system matrix" defined by

$$P = \begin{bmatrix} D & -N \\ Q & W \end{bmatrix} \qquad (14)$$

---

[6]The notation A + B is used here for the "disjoint union" of sets A and B, and C - D for the complement of C in D (when $D \subset C$). Note that definition (13) is consistent, because $\{\text{i.o.d.z.}\} \subset \{\text{i.d.z.}\} \cap \{\text{o.d.z.}\}$.

Hence, the invariant zeros of $\mathcal{D}$ are the invariant roots of P.

3.3.2 Definition: The *transmission zeros* of $\mathcal{D}$ are the invariant zeros of its controllable-and-observable dynamics, or, more precisely, are the invariant roots of $(\Phi \cap [y, u])/(\Phi \cap [y])$.

Consider the equation (10) of $\Phi \cap [y, u]$. The equation of $(\Phi \cap [y, u])/(\Phi \cap [y])$ is

$$N_0(\frac{d}{dt})\, \mathring{u} = 0$$

where the meaning of $\mathring{u}$ is obvious. As a result, the transmission zeros are the invariant roots of N_0.

Now, consider the Smith-McMillan form (12) of the transfer matrix G: the transmission zeros are also the roots over $\mathbb{C}$ of the "transmission zero polynomial" $\varepsilon(s) = \varepsilon_1(s)... \varepsilon_\rho(s)$ (§ 3.1.5).

Definition: $\mathcal{D}$ is said to be *minimum phase* iff all its transmission zeros are in the OLHP.

3.3.3. The *blocking zeros* of $\mathcal{D}$ are the roots over $\mathbb{C}$ of $\varepsilon_1(s)$.

Note that ε is the greatest common divisor (GCD) of the minors of order ρ of N_0, whereas ε_1 is the GCD of the elements of N_0. Obviously,

$$\{\text{blocking zeros}\} \subset \{\text{transmission zeros}\} \quad (15)$$

3.4. Relationships between various poles and zeros

3.4.1 In this Section, the following properties will be useful and are recalled for the convenience of the reader [see, e.g., (MacLane and Birkhoff, 1968)]:

i) Let M, M', M" be modules such that $M'' \subset M' \subset M$. Then

$$M/M' \cong (M/M'')/(M'/M'') \quad (16)$$

ii) Let M, M', M", N', N" be modules such that $M = M' \oplus M''$, $N' \subset M'$, $N'' \subset M''$. Then

$$(M' \oplus M'')/(N' \oplus N'') \cong (M'/N') \oplus (M''/N'') \quad (17)$$

iii) Let T_1 and T_2 be two torsion modules, with $T_2 \subset T_1$; then

$$\{\text{invariant roots of } T_1\} = \{\text{invariant roots of } T_2\} + \{\text{invariant roots of } T_1/T_2\} \quad (18)$$

3.4.2. As [u] is free, one has by (17)

$$\mathcal{D}/[u] = (T \oplus \Phi)/[u] \cong T \oplus \Phi/[u]$$

Therefore,

$$\{\text{system poles}\} = \{\text{i.d.z.}\} + \{\text{controllable poles}\} \quad (19)$$

3.4.3. One has by (17)

$$\mathcal{D}/[y, u] \cong (\mathcal{D}/[u])/([y, u]/[u])$$

so that by (18)

$$\{\text{system poles}\} = \{\text{o.d.z.}\} + \{\text{observable poles}\} \quad (20)$$

3.4.4. One has by (1) and (17)

$$[y, u]/[u] \cong (\Phi \cap [y, u])/[u] \oplus (T \cap [y, u]) \quad (21)$$

Now, by definition of the i.o.d.z' s and by (18)

$$\{\text{invariant roots of } T \cap [y, u]\} = \{\text{i.d.z.}\} - \{\text{i.o.d.z.}\} \quad (22)$$

Hence, by (21)

$$\{\text{observable poles}\} = \{\text{transmission poles}\} + \{\text{i.d.z.}\} - \{\text{i.o.d.z.}\} \quad (23)$$

3.4.5. From (20), (23) and (13), we obtain

$$\{\text{system poles}\} = \{\text{transmission poles}\} + \{\text{hidden modes}\} \quad (24)$$

3.4.6. One has by (17)

$$\mathcal{D}/[y] = (T \oplus \Phi)/[y] \cong T/(T \cap [y]) \oplus \Phi/(\Phi \cap [y]) \quad (25)$$

Now, $(\Phi \cap [y, u])/(\Phi \cap [y]) \subset \Phi/(\Phi \cap [y])$ and $T/(T \cap [y, u]) \subset T/(T \cap [y])$[7], so that

$$\{\text{transmission zeros}\} + \{\text{i.o.d.z.}\} \subset \{\text{invariant zeros}\} \quad (26)$$

Similarly, one can prove the following results:

3.4.7. Assume that $\rho = p$, i.e. G is right-invertible. Then,

$$\{\text{transmission zeros}\} + \{\text{i.d.z.}\} \subset \{\text{invariant zeros}\} \quad (27)$$

Note that in this case, the set of o.d.z. is generally not included in the set of invariant zeros.

3.4.8. Assume that $\rho = m$, i.e. G is left-invertible. Then,

$$\{\text{transmission zeros}\} + \{\text{o.d.z.}\} \subset \{\text{invariant zeros}\} \quad (28)$$

Note that in this case, the set of i.d.z. is generally not included in the set of invariant zeros.

[7] Isomorphic modules are identified throughout this Section.

3.4.9. Assume that G is invertible (i.e. p = m = ρ). Then,

$$\{\text{transmission zeros}\} + \{\text{hidden modes}\} = \{\text{invariant zeros}\} \quad (29)$$

3.4.10. In the general case, the right-hand member of the above equality was defined by Rosenbrock (1974) as the set of system zeros. These zeros can directly be calculated by taking the GCD of specific minors of the system matrix (14). In the case where G is invertible, of course, the system zeros are nothing but the invariant zeros. We have in the general case

$$\{\text{invariant zeros}\} \subset \{\text{system zeros}\} \quad (30)$$

4. CONCLUDING REMARKS

In this paper, we defined poles and zeros and we studied their relationships in an "invariant approach". In our opinion, this viewpoint clarifies these notions; it also makes it possible to establish the fundamental relationships between various poles and zeros without using tedious polynomial matrix calculations.

The results (24), (29) and (30) were first established by Rosenbrock (1974) in the PMD framework, and the results (27), (28) by Rosenbrock (1976). To the best of our knowledge, the inclusion (26), which holds in the general case, is new (however, it is well known that {transmission zeros} ⊂ {invariant zeros}).

Discrete time may be treated along the same lines with the formalism of (Fliess, 1992b).

5. REFERENCES

Blomberg, H., and Ylinen, R., 1983, *Algebraic Theory for Multivariable Linear Systems* (London: Academic Press).

Bourlès, H., 1994, Semi-cancellable fractions in system theory. IEEE Trans. on Autom. Control, **38**, 2148-2153.

Conte, G., and Perdon, A. M., 1985, Zero modules and factorization problems. *Contemporary Mathematics: Linear Algebra and its Role in Systems Theory* (Providence: American Mathematical Society), **47**, 81-94.

Conte, G., and Perdon, A. M., 1986, Zeros of cascade compositions. *Frequency Domain and State Space Methods for Linear Systems*, Byrnes, C. I., and Lindquist, A., Eds. (Amsterdam: North-Holland).

Ferreira, P. G., Bhattacharyya, S. P., 1977, On blocking zeros, IEEE Trans. on Autom. Control, **22**, 258-259.

Fliess, M., 1990, Some basic structural properties of generalized linear systems, Systems Control Lett., **15**, 391-396.

Fliess, M., 1991, A simple definition of hidden modes, poles and zeros. Kybernetika, **27**, 186-189.

Fliess, M., 1992a, A remark on Willems' trajectory characterization of linear controllability. Systems Control Lett., **19**, 43-45.

Fliess, M., 1992b, Reversible Linear and Nonlinear Discrete-Time Dynamics. IEEE Trans. on Autom. Control, **37**, 1144-1153.

Fliess, M., 1994, Une interprétation algébrique de la transformation de Laplace et des matrices de transfert. Linear Algebra Applications, **203-204**, 429-442.

Fliess, M., and Glad, S. T., 1993, An algebraic approach to linear and nonlinear control. *Essays on Control: Perspectives in the Theory and its Applications*, Trentelman H. and Willems J. C., Eds (Boston: Birkhaüser), 223-267.

Fliess, M., and Mounier, H., 1994, Quelques propriétés structurelles des systèmes linéaires à retards constants. C. R. Acad. Sc. Paris, t. 319, I, 289-294.

Freudenberg, J. S., and Looze, D. P., *Frequency Domain Properties of Scalar and Multivariable Feedback Systems* (Berlin: Springer Verlag).

Glover K., and Doyle J. C., 1988, State-space formulae for all stabilizing controllers that satisfy an H_∞-norm bound and relations to risk sensitivity. Systems Control Lett., **11**, 167-172.

Kailath, T., 1980, *Linear Systems* (Englewood Cliffs, New Jersey: Prentice-Hall).

MacFarlane, A. G. J., and Karkanias, N., 1976, Poles and zeros of linear multivariable systems: a survey of the algebraic, geometric and complex-variable theory. International Journal of Control, **24**, 33-74.

MacLane, S., and Birkhoff, G., 1968, *Algebra* (New-York: The MacMillan Company).

Schrader, C. B., and Sain, M., 1989, Research on systems zeros: a survey. International Journal of Control, **50**, 1407-1433.

Rosenbrock, H. H., 1970, *State-space and Multivariable Theory* (New-York: Wiley).

Rosenbrock, H. H., 1974, Corrections to "The zeros of a system". International Journal of Control, **20**, 525-527.

Rosenbrock, H. H., 1977, Comments on "Poles and zeros of linear multivariable systems: a survey of the algebraic, geometric and complex-variable theory". International Journal of Control, **26**, 157-161.

Willems, J. C., 1983, Input-output and state-space representations of finite dimensional linear time invariant systems. Linear Algebra Applications, **50**, 581-608.

Willems, J. C., 1991, Paradigms and puzzles in the theory of dynamical systems. IEEE Trans. on Autom. Control, **36**, 259-294.

Wonham, W. M., 1984, *Linear Multivariable Control: a Geometric Approach* (Berlin: Springer Verlag).

Wyman, B. F., and Sain, M. K., 1981, Exact sequences for pole-zero cancellations. Proc. of *International Symposium on the Mathematical Theory of Networks and Systems*, 278-280.

Wyman, B. F., and Sain, M. K., 1987, Module theoretic zero structures for system matrices. SIAM J. Control and Optimization, **25**, 87-99.

Youla, D. C., Bongiorno, J. J., Lu, Jr. and C. N., 1974, Single-loop feedback stabilization of linear multivariable plants. Automatica, **10**, 159-173.

DUALITY FOR CONTROL SYSTEMS REVISITED

J.F. POMMARET
Centre d'Enseignement et de Recherche en Mathématiques, Informatique et Calcul Scientifique (CERMICS)
Ecole Nationale des Ponts et Chaussées (ENPC)
La Courtine, 93167 Noisy-le-Grand Cedex, France

Abstract. The purpose of this communication is to revisit the concept of duality existing in classical control theory while extending it to partial differential control theory in order to test controllability in this new framework. We prove in particular that a linear control system is controllable if and only if the corresponding differential operator expresses the compatibility conditions of a preceding parametrizing differential operator that we construct in an effective way. In the differential algebraic case, these results are extended to nonlinear control systems and show that controllability is a "built-in" property of a control system that does not depend on the choice of inputs and outputs. Many explicit examples will illustrate this new approach.

Résumé. Le but de cette communication est de revisiter le concept de dualité existant en théorie du contrôle classique tout en l'étendant au cas du contrôle aux dérivées partielles, pour pouvoir tester la contrôlabilité dans ce nouveau cadre. De nombreux exemples illustrent cette nouvelle approche.

Key Words. Control theory, controllability, differential operator, differential sequence, adjoint operator, duality theory.

At the beginning of this paper we recall two problems, one in control theory and one in differential geometry, which do not seem to be connected at first sight. The use of duality theory for differential sequences will not only show out the surprising link existing between these two problems, but also solve them by means of *totally new effective techniques* in computer algebra (Pommaret, 1994).

In classical control theory, with standard notations in that domain, let us consider a linear input/state/output control system:

$$\dot{x} = Ax + Bu \quad , \quad y = Cx + Du$$

with a state x of dimension n. We know that such a system is *controllable* if and only if:

$$rank(B, AB, ..., A^{n-1}B) = n$$

by exhibiting the *controllability matrix* and *observable* if and only if:

$$rank(\tilde{C}, \tilde{A}\tilde{C}, ..., \tilde{A}^{n-1}\tilde{C}) = n$$

by exhibiting the *observability matrix* where transposed matrices are denoted with a tilde sign. As a byproduct, from elementary properties of matrices, the *dual system*, or, as we shall see later on, in a more appropriate way, the *adjoint system*:

$$\dot{x}_a = -\tilde{A}x_a - \tilde{C}u_a \quad , \quad y_a = \tilde{B}x_a + \tilde{D}u_a$$

$$nb(u_a) = nb(y), nb(x_a) = nb(x), nb(y_a) = nb(u)$$

with dual correspondence:

$$u \longrightarrow x \longrightarrow y$$

$$y_a \longleftarrow x_a \longleftarrow u_a$$

is controllable (observable) if and only if the initial system is observable (controllable).

Nevertheless, despite a few recent works on the subject (Crouch et al., 1993; van der Schaft, 1991), the preceding definition still seems to be quite artificial as one cannot avoid introducing the state.

Exactly like ordinary differential equations (ODE) are examples of partial differential equations (PDE), a PD control system will be defined by any number of independent variables, inputs and outputs, together related by a system of (nonlinear) PDE of any order. Of course, in such a framework, the usual *functional definition* of controllability, in terms of a property of the trajectories, cannot exist any longer, even if, at least for linear systems as we saw and for certain nonlinear systems (Nijmeijer, 1990), it can be checked by means of a

formal test.
In Pommaret (1986), we gave for the first time a *formal definition* of controllability, based on the concept of *differential algebraic closure* in differential algebra and valid for both (linear or nonlinear) ODE or PDE as well, that we now recall.
By an *observable*, we mean a function of the control variables (inputs and outputs solutions of the control system) and their derivatives. Then there are only two possibilities:
• The observable is *free*, that is does not satisfy any ODE or PDE by itself.
• The observable is *not free*, or *constrained*, that is must satisfy at least one ODE or PDE by itself.

DEFINITION: A control system is *controllable* in our sense if and only if any observable is free.

Unhappily, even if we have proved that such a definition is perfectly coherent with any known classical test, *it is not possible to test it in general* and only particular cases have been considered in Pommaret (1992, 1993).

REMARK: The mono-input/mono-output differential algebraic system $u\ddot{y} - \dot{u} = a = cst$ is controllable if and only if $a \neq 0$ but the nonalgebraic observable $z = \dot{y} - logu$ nevertheless satisfies $\dot{z} = 0$. Hence differential algebra can only be a guide towards controllability (Pommaret, 1992).

In the case of linear control systems, as noticed by Fliess (1991), the concept of differential algebraic closure is replaced by the concept of *torsion submodule* of the residual differential module obtained by considering the module of all linear combinations of inputs and outpts and their derivatives modulo the submodule generated by the given equations of the control system and all their derivatives. We recall that the torsion submodule of a module M over a ring A is made by all the *torsion elements* $m \in M$ such that $\exists a \in A, am = 0$. In the ordinary differential case with constant coefficients, the ring A becomes a principal ideal domain (each ideal is generated by a single element) and, *in that very specific case only*, a module is *torsionless* if and only if it is *free*, that is admits a *basis* or equivalently can be *parametrized* by certain free elements, themselves in the module (Lang, 1971). Such a property has led to the definition of *flat* systems by extending the concept of parametrization to the nonlinear situation (Fliess et al., 1992).

Unhappily, in the partial differential case, nothing is left from this result and we shall discover that the parametrization property is much more important than the freedom property in characterizing a torsionless (partial differential) module. As an easy counterexample, the *div* operator is parametrized by the *rot* operator and the corresponding module is therefore torsionless though not free.

Recapitulating the results so far obtained, we discover that a PD linear control system can be considered as a set of PDE under the form $\mathcal{D}_1\eta = 0$ where $\mathcal{D}_1$ is a linear PD operator and η is describing *all* the control variables, namely *both* inputs and outputs. Hence, what we are looking for is a controllability test for $\mathcal{D}_1$, that is a test for the torsionless property of the corresponding differential module, that should give back, of course, the classical ranking test of the controllability matrix for the ordinary differential case in Kalman form. In the nonlinear situation, one should adapt this result to the linearized system ... whenever a sufficiently generic linearization procedure can be exhibited, for example in the differential algebraic case.

Meanwhile, taking into account the parametrization aspect, we are led to question about the *existence* and *construction* of a *differential sequence* (Pommaret, 1994):

$$E \xrightarrow{\mathcal{D}} F_0 \xrightarrow{\mathcal{D}_1} F_1$$

for convenient vector bundles, such that, not only $\mathcal{D}_1 \circ \mathcal{D} \equiv 0$ but also $\mathcal{D}_1$ should represent/generate *all* the compatibility conditions for $\mathcal{D}$, that is $\mathcal{D}\xi = \eta \Longrightarrow \mathcal{D}_1\eta = 0$ and conversely any solution of $\mathcal{D}_1\eta = 0$ is of the form $\mathcal{D}\xi = \eta$ for certain ξ.
In fact, though surprising it could be, the main result of this paper is a constructive proof of the following theorem:

THEOREM: A linear partial differential control system is controllable if and only if it is parametrizable.

We shall only sketch the proof of this theorem and refer the reader for the details to a forthcoming book (Pommaret, 1994). Rather we shall focus on the effectiveness of such a result and its important consequences in control theory, namely:

• *Controllability admits a formal definition.*
• *Controllability admits a formal test.*
• *Controllability does not depend on the separation of the control variables between input and output.*
• *Controllability does not depend on the state or even any representation.*
• *Controllability is coherent with the classical definition through the inclusion* ODE⊂PDE .

The key piece of machinery will be the *duality theory for differential sequences*, which, surprisingly too, is hard to find in the litterature (Vinogradov,

1984). For this, if E is a vector bundle over a manifold of dimension n, we shall introduce the *dualizing* vector bundle $\tilde{E} = \bigwedge^n T^* \otimes E^*$ where $T^* = T^*(X)$ is the cotangent bundle of X, $\bigwedge$ denotes exterior roduct and $*$ denotes the usual duality of tensor calculus. Now, if $\mathcal{D} : E \longrightarrow F$ is a linear differential operator of order q, we shall denote by $ad(\mathcal{D}) = \tilde{\mathcal{D}} : \tilde{F} \longrightarrow \tilde{E}$ the *adjoint* of $\mathcal{D}$ in the formal sense which is defined uniquely by the three following rules:

* The adjoint of any matrix of coefficients (zero order operator) is the transposed matrix.

** The adjoint of ∂_i is $-\partial_i$.

*** For any couple of operators that can be composed, we have the property:

$$ad(\mathcal{P} \circ \mathcal{Q}) = ad(\mathcal{Q}) \circ ad(\mathcal{P})$$

In that case, one has the following useful consequence on test sections (replacing test functions for vector valued operators) where d is the exterior derivative and $<>$ denotes the usual duality contraction instead of using column/line vector notation:

$$< \mathcal{D}\xi, \mu > = < \xi, \tilde{\mathcal{D}}\mu > + d()$$

This technical result proves that the adjoint operator can also be constructed inductively through integration by part.

First of all, let us suppose that $\mathcal{D}_1\eta = 0$ represents all the compatibility conditions of the linear inhomogeneous system $\mathcal{D}\xi = \eta$. Using the fact that a linear differential ideal is always prime, we may use at once basic results on the differential transcendence degree in differential algebra whenever the coefficients of $\mathcal{D}$ and thus also $\mathcal{D}_1$ are in a differential field K containing the rational numbers. We have the inclusions $K \subset K < \eta > \subset K < \xi >$ and obtain successively:

$$\begin{aligned} dimE &= difftrdK < \xi > /K \\ &= difftrdK < \xi > /K < \eta > \\ &\quad + difftrdK < \eta > /K \\ &= difftrd\mathcal{D} + difftrd\mathcal{D}_1 \end{aligned}$$

Hence, setting $diffrk\mathcal{D} = dimE - difftrd\mathcal{D}$, we get the important formula;

$$diffrk\mathcal{D} + diffrk\mathcal{D}_1 = dimF_0$$

which generalizes the ranking condition existing for the linear maps of exact sequences in homological algebra.

Similarly, one can prove the equality:

$$diffrk\mathcal{D} = diffrk\tilde{\mathcal{D}}$$

which generalizes the well known ranking property of matrices.

The scheme of the proof is based on the following *dual differential sequences*;

$$\begin{array}{cccccccc} & & & & & \xrightarrow{\mathcal{D}_1'} & F_1' & 5 \\ 4 & E & \xrightarrow{\mathcal{D}} & F_0 & & \xrightarrow{\mathcal{D}_1} & F_1 & 1 \\ 3 & \tilde{E} & \xleftarrow{\tilde{\mathcal{D}}} & \tilde{F}_0 & & \xleftarrow{\tilde{\mathcal{D}}_1} & \tilde{F}_1 & 2 \end{array}$$

where numbers aside indicate the successive constructions, $\tilde{\mathcal{D}}$ represents *all* the compatibility conditions of $\tilde{\mathcal{D}}_1$, $\mathcal{D}$ is the adjoint of $\tilde{\mathcal{D}}$ and $\mathcal{D}_1'$ represents all the compatibility conditions of $\mathcal{D}$.

Counting the differential transcendence degrees and dimensions, we get:

$$\begin{aligned} dimF_0 &= diffrk\mathcal{D} + diffrk\mathcal{D}_1' \\ dim\tilde{F}_0 &= diffrk\tilde{\mathcal{D}} + diffrk\tilde{\mathcal{D}}_1 \\ &= diffrk\mathcal{D} + diffrk\mathcal{D}_1 \end{aligned}$$

and thus $diffrk\mathcal{D}_1 = diffrk\mathcal{D}_1'$ though in general $\mathcal{D}_1 \leq \mathcal{D}_1'$.

It remains to prove that $\mathcal{D}_1$ is controllable if and only if the equality holds, that is $\mathcal{D}_1 = \mathcal{D}_1'$.

Indeed, setting $\mathcal{D}_1\eta = \zeta$ and $\mathcal{D}_1'\eta = \zeta'$, we have the inclusions:

$$K \subset K < \zeta > \subseteq K < \zeta' > \subset K < \eta >$$

and therefore:

$$\begin{aligned} & difftrdK < \zeta' > /K < \zeta > \\ = \; & difftrdK < \eta > /K < \zeta > \\ & - difftrdK < \eta > /K < \zeta' > \\ = \; & difftrd\mathcal{D}_1 - difftrd\mathcal{D}_1' \\ = \; & diffrk\mathcal{D}_1' - diffrk\mathcal{D}_1 \\ = \; & 0 \end{aligned}$$

Hence, whenever η is a (generic) solution of the control system $\mathcal{D}_1\eta = 0$, it follows that any new compatibility condition ζ', *if there is one*, determines an observable which is differentially constrained over K, a contradiction whenever $\mathcal{D}_1$ is assumed to be controllable.

Conversely, when $\mathcal{D}_1$ is (at least formally) parametrized by $\mathcal{D}$, any observable in $K < \eta >$ being expressed as an element in $K < \xi >$ through $\mathcal{D}\xi = \eta$, is therefore free because K is differentially algebraically closed in the purely differentially transcendental differential extension $K < \xi >$ of K and $\mathcal{D}_1$ is thus controllable.

Similarly, one can prove:

COROLLARY: A controllable $\mathcal{D}_1$ determines a free differential module if an ony if $\mathcal{D}$ is injective when $\tilde{\mathcal{D}}$ is surjective.

An additional striking consequence for classical control theory (n=1) is the following result providing a new test for controllability which relates such a property to formal integrability without refering to any state representation. We recall that a system of PDE of order q is formally integrable whenever all the resulting PDE of order $q+r$ are obtained after only r differentiations, $\forall r \geq 0$.

PROPOSITION: A surjective linear (ordinary differential) control operator $\mathcal{D}_1$ is controllable if and only if its adjoint is injective.

Proof: We give two proofs:
1) When the control system can be written in the Kalman form $\dot{y} = Ay + Bu$, the kernel of the dual operator is easily seen to be defined by the two equations $\dot{\lambda} + \lambda A = 0$ and $\lambda B = 0$. Such a system is clearly not formally integrable as one may obtain the additional zero order equations $\lambda AB = 0, \lambda A^2 B = 0, \ldots$, a result directly linking the rank of the controllability matrix to the defect of formal integrability of the dual system.
2) If $\mathcal{D}_1$ were not injective, let us use a test section of $\tilde{F}_1$ that is killed by $\tilde{\mathcal{D}}_1$. Then $< \mathcal{D}_1 \eta, \lambda >$ is the total differential of an observable which is therefore noy free as it is a "constant of motion" when η is a solution of he control system.Conversely, it is known that any differential expresssion, for example the one constraining an observable, admits an integrating factor, a result contradicting the injectivity of $\tilde{\mathcal{D}}_1$.

REMARK 1: We want to stress out a *basic confusion* done in the recent existing litterature on duality (Crouch et al., 1993; van der Schaft, 1991). Indeed, let us suppose that we have the control system :

$$\dot{x} = Ax + Bu \quad , \quad y = x$$

as before. We should obtain the dual system:

$$\dot{x}_a + \tilde{A}x_a = u_a \quad , \quad \tilde{B}x_a = y_a$$

and we discover that what the people use to call the dual of $\mathcal{D}_1$ is in fact ... $\tilde{\mathcal{D}}$!.

REMARK 2: It is well known in 2-dimensional elasticity theory that, whenever $\mathcal{D}$ defines the (small) strain tensor:

$$\epsilon_{11} = \partial_1 \xi^1, \epsilon_{12} = \frac{1}{2}(\partial_1 \xi^2 + \partial_2 \xi^1), \epsilon_{22} = \partial_2 \xi^2$$

then $\mathcal{D}_1 = \mathcal{D}_1'$ is defined by the unique compatibility condition:

$$\partial_{11}\epsilon_{22} + \partial_{22}\epsilon_{11} - 2\partial_{12}\epsilon_{12} = 0$$

Caring about duality for the stress tensor by noticing that:

$$\sigma^{ij}\epsilon_{ij} = \sigma^{11}\epsilon_{11} + 2\sigma^{12}\epsilon_{12} + \sigma^{22}\epsilon_{22}$$

we discover that $\tilde{\mathcal{D}}_1$ just expresses the way to define the stress by means of an Airy function:

$$\sigma^{11} = \partial_{22}\lambda, \sigma^{12} = -\partial_{12}\lambda, \sigma^{22} = \partial_{11}\lambda$$

while $\tilde{\mathcal{D}}$ just expresses the stress equations:

$$\partial_1\sigma^{11} + \partial_2\sigma^{21} = 0, \partial_1\sigma^{12} + \partial_2\sigma^{22} = 0$$

The reader must understand through this example how duality for differential sequences can be used in other domains of engineering sciences, even though they look very far away from control theory (Pommaret, 1994).

EXAMPLE: Let us consider the following linear control system:

$$\dot{y}^1 = y^2 + u \quad , \quad \dot{y}^2 = ay^1 + u$$

where a is a constant parameter. We could observe this system by means of the single output $y = y^1$ and study equivalently the corresponding mono-input/mono-output system:

$$\ddot{y} - ay - u - \dot{u} = 0$$

Conversely, in order to decide about the controllability of such a system, we should transform it back to the preceding Kalman form and use the classical test:

TEST 1: We have:

$$A = \begin{pmatrix} 0 & 1 \\ a & 0 \end{pmatrix}, B = \begin{pmatrix} 1 \\ 1 \end{pmatrix}$$

$$det(B, AB) = det \begin{pmatrix} 1 & 1 \\ 1 & a \end{pmatrix} = a - 1$$

and therefore the control system is controllable if and only if $a \neq 1$. If $a = 1$, substracting the first ODE from the second gives the uncontrollable part of the system under the form $\dot{z} + z = 0$ with $z = y^2 - y^1 = \dot{y} - y - u$.
We now study the module theoretic approach:

TEST 2: Up to a constant factor, any torsion element can be written $z = \dot{y} + \lambda y + \mu_0 u + \mu_1 \dot{u} + \ldots$. Substituting into a constraint equation of the form $\dot{z} + \nu z = 0$ brings the conditions $\mu_1 = \ldots = 0$ and, setting $\mu_0 = \mu$, we are left with the ODE:

$$\ddot{y} + (\lambda + \nu)\dot{y} + \lambda\nu y + \mu\nu u + \mu\dot{u} = 0$$

Identifying with the control ODE, we get:

$$\lambda + \nu = 0, \lambda\nu = -a, \mu\nu = -1, \mu = -1$$

$$\Longrightarrow \lambda = -1, \mu = -1, \nu = 1 \Longrightarrow a = 1$$

TEST 3: Setting now $\eta = (\eta^1, \eta^2)$ with $\eta^1 = u, \eta^2 = y$ in order to really mix together input and output, we may rewrite the control system in the form:

$$\mathcal{D}_1\eta \equiv \ddot{\eta}^2 - a\eta^2 - \eta^1 - \dot{\eta}^1 = 0$$

Multiplying by a test function λ and integrating by part, we obtain:

$$\tilde{\mathcal{D}}_1\lambda = \mu = (\mu_1, \mu_2)$$

$$\dot{\lambda} - \lambda = \mu_1, \ddot{\lambda} - a\lambda = \mu_2$$

and the kernel of $\tilde{\mathcal{D}}_1$ is defined by the nonformally integrable second order system:

$$\ddot{\lambda} - a\lambda = 0, \dot{\lambda} - \lambda = 0 \Longrightarrow (a-1)\lambda = 0$$

It follows that $\tilde{\mathcal{D}}_1$ is injective if and only if $a \neq 1$. Now, if $a = 1$, a solution may be $\lambda = e^t$ and we obtain at once the relation:

$$\dot{y} - y - u = z$$

$$(\ddot{y} - y - u - \dot{u})e^t = \frac{d}{dt}(e^t z)$$

exhibiting the link with the preceding test.

TEST 4: From the preceding test, we notice the following compatibility condition when $a = 1$:

$$\nu = \tilde{\mathcal{D}}\mu = \dot{\mu}_1 + \mu_1 - \mu_2 = 0$$

Multiplying by ξ and integrating by part, we get:

$$\mathcal{D}\xi = \eta \Longleftrightarrow -\dot{\xi} + \xi = \eta^1, -\xi = \eta^2$$

and therefore the *first order* compatibility condition:

$$\zeta' = \mathcal{D}_1'\eta \equiv \dot{\eta}^2 - \eta^2 - \eta^1 = 0$$

leading to $\dot{\zeta}' + \zeta' - \zeta = 0$. Hence, when $\zeta = 0$, we find back the constrained observable $z = \zeta'$.

We now present an example showing how the preceding techniques may even aply to control systems with variable coefficients.

EXAMPLE: When $n = 3$, let us define $\mathcal{D}_1$ by the system:

$$A \equiv \partial_{222}\eta^1 - \partial_{233}\eta^2 + x^2\partial_{112}\eta^2 + 3\partial_{11}\eta^2 = 0$$

Multiplying by a test function λ and integrating by part *with care*, we obtain for $\tilde{\mathcal{D}}_1$ the operator:

$$-\partial_{222}\lambda = \mu_1, \partial_{233}\lambda - x^2\partial_{112}\lambda + 2\partial_{11}\lambda = \mu_2$$

and for $\tilde{\mathcal{D}}$ the only compatibility condition:

$$\partial_{33}\mu_1 - x^2\partial_{11}\mu_1 + \partial_{22}\mu_2 = 0$$

Accordingly, we obtain for $\mathcal{D}$ the operator:

$$\partial_{33}\xi - x^2\partial_{11}\xi = \eta^1, \partial_{22}\xi = \eta^2$$

which has been studied with details in (Pommaret et al., 1991; Pommaret, 1994). In particular, we know from computer algebra that $\mathcal{D}_1'$ contains one additional 6^{th}order compatibility condition $B = 0$ that determines an observable constrained by the PDE $\partial_2 B = 0$ whenever $A = 0$ and $\mathcal{D}_1$ is therefore not controllable, a result not evident at all at first sight in view of the difficulty to work out B explicitly.

We end this paper with an example showing how to apply the duality technique on an arbitrary nonlinear control system.

EXAMPLE: In the case of the nonlinear differential algebraic control system $u\ddot{y} - \dot{u} = a = cst$ already considered in a remark, we define it over a ground differential field K containing a and get an input/output differential extension N as in Pommaret (1993). Though K is differentially algebraically closed in N for any value of a, we can linearize this system over N (*care*) in order to get, with capital letters for the variations, the following linear control system over N:

$$u\ddot{Y} + \ddot{y}U - \dot{U} = 0$$

apparently no longer depending on a. However, N/K *effectively* depends on a and, using the test 3 of a preceding example, we obtain:

$$u\ddot{\lambda} + 2\dot{u}\dot{\lambda} + \ddot{u}\lambda = 0, \dot{\lambda} + \ddot{y}\lambda = 0$$

leading to $a\dot{\lambda} = 0$ and thus to $a\lambda = 0$ when taking into account the initial control system and its first prolongation. Hence the linearized control system is controllable if an only if $a \neq 0$ in a coherent way with a direct approach (Pommaret, 1992).

The final example will establish a link between these new techniques and modern geometric ordinary control theory as it can be found in Isidori (1985) or Nijmeijer et al. (1990).

EXAMPLE: Let us consider the nonlinear control system $\dot{x} = f(x) + g_r(x)u^r$ without indicating all the indices for simplicity. Using capital letters for the variations, the linearized system will be defined by the ODE:

$$\dot{X} = \frac{\partial f(x)}{\partial x}X + \frac{\partial g_r(x)}{\partial x}u^r X + g_r(x)U^r$$

The kernel of the dual system will be defined by the ODE:

$$\dot{\lambda} + \lambda\frac{\partial f(x)}{\partial x} + \lambda\frac{\partial g_r(x)}{\partial x}u^r = 0, \qquad \lambda g_r(x) = 0$$

This system is not formally integrable and one must introduce the new ODE:

$$\lambda([f, g_r]) + u^s[g_s, g_r]) = 0$$

with rank equal to $rk([f, g_r], [g_s, g_r])$ because u is an indeterminate,... and so on. Hence, we find back at once that the above nonlinear system is (formally) controllable if and only if the corresponding adjoint system admits only the zero solution, that is to say if and only if the so-called *strong controllability distribution* $(g, [f, g], [g, g], ...)$ has maximum (generic) rank equal to dim x.

BIBLIOGRAPHY

CROUCH, P.E., F. LAMNABHI-LAGARRIGUE, and A.J. van der SCHAFT (1993). Ajjoint and Hamiltonian input-output differential equations, Memorandum 1187, Twente University, *IEEE Trans. Aut. Control*, 40, 4, to appear.

FLIESS, M.(1991). Controllability revisited, in "The influence of R.E. Kalman", A.C. Antoulas ed., Springer, 463-474.

FLIESS, M., J. LEVINE, P. MARTIN, and P. ROUCHON (1992). Sur les systèmes linéaires différentiellement plats, C. R. Acad. Sc. Paris, 315, I, 619-624.

ISIDORI, A. (1985). Nonlinear control systems : an introduction, Lecture Notes in Control and Information Sciences 72, Springer, Berlin, 297 pp.

LANG, S. (1971). Algebra, Addison-Wesley.

NIJMEIJER, H., and A.J. van der SCHAFT (1990). Nonlinear dynamical control systems, Springer, 470p.

POMMARET, J.F. (1986). Géométrie différentielle algébrique et théorie du contrôle, C. R. Acad. Sc. Paris, 302, I, 547-550.

POMMARET, J.F. (1992). New perspectives in control theory for partial differential equations, *IMA J. Math. Control and Information*, 9, 305-330.

POMMARET, J.F. (1993). Structure at infinity revisited, *Systems and Control Letters*, 20, 327-333.

POMMARET, J.F. (1994). Partial differential equations and group theory: new perspectives for applications, Kluwer, Dordrecht, 480p, to appear.

POMMARET, J.F., and A. HADDAK (1991). Effective methods for systems of algebraic PDE, in "Effective methods in algebraic geometry", T. Mora, C. Traverso eds.,Birkhäuser, 411-426.

SCHAFT, A.J. van der (1991). Duality for linear systems, in "Analysis of controlled dynamical systems", B. Bernard, B. Bride, J.P. Gauthier, I. Krupka eds., Progress in systems and control theory, Birkhäuser, 393-403.

VINOGRADOV, A.M. (1984). The *C*-spectral sequence, lagrangian formalism and conservation laws, *J. of Math. Anal. and Applications*, 100, 1, 1984, 1-40.

INVITATION TO DISSYMMETRY — SYSTEM INVARIANCE AND BASIC PROPERTIES OF LINEAR SYSTEMS —

R. TANAKA, S. SHIN and N. SEBE

Department of Mathematical Engineering and Information Physics, Faculty of Engineering, the University of Tokyo, Hongo Bunkyo-ku, Tokyo 113, Japan

Abstract. In this paper, a particular symmetry is defined for linear systems as an invariance under state transformations. It is shown the basic properties of the systems possessing this symmetry. It results in revealing that symmetric systems are uncontrollable, although structural symmetry is advantageous to simplify design and analysis of the systems. Moreover, as systems become larger and more complex, their reliability acquires greater importance in control problems. Then this paper shows two examples where certain failure patterns induce symmetry in the systems, so that they become uncontrollable. Investigations into these failures will serve for evaluation of fault-tolerance.

Résumé. Dans ce document, un type particulier de symétrie des systèmes linéaires est défini comme une invariance par certaines transformations d'état. On décrit les propriétés de base des systèmes qui possèdent cette symétrie. Il en résulte que de tels systèmes sont incontrôlables bien que la symétrie de leur structure soit avantageuse pour simplifier leur forme ainsi que leur analyse. D'autre part, plus les systèmes deviennent grands et complexes, plus leur fiabilité acquiert une grande importance dans les problèmes de contrôle. Aussi ce document présente deux cas où des arrangements défectueux des sous-systèmes font apparaître la symétrie considérée et rendent ainsi le système incontrôlable. Les recherches sur ces arrangements des défauts serviront à l'évaluation de la tolérance aux erreurs.

Key Words. Controllability; decentralized systems; fault tolerance; invariance.

1. INTRODUCTION

Dissymmetry, the state which comes after the rupture of a balance or of a symmetry. It appears clearly as an intervention, necessitated subversion or premediated modification (Caillois, 1976). This idea may be true of control problems in a sense. Symmetry seems to be favorable for systems, but it should be destroyed for the controllability. The contribution of this paper is to reveal a disadvantage of the symmetry in control systems.

Fig. 1. Some traditional Japanese crests. Each crest has geometrical symmetry.

A noticeable advantage of symmetry is found in some Japanese traditional crests (Fig.1), which is recognized as a cultural heritage (Inui, 1976). Thanks to the apparent geometric symmetry, drawing or analysis of the minimum unit can be applied to the symmetric parts. It facilitates drawing or analysis of each of the crests. The same is true of engineering systems. As systems become larger, homogeneity or symmetry are often introduced in system structures to simplify their designs and analyses. In practical point of view, these characteristics are suitable for mass production.

With such an idea, many studies have been made for systems with *symmetrically interconnected subsystems* (for example, see Lunze, 1989; Sundareshan and Elbanna, 1991; Xiaoping, 1992), and for systems conerning to symmetry (for example, see Grizzle, 1985; Nijemeijer, 1985). In this paper, how-

ever, symmetry is differentiated from homogeneity. Homogeneity, is defined by the identical nature of the operational characteristic of the subsystems. On the other hand, symmetry discussed in this paper is defined with respect to a system invariance under state transformation. This definition enables us to deal with symmetry for broader types of systems. As for linear systems with the symmetry in this sense, basic properties are clarified especially from the viewpoint of controllability. It results in revealing the disadvantage of the symmetry.
For the sake of illustrating this characteristic, a familiar example of symmetry — mirror images shall mentioned here. Imagine a person standing in front of a mirror. Naturally, the person can move freely as he wants, while his image in the mirror never moves independently of his movement. They move exactly in the same way on both sides of the mirror.

Example 1.
Let us consider a control system that behaves similarly to this phenomenon (Fig.2). Both of the person and his image are compared to two symmetric plants in the system, say, Plant 1 and Plant 2, respectively. And his intention functions as the control input to Plant 1. In this correspondence, the above description is restated as follows: the states of Plant 1 are controlled freely by the input. As for Plant 2, its states have to change just as those of Plant 1 and can not change independently of the ones of Plant 1.

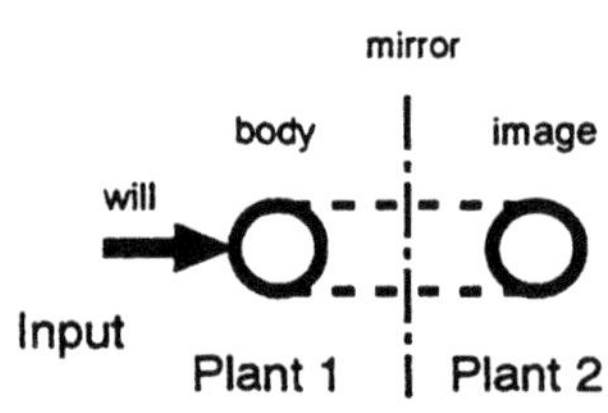

Fig. 2. A model of a person in front of a mirror. His intention (input) controls the movement of his body (Plant 1). And his image (Plant 2) moves in the same way.

This statement directly implies that the states of Plant 2 are uncontrollable by the control input. Consequently, the system is uncontrollable as a whole. □

In the second part of this paper, this result is applied to ADSs (Autonomous Decentralized Systems), especially those with homogeneity of subsystems. Since systems utilized in industry become larger and more complex, their reliability acquires greater importance in control problems. For several years, many studies have been done for ADSs (for example, see Ito *et al.*, 1994; Shin and Ikeda, 1993) as systems with fault-tolerance, and has been applied in various fields such as robotics (for example, see Habib *et al.*, 1992) and computer networks (for example, see Ihara and Mori, 1984). Specifically, controllability for ADSs is defined and investigated as that of an entire system in spite of failures in some subsystems (Tanaka *et al.*, 1994). This paper gives some examples of special failure patterns which induce symmetry, so that the system becomes uncontrollable. By the use of the symmetry, it becomes easy to analyze uncontrollability induced by failures. Moreover, investigation into these failure patterns will serve for evaluation of fault-tolerance.

Throughout this paper, O_k and I_k denote the zero matrix and the unit matrix of the size k, respectively.

2. DEFINITION OF SYMMETRY

Consider a time-invariant linear system $\mathcal{S}$ described by state-space equations

$$\dot{\boldsymbol{x}}(t) = A\boldsymbol{x}(t) + B\boldsymbol{u}(t) \tag{1}$$

and

$$\boldsymbol{y}(t) = C\boldsymbol{x}(t), \tag{2}$$

where $\boldsymbol{x}(t) \in \mathbf{R}^n$, $\boldsymbol{u}(t) \in \mathbf{R}^r$ and $\boldsymbol{y}(t) \in \mathbf{R}^l$ are the state, the input and the output of the system, respectively. Here, the dimensions of the matrices A and B are defined consistently, and the argument t is omitted in the following. Moreover, as for $\mathcal{S}$, a state transformation described by a nonsingular matrix $T \in \mathbf{R}^{n\times n}$ changes state variables, i.e.,

$$\boldsymbol{z} = T\boldsymbol{x}, \tag{3}$$

so that (1) and (2) are modified into

$$\dot{\boldsymbol{z}} = TAT^{-1}\boldsymbol{z} + TB\boldsymbol{u} \tag{4}$$

and

$$\boldsymbol{y} = CT^{-1}\boldsymbol{z}, \tag{5}$$

respectively.

The symmetry of a system $\mathcal{S}$ is now defined with an invariance of state transition equations (1) and (4) under state transformations.

Definition 1.
A system $\mathcal{S}$ is said to be *symmetric* if there

exists a nonsingular matrix $T(\neq I)$ which satisfies the following equations

$$A = TAT^{-1} \tag{6}$$

and

$$B = TB. \tag{7}$$

□

It must be noted here that the matrix T employed above allows any kinds of state transformations. Consequently, this definition permits us to deal with symmetry for broader types of systems. The following example exaplains this definition.

Example 2.
Referring to Example 1, let x_1 and x_2 denote the states of Plant 1 and Plant 2, respectively. The states of the entire system is $x := (x_1, x_2)$, with $x_1 = x_2$, if the effect of the initial states is invariant. Consider the permutation of x_1 and x_2 as a state transformation. Then the transformed state is $z = (x_2, x_1)$. Because of the equality of x_1 and x_2, the state transition equations (1) and (4) are invariant, i.e., (6) and (7) hold.

In this example, a permutation of corresponding parts of states is employed as a special case of state transformations, while T is allowed to any nonsingular matrices in Definition 1.

3. UNCONTROLLABILITY OF SYMMETRIC SYSTEMS

As shown in Example 1, the symmetry of systems is related to uncontrollability. This relationship is summarized in the following theorem. The well-known criteria (Kailath, 1980) to test the controllability will be used.

Lemma 1.
The system $\mathcal{S}$ is controllable if and only if the controllability matrix defined as

$$R(A,B) = [B\ AB\ A^2B \cdots A^{n-1}B] \in \mathbf{R}^{n\times nr} \tag{8}$$

is full row rank. □

Lemma 2.
The system $\mathcal{S}$ is uncontrollable if and only if there exist a nonzero vector x and a number λ which satisfy

$$x^{\mathrm{T}}A = \lambda x^{\mathrm{T}} \text{ and } x^{\mathrm{T}}B = \mathbf{o}\,. \tag{9}$$

□

Apparently from (8),

$$R(TAT^{-1}, TB) = TR(A,B) \tag{10}$$

holds for any nonsingular matrices T.

Theorem 1.
A system $\mathcal{S}$ is symmetric *if and only if* $\mathcal{S}$ is uncontrollable. □

Proof.
The sufficiency is shown.
Suppose that $\mathcal{S}$ is symmetric, so that there exists a matrix $T(\neq I)$ which satisfies (6) and (7) from Definition 1. Hence,

$$R(TAT^{-1}, TB) = R(A,B) \tag{11}$$

holds. It follows from (10) and (11) that

$$(T-I)R(A,B) = 0. \tag{12}$$

Since $T \neq I$, (12) implies that $R(A,B)$ is not of full row rank, that is, $\mathcal{S}$ is uncontrollable.

Next, the necessity is shown.
Suppose that $\mathcal{S}$ is uncontrollable. For x and λ that satisfy (9), there exists an n-dimensional vector y such that

$$Ay = \lambda y\,. \tag{13}$$

It follows from (9)–(13) that the equations (6) and (7) hold for T defined as

$$T = I_n + y\,x^{\mathrm{T}}. \tag{14}$$

Therefore, $\mathcal{S}$ is symmetric from Definition 1.
■

This theorem declares the important fact that symmetric systems fail to be controllable. Generally, symmetry is considered to be favorable. This theorem, however, calls attention to its potential disadvantage from the viewpoint of controllability, which is a classic and fundamental characteristic in control problems.

Let us seek the general formulation of T which satisfies (6) and (7) for $\mathcal{S}$. Suppose that A is decomposed by a nonsingular matrix $S \in \mathbf{R}^{n\times n}$ into a Jordan canonical form:

$$S^{-1}AS = \begin{bmatrix} J_1 & & & \\ & J_2 & & \\ & & \ddots & \\ & & & J_g \end{bmatrix}, \tag{15}$$

where the matrix J_i ($i = 1, 2, \cdots g$) is a k_i-dimensional Jordan block for an eigenvalue α_i:

$$J_i = \begin{bmatrix} \alpha_i & 1 & 0 & \cdots & 0 \\ 0 & \ddots & \ddots & \ddots & \vdots \\ \vdots & \vdots & \ddots & \ddots & 0 \\ \vdots & & \ddots & \ddots & 1 \\ 0 & \cdots & \cdots & 0 & \alpha_i \end{bmatrix} \in \mathbf{R}^{k_i \times k_i}, \quad (16)$$

with $\sum_{i=1}^{g} k_i = n$. According to the dimension k_i of each Jordan block J_i, S^{-1} and S are segmentalized as

$$S^{-1} = \begin{bmatrix} P_1 \\ P_2 \\ \vdots \\ P_g \end{bmatrix} \quad \text{and} \quad S = [Q_1 \; Q_2 \; \cdots \; Q_g],$$

where $P_i \in \mathbf{R}^{k_i \times n}$ and $Q_i \in \mathbf{R}^{n \times k_i}$ are set of row vectors and of column vectors, respectively. Consequently, the following theorem derives from some fundamental calculations.

Theorem 2. The equations (6) and (7) hold if T is written in the form of

$$T = I_n + \sum_{i=1}^{g} h_i Q_i P_i \quad (17)$$

with arbitrary coefficients h_i ($1 \leq i \leq g$), where

$$h_i P_i B = 0. \quad (18)$$

□

The same discussion is true of unobservability owing to the duality of linear systems.

4. SYMMETRY INDUCED BY FAILURES

The discussion above has shown the uncontrollability of systems with the symmetry defined in this paper. It evokes us a disadvantageous aspect of symmetry. This result is applied here to ADSs, especially to homogeneous systems which have homogeneity of subsystems. As the reliability of systems acquires greater importance, it is required to consider failures in a system. In this section, it is illustrated that special failure patterns induce symmetry, so that the system becomes uncontrollable.

An ADS considered here (Fig. 3) is formalized as follows. It consists of m subsystems $\{\mathcal{S}_1, \mathcal{S}_2, \cdots, \mathcal{S}_m\}$. All the subsystems are supposed to collaborate with each other and thus the entire system is organized. Each subsystem $\mathcal{S}_i$ (i=1, 2, $\cdots, m$) is described by a state transition equation

$$\dot{x}_i(t) = A_{ii}x_i(t) + B_{ii}u_i(t) + \sum_{j \neq i} A_{ij}x_j(t) + \sum_{j \neq i} B_{ij}u_j(t). \quad (19)$$

where $\boldsymbol{x}_i(t) \in \mathbf{R}^{n_i}$ and $\boldsymbol{u}_i(t) \in \mathbf{R}^{r_i}$ denote the state and the input of the subsystem $\mathcal{S}_i$, respectively, and A_{ij} and B_{ij} are matrices with consistent dimensions.

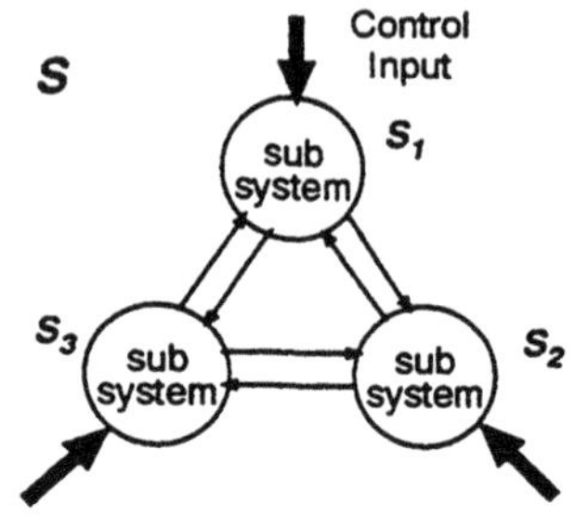

Fig. 3. Autonomous Decentralized System with three subsystems (m=3).

The state and the input of the entire system $\mathcal{S}$ are denoted as $\boldsymbol{x}(t) \in \mathbf{R}^n$ and $\boldsymbol{u}(t) \in \mathbf{R}^r$, and obtained by

$$\boldsymbol{x}(t) = (\boldsymbol{x}_1(t)^{\mathrm{T}} \; \boldsymbol{x}_2(t)^{\mathrm{T}} \cdots \boldsymbol{x}_m(t)^{\mathrm{T}})^{\mathrm{T}} \quad (20)$$

and

$$\boldsymbol{u}(t) = (\boldsymbol{u}_1(t)^{\mathrm{T}} \; \boldsymbol{u}_2(t)^{\mathrm{T}} \cdots \boldsymbol{u}_m(t)^{\mathrm{T}})^{\mathrm{T}}, \quad (21)$$

respectively. Then the system $\mathcal{S}$ is described by a state transition equation (1), where

$$A = \begin{bmatrix} A_{11} & A_{12} & \cdots & A_{1m} \\ A_{21} & A_{22} & \cdots & A_{2m} \\ \vdots & \vdots & \ddots & \vdots \\ A_{m1} & A_{m2} & \cdots & A_{mm} \end{bmatrix} \text{ and } B = \begin{bmatrix} B_{11} & B_{12} & \cdots & B_{1m} \\ B_{21} & B_{22} & \cdots & B_{2m} \\ \vdots & \vdots & \ddots & \vdots \\ B_{m1} & B_{m2} & \cdots & B_{mm} \end{bmatrix}.$$

The diagonal blocks of A and B stand for the effects of each subsystem $\mathcal{S}_i$ on its state $\boldsymbol{x}_i$. On the other hand, the off-diagonal blocks imply the interactions among subsystems. An ADS is said to be *homogeneous*, if

$$n_i = n_0, \quad r_i = r_0, \quad A_{ii} = A^\circ \text{ and } B_{ii} = B^\circ$$

for $1 \leq \forall i \leq m$, where $n_0 = \frac{n}{m}$, $r_0 = \frac{r}{m}$, and A° and B° are arbitrary specified matrices of the size $n_0 \times n_0$ and $n_0 \times r_0$, respectively.
In this paper, failure of a subsystem is restricted to that of its controller (Mori, 1981).

In other words, it is said that failure occurs in a subsystem $\mathcal{S}_i$ at time t_0 if

$$u_i\ (t) = 0 \qquad \text{for} \quad \forall t \geq t_0. \qquad (22)$$

This model of the failure is often adopted and is quite general for practical use. That is because systems are usually equipped to make the controller stop when failures occur in the subsystem. For later use, a failure matrix F is defined as follows with reference to a failure pattern.

$$F = \text{block-diag}\{F_1, F_2, \cdots, F_m\}, \qquad (23)$$

where

$$F_i = \begin{cases} O_{r_i} (\mathcal{S}_i : \text{ out of order}), \\ I_{r_i} \text{ (otherwise)}. \end{cases} \qquad (24)$$

The matrix BF after the failure has zero column blocks corresponding to all the subsystems in the outage.

Let us now consider failures in special two types of homogeneous systems. The following examples demonstrate that failures in some subsystems may cause the system to become symmetric, i.e. uncontrollable.

Example 3.
Consider a ring-type homogeneous system consisting of four subsystems. Suppose that the matrices A and B of the system is given as

$$A = \begin{bmatrix} a & g & 0 & 0 \\ g & a & g & 0 \\ 0 & g & a & g \\ 0 & 0 & g & a \end{bmatrix} \text{ and } B = \begin{bmatrix} b & h & 0 & 0 \\ h & b & h & 0 \\ 0 & h & b & h \\ 0 & 0 & h & b \end{bmatrix},$$

where a, b, g and h are independent parameters. If failures occur in the two subsystems $\mathcal{S}_1$ and $\mathcal{S}_3$ as shown in Fig.4,

$$BF = \begin{bmatrix} 0 & h & 0 & 0 \\ 0 & g & 0 & 0 \\ 0 & h & 0 & h \\ 0 & 0 & 0 & g \end{bmatrix}. \qquad (25)$$

The equations (6) and (7) hold with T defined as

$$T = \begin{bmatrix} 0 & 0 & 1 & 0 \\ 0 & 1 & 0 & 0 \\ 1 & 0 & 0 & 0 \\ 0 & 0 & 0 & 1 \end{bmatrix}. \qquad (26)$$

Hence the symmetry, i.e., the uncontrollability of the system after the failure. □

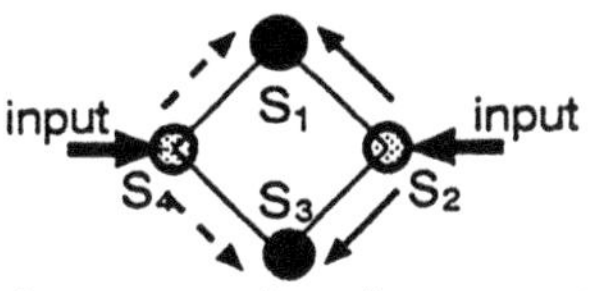

Fig. 4. An example of symmetry induced by failures in a homogeneous system. Failures occur in the subsystems represented by bullets(•). An arrow (⇒) represents an effective input. Then this system is symmetric, i.e., uncontrollable.

Example 4.
Consider a string-type homogeneous system consisting of five subsystems. Suppose that the matrices A and B of the system is given as

$$A = \begin{bmatrix} a & g & 0 & 0 & 0 \\ g & a & g & 0 & 0 \\ 0 & g & a & g & 0 \\ 0 & 0 & g & a & g \\ 0 & 0 & 0 & g & a \end{bmatrix} \text{ and } B = \begin{bmatrix} b & h & 0 & 0 & 0 \\ h & b & h & 0 & 0 \\ 0 & h & b & h & 0 \\ 0 & 0 & h & b & h \\ 0 & 0 & 0 & h & b \end{bmatrix},$$

where a, b, g and h are independent parameters. If failures occur in the three subsystems $\mathcal{S}_1$, $\mathcal{S}_3$ and $\mathcal{S}_5$ as shown in Fig.5,

$$BF = \begin{bmatrix} 0 & h & 0 & 0 & 0 \\ 0 & g & 0 & 0 & 0 \\ 0 & h & 0 & h & 0 \\ 0 & 0 & 0 & g & 0 \\ 0 & 0 & 0 & h & 0 \end{bmatrix}. \qquad (27)$$

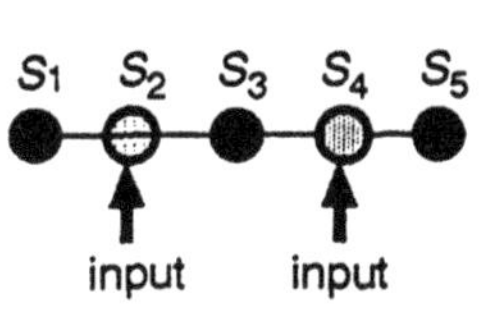

Fig. 5. Another example of symmetry induced by failures in $\mathcal{S}_1$, $\mathcal{S}_3$ and $\mathcal{S}_5$ of a homogeneous system. This system is symmetric, i.e., uncontrollable.

Let us find T that satisfies (6) and (7) according to the procedure shown in Proof of Theorem 1.

Some calculations bring

$$\boldsymbol{x} = \boldsymbol{y} = \begin{bmatrix} 1 & 0 & -1 & 0 & 1 \end{bmatrix}^{\mathrm{T}} \qquad (28)$$

that satisfy (9) and (13). It follows from (14) that

$$T = \begin{bmatrix} 2 & 0 & -1 & 0 & 1 \\ 0 & 1 & 0 & 0 & 0 \\ -1 & 0 & 2 & 0 & 1 \\ 0 & 0 & 0 & 1 & 0 \\ 1 & 0 & -1 & 0 & 2 \end{bmatrix}, \qquad (29)$$

which satisfies the equations (6) and (7). Hence the symmetry, i.e., the uncontrollability of the system after the failure. □

In the examples above, there is no need to calculate the controllability matrix. This investigation of uncontrollability is based merely on structural information of systems. Therefore, every system in a set of ring-type homogeneous systems consisting of four subsystems becomes uncontrollable by aforementioned failures. The same is true of a set of string-type homogeneous systems consisting of five subsystems. Note that structural information utilized above is based on the homogeneity of subsystems and of interconnections among subsystems, so that structural zeros (Van der Woude, 1991) are adopted. The relationthip with the study of structured systems (for example, see Lin, 1974; Dion and Commault, 1993) is a topic of further investigation.

5. CONCLUSION

In this paper, a symmetry of systems is defined as an invariance under state transformations. With this definition, it is shown that symmetric systems are uncontrollable. Structural symmetry has an advantage to simplify design and analysis of the systems. But when systems have the symmetry, *symmetry may prevent developments, and become an obstacle which should be broken when the right time comes*(Caillois, 1976). Consequently, It is required to introduce a kind of dissymmetry. Similarly, homogeneity of systems is normally favorable, especially for analysis of large complex systems. However, it is illustrated here that special failure patterns in homogeneous systems induce symmetry, so that the systems become uncontrollable. Concerning to this, failure patterns which induce symmetry in homogeneous systems have been further investigated by Tanaka *et al.*(1995).

The authors wish to express our gratitude to Mr. S. Iwata of Kyoto University and Mr. Y. Oishi of the University of Tokyo for their effective suggestion in the course of this research.

6. REFERENCES

Caillois, R. (1976). La Dissymétrie. Gallimard, Paris.

Dion, J. M., and C. Commault (1993). Feedback Decoupling of Structured Systems. *IEEE Trans. Aut. Cont.*, **38**, 1132–1135.

Grizzle, J. W., and S. I. Marcus (1985). The Structure of Nonlinear Control Systems Possessing Symmetries. *IEEE Trans. Aut. Cont.*, **30**, 248–257.

Habib, M. K., H. Asama, Y. Ishida, A. Matsumoto and I. Endo (1992). Simulation Enviroment for An Autonomous and Decentralized Multi-Agent Robot System. *IEEE Proc. of Int. Conf. on Robotics & Automation*, 1550–1557.

Ihara, H. and K. Mori (1984). A. D. Computer Control Systems. *IEEE COMPUTER*, **17**, 57–66.

Inui, T., Y. Tanabe, and Y. Onodera (1976). Applied Group Theory. Shokabo, Tokyo.

Ito, M., *et al.* (1994). Autonomous Decentralized Systems. *Technical Report of Scientific Foundation of Ministry of Education*, Japan.

Kailath, T. (1980). Linear Systems. Printice-Hall, Inc.

Lin, C-T. (1974). Structural Controllability. *IEEE Trans. Aut. Cont.*, **19**, 201–208.

Lunze, J. (1989). Stability Analysis of Large-scale Systems Composed of Strongly Coupled Similar Subsystems. *Automatica*, **25**, 561–570.

Mori, K., K. Sano, and H. Ihara (1981). Autonomous Controllability of Decentralized System Aiming at Fault-tolerance. *Proc. eighth IFAC World Congress*, 1833–1839.

Nijemeijer, H., and A. J. van der Schaft (1985). Partial Symmetries for Nonlinear Systems. *Mathematical Systems Theory*, **18**, 79–96.

Shin, S., and K. Ikeda (1993). Autonomous decentralized adaptive control. *Proc. second IEEE Int. Workshop on Emerging Technologies and Factory Automation*, 204–208.

Sundareshan, M.K., and R. M. Elbanna (1991). Qualitative Analysis and Decentralized Controller Synthesis for a Class of Large-scale Systems with Symmetrically Interconnected Subsystems. *Automatica*, **27**, 383–388.

Tanaka, R., S. Shin, and N. Sebe (1994). Controllability of Autonomous Decentralized Systems. *Proc. third Seiken/IEEE Symposium on Emerging Technologies and Factory Automation*, 265–272.

Tanaka, R., S. Iwata, S. Shin, and N. Sebe (1995). Fault-tolerance Evaluation of Homogeneous Systems by Autonomous Controllability. Submitted to CDC '95.

Van der Woude, J. W. (1991). On the structure at infinity of a structured system. *Linear Algebra and its Applications*, **148**, 145-169.

Xiaoping, L. (1992). Output Regulation of Strongly Coupled Symmetric Composite Systems. *Automatica*, **28**, 1037–1041.

STRICTLY DOUBLY COPRIME FACTORIZATIONS: THE DISCRETE TIME CASE

P. HIPPE

Institut für Regelungstechnik, Universität Erlangen-Nürnberg, Cauerstr. 7, D-91058 Erlangen, Germany

Abstract. Doubly coprime factorizations (DCF) are defined as stable proper rational matrices, and they are used in the fractional approach to all stabilizing compensators. The DCFs related to reduced order observers are partly improper, a problem which can be overcome by introducing artificial stable dynamics, the so-called identity elements. These cancelling factors are contrary to the usual notion of coprimeness. In an earlier paper, strictly doubly coprime factorizations (SDCF) were introduced, which do not contain identity elements, and which are consequently not all proper. It was proven, however, that these SDCFs can equally be used in the parametrization of all stabilizing compensators by a simple modification of the parametrizing matrix. Here, the SDCFs for discrete time systems are investigated.

Keywords. Linear systems, discrete time systems, reduced order observers, strictly doubly coprime factorizations, multivariable systems.

1. INTRODUCTION

Transfer matrices can either be represented in matrix fraction descriptions (MFD) or in doubly coprime fractional representations (DCF). While the MFDs use mutually coprime pairs of polynomial matrices, the DCFs incorporate eight stable rational quantities satisfying the Bezout identity. Following the lines of Nett et al. (1984), the DCFs can directly be related to the observer based compensation problem. Hippe (1989b) showed that an extention of these ideas to the reduced order observer case is possible. This yields, however, partly improper factorizations, which is contrary to the usual definition of DCFs, given e.g. by Vidyasagar (1985).

By an introduction of additional stable factors, so-called identities in the ring of stable rational proper functions, also the DCFs related to reduced order observers can be made stable and proper (Telford and Moore, 1989; Antsaklis, 1986). These cancelling factors have no meaning in an observer based compensation scheme, they are contrary to the usual notion of coprimeness, and they are superfluous. This has been demonstrated in Hippe (1992) and Hippe (1995), where strictly doubly coprime factorizations (SDCF) without identities have been defined, and where it was shown, that also the possibly improper SDCFs can be used in the parametrization of all stabilizing compensators.

Though the continuous and the discrete time cases are closely related, the continuous time results presented in Hippe (1990) cannot directly be carried over to the discrete time case. Beyond that, the use of an innovated state estimate $\hat{x}^+$ for state feedback has no correspondence in the continuous time.

First, the notations for the observer based compensation scheme in discrete time are introduced, and then the SDCFs related to reduced order observer based compensators are presented for both cases, namely when the one step prediction estimate, and when the innovated state estimate are used for state feedback.

2. STATE FEEDBACK AND REDUCED ORDER OBSERVERS

We consider completely controllable and observable linear time invariant discrete time systems

$$\begin{aligned} x(k+1) &= Ax(k) + Bu(k) \\ y(k) &= Cx(k) \end{aligned} \qquad (1)$$

with $x\epsilon\mathbb{R}^n$, $y\epsilon\mathbb{R}^p$, $u\epsilon\mathbb{R}^m$. Joining the κ outputs to be used for the reconstruction of x in y_2, the output vector y is subdivided into

$$y(k) = \begin{bmatrix} y_1(k) \\ y_2(k) \end{bmatrix} = \begin{bmatrix} C_1 \\ C_2 \end{bmatrix} x(k) \tag{2}$$

with $y_1\epsilon\mathbb{R}^{p-\kappa}$, $y_2\epsilon\mathbb{R}^{\kappa}$. A reduced order observer of order $n-\kappa$

$$\zeta(k+1) = F\zeta(k)+[H_1 \ H_2]\begin{bmatrix} y_1(k) \\ y_2(k) \end{bmatrix}+TBu(k) \tag{3}$$

yields $\zeta(k) = Tx(k)$ in steady state, if

$$TA - FT = [H_1 \ H_2] \begin{bmatrix} C_1 \\ C_2 \end{bmatrix} \tag{4}$$

holds (Luenberger, 1971). If C_2 and T are linearly independent the state estimate $\hat{x}$ is given by

$$\hat{x} = \begin{bmatrix} C_2 \\ T \end{bmatrix}^{-1} \begin{bmatrix} y_2 \\ \zeta \end{bmatrix} = [\Psi_2 \ \Theta] \begin{bmatrix} y_2 \\ \zeta \end{bmatrix} \quad . \tag{5}$$

In the sequel it will be of importance that

$$\begin{bmatrix} C_2 \\ T \end{bmatrix} [\Psi_2 \ \Theta] = \begin{bmatrix} I_\kappa & 0 \\ 0 & I_{n-\kappa} \end{bmatrix} \tag{6}$$

and

$$\Psi_2 C_2 + \Theta T = I_n \tag{7}$$

are implied by (5). With L_1 such that $TL_1 = H_1$ the observer (3) can equally be represented by

$$\begin{aligned} \zeta(k+1) &= T(A - L_1C_1)\Theta\zeta(k) \\ &+ \Big[TL_1 \vdots T(A-L_1C_1)\Psi_2\Big]\begin{bmatrix} y_1(k) \\ y_2(k) \end{bmatrix}+TBu(k). \end{aligned} \tag{8}$$

In the continuous time case (Hippe, 1990) a choice of $L_1 = \Theta H_1$, yielding $C_2L_1 = 0$, was motivated by the optimal reduced order estimation scheme (reduced order Kalman filter (Hippe, 1989c)). However, for discrete time systems the optimal reduced order estimation scheme does not imply $C_2L_1 = 0$, (Hippe and Wurmthaler, 1990) and consequently the following developments differ from the continuous time case. Substituting in the state feedback law $u = -Kx$ the state x by the (one step prediction) estimate $\hat{x}$ (5), the compensator transfer matrix $u(z) = -F_C(z)y(z)$, is given by

$$\begin{aligned} F_C(z) &= K\Theta\,[zI - T(A - L_1C_1 - BK)\Theta]^{-1} \\ &\Big[TL_1 \vdots T(A - L_1C_1 - BK)\Psi_2\Big] + \Big[0 \vdots K\Psi_2\Big] \ . \end{aligned} \tag{9}$$

This transfer matrix can be represented in a left coprime matrix fraction description (MFD) $F_C(z) = d_C^{-1}(z)n_C(z)$ or in a right coprime MFD $F_C(z) = \bar{n}_C(z)\bar{d}_C^{-1}(z)$. Similarly the system transfer matrix $F(z) = C(zI-A)^{-1}B$ can be represented in a right coprime MFD $F(z) = n(z)d^{-1}(z)$ or in a left coprime MFD $F(z) = \bar{d}^{-1}(z)\bar{n}(z)$. Correspondingly the system and the compensator transfer matrices $F(s)$ and $F_C(s)$ can be represented in doubly coprime fractional representations

$$F(z) = N(z)D^{-1}(z) = \bar{D}^{-1}(z)\bar{N}(z) \tag{10}$$

and

$$F_C(z) = D_C^{-1}(z)N_C(z) = \bar{N}_C(z)\bar{D}_C^{-1}(z) \tag{11}$$

where the eight rational matrices N, D, $\bar{N}$, $\bar{D}$, N_C, D_C, $\bar{N}_C$, $\bar{D}_C$ have stable denominator polynomials (with roots inside the unit circle), and the Bezout identity

$$\begin{bmatrix} D_C & N_C \\ -\bar{N} & \bar{D} \end{bmatrix} \begin{bmatrix} D & -\bar{N}_C \\ N & \bar{D}_C \end{bmatrix} = \begin{bmatrix} I_m & 0 \\ 0 & I_p \end{bmatrix} \tag{12}$$

holds.

3. STRICTLY DOUBLY COPRIME FACTORIZATIONS RELATED TO REDUCED ORDER OBSERVERS

A natural way to obtain the eight rational matrices in the DCF is to relate the polynomial matrices of the plant and of the compensator MFDs to the (stable) polynomial matrices characterizing the controlled plant and the observer dynamics (Hippe, 1989a and 1989b). In Nett et al. (1984) the case of identity observers was considered and consequently, all eight quantities in (10) and (11) were proper, since here, plant and observer/compensator orders coincide.

When using reduced order observer based compensators, the rational matrix $\bar{D}$ (which in the scalar case is the plant denominator polynomial divided by the characteristic polynomial of the reduced order observer) is no longer proper (Hippe, 1989b and 1990). Therefore, attempts have been made to obtain proper factorizations also for reduced order observers (Telford and Moore, 1989; Antsaklis, 1986), and this was achieved by introducing cancelling (stable) factors, the so-called identities in the ring of proper stable rational matrices. The definition of DCFs allows for such factors (Vidyasagar, 1985), but they are definitely contrary to the usual notion of coprimeness, and they are superfluous. The continuous time factorizations introduced in (Hippe, 1989b and 1990) related to reduced order observers do not contain identities, and they are consequently coprime in the usual sense. Therefore, they have been named strictly doubly coprime factorizations (SDCF) in Hippe (1992). Though not all SDCFs are proper, they can equally be used in the parametrization of all stabilizing compensators (Hippe, 1995).

Here the continuous time results of Hippe (1990) shall be generalized to the discrete time case. The block diagrams of Hippe (1990) remain valid, if $s^{-1}I$ is replaced by $z^{-1}I$. However, the actual results of Hippe (1990) cannot simply be transferred to the discrete time case by substituting s by z, because of the fact $C_2L_1 \neq 0$ (see above). Following the lines of Hippe (1990), but assuming $C_2L_1 \neq 0$, the strictly doubly coprime factorizations related to discrete time observers of arbitrary orders are given by Theorem 1.

Theorem 1. Suppose $F(z) = C(zI - A)^{-1}B$ describes an nth order completely controllable and observable system with m inputs and p outputs, and $F_C(z)$ as defined by (9) describes a reduced order compensator of order $n - \kappa$ for this system. With K and $F = TA_1\Theta$, where $A_1 = A - L_1C_1$, such that $A - BK$ and F have eigenvalues inside the unit circle, define

$$N(z) = C(zI - A + BK)^{-1}B \tag{13}$$
$$D(z) = I - K(zI - A + BK)^{-1}B \tag{14}$$

$$D_C(z) = I + K\Theta(zI - F)^{-1}TB \tag{15}$$
$$N_C(z) = K\Theta(zI - F)^{-1}\left[TL_1 \vdots TA_1\Psi_2\right] + \left[0 \vdots K\Psi_2\right] \tag{16}$$

$$\bar{D}(z) = \begin{bmatrix} I - C_1\Theta(zI - F)^{-1}TL_1 & \vdots & -C_1\left[I + \Theta(zI - F)^{-1}TA_1\right]\Psi_2 \\ -C_2\left[I + A_1\Theta(zI - F)^{-1}T\right]L_1 & \vdots & C_2\left[zI - A_1 - A_1\Theta(zI - F)^{-1}TA_1\right]\Psi_2 \end{bmatrix} \tag{17}$$

$$\bar{N}(z) = \begin{bmatrix} C_1\Theta(zI - F)^{-1}TB \\ C_2\left[I + A_1\Theta(zI - F)^{-1}T\right]B \end{bmatrix} \tag{18}$$

$$\bar{N}_C(z) = K(zI - A + BK)^{-1}\left[L_1 \vdots \Psi_2\right] \tag{19}$$

$$\bar{D}_C(z) = C(zI - A + BK)^{-1}\left[L_1 \vdots \Psi_2\right] + \begin{bmatrix} I_{p-\kappa} & 0 \\ 0 & 0_\kappa \end{bmatrix}. \tag{20}$$

Then

(i) all eight matrices (13) – (20) have stable denominator polynomials

(ii) $D(z)$, $\bar{D}(z)$, $D_C(z)$ and $\bar{D}_C(z)$ are nonsingular

(iii) $F(z) = N(z)D^{-1}(z) = \bar{D}^{-1}(z)\bar{N}(z)$

(iv) $F_C(z) = D_C^{-1}(z)N_C(z) = \bar{N}_C(z)\bar{D}_C^{-1}(z)$

(v) $\begin{bmatrix} D_C(z) & N_C(z) \\ -\bar{N}(z) & \bar{D}(z) \end{bmatrix}\begin{bmatrix} D(z) & -\bar{N}_C(z) \\ N(z) & \bar{D}_C(z) \end{bmatrix} = \begin{bmatrix} I_m & 0 \\ 0 & I_p \end{bmatrix}.$

Proof of Theorem 1: Part (i) follows from the fact that the zeros of $\det(zI - A + BK)$ and of $\det(zI - F) \hat{=} \det\left[zI - T(A - L_1C_1)\Theta\right]$ lie within the unit circle of the z-plane.

Part (ii): the matrices $D(z)$ and $D_C(z)$ are nonsingular because of the identity matrix appearing in both expressions. Because of $C_2\Psi_2 = I$ the polynomial parts of $\bar{D}(z)$ and $\bar{D}_C(z)$ have nonvanishing determinants and consequently, also these matrices are nonsingular. The proofs for (iii) through (v) can be developed in a similar fashion as in Hippe (1990) for the continuous time case, and they have to be omitted here for brevity.

The main difference between the continuous and the discrete time cases stems from the fact that $C_2L_1 = 0$ does not hold in the discrete time case. The optimal linear estimator (Kalman Filter) in the continuous time case (Hippe, 1989c) is characterized by an optimal feedback matrix L_1 which meets the condition $C_2L_1 = 0$. In Hippe and Wurmthaler (1990) it was shown, that this is not the case for the optimal discrete time estimator.

Though the rational matrix $\bar{D}(z)$ is no longer proper, the above factorizations (13) – (20) can equally be used in the parametrization of all stabilizing compensators by a simple modification of the parametrizing matrix (Hippe, 1992 and 1995).

4. DOUBLY COPRIME FRACTIONAL REPRESENTATIONS RELATED TO THE UPDATED STATE ESTIMATE

In the discrete time case the optimal feedback matrix L_1 of the reduced order Kalman filter is such that

$$L_1 = A\lambda_1 \tag{21}$$

with $C_2\lambda_1 = 0$ holds (Hippe and Wurmthaler, 1990).

Instead of $\hat{x}$ as defined above the updated state estimate

$$\hat{x}^{+}(k) = \hat{x}(k) + \lambda_1 [y_1(k) - C_1\hat{x}(k)] \tag{22}$$

is often used (Gelb, 1974). Substituting this state estimate in the state feedback $u = -Kx$, the compensator transfer matrix is given by

$$\begin{aligned} F_C^{+}(z) &= K(I-\lambda_1 C_1)\Theta\,[zI - T(A-BK) \\ &(I-\lambda_1 C_1)\Theta]^{-1} T(A-BK)\Big[\lambda_1 \vdots (I-\lambda_1 C_1)\Psi_2\Big] \\ &+ \Big[K\lambda_1 \vdots K(I-\lambda_1 C_1)\Psi_2\Big]. \end{aligned} \tag{23}$$

The corresponding DCFs are then given by the following theorem.

Theorem 2. Given an nth order completely controllable and observable linear system with m inputs and p outputs, with its discrete transfer matrix $F(z) = C(zI-A)^{-1}B$, and an $(n-\kappa)$th order (optimal) observer whose updated state estimate is fed back via $u = -K\hat{x}^{+}$ yielding the reduced order compensator transfer matrix (23). Now select K and $F = TA_\lambda\Theta$ with $A_\lambda = A(I-\lambda_1 C_1)$ such that the eigenvalues of $A - BK$ and F lie within the unit circle of the z-plane.

Define

$$N^{+}(z) = C(zI - A + BK)^{-1}B \tag{24}$$

$$D^{+}(z) = I - K(zI - A + BK)^{-1}B \tag{25}$$

$$D_C^{+}(z) = I + K(I-\lambda_1 C_1)\Theta(zI-F)^{-1}TB \tag{26}$$

$$\begin{aligned} N_C^{+}(z) &= K(I-\lambda_1 C_1)\Theta(zI-F)^{-1} \\ &\Big[TA\lambda_1 \vdots TA_\lambda\Psi_2\Big] + K\Big[\lambda_1 \vdots (I-\lambda_1 C_1)\Psi_2\Big] \end{aligned} \tag{27}$$

$$\bar{D}^{+}(z) = \begin{bmatrix} I - C_1\Theta(zI-F)^{-1}TA\lambda_1 & \vdots & -C_1\left[I+\Theta(zI-F)^{-1}TA_\lambda\right]\Psi_2 \\ -C_2\left[A + A_\lambda\Theta(zI-F)^{-1}TA\right]\lambda_1 & \vdots & C_2\left[zI - A_\lambda - A_\lambda\Theta(zI-F)^{-1}TA_\lambda\right]\Psi_2 \end{bmatrix} \tag{28}$$

$$\bar{N}^{+}(z) = \begin{bmatrix} C_1\Theta(zI-F)^{-1}TB \\ C_2\left[I + A_\lambda\Theta(zI-F)^{-1}T\right]B \end{bmatrix} \tag{29}$$

$$\begin{aligned} \bar{N}_C^{+}(z) &= K(zI-A+BK)^{-1}\Big[(A-BK)\lambda_1 \vdots \Psi_2\Big] \\ &\quad + \Big[K\lambda_1 \vdots 0\Big] \end{aligned} \tag{30}$$

$$\begin{aligned} \bar{D}_C^{+}(z) &= C(zI-A+BK)^{-1}\Big[(A-BK)\lambda_1 \vdots \Psi_2\Big] \\ &\quad + \begin{bmatrix} I_{p-\kappa} & 0 \\ 0 & 0_\kappa \end{bmatrix}. \end{aligned} \tag{31}$$

Then

(i) all eight matrices (24)–(31) have stable denominator polynomials

(ii) $D^{+}(z)$, $\bar{D}^{+}(z)$, $D_C^{+}(z)$ and $\bar{D}_C^{+}(z)$ are nonsingular

(iii) $F(z) = N^{+}(z)D^{+-1}(z) = \bar{D}^{+-1}\bar{N}^{+}(z)$

(iv) $F_C^{+}(z) = D_C^{+-1}(z)N_C^{+}(z) = \bar{N}_C^{+}(z)\bar{D}_C^{+-1}(z)$

(v)
$$\begin{bmatrix} D_C^{+}(z) & N_C^{+}(z) \\ -\bar{N}^{+}(z) & \bar{D}^{+}(z) \end{bmatrix}\begin{bmatrix} D^{+}(z) & -\bar{N}_C^{+}(z) \\ N^{+}(z) & D_C^{+}(z) \end{bmatrix} = \begin{bmatrix} I_m & 0 \\ 0 & I_p \end{bmatrix}.$$

The proof of this Theorem goes along the same lines as the proof for Theorem 1.

5. CONCLUSIONS

Starting from the continuous time results, strictly doubly coprime representations have been presented for discrete time systems. They cover the case of arbitrary observer orders $n-\kappa$ within the range $0 < \kappa < p$. For $\kappa = 0$ observer and system orders coincide and consequently, all factorizations are proper and stable. For $\kappa > 0$ some fractional quantities become strictly proper and others (in corresponding parts) become improper. This, however, does not cause any problems in the parametrization of all stabilizing compensators, when the parametrizing matrix is modified in an appropriate manner (Hippe, 1992 and 1995). Compared to DCFs containing identities, such SDCFs have the following benefits, as they

- are directly related to the two dual closed loop representations of observer based compensation schemes (Hippe, 1991)
- directly appear in the frequency domain design of observer based compensators (Hippe, 1991)
- contain polynomials of smallest possible degrees, which is desirable, as numerical problems rapidly increase with polynomial degree.

Therefore they should always be preferred to factorizations containing identities.

6. REFERENCES

Antsaklis, P.J. (1986). Proper stable transfer matrix factorizations and internal system descriptions. *IEEE Trans. Automat. Control 31*, 634–638.

Gelb, A. (1974). Applied Optimal Estimation. Cambridge, Mas.: The MIT Press.

Hippe, P. (1989a). The computation of doubly coprime fractional representations directly in the frequency domain. *IEEE Conference on Control and Applications - ICCON '89*, Jerusalem, April 3–6, RA–2–5.

Hippe, P. (1989b). Modified doubly coprime fractional representations related to the reduced order observer, *IEEE Trans. Autom. Control.*, AC–34, 573–576.

Hippe, P. (1989c). Design of reduced order optimal estimators directly in the frequency domain. *Int. J. Control*, 50, 2599–2614.

Hippe, P. (1990). A nonminimal representation of reduced order observers. *Automatica 26*, 405–409.

Hippe, P. (1991). Design of observer based compensators: The polynomial approach. *Kybernetika*, 27, 125–150.

Hippe, P. (1992). Strictly doubly coprime factorizations related to reduced order observers. *2nd IFAC Workshop on System Structure and Control*, 3.–5.9.92, Prague.

Hippe, P. (1995). Strictly doubly coprime factorizations and all stabilizing compensators related to reduced order observers. *Automatica*, 31, to appear.

Hippe, P., and Wurmthaler, Chr. (1990). Optimal reduced order estimators in the frequency domain: The discrete time case. *Int. J. Control 52*, 1051–1064.

Luenberger, D.G. (1971). An introduction to observers. *IEEE Trans. Automat. Contr.* AC–16, 596–603.

Nett, C.N., Jacobson C.A. and Balas, M.J. (1984). A Connection between state–space and doubly coprime fractional representations. *IEEE Trans. Automat. Contr.*, AC–29, 831–832.

Telford, A.J. and Moore, J.B. (1989). Doubly coprime factorizations, reduced order observers and dynamic state estimate feedback. *Int. J. Control 50*, 2583–2597.

Vidyasagar, M. (1985). *Control system synthesis: A factorization approach.* Cambridge, Mass.

IMAGE REPRESENTATION OF CONTROLLABLE BEHAVIORS OVER A PRINCIPAL IDEAL DOMAIN

FABIO FAGNANI* and **SANDRO ZAMPIERI****

**Scuola Normale Superiore, Piazza dei Cavalieri 7, 56126 Pisa, Italy*

***Dipartimento di Elettronica ed Informatica, Università di Padova, via Gradenigo 6/a, 35131 Padova, Italy*

Abstract. In this paper we study systems over rings using the behavioral approach. Attention is here mainly focused on the case the base ring is a principal ideal domain. In this case a systematic study of matrix shift operators is presented and it is then used to obtain results on the structure and representation of controllable systems.

Résumé. On montre que un système commandable sur un domain à ideaux principaux admet une répresentation image de type injectif. Ce résultat pent être obtenu à l'aide de qualques proprietés des opérateurs de translation.

Key Words. Behavioral approach; controllability; convolutional codes; Laurent polynomials; principal ideal domains; shift operators; symbolic dynamics; systems over rings

1. INTRODUCTION

The study of the dynamical systems whose signals take value over groups, rings, and modules has been the object of a great attention for various reasons. First, since it seemed to be the most promising way to develop some important extensions of the classical linear systems theory, such as the theory of continuous linear systems with delays (Kamen, 1975), the theory of parametrized systems (Hautus and Sontag, 1985), and the theory of distributed parameter systems (Brockett and Willems, 1974).

Moreover, in the early 70's, with the introduction of the convolutional codes, the importance of linear systems theory became apparent also for the coding theory. In this framework the encoder could be interpreted as an input/output dynamical system defined over a finite field. Many of the results that were available for linear systems over $\mathbb{R}$ or $\mathbb{C}$ could also be applied in the theory of convolutional codes. Viceversa, some results, such as canonical state realizations based on minimal bases representations (Forney, 1970), that have been found by researchers working on convolutional codes, gave a great contribution to linear systems theory. However, in coding theory finite fields are sometimes too restrictive and there has been the necessity to extend the theory of convolutional codes to less structured alphabet sets. This was another motivation to the development of a theory of linear systems over rings. Rather complete surveys of this topic can be found in (Brewer *et al.*, 1986)

In the last years an alternative approach to linear systems theory, called the behavioral approach, has been proposed by Willems (Willems, 1988). According to this approach, a linear system is characterized by a time set T (which can be either $\mathbb{Z}$ or $\mathbb{R}$), a signal alphabet W which is a finite-dimensional vector space over a field k and by a behavior $\mathcal{B}$ that is a k-subspace of W^T and that specifies the dynamics of the system simply listing what are the allowed trajectories. Usually, the behavior $\mathcal{B}$ is assumed to be invariant with respect to the standard backward shift operator σ (Willems, 1988).

The behavioral approach appeared to be a nice framework where quite general problems of modelling and identification can be formulated and solved (Willems, 1986). One of the key feature of this approach is the fact that, differently from the classical one, there is no *a priori* distinction between input and output signals. From a system theoretic point of view this has the advantage of allowing a more general type of interaction of a system with the external environment and a wider range of system interconnections: this is of concrete utility in the analysis of electrical networks. On the other hand, this feature is also one of the reasons why this approach has been analyzed and appreciated also by the coding theory community. Indeed a code can be seen as a behavior and the language of the behavioral approach seems to be the more natural in the formulation and solutions of convolutional codes problems, too (Forney and Trott, 1993; Loeliger *et al.*, 1994; Fornasini and Valcher, 1994).

For the same reasons mentioned above, researchers considered interesting to extend the behavioral approach to systems over more general alphabets. This is in fact what has been done in (Forney and Trott, 1993; Loeliger *et al.*, 1994; Zampieri and Mitter, 1994), that constitute the first attempts to extend the properties that hold for linear systems over fields to systems over groups and to systems over rings. In (Forney and Trott, 1993) it is shown that the concept of canonical state space can be introduced even for systems over noncommutative groups and that many properties hold true in this setup. In (Zampieri and Mitter, 1994) linear systems over Noetherian rings are analyzed. It is shown that two important categories of systems, i.e. autonomous systems and controllable systems, maintain some nice properties.

Finally, the behavioral approach to systems over groups has shown surprising contacts with symbolic dynamics over groups (Kitchens and Schmidt, 1989). This relationship has been analyzed in (Fagnani, 1994*b*; Fagnani, 1994*a*).

1.1. *Notation and preliminaries*

We now introduce all the basic concepts and notation. If A is a commutative ring, denote by $A^{l\times q}$ the A-module of $l \times q$ matrices with entries in A and by A^* the multiplicative group of units of A.

In the sequel R will always denote a commutative ring with identity and V, V_i, W will always denote finitely generated R-modules. $R[u,u^{-1}]$ is the ring of Laurent polynomials with coefficients in R. $V[u,u^{-1}] := V \otimes_R R[u,u^{-1}]$ is the $R[u,u^{-1}]$-module of Laurent polynomials with coefficients in V. $V^{\mathbb{Z}}$ denotes the R-module of bi-infinite sequences over V. On $V^{\mathbb{Z}}$ we can introduce a module structure over the ring $R[u,u^{-1}]$ by defining, for $v \in V^{\mathbb{Z}}$,

$$(u \cdot v)(t) := (\sigma v)(t) := v(t+1) \quad \forall t \in \mathbb{Z}. \qquad (1)$$

σ is called the *shift operator*. From now on we will always assume that any finitely generated R-module V is equipped with the discrete topology and $V^{\mathbb{Z}}$ with the corresponding product topology. It is clear that the topology just introduced on $V^{\mathbb{Z}}$ is metrizable and a sequence $w_n \in V^{\mathbb{Z}}$ converges to $w \in V^{\mathbb{Z}}$ if and only if $w_n(t) \to w(t)$ for all $t \in \mathbb{Z}$: for this reason this topology is called *pointwise convergence topology*. An *R-behavior* on V is any closed $R[u,u^{-1}]$-submodule $\mathcal{B} \subseteq V^{\mathbb{Z}}$. $V^{\mathbb{Z}}$ is called the *full* R-behavior over V. Closure in the pointwise convergence topology is equivalent to the important system theoretic notion of completeness (Willems, 1988).

Let $\mathcal{B} \subseteq V^{\mathbb{Z}}$ be an R-behavior. $\mathcal{B}$ is said to be *controllable* (Willems, 1988) if for all $v_1, v_2 \in \mathcal{B}$, there exists $n \in \mathbb{N}$ and $v \in \mathcal{B}$ with

$$v(t) = v_1(t) \ \forall t < 0, \quad (\sigma^n v)(t) = v_2(t) \ \forall t \geq 0.$$

It is shown in (Zampieri and Mitter, 1994) that, if V is Noetherian, then, every controllable $\mathcal{B}$ over V satisfies a stronger notion of controllability in the sense that the number $n \in \mathbb{N}$ in the previous definition can be chosen a priori for all pairs $v_1, v_2 \in \mathcal{B}$.

A weaker notion of controllability has been introduced in (Fagnani, 1994*a*) as follows. Let $\mathcal{B}$ be an R-behaviour and let $\mathcal{B}_f$ be the R-submodule of $\mathcal{B}$ consisting of finite support trajectories in $\mathcal{B}$. Then $\mathcal{B}$ is called *weakly controllable* if $\overline{\mathcal{B}_f} = \mathcal{B}$. It is easy to see that controllability indeed implies weak controllability. We will see in this paper that these two concepts coincide in the case R is a PID.

If $I \subseteq \mathbb{Z}$, denote by $\mathcal{B}_{|I}$ the R-module of restrictions to I of the bi-infinite sequences in $\mathcal{B}$. We recall now the concept of memory (Willems, 1988). $\mathcal{B}$ is said to have *memory* $n \in \mathbb{N}$ if

$$v \in V^{\mathbb{Z}} \text{ and } v_{|[t,t+n]} \in \mathcal{B}_{|[t,t+n]} \ \forall t \in \mathbb{Z} \Rightarrow v \in \mathcal{B}.$$

$\mathcal{B}$ is said to have *finite memory* (or to be of *finite type*), if it has memory n for some $n \in \mathbb{N}$. If V is Artinian, every R-behavior over V has automatically finite memory (Zampieri and Mitter, 1994).

Two R-behaviors $\mathcal{B}_1 \subseteq V_1^{\mathbb{Z}}$ and $\mathcal{B}_2 \subseteq V_2^{\mathbb{Z}}$ are said to be *R-equivalent* (and we write $\mathcal{B}_1 \simeq \mathcal{B}_2$) if there exists an invertible $R[u,u^{-1}]$-homomorphism $\phi : \mathcal{B}_1 \to \mathcal{B}_2$ which is also a homeomorphism. Such a ϕ will be called an *R-equivalence*.

We have the following result.

Proposition 1 Let V, W be R-modules and let $\mathcal{B} \subseteq V^{\mathbb{Z}}$ be an R-behavior. A map $\phi : \mathcal{B} \to W^{\mathbb{Z}}$ is a continuous $R[u,u^{-1}]$-homomorphism if and only if there exists numbers $h, k \in \mathbb{N}$ and $\zeta \in \mathrm{Hom}_R(\mathcal{B}_{|[0,h+k]}, W)$ such that

$$\phi(v)(t) = \zeta(v_{|[t-h,t+k]}) \quad \forall v \in \mathcal{B},\ t \in \mathbb{Z}. \qquad (2)$$

Remark Operators like ϕ in (2) are also called *sliding block operators* in the coding theory literature and in symbolic dynamics since if $v \in \mathcal{B}$ then the value of $(\phi v)(t)$ depends on the values of v only in the finite block $[t-h, t+k]$ which slides with t.

It follows from (1) that a polynomial $p \in R[u, u^{-1}]$ can be interpreted as a $R[u, u^{-1}]$-homomorphism

$$p(\sigma, \sigma^{-1}) : R^{\mathbb{Z}} \to R^{\mathbb{Z}}. \tag{3}$$

In an analogous way, if V, W are R-modules, every $M \in \mathrm{Hom}_R(V, W)[u, u^{-1}]$ naturally induces a $R[u, u^{-1}]$-homomorphism

$$M(\sigma, \sigma^{-1}) : V^{\mathbb{Z}} \to W^{\mathbb{Z}}. \tag{4}$$

The homomorphisms defined in this way are called *shift operators*. In case $V = R^q$ and $W = R^l$ we identify $\mathrm{Hom}_R(V, W)[u, u^{-1}]$ with $R[u, u^{-1}]^{l \times q}$: the $l \times q$ matrices with coefficients in $R[u, u^{-1}]$. In this case, corresponding shift operators are also called *matrix shift operators*. A straightforward consequence of Proposition 1 is that all the continuous $R[u, u^{-1}]$-homomorphisms between full R-behaviors are given by shift operators.

It is easy to see that if $\mathcal{B}$ is the image through a continuous $R[u, u^{-1}]$-homomorphism of a controllable R-behavior, then also $\mathcal{B}$ is controllable. In particular, $\mathcal{B}$ is controllable if it is the image of a shift operator. The converse was proven in (Zampieri and Mitter, 1994) when V is a Noetherian module: in this case every controllable R-behavior over V is necessarily the image of a shift operator which is called an *image representation* for $\mathcal{B}$. In general, it is not possible to find an injective image representation as it is shown in (Fagnani, 1994*b*). On the other hand, it is shown in (Willems, 1988) that for linear systems ($R = k$ a field and V a finite-dimensional vector space) controllability is the necessary and sufficient condition to have that the behavior admits an injective image representation. Moreover, in this case, the inverse is automatically continuous and hence, a sliding block operator. A question that can be posed is thus under which conditions on the ring R and on the R-module V, we have that every controllable R-behavior over V admits an injective image representation or, even more, is R-equivalent to a full R-behavior. A related complementary question is, given R and V, which are the conditions, beside controllability, which characterize the R-behaviors over V which are R-equivalent to a full R-behavior. This second question has been answered in (Fagnani, 1994*b*) for $R = \mathbb{Z}$ and V a torsion R-module.

In the language of coding theory, a behavior is a code and when the behavior is the image of an operator, this can be interpreted as an encoder: it codes unconstrained sequences into the constrained sequences of a code. Encoders with sliding block structure are important in applications since the fact that they have to keep track of only a finite portion of the encoding sequence makes them suitable for concrete implementations. Injectivity of the encoder is of course a necessity for decoding. The inverse of the encoder, if it exists, is called the decoder and also for this, sliding block structure, and hence continuity, is particularly suitable. This should make clear the importance of the above questions in coding theory.

One more interpretation. By the point of view of topological dynamics, the pair $(\mathcal{B}, \sigma_{|\mathcal{B}})$ is a topological dynamical system with an additional R-module structure on the space $\mathcal{B}$. R-equivalence then is the usual topological conjugacy with the additional requirement of preserving the R-module structure. Previous questions can be seen then as problems of classifications of these topological dynamical systems up to this algebraic-topological conjugacy (Kitchens and Schmidt, 1989; Fagnani, 1994*b*; Fagnani, 1994*a*).

In this paper we focus on the case when R is a principal ideal domain (PID). In Section 2 we provide the characterization of some properties of matrix shift operators to algebraic properties of the corresponding inducing polynomial matrices. The results of Section 2, beside their intrinsic interest, are used, in Section 3, to prove Theorem 11. This is one of our main results: it asserts that if V is a finitely generated free R-module, then any controllable R-behavior over V is R-equivalent to a full R-behavior. After that, we prove an interesting result showing that controllability and weak controllabilty coincide for R-behaviors in the case R is a PID, and they both imply finite memory. The proofs of all the results presented here can be found in (Fagnani and Zampieri, 1994).

The most interesting case, for coding theory and for topological dynamics, is probably given by $R = \mathbb{Z}$. On the other hand, we believe that the case $R = \mathbb{R}[x]$ could be of potential interest to develop a behavioral approach to the theory of parametrized linear systems. These issues will be considered elsewhere. We are aware of the fact that there are many more systems theoretic issues to be studied even in the case when R is a PID. In particular, we have not considered here kernel representations which are nevertheless of central importance in the behavioral approach to systems. These issues will be treated in a forthcoming paper.

2. Matrix shift operators

In this section we will analyze the properties of matrix shift operators when R is a principal ideal domain. We start from the scalar case.

We first need to introduce some more notation. Let R be a unique factorization domain (UFD).

The rank of $M \in R^{l\times q}$ is defined as the usual rank with respect to the field of fractions of R.

A multiplicatively closed set is a subset S of R that is closed under multiplication and such that $1 \in S$ and $0 \notin S$. A typical example is given by $R_0 := R \setminus \{0\}$ which is actually the largest multiplicatively closed subset of R. A multiplicatively closed subset $S \subseteq R$ is said to be saturated if for all $a, b \in R$ we have that $ab \in S \Leftrightarrow a, b \in S$. Note that a saturated multiplicatively closed set contains the set R^* of the invertible elements of R. Let $S \subseteq R$ be a saturated multiplicatively closed set. Define

$$\tilde{S} := \{a \in R | (b \notin R^*,\ b \text{ divides } a) \Rightarrow b \notin S\}. \quad (5)$$

It is easy to see that $\tilde{S}$ is also a saturated multiplicatively closed set in R and $\tilde{\tilde{S}} = S$. Moreover, $S \cap \tilde{S} = R^*$ and $R_0 = S\tilde{S} := \{st | s \in S,\ t \in \tilde{S}\}$.

Let $S \subseteq R$ be a multiplicatively closed subset. A full column rank matrix $M \in R^{l\times q}$ is said to be *right S-factor prime* if $M = M'F$ with $M' \in R^{l\times q}$ and $F \in R^{q\times q}$ yields $\det F \in S$. Similarly, a matrix M is said to be *left S-factor prime*, if M^T is right S-factor prime. It is easy to see that every full column rank $M \in R^{l\times q}$ can be factored as $M = M'\Lambda$ where $M' \in R^{l\times q}$ is right S-factor prime and $\Lambda \in R^{q\times q}$. In what follows we will need several results on left and right prime Laurent polynomial matrices which are gathered in the Appendix. These results have appeared in a slightly less general form in (Lèvy, 1981).

Let R be a UFD. Then also the ring $R[u, u^{-1}]$ is a UFD. Let $p = \sum_{i=l}^{L} p_i u^i \in R[u, u^{-1}]$ and suppose that p_l, p_L are nonzero. If $p_l, p_L \in R^*$ we say that p is *bimonic*. Denote by S_b the set of all the bimonic polynomials. It is easy to see that S_b is a saturated multiplicatively closed subset of $R[u, u^{-1}] \setminus \{0\}$. As defined above, Consider moreover the saturated multiplicatively closed set $\tilde{S}_b$ that is obtained from S_b as in (5). We have the follwing result.

Proposition 2 Let R be a Noetherian UFD. Consider a polynomial $p \in R[u, u^{-1}]$. Then $p(\sigma, \sigma^{-1})$ is an injective shift operator if and only if $p \in \tilde{S}_b$

Example: Let $R = \mathbb{Z}$ and $p = 2u - 1$. Consider $w \in \mathbb{Z}^{\mathbb{Z}}$ such that $p(\sigma, \sigma^{-1})w = 0$. Then, for all $t \in \mathbb{Z}$,

$$2w(t+1) = w(t).$$

Consequently

$$\begin{array}{ll} w(0) = 2w(1) & \\ w(1) = 2w(2) & w(0) = 4w(2) \\ w(2) = 2w(3) & w(0) = 8w(3) \\ \cdots & \cdots \end{array}$$

and so we argue that 2^t devides $w(0)$ for all $t \in \mathbb{N}$. This implies that $w(0) = 0$ and so that $w = 0$.

Define moreover Σ to be the saturated multiplicatively closed set of Laurent polynomials which induce surjective shift operators. It is easy to see that $S_b \subseteq \Sigma$. We have equality in some special cases.

Proposition 3 Let R be a semisimple (the intersection of its maximal ideals is 0) Noetherian UFD. Let $p \in R[u, u^{-1}]$. Then

1. $p(\sigma, \sigma^{-1})$ is a surjective shift operator if and only if $p \in S_b$
2. $p(\sigma, \sigma^{-1})$ is a bijective shift operator if and only if $p \in R[u, u^{-1}]^*$, i.e., p is a monomial with coefficient in R^*.

In the following example we show that Proposition 3 does not hold in general without the assumption of semisemplicity.

Example: Let $R = \mathbb{R}[[x]]$ and consider $p := x + u$. It follows from Proposition 2 that $p(\sigma, \sigma^{-1})$ is injective. We now prove that it is also surjective, in spite of the fact that p is not bimonic. Consider the inverse of p in the ring $R[[x]][[u, u^{-1}]]$. We obtain

$$q(u^{-1}) = (x+u)^{-1} = \sum_{i=0}^{+\infty} (-1)^i x^i u^{-(i+1)}. \quad (6)$$

$q(u^{-1})$ induces an $R[u, u^{-1}]$-homomorphism from $R^{\mathbb{Z}}$ to itself by

$$(q(\sigma^{-1})w)(t) := \sum_{i=0}^{+\infty} (-1)^i x^i w(t-i-1). \quad (7)$$

It is clear that the right hand side of (7) is well-defined in $R[[x]]$. It follows by the construction that $q(\sigma^{-1})$ is the inverse of $p(\sigma, \sigma^{-1})$. This proves our claim. Notice, moreover, that $q(\sigma^{-1})$ is not continuous. This shows that in this case there are continuous invertible scalar shift operators whose inverse is not continuous.

Now, in order to complete the study of scalar shift operators we give the characterization of closed range shift operators and of open shift operators.

Proposition 4 Let R be a PID. Consider a polynomial $p \in R[u, u^{-1}]$. Then $p(\sigma, \sigma^{-1})$ has closed range if and only if $p \in R\Sigma := \{as | a \in R,\ s \in \Sigma\}$

Proposition 5 Let R be a Noetherian UFD. Consider a polynomial $p \in R[u, u^{-1}]$. Then $p(\sigma, \sigma^{-1})$ is an open mapping if and only if $p \in S_b$

Consider now the general vector case. In the sequel we will give complete characterizations of four important classes of matrix shift operators: injective matrix shift operators (Theorem 6), surjective matrix shift operators (Theorem 7), closed range matrix shift operators (Theorem 8) and open matix shift operators (Theorem 10). Note that these characterizations present nice analogies. We begin by giving the characterization of the injective matrix shift operators.

Theorem 6 Let R be a PID and let $M \in R[u,u^{-1}]^{l\times q}$. Then the following facts are equivalent

1. $M(\sigma,\sigma^{-1})$ is an injective shift operator.
2. There exists a polynomial matrix $X \in R[u,u^{-1}]^{q\times l}$ such that
$$XM = aI, \tag{8}$$
where $a \in \tilde{S}_b$.
3. M is right $\tilde{S}_b$-factor prime.

Theorem 7 Let R be a PID and let $M \in R[u,u^{-1}]^{l\times q}$. Then the following facts are equivalent

1. $M(\sigma,\sigma^{-1})$ is a surjective shift operator.
2. There exists $X \in R[u,u^{-1}]^{q\times l}$ such that
$$MX = aI, \tag{9}$$
where $a \in \Sigma$.
3. M is left Σ-factor prime.

We are now ready to characterize closed range shift operators. We first introduce another notion of primeness which is a generalization of previous ones. Let $S \subseteq R_0$ be a multiplicatively closed set and let $M \in R[u,u^{-1}]^{l\times q}$ be a matrix of rank r. M is said to be S-factor prime if given any factorization of type $M = M_1\Lambda M_2$ where $M_1 \in R[u,u^{-1}]^{l\times r}$, $\Lambda \in R[u,u^{-1}]^{r\times r}$, and $M_2 \in R[u,u^{-1}]^{r\times q}$, we have that $\det\Lambda \in S$. It is easy to see that if $r = l$ (resp. $r = q$), then M is S-factor prime if and only if it is left (resp. right) S-factor prime.

Theorem 8 Let R be a PID and let M be a rank $r > 0$ matrix in $R[u,u^{-1}]^{l\times q}$. Then the following facts are equivalent

1. $M(\sigma,\sigma^{-1})$ has a closed range.
2. There exist matrices $X \in R[u,u^{-1}]^{r\times l}$, $Y \in R[u,u^{-1}]^{q\times r}$ such that
$$XMY = aI, \tag{10}$$
where $a \in R_0\Sigma$.
3. M is $R_0\Sigma$-factor prime.

An interesting consequence of this theorem is provided by the following proposition which provides a nice characterization the closure of the image of a shift operator.

Proposition 9 Let R be a PID and let M be a rank r matrix in $R[u,u^{-1}]^{l\times q}$. Then
$$\overline{\text{Im } M(\sigma,\sigma^{-1})} = \text{Im } M_1(\sigma,\sigma^{-1}), \tag{11}$$
where $M = M_1M_2$ and $M_1 \in R[u,u^{-1}]^{l\times r}$ is right R_0-factor prime and $M_2 \in R[u,u^{-1}]^{r\times q}$ is left $\tilde{R}_0$-factor prime.

Finally we give the characterization of open matrix shift operators.

Theorem 10 Let R be a PID and let $M \in R[u,u^{-1}]^{l\times q}$. Then the following facts are equivalent

1. $M(\sigma,\sigma^{-1})$ is open.
2. There exists $X \in R[u,u^{-1}]^{q\times l}$ such that
$$MX = aI, \tag{12}$$
where $a \in S_b$.
3. M is left S_b-factor prime.

3. Controllable behaviors

In this section R will always denote a PID. In this section we will apply the results of Section 2 to characterize weakly controllable R-behaviors over finitely generated free R-modules. The first result is provided by the following theorem.

Theorem 11 Let R be a PID and let $\mathcal{B} \subseteq (R^l)^{\mathbb{Z}}$ be a weakly controllable R-behavior. Then there exists a $q \in \mathbb{N}$ such that $\mathcal{B} \simeq (R^q)^{\mathbb{Z}}$. In other words, there exists an injective shift operator $M \in R[u,u^{-1}]^{l\times q}$ such that
$$\mathcal{B} = \text{Im } M(\sigma,\sigma^{-1}) \tag{13}$$
and such that the inverse of $M(\sigma,\sigma^{-1})$ on the image $\mathcal{B}$ is continuous.

Remark Note that Theorem 11 shows that a weakly controllable R-behavior $\mathcal{B} \subseteq (R^l)^{\mathbb{Z}}$ is controllable and so that for R-behaviors in $(R^l)^{\mathbb{Z}}$ the notions of controllability and weak controllability are equivalent. Moreover, it also follows from Theorem 11 that every controllable $\mathcal{B} \subseteq (R^l)^{\mathbb{Z}}$ is automatically of finite memory. These results will be generalized in the sequel.

Remark According to Theorem 11, any controllable R-behavior $\mathcal{B} \subseteq (R^l)^{\mathbb{Z}}$ can be suitable to be a code. Actually, not only it admits an image representation and so there exists an encoder for it, but there exists also the decoder that is a suitable continuous operator that provides the original trajectory from the encoded one. As shown in the previous theorem, there exists a right $(R_0)R[u,u^{-1}]^*$-factor prime polynomial matrix $M \in R[u,u^{-1}]^{l\times q}$ such that $\mathcal{B} = \text{Im } M(\sigma,\sigma^{-1})$. Therefore it can be showed (Fagnani and Zampieri, 1994) that there exists $X \in R[u,u^{-1}]^{q\times l}$ such that $XM = aI$, with $a \in R_0$. Therefore the decoder is given by the matrix shift operator $X(\sigma,\sigma^{-1})$ with an additional division of the output by the constant a as shown in Figure 1.

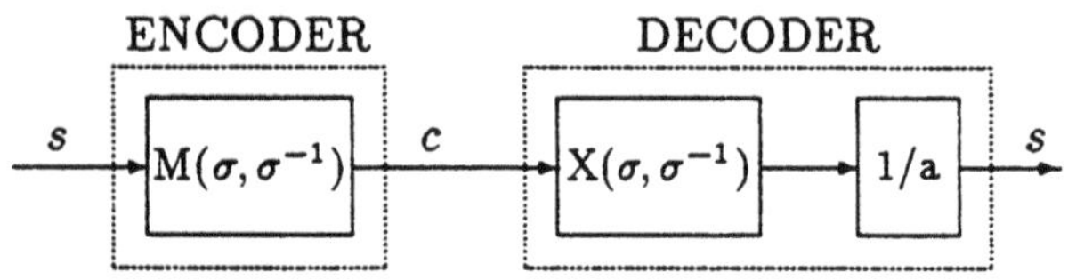

s: signal to code
c: coded signal

Fig. 1. Scheme of the encoder and the decoder

The following corollary provides a classification result about controllable R-behaviors over finitely generated free R-module. More precisely it shows that an R-behavior $\mathcal{B} \subseteq V^{\mathbb{Z}}$, where V is finitely generated free R-module, is controllable if and only if $\mathcal{B}$ is R-equivalent to a full R-behavior $H^{\mathbb{Z}}$, where H is a finitely generated free R-module.

It is easy to see that, given two finitely generated R modules H and H' we have that $H^{\mathbb{Z}} \simeq H'^{\mathbb{Z}}$ if and only if H and H' are isomorphic R-modules. Hence, it follows that we have established a one-to-one correspondence between R-equivalence classes of controllable R-behaviors over free finitely generated R-modules and R-isomorphism classes of finitely generated free R-modules.

4. REFERENCES

Brewer, W., J.W. Bunce and F.S. Van Vleck (1986). *Linear systems over commutative rings.* Dekker.

Brockett, R. and J.L. Willems (1974). Discretized partial differential equations: examples of control systems defined on modules. *Automatica* **10**, 507–515.

Fagnani, F. (1994*a*). Shifts on compact and discrete Lie groups: algebraic-topological invariants and classification problems. *To appear in Advances in Mathematics.*

Fagnani, F. (1994*b*). Some results on the structure of abelian group subshifts. *To appear to Ergodic Theory and Dynamical Systems.*

Fagnani, F. and S. Zampieri (1994). Matrix shift operators and controllable systems over a principal ideal domain. Preprints di Matematica 36. Scuola Normale Superiore di Pisa.

Fornasini, E. and M.E. Valcher (1994). Algebraic aspects of 2D convolutional codes. *IEEE Trans. Information Theory* **IT-40**, 1068–1082.

Forney, G.D. (1970). Convolutional codes I: Algebraic structure. *IEEE Trans. on Infornation Theory* **IT-16**, 720–738.

Forney, G.D. and M.D. Trott (1993). The dynamics of group codes: State spaces, trellis diagramns and canonical encoders. *IEEE Trans. on Infornation Theory* **IT-39**, 1491–1513.

Hautus, M.L.J. and E.D. Sontag (1985). New results on pole shifting for parametrized families of systems. *Journal Pure Appl. Algebra* **40**, 229–244.

Kamen, E. (1975). On an algebraic theory of systems defined by convolutional operators. *Math. System Theory* **9**, 57–74.

Kitchens, B.P. and K. Schmidt (1989). Automorphisms of compact groups. *Ergodic Theory and Dynamical Systems* **9**, 691–735.

Lèvy, B. (1981). 2D polynomial and rational matrices, and their applications for the modelling of 2D dynamical systems. Technical Report M735-11. Stanford Electronics Lab.

Loeliger, H.A., G.D. Forney, T. Mittelholzer and M.D. Trott (1994). Minimality and observability of groups systems. *Linear Algebra and its Applications* **205-206**, 937–963.

Willems, J.C. (1986). From time series to linear system, part II: Exact modelling. *Automatica* **22**, 675–684.

Willems, J.C. (1988). Models for dynamics. *Dynamics reported* **2**, 171–269.

Zampieri, S. and S.K. Mitter (1994). Linear systems over noetherian rings in the behavioural approach. *Submitted to Journal of Mathematical Systems, Estimation, and Control.*

LINEAR SYSTEMS DEFINED OVER A FINITE FIELD, LQ THEORY AND CONVOLUTIONAL CODES

J. Rosenthal* and E.V. York †

Department of Mathematics, University of Notre Dame
Notre Dame, IN 46556, USA

Abstract. In this note, we describe several different representations of a convolutional code as a linear system. We then show how this leads one to a natural interpretation of one of the the main problems in coding theory in systems theoretic terms. In particular, we show how the computation of the free distance of a code can be viewed as a modified LQ problem.

Résumé. Dans ce papier nous faisons une description des représentations vairiées d'un code de convolution. Ensuite nous expliquous un problème principal dans la théorie de codage en user un langage d'analyse du système. En particulier nous faisons la démonstration que l'évaluation de la distance d'un code de convolution est un problème modifié du type LQ.

Keywords. Convolutional code, linear quadratic regulator, free distance, autoregressive systems.

1. Introduction

Since the early seventies it has been widely known that convolutional codes and linear systems which are defined over a finite field are essentially the same objects. More recently there has been a new and increased interest in the connection of linear systems theory and convolutional coding theory and we refer e.g. to the recent article (Loeliger *et al.* 1994). In this more recent approach a convolutional code is viewed as a linear time invariant behavior in the sense of (Willems 1991). Since convolutional codes are easily implemented and have good error correcting properties, they are among the most widely used codes in data transmission (refer to (Lin and Costello 1983; Chapter 17) to underline this point). The main purpose of this paper is to show that one of the major unsolved problems in coding theory can be viewed as the solution of a discrete time LQ problem having a particular cost function.

In coding theory *convolutional encoders* are typically described either through a generator matrix or through a state respectively a *trellis diagram.* The dynamics of the trellis representation can be described as a finite state machine via a traditional A, B, C, D representation. This result is not new, in fact this viewpoint was expressed by Massey and Sain as early as 1967.

In this note, we will generalize this viewpoint and explore some of its advantages. In particular, we show that the computation of the

*This author was supported in part by NSF grant DMS-94-00965. The research for this paper was carried out in part while both authors were visiting at CWI. *e-mail:* Joachim.Rosenthal@nd.edu

† *e-mail:* Eric.V.York@nd.edu

free distance of a convolutional code naturally describes a tracking problem where some 'quadratic cost function' has to be minimized. We conclude the paper with several examples.

2. Convolutional codes viewed as linear systems.

In the sequel we explain some basic properties of convolutional codes from a systems theory point of view. For a detailed treatment of this active research area we refer to the excellent monograph of (Piret 1988).

From a systems theory point of view, convolutional codes are essentially the behavior of a discrete-time linear system defined over a finite field. To be precise, let $\mathbb{F} = \mathbb{F}_q$ be the Galois field with q elements and consider a $(m+p) \times m$ matrix $G(\mathcal{D})$ defined over the polynomial ring $\mathbb{F}[\mathcal{D}]$. $G(\mathcal{D})$ generates a rate $\frac{m}{m+p}$ convolutional code through:

$$C := \{w(\mathcal{D}) \mid w(\mathcal{D}) = G(\mathcal{D})\ell(\mathcal{D})\}. \quad (1)$$

From a systems theory point of view (1) describes a MA-representation, the m-vectors $\ell(\mathcal{D})$ describe the set of latent variables and the set of $(m+p)$-vectors $w(\mathcal{D})$ describe the so called behavior. Every MA representation is by definition always 'controllable' but not necessarily 'observable'. By definition system (1) is observable if, and only if the $m \times m$ full size minors of the generator matrix $G(\mathcal{D})$ are coprime.

The notion of observability corresponds in the coding literature to the notion of *non-catastrophic*. In other words one says the convolutional encoder is *catastrophic* (Piret 1988) if the system (1) is not observable, i.e. if the full size minors of $G(\mathcal{D})$ have a nontrivial greatest common divisor. As the name indicates it is most important to have non-catastrophic convolutional encoders. In the coding literature the expression 'basic encoder' (Piret 1988, Johannesson and Wan 1992) is also used for a non-catastrophic encoder but we consider the expression 'observable' very appealing also in the coding context.

Equivalently, a convolutional code can be given by a parity check matrix $H(\mathcal{D})$, where $H(\mathcal{D})$ is a $p \times (m+p)$ full rank polynomial matrix having the property that $H(\mathcal{D})G(\mathcal{D}) = 0$. $H(\mathcal{D})$ gives an AR representation for the convolutional code C introduced in (1) through:

$$C = \{w(\mathcal{D}) \mid H(\mathcal{D})w(\mathcal{D}) = 0\}. \quad (2)$$

We say $H(\mathcal{D})$ is minimal if the 'high order coefficient matrix' of $H(\mathcal{D})$ has full row rank and we say the representation is *controllable* if $p \times p$ full size minors of the matrix $H(\mathcal{D})$ are coprime. Note the use of language: An encoder should be observable and a parity check matrix (decoder) should be controllable.

3. Representations of convolutional codes.

In the sequel we will assume for simplicity that $G(\mathcal{D})$ is observable. We also will assume that $G(\mathcal{D})$ is minimal (Piret 1988, Johannesson and Wan 1992) having column indices (constraint length of the i-input) $\nu_1, \ldots, \nu_k$ and overall constraint length $c := \sum_{i=0}^{k} \nu_i$.

Under those assumptions we have the following equivalent first order description of the encoder $G(\mathcal{D})$ (Kuijper and Schumacher 1990):

Theorem 1 There exist $(m+c) \times c$ matrices K, L and a $(m+c) \times (m+p)$ matrix M (all defined over $\mathbb{F}$) such that (1) is equivalently described through

$$\{w(\mathcal{D}) \mid \exists x(\mathcal{D}) : Kx_{t+1} + Lx_t = Mw_t\}. \quad (3)$$

where $w(\mathcal{D}) = \sum w_t \mathcal{D}^t \in \mathbb{F}^{m+p}[\mathcal{D}]$, and $x(\mathcal{D}) = \sum x_t \mathcal{D}^t \in \mathbb{F}^c[\mathcal{D}]$.

First order representations of the form (3) can be found in (Willems 1991; p. 269) but as explained in (Ravi and Rosenthal 1995) such representations arise also naturally in the algebraic geometry literature. In the systems theory literature the c-vector $x(\mathcal{D})$ describes the state of the MA-system and it is not difficult to verify that the same interpretation can be given in the coding context.

We can associate the following minimal conditions to the first order representation given in (3):

M1: K has full column rank.
M2: $[K \;\; M]$ has full row rank.

Conditions (M1) and (M2) reduce system (3) to an 'input free' A, B, C, D representation (i.e. a representation in terms of the behavior of the system).

Representations of the form (3) also exist if the generator matrix $G(\mathcal{D})$ is not observable. Moreover they are in general not unique. Indeed if $T \in Gl_{m+c}$ and $S \in Gl_c$ one has natural equivalent descriptions given through the equivalence relation:

$$(K, L, M) \sim (TKS^{-1}, TLS^{-1}, TM).$$

On the other hand one has the interesting result that the equivalence classes of so called admissible triples K, L, M describe the generator matrix $G(\mathcal{D})$ uniquely up to post multiplication by a $m \times m$ unimodular matrix.

MA-representations and AR-representations are dual in a certain sense and, as there are canonical first order representations for MA-representations, there are also canonical first order representations for AR-representations. The following Theorem clarifies this.

Theorem 2 There exist $c \times (c+p)$ matrices P, Q and a $(m+p) \times (c+p)$ matrix R (all defined over $\mathbb{F}$) such that (2) is equivalently described through

$$\{w_t = Rz_t \mid Pz_{t+1} + Qz_t = 0\}. \qquad (4)$$

where $z(D) = \sum_t z_t D^t \in \mathbb{F}^{c+p}[\mathcal{D}]$.

In this representation the $(c+p)$ vector $z(D)$ describes the set of 'internal variables' and $w(\mathcal{D})$ describes again the set of 'external variables', i.e. the behavior of the system. The matrices P, Q are linear maps from the space of internal variables to the space of state variables. Corresponding to change of coordinates one has a natural equivalence among pencil representations

$$(P, Q, R) \sim (SPT^{-1}, SQT^{-1}, RT^{-1}),$$

where $S \in Gl_c$ and $T \in Gl_{c+p}$. As in the case of MA-representations every equivalence class describes a parity check matrix $H(\mathcal{D})$ uniquely up to pre multiplication by a $m \times m$ unimodular matrix.

Similar to the minimality conditions that we imposed on (3), we have the following conditions related to (4):

M1': P has full column rank.
M2': $[P^t \;\; R^t]^t$ has full column rank.

In this case, condition (M1') reduces the first order representation given in (4) to a traditional A, B, C, D representation of a code as considered by Massey and Sain, while both conditions (M1') and (M2') reduce system (4) to an A, B, C, D representation in terms of the behavior of the code.

We would like to remark that the dynamics described through

$$Pz_{t+1} = Qz_t; \quad w_t = Rz_t. \qquad (5)$$

is usually characterized in the coding literature through graphical methods - e.g. a *trellis diagram* (Piret 1988) and the description (4) seems to be new in the coding context.

4. The free distance of a code and state graphs.

One of the most important parameters of a given convolutional code is the *free distance* which we now define. Let $v = (v_1, v_2, \ldots v_s) \in \mathbb{F}^s$ and define $S_v = \{v_i \mid v_i \neq 0\}$. We define the *weight* of v by:

$$\mathrm{wt}(v) := \mid S_v \mid .$$

Given $y(\mathcal{D}) \in \mathbb{F}^s[\mathcal{D}]$ we define the weight of $y(\mathcal{D})$ by:

$$\mathrm{wt}(y(\mathcal{D})) := \sum_0^\infty \mathrm{wt}(y_t)$$

Finally, given a convolutional code C, we define the *free distance* of C (denoted d_f) by:

$$d_f := \min\{\mathrm{wt}(w(\mathcal{D})) \mid 0 \neq w(\mathcal{D}) \in C\}$$

The free distance is the parameter of a convolutional code that measures the ability of the code to perform error correction when used in conjunction with certain types of channels and decoders (Lin and Costello 1983).

A key problem in convolutional coding theory has been to find a method for effectively characterizing the free distance of a given convolutional code. Very much related to this problem is the task of designing codes of a given rate and complexity with good free distance. At present, the most effective techniques for doing this has been to make an exhaustive search of the class of codes determined by the rate and complexity,and compute the free distance of encoders in this class, until one with maximal or *near* maximal free distance is found. To date, there have been no significant construction techniques developed for convolutional codes. In the following, we will explain how convolutional codes are often viewed in the coding literature.

Let C be the rate $\frac{1}{2}$ convolutional code with generator matrix $G(\mathcal{D}) = (\mathcal{D}^2{+}1, \mathcal{D}^2{+}\mathcal{D}{+}1)^t$. In the coding literature it is usual to depict the dynamics of an encoder/decoder in terms of a shift register diagram composed of adders ($\oplus$) and delay elements ($\square$). For C, this diagram is of the form:

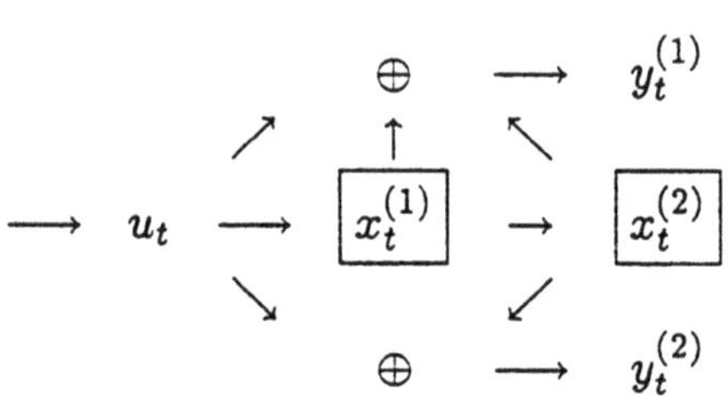

At time t, the input to the system is u_t, the state s_t is given by $(x_t^{(1)}, x_t^{(2)})$, and the output $(y_t^{(1)}, y_t^{(2)})$ is obtained by simultaneously performing the indicated operations. At time $t+1$, u_{t+1} is inputted into the system and the state shifts to $(u_t, x_t^{(1)})$ and $x_t^{(2)}$ becomes obsolete.

To each shift register diagram we can associate a weighted, directed graph S. Define the vertices of S to be the q^c possible states of the code. Let $s_i, s_j \in S$, we associate and edge $\mathcal{E}$ from s_i to s_j if and only if there exists an input vector u that drives the state from s_i to s_j, furthermore, we associate to $\mathcal{E}$ the output vector w obtained from the shift register when it is in the state s_i and the input vector u is applied.

In this way, we obtain a directed graph that models the dynamics of the convolutional code. In the coding literature it is common to associate to each edge $\mathcal{E}$, the weight of the output vector w. With this modification of the graph S, we can view the computation of the free distance of a code as the equivalent problem of determining the least weight nontrivial closed path initiating from the zero state. Below, we present a different interpretation of this problem in system theoretic terms.

5. Characterization of d_f as an LQ problem.

In this section we will show that the determination of the free distance of a convolutional code can be viewed as a discrete time LQ problem having a particular cost function.

Consider the system given in (4) with behavior partitioned as:

$$w(\mathcal{D}) = \begin{pmatrix} y(\mathcal{D}) \\ u(\mathcal{D}) \end{pmatrix}.$$

Where $y(\mathcal{D}) \in \mathbb{F}^m[\mathcal{D}]$ and $u(\mathcal{D}) \in \mathbb{F}^p[\mathcal{D}]$. The minimality conditions (M1') and (M2') reduce P, Q, R in (4) to the form:

$$\begin{aligned} P &= (I \;\; 0) \\ Q &= (A \;\; B) \\ R &= \begin{pmatrix} C & E \\ 0 & I \end{pmatrix} \end{aligned}$$

Fom this we obtain the representation:

$$\begin{aligned} x_{t+1} &= Ax_t + Bu_t \\ y_t &= Cx_t + Du_t \end{aligned} \qquad (6)$$

Note that the input u_t and the output y_t of (6) make up the code word $w(\mathcal{D})$. We will define the cost (weight) of $w(\mathcal{D})$ as given by:

$$\text{cost} := \sum_{0}^{\infty} \text{wt}(y_t) + \text{wt}(u_t). \qquad (7)$$

We can now express the the problem of determining the free distance of a convolutional code as follows:

> Given the system determined by (6), find the input sequence $\{u_t\}$ such that the cost function determined by (7) is minimal.

This is exactly the discrete time linear quadratic regulator problem (Delchamps 1988; Chapter 26) over a finite field. Techniques (i.e. Riccati equations) for solving these types of problems exist in the systems community and it is our belief that some of these methods may provide one with algebraic information concerning the free distance of a convolutional code. These and related ideas are currently under investigation.

6. Viterbi Decoding

The main method used for the decoding of convolutional codes has been the Viterbi Algorithm (Viterbi 1967). The Viterbi Algorithm is a maximum likelihood decoding algorithm for convolutional codes (Forney 1974).

The decoding of a convolutional code via the Viterbi decoding algorithm can also be viewed as an LQ problem, and we explain this connection as well. Let $\hat{w}(\mathcal{D}) \in \mathbb{F}^{n+p}[\mathcal{D}]$ be the transmitted vector and partition $\hat{w}(\mathcal{D})$ as:

$$\hat{w}(\mathcal{D}) = \begin{pmatrix} \hat{y}(\mathcal{D}) \\ \hat{u}(\mathcal{D}) \end{pmatrix}.$$

Then our objective is to find the codeword $w(\mathcal{D})$ such that the cost function:

$$\text{cost} := \sum_{t=0}^{\infty} \text{wt}(y_t - \hat{y}_t) + \text{wt}(u_t - \hat{u}_t). \quad (8)$$

is minimal. The Viterbi decoding algorithm computes the code word $w(\mathcal{D})$ which minimizes the cost (8) by applying the principle of dynamic programming. This interpretation itself has been known (Omura 1969) but we want to note that the cost function (8) has a quadratic nature and the problem therefore very much resembles the well known quadratic regulator problem well studied in the systems literature. Those intriguing connections are currently under investigation.

7. Examples

Consider the (controllable) parity check matrix

$$H(\mathcal{D}) := (\mathcal{D}^2 + \mathcal{D} + 1, \mathcal{D}^2 + 1) \quad (9)$$

defined over $\mathbb{F}_2$.

Example 3 A pencil representation of the form (2) can easily be computed and is given through e.g.:

$$\begin{pmatrix} P + \mathcal{D}Q \\ R \end{pmatrix} = \begin{pmatrix} 1 & 0 & \mathcal{D} \\ \mathcal{D} & 1 & 0 \\ \hline 1 & 1 & 1 \\ 0 & 1 & 1 \end{pmatrix}.$$

Example 4 Partitioning the external variables in state and input variables results in a description of the trellis dynamics:

$$\begin{aligned} x_{t+1} &= \begin{pmatrix} 0 & 0 \\ 1 & 0 \end{pmatrix} x_t + \begin{pmatrix} 1 \\ 0 \end{pmatrix} u_t \\ w_t &= \begin{pmatrix} 1 & 1 \\ 0 & 1 \end{pmatrix} x_t + \begin{pmatrix} 1 \\ 1 \end{pmatrix} u_t \end{aligned} \quad (10)$$

Example 5 The behavioral representation of (9) is given through:

$$\begin{aligned} x_{t+1} &= \begin{pmatrix} 0 & 1 \\ 1 & 0 \end{pmatrix} x_t + \begin{pmatrix} 1 \\ 0 \end{pmatrix} u_t \\ y_t &= \begin{pmatrix} 1 & 0 \end{pmatrix} x_t + (1) u_t \end{aligned} \quad (11)$$

The input sequence that minimizes the cost function given in (7) is given by $u_0 = u_2 = 1$ and $u_i = 0$ for all $i \neq 0, 2$.

REFERENCES

Delchamps, D. F. (1988). *State Space and Input-Output Linear Systems.* Springer Verlag. New York.

Forney, G. D. (1970). 'Convolutional codes I: Algebraic structure'. *IEEE Trans. Inform. Theory* **IT-16**(5), 720–738.

Forney, G. D. (1974). 'Convolutional codes II: Maximum likelihood decoding'. *Inform. Control* **25**, 222–266.

Johannesson, R. and Z. Wan (1992). 'A linear algebra approach to minimal convolutional encoders'. *IEEE Trans. Inform. Theory* **IT-39**(4), 1219–1233.

Kuijper, M. (1994). *First-Order Representations of Linear Systems*. Birkhäuser. Boston.

Kuijper, M. and J. M. Schumacher (1990). 'Realization of autoregressive equations in pencil and descriptor form'. *SIAM J. Control Optim.* **28**(5), 1162–1189.

Lin, S. and D. Costello (1983). *Error Control Coding: Fundamentals and Applications*. Prentice-Hall. Englewood Cliffs, NJ.

Loeliger, H. A., G. D. Forney, T. Mittelholzer and M. D. Trott (1994). 'Minimality and observability of group systems'. *Linear Algebra Appl.* **205/206**, 937–963.

Omura, J. K. (1969). 'On the Viterbi decoding algorithm'. *IEEE Trans. Inform. Theory* **IT-15**, 177–179.

Piret, P. (1988). *Convolutional Codes, an Algebraic Approach*. MIT Press. Cambridge, MA.

Ravi, M. S. and J. Rosenthal (1995). 'A general realization theory for higher order linear differential equations'. *Systems & Control Letters*. To appear.

Viterbi, A. J. (1967). 'Error bounds for convolutional codes and an asymptotically optimum algorithm,'. *IEEE Trans. Inform. Theory* **IT-13**, 260–269.

Willems, J. C. (1991). 'Paradigms and puzzles in the theory of dynamical systems'. *IEEE Trans. Automat. Control* **AC-36**(3), 259–294.

ON THE RELATIONSHIPS BETWEEN GLOBAL AND LOCAL STRUCTURAL PROPERTIES OF 2D SYSTEMS

MAURO BISIACCO*

* *Department of Electronics and Computer Science, University of Padova, Via Gradenigo 6/A, 35131 Padova, Italy*

Abstract. In this paper various definitions of global controllability and reachability are considered in the 2D context. It is shown how all these different concepts can be treated in terms of the same 2D reachability and PBH controllability matrices, whose rank-singularities sets structure completely characterize these structural properties. A comparison between these concepts and others already existent in the literature is also performed.

Résumé. Pour les systèmes bidimensionnels, comme l'on soit, il y a plusieurs définitions de commandabilité et de atteignabilité. Dans ce travail nous montrous que ces proprietés peuvent être analysées à l'aide de la matrice de commandabilité et de la matrice du critère de Popov, qui permettent de characterizer complètement la structure d'un couple entrée-état donné. Comme conclusion, nous soulignons quelques differences et analogies avec des résultats déja parus dans la litérature.

Key words. 2D systems, systems over rings, modules over $\mathbf{R}[z, z^{-1}]$, constrained global reachability, constrained global controllability.

1. INTRODUCTION

In the 2D context, reachability, controllability and stabilizability concepts are subject to various interpretations, because of the distinction between global and local states which reflects into a variety of possible definitions, often corresponding to uncorrelated properties. For instance, local reachability and local controllability are uncorrelated each other (Bisiacco,1985), while on the other hand global reachability implies global controllability. Moreover, global properties are sometimes stronger and sometimes weaker than the corresponding local ones (local controllability implies global controllability, while global reachability implies local reachability). A further complicating factor, from the control and stabilization point of view, is the possibility of introducing constraints on the control input and on the initial global state, which reflect into radical changes in the problem solution.

The introduction of suitable constraints is a standard fact in 2D system theory. For instance, boundedness of the initial condition is required for obtaining a consistent definition of asymptotic stability (Fornasini and Marchesini, 1980), membership of both input and state to ℓ_2-spaces is needed for formulating and solving a well-defined optimal control problem (Bisiacco and Fornasini, 1990), and casual dependence of the control input from the local state is a necessary and sufficient condition for implementability of the control via a feedback scheme based on a controller having a 2D system structure itself (Bisiacco, 1985). Moreover, resorting to unbounded inputs is quite unrealistic and useless from the practical point of view. This motivates the restriction of the analysis to bounded inputs, as happens when studying ℓ_∞-stabilization questions (Bisiacco and Fornasini, 1990). Therefore the investigation of the rôle played by constraints such as membership to ℓ_1, ℓ_2 or ℓ_∞-spaces of inputs and states, or by stronger requirements such as causal dependence of the input from the local state, is needed for developing a consistent theory.

In this paper an unified treatment of various controllability and reachability properties will be developed, leading to a complete characterization of these structural properties and of their relationships. This will be done by showing how reachability, controllability and stabilizability properties can be expressed in terms of simple polynomial matrices, whose rank-singularities sets structure depends on the particular constraint. Rank-singularities analysis will be proved to be the basic tool for deciding about reachability and controllability, and will permit us to simply perform a comparison between the various properties.

Roughly speaking, the original 2D system will be embedded in a 1D system over the ring $\mathbf{R}[\xi, \xi^{-1}]$. However, standard results about systems over rings doesn't apply to our problem, since a system over a ring R is usually defined in terms of state and input spaces of the form R^n and R^m respectively. Actually our context is more general, since it requires to deal with state and input spaces of the form M^n and M^m, where M constitutes a suitable module over R. This makes the analysis more difficult with respect to the standard case, and justify the development of a new theory for systems over rings which covers our situation.

Finally, a comparison with already known results, in particular with local control and stabilization and with weakly-causal ℓ_∞-stabilization will be performed in the last section, where a hierarchic diagram explaining the connections between the considered various properties is proposed.

2. GLOBAL REACHABILITY

A 2D system is described by the following linear model

$$\begin{aligned} x(h+1,k+1) &= A_1 x(h,k+1) + A_2 x(h+1,k) \\ &\quad + B_1 u(h,k+1) + B_2 u(h+1,k) \end{aligned} \tag{2.1}$$

where $x(h,k) \in \mathbf{R}^n$ and $u(h,k) \in \mathbf{R}^m$ are the local state and the input value at $(h,k) \in \mathbf{Z}x\mathbf{Z}$ respectively and A_1, A_2, B_1 and B_2 are real matrices of suitable size. By introducing a formal power series representation for the so-called global states and inputs (Fornasini and Marchesini, 1982)

$$\begin{aligned} \mathcal{X}_k(\xi) &:= \sum_{i=-\infty}^{+\infty} x(k-i,i)\xi^i \\ \mathcal{U}_k(\xi) &:= \sum_{i=-\infty}^{+\infty} u(k-i,i)\xi^i \end{aligned} \tag{2.2}$$

which are associated with the restrictions of state and input to the k-th separation set

$$\mathcal{C}_k := \{(i,j) \in \mathbf{Z}x\mathbf{Z} : i+j=k\} \tag{2.3}$$

we can rewrite (2.1) as

$$\mathcal{X}_{k+1}(\xi) = F(\xi)\mathcal{X}_k(\xi) + G(\xi)\mathcal{U}_k(\xi) \tag{2.4}$$

where $F(\xi):=A_1+A_2\xi$ and $G(\xi):=B_1+B_2\xi$. Introducing the "k-steps reachability matrix" $\mathcal{R}_k(\xi)$ given by

$$\mathcal{R}_k(\xi) := \left[G(\xi) \middle| F(\xi)G(\xi) \middle| \cdots \middle| F^{k-1}(\xi)G(\xi)\right] \tag{2.5}$$

which fully corresponds, in a pointwise sense, to the standard k-steps 1D reachability matrix of the ξ-dependent system given by matrices $F(\xi)$ and $G(\xi)$, and assuming $\mathcal{X}_0(\xi)=0$, we have from (2.4)

$$\mathcal{X}_k(\xi) = \mathcal{R}_k(\xi) \begin{bmatrix} \mathcal{U}_{k-1}(\xi) \\ \vdots \\ \mathcal{U}_0(\xi) \end{bmatrix} \tag{2.6}$$

which expresses the dependence of the k-th global state on the sequence of global inputs $\mathcal{U}_0(\xi)$, ..., $\mathcal{U}_{k-1}(\xi)$. A recursive argument based on the application of the Cayley-Hamilton theorem to $F(\xi)$ in a pointwise sense permits us to express also any k-steps reachability matrix, with $k \geq n$, in terms of the n-step reachability matrix $\mathcal{R}(\xi) := \mathcal{R}_n(\xi)$

$$\mathcal{R}_k(\xi) = \mathcal{R}(\xi)\left[I_{nm} \middle| P_{k-n}(\xi)\right], \qquad k \geq n \tag{2.7}$$

where $P_i(\xi)$ denotes a suitable polynomial matrix. Recalling the row-Hermite form of a polynomial matrix, we can introduce a unimodular matrix $T(\xi)$ such that

$$\begin{aligned} \mathcal{R}(\xi) &= \left[\begin{array}{ccc|c} a_{11}(\xi) & & 0 & 0 \\ \vdots & \ddots & & \vdots \\ a_{n1}(\xi) & \cdots & a_{nn}(\xi) & 0 \end{array}\right] T(\xi) \\ &:= \left[H(\xi) \middle| 0\right] T(\xi) \end{aligned} \tag{2.8}$$

where all the considered matrices are polynomial ones. Starting from (2.6) and using (2.7) and (2.8), we can express $\mathcal{X}_k(\xi)$ as

$$\mathcal{X}_k(\xi) = \left[H(\xi) \middle| 0\right] \begin{bmatrix} u_1(\xi) \\ \vdots \\ u_n(\xi) \\ \hline u_{n+1}(\xi) \\ \vdots \\ u_{nm}(\xi) \end{bmatrix} = H(\xi) \begin{bmatrix} u_1(\xi) \\ \vdots \\ u_n(\xi) \end{bmatrix} \tag{2.9}$$

where the scalars $u_i(\xi)$'s and the vector inputs $\mathcal{U}_0(\xi)$, ..., $\mathcal{U}_{k-1}(\xi)$ are related by polynomials operations. Only the first components $u_1(\xi),\ldots,u_n(\xi)$ gives a real contribute to the computation of the k-steps reached state $\mathcal{X}_k(\xi)$. If $x_1(\xi),\ldots,x_n(\xi)$ denote the components of the vector $\mathcal{X}_k(\xi)$, we can rewrite (2.9) as

$$\begin{cases} a_{11}(\xi)u_1(\xi) = x_1(\xi) \\ a_{22}(\xi)u_2(\xi) = x_2(\xi) - a_{21}(\xi)u_1(\xi) \\ \vdots \\ a_{nn}(\xi)u_n(\xi) = x_n(\xi) - \ldots - a_{nn-1}(\xi)u_{n-1}(\xi) \end{cases} \tag{2.10}$$

whose solvability w.r. to the unknowns $u_i(\xi)$'s has to be analyse, in order to understand when an arbitrary global state $\mathcal{X}_k(\xi)$ is reachable by using suitable inputs.

Let us note that in deriving all previous equations no constraints have been imposed on inputs and states, so that they are in general vectors with entries in the set of real-valued bilateral sequences $\mathbf{R}_b((\xi))$. In fact all the previous relations involve only sums and products by polynomials, so they always make sense by being well-defined the product between any sequence in $\mathbf{R}_b((\xi))$ and any polynomial. However it is clear that these relations mantain their validity even if some constraints are imposed (for instance, if global states and inputs are restricted to exhibit a "bounded shape"), since they simply correspond to assume the entries of $\mathcal{X}_i(\xi)$'s and $\mathcal{U}_i(\xi)$'s belonging to suitable subsets of $\mathbf{R}_b((\xi))$. In fact the Cauchy-product between elements in $\mathbf{R}_b((z))$ is in general undefined, so that $\mathbf{R}_b((z))$ is not a ring but only an abelian group w.r. to the usual sum. However we have already seen that bilateral series can be multiplied by polynomials, and in fact $\mathbf{R}_b((z))$ becomes a module over the ring $\mathbf{R}[z,z^{-1}]$ of bilateral polynomials w.r. to this product.

We can introduce some submodules of $\mathbf{R}_b((z))$, such as $\mathbf{R}_\infty((z))$, $\mathbf{R}_2((z))$ and $\mathbf{R}_1((z))$, respectively the sets of bounded, square summable and summable bilateral sequences, and also the submodule $\mathbf{R}_e((z))$ of exponentially convergent series. They satisfy the following inclusions-chain

$$\mathbf{R}_b((z)) \supseteq \mathbf{R}_\infty((z)) \supseteq \mathbf{R}_2((z)) \supseteq \supseteq \mathbf{R}_1((z)) \supseteq \mathbf{R}_e((z)) \supseteq \mathbf{R}[z,z^{-1}] \tag{2.11}$$

An important feature of these submodules is the possibility of discussing convergence properties for their elements. In fact series in $\mathbf{R}_2((z))$ converge for almost all z on the unit circle, as a consequence of the theory of Fourier-transforms, once z is replaced with $e^{i\omega}$. Moreover, elements in $\mathbf{R}_e((z))$ can be viewed as Cauchy-Laurent expansions of analytical functions defined on a suitable annulus including the unit circle. Finally, bilateral polynomials uniquely define analytical functions on the whole $\mathbf{C}$, except at most for the point $z=0$ and for the improper point $z=\infty$. Some of these submodules are also rings with the same Cauchy product. This happens for $\mathbf{R}_1((z))$ and $\mathbf{R}_e((z))$, besides obviously to $\mathbf{R}[z,z^{-1}]$, and any submodule of the chain (2.11), except for $\mathbf{R}_b((z))$, can be viewed also as a module over any ring which is included in it. So, for instance, a product between bounded sequences in $\mathbf{R}_\infty((z))$ and exponentially convergent sequences is well-defined, and the result of the product is always a bounded sequence itself.

Given a module M over a ring R such that $M \supseteq R$, let us say that $r \in R$ is invertible in M if $m \in M$ exists such that $rm=1 \in R \subseteq M$. Clearly, if M=R our definition collapses into the notion of unit (or invertible) elements of a ring. For instance, $p \in \mathbf{R}[z,z^{-1}]$ is invertible in $\mathbf{R}[z,z^{-1}]$ if and only if $p(z)=h\,z^k$ for some $h \neq 0$ and $k \in \mathbf{Z}$, while is invertible in $\mathbf{R}_e((z))$ if and only if $p(e^{i\omega}) \neq 0$ for any $\omega \in [0,2\pi]$. This is in fact a necessary and sufficient condition for $1/p(z)$ being an analytical function in a suitable annulus including the unit circle, and therefore corresponding to some element in $\mathbf{R}_e((z))$. Moreover, p is invertible in $\mathbf{R}_b((z))$ if and only if $p \neq 0$. For this, it suffices to note that an inverse of p always exists in $\mathbf{R}((z)) \subseteq \mathbf{R}_b((z))$ if and only if $p \neq 0$, since the set $\mathbf{R}((z))$ of unilateral series is actually a field. Let us note that since the same reasoning can be performed on $\mathbf{R}((z^{-1}))$, we have that inverses can be not uniquely determined.

Given an element $r \in R$, we can define a linear map f_r from M into itself by letting

$$f_r(m) := r\,m \tag{2.12}$$

and the properties of this map when $R = \mathbf{R}[z,z^{-1}]$ and M varies on the submodules-sequence described by (2.11) are expressed by the following two lemmas (Bisiacco, 1995).

Lemma 1: Let $p \in \mathbf{R}[z,z^{-1}]$. Then $f_p : \mathbf{R}_b((z)) \to \mathbf{R}_b((z))$ is onto if and only if p is invertible in $\mathbf{R}_b((z))$, $f_p : \mathbf{R}_i((z)) \to \mathbf{R}_i((z))$ $(i = \infty, 2, 1, e)$ is onto if and only if p is invertible in $\mathbf{R}_e((z))$, and $f_p : \mathbf{R}[z,z^{-1}] \to \mathbf{R}[z,z^{-1}]$ is onto if and only if p is invertible in $\mathbf{R}[z,z^{-1}]$. ■

Lemma 2: Let $p \in \mathbf{R}[z,z^{-1}]$. Then $f_p : \mathbf{R}_b((z)) \to \mathbf{R}_b((z))$ is one-to-one if and only if p is invertible in $\mathbf{R}[z,z^{-1}]$, $f_p : \mathbf{R}_\infty((z)) \to \mathbf{R}_\infty((z))$ is one-to-one if and only if p is invertible in $\mathbf{R}_e((z))$, and $f_p : \mathbf{R}_i((z)) \to \mathbf{R}_i((z))$ $(i = 2, 1, e)$ and $f_p : \mathbf{R}[z,z^{-1}] \to \mathbf{R}[z,z^{-1}]$ are one-to-one if and only if p in invertible in $\mathbf{R}_b((z))$. Moreover, in all cases in which f_p is not one-to-one, ker f_p results to be a finite-dimensional vector space over $\mathbf{R}$. ■

The previous Lemmas are useful in studying solvability of equations of the form

$$p\,x = s \tag{2.13}$$

with $p \in \mathbf{R}[z,z^{-1}]$ and x,s belonging to some submodule of $\mathbf{R}_b((z))$. In fact the **existence** of the solution x for any s depends on the **onto** property of f_p, while the **unicity** of the solution depends on the **one-to-one** property. Let us note that invertibility of p in a module M is equivalent to solvability of $px=1$, so it is implied by the onto property, but the converse is not always true. For instance, if $p=1-z$, we know that (2.13) is not solvable in $\mathbf{R}_\infty((z))$ for any s. However, for s=1 we have a solution of the form $x(z) = 1 + z + z^2 + ..$ which belongs to $\mathbf{R}_\infty((z))$.

Now we can define reachability concepts in terms of the particular $\mathbf{R}[\xi,\xi^{-1}]$-module to whom inputs and states belong. When solving (2.6), we will speak about global reachability tout-court if no constraints are imposed on the entries of $\mathcal{X}_k(\xi)$'s and $\mathcal{U}_k(\xi)$'s, while we speak about bounded (or ℓ_∞-) global reachability, ℓ_2-global reachability, ℓ_1-global reachability, exponential reachability and polynomial reachability if they are assumed to belong to $\mathbf{R}_\infty((\xi))$, $\mathbf{R}_2((\xi))$, $\mathbf{R}_1((\xi))$, $\mathbf{R}_e((\xi))$ and $\mathbf{R}[\xi,\xi^{-1}]$ respectively. For instance, ℓ_2-global reachability means the possibility of reaching **any** ℓ_2-global state by using **only** ℓ_2-inputs, and analogously for the other properties. By being any module over $\mathbf{R}[\xi,\xi^{-1}]$ invariant w.r. to products by polynomials, the polynomial relationships between the $u_i(\xi)$'s and the $\mathcal{U}_i(\xi)$'s doesn't modify the module they belong to, so that reachability in any sense reduces to solvability of (2.10) for any $x_i(\xi)$'s belonging to some module M in the unknown $u_i(\xi)$'s belonging to the same module M. Therefore the only thing we need is to analyse solvability of (2.10) with $x_i(\xi)$ and $u_i(\xi)$ constrained to belong to a given module M over $\mathbf{R}[\xi,\xi^{-1}]$.

Recalling the previous Lemmas 1 and 2, we can easily derive the following theorem, which generalizes the results expressed in (Fornasini and Marchesini, 1982), and whose proof can be found in (Bisiacco, 1995).

Theorem 1: A 2D system is globally-reachable if and only if

i) reachability tout-court $(\mathbf{R}_b((\xi))) \Leftrightarrow a_{ii}(\xi) \neq 0$ $\forall i = 1,2,\dots,n$;

ii) bounded, ℓ_2-, ℓ_1-, exponential reachability $(\mathbf{R}_\infty((\xi)),\ \mathbf{R}_2((\xi)),\ \mathbf{R}_1((\xi)),\ \mathbf{R}_e((\xi)))$ $\Leftrightarrow a_{ii}(e^{i\omega}) \neq 0$ for any i=1,2,...,n and for any $\omega \in [0,2\pi]$;

iii) polynomial $(\mathbf{R}[\xi,\xi^{-1}])$ reachability $\Leftrightarrow a_{ii}(\xi) = h_i z^{k_i}$ for any i=1,2,..,n, with $h_i \neq 0$ and $k_i \in \mathbf{Z}$. ■

Let us note that polynomial reachability corresponds to the possibility of reaching any finite-support global state by using only finite-support inputs. The previous theorem can be expressed in term of the set of rank-singularities of $\mathcal{R}(\xi)$:

$$\begin{aligned}\mathcal{V}(\mathcal{R}(\xi)) &:= \{\xi \in \mathbf{C}: \mathcal{R}(\xi) \text{ is not full - rank}\} \\ &= \{\xi \in \mathbf{C}: \prod_{i=1}^{n} a_{ii}(\xi) = 0\}\end{aligned} \tag{2.14}$$

In fact if $\mathcal{V}(\mathcal{R}(\xi)) = \mathbf{C}$ the system is unreachable in any sense, if $\mathcal{V}(\mathcal{R}(\xi))$ is a finite set it is globally reachable, while if $\mathcal{V}(\mathcal{R}(\xi)) \cap \gamma_1 = \phi$, where $\gamma_1 := \{z \in \mathbf{C}: |z| = 1\}$ is the unit circle, it is globally reachable and also reachable in the bounded, ℓ_2, ℓ_1 and exponential senses. Finally, if $\mathcal{V}(\mathcal{R}(\xi)) \cap (\mathbf{C} - \{0\}) = \phi$ the system is globally reachable in any sense. Let us note that the third situation implies the second one, since there are no other possibilities for $\mathcal{V}(\mathcal{R}(\xi))$ besides to coinciding with **C** and having finite cardinality, and the first situation is clearly excluded by the third one. Let us also note that ℓ_∞, ℓ_1, ℓ_2 and exponential reachability are completely equivalent properties.

The previous result can be restated in terms of pointwise reachability of the ξ-dependent 1D pair $(F(\xi), G(\xi))$. In fact polynomial reachability means that $(F(\xi), G(\xi))$ is reachable in the 1D-sense for any $\xi \in \mathbf{C}$, except possibly for $\xi=0$ or, equivalently, $(z_1A_1+z_2A_2,\ z_1B_1+z_2B_2)$ is reachable in the 1D-sense for any $(z_1,z_2) \in \mathbf{C}\times\mathbf{C}-\mathbf{X}_1-\mathbf{X}_2$, with $\mathbf{X}_1 := \{(0,z_2): z_2 \in \mathbf{C}\}$ and $\mathbf{X}_2 := \{(z_1,0): z_1 \in \mathbf{C}\}$. Analogously, ℓ_∞, ℓ_2, ℓ_1 and exponential reachability corresponds to the reachability of the frequency-dependent 1D-system $(F(e^{i\omega}), G(e^{i\omega}))$ for any $\omega \in [0,2\pi]$, while global reachability admits the possibility of a finite set of frequencies where the 1D system results to be unreachable.

Example 1: Let $A_1=A_2=0$, $B_1=1$, $B_2=-1$. We have $\mathcal{R}(\xi)=(1-\xi)$ which is not full-rank at the only point $\xi = 1 \in \gamma_1$. Therefore this system is globally reachable but not reachable in any other sense. If the one-step state to reach is the following

$$\mathcal{X}_1(\xi) = \sum_{i=-\infty}^{+\infty} \xi^i \tag{2.15}$$

we have to solve this simple equation w.r. to $\mathcal{U}_0(\xi) := \sum_{i=-\infty}^{+\infty} u_i \xi^i$

$$\sum_{i=-\infty}^{+\infty} \xi^i = (1-\xi)\,\mathcal{U}_0(\xi) \tag{2.16}$$

which is equivalent to $u_i - u_{i-1} = 1$ for any $i \in \mathbf{Z}$. The general solution of this equation is the following

$$u_i = u_0 + i \qquad \text{for any } i \in \mathbf{Z} \tag{2.17}$$

which is clearly an unbounded control input.

Example 2: Let $A_1=A_2=0$, $B_1=1$, $B_2=2$. In this case $\mathcal{R}(\xi)=(1+2\xi)$ exhibits the only rank-singularity $\xi = -1/2$, which doesn't belong to γ_1. Therefore the system is globally reachable in any sense except for the polynomial one. In fact the local state $\mathcal{X}_1(\xi)=1$ cannot be reached by a finite-support input, since $1=(1-\xi)\mathcal{U}_0(\xi)$ doesn't admit polynomial solutions.

3. GLOBAL CONTROLLABILITY

Let us focus an alternate notation for expressing the solution of (2.1) which appears to be more interesting from the point of view of controllability problems. Let us denote with

$$\begin{aligned}\mathcal{X}_k(\eta,\xi) &:= \sum_{i=0}^{k} \mathcal{X}_i(\xi)\eta^i \in \mathbf{R}_b^n((\xi))[\eta] \\ \mathcal{X}(\eta,\xi) &:= \sum_{i=0}^{+\infty} \mathcal{X}_i(\xi)\eta^i \in \mathbf{R}_b^n((\xi))[[\eta]]\end{aligned} \tag{3.1}$$

the restrictions of the state-evolution to the strip $\bigcup_{i=0}^{k} \mathcal{C}_i$ and to the half-plane $\bigcup_{i=0}^{+\infty} \mathcal{C}_i$ respectively, and use a similar notation for inputs. It is easy to verify that

$$\mathcal{X}_0(\xi) - \eta^{k+1}\mathcal{X}_{k+1}(\xi) = \left[I - \eta F(\xi) \middle| \eta G(\xi)\right]\left[\mathcal{X}_k^T(\eta,\xi) \middle| -\mathcal{U}_k^T(\eta,\xi)\right]^T \quad (3.2)$$

so that, in the particular case $\mathcal{X}_{k+1}(\xi)=0$, we have that use of the η-polynomial input $\mathcal{U}(\eta,\xi) := \mathcal{U}_k(\eta,\xi)$ leads to a η-polynomial evolution $\mathcal{X}(\eta,\xi) = \mathcal{X}_k(\eta,\xi)$, and (3.2) becomes

$$\mathcal{X}_0(\xi) = \left[I - \eta F(\xi) \middle| \eta G(\xi)\right]\left[\frac{\mathcal{X}(\eta,\xi)}{-\mathcal{U}(\eta,\xi)}\right] \quad (3.3)$$

which expresses, with $\mathcal{X}(\eta,\xi)$ and $\mathcal{U}(\eta,\xi)$ polynomial vectors w.r. to η, the controllability property of the initial global state $\mathcal{X}_0(\xi)$ in a number of steps depending on the degrees of such polynomials w.r. to η.

In other words, solvability of (3.3) for any given $\mathcal{X}_0(\xi) \in \mathbf{R}_b^n((\xi))$ in the unknowns $\mathcal{X}(\eta,\xi) \in \mathbf{R}_b^n((\xi))[\eta]$ and $\mathcal{U}(\eta,\xi) \in \mathbf{R}_b^m((\xi))[\eta]$ is necessary and sufficient for global controllability of the given 2D system and therefore the following matrix

$$C(\eta,\xi) := \left[I - \eta F(\xi) \middle| \eta G(\xi)\right] \quad (3.4)$$

becomes crucial in the controllability property analysis. We will refer to (3.4) as to the 2D PBH-controllability matrix since, analogously to what we told about reachability, it can be viewed as the standard PBH-controllability matrix of the ξ-dependent 1D system $(F(\xi), G(\xi))$. Moreover, letting $z_1 := \eta$ and $z_2 := \eta\xi$, we can rewrite (3.4) as

$$\begin{cases} \mathcal{C}(z_1,z_2) := \left[I - A_1 z_1 - A_2 z_2 \middle| B_1 z_1 + B_2 z_2\right] \\ C(\eta,\xi) = \mathcal{C}(\eta,\eta\xi) \end{cases} \quad (3.5)$$

The matrix $\mathcal{C}(z_1,z_2)$ has already been introduced in (Bisiacco, 1985) for studying local controllability, and the analogies with its 1D counterpart have been there discussed. Denoting by $m_1(\eta,\xi),\ldots,m_r(\eta,\xi)$ the maximal order minors in $C(\eta,\xi)$, and by $p(\eta,\xi)$ their greatest common divisor, so that

$$m_i(\eta,\xi) = p(\eta,\xi) r_i(\eta,\xi) \quad i = 1,2,\ldots,r \quad (3.6)$$

we know that the algebraic variety $\mathcal{N} := \mathcal{V}(r_1, r_2, \ldots, r_r)$ is a (possibly empty) finite set, since the r_i's are coprime polynomials, and moreover the set of rank-singularities satisfies

$$\mathcal{V}(C(\eta,\xi)) := \{(\eta,\xi) \in \mathbf{C}\times\mathbf{C}: \operatorname{rank} C(\eta,\xi) < n\} = \mathcal{V}(p) \cup \mathcal{N} \quad (3.7)$$

Now, performing a reasoning on $\mathbf{R}(\xi)[\eta]$ permits us to obtain the row-Hermite form of $C(\eta,\xi)$

$$C(\eta,\xi)S(\eta,\xi) = \left[\begin{array}{ccc|c} a_{11}(\eta,\xi) & & 0 & 0 \\ \vdots & \ddots & & \vdots \\ a_{n1}(\eta,\xi) & \cdots & a_{nn}(\eta,\xi) & 0 \end{array}\right] \quad (3.8)$$

where $\det S(\eta,\xi) \in \mathbf{R}(\xi)$ is different from zero and both $S(\eta,\xi)$ and $a_{ij}(\eta,\xi)$ are polynomial w.r. to η but rational w.r. to ξ. If $\mathcal{M}$ denotes the finite set of ξ-values where at least one denominator in (3.8) or $\det S(\eta,\xi)$ vanishes, we have that $\xi \in \mathbf{C}-\mathcal{M}$ implies

$$\operatorname{rank} C(\eta,\xi) = n \Leftrightarrow \prod_{i=1}^{n} a_{ii}(\eta,\xi) \neq 0 \quad (3.9)$$

Starting from (3.3), we can now define global controllability, bounded (or ℓ_∞-) controllability, ℓ_2-controllability, ℓ_1-controllability, exponential controllability and polynomial controllability in terms of the module M which the components of $\mathcal{X}_0(\xi)$ and of the η-polynomial vectors $\mathcal{X}(\eta,\xi)$ and $\mathcal{U}(\eta,\xi)$ belong to. The main result, which can be found in (Bisiacco, 1995), is the following.

Theorem 2: A 2D system is globally controllable if and only if $p(\eta,\xi)=1$, i.e. if and only if $C(\eta,\xi)$ is a left-coprime matrix, i.e. if and only if the rank-singularities $\mathcal{V}(C(\eta,\xi))$ constitute a finite set. Moreover, it is globally controllable in the bounded, ℓ_2-, ℓ_1-, or exponential sense if and only if $C(\eta,\xi)$ is full-rank $\forall\eta\in\mathbf{C}$ and $\forall\xi\in\mathbf{C}:|\xi|=1$, while it is polynomially controllable if and only if $C(\eta,\xi)$ is full-rank for any $(\eta,\xi)\in\mathbf{C}\times\mathbf{C}$ with $\xi\neq0$. ■

Analogously to the reachability-case, we can restate the previous theorem in terms of pointwise controllability of a ξ-dependend 1D system. In fact we have that global controllability fully corresponds to controllability of the 1D-system $(F(e^{i\omega}), G(e^{i\omega}))$ for almost all ω's in $[0,2\pi]$, bounded, ℓ_2-, ℓ_1- and exponential controllability to controllability of $(F(e^{i\omega}), G(e^{i\omega}))$ for any $\omega\in[0,2\pi]$, and eventually polynomial controllability requires $(F(\xi), G(\xi))$ controllable for any $\xi\neq0$ or, equivalently, $(z_1A_1 + z_2A_2, z_1B_1 + z_2B_2)$ controllable for any $(z_1,z_2) \in \mathbf{C}\times\mathbf{C} - \mathbf{X}_1 - \mathbf{X}_2$. Recalling the results about causal control expressed in (Bisiacco, 1985), we see that local controllability is stronger than polynomial controllability, since it requires controllability of the pair $(z_1A_1 + z_2A_2, z_1B_1 + z_2B_2)$ for any $(z_1,z_2) \in \mathbf{C}\times\mathbf{C}$. In fact, when a system is polynomially but not locally controllable, a local state requires in general a weakly-causal control for driving the state to zero, as shown in the following example.

Example 3: Let $A_1=B_2=1$, $A_2=B_1=0$, so that $\mathcal{C}(z_1,z_2) = [1-z_1|z_2]$ is not full-rank at

$(z_1,z_2) = (1,0) \in \mathcal{P}_1 := \{(z_1,z_2) \in \mathbf{C} \times \mathbf{C}: |z_1| \le 1$ and $|z_2| \le 1\}$. Hence the system is neither local stabilizable nor locally controllable (Bisiacco, 1985) and no causal input (with support in the quarter plane h≥0, k≥0) is able of driving to zero the initial local state $\mathcal{X}_0(\xi)=1$, neither in a finite number of steps nor asymptotically.

However we have $C(\eta,\xi) = [1-\eta | \eta\xi]$ and $C(\eta,\xi)[\mathcal{X}^T(\eta,\xi) \; -\mathcal{U}^T(\eta,\xi)]^T = 1$ for $\mathcal{X}(\eta,\xi)=1$ and $\mathcal{U}(\eta,\xi)=-\xi^{-1}$. Therefore the "weakly-causal" control law $u(h,k) = -x(h-1, k+1)$ is able to control to zero any (polynomial or not) global state $\mathcal{X}_0(\xi)$ in exactly one step, by effects superposition.

4. PROPERTIES COMPARISON

At this point we know a lot of reachability and controllability concepts. Recalling other concepts from literature, such as local reachability (Fornasini and Marchesini, 1978), weak reachability (Fornasini, 1978), local controllability (Bisiacco, 1985), and also some stabilizability concepts such as ℓ_∞- and ℓ_2- ones (Bisiacco and Fornasini, 1990) and local one (Bisiacco, 1985), we can try to describe all these properties by means of the same objects (i.e. the reachability and PBH-controllability matrices) in order to investigate their relationships. Except for the local and the weak reachability concepts, which cannot be defined in terms of these matrices, all other concepts can, and more precisely all reachability concepts are expressible in terms of rank-properties of $\mathcal{R}(\xi)$, while on the other hand all controllability and stabilizability concepts can be referred to the previously defined matrix $\mathcal{C}(z_1,z_2)$. The only think that changes is the region when these matrices can have not full-rank. More precisely, let us summarize the various possibilities for the set $\mathcal{V}(\mathcal{C})$ of rank-singularities of $\mathcal{C}(z_1,z_2)$:

$\mathcal{V}(\mathcal{C}) = \phi \quad \Leftrightarrow$ local controllability

$\mathcal{V}(\mathcal{C}) \cap \mathcal{P}_1 = \phi \quad \Leftrightarrow$ local stabilizability

$\mathcal{V}(\mathcal{C}) \cap \mathcal{M} = \phi \quad \Leftrightarrow \ell_\infty(\ell_1, \ell_2,$ exp)-controllability

$\mathcal{V}(\mathcal{C}) \cap \mathcal{M} \cap \mathcal{P}_1 = \phi \quad \Leftrightarrow \ell_\infty$-stabilizability (or ℓ_2)

$\mathcal{V}(\mathcal{C}) =$ finite set $\quad \Leftrightarrow$ global controllability

$\mathcal{V}(\mathcal{C}) \cap \mathbf{X}_1 \cap \mathbf{X}_2 = \phi \quad \Leftrightarrow$ polynomial controllability

where $\mathcal{M} := \{(z_1,z_2) \in \mathbf{C} \times \mathbf{C}: |z_1| = |z_2|\}$ and $\mathcal{P}_1$, $\mathbf{X}_1$ and $\mathbf{X}_2$ have been previously defined. Analogously, if $\mathcal{V}(\mathcal{R})$ is the set of $\mathcal{R}(\xi)$-rank-singularities, we have that

$\mathcal{V}(\mathcal{R}) \subseteq \{0\} \quad \Leftrightarrow$ polynomial reachability

$\mathcal{V}(\mathcal{R}) \cap \gamma_1 = 0 \quad \Leftrightarrow \ell_\infty(\ell_1, \ell_2,$ exp)-reachability

$\mathcal{V}(\mathcal{R}) =$ finite set $\quad \Leftrightarrow$ global reachability

where γ_1 has been previously defined. Since controllability and reachability properties can be expressed in terms of pointwise properties of the 1D system $(F(\xi), G(\xi))$, we can obtain a relationships-diagram of the following form

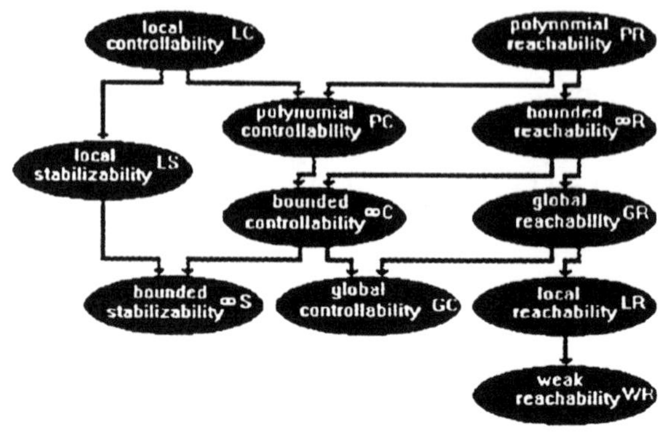

Fig.1: *Reachability and controllability relationships diagram.*

where "⇒" means "implies and is not implied by", and the absence of connections means "doesn't imply and is not implied by". We see that in this hierarchic diagram the "stronger" properties are local controllability and polynomial reachability, but none of these is stronger of the other, while the "weaker" properties are ℓ_∞-stabilizability, global controllability and weak-reachability, which are on the other hand uncorrelated one each other.

5. REFERENCES

Bisiacco, M. (1985). State and output feedback stabilizability of 2-D systems. *IEEE Trans. on CAS, vol. CAS-32, n.12.*

Bisiacco, M. (1995). On some reachability and controllability properties of 2D systems. *To appear.*

Bisiacco, M. and Fornasini, E. (1990). Optimal control of two-dimensional systems. *SIAM J. on Control and Optimization, vol.28, n.3.*

Fornasini, E. (1978). On the relevance of noncommutative power series in spatial filters realization. *IEEE Trans. on CAS, vol. CAS-25, n. 5.*

Fornasini, E. and Marchesini, G. (1978). Doubly-indexed dynamical systems: state-space models and structural properties. *Mathematical System Theory, n. 12.*

Fornasini, E. and Marchesini, G. (1980). Stability analysis of 2D systems. *IEEE Trans. on CAS, vol. CAS-27, n. 12.*

Fornasini, E. and Marchesini, G. (1982). Global Properties and duality in 2-D systems. *Systems and Control Letters, vol.2, n.1.*

SINGLE-LOOP ROBUST MULTIDIMENSIONAL FEEDBACK STRUCTURES

N. K. BOSE

The Pennsylvania State University, Department of Electrical Engineering, Spatial and Temporal Signal Processing Center, University Park, PA 16802, U.S.A.

Abstract. This article focusses attention on the problems encountered in the robust design of multidimensional feedback control systems subject to the constraints of robust stability as well as robust performance. Vertex results, edge results, as well as a transform domain approach to the evaluation of H_2 performance measure over an uncertainty set are considered in relation to other developments in the last five years and some open problems are delineated.

Key Words. Uncertain dynamic systems; robust stability; multidimensional systems

1. INTRODUCTION

The results of research during the last decade in the area of stability of dynamic systems under parametric uncertainty have been documented in several monographs (Ackermann *et al.* (1993), Barmish (1994), Bhattacharyya *et al.* (1995)). A significant amount of literature exists on the subclass of uncertain systems called interval systems. Those systems are analytically tractable because the variations of the parameters are independent of one another. The underlying mathematical structure supporting the topic of interval systems is quite extensive (Neumaier (1990)). There are many situations where the interval property in the uncertainty structure is either lost or not present. For example, the set of polynomials resulting from the application of bilinear transformation on an interval polynomial is not an interval set. In feedback systems with interval plants, only very special types of compensators in the feedback loop preserve the interval property in the closed-loop characteristic polynomial.

Consider a single-input single-output continuous-time plant characterized by the interval transfer function

$$h(s,\boldsymbol{a}_m,\boldsymbol{b}_n) = \frac{a(s,\boldsymbol{a}_m)}{b(s,\boldsymbol{b}_n)}, \qquad (1)$$

where $a(s,\boldsymbol{a}_m)$ and $b(s,\boldsymbol{b}_n)$ are, respectively, interval polynomials of degrees m and n. Consider a fixed proper controller in the simple feedback loop. This controller is characterized by the transfer function

$$f(s) = \frac{c(s)}{d(s)}, \qquad (2)$$

where $c(s)$ and $d(s)$ are polynomials. The closed-loop characteristic polynomial set,

$$g(s,\boldsymbol{q}_m,\boldsymbol{q}_n) = c(s)a(s,\boldsymbol{q}_m) + d(s)b(s,\boldsymbol{q}_n) \qquad (3)$$

is not, in general, an interval polynomial. However, the parameter dependency of $g(s,\boldsymbol{q}_m,\boldsymbol{q}_n)$ is affine on the uncertain parameters a_i, b_j in the interval polynomials,

$$a(s,a_m) = \sum_{k=0}^{m} a_k s^k,\ a_k \in \left[a_k^-, a_k^+\right]$$
$$b(s,b_n) = \sum_{k=0}^{n} b_k s^k,\ b_k \in \left[b_k^-, b_k^+\right]. \qquad (4)$$

Kharitonov-type extreme point (vertex) results are known to occur with fixed first order compensators,

$$f(s) = \frac{c_0 + c_1 s}{d_0 + s}. \qquad (5)$$

The proof is known to be complicated (Ackermann *et al.* (1993)). The affine dependency of the closed-loop characteristic polynomial always permits the application of the *edge theorem*. In general, the number of edges to be tested increases exponentially with the dimension of the uncertainty box. This dimension being equal to the number of uncertain parameters may be as high as $m + n + 2$. However, in the case under consideration, a special type of affine dependency reduces substantially the complexity of the test. It suffices only to check no more than 32 edge polynomials (*Kharitonov segments*) for stability.

2. MULTIDIMENSIONAL FEEDBACK SYSTEMS

Kharitonov-type vertex results hold in the case of multivariate polynomials when the property of interest is the scattering Hurwitz property (Bose (1988), Basu (1990)). This property provides the isomorphism between the class of multivariate scattering Hurwitz polynomials and the class of multivariate rational functions characterized by the reactance property. In this spirit, the scattering Hurwitz property is the desired generalization of the strict Hurwitz property associated with a class of univariate polynomials for which Kharitonov developed his original results (Kharitonov (1979), Bose (1985)). The one-dimensional results summarized in the Introduction may be generalized to obtain either vertex or edge results in parameter space for the generic single-loop feedback system comprised of a linear shift-invariant uncertain multidimensional interval rational plant and a fixed rational compensator when the property of interest is scattering Hurwitz. In the n-dimensional (n-D) case, the compensators of the type,

$$g(s_1,s_2,\ldots s_n) = \frac{\ell}{s_1^{k_1} s_2^{k_2} \ldots s_n^{k_n}}, \qquad (6)$$

where ℓ is a real constant, and $k_1, k_2, \ldots k_n$ are fixed integers, preserve the interval property in the closed-loop characteristic polynomial of the uncertain dynamic feedback system under consideration here and, therefore, vertex results with respect to the scattering Hurwitz property hold. Extension to a wider class of fixed compensators of low order for which a Kharitonov-like result (i.e. a result that yields a testing set with a fixed number of elements, independent of the number of uncertain parameters) applies is *an open problem*.

Affine dependency of a special type hold for n-D uncertain dynamic feedback systems of the generic type comprised of an interval rational plant and a fixed rational compensator. Therefore, edge results again apply. The *challenge* is to discover the reduced set in the n-D case, because of the special nature of the affine dependency, that would be a true generalization of the 1-D results, where the cardinality of the determining set of edges or segments is 32.

3. PERFORMANCE AND ROBUSTNESS

Stability as well as performance are important measures for good design of control systems. Performance is often expressed through H_2 norm constraints while robust stability is expressed in the H_∞ norm. Recently, considerable interest has been shown in the approach whose objective is to minimize the nominal H_2 norm (guarantees nominal performance) under the constraint of an H_∞ norm bound (guarantees robust stability) (Stoorvogel (1993)). Consider a linear time-invariant closed loop discrete-time system which is characterized by the transfer function,

$$H_{m,n}(z,\boldsymbol{q}) \triangleq \frac{B_m(z,\boldsymbol{q})}{A_n(z,\boldsymbol{q})} = \frac{\sum_{k=0}^{m} b_{m,q} z^{-k}}{\sum_{\ell=0}^{n} a_{n,\ell} z^{-\ell}}, \tag{7}$$

where $q \epsilon Q$ is a vector of uncertain parameters. The square of the H_2 norm is defined to be a specialization of

$$I_{m,n,p}(\boldsymbol{q}) = \oint_{|z|=1} \frac{B_m(z,\boldsymbol{q}) B_m(z^{-1},\boldsymbol{q})}{A_n(z,\boldsymbol{q}) A_n(z^{-1},\boldsymbol{q})} z^{p-1} dz, \tag{8}$$

corresponding to $p = 0$. The main problem of calculating the supremum of $I_{m,n,o}(\boldsymbol{q})$ as q varies in Q has recently been successfully tackled in the special case of a single uncertain parameter with affine constraints on the state-space matrices characterizing the uncertain dynamic feedback system under consideration (Friedman *et al.* (1994)). The method is closely related to the bisection method for computation of the H_∞ norm of linear systems. It is suggested that in the multiparameter case, the problem is linked to tests for positivity of multivariate polynomials that have been developed in multidimensional systems theory (Bose (1982)). Another recent approach considers the robust performance in H_∞ with multiple scalar perturbations (Bamieh *et al.* (1994)). The worst-case H_∞ performance is regarded as an H_∞ norm of a function of several complex variables. However, it is admitted that methods for computing the H_∞ norm of such functions are still lacking.

In this paper, it is pointed out that $I_{m,n,o}(\boldsymbol{q})$ can be directly calculated in the transform domain, unlike the state-space approach of (Friedman *et al.* (1994)), using a generalization of the fast evaluation procedure developed for the case when the $b_{m,k}$'s and $a_{n,\ell}$'s are constants (Chun and Bose (1995)). In fact, the one term recursion on p for arbitrary but fixed nonnegative integers m and n, developed in (Chun and Bose (1995)) will allow the computation of performance measures $I_{m,n,p}(\boldsymbol{q})$, for $p > 0$. The generalization is clearly possible because it involves only the basic arithmetic operations on the coefficients $b_{m,k}(\boldsymbol{q})$ for $0 \leq k \leq m$ and $a_{n,\ell}(\boldsymbol{q})$ for $0 \leq \ell \leq n$ in the method of Chun and Bose (1995), where the parameter vector q does not occur. The remaining task of computing the supremum of $I_{m,n,p}(\boldsymbol{q})$ over the multiparameter uncertainty set Q containing q may be done using techniques in multidimensional systems theory when the uncertainty is rational in q (in which case the coefficients $b_{m,k}(\boldsymbol{q})$ and $a_{n,\ell}(\boldsymbol{q})$ are rational functions in the parameters of q).

4. CONCLUSIONS

During the last five years, robustness analysis results for single-loop feedback structures have been introduced. Consider an uncertain plant, characterized by an interval transfer function (ratio of two interval polynomials) with a fixed proper (controller) transfer function in the standard feedback loop. It is easily verified that the characteristic polynomial of the closed-loop system is not an interval polynomial. Only very special types of compensators in the feedback loop preserve the interval property of the closed-loop characteristic polynomial. As a consequence, for one-dimensional systems there exists a Kharitonov-like *vertex* result for a first order compensator and for a compensator with arbitrary order one can only get a special *edge* theorem. The edge polynomials are called *Kharitonov segments* and results on stability of convex combinations of stable polynomials may then be applied to each of the 32 Kharitonov segments. Another result on stabilization of an interval plant by using a nonlinear controller requires test on 16 fixed Kharitonov plants each with a similar nonlinear controller (Mori and Barnett (1988)). In this paper, counterparts of some of the results enumerated above for 1-D systems are identified for n-D systems. Subsequently, an approach towards mixed H_2/H_∞ feedback control design problem in the presence of multiparameter uncertainty is introduced. The scopes for generalization of this approach to multidimensional feedback systems also exist. For the sake of brevity, proofs are not included here. More complete results with proofs in this fertile arena of research activity in robust multidimensional control system design will be reported in due course.

5. REFERENCES

Ackermann, J., A. Bartlett, D. Kaesbauer, W. Sienel, and R. Steinhauser (1993). *Robust Control: Systems with Uncertain Physical Parameters*. Springer-Verlag, London.

Bamieh, B., M. Dahleh and P. Voulgaris (1944). On robust performance in H^∞. *Proceedings Amer. Control Conference*, Baltimore, MD 1588-1592.

Basu, S. (1990). On boundary implications of stability and positivity properties of multidimensional systems. Proceedings of the IEEE, **78**, 614-626.

Bhattacharyya, S. P., L. H. Keel, and H. Chapellat (1995). *Robust Control: The Parametric Approach*. Prentice-Hall, Englewood Cliffs, NJ.

Bose, N. K. (1982). *Applied Multidimensional Systems Theory*. Van Nostrand Reinhold, New York.

Bose, N. K. (1985). A system-theoretic approach to stability of sets of polynomials. *Contemporary Mathematics*, **47**, 25-34.

Bose, N. K. (1988). Robust multivariate scattering Hurwitz interval polynomials. *Linear Algebra and Its Applications*, **98**, 123-136.

Chun, J. and N. K. Bose (1995). Fast evaluation of an integral occurring in digital filter applications. *IEEE Transactions Signal Processing*, **43**, scheduled to appear.

Friedman, J. H., P. T. Kamba, and P. P. Khargonekar, (1994). Worst-case and average performance analysis against real constraint parametric uncertainty. *Proceedings of Amer. Control Conference*, Baltimore, MD, 2406-2410.

Kharitonov, V. L. (1979). Asymptotic stability of an equilibrium position of a family of systems of linear differential equations. *Differential Equations*, **14**, 1483-1485.

Mori, T. and S. Barnett (1988). On stability tests for some classes of dynamical systems with perturbed coefficients. *IMA Journal of Mathematical Control and Information*, **5**, 117-123.

Neumaier, A. (1990). *Interval Methods For Systems of Equations*. Cambridge University Press, Cambridge.

Stoorvogel, A. A. (1993). The robust H_2 control problem. *IEEE Transactions on Automatic Control*, **38**, 1358-1370.

COMBINATORIAL PROPERTIES OF NONNEGATIVE MATRIX PAIRS AND BEHAVIOR OF 2D SYSTEMS

E. FORNASINI* and M.E. VALCHER *

*Dip. di Elettronica ed Informatica, Univ. di Padova, via Gradenigo 6a, 35131 Padova, ITALY

Abstract. The paper aims to connect the spectral properties of a nonnegative matrix pair with the combinatorial structure of the associated digraphs. Dynamical features, such as finite memory, separability and structural linearity, are reduced to the existence of particular canonical forms for the state updating matrices. Finally, 2D positive systems whose characteristic polynomial has support included in a straight line are investigated, and the combinatorial nature of the associated pairs is enlightened.

Résumé. Dans ce travail, nous établissons quelques connexions entre les proprietés spectrales des couples de matrices d'un systrhe 2D positif et la structure combinatoire des graphes de transfert. Nous ramenons plusieurs caractfistiques, telles que la mémoire finie, la séparabilité et la linéarité structurelle à l'existence de formes canoniques pour les couples de matrices qui jouent dans l'équation d'état. On s'adresse enfin à la classe des systrhes 2D douès d'un polynôme caractéristique dont le support est inclus dans une droite, et l'on analyse quelques aspects combinatoires des matrices correspondantes.

Key Words. Reducible/irreducible matrices, finite memory, separability, property L

1. INTRODUCTION

A 2D system is the mathematical model of a process whose dynamics depends on two independent variables. In the last twenty years the theory of 2D systems mainly developed in a linear vector spaces context, where 2D signals take values on the whole real line. Quite recently, however, an increasing interest has been conveyed toward positive 2D systems, in which input, state and output variables are always positive (or at least nonnegative) in value. This is mainly due to the fact that they constitute an appropriate setting when the physical systems one tries to model involve quantities that are intrinsically nonnegative, such as pressures, concentrations, population levels, etc.

In this paper we consider homogeneous positive 2D systems, described by the equation

$$x(h+1,k+1) = A\,x(h,k+1)+B\,x(h+1,k), \quad (1.1)$$

where the doubly indexed local state sequence $x(\cdot,\cdot)$ takes values in the positive cone

$$\mathbb{R}^n_+ := \{x \in \mathbb{R}^n : x_i \geq 0,\ i = 1,2,...,n\},$$

A and B are nonnegative $n \times n$ matrices, and the initial conditions are assigned by specifying the nonnegative values of the local states on the *separation set* $\mathcal{C}_0 := \{(i,-i) : i \in \mathbb{Z}\}$.
The aim of this contribution is to show how some spectral properties, which characterize interesting classes of nonnegative matrix pairs (A, B), are induced by the combinatorial structure of the associated influence graphs, and hence are invariant with respect to structured perturbations of the pairs. As a consequence, the system features that depend on the characterystic polynomial of the pair are robust and remain unchanged as the nonzero entries of A and B vary over $\mathbb{R}_+$. This is particularly relevant when the information available on some physical process allows to assume that no interaction exists among certain variables, and, consequently, some entries of matrices A and B are exactly 0, whereas the others are nonnegative and known with some level of uncertainty.

A typical example is provided by compartmental models, where nonzero entries correspond to the existence of flows between different compartments, and physical or biological reasons guarantee that some pairs of compartments have no direct interaction at all. In this setting the situation when all uncertain values are positive represents the "worst case", and, therefore, if such properties are verified in the worst case, they are preserved under all perturbations of the nonnegative entries.

Before proceeding, we introduce some notation. If $M = [m_{ij}]$ is a matrix (in particular, a vector), we write $M \gg 0$ (*M strictly positive*), if $m_{ij} > 0$ for all i, j; $M > 0$ (*M positive*), if $m_{ij} \geq 0$ for all i, j, and $m_{hk} > 0$ for at least one pair (h, k);

$M \geq 0$ (*M nonnegative*), if $m_{ij} \geq 0$ for all i, j. The positive matrix whose (i, j)th entry is 1, while all others are 0, is denoted by E_{ij}. To every $n \times n$ nonnegative matrix M we associate (Minc, 1988) a *digraph* (directed graph) $D(M)$ of order n, with vertices indexed by $1, 2, ..., n$. There is an arc $\alpha = (i, j)$ from i to j if and only if $m_{ij} > 0$. Two $n \times n$ nonnegative matrices $M = [m_{ij}]$ and $N = [n_{ij}]$ have the same *zero pattern* if $m_{ij} = 0$ implies $n_{ij} = 0$ and vice versa. M and N have the same zero pattern if and only if $D(M) = D(N)$. An $n \times n$ nonnegative matrix M is said to be *cogredient* to a nonnegative matrix $\bar{M}$ if there exists a permutation matrix P such that $\bar{M} = P^T M P$. Two matrices are cogredient if and only if they have isomorphic digraphs. In some cases, it will be convenient to denote the (i, j)th entry of a matrix M as $[M]_{ij}$. The symbol $*$ represents the *Hadamard product* : if A and B are $m \times n$ (nonnegative) matrices, then $A * B$ is the $m \times n$ matrix whose entries are $[A * B]_{ij} = [A]_{ij}[B]_{ij}$. The *characteristic polynomial* of a pair (A, B) of (not necessarily nonnegative) matrices is

$$\Delta_{A,B}(z_1, z_2) := \det(I - Az_1 - Bz_2). \qquad (1.2)$$

Given an alphabet $\Xi = \{\xi_1, \xi_2\}$, the free monoid Ξ^* is the set of all words $w = \xi_{i_1}\xi_{i_2}\cdots\xi_{i_m}$, $m \in \mathbb{N}$, $\xi_{i_h} \in \Xi$. The integer m is called the length of w and denoted by $|w|$, while $|w|_i$ represents the number of occurencies of ξ_i in w, $i = 1, 2$. For each pair $A, B \in \mathbb{C}^{n \times n}$, the map ψ defined on $\{1, \xi_1, \xi_2\}$ by the assignements $\psi(1) = I_n$, $\psi(\xi_1) = A$ and $\psi(\xi_2) = B$, uniquely extends to a monoid morphism of Ξ^* into $\mathbb{C}^{n \times n}$. The ψ-image of $w \in \Xi^*$ is denoted by $w(A, B)$.
The Hurwitz products of two square matrices A and B are inductively defined as $A^i \sqcup\!\sqcup^0 B = A^i$, $i > 0$ and $A^0 \sqcup\!\sqcup^j B = B^j$ $j > 0$ and, when i and j are both greater than zero, $A^i \sqcup\!\sqcup^j B = A(A^{i-1} \sqcup\!\sqcup^j B) + B(A^i \sqcup\!\sqcup^{j-1} B)$. Clearly

$$A^i \sqcup\!\sqcup^j B = \sum_{\substack{|w|_1=i,\ |w|_2=j \\ w\ \in\ \Xi^*}} w(A, B), \qquad (1.3)$$

namely, the (i, j)-th Hurwitz product is the sum of all matrix products that include the factors A and B, i and j times respectively. Assuming in (1.1) zero initial conditions on $\mathcal{C}_0$, except at $(0, 0)$, $x(h, k)$ can be expressed as $x(h, k) = A^h \sqcup\!\sqcup^k B\ x(0, 0)$, $\forall h, k \geq 0$. Linearity and shift-invariance allow to compute $x(h, k)$ for arbitrary initial conditions on $\mathcal{C}_0$.

2. FINITE MEMORY, SEPARABILITY AND PROPERTY L

When dealing with a generic 2D system, some assumptions on the structure of the characteristic polynomial allow to single out important classes of systems, whose dynamical behaviour exhibits some distinguishing features. As an example, finite memory systems (Bisiacco, 1985), i.e. 2D systems whose state evolution goes to zero in a finite number of steps, are characterized by the condition $\Delta_{A,B}(z_1, z_2) = 1$. On the other hand, separable systems (Fornasini *et al.*, 1994), that are usually viewed as the simplest examples of IIR 2D systems, are those which satisfy the factorization property $\Delta_{A,B}(z_1, z_2) = r(z_1)s(z_2)$, for suitable polynomials $r(z_1) \in \mathbb{R}[z_1]$ and $s(z_2) \in \mathbb{R}[z_2]$.

A more general class, which encompasses both finite memory and separable systems, refers to the so-called property L (Motzkin and O., 1952; Motzkin and O., 1955) and corresponds to the possibility of factorizing $\Delta_{A,B}(z_1, z_2)$ into the product of first degree polynomials with complex coefficients

$$\Delta_{A,B}(z_1, z_2) = \prod_{i=1}^{n}(1 - \lambda_i z_1 - \mu_i z_2). \qquad (2.1)$$

Equivalently, the pair (A, B) has property L if and only if the eigenvalues of A and B can be ordered into two n-tuples

$$\Lambda(A) = (\lambda_1, ..., \lambda_n), \ \Lambda(B) = (\mu_1, ..., \mu_n) \qquad (2.2)$$

such that, for all α, β in $\mathbb{C}$, the spectrum of $\alpha A + \beta B$ is given by

$$\Lambda(\alpha A + \beta B) = (\alpha\lambda_1 + \beta\mu_1, ..., \alpha\lambda_n + \beta\mu_n). \qquad (2.3)$$

It turns out that L-pairs are so special to provide a basis for a rich and interesting theory, but also general enough to include several models of practical importance, such as systems described by triangular or commutative matrix pairs.

The nonnegativity assumption introduces further constraints on the structure of the above systems, we will explore in some detail. In the finite memory and in the separable cases, the combinatorial nature and the spectral features of the pair (A, B) are clarified by Propositions 2.1 and 2.2 below.

Proposition 2.1 (Fornasini and M.E., 1995*b*) *For a pair of $n \times n$ nonnegative matrices (A, B), the following statements are equivalent*

i) $\Delta_{A,B}(z_1, z_2) = 1$;
ii) $A + B$ *is a nilpotent (and, a fortiori, a reducible) matrix;*
iii) $A^i \sqcup\!\sqcup^j B$ *is nilpotent, for all* $(i, j) \neq (0, 0)$;
iv) $w(A, B)$ *is nilpotent, for all* $w \in \Xi^* \setminus \{1\}$;
v) *there exists a permutation matrix P such that both P^TAP and P^TBP are upper triangular nilpotent matrices.* ■

Proposition 2.2 (Fornasini and M.E., 1995*b*) *For a pair of $n \times n$ positive matrices A and B, the following statements are equivalent:*

i) $\Delta_{A,B}(z_1, z_2) = r(z_1)s(z_2)$;
ii) $\det[I-(A+B)z] = \det[I-Az]\det[I-Bz]$;
iii) $A^i \sqcup\!\sqcup^j B$ *is nilpotent for all* (i,j) *with* $i,j>0$;
iv) $w(A,B)$ *is nilpotent, for all* $w \in \Xi^*$ *such that* $|w|_i > 0$, $i=1,2$;
v) *there exists a complex valued nonsingular matrix* T *such that* $\hat{A} = T^{-1}AT$ *and* $\hat{B} = T^{-1}BT$ *are upper triangular matrices, and* $[\hat{A}]_{hh} \neq 0$ *implies* $[\hat{B}]_{hh} = 0$;
vi) *there exists a permutation matrix* P *such that* P^TAP *and* P^TBP *are conformably partitioned into block triangular matrices*

$$P^TAP=\begin{bmatrix} A_{11} & * & * \\ & \ddots & * \\ & & A_{tt}\end{bmatrix},\ P^TBP=\begin{bmatrix} B_{11} & * & * \\ & \ddots & * \\ & & B_{tt}\end{bmatrix}$$

where the nonzero diagonal blocks are irreducible and $A_{ii} \neq 0$ *implies* $B_{ii} = 0$. ∎

Statement *v)* in Proposition 2.1 and *vi)* in Proposition 2.2 show that, for a nonnegative matrix pair, finite memory and separability properties depend on its zero pattern only. This is no more true when property L is considered. To see that property L cannot be deduced from the structure of the directed graphs, $D(A)$ and $D(B)$, of the nonnegative matrices A and B, just consider

$$(A_1, B_1) = \left(\begin{bmatrix}1 & 1\\0 & 0\end{bmatrix}, \begin{bmatrix}0 & 0\\1 & 1\end{bmatrix}\right)$$

and

$$(A_2, B_2) = \left(\begin{bmatrix}1 & 2\\0 & 0\end{bmatrix}, \begin{bmatrix}0 & 0\\1 & 1\end{bmatrix}\right).$$

The pairs (A_1, B_1) and (A_2, B_2), of course, have the same zero pattern, but only the first one is endowed with property L. So, it seems natural to ask under what conditions a nonnegative pair has property L, *independently of the values of its nonzero entries*. When so, property L turns out to be a feature which depends only on the combinatorial structure of the matrices, and therefore will be called *structural linearity* (property SL, for short).

Definition A pair of $n \times n$ nonnegative matrices (A,B) has property SL if $(A*M, B*N)$ has property L for all $n \times n$ nonnegative matrices M and N.

Two matrices with property SL can be simultaneously reduced to a very particular block triangular form by a suitable cogredience transformation. To obtain this result, we need the following lemma.

Lemma 2.3 *Let* (A,B) *be a pair of* $n\times n$ *nonnegative matrices, with* $n>1$. *If* (A,B) *has property SL, then either* $A+B$ *is reducible or one of the two matrices is zero.*

PROOF As a preliminary step, we prove that if one of the matrices, say A, is irreducible, the other is zero. Assume, by contradiction, that B has an element $b_{ij} > 0$, and suppose first $i \neq j$. As A is irreducible, we can find in $D(A)$ a minimal path γ, of length $m-1$, which connects vertex j to vertex i.
Possibly after relabelling the vertices, we can assume $i=1$, $j=m$ and $\gamma = \{(m, m-1), (m-1, m-2), \dots, (2,1)\}$. This implies the existence of nonnegative matrices M and N which reduce A and B to $\hat{A} := A*M = \sum_{i=1}^{m-1} E_{i+1,i}$ and $\hat{B} := B*N = E_{1,m}$. By the SL property of (A,B), the pair $(\hat{A}, \hat{B})$ should be endowed with property L, which is impossible since

$$\det\begin{bmatrix} 1 & & & & -z_2 & \\ -z_1 & 1 & & & & \\ \vdots & -z_1 & \ddots & & \vdots & 0 \\ & & \ddots & 1 & & \\ 0 & 0 & \dots & -z_1 & 0 & \\ & & \mathbf{0} & & & \mathbf{I}\end{bmatrix} = 1 - z_1^{m-1}z_2$$

does not factor into linear factors, as $m-1$ is positive.
On the other hand, if b_{ii} were nonzero for some i, by the irreducibility assumption on A, there would be in $D(A)$ a circuit γ, of length $m-1>1$, connecting vertex i to itself and, as before, we can assume $i=1$ and $\gamma = \{(1,m), (m, m-1), \dots, (2,1)\}$. Hence, there exist nonnegative matrices M and N such that

$$\bar{A} := A*M = E_{1,m} + \sum_{i=1}^{m-1} E_{i+1,i} \quad \bar{B} := B*N = E_{1,1}.$$

The characteristic polynomial of $(\bar{A}, \bar{B})$ is $1 - z_2 - z_1^m$, which cannot factor into linear terms as $m>1$, thus contradicting the SL assumption on (A,B). So, every pair of positive matrices with property SL consists of reducible matrices.
It remains to prove that also their sum, $A+B$, is reducible. Suppose, by contradiction, that it is not. As the irreducibility of $A+B$ only depends on its zero pattern, when both a_{ij} and b_{ij} are nonzero we may replace one of them with zero without destroying the irreducibility of $A+B$ and the property SL. Since both A and B are still nonzero (otherwise $A+B$ would coincide either with A or with B, and hence would be reducible), and $A*B=0$, there exist nonzero entries a_{ij} and b_{hk} with $(i,j)\neq(h,k)$. By the irreducibility of $A+B$, two directed paths can be found in $D(A+B)$ connecting j with h and k with i, thus producing a closed walk τ with distinct edges, including arcs of both $D(A)$ and $D(B)$. Among the elementary circuits of τ, either there is a circuit γ including arcs of both $D(A)$ and $D(B)$, or there are two circuits, γ_A in $D(A)$ and γ_B in $D(B)$, with a common vertex. In the first case, by resorting to reasonings of the same kind as in the first part of the proof, we obtain a matrix pair $(\tilde{A}, \tilde{B}) := (A*M, B*N)$ whose characteristic polynomial

$$\Delta_{\tilde{A},\tilde{B}}(z_1, z_2) = 1 - z_1^r z_2^s \qquad r,s \geq 1$$

does not split into linear factors. In the second case, by relabelling the vertices of the digraph $D(A+B)$, we can assume that the common vertex of γ_A and γ_B is 1 and

$$\gamma_A = \{(1,2),(2,3),...,(m-1,m)\}$$
$$\gamma_B = \{(1,m+1),(m+1,m+2),...,(m+n-1,1)\}.$$

By suitably choosing M and N, we obtain once more a pair (A', B'), with $A' := A * M = E_{m,1} + \sum_{i=1}^{m-1} E_{i,i+1}$ and $B' := B * N = E_{m+n-1,1} + E_{1,m+1} + \sum_{i=1}^{n-2} E_{m+i,m+1+i}$, whose characteristic polynomial $\Delta_{A',B'}(z_1,z_2) = 1-z_1^m-z_2^n$, $n+m>2$ does not factor into linear factors. Property SL, however, would imply that (A', B') has property L, a contradiction. ■

Proposition 2.4 *Let (A,B) be a pair of $n \times n$ nonnegative matrices with property SL. Then there exists a permutation matrix P such that P^TAP and P^TBP are conformably partitioned into block triangular matrices*

$$P^TAP = \begin{bmatrix} A_{11} & * & * \\ & \ddots & * \\ & & A_{tt} \end{bmatrix}, \; P^TBP = \begin{bmatrix} B_{11} & * & * \\ & \ddots & * \\ & & B_{tt} \end{bmatrix}$$

where the diagonal pairs (A_{ii}, B_{ii}) of dimension greater than 1 consist of an irreducible and a zero matrix.

PROOF The case $n=1$ is trivial. If $n>1$ and $A+B$ is irreducible, either A or B is zero, and we can reduce the other matrix to Frobenius normal form (Minc, 1988). If $A+B$ is reducible, we can reduce it to Frobenius normal form by using a cogredience transformation, and apply to each irreducible diagonal block the previous arguments. ■

As a consequence of Proposition 2.4, the zero pattern of a nonnegative matrix A completely characterizes the class of nonnegative matrices B such that (A,B) has property SL. A more difficult problem is that of obtaining, for a given (nonnegative) matrix A, all nonnegative pairs (A,B) with property L. In the case of a diagonal matrix A, however, a complete solution is available, which sheds ligth on some interesting connections between properties L and SL.

Proposition 2.5 (Fornasini and M.E., 1995*b*) *Let $A = \mathrm{diag}\{a_1 I_{\nu_1},\ a_2 I_{\nu_2},\ \ldots, a_k I_{\nu_k}\}$ be a nonnegative (block) diagonal matrix, with $a_i \neq a_j$ if $i \neq j$, and let $B \geq 0$ be partitioned conformably with the partition of A, as follows*

$$B = \begin{bmatrix} B_{11} & B_{12} & \ldots & B_{1k} \\ B_{21} & B_{22} & \ldots & B_{2k} \\ \vdots & & \ddots & \vdots \\ B_{k1} & B_{k2} & \ldots & B_{kk} \end{bmatrix}.$$

The following statements are equivalent:

i) *(A,B) has property L;*
ii) $\det(zI_n - B) = \prod_{i=1}^{k} \det(zI_{\nu_i} - B_{ii})$;
iii) *there exists a permutation matrix P such that*

$$P^TAP = \begin{bmatrix} \hat{A}_{11} & & \\ & \ddots & \\ & & \hat{A}_{pp} \end{bmatrix}, \; P^TBP = \begin{bmatrix} \hat{B}_{11} & * & \hat{B}_{1p} \\ & \ddots & * \\ & & \hat{B}_{pp} \end{bmatrix}$$

where the $\hat{A}_{ii}$'s are scalar matrices. Moreover, each $\hat{B}_{ii}$ is a diagonal block of the Frobenius normal form of B_{jj}, for some j. ■

3. CHARACTERISTIC POLYNOMIALS WITH SINGLE LINE SUPPORTS

As a consequence of the quarter plane causality underlying the state updating in (1.1), each nonzero local state in $(0,0)$ produces a free state evolution which generally extends to the whole positive orthant $\{(h,k) \in \mathbb{Z} \times \mathbb{Z} : h,k \geq 0\}$. A noteworthy exception is represented by the class of 2D systems whose free evolution is identically zero except on a "strip" that includes a straight line passing through the origin. This kind of systems exhibits several features which strictly resemble those of 1D systems. Furthermore, no matter how far (h,k) is from the set $\mathcal{C}_0 = \{(i,-i) : i \in \mathbb{Z}\}$ where the initial conditions are given, the local state in (h,k) is determined only by a finite subset of the initial global state, whose cardinality does not exceed a fixed integer $M > 0$.

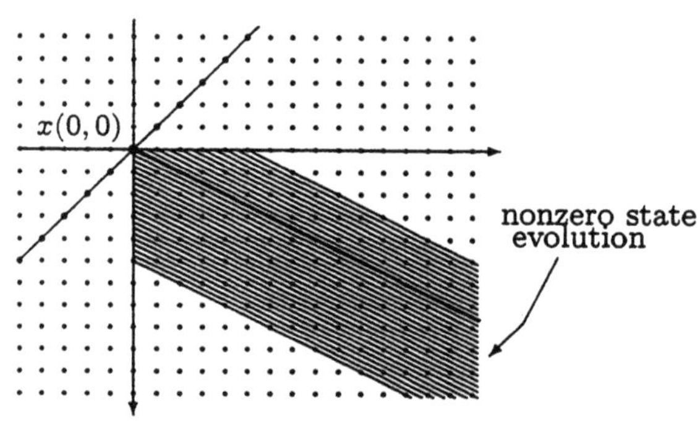

FIG.1

Like property L, the above property can be expressed as a condition on the spectra of the linear combinations $\alpha A + \beta B$, $\alpha, \beta \in \mathbb{C}$, or, equivalently, as a constraint on the support of the characteristic polynomial of the pair (A,B), which has to be a subset of a straight line through the origin in $\mathbb{Z} \times \mathbb{Z}$, i.e. there exists $(\ell, m) \neq (0,0)$ in $\mathbb{N} \times \mathbb{N}$ such that $\mathrm{supp}(\Delta_{A,B}(z_1,z_2)) \subset \{(k\ell, km), k \in \mathbb{N}\}$.The main results on these systems have been presented in (Fornasini and M.E., 1995*a*) and are summarized by the following proposition.

Proposition 3.1 *Let (A,B) be a pair of $n \times n$ matrices with entries in $\mathbb{R}$ and $\Delta_{A,B}(z_1,z_2)$ its* *characteristic polynomial. Assume moreover that (ℓ, m) is a pair of nonnegative integers and introduce the two sets*

$$\mathcal{R} := \{(k\ell, km) : k \in \mathbb{N}\}$$
$$\mathcal{S} := \{(i,j) \in \mathbb{N}^2 : -\ell(n-1) \leq mi - \ell j < m(n-1)\}.$$

The following statements are equivalent:

i) $\Delta_{A,B}(z_1, z_2) = 1 - \sum_{h=1}^{r} d_h (z_1^\ell z_2^m)^h$;

ii) there exist $c_1, c_2, \ldots, c_n$ in $\mathbb{C}$ such that, for every $(\alpha, \beta) \in \mathbb{C} \times \mathbb{C}$ and every $(\ell+m)$th root of $\alpha^\ell \beta^m$, $\Lambda(\alpha A + \beta B) = \left(c_1(\alpha^\ell\beta^m)^{\frac{1}{\ell+m}}, \ldots, c_n(\alpha^\ell\beta^m)^{\frac{1}{\ell+m}}\right)$;

iii) the trace $\mathrm{tr}\,(A^i ⧢^j B)$ *is zero for every $(i,j) \notin \mathcal{R}$;*

iv) for all $\alpha, \beta \in \mathbb{C}$ and suitable $b_k \in \mathbb{C}$,

$$\mathrm{tr}(\alpha A + \beta B)^k = \begin{cases} b_k(\alpha^\ell \beta^m)^\nu & \text{if } k = (\ell+m)\nu \\ 0 & \text{otherwise;} \end{cases}$$

v) $(i,j) \notin \mathcal{S}$ implies $A^i ⧢^j B = 0$. ■

An immediate consequence of property *iii)* in the above proposition is the following

Corollary 3.2 *If* $\mathrm{supp}(\Delta_{A,B}(z_1,z_2))$ *is a subset of a straight line, different from the coordinate axes, then both A and B are nilpotent.* ■

The nonnegativity assumption on A and B allows to translate conditions on the matrices $A^i ⧢^j B$ into conditions on the matrices $w(A,B)$, $w \in \Xi^*$, involved in each Hurwitz product, thus extending the list of equivalent conditions given in Proposition 3.1.

Proposition 3.3 *Let (A, B) be a pair of $n \times n$ nonnegative matrices. Then properties (i) ÷ (vi) in Proposition 3.1 are equivalent to anyone of the following statements:*

vi) for all $w \in \Xi^*$ such that $(|w|_1, |w|_2) \notin \mathcal{R}$, $\mathrm{tr}(w(A,B)) = 0$;

vii) for all $w \in \Xi^*$ such that $(|w|_1, |w|_2) \notin \mathcal{R}$, $w(A,B)$ is nilpotent;

viii) $(|w|_1, |w|_2) \notin \mathcal{S}$ implies $w(A,B) = 0$.

PROOF *(iii)* ⇔ *(vi)* As a consequence of (1.3), we have

$$\mathrm{tr}(A^i ⧢^j B) = \sum_{\substack{|w|_1 = i,\ |w|_2 = j \\ w \in \Xi^*}} \mathrm{tr}(w(A,B)),$$

and the nonnegativity assumption implies that $\mathrm{tr}(A^i ⧢^j B)$ is zero if and only if $\mathrm{tr}(w(A,B))$ is zero for all w's satisfying $|w|_1 = i, |w|_2 = j$.

(vi) ⇔ *(vii)* The nilpotency of $w(A,B)$ implies $\mathrm{tr}(w(A,B)) = 0$, which gives *(viii)* ⇒ *(vii)*. To prove the converse, note that if $(|w|_1, |w|_2)$ does not belong to $\mathcal{R}$, the same holds true for all powers w^i, $i > 1$. Consequently assumption (vii) gives $\mathrm{tr}(w^i(A,B)) = 0$, $\forall i \in \mathbb{N}_+$, which implies the nilpotency of $w(A,B)$.

(viii) ⇔ *(v)* Same argument as in *(iii)* ⇔ *(vi)*. ■

The above proposition makes it clear that, under the nonnegativity assuption, the situation when the characteristic polynomial $\Delta_{A,B}(z_1, z_2)$ has a single line support represents a structural property of the pair (A, B), and does not depend on the particular values taken by the nonnegative entries of A and B.

Proposition 3.4 *Let (A, B) be a pair of $n \times n$ nonnegative matrices. The characteristic polynomial of the pair, $\Delta_{A,B}(z_1, z_2)$, belongs to $\mathbb{R}[z_1^\ell z_2^m]$, ℓ and $m \in \mathbb{N}$, if and only if each circuit in $D(A+B)$ consists of $k\ell$ arcs of $D(A)$ and km arcs of $D(B)$, k a positive integer.*

PROOF As a consequence of *viii)* of Proposition 3.3 and the nonnegativity assumption, the characteristic polynomial of the pair (A, B) is in $\mathbb{R}[z_1^\ell z_2^m]$ if and only if $[w(A,B)]_{ii} = 0$ for every $i = 1, 2, \ldots, n$ and every $w \in \Xi^*$, $(|w|_1, |w|_2) \notin \mathcal{R}$. But this amounts to say that no circuit can be found in $D(A+B)$ except, possibly, for those that include $k\ell$ arcs of $D(A)$ and km arcs of $D(B)$, for some positive integer k. ■

As each circuit in $D(A+B)$ consists of no more than n arcs, n being the dimension of the nonnegative matrices A and B, every statement involving Hurwitz and matrix products given in Propositions 3.1 and 3.3 can be restricted to index pairs (i, j) with $i + j \leq n$ and words $w \in \Xi^*$ with $|w| \leq n$.

When restricting ourselves to matrix pairs (A, B) with irreducible sum $A + B$, the graph interpretation provided in Proposition 3.4 allows for deeper insights into the structure of the matrix pairs, and leads to the following result.

Proposition 3.5 *Let A and B be $n \times n$ nonnegative matrices with irreducible sum $A+B$ and suppose that their characteristic polynomial $\Delta_{A,B}(z_1, z_2)$ is in $\mathbb{R}[z_1^\ell z_2^m]$, ℓ and $m \in \mathbb{N}$. Then $A * B = 0$. Moreover, if $\Delta_{A,B}(z_1, z_2)$ is in $\mathbb{R}[z_1^\ell z_2^m]$ but does not belong to $\mathbb{R}[z_1^{\ell d} z_2^{md}]$, for any positive integer d different from 1, then the imprimitivity index of $A+B$ coincides with $\ell + m$.*

PROOF Assume $A_{ij} > 0$. We aim to show that B_{ij} is zero. As $A + B$ is an irreducible matrix, there exists a directed path in $D(A+B)$ from j to i, and hence a word $w \in \Xi^*$ such that $[w(A,B)]_{ji} > 0$. Consequently, $[A\ w(A,B)]_{ii}$ is positive, and, by the assumption on the characteristic polynomial of (A, B), one gets $(|w|_1, |w|_2) =$

$(k\ell-1, km)$ for some positive integer k. If B_{ij} were nonzero, then also $[B\ w(A, B)]_{ii}$ would be strictly positive, and hence it should be $(|w|_1, |w|_2) = (k'\ell, k'm - 1)$ for some positive integer k'. But the two conditions on $(|w|_1, |w|_2)$ cannot be simultaneously verified.

As well known (Minc, 1988), the imprimitivity index of a matrix coincides with the g.c.d. of the lengths of all circuits appearing in its digraph. By the assumptions on the pair, every circuit in the digraph $D(A + B)$ has length $k(\ell + m)$ for some $k \in \mathbb{N}$ and therefore $\ell + m$ clearly divides h.
On the other hand, as $\Delta_{A,B}(z_1, z_2)$ is in $\mathbb{R}[z_1^\ell z_2^m]$, $\ell, m \in \mathbb{N}$, but does not belong to $\mathbb{R}[z_1^{\ell d} z_2^{md}]$, for any positive integer $d > 1$, two words w' and w'' can be found in Ξ^* such that $\mathrm{tr}(w'(A, B)) \neq 0$, $\mathrm{tr}(w''(A, B)) \neq 0$, and $(|w'|_1, |w'|_2) = (k_1\ell, k_1 m)$, $(|w''|_1, |w''|_2) = (k_2\ell, k_2 m)$, with k_1 and k_2 coprime positive integers. So, in $D(A+B)$ there are two circuits of lengths $k_1(\ell+m)$ and $k_2(\ell+m)$, respectively, and therefore $h|k_1(\ell+m)$ and $h|k_2(\ell+m)$ together imply $h|(\ell + m)$, which proves the result. ∎

4. REFERENCES

Bisiacco, M. (1985). State and output feedback stabilizability of 2D systems. *IEEE Trans. on Circ. and Sys* **CAS 32**, 1246–1249.

Fornasini, E. and Valcher M.E. (1995*a*). Matrix pairs in 2D systems: an approach based on trace series and Hankel matrices. *SIAM J. of Contr. and Optim.* **33, no.4**, 511–534.

Fornasini, E. and Valcher M.E. (1995*b*). On the spectral and combinatorial structure of 2D positive systems. *to appear in Lin. Alg. and Appl.*

Fornasini, E., Marchesini G. and Valcher M.E. (1994). On the structure of finite memory and separable 2D systems. *Automatica* **29**, 347–350.

Minc, H. (1988). *Nonnegative Matrices.* J.Wiley.

Motzkin, T.S. and Taussky O. (1952). Pairs of matrices with property L. *Trans. Amer. Math.Soc.* **73**, 108–114.

Motzkin, T.S. and Taussky O. (1955). Pairs of matrices with property L (II). *Trans. Amer. Math.Soc.* **80**, 387–401.

NEW 2D SYSTEMS MODELS FOR REPETITIVE PROCESSES

K Galkowski[+], E Rogers[++] and DH Owens[+++]

+ *Institute of Telecommunication & Accoustics, Technical University of Wroclaw, Poland.*
++ *Image, Speech & Intelligent Systems Research Group, Dept of Electronics & Computer Science, University of Southampton, UK.*
+++ *Centre for Systems and Control Engineering, School of Engineering, University of Exeter, UK.*

Resume: Cet article developpe une nouvelle classe de modeles des processus lineaires, discrets et repetitifs. L'objectif est du'utiliser conjointement ces modeles et la theorie classique des systemes 2D afin der examiner, pour cette classe de systemes repetitifs, des notions de base la theorie des systemes. A cette fin, il est montreque les theories de la stabili te sont equivalentes. Une matrice de transition est definie, qui nous permet de presenter quelques resultats de base concernant latteignab ilite.

Abstract: This paper develops new 2D systems state space models for discrete linear repetitive processes. The overall aim is to use these models and well established 2D systems theory to address basic systems theoretic questions for this subclass of repetitive processes. To this end, it is shown that the stability theories are equivalent and a state transition matrix is developed and used to present some basic results on reachability.

Keywords: 2D Systems, Stability theory, Reachability.

1. INTRODUCTION

Repetitive, or multipass processes are uniquely characterised by a recursive action with interaction between successive outputs or pass profiles. To introduce a formal definition, suppose that the necessarily finite pass length α is constant. Then in a repetitive process the output, or pass profile, $Y_k(t)$, $0 \leq t \leq \alpha$, produced on the kth pass acts as a forcing function on, and hence contributes to, the new pass profile $Y_{k+1}(t)$, $0 \leq t \leq \alpha$, $k \geq 0$.

Industrial examples include long-wall coal cutting operations, metal rolling and certain classes of vehicle convoys. Interest in the dynamic behaviour of repetitive processes has recently been increased by the recognition of the fact that they from a natural basis for the analysis and design of iterative learning control schemes. Basically (see, for example Amann etal [1994]) such schemes learn the input necessary to generate a desired output by a series of trials where the input to the current trial is a function of the previous outputs and inputs and (possibly) the causal part of the current output.

Suppose now that $\{Y_k\}_{k\geq 1}$ denotes the output sequence produced by a repetitive process. Then the unique control problem for these processes is that $\{Y_k\}_{k\geq 1}$ may contain oscillations which increase in amplitude in the pass to pass direction. Such behaviour is easily generated both in simulation studies and in experiments on scaled models of industrial examples.

(Smyth [1992] gives a detailed treatment of this area).

Consider now the case of linear dynamics and a constant finite pass length where this sub-class contains a wide range of known examples as special cases. Then a rigorous stability theory for this subclass has been developed by Rogers and Owens [1992] based, in effect, on an abstract model in a Banach space setting which includes all such examples as special cases. Also the results of applying the resulting conditions to a wide range of special cases are known and one of these is the case of so-called discrete processes which are the subject of this paper.

These processes exhibit strong structural links with 2D linear time invariant systems described by the Roesser model [1975]. This raises the possibility of using well established 2D theory to answer basic systems theoretic (and controller design) questions for discrete linear repetitive processes. One such area is reachability, controllability and observability and related problems.

Previous work (Rocha etal [1995] for example) has shown that stability tests can be interchanged between these two (apparently distinct) areas. This, however, does not provide a suitable basis on which to study the application of well established 2D theory to currently open systems theoretic questions for discrete linear repetitive processes. The main objective of this paper is to remove this basic difficulty by showing that several equivalent 2D systems representations of the dynamics of these repetitive processes exist. Some immediate consequences of the existence of these representations are also reported. These are (i) a proof that the stability theories are equivalent, (ii) a state transition matrix, and (iii) some basic results on reachability.

2. Background

A detailed treatment of the abstract model based stability theory can be found in Rogers and Owens [1992]. In this paper the interest is in the sub-class of so-called discrete unit memory linear repetitive processes with state space model

$$X(k+1, p+1) = AX(k+1, p) + BU(k+1, p) + B_o Y(k,p)$$
$$Y(k+1, p) = CX(k+1, p) + DU(k+1, p) + D_o Y(k, p) \qquad (1)$$

Here on pass k, $X(k, p)$ is the $n \times 1$ state vector at sample p along a pass of length $\alpha < +\infty$, $Y(k, p)$ is the $m \times 1$ vector pass profile, and $U(k, p)$ is the $l \times 1$ vector of control inputs.

Two essentially different concepts of stability for linear dynamics constant pass length repetitive processes exist. These are termed asymptotic stability and stability along the pass respectively where the former is a necessary condition for the latter, which (see Smyth [1992] for a detailed discussion of this key point) is required in the vast majority of practical cases. The following (Rogers and Owens [1992] theorem 3.3.8) is one of several equivalent sets of necessary and sufficient conditions for stability along the pass of processes described by (1).

Theorem 1: Suppose that the pair $(A, B_o\}$ is controllable and the pair $\{C, A\}$ is observable. Then discrete unit memory linear repetitive processes described by (1) are stable along the pass if, and only if,

(i) D_o and A are stable matrices in the standard sense (i.e. all eigenvalues lie inside the unit circle in the complex plane); and

(ii) all eigenvalues of the transfer function matrix

$$G(z) := C(zI_n - A)^{-1} B_o + D_o \qquad (2)$$

have modulus strictly less than unity $\forall \ |z| = 1$.

Note that the condition on the matrix D_o is the necessary and sufficient condition of asymptotic stability. Under this condition, the output sequence converges to the so-called limit profile which is described by a standard linear systems state space model. The stronger condition of stability along the pass is, however, required to ensure that this limit profile has 'acceptable'' along the pass dynamics, the basic requirement of which is stability (in the standard sense).

A key feature of these conditions is that they can, in effect, be tested by direct application of standard linear systems tests. Smyth [1992] gives a detailed treatment of this key area. They also provide the first links with the class of 2D linear time invariant systems recursive in the positive quadrant described by the well known Roesser model [1975] (or equivalents).

The Roesser model (omitting the output equation which has no role in this work) has the following state space structure

$$X_h(k, p+1) = A_1 X_h(k, p) + A_2 X_v(k, p) + B_1 U(k, p)$$
$$X_v(k+1, p) = A_3 X_h(k, p) + A_4 X_v(k, p) + B_2 U(k, p) \qquad (3)$$

Here p and k are the (integer valued) horizontal and vertical coefficients, X_h is the $n \times 1$ vector of 'horizontally' transmitted information, X_v is the $m \times 1$ vector of 'vertically' transmitted information, and U is the $l \times 1$ vector of control inputs.

Numerous equivalent sets of conditions for bounded-input/bounded-output (BIBO) stability of processes described by (3) (or equivalents) are known. Applying one such set leads (see theorem 3.4.4 in Rogers and Owens [1992] for example) to the following result.

Theorem 2: The same set of conditions result from the following operations:

(i) applying the conditions for stability along the pass to processes described by (1).

(ii) applying any set of necessary and sufficient conditions for BIBO stability of the Roesser model with the matrices A_1, A_2, A_3 and A_4 replaced by A, B_o, C and D_o respectively.

This result has enabled stability tests (see Rocha etal [1995] for example) to be interchanged to great effect between these two areas. It has not, however, provided a suitable basis on which to employ well established 2D concepts and results in answering basic systems theoretic questions for (1), such as what is meant by controllability and observability. The removal of this basic difficulty is the subject of the remainder of this paper.

2D MODELS - BASICS

The premise that (1) is a Roesser model is based on interpreting the state vector $\{X\}$ as 'horizontally' transmitted information and the pass profile $\{Y\}$ as 'vertically' transmitted information. Theorem 2 does not, however, show that the linear repetitive process (1) has the structure of a Roesser model, but only that the same set of stability conditions result from applying the corresponding stability criteria to each case. In fact, a number of structural differences exist between this repetitive process state space model and the Roesser state space model of (3).

These structural differences are discussed fully in Galkowski etal [1995]. The main conclusion is that an appropriate 2D model for the dynamics of (1) is of the Fornasini Marchesini [1978] type. In such models, the state is not split into horizontal and vertical components. Here again the output equation is not required and, with $Z(k, p)$ denoting the state vector at point (k, p), $k \geq 0$, $p \geq 0$,the basic model of this type has the structure

$$Z(k+1, p+1) = A_5 Z(k, p+1) + A_6 Z(k+1, p) + A_7 Z(k, p) + B_3\, U(k+1, p) + B_4\, U(k, p+1) \quad (4)$$

As before U is the (appropriately dimensioned) vector of control inputs.

Suppose now that I_p denotes the $p \times p$ identity matrix and O_{pq} the null matrix of dimension $p \times q$. Also define the vector

$$Z(k, p) = [X^T(k, p), Y^T(k, p)]^T \quad (5)$$

Then it follows that the discrete unit memory linear repetitive process state space model (1) can be written in the form

$$E\, Z(k+1, p+1) = A_8\, Z(k+1, p) + A_9\, Z(k, p) + B_5\, U(k+1, p) \quad (6)$$

where

$$E = \begin{bmatrix} I_n & O_{nm} \\ O_{mn} & O_{mm} \end{bmatrix}, \quad A_8 = \begin{bmatrix} A & O_{nm} \\ C & -I_m \end{bmatrix}$$

$$A_9 = \begin{bmatrix} O_{nn} & B_0 \\ O_{mn} & D_0 \end{bmatrix}, \quad B_5 = \begin{bmatrix} B \\ D \end{bmatrix} \quad (7)$$

i.e. the singular version of the model of (4). (For background on the structure and control of 2D linear systems described by singular state space models see, for example, Kaczorek [1985] and numerous publications since then).

As in standard 2D linear systems theory (see, for example, Lewis [1992]) it is possible to write (7) as a singular Roesser model. To detail this, introduce the new state variable

$$\zeta(k, p) = E\, Z(k, p+1) - A_8\, Z(k, p) \quad (8)$$

and hence, using (6) and (8),

$$\begin{aligned}\zeta(k+1, p) &= E\, Z(k+1, p+1) - A_8\, Z(k+1, p) \\ &= A_9\, Z(k, p) + B_5\, U(k+1, p)\end{aligned} \quad (9)$$

These equations now define a Roesser model since

$$E\, Z(k, p+1) = A_8\, Z(k, p) + \zeta(k, p)$$

$$\zeta(k+1, p) = A_9\, Z(k, p) + B_5\, U(k+1, p) \quad (10)$$

In this Roesser model both the state vector $\{X\}$ and the pass profile $\{Y\}$, which define the vector Z, play the role of horizontally transmitted information and the new artificial variable ζ plays the role of vertically transmitted information. Note also that it has a time (spatial) shift in the input to the second equation. This problem can be removed by employing the transformation

$$\zeta(k,p) = \varphi(k,p) + B_5 U(k,p) \quad (11)$$

to yield (in matrix form)

$$E_1 \begin{bmatrix} Z(k,p+1) \\ \varphi(k+1,p) \end{bmatrix} = \begin{bmatrix} A_8 & I_{n+m} \\ A_9 & O_{n+m,n+m} \end{bmatrix} \begin{bmatrix} Z(k,p) \\ \varphi(k,p) \end{bmatrix} + B_6 U(k,p) \quad (12)$$

where

$$E_1 = \begin{bmatrix} E & 0 \\ 0 & I_{n+m} \end{bmatrix}, \quad B_6 = \begin{bmatrix} B_5 \\ 0 \end{bmatrix} \quad (13)$$

Note that the (state) dimension of this model is $2(n + m)$ and it is singular. It is, however, possible to reduce this dimension to $2n + m$ and hence remove (some of) its inherent redundancy. To do this, return to (1) and define the new state variable

$$\omega(k,p) = X(k+1,p) \quad (14)$$

Then this reptitive process state space model can be written in the form

$$\begin{bmatrix} X(k+1,p) \\ Y(k+1,p) \\ \omega(k,p+1) \end{bmatrix} = \begin{bmatrix} 0 & 0 & I \\ 0 & D_0 & C \\ 0 & B_0 & A \end{bmatrix} \begin{bmatrix} X(k,p) \\ Y(k,p) \\ \omega(k,p+1) \end{bmatrix} + \begin{bmatrix} 0 \\ D \\ B \end{bmatrix} U(k+1,p) \quad (15)$$

Here X and Y play the role of 'vertically' transmitted information and the new variable ω plays the role of 'horizontally' transmitted information.

If it is desirable to have a $U(k, p)$ term in this last model, this can be achieved by using the transformation

$$Y(k,p) = W(k,p) + D\,U(k,p) \quad (16)$$

and hence

$$\begin{bmatrix} X(k+1,p) \\ W(k+1,p) \\ \omega(k,p+1) \end{bmatrix} = \begin{bmatrix} 0 & 0 & I \\ 0 & D_0 & C \\ 0 & B_0 & A \end{bmatrix} \begin{bmatrix} X(k,p) \\ W(k,p) \\ \omega(k,p+1) \end{bmatrix} + \begin{bmatrix} 0 \\ D_0 D \\ B_0 D \end{bmatrix} U(k,p) + \begin{bmatrix} 0 \\ 0 \\ B \end{bmatrix} U(k+1,p) \quad (17)$$

Note, however, that the shift k+1 in the input vector cannot be avoided. Also the standard nonsingular Roesser model results from (17) on setting $B = 0$.

It is also possible to follow the 2D systems literature (see, for example, Lewis [1992]) and develop yet more 2D models for (1). For example, the so-called strictly causal nonsingular Roesser model description could easily be obtained from (17). These are omitted here (see Galkowski etal [1995] for a full treatment) since the analysis of the next section only requires the models of (17) and (6) - (7).

STABILITY, STATE TRANSITION AND REACHABILITY

Theorem 3 below is the first major result of this section and (formally) establishes that repetitive processes described by (1) are stable along the pass if, and only if, the corresponding Roesser model interpretation is BIBO stable. This requires the following preliminary result giving a set of necessary and sufficient conditions for BIBO stability of the Roesser model of (17).

Lemma 1: 2D linear time invariant systems described by the Roesser model of (17) are BIBO stable, if and only if,

a) D_o and A are stable matrices in the standard sense; and
b) all eigenvalues of the transfer function matrix

$$Q(z_1^{-1}) := C(z_1^{-1} I_n - A)^{-1} B_o + D_o \quad (18)$$

have modulus strictly less than unity $\forall \; |z_1| = 1$.

Proof: This follows identical steps to that of theorem 3.4.3 in Rogers and Owens [1992] and hence the details are omitted.

Theorem 3: Regard repetitive processes described by the state space model (1) as 2D linear systems described by Roesser model of (17). Then such repetitive processes are stable along the pass if, and only if, they are BIBO stable in the sense of lemma 1.

Proof: In effect, this consists of showing that the conditions of theorem 1 and lemma 1 are equivalent. Suppose first therefore that the Huang [1972] conditions for BIBO stability are applied to the characteristic polynomial of (17). Then the set of conditions which result are precisely those of theorem 1 for stability along the pass. Conversely, suppose that the conditions of lemma 1 hold. Then the proof that these imply stability along the pass is identical to that of theorem 1 and hence the details are omitted.

Note: In studying BIBO stability of the Roesser model, the existence of so-called non-essential singularities of the second kind must be excluded. This is not necessary in the repetitive process problem studied here since it can be shown that any such singularities are unstable along the pass.

The remainder of this paper, uses the singular 2D model of (6) - (7) to develop some basic results on pass reachability. Some areas for further work are also noted. A basic building block for this analysis is a transition matrix which is now introduced with a full treatment in Galkowski etal [1995].

Kaczorek [1992] has developed a recursive formula for computing the transition matrix, denoted $T_{p,q}$, for the general singular Fornasini Marchesini model. The case under consideration here has special structure which simplifies the computation of the transition matrix. In particular, the index of this matrix, i.e. the minimum integers β and γ such that $T_{\beta,\gamma}$ is not zero, are given by

$$\beta = \gamma = -1 \tag{19}$$

Hence the so-called initial matrix for the recursive formula in this particular case is

$$T_{0,-1} = \begin{bmatrix} 0 & 0 \\ 0 & I_m \end{bmatrix} \tag{20}$$

Further, to obtain the general element $T_{p,q}$, $p, q \geq 1$ solve the following two equations for $T_{p,q}$

$$T_{p,q}{}^{1j} = AT_{p,q-1}{}^{1j} + B_o T_{p-1,q-1}{}^{2j}, j = 1, 2 \tag{21}$$

and the following two equations for $T_{p,q+1}$

$$T_{p,q}{}^{2j} = CT_{p,q}{}^{1j} + D_1 T_{p-1,q}{}^{2j}, j = 1, 2 \tag{22}$$

Here the superscript ij denotes the corresponding element in T.

Before proceeding to consider the property of reachability of processes described by (1) written in the form (6) - (7) it is necessary to investigate the problem of acceptability of the inputs for a given set of initial conditions. This is because [see Kaczorek [1993] for a detailed treatment) for singular 2D systems it is possible that a state trajectory does not exist. In the case considered here the following result, proved in Galkowski etal [1995], provides a partial answer.

Lemma 2: The linear repetitive process of (2) accepts all input sequences on the rectangle

$$[0, \zeta] \times [0, Y] := \{k, p\} : 0 \leq k \leq \zeta, 0 \leq Y \leq p \tag{23}$$

provided at least one of the matrices E, A_8 A_9 of (6) - (7) has full row rank.

The condition of lemma 2 means that the state matrix A of (1) must have full row rank. It is, however, possible that this condition does not hold. In which case an acceptable input sequence can be constructed by, in effect, modifying the procedure given by Kaczorek [1993].

The following definition of reachability for processes described by (1) can now be introduced.

Definition 1: The discrete linear repetitive process state space model of (1) is locally reachable in the rectangle $[0, \zeta] \times [0, Y]$ if for admissible boundary conditions $z(k, 0) \in R^{n+m}$, $k = 1, 2, \ldots$ and $z(0, p) \in R^{n+m}$, $p = 0, 1, 2, \ldots$, α and any vector $z_f \in R^{n+m}$ there exists a system of input vectors $U(k, p)$ on $(0, 0) \leq (k, p) \leq (\zeta + \beta + 1, Y + \gamma + 1)$ such that $z(\zeta, Y) = z_f$.

Conditions for local reachability of processes described by (1) follow immediately from those of Kaczorek [1992] for the general singular 2D model. In terms of applications, however, a more relevant question is under what conditions can a specified complete pass profile be reached, i.e.

$$Z^*(\zeta) := [z^*(\zeta, 0), z^*(\zeta, 1), \ldots, z^*(\zeta, \alpha)], \zeta \geq 1 \tag{24}$$

where, without loss of generality, $Z^*(0) = 0$. The following results (see again Galkowski etal [1995] for the proofs) answer this question.

Theorem 4: The discrete linear repetitive process state space model of (1) is reachable in all rectangles

$$[0, \zeta] \times [0, 0], [0, \zeta] \times [0, 1], \ldots, [0, \zeta] \times [0, \alpha]$$

of pass ζ if the matrices D and B are square and non singular.

This condition is too strong and can be relaxed to the following.

Theorem 5: The discrete linear repetitive process state space model of (1) is reachable in all rectangles

$$[0, \zeta] \times [0, 0], [0, \zeta] \times [0, 1], \ldots, [0, \zeta] \times [0, \alpha]$$

of pass ζ if

$$\text{rank } W(\zeta) = m \tag{25}$$

and

$$\text{rank}\,[B_o\, W(\zeta - 1), \text{B}] = n \qquad (26)$$

where

$$W(\zeta) = [D_o^{\zeta-1}D, D_o^{\zeta-2}D, \ldots, D] \qquad (27)$$

Note also that a similar analysis can be employed to define and develop conditions for local controllability and observability. The details are again in Galkowski etal [1995]. Current work is concentrating on the role of these properties in terms of the basic dynamics and control of processes described by (1). Results from this work will be reported in due course.

CONCLUSIONS

This paper has developed new 2D systems models for discrete linear repetitive processes to serve as a strating point for addressing currently open control theoretic research problems for them. The main results, in addition to the models themselves, are a proof of stability equivalence, a state transition matrix, and some basic results on reachability. In depth development work is continuing on these and other areas and will be reported in due course.

REFERENCES

Amann, N, Owens, DH and Rogers, E. 1994. New Results in Iterative Learning Control, IEE International Conference Control 94, Vol1, pp640-645.

Lewis, FL. 1992. A Review of 2D Implicit Systems, Automatica, Vol28, No2, pp345-354.

Fornasini, E and Marchesini, G. 1978. Doubly Indexed Dynamical Systems; State Space Models and Structural Properties, Maths System Theory, Vol12, pp59-72.

Galkowski, K, Rogers, E and Owens, DH. 1995. 2D Models and a Transition Matrix for Discrete Linear Repetitive Processes, Submitted to Multidimensional Systems and Signal Processing.

Huang, TS. 1972. Stability of Two-Dimensional and Recursive Filters, IEEE Trans Audio and Electroacoustics, Vol AU-20, 2, 2, pp158-163.

Kaczorek, T. 1985. Two-Dimensional Linear Systems, Springer Verlag LNCIS, Vol 68, Berlin.

Kaczorek, T. 1992. Local Controllability, Reachability of the General Singular Model of 2D Systems, IEEE Trans Auto Control, VolAC-37, No10, pp1527-1530.

Kaczorek, T. 1993. Acceptable Input Sequences for Singular 2D Systems, IEEE Trans Auto-Control. VolAC-38, No9, pp600-602.

Rocha, P, Rogers, E and Owens, DH. 1995. Stability of Discrete non-unit Memory Linear Repetitive Processes - a 2D Systems Interpretation, Int. J. Control, To appear.

Roesser, RP. 1975. A Discrete State Space Model for Linear Image Processing, IEEE Trans Auto Control, Vol AC-20, No1, pp1-10.

Rogers E and Owens DH. 1992. Stability Analysis for Linear Repetitive Processes, Springer Verlag LNCIS, Vol175. Berlin.

Smyth, KJ. 1992. Computer Aided Analysis for Linear Repetitive Processes, PhD Thesis, University of Strathclyde, Glasgow, UK.

GENERALIZED 2-D CONTINUOUS-DISCRETE LINEAR SYSTEMS WITH DELAYS

T.KACZOREK

Warsaw University of Technology, Faculty of Electrical Engineering, Institute of Control and Industrial Electronics, 00-662 Warszawa, ul. Koszykowa 75, POLAND

Abstract. A general Kurek (K) type model and a general Roesser (R) type model of 2-D continuous-discrete linear systems with deviating arguments are introduced. A solution and the general response formula for the regular general type model of 2-D continuous-discrete linear systems with delays are derived. Necessary and sufficient conditions for the local reachability and the local controllability of the regular general K type model are established. The minimum energy control of the regular model is solved.

Key Words. Generalized 2-D continuous-discrete systems delays local reachability, local controllability, minimum energy control.

1. INTRODUCTION

The most popular models of two-dimensional (2-D) linear systems are the discrete state space models introduced by Roesser (1975), Fornasini and Marchesini (1976, 1978) and Kurek (1985). Generalized multidimensional linear systems with deviating arguments (specially with delays) have been investigated by Kaczorek (1993a,b). Recently by Kaczorek (1994) singular 2-D continuous-discrete models of linear systems have been introduced. In 2-D continuous - discrete systems one independent variable is continuous and the second independent variable is discrete. Such continuous-discrete models appear for example in the iterative learning control synthesis (Kurek and Zaremba, 1993) and the repetitive processes analysis (Rogers and Owens, 1992).In this paper a general Kurek (K) type model of 2-D continuous-discrete systems with deviating arguments and a general Roesser (R) type model of 2-D continuous-discrete systems with deviating arguments will be introduced. The general response formula for the regular general K type model of 2-D continuous-discrete linear systems with delays will be derived and necessary and sufficient conditions for the local reachability and local controllability will be established. The minimum energy control problem for the general K type model will be solved.

2. MODELS OF 2-D CONTINUOUS-DISCRETE LINEAR SYSTEMS.

Consider a general type model of 2-D continuous-discrete model linear system with deviating arguments described by the equations

$$\begin{aligned}E\dot{x}(t,k+1) &= A_0x(t,k)+A_1\dot{x}(t,k)+\\ &A_2x(t,k+1)+B_0x(t-d_1,k-d_2)+ \qquad (1a)\\ &B_1\dot{x}(t-d_1,k-d_2)+B_2x(t-d_1,k-d_2+1)+\\ &+C_0u(t,k)+C_1\dot{u}(t,k)+C_2u(t,k+1)\end{aligned}$$

$t,d_1 \in R_+,\ k,d_2 \in Z_+$

$$y(t,k) = D_0x(t,k)+D_1u(t,k) \qquad (1b)$$

where $\dot{x}(t,k)=\dfrac{\partial x(t,k)}{\partial t}$, $x(t,k)\in R^n$ is the semistate vector, $u(t,k)\in R^m$ is the input vector, $y(t,k)\in R^p$ is the output vector, $E\in R^{q\times n}, A_i \in R^{q\times n}$, $B_i \in R^{q\times n}$, $C_i \in R^{q\times m}$, i=0,1,2, $D_0 \in R^{p\times n}$, $D_1 \in R^{p\times m}$, and $R^{q\times n}$ is the set of $q\times n$ real matrices, R_+ and Z_+ is the set of nonnegative real numbers and integers, respectively. If $q\neq n$ or $\det E=0$ when $q=n$, the model (1) is called singular (implicit). If $q=n$ and $\det E\neq 0$ the model will be called standard. In this case premultiplying (1a) by E^{-1} we obtain a model with $E=I$ (the identity matrix). If $q=n$ and $\det E=0$ but

$$\det\left[Esz - A_0 - A_1s - A_2z - B_0e^{-sd_1}z^{-d_2} - B_1se^{-d_1}z^{-d_2} - B_2e^{-d_1s}z^{1-d_2}\right]\neq 0$$

for some $s\in$C (the field of complex numbers) the model (1) will be called regular. If $d_1>0$ and $d_2>0$ then (1) will be called a model with delays (retarded arguments) and if $d_1 < 0$ and $d_2 < 0$ then it will be called a model with advanced arguments. When all or some entries of the matrices E, A_i, B_i, C_i, $i = 0,1,2$, D_0 *and* D_1 depend on t *and* k then (1) will be the called the model with variable coefficients. From (1) for $C_1 = 0$ and $C_2 = 0$ we obtain the first generalized Fornasini-Marchesini model of 2-D continuous-discrete linear systems with deviating arguments. Similarily, from (1)for $A_0 = 0, B_0 = 0$ *and* $C_0 = 0$ we obtain the second generalized Fornasini-Marchesini model of 2-D continuous-discrete linear systems with deviating arguments.The general type model of 2-D continuous-discrete linear systems with deviating arguments has the form

$$E\begin{bmatrix}\dot{x}^h(t,k)\\ x^v(t,k+1)\end{bmatrix}=\begin{bmatrix}A_{11} & A_{12}\\ A_{21} & A_{22}\end{bmatrix}\begin{bmatrix}x^h(t,k)\\ x^v(t,k)\end{bmatrix}+ \tag{2a}$$

$$+\begin{bmatrix}B_{11} & B_{12}\\ B_{21} & B_{22}\end{bmatrix}\begin{bmatrix}x^h(t-d_1,k-d_2)\\ x^v(t-d_1,k-d_2)\end{bmatrix}+\begin{bmatrix}C_{10}\\ C_{20}\end{bmatrix}u(t,k)$$

$$y(t,k)=\begin{bmatrix}D_{01} & D_{02}\end{bmatrix}\begin{bmatrix}x^h(t,k)\\ x^v(t,k)\end{bmatrix}+D_1u(t,k) \tag{2b}$$

where

$x^h(t,k) \in R^{\bar{n}_1}$ is the horizontal semistate vector,

$x^v(t,k) \in R^{\bar{n}_2}$ is the vertical semistate vector,

$u(t,k) \in R^m$ is the input vector,

$y(t,k) \in R^p$ is the output vector, and $A_{11}, B_{11} \in R^{q_1\times\bar{n}_1}, A_{12}, B_{12} \in R^{q_1\times\bar{n}_2}$, $A_{21}, B_{21} \in R^{q_2\times\bar{n}_1}, A_{22}, B_{22} \in R^{q_2\times\bar{n}_2}, C_{10} \in R^{q_1\times m}$, $C_{20} \in R^{q_2\times m}, D_{01} \in R^{p\times\bar{n}_1}, D_{02} \in R^{p\times\bar{n}_2}$ $D_1 \in R^{p\times m}$

Similarily as for (1) the generalized Roesser model (2) will be called singular if $q_1 \neq \bar{n}_1, q_2 \neq \bar{n}_2$ or $\det E = 0$ when $q_1 = \bar{n}_1, q_2 = \bar{n}_2$. If E is square and det $E \neq 0$ then the model (2) will be called standard. It can be shown that the generalized Roesser model (2) is a special case of the model (1).

3. SOLUTION TO THE REGULAR GENERAL MODEL.

To simplify the notion we write (1a) in the form

$$E\dot{x}(t,k+1) = A_0x(t,k) + A_1\dot{x}(t,k) +$$
$$+A_2x(t,k+1) + B_0x(t-d_1,k-d_2) + \tag{3a}$$
$$B_1\dot{x}(t-d_1,k-d_2) + B_2x(t-d_1,k-d_2+1) + f(k)$$

where

$$f(t,k) := C_0u(t,k) + C_1\dot{u}(t,k) + C_2u(t,k+1) \tag{3b}$$

It is assumed that $q = n, \det E = 0$ and $\det[Es - A_2] \neq 0$ for some $s \in \mathbb{C}$ (the field of complex numbers). (4)

It is well-known (Kaczorek 1993) that if (4) holds then there exist nonsingular matrices $P, Q \in R^{n\times n}$ such that

$$P[Es - A_2]Q = \begin{bmatrix}I_{n_1}s - A_{21} & 0\\ 0 & Ns - I_{n_2}\end{bmatrix} \tag{5}$$

where n_1 is the degree of det $[Es - A_2]$, $n_2 := n - n_1$, $A_{21} \in R^{n_1\times n_1}$, $N \in R^{n_2\times n_2}$ is a nilpotent matrix with index ν $(N^{\nu-1} \neq 0$ *and* $N^\nu = 0)$. Premultiplying (3a) by P, introducing the new vector

$$Q^{-1}x(t,k) := \begin{bmatrix}x_1(t,k)\\ x_2(t,k)\end{bmatrix},\ x_1(t,k) \in R^{n_1},$$

$x_2(t,k) \in R^{n_2}$ and using (5) we obtain

$$\dot{x}_1(t,k+1) = A_{01}x_1(t,k) + A_{02}x_2(t,k) +$$
$$+A_{11}\dot{x}_1(t,k) + A_{12}\dot{x}_2(t,k) + A_{21}x_1(t,k+1) +$$
$$+B_{01}x_1(t-d_1,k-d_2) + B_{02}x_2(t-d_1,k-d_2)$$
$$+B_{11}\dot{x}_1(t-d_1,k-d_2) + B_{12}\dot{x}_2(t-d_1,k-d_2) +$$
$$+B_{21}x_1(t-d_1,k-d_2+1) + B_{22}x_2(t-d_1,k-d_2+1)$$
$$f_1(t,k) \tag{6a}$$

$$N\dot{x}_2(t,k+1) = A_{03}x_1(t,k) + A_{04}x_2(t,k) + A_{13}\dot{x}_1(t,k)$$
$$+A_{14}\dot{x}_2(t,k) + x_2(t,k+1) + B_{03}x_1(t-d_1,k-d_2)$$
$$+B_{04}x_2(t-d_1,k-d_2) + B_{13}\dot{x}_1(t-d_1,k-d_2) +$$
$$B_{14}\dot{x}_2(t-d_1,k-d_2) + B_{23}x_1(t-d_1,k-d_2+1) +$$
$$+B_{24}x_2(t-d_1,k-d_2+1) + f_2(t,k) \quad for\ t \in R_+, k \in Z_+ \tag{6b}$$

where

$$PA_kQ = \begin{bmatrix}A_{k1} & A_{k2}\\ A_{k3} & A_{k4}\end{bmatrix},\ k = 0,1,\quad PB_jQ = \begin{bmatrix}B_{j1} & B_{j2}\\ B_{j3} & B_{j4}\end{bmatrix},$$

$$j = 0,1,2,\quad Pf(t,k) = \begin{bmatrix}f_1(t,k)\\ f_2(t,k)\end{bmatrix}$$

The submatrices A_{ki}, B_{ji}, $k = 0,1$; $i = 1,2,3,4$; $j = 0,1,2$ and the vectors $f_1(t,k), f_2(t,k)$ have dimensions compatible with the dimensions of x_1 and x_2, respectively.

Let the boundary conditions for (3a) be given by

$$x(t,k)\ and\ \dot{x}(t,k)\ for\ t \geq -d_1, -d_2 \leq k \leq 0$$
$$and\ -d_1 \leq t \leq 0,\ k > 0 \tag{7}$$

It is assumed that the boundary conditions and $u(t,k)$ are $(\nu+1)$ - times differentable with respect to t. The boundary conditions (7) are called admissible if for a given $u(t,k)$ there exists a solution $x(t,k)$ to (3a).

Knowing (7) we may find the boundary conditions for (6)

$$\begin{bmatrix} x_1(t,k) \\ x_2(t,k) \end{bmatrix} = Q^{-1}x(t,k) \; and \; \begin{bmatrix} \dot{x}_1(t,k) \\ \dot{x}_2(t,k) \end{bmatrix} = Q^{-1}\dot{x}(t,k) \quad (8)$$

for $t \geq -d_1, -d_2 \leq k \leq 0$ and $-d_1 \leq t \leq 0,\ k > 0$

Note that the equations (6) are coupled by the matrices $A_{i2}, A_{i3}, i = 0,1, B_{j2}, B_{j3}, j = 0,1,2$ and if at least one of the matrices is nonzero the equations can not be solved independently.To find the solutions $x_1(t,k)$ and $x_2(t,k)$ to (6) with boundary conditions (8) let us consider the equations for $k = 0$

$$\dot{x}_1(t,1) = A_{21}x_1(t,1) + F_{10}(t) \quad (9a)$$

$$N\dot{x}_2(t,1) = x_2(t,1) + F_{20}(t) \quad (9b)$$

where

$$\begin{aligned} F_{10}(t) &:= A_{01}x_1(t,0) + A_{02}x_2(t,0) + A_{11}\dot{x}_1(t,0) + \\ &A_{12}\dot{x}_2(t,0) + B_{01}x_1(t-d_1,-d_2) + B_{02}x_2(t-d_1,-d_2) \\ &+B_{11}\dot{x}_1(t-d_1,-d_2) + B_{12}\dot{x}_2(t-d_1,-d_2) + \quad (10a) \\ &+B_{21}x_1(t-d_1,1-d_2) + B_{22}x_2(t-d_1,1-d_2) + f_1(t,0) \end{aligned}$$

$$\begin{aligned} F_{20}(t) &:= A_{03}x_1(t,0) + A_{04}x_2(t,0) + A_{13}\dot{x}_1(t,0) + \\ &A_{14}\dot{x}_2(t,0) + +B_{03}x_1(t-d_1,-d_2) + B_{04}x_2(t-d_1,-d_2) \\ &+B_{13}\dot{x}_1(t-d_1,-d_2) + B_{14}\dot{x}_2(t-d_1,-d_2) + \quad (10b) \\ &B_{23}x_1(t-d_1,1-d_2) + B_{24}x_2(t-d_1,1-d_2) + f_2(t,0) \end{aligned}$$

are known for given (8), $f_1(t,k)$ *and* $f_2(t,k)$

The solutions $x_1(t,1)$ *and* $x_2(t,1)$ to (9a) and (9b) have the form[5,6]

$$x_1(t,1) = e^{A_{21}t}x_1(0,1) + \int_0^t e^{A_{21}(t-\tau)}F_{10}(\tau)d\tau,\ t \in R_+ \quad (11)$$

$$x_2(t,1) = -\sum_{i=0}^{\nu-1} N^i F_{20}^{(i)}(t) \quad (12)$$

If $d_2 \geq 2$ then substituting $k = 1$ into (6) and using (11) and (12) we obtain

$$\dot{x}_1(t,2) = A_{21}x_1(t,2) + F_{11}(t) \quad (13)$$

$$N\dot{x}_2(t,2) = x_2(t,2) + F_{21}(t) \quad (14)$$

where

$$\begin{aligned} F_{11}(t) &:= \overline{A}_1\left[e^{A_{21}t}x_1(0,1) + \int_0^t e^{A_{21}(t-\tau)}F_{10}(\tau)d\tau\right] + \\ &A_{11}F_{10}(t) - \sum_{i=0}^{\nu-1}\left[A_{02}N^iF_{20}^{(i)}(t) + A_{12}N^iF_{20}^{(i+1)}(t)\right] + \\ &+B_{01}x_1(t-d_1,1-d_2) + B_{02}x_2(t-d_1,1-d_2) + \\ &+B_{11}\dot{x}_1(t-d_1,1-d_2) + B_{12}\dot{x}_2(t-d_1,1-d_2) + \\ &+B_{21}x_1(t-d_1,2-d_2) + B_{22}x_2(t-d_1,2-d_2) + f_1(t,1), \end{aligned}$$

$$\begin{aligned} F_{21}(t) &:= \overline{A}_2\left[e^{A_{21}t}x_1(0,1) + \int_0^t e^{A_{21}(t-\tau)}F_{10}(\tau)d\tau\right] + \\ &A_{13}F_{10}(t) - \sum_{i=0}^{\nu-1}\left[A_{04}N^iF_{20}^{(i)}(t) + A_{14}N^iF_{20}^{(i+1)}(t)\right] + \\ &+B_{03}x_1(t-d_1,1-d_2) + B_{04}x_2(t-d_1,1-d_2) + \\ &+B_{13}\dot{x}_1(t-d_1,1-d_2) + B_{14}\dot{x}_2(t-d_1,1-d_2) + \\ &+B_{23}x_1(t-d_1,2-d_1) + B_{24}x_2(t-d_1,2-d_2) + f_2(t,1), \end{aligned}$$

$$\overline{A}_1 := A_{01} + A_{11}A_{21},\ \overline{A}_2 := A_{03} + A_{13}A_{21}$$

The solutions $x_1(t,2)$, $x_2(t,2)$ to (13) and (14) have the form

$$x_1(t,2) = e^{A_{21}t}x_1(0,2) + \int_0^t e^{A_{21}(t-\tau)}F_{11}(\tau)d\tau$$

$$t \in R_+$$

$$x_2(t,2) = -\sum_{i=0}^{\nu-1} N^i F_{21}(t)$$

If $d_2 < 2$ we have also to substitute (11) and (12) instead of $x_1(t-d_1,2-d_2)$ and $x_2(t-d_1,2-d_2)$ into (13) and (14).

Continuing this procedure after k steps we obtain

$$x_1(t,k) = e^{A_{21}t}x_1(0,k) + \int_0^t e^{A_{21}(t-\tau)}F_{1,k-1}(\tau)d\tau \quad (15a)$$

$$x_2(t,k) = -\sum_{i=0}^{\nu-1} N^i F_{2,k-1}^{(i)}(t) \quad (15b)$$

where $F_{1,k-1}(t)$, $F_{2,k-1}(t)$ for $k \geq d_2$ satisfy the equations

$$\begin{aligned} F_{1,k}(t) &= \overline{A}_1\left[e^{A_{21}t}x_1(0,k) + \int_0^t e^{A_{21}(t-\tau)}F_{1,k-1}(\tau)d\tau\right] + \\ &A_{11}F_{1,k-1}(t) + -\sum_{i=0}^{\nu-1}\left[A_{02}N^iF_{2,k-1}^{(i)}(t) + A_{12}N^iF_{2,k-1}^{(i+1)}(t)\right] \\ &+B_{01}x_1(t-d_1,k-d_2) + B_{02}x_2(t-d_1,k-d_2) + \\ &+B_{11}\dot{x}_1(t-d_1,k-d_2) + B_{12}\dot{x}_2(t-d_1,k-d_2) + \\ &+B_{21}x_1(t-d_1,k-d_2+1) + B_{22}x_2(t-d_1,k-d_2+1) + \\ &f_1(t,k) \end{aligned} \quad (16a)$$

$$F_{2,k}(t)=\overline{A}_2\left[e^{A_{21}t}x_1(0,k)+\int_0^t e^{A_{21}(t-\tau)}F_{1,k-1}(\tau)d\tau\right]+$$
$$+A_{13}F_{1,k-1}(t)-\sum_{i=0}^{\nu-1}\left[A_{04}N^iF_{2,k-1}^{(i)}(t)+A_{14}N^iF_{2,k-1}^{(i+1)}(t)\right]$$
$$+B_{03}x_1(t-d_1,k-d_2)+B_{04}x_2(t-d_1,k-d_2)$$
$$+B_{13}\dot{x}_1(t-d_1,k-d_2)+B_{14}\dot{x}_2(t-d_1,k-d_2)+$$
$$+B_{23}x_1(t-d_1,k-d_2+1)+B_{24}x_2(t-d_1,k-d_2+1)$$
$$+f_2(t,k) \qquad (16b)$$

Let $P_t:=\begin{bmatrix}P_{11} & P_{12}\\ P_{21} & P_{22}\end{bmatrix}$ be a linear operator (map) defined as follows

$$P_{11}F_1(t):=\overline{A}_1\int_0^t e^{A_{21}(t-\tau)}F_1(\tau)d\tau+A_{11}F_1(t)$$
$$P_{12}F_2(t):=-\sum_{i=0}^{\nu-1}\left[A_{02}N^iF_2^{(i)}(t)+A_{12}N^iF_2^{(i+1)}(t)\right]$$
$$P_{21}F_1(t):=\overline{A}_2\int_0^t e^{A_{21}(t-\tau)}F_1(\tau)d\tau+A_{13}F_1(t) \qquad (17)$$
$$P_{22}F_2(t):=-\sum_{i=0}^{\nu-1}\left[A_{04}N^iF_2^{(i)}(t)+A_{14}N^iF_2^{(i+1)}(t)\right]$$

and $P^k:=P\circ P\circ\ldots\circ P$ is the k multiple composition of P (by definition $P^o:=I$ (the identity operator)).Using (17) and

$$h_1(t,k):=\overline{A}_1e^{A_{21}t}x_1(0,k)+B_{01}x_1(t-d_1,k-d_2)+$$
$$+B_{02}x_2(t-d_1,k-d_2)+B_{11}\dot{x}_1(t-d_1,k-d_2)+$$
$$+B_{12}\dot{x}_2(t-d_1,k-d_2)+B_{21}x_1(t-d_1,k-d_2+1)$$
$$+B_{22}x_2(t-d_1,k-d_2+1)+f_1(t,k) \qquad (18)$$
$$h_2(t,k):=\overline{A}_2e^{A_{21}t}x_1(0,k)+B_{03}x_1(t-d_1,k-d_2)+$$
$$+B_{04}x_2(t-d_1,k-d_2)+B_{13}\dot{x}_1(t-d_1,k-d_2)+$$
$$+B_{14}\dot{x}_2(t-d_1,k-d_2)+B_{23}x_1(t-d_1,k-d_2+1)+$$
$$+B_{24}x_2(t-d_1,k-d_2+1)+f_2(t,k)$$

we may write the equations (16) in the form

$$\begin{bmatrix}F_{1,k}(t)\\ F_{2,k}(t)\end{bmatrix}=P_t\begin{bmatrix}F_{1,k-1}(t)\\ F_{2,k-1}(t)\end{bmatrix}+\begin{bmatrix}h_1(t,k)\\ h_2(t,k)\end{bmatrix} \quad t\in R_+,\ k\in Z_+ \qquad (19)$$

It is easy to show that the solution to (19) has the form

$$\begin{bmatrix}F_{1,k}(t)\\ F_{2,k}(t)\end{bmatrix}=P_t^k\begin{bmatrix}F_{10}(t)\\ F_{20}(t)\end{bmatrix}+\sum_{i=0}^{k-1}P_t^{k-i-1}\begin{bmatrix}h_1(t,i+1)\\ h_2(t,i+1)\end{bmatrix}$$
$$k>0 \qquad (20)$$

where $F_{10}(t),F_{20}(t)$ are defined by (10).

In a similar way we may defined the operator P and the vectors $h_1(t,k),\ h_2(t,k)$ for $k<d_2$. Therefore, the following theorem has been proved.

Theorem 1. If the assumption (4) holds then the solution $x(t,k)$ to (3a) with (7) has the form

$$x(t,k)=Q\begin{bmatrix}e^{A_{21}t}x_1(0,k)+\int_0^t e^{A_{21}(t-\tau)}F_{1,k-1}(\tau)d\tau\\ -\sum_{i=0}^{\nu-1}N^iF_{2,k-1}^{(i)}(t)\end{bmatrix},$$
$$t\in R_+,\ k\in Z_+ \qquad (21)$$

where $F_{1,k}(t),\ F_{2,k}(t)$ are given by (20).

Substitution of (21) into (1b) yields the desired general response formula, which able us to find $y(t,k)$ for given $u(t,k)$ and admissible boundary conditions (7).

4. LOCAL REACHABILITY AND LOCAL CONTROLLABILITY.

In this section necessary and sufficient conditions for the local reachability and local controllability of the regular system (3a) with $f(t,k)=Cu(t,k)$ will be established.

Definition 1. The system (3a) with $f(t,k)=Cu(t,k)$ is called locally reachable in the rectangle
$$[h,r]:=\{(t,k)\in R_+\times Z_+:0\le t\le h,0<k\le r\} \qquad (22)$$
if for any admissible boundary conditions (7) and every vector $x_f\in R^n$ there exists $u(t,k)$ for $0\le t\le h,0<k\le r-1$ such that $x(h,r)=x_f$.

Using

$$P_t^i:=\begin{bmatrix}P_{1,t}^i\\ P_{2,t}^i\end{bmatrix},i\in Z_+,\ F_0(t):=\begin{bmatrix}F_{10}(t)\\ F_{20}(t)\end{bmatrix},$$
$$h(t,k):=\begin{bmatrix}h_1(t,k)\\ h_2(t,k)\end{bmatrix},\ t\in R_+,\ k\in Z_+$$

we may write (20) as

$$F_{1,k}(t)=P_{1,t}^kF_0(t)+\sum_{i=0}^{k-1}P_{1,t}^{k-i-1}h(t,i+1)$$
$$k>0 \qquad (23)$$
$$F_{2,k}(t)=P_{2,t}^kF_0(t)+\sum_{i=0}^{k-1}P_{2,t}^{k-i-1}h(t,i+1)$$

From (21) we have

$$x(t,k)=Q_1\left[e^{A_{21}t}x_1(0,k)+\int_0^t e^{A_{21}(t-\tau)}F_{1,k-1}(\tau)d\tau\right]+$$

$$-\sum_{i=0}^{v-1}Q_2N^iF_{2,k-1}^{(i)}(t) \qquad (24)$$

where
$Q=[Q_1,Q_2]$, $Q_1\in R^{n\times n_1}$ *and* $Q_2\in R^{n\times n_2}$.

Substitution of (23) into (24) yields

$$x(t,k)=Q_1e^{A_{21}t}x_1(0,k)+\int_0^t Q_1e^{A_{21}(t-\tau)}P_{1,\tau}^{k-1}F_0(\tau)d\tau+$$
$$+\sum_{i=0}^{k-2}\int_0^t Q_1e^{A_{21}(t-\tau)}P_{1,\tau}^{k-i-2}h(\tau,i+1)d\tau+$$
$$\sum_{i=0}^{v-1}Q_2N^i\left[P_{2,t}^{k-1}F_0(t)\right]^{(i)}+$$
$$-\sum_{i=0}^{v-1}\sum_{j=0}^{k-2}Q_2N^i\left[P_{2,t}^{k-j-2}h(t,j+1)\right]^{(i)} \qquad (25)$$
$$k>1$$

From (10) and (18) for $f(t,k)=Cu(t,k)$ we have

$$F_0(t)=F_{bc}(t)+PCu(t,0) \qquad (26a)$$
$$h(t,k)=h_{bc}(t,k)+PCu(t,k) \qquad (26b)$$

where

$$F_{bc}(t):=PA_0x(t,0)+PA_1\dot{x}(t,0)+PB_0x(t-d_1,-d_2)+$$
$$+PB_1\dot{x}(t-d_1,-d_2)+PB_2x(t-d_1,1-d_2)$$
$$h_{bc}(t,k):=\begin{bmatrix}\bar{A}_1\\ \bar{A}_2\end{bmatrix}e^{A_{21}t}x_1(0,k)+PB_0x(t-d_1,k-d_2)+$$
$$+PB_1\dot{x}(t-d_1,k-d_2)+PB_2x(t-d_1,k-d_2+1)$$

Let assume that

$$u(t,k):=u_k \quad for \quad 0\le t\le h,\ 0\le k\le r \qquad (27)$$

where u_k is independent of t.
Taking into account (26) and (27) we may write(25) in the form

$$x(t,k)=x_{bc}(t,k)+\int_0^t Q_1e^{A_{21}(t-\tau)}P_{1,\tau}^{k-1}PCd\tau u_0+$$
$$+\sum_{i=0}^{k-2}\int_0^t Q_1e^{A_{21}(t-\tau)}P_{1,\tau}^{k-i-2}PCd\tau u_{i+1}+$$
$$-\sum_{i=0}^{v-1}Q_2N^i\left[P_{2,t}^{k-1}\right]^{(i)}PCu_0+ \qquad (28)$$
$$-\sum_{i=0}^{v-1}\sum_{j=0}^{k-2}Q_2N^i\left[P_{2,t}^{k-j-2}\right]^{(i)}PCu_{j+1}$$

where

$$x_{bc}(t,k):=Q_1e^{A_{21}t}x_1(0,k)+\int_0^t Q_1e^{A_{21}(t-\tau)}P_{1,\tau}^{k-1}F_{bc}(\tau)d\tau+$$
$$+\sum_{i=0}^{k-2}\int_0^t Q_1e^{A_{21}(t-\tau)}P_{1,\tau}^{k-i-2}h_{bc}(\tau,i+1)+$$
$$-\sum_{i=0}^{v-1}Q_2N^i\left[P_{2,t}^{k-1}F_{bc}(t)\right]^{(i)}+ \qquad (29)$$
$$-\sum_{i=0}^{v-1}\sum_{j=0}^{k-2}Q_2N^i\left[P_{2,t}^{k-j-2}h_{bc}(t,j+1)\right]^{(i)}$$

Theorem 2
The system (3a) with $f(t,k)=Cu(t,k)$ is locally reachable in the rectangle (22) if and only if

$$rank[R_0,R_1,\ldots,R_{r-1}]=n \qquad (30)$$

where

$$R_{r-i-1}=R(h):=\int_0^h Q_1e^{A_{21}(h-\tau)}P_{1,\tau}^iPCd\tau+$$
$$-\sum_{j=0}^{v-1}Q_2N^j\left[P_{2,h}^i\right]^{(j)}PC \qquad (31)$$
$$i=0,1,\ldots,r-1$$

Proof
Using (28) for $t=h,k=r$, (29) and $x(h,r)=x_f$ we obtain

$$x_f-x_{bc}(h,r)=[R_0,R_1,\ldots,R_{r-1}]u \qquad (32)$$

where $u:=\left[u_0^Tu_1^T,\ldots,u_{r-1}^T\right]^T$
From (31) it follows that for any admissible boundary conditions (7) and every vector x_f there exists a sequence $u_0,u_1,\ldots,u_{r-1}$ if and only if (28) holds.□

Definition 2. The system (3a) with $f(t,k)=Cu(t,k)$ is called locally controllable in the rectangle (22) if for any admissible boundary conditions (7) there exists $u(t,k)$ for $0\le t\le h,\ 0\le k\le r-1$ such that $x(h,r)=0$.

Theorem 3
The system (3a) with $f(t,k)=Cu(t,k)$ is locally controllable in the rectangle (22) if and only if the condition (30) or (30) holds.

Proof. Using (28) for $t=h,\ k=r$, (29) and $x(h,r)=0$ we obtain (32) for $x_f=0$. From (29) for $t=h,\ k=r$ it follows that for any $x_1(0,k)$ the term $Q_1e^{A_{21}h}x_1(0,r)$ is an arbitrary n-dimensional vector. Therefore, there exists a sequance $u_0, u_1, \ldots, u_{r-1}$ satisfying (32) for $x_f = 0$ if and only if (30) holds. □

From theorem 2 and theorem 3 it follows that for 2-D continuous-discrete regular system (3a) the local controllability is equivalent to its local reachability.

5. MINIMUM ENERGY CONTROL

Consider the system (3a) with $f(t,k) = Cu(t,k)$ and the performance index

$$I(u) := \sum_{i=0}^{r-1} u_i^T Q u_i \qquad (33)$$

where Q is the $m \times m$ symmetric and positive definite weighting matrix. The minimum energy control problem for the system (3a) can be stated as follows. Given the matrices $A_i, B_i, i = 0,1,2, C$ of (3a), the weighting matrix Q, the numbers h, r and the boundary conditions (7), find a sequence $u_0, u_1, \ldots, u_{r-1}$ defined by (27) which transfers the system to the desired final state x_f, $x(h,r) = x_f$ and and minimizes the performance index (33). To solve the problem we define the matrix

$$W_Q := \sum_{i=0}^{r-1} R_i Q^{-1} R_i^T \qquad (34)$$

where R_i is defined by (31).

It is easy to show that the matrix (34) is nonsingular if and only if the system (3a) is reachable in (22).

We may define the input sequence

$$\hat{u}_k := Q^{-1} R_k^T W_Q^{-1} (x_f - x_{bc}),\ k = 0,1,\ldots,r-1 \qquad (35)$$

where x_{bc} is given by (29) for $t = h$, $k = r$.

Theorem 4.

Let us assume that

i) the system (3a) with $f(t,k) = Cu(t,k)$ is reachable in the rectangle (22),

ii) $\bar{u}_k$, $k \in [0, r-1]$ is any input sequence which transfer the system to x_f.

Then the input sequence (35) accomplishes the same task and

$$I(\hat{u}) < I(\bar{u}) \qquad (36)$$

Moreover, the minimum value of (33) is given by

$$I(\hat{u}) = (x_f - x_{bc})^T W_Q^{-1} (x_f - x_{bc}) \qquad (37)$$

The proof is similar to the one in (Klamka 1991, Kaczorek 1993).

6. CONCLUSIONS.

The general Kurek type and the general Roesser type model of 2-D continuous-discrete linear systems with deviating arguments have been introduced. The solution (21) to the regular model (3a) satisfying the condition (4) with admissible boundary conditions (7) has been derived. Necesary and sufficient conditions for the local reachability (theorem 2) and the local controllability (theorem 3) of the regular model (3a) have been established. It has been shown that for regular 2-D continuous-discrete linear systems with delays the local controllability is equivalent to their local reachability. The minimum energy control problem for the regular 2-D continuous-discrete linear systems with delays has been solved (theorem 4). An extension of the theorems for the regular model (1) satisfying the condition (4) and n-D (n>2) case is straightforward.

REFERENCES

Fornasini, E. and Marchesini, G. (1976), *State space realization of two- dimensional filters*, IEEE Trans. Autom. Control, AC-21, 484-491.

Fornasini, E. and Marchesini, G. (1978), *Doubly indexed dynamical systems: State space models models and structural properties*, Math. Syst. Theory 12.

Kaczorek, T. (1993), *Generalized multidimensional linear systems with deviating arguments*, Arch. Control Sciences, vol. 2, No 2-4, 165-183

Kaczorek, T. (1993), *Linear two-dimensional systems with deviating arguments*, Differential Equations, Dynamical Systems and Control Science, vol. 152, 623-640.

Kaczorek, T.(1993), *Linear Control Systems*, vol. II, Research Studies Press LTD. J.Wiley, New York.

Kaczorek, T. (1994), *Singular 2-D continuous-discrete linear systems*, Bull. Pol. Acad. Techn. Sci., vol. 42, No 1.

Klamka, J. (1991), *Controllability of Dynamical Systems*, PWN-Kluwer Academie Publishers, Warszawa-Dordrecht-London.

Kurek, J. (1985), *The general state-space model for two-dimensional linear digital system*, IEEE Trans. Autom. Control, AC-30, 600-602.

Kurek, J. and Zaremba, M.B. (1993), *Iterative learning control synthesis on 2-D system theory*, IEEE Trans.Autom.Contr.,vol.38, No 1, 121-125.

Roesser, R.P. (1975), *A discrete state-space model for linear image processing*, IEEE Trans. Autom. Control,AC-21,1-10.

Rogers, E. and Owens, D.H. (1992), *Stability Analysis for Linear Repetitive Processes*. Springer -Verlag.

SYSTEMS STRUCTURE IN ITERATIVE LEARNING CONTROL

D. H. OWENS*, N. AMANN* and E. ROGERS**

* *Centre for Systems and Control Engineering, University of Exeter, Exeter EX4 4QF, U.K.*
** *Electronics and Computer Science, University of Southampton, Southampton S09 5NH, U.K.*

Contenu: L'article fait part de récents résultats dans le domaine du contrôle d'apprentissage par itérations, avec un intérêt particulier pour sa structure 2D et l'effet des propriétés liées aux systèmes sur la convergence. Une attention particulière est donnée aux concepts de gain élevé et les approches d'optimisation de H_∞ et linéaire quadratique.

Abstract: The paper reviews recent results in the area of Iterative Learning Control with particular emphasis on its 2D structure and the effect of systems structural properties on convergence. Particular attention is focussed on high-gain concepts and H_∞ and LQ optimisation approaches.

Keywords: iterative learning control, adaptive control, learning systems, optimisation, poles and zeros

1 Introduction

Iterative Learning Control is a new area of Intelligent Control where the central problem is that knowledge is incomplete and the learning of the appropriate information must be done in the presence of a specified dynamical system. The dynamics of the process can interact with the learning in a destructive manner, destroying stability of the combined system. The rigourous analysis of the behaviour of the combined learning and dynamical system is only now emerging (Rogers and Owens 1992) and is dependent on the structure of the system in a subtle way. This paper describes recent results obtained at the University of Exeter supported by a collaborative UK EPSRC contract. The paper includes a survey of work in this area and highlights new results that extends previous work (Amann et. al. 1994).

Iterative Learning Control can be applied to systems that constantly repeat the same task. One economically important and often studied example is that of robot manipulators in the production industry (Arimoto 1990). These would perform one task, e.g. following a geometrical trajectory, repetitively. Using normal control, they would exhibit the same performance (performance designates the difference between the desired and actual output) at each repetition/trial. Motivated by human learning, Iterative Learning Control continuously uses information from previous trials in order to *improve* performance (Arimoto et. al. 1984). Mathematically expressed, this means that the control input $u_{k+1}(t)$ at the $(k+1)^{\text{th}}$ trial to the plant is given as a function of previous inputs and errors (Arimoto 1990)

$$u_{k+1} = f(u_k, e_k) \quad , \tag{1}$$

where e_k denotes the tracking error signal on trial k.

The intuitive notion of "improving performance progressively" can be refined to a convergence condition on the error, i.e. (in some norm topology)

$$\lim_{k\to\infty} \|e_k(\cdot)\| = 0 \quad . \tag{2}$$

This is a stability problem on a 2D-product space, typically of the form $\mathbb{N} \times L_2[0,T]$. This places the analysis of Iterative Learning Control systems firmly outside of the traditional realm of control theory although it is illustrated below how "classical" control theory is a valuable tool in the area.

Iterative Learning Control makes it possible for precise tracking to be achieved. It is relevant to practice however to consider the issue of whether or not a zero limit error can be achieved, even theoretically. In the "difficult" cases where it cannot, the aim is to achieve the smallest possible limit error or an error within a specified tolerance.

As the algorithms are implemented through repeated experimentation (real or simulated), the saving of time is essential. This requirement is equivalent to the additional goal of achieving a rapid reduction of the magnitude of the error from trial to trial (high convergence rate).

Previous work can be divided into two classes. In the larger class, nonlinear systems and particularly the specific nonlinear class of mechanical systems, as found in robotics, are studied. This paper considers linear systems. This permits a very general analysis in principle and, in particular, it enables the use of well-known analysis and design methods from "classical" control theory. For example, frequency domain methods (Owens and Neuffer 1992, Padieu and Su 1990) allow the derivation of convergence conditions as norm bounds on the operator that relates the error from trial k to trial $k+1$ (the so-called "error transition operator").

In contrast to other papers in this area, this paper studies Iterative Learning Control schemes based on a learning law that takes the error of the current trial into account, as in the functional algorithm

$$u_{k+1} = u_k + f(e_{k+1}) , \qquad (3)$$

as opposed to the normally used scheme (1) which includes the previous trial error only. A discussion of algorithms of the form (1) can be found in Moore (1993). The use of the previous error in the control law corresponds to "feedforward" (trial to trial) control while the algorithm (3) uses a "feedback" (current trial) control. The use of feedback control enables convergence for a wider class of systems (Neuffer 1992) and has besides that the usual advantages of feedback control, e.g. stabilisation of the closed loop and the potential for increased robustness.

The outline of the next sections is as follows: in section 2 high gain controllers are studied, connecting learning control to the area of universal adaptive stabilisation (Ilchmann 1993, Owens et. al. 1987). Section 3 regards Iterative Learning Control as a form of 2D-systems theory. In section 4, frequency domain convergence conditions are reviewed and based on this, a specific notion of optimality is introduced. The comparison of the algorithms addresses the issues of weak versus strong convergence, limitations of Iterative Learning Control due to plant properties and the form of the limit error. The paper concludes with a brief section describing Iterative Learning Control design based on a LQR in traditional optimal control. In this context stabilisation is no problem (intuitively because full state feedback and hence full information is used) but systems structure can still have a devastating effect on the rate of convergence.

2 High Gain Feedback and Adaptive Control

High gain feedback is well-known in classical control for its ability to reduce errors, reject disturbances and reduce sensitivity to modelling errors. This idea is studied in this section for Iterative Learning Control in both non-adaptive and adaptive cases. The adaptive approach is a generalisation of the concepts of universal adaptive stabilisation (Byrnes and Willems 1984, Ilchmann 1991, 1993) to Iterative Learning Control systems. It is assumed that the plant

$$\begin{aligned} \dot{x}_k(t) &= A\,x_k(t) + B\,u_k(t) \qquad 0 \le t \le T < \infty \\ e_k(t) &= r - C\,x_k(t) \\ & x_k \in \mathbb{R}^n, u_k \in \mathbb{R}^m, e_k \in \mathbb{R}^m, r \in \mathbb{R}^m \qquad (4) \end{aligned}$$

is unspecified, but satisfies the following *structural* properties:

- It is minimum phase.
- CB is square, nonsingular and has spectrum in the open right half complex plane.

The learning law considered has the form:

$$u_{k+1}(t) = u_k(t) + K e_{k+1}(t) \; . \qquad (5)$$

Under the above conditions and if the scalar, positive gain K is sufficiently large, it has been shown that the error sequence has the following properties:

1. The error satisfies a *slow change condition*:
$$\sum_{k=0}^{\infty} \|e_{k+1} - e_k\|^2_{L_2^m[0,T]} < \infty \qquad (6)$$
2. It is a *monotonically* non-increasing sequence:
$$\|e_{k+1}\|^2_{L_2^m[0,T]} \le \|e_k\|^2_{L_2^m[0,T]} \qquad (7)$$
3. $e_{k+1} - e_k$ converges in norm to zero as $k \to \infty$ in $L_2(0,\infty) \cap L_\infty(0,\infty)$.
4. The error sequence $\{e_k\}$ is uniformly bounded and it converges to zero in the *weak* topology on $L_2^m[0,T]$, i.e.
$$\langle f, e_k \rangle = 0 \quad \forall f \in L_2^m[0,T] \qquad (8)$$

The proof is given in Owens (1992) and depends crucially on the structural assumptions on the plant. Note that the structural assumptions are precisely those familiar properties that produce excellent performance from classical high gain feedback systems, i.e. some of the design properties and problems of classical design carry over with little change to the Iterative Learning Control case.

The difficulty of this scheme for situations of high uncertainty is to find a large enough gain (which depends upon the system parameters in a complex manner). This problem was eased through recent extensions by one of the authors through adapting the control gains (Owens 1993). At each trial, a gain K_{k+1} is employed. This gain is adapted between trials according to the typical update law

$$K_{k+1} = K_k + \gamma \|e_{k+1} - e_k\|^2_{L_2^m[0,T]} \; . \qquad (9)$$

where γ represents the sensitivity of the gain to observed data. Intuitively, as γ increases, convergence rates will increase. This is observed in simulations. Under the same conditions as above, in addition to the above properties of $\{e_k\}$, the gain sequence $\{K_{k+1}\}$ converges to a limit gain K_∞ if the error sequence $\{e_k\}$ is bounded. Sufficient conditions for boundedness of the error sequence is either that the system is SISO and positive real. Alternatively, boundedness can be guaranteed if the initial choice K_1 is sufficiently large.

The adaptation of K_{k+1} parallels the basic control $\dot{k} = y^2$ introduced by Byrnes and Willems (1984) in adaptive stabilisation theory. One advantage of this method is that the plant may be unknown as long as it belongs to the above defined class. It is however disappointing that only weak convergence is achieved. This means that all Fourier-coefficients of the limit error go to zero, but it could still be the case that the error does not go pointwise to zero as $k \to \infty$. Of course this possibility is not seen in simulations and, if a physicists view is taken, it will not occur in practice - this is, however, unproven!

3 2D Systems Theory and Iterative Learning Control

As a 2D systems problem, Iterative Learning Control has a causality that stresses the repetitive action of the system. The problem of convergence in the k-direction is in this light a stability problem. In Edwards and Owens (1982) and Rogers and Owens (1992), the stability theory for repetitive systems (originally derived by Owens) is stated and extended to provide computational design tools. In the following, a general Iterative Learning Control system is regarded as such a "linear repetitive system" so that the stability theory described in [7] and [18] can be applied.

The learning law of the Iterative Learning Control system considered by the Exeter group uses, in its most general form, information from several (n say) previous trials, thus making up what can be called an n-th order Iterative Learning Control system. It also incorporates relaxation effects in the form of *forgetting factors* α_i so that the inputs of previous trials are weighted. Relaxation has a well defined effect on stability (independent of the systems structure) and also increases any natural robustness of the algorithm.

The learning law proposed is as follows:

$$\begin{aligned} u_{k+1}(t) &= \sum_{i=1}^{n} \alpha_i u_{k+1-i}(t) \\ &\quad + \sum_{i=1}^{n} K_i[e_{k+1-i}](t) \\ &\quad + K_0[e_{k+1}](t) \end{aligned} \qquad (10)$$

Regarded as a linear repetitive process, the error is the "pass profile" and the reference is the pass disturbance. The system equation for $e_{k+1}(t)$ follows from (10) and the plant equation $e_{k+1}(t) = r(t) - G[u_{k+1}](t)$ as

$$\begin{aligned} e_{k+1} &= (I + GK_0)^{-1} \Big\{ \sum_{i=1}^{n} (\alpha_i I - GK_i) e_{k+1-i} \\ &\quad + (1 - \sum_{i=1}^{n} \alpha_i) r \Big\} . \end{aligned} \qquad (11)$$

Defining a supervector $\hat{e}_{k+1}(t)$ consisting of the errors appearing in (11) is comparable to introducing the (global) state of the 2D-system. The error supervector is given by

$$\hat{e}_k(t) = \begin{bmatrix} e_{k+1-n}(t) \\ \vdots \\ e_{k-1}(t) \\ e_k(t) \end{bmatrix} . \qquad (12)$$

The notation is further streamlined by introducing the compact notation for the operators $E_i[\cdot](t)$ as $E_0 = (I + GK_0)^{-1}$ and $E_i = \alpha_i I - GK_i$ for $i = 1, \ldots, n$. This yields the final familiar repetitive form of the system:

$$\hat{e}_{k+1} = L_T \hat{e}_k + b_k \qquad (13)$$

with

$$L_T = \begin{bmatrix} 0 & 1 & 0 & \cdots & 0 \\ 0 & 0 & 1 & \cdots & 0 \\ \vdots & \vdots & \vdots & \ddots & \vdots \\ 0 & 0 & 0 & \cdots & 1 \\ E_0 E_n & E_0 E_{n-1} & E_0 E_{n-2} & \cdots & E_0 E_1 \end{bmatrix} \qquad (14)$$

and

$$b_k(t) = \begin{bmatrix} 0 \\ 0 \\ \vdots \\ 0 \\ 1 - \sum_{i=1}^{n} \alpha_i \end{bmatrix} r(t) . \qquad (15)$$

This is a linear repetitive process with operator L_T acting on the state (the error supervector) $\hat{e}_{k+1}$ and a trial independent disturbance $b(t)$. The stability theory due to Owens and described in [7] and [18] shows that this system is asymptotically stable (in a precisely defined robust sense) if and only if the spectral radius of L_T is $\rho(L_T) < 1$. The surprising result for a finite trial length $T < \infty$ is that the spectral radius $\rho(L_T)$ is given by the maximal eigenvalue of D_{L_T}

$$\rho(L_T) = \max_i \lambda_i(D_{L_T}) \qquad (16)$$

where D_{L_T} is the direct feedthrough/feedforward matrix of any state space realization of the system L_T. If the underlying plant is strictly proper, D_{L_T} is given by

$$D_{L_T} = \begin{bmatrix} 0 & 1 & 0 & \cdots & 0 \\ 0 & 0 & 1 & \cdots & 0 \\ \vdots & \vdots & \vdots & \ddots & \vdots \\ 0 & 0 & 0 & \cdots & 1 \\ \alpha_n & \alpha_{n-1} & \alpha_{n-2} & \cdots & \alpha_1 \end{bmatrix} \qquad (17)$$

and λ_i are the roots of its characteristic polynomial

$$\lambda^n - \alpha_1 \lambda^{n-1} - \cdots - \alpha_{n-1}\lambda - \alpha_n = 0 . \qquad (18)$$

For example, for a first order system ($n = 1$), this requires that the solution of $\lambda - \alpha = 0$ has modulus less than unity, i.e. $|\alpha| < 1$ is a necessary and sufficient condition for (asymptotic) stability along the k-axis.

The counter-intuitive result is that stability is largely independent of the plant and the controllers! This is a direct result of the assumption of a finite trial endtime T because on a finite time interval, a linear system can only produce a bounded output, even if it is unstable. In the stability notion considered here these unstable outputs are still "acceptable". The problem with this result is thus that even if $\{e_k(t)\}$ is guaranteed to converge to a limit $e_\infty(t)$, this terminal error profile might be unstable and/or it might be worse than $e_1(t)$, i.e. that learning did not result in an improvement of the error. For an acceptable $e_\infty(t)$, additional conditions on E_i and α, taking the structure of the plant and the learning structure into account, have to be satisfied. To illustrate the ideas, consider the case where $n = 1$. A simple analysis then gives the following expression for limiting error e_∞

$$e_\infty = (I + \frac{1}{1-\alpha} GK)^{-1} r \qquad (19)$$

and hence the tracking error converges to the error of a unity feedback system for the plant G with the *effective controller*

$$K_{\text{eff}} = \frac{1}{1-\alpha}K. \tag{20}$$

For small terminal error, it is necessary (but not sufficient) that the effective controller has a high gain i.e. α is very close to $1-$. The fact that this is not sufficient is easily seen in the single-input/single-output case using classical root-locus concepts for the system GK. Omitting details for brevity, the essential conclusions are that:

(1). The pole-zero excess/relative degree of GK should be ≤ 2.

(2). GK must be minimum-phase.

These results can easily be seen using simple partial fraction analysis. This analysis also shows that systems structure has a significant effect -

(a). If the relative degree $\partial(GK)$ is unity, then, as $\alpha \to 1-$, $e_\infty \to 0$ uniformly on $[0, T]$.

(b). If the relative degree $\partial(GK)$ is two, then, as $\alpha \to 1-$, $e_\infty \to 0$ on $[0, T]$ but consists of an oscillation of decreasing magnitude but increasing frequency i.e. for values of α close to unity the Iterative Learning Control algorithm converges to a "jitter".

(c). If the relative degree $\partial(GK)$ is ≥ 3 and $|\alpha| < 1$, the Iterative Learning Control algorithm converges but to an unstable transient, i.e. although stable in the repetitive systems sense, the algorithm has no stability in a practical sense.

The forgetting factor α and systems structure hence have a substantial effect on stability and terminal errors. Rate of convergence depends on the system in a complex way. Use of repetitive systems techniques gives bounds of the form

$$\|e_k - e_\infty\| \leq M_0(\epsilon)(\alpha+\epsilon)^k \left\{ \|e_0\| + \frac{1-\alpha}{1-\alpha-\epsilon}\|r\| \right\} \tag{21}$$

for any choice of $\epsilon > 0$ satisfying $\alpha + \epsilon < 1$ and some $M_0(\epsilon) > 0$. Only if GK is positive-real can we choose $\epsilon = 0$ and $M_0(\epsilon) = 1$. Fast convergence is hence obtained (but not exclusively) for small values of α. There is hence a trade-off between fast convergence (small α) and terminal accuracy ($\alpha \to 1-$).

4 H_∞ Optimal Iterative Learning Control

With relaxation, convergence can always be achieved at the cost of terminal errors although this convergence may not be acceptable. The desire to improve algorithm performance is being addressed through the use of optimal control methods. Optimality in this context is desired with respect to the "convergence rate". Iterative Learning Control normally operates on a finite time interval $[0, T]$. Frequency domain techniques implicitly use the infinite interval $[0, \infty)$. As is known from Padieu and Su (1990) and Owens and Neuffer (1992), the convergence condition in the frequency domain for a learning system (without relaxation effects) on $[0, \infty)$ is that the error transition operator $E(s)$ which relates the error in one trial to the error of the next trials and which is defined by

$$e_{k+1}(s) = E(s)e_k(s) \tag{22}$$

has H_∞-norm less than or equal one. It is only a sufficient condition on finite time intervals however. Furthermore, if it has H_∞-norm γ with $\gamma < 1$, the learning rate is exponential and the following estimate for the error norm holds pointwise:

$$|e_{k+1}(j\omega)| = \gamma^{k+1}|e_0(j\omega)| \,. \tag{23}$$

In order to achieve the maximum reduction of the error norm from trial to trial, it is hence sufficient to have a small value of γ. The design strategy suggested by this discussion is to construct learning controllers $K(s)$ such that $\|E(s)\|_\infty$ is minimised. For a learning control law of the form

$$u_{k+1}(s) = u_k(s) + K(s)e_{k+1}(s) \tag{24}$$

the error transition operator $E(s)$ is equal to the sensitivity matrix $S(s) = (I + G(s)K(s))^{-1}$. This simple form allows the use of standard algorithms for the design of the H_∞ optimal controller (Doyle et. al. 1989). To satisfy regularity conditions, a mixed sensitivity problem of the form $\|S, KS\|_\infty$ is solved where the inclusion of the term KS can be justified by the goal to achieve robustness of the closed loop of stability with regard to the time axis and additive perturbations to the plant. A frequency dependent weight is usually included to shape the sensitivity function. In Owens and Rogers (1992), this idea was introduced for repetitive systems and, within this context, the relationship of H_∞-minimizes controllers to high gain feedback was discussed.

There are two important issues about H_∞-optimality. The first is the possible conservativeness of the H_∞-convergence criterion for Iterative Learning Control on a *finite* trial length. All systems that are stable on an infinite time interval are stable for a finite duration of the trial as well, but not vice versa! It is questionable whether the strong convergence condition in the frequency domain $\|E(s)\|_\infty < 1$ asks too much from a system on finite trial length. This is certainly true in principle but, if T is long relative to the time constants of the process, it is essentially "infinite" (particularly if the target trajectory r is more or less constant over its terminal portion).

The second question is the necessary shape for the weight function in the H_∞-optimization problem. Taking the norm of (22) suggests that $\bar{\sigma}(E(j\omega))$ should be as small as possible for a frequency range as large as possible to achieve a fast reduction of the error. When the reference signal is taken into account, this objective can be refined a little, as explained below.

The H_∞-optimization approach, as becomes evident even in the SISO case, can only be successful for a restricted class of systems. It is well-known that for non-minimum phase plants and plants of relative degree greater than one the sensitivity function always has peaks in $\bar{\sigma}(S(j\omega)) > 1$ (Maciejowski 1989). The studies undertaken showed however that even in these cases the use of iterative learning control can be of practical advantage. This is because even if the error e_k in the limit $k \rightarrow \infty$ may become large or infinite, it can decrease substantially during the first few trials. Hence, even with Iterative Learning Control schemes that violate the convergence condition $\|S(s)\|_\infty < 1$ the error $|e_k(j\omega)|$ is often reduced substantially during initial trials before any instability sets in. This situation represents a kind of "practical convergence" where during a limited number of trials $|e_k(j\omega)|$ decreases and only then does the expected divergence appear. This leaves open the possibility that the algorithm can be terminated at that point with small tracking error.

To understand this effect and the importance of the frequency content of the reference r, consider the magnitude of the involved signals and the maximal singular value of the error transition operator. The reference signal has typically large magnitude for low frequencies and small magnitude for high frequencies. This is because the reference signal must have a higher "roll-off" for high frequencies than the plant for a solution to exist. In other words, the reference magnitude $|r(j\omega)|$ must decrease faster than a rate of 20 dB/decade times the relative degree of the plant for $\omega \rightarrow \infty$.

Using weighting functions, the error transition operator can normally be forced to have an amplification of much less than one for frequencies below the bandwidth. If the bandwidth can be made large enough (this depends on the plant structure, e.g. the location of non-minimum phase zeros) so that the reference signal is already small at the frequency where $\bar{\sigma}(S(j\omega)) > 1$, then the application of Iterative Learning Control will reduce $|e_0(j\omega)| = |r(j\omega)|$ (with $u_0(s) \equiv 0$) and the next few errors as well. Only after a (possibly large) number of trials will $|e_k(j\omega)|$ be noticeably large at the frequencies where it is amplified.

The conclusions of these considerations is that Iterative Learning Control in the described scenario may diverge but it may be used under certain conditions until the error norm is below a desired tolerance ε or until it increases. Depending on achieving a large enough bandwidth, the achievable tolerance can be set quite small. As a guideline, it is suggested that the learning controller is designed such that it has a large bandwidth, a small gain especially at low frequencies and no or only a small peak of $\bar{\sigma}(S(j\omega)) > 1$. These are essentially the same "rules of thumb" for the shape of $\bar{\sigma}(S(j\omega))$ as in classical controller design.

5 Optimality and State Feedback

Recent results (submitted) on the use of optimal control techniques in Iterative Learning Control indicate that state feedback and previous trial feedforward can guarantee theoretical convergence for any linear (even time-varying) plant. Systems structure does still however have an impact on the practical convergence rate. In particular, the presence of right-half-plane zeros can introduce a severe property of slow convergence that has been interpreted by Amann and Owens (1994) in a recent paper.

The basic idea behind the analysis is that the aim of the Iterative Learning Control algorithm is to minimise $\|e\|^2$ subject to the dynamic constraints represented by the systems dynamical equations. The problem is a singular optimal control problem that is naturally solved iteratively and hence, provided the causality of the solution permits implementation as an Iterative Learning Control algorithm, fairly standard optimal control numerical schemes may be used. The algorithm chosen generates the input u_{k+1} for the $(k+1)$th trial as the solution of the Hilbert space optimisation problem (interpreted as a modified gradient method)

$$J_{k+1} = \|e_{k+1}\|^2 + \|u_{k+1} - u_k\|^2 \qquad (25)$$

which has the formal and apparently anti-causal solution

$$u_{k+1} = u_k + G^* e_{k+1} \qquad (26)$$

where G^* is the adjoint operator to the plant operator G. The most interesting case is when the norm chosen on the underlying space is the usual LQR performance index with weighting matrices Q on the error, R on the input change and F on the terminal error. Under these conditions, it has been shown that the solution can be expressed as a familiar Riccati state feedback of the *current* trial state vector plus an additional feedforward term generated by the *previous* trial data including, depending on the realisation chosen, the state, the input and the error. Given access to the state, realisation is, in principle, no problem.

It has been shown that the algorithm always converges to a zero limit error, the convergence being geometric under well defined invertibility conditions on the plant. The convergence is also monotone in the sense that

$$\|e_{k+1}\| \leq \|e_k\|, \quad \forall k \geq 0 \qquad (27)$$

The algorithm convergence rate is firstly governed by the choice of weight Q, R and F with the simple (high gain) rule that R should be small for rapid convergence. Convergence is then rapid in the case of minimum-phase systems but is slow for non-minimum-phase systems as the right-half-plane zero creates a direction in function space of great sensitivity. This sensitivity intuitively reduces the step length that can be tolerated and hence increases the number of iterations/trials required (Amann and Owens 1994).

More recent work shows that complete control over convergence, even in the case of non-minimum-phase systems can be introduced using a form of predictive control algorithm which, in its infinite horizon form, creates convergence at an exponential rate γ^k to zero for any choice of γ. The price to be paid for this is in the complexity of the algorithm but the theory defines a benchmark solution against which all others can be compared. This work will be published separately.

6 Conclusions

The paper has reviewed the current state of development of new ideas in Iterative Learning Control with particular attention focussed on the effect of systems structure on its 2D dynamical properties. It is seen that, for the class of algorithms considered, the familiar properties of relative degree and non-minimum-phase structure are major limitations to convergence and/or convergence to an acceptable error. These ideas are made particularly clear through the use of high gain techniques and optimisation methodologies. Only in the case of LQ-type optimal Iterative Learning Control is convergence achieved independent of the details of the systems structure although it is noted that non-minimum-phase structure has a serious effect on convergence rates.

Acknowledgements

Current work is supported by the UK Science and Engineering Research Council under contract No. GR/H/48286. The preliminary calculations in high gain adaptive Iterative Learning Control were obtained by Dieter Neuffer at Exeter. The preliminary computational results for the application of H_∞ to Iterative Learning Control were obtained with the help of Anja Wahl, an ERASMUS student from the Institut für Systemdynamik und Regelungstechnik, Universität Stuttgart.

References

[1] Amann, N. and Owens, D. H. (1994). Non-minimum phase plants in iterative learning control. In *Proc. 2nd Int. Conf. on Intelligent Systems Engineering*, Hamburg-Harburg.

[2] Amann, N., Owens, D. H., and Rogers, E. (1994). New results in iterative learning control. In *Proc. IEE Int. Conf. Control '94*, pages 640–645, Coventry.

[3] Arimoto, S. (1990). Learning control theory for robotic motion. *Int. J. of Adaptive Control and Signal Processing*, 4:543–564.

[4] Arimoto, S., Kawamura, S., and Miyazaki, F. (1984). Bettering operations of robots by learning. *J. Robotic Systems*, 1(2):123–140.

[5] Byrnes, C. I. and Willems, J. C. (1984). Adaptive stabilization of multivariable linear systems. In *Proc. 23rd IEEE Conf. on Decision and Control*, pages 1574–1577, Las Vegas, Nevada.

[6] Doyle, J. C., Glover, K., Khargonekar, P. P., and Francis, B. A. (1989). State-space solutions to standard H_2 and H_∞ control problems. *IEEE Trans. on Automatic Control*, 34(8):831–847.

[7] Edwards, J. B. and Owens, D. H. (1982). *Analysis and Control of Multipass Processes.* Research Studies Press (John Wiley), Chicester.

[8] Ilchmann, A. (1991). Non-identifier based control of dynamical systems – a survey. *IMA J. Math. Control and Inf.*, 8:321–366.

[9] Ilchmann, A. (1993). *Non-Identifier-Based High-Gain Adaptive Control*, volume 189 of *Lecture Notes in Control and Information Sciences.* Springer-Verlag, London.

[10] Maciejowski, J. M. (1989). *Multivariable Feedback Design.* Addison-Wesley, Wokingham.

[11] Moore, K. L. (1993). *Iterative Learning Control for Deterministic Systems.* Advances in Industrial Control Series. Springer-Verlag, London.

[12] Owens, D. H. (1992). Iterative learning control – convergence using high gain feedback. In *Proc. 31st IEEE Conf. on Decision and Control*, pages 2545–2546, Tucson, AZ.

[13] Owens, D. H. (1993). Iterative learning control using adaptive high gain feedback. In *Proc. 2nd European Control Conf.*, pages 195–198, Groningen.

[14] Owens, D. H. and Neuffer, D. (1992). Theoretical and computational studies in iterative learning control. Internal report series of the centre for systems and control engineering, University of Exeter. Report No. 92/02.

[15] Owens, D. H., Prätzel-Wolters, D., and Ilchmann, A. (1987). Positive real structure and high gain adaptive stabilization. *IMA J. Math. Control and Inf.*, 4:167–181.

[16] Owens, D. H. and Rogers, E. (1992). H-infinity-norm minimisation and the stabilisation of systems with repetitive dynamics. *Trans. Inst. Measurement and Control*, 14(3):126–129.

[17] Padieu, F. and Su, R. (1990). An H_∞ approach to learning control systems. *Int. J. of Adapt. Control and Signal Processing*, 4:465–474.

[18] Rogers, E. and Owens, D. H. (1992). *Stability Analysis for Linear Repetitive Processes*, volume 175 of *Lecture Notes in Control and Information Sciences*. Springer-Verlag, Berlin.

Behavioral Interpretation of Weak and Strong Controllability for Delay-Differential Systems

Paula Rocha

Department of Mathematics; University of Aveiro; 3800 Aveiro;Portugal.

Résumé. On généralise les propriétés de commandabilité faible et forte pour des systèmes de la forme $R(D, \Delta)w = 0$, où D désigne l'opérateur de différentiation, Δ l'opérateur de retard et $R(z_1, z_2)$ est une matrice polynomiale 2D. Ces concepts ont été introduits comme des conditions sur la charactéristique de certaines matrices. Ici on donne une charactérisation altérnative en termes des trajectoires du système.

1 Introduction

In spite of their one dimensional nature, delay-differential systems are often considered within an ND setting, since they can be analysed by means of multidimensional techniques. Consider a system Σ described by the following pseudo-state equations.

$$\begin{cases} Dx &= A(\Delta)x + B(\Delta)u \\ y &= C(\Delta)x + D(\Delta)u \end{cases} \qquad (1)$$

where D is the differentiator and Δ is the unit delay operator defined for every function f of real variable as $\Delta f(t) = f(t-1)$ $\forall t \in I\!R$; $A(z_2)$, $B(z_2)$, $C(z_2)$ and $D(z_2)$ are polynomial matrices in the indeterminate z_2. Denoting $w := col(x, u, y)$, (1) can be written as:

$$R(D, \Delta)w = 0 \qquad (2)$$

where $R(z_1, z_2) := col([z_1 - A(z_2) \mid -B(z_2) \mid 0], [C(z_2) \mid D(z_2) \mid -I])$ is a 2D polynomial matrix. This suggests an analogy with 2D systems that can be described by:

$$R(\sigma_1, \sigma_2)w = 0 \qquad (3)$$

with σ_1 and σ_2 denoting either the 2D shifts or the partial differentiators. Note, however, that while σ_1 and σ_2 act independently along two different directions, the same does not apply to the differentiator D and the delay Δ - for instance, $Dw = 0$ implies that $(\Delta - 1)w = 0$, showing that D and Δ are not independent.

Now, since the description matrix $R(z_1, z_2)$ completely specifies the delay - differential system Σ, the different structural properties of Σ will correspond to certain properties of R. In this paper we are concerned with the characterization of different kinds of controllability properties for delay-differential systems in terms of their description matrices.

Within a behavioral framework (which essentially views a system as a set of trajectories) the natural concept of controllability corresponds to the possibility of making compatible the past of a given trajectory with the (sufficiently far) future of any other one. It has been recently shown in Rocha and Willems (1995) that a delay-differential system with a description $R(D, \Delta)w = 0$, where $R(z_1, z_2)$ is an arbitrary 2D polynomial matrix of rank r, is behaviorally controllable if and only if:

$$rank R(\lambda, e^{-\lambda}) = r, \ \forall \lambda \in \mathbb{C}. \quad (4)$$

When applied to systems with a description as given in (1), (4) amounts to the condition of spectral controllability introduced in Manitius (1981). This gives a behavioral interpretation of that controllability property, which had previously only been defined as a rank condition.

Other interesting properties are weak and strong controllability, which have been considered in Morse (1976) and Levy (1981) within an algebraic approach. Similar to spectral controllability, these have been defined as rank conditions. Here we will provide these properties with the missing system theoretic interpretation, and relate them with the other existing controllability notions.

2 Weak and strong controllability

The properties of weak and strong controllability have been defined in Morse (1976) and Levy (1981) for pseudo-state equations of the form (1) in the following way.

Definition 1 *A system of the form (1) is siad to be weakly controllable if* $[A(z_1)^{n-1}B(z_1) \mid A(z_1)^{n-2}B(z_1) \mid B(z_1)]$ *has full row rank over* $\mathbb{R}[z_1]$.
The system is said to be strongly (or strictly) controllable if $[A(\lambda)^{n-1}B(\lambda) \mid A(\lambda)^{n-2}B(\lambda) \mid \ldots \mid B(\lambda)]$ *has full row rank for all* $\lambda \in \mathbb{C}$.

These controllability properties can also be stated in terms of primeness properties for the 2D matrix $C(z_1, z_2) = [z_1 - A(z_2) \mid B(z_2)]$ associated with (1).

Proposition 1 Levy (1981)
(i) (1) is weakly controllable iff $C(z_1, z_2)$ *is left-prime.*
(ii) (1) is strongly controllable iff $C(z_1, z_2)$ *is left-zero-prime, i.e.* $C(\lambda_1, \lambda_2)$ *has full rank for all* $\lambda_1, \lambda_2 \in \mathbb{C}$.

Note that the primeness properties of $C(z_1, z_2)$ in Proposition 1 are equivalent to the same primeness properties for the matrix $R(z_1, z_2)$ associated with (2). So, (1), or equivalently (2), is weakly/strongly controllable if and only if $R(z_1, z_2)$ is left-prime/zero-prime. This suggests that the notions of weak and strong controllability could be extended for delay-differential systems Σ given by $R(D, \Delta)w = 0$, where $R(z_1, z_2)$ is now a general 2D polynomial matrix, in the following way. We would say that Σ is *weakly controllable* if $R(z_1, z_2)$ is left-prime and that it is *strongly controllable* if $R(z_1, z_2)$ is left-zero-prime. However, this does not provide a system theoretic interpretation of the defined controllability properties. In order to give such an interpretation, we introduce next some fundamental concepts.

Let Σ be the system described by the delay-differential equation

$$R(D, \Delta)w = 0,$$

where $R(z_1, z_2)$ is a 2D polynomial matrix and $R(D, \Delta)$ is regarded as an operator acting on C^∞ trajectories, and denote the system behavior by $\mathcal{B} := ker R(D,\Delta)$. A system $\bar{\Sigma}$ is said to be a *subsystem* of Σ if $\bar{\Sigma}$ is described by a polynomial delay-differential equation $\bar{R}(D,\Delta)w = 0$ and $\bar{\mathcal{B}} := ker\bar{R}(D,\Delta) \subseteq \mathcal{B}$. A subsystem $\bar{\Sigma}$ of Σ is said to be an *absorbing subsystem* of Σ if

$$(w \in \mathcal{B}, w|_{(-\infty,0)} \in \bar{\mathcal{B}}|_{(-\infty,0)}) \Rightarrow (w \in \bar{\mathcal{B}})$$

This means that the system trajectories which in the past satify the laws of the subsystem $\bar{\Sigma}$ must continue to satisfy these laws also in their future evolution, i.e. they are never allowed to leave $\bar{\mathcal{B}}$.

Definition 2 *We will call Σ a weakly controllable system if it does not allow for any proper absorbing subsystem.*

This corresponds to the desired definition of weak controllability in system theoretic terms. Indeed,

Theorem 1 *A delay-differential system Σ described by $R(D,\Delta)w = 0$ is weakly controllable if and only if $R(z_1,z_2)$ is left-prime.*

Proof:
If $R(z_1,z_2)$ is not left-prime, there exist 2D polynomial matrices $F(z_1,z_2)$ with full column rank and $\bar{R}(z_1,z_2)$ with full row rank such that:

$$R(z_1,z_2) = F(z_1,z_2)\bar{R}(z_1,z_2),$$

and where moreover $F(z_1,z_2)$ is a nontrivial factor. This implies that:

$$\bar{R}(D,\Delta)(\mathcal{B}) = ker F(D,\Delta),$$

where, as before, $\mathcal{B} := ker R(D,\Delta)$. Now, since $F(z_1,z_2)$ has full column rank, $ker F(D,\Delta)$ has the following property:

$$(v \in ker F(D,\Delta), v|_{(-\infty,0)} = 0) \Rightarrow (v \equiv 0).$$

Therefore, if w is a trajectory in $\mathcal{B}$ and $(\bar{R}(D,\Delta)w)|_{(-\infty,0)} = 0$ then $\bar{R}(D,\Delta)w \equiv 0$. In other words, if $w \in \mathcal{B}$ and $w|_{(-\infty,0)} \in \bar{\mathcal{B}}|_{(-\infty,0)}$ then $w \in \bar{\mathcal{B}}$, where $\bar{\mathcal{B}} := ker\bar{R}(D,\Delta)$. So, the system associated with $\bar{\mathcal{B}}$ is an absorbing proper subsystem of Σ, meaning that Σ is not weakly controllable. Thus, the left-primeness of $R(z_1,z_2)$ is a necessary condition for the weak controllability of Σ.

In order to show the converse implication, suppose that Σ is not weakly controllable, allowing therefore for a proper absorbing subsytem $\bar{\Sigma}$. Then there exist 2D polynomial matrices $R'(z_1,z_2)$, $\bar{R}(z_1,z_2)$ (of full row rank) and $F(z_1,z_2)$ such that Σ is associated with $R'(z_1,z_2)$, $\bar{\Sigma}$ is associated with $\bar{R}(z_1,z_2)$ and $R' = F\bar{R}$. Taking into account that $\bar{\Sigma}$ is a proper absorbing subsystem of Σ, it is possible to show that $F(z_1,z_2)$ is a nontrivial factor with full column rank. Consequently R' is not left-prime. This implies that also R is not left-prime, proving that the left-primeness of $R(z_1,z_2)$ is also sufficient condition for the weak controllability of Σ.□

As for the strong controllability property, we will define it as the possibility of (via a transformation) separating the system variables into two cathegories, namely: completely free variables and variables which are required to remain zero.

We will say that the system Σ in q variables is *rectified* if the set B of its admissible trajectories can be decomposed as $\mathcal{B} = \mathcal{B}_0 \text{ x } \mathcal{B}_f$, where $\mathcal{B}_0$ and $\mathcal{B}_f$ are sets of trajectories in q_1 resp. q_2 variables (with $q_1+q_2 = q$) such that $\mathcal{B}_0 = \{w_1 : \mathbb{R} \rightarrow \mathbb{R}^{q_1} : w_1(t) = 0 \ \forall t \in \mathbb{R}\}$ and $\mathcal{B}_f = \{w_2 : \mathbb{R} \rightarrow \mathbb{R}^{q_2}\}$. We will say that Σ is *rectifiable* if there exists an invertible polynomial delay-differential operator $V(D,\Delta)$ such that $V(\mathcal{B})$ is rectified.

Definition 3 *Σ is said to be strongly controllable if it is rectifiable.*

The next result shows that this corresponds to the desired system theoretic definition of strong controllability.

Theorem 2 *A delay-differential system Σ described by $R(D,\Delta)w = 0$ is strongly controllable if and only if $R(z_1, z_2)$ is left-zero-prime.*

Proof:
If $R(z_1, z_2)$ is left-zero-prime, there exist unimodular matrices $U(z_1, z_2)$ and $W(z_1, z_2)$ such that:

$$WRU = \begin{bmatrix} I & 0 \\ 0 & 0 \end{bmatrix}.$$

Therefore

$$Rw = 0 \Leftrightarrow [I \;\; 0]Vw = 0,$$

where $V(z_1, z_2)$ is the inverse of $U(z_1, z_2)$. This implies that $V(D,\Delta)(\mathcal{B}) = ker[I \;\; 0]$, and so Σ is rectifiable, i.e. it is strongly controllable.
Suppose now that Σ is strongly controllable. Then there exists an invertible operator $V(D,\Delta)$ such that $V(D,\Delta)(kerR) = ker[I \;\; 0]$. Consequently, $kerR = ker[I \;\; 0]V$ and as $[I \;\; 0]V(z_1, z_2)$ is left-zero-prime the same holds for $R(z_1, z_2)$.□

3 Comparison with other controllability concepts

Taking into account their characterizations in terms of the description matrix $R(z_1, z_2)$, it is now possible to relate the different controllability properties. It is not difficult to see that:

strong controllability $\Rightarrow$
$\Rightarrow$ behavioral controllability $\Leftrightarrow$ spectral controllability $\Rightarrow$
$\Rightarrow$ weak controllability.

The examples below show that weak controllability does not imply behavioral controllability, which in turn does not imply strong controllability.

Example 1 The delay-differential system

$$[\Delta - 1 \;\; D]w = 0,$$

has as associated 2D polynomial matrix

$$R(z_1, z_2) = [z_2 - 1 \;\; z_1],$$

which is left-prime, and hence it is weakly controllable. However,

$$R(\lambda, e^{-\lambda}) = [e^{-\lambda} - 1 \;\; \lambda],$$

drops in rank for $\lambda = 0$, and therefore the system is not behaviorally controllable.

Example 2 The system $R(D,\Delta)w = 0$ with

$$R(z_1, z_2) = [z_2 + 1 \;\; z_1]$$

is behaviorally controllable since

$$rank[e^{-\lambda} + 1 \;\; \lambda] = 1 \; \forall \lambda \in \mathbb{C}.$$

But, because $R(\lambda_1, \lambda_2)$ drops in rank for $(\lambda_1, \lambda_2) = (1, -1)$, the system is not strongly controllable.

If we consider the discrete 2D system $R(\sigma_1, \sigma_2)w = 0$ associated with $R(z_1, z_2)$, where w is now a trajectory over $\mathbb{Z}^2$ and σ_1 and σ_2 denote the 2D shifts, we have that the weak resp. strong controllability of the delay-differential system corresponds to the weak resp. strong controllability of the associated 2D system (see Rocha and Willems (1991a) and Rocha and Willems (1991b)). Moreover, in the 2D case weak and behavioral controllability are equivalent.
As for the 1D case, all the aforementioned controllability properties turn out to be equivalent.

4 References

Lévy, B.C.(1981).
'2-D Polynomial and Rational Matrices and their Applications for the Modelling of 2-D Dynamical Systems'. Ph.D. Thesis, Stanford University, June 1981.

Manitius, A.(1981).
'Necessary and Sufficient Conditions of Approximate Comtrollability for General Linear Retarded Systems'. *SIAM J. Control and Optimization*, vol. 19, n. 4, pp. 516-532, 1981.

Morse, A. S.(1976).
'Ring Models for Delay-Differential Systems'. *Automatica*, vol. 12, pp. 529-531, 1976.

Rocha, P. and J.C. Willems (1991a).
Controllability of 2D Systems. *IEEE Trans. Aut. Cont.*, vol. AC-36, n. 4, pp. 413-423, 1991.

Rocha, P. and J.C. Willems (1991b).
On the Representation of 2D Systems. *Kybernetika*, vol. 27, n.3, pp. 225-230, 1991.

Rocha, P. and J.C. Willems (1995).
'Behavioral Controllability of Delay-Differential Systems'.Submitted. 1995.

ON $\mathcal{H}_\infty$ PROBLEMS FOR 2-D SYSTEMS

M. ŠEBEK* and F. J. KRAUS

Automatic Control Laboratory, Swiss Federal Institute of Technology, Physikstr. 3, 8092 Zürich, Switzerland; e-mail: msebek@aut.ee.ethz.ch and kraus@aut.ee.ethz.ch

Abstract. Robust $\mathcal{H}_\infty$ control problem for 2-D systems is attacked by polynomial techniques. The design procedure is shown to consist of the same major steps as its 1-D counterpart. For the particular case of the 2-input 2-output considered here this include one scalar polynomial spectral factorization and one matrix polynomial J-spectral factorization. However, the both operation are much more difficult to perform in 2-D.

Resumé. La régulation robuste par méthode $\mathcal{H}_\infty$ appliquée à des systèmes 2-D est résolue au moyen de techniques polynômiales. Le procédé de synthèse se compose des mêmes étapes principales que dans le cas 1-D. Le cas particulier de systèmes à deux entrées et deux sorties presenté ici comprend une factorisation spectrale d'un polynôme scalaire et une factorisation J-spectrale d'un polynôme matriciel. Cependant les deux opérations sont beaucoup plus difficiles à réaliser dans le cas 2-D.

Key Words. 2-D systems; polynomial techniques; robust control; $\mathcal{H}_\infty$ control.

1. INTRODUCTION

Polynomial techniques have become very popular in the last decade. They were applied to solve standard tasks in communication and control. In addition, they have been employed recently to cope with some new problems of robustness as $\mathcal{H}_\infty$ optimization for standard (1-D) linear systems. On the other hand, polynomial techniques have been generalized to cover standard problems for some non-standard classes of systems. This namely comprehends two-dimensional (2-D) linear systems and filters, which are now of rising importance in many areas of engineering and science. Influenced by both the trends mentioned above, this paper is attempts to generalize polynomial methods for robust $\mathcal{H}_\infty$ problems in control of 2-D systems.

It is shown that, as expected, all the basic steps of the design procedure remain the same as in 1-D case. However, their content is different and, namely, completely new numerical algorithms are required to perform every particular step of the design. In fact, the same phenomenon was encountered when studying 2-D version of $\mathcal{H}_2$ optimization (Šebek, 1992; Šebek and Kraus, 1995) and other 2-D control problems.

All signals considered throughout the paper are two-dimensional discrete sequences (Dudgeon and Mersereau, 1984). As usually, such 2-D a sequence s is described by means of the formal power series $s(z_1, z_2)$ in horizontal and vertical delay operators z_1 and z_2, respectively (Šebek, 1983; Šebek, Fornasini and Bisiacco, 1988). A 2-D sequence s is called *causal* if it is zero outside the first quadrant of the plane $\mathcal{Z} \times \mathcal{Z}$ and *bicausal* if, moreover, $s(0,0) \neq 0$. Further a causal sequence $s(z_1, z_2) = \sum_{i,j \geq 0} s_{ij} z_1^i z_2^j$ is called *absolutely summable* if $\sum |s_{ij}| < \infty$ and *minimum phase* if, moreover, its inverse s^{-1} is absolutely summable as well. A *conjugate* sequence to s is denoted by s^* and defined by $s^*(z_1, z_2) = s(z_1^{-1}, z_2^{-1})$.

A finite-extent causal 2-D sequence p is called (2-D) *polynomial* (in the indeterminates z_1 and z_2). In accordance with 1-D theory, the polynomial

*The first author was on leave from the Institute of Information Theory and Automation, Prague, CZ (msebek@utia.cas.cz). His stay in Switzerland was funded by the Swiss National Science Foundation Grant 21-37465.

$p(z_1, z_2)$ is called *causal* if it is a bicausal 2-D sequence, that is, if $p(0,0) \neq 0$. It is *stable* if $p(z_1, z_2) \neq 0$ on the *closed unit bidisc*

$$\{(z_1, z_2) \in C \times C \mid |z_1| \leq 1, |z_2| \leq 1\}. \quad (1)$$

A 2-D sequence is called *recurrent* if it is described by a 2-D polynomial fraction $r(z_1, z_2) = n(z_1, z_2)/d(z_1, z_2)$. Throughout the paper, all polynomial fraction will possess causal denominator and, hence, they will be well defined.

Finally, the ∞-norm of the recurrent sequence r with stable denominator d is defined via

$$\begin{aligned} ||r(z_1, z_2)||_\infty &= \sup_{|z_1|=1, |z_2|=1} |r(z_1, z_2)| \\ &= \sup_{\omega_1, \omega_2} |r(e^{j\omega_1}, e^{j\omega_2})|. \end{aligned} \quad (2)$$

In particular, for a 1-D sequence r this reads

$$||r(z_1)||_\infty = \sup_{|z_1|=1} |r(z_1)| = \sup_{\omega_1} |r(e^{j\omega_1})|. \quad (3)$$

2. STANDARD SOLUTION: THE 2-INPUT 2-OUTPUT CASE

As a starting point, let us consider the standard 1-D problem which is defined by the well-known configuration of Fig 1. The block marked "G" is the "plant," that is, the system to be controlled. The signal w represents external, uncontrollable inputs. The signal u is the control input. The output z has the meaning of control error, which ideally should be zero. The signal y, finally, is the measured output, available for feedback via the controller, which is the block marked "K."

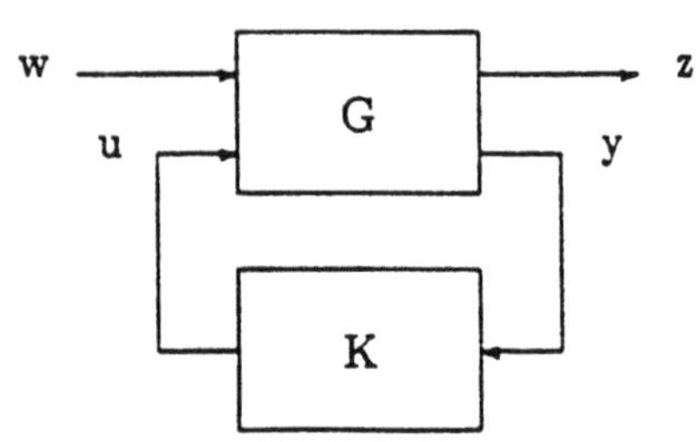

Figure 1: The standard $\mathcal{H}_\infty$ problem

Our show introduces the results of Kwakernaak (1991) but in the simplest form with all four signals u, w and y, z scalar. Such a simplification makes practical application limited, of course. On the other hand, the differences between 1-D and 2-D solutions become more transparent.

Hence, the plant is governed by the equation

$$\begin{bmatrix} z \\ y \end{bmatrix} = \underbrace{\begin{bmatrix} g_{11} & g_{12} \\ g_{21} & g_{22} \end{bmatrix}}_{G} \begin{bmatrix} w \\ u \end{bmatrix} \quad (4)$$

with 2×2 rational transfer matrix G. Hence, all the g_{ij}'s are scalars and, in addition, it is assumed that $g_{12} \neq 0$ and $g_{21} \neq 0$. As it is typical for polynomial approach, the rational matrix G is expressed as a coprime fraction of polynomial matrices. The fraction

$$\begin{bmatrix} z \\ y \end{bmatrix} = \begin{bmatrix} a_z & 0 \\ 0 & a_y \end{bmatrix}^{-1} \begin{bmatrix} d_z & c_z \\ e_y & b_y \end{bmatrix} \begin{bmatrix} w \\ u \end{bmatrix} \quad (5)$$

arises quite naturally. Here all the entries are scalar polynomials and, to make the matrix fraction left coprime, the both triples a_z, c_z, d_z and a_y, b_y, e_y are taken coprime.

Similarly, the controller is described by the equation

$$u = ky = \frac{m}{n} y. \quad (6)$$

Its transfer function k is a fraction of coprime polynomials n, m with $n \neq 0$.

In the configuration of Fig 1, the closed-loop transfer function h from the external input w to the control error z is easily found to be given by

$$\begin{aligned} h &= \frac{g_{11} + (g_{12}g_{21} - g_{11}g_{22})}{1 - kg_{22}} \\ &= \frac{n(d_z a_m + e_m c_z - d_z b_y)}{a_z(a_y n - b_y m)} \end{aligned} \quad (7)$$

and the standard $\mathcal{H}_\infty$-optimal regulation problem is defined (Kwakernaak, 1991) as follows:

Definition - Standard $\mathcal{H}_\infty$-optimal problem. Determine the controller k such that

- the closed-loop system is stable,
- $||h||_\infty$ is minimal. $(\star)$

Internal stability of the structure from Fig 1 hinges, of course, on the stability of the whole characteristic polynomial

$$\det \begin{bmatrix} a_z & c_z m \\ 0 & a_y n - b_y m \end{bmatrix} = a_z(a_y n - b_y m) \quad (8)$$

For practical reasons, however, the controller is called stabilizing in Kwakernaak (1991) whenever just the polynomial $\bar{a}_y n - \bar{b}_y m$ is stable where $\bar{a}_y, \bar{b}_y$ are coprime polynomials such that $\bar{b}_y/\bar{a}_y = b_y/a_y$.

It is well known that the solution of $\mathcal{H}_\infty$-optimal problem consists of a chain of successive solutions of $\mathcal{H}_\infty$-sub-optimal problems with λ approaching the optimal value $\lambda_{\min}$. In every step of iteration, the following problem solution is exercised:

Definition - Sub-optimal problem. For a given $\lambda \geq 0$ (and, in fact, $\lambda \geq \lambda_{\min}$) determine the controller k_λ such that

- the closed-loop system is stable,

- $||h_\lambda||_\infty \leq \lambda$ $(\star\star)$

Besides its repeated appearance in the $\mathcal{H}_\infty$-optimum search, the sub-optimal solution itself is known to guarantee certain level of robustness (Kwakernaak, 1991). This problem has recently been solved both via state space methods (Francis, 1987) and polynomial techniques (Kwakernaak, 1991). We now briefly review the 2 input-2 output version of the polynomial solution of this problem developed by Kwakernaak (1991).

The underlaying philosophy is to transform the original problem to the solution of the rational matrix inequality

$$\begin{bmatrix} n^* & m^* \end{bmatrix} \Pi_\lambda \begin{bmatrix} n \\ m \end{bmatrix} \geq 0 \tag{9}$$

on the imaginary axis. Its solution parameterizes the class of all controllers (6) that warrant $(\star\star)$. In our simple 2×2 case,

$$\begin{aligned} \Pi_\lambda \; = \; & \begin{bmatrix} 0 & a_y^* \\ -c_z^* & -b_y^* \end{bmatrix} \times \\ & \begin{bmatrix} d_z^* d_z - \lambda^2 a_z a_z^* & d_z e_y^* \\ e_y d_z^* & e_y e_y^* \end{bmatrix}^{-1} \times \\ & \begin{bmatrix} 0 & -c_z \\ a_y & -b_y \end{bmatrix}. \end{aligned} \tag{10}$$

After some algebraic manipulations, one gets

$$\Pi_\lambda = \frac{P}{\pi} \tag{11}$$

with the matrix polynomial numerator

$$\begin{aligned} P \; = \; & \begin{bmatrix} 0 & a_y^* \\ -c_z^* & -b_y^* \end{bmatrix} \times \\ & \begin{bmatrix} -\frac{1}{\lambda^2} e_y e_y^* & \frac{1}{\lambda^2} d_z e_y^* \\ \frac{1}{\lambda^2} e_y d_z^* & a_z a_z^* - \frac{1}{\lambda^2} d_z^* d_z \end{bmatrix} \times \\ & \begin{bmatrix} 0 & -c_z \\ a_y & -b_y \end{bmatrix} \end{aligned} \tag{12}$$

and the scalar polynomial denominator

$$\pi = e_y^* a_z^* a_z e_y. \tag{13}$$

The next step is to find a *rational* symmetric factorization of Π_λ which, in our case, is accomplished by a scalar *polynomial* spectral factorization of the denominator

$$\pi = q^* q \tag{14}$$

resulting in a stable polynomial q, and a matrix polynomial factorization of the numerator

$$P = \Gamma_\lambda^* J \Gamma_\lambda. \tag{15}$$

which results in a stable polynomial matrix Γ and a real diagonal matrix of plus and minus ones denoted by J.

Numerical algorithms to perform (14) are standard (Kučera (1980) and references therein). On the other hand, the matrix P is always indefinite on the unit circle (because Π_λ is so) and this fact makes the factorization (15) numerically involved. Anyway, this problem, which is called *J-spectral factorization*, has received much attention recently and workable algorithms have been already derived (Kwakernaak and Šebek, 1995).

Having computed (14) and (15), we receive the rational factorization of Π_λ

$$\Pi_\lambda = \left(\frac{\Gamma_\lambda}{q}\right)^* J \left(\frac{\Gamma_\lambda}{q}\right) \tag{16}$$

Then, according to Kwakernaak (Kwakernaak, 1991), the class of all controllers (6) that guarantee $(\star\star)$ is given by

$$k = \frac{\tilde{m}}{\tilde{n}}, \tag{17}$$

where rational functions $\tilde{n}, \tilde{m}$ result from

$$\begin{bmatrix} \tilde{n} \\ \tilde{m} \end{bmatrix} = q\,\Gamma_\lambda^{-1} \begin{bmatrix} 1 \\ r \end{bmatrix} \tag{18}$$

with an arbitrary rational function r such that

$$||r||_\infty \leq 1. \tag{19}$$

To result in a stabilizing controller, a stable r should be chosen in (18). Denoting further

$$\operatorname{adj} \Gamma_\lambda = \begin{bmatrix} \gamma_{11} & \gamma_{12} \\ \gamma_{21} & \gamma_{22} \end{bmatrix}, \tag{20}$$

and

$$\det \Gamma_\lambda = \gamma, \tag{21}$$

then (18) reads

$$\begin{bmatrix} \tilde{n} \\ \tilde{m} \end{bmatrix} = \frac{q}{\gamma} \begin{bmatrix} \gamma_{11} + \gamma_{12} r \\ \gamma_{21} + \gamma_{22} r \end{bmatrix} \tag{22}$$

and, finally, the desired solution is parameterized by

$$\frac{m}{n} = \frac{\gamma_{21} + \gamma_{22} r}{\gamma_{11} + \gamma_{12} r} \tag{23}$$

with all the γ_{ij}'s polynomials from (20) and r stable rational function satisfying (19). Typically, the *central controller* is chosen by taking $r = 0$, for which

$$\frac{m}{n} = \frac{\gamma_{21}}{\gamma_{11}}. \tag{24}$$

Although Kwakernaak derived his polynomial solution for continuous-time systems (Kwakernaak, 1991), it applies, *mutatis mutandis* for discrete-time systems as well.

3. 2-D SCALAR CASE

Let us consider the standard structure again but now all the signals u, w and z, y are scalar 2-D causal sequences and all the g_{ij}'s in (4) are fractions of 2-D polynomials with causal denominators. Consequently, a_y, a_z, d_z, c_z, e_y and b_y are 2-D polynomials with $a_y(0,0) \neq 0$ and $a_z(0,0) \neq 0$. The both triples a_z, c_z, d_z and a_y, b_y, e_y must be factor coprime.

The controller (6), in general, is expected with m, n ranging 2-D causal sequences with n bicausal. It is stabilizing provided that

$$(\bar{a}_y n - \bar{b}_y m) \qquad (25)$$

is a minimum phase 2-D sequence (or, in particular, a stable 2-D polynomial).

It can be shown that the solution of the 2-D problem follows the lines of the 1-D solution: At first, the scalar spectral factorization (14) should be performed to find a bicausal minimum phase 2-D sequence $q(z_1, z_2)$ such that

$$\pi = q^*(z_1, z_2) q(z_1, z_2). \qquad (26)$$

As the second step, 2×2 matrix 2-D J-spectral factorization must be computed to get 2×2 2-D sequential matrix $\Gamma_\lambda(z_1, z_2)$ such that det Γ_λ is minimum phase and

$$P(z_1, z_2) = \Gamma_\lambda^*(z_1, z_2)\, J\, \Gamma_\lambda(z_1, z_2). \qquad (27)$$

Example: Consider the standard structure with

$$G(z_1, z_2) = \begin{bmatrix} 1 & \dfrac{1}{2+z_1} \\ \dfrac{2+z_1}{2+z_2} & \dfrac{z_1 z_2}{2+z_2} \end{bmatrix}$$

and calculate the class of robust sub-optimal controllers that guarantee

$$||h||_\infty \leq \lambda = 1.$$

The given data give rise to the polynomials

$$\begin{array}{ll} a_y = 2 + z_2 & a_z = 2 + z_1 \\ d_z = 2 + z_1 & c_z = 1 \\ e_y = 2 + z_1 & b_y = z_1 z_2 \end{array}$$

and, hence, $\lambda = 1$ yields (12)

$$\begin{aligned} P(z_1, z_2) &= (2 + z_1^{-1})(2 + z_1) \times \\ &\begin{bmatrix} 0 & 2 + z_2^{-1} \\ -1 & -z_1^{-1} z_2^{-1} \end{bmatrix} \times \\ &\begin{bmatrix} -1 & 1 \\ 1 & 0 \end{bmatrix} \begin{bmatrix} 0 & -1 \\ 2 + z_2 & -z_1 z_2 \end{bmatrix} \end{aligned}$$

and (13)

$$\pi(z_1, z_2) = (2 + z_2^{-1})^2 (2 + z_2)^2.$$

In this particular example, the both spectral factorizations happen to be polynomials. A simple calculation results in stable polynomials

$$q(z_1, z_2) = (2 + z_2)^2$$

and

$$\begin{aligned} \Gamma(z_1, z_2) &= (2 + z_2) \begin{bmatrix} \frac{3}{4} & \frac{1}{2} \\ -\frac{5}{4} & \frac{1}{2} \end{bmatrix} \times \\ &\begin{bmatrix} 0 & -1 \\ 2 + z_2 & -z_1 z_2 \end{bmatrix} \\ &= \frac{2 + z_2}{2} \begin{bmatrix} 2 + z_2 & -\frac{3}{2} - z_1 z_2 \\ 2 + z_2 & \frac{5}{2} - z_1 z_2 \end{bmatrix} \end{aligned}$$

with

$$J = \begin{bmatrix} 1 & 0 \\ 0 & -1 \end{bmatrix}.$$

Since

$$\text{adj } \Gamma = \begin{bmatrix} \frac{5}{2} - z_1 z_2 & \frac{3}{2} + z_1 z_2 \\ -2 - z_2 & 2 + z_2 \end{bmatrix},$$

we obtain

$$\frac{m}{n} = \frac{-2 - z_2 + (2 + z_2) r}{\frac{5}{2} - z_1 z_2 + (\frac{3}{2} + z_1 z_2) r}$$

and, for example, the central controller reads

$$\frac{m}{n} = \frac{-2 - z_2}{\frac{5}{2} - z_1 z_2}.$$

This controller guarantees

$$||h||_\infty \leq 1$$

and gives rise to the closed-loop characteristic polynomial (8)

$$\frac{5}{2}(2 + z_1)(2 + z_2)$$

which is stable, indeed.

4. SPECTRAL FACTORIZATION

The main difference between 1-D and 2-D solutions emerges in the operation of spectral factorization. So the scalar factorization (14)

$$\pi(z_1, z_2) = q^*(z_1, z_2) q(z_1, z_2) \qquad (28)$$

exists whenever $a_z(z_1, z_2)$ and $e_y(z_1, z_2)$ (and, hence, $\pi(z_1, z_2)$) are nonzero on the *unit torus*

$$\{(z_1, z_2) \in C \times C \mid |z_1| = 1, |z_2| = 1\} \qquad (29)$$

but, in general, it results in a factor $q(z_1, z_2)$ with an infinite number of terms. Hence, $q(z_1, z_2)$ is an infinite 2-D sequence rather than a 2-D polynomial (Dudgeon and Mersereau, 1984). As an example, for $\pi = z_1^{-1} + z_2^{-1} + 5 + z_1 + z_2$ it is not possible to find a 2-D polynomial $q(z_1, z_2)$ to satisfy (14). An algorithm to compute $q(z_1, z_2)$ has been published (Dudgeon and Mersereau, 1984) but in practice, it always ends in a truncated (i.e., polynomial) version $\bar{q}(z_1, z_2)$ of the sequence $q(z_1, z_2)$. Of course, the truncation must retain the stability of $\bar{q}(z_1, z_2)$ which is another problem.

The second J-spectral factorization (15), i.e.,

$$P = \Gamma_\lambda^* J \Gamma_\lambda, \tag{30}$$

combines both indefiniteness and two-dimensionality and remains unsolved thus far. It will be a subject of further research.

5. CONCLUSIONS

The problem of $\mathcal{H}_\infty$ robust control has been discussed for 2-D systems with scalar standard structure. It was shown that the 1-D polynomial design procedure (Kwakernaak, 1991) fits but its basic steps (spectral and J-spectral factorizations) are to be replaced by their 2-D versions. While the problem of scalar 2-D spectral factorization is now partly solved, the problem of 2-D matrix J-spectral factorization is a subject of further research.

6. REFERENCES

Dudgeon, D. E. and Mersereau, R. M. (1984). *Multidimensional Digital Signal Processing*. Prentice-Hall Signal Processing Series. Prentice Hall. Englewood Cliffs.

Francis, B. A. (1987). *A Course in $\mathcal{H}_\infty$ Control Theory*. Lecture Notes in Control and Information Sciences. Springer Verlag. Berlin.

Kučera, V. (1980). *Discrete Linear Control: The Polynomial Approach*. Wiley. Chichester.

Kwakernaak, H. (1991). The polynomial approach to $\mathcal{H}_\infty$-optimal regulation. In: *$\mathcal{H}_\infty$-Control Theory* (E. Mosca and L. Pandolfi, Eds.). Lecture Notes in Mathematics. Springer-Verlag. Berlin.

Šebek, M. (1983). 2-D exact model matching. *IEEE Transactions on Automatic Control* **AC-28**(2), 215–217.

Šebek, M. (1992). Polynomial solution of 2-D Kalman-Bucy filtering problem. *IEEE Transactions on Automatic Control* **AC-37**(10), 1530–1533.

Šebek, M., Fornasini, E. and Bisiacco, M. (1988). Controllability and reconstructibility conditions for 2-D systems. *IEEE Transactions on Automatic Control* **AC–33**(5), 496–499.

Šebek, M. and Kraus, F. J. (1995). Stochastic LQ-optimal control for 2-D systems. *Multidimensional Systems and Signal Processing*. To appear.

Kwakernaak H. and Šebek, M. (1994). Polynomial J-Spectral Factorization. *IEEE Transactions on Automatic Control* **AC–39**(2), 315–328.

IDENTIFIABILITY OF 2D SCALAR BEHAVIORS

SANDRO ZAMPIERI*

Dipartimento di Elettronica ed Informatica, Università di Padova, via Gradenigo 6/a, 35131 Padova, Italy

Abstract. A modelling procedure is an algorithm that analyzing a set of data provides a mathematical description of a phenomenon, that is called model. In some cases a dynamical description is more reasonable and in these cases dynamical systems are the models that are considered. In this paper the particular case when the data have a two-dimensional support and are noise free is considered. The model class is constituted by systems of two-dimensional difference equations. In this setup two indices describing the identifiability of the model in the model class are introduced. They provide an indication on the amount of data that is needed to identify a model.

Résumé. On propose deux procédés pour construire un modèle 2D quand les donnés sont definis sur un reseau bidimensionnel. Puis on étudie les proprietés d'identifiabilité des systèmes 2D en utilisant les techniques introduites dans la première partie.

Key Words. Behavioral approach; modelling; system identification; identifiability indices; 2D systems; 2D Padè approximation; Gröbner bases

1. INTRODUCTION

The problem of modelling real phenomena is a central issue in dynamical systems theory. A modelling procedure can be considered as a technique that allows to construct a mathematical description, called model, of a phenomenon starting from some data that come from measurements. When the phenomenon evolves in time, the model class that is the most suitable to its description is given by the class of 1D dynamical systems and most of the common interest situations are covered by the class of 1D linear time-invariant systems. When the phenomenon lives on multidimensional domains, such as space-time domains, the class 1D dynamical systems can be sometimes to restrictive. Therefore the development of modelling procedures for the description of phenomena living on multidimensional domains is an essential step for the application of all the methodologies developed for multidimensional systems.

Every modelling procedure can be described by specifying:

1. The structure of the data: The data set is characterized by the time set on which the data are supported (unidimensional data or multidimensional data), by the subset of the time set in which the data are effectively given (they can be known on the whole time set or only on a part of it), and finally by a description of the noise which corrupts the data.
2. The class of admissible models: This choice is very delicate since a too rich class may cause inefficiency and problems of identifiability. On the other hand, if the class is too small, the procedure may provide meaningless models.
3. The algorithm that associates the most suitable model to the available data: In this case the accuracy in the identification and the computational complexity are conflicting requirements and the right equilibrium between them depends on the particular application.

This paper deals with the modelling problem in the case when the data are 2D scalar sequences, i.e. data characterized by the time set $\mathbb{N}^2$ that may be known completely or only partially. More precisely the data will be in general a finite family $w_1, \ldots, w_l$ of sequences in $\mathbb{R}^I$ (the set of functions from I to $\mathbb{R}$), where $I = \mathbb{N}^2$ or I is a finite subset of $\mathbb{N}^2$ with particular structure. The first case corresponds to the idealistic case when the data are known on the whole domain, while the second corresponds to the more realistic case when the known data are finite. In both cases the data will be supposed noise free.

The class of model considered in this paper is given by the set of all linear shift-invariant systems as defined in the behavioral approach. In general in this framework a dynamical system is described by a triple $\Sigma = (T, W, \mathcal{B})$, where T is the time set, W is the signal set and $\mathcal{B}$ is a subset of

W^T, called behavior, which describes which trajectories of W^T can occur. A 2D discrete linear shift-invariant and complete system corresponds to a dynamical system with $\mathbb{N}^2$ as the time set, $\mathbb{R}^q$ as the signal alphabet and a behavior $\mathcal{B}$ which is a linear subspace of W^T, invariant w.r. to both the operators σ_1, σ_2, that are the backward shift operators in the two directions of the grid, and with the property that a trajectory $w \in \mathcal{B}$ iff $w \in \mathcal{B}_{|I}$ for all finite subset I of $\mathbb{N}^2$ (see (Rocha, 1990)). It can be seen (Rocha, 1990) that the behavior of systems of this kind coincides with the set of solutions of a linear difference equation and for this reason they are called 2D autoregressive (AR) systems. More precisely, given a 2D AR system, there exists a difference operator $R \in \mathbb{R}[\sigma_1, \sigma_2]^{l \times q}$

$$R = \sum R_{ij}\sigma_1^i\sigma_2^j, \quad R_{ij} \in \mathbb{R}^{l \times q},$$

where the sum is finite, such that $\mathcal{B} = \ker R$, i.e.,

$$\mathcal{B} = \{w \in (\mathbb{R}^q)^{\mathbb{N}^2} \; : \; \sum R_{ij}w(h+i, k+j) = 0, \\ \forall (h,k) \in \mathbb{N}^2\}.$$

In this way the behavior $\mathcal{B}$ is specified by the 2D difference equation associated to the polynomial matrix R. It can be proved (Rocha, 1990) that two 2D AR systems $\Sigma_1 = (\mathbb{N}^2, \mathbb{R}^q, \mathcal{B}_1)$ and $\Sigma_2 = (\mathbb{N}^2, \mathbb{R}^q, \mathcal{B}_2)$, with $\mathcal{B}_1 = \ker R_1$ and $\mathcal{B}_2 = \ker R_2$, coincide if and only if there exist polynomial matrices X_1, X_2 of suitable dimensions such that $R_1 = X_2R_2$ and $R_2 = X_1R_1$.

In this paper only scalar 2D AR systems will be considered. For the representation of this systems column matrices are needed and so the behaviors will take the following form

$$\mathcal{B} = \ker(p_1, \ldots, p_n)^T,$$

where $p_1 \ldots, p_n \in \mathbb{R}[\sigma_1, \sigma_2]$. In this particular case two scalar 2D AR systems $\Sigma_1 = (\mathbb{N}^2, \mathbb{R}, \mathcal{B}_1)$ and $\Sigma_2 = (\mathbb{N}^2, \mathbb{R}, \mathcal{B}_2)$, with $\mathcal{B}_1 = \ker(p_1, \ldots, p_n)^T$ and $\mathcal{B}_2 = \ker(q_1, \ldots, q_m)^T$, coincide if and only if the ideal generated by the polynomials $p_1, \ldots, p_n$ and the ideal generated by the polynomials $q_1, \ldots, q_m$ coincide.

Example Consider the scalar 2D AR system $\Sigma = (\mathbb{N}^2, \mathbb{R}, \mathcal{B})$, where

$$\mathcal{B} = \ker \begin{bmatrix} \sigma_1 - 1 \\ \sigma_2 - 1 \end{bmatrix}.$$

It is easy to see that in this case the behavior $\mathcal{B}$ is constituted by the trajectories $w \in \mathbb{R}^{\mathbb{N}^2}$ that are constant on all the domain $\mathbb{N}^2$.

Finally the modelling algorithms that are proposed in this paper are the natural extensions to the 2D case of analogous algorithms proposed in the 1D case in (Willems, 1986; Heij, 1992). They are based on the concept of unfalsified model. In our specific case, given a set $I \subseteq \mathbb{N}^2$ and a family of 2D scalar arrays $v_1, \ldots, v_l \in \mathbb{R}^I$, a scalar 2D AR system $\Sigma = (\mathbb{N}^2, \mathbb{R}, \mathcal{B})$ is an unfalsified model of $v_1, \ldots, v_l$ if there exists $w_1, \ldots, w_l \in \mathcal{B}$ such that $w_{i|I} = v_i$. The unfalsified models are the systems that agree with the observations and so that can be accepted to be models of the data.

In general a modelling procedure must choose an unfalsified model starting from the data $v_1, \ldots, v_l$. Therefore, since there are infinitely many unfalsified models, a recipe allowing to compare different unfalsified models is necessary. Between two unfalsified models $\Sigma_1 = (\mathbb{N}^2, \mathbb{R}, \mathcal{B}_1)$ and $\Sigma_2 = (\mathbb{N}^2, \mathbb{R}, \mathcal{B}_2)$ such that $\mathcal{B}_1 \subseteq \mathcal{B}_2$, the model Σ_1 can be considered better than Σ_2, since it is more falsifiable and so more powerful in describing the system generating $v_1, \ldots, v_l$. Moreover it has more predictive power since it allows a smaller number of trajectories. The concept of undominated unfalsified model of $v_1, \ldots, v_l$ can be introduced following this idea. Actually, a system $\Sigma = (\mathbb{N}^2, \mathbb{R}, \mathcal{B})$ is an undominated unfalsified model of $v_1, \ldots, v_l$, if it is one of the minimal elements in the set of all the unfalsified models w.r. to the partial ordering $\subseteq$ or, in other words, if for every unfalsified model $\Sigma' = (\mathbb{N}^2, \mathbb{R}, \mathcal{B}')$ such that $\mathcal{B}' \subseteq \mathcal{B}$, then $\mathcal{B}' = \mathcal{B}$. Finally $\Sigma = (\mathbb{N}^2, \mathbb{R}, \mathcal{B})$ is said to be the most powerful unfalsified model, if it is the smallest unfalsified model, i.e. if every unfalsified model $\Sigma' = (\mathbb{N}^2, \mathbb{R}, \mathcal{B}')$ is such that $\mathcal{B} \subseteq \mathcal{B}'$.

In general if there exists the most powerful unfalsified model, then this can be considered the best candidate to be the model of the data. This is the case when $I = \mathbb{N}^2$, and so in this situation the most reasonable strategy is to choose always the most powerful unfalsified model to be the model of the data. The procedure based on this strategy will be called the exact modelling procedure and the model it provides, that is the most powerful unfalsified model, will be denoted by the symbol $\Sigma^*(v_1, \ldots, v_l)$.

When I is a finite subset of $\mathbb{N}^2$, the situation is much more complicate, since the most powerful unfalsified model does not exist. The algorithm proposed in this paper is based of a 2D Padé approximation algorithm developed by Sakata in (Sakata, 1988; Sakata, 1990). This algorithm allows to obtain all the minimal order recurring relations that exist in the data when the support I has the following particular structure:

$$I = \{s = (s_1, s_2) \in \mathbb{N}^2 : s_1 + s_2 \leq k\}, \qquad (1)$$

This subset of $\mathbb{N}^2$ depends on k and will be denoted as I_k. A first attempt to construct a modelling procedure in this setup can be found in (Zampieri, 1994; Zampieri, 1992), where an algorithm with nice properties have been proposed. The only serious drawback of the method proposed there is the fact that a large class of scalar 2D AR systems are not identifiable w.r. to this procedure. The modelling procedure proposed in this paper overcome this problem by rejecting the recurring relations whose order are too big with respect to the dimension of the support I of the data. The result of this strategy is that in general this procedure does not provide an undominated unfalsified model. However the advantage is that by this procedure every scalar 2D AR system is identifiable. The following 1D example will clarify the motivations of this strategy.

Example Suppose that a 1D finite array $v \in \mathbb{R}^I$, where $I = [0, 2n+1] \subset \mathbb{N}$ is given and fix the class of the linear shift-invariant complete systems over the time set $\mathbb{N}$ to be the family of admissible models. For every model $\Sigma = (\mathbb{N}, \mathbb{R}, \mathcal{B})$, such that $\mathcal{B} \neq \mathbb{R}^{\mathbb{N}^2}$, there exists a 1D polynomial $p \in \mathbb{R}[\sigma]$ such that $\mathcal{B} = \ker p$ and the degree of p coincides with the dimension of the finitely generated vector space $\mathcal{B}$. If the model for v is chosen in the class of the unfalsified models with smallest dimension, then such a model is undominated and provides an optimal representation of the data. However this representation is effectively efficient if and only if the dimension is less than or equal to n, since n parameters are needed for fixing the polynomial p in the representation $\ker p$ of the model, and moreover $v_{|[0,n-1]}$ must be provided in order to be able to reconstruct v from the model. Consequently only models $\Sigma = (\mathbb{N}, \mathbb{R}, \mathcal{B})$, with $\dim \mathcal{B} \leq n$ are really meaningful, while the others must be excluded in advance. Their exclusion is motivated by the fact that they provide no data compression.

Since this procedure is developed for noise free and partially known data, then it will be called partial exact modelling procedure and the model it provides will be denoted by the symbol $\Sigma_I^*(v_1, \ldots, v_l)$. The detailed description of the partial exact modelling procedure will be given in the next section.

2. PARTIAL EXACT MODELLING PROCEDURE

In order to describe and motivate the partial exact modelling procedure some preliminary definitions and results are needed. The first considered is the concept of annihilating polynomial. Given any subset I of $\mathbb{N}^2$ and denote by $\mathbb{R}[z_1, z_2]_I$ the subset of all polynomials in $\mathbb{R}[z_1, z_2]$ with support in contained in I. Consider the following bilinear map

$$\langle \cdot, \cdot \rangle_I \; : \; \mathbb{R}[z_1, z_2]_I \times \mathbb{R}^I \longrightarrow \mathbb{R}$$

defined for every $v \in \mathbb{R}^I$ and every polynomial

$$p = \sum_{(i,j) \in I} p_{ij} z_1^i z_2^j$$

as follows

$$\langle p, v \rangle_I := \sum_{(i,j) \in I} p_{ij} v(i,j).$$

Define moreover for any $v \in \mathbb{R}^I$, the set of polynomials

$$\mathcal{A}_I(v) := \{p \in \mathbb{R}[z_1, z_2] : \forall (i_1, i_2) \in \mathbb{N}^2, \; z_1^{i_1} z_2^{i_2} p \in \mathbb{R}[z_1, z_2]_I \;\Rightarrow\; \langle z_1^{i_1} z_2^{i_2} p, v \rangle_I = 0\}.$$

Given $v_1, \ldots, v_l \in \mathbb{R}^I$, define

$$\mathcal{A}_I(v_1, \ldots, v_l) := \bigcap_{i=1}^{l} \mathcal{A}_I(v_i). \qquad (2)$$

The polynomials in $\mathcal{A}_I(v_1, \ldots, v_l)$ are called annihilators of the sequences $v_1, \ldots, v_l$. Note that if $I = \mathbb{N}^2$ the set of annihilators $\mathcal{A}_I(v_1, \ldots, v_l)$ is an ideal. There exists a strong relation between unfalsified models and the set of annihilators as shown in the following proposition, whose proof is easy consequence of analogous results that appear in (Zampieri, 1994).

Proposition 1 Let I be any subset of $\mathbb{N}^2$, $p_1, \ldots, p_n$ polynomials in $\mathbb{R}[z_1, z_2]$ and $v_1, \ldots, v_l$ 2D arrays in $\mathbb{R}^I$. Then

1. The system $\Sigma = (\mathbb{N}^2, \mathbb{R}, \mathcal{B})$, where $\mathcal{B} = \ker(p_1, \ldots, p_n)^T$ is an unfalsified model of $v_1, \ldots, v_l$ if and only if

$$\mathcal{I} \subseteq \mathcal{A}_I(v_1, \ldots, v_l),$$

   where $\mathcal{I}$ is the ideal generated by $p_1, \ldots, p_n$.

2. The system $\Sigma = (\mathbb{N}^2, \mathbb{R}, \mathcal{B})$, where $\mathcal{B} = \ker(p_1, \ldots, p_n)^T$ is an undominated unfalsified model of $v_1, \ldots, v_l$ if and only if $\mathcal{I}$ is a maximal ideal in the class of the ideals contained in $\mathcal{A}_I(v_1, \ldots, v_l)$.

■

This proposition shows how the most powerful unfalsified model of a family of scalar 2D sequences $v_1, \ldots, v_l \in \mathbb{R}^{\mathbb{N}^2}$ can be determined at least in principle. Actually, in this case $\mathcal{A}_I(v_1, \ldots, v_l)$ is an ideal. Therefore it is easy to argue that, if $\{p_1, \ldots, p_n\}$ is a set of its generators, then

$$\mathcal{B}^* = \ker(p_1, \ldots, p_n)^T$$

provide the behavior of the most powerful unfalsified model.

On the other hand the previous proposition cannot be directly used for verifying if a model is unfalsified by a family of sequences $v_1, \ldots, v_l \in \mathbb{R}^I$, when $I \neq \mathbb{N}^2$. In order to obtain an effective verification method the freedom in the choice of the set I must be restricted. Let $<_T$ be any admissible ordering in $\mathbb{N}^2$, that is a total ordering compatible with the monoid structure of $\mathbb{N}^2$ and such that $(0,0)$ is the smallest element in $\mathbb{N}^2$ w.r. to this ordering (Buchberger, 1985). If there exists $q \in \mathbb{N}^2$ such that the subset I can be described as

$$I = \{s \in \mathbb{N}^2 : s <_T q\}, \tag{3}$$

then the previous proposition can be simplified and, when I is a finite set, it allows to verify effectively if a model is unfalsified, through the computation of a suitable Gröbner basis.

Proposition 2 Let $\{g_1, \ldots, g_m\}$ be a Gröbner basis w.r. to the admissible ordering $<_T$. Let moreover $I = \{s \in \mathbb{N}^2 : s <_T q\}$ for some $q \in \mathbb{N}^2$ and $v_1, \ldots, v_l \in \mathbb{R}^I$. Then the system $\Sigma = (\mathbb{N}^2, \mathbb{R}, \mathcal{B})$, where $\mathcal{B} = \ker(g_1, \ldots, g_m)^T$ is an unfalsified model of $v_1, \ldots, v_l$ if and only if

$$\{g_1, \ldots, g_m\} \subseteq \mathcal{A}_I(v_1, \ldots, v_l).$$

■

The previous proposition suggests a constructive procedure verifying if a model $\Sigma = (\mathbb{N}^2, \mathbb{R}, \mathcal{B})$, $\mathcal{B} = \ker(p_1, \ldots, p_n)^T$ is unfalsified by $v_1, \ldots, v_l \in \mathbb{R}^I$ in this case. Actually the first step is to compute a Gröbner basis $\{g_1, \ldots, g_m\}$ of the ideal generated by $p_1, \ldots, p_n$. Since, as mentioned above, $\ker(p_1, \ldots, p_n)^T = \ker(g_1, \ldots, g_m)^T$, the second step is to verify that all g_i are annihilators of $v_1, \ldots, v_l$. This last condition can be tested in a finite number of steps if I is a finite set.

In this paper a procedure providing a model of a family of arrays $v_1, \ldots, v_l \in \mathbb{R}^I$ is presented when $I = I_k$, for some $k \in \mathbb{N}$, where I_k is a finite subset of $\mathbb{N}^2$ defined in (1). This kind of subsets of $\mathbb{N}^2$ can be easily seen as particular cases of subsets like (3), when the selected admissible ordering is the total degree ordering (Buchberger, 1985; Sakata, 1988). This can be defined in the following way:
Let $r = (r_1, r_2), s = (s_1, s_2) \in \mathbb{N}^2$.

$$r <_D s \overset{\text{def}}{\Leftrightarrow} \begin{array}{l} r_1 + r_2 < s_1 + s_2 \text{ or} \\ r_1 + r_2 = s_1 + s_2 \text{ and } r_1 < s_1, \end{array}$$

Therefore, the characterization of the unfalsified models described in proposition 2 holds true when $I = I_k$.

The partial exact modelling procedure is based on proposition 2 and on the concept of minimal polynomial set of a family of 2D arrays that has been introduced by Sakata in (Sakata, 1988). This can be defined as follows.

Consider a subset $I = \{s \in \mathbb{N}^2 : s <_D q\}$ and the 2D arrays $v_1, \ldots, v_l \in \mathbb{R}^I$. Consider the set of annihilators $\mathcal{A}_I(v_1, \ldots, v_l)$ and let deg $\mathcal{A}_I(v_1, \ldots, v_l)$ be the set of the degrees of all the polynomials in $\mathcal{A}_I(v_1, \ldots, v_l)$, where the degree of a polynomial is defined as the greatest element in the support of the polynomial w.r to the ordering $<_D$. Consider moreover the set $Min(\text{deg } \mathcal{A}_I(v_1, \ldots, v_l))$ constituted by the minimal elements of deg $\mathcal{A}_I(v_1, \ldots, v_l)$ w.r. to the usual partial ordering in $\mathbb{N}^2$. This coincides with the smallest subset S of $\mathbb{N}^2$ such that

$$S + \mathbb{N}^2 = \text{deg } \mathcal{A}_I(v_1, \ldots, v_l).$$

A subset $\mathcal{F}$ of $\mathcal{A}_I(v_1, \ldots, v_l)$ is said to be a minimal polynomial set of $v_1, \ldots, v_l$ if

$$\text{deg } \mathcal{F} \supseteq Min(\text{deg } \mathcal{A}_I(v_1, \ldots, v_l)).$$

A minimal polynomial set is called reduced minimal polynomial set, if every $f \in \mathcal{F}$ is in normal form modulo $\mathcal{F} \setminus \{f\}$ (see (Buchberger, 1985)), i.e. if every $f \in \mathcal{F}$ is monic and satisfy the following condition

$$(\text{supp } f) \cap \bigcup_{f' \in \mathcal{F} \setminus \{f\}} (\text{deg } f' + \mathbb{N}^2) = \emptyset.$$

It is easy to see that a reduced minimal polynomial set of a family of arrays $v_1, \ldots, v_l \in \mathbb{R}^I$ always exists, but that it is not unique in general. The following proposition provides a condition ensuring the uniqueness of the reduced minimal polynomial set.

Proposition 3 Let $v_1, \ldots, v_l \in R^I$, where $I = \{s \in \mathbb{N}^2 : s <_D q\}$, and let

$$\Delta := \bigcup_{i=1}^{l} (\mathbb{N}^2 \setminus \text{deg} \mathcal{A}_I(v_i)).$$

Suppose that $\mathcal{F}$ and $\mathcal{F}'$ are reduced minimal polynomial sets of $v_1, \ldots, v_l$. If $f \in \mathcal{F}$, $f' \in \mathcal{F}'$ are such that $\text{deg} f = \text{deg} f'$ and $\text{deg} f + \Delta \subseteq I$, then $f = f'$.

In (Sakata, 1990) an efficient algorithm that is able to provide a reduced minimal polynomial set in this context is proposed. Note the strong similarity between the definitions of the minimal polynomial set and reduced minimal polynomial set with the definitions of Gröbner basis and reduced

Gröbner basis. In fact, if $\mathcal{A}_I(v_1, \ldots, v_l)$ is an ideal, then a minimal polynomial set is a Gröbner basis.

Given the arrays $v_1, \ldots, v_l \in \mathbb{R}^I$, where $I = I_k$ for some $k \in \mathbb{N}$, the partial exact modelling procedure proceeds in the following way:

1. Compute a reduced minimal polynomial set $\mathcal{F}$ of $v_1, \ldots, v_l$.
2. Let $\mathcal{F}'$ be the subset of $\mathcal{F}$ defined as follows

$$\mathcal{F}' := \{f \in \mathcal{F} \ : \ 3\deg\ f \in I\},$$

   i.e. $\mathcal{F}'$ is the biggest subset of $\mathcal{F}$ such that $3\deg\ \mathcal{F}' := \{3\deg\ f \ : \ f \in \mathcal{F}\} \subseteq I$.
3. If $\mathcal{F}' = \{f_1, \ldots, f_m\}$ is a Gröbner basis w.r. to the admissible ordering $<_T$, then

$$\Sigma_I^*(v_1, \ldots, v_l) = (\mathbb{N}^2, \mathbb{R}, \mathcal{B}^*),$$

   where $\mathcal{B}^* = \ker(f_1, \ldots, f_m)^T$. If $\mathcal{F}' = \{f_1, \ldots, f_m\}$ is not a Gröbner basis, then the algorithm proposes $\Sigma_I^*(v_1, \ldots, v_l) = (\mathbb{N}^2, \mathbb{R}, \mathbb{R}^{\mathbb{N}^2})$ as the model of the data. In this case the procedure is not able to choose a really meaningful model from the data and therefore it chooses the most conservative one.

Note that in any case this procedure provides an unfalsified model of the data that is in general not undominated.

3. IDENTIFIABILITY OF AR SYSTEMS WITH RESPECT TO THE MODELLING PROCEDURES

In this section the concept of identifiability of a model with respect to a modelling procedure will be specified and the identifiability characteristics of the procedures that have been proposed above will be given. A modelling procedure can be seen as an operator P that maps a family of data $d_1 \ldots, d_l$ to a model M in a certain model set $\mathbb{M}$. In general, given the modelling procedure P, the model M is said to be identifiable w.r. to this procedure if there exists a finite family of data $d_1 \ldots, d_l$ such that $P(d_1 \ldots, d_l) = M$, i.e. M is exactly the model proposed by the procedure from the given data. In other words M is identifiable if it is in the codomain of P.

Considering the particular case treated in this paper it can be shown that all scalar 2D AR systems are identifiable w.r to both the modelling procedures. More specifically, in case when $I = \mathbb{N}^2$ the following property holds. Given a scalar 2D AR system $\Sigma = (\mathbb{N}^2, \mathbb{R}, \mathcal{B})$, there exists a identifiability index $i_\Sigma \in \mathbb{N}$ with the following property:
there exist $w_1, \ldots, w_l \in \mathbb{R}^{\mathbb{N}^2}$ such that $\Sigma^*(w_1, \ldots, w_l) = \Sigma$ if and only if $l \geq i_\Sigma$.

More intuitively if the number of trajectories is large enough, the exact modelling procedure is able to identify all scalar 2D AR systems.

Also the partial exact modelling procedure enjoys an analogous property. More precisely suppose that $I = I_k$ for some $k \in \mathbb{N}$ as in (1). Given a scalar 2D AR system $\Sigma = (\mathbb{N}^2, \mathbb{R}, \mathcal{B})$, there exists another identifiability index $j_\Sigma \in \mathbb{N}$ satisfying the following property:
If $l \geq i_\Sigma$ and $k \geq j_\Sigma$, then there exist $w_1, \ldots, w_l \in \mathbb{R}^I$ such that $\Sigma_I^*(w_1, \ldots, w_l) = \Sigma$.
Therefore also in this case all scalar 2D AR systems are identifiable.

It is clear that these modelling procedures has not much practical value. However these idealistic modelling methods are interesting since they give theoretical bounds on the performances of any realistic modelling procedure. Actually, since in order to identify a system Σ at least i_Σ noise free and completely known trajectories are needed, then i_Σ can be considered a lower bound on the number of trajectories that are needed in the identification of Σ also in the more realistic case in which the trajectories are noisy and only partially known. Also the index j_Σ provides an useful information on the dimension of the support of the data that can be considered sufficiently large to be meaningful for the identification of the system Σ.

Note that the identifiability condition connected with the index i_Σ is necessary and sufficient:
If $l \geq i_\Sigma$, then there exist $v_1, \ldots, v_l \in \mathbb{R}^{\mathbb{N}^2}$ such that $\Sigma^*(v_1, \ldots, v_l) = \Sigma$, while if $l < i_\Sigma$, then such trajectories do not exist. On the contrary, the identifiability condition related with the partial exact modelling procedure is only sufficient. In other words, if $l \geq i_\Sigma$, there could exist $v_1, \ldots, v_l \in \mathbb{R}^I$, $I = I_k$, such that $\Sigma_I^*(v_1, \ldots, v_l) = \Sigma$ even if $k < j_\Sigma$. Actually, it is reasonable to expect that the dimension of the support of the arrays that are needed should be connected not only with the system but also with the number of arrays that are available. More precisely, if more arrays are available, then smaller supports should be needed. In order to obtain a necessary and sufficient condition a more refined modelling procedure should be found. However this seems to be a hard problem even in the 1D case.

4. COMPUTATION OF THE IDENTIFIABILITY INDICES

This last section will be devoted to the computation the identifiability indices i_Σ and j_Σ starting from any AR representation of the 2D AR system Σ. The proof of the results mentioned below will be presented in the paper (Zampieri, 1995)

Given an 2D AR system $\Sigma = (\mathbb{N}^2, \mathbb{R}, \mathcal{B})$, where $\mathcal{B} = \ker(p_1, \dots, p_n)^T$, $p_1 \dots, p_n \in \mathbb{R}[\sigma_1, \sigma_2]$. The algorithm that computes i_Σ consists in the following three steps

1. First compute the greatest common divisor p of $p_1, \dots, p_n$ and consider $\bar{p}_1, \dots, \bar{p}_n$, where $\bar{p}_i := p_i/p$.
2. Consider the behavior $\bar{\mathcal{B}} = \ker(\bar{p}_1, \dots, \bar{p}_n)^T$. As shown in (Fornasini *et al.*, 1993), $\bar{\mathcal{B}}$ is finite dimensional subspace of $\mathbb{R}^{\mathbb{N}^2}$ and moreover, if k is its dimension, then there exist matrices $C \in \mathbb{R}^{1\times k}, M_1, M_2 \in \mathbb{R}^{k\times k}$ such that
$$\mathcal{B} = \{w \in \mathbb{R}^{\mathbb{N}^2} : w(i,j) = CM_1^i M_2^j x,\ x \in \mathbb{R}^k\}$$
In (Fornasini *et al.*, 1993) it is also proposed an algorithm that provides such matrices through the computation of the Gröbner basis of the ideal generated by $\bar{p}_1, \dots, \bar{p}_n$.
3. Compute the number
$$lrk(M_1, M_2) := k - \min_{(\lambda_1,\lambda_2)\in\mathbb{C}^2} \operatorname{rank}\,[\lambda_1 I - M_1 \mid \lambda_2 I - M_2].$$

It can be shown that
$$i_\Sigma = lrk(M_1, M_2).$$

Also the computation of the identifiability index j_Σ is based on the computation of a Gröbner basis. Given an 2D AR system $\Sigma = (\mathbb{N}^2, \mathbb{R}, \mathcal{B})$, where $\mathcal{B} = \ker(p_1, \dots, p_n)^T$, $p_1 \dots, p_n \in \mathbb{R}[\sigma_1, \sigma_2]$. Compute a Gröbner basis $\{g_1, \dots, g_m\}$ w.r. to the total degree ordering $<_D$ of the ideal generated by $p_1 \dots, p_n$. It can be shown that
$$j_\Sigma = \max\{h_i + k_i : (h_i, k_i) = \deg g_i,\ i = 1, \dots, m\}.$$

Through a more careful analysis of the identifiability properties of the modelling procedure proposed in this paper it could be noticed the following. As mentioned above, given a scalar 2D AR system $\Sigma = (\mathbb{N}^2, \mathbb{R}, \mathcal{B})$, there exists a identifiability index $i_\Sigma \in \mathbb{N}$ with the following property:
if $l \geq i_\Sigma$, then there exist $w_1, \dots, w_l \in \mathcal{B}$ such that $\Sigma^*(w_1, \dots, w_l) = \Sigma$. In fact it can be shown that the families of trajectories $w_1, \dots, w_l \in \mathcal{B}$ such that $\Sigma^*(w_1, \dots, w_l) = \Sigma$ form a generic subset of $\mathcal{B}^l$. Moreover it can be proven not only that if $I = I_k$ for some $k \in \mathbb{N}$ and if $l \geq i_\Sigma$ and $k \geq j_\Sigma$, then there exist $w_1, \dots, w_l \in \mathcal{B}_{|I}$ such that $\Sigma_I^*(w_1, \dots, w_l) = \Sigma$, but also that in those hypothesis the families of trajectories $w_1, \dots, w_l \in \mathcal{B}_{|I}$ such that $\Sigma_I^*(w_1, \dots, w_l) = \Sigma$ form a generic subset of $\mathcal{B}_{|I}^l$.

These are very interesting facts since they show that generically the modelling procedure proposed in this paper provide the right model. Actually, if we suppose that the data have been really generated by an AR model, under the condition that enough data are available, then we can be almost sure that the modelling procedure proposed here are able to reconstruct the right model.

5. REFERENCES

Buchberger, B. (1985). Gröbner basis: An algorithmic method in polynomial ideal theory. In: *Multidimensional Systems Theory* (N. K. Bose, Ed.). pp. 184–232. D. Reidel.

Fornasini, E., P. Rocha and S. Zampieri (1993). State space realization of 2D finite dimensional behaviours. *SIAM J. of Control and Optimization* **31**, 1502–1517.

Heij, C. (1992). Exact modelling and identifiability of linear systems. *Automatica* **28**, 325–344.

Rocha, P. (1990). Structure and Representation of 2-D Systems. PhD thesis. University of Groningen, The Netherlands.

Sakata, S. (1988). Finding a minimal set of linear recurring relations capable of generating a given two-dimensinal array. *Journal of symbolic computation* **5**, 321–337.

Sakata, S. (1990). A Gröbner basis and a minimal polynomial set of a finite nD array. In: *Lecture Notes in Computer Science* (S. Sakata, Ed.). Vol. 508. pp. 280–291. Springer and Verlag.

Willems, J.C. (1986). From time series to linear system, part II: Exact modelling. *Automatica* **22**, 675–684.

Zampieri, S. (1992). Exact modelling of finite 2D arrays. In: *Proc. of IFAC Conf. on System Structure and Control.* Praga. pp. 428–431.

Zampieri, S. (1994). Exact modelling of scalar 2D arrays. *Kybernetica* **30**, 129–140.

Zampieri, S. (1995). Identifiability of 2D scalar systems. In preparation.

GROWTH CONDITIONS FOR GLOBAL STABILIZATION OF CASCADE NONLINEAR SYSTEMS

A. FERFERA* and M.A. HAMMAMI*

*INRIA Lorraine (CONGE Project) & University of Metz.
4, rue Marconi, 57 070 METZ - FRANCE.
e-mail: {ferfera,hammami}@ilm.loria.fr

Abstract. In this paper we deal with the problem of the global stabilization for a class of cascade nonlinear control systems. It is well known that, in general, the global asymptotic stability of the cascaded subsystems does not imply the global asymptotic stability of the composite closed-loop system. Using stability properties of triangular systems, additional sufficient conditions for the global stabilization of a cascade nonlinear system are given.

Résumé. On s'intéresse dans ce travail, au problème de la stabilisation globale de certains systèmes non linéaires en cascade. Celle-ci ne se déduisant pas, en général, des propriétés de stabilité asymptotique globale des sous-systèmes mis en cascade, on donne ici des conditions supplémentaires suffisantes pour assurer la stabilité asymptotique globale du systéme composé bouclé.

Key Words. Cascade nonlinear systems, stabilization, Lyapunov function, feedback.

Mots-Clés. Systèmes non linéaires en cascade, stabilisation, fonction de Liapounov, feedback.

1. INTRODUCTION

In this paper we propose sufficient conditions for global stabilization of cascade nonlinear control systems of the form

$$\begin{cases} \dot{x} = f(x,y) \\ \dot{y} = F(y,u) \end{cases} \tag{1}$$

by means of a state feedback $u(y)$ depending only on the state y of the subsystem

$$\dot{y} = F(y,u)$$

Many recent papers (Kokotovic and Sussmann, 1989; Kokotovic and Sussmann, 1991; Saberi *et al.*, 1990) and references therein, addressed the problem of the stabilization of partially linear composite systems of the form

$$\begin{cases} \dot{x} = f(x,y) \\ \dot{y} = Ay + Bu \end{cases} \tag{2}$$

where A and B are constant matrices and f is a smooth vector field such that $x = 0$ is a globally asymptotically stable (GAS) equilibrium point of

$$\dot{x} = f(x,0) \tag{3}$$

In (Saberi *et al.*, 1990), by assuming the global exponential stability of (3) and an appropriate growth condition on the nonlinear part of the system (2), it is shown that this system is stabilizable by any linear feedback $u = Ky$ which stabilizes the linear subsystem.

In this paper, using stability properties of triangular systems (Seibert and Suarez, 1991; Vidyasagar, 1980), we give, for more general systems of the form (1), sufficient growth conditions on $f(x,y)$ under which the system (1) is globally asymptotically stabilizable by a feedback $u(y)$ stabilizing

$$\dot{y} = F(y,u)$$

2. SUFFICIENT GROWTH CONDITIONS FOR STABILIZATION

Consider a cascade nonlinear control system of the form

$$\begin{cases} \dot{x} = f(x,y) \\ \dot{y} = F(y,u) \end{cases} \tag{4}$$

where $x \in \mathbb{R}^n$, $y \in \mathbb{R}^m$, $u \in \mathbb{R}^p$ and f and F are smooth vector fields such that $f(0,0) = 0$ and $F(0,0) = 0$

Notice that, due to the smoothness property, it is always possible to decompose $f(x,y)$ in the form

$$f(x,y) = f(x,0) + g(x,y)y \tag{5}$$

where

$$g(x,y) = \int_0^1 \frac{\partial f}{\partial y}(x,ty)dt$$

Assume that the system (4) satisfies the following assumptions:

(H_1) The equilibrium $x = 0$ of

$$\dot{x} = f(x,0)$$

is GAS and there exists a proper smooth Lyapunov function

$$V(x) > 0, \; x \neq 0, \; V(0) = 0$$

such that

$$\nabla V(x) f(x,0) < 0, \; \forall x \neq 0 \tag{6}$$

and

$$||\nabla V(x)|| \leq \alpha(1 + V(x)), \quad \alpha > 0 \tag{7}$$

(H_2) There exists a nondecreazing scalar function $\Theta(||y||) \geq 0$, bounded for all bounded y, such that for any $x \in \mathbb{R}^n$ and any $y \in \mathbb{R}^m$

$$||g(x,y)|| \leq \Theta(||y||) \tag{8}$$

(H_3) There exists a feedback law $u(y)$ such that $y = 0$ is a globally exponentially stable (GES) equilibrium point of the system

$$\dot{y} = F(y, u(y)) \tag{9}$$

Then one has :

Theorem 1 Under the assumptions (H_1), (H_2) and (H_3), $(x,y) = (0,0)$ is a GAS equilibrium point of the composite closed-loop system

$$\begin{cases} \dot{x} = f(x,0) + g(x,y)\,y \\ \dot{y} = F(y, u(y)) \end{cases} \tag{10}$$

Proof: It is known (Seibert and Suarez, 1991) that if

$$\dot{x} = f(x,0)$$

and (9) are both GAS, and all orbits of (10) are bounded for $t \geq 0$, then (10) is GAS. According to this fact, it suffices to prove the boundedness of the component $x(t)$ of any trajectory $(x(t), y(t))$, $t \geq 0$, of (10).

On the one hand, by the assumption (H_3), the component $y(t)$ of any trajectory of the system (10) satisfies

$$||y(t)|| \leq k_1 e^{-a_1 t} ||y(0)||, \quad \forall t \geq 0 \tag{11}$$

with $k_1 > 0$ and $a_1 > 0$.

On the other hand, the derivative of the Lyapunov function V given in (H_1) with respect to the time-varying equation

$$\dot{x} = f(x,0) + g\big(x, y(t)\big)\, y(t)$$

is given by

$$\begin{aligned} \dot{V}(x) \;&=\; \nabla V(x) f(x,0) \\ &\quad + \nabla V(x)\, g\big(x, y(t)\big)\, y(t) \end{aligned}$$

Tacking into account (6), (7), (8) and (11), one can deduce that

$$\dot{V}(x) \leq \nu\, e^{-a_1 t} \big(1 + V(x)\big) \tag{12}$$

where

$$\nu = \alpha\, k_1 \Theta\big(||y(0)||\big)$$

It follows that $\ln\big(1 + V(x)\big)$ is upper bounded by a positive constant, and so $x(t)$ is bounded. ∎

Now, if one replaces the assumptions (H_1), (H_2) and (H_3) respectively by:

(H_1') The equilibrium $x = 0$ of

$$\dot{x} = f(x,0)$$

is GES.

(H_2') There exists a nondecreazing scalar function $\Theta(||y||) \geq 0$, bounded for all bounded y, such that for any $x \in \mathbb{R}^n$ and any $y \in \mathbb{R}^m$

$$||g(x,y)|| \leq \Theta(||y||)||x|| \tag{13}$$

(H_3') There exists a feedback law $u(y)$ such that $y = 0$ is a GAS equilibrium point of the system

$$\dot{y} = F(y, u(y)) \tag{14}$$

then one can state :

Theorem 2 If the assumptions (H_1'), (H_2') and (H_3') hold, then the system (10) is GAS.

Proof: Indeed, using the same idea as in the proof of the above Theorem, it suffices to show the boundedness of the component $x(t)$ of any trajectory $(x(t), y(t))$, $t \geq 0$ of (10).

By assumption (H_1'), there exists a proper smooth Lyapunov function

$$V(x) > 0,\ \forall x \neq 0,\ V(0) = 0$$

and some positive constants α_0, α_1, α_2, α_3 such that $V(x)$ satisfies, for all x in $\mathbb{R}^n$ (Hahn, 1967)

$$\alpha_1\|x\|^2 \leq V(x) \leq \alpha_2\|x\|^2 \tag{15}$$

$$\|\nabla V(x)\| \leq \alpha_3\|x\| \tag{16}$$

and

$$\dot{V}(x) = \nabla V(x)\, f(x,0) \leq -\alpha_0 V(x) \tag{17}$$

Furthemore, by the assumption (H_3'), there exists a real-valued function $\beta(r,s)$, defined for $r \geq 0$ and $s \geq 0$, which belongs to class $\mathcal{KL}$ (Hahn, 1967), such that the component $y(t)$ of any trajectory of the system (10) satisfies

$$\|y(t)\| \leq \beta(\|y(0)\|, t), \quad \forall t \geq 0. \tag{18}$$

The derivative of V with respect to the time-varying equation

$$\dot{x} = f(x,0) + g(x, y(t))\, y(t)$$

is given by

$$\begin{aligned}\dot{V}(x) &= \nabla V(x)\, f(x,0)\\ &\quad + \nabla V(x)\, g(x, y(t))\, y(t)\end{aligned}$$

Tacking into account (13), (15), (16), (17) and (18), one can deduce that

$$\begin{aligned}\dot{V}(x(t)) &\leq -\alpha_0\, V(x(t))\\ &\quad + \frac{\alpha_3\Theta(\beta(\|y(0)\|, t))}{\alpha_1} \times\\ &\qquad \beta(\|y(0)\|, t)\, V(x(t))\end{aligned}$$

that is

$$\begin{aligned}\dot{V}(x(t)) &\leq -\alpha_0\, V(x(t))\\ &\quad + \mu\, \beta(\|y(0)\|, t)\, V(x(t))\end{aligned}$$

where

$$\mu = \frac{\alpha_3\Theta(\beta(\|y(0)\|, 0))}{\alpha_1} > 0$$

Since $\beta \in \mathcal{KL}$ then $\beta(\|y(0)\|, t)$ decreases to zero as t increases, and it follows that

$$\exists T > 0,\ \forall t \geq T,\ \beta(\|y(0)\|, t) < \frac{\alpha_0}{2\mu}$$

Hence

$$\dot{V}(x(t)) \leq -\frac{\alpha_0}{2} V(x(t)), \quad \forall t \geq T$$

and

$$V(x(t)) \leq k\, \exp(-\frac{\alpha_0}{2} t), \quad \forall\, t \geq T$$

with

$$k = V(x(T))\, \exp(\frac{\alpha_0}{2} T)$$

This allows to state the boundedness of $x(t)$, $t \geq 0$, and completes the proof. ∎

Remark 1: Notice that if one assumes in the theorem 2 that the system (14) is GES then, the global exponential stability of the system (10) is got, and an estimate

$$\|x(t)\| \leq k_2\, e^{-a_2 t}\|x(0)\|, \quad \forall t \geq 0$$

holds for some positive constants k_2, a_2.

Remark 2: If, in (H_1), one replaces (6) by an estimate of the form

$$\nabla V(x) f(x,0) \leq -\eta\, V(x), \quad \eta > 0 \tag{19}$$

and if (H_1), (H_2) and (H_3') hold, then the inequality (12) becomes

$$\begin{aligned}\dot{V}(x) &\leq -\eta\, V(x)\\ &\quad + \nu\, \beta(\|y(0)\|, t)\,(1 + V(x))\end{aligned}$$

where

$$\nu = \alpha\, \Theta(\beta(\|y(0)\|, 0))$$

So, under these assumptions, the system (10) is still GAS.

Example: Set $x = (x_1, x_2)^T$, $y = (y_1, y_2)^T$, and consider the following partially bilinear system which evolves on $\mathbb{R}^4$

$$\begin{cases} \dot{x} = f(x,0) + y_1 g(x,y) \\ \dot{y} = Ay + uBy \end{cases} \tag{20}$$

where $f(x,0) = (f_1(x,0), f_2(x,0))^T$ with

$$\begin{aligned} f_1(x,0) &= -2x_1 + 2x_2 + x_1^2 x_2^2 \\ &\quad - \frac{1}{2}x_1^3 - x_1^4 - \frac{1}{2}x_1^5 \\ f_2(x,0) &= -2x_1 - 2x_2 + x_1^2 x_2 - \frac{1}{2}x_2^3 \end{aligned}$$

and

$$g(x,y) = \begin{pmatrix} \sin(y_1)\cos(x_1) \\ \sin(y_2)\cos(x_1) \end{pmatrix}$$

$$A = \begin{pmatrix} 3 & 3 \\ -2 & 3 \end{pmatrix}, \quad B = \begin{pmatrix} 1 & 0 \\ 0 & -1 \end{pmatrix}$$

The system

$$\dot{x} = f(x,0)$$

is GAS. Indeed, the derivative $\dot{V}(x)$ of the homogeneous Lyapunov function

$$V(x) = ||x||^2$$

along the trajectories of

$$\dot{x} = f(x,0)$$

satisfies

$$\begin{aligned} \dot{V}(x) &= -4V(x) - \left(x_1^2 + x_1^3 - x_2^2\right)^2 \\ &\leq -4V(x) \end{aligned}$$

and so (19) is satisfied. Furthermore, since the Liapunov function V is homogeneous, then (7) is also satisfied. Moreover, one can verify that $g(x,y)$ satisfies (8). Besides, the homogeneous feedback law of degree zero given by $u(0) = 0$ and, $\forall y \neq 0$,

$$u(y) = \frac{-14y_1^2 + 14y_1 y_2 + 21y_2^2}{2y_1^2 + 3y_2^2}$$

stabilizes the bilinear subsystem

$$\dot{y} = Ay + uBy$$

which is not stabilizable by any continuous at the origin feedback because of

$$\mathrm{Tr}(A + \lambda B) = \mathrm{Tr}A > 0, \ \forall \lambda \in \mathbb{R}$$

Indeed, let $W(y)$ be the Lyapunov function given by

$$\begin{aligned} W(y) &= 4y_1^2 y_2^2 \\ &\quad + 2y_1 y_2(-2y_1^2 + 6y_1 y_2 + 3y_2^2) \\ &\quad + (-2y_1^2 + 6y_1 y_2 + 3y_2^2)^2 \end{aligned}$$

A simple computation shows that the derivative of $W(y)$ with respect to the closed-loop bilinear subsystem

$$\dot{y} = Ay + u(y)By \tag{21}$$

is

$$\dot{W}(y) = -2W(y)$$

and so $y = 0$ is a GAS equilibrium point of (21). Hence, the assumptions (H_1), (H_2) and (H_3') hold, and, following the Remark 2, one can deduce that $u(y)$ is a stabilizing feedback law for the cascade system (20)

3. REFERENCES

Hahn, W. (1967). *Stability of motion.* Springer-Verlag. New York.

Kokotovic, P.V. and H.J. Sussmann (1989). A positive real condition for global stabilization of nonlinear systems. *Systems & Control Letters* **13**, 125–133.

Kokotovic, P.V. and H.J. Sussmann (1991). The peaking phenomenon and the global stabilization of nonlinear systems. *IEEE Trans. Automat. Control* **36**, 424–440.

Saberi, A., P.V. Kokotovic and H.J. Sussmann (1990). Global stabilization of partially linear composed systems. *SIAM J. Control and Optimisation* **28**, 1491–1503.

Seibert, P. and R. Suarez (1991). Global stabilization of a certain class of nonlinear systems. *Systems & Control Letters* **16**, 17–23.

Vidyasagar, M. (1980). Decomposition techniques for large-scale systems with nonadditive interactions stability and stabilizability. *IEEE Trans. Automat. Control* **25**, 773–779.

ABSOLUTE STABILIZABILITY OF DISCRETE-TIME UNCERTAIN NON-LINEAR SYSTEMS

Andrey V. Savkin* and Ian R. Petersen**

* Department of Electrical and Electronic Engineering and Cooperative Research Center for Sensor Signal and Information Processing, University of Melbourne, Parkville, Victoria 3052, Australia, email: savkin@ee.mu.oz.au.
** Department of Electrical Engineering, Australian Defence Force Academy, Campbell, 2600, Australia, email: irp@ee.adfa.oz.au

Abstract. L'article donne une condition nècessaire et suffisant pour la stabilisation absolu d'un systéme incertain non-linèaire. L'article fonde le rèsultat principal avec un rèultat "S-procèdure".

Keywords. Absolute stabilizability; nonlinear systems; discrete-time uncertain systems

1. INTRODUCTION

The development of linear H^∞ control theory such as in (Petersen 1987) and (Doyle *et al.* 1989) has had an important impact on the field of Robust Control. In particular, results on linear H^∞ control theory can be applied to give necessary and sufficient conditions for the robust stabilization of uncertain linear systems; e.g., see (Khargonekar *et al.* 1990, Savkin and Petersen 1994*b*). Furthermore, by combining H^∞ control theory with the recently published "S-procedure" Theorem of (Megretsky and Treil 1993), necessary and sufficient conditions can be obtained for the robust stabilization via linear control of uncertain linear systems containing structured uncertainty.

In practice however, it is often the case that the system to be controlled is essentially nonlinear and it is not reasonable to regard all of the nonlinearity as uncertainty. This fact has motivated the development of nonlinear H^∞ control theory such as in (van der Schaft 1991, van der Schaft 1992, Isidori and Astolfi 1992*a*, Isidori and Astolfi 1992*b*, Baras and James 1993). In these papers, the Riccati equation approach to linear H^∞ control is extended to give a dynamic programming approach to nonlinear H^∞ control. Furthermore in (Astolfi and Guzzella 1993), nonlinear H^∞ control results are applied to a robust control problem for nonlinear systems. However, the results of (Astolfi and Guzzella 1993) give only sufficient conditions and they are based on the sufficient conditions for nonlinear H^∞ control given in (Isidori and Astolfi 1992*a*).

In this paper, we consider the problem of robustly stabilizing an uncertain nonlinear system containing structured uncertainty. The launching point for this paper is the recent paper (Baras and James 1993) which contains necessary and sufficient conditions for the solution of a discrete-time nonlinear H^∞ control problem. These results are given in terms of the existence of suitable solutions to a dynamic programming equation. Using these results, we are able to develop our main result which is a necessary and sufficient condition for the robust stabilizability (via state feedback) of a discrete-time uncertain nonlinear system containing structured uncertainties.

In order to obtain our necessary and sufficient conditions for robust stabilizability, we consider a new class of discrete-time uncertain nonlinear systems. The class of discrete-time uncertain nonlinear systems being considered, involves uncertainties satisfying a certain "Sum Constraint". This notion extends the standard (continuous-time) Integral Quadratic Constraint which originated in the work of Yakubovich (Yakubovich 1973, Yakubovich 1988*b*, Yakubovich 1988*a*); see also, (Savkin and Petersen 1994*b*, Savkin and Petersen 1995*b*, Savkin and Petersen 1995*a*, Savkin and Petersen 1994*a*, Savkin and Petersen n.d.). This uncertainty class is extremely rich and allows for nonlinear time-varying dynamic structured uncertainties. Also central to our results is a new definition of absolute stabilizability for discrete-time uncertain nonlinear systems. This definition extends to discrete-time uncertain nonlinear systems, the notion of absolute stabilizability introduced in (Savkin and Petersen 1994*b*, Savkin and Petersen 1995*b*, Savkin and Petersen 1994*a*) for the corresponding notion of absolute stability.

An important feature of the main result of this paper is that a necessary and sufficient condition has been given for the absolute stabilizability of uncertain nonlinear systems containing *structured uncertainty*. In recent years, a number of similar results have appeared for various classes of uncertain linear systems ; e.g., see (Khammash and Pearson 1993, Shamma 1992, Megretsky 1993). In each of these results, the uncertainty structure is exploited by introducing a set of scaling parameters. The use of such scaling parameters is an important tool in the theory of robust control and originated in the work of Doyle (Doyle 1982) and Safonov (Safonov 1982). The key idea is to use an extension of the S-procedure Theorem (also known as Finsler's Theorem); e.g., see (Yakubovich 1971, Uhlig 1979). Such a result enables the problem of robust stability of an uncertain system containing structured uncertainty to be converted into a problem of robust stability of an uncertain system containing unstructured uncertainty.

In all of the results contained in (Khammash and Pearson 1993, Shamma 1992, Megretsky 1993), the underlying uncertain system is required to be linear. One of the main contributions of this paper is to show that these ideas can be extended into the realm of nonlinear systems. That is, we show using a nonlinear version of the S-procedure result of (Megretsky and Treil 1993), that the problem of absolute stabilizability for a nonlinear uncertain system with structured uncertainty is equivalent to the existence of a set of real scaling parameters such that a corresponding nonlinear H^∞ control problem system has a solution.

We can summarize our approach as a *three step procedure* containing the following three steps:

- The nonlinear S-procedure Theorem;
- Nonlinear H^∞ Control;
- A nonlinear version of the Bounded Real Lemma.

A linear version of this three step method for robust stabilizability can also be found in the approach of (Savkin and Petersen 1995*b*).

The remainder of this paper proceeds as follows: In Section 2, we introduce the class of uncertain systems under consideration and define our corresponding notion of "Absolute Stabilizability". In Section 3, we present our new S-procedure result which enables us to deal with uncertain nonlinear systems with structured uncertainty. Section 4 presents the main result of the paper which is a necessary and sufficient condition for absolute stabilizability via state feedback. This result is given in terms of the existence of a suitable solution to a parameterized dynamic programming equation. If such a solution exists, then a corresponding dynamic programming inequality can be used to construct a non-dynamic state feedback absolutely stabilizing controller. Such dynamic programming inequalities are closely related to the dissipativenss theory of nonlinear systems developed in the work of Willems (Willems 1972) and Hill and Moylan (Hill and Moylan 1976, Hill and Moylan 1977); see also (James 1993). A corollary to our main result is the fact that if an uncertain system can be absolutely stabilized via a dynamic (possibly infinite dimensional) state feedback controller then it can be absolutely stabilized via a static state feedback controller. The main result of the paper is proved by converting the absolute stabilization problem into an equivalent nonlinear H^∞ control problem which can be solved using the methods of (Baras and James 1993).

2. PROBLEM STATEMENT

Consider the following discrete-time uncertain nonlinear system defined on the infinite time interval $t = 0, 1, 2, \ldots$:

$$\begin{aligned} x(t+1) &= b(x(t), u(t), \xi_0(t), \ldots, \xi_k(t)); \\ z_0(t) &= l_0(x(t), u(t)); \\ &\vdots \\ z_k(t) &= l_k(x(t), u(t)) \end{aligned} \tag{1}$$

where $x(t) \in \mathbf{R}^n$ is the *state*, $u(t) \in \mathbf{R}^m$ is the *control input*, $z_0(t) \in \mathbf{R}^{h_0}$, $z_1(t) \in \mathbf{R}^{h_1}, \ldots, z_k(t) \in \mathbf{R}^{h_k}$ are the ***uncertainty outputs***, $\xi_0(t) \in \mathbf{R}^{r_0}$, $\xi_1(t) \in \mathbf{R}^{r_1}, \ldots, \xi_k(t) \in \mathbf{R}^{r_k}$ are the ***uncertainty inputs***, and $b(\cdot,\cdot,\cdot,\ldots,\cdot), l_0(\cdot,\cdot), \ldots, l_k(\cdot,\cdot)$ are continuous functions.

System Uncertainty The uncertainty in the above system is described by a set of equations of the form:

$$\begin{aligned} \xi_0(t) &= \phi_0(t, z_0(\cdot)|_0^t) \\ &\vdots \\ \xi_k(t) &= \phi_k(t, z_k(\cdot)|_0^t). \end{aligned} \tag{2}$$

Alternatively, the uncertainty inputs and outputs may be collected together into two vectors. That is, we define $\xi(t) \triangleq [\xi_0(t)' \ \xi_1(t)' \ \ldots \ \xi_k(t)' \ ]'$ and $z(t) \triangleq [z_0(t)' \ z_1(t)' \ \ldots \ z_k(t)' \ ]'$. Then (2) can be re-written in the more compact form

$$\xi(t) = \Phi(t, z(\cdot)|_0^t). \tag{3}$$

The uncertainty described by the above equations is required to satisfy a certain "Sum Constraint"; see also (Yakubovich 1973, Yakubovich 1988*b*, Yakubovich 1988*a*, Savkin and Petersen 1994*b*, Savkin and Petersen 1995*b*, Savkin and Petersen 1995*a*, Savkin and Petersen 1994*a*, Savkin and Petersen n.d.). This uncertainty description allows for non-linear, time-varying, dynamic, non-casual uncertainties.

Definition 1 (Sum Constraint) An uncertainty of the form (2) is an admissible uncertainty for the system (1) if the following conditions hold: Given any control input $u(\cdot)$ and any corresponding solution to equations (1), (2), then there exists a sequence $\{t_i\}_{i=1}^\infty$ and constants $d_0 \geq 0, d_1 \geq 0, \ldots, d_k \geq 0$ such that $t_i \to \infty$, $t_i \geq 0$ and

$$\sum_{t=0}^{t_i} \|\xi_s(t)\|^2 \leq d_s + \sum_{t=0}^{t_i} \|z_s(t)\|^2 \tag{4}$$

for all i and for all $s = 0, 1, \ldots, k$. Here $\|\cdot\|$ denotes the standard Euclidean norm. Also, note that t_i may be equal to infinity.

Remarks Given an admissible uncertainty of the form (2), then for every solution to the equations (1), (2), there will be corresponding parameters $d_s \geq 0$ such that condition (4) is satisfied. The values of the parameters d_s are a 'measure of mismatch' between the resulting uncertainty input $\xi(\cdot)$ and the following *Homogeneous Sum Constraint:*

$$\sum_{t=0}^{t_i} \|\xi_s(t)\|^2 \leq \sum_{t=0}^{t_i} \|z_s(t)\|^2 \quad \forall i.$$

Note that the uncertainty description given above allows for $\xi(t)$ to depend dynamically on $z(t)$. Indeed in this case, one could interpret the 'mismatch' mentioned above as being due to a non-zero initial condition on the uncertainty dynamics. Also, it is clear that the uncertain system (1), (2) allows for structured uncertainty satisfying norm bound conditions. In this case, the uncertain system would be described by the state equations

$$x(t+1) = A(x(t), u(t)) +$$
$$\sum_{s=0}^{k} B_s(x(t), u(t))\Delta_s(t)C_s(x(t), u(t));$$

where $\Delta_s(t)$ are the uncertainty matrices, $\|\Delta_s(t)\| \leq 1$ and $\|\cdot\|$ denotes the standard induced matrix norm. To verify that such uncertainty is admissible for the uncertain system (1), (2), let $\xi_s(t) \triangleq \Delta_s(t)C_s(x(t), u(t))$ where $\|\Delta_s(t)\| \leq 1$ for all $t \geq 0$ and $z_s(t) \triangleq C_s(x(t), u(t))$. Then $\xi(\cdot)$ satisfies condition (4) with any t_i and with $d_s = 0$.

We will consider the problem of absolute stabilization via an infinite dimensional nonlinear dynamic state feedback controller of the form:

$$\begin{aligned} x_c(t+1) &= \Lambda(x_c(t), x(t)); \\ u(t) &= \lambda(x_c(t), x(t)) \end{aligned} \quad (5)$$

where $\Lambda(\cdot,\cdot)$ and $\lambda(\cdot,\cdot)$ are vector functions such that $\Lambda(0,0) = 0$ and $\lambda(0,0) = 0$. Here $x_c(t) \in \mathbf{B}$ where $\mathbf{B}$ is a Banach space. We will restrict attention to controllers of the form (5) such that the resulting closed loop system satisfies the following local controllability type assumption with respect to the uncertainty input:

Assumption 2.1 For any $\varepsilon > 0$ there exists a $\delta > 0$ such that the following condition holds: for any initial condition $[x_0,\ x_{c0}] \in \{[x_0,\ x_{c0}] \in \mathbf{R}^n \times \mathbf{B} : \|x_0\| + \|x_{c0}\| \leq \delta\}$, there exists an integer N and an uncertainty input $\xi(\cdot)$ such that the corresponding solution to the closed loop system (1), (5) satisfies $x(N) = 0,\ \ x_c(N) = 0$ and

$$\sum_{t=0}^{N-1} (\|x(t)\|^2 + \|x_c(t)\|^2 + \|u(t)\|^2$$
$$+\|z(t)\|^2 + \|\xi(t)\|^2) \leq \varepsilon.$$

Remark The above assumption is a weak technical assumption required for the proof of our main results. Indeed, given any system of the form (1) and controller of the form (5), an arbitrarily small perturbation of the system and controller will lead to the satisfaction of this assumption.

Notation Throughout this paper, $\|\cdot\|$ denotes the norm in the corresponding Banach space or the standard Euclidean norm in $\mathbf{R}^n$, $\mathbf{l}_2$ denotes the Hilbert space of vector sequences $\{h(t)\}_{t=0}^{\infty}$ such that $\sum_{t=0}^{\infty} \|h(t)\|^2 < \infty$. Also, $\|h(\cdot)\|_2^2 \triangleq \sum_{t=0}^{\infty} \|h(t)\|^2$.

We now introduce a notion of absolute stabilizability for the uncertain system (1), (2). This definition extends to the case of nonlinear discrete-time uncertain systems, the definition of absolute stabilizability given in (Savkin and Petersen 1994*b*, Savkin and Petersen 1995*b*).

Definition 2 The uncertain system (1), (4) is said to be ***absolutely stabilizable via nonlinear state feedback control*** if there exists a controller of the form (5) satisfying Assumption 2.1, a function $W(x_0, x_{c0}) \geq 0$ and a constant $c > 0$ such that the following conditions hold:

(i) $W(0,0) = 0$ and the function $W(\cdot,\cdot)$ is continuous at the point $x_0 = 0,\ \ x_{c0} = 0$.

(ii) For any initial condition $[x(0)\ x_c(0)] = [x_0\ x_{c0}]$ and any admissible uncertainty $\xi(\cdot) = \left[\ \xi_0(\cdot)\ \ \ldots\ \ \xi_k(\cdot)\ \right]$, then $\left[\ x(\cdot)\ \ x_c(\cdot)\ \ u(\cdot)\ \ z(\cdot)\ \ \xi(\cdot)\ \right] \in \mathbf{l}_2$, and

$$\|x(\cdot)\|_2^2 + \|x_c(\cdot)\|_2^2 + \|u(\cdot)\|_2^2 + \|z(\cdot)\|_2^2$$
$$+\|\xi(\cdot)\|_2^2 \leq W(x_0, x_{c0}) + c\sum_{s=0}^{k} d_s. \quad (6)$$

3. S-PROCEDURE FOR NONLINEAR SYSTEMS

In this section, we present a result which extends the "S-Procedure" of (Megretsky and Treil 1993). The main result of this section applies to an infinite dimensional, nonlinear, time-invariant system of the form

$$h(t+1) = \Pi(h(t), \psi(t)) \quad (7)$$

where $h(t) \in \mathbf{B}_h$ is the *state*, $\psi(t) \in \mathbf{B}_\psi$ is the *input* and $\mathbf{B}_h$ and $\mathbf{B}_\psi$ are Banach spaces. Associated with the system (7) is the following set of functionals:

$$f_s\,(h(\cdot), \psi(\cdot)) = \sum_{t=0}^{\infty} \nu_s(h(t), \psi(t))$$

for $s = 0, 1, \ldots, k$.

Assumptions The system (7) and associated set of functionals satisfy the following assumptions:

3.1 For all $\psi(\cdot) \in \mathbf{l}_2$ and all initial conditions $h(0) \in \mathbf{B}_h$, the corresponding solution $h(\cdot)$ belongs to $\mathbf{l}_2$ and the corresponding quantities $f_0\,(h(\cdot), \psi(\cdot))$, $\ldots$, $f_k\,(h(\cdot), \psi(\cdot))$ are finite.

3.2 For any $\varepsilon > 0$, there exists a $\delta > 0$ such that the following condition holds: for any initial condition $h_0 \in \{h_0 \in \mathbf{B}_h : \|h_0\| \leq \delta\}$, there exists an integer N and an input $\psi_0(\cdot)$ such that the corresponding solution the system (7) satisfies

$$h_0(N) = 0, \quad |\sum_{t=0}^{N-1} \nu_s(h_0(t), \psi_0(t))| \le \varepsilon \quad (8)$$

for $s = 0, 1, \ldots, k$.

Note, Assumption 3.2 is a local controllability type assumption on the system (7).

Notation For the system (7) satisfying the above assumptions, we define the set $\Omega \subset \mathbf{l}_2$ as follows: Ω is the set of $\{h(\cdot), \psi(\cdot)\}$ such that $\psi(\cdot) \in \mathbf{l}_2$ and $h(\cdot)$ is the corresponding solution to (7) with initial condition $h(0) = 0$.

Theorem 1 Consider the system (7) and associated functionals and suppose the Assumptions 3.1 – 3.2 are satisfied. If $f_0(h(\cdot), \psi(\cdot)) \ge 0$ for all $\{h(\cdot), \psi(\cdot)\} \in \Omega$ such that $f_1(h(\cdot), \psi(\cdot)) \ge 0, \ldots, f_k(h(\cdot), \psi(\cdot)) \ge 0$, then there exist constants $\tau_0 \ge 0, \tau_1 \ge 0, \ldots, \tau_k \ge 0$ such that $\sum_{s=0}^{k} \tau_s > 0$ and

$$\tau_0 f_0(h(\cdot), \psi(\cdot)) \ge \sum_{s=1}^{k} \tau_s f_s(h(\cdot), \psi(\cdot)) \quad (9)$$

for all $\{h(\cdot), \psi(\cdot)\} \in \Omega$.

In order to prove this theorem, we will use the following convex analysis result, the proof of which was given in (Savkin and Petersen 1995*b*). However, we first introduce some notation.

Notation Given $S \subset \mathbf{R}^n$ and $T \subset \mathbf{R}^n$ and $\lambda \in \mathbf{R}$, then $S + T \triangleq \{x + y : x \in S, y \in T\}$. Also, cl$(S)$ denotes the closure of the set S.

Lemma 1 (See (Savkin and Petersen 1995*b*) for proof) Consider a set $M \subset \mathbf{R}^{k+1}$ with the property that $a + b \in \text{cl}(M)$ for all $a, b \in M$. If $x_0 \ge 0$ for all vectors $\begin{bmatrix} x_0 & \cdots & x_k \end{bmatrix}' \in M$ such that $x_1 \ge 0, \ldots x_k \ge 0$, then there exist constants $\tau_0 \ge 0, \ldots, \tau_k \ge 0$ such that $\sum_{s=0}^{k} \tau_s > 0$ and $\tau_0 x_0 \ge \tau_1 x_1 + \tau_2 x_2 + \cdots + \tau_k x_k$ for all $\begin{bmatrix} x_0 & \cdots & x_k \end{bmatrix}' \in M$.

Proof of Theorem 1 Suppose $f_0(h(\cdot), \psi(\cdot)) \ge 0$ for all $\{h(\cdot), \psi(\cdot)\} \in \Omega$ such that $f_1(h(\cdot), \psi(\cdot)) \ge 0, \ldots, f_k(h(\cdot), \psi(\cdot)) \ge 0$ and let the set M be defined by

$$M \triangleq \{\begin{bmatrix} f_0(h(\cdot), \psi(\cdot)) & \ldots & f_k(h(\cdot), \psi(\cdot)) \end{bmatrix} : \{h(\cdot), \psi(\cdot)\} \in \Omega\}.$$

It follows from the assumption on the set Ω that $x_0 \ge 0$ for all vectors $\begin{bmatrix} x_0 & \cdots & x_k \end{bmatrix}' \in M$ such that $x_1 \ge 0, \ldots, x_k \ge 0$. Now let $\{h_a(\cdot), \psi_a(\cdot)\} \in \Omega$ and $\{h_b(\cdot), \psi_b(\cdot)\} \in \Omega$ be given. We will prove that for any $\epsilon_0 > 0$ there exists a $\{h_{\epsilon_0}(\cdot), \psi_{\epsilon_0}(\cdot)\} \in \Omega$ such that

$$|f_s(h_{\epsilon_0}(\cdot), \psi_{\epsilon_0}(\cdot)) - f_s(h_a(\cdot), \psi_a(\cdot)) - f_s(h_b(\cdot), \psi_b(\cdot))| \le \epsilon_0 \quad (10)$$

for $s = 0, 1, \ldots, k$.

Indeed, let $\delta > 0$ be the constant defined as in Assumption 3.2 corresponding to $\varepsilon = \frac{\epsilon_0}{2}$. Since $\{h_a(\cdot), \psi_a(\cdot)\} \in \mathbf{l}_2$, there exists an integer i such that $\|h_a(i)\| \le \delta$ and furthermore, Assumption 3.1 implies that i can be chosen so that

$$|\sum_{t=i}^{\infty} \nu_s(h_a(t), \psi_a(t))| \le \varepsilon \quad \forall s = 0, 1, \ldots, k. \quad (11)$$

Now given the initial condition $h_0(0) = h_a(i)$, let the integer N and the input $\psi_0(\cdot)$ be defined as in Assumption 3.2 so that condition (8) holds for the corresponding solution to the system (7). Also, consider $\{h_{\epsilon_0}(\cdot), \psi_{\epsilon_0}(\cdot)\} \in \Omega$ defined so that

$$\psi_{\epsilon_0}(t) = \begin{cases} \psi_a(t) & t = 0, 1, \ldots, i-1; \\ \psi_0(t-i) & i + N - 1 \ge t \ge i; \\ \psi_b(t-i-N) & t \ge i + N. \end{cases}$$

It follows from the time invariance of the system (7) and condition (8) that

$$\begin{aligned} f_s(h_{\epsilon_0}(\cdot), \psi_{\epsilon_0}(\cdot)) &= \sum_{t=0}^{\infty} \nu_s(h_{\epsilon_0}(t), \psi_{\epsilon_0}(t)) \\ &= \sum_{t=0}^{i-1} \nu_s(h_a(t), \psi_a(t)) \\ &+ \sum_{t=i}^{i+N-1} \nu_s(h_0(t-i), \psi_0(t-i)) \\ &+ \sum_{t=i+N}^{\infty} \nu_s(h_b(t-i-N), \psi_b(t-i-N)) \\ &= f_s(h_a(\cdot), \psi_a(\cdot)) + f_s(h_b(\cdot), \psi_b(\cdot)) - \\ &\sum_{t=i}^{\infty} \nu_s(h_a(t), \psi_a(t)) + \sum_{t=0}^{N-1} \nu_s(h_0(t), \psi_0(t)). \end{aligned}$$

Using the fact that $\varepsilon = \frac{\epsilon_0}{2}$, conditions (8) and (11) now imply (10).

From the above, it follows that the set M has the property that $a + b \in \text{cl}(M)$ for all $a, b \in M$. Hence, Lemma 1 implies that there exist constants $\tau_0 \ge 0, \ldots, \tau_1 \ge 0$ such that $\sum_{s=0}^{k} \tau_s > 0$ and $\tau_0 x_0 \ge \tau_1 x_1 + \cdots + \tau_k x_k$ for all $\begin{bmatrix} x_0 & \cdots & x_k \end{bmatrix}' \in M$. That is, condition (9) is satisfied. □

4. THE MAIN RESULT

In this section, we present the main result of this paper which establishes a necessary and sufficient condition for the existence of an absolutely stabilizing nonlinear state feedback control of the form (5) for the nonlinear uncertain system (1), (2). This condition is given in terms of the existence of a suitable solution to a parameter dependent dynamic programming equation. The equation under consideration is defined as follows: Let $\tau_0 \ge 0, \ldots, \tau_k \ge 0, \delta > 0$ be given constants and consider the equation

$$V_\delta(x) = \inf_u \sup_{\xi_0, \ldots, \xi_k} \{V_\delta(b(x, u, \xi_0, \ldots, \xi_k)) + G_\delta(x, u, \xi_0, \ldots, \xi_k)\} \quad (12)$$

where

$$G_\delta(x,u,\xi_0,\ldots,\xi_k) \triangleq \sum_{s=0}^{k} \tau_s(\|z_s\|^2 - \|\xi_s\|^2) + \delta(\|x\|^2 + \|u\|^2 + \sum_{s=0}^{k} \|z_s\|^2). \quad (13)$$

We will construct a static state feedback controller of the form

$$u = u^*(x) \quad (14)$$

so that the following dynamic programming inequality is satisfied:

$$F_\delta(x) \geq \sup_{\xi_0,\ldots,\xi_k} \{F_\delta(b(x,u^*(x),\xi_0,\ldots,\xi_k)) + G_\delta(x,u^*(x),\xi_0,\ldots,\xi_k)\}. \quad (15)$$

Our main result requires that the uncertain system (1), (4) satisfies the following assumptions.

Assumptions

4.1 $b(0,0,0,\ldots,0) = 0$, $l_0(0,0) = 0,\ldots, l_k(0,0) = 0$.

4.2 There exist continuous functions $\xi_0^*(x),\ldots,\xi_k^*(x)$ such that $\xi_0^*(0) = 0,\ldots,\xi_k^*(0) = 0$ and $b(0,0,\xi_0^*(x),\ldots,\xi_k^*(x)) = x$ for all $x \in \mathbf{R}^n$.

4.3 For any $\varepsilon > 0$, there exists a $\delta > 0$ such that the following condition holds: for any $x_0 \in \{x_0 \in \mathbf{R}^n : \|x_0\| \leq \delta\}$, $u_0 \in \{u_0 \in \mathbf{R}^m : \|u_0\| \leq \delta\}$ and $x_* \in \{x_* \in \mathbf{R}^n : \|x_*\| \leq \delta\}$ there exists a vector $\xi^* = [\xi_0^*,\ldots,\xi_k^*]$ such that $b(x_0,u_0,\xi_0^*,\ldots,\xi_k^*) = x_*$ and $\|\xi^*\| \leq \varepsilon$.

Note, for a linear discrete-time system of the form

$$x(k+1) = Ax(k) + Bu(k) + D\xi(k),$$

Assumption 4.2 amounts to the requirement that the matrix D is of full rank. For linear systems, such an assumption can be made without loss of generality. Assumption 4.2 in this paper extends this condition to the case of nonlinear systems.

Assumption 4.3 is a related assumption which holds in the neighbourhood of the point $x = 0$, $u = 0$.

We are now in a position to present our main result on the problem of absolute stabilizability via state feedback.

Theorem 2 Consider the uncertain system (1), (2) and suppose that Assumptions 4.1-4.3 are satisfied. Then the following statements are equivalent:

(i) The uncertain system (1), (2) is absolutely stabilizable via an infinite dimensional nonlinear state feedback controller of the form (5).

(ii) There exist constants $\tau_0 \geq 0,\ldots,\tau_k \geq 0$ and $\delta_* > 0$ such that for all $\delta \in [0,\delta_*]$, the dynamic programming equation (12) has a solution $V_\delta(x)$ such that $V_\delta(x) \geq 0$ for all $x \in \mathbf{R}^n$, $V_\delta(0) = 0$ and $V_\delta(\cdot)$ is continuous at the origin. Furthermore if $x = 0$, the infimum in (12) is achieved at $u = 0$.

Moreover, if condition (ii) holds, then for any $\delta \in (0,\delta_*)$, there exists a static state feedback control of the form (14) and a function $F_\delta(x) \geq 0$ such that $u^*(0) = 0$, $u^*(\cdot)$ is continuous at the origin, $F_\delta(0) = 0$, $F_\delta(\cdot)$ is continuous at the origin and inequality (15) is satisfied. Furthermore, any such control absolutely stabilizes the uncertain system (1), (2).

Acknowledgement. This work was supported by the Australian Research Council.

REFERENCES

Astolfi, A. and L. Guzzella (1993). Robust control of nonlinear systems an H_∞-approach. In 'Proceedings of the 12th IFAC World Congress'. Vol. 9. Sydney. pp. 281–284.

Baras, J. S. and M. R. James (1993). Robust output feedback control for discrete-time nonlinear systems: The finite time case. In 'Proceedings of the 32nd IEEE Conference on Decision and Control'. San Antonio, Tx. pp. 51–55.

Doyle, J. C. (1982). 'Analysis of feedback systems with structured uncertainty'. *IEE Proceedings Part D* **129**(6), 242–250.

Doyle, J. C., K. Glover, P. P. Khargonekar and B. Francis (1989). 'State-space solutions to the standard H_2 and H_∞ control problems'. *IEEE Transactions on Automatic Control* **34**(8), 831–847.

Hill, D. and P. Moylan (1976). 'The stability of nonlinear dissipative systems'. *IEEE Transactions on Automatic Control* **AC-21**, 708–7811.

Hill, D. and P. Moylan (1977). 'Stability results for nonlinear feedback systems'. *Automatica* **13**, 377–382.

Isidori, A. and A. Astolfi (1992*a*). 'Disturbance attenuation and H_∞-control via measurement feedback'. *IEEE Transactions on Automatic Control* **37**, 1283–1293.

Isidori, A. and A. Astolfi (1992*b*). 'Nonlinear H_∞-control via measurement feedback'. *Journal of Mathematical Systems, Estimation and Control* **2**(1), 31–44.

James, M. R. (1993). 'A partial differential inequality for dissipative nonlinear systems'. *Systems and Control Letters* **21**(4), 315–320.

Khammash, M. and J.B. Pearson (1993). 'Analysis and design for robust performance with structured uncertainty'. *Systems and Control Letters* **20**(3), 179–188.

Khargonekar, P. P., I. R. Petersen and K. Zhou (1990). 'Robust stabilization of uncertain systems and H^∞ optimal control'. *IEEE Transactions on Automatic Control* **AC-35**(3), 356–361.

Megretsky, A. (1993). 'Necessary and sufficient conditions of stability: A multiloop generalization of the circle criterion'. *IEEE Transactions on Automatic Control* **AC-38**(5), 753–756.

Megretsky, A. and S. Treil (1993). 'Power distribution inequalities in optimization and robustness of uncertain systems'. *Journal of Mathematical Systems, Estimation and Control* **3**(3), 301–319.

Petersen, I. R. (1987). 'Disturbance attenuation and H^∞ optimization: A design method based on the algebraic Riccati equation'. *IEEE Transactions on Automatic Control* **AC-32**(5), 427–429.

Safonov, M. G. (1982). 'Stability margins of diagonally perturbed multivariable feedback systems'. *IEE Proceedings Part D* **129**(6), 251–256.

Savkin, A. and I.R. Petersen (1994*a*). Time-varying problems of H^∞ control and absolute stabilizability of uncertain systems. In 'Proceedings of the 1994 American Control Conference'. Baltimore, MD.

Savkin, A. V. and I. R. Petersen (1994*b*). 'A connection between H^∞ control and the absolute stabilizability of uncertain systems'. *Systems and Control Letters* **23**(3), 197–203.

Savkin, A. V. and I. R. Petersen (1995*a*). Minimax optimal control of uncertain systems with structured uncertainty. Accepted for publication in the International Journal of Robust and Nonlinear Control.

Savkin, A. V. and I. R. Petersen (1995*b*). Nonlinear versus linear control in the absolute stabilizability of uncertain linear systems with structured uncertainty. Accepted for publications in the IEEE Transactions on Automatic Control.

Savkin, A. V. and I. R. Petersen (n.d.). Structured dissipativeness and absolute stability of nonlinear systems. Accepted for publication in the *International Journal of Control.*

Shamma, J. S. (1992). Robustness analysis for time-varying systems. In 'Proceedings of the 31st IEEE Conference on Decision and Control'. Tuscon, Arizona.

Uhlig, F. (1979). 'A recurring theorem about pairs of quadratic forms and extensions: A survey'. *Linear Algebra and its Applications* **25**, 219–237.

van der Schaft, A. (1991). 'On a state space approach to nonlinear H_∞ control'. *Systems and Control Letters* **16**(1), 1–8.

van der Schaft, A. J. (1992). 'L_2-gain analysis of nonlinear systems and nonlinear state feedback control'. *IEEE Transactions on Automatic Control* **37**(6), 770–784.

Willems, J. C. (1972). 'Dissipative dynamical systems – part I: General theory'. *Archive of Rational Mechanics and Analysis* **45**, 321–351.

Yakubovich, V. A. (1971). 'S-procedure in nonlinear control theory'. *Vestnik Leningrad University, Series 1* **13**(1), 62–77.

Yakubovich, V. A. (1973). 'Minimization of quadratic functionals under the quadratic constraints and the necessity of a frequency condition in the quadratic criterion for absolute stability of nonlinear control systems'. *Soviet Mathematics Doklady* **14**, 593–597.

Yakubovich, V. A. (1988*a*). 'Absolute stability of nonlinear systems with a periodically nonstationary linear part'. *Soviet Physics Doklady* **32**(1), 5–7.

Yakubovich, V. A. (1988*b*). 'Dichotomy and absolute stability of nonlinear systems with periodically nonstationary linear part'. *Systems and Control Letters* **11**(3), 221–228.

MECHANICAL SYSTEM STABILIZATION VIA DIFFERENTIAL OBSERVER

Ilya V. BURKOV

Dept. of Higher Mathematics, St. Petersburg Technical University
Hlopina str. 9-3-67, 194021 St. Petersburg, Russia
Fax (812)5526086, ph. 5344916, tlx. 121187 LSUT SU
E-mail root@fmf.stu.spb.su

Abstract. Problem of tracking the prescribed path by nonlinear controlled mechanical system, descrided by Lagrange equations, is considered. It is proposed dynamic feedback not using the measurements of the generelised velocities. During the control it is necessarry to solve linear nonstationarly differential equations. The results may be applied for stabilization of the manipulators with absolutely rigid links or for ones with elastic joints. Proofs of the stability of proposed regulators are based on application of the theory of singularly perturbed differential equations.

Résumé. Dans cet article on étudie le probleme de la stabilisation (par retour) des trajectoires desirées des systemes du contrôle Lagrangiennes non linéaires. On propose le retour dynamique qui n' utilise pas la mesure des vitesses généralisées. En cours de processsus du contrôle il faut intégrer numériquement des equations lineaires nonstationaires differentielles. On peut appliquer les resultats obtenus pour guidage des robots avec charnières absolument rigides ou élastiques. La démonstration de stabilité au sens de Liapounoff repose pour l'essentiel sur la thèorie des equations singulières (à paramètre petit).

Keywords. Robots, singular perturbations, stability, dynamic feedback, nonlinear systems.

INTRODUCTION

The problem of asymptotic stabilization of nonlinear Lagrangian system has been studied extensively. Velocities information is required by most of control schemes. The required information may be obtained by numerical differentiation, but this leads to noise added to velocities information. If the program trajectory is constant vector then the problem of asymptotic stabilization without velocity measurements has been succesively solved in papers by Berghius and Nijmeijer (1993); Kelly (1993), Burkov (1993). It may be noticed that in paper by Kelly there is a easy correctable error: author wrongly considers that the product of two positve definite matrices is the positive definite matrix. To solve the problem of trajectory stabilization Berghius and Nijmeijer (1991) Canudas de Wit , Åström , Fixot (1990) have proposed nonlinear, diffucult in implementation dynamic feedbacks. The solution by Nicosia and Tomei is unsatisfactory, because in their paper apriori boundness of some components of phase vector is assumed. The problem of stabilization of the program trajectory which is prescribed vector function of time is essentially more diffucult than the problem of stabilization of system in constant program position. The closed-loop system of differential equations which describes such systems is as a rule essentially nonstationarly. The result of this paper is that the program trajectory of nonlinear Lagrangian system may be asymptotically stabilized via very simple dynamic feedback, which need not velocity measurements.

STABILIZATION OF THE MANIPULATOR WITH ABSOLUTELY RIGID LINKS

The dynamics of the system is described by Lagrange equation

$$(\partial K/\partial \dot{q})^{\cdot} - \partial K/\partial q + g(q) = M,$$

which is equivalent to the following one

$$A(q)\ddot{q} + b(q,\dot{q}) + g(q) = M, \qquad (1.1)$$

where $q \in R^n$ is coordinate vector, K is kinetic energy, $g(q)$ is vector of potential forces, $A(q)$ is positive definite inertia matrix, $b(q,\dot{q})$ is vector of quadratic terms in reference to $\dot{q}$, M is vector of controlling generalized forces. Let $q_p(t) \in C^2[0,\infty)$ denote the program motion. Let matrix $A^{-1}(q)$, vector $b(q,\dot{q})$ and their first and second partial derivatives be bounded in some neighbourhood of the program motion.

To stabilize the program motion it may be proposed the following regulator

$$M = A(q)[\ddot{q}_p - K_1 x_1 - K_2 y] + b(q, y + \dot{q}_p) + g(q) \qquad (1.2)$$

where $x_1 = q - q_p$, $y = xK_3 q - \hat{q}$, and the differential observer ($\hat{q}$ is estimate of $xK_3 q$)

$$\dot{\hat{q}} = xK_3(xK_3 q - \hat{q}) + xK_3 \dot{q}_p(t). \qquad (1.3)$$

Here K_1, K_2, K_3 are positive definite constant matrices.

Theorem 1. The closed loop-system (1.1), (1.2), (1.3) is globally asymptotically stable in reference to $q - q_p$, $\dot{q} - \dot{q}_p$, y for all sufficiently large gain coefficients x.

Proof. The equation of observer is equivalent to the following ones

$$\dot{\hat{q}} - xK_3\dot{q} = xK_3(xK_3 q - \hat{q}) + xK_3\dot{q}_p - xK_3\dot{q}$$

$$-\dot{y} = xK_3 y - xK_3 x_2, \quad x_2 = \dot{q} - \dot{q}_p$$

$$\varepsilon\dot{y} = -K_3 y + K_3 x_2, \quad \varepsilon = 1/x.$$

The latter equation degenerates to the algebraic one: $y = x_2$, if small parameter ε becomes zero.

The equation of the dynamics of mechanical system may be rewritten in the following form

$$\dot{x}_2 + K_1 x_1 + K_2 y + A^{-1}(q)[b(q,\dot{q}) - b(q, y + \dot{q}_p)] = 0.$$

The dynamics of the linear nonstationarly approximation of the closed-loop system when small parameter ε degenerates to zero is described by the second order vector differential equation:

$$\dot{x}_2 + K_1 x_1 + K_2 y = 0, \quad y = x_2.$$

Evidently, the latter system is uniformly asymptotically stable in reference to the initial time. According to the Klimushev theorem (see appendix) the nonlinear closed loop-system is asymptotically stable for all sufficiently large parameters x.

Theorem 2 from the paper of Hoppensteadt (1974) allows to establish *global* asymptotic stability of the closed-loop system. The key condition of Hoppensteadt theorem is reduced to the fact that the system (1.1) controlled by the well-known input $M = A(q)[\ddot{q}_p - K_1 x_1 - K_2 x_2] + b(q,\dot{q}) + g(q)$ is *globally* asymptotically stable in reference to the variables x_1, x_2.

Note, that proposed control scheme need not measuring generalized velocity $\dot{q}$. Another advantage of this control scheme is that the observer (1.3) is linear.

To exclude on-line computation of the matrix $A(q)$ and the vectors $b(q, y + \dot{q}_p)$, $g(q)$ the following regulator may be proposed (Burkov, 1995)

$$M = M_p(t) - \gamma(K_0(q - q_p) + \upsilon K_1 y), \qquad (1.4)$$

where $M_p(t) = A(q_p)\ddot{q}_p + b(q_p, \dot{q}_p) + g(q_p)$, K_0, K_1 are positive definite diagonal matrices and the differential observer (1.3). It is proved that the closed-loop system (1.1),(1.3),(1.4) is asymptotically stable in reference to $q-q_p$, $\dot{q}-\dot{q}_p$, y for all sufficiently large gain coefficients x, γ, υ.

The control (1.4) is simpler in practical implementation than (1.2), but the conditions of stability are more restrictive.

CONTROL OF THE MANIPULATOR WITH ELASTIC JOINTS

Dynamics of elastic n-link plane manipulator may be described by the following equations (Spong, 1987; Burkov, 1987)

$$A(q_1)\ddot{q}_1+B(q_1,\dot{q}_1)+k(q_1-q_2)=0$$
$$J\ddot{q}_2-k(q_1-q_2)=M, \quad (2.1)$$

where $q_1 \in R^n$ is vector of coordinates of links, $q_2 \in R^n$ is vector of coordinates of drive rotors, $A(q_1)$ is positive definite symmetric matrix of inertia, $B(q_1,\dot{q}_1) \in R^n$ is vector of Coriolis and potential generalized forces, J is diagonal matrix of rotors inertiae, its diagonal elements are positive, k is diagonal matrix of joints stiffness, its diagonal elements are positive, $M \in R^n$ is vector of controlling forces.

Let $q_{1p}(t) \in C^4([0,\infty) \to R^n)$ denotes the program motion, it is assumed that its derivatives from first to fourth are bounded on semi-infinite interval.

It may be defined the following vector functions of time

$$q_{2p}(t)=k^{-1}(A(q_{1p})\ddot{q}_{1p}+B(q_{1p},\dot{q}_{1p})+kq_{1p}),$$

$$M_p(t)=J\ddot{q}_{2p}-k(q_{1p}-q_{2p}).$$

Note, that functions $q_{2p}(t)$ and $M_p(t)$ may be computed off-line and remembered, for example by means of a spline. It may be introduced the following variables

$$x=q_1-q_{1p}, \quad v=\dot{q}_1-\dot{q}_{1p}, \quad w=q_2-q_{2p}, \quad z=\dot{q}_2-\dot{q}_{2p}.$$

Consider the regulator

$$M=M_p(t)-K_z z-K_w w-K_v v-K_x x, \quad (2.2)$$

where K_z, K_w, K_v, K_x are diagonal matrices with positive diagonals elements.

Physical meanings of matrices K_z, K_w, K_v, K_x are gain coefficients, and physical meaning of $M_p(t)$ is the solution of inverse dynamics problem.

It is proved (Burkov, 1992) that the closed-loop system (2.1),(2.2) is uniformly (with respect to the initial time) asymptotically stable in reference to the variables x,v,w,z for all sufficiently large matrices K_z, K_w, K_v, K_x.

To exclude measurements of the velocities it may be proposed simple dynamic feedback

$$M=M_p(t)-K_z\bar{z}-K_w w-K_v\bar{v}-K_x x, \quad (2.3)$$

$$\dot{\hat{q}}_1=xK_4(xK_4q_1-\hat{q}_1)+xK_4\dot{q}_{1p}, \quad \bar{v}=xK_4q_1-\hat{q}_1$$
$$\dot{\hat{q}}_2=xK_5(xK_5q_2-\hat{q}_2)+xK_5\dot{q}_{2p}, \quad \bar{z}=xK_5q_2-\hat{q}_2 \quad (2.4)$$

Theorem 2.1. Assume, that matrices $A(q_1)$, $A^{-1}(q_1)$ and vector $B(q_1,\dot{q}_1)$ are twice differentiable in reference to $q_1, \dot{q}_1$ they and their derivatives are upper bounded is some neighbourhood of the program motion. Assume also, that $|\dot{q}_1^* A(q_1)\dot{q}_1| \geq const|\dot{q}_1|^2$ $\forall \dot{q}_1$ (const>0). Then the closed-loop system (2.1),(2.3),(2.4) is asymptotically stable in reference to the variables $z,w,v,x,\bar{z},\bar{v}$ for all sufficiently large $x>0$, for all sufficiently large matrices K_z, K_w, K_v, K_x and for all positive definite matrices K_4, K_5.

Sketch of proof. Applying Klimushev theorem in reference to the small parameter $\varepsilon=1/x$ the fact of stability of the closed-loop system (2.1),(2.3),(2.4) may be reduced to the fact of stability of the system (2.1),(2.2). The latter is proven in Burkov, 1992.

Remark 2.1. The transition process may be made arbitrary fast by increasing gain coefficients x, K_z, K_w, K_v, K_x; this is directly deduced from the theory of singular perturbations (Hoppensteadt, 1971, 1974).

Remark 2.2. In real system there is no possibility of unlimited increasing gain coefficients, because there is a noise in measurements of vector of states and this noise will be multiplied by large gain coefficients; beside this, large position gain coefficient may induce high-frequent oscillations, which may resonate with high-frequent elastic dynamics (unmodelled here explicitly).

CONCLUSION

Application of the theory of singular perturbations allowed to demonstrate the stability of the closed-loop systems very simply. The shortage of such approach is absence of precise information on gain coefficients.

APPENDIX

The following theorem was formulated and proved by Klimushev (1973).

Consider system of differential equations of the following kind

$$\dot{x}=(A(t)+\mu A_1(t))x+(B(t)+\mu B_1(t))y+X(t,x,y)$$

$$\mu\dot{y}=(C(t)+\mu C_1(t))x+(D(t)+\mu D_1(t))y+Y(t,x,y),$$

where $\mu>0$ - small positive parameter $x,X(t,x,y)\in R^n$, $Y(t,x,y)\in R^m$; A,A_1,B,B_1,C,C_1,D,D_1 matrices of appropriate dimensions. Assume these matrices with their derivatives continuously-differentiable and bounded when $t\in[0,\infty)$. Assume also that $|\det D(t)|>const>0$, and functions $X(t,x,y)$, $Y(t,x,y)$ in region $|x|\le H$, $|y|\le H$, $(H>0)$ $t\in[0,\infty)$ satisfy the conditions $X(t,0,0)=0$, $Y(t,0,0)=0$ and are expanded in Taylor series beginning from the second power of x,y.

System $\dot{x}=A(t)x+B(t)y$, $0=C(t)x+D(t)y$, which equivalent to $\dot{x}=R(t)x$, is called reduced system of linear approximation. System $\dot{y}=D(t)y$ is called shortened system of linear approximation.

Theorem . If shortened and reduced systems are uniformly asymptotically stable (with respect to the initial time) then initial system is also uniformly asymptotically for all sufficiently small $\mu>0$.

This formulation is strengthened in some way in comparison with Klimushev (1973), but proof is principally the same.

ACKNOWLEDGEMENT

This work has been partially supported by Russian Fund of Fundamental Researches

REFERENCES

Berghuis H., Löhnberg P., Nijmeijer H. (1991) Tracking control of robots using only position measurements. *30th Conf. on Decision and Control. 1.* P.1039-1040.

Berghuis H., Nijmeijer H. (1993). Global regulation of robots using only position measurements. *Syst. Contr. Lett. 21.*

Burkov I.V., Zaremba A.T. (1987). Dynamics of elastic manipulator with electric drives. *Izv. AN SSSR. Mekh. Tverd. Tela*, no. 1, p.57-64 . Engl. transl. in *Mechanics of Solids.*

Burkov I.V. (1992). Tracking prescribed path by manipulator with elastic joints. *Izv. Akad. Nauk. Tekhn. Kibern.* no. 4. P. 169-174. Engl. tranl. in *J. Comput. Systems, Sci., Intenat.* 1993, no. 6.

Burkov I.V. (1993) Asymptotic stabilization of nonlinear Lagrangian systems without measuring velocities. *Active Control in Mechanical Engineering. Proc. Int. Sympos. (Lyon, France, 1993).* Lyon: Association MV2. Vol. 2. Reprinted by Hermès, Paris. 1995. Ed.: L. Jézéquel.

Burkov I.V. (1995). Stabilization of a natural mechanical system without measuring its velocities. *Prikl. mat. i mekh.* (in print). Engl. transl. in *J. Appl. Math. Mech.*

Butuzov V.F. (1963). Asymptotic formulae for the solution of a system of differential equations with small parameter multiplying derivatives on semi-infinite interval. *Vestnik Moskovsk. Univ.*, Ser. 1. no.1, p.3-14. In Russ.

Canudas de Wit C., Åström K. J., Fixot N. (1990) Computed torque control via non-linear observer. *Int. J. Adapt. Contr. Sign. Process. 4.* P. 443-452.

Hoppensteadt F. (1971). Properties of solutions of ordinary diffferential equations with small parameters. *Comm. Pure Appl. Math., 24,* no.6, p.807-840.

Hoppensteadt F. (1974) Asymptotic stability in singular perturbation problems. *J. Diff. Eq., 15,* no.3, p.510-521.

Kelly R. (1993). A simple set-point robot controller by using only position measurement. *Proc. 12th IFAC World Congress. 6.*

Kelly R., Ortega R., Ailon A., Loria A. (1994) Global regulation of flexible joints robots using approximate differentiation. *IEEE Trans. Aut. Contr. 39,* no. 6. P. 1222-1224.

Klimushev A.I. (1973). Stability of nonlinear systems of differential equations with small parameter multiplying the derivatives. *Trudy Uralskogo Politekhn. Inst.*, no.211, p.44-54 . In Russ.

Nicosia S., Tomei P.(1992) State observers for rigid and elastic joint robots. *Robotics and CIM. 9,* no. 2. P. 113-120.

Spong M.W. (1987). Modelling and control of robots with elastic joints. *J. Dyn. Syst. Meas, Control, 109.* P.310-319.

NEW LYAPUNOV METHOD BASED METHODOLOGY FOR ASYMPTOTIC STABILITY

Ljubomir T. GRUJIĆ *

* Ecole Nationale d'Ingénieurs de Belfort, 525, 90016 Cedex, FRANCE.

Abstract. A new Lyapunov method based methodolgy for testing asymptotic stability of the zero state of a nonlinear system results from new conditions that are both necessary and sufficient for the asymptotic stability and which are established in the paper. Besides, the necessary and sufficient conditions are determined for both an exact construction of a system Lyapunov function and for a set to be the exact domain of the asymptotic stability.

Key Words. Lyapunov method, Lyapunov functions, Lyapunov stability, stability domains

Résumé. Une nouvelle méthodologie basée sur la méthode de Lyapounov est établie pour essayer la stabilité asymptotique de l'équilibre zéro d'un système non-linéaire en utilisant les nouvelles conditions qui sont nécessaires et suffisantes pour la stabilité asymptotique et qui sont établies dans ce papier. D'ailleurs, les conditions nécessaires et suffisantes sont déterminées pour les deux, pour la construction exacte de la fonction de Lyapounov du système et pour qu'un ensemble soit le domaine exacte de la stabilité asymptotique.

1. INTRODUCTON

The (second, direct) Lyapunov method (1892) is characterized by analysing stability properties of dynamical systems via testing positive definiteness property of a function v and (semi-) definiteness of its total time derivative D^+v along system motions. The well known methodology how to cary out the stability for nonlinear system, consists *first* in a choice of a positive definite function v and, *afterthat,* in verification of an appropriate (semi-) definiteness property of D^+v. Such a methodology results from the essence of Lyapunov's original stability conditions [Lyapunov (1892)] and from all those based on them or following them : for x=0 of a smooth nonlinear system $x^{(1)}=f(x)$ to be asymptotically stable it is both necessary and sufficient that there **exists** a positive definite v with negative definite D^+v. Those criteria and the resulting metholology do not clarify how to chose, construct v for a given system. Therefore, if D^+v is not negative (definite) for chosen v, then another v is to be selected. Since there is not a guideline for such a selection, the process can be endless. This explains why the problem of the Lyapunov function v construction has been left unsolved and why it is the fundamental problem of the Lyapunov stability theory. Another related fundamental problem is that of the exact determination of the asymptotic stability domain. Probably the best contribution to solving these problems in the framework of such a metholology is due to Zubov (1964. It suffers from the same drawbacks : the solution is expressed via the requirement for **existence** of two functions (of v and another function φ) with appropirate features. Howere, no gideline is given how to construct one of them (either v or φ) in order to calculate another one for a given nonlinear system. A number of other approahes to solving the problem of a Lyapunov function generation, or the problem of the exact determination of the asymptotic stabiliy domain, are also based on **existence** of the function.

The paper further develops a new Lyapunov method based methology established [Grujić (1989 - 1995)] that enables to solve the problems of the necessary and sufficient conditions for asymptotic stability and for an one-shot exact construction of a system Lyapunov function, and the problem of the necessary and sufficient conditions for an accurate on-shot determination of the asymptotic stability domain.

2. NOTATION

ClB_δ and ∂B_δ denote the closure and boundary of $B_\delta = \{x: \|x\| < \delta\}$ for $\|x\| = (x^T x)^{1/2}$ and $x \varepsilon R^n$. The empty set will be denoted by Φ. An open neighbourhood of $x = 0$ will be designated by S, and its open connected neighbourhood by A or N with In N used for the interior of N. The letters D_a, D_S and D will be used for the domain of attraction, of stability and of asymptotic stability, respecively. For their definitions and properties [see Grujić (1975, 1979, 1987, 1989 - 1995), Grujić et al. (1987) and Grujić and Pavella (1979)].

A motion of the system is denoted by $\underline{x}(.\ ;x_o)$ with $\underline{x}(.\ ;x_o) \equiv x_o$. If $v : R^n \to R$ is continuous then $D^+v(x) = \lim\sup\{[v[\underline{x}(\theta;x)] - v(x)]\theta^{-1} : \theta \to 0^-\}$.

3. MATHEMATICAL MODEL OF A SYSTEM

Let $f(.) : R^n \to R^n$ determine a mathematical model of a physical system via (1),

$$\frac{dx}{dt} = f(x),$$

which will be called the system. It will be required to posses one of the net properties :

Weak Smootthness Property :

(i) there is an open connected neighbourhood S of x = 0 such that for every $x_o \in S$:

(a) the system (1) has the unique solution $\underline{x}(.\ ;x_o)$, and

(b) the solution $\underline{x}(.\ ;x_o)$ is defined and continuous on $I_o \times S$, $I_o \subseteq R_+$, $I_o \neq \Phi$ and $I_o = I_o(x_o)$.

(ii) For every $x_o \in (R^n - S)$ every motion $\underline{x}(.\ ;x_o)$ of system (1) is continuous in $t \in I_o$.

Strong Smoothness Property

(i) System (1) has the Weak Smoothness Property.

(ii) If the boundary ∂S of S is non-empty then every motion of the system (1) passing through $x_o \in \partial S$ obeys:

$$\inf\left[\|x(t, x_o)\| : t \in I_o\right] > 0 \text{ for every } x_0 \in \partial S.$$

4. THEOREMS

A function $v(.) : R^n \to R$ is positive definite if and only if there is an open connected neighbourhood A of x = 0 such that

1) v(.) is continuous on A : $v(x) \in C(A)$,
2) $v(0) = 0$ and
3) $v(x) > 0$ for every $(x \neq 0) \in A$.

In order to construct a suitable function v(.) we introduce the next definition.

Definition 1. Q(s;f) is the family of all the functions $q(.) : R^n \to R$ with the folowing properties :

1) q(.) is continuous on S,
2) for any $\alpha > 0$ for which the set S is a neighbourhood of B_α there is $\beta > 0$, $\beta = \beta(\alpha ;q ;S)$ obeing $\inf[q(x) : x \in (S - B_\alpha)] = \beta$,
3) there is $\mu \in]0,+\infty[$, $\mu = \mu(q, f)$, such that there exists a solution to equations (2) with (2a) determined along motions of the system (1),

$$D^+v(x) = -[1 - v(x)]q(x) \qquad (2a)$$

$$v(0) = 0 \qquad (2b)$$

which is defined and continuous on Cl B_μ.

Notice that Definition 1 does allow the function $q(.) \in Q(S ;f)$ and a solution function v(.) to equations (2) to be indefinite.

Theorem 1. In order for the state x = 0 of system (1) with the Strong Smoothness Property to have the domain of asymptotic stability D and for a set N, $N \subseteq R^n$ *, to be the domain of ite asymptotic stability, N =D, it is both necessary and sufficient that*

1) the set N is open connected neighbourhood of x = 0 and $N \subseteq R^n$,

2) f(x) = 0 for $x \in N$ *if and only if* x =

0,

and,

3) *for arbitrarily chosen positive definite* q(.) *obeying* $q(.) \in Q(S;f)$, *the equations* (2) *have the unique solution function* v(.) *on* N *with the following properties* ;

(i) v(.) *is positive definite on* N,

(ii) *if the bondary* ∂N *of* N *is non-empty then* $v(x) \to +1$ *as* $x \to \partial N$, $x \in N$.

Proof. Omitted due to the space limitation [see Grujić (1989-1995)] of the proofs of analogous results. If the system possesses the Weak rather than the Strong Smoothness Propety then the criterion takes the next form :

Theorem 2. For the state x = 0 of system (1) possessing the Weak Smoothness Property to have the domain of asymptotic stability D and that a subset N of S, $N \subseteq R^n$, *aquales* D : N=D, *it is both necessary and sufficient that:*

1) *the set* N *is open connected neighourbhood of* $x = 0$,

2) $f(x) = 0$ *for* $x \varepsilon N$ *if and* $x = 0$, *and*

3) *for arbitrarily chosen positive definite function* q(.) *obeying* $q(.) \in Q(R^n;f)$, *equations* (2) *have the unique solution function* v(.) *on* N *with the following properties* :

(i) v(.) *is positive definite on* N, *and*

(ii) *if the boundary* ∂N *of* N *is non-empty then* $v(x) \to +1$ *as* $x \to \partial N$, $x \in N$.

5. EXAMPLES

Example 1. Let the second order system (1) take the following specific form :

$$\frac{dx_1}{dt} = -(20-|x_1|)(|x_1|+|x_2|)x_1 = f_1(x_1,x_2) = f_1(x), \qquad (3,1)$$

$$\frac{dx_{12}}{dt} = -(50-|x_2|)(|x_1|+|x_2|)x_2 = f_2(x_1,x_2) = f_2(x), \qquad (3,2)$$

$$f(.) = [f_1(.) \quad f_2(.)]^T, \; x = (x_1 \quad x_2)^T \qquad (3,3)$$

The system has the unique motion $\underline{x}(.\,;x_o)$ for every $x_o \varepsilon R^2$, wich is defined and continuous in $(t, x_o) \in R_+ \times R^2$. The set S_e of its equilibrium states is found as

$$S_e = \{x : x \in R^2, (x=0) \text{ or } (|x_1| = 20 \text{ and } |x_2| = 50)\}.$$

The set S_e suggests the set S as defined by

$$S = \{x ; x \varepsilon R^2, |x_1| < 20 \text{ and } |x_2| < 50\}. \qquad (4)$$

Since the set S_e is positive invariant set of the system (3) and $S_e \cap S = \{0\}$ then it is easy to verify that the system possesses the Strong Smoothness Property on set S (4). Let

$$q(x) = (|x_1| + |x_2|)^2 \qquad (5)$$

which yields the solution v(.) to equations (2) in the next form :

$$v(x) = 1 - \frac{(20-|x_1|)(50-|x_2|)}{1000} \qquad (6)$$

The function v(.) is well defined on ClB_μ for $\mu \varepsilon \,]0, 20[$. Since $\inf [q(x) : x \in (S - B_\alpha)] = \beta > 0$ for any $\alpha \in \,]0, 20[$, then we conclude that the positive definite function q(.) (5) obeys Definition 1, that is that $q \in Q(S, f)$ for S and f given by (4) and (3), respectively. The function v(.) obeys all the conditions of Theorem 1 on S. Since x = 0 is the unique equilibrium state of system (3) in S then set N = S is domain D of asymptotic stability of x = 0 of the system,

$$D = \{x : x \in R^2, |x_1| < 20 \text{ and } |x_2| < 50\}.$$

Disussion 1. If we follow the existing metholology we can accept a tentative Lyapunov function as

A) $v(x) = \frac{1}{2}(x_1^2 + x_2^2)$, or

B) $v(x) = |x_1| + |x_2|$.

A) the derivative of $v(x) = \frac{1}{2}(x_1^2 + x_2^2)$ along motions of the systems (3) is obtained as

$$\dot{v}(x) = -[(20-|x_1|)x_1^2 + (50-|x_2|)x_2^2](|x_1| + |x_2|)$$

Hence,

$$\dot{v}(x) < 0, \quad \forall (x \neq 0) \in S$$

and

$$\dot{v}(x) \leq 0, \quad \forall x \in \partial S$$

The existing theorems enable to conclude that $E = \{x : x \in R^2, |x_1| < 20\}$ is an estimate of domain D of asymptotic stability of x = 0. The same result can be deduced classically by using the function v(.) under B). In both cases $E \subset D$, $E \neq D$

Simulations were carried out for two different initial conditions :

a) Fig. 1 : the initial state x(o) was in the domain D asymptotic stabilty of x = 0 so that motion $x(. ; x_o)$ converges to x = 0.

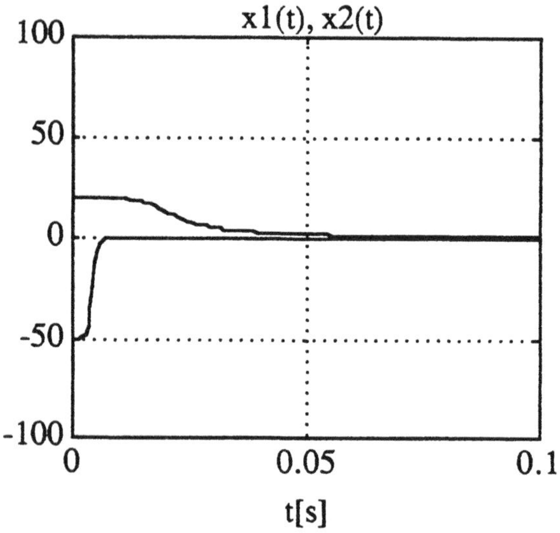

Fig. 1: x(0)= (19.9999 - 49.9999)$^T \in$ D.

b) Fig. 2 : the initial state x(o) was out of the asymptotic stability domain. The notion $x(. ; x_o)$ diverges vers infinity.

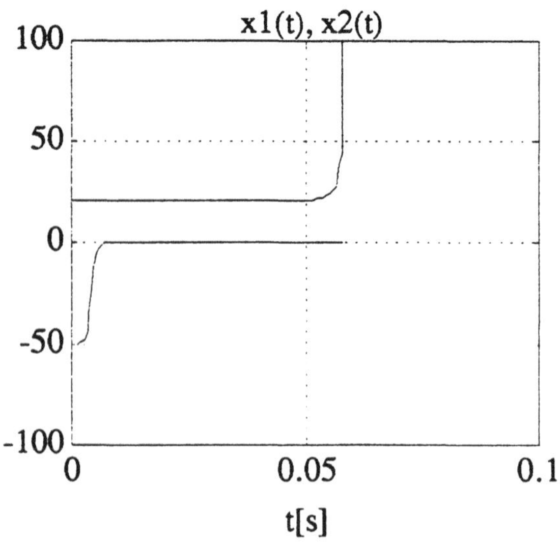

Fig. 2: x(0)= (20.0000000001 - 49.9999)$^T \notin$ D.

Discussion 2. The theorems can be given the following physical interpretation in case model (1) describes a physical system . Let $P(.) : R^n \to R$ and $E(.) : R^n \to R$ be system power and energy, respectively. If they are both definited and continuous on B_μ for some $\mu \in R_+$ and if the power P(.) obeys the condition 2) of Definition 1 then the functions q(.) and v(.) defined by

$$q(x) = -P(x),$$

and

$$v(x) = 1 - \exp[-E(x)],$$

obey completely the Definition 1. The conditions 3) of Theorem 1 (and Theorem 2) reduce to the following, respectively :

3') if the power P(.) is negative definite on N (on R^n) then :

i) the energy E(x) is to be positive definite on N , and

(ii) if the boundary ∂N of N is non-empty then $E(x) \to \infty$ as $x \to \partial R$, $x \in N$

6. CONCLUSION

The necessary and sufficient conditions for system Lyapunov function generation and determination of the asymptotic stability domain enable single step construction of the Lyapuov function and single asymtotic stability test : x = 0 of the system is not asymptoticaly stable if the function v constructed via Theorem 1 or Theorem 2 is not positive definite. There is not any sense to try to generate another v(.)- function. If all the conditions of either theorem hold then we exactly algorithmically solve the problems of the Lyapunov function generation and determination of the asymptotic stability domain. In case of the asymptotic stability in the whole the function v(.) generated via Theorem 1 or Theorem 2 can be, but need not be, radially unbounded.

The new methodolgy results from theorems. After verifying their conditions (i) and (ii), we should accept a positive definite q(.)

from Q(S , f) (Definition 1) *at first*, and *afterthat* we should verify properties of the solution v(.) to (2). This order of the steps is inverse to that of the well known methodology resulting from Lyapunov's results.

7. REFERENCES

Lj.T. Grujić, 1975, "Novel development of Lyapunov stability motion" *Int. J Control*, vol. 22, N°. 4, pp. 525 - 549.

Lj.T. Grujić, 1979, "Sets and singularly perturbed systems", *Systems Science*, Wroclaw (Poland) vol. 5, N°. 4, pp. 327 - 338.

Lj.T. Grujić, 1981, "On Absolute Stability and Aizerman Conjecture", *Automatica*, Vol. 17, N°. 2, pp. 335 - 349.

Lj.T. Grujić, 1987, Stability Domains of General and Large-Scale Stationary Systems", *Applied Modelling and Simulation of Technological Systems*", Ed. P. Borne and S. Y. Tzafestas, Elsevier Science Publishers, B. V. (North Holland) IMACS, pp. 317 - 327.

Lj.T. Grujić, 1989, "On Determination of Lyapunov Functions and the Asymptotic Stability Domain", (in Serbian), *Proc. of the 3rd Conference SAUM (on Systems, Automatic Control and Measurement)*, Vrnjachka Banja, Serbia, Yugoslavia, October 18-20, 1989, pp. 12 - 20.

Lj.T. Grujić, 1990, "Solutions to Lyapunov Stability Problems : Nonlinear Systems with Globally Differentiable Motions", in *"THE LYAPUNOV functions method and applications"*, ed. P. Borne and V. Matrosov, IMACS 1990, J.C. Baltzer Scientific Publishing Co., Basel, pp. 19 - 27.

Lj.T. Grujić, 1991, "Solutions to Lyapunov Stability Problems : Nonlinear Systems with Differentiable Motions", *Proc. of the 13th IMACS World Congress on Computation and Applied Mathematics*, Ed. R. Vichnevetsky and J.J.H. Miller, Dublin, Ireland, July 22-26, 1991, Vol. 3, pp. 1228 - 1231.

Lj.T. Grujić, 1991, "The Necessary and Sufficient Conditions for the Exact Construction of a Lyapunov Function and the Asymptotic Domain", *Proc. of the 30th Conference on Decision and Control*, Brighton, England, 11-13 December 1991, Vol. 3, pp. 2885 -2888.

Lj.T. Grujić, 1992, "On solutions to Lyapunov Stability Problems", *Facta Universitatis, Series: Mechanics, Automatic Control and Robotics*, University of Nish, Nish, Serbia, Vol. 1, No. 2, pp. 121 - 138.

Lj.T. Grujić, 1992, "Exact determination of a Lyapunov function and the asymptotic stability domain", *Int. J. Systems Sci.*, Vol. 23, No. 11, pp. 1871-1888.

Lj.T. Grujić, 1992, "Solutions to Lyapunov stability problems: nonlinear systems with differentiable motions", in *COMPUTATIONAL AND APPLIED MATHEMATICS II : DIFFERENTIAL EQUATIONS*, ed. W.F. Ames and P. J. van der Houwen, Elsevier Sci. Publ., Amsterdam, The Netherlands, submited for publication 24 February 1992.

Lj.T. Grujić, 1993, " Exact Both Construction of Lyapunov Function and Asymptotic Stability Domain Determination", *IEEE Inernational Conference on Systems, Man and Cybernetics*, Le Touquet, France, Vol. 1, pp. 331-336.

Lj.T. Grujić, 1994, "Solutions to Lyapunov Stability Problems of Sets : Nonlinear Systems with Differentiable Motions", *Int. Journ. of Mathematics & Mathematical Sciences*, Vol. 17, No. 1, pp. 103-

112.

Lj.T. Grujić, 1994, "Complete Exact Solution to the Lyapunov Stability Problem : Time-Varying Nonlinear Systems with Differentiable Motions", *Nonlinear Analysis:Theory, Methods & Applications*, Vol. 22., No. 8, pp. 971-981.

Lj.T. Grujić, 1994, "Solutions to Lyapunov Stability Problems : Time-Invariant Systems", *Proc. 14th IMACS World Congress*, Atlanta, Georgia, USA, Vol. 1, pp. 203-205.

Lj.T. Grujić, 1994, "Solutions to Lyapunov Stability Problems : Nonlinear Systems with Continuous Motions", *Int. Journ. of Mathematics & Mathematical Sciences*, Vol. 17, No. 3, pp. 587-596.

Lj.T. Grujić, 1995, "Exact Solutions for Asymptotic Stability: Nonlinear Systems", *Int. Journ. Non-Linear Mechanics*, Vol. 30, No. 1, pp. 45-56.

Lj.T. Grujić, 1995, "New-Lyapunov method based methodology for asymptotic stability", this proceedings.

Lj.T. Grujić, A. A. Martynyuk and M. Ribbens-Pavella 1987, "*Large-Scale Systems Stability under Structural and Singular Perturbations*", Springer Verlag, Berlin, (Russian ed. 1984, Naukova Dumka, Kiev).

Lj.T. Grujić and M. Ribbens-Pavella, 1979, "Asymptotic stability of large-scale systems with application to power systems. Part 1 : domain estimation", *Electrical Power and Energy Systems*, Vol. 1, N°. 63, pp. 151 -157.

A.M. Lyapunov , 1892 "*The General Problem of the stability of Motion* (in Russian), Kharkov Mathematical Society, Kharkov, English translation : 1992, *Int. J. C.* Control, Vol. 55, pp.531 -773, also the book, Taylor & Francis, London.

V.I. Zubov, 1964, "*Methods of A.M. Lyapunov and Their Application*", Nordhoff, Groningen.

GLOBAL STABILIZATION OF UNCERTAIN BILINEAR SYSTEMS BY DISCONTINUOUS FEEDBACK CONTROLS

J. WANG and D.P. GOODALL

Control Theory & Applications Centre, Coventry University, Priory Street, Coventry CV1 5FB, U.K.

Abstract. A class of uncertain dynamical control systems is considered, which is based on a bilinear prototype and modelled as a controlled differential inclusion system. Using Lyapunov-based theory, global asymptotic stability of a compact set, containing the system state origin, is investigated. A class of generalized state-feedbacks, formulated as set-valued maps, is developed which guarantees the stability property for such a class of bilinear uncertain systems. Practical controls are any particular selections, chosen from a proposed generalized state-feedback, and these feedback controls are discontinuous in nature.

Résumé. Une classe de systèmes dynamiques de contrôle avec présence d'incertitude est présentée. Cette classe est basée sur un prototype bilinéaire et modélisée tel un système commandé à inclusion différentielle. En utilisant la théorie de Lyapunov, la stabilité asymptotique globale d'un ensemble compacte contenant l'origine est étudiée. Une classe de rétroaction d'états généralisés, exprimée sous la forme de multi-fonctions est développée qui assure la stabilité de cette classe de systèmes bilinéaires incertains. Le choix pratique d'une commande peut se faire à partir de n'importe quel élément issu d'une multi-fonction à condition d'être sélectioné parmi la rétroaction d'états généralisés proposée. Ces commandes à rétroactions sont discontinues par nature.

Key Words. Bilinear systems; controlled differential inclusions; discontinuous feedbacks; global stabilization; uncertain systems.

1. INTRODUCTION

Almost all practical industrial systems exhibit nonlinear characteristics and are also subject to various forms of disturbances. Mathematical models of real systems always contain uncertain elements which model the designer's lack of knowledge about parameter values, disturbances and the imprecision in the model. Uncertainties, arising from imperfect knowledge of system inputs and inaccuracies in the mathematical modelling itself, may contribute to performance degradation of the feedback control system. Of particular interest to a control designer is the problem of synthesizing a controller which guarantees stability properties for the system in the presence of uncertainties.

To obtain desired system response, two approaches have been widely used: stochastic and deterministic control. Using a deterministic approach, based on Lyapunov stability theory, numerous papers dealing with the control of continuous-time systems containing uncertainties have been published (see, for example, Corless and Leitmann (1981), Ryan and Corless (1984), Qu (1993); for overviews, see Corless (1993), Leitmann (1993) and Zinober (1994); see, also, the bibliographies therein).

In this paper, a deterministic control design method is formulated for a class of uncertain systems which is based on a known bilinear prototype system. Bilinear systems are an important subclass of nonlinear systems which have numerous applications to engineering, biology, and economics. A general time-invariant bilinear system can be modelled as

$$\dot{x}(t) = Ax(t) + \sum_{i=1}^{m} u_i(t)B_i x(t) + Cu(t) \qquad (1)$$

where $x(t) \in \mathbb{R}^n$, $u(t) = [u_1(t) \ldots u_m(t)]^{\mathrm{T}} \in \mathbb{R}^m$, $m \leq n$, $A, B_i \in \mathbb{R}^{n\times n}$ and $C \in \mathbb{R}^{n\times m}$. The bilinear system is called *homogeneous-in-the-state* if $C = 0$, *homogeneous-in-the-input* if $A = 0$ and *strictly bilinear* if $A = 0$ and $C = 0$ (see Bacciotti (1992)). Such a class of bilinear systems (1), with $C \neq 0$, is widely adopted to model practical industrial systems (see, for example, Mohler (1971)).

The prototype system (1) is assumed to be subject to uncertainties, modelled additively by the addition of unknown functions and nonempty multifunctions (i.e. set-valued maps) which comprise all possible uncertainties in the system description together with any known time-dependent or nonlinear elements. More specifically, the uncertain system is assumed to be modelled by the controlled differential inclusion system:

$$\begin{aligned}\dot{x}(t) \in\ & Ax(t) \\ &+ \sum_{i=1}^{m}[u_i(t) + q_i(t, x(t), u(t))]B_i x(t) \\ &+ C[u(t) + \mathcal{G}(t, x(t), u(t))] \\ &+ \mathcal{H}(t, x(t), u(t)) \qquad (2)\end{aligned}$$

where the functions $q_i : \mathbb{R} \times \mathbb{R}^n \times \mathbb{R}^m \to \mathbb{R}$ are continuous and model uncertainties which are 'matched' with respect to B_i, the multifunction $\mathcal{G} : \mathbb{R} \times \mathbb{R}^n \times \mathbb{R}^m \rightsquigarrow \mathbb{R}^m$ (i.e. $\mathcal{G}$ maps $\mathbb{R} \times \mathbb{R}^n \times \mathbb{R}^m$ into subsets of $\mathbb{R}^m$) models uncertainties which are 'matched' with respect to C (assumed to be full rank), and $\mathcal{H} : \mathbb{R} \times \mathbb{R}^n \times \mathbb{R}^m \rightsquigarrow \ker(C^T)$ models 'residual' uncertainty. The known multifunctions $\mathcal{G}$ and $\mathcal{H}$ are assumed to be nonempty and have properties of upper semicontinuity, convexity and compactness of their values.

Remark: A multifunction $\mathcal{G} : \mathbb{R}^p \rightsquigarrow \mathbb{R}^q$, with compact values, is upper semicontinuous at $\alpha \in \mathbb{R}^p$ iff, for each $\varepsilon > 0$, there exists a $\delta > 0$ such that $\mathcal{G}(\bar{\alpha}) \subset \mathcal{G}(\alpha) + \varepsilon \mathbb{B}_q$, where $\mathbb{B}_q$ denotes the open unit ball centred at the origin in $\mathbb{R}^q$, for all $\bar{\alpha} \in \alpha + \delta \mathbb{B}_p$.

For uncertain, homogeneous-in-the-state, bilinearly coupled systems, the stability problem has been investigated by Goodall and Ryan (1991) and the problem of practical stabilization discussed in Goodall and Wang (1993). In this paper, a class of generalized state-feedback controls, with practical analogues in the form of discontinuous feedbacks, has been synthesized to globally uniformly asymptotically stabilize the class of uncertain bilinear systems for the case $C \neq 0$ and a further discussion on the case $C = 0$ is included.

2. PROBLEM FORMULATION

It is assumed that the ensuing hypothesis holds for the bilinear prototype system.

H1: (A, C) is a stabilizable pair.

As a consequence of H1, there exists a feedback matrix $K \in \mathbb{R}^{m \times n}$ such that $(A + CK)$ is asymptotically stable. Thus, given a symmetric $Q > 0$, there exists a unique, real, symmetric matrix $P > 0$ that satisfies the Lyapunov equation

$$P(A + CK) + (A + CK)^T P + Q = O \ . \qquad (3)$$

Hypotheses characterizing the uncertainty in system (2) are prescribed below.

For a nonempty compact set $\mathcal{A} \subset \mathbb{R}^n$, or $\mathbb{R}^m$ (as appropriate), $\xi(\mathcal{A}) := \max\{\|a\| : a \in \mathcal{A}\}$; moreover, if $\mathcal{A} = \emptyset$, then $\xi(\mathcal{A}) := 0$. Define the multifunctions $\mathcal{P}_1 : \mathbb{R}^n \rightsquigarrow \mathbb{R}^m$ and $\mathcal{P}_2 : \mathbb{R}^n \rightsquigarrow \mathbb{R}^n$ by

$$x \mapsto \mathcal{P}_1(x) := \{v \in \mathbb{R}^m : \langle v, C^T P x\rangle \geq 0\}$$

and

$$x \mapsto \mathcal{P}_2(x) := \{w \in \mathbb{R}^n : \langle w, Px\rangle \geq 0\} \ .$$

H2:

(a) There exist real constants $\alpha_1, \alpha_3, \alpha_4 \geq 0$ and $0 \leq \alpha_2 < 1$ such that, for all (t, x, u) and for all i,

$$|q_i(t, x, u)| \leq \alpha_1 + \alpha_2|u_i| + \alpha_3\|x\| + \alpha_4|\langle x, PB_i x\rangle| .$$

(b) $\mathcal{G}(t, x, u) = \{[w_1 \ldots w_m]^T \in \mathbb{R}^m : w_i \in \mathcal{G}_i(t, x, u)\}$ and there exist real constants $\beta_1, \beta_2, \beta_3 \geq 0$ such that, for all i and (t, x, u),

$$\xi\left(\mathcal{G}_i(t, x, u) \cap \mathcal{P}_1(x)\right) \leq \beta_1 + \beta_2|u_i| + \beta_3\|x\| \ .$$

(c) There exists a real constants $\gamma_1, \gamma_2 \geq 0$ such that, for all (t, x, u),

$$\xi\left(\mathcal{H}(t, x, u) \cap \mathcal{P}_2(x)\right) \leq \gamma_1\|x\| + \gamma_2 \sum_{i=1}^{m} |\langle x, PB_i x\rangle| \ .$$

The principle objective is to determine a class of feedback controls which globally uniformly asymptotically stabilize the uncertain system (2). The admissible feedbacks are embedded in a class of *generalized state-feedbacks*.

Definition 1 A multifunction $\mathcal{F} : \mathbb{R}^n \rightsquigarrow \mathbb{R}^m$ is a *generalized state-feedback* if

(*i*) $\mathcal{F}$ is upper semicontinuous with nonempty, convex and compact values;

(*ii*) $\mathcal{F}$ is singleton-valued except on a set $\Sigma_{\mathcal{F}}$ of Lebesgue measure zero in $\mathbb{R}^n$.

Let $\overline{\mathbb{B}}_n$ denote the closure of $\mathbb{B}_n$ and let $\mathcal{A} \supset \{0\}$ be compact. Define $\rho(a, \mathcal{A}) := \inf\{\|a - \alpha\| : \alpha \in \mathcal{A}\}$ to be the distance from a point $a \notin \mathcal{A}$ to the set $\mathcal{A}$ and define $\mathcal{N}(\mathcal{A}, \varepsilon) := \{x \in \mathbb{R}^n : \rho(a, \mathcal{A}) < \varepsilon\}$ to be the ε-neighbourhood of $\mathcal{A}$.

Definition 2 A compact set $\mathcal{A}$, containing the state origin of system (2), is globally uniformly asymptotically stable if the following conditions hold.

(a) *Existence and indefinite continuation of solutions.* For each $t_0 \in \mathbb{R}$ and $x^0 \in \mathbb{R}^n$, there exists a local solution $x : [0, \tau) \to \mathbb{R}^n$ (i.e. an absolutely continuous function satisfying (2) a.e. and $x(t_0) = x^0$), and every such solution can be extended into a solution on $[t_0, \infty)$.

(b) *Uniform boundedness of solutions.* For each $\delta > 0$, there exists $d(\delta) > 0$ (independent of t_0) such that $x(t) \in d(\delta)\overline{\mathbb{B}}_n$ for all $t \geq t_0$ on every solution $x(\cdot)$ of (2) with $x^0 \in \delta\overline{\mathbb{B}}_n$.

(c) *Uniform stability of $\mathcal{A}$.* For each $\varepsilon > 0$ and $t_0 \in \mathbb{R}$, there exists $r(\varepsilon) > 0$ (independent of t_0) such that $x(t) \in \mathcal{N}(\mathcal{A}, \varepsilon)$ for all $t \geq t_0$ on every solution $x(\cdot)$ of (2) with $x(0) \in \mathcal{N}(\mathcal{A}, r(\varepsilon))$.

(d) *Global uniform attractivity of $\mathcal{A}$.* For each $\delta > 0$ and $\varepsilon > 0$, there exists $T(\delta, \varepsilon) \geq 0$ (independent of t_0) such that $x(t) \in \mathcal{N}(\mathcal{A}, \varepsilon)$ for all $t \geq t_0 + T(\delta, \varepsilon)$ on every solution $x(\cdot)$ of (2) with $x^0 \in \mathcal{N}(\mathcal{A}, \delta)$ and these conditions hold for δ arbitrarily large.

To show that the differential inclusion system (2) has indefinite continuation of solutions, the following proposition (see Ryan (1990)) is required. For existence of a local solution see Aubin and Cellina (1984).

Proposition 1 For each $x^0 \in \mathbb{R}^n$ and continuous $u : \mathbb{R} \to \mathbb{R}^m$, there exists a local solution of (2) which can be extended into a maximal solution $x : [0, \tau) \to \mathbb{R}^n$ and if $x(\cdot)$ is bounded, then $\tau = \infty$.

3. CLASS OF GENERALIZED STATE-FEEDBACKS

The proposed class of generalized state-feedbacks consist of multifunctions $\mathcal{F} = [\mathcal{F}_1 \dots \mathcal{F}_m]^{\mathrm{T}} : \mathbb{R}^n \rightsquigarrow \mathbb{R}^m$, with convex and compact values, having the following structure:

$$\begin{aligned} x \mapsto \mathcal{F}_i(x) &:= k_i x - \mu\langle x,\ PB_i x\rangle \\ &- \phi_i(x)\mathcal{S}(c_i^{\mathrm{T}} Px) \\ &- \psi_i(x)\mathcal{S}(\langle x,\ PB_i x\rangle)\ , \end{aligned} \tag{4}$$

where k_i denotes the ith row of $K \in \mathbb{R}^{n\times m}$, c_i denotes the ith column of $C \in \mathbb{R}^{n\times m}$, $\mathcal{S} : \mathbb{R} \rightsquigarrow \mathbb{R}$ is defined by

$$z \mapsto \mathcal{S}(z) := \begin{cases} \{|z|^{-1}z\}\ , \text{ if } \ z \neq 0 \\ [-1,\ 1]\ , \text{ if } \ z = 0 \end{cases}$$

and, for each i, $\phi_i, \psi_i : [0, \infty) \to [0, \infty)$ are continuous functions specified, by the design procedure, as

$$\phi_i(x) := \bar{\phi}_1 + \bar{\phi}_2\|x\| + \bar{\phi}_3|\langle x,\ PB_i x\rangle| \tag{5}$$

$$\psi_i(x) := \bar{\psi}_1 + \bar{\psi}_2\|x\| + \bar{\psi}_3|c_i^{\mathrm{T}} Px|\ , \tag{6}$$

$\bar{\phi}_i,\ \bar{\psi}_i \geq 0$ for all i. By definition, $\mathcal{S}$ is upper semicontinuous and has convex and compact values and, hence, each $\mathcal{F}_i$ qualifies as a generalized state-feedback.

4. GLOBAL STABILIZATION

Consider the differential inclusion system:

$$\dot{x}(t) \in \mathcal{M}(t, x(t))\ , \qquad x(t_0) = x^0\ , \tag{7}$$

where $\mathcal{M} : \mathbb{R} \times \mathbb{R}^n \rightsquigarrow \mathbb{R}^n$ is defined by

$$\mathcal{M}(t,x) := \Bigg\{ Ax + \sum_{i=1}^{m}[u_i + q_i(t,x,u)]B_i x + C[u + \mathcal{G}(t,x,u)] + \mathcal{H}(t,x,u)\ :\ u \in \mathcal{F}(x) \Bigg\}\ .$$

The multifunction $(t, x) \mapsto \mathcal{M}(t, x)$ is upper semicontinuous with nonempty, convex and compact values and, hence, for each x^0, the initial value problem (7) admits a maximal solution $x : [0, \tau) \to \mathbb{R}^n$. A Lyapunov function candidate, $v : \mathbb{R}^n \to [0, \infty)$, is chosen to be

$$x \mapsto v(x) := \langle x,\ Px\rangle\ .$$

A straightforward calculation shows that, along each maximal solution $x(\cdot)$ of (7),

$$\begin{aligned} \dot{v}(x(t)) \leq &- \left(\tfrac{1}{2}\sigma_{\min}(Q) - \gamma_1\sigma_{\max}(P)\right)\|x\|^2 \\ &- (\nu_1 - \beta_1)\sum_{i=1}^{m}|c_i^{\mathrm{T}}Px| \\ &- (\nu_2 - \beta_2\|K\| - \beta_3)\|x\|\sum_{i=1}^{m}|c_i^{\mathrm{T}}Px| \\ &+ \bar{\psi}_3(1+\beta_2)\sum_{i=1}^{m}|c_i^{\mathrm{T}}Px|^2 \\ &- (\nu_3 - \mu(1+\beta_2))\sum_{i=1}^{m}|\langle x,\ PB_i x\rangle|\,|c_i^{\mathrm{T}}Px| \\ &- (\nu_4 - \alpha_1)\sum_{i=1}^{m}|\langle x,\ PB_i x\rangle| \\ &- (\nu_5 - \alpha_3 - (1+\alpha_2)\|K\| \\ &\qquad -\gamma_2\sigma_{\max}(P))\|x\|\sum_{i=1}^{m}|\langle x,\ PB_i x\rangle| \\ &- (\nu_6 - \alpha_4)\sum_{i=1}^{m}|\langle x,\ PB_i x\rangle|^2\ , \end{aligned}$$

where

$$\begin{aligned} \nu_1 &:= \bar{\phi}_1(1-\beta_2) - \bar{\psi}_1(1+\beta_2) \\ \nu_2 &:= \bar{\phi}_2(1-\beta_2) - \bar{\psi}_2(1+\beta_2) \\ \nu_3 &:= \bar{\psi}_3(1-\alpha_2) + \bar{\phi}_3(1-\beta_2) \\ \nu_4 &:= \bar{\psi}_1(1-\alpha_2) - \bar{\phi}_1(1+\alpha_2) \\ \nu_5 &:= \bar{\psi}_2(1-\alpha_2) - \bar{\phi}_2(1+\alpha_2) \\ \nu_6 &:= \mu(1-\alpha_2) - \bar{\phi}_3(1+\alpha_2)\ . \end{aligned}$$

If the design parameters μ, $\bar{\psi}_1$, $\bar{\psi}_2$, $\bar{\psi}_3$ are chosen to be

$$\begin{aligned}\mu &:= (1-\alpha_2)^{-1}[\bar{\phi}_3(1+\alpha_2)+\alpha_4]\\ \bar{\psi}_1 &:= (1-\alpha_2)^{-1}[\bar{\phi}_1(1+\alpha_2)+\alpha_1]\\ \bar{\psi}_2 &:= (1-\alpha_2)^{-1}[\bar{\phi}_2(1+\alpha_2)+\alpha_3\\ &\qquad + (1+\alpha_2)\|K\| + \gamma_2\sigma_{\max}(P)]\\ \bar{\psi}_3 &:= (1-\alpha_2)^{-2}[2\bar{\phi}_3(\alpha_2+\beta_2)+\alpha_4(1+\beta_2)]\end{aligned}$$

then, along all maximal solutions to (7) and for almost all t,

$$\begin{aligned}\dot{v}(x(t)) &\le -\left(\tfrac{1}{2}\sigma_{\min}(Q) - \gamma_1\sigma_{\max}(P)\right)\|x\|^2\\ &+(\bar{\nu}_1(1-\alpha_2)^{-1}+\beta_1)\sum_{i=1}^{m}|c_i^{\mathrm{T}}Px|\\ &+((1-\alpha_2)^{-1}[\bar{\nu}_2+(1+\beta_2)((1+\alpha_2)\|K\|\\ &+\gamma_2\sigma_{\max}(P))]+\beta_2\|K\|+\beta_3)\|x\|\sum_{i=1}^{m}|c_i^{\mathrm{T}}Px|\\ &+\bar{\nu}_3\frac{(1+\beta_2)}{(1-\alpha_2)^2}\sum_{i=1}^{m}|c_i^{\mathrm{T}}Px|^2\ ,\end{aligned}$$

where

$$\begin{aligned}\bar{\nu}_1 &:= 2\bar{\phi}_1(\alpha_2+\beta_2)+\alpha_1(1+\beta_2)\\ \bar{\nu}_2 &:= 2\bar{\phi}_2(\alpha_2+\beta_2)+\alpha_3(1+\beta_2)\\ \bar{\nu}_3 &:= 2\bar{\phi}_3(\alpha_2+\beta_2)+\alpha_4(1+\beta_2)\ .\end{aligned}$$

The design parameters $\bar{\phi}_1$, $\bar{\phi}_2$, $\bar{\phi}_3$ are, as yet, unspecified and, hence, their values may be chosen as $\bar{\phi}_1 = \bar{\phi}_2 = \bar{\phi}_3 = 0$. This has the advantage of producing a controller with a more simple structure than that given by (4)-(6). Thus, the generalized state-feedbacks have the form :

$$\begin{aligned}x \mapsto \mathcal{F}_i(x) &:= k_i x - \mu\langle x,\, PB_i x\rangle\\ &\quad - \psi_i(x)\mathcal{S}(\langle x,\, PB_i x\rangle)\ ,\end{aligned}\tag{8}$$

where

$$\mu := (1-\alpha_2)^{-1}\alpha_4 \tag{9}$$

$$\bar{\psi}_1 := (1-\alpha_2)^{-1}\alpha_1 \tag{10}$$

$$\begin{aligned}\bar{\psi}_2 &:= (1-\alpha_2)^{-1}[\alpha_3+(1+\alpha_2)\|K\|\\ &\qquad + \gamma_2\sigma_{\max}(P)]\end{aligned}\tag{11}$$

$$\bar{\psi}_3 := (1-\alpha_2)^{-2}\alpha_4(1+\beta_2)\ . \tag{12}$$

A simple calculation shows that

$$\dot{v}(x(t)) \le -\tfrac{1}{2}(\sigma_{\min}(Q)-\eta)\|x\|^2 + \zeta\|x\|\ , \tag{13}$$

where

$$\begin{aligned}\eta &:= 2\left[\gamma_1\sigma_{\max}(P)+\bar{\eta}_1+\bar{\eta}_2+\bar{\eta}_3\right] > 0\\ \bar{\eta}_1 &:= m\alpha_4\left(\frac{1+\beta_2}{1-\alpha_2}\right)^2\|C^{\mathrm{T}}P\|^2\\ \bar{\eta}_2 &:= m[\alpha_3+(1+\alpha_2)\|K\|\\ &\qquad +\gamma_2\sigma_{\max}(P)]\left(\frac{1+\beta_2}{1-\alpha_2}\right)\|C^{\mathrm{T}}P\|\\ \bar{\eta}_3 &:= m(\beta_2\|K\|+\beta_3)\|C^{\mathrm{T}}P\|\\ \zeta &:= m\left(\alpha_1\left(\frac{1+\beta_2}{1-\alpha_2}\right)+\beta_1\right)\|C^{\mathrm{T}}P\| \ge 0\ .\end{aligned}$$

If the design parameter Q can be chosen to satisfy $\sigma_{\min}(Q) > \eta$ then, as a consequence of Proposition 1, every maximal solution can be extended into a solution on $[0,\infty)$, since finite escape times are precluded by (13). Moreover, by (13), properties of uniform boundedness, stability and attractivity hold and, hence, any compact set containing the ellipsoid

$$\mathcal{E} := \left\{x \in \mathbb{R}^n : \langle x,\, Px\rangle \le \frac{\sqrt{2}\zeta\sigma_{\max}(P)}{\eta[\sigma_{\min}(P)]^{\frac{1}{2}}}\right\}$$

is globally uniformly asymptotically stable.

Theorem 1 Under hypotheses H1-H2, and with $u = \mathcal{F}(x)$, given by (8), (9)-(12), if $\sigma_{\min}(Q) > \eta$, then any compact set containing the ellipsoid $\mathcal{E}$ is globally uniformly asymptotically stable for the differential inclusion system (7).

Noting that $\zeta = 0$ when $\alpha_1 = \beta_1 = 0$, the following corollary may be deduced.

Corollary 1 Under hypotheses H1-H2, and with $u = \mathcal{F}(x)$, given by (8), (9)-(12), if $\alpha_1 = \beta_1 = 0$ and $\sigma_{\min}(Q) > \eta$, then $\{0\}$ is globally uniformly asymptotically stable under the dynamics of (7).

5. HOMOGENEOUS-IN-THE-STATE

Suppose $C \equiv O$ in the bilinear prototype, modelled by (1). Then the prototype system is homogeneous-in-the-state and the associated system with uncertainty is modelled as

$$\begin{aligned}\dot{x}(t) &\in Ax(t)\\ &+\sum_{i=1}^{m}[u_i(t)+q_i(t,x(t),u(t))]B_i x(t)\\ &+\mathcal{H}(t,x(t),u(t))\end{aligned}\tag{14}$$

The bilinear prototype is assumed to be subject to the following hypotheses.

H3:

(a) A is asymptotically stable, that is $\sigma(A) \in \mathbb{C}^-$.
(b) There exists a nonempty set $\Omega \subset \mathbb{R}^n \setminus \{0\}$ and, for each $x \in \Omega$, there exists $k \in \mathbb{N}$ such that the dimension of

$$\mathrm{span}\{Qx\ Ax\ ad^0(A,B_i)x \ldots ad^k(A,B_i)x :\ i = 1,\ldots m\}$$

is n, where the operators $ad^j(\bullet,\bullet)$ are defined recursively as :

$$\begin{aligned} ad^0(A,B_i) &:= B_i \\ ad^1(A,B_i) &:= ad(A,B_i) := AB_i - B_iA \\ ad^j(A,B_i) &:= ad(A, ad^{j-1}(A,B_i)) \ , \end{aligned}$$

with $j = 2, 3, \ldots$.

Remarks:
(i) As a consequence of H3(a), given any symmetric $Q > 0$, there exists a unique, symmetric $P > 0$ which satisfies the Lyapunov equation

$$PA + A^{\mathrm{T}}P + Q = O \ .$$

(ii) H3(b) is essentially a controllability hypothesis for the bilinear prototype (1) with $C = O$. See Jurdjevic and Quinn (1978) and Bacciotti (1992) for more details.

Assumptions on the 'matched' uncertainty (with respect to B_i) and the 'residual' uncertainty are as stated in H2(a) and (c), respectively. The generalized state-feedbacks are designed as

$$x \mapsto \mathcal{F}_i(x) := -\bar{\mu}\langle x,\ PB_ix\rangle - \theta_i(x)\mathcal{S}(\langle x,\ PB_ix\rangle) \ , \quad (15)$$

where $\theta_i(x) = \bar{\theta}_1 + \bar{\theta}_2\|x\|$ and $\bar{\theta}_1,\ \bar{\theta}_2 \geq 0$ are design parameters. Choosing

$$\mu \geq \alpha_4(1-\alpha_2)^{-1} \ , \quad (16)$$

$$\bar{\theta}_1 \geq \alpha_1(1-\alpha_2)^{-1} \ , \quad (17)$$

$$\bar{\theta}_2 \geq (\alpha_3 + \gamma_2\sigma_{\max}(P))(1-\alpha_2)^{-1} \ , \quad (18)$$

one may deduce, using Lyapunov's direct method, that for almost all t

$$\dot{v}(x(t)) \leq -\left(\tfrac{1}{2}\sigma_{\min}(Q) - \gamma_1\sigma_{\max}(P)\right)\|x\|^2 \ ,$$

along all maximal solutions to (14). Thus, the following theorem is obtained.

Theorem 2 Suppose hypotheses H2(a), H2(c), H3 hold and $u = \mathcal{F}(x)$ is given by (15)-(18). If the design parameter Q can be chosen so that $\sigma_{\min}(Q) > 2\gamma_1\sigma_{\max}(P)$, then $\{0\}$ is globally uniformly asymptotically stable for the feedback-controlled differential inclusion system (14).

6. EXAMPLE

To illustrate the above theory, the dynamical system

$$\dot{x}(t) \in Ax(t) + [u(t) + q(t,x(t),u(t))]Bx(t) + C[u(t) + \mathcal{G}(t,x(t),u(t))] + \mathcal{H}(t,x(t),u(t)) \ , \quad (19)$$

where $x(t) \in \mathbb{R}^2$, $u(t) \in \mathbb{R}$,

$$A = \begin{bmatrix} 0 & 1 \\ -5 & -5 \end{bmatrix}, \quad C = \begin{bmatrix} 0 \\ 1 \end{bmatrix}, \quad K = [0 \ \ 0] \ ,$$

is adopted. Given

$$Q = \begin{bmatrix} 1 & 0 \\ 0 & 1 \end{bmatrix},$$

the solution to the Lyapunov equation (3) is

$$P = \frac{1}{50}\begin{bmatrix} 55 & 5 \\ 5 & 6 \end{bmatrix} .$$

Then

$$\sigma_{\max}(P) = \frac{1}{100}\left(61 + \sqrt{2501}\right)$$

and

$$C^{\mathrm{T}}P = \begin{bmatrix} \frac{1}{10} & \frac{3}{25} \end{bmatrix} .$$

It is assumed that, for all (t,x,u),

$$|q(t,x,u)| \leq \alpha_1 + \alpha_2|u| + \alpha_3\|x\| + \alpha_4|\langle x,\ PBx\rangle| ,$$

$$\xi(\mathcal{G}(t,x,u) \cap \mathcal{P}_1(x)) \leq \beta_1 + \beta_2|u| + \beta_3\|x\| \ ,$$

$$\xi(\mathcal{H}(t,x,u) \cap \mathcal{P}_2(x)) \leq \gamma_1\|x\| + \gamma_2|\langle x,\ PBx\rangle| \ ,$$

where the uncertainty parameters have the following values :

$$\alpha_1 = 0 \quad \alpha_2 = 0.1 \quad \alpha_3 = 0.4 \quad \alpha_4 = 0.5$$

$$\beta_1 = 0 \quad \beta_2 = 0.2 \quad \beta_3 = 0.5 \quad \gamma_1 = 0.2 \quad \gamma_2 = 0.4 \ .$$

For these values $\eta < 0.996$ and, therefore, the inequality $\sigma_{\min}(Q) > \eta$ is satisfied. The design parameters μ, $\bar{\psi}_1$ $\bar{\psi}_2$ $\bar{\psi}_3$ are chosen to have the values

$$\mu = \frac{5}{9} \qquad \bar{\psi}_1 = 0 \quad (20)$$

$$\bar{\psi}_2 = \frac{4}{9}(1 + \sigma_{\max}(P)) \qquad \bar{\psi}_3 = \frac{2}{3} \quad (21)$$

and the generalized state-feedback (8) becomes

$$\mathcal{F}(x) = -\mu\langle x,\ PBx\rangle - (\bar{\psi}_2\|x\| + \bar{\psi}_3|C^{\mathrm{T}}Px|)\mathcal{S}(\langle x,\ PBx\rangle) \ . \quad (22)$$

Thus, with $\mathcal{F}$ given by (20)-(22), $\{0\}$ is globally uniformly asymptotically stable under the dynamics of the uncertain system (19).

7. REFERENCES

Aubin, J.-P., and A. Cellina (1984). *Differential Inclusions.* Springer-Verlag, New York.

Bacciotti, A. (1992). *Local Stabilizability Of Nonlinear Control Systems.* World Scientific, Singapore.

Corless, M. (1993). Control of uncertain nonlinear systems. *ASME J. Dyn. Sys., Meas., and Contr.*, **115**, 362-372.

Corless, M. and G. Leitmann (1981). Continuous state feedback guaranteeing uniform ultimate boundedness for uncertain dynamic systems. *IEEE Trans. Aut. Contr.*, **AC-26**, 1139-1144.

Goodall, D.P. and E.P. Ryan (1991). Feedback stabilization of a class of nonlinearly coupled uncertain dynamical systems. *IMA J. Math. Contr. and Inf.*, **8**, 81-92.

Goodall, D.P. and J. Wang (1993). Practical stabilization of a class of uncertain composite control systems. *IEE Colloquium on Advances in the Application of Robust Controllers*, IEE Digest No. **1993/220**.

Jurdjevic, V. and J.P. Quinn (1978). Controllability and stability. *J. Diff. Eqns.*, **28**, 381-389.

Leitmann, G. (1993). On one approach to the control of uncertain systems. *ASME J. Dyn. Syst. Meas. Contr.*, bf 115, 372-382.

Mohler, R.P. (1973). *Bilinear Control Processes.* Academic Press, New York.

Qu, Z. (1993). Robust control of nonlinear uncertain systems under generalized matching conditions. *Automatica*, **29**, 985-998.

Ryan, E.P. (1990). Discontinuous feedback and universal adaptive stabilization. In *Control Of Uncertain Dynamical Systems* (eds. Hinrichsen and Mårtensson), Birkhauser, Boston.

Ryan, E.P. and M. Corless (1984). Ultimate boundedness and asymptotic stability of a class of uncertain dynamical systems via continuous and discontinuous feedback control. *IMA J. Math. Contr. and Inf.*, **1**, 223-242.

Zinober, A.S.I. (1994). *Variable Structure And Lyapunov Control.* Springer-Verlag, London.

ASYMPTOTIC STABILITY VIA ENERGY AND POWER. PART I : NEW LYAPUNOV METHODOLOGY FOR NONLINEAR SYSTEMS

LJUBOMIR T. GRUJIĆ* and GENEVIÈVE DAUPHIN-TANGUY**

* Laboratoire de Mécanique et Productique, Ecole Nationale d'Ingénieurs de Belfort, BP 525, 90016 Belfort -Cedex, FRANCE
** Laboratoire d'Automatique et informatique de Lille ; CNRS, Ecole centrale de Lille, BP 48, Cité Scientifique, 59650 Villeneuve d'Ascq - Cedex, FRANCE

Abstract. A large class of nonlinear dynamical physical systems is considered. Complete solutions for asymptotic stability of a state and for a set to be the asymptotic stability domain are discovered in terms of the stystem power and energy. The power is permitted to be semi-definite. The necessary and sufficient conditions are imposed on the energy.This reflects a new methodology of testing the asymptotic stability. It is essentially inverse relative to the widely used methodology of the Lyapunov method application to nonlinear systems. The first part of this two-part paper presents the theoretical foundation of the new methodology and criteria.

Key Words. Lyapunov method, Lyapunov functions, asymptotic stability and domains.

Résumé. Une classe large des systèmes physiques non-linéaires est considérée. Les solutions complètes pour la stabilité asymptotique de l'état et pour qu'un ensemble soit le domaine de la stabilité asymptotique sont découvertes en terme de l'énergie et de la puissance du système. La puissance peut-être demi-finie. Les conditions nécessaires et suffisantes sont imposées sur l'énergie. Cela reflète une nouvelle méthodologie de l'essai de la stabilité asymptotique. Elle est essentiellement inversée par rapport à la méthodologie utilisée, largement de l'application de la méthode de Lyapounov aux systèmes non-linéaires. La première partie de ce papier comportant deux parties, présente le fondement théorique de la nouvelle méthodologie et des critères.

1.INTRODUCTION

It has been well known that the system energy is its Lyapunov function as soon as the system is conservative. Hopfield (1984) showed that energy of a symmetric neural dynamical (Hopfield) network with output feedbacks is the network Lyapunov function. Such results bear special significance. They give physical meaning to a system Lyapunov function. Moreover, they avoid very complex mathematical problem of an exact construction of a system Lyapunov function. Besides, they provide both physical and mathematical basis for stabilizing control synthesis.

Relationships between the system energy, asymptotic stability and the Lyapunov method [Lyapunov (1892)] have not been clarified in general. This means that conditions imposed on the energy, which are both necessary and sufficient for asymptotic stability of a state of a nonlinear system, have not been known in general or at least for wide classes of nonlinear dynamical physical systems.

In what follows, general qualitative properties of the systems are specified framing the class of the systems for which Lyapunov-like conditions are imposed on the system energy. The conditions are both necessary and sufficient for asymptotic stability of a state.

La Salle and Lefschetz(1961) pointed out that the Lyapunov stability concept can have only academic sense if the domain of a requested stability property, or at least its good estimate is not determined.

Zubov (1964) and Vanelli and Vidyasagar (1985) contributed essentially to solving the problem of an exact determination of the attraction domain. Their conditions for a set to be the domain are expressed in the classical Lyapunov-like form via *existence* of Lyapunov functions and of comparison functions related to the derivatives of the former. Such a methodology is characterized by starting with a tentative Lyapunov function v(·), verifying its properties and, provided they are satisfactory, by testing features of their derivatives relative to existence of the comparison functions. If such functions are not found, the procedure should be repeated and can be endless. This characterizes Lyapunov's methodology for nonlinear systems.

Another methodology has been also developed [Grujić (1989-1995)]. It is also based on Lyapunov's idea to test a stability property without solving a system mathematical model, but by using a scalar function v(·) and its derivative $v^{(1)}(\cdot)$. The new methodology starts with **arbitrary choice** of a scalar function *p(·)* from a specificied class and with the positive definite property. Afterthat, the solution function v(·) to $v^{(1)}(\cdot) = -p(\cdot)$ {or to $v^{(1)}(\cdot) = -p(\cdot)[1-v(\cdot)]$} is to be tested. The necessary and sufficient conditions on v(·) for asymptotic stability of x=o and for a set to be its domain are to be verified. The result of the procedure gives the final solution to the problem whether the state is asymptotically stable and whether the set is its domain.

A study of a class of technical systems considered in Part II faces us with reality of the semi-definite system power.

A question rises whether the p(·)- function may be positive semi-definite, what are then necessary and sufficient properties of the v(·)-function and what is energy-power meaning to all that. This question will be completely replied in the sequel.

The problem of testing the asymptotic stability via system energy E(·) in one shot when system power P(·) is semi-definite is solved in this part of the paper for a large class of non-linear dynamical systems and effectively applied to a technical system in Part II.

2. SYSTEM DESCRIPTION

A system to be studied is described by (1),

$$\frac{dx}{dt} = f(x), \qquad (1)$$

where $f(.) : R^n \to R^n$, $t \in R$ and $x \in R^n$. Function f(.) and initial state $x_o = x(o)$ at initial time $t_o = o$ determine the system internal dynamics. Function f(.) is assumed such that the system possesses the following properties related to some $x^* \in R^n$ of our interest.

Property 1. There is open connected neighbourhood M of $x^*, M = M(x^*)$, such that for every $x_o \in M$:

a) system (1) has a unique motion $\underline{x}(\cdot\,;x_o)$ through x_o at t = 0, $\underline{x}(t;x_o) \equiv x_o$, with the largest interval of existence and definition of $\underline{x}(\cdot\,;x_o)$ denoted by $J_o = J_o(x_o) \subseteq R$, and

b) $\underline{x}(t;x_o)$ is continuous and continuously differentiable in $(t;x_o) \in J_{o+} \times M$ where $J_{o+} = J_o \cap R_+, J_o = J_o(x_o\,;\,\underline{x})$ is the largest interval of existence and definition of $\underline{x}(.\,;x_o)$ and $R_+ = [0 + \infty[$.

Property 2. For every $x_o \in (R^n - M)$ every motion $\underline{x}(\cdot\,;x_o)$ of the system through x_o at t = 0 is at least continuous in $t \in J_{o+}(x_o;\underline{x})$. It is permitted that $J_o(x_o;\underline{x}) = \emptyset$ for any $x_o \in (R^n - M)$, where $\emptyset$ is the empty set.

Property 3. There is open connected neighbourhood E of x^*, $E = E(x^*)$, such that system energy $E(\cdot)$ is continuously differentiable function in $x \in \bar{E}$: $E(x) \in C^{(1)}(\bar{E})$, where diameter $d(\bar{E})$ of $\bar{E}$, $d(\bar{E}) = \sup\{\|x-y\| : x, y \in \bar{E}\}$, $\|x\| = (x^T x)^{1/2}$, may be arbitrarily small but positive, and $\bar{E}$ is the closure of E. An open hyperball $B_\rho(x^*) = \{x : \|x - x^*\| < \rho\}$ may be accepted for E, and its diameter $2\rho > 0$ may be arbitrarily small.

For a class of technical systems considered in Part II the energy is differentiable on a neighbourhood of chosen x*. This is a reason to accept the validity of Property 3.

3. ASYMPTOTIC STABILITY CRITERION

Asymptotic stability criterion to follow will be expressed in terms of a system power P(.) and (corresponding) energy E(.). For the class of technical systems considered in Part II, power P(.) can be equal to zero out of the system equilibrium state. More precisely, every neighbourhood of equilibrium state x_e has at least one point $y \neq x_e$ such that $P(y) = 0$. This is a reason to introduce

$$P_o = P_o(A) = \{x : x \in A, P(x) = o\}, A \subset R^n.$$

Power P(.) will be called **negative semi-definite with respect to x* on *A*** if and only if : 1) A is open connected neighbourhood of x*, 2) $P(x) \in C(A)$, 3) $P(x^*) = 0$ and 4) $P(x) \leq 0$ for all $x \in A$.
Furthermore, energy E(.) is **positive definite with respect (or, relative) to x* on S** if and only if : 1) S is open connected neighbourhood of x*, 2) $E(x) \in C(S)$, 3) $E(x^*) \in R$ and 4) $[E(x) - E(x^*)] > 0$ for all $(x \neq x^*) \in S$. Notice that if we use state deviation $x = x - x^*$ and define $v(x) = [E(x^* + x) - E(x^*)]$ then $v(.)$ is positive definite on $S_o = \{x : x = x - x^*, x \in S\}$ if and only if E(.) is positive definite relative to x^* on S. Evidently, $v(o) = o$.

A set S will be called as usual **positively invariant set** if and only if $x_o \in S$ implies $\underline{x}(t; x_o) \in S$ for all $t \in R_+$, where $x_o \in S$ means that S is non-empty. A set L_+ **is the largest positively invariant set** if and only if it is positively invariant and any other positively invariant set S_+ is its subset : $\forall S_+ \subseteq L_+$.

Theorem 1. **Let system (1) possess Properties 1-3. Let system power P(.) be negative semi-definite relative to x^* on $A \subseteq R^n$ and let singleton $\{x^*\}$ be the largest positive invariant set of the system in set $P_o(A)$. In order for x^* to be asymptotically stable it is both necessary and sufficient that there exists an open connected neighbourhood S of x^* such that**

(i) $S \subseteq [A \cap E \cap M]$

(ii) **$f(x) = o$ for $x \in S$ if and only if $x = x^*$, and**

(iii) **energy E(.) is positive definite relative to x^* on S and differentiable on S.**

Proof. Omitted due to the space limitation.

Comment 5. For the sake of a comparison of Theoriem 1 with other results let us remind first of the latter. Lyapunov's original results (1892) and all those following them, which will be called **classical**, start with a choice of a function $v(.)$ with appropriate properties such as positive definiteness (and often differentiability) in the framework of nonlinear systems. **Afterthat,** properties of its derivative $v^{(1)}(.)$ are to be tested. If $v^{(1)}(.)$ is semi-definite or definite then the classical results resolve the problem. If $v^{(1)}(.)$ is not semi-definite then they do not povide any solution. The procedure should be repeated from the begining and can be endless. If fortunately a $v(.)$-function was found to satisfy, together with $v^{(1)}(.)$, all the conditions of the classical results, there still stays open problem whether such a choice of $v(.)$- function is the best. The classical results do not provide solution to this problem.

Differently than the classical results, in fact inversely to them, Theorem 1 **presents a new methodology** for using the Lyapunov second method itself in the framework of nonlinear systems. It is characterized by verifying certain system features (Properties 1-3), semi-definiteness of power $P(.)$ and the property of set $P_o(A)$ **at the begining. Afterthat,** the conditions in (iii) imposed on system energy E(.) are to be tested together with conditions (i) and (ii). They are both necessary and sufficient for asymptotic stability of x^*. If they are satisfied then x^* ***IS asymptotically stable*** (sufficiency) and there is not a better Lyapunov function than $v(.) = E(.) - E(x^*)$ with $v^{(1)}(.) = P(.)$ (necessity). If any of them is not satisfied then x^* is **not asymptotically stable** (necessity), and there is not a sense to try finding "a better function $v(.)$" because it does not exist. Hence, Theorem 1 provides an one-shot resolution to the problem.

4. ASYMPTOTIC STABILITY DOMAIN CONDITIONS

For the definition of the asymptotic stability domain see Grujić (1975, 1981,1987,1989 1995) and Grujić et al. (1979, 1987).

Power P(.) of the system is considered as (negative) semi-definite function of state x rather than (negative) definite. This is accepted because the whole class of nonlinear technical systems considered in Part II is characterized by negative semi-definite power P(.).

If set $M = R^n$ then a neighborhood N of M, $N = N(M)$, equals $R^n: N = R^n$, in the sequel.

Theorem 2. **Let system (1) possess Properties 1-3, its power be continuous on a neighbourhood N of $M: P(x) \in C(N)$ and negative semi-definite relative to x^* on $A \subseteq N$. Let an open connected neighbourhood S of x^*, $S = S(x^*)$, obey (a)-(d),**

(a) $S \subseteq (A \cap M)$

(b) **singleton $\{x^*\}$ is the largest positive invariant set in $P_o(S)$;**

(c) **if boundary ∂S of S is non-empty then $x \to \partial S$ implies $|E(x) - E(x^*)| \to \infty$,**

(d) **if set S is not bounded then $|E(x) - E(x^*)| \to \infty$ as soon as $\|x - x^*\| \to \infty$, $x \in S$.**

Then for set S to be domain $D(x^*)$ of asymptotic stability of x^* it is both necessary and sufficient that (i) and (ii) hold :

(i) **$f(x) = o$ for $x \in S$ if and only if $x = x^*$,**

(ii) **system energy E(.) is positive definite with respect to x^* on S and continuously differentiable on S.**

Proof : Omitted due to the space limitation.

In case of the global asymptotic stability analys, set S should be obviously the whole state space,

$$S = D(x^*) = R^n$$

so that Theorem 2 then becomes the following :

Theorem 3. **Let system (1) possess Properties 1-3 with $M = R^n$ and its power $P(.)$ be negative semi-definite relative to x^* on R^n. If**

(a) **singleton $\{x^*\}$ is the largest positively invariant set in $P_o(R^n)$, and**

(b) $[E(x) - E(x^*)] \to \infty$ **as** $\|x - x^*\| \to \infty$,

then for x^* to be globally asymptotically stable it is both necessary and sufficient that both (i) and (ii) hold,

(i) $f(x) = o$ **if and only if** $x = x^*$,

(ii) **system energy $E(.)$ is positive definite with respect to x^* on R^n and continuously differentiable on R^n.**

The proof of this theorem is just a repetition of the proof of Theorem 2 with $A = M = N = R^n$

5. CONCLUSION

In order to use the system energy for an one-shot test of asymptotic stability of a state and/or its domain, properties of the system (Properties 1-3) and its power should be verified at first. The power may be semi-definite rather than definite function of the system state {if it is indefinite it can be modified by "corrective" "power" $P_c(.)$ so that "modified power" $P_M(.) = [P(.) + P_c(.)]$ replaces P(.) everywhere, and "modified energy" $E_M(.) = [E(.) + E_c(.)]$ **replaces energy** E(.) everywhere, with $E_c^{(1)}(.) = P_c(.)$}. **Afterthat,** properties of energy E(.) are to be tested, among which positive definiteness is crucial. The state ***is asymptotically stable and the set is its domain*** if the energy has all requested properties. Otherwise, the state **is not** asymptotically stable and/or the set **is not** its domain. The analysis of the asymptotic stablility terminates after the first test of the energy properties.

Knowing the system power and energy, with power P(.) possessing the specified features, we resolve the asymptotic stability problem without solving the complex problem of the mathematical construction of a system Lyapunov function.

This part of the paper emphasizes the need for an effective determination of the system power and energy in order to bridge the problem ot the mathematical determination of E(.) from $E^{(1)}(.) = P(.)$ for known P(.). Such a bridge is built in Part II of the paper.

6. REFERENCES

Lj.T. Grujić, 1975, "Novel development of Lyapunov stability motion" *Int. J Control,* vol. 22, N°. 4, pp. 525 - 549.

Lj.T. Grujić, 1979, "Sets and singularly perturbed systems", *Systems Science,* Wroclaw (Poland) vol. 5, N°. 4, pp. 327 - 338.

Lj.T. Grujić, 1981, "On Absolute Stability and Aizerman Conjecture", *Automatica,* Vol. 17, N°. 2, pp. 335 - 349.

Lj.T. Grujić, 1987, Stability Domains of General and Large-Scale Stationary Systems", *Applied Modelling and Simulation of Technological Systems",* Ed. P. Borne and S. Y. Tzafestas, Elsevier Science Publishers, B. V. (North Holland) IMACS, pp. 317 - 327.

Lj.T. Grujić, 1989, "On Determination of Lyapunov Functions and the Asymptotic Stability Domain", (in Serbian), *Proc. of the 3rd Conference SAUM (on Systems, Automatic Control and Measurement),* Vrnjachka Banja, Serbia, Yugoslavia, October 18-20, 1989, pp. 12 - 20.

Lj.T. Grujić, 1990, "Solutions to Lyapunov Stability Problems : Nonlinear Systems with Globally Differentiable Motions", in *"THE LYAPUNOV functions method and applications",* ed. P. Borne and V. Matrosov, IMACS 1990, J.C. Baltzer Scientific Publishing Co., Basel, pp. 19 - 27.

Lj.T. Grujić, 1991, "Solutions to Lyapunov Stability Problems : Nonlinear Systems with Differentiable Motions", *Proc. of the 13th IMACS World Congress on Computation and Applied Mathematics*, Ed. R. Vichnevetsky and J.J.H. Miller, Dublin, Ireland, July 22-26, 1991, Vol. 3, pp. 1228 - 1231.

Lj.T. Grujić, 1991, "The Necessary and Sufficient Conditions for the Exact Construction of a Lyapunov Function and the Asymptotic Domain", *Proc. of the 30th Conference on Decision and Control*, Brighton, England, 11-13 December 1991, Vol. 3, pp. 2885 -2888.

Lj.T. Grujić, 1992, "On solutions to Lyapunov Stability Problems", *Facta Universitatis, Series: Mechanics, Automatic Control and Robotics*, University of Nish, Nish, Serbia, Vol. 1, No. 2, pp. 121 - 138.

Lj.T. Grujić, 1992, "Exact determination of a Lyapunov function and the asymptotic stability domain", *Int. J. Systems Sci.*, Vol. 23, No. 11, pp. 1871-1888.

Lj.T. Grujić, 1992, "Solutions to Lyapunov stability problems: nonlinear systems with differentiable motions", in *COMPUTATIONAL AND APPLIED MATHEMATICS II : DIFFERENTIAL EQUATIONS*, ed. W.F. Ames and P. J. van der Houwen, Elsevier Sci. Publ., Amsterdam, The Netherlands, submited for publication 24 February 1992.

Lj.T. Grujić, 1993, " Exact Both Construction of Lyapunov Function and Asymptotic Stability Domain Determination", *IEEE Inernational Conference on Systems, Man and Cybernetics*, Le Touquet, France, Vol. 1, pp. 331-336.

Lj.T. Grujić, 1994, "Solutions to Lyapunov Stability Problems of Sets : Nonlinear Systems with Differentiable Motions", *Int. Journ. of Mathematics & Mathematical Sciences*, Vol. 17, No. 1, pp. 103-112.

Lj.T. Grujić, 1994, "Complete Exact Solution to the Lyapunov Stability Problem : Time-Varying Nonlinear Systems with Differentiable Motions", *Nonlinear Analysis:Theory, Methods & Applications*, Vol. 22., No. 8, pp. 971-981.

Lj.T. Grujić, 1994, "Solutions to Lyapunov Stability Problems : Time-Invariant Systems", *Proc. 14th IMACS World Congress*, Atlanta, Georgia, USA, Vol. 1, pp. 203-205.

Lj.T. Grujić, 1994, "Solutions to Lyapunov Stability Problems : Nonlinear Systems with Continuous Motions", *Int. Journ. of Mathematics & Mathematical Sciences*, Vol. 17, No. 3, pp. 587-596.

Lj.T. Grujić, 1995, "Exact Solutions for Asymptotic Stability: Nonlinear Systems", *Int. Journ. Non-Linear Mechanics*, Vol. 30, No. 1, pp. 45-56.

Lj.T. Grujić, 1995, "New-Lyapunov method based methodology for asymptotic stability", this proceedings.

Lj.T. Grujić, A. A. Martynyuk and M. Ribbens-Pavella 1987, "*Large-Scale Systems Stability under Structural and Singular Perturbations*", Springer Verlag, Berlin, (Russian ed. 1984, Naukova Dumka, Kiev).

Lj.T. Grujić and M. Ribbens-Pavella,1 979, "Asymptotic stability of large-scale systems with application to power systems. Part 1 : domain estimation", *Electrical Power and Energy Systems*, Vol. 1, N°. 63, pp. 151 -157.

J.J. Hoppfield, 1984, "Neurons with graded reponses have collective computational properties like those of two-state neurons" *Proc. Natl. Acad. Sci. U.S.A.*, Vol. 81, pp. 3088 - 3092.

J.P La Salle , 1976, "*The stability of Dynamical Systems*", Philadelphia, SIAM.

J.P La Salle and S. Lefschetz, 1961, *Stability by Lyapunov's Direct Method*", Academic Press, New-York.

A.M. Lyapunov , 1892 "*The General Problem of the stability of Motion* (in Russian), Kharkov Mathematical Society, Kharkov, English translation : 1992, *Int. J. C.* Control, Vol. 55, pp.531 -773, also the book, Taylor & Francis, London.

A. Vannelli and M. Vidyasagar, 1985, "Maximal Lyapunov Functions and Domains of Attraction for Autonomous Nonlinear Systems" *Automatica*, Vol. 21, N°. 1, pp. 69 -80.

V.I. Zubov, 1964, "*Methods of A.M. Lyapunov and Their Application*", Nordhoff, Groningen.

MULTIRATE PREDICTIVE CONTROL WITH MULTIPLE REFERENCE MODEL APPLICATION ON THERMIC PROCESS

V. KONCAR[*], C. VASSEUR[*], P. BRUNIAUX[*] and D. PINCHON[+]

[*]*Ecole Nationale Superieure des Arts et Industries Textiles, GEMTEX*
2, Place des Martyrs de la Resistance, 59070 Roubaix CEDEX, FRANCE
Phone: 20.25.64.64. Fax: 20.24.84.06.

[+]*Centre d'Automatique de Lille*
BAT P2, Cité Scientifique, 59650 Villeneuve d'Ascq, FRANCE
Phone: 20.43.48.76.

Abstract. The Multirate Predictive Control (Sliding Criterion J_m) with Multiple Reference Model has been used to control a thermic process including important non-linearities and dead time. The thermic process is controlled by pulse width modulation and it has been modelized by simple linear transfer function including only heating part. This linearized model has been used to design the polynomial multirate predictive controller with modified sliding criterion. The high frequencies measurement disturbances have been partially rejected by implementation of an IIR digital filter.

Key Words. Multirate; Predictive Control; Thermic process; Filters

Résumé. La Commande Prédictive Multiéchantillonnée (Critère à glissement J_m) avec Modèles de Référence Multiples a été utilisée pour commander un processus thermique incluant de fortes non-linéarités et un retard important. Le processus thermique est commandé par la modulation d'impulsion en largeur et est modélisé par un modèle linéaire concernant uniquement sa partie de chauffage. Ce même modèle est utilisé pour la synthèse du régulateur prédictif. Les perturbations dues à la mesure sont partiellement rejetées par l'utilisation d'un filtre à RII.

Mots clés. Multiéchantillonnage; Commande Prédictive; Processus thermique; Filtres

1. INTRODUCTION

Digital computers were originally used as components in complicated process control systems. However, because of their small size and low price, digital computers are now also being used in regulators for individual control loops. Because of these developments, the approach to analysis, design and implementation of control systems are changing drastically. Originally it was only a matter of translating the analog design into the new technology (digital PID and State feedback redesign). However it has been realized that there is much to be gained by exploiting the full potential of the new technology. New control theories have also been developed giving the advanced control methods which can be implemented even for basic application such as thermic process as shown in this paper.

In the last decade many papers have been published on predictive control methods. These methods promises good control proprieties and robust behaviour, Richalet *and al.* and Clarke *and al.* In some cases the measurement errors and large load disturbances can induce important mathematical model errors and then predictive control gives poor control properties. Because of this, digital filters have to be used to eliminate measurement disturbances.

The one step ahead predictive control algorithms based on non parametric models have been proposed by R.Bars (1992). The linear system output is predicted on the basis of the non-parametric discrete weighting model.

In general, the predictive control method is based on the computation of series of the control signal by minimizing a quadratic deviation of reference signal and the predicted output in a given prediction horizon. This technique is very effective especially in the case of plants with important dead time (thermic process). Adaptive generalized predictive control method with multiple reference model has been developed by Irving and al. (1986). This control method based on two reference model tracking (control reference model and output reference model), enables the resolution of the tracking problem and the regulation problem (disturbances rejection). The first reference model is used to impose the output dynamics and the second reference model to impose the control dynamics. This control scheme leads to the minimizing of the quadratic criterion in a given prediction horizons. M. Duque proposed in 1989 the algorithm for automatic setting of the output horizon N_y and the control weighting parameter λ.. The control horizon N_u is usually set on one (N_u=1).

V. Koncar and C. Vasseur (1994) proposed the modified quadratic criterion for the general predictive control with multiple reference model. The prediction horizons N_y and N_u are defined with the sampling period T_e and on the other side the quadratic criterion J_m is computed with the sampling period t_e ($t_e<T_e$), then the control vector obtained by minimizing J_m is applied with the sampling period t_e. This control scheme with sliding criterion J_m permits better control performances and improvement of high frequencies disturbances rejection. Moreover, the problem of oversampling when computing the discrete transfer function, which may occur with t_e sampling period, has been eliminated. An improvement of control robustness is also obtained. High frequency measurement errors will be attenuated by the regulator dynamics.

Multirate sampling has already been used in many control schemes e.g. by B. Lenartson (1989) in multirate sampled data control of two scale systems, D. Eckardt (1989) in the design of finite impulse response controller by pole assignment or by V. Koncar (1990) in digital state control by oversampling.

P. Boucher and D. Dumur (1992) have developed a multirate version of the generalized predictive control law performed within a polynomial cascade structure framework. This method has been tested on a brushless motor MASAP Télémécanique.

In this paper the Multirate Predictive Control (Sliding Criterion J_m) with Multiple Reference Model has been used to control a thermic process including important nonlinearities and dead time.

The thermic process is composed of following elements:

- tank containing the liquid,

- heater fixed in the tank controlled by PC computer,

- cooling system, controlled by PC computer, including liquid pump and radiator with fun,

- shaker used to keep the liquid temperature homogeneous

- sensors for liquid temperature measurement.

The Thermic process is controlled by pulse width modulation and it has been modelized by simple linear transfer function including only heating part. This linearized model has been used to design the polynomial multirate predictive controller with modified sliding criterion. The high frequencies measurement disturbances have been partially rejected by implementation of an IIR digital filter.

2. BASIC THEORY

Before starting the problem of the thermic process control it is reasonable to introduce the Multirate Predictive Control with Multiple Reference Model.

Multirate Predictive Control With Multiple Reference Model

Controlled system model is given by the equation (1),

$$A(q^{-1})\,y(t) = B(q^{-1})\,u(t) + w(t) \tag{1}$$

with

$$A(q^{-1}) = 1 + \sum_{i=1}^{n} a_i q^{-i}\ ,\ B(q^{-1}) = B'(q^{-1})\,q^{-d}\ , \tag{2}$$

$$B'(q^{-1}) = \sum_{i=1}^{n} b_i q^{-i} \tag{3}$$

w(t) is a white noise,

d is a dead time,

$u(t)$ is a control vector,

$y(t)$ is a system output,

and the sampling period is T_e.

For the pole placement based on input-output models, the servo specifications are expressed in terms of a model that gives the desired response to control signal. The desired close loop pulse transfer function is given by :

$$H_m(z)=\frac{B_r(z)}{A_r(z)} \tag{4}$$

The global system dynamics is given by :

$$A_r(q^{-1})\,y(t)=B_r(q^{-1})\,y_{cc}(t) \tag{5}$$

Control scheme is given in Fig. 1 with deg T ≤ deg S and deg R ≤ deg S (causality). Therefore, the system controller is given by the linear equation

$$S(q^{-1})\,u(t)=-R(q^{-1})\,y(t)+T(q^{-1})\,y_{cc}(t) \tag{6}$$

with :

$$S(q^{-1})=1+\sum_{i=1}^{ns} s_i\,q^{-i}\ , \tag{7}$$

$$R(q^{-1})=\sum_{i=0}^{n_R} r_i\,q^{-i}\ ,\ T(q^{-1})=\sum_{i=0}^{n_T} t_i\,q^{-i} \tag{8}$$

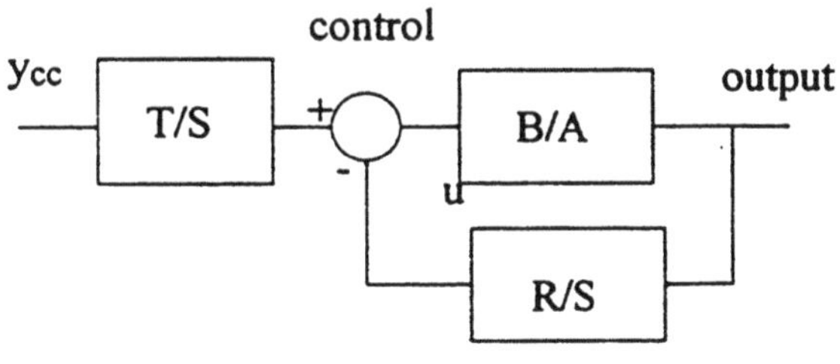

Fig.1.Control scheme

Double reference model permits the tracking problem resolution and the regulation problem resolution.

Output reference model :

$$A_r(q^{-1})\,y_m(t)=B(q^{-1})\,T(q^{-1})\,y_{cc}(t) \tag{9}$$

Control reference model :

$$A_r(q^{-1})\,u_m(t)=A(q^{-1})\,T(q^{-1})\,y_{cc}(t) \tag{10}$$

The polynomial $A_r(q^{-1})$ is imposed by user.

Then the modified sliding criterion to minimize every ($t_e<T_e$)second is defined by:

$$J_m(lt_e,N_y,N_u)=E\left[\sum_{i=lt_e}^{lt_e+(N_y-1)T_e}\varepsilon_{sy}^2(i+d)+\lambda(lt_e)\sum_{i=lt_e}^{lt_e+(N_u-1)T_e}\varepsilon_{su}^2(i)\right]$$

Errors are Now :

$$\varepsilon_{sy}(lt_e)=A_r(q^{-1})\,y(lt_e)-B(q^{-1})\,T(q^{-1})\,y_{cc}(lt_e)$$

$$\varepsilon_{su}(lt_e)=A_r(q^{-1})\,u(lt_e)-A(q^{-1})\,T(q^{-1})\,y_{cc}(lt_e)$$

In the case of the multirate general predictive control, the sliding criterion J_m is minimized and the system is controlled with the sampling period te. The control vector is computed by the expression:

$$u(lt_e)=\frac{T(q^{-1},\lambda)}{S(q^{-1},\lambda)}\,y_{cc}(lt_e)-\frac{R(q^{-1},\lambda)}{S(q^{-1},\lambda)}\,y\,(lt_e) \tag{11}$$

3. THERMIC PROCESS

The multirate predictive control with multiple reference model has been used to control the thermic process given in Fig. 2.

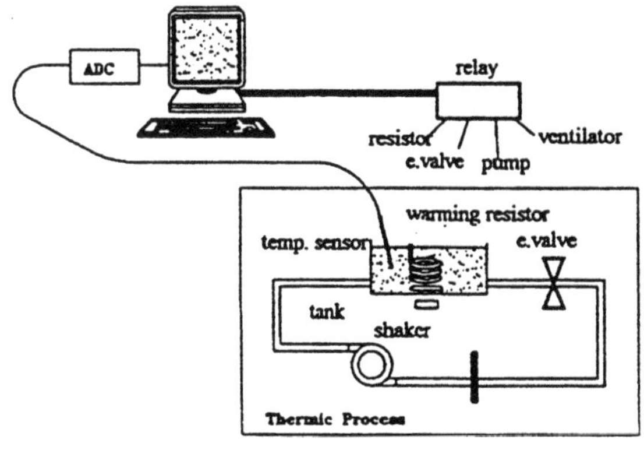

Fig.2. Thermic process

This process is modelized by the linear transfer function given bellow.

$$H(q^{-1}) = \frac{B(q^{-1})}{A(q^{-1})} = \frac{\beta q^{-1}}{1 + \alpha_1 q^{-1} + \alpha_2 q^{-2}}$$

with

$\alpha_1 = -1.6065$

$\alpha_2 = 0.6065$

$\beta_1 = 0.3935$

The desired dynamics is defined by:

$A_r(q^{-1}) = 1 + a_1 q^{-1} + a_2 q^{-2}$

with:

$a_1 = 0.84$

$a_2 = 0.1864$

Then the global discrete control system roots are:

$R_{1,2} = -0.42 \pm 0.1j$

The control and output horizons are:

$N_u = 2$

$N_y = 3$

the dead time is supposed to be d=0

Sampling period $T_e = 10\,s.$ and $t_e = 2\,s.$

Error is now:

$$\varepsilon_{su}(lt_e) = (1+0.84q^{-1}+0.1864q^{-2})u(lt_e) + \\ +(1-1.6065q^{-1}+0.6065q^{-2})T(q^{-1})y_{cc}(lt_e)$$

After resolving the Diophantine equations, we obtained the controller design:

$$u(lt_e) = \frac{-RQ}{V\,(1+a_1q^{-1}+a_2q^{-2})} y_{cc}(lt_e) - \\ -\frac{X+Yq^{-1}+Zq^{-2}+Wq^{-3}}{(1+a_1q^{-1}+a_2q^{-2})} y(lt_e) \qquad (12)$$

with

$$V = (\beta^2+\alpha)\,(\beta^2+\alpha_1^2\beta^2+\lambda) - \alpha_1^2\beta^4$$

$$X = (\beta^2+\lambda)\beta\alpha_1 + \alpha_1\beta\lambda(\alpha_1^2-\alpha_2^2))/V$$

$$Y = ((\beta^2+\lambda)\,\beta\,\alpha_2 + \alpha_1(\beta^2+\lambda)\,\beta\,\alpha_1 + \alpha_1^2\,\alpha_2\,\beta\,\lambda + \alpha_1\,a_1\,\beta\,\lambda\,(\alpha_1^2-\alpha_2))/V$$

$$Z = (\alpha_1 a_2\beta(\beta^2+\lambda) - \alpha_2 a_1\beta(\beta^2+\lambda) + \alpha_1^2 a_2 a_1\beta\lambda + \alpha_1\beta\lambda a_2(\alpha_1^2-\alpha_2))/V$$

$$W = (\alpha_2 a_2\beta(\beta^2+\lambda) + \alpha_1^2\alpha_2 a_2\beta\lambda)/V$$

$$Q = (\beta^2+\lambda)\beta^2(\alpha_1+\alpha_2) + \alpha_1\beta^2\lambda(\alpha_1+\alpha_2) - \alpha_1^2\beta^2\lambda(1+\alpha_1+\alpha_2) - V(1+\alpha_1+\alpha_2)$$

$$R = V(X+Y+Z+W)/Q$$

so the polynomials R, S and T are deined as:

$$\frac{T(q^{-1},\lambda)}{S(q^{-1},\lambda)} = \frac{-RQ}{V\,(1+a_1q^{-1}+a_2q^{-2})} \qquad (13)$$

and

$$\frac{R(q^{-1},\lambda)}{S(q^{-1},\lambda)} = \frac{X+Yq^{-1}+Zq^{-2}+Wq^{-3}}{(1+a_1q^{-1}+a_2q^{-2})} \qquad (14)$$

4. FILTER

The high frequencies measurement disturbances generated by the non homogeneity of the liquid temperature and by the measurement system has been partially eliminated by the use of the low pass filters.

The IIR filter defined by the equation (15) has been used.

$$S(kt_e) = 0.2E(kt_e) + 0.1E((k-1)t_e) + \\ 0.2E((k-2)t_e) + 0.6S((k-1)t_e) - 0.1S((k-2)t_e) \qquad (15)$$

with S the filter output signal and E the filter input signal.

5. PERFORMANCES

The control synthesis is realised for $T_e = 10\,s.$ and $t_e = 2\,s.$ and the weighting parameter is initialized by $\lambda = 0.01$.

The liquid temperature, filtered liquid temperature and the control signal are given in the Fig.3. In this case the setting point $y_{cc}(lt_e)$ is constant. $y_{cc}(lt_e)$ = 42°C and initial temperature is 36°C.

In the Fig.4. the $y_{cc}(lt_e)$ varies.

Finally in the Fig.5. and the Fig.6. we can see the thermic process when the cooling part is controlled (setting point < initial liquid temperature). In this case the thermic process, modelized by simple linear transfer function including only heating part, is controlled with the controller based on this simple transfer function.

The performances of the Predictive Control with Multiple Reference Model are shown in the Fig.5.

The modified sliding criterion is introduced and the results obtained by the use of the Multirate Predictive Control with Multiple Reference Model are given in Fig.6.

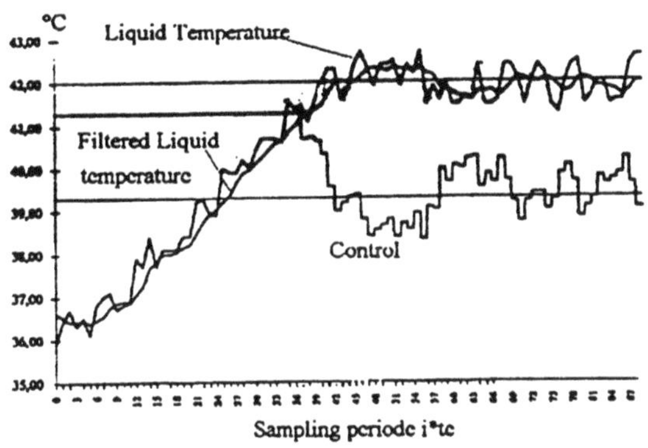

Fig.3. Multirate Predictive Control, thermic process

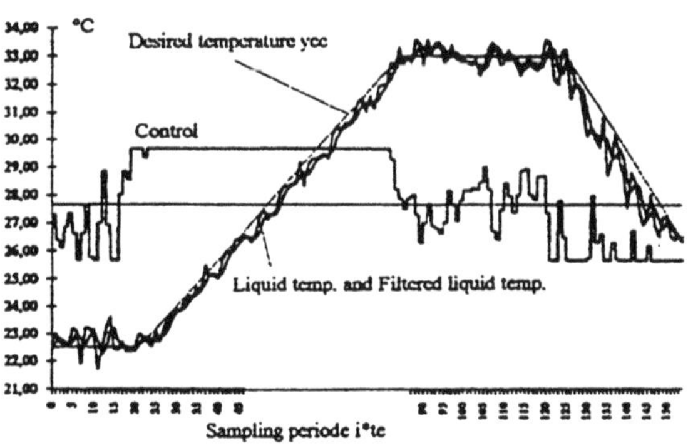

Fig.4. Multirate Predictive Control, thermic process

In the case where the traditional P.C. with M.R.M. is used (Fig.5.) the control results are not satisfying. This can be explained by the controller designed from the simple heating part model and applied to control thermic process when cooling. In the second case, the introduction of the modified sliding criterion J_m enables better control performances and improves a control robustness. The controller designed only from the simple heating part model assures good control properties.

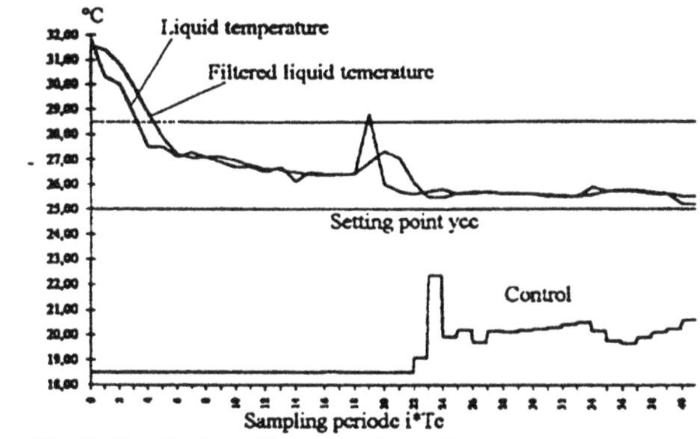

Fig.5. Predictive Control, thermic process

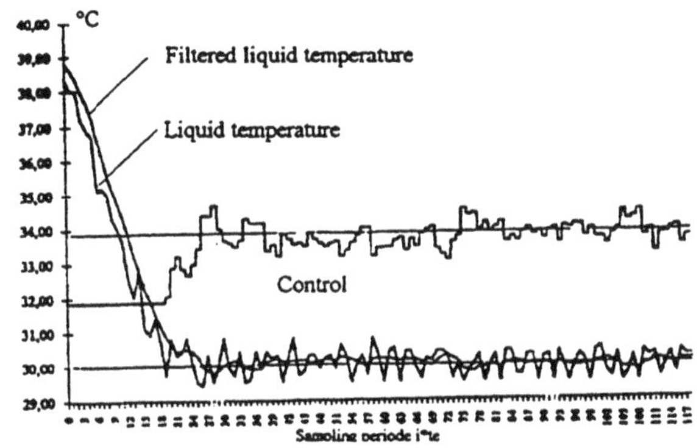

Fig.6. Multirate Predictive Control, thermic process

Moreover, the sliding criterion J_m permits better control of high frequencies disturbances rejection. Sensitivity function of the global control system is improved as shown by V. Koncar (1994).

6. CONCLUSION

After the elaboration of the multirate predictive control law with Multiple reference model and the description of advantages in term of robustness, the multirate aspect of predictive control has been adapted to a thermic process including two different dynamics corresponding to heating and cooling parts.

The control results obtained with MPC with MRM are excellent and shows good control robustness. Therefore, the MPC controller based on the linearized heating model without dead time enables good control properties of the non linear thermic process with dead time, which is in reality composed of heating and cooling parts.

This structure seems very interesting and well adapted in the area of complex systems with approximate models.

7. REFERENCES

Bars R. (1992). "One Step Ahead Predictive Control Algorithms based on Nonparametric Models", *MATE Scientific Society of Measurement and Automation,* Budapest, Hungary, 18-19 Feb. 1992.

Boucher P., Dumur D. (1992)., "Multirate Polynomial Predictive Cascade Control", Second IEEE Conference on Control Application, September 13-16, Vancouver, B.C.

Clarke D. W. Mohtadi C. Tuffs P. S. "Generalized Predictive Control", Part I.,II., *Automatica,* Vol. 23., N. 2., pp. 137-160.

Duque M., (1989). "Contribution à la mise en oeuvre de la commande adaptative", *Thèse de Doctorat de l'INPG*, Grenoble, France, 1989.

Eckardt D. (1989). "Design of Finite Impulse Response Controllers by Pole Assignment in Multirate Sampled Data Systems", *Int. J. Control,* N. 4, pp. 1185-1193, 1989.

Irving E., Falinower C. M., and Fonte C. (1986). "Adaptive Generalized Predictive Control with Multiple Reference Model", *Proc. of the 2nd IFAC workshop on adaptive systems in control and signal processing,* Lund, Sweden, 1986.

Lenartson B. (1989). "Multirate Sampled Data Control of Two Time Scale Systems", *IEEE Trans. Automatic Control.*, Vol. 34., pp. 624-644, June 1989.

Koncar V., Vasseur C. (1990). "Commande numerique par surechantillonnage, Application la realisation d'une carte d'axe", *ISMM,* Lugano, Switzerland, June 1990.

Koncar V., Vasseur C. (1994). "Multirate Predictive Control with Multiple Reference Model, Application on DC Motor", *3rd Int. IEEE Workshop on Advanced Motion Control,* Berkeley, Cal, U.S.A. March 20-23, 1994.

Richalet. J. , Rault A. , Testud J. L. , Papon J. "Model Predictive heuristic control An Application to industrial processes", *Automatica,* Vol. 14, p. 413-428.

EIGENVALUE PROBLEMS ARISING IN THE CONTROL OF A DISTRIBUTED PARAMETER BIOREACTOR

M.T. NIHTILÄ* and J.P. BABARY**

**University of Kuopio, Faculty of Natural and Environmental Sciences, Department of Computer Science and Applied Mathematics, P.O.Box 1627, FIN-70211 Kuopio, FINLAND, e-mail: nihtila@cs.uku.fi*

***Centre National de la Recherche Scientifique, Laboratoire d'Analyse et d'Architecture des Systèmes, 7, avenue du Colonel Roche, F-31077 Toulouse Cédex, FRANCE, e-mail: babary@laas.fr*

Résumé. L'objet de cette communication concerne l'étude d'un modèle à paramètres répartis d'un bioréacteur à lit fixe fonctionnant en mode continu. L'accent est porté sur une propriété structurelle du modèle du procédé formulé sous la forme d'un système aux dérivées partielles. Dans le cas de systèmes à paramètres localisés, cette propriété, relative à *un déphasage non-minimal*, est à l'origine de certaines difficultés dans la conception de lois de commande. Dans le cas d'un modèle à paramètres répartis, la loi de commande est obtenue après semi-discrétisation. La seule variable d'espace de ce modèle est discrétisée à l'aide de la méthode des éléments finis de Galerkin. Pour certaines valeurs paramétriques du modèle aux dérivées partielles, la commande linéarisante du modèle semi-discrétisé engendre un comportement instable dans le sens où la dynamique des zéros du système est instable. Ce comportement instable avait déjà été observé quand la discrétisation est effectuée à l'aide de la méthode de collocation orthogonale. Les relations existant entre la position des zéros du modèle aux dérivées partielles linéarisé autour de sa solution en régime permanent et les propriétés de stabilité/instabilité de la commande linéarisante du modèle semi-discrétisé sont discutées de façon analogue au cas des systèmes à paramètres localisés.

Mot-clés. Systèmes à paramètres répartis; Approximation par collocation orthogonale; Approximation aux éléments finis; Commande linéarisante; Déphasage non-minimal; Dynamique des zéros; Fonction de transfert linéarisée.

Abstract. A distributed parameter model of a continuous flow fixed-bed reactor is studied. The main emphasis lies on a structural property of the partial differential equation (PDE) system model. This property, which in lumped parameter systems, is called *nonminimum phase property* has certain implications in the controller design. The controller design for the PDE model is constructed via semidiscretisation. The only space variable of the PDE model is discretised by using Galerkin's finite element method (FEM). For some parameter values of the PDE model the linearising control of the semidiscretised model resulted in unstable behaviour in the sense that the zero dynamics of the model was unstable. This same unstable behaviour was also earlier observed in using the orthogonal collocation for the semidiscretisation. Connections between the location of the zeros of the original PDE model linearised around its steady state solution and the stability/instability properties of the linearising control of the semidiscrete model are discussed in relation to the same issues in lumped parameter differential system models.

Keywords. Distributed parameter systems; Orthogonal collocation approximation; Finite element method approximation; Linearizing control; Nonmimimum phase property; Zero dynamics; Linearized transfer function.

1. INTRODUCTION

The system studied is a continuous flow fixed-bed bioreactor. It is aimed at removing harmful nitrogen (and carbon) compounds from the influent waste water. The goal is to decrease the overall concentrations at the outlet of the reactor under some prescribed allowable limits. Most conventional wastewater plants are based on the activated sludge process. The new type of biological wastewater treatment plants considered here are also called biofilter processes. Their main advantages are the high compacity and efficiency for biological epuration (nitrogen and carbon compounds removal), their good integration in the environment (no noise, no smell), their low energy consumption and sludge production and the fact that they do not need secondary clarifiers.

A submerged granular fixed-bed biofilter can be considered as a continuous flow distributed parameter bioreactor, in which the wastewater flows through the tubular reactor. The microorganisms fixed in the reactor absorb these harmful nutrients in such a way that the substrate concentrations has been decreased in the outflowing water.

Due to the clearly distributed nature of the bio-reactor it is modelled by a set of two partial differential equations of parabolic type. The model is nonlinear due to the dependence of the specific growth rate of the biomass of the microorganisms on the concentrations of nitrogen and biomass. The model has two independent variables: the time-to-go and one space variable, the distance from the inlet of the reactor. Some basic assumptions in modelling have been applied, such as; the biomass and nitrogen concentrations are both described by one state variable; the gaseous phase is not taken into account because the gaseous hold-up may be neglected as compared to the liquid one; the temperature and acidity (pH-value is close to 7) are constant. The modelling is based on mass balances and on the specific growth rate model of Contois for the biomass.

The key technique in studying the behaviour of and designing a controller for the system is to approximate the infinite dimensional PDE model by a nonlinear finite dimensional ordinary differential equation (ODE) model. A method of orthogonal collocation (Tali-Maamar *et al.*, 1993) and a finite element method (FEM) of Galerkin type (Nihtilä *et al.*, 1994) have been applied to construct the approximations.

Direct application of the linearizing control technique in the controller design of the approximated (semidiscretised) systems results, however, in the unstable behaviour of the internal modes of the controlled system for some values of the process parameters (c.f. Tali-Maamar, 1994).

It is a known fact that in nonlinear systems which are of nonminimum phase character, i.e. their zero dynamics is unstable, the application of linearizing control techniques leads into internally unstable closed-loop systems (Isidori, 1989). The origin of this study is consequently in the question:
Is the nonminimum phase property of the ODE systems obtained via semidiscretisations also the property of the original infinite dimensional PDE system or is it only due to the approximation methods applied ?

The organisation of this paper is the following. The system models - the original PDE system and its approximation obtained via FEM-semidiscretisation - are presented in Ch.2. The linearising control idea and the algorithm for the semidiscretised model with some simulation results are given in Ch.3. The transfer function of the PDE system model linearised around the steady state solution function is presented and discussed in Ch.4. Some concluding remarks complete the paper.

2. SYSTEM MODELS

The bioreactor of the denitrification process is modelled by using partial differential equations due to the clear distributed nature of the system. Different variants of the basic model structure, which include compartments for the biomass, which is fixed, and for the substrates, which are moving with the moving liquid medium, have been developed. A four-dimensional model is studied in Babary *et al.* (1995). Two-dimensional models with and without diffusion were considered in Dochain *et al.* (1992) and Nihtilä *et al.* (1992). Here the system is described by a two-dimensional model with one state varible for the biomass and the other for the substrate.

The overall model is given by the following equations.

$$\frac{\partial u_1}{\partial t}(z,t) = [\mu(u_1,u_2) - k_d]u_1 \tag{2.1}$$

$$\frac{\partial u_2}{\partial t}(z,t) = D\frac{\partial^2 u_2}{\partial z^2} - c(t)\frac{\partial u_2}{\partial z} - k_1\mu(u_1,u_2)u_1 \tag{2.2}$$

$$t \in]0,T[\quad ; \quad z \in G \triangleq]0,1[$$

$$\mu(u_1,u_2) = \mu_m u_2/(k_2 u_1 + u_2) \tag{2.3}$$

in which u_1 and u_2 denote the concentrations of the biomass and nitrogen, respectively. The specific growth rate model Eq.(2.3) is the model of Contois (1959). The boundary conditions are given by

$$\frac{\partial u_2}{\partial z}(0,t) = (c(t)/D)[u_2(0,t) - s_a(t)] \tag{2.4}$$

$$\frac{\partial u_2}{\partial z}(1,t) = 0 \tag{2.5}$$

Functions $c(t)$ and $y(t) = u_2(1,t)$ are the control input, defined as the flow rate per the cross-sectional area of the bioreactor, and the output, correspondingly. The initial condition functions which are obtained for constant values of the control and the disturbance variable $s_a(t)$ are chosen as the steady state solutions of the problem.

$$u_1(z,0) = \bar{u}_1(z) \tag{2.6}$$

$$u_2(z,0) = \bar{u}_2(z) \tag{2.7}$$

For the chosen specific growth rate model of Contois (2.3) these are given by the equations

$$\bar{u}_2(z) = \{q\cosh[q(1-z)] + p\sinh[q(1-z)]\}\bar{s}_a\exp(pz)/[q\cosh q + (a+p)\sinh q] \tag{2.8}$$

$$\bar{u}_1(z) = \bar{u}_2(z)(\mu_m - k_d)/(k_2 k_d) \tag{2.9}$$

The coefficients p and q are defined by

$$P = \bar{c}/D \quad ; \quad p = P/2 \tag{2.10}$$

$$a = k_1(\mu_m - k_d)/(k_2\bar{c}) \tag{2.11}$$

$$q = \sqrt{P^2/4 + Pa} \tag{2.12}$$

A firm mathematical basis for the problem is given in the form of the following existence theorem.

Theorem 1. Suppose that c and s_a are positive $C^2([0,\infty[\,)$ functions with the property

$$c(0) = \bar{c} \quad ; \quad s_a(0) = \bar{s}_a \tag{2.13}$$

then the problem of the equations (2.1-.3) with the boundary and initial values of (2.4-.9) has a unique classical solution $u = (u_1, u_2)$ such that

$$u_1 \in C^1(]0,T[,C(\bar{G})) \cap C([0,T[,C(\bar{G}))$$

$$u_2 \in C^1(]0,T[,C(\bar{G})) \cap C([0,T[,C(\bar{G})) \cap C(]0,T[,C^2(\bar{G}))$$

The proof is based on the abstract formulation of the problem in the form

$$dv/dt + A(t)v = \tilde{F}(t,v) \quad ; \quad v(0) = v_0 \tag{2.14}$$

where $v = (v_1, v_2) = (u_1, u_2 - s_a)$ is in a suitably defined Sobolev space. The operator $A(t)$, which generates an analytic semigroup for any fixed t, and $\tilde{F}$ is a uniformly Lipschitz continuous function with respect to v. It is obtained from the Contois model (2.3). The proof is outlined in Nihtilä *et al.* (1994) and given in detail in Goldstein & Tervo (1995).

2.1 Semidiscretisation

The input-output behaviour of the PDE system was simulated in the open-loop and closed-loop cases. Galerkin's finite element method was applied to convert the PDE system into a set of nonlinear ODEs. Details of the semidiscretisation are presented in Nihtilä *et al.* (1994).

Finite elemet methods are mathematically well proved and justified spatial discretization methods for PDE systems (Thomee *et al.*, 1975 & 1984). Their error estimates and convergence proofs as a function of the number of discretisation points are based on the abstract theory of the solutions in a weak sense in suitable Sobolev spaces.
Here, the solutions of the initial/boundary value problem are approximated by finite series

$$u_1(z,t) \approx \tilde{u}_1(z,t) = \textstyle\sum_{i=1}^{N} x_i\phi_i(z)$$
$$u_2(z,t) \approx \tilde{u}_2(z,t) = \textstyle\sum_{i=1}^{N} x_{N+i}\phi_i(z) \tag{2.15}$$

in which the base functions ϕ_i are triangular, positive, piecewise linear functions. Application of the series (2.15) in the weak form of the PDE system results in the ODE system

$$M\dot{x} = -DKx - c(t)[Rx + w(t,x)] + F(x)$$
$$x(0) = [\alpha^T \ \beta^T]^T \tag{2.16}$$

where the $2N$-vector $\boldsymbol{x}$ and the coefficient matrices $\boldsymbol{M}$, $\boldsymbol{K}$, and $\boldsymbol{R}$, and the vector-valued functions $\boldsymbol{w}$ and $\boldsymbol{F}$ can be found in Nihtilä *et al.* (1994).

The orthogonal collocation method, studied e.g. by Villadsen & Michelsen (1978) offers another alternative for spatial discretisation. Then the base functions ϕ_i are obtained via Lagrange interpolation polynomials. Details of the application can be found in Tali-Maamar *et al.* (1993, 1994).

The system (2.16) can be put into the basic control theoretic form in the case of the constant disturbance $s_a(t)$, i.e.

$$\dot{x} = f(x) + g(x)c(t) \tag{2.17}$$

$$y = h(x) \tag{2.18}$$

where the output mapping is given by

$$h(x) = [0 \dots 0\ 1]x \tag{2.19}$$

3. LINEARISING CONTROL OF THE SEMIDISCRETE PROBLEM

The relative degree of the model (2.17-.18) is obtained by differentiating the output function $\boldsymbol{h(x)}$ with respect to the time. Then by taking into account (2.17) the control $\boldsymbol{c(t)}$ appears in the diffentiated output giving the relative degree $\boldsymbol{r = 1}$.

The asymptotic output tracking of the reference input y_R is obtained by applying the linearising control law (Isidori, 1989; p.191, Eq.(5.2))

$$c(t) = \frac{1}{L_g h(x)}\{-L_f h(x) + \dot{y}_R - c_0[h(x) - y_R]\} \tag{3.1}$$

Then the closed-loop system (from y_R to y) has the transfer function

$$G_{CL}(\lambda) = 1/(\lambda + c_0) \tag{3.2}$$

In order that the linearising control law results in (also) internally stable system the zero dynamics of the model (2.17-.18), with respect to the (constant) reference input y_R, must be stable.

Typical parameter values, in the case $N = 4$, which gave for the semidiscrete (ODE) system the stable linearising control are given by

$$k_1 = 0.4 \qquad k_2 = 0.4$$
$$k_d = 0.05 \qquad \mu_m = 0.35$$
$$D = 0.005 \qquad \bar{c} = 0.1$$

A simulation run is depicted in Fig.1.[#] The disturbance $s_a(t)$ changed from **5.0** to **4.0** at t = **26.7**, and the reference $y_R(t)$ from **0.3487**, that corresponds to the value **0.1** of the constant control, to **0.4** at t = **53.3**.

[#] The simulations and the control law calculations were expertly carried out by Dr. Jouko Tervo, which is greatly acknowledged.

4. ON THE TRANSFER FUNCTION OF THE LINEARISED PDE

A possible way to study the qualitative behaviour of the linearising control of the semidiscretised system is to study zeros of the (transcendental) transfer function of the linearised system. Now, the linearisation of the PDE system means linearisation around the *steady-state function* $\bar{u} = (\bar{u}_1, \bar{u}_2)$ given by Eqs.(2.8-.9). The transfer function (as a function of the Laplace variable λ) of the system linearised for the small changes, say $\delta c(t)$, of the input on the small changes, say $\delta y(t)$ of the output, is formally given by the equation

$$\frac{\delta Y(\lambda)}{\delta C(\lambda)} = K\frac{B(r(\lambda))}{A(r(\lambda))} \tag{4.1}$$

After some tedious calculations, including a.o. the solving of the non-constant coefficient linear ODEs, the nominator and the denominator functions are obtained in the following form

$$A(r) = [(p^2+r^2)\sinh r + 2pr\cosh r](r^2-q^2) \tag{4.2}$$

$$B(r) = b_1 r\cosh r + b_2\sinh r + r^3 - b_3 r \tag{4.3}$$

$$\begin{aligned} b_1 &= 2pq/\sinh q \\ b_2 &= 2p^2q/\sinh q \\ b_3 &= q^2+2p^2+2pq\cosh q/\sinh q \\ K &= (a/D)\bar{s}_a e^p/[q\cosh q + (a+p)\sinh q] \end{aligned} \tag{4.4}$$

in which the inner function r is given by the equation

$$r(\lambda)^2 = (\lambda^2+d_1\lambda+d_2)/[D(\lambda+a_1)] \tag{4.5}$$

$$\begin{aligned} d_1 &= a_1 + k_1 a_2 + \bar{c}^2/(4D) \\ d_2 &= k_1 k_d a_2 + a_1\bar{c}^2/(4D) \\ a_1 &= (1-k_d/\mu_m)k_d \\ a_2 &= (1-k_d/\mu_m)(\mu_m - k_d)/k_2 \end{aligned} \tag{4.6}$$

It has to be noted that when considering the relation between the location of the zeros of $B(r(\lambda))$ on the complex λ-plane and the stability of the corresponding transfer function $1/B(r(\lambda))$ the original overall transfer function can also be given by

$$\frac{\delta Y(\lambda)}{\delta C(\lambda)} = K\frac{\bar{B}(r(\lambda))}{\bar{A}(r(\lambda))} \tag{4.7}$$

$$\bar{A}(r) = [p^2 f(r) + r\sinh r + 2p\cosh r](r^2-q^2) \tag{4.8}$$

$$\bar{B}(r) = b_1\cosh r + b_2 f(r) + r^2 - b_3 \tag{4.9}$$

$$f(r) = \sinh r/r \tag{4.10}$$

By considering the power series expansions of the function $f(r)$ and $\cosh r$

$$\frac{\sinh r}{r} = 1 + \frac{r^2}{3!} + \frac{r^4}{5!} + \frac{r^6}{7!} + \ldots \tag{4.11}$$

$$\cosh r = 1 + \frac{r^2}{2!} + \frac{r^4}{4!} + \frac{r^6}{6!} + \ldots \tag{4.12}$$

it is seen that $\bar{B}(r)$ in fact depends only on r^2 but not on r itself. The Riemannian cutting of the complex plane, due to the square root in the expression of r in calculating the inverse transforms may then be avoided. These issues need, however, a more detailed and careful analysis, especially to be able, if it is at all possible, to connect the location of the zeros and the stability/instability of the linearising control. Furthermore, the nominator and denominator functions $A(r(\lambda))$ and $B(r(\lambda))$ as functions of λ have an essential singularity at the singularity of the function $r(\lambda)$, i.e. at the point

$$\lambda = -a_1 \quad (<0)$$

This essential singularity cannot be removed by multiplying the nominator and denominator functions by some suitable power

$$p_k(\lambda) = (\lambda + a_1)^k$$

as it is evident by considering the substitution of the power series (4.11-.12) to the equations (4.8-.9).

The parameter values which gave for the semi-discrete system the stable linearising control and which were applied in calculating system zeros are given by

$k_1 =$	0.4	$k_2 =$	0.4
$k_d =$	0.05	$\mu_m =$	0.35
$D =$	0.005	$\bar{c} =$	0.2
$\bar{s}_a =$	5.0		

A specific feature of the transfer function is that the functions $A(r)$ and $B(r)$ have three common zeros in the r-plane: $r_0 = 0$, $r_{1,2} = \pm q$. These correspond to the four zeros in the λ-plane obtained from the equations (c.f. (4.5))

$$\lambda^2 + d_1\lambda + d_2 = 0 \tag{4.13}$$

$$\lambda^2 + (d_1 - q^2 D)\lambda + d_2 - q^2 D a_1 = 0 \tag{4.14}$$

The former equation, due to the fact that the coefficients d_1 and d_2 are strictly positive, gives the result that the both roots have negative real parts. The latter reduces to the equation

$$\lambda[\lambda + (1 - k_1/k_2)a_2] = 0 \tag{4.15}$$

which has the zeros

$$\lambda_1 = 0; \quad \lambda_2 = (1 - k_1/k_2)a_2$$

However, based on these singular zeros, for $k_1 > k_2$, it cannot be concluded anything about the stability of the FEM-approximated semidiscrete system. The above-mentioned common pole-zero group seems to be the result of the uncontrollability of the output of the original PDE system. It is clear that the output always stays on the following interval

$$u_2(1,t) \in [0, \bar{s}_a]$$

for a constant disturbance showing a kind of output uncontrollability property.

A subset of the ordinary zeros in the r-plane and the corresponding λ-plane are given in Table 1. Furthermore, it can be concluded that for the sequnce of r-zeros, which are pairwise r_i and $\bar{r}_i$,

$$\lim_{i\to+\infty} |r_i| = \infty \tag{4.16}$$

The corresponding λ-zeros are obtained from the equations

$$r_i^2 = \frac{\lambda^2 + d_1\lambda + d_2}{D(\lambda + a_1)} \tag{4.17}$$

$$\bar{r}_i^2 = \frac{\lambda^2 + d_1\lambda + d_2}{D(\lambda + a_1)} \tag{4.18}$$

Due to the infinite limit (4.16) at least two of the corresponding four λ-zeros (pairwise complex conjugate values) have the limits

$$\lim_{i \to +\infty} \lambda_i = \lim_{i \to +\infty} \bar{\lambda}_i = -a_1$$

as it is evident from the root locus principle. Actually, this is seen also in Table 1. because on the last row there are

$$Re(\lambda_i) = -0.04289\,;\; Im(\lambda_i) \approx 0 \;\; (a_1 = 0.04286)$$

5. SOME SPECULATIONS, OPEN QUESTIONS AND CONCLUDING REMARKS

The infinite dimensional nature of distributed parameter systems is the main obstacle in the efforts to transfer concepts and ideas directly usable in finite-dimensional, even if nonlinear, ODE systems. The concept of the linearising control may have its counterpart in PDE systems. Especially, the concept of zero dynamics can be formulated (to be done later on) as the inverse problem of the original control problem of Eqs.(2.1-.7). On the other hand, boundedness of the solution of the inverse problem as the time goes to infinity is not necessarily directly related to the location of the zeros of the original PDE system linearised around its steady state solution as it is the case in finite-dimensional ODE systems.

The (transcendental) transfer function of the linearised PDE system has four common pole-zero pairs. This is seemingly due to the evident uncotrollability of the output, which corresponds to the analogous case in linear ODE systems.

The practical treatment, e.g. the controller design for PDE systems having the pointwise boundary control and output, can be done based on some semidiscretisation, i.e. spatial, procedure. Then the system model is approximated by an ODE system. Several theoretical questions remained, however, unanswered in this paper: What is the relation of the instability of the linearising control in the ODE approximations obtained via the orthogonal collocation and FEM for some parameter values and the location of zeros of the linearised transfer function of the original PDE system ? Is it possible to formulate a general theory of linearising control for (certain type of) PDE systems ?

REFERENCES

Babary, J.P., Bourrel, S.V., Nihtilä, M.T., and Dochain, D. (1995). Sur la représentation d'etat des systèmes á paramètres répartis, in the book: *Les Systèmes de Régulation*, A.Rachid, ed., Scientifika, France, 23 pp. (in print)

Contois, D. (1959). Kinetics of bacterial growth relationship between population density and specific growth rate of continuous cultures, *Journal of General Microbiology* **21**, 40.

Dochain, D., Babary, J.-P., and Tali-Maamar, N. (1992). Modelling and adaptive control of non-linear distributed parameter bioreactors via orthogonal collocation. *Automatica*, **28**(9), 873-883.

Goldstein, J.A. and Tervo, J. (1995). Existence results of solutions for a system of nonlinear partial differential equations related to bio-reactor. Department of Computer Science and Applied Mathematics, University of Kuopio, 15 pp. (in preparation).

Isidori, A. (1989). *Nonlinear Control Systems*, 2nd Ed., Berlin: Springer.

Nihtilä, M., Tali-Maamar, N., and Babary, J.P. (1992). Controller design for a nonlinear distributed parameter system via a collocation approximation. *Proceedings. IFAC Symposium on Nonlinear Control System Design, NOLCOS'92*, Bordeaux, France, June 24-26, 1992, M.Fliess, ed., pp.447-452.

Nihtilä, M., Tervo, J., and Kaipio, J. (1994). Simulation of a nonlinear distributed parameter bioreactor by FEM approach. Reports, A5, Department of Computer Science and Applied Mathematics, University of Kuopio, 19 pp.

Tali-Maamar, N. (1994). Modèlisation, Analyse et Commande d'un Procédé Biotechnologique à Gradient Spatial de Concentration. Thèse de Docteur de l'Université Paul Sabatier de Toulouse, Rapport LAAS n:o 94079, 106 pp.

Tali-Maamar, N., Damak, T., Babary, J.P., and Nihtilä, M.T. (1993). Application of a collocation method for simulation of distributed parameter bioreactors. *Mathematics and Computers in Simulation*, **31**(11), 303-319.

Thomee, V. and Wahlbin, L. (1975). On Galerkin methods in semilinear parabolic problems. *SIAM J. Numer. Anal.* **12**, 378-389.

Thomee, V. (1984). *Galerkin Finite Element Methods for Parabolic Problems*. Lecture Notes in Mathematics. **1054**, Berlin: Springer.

Villadsen, J. and Michelsen, M.L. (1978). *Solution of Differential Equation Models by Polynomial Approximation*. London: Prentice Hall.

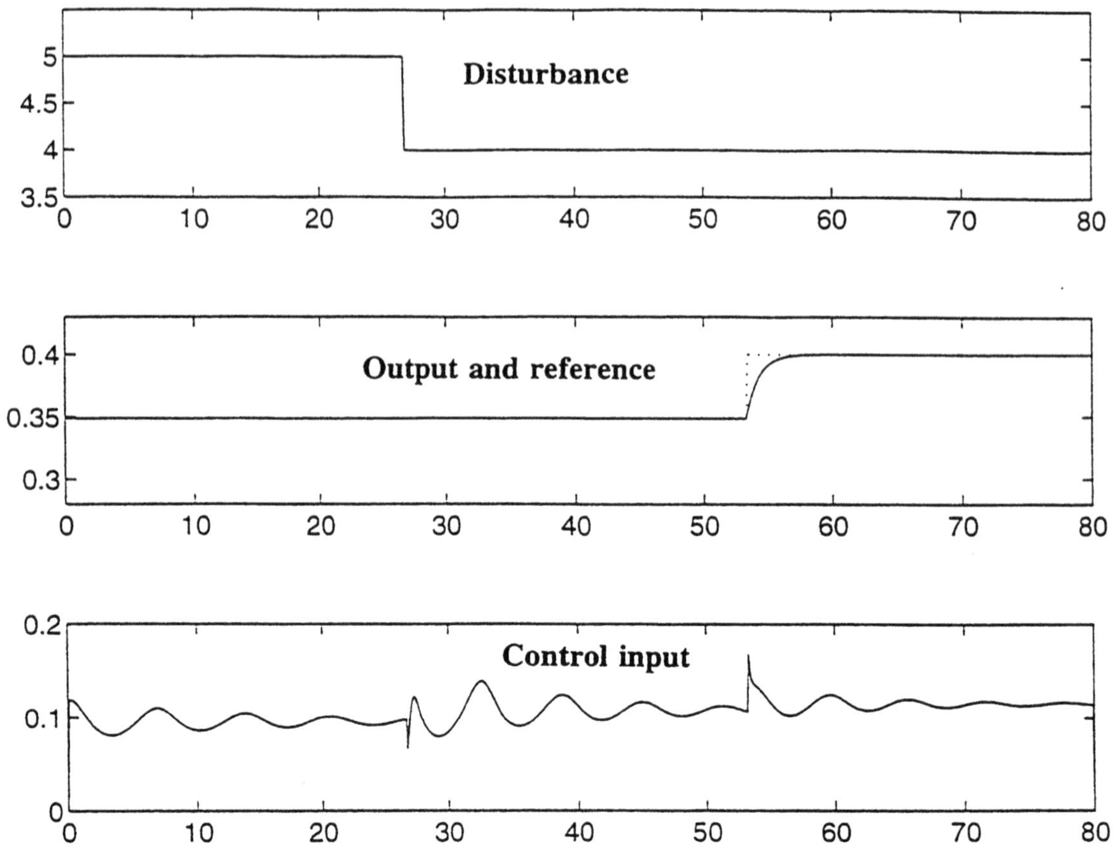

Fig.1. Simulation of step responses for the steps in the disturbance s_a and in the reference y_R for $c_0 = 1$.

Table 1. Some zeros of the linearised transfer function from the infinite set.

- r-zeros: $r_k = r_1 + i r_2$; $\bar{r}_k = r_1 - i r_2$
- λ-zeros: $\lambda_{2k-1} = \lambda_1 + i\lambda_2$; $\bar{\lambda}_{2k-1} = \lambda_1 - i\lambda_2$
  $\lambda_{2k} = \lambda_3 + i\lambda_4$; $\bar{\lambda}_{2k} = \lambda_3 - i\lambda_4$

r_1	r_2	λ_1	λ_2	λ_3	λ_4
-21.4476109	.0000000	.0000000	-.0000745	.0000000	.0000745
-21.5126419	6.2590432	-.0429539	.0013559	-.1389555	-1.3478414
-21.6811600	12.5060987	-.0430097	.0006409	-.6886393	-2.7121081
-21.9022617	18.7453690	-.0430066	.0003902	-1.6153923	-4.1060500
-22.1372890	24.9867039	-.0429933	.0002611	-2.9283857	-5.5316396
-22.3663654	31.2357998	-.0429778	.0001834	-4.6341267	-6.9864960
-22.5817642	37.4939537	-.0429626	.0001334	-6.7363396	-8.4669294
-22.7816715	43.7602692	-.0429488	.0000996	-9.2368336	-9.9694204
-22.9666996	50.0331573	-.0429368	.0000757	-12.1363010	-11.4910402
-23.1382580	56.3110428	-.0429268	.0000586	-15.4348450	-13.0294533
-23.2978764	62.5925789	-.0429177	.0000464	-19.1322803	-14.5827875
-23.4469700	68.8767166	-.0429107	.0000367	-23.2282982	-16.1495399
-23.5867672	75.1626663	-.0429049	.0000295	-27.7225494	-17.7284737
-23.7183132	81.4498291	-.0428993	.0000250	-32.6146812	-19.3185501
-23.8424950	87.7377930	-.0428943	.0000208	-37.9043808	-20.9188995
-23.9600658	94.0262299	-.0428910	.0000174	-43.5913429	-22.5287647
-24.0716686	100.3149414	-.0428866	.0000153	-49.6753235	-24.1474953

SHAPE IDENTIFICATION AND BOUNDARY HEAT FLUX DETERMINATION. APPLICATION TO A PHASE CHANGE APPARATUS.

S. PÉNEAU*, J.P. HUMEAU*, Y. JARNY** and A. SARDA**

* Laboratoire d'Automatique de Nantes, Ecole Centrale de Nantes, 1 rue de la noe, 44072 Nantes cedex 03, FRANCE, Tel: (33) 40.37.16.00, FAX: (33) 40.37.25.22

** Laboratoire de Thermocinétique, ISITEM, La chantrerie, 44087 Nantes cedex 03, FRANCE, Tel: (33) 40.68.31.15. FAX: (33) 40.68.31.41

Abstract. In order to solve an inverse heat conduction problem, a shape identification method for an elliptic problem is devised : the shape of an isothermal part of the boundary of a solid material and the heat flux through this boundary are determined from temperature measurements in the domain or on its boundary. The problem is formulated as a minimization problem; several descent methods are presented and the gradient is computed either by direct calculation or by using an adjoint state. Simulated data, noisy simulated data and experimental data allow a validation of the method and show its robustness.

Résumé. Dans le but de résoudre un problème inverse de conduction thermique, une méthode d'optimisation de forme pour une équation elliptique est développée : la forme d'une partie isotherme de la frontière d'un domaine est déterminée à partir de mesures de temperatures à l'intérieur ou sur la frontière du domaine. Le problème est formulé comme un problème de minimisation. Plusieurs méthodes de descente sont présentées, le gradient est calculé par une méthode directe ou en fonction d'un état adjoint. Des données issues de simulation, puis bruitées et enfin des données expérimentales permettent de valider la méthode et de montrer sa robustesse.

Keywords. Distributed parameters systems, shape identification, Pontryagin's principle, phase change apparatus.

1. INTRODUCTION

The aim is to solve an inverse heat conduction problem which consists in the determination of the shape of a domain, from temperature measurements in the domain and/or on its boundary. More precisely, the unknown shape is defined by an isothermal part of the boundary of a solid material, especially the phase change isotherm.

The following method is rather more general and allows to solve shape identification problems in system governed by elliptic partial differential equations. The domain is bounded by a fixed boundary and by a free-boundary. First the state of the system (temperature in the application) is determined with a known free boundary. This is called the direct problem.

Then a criterion J is introduced, to characterize the deviation between temperature measurements and calculated temperature depending on the shape of the domain.The identification problem is solved as an optimization problem, *find the free boundary which minimizes J*.

A direct and a Lagrangian method are proposed to calculate the gradient of the criterion.

An improvement based on the penalization of the first derivative of the free boundary is carried out. The method is applied to a phase change system with simulated data and noisy data. An apparatus designed to study fundamental heat-transfer mechanisms arising in fusion/solidification processes allows to test the shape identification method with experimental data. This process, which allows to obtain a steady-state one-phase Stefan problem consists in a parallelepipedic cell which can be accurately described by 2D-model presented hereafter.

2. DIRECT PROBLEM

2.1 Notations

L is a real strictly positive number. $X = (x_1, x_2)$ is an element of $\mathbb{R}^2$, $\gamma_S : \mathbb{R} \mapsto \mathbb{R}$ is a continuous function parametrized by $[S_i]_{i=0}^{n-1} \in \mathbb{R}^n$. $\Omega(S)$ is an open subset of $\mathbb{R}^2$ bounded by $\Gamma_1 \cup \Gamma_2 \cup \Gamma_3(S) \cup \Gamma_4$. $\Omega(S)$ and $\Gamma_3(S)$ are definied by

$$\Omega(S) = \{X : 0 < x_1 < L; 0 < x_2 < \gamma_S(x_1)\},$$
$$\Gamma_3(S) = \{X : 0 < x_1 < L; x_2 = \gamma_S(x_1)\}.$$

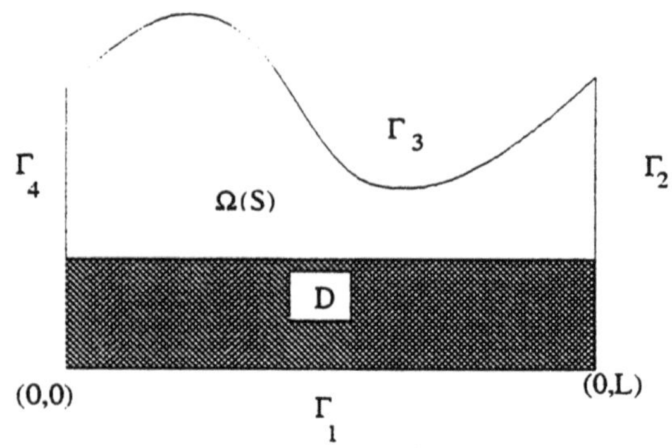

Fig. 1. domain

$\Gamma_3(S)$ is the unknown boundary to be determined. The measurement area $\mathcal{D}$ is a fixed subdomain of $\Omega(S)$.
γ_S is taken piecewise linear such that for i=0...n-1

$$\gamma_S(x_i) \quad = \quad S_i \text{ with } x_i = \frac{i.L}{n-1}.$$

2.2 State equations

In this section, S is assumed to be known, so are $\Omega(S)$ and $\gamma(S)$. $\Gamma_2 \cup \Gamma_3(S) \cup \Gamma_4$ is denoted Γ_0 and $\frac{\partial u}{\partial n}$ is the normal derivative on Γ_1. u_Γ and ϕ_Γ are respectively the Dirichlet condition on Γ_0 and the Neumann condition on Γ_1.
Then the temperature u is the solution of the direct problem $DP(S)$.

$$DP(S) \left\{ \begin{array}{rcll} -\Delta u & = & 0 & \text{in} \quad \Omega(S) \\ u & = & u_\Gamma & \text{on} \quad \Gamma_0 \\ \frac{\partial u}{\partial n} & = & \phi_\Gamma & \text{on} \quad \Gamma_1 \end{array} \right. \qquad (1)$$

2.3 Varitional formulation

The direct problem is solved by a finite element method, a variational formulation (VF) of the direct problem is introduced.

$$VF \left\{ \begin{array}{l} \text{Find u} \in H^1(\Omega(S)) \text{ such that} \\ \displaystyle\int_{\Omega(S)} \nabla u . \nabla v \, dX = \int_{\Gamma_1} \phi_\Gamma v \, dX, \\ \text{on } \Omega(S), \ \forall v \in H^1(\Omega(S)), \\ u = u_\Gamma \text{ on } \Gamma_0, \end{array} \right. \qquad (2)$$

avec
$H^1(\Omega) = \left\{ v \in L^2(\Omega) ; \frac{\partial u}{\partial x_i} \in L^2(\Omega) \ i = 1, 2 \right\}$,
$V = \left\{ v \in H^1(\Omega) ; v|_{\Gamma_0} = 0 \right\}$.

2.4 Finite element method

The grid is built, as shown on figure 2. A moving grid, depending on S, has been chosen.

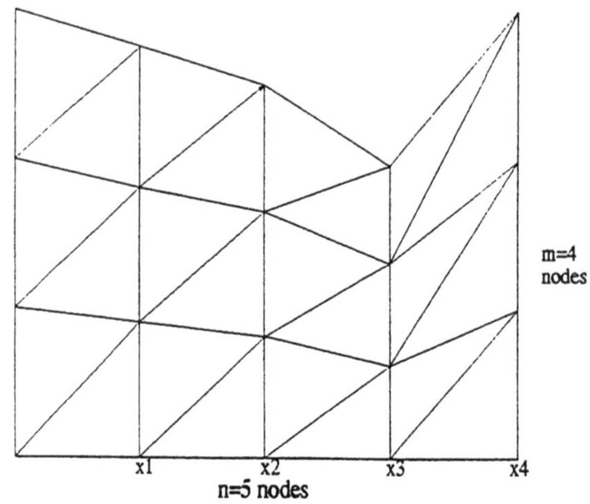

Fig. 2. moving grid

n (respectively m) is the number of nodes on Γ_1 and Γ_3 (resp. Γ_1 and Γ_3). $nf (= (n-2).(m-1))$ is the number of unknown values.Nodes define a set of triangles T_h. V is approached by the following space
$V_h = \{v \in V$ piecewise linear on each triangles$\}$
Then $u(S)$ is the solution of the following u-linear equation

$$D(S).u(S) = E(S) \qquad (3)$$

Similar results are obtained if the state equation "$\Delta u = 0$" is a more general elliptic equation such that

$$-\sum_{ij=1}^{2} \frac{\partial}{\partial x_i} \left(a_{ij} . \frac{\partial u}{\partial x_j} \right) + a_0 . u = f$$

A non-linear application $S \rightarrow u(S)$ is defined by equation 3.

3. IDENTIFICATION PROBLEM

The identification problem (IP) is formulated by introducing a criterion $J(S)$. J characterizes the deviation between temperature measurements and calculated temperatures. Let $[x_s(i); y_s(i)]_{i=1}^{ns}$ and $[u_d(i)]_{i=1}^{ns}$ be respectively the coordinates of the sensors and the temperature measurements at the sensors. The identification problem is to find S^* which minimizes :

$$J(S) = ||C(S).u(S) - u_d||^2 \qquad (4)$$

The node positions depends on S so the observation matrix $C(S)$ depends also on S.

The array $C(S).u(S)$ gives the temperature at $[x_s(i); y_s(i)]_{i=1}^{ns}$ for a temperature field $u(S) \in \mathbb{R}^{nf}$. To get a simpler form for $C(S)$, the sensors are located between two nodes on the same vertical line of the mesh.

4. CALCULATION OF THE GRADIENT

4.1 Direct method

Let $(\ ,\ )_{\mathbb{R}^{ns}}$ denote the scalar product in $\mathbb{R}^{ns}$. Differentiation of 4 according to S_i, $i = 1, \ldots n-1$ leads to

$$\frac{\partial J(S)}{\partial S_i} = 2.\left(Cu - u_d, \frac{\partial C}{\partial S_i}.u + C\frac{\partial u}{\partial S_i}\right)_{\mathbb{R}^{ns}} \quad (5)$$

The differentiation of $C(S)$ is developed in Péneau *et al.*(1994)
To calculate the sensitivity of temperature $\frac{\partial u}{\partial S_i}$ due to a variation of the shape of the boundary, equation 3 is differentiated. For $i = 1, \ldots, n-1$

$$D(S)\frac{\partial u(S)}{\partial S_i} = \frac{\partial E(S)}{\partial S_i} - \frac{\partial D(S)}{\partial S_i}.u(S) \quad (6)$$

The main difficulty lies in the calculation of the sensitivity of the matrices D, E and C. The assembled matrices in a finite element method depend on the geometry of the domain. This dependance is explicitely written, then elementary matrices(before assemblying) are differentiated with respect to S_i.
The n linear equations 6 are solved then 5 allows to calculate $\frac{\partial J(S)}{\partial S_i}$.
The resolution of $(n+1)$ linear systems is needed by the direct method.

4.2 Adjoint method

The calculation of the gradient can be done by using a Lagrangian and the Pontryagin's principle. Let introduce a functionnal $\mathcal{L}$ from the criterion's definition 4 and the constraint equation 3

$$\mathcal{L}(S, u, p) = J(S) + (p, D(S).u - E(S))_{\mathbb{R}^{nf}}$$

$\mathcal{L}$ depends on S, u and p, so

$$\frac{d\mathcal{L}(S,u,p)}{dS}.\delta S = \frac{\partial \mathcal{L}}{\partial S}.\delta S + \frac{\partial \mathcal{L}}{\partial u}.\frac{\partial u}{\partial S}.\delta S + \frac{\partial \mathcal{L}}{\partial p}.\frac{\partial p}{\partial S}.\delta S$$

Let denote P^* the following adjoint problem:
find $p \in \mathbb{R}^{nf}$ *such that*

$$D^*.p = -2C^*(Cu - u_d) \quad (7)$$

If p is the solution of P^* and u is the solution of 3, $\frac{\partial \mathcal{L}}{\partial p}.\delta p = 0$ and then

$$\frac{d\mathcal{L}(S,u,p)}{dS}.\delta S = \frac{\partial \mathcal{L}}{\partial S}.\delta S (\nabla J(S), \delta S)_{\mathbb{R}^n}$$

The calculation of $\frac{\partial L}{\partial S}.\delta S$ gives

$$\frac{\partial L}{\partial S}.\delta S = 2.\left(C(S)u - u_d, \frac{\partial}{\partial S}(C(S)u - u_d).\delta S\right)_{\mathbb{R}^{nf}} + .\left(p, \frac{\partial}{\partial S}(D(S)u - E(S)).\delta S\right)_{\mathbb{R}^{nf}}$$

and then for $k = 0, \ldots, n-1$

$$\frac{\partial J(S)}{\partial S_k} = 2.u^T.\left(\frac{\partial C(S)}{\partial S_k}\right)^T (C(S)u - u_d) + .\left[u^T.\left(\frac{\partial C(S)}{\partial S_k}\right)^T - \left(\frac{\partial E(S)}{\partial S_k}\right)^T\right].p$$

The calculation of $\nabla J(S)$ requires then the resolution of two linear systems 3 and 7 (instead of $n+1$ linear systems by the direct method)

4.3 CPU time comparison

Figure 3 shows the CPU time needed by both methods, the time needed by the resolution of the linear systems explain the difference between the methods, so the ratio increases with the number of nodes

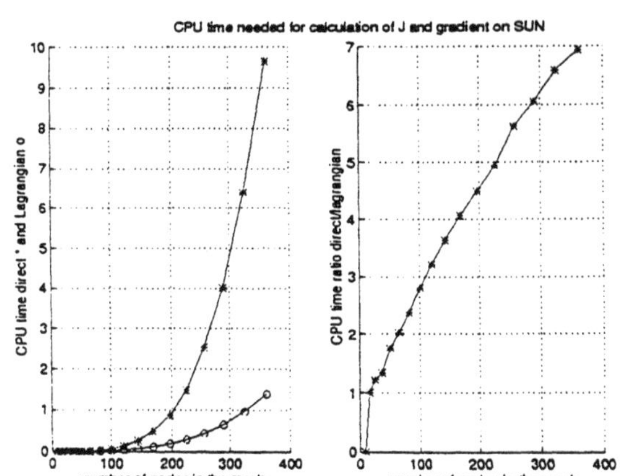

Fig. 3. Calculation of ∇J. CPU time on Sun

5. IDENTIFICATION WITH SIMULATED DATA

Let u_1 be a quadratic field of temperature

$$u_1(x_1, x_2) = a^2 . \left[(x_1 - x_0)^2 + (x_2 - y_0)^2\right] - 1$$

Simulations are done with :
$L = 1, a = 0.4, y_0 = 1 - \sqrt{\frac{41}{10}}, x_0 = \frac{1}{2}$.
The identification problem is an *ill-posed problem* so a cost function is added to the criterion to take into account the smoothness of the free boundary. ρ being a real positive constant, a new criterion $\tilde{J}$ is introduced

$$\tilde{J}(s) = J(S) + \rho \int_0^L \left(\frac{\partial \gamma_S(\sigma)}{\partial X}\right)^2 d\sigma$$

5.1 Optimisation methods

Several descent methods has been implemented. The results from a steepest descent method and quasi-Newton method are presented. The parameters are ns=5 sensors, ρ=0.01, for $i = 1, \ldots, n-1$). The initial guess is a random initialization. On figure 4 and 5 are represented:

- the value of criterion $\tilde{J}$ and J versus the iteration number.
- the maximum erreur on S, $\max_{0 \le i < n} |S_i - S_i^*|$ versus the iteration number.
- the random initial guess, the desired value S^* and the final final shape of the unknown boundary.

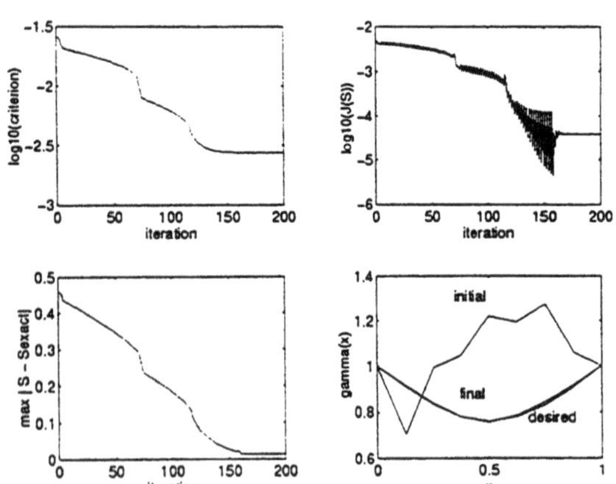

Fig. 4. steepest descent optimization

Convergence is obtained after 150 iterations for steepest descent and 20 iterations for Quasi-Newton, so figure 6 shows CPU time needed for the convergence

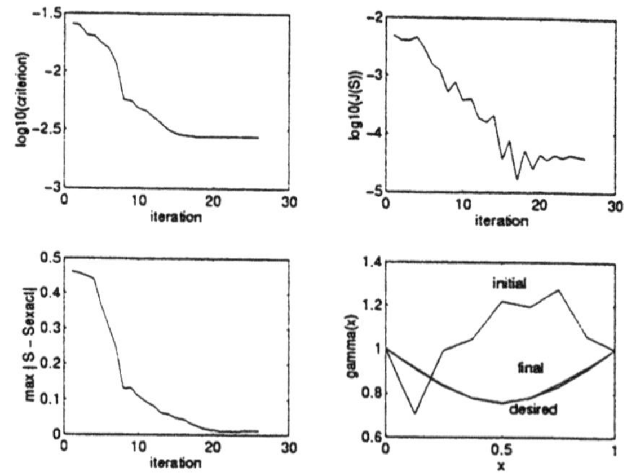

Fig. 5. quasi-Newton optimization

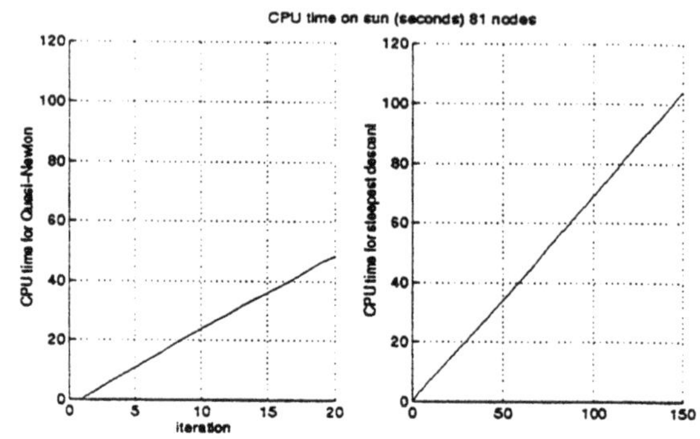

Fig. 6. CPU time quasi-Newton and steepest descent

6. IDENTIFICATION WITH NOISY SIMULATED DATA

The temperature measurements $u_d(i)$ are perturbed by a uniform random noise. $u_d(i)$ being the exact simulated value, for each sensor perturbed measurement is

$$\tilde{u}_d(i) = u_d(i).(1 + noise).$$

The method is able to find the shape of the zero isotherm with an error on the measurement equals to 7%. The shapes of the isotherm is presented on figure 7 (7 sensors, ρ=0.01 for the penalization, adjoint state for calculation of the gradient, quasi-newton optimisation) and for differents amplitudes of noise. Only the 10% case gives a shape far from the real one.

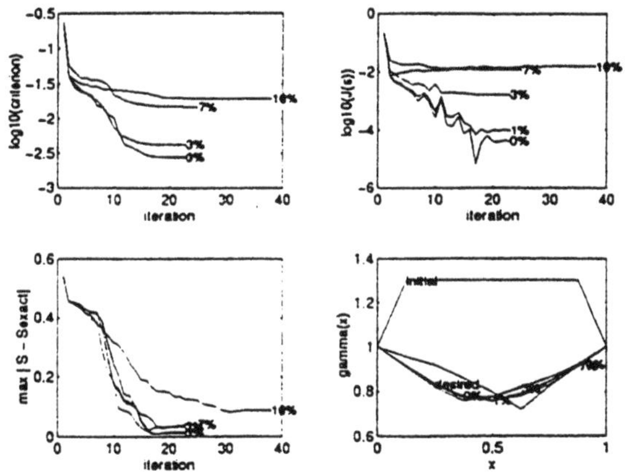

Fig. 7. Noisy data

7. IDENTIFICATION WITH EXPERIMENTAL DATA

The apparatus is based on a vertical Bridgman configuration, a schematic description is given on figure 8. The parallelepipedic cell, containing a pure phase change material (an hexadecane paraffin), has been designed to study bidimensional heat transfer phenomenon.

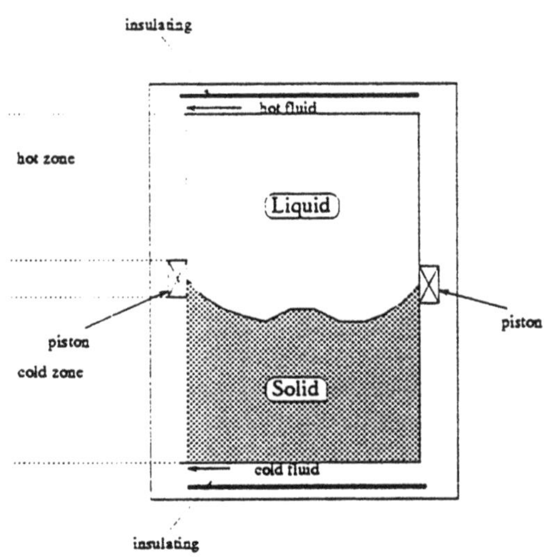

Fig. 8. apparatus

A grid of 64 thermocouples 9, streched in the cell, gives temperature measurements in half of the cell. The domain is assumed to be symmetrical.

In steady-state, heat transfers in the liquidus are combination of conduction and convection phenomena. The aim is to find the shape of the phase change isotherm with the preceding method, by considering heat transfer only in the solidus (without convection). Sensors on the boundaries of the hachured domain of the figure 9 are used to give boundaries conditions on $\Gamma_2 \cup \Gamma_4$ and temperature measurements on Γ_1. The calculated shapes of the 18 (phase change temperature) and 10 degrees cel-

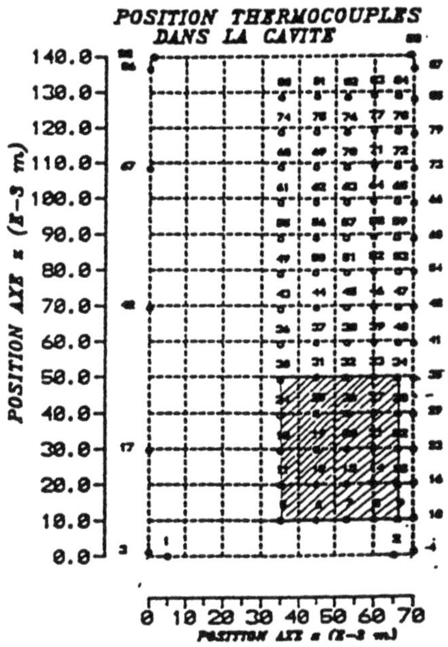

Fig. 9. sensors location

sius isotherms are compared to the experimental values given by the whole grid of thermocouples.

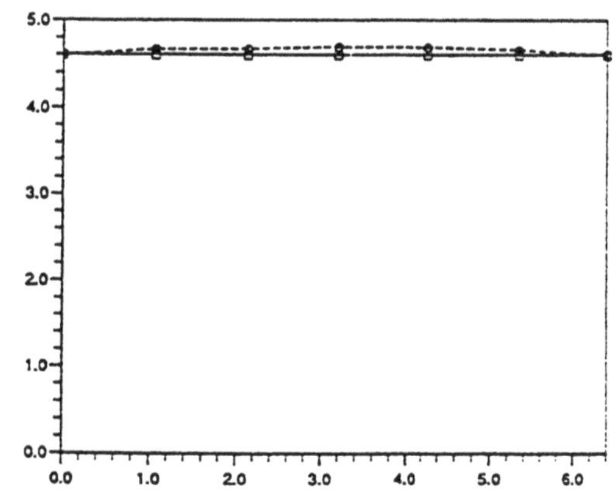

Fig. 10. Phase change isotherm: initial and final shape

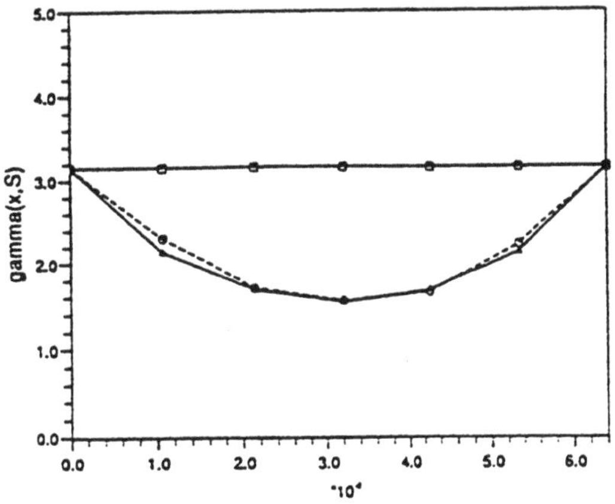

Fig. 11. 10^oC isotherm: initial, final and desired shape

Heat flux measurement is difficult to perform, especially on an unknown free-boundary. As a result of the shape identification, the temperature field u is known in the domain, so the heat flux is determined by calculation of the normal derivative of u on the free boundary. The length of the arrows is proportional to the normal heat flux values (given in the table).

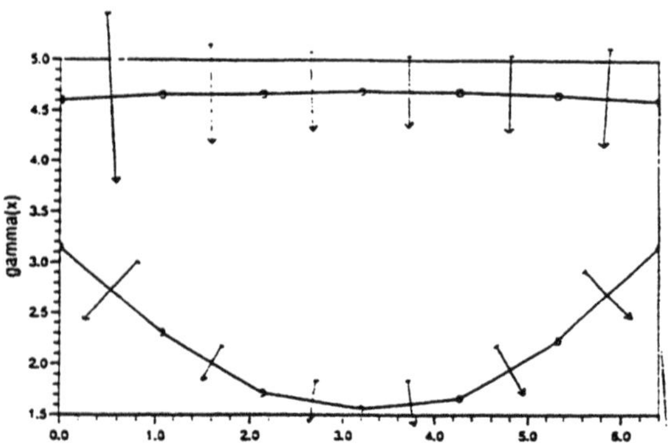

Fig. 12. heat flux on the isotherms

heat flux density $(\times(-1))$ in $W.m^{-2}$

18°	283.4	166.6	129.7	117.9	129.9	165.2
10°	99.1	54.6	67.1	72.8	81.2	80.2

8. CONCLUSIONS

The main difficulty in the implementation of the algorithm lies in the calculation of the matrices $C(S)$, $D(S)$ et $E(S)$ and their sensitivity. Simulated, noisy data and temperatures measurements from an experimental apparatus are used to check the method and show its robustness. The results presented above show that the method allows to compute the heat flux on an unknown boundary for a steady-state one phase Stefan problem. It allows then to determine the convective heat exchange on this free boundary. The non-steady state case (moving free boundary) presents a great practical interest especially in thermal processus.

REFERENCES

Sarda A. and Delaunay D., "Experimental study of a vertical Bridgman configuration", Eurotherm n30, Oct 1993.

Péneau S., Jarny Y. and Sarda A., "Isotherm shape identification for a tow dimension heat conduction problem", Inverse Problem in Engineering Mechanics, ISIP'94, Paris, Nov 1994.

Banks H.T. and Fumio Kojima, "Boundary shape identification problems in two dimensional domains related to thermal testing of materials", Quartely of applied Math., vol.XLVII n2, June 1989.

Certes H., Knopf-Lenoir C., Touzot G., Yvon J.P., "Optimisation de forme d'une pièce mécanique", 1er colloque international sur les méthodes numériques dans les sciences de l'ingénieur, Nov. 1978.

Lions J.L., "Contrôle optimal de systèmes gouvernés par des équations aux dérivées partielles"; Dunod , Paris 1968.

Raviart P.A., Thomas J.M., "Introduction à l'analyse numérique des équations aux dérivées partielles"; Masson; 1983.

Pironneau O., "Optimal shape design for elliptic systems"; Springer Verlag; New-York; 1984.

Sarda A., " Construction experimentale en vue de valider des algorithmes d'identification et de commandes de frontière mobiles, application la solidification dirigée"; Thèse de Doctorat; Université de Nantes, Ecole Centrale de Nantes; 1992

NONLINEAR CONTROLLERS FOR INDUCTION MOTORS

JOHN CHIASSON

348 Benedum Hall, Department of Electrical Engineering, University of Pittsburgh, Pittsburgh, PA 15261, USA, chiasson@ee.pitt.edu

Abstract. Tremendous breakthroughs in the techniques of nonlinear control have been achieved in the last twenty years or so. These methods include the differential-geometric techniques, differential-algebraic methods, variable structure control, back-stepping, etc. This paper specifically considers some of the differential-geometric techniques in terms of their suitability for high-performance induction motor control. It is shown that both dynamic feedback linearization and input-output linearization are applicable to induction motor control and a comparison of the two is made. The particular effectiveness of the input-output linearization controller is demonstrated with experimental results.

Key Words. Induction Motor, Dynamic Feedback Linearization, Input-Output Linearization, Nonlinear Observers, Experimental Results

1 INTRODUCTION

With the incredible capabilities of today's computer technology, ever increasing performance is being demanded of modern electromechanical systems. These systems include such devices as induction, synchronous and variable-reluctance motors as well as magnetic bearings, all of which have dynamic models that are nonlinear. The classical (and still important) method of nonlinear control is linearization. That is, the use of an approximating linear model valid about an operating point to design a linear controller and then gain scheduling based on the operating condition. However, this methodology just cannot provide the performance demanded of electromechanical systems. Interestingly, over the last 75 years or so, electric machinery engineers have made some ad-hoc discoveries of nonlinear techniques that are still extremely useful. For example, the direct-quadrature (dq) transformation of electric machine theory is a feedback linearizing transformation for synchronous machines. However, the recently developed nonlinear theories have not only provided formal systematic approaches to such existing ad-hoc methods (e.g., to determine if a feedback linearizing transformation exists for any given nonlinear system), but have actually developed approaches that hitherto had not been considered by application oriented engineers (e.g., dynamic feedback linearization, input-output linearization, differential-algebraic methods, etc.). This has lead to a renaissance of interest in the control of electromechanical systems from the viewpoint of these new methods [39] which include the differential-geometric approach [24][25], back-stepping methods [27], passivity theory [40], differential-algebraic methods [11], and variable structure techniques [9].

Electromechanical systems are good candidates for the new nonlinear methodologies as they typically have well-defined nonlinear mathematical models. In fact, by appropriate changes of coordinates, these models are often either a polynomial or rational function of the state variables. Because of their well-defined nonlinearities, one expects that the performance of electromechanical systems would be significantly enhanced with the incorporation of such knowledge in the design of the controller. This expectation is being fulfilled especially because these new nonlinear control methods turn out to be particularly accommodating for the nonlinear models characterizing electromechanical systems [39]. As a sampling of specific applications of these techniques to electromechanical systems, the method of feedback linearization (static/dynamic) has been applied to stepper, variable reluctance and induction motors in [18][19][33][38], input-output linearization has been used to obtain high-performance control of induction motors in [4][6][7][8][19], the back-stepping method was used to design an output feedback controller for an induction motor in [26] and for a dc motor in [13], a passivity-based technique to estimate the rotor flux in induction motors is given in [14] while controllers based on this concept are given in [21][23], and finally, variable structure (sliding mode) control methods have been formulated for use with dc, synchronous and induction motors in [9][10]. It is also quite fortuitous that, in addition to these theoretical advances in nonlinear control, there continues to be advances in the performance of high-speed digital signal processors. That is, the computational capability is now available to cheaply implement these sophisticated nonlinear control algorithms.

In this paper, the induction motor is chosen as a practical yet important example for which the advances in nonlinear control can be exploited. It is the motor of choice in many industrial applications due to its reliability, ruggedness, and relatively low cost. The mechanical reliability of the induction motor is due to the fact the current commutation is done electrically rather than mechanically (e.g., there are no brushes nor commutator to wear out as in a DC motor). The relatively low cost of an induction motor is a result of permanent magnets not being used as in many synchronous and dc machines as well as to the simple construction of the (squirrel cage) rotor making it cheap to manufacture. However, the induction motor presents a very challenging control problem which is primarily due to three issues: (i) The dynamic model of the system is nonlinear (with or without magnetic saturation) (ii) Two of the state variables (rotor fluxes/currents) are not usually measurable (iii) Due to Ohmic heating, the rotor resistance can vary considerably with a significant impact on the system dynamics. These challenges make the induction motor a particularly interesting choice as a control problem. It is the aim of this paper to consider the differential-geometric techniques in terms of providing control algorithms with the capability of making the induction motor into a high-performance machine.

The paper is outlined as follows: Section 2 presents a brief derivation of the accepted model of the induction motor. In Section 3, the field-oriented model of the induction motor as well as the method of field-oriented control are described. Section 4 considers two different dynamic feedback linearization controllers derived from the field-oriented model of the induction motor. In Section 5 the input-output linearization approach to induction motor control is also derived based on the field-oriented model and it is shown how it provides the decoupling of flux and speed. Section 6 considers the design of nonlinear state/disturbance observers for induction motors. Finally, experimental results using the input-output linearization controller are presented in Section 7 with some brief concluding remarks given in Section 8.

2 Mathematical Model

The mathematical model of a sinusoidally wound, two-phase, n_p pole-pair induction motor is now briefly outlined. It is standard practice, and validated experimentally in [15], to also use this same model for induction motors with squirrel cage rotors. Let i_{Sa} and i_{Sb} denote the current in phases a and b of the stator, i_{Ra} and i_{Rb} denote the currents in phases a and b of the rotor, θ be the rotor position, ω be the rotor speed and, u_{Sa} and u_{Sb} be the applied voltages to phases a and b of the stator. These quantities are indicated in Figure 1 with $n_p = 1$ (adapted from Figure 4.1.7a of [3]) where the single coil in the figure for each phase represents a multi-turn sinusoidally wound phase.

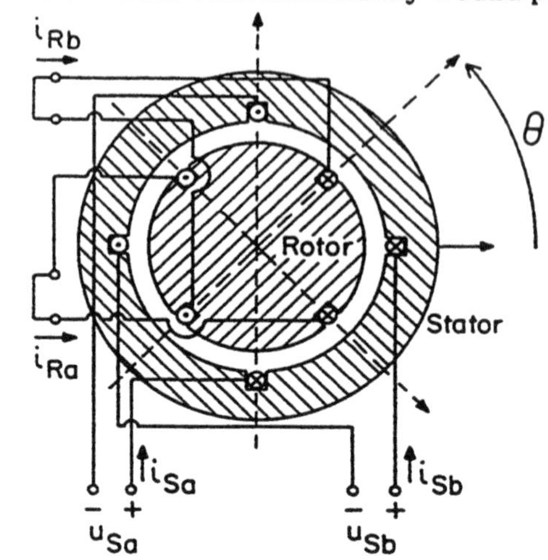

Figure 1 – Induction Motor

The total flux linkage in stator phases a and b are

$$\begin{aligned}\lambda_{Sa} &= L_S i_{Sa} + M(i_{Ra}\cos(n_p\theta) - i_{Rb}\sin(n_p\theta)) \\ \lambda_{Sb} &= L_S i_{Sb} + M(i_{Ra}\sin(n_p\theta) + i_{Rb}\cos(n_p\theta))\end{aligned} \quad (1)$$

respectively, where L_S is the self-inductance of each stator phase and M is the coefficient of mutual inductance between the stator and rotor windings (phases). Similarly, the total flux linkage in each of the rotor phases a and b is, respectively

$$\begin{aligned}\lambda_{Ra} &= L_R i_{Ra} + M(i_{Sa}\cos(n_p\theta) + i_{Sb}\sin(n_p\theta)) \\ \lambda_{Rb} &= L_R i_{Rb} + M(-i_{Sa}\sin(n_p\theta) + i_{Sb}\cos(n_p\theta))\end{aligned} \quad (2)$$

where L_R is the self-inductance of each rotor phase. These expressions for the fluxes can be derived as shown in [1]. However, by studying Figure 1, the reader should be able to see that they are at least plausible. The simplicity of the expressions (1), (2) for the fluxes is due to the assumption of the rotor and stator phases being sinusoidally wound.

By applying Faraday's and Ohm's Laws to the stator phases, it follows that the electrical equations for these two phases are

$$\begin{aligned}u_{Sa} - d\lambda_{Sa}/dt - R_S i_{Sa} &= 0 \\ u_{Sb} - d\lambda_{Sb}/dt - R_S i_{Sb} &= 0\end{aligned} \quad (3)$$

and similarly, the electrical equations of the rotor phases are

$$\begin{aligned} -d\lambda_{Ra}/dt - R_R i_{Ra} &= 0 \\ -d\lambda_{Rb}/dt - R_R i_{Rb} &= 0 \end{aligned} \tag{4}$$

where the rotor phases are assumed to be shorted. An expression for the electromagnetic torque produced by the motor is [1]

$$\tau = n_p\left(i_{Sb}\lambda_{Sa} - i_{Sa}\lambda_{Sb}\right) \tag{5}$$

Substituting the expressions (1) (2) for the stator and rotor fluxes into (3)(4)(5), a model of the induction motor is given by system of equations [1]

$$\begin{aligned} u_{Sa} &= R_S i_{Sa} + L_S \tfrac{d}{dt} i_{Sa} \\ &\quad + M\tfrac{d}{dt}(i_{Ra}\cos(n_p\theta) - i_{Rb}\sin(n_p\theta)) \\ u_{Sb} &= R_S i_{Sb} + L_S \tfrac{d}{dt} i_{Sb} \\ &\quad + M\tfrac{d}{dt}(i_{Ra}\sin(n_p\theta) + i_{Rb}\cos(n_p\theta)) \\ 0 &= R_R i_{Ra} + L_R \tfrac{d}{dt} i_{Ra} \\ &\quad + M\tfrac{d}{dt}(i_{Sa}\cos(n_p\theta) + i_{Sb}\sin(n_p\theta)) \\ 0 &= R_R i_{Rb} + L_R \tfrac{d}{dt} i_{Rb} \\ &\quad + M\tfrac{d}{dt}(-i_{Sa}\sin(n_p\theta) + i_{Sb}\cos(n_p\theta)) \\ J d\omega/dt &= n_p M(i_{Sb}(i_{Ra}\cos(n_p\theta) - i_{Rb}\sin(n_p\theta)) \\ &\quad - i_{Sa}(i_{Ra}\sin(n_p\theta) + i_{Rb}\cos(n_p\theta))) \\ &\quad - \tau_L \\ d\theta/dt &= \omega \end{aligned} \tag{6}$$

The control problem is to choose u_{Sa}, u_{Sb} in such a way as to force ω and/or θ to track a given trajectory/reference. Measurements of i_{Sa}, i_{Sb} and θ are usually available for feedback control while measurements of i_{Ra} and i_{Rb} are typically *not* available. In fact, a common type of induction motor is the so-called squirrel cage motor, where rotor currents are distributed on the surface of the rotor so that the total equivalent current cannot be measured directly. Though the resulting flux can be measured through the use of Hall sensors placed in the air-gap, such sensors are not economical and diminish the overall reliability of the control system.

The particular set of nonlinear differential equations (6) describing the induction motor is complicated and a control strategy is by no means self-evident. The solution lies in finding an *equivalent* set of equations which are simpler in form for which the control design becomes apparent. As a first step toward simplifying the above equations, the dynamic equations are rewritten in terms of an equivalent rotor flux pair (ψ_{Ra}, ψ_{Rb}) resulting in a mathematical model for which the $\cos(n_p\theta)$ and $\sin(n_p\theta)$ expressions are eliminated. That is, one defines an equivalent set of rotor fluxes as [1]

$$\begin{aligned} \psi_{Ra} &= \cos(n_p\theta)\lambda_{Ra} - \sin(n_p\theta)\lambda_{Rb} \\ &= L_R(\cos(n_p\theta)i_{Ra} - \sin(n_p\theta)i_{Rb}) + M i_{Sa} \\ \psi_{Rb} &= \sin(n_p\theta)\lambda_{Ra} + \cos(n_p\theta)\lambda_{Rb} \\ &= L_R(\sin(n_p\theta)i_{Ra} + \cos(n_p\theta)i_{Rb}) + M i_{Sb} \end{aligned}$$

Then, in terms of the state variables θ, ω, ψ_{Ra}, ψ_{Rb}, i_{Sa} and i_{Sb}, the system (6) is transformed to the so-called $\alpha-\beta$ model given by

$$\begin{aligned} d\theta/dt &= \omega \\ d\omega/dt &= \frac{n_p M}{J L_R}(i_{Sb}\psi_{Ra} - i_{Sa}\psi_{Rb}) - \tau_L/J \\ d\psi_{Ra}/dt &= -\frac{R_R}{L_R}\psi_{Ra} - n_p\omega\psi_{Rb} + \frac{M R_R}{L_R} i_{Sa} \\ d\psi_{Rb}/dt &= -\frac{R_R}{L_R}\psi_{Rb} + n_p\omega\psi_{Ra} + \frac{M R_R}{L_R} i_{Sb} \\ u_{Sa} &= R_S i_{Sa} + \sigma L_S d i_{Sa}/dt + \frac{M}{L_R} d\psi_{Ra}/dt \\ u_{Sb} &= R_S i_{Sb} + \sigma L_S d i_{Sb}/dt + \frac{M}{L_R} d\psi_{Rb}/dt \end{aligned} \tag{7}$$

or, in state-space form (cf., [4])

$$\begin{aligned} d\theta/dt &= \omega \\ d\omega/dt &= \mu(\psi_{Ra} i_{Sb} - \psi_{Rb} i_{Sa}) - \tau_L/J \\ d\psi_{Ra}/dt &= -\eta\psi_{Ra} - n_p\omega\psi_{Rb} + \eta M i_{Sa} \\ d\psi_{Rb}/dt &= -\eta\psi_{Rb} + n_p\omega\psi_{Ra} + \eta M i_{Sb} \\ d i_{Sa}/dt &= \eta\beta\psi_{Ra} + \beta n_p\omega\psi_{Rb} - \gamma i_{Sa} + \frac{u_{Sa}}{\sigma L_S} \\ d i_{Sb}/dt &= \eta\beta\psi_{Rb} - \beta n_p\omega\psi_{Ra} - \gamma i_{Sb} + \frac{u_{Sb}}{\sigma L_S} \end{aligned} \tag{8}$$

with $\eta = R_R/L_R$, $\beta = M/(\sigma L_R L_S)$, $\mu = n_p M/(J L_R)$, $\gamma = M^2 R_R/(\sigma L_R^2 L_S) + R_S/(\sigma L_S)$. In practice, the usual case is that ψ_{Ra}, ψ_{Rb} (or, equivalently i_{Ra}, i_{Rb}) are *not* available for feedback. Further, if the motor has a position sensor, then it typically does not have a speed sensor.

3 Field-Oriented Control

The present state of the art in induction motor control is the so-called field-oriented (or, vector control) technique introduced by Blaschke [1]. This method consists in rewriting the dynamic equations of the induction motor in a reference frame that rotates with the rotor flux "vector" (i.e., a time-varying complex phasor representation of the fluxes in the rotor windings). In this new coordinate system, it is observed that by holding the magnitude of the rotor flux constant, there is a *linear* relationship between a control variable and the speed. To outline this approach, consider an equivalent model of (8) found by using the following nonlinear state-space transformation [4]

$$\begin{aligned}\omega &= \omega\\ \rho &= \tan^{-1}(\psi_{Rb}/\psi_{Ra})\\ i_d &= \cos(\rho)i_{Sa}+\sin(\rho)i_{Sb}\\ i_q &= -\sin(\rho)i_{Sa}+\cos(\rho)i_{Sb}\\ \psi_d &= \cos(\rho)\psi_{Ra}+\sin(\rho)\psi_{Rb}=\sqrt{\psi_{Ra}^2+\psi_{Rb}^2}\\ \psi_q &= -\sin(\rho)\psi_{Ra}+\cos(\rho)\psi_{Rb}\equiv 0\end{aligned} \tag{9}$$

along with defining a new set of inputs

$$\begin{aligned}u_d &= \cos(\rho)u_{Sa}+\sin(\rho)u_{Sb}\\ u_q &= -\sin(\rho)u_{Sa}+\cos(\rho)u_{Sb}.\end{aligned}$$

to represent the system in the *rotor-flux field-oriented* coordinates.

Note that ψ_d and ρ are just the polar coordinates for the ordered pair (ψ_{Ra}, ψ_{Rb}). The term *field-oriented* refers to this new rotating coordinate system whose angular position is ρ so that $\psi_q \equiv 0$. That is, this coordinate system is aligned with the field flux whose position is ρ and magnitude is ψ_d. This transformation of the currents/fluxes/voltages is referred to as the direct-quadrature or dq transformation. In these new coordinates, the system equations become

$$\begin{aligned}d\omega/dt &= \mu\psi_d i_q-\tau_L/J\\ d\psi_d/dt &= -\eta\psi_d+\eta M i_d\\ di_d/dt &= -\gamma i_d+\eta\beta\psi_d+n_p\omega i_q\\ &\quad +\eta M i_q^2/\psi_d+u_d/(\sigma L_S)\\ di_q/dt &= -\gamma i_q-\beta n_p\omega\psi_d-n_p\omega i_d\\ &\quad -\eta M i_q i_d/\psi_d+u_q/(\sigma L_S)\\ d\rho/dt &= n_p\omega+\eta M i_q/\psi_d\end{aligned} \tag{10}$$

Note that the electromagnetic torque $\tau = J\mu\psi_d i_q$ is now just proportional to the product of two state variables. On the other hand, the differential equations for i_d and i_q still contain quite complicated nonlinearities. One possibility to simplify these current dynamics is to use feedback cancellation, that is, let

$$\begin{aligned}u_d &= -(\sigma L_S)\left(-\gamma i_d+\eta\beta\psi_d+n_p\omega i_q\right.\\ &\quad \left.+\eta M i_q^2/\psi_d\right)+(\sigma L_S)v_d\\ u_q &= -(\sigma L_S)\left(-\gamma i_q-\beta n_p\omega\psi_d-n_p\omega i_d\right.\\ &\quad \left.-\eta M i_q i_d/\psi_d\right)+(\sigma L_S)v_q\end{aligned} \tag{11}$$

so that (10) becomes

$$\begin{aligned}d\omega/dt &= \mu\psi_d i_q-\tau_L/J\\ d\psi_d/dt &= -\eta\psi_d+\eta M i_d\\ di_d/dt &= v_d\\ di_q/dt &= v_q\\ d\rho/dt &= n_p\omega+\eta M i_q/\psi_d\end{aligned} \tag{12}$$

However, the feedback (11) requires accurate knowledge of the motor parameters and the state variables. The motor parameters R_R, R_S can vary significantly due to Ohmic heating while L_R, L_S can also vary due to magnetic saturation. Furthermore, the state variables ρ, ψ_d have to be estimated leaving uncertainty not only in their values, but in currents i_d, i_q as the value of ρ is needed to calculate them. It turns out that such a computation is not needed because the effect of these nonlinear terms can be easily eliminated by forcing the system into *current – command* mode using high-gain feedback [2]. That is, one uses PI current loops of the form

$$\begin{aligned}u_d &= K_{dI}\int_0^t (i_{dr}-i_d)dt+K_{dP}(i_{dr}-i_d)\\ u_q &= K_{qI}\int_0^t (i_{qr}-i_q)dt+K_{qP}(i_{qr}-i_q)\end{aligned} \tag{13}$$

to force i_d and i_q to track their corresponding references i_{dr} and i_{qr}, respectively. These PI current loops result in fast responses for the currents by using large feedback gains. As a result, i_{dr} and i_{qr} can then be considered as the new inputs and the system equations may be simplified (approximated) as

$$\begin{aligned}d\omega/dt &= \mu\psi_d i_{qr}-\tau_L/J\\ d\psi_d/dt &= -\eta\psi_d+\eta M i_{dr}\\ d\rho/dt &= n_p\omega+\eta M i_q/\psi_d\end{aligned} \tag{14}$$

Field-oriented control then consists of choosing

$$i_{dr} = K_P(\psi_{dr}-\psi_{do})+K_I\int_0^t(\psi_{dr}-\psi_{do})dt$$

to force ψ_d to $\psi_{do} = Mi_{do}$ (constant) [1]. As a consequence, there is a *linear* relationship between the input i_{qr} and the speed ω. The control of speed and/or position is then done through the input i_{qr} by choosing, for example,

$$\begin{aligned}i_{qr} &= \Big(K_0'\textstyle\int_0^t(\theta_r-\theta)dt+K_1'(\theta_r-\theta)\\ &\quad +K_2'(\omega_r-\omega)+\alpha_r+\tau_L/J\Big)/\mu\psi_{do}\end{aligned}$$

As ψ_{do} is constant, it is standard practice to incorporate $\mu\psi_{do}$ directly into the controller gains to obtain

$$\begin{aligned}i_{qr} &= K_0\textstyle\int_0^t(\theta_r-\theta)dt+K_1(\theta_r-\theta)\\ &\quad +K_2(\omega_r-\omega)+\bar{\alpha}_r+\tau_L/J\end{aligned} \tag{15}$$

where $K_0 = K_0'/\mu\psi_{do}$, $K_1 = K_1'/\mu\psi_{do}$, $K_2 = K_2'/\mu\psi_{do}$ and $\bar{\alpha}_r = \alpha_r/\mu\psi_{do}$.

As shown above, the field-oriented controller requires that the magnitude of the rotor flux be regulated to a constant value, that is, the rotor speed is only *asymptotically* decoupled from the rotor flux [4]. However, it is often necessary to decrease the rotor flux in order for the motor to achieve higher speeds with a limited voltage supply. Specifically, recall that the torque is controlled by the quadrature current i_q and inspection of the fourth equation in (10) reveals that to maintain or increase i_q, the voltage u_q must be chosen such that $di_q/dt \geq 0$, that is,

$$\begin{aligned} u_q \;\geq\; & \sigma L_S\gamma i_q + \sigma L_S\beta n_p\omega\psi_d \\ & +\sigma L_S n_p\omega i_d + \sigma L_S\eta M i_q i_d/\psi_d \end{aligned}$$

Note that $\sigma L_S\gamma = M^2R_R/L_R^2 + R_S \approx R_R + R_S$ and $\sigma L_S\beta n_p\omega\psi_d = Mn_p\omega\psi_d/L_R \approx n_p\omega\psi_d$. Further, in steady-state $\psi_d = Mi_d$ so that the last term becomes $\sigma L_S\eta i_q$ in steady-state. Consequently, as the quantity σL_S is quite small, it is clear that $\sigma L_S\beta n_p\omega\psi_d \approx n_p\omega\psi_d$ is the dominant term on the right-hand side of this inequality. Thus, taking the motor up to higher speeds for a given (constant) ψ_d requires increasing the available input voltage. However, if the flux ψ_d is decreased as the speed ω increases, then the amount of input voltage required is reduced. In particular, the input i_{dr} is often chosen to force ψ_d to vary as $1/\omega$ at higher speeds so that $n_p\omega\psi_d$ is constant, thus limiting the amount of input voltage required to overcome it. This method of reducing ψ_d at higher speeds is referred to as *field weakening*. The trade-off here is that the torque produced by the machine is less due to the decreased flux level.

4 Dynamic Feedback Linearization

The concept behind feedback linearization is to find a coordinate system transformation such that the nonlinear dynamics represented in the new coordinate system can be canceled out by state feedback. This is a quite appealing approach when such a transformation exists. Unfortunately, it happens that the system (8) (or equivalently, the model (10)) of the induction motor is *not* feedback linearizable [4]. Furthermore, the reduced-order model (14) with inputs i_{dr}, i_{qr} and state variables ω, ψ_d, ρ is also *not* feedback linearizable. However, a quite interesting result is that for some dynamic systems which are not feedback linearizable, the addition of integrators in some of the inputs can result in the (higher-order) system being feedback linearizable [28][29][30]. This is a phenomenon of multi-input systems in that a single-input system is dynamically feedback linearizable if and only if it is (statically) feedback linearizable [30].

As the induction motor is a multi-input system which is not feedback linearizable, it is natural to ask if it is dynamically feedback linearizable. The model (8) was considered in [33] where it was shown that a dynamic feedback linearizing transformation existed as long at the torque $\tau = J\mu(\psi_{Ra}i_{Sb} - \psi_{Rb}i_{Sa})$ was nonzero. However, the feedback structure in [33] required switching between two computationally complex controllers to avoid singularities. On the other hand, by considering the model (10) and adding an integrator to either the d- or q-axis input, a much simpler controller results. Furthermore, in the case of the d-axis input, the singularity condition is less restrictive.

4.1 Addition of an Integrator to the d-axis Input

Consider the addition of an integrator in the d-axis of (14). That is, let $x_1 = \omega$, $x_2 = \psi_d$, $x_3 = \rho$ and $x_4 = i_{dr}$, $dx_4/dt = v_1$, $v_2 = i_{qr}$ so that (14) becomes

$$\begin{aligned} dx_1/dt &= \mu x_2 v_2 - \tau_L/J \\ dx_2/dt &= -\eta x_2 + \eta M x_4 \\ dx_3/dt &= n_p x_1 + \eta M v_2/x_2 \\ dx_4/dt &= v_1 \end{aligned} \qquad (16)$$

More compactly, (16) may be written as

$$\frac{dx}{dt} = f(x) + g_1 v_1 + g_1 v_2$$

with $f(x) = \begin{bmatrix} -\tau_L/J \\ -\eta x_2 + \eta M x_4 \\ n_p x_1 \\ 0 \end{bmatrix}$, $g_1 = \begin{bmatrix} 0 \\ 0 \\ 0 \\ 1 \end{bmatrix}$, $g_2 = \begin{bmatrix} \mu x_2 \\ 0 \\ \eta M/x_2 \\ 0 \end{bmatrix} \in \Re^4$.

The conditions for feedback linearization are well-known [24][25] and it is straightforward to check that the system (16) does indeed satisfy these conditions. The set of software programs CONDENS [31] written in MACSYMA [32] can be used to easily check the conditions. The Kronecker indices are $\kappa_1 = 2, \kappa_2 = 2$. Although CONDENS is unable to construct the feedback linearizing transformation, it is straightforward to do this (see [24][25]) by solving for $h_1(x)$, $h_2(x)$ satisfying

$$dh_i \begin{bmatrix} g_1 & ad_f g_1 & g_2 & ad_f g_2 \end{bmatrix} =$$

$$\begin{bmatrix} 0 & < dh_i, ad_f g_1 > & 0 & < dh_i, ad_f g_2 > \end{bmatrix}$$

for $i = 1, 2$ and

$$\det \begin{bmatrix} < dh_1, ad_f g_1 > & < dh_1, ad_f g_2 > \\ < dh_2, ad_f g_1 > & < dh_2, ad_f g_2 > \end{bmatrix} \neq 0$$

In this case, a solution set is

$$\begin{array}{rcccl} y_1 & = & h_1(x) & = & x_2 \\ y_2 & = & h_2(x) & = & x_1 - \mu x_2^2 x_3/\eta M \end{array}$$

The feedback linearizing transformation is then

$$\begin{array}{rcccl} z_1 & = & h_1(x) & = & x_2 \\ z_2 & = & \mathcal{L}_f h_1(x) & = & -\eta x_2 + \eta M x_4 \\ z_3 & = & h_2(x) & = & x_1 - \mu x_2^2 x_3/\eta M \\ z_4 & = & \mathcal{L}_f h_2(x) & = & -\frac{2\mu}{\eta M} x_2 x_3(-\eta x_2 + \eta M x_4) \\ & & & & -\frac{\mu n_p}{\eta M} x_2^2 x_1 - \tau_L/J \end{array} \tag{17}$$

where $\mathcal{L}_f(\cdot)$ denotes the Lie derivative with respect to f. In these new coordinates, the system equations become

$$\begin{array}{rcl} dz_1/dt & = & z_2 \\ dz_2/dt & = & \mathcal{L}_f^2 h_1(x) + v_1 \mathcal{L}_{g_1}\mathcal{L}_f h_1 + v_2 \mathcal{L}_{g_2}\mathcal{L}_f h_1 \\ dz_3/dt & = & z_4 \\ dz_4/dt & = & \mathcal{L}_f^2 h_2(x) + v_1 \mathcal{L}_{g_1}\mathcal{L}_f h_2 + v_2 \mathcal{L}_{g_2}\mathcal{L}_f h_2 \end{array}$$

where

$$\begin{array}{rcl} \mathcal{L}_f^2 h_1(x) & = & -\eta(-\eta x_2 + \eta M x_4) \\ \mathcal{L}_f^2 h_2(x) & = & -2\eta M \mu x_3 x_4^2 + 6\eta\mu x_2 x_3 x_4 \\ & & -4\mu n_p x_1 x_2 x_4 \\ & & + (-4\eta\mu x_2^2 x_3 + 4\mu n_p x_1 x_2^2)/M \\ & & +\tau_L \mu n_p x_2^2/(\eta M J) \end{array}$$

and

$$\begin{array}{rcl} \mathcal{L}_{g_1}\mathcal{L}_f h_1 & = & \eta M \\ \mathcal{L}_{g_1}\mathcal{L}_f h_2 & = & -2\mu x_2 x_3 \\ \mathcal{L}_{g_2}\mathcal{L}_f h_1 & = & 0 \\ \mathcal{L}_{g_2}\mathcal{L}_f h_2 & = & -2\mu(-\eta x_2 + \eta M x_4) - \frac{\mu^2 n_p}{\eta M} x_2^3 \end{array}$$

Application of the feedback

$$\begin{bmatrix} v_1 \\ v_2 \end{bmatrix} = \begin{bmatrix} \mathcal{L}_{g_1}\mathcal{L}_f h_1 & \mathcal{L}_{g_2}\mathcal{L}_f h_1 \\ \mathcal{L}_{g_1}\mathcal{L}_f h_2 & \mathcal{L}_{g_2}\mathcal{L}_f h_2 \end{bmatrix}^{-1} \begin{bmatrix} u_1 - \mathcal{L}_f^2 h_1(x) \\ u_2 - \mathcal{L}_f^2 h_2(x) \end{bmatrix}$$

results in two decoupled second-order *linear* systems

$$\begin{array}{rcl} dz_1/dt & = & z_2 \\ dz_2/dt & = & u_1 \\ dz_3/dt & = & z_4 \\ dz_4/dt & = & u_2 \end{array} \tag{18}$$

The controller is singular when

$$\det \begin{bmatrix} \mathcal{L}_{g_1}\mathcal{L}_f h_1 & \mathcal{L}_{g_2}\mathcal{L}_f h_1 \\ \mathcal{L}_{g_1}\mathcal{L}_f h_2 & \mathcal{L}_{g_2}\mathcal{L}_f h_2 \end{bmatrix} = 0$$

or, explicitly

$$-2\eta M\mu(-\eta x_2 + \eta M x_4) - \mu^2 n_p x_2^3 = 0$$

Rewriting this singularity condition in the dq coordinates as

$$-(2\eta M\mu) d\psi_d/dt - \mu^2 n_p \psi_d^3 = 0$$

it is seen to be easily avoided by choosing, for example, the input u_1 to regulate ψ_d to a positive constant (so that $d\psi_d/dt = 0$). On the other hand, during field-weakening when ψ_d is reduced, this condition just restricts how fast ψ_d can be decreased, i.e.,

$$0 \geq d\psi_d/dt > -(\frac{\mu n_p}{2\eta M})\psi_d^3 = -(\frac{n_p^2}{2JR_R})\psi_d^3.$$

At start-up of the motor, when the flux is brought up from zero to its nominal value, the singularity is not encountered as $d\psi_d/dt > 0$.

Remarks Note from (17) that none of the state variables z_i are linearly related to the speed ω. Consequently, $d\theta/dt = \omega$ is *not* a linear function of the z_i so that the position cannot be appended to the system (18) without losing linearity of the system. That is, if $d\theta/dt = \omega = x_1$ is appended to the system (16), then it is straightforward to show that the resulting system is *not* feedback linearizable (a similar situation arises in the case of a shunt-connected dc motor [34]). This situation also complicates the speed control design as the speed is indirectly controlled through the input u_2 vis-a-vis $z_3 = x_1 - \mu x_2^2 x_3/\eta M = \omega - \mu\psi_d^2\rho/\eta M$. For example, if a constant speed ω_0 is desired, $z_3 = \omega_0 - \mu\psi_d^2\rho/\eta M$ must be tracked using the input u_2. This is sensitive to errors in the parameter values as well as errors in the estimation of the state variables ψ_d and ρ.

4.2 Addition of an Integrator to the q-axis Input

It is quite natural to consider adding an integrator in the q-axis of (14) rather than the d-axis for comparison. To proceed, let $x_1 = \omega$, $x_2 = \psi_d$, $x_3 = \rho$ and $x_4 = i_{qr}$, $dx_4/dt = v_2, v_1 = i_{dr}$ so that (14) becomes

$$\begin{array}{rcl} dx_1/dt & = & \mu x_2 x_4 - \tau_L/J \\ dx_2/dt & = & -\eta x_2 + \eta M v_1 \\ dx_3/dt & = & n_p x_1 + \eta M x_4/x_2 \\ dx_4/dt & = & v_2 \end{array} \tag{19}$$

More compactly, (19) may be written as

$$\frac{dx}{dt} = f(x) + g_1 v_1 + g_1 v_2$$

with $f(x) = \begin{bmatrix} \mu x_2 x_4 - \tau_L/J \\ -\eta x_2 \\ n_p x_1 + \eta M x_4/x_2 \\ 0 \end{bmatrix}$, $g_1 = \begin{bmatrix} 0 \\ \eta M \\ 0 \\ 0 \end{bmatrix}$,

$g_2 = \begin{bmatrix} 0 \\ 0 \\ 0 \\ 1 \end{bmatrix} \in \Re^4.$

Again using CONDENS [31], it is easy to see that the system (19) is feedback linearizable with Kronecker indices $\kappa_1 = 2, \kappa_2 = 2$. In this case, a solution set is easily found to be

$$\begin{array}{rcccl} y_1 & = & h_1(x) & = & x_1 \\ y_2 & = & h_2(x) & = & x_3 \end{array}$$

The feedback linearizing transformation is then

$$\begin{array}{rcccl} z_1 & = & h_1(x) & = & x_1 \\ z_2 & = & \mathcal{L}_f h_1(x) & = & \mu x_2 x_4 - \tau_L/J \\ z_3 & = & h_2(x) & = & x_3 \\ z_4 & = & \mathcal{L}_f h_2(x) & = & n_p x_1 + \eta M x_4/x_2 \end{array} \quad (20)$$

In these new coordinates, the system equations become

$$\begin{array}{rcl} dz_1/dt & = & z_2 \\ dz_2/dt & = & \mathcal{L}_f^2 h_1(x) + v_1 \mathcal{L}_{g_1}\mathcal{L}_f h_1 + v_2 \mathcal{L}_{g_2}\mathcal{L}_f h_1 \\ dz_3/dt & = & z_4 \\ dz_4/dt & = & \mathcal{L}_f^2 h_2(x) + v_1 \mathcal{L}_{g_1}\mathcal{L}_f h_2 + v_2 \mathcal{L}_{g_2}\mathcal{L}_f h_2 \end{array}$$

where

$$\begin{array}{rcl} \mathcal{L}_f^2 h_1(x) & = & -\eta\mu x_2 x_4 \\ \mathcal{L}_f^2 h_2(x) & = & \mu n_p x_2 x_4 + \eta^2 M(x_4/x_2) - n_p \tau_L/J \end{array}$$

and

$$\begin{array}{rclrcl} \mathcal{L}_{g_1}\mathcal{L}_f h_1 & = & \eta M \mu x_4 & \mathcal{L}_{g_2}\mathcal{L}_f h_1 & = & \mu x_2 \\ \mathcal{L}_{g_1}\mathcal{L}_f h_2 & = & -\frac{\eta^2 M^2 x_4}{x_2^2} & \mathcal{L}_{g_2}\mathcal{L}_f h_2 & = & \frac{\eta M}{x_2} \end{array}$$

Application of the feedback

$$\begin{bmatrix} v_1 \\ v_2 \end{bmatrix} = \begin{bmatrix} \mathcal{L}_{g_1}\mathcal{L}_f h_1 & \mathcal{L}_{g_2}\mathcal{L}_f h_1 \\ \mathcal{L}_{g_1}\mathcal{L}_f h_2 & \mathcal{L}_{g_2}\mathcal{L}_f h_2 \end{bmatrix}^{-1} \begin{bmatrix} u_1 - \mathcal{L}_f^2 h_1(x) \\ u_2 - \mathcal{L}_f^2 h_2(x) \end{bmatrix}$$

results in two decoupled second-order systems as in the d-axis case. This controller is singular when

$$\begin{array}{rcl} \det \begin{bmatrix} \mathcal{L}_{g_1}\mathcal{L}_f h_1 & \mathcal{L}_{g_2}\mathcal{L}_f h_1 \\ \mathcal{L}_{g_1}\mathcal{L}_f h_2 & \mathcal{L}_{g_2}\mathcal{L}_f h_2 \end{bmatrix} & = & 2\eta^2 M^2 \mu (x_4/x_2) \\ & = & 0 \end{array}$$

This singularity condition is comparable to the condition that the torque $\tau = J\mu\psi_d i_q = J\mu x_2 x_4 \neq 0$ given in [33]. However, in contrast to [33], the approach here requires only a single transformation and the resulting controller requires relatively little feedback computation. A drawback here is the requirement that the quadrature current be nonzero ($\psi_d > 0$ is straightforward to maintain and is common to other controllers [1][8][4]). Consequently, to use this controller would require switching to another control strategy when it is desired to change the sign of the torque. Note that, in contrast to the case where the integrator was in the d-axis, one of the transformed state variables is the speed which is easily controlled using the input u_1.

By appending $d\theta/dt = \omega = x_1$ to (19), it is straightforward to check that the fifth-order system is feedback linearizable. In fact, adding the state variable $z_0 = \theta$ to the transformation (20) gives a feedback linearizing controller for *position* control. However, to avoid the singularity of the controller, it is still required that $i_q \neq 0$.

5 Input-Output Linearization

In the field-oriented control scheme presented in Section 3, the speed ω and flux ψ_d are only *asymptotically* decoupled [4]. That is, the speed is linearly related to i_{qr} only *after* ψ_d is constant. However, it is often required to vary the flux (e.g., field weakening) in which case the dynamics of ψ_d interfere with the dynamics of ω and is especially troublesome when the speed varies rapidly. The dynamic feedback linearization controllers presented in Section 4 can provide decoupled control of flux and speed. However, as pointed out there, the d-axis controller does not have speed as a transformed variable nor can it do position control while the q-axis controller requires the quadrature current be non-zero to avoid singularities.

A quite effective way to achieve decoupling of the flux and speed dynamics is the so-called *input-output linearizing* controller [4][6][7][8][10][19]. Specifically, let $i_{dr} = u_1$ and $i_{qr} = u_2/(\mu\psi_d)$ so that the system (14) becomes

$$\begin{array}{rcl} d\omega/dt & = & u_2 - \tau_L/J \\ d\psi_d/dt & = & -\eta\psi_d + \eta M u_1 \\ d\rho/dt & = & n_p\omega + \eta M i_q/\psi_d \end{array} \quad (21)$$

The flux dynamics are now decoupled from the speed dynamics, that is, with u_1, u_2 the inputs and ω, ψ_d the outputs, the system (21) is *linear* from these new inputs to the outputs so that *input-output linearization* has been achieved. The overall system is still nonlinear since the dynamics of ρ are nonlinear. However, the stability of ρ is guaranteed since it is an angle which is simply reset to 0 every $\pm 2\pi$.

With $\psi_{dr}(t)$ an *arbitrary* flux reference, the input u_1 is chosen to force the motor's flux to track this reference by setting [1]

$$\begin{array}{rcl} u_1 & = & i_{dr} \\ & = & K_{P\psi}(\psi_{dr} - \psi_d) + K_{I\psi}\int_0^t (\psi_{dr} - \psi_d)dt \end{array} \quad (22)$$

Furthermore, with $\theta_r(t)$, $\omega_r(t) = d\theta_r/dt$ and $\alpha_r(t) = d\omega_r/dt$ the mechanical reference trajectory, the input u_2 is chosen as

$$\begin{array}{rcl} u_2 & = & K_0 \int_0^t (\theta_r - \theta)dt + K_1(\theta_r - \theta) \\ & & + K_2(\omega_r - \omega) + \alpha_r + \tau_L/J \end{array} \quad (23)$$

to force $\theta \to \theta_r$ and $\omega \to \omega_r$ as $t \to \infty$.

In Section 7, experimental results are presented that clearly demonstrate not only the decoupling of the flux and speed control, but how this capability is exploited so that the motor is able to execute fast point-to-point moves.

Remarks The full-order (voltage-command) system (12) can also be input-output linearized (see [4] where the linearization was done directly on the model (8)). Specifically, the transformation

$$\begin{aligned}\omega &\triangleq \omega\\ \alpha &\triangleq \mu\psi_d i_q\\ \psi_d &\triangleq \psi_d\\ \psi_d' &\triangleq -\eta\psi_d+\eta M i_d\\ \rho &\triangleq \rho\end{aligned}$$

is applied to (12) resulting in

$$\begin{aligned}d\omega/dt &= \alpha-\tau_L/J\\ d\alpha/dt &= \mu\psi_d' i_q+(\mu\psi_d)v_q\\ d\psi_d/dt &= \psi_d'\\ d\psi_d'/dt &= -\eta\psi_d'+(\eta M)v_d\\ d\rho/dt &= n_p\omega+\eta M\alpha/(\mu\psi_d^2).\end{aligned}$$

Then, application of the feedback

$$\begin{aligned}v_d &= \psi_d'/M+w_d/(\eta M)\\ v_q &= -\mu\psi_d' i_q/(\mu\psi_d)+w_q/(\mu\psi_d)\end{aligned}\tag{24}$$

results in the input-output linearized system

$$\begin{aligned}d\omega/dt &= \alpha-\tau_L/J\\ d\alpha/dt &= w_q\\ d\psi_d/dt &= \psi_d'\\ d\psi_d'/dt &= w_d\\ d\rho/dt &= n_p\omega+\eta M\mu\alpha/\psi_d^2.\end{aligned}$$

A question arises as to whether the voltage-command approach has any advantages over the current-command approach. The current-command PI controller (13) works well as long as the current dynamics are much faster that the mechanical dynamics. When a machine has a large number of pole-pairs, this may not be the case as in [18] where a 50 pole-pair stepper motor was run at high-speed. There, the stator impedances were $R_S+jn_p\omega L_S$ and, as $n_p=50$ is large, it takes a significant amount of time (relative to the mechanical motion of the rotor) to build up the current in the phase. Consequently, it was necessary to use voltage feedback (cancellation) similar to (11) and (24) to account for the current dynamics rather than swamping out their effect with high-gain feedback [18]. In regards to the induction motor, the stator impedances are also (approximately - neglecting leakage) given by $R_S+jn_p\omega L_S$. However, induction machines rarely have a large number of pole-pairs and, even if they do (say $n_p=12$), such machines are not run at high speeds. Thus, in practice, the performance of PI current controllers is excellent.

6 Observers

The controllers presented above require full state feedback as well as knowledge of the load-torque τ_L. This, of course, is a problem that needs to be addressed since usually neither of them are completely available. The stator currents i_{Sa}, i_{Sb} are easily measured and this will be assumed to be the case here. As the primary interest here is position control, the rotor angle θ will also be assumed to be measured. However, in such a case, a separate sensor for the speed ω is usually *not* provided. Consequently, in what follows, only measurements of the stator currents and rotor position will be assumed with the problem being the estimation of the rotor flux pair (ψ_{Ra}, ψ_{Rb}), the rotor speed ω and the load-torque τ_L.

A speed estimate is usually obtained by just numerically differentiating (backward difference) the position measurement. However, a smoother estimate of the speed along with an estimate of the rotor fluxes and load-torque can be obtained using a state observer. As in the case of the control problem, the key to developing such an estimator is to find the appropriate coordinate system to do the design and analysis. In this case, the rotor coordinate system is the appropriate one. Specifically, as in [15], consider the transformation

$$\begin{aligned}i_{Sx} &\triangleq \cos(n_p\theta)i_{Sa}+\sin(n_p\theta)i_{Sb}\\ i_{Sy} &\triangleq -\sin(n_p\theta)i_{Sa}+\cos(n_p\theta)i_{Sb}\\ \psi_{Rx} &\triangleq \cos(n_p\theta)\psi_{Ra}+\sin(n_p\theta)\psi_{Rb}\\ \psi_{Ry} &\triangleq -\sin(n_p\theta)\psi_{Ra}+\cos(n_p\theta)\psi_{Rb}\end{aligned}\tag{25}$$

The currents i_{Sx}, i_{Sy} are *known* as the angle θ is measured. Note also that $\psi_{Rx}=\lambda_{Ra}$, $\psi_{Ry}=\lambda_{Rb}$ as given in (2). In order to estimate the load-torque, it is modeled as a constant, that is, its dynamic equation is taken to be $d(\tau_L/J)/dt=0$. Transforming the flux, speed and position dynamics of (8) by (25) and appending the load-torque equation, one obtains

$$\begin{aligned}\frac{d\psi_{Rx}}{dt} &= -\frac{R_R}{L_R}\psi_{Rx}+\frac{MR_R}{L_R}i_{Sx}\\ \frac{d\psi_{Ry}}{dt} &= -\frac{R_R}{L_R}\psi_{Ry}+\frac{MR_R}{L_R}i_{Sy}\\ \frac{d\theta}{dt} &= \omega\\ \frac{d\omega}{dt} &= \mu(i_{Sy}\psi_{Rx}-i_{Sx}\psi_{Ry})-\tau_L/J\\ \frac{d(\tau_L/J)}{dt} &= 0\end{aligned}\tag{26}$$

The key point here is that in this state-space representation, the (unmeasured) speed ω has been *eliminated* from the flux equations. An estimator for the fluxes, speed and load-torque is then defined by

$$
\begin{aligned}
d\hat{\psi}_{Rx}/dt &= -\tfrac{R_R}{L_R}\hat{\psi}_{Rx} + \tfrac{MR_R}{L_R} i_{Sx} \\
d\hat{\psi}_{Rx}/dt &= -\tfrac{R_R}{L_R}\hat{\psi}_{Ry} + \tfrac{MR_R}{L_R} i_{Sy} \\
d\theta/dt &= \hat{\omega} + \ell_1(\theta - \hat{\theta}) \\
d\omega/dt &= \mu(i_{Sy}\hat{\psi}_{Rx} - i_{Sx}\hat{\psi}_{Ry}) \\
&\quad -\hat{\tau}_L/J + \ell_2(\theta - \hat{\theta}) \\
d(\hat{\tau}_L/J)/dt &= 0 + \ell_3(\theta - \hat{\theta})
\end{aligned} \tag{27}
$$

Subtracting (27) from (26) results in the error system

$$
\begin{aligned}
\frac{de_{Rx}}{dt} &= -\frac{R_R}{L_R} e_{Rx} \\
\frac{de_{Ry}}{dt} &= -\frac{R_R}{L_R} e_{Ry} \\
\frac{de_\theta}{dt} &= e_\omega - \ell_1(\theta - \hat{\theta}) \\
\frac{de_\omega}{dt} &= \mu(i_{Sy}e_{Rx} - i_{Sx}e_{Ry}) - e_{\tau_L} - \ell_2(\theta - \hat{\theta}) \\
\frac{de_{\tau_L}}{dt} &= 0 - \ell_3(\theta - \hat{\theta})
\end{aligned} \tag{28}
$$

The flux error is stable as $e_{Rx}(t) = e_{Rx}(0)e^{-t/T_R}$, $e_{Ry}(t) = e_{Ry}(0)e^{-t/T_R}$ go to 0 as $t \to \infty$. As long as the currents i_{Sx}, i_{Sy} are bounded (consistent with the assumed current-command operation), the perturbation term $\mu(i_{Sy}e_{Rx} - i_{Sx}e_{Ry})$ due to the error in the estimate of motor torque dies out, that is, $|\mu(i_{Sy}e_{Rx} - i_{Sx}e_{Ry})| \le Ce^{-t/T_R}$ for some $C > 0$. By choosing the gains ℓ_1, ℓ_2, ℓ_3 appropriately, the subsystem of (28) consisting of e_θ, e_ω, e_{τ_L} is a stable linear time-invariant system driven by the exponentially decaying input $\mu(i_{Sy}e_{Rx} - i_{Sx}e_{Ry})$. Consequently, $e_\theta \to 0$, $e_\omega \to 0$, $e_{\tau_L} \to 0$ as $t \to \infty$ showing convergence of the estimator (27). However, it is clear from the above analysis that the rate of convergence of the error dynamics of this observer is limited by the rotor time constant $T_R = L_R/R_R = 1/\eta$ [14].

There are important numerical considerations in the implementation of the estimator (27). This system is driven by the stator currents i_{Sx}, i_{Sy} which vary at the slip frequency ω_s. These currents are (at least, in steady-state) sinusoidal functions of time which can oscillate quite rapidly at high slip according to $\omega_s t$. Consequently, a very small step size may be required to accurately (numerically) integrate these sinusoidal functions. To get around this numerical problem, the estimator is implemented in the field-oriented (dq) coordinate system as

$$
\begin{aligned}
d\hat{\rho}/dt &= n_p\hat{\omega} \\
&\quad +\eta M(-i_{Sa}\sin(\hat{\rho}) + i_{Sb}\cos(\hat{\rho}))/\hat{\psi}_d \\
d\hat{\psi}_d/dt &= -\eta\hat{\psi}_d + \eta M(i_{Sa}\cos(\hat{\rho}) + i_{Sb}\sin(\hat{\rho})) \\
d\hat{\theta}/dt &= \hat{\omega} + \ell_1(\theta - \hat{\theta}) \\
d\hat{\omega}/dt &= \mu\hat{\psi}_d(-i_{Sa}\sin(\hat{\rho}) + i_{Sb}\cos(\hat{\rho})) \\
&\quad -\hat{\tau}/J + \ell_2(\theta - \hat{\theta}) \\
d(\hat{\tau}/J)/dt &= 0 + \ell_3(\theta - \hat{\theta})
\end{aligned} \tag{29}
$$

More details on this approach are given in [20].

Remarks As noted, the rate of convergence of the estimation scheme (27) (or, 29) is limited by the rotor time constant $T_R = L_R/R_R = 1/\eta$. However, given that the speed is known, an ingenious approach has been given by Verghese and Sanders [14] that allows one to specify an *arbitrary* rate of convergence.

As in any model-based scheme, this estimation scheme is sensitive to parameter error. In this case, the significant parameter is the rotor time constant T_R as its value can vary as the rotor resistance heats up or cools down. An analysis of this sensitivity is given in [14]. An approach for estimating the rotor fluxes with adaptation to the rotor time constant has recently been reported in [5].

7 Experimental Results

The current-command input-output linearization controller presented in Section 5 provides an extremely effective means for obtaining high-performance control of the induction motor. Not only does it allow independent speed and flux control, but its only significant computational requirement over the standard field-oriented controller is the computation of $1/\psi_d$ (division) in real-time. Note also that since the state variables are in the field-oriented (dq) coordinate system, they vary slowly compared to the electrical frequency making it easy to numerically integrate the controller/observer equations.

To illustrate the performance of the input-output linearization controller, experimental results are now presented [19]. The motor chosen is a 6-pole 1/12 horsepower two-phase induction motor with a squirrel cage rotor. This small motor is rated for 2.4 A (continuous) and 60 V and is intended for servo use. The hardware setup for controlling the motor consists of a Motorola DSP56001 digital signal processing system, two PWM amplifiers (± 80 V and ± 6 A), A/D's for the stator currents, D/A's for stator voltages, an electronic interface board and a 2000 pulse/rev encoder (resolution of $360^o/2000 = 0.18^o$) for the position measurement. The load-torque was modeled as viscous-friction so that $\tau_L = -f\omega$. The parameters were identified using the technique given in Stephan *et al* [15], and were found to be $M = 0.011$ H, $R_R = 3.9\,\Omega$, $R_S = 1.7\,\Omega$, $L_R = 0.014$ H, $L_s = 0.014$ H, $f = .00014$ N-m/rad/sec and $J = 0.00011$ kg-m^2.

An experiment requiring a point-to-point position move was carried out in which the motor was brought up from zero to a speed of 8000 rpm (837.8 radian/sec) in 0.38 seconds and brought down from 8000 rpm to 0 rpm in 0.265 seconds (see Figure 4). The PI current gains in (13) were set at $K_{dI} = 9000$, $K_{dP} = 15$, $K_{qI} = 9000$, $K_{qP} = 15$, the flux gains in (22) set as $K_{P\psi} = 1600$, $K_{I\psi} = 23000$, while the PID gains for the tracking error in (23) were chosen as $K_0 = 3.0 \times 10^5$, $K_1 = 5.5 \times 10^4$, $K_2 = 125$. The flux/speed observer (29) was used with the gains chosen as $\ell_1 = 1.8 \times 10^3$ and $\ell_2 = 8 \times 10^5$ (the term $\hat{\tau}_L/J$ was replaced by $-(f/J)\hat{\omega}$ in (29) making the last equation in (29) superfluous). The sample rate was 4 kHz.

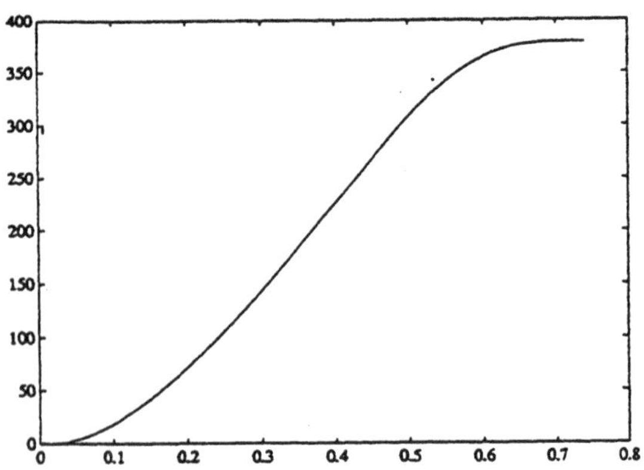

Figure 2 – θ and θ_{ref} in radians vs time in secs

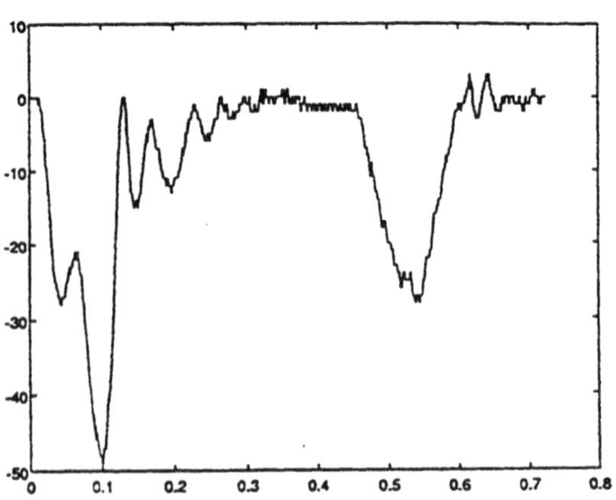

Figure 3 – Error $\theta_{ref} - \theta$ in counts vs time in secs

Figure 2 shows the tracking of the measured position to its reference (which appear to be right on top of each other) while the position tracking error *in encoder counts* is shown in Figure 3. Note in Figure 3 that the final position error is zero at the end of the run as required for a point-to-point move. Comparing with Figure 6 which gives the torque reference (dashed curve), it is seen that the maximum position errors occur when the acceleration or deceleration are at a maximum. As seen in Figure 6, this is especially true at about 0.1 second where in addition the torque is not smooth (i.e., the jerk is discontinuous). This position error can be brought down significantly (less than 3-4 counts for the entire move) if one either provides an amplifier capable of putting out more voltage (i.e., than the 80 Volts here) or the stringent trajectory requirements are reduced (e.g., take the motor up to only 7000 rpm in the same amount of time).

Figure 4 is a plot of the estimated speed and the reference speed in which the two are barely discernible. The excellent speed tracking ensues despite the time-varying flux shown in Figure 5. This vividly illustrates the capability of the input-output controller to provide such independent tracking of speed and flux. The flux reference shown in Figure 5 (dashed line) was chosen to optimize (maximum for acceleration and minimum for deceleration) the motor torque $\tau = J\mu\psi_d i_q$ subject to $i_d^2 + i_q^2 \leq I_{\max}^2$, $u_d^2 + u_q^2 \leq V_{\max}^2$ ($I_{\max} = 6$ A, $V_{\max} = 80$ V) [16][17][36].

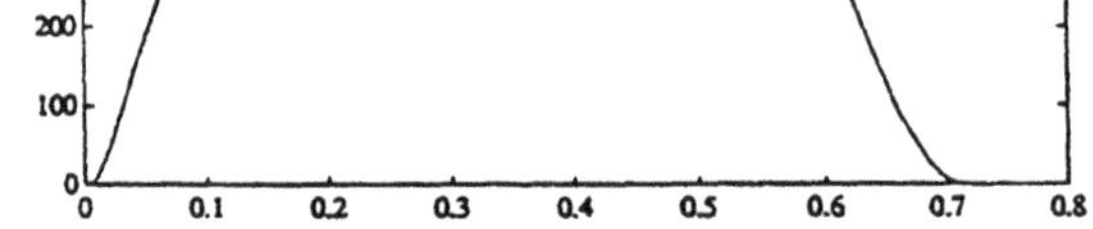

Figure 4 – ω and ω_{ref} in rad/sec vs time in sec

The dashed line in Figure 6 is a plot of the reference torque corresponding to the position and speed references given in Figures 2 and 4, respectively. The solid line is a plot of the optimum torque achievable based on the results in [16][17][36]. Note the closeness of the reference to the optimum.

These high-speed experiments have also been carried out using a field-oriented controller (including field weakening) and it was found that tracking could not be achieved due to saturation of

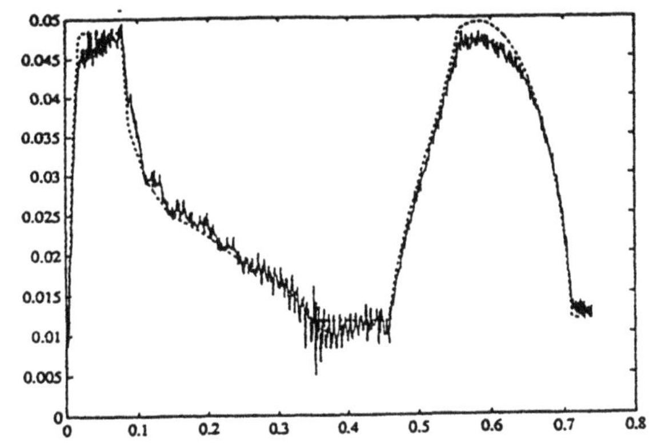

Figure 5 – ψ_d and ψ_{dref} in Webers vs time in secs

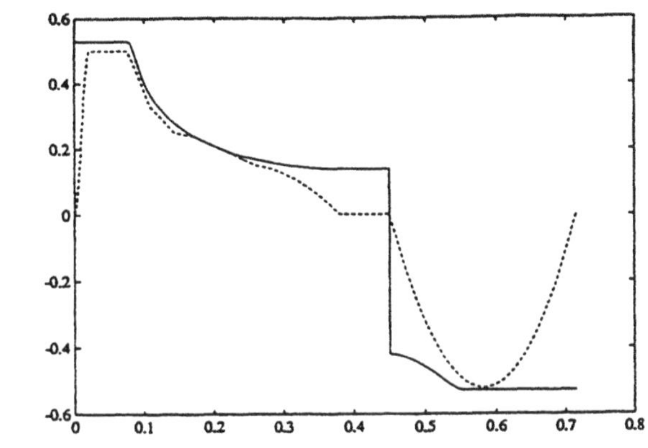

Figure 6 – τ_{ref} and $\tau_{optimum}$ in N-m vs time in secs

the phase voltages [19]. Consequently, the input-output linearization controller is essential for obtaining such high-performance.

8 Concluding Remarks

This paper has only touched on some of the modern nonlinear techniques in terms of their applicability to induction motor control. The two approaches of dynamic feedback linearization and input-output linearization were considered as alternatives to the standard field-oriented controller. The current-command input-output linearization controller is seen to be both theoretically and experimentally able to achieve high-performance motion control, that is, the precise tracking of demanding point-to-point moves. This performance was directly due to it providing a means of decoupling the speed and flux dynamics of the machine. That is, a time-varying flux trajectory was chosen in order to be able to extract nearly optimum torque from the machine. Simultaneously, while following this flux reference, a time-varying speed reference which required nearly optimum torque was tracked in order to carry out the point-to-point move.

As stated in the introduction, there are many other approaches and the two considered here were chosen based on the author's own interests and work. Through the kind correspondence of Pierre Rouchon, the recent work in [37] has just come to the author's attention. There, the demonstration of the dynamic feedback linearizability of the induction motor has been simplified using the notion of flatness [12]. Such work clearly shows that electromechanical systems will continue to be a source of real-world problems that challenge existing theory as well as provide control theorists and practitioners with fruitful areas of collaboration.

Acknowledgments

First of all, the author would like to thank the program committee for providing him with the opportunity to present this work. It is sincerely appreciated.

Much of the authors' research in the control of electric machines (including many results presented here) has been done in close collaboration with his colleague (and, more importantly, friend) Marc Bodson along with his now former graduate student (and hopefully, still friend) Bob Novotnak. Thank you both for such an enjoyable and productive collaborative effort.

I am very grateful to Professor Werner Leonhard of the Technical University of Braunschweig for his encouragement of my work and for writing such a beautiful book that stimulated my interest in induction motor control. I am also grateful to Professor E.W. Kamen, past director of the Center for Motion Control Research, and to Mr. Stephen Botos, President of Aerotech Inc. for supporting the early stages of much of my work.

The indispensable help of Michael Aiello of Aerotech, Inc. including his design of the interface hardware & amplifiers as well as the donation of the DSP56001 Development System by the Motorola Corporation are both sincerely appreciated.

References

[1] Leonard, W., *Control of Electrical Drives*. New York: Springer-Verlag, 1985.

[2] Lessmeier, R., W. Schumacher and W. Leonhard, "Microprocessor-controlled AC-servo drives with synchronous or induction motors: Which is Preferable?" *IEEE Transactions on Industry Applications*, Vol. IA-22, No. 5, Sept/Oct 1986.

[3] Woodson, H.H., and J.R. Melcher, *Electromechanical Dynamics: Part 1 Discrete Systems*, New York: Wiley, 1968.

[4] Marino, R., S. Peresada, and P. Valigi, "Adaptive input-output linearizing control of induction motors", *IEEE Transactions on Automatic Control*, Vol. 38, No. 2, February 1993.

[5] Marino, R., S. Peresada and P. Tomei, "Adaptive observer for induction motors with unknown rotor resistance", *Proceedings of the 33rd Conference on Deci-*

sion and Control, pp. 696-697, Lake Buena Vista FL December 1994.

[6] Kim, D.I., I.J. Ha, and M.S. Ko, "Control of induction motors via feedback linearization with input-output decoupling," *International Journal of Control*, Vol. 51, No. 4, pp. 863-883, 1990.

[7] Kim, G.S., I.J. Ha, and M.S. Ko, "Control of induction motors for both high dynamic performance and high-power efficiency", *IEEE Transactions on Industrial Electronics*, Vol. 39, No. 4, pp. 323-333, August 1992.

[8] Krzeminski, Z., "Nonlinear control of the induction motor," in *10th IFAC World Congress*, pp. 349-354, Munich, 1987.

[9] Utkin, V., *Sliding Modes in Control Optimization*, Springer-Verlag, 1992.

[10] Sabanovic, A., and D.B. Izosimov, "Application of sliding modes to induction motor control", *IEEE Transactions on Industry Applications*, Jan/Feb 1981.

[11] Fliess, M., "Generalized controller canonic forms for linear and nonlinear dynamics," *IEEE Transactions on Automatic Control*, Vol. 35, No. 9, pp. 994-1001, September 1990.

[12] Fliess, M., J. Lévine, P. Martin, and P. Rouchon, "Flatness and defect of nonlinear systems: introductory theory and applications", to appear in *International Journal of Control*.

[13] Dawson, D.M., J.H. Carroll and M. Schneider, "Integrator backstepping control of a brush dc motor turning a robotic load", *IEEE Transactions on Control Systems Technology*, Vol. 2, No. 3, September 1994.

[14] Verghese, G.C., and S.R. Sanders, "Observers for flux estimation in induction machines," *IEEE Transactions on Industrial Electronics*, vol. 35, pp. 85-94, Feb. 1988.

[15] Stephan, J., M. Bodson and J. Chiasson," Real-time estimation of induction motor parameters, *IEEE Transactions on Industry Applications*, Vol. 30, No. 3, May/June 1994.

[16] Bodson, M., and J. Chiasson, "A systematic approach to selecting optimal flux references in induction motors", *Conference Record IEEE IAS Annual Meeting*, Houston, Oct 1992.

[17] Bodson, M., J. Chiasson and R. Novotnak, "A systematic approach to selecting flux references for torque maximization in induction motors", Submitted for publication.

[18] Bodson, M., J. Chiasson, R. Novotnak, R. Rekowski, High performance nonlinear control of a permanent magnet stepper motor, *IEEE Transactions Control Systems Technology*, vol. 1, no. 1, pp. 5-14, March 1993.

[19] Bodson, M., J. Chiasson and R. Novotnak, "High-performance induction motor control via input-output linearization", *IEEE Control Systems*, Vol. 14, No. 4, pp. 25-33, August 1994.

[20] Bodson, M., J. Chiasson and R. Novotnak, "Nonlinear speed observer for high-performance induction motor control," to appear in the *IEEE Transactions on Industrial Electronics*.

[21] Ortega, R., and G. Espinosa, "Passivity properties of induction motors", *Proceedings of the 1991 Conference on Industry Applications*, Dearborn, MI.

[22] Ortega, R., C. Canudas, and S.I. Seleme, "Nonlinear control of induction motors: Torque tracking with unknown load disturbance", *IEEE Transactions on Automatic Control*, Vol. 38, No. 11, pp. 1675-1680, November 1993.

[23] Ortega, R., P.J.H. Nicklasson and G. Espinosa-Pérez, "On speed control of induction motors", preprint.

[24] Nijmeijer, H., and A.J. van der Schaft, *Nonlinear Dynamical Control Systems*. New York: Springer-Verlag, 1990.

[25] Isidori, A., *Nonlinear Control Systems*, Second Edition, New York: Springer-Verlag, 1989.

[26] Kanellakopoulos, I., P.T. Krein, and F. Disilvestro, "Nonlinear flux-observer-based control of induction motors," in *Proceedings of the 1992 American Control Conference*, June 26-28, Chicago, IL.

[27] Kokotovic, P. V., "The joy of feedback: Nonlinear and adaptive", 1991 Bode Prize Lecture, reprinted in the *IEEE Control Systems Magazine*, Vol. 12, No. 3, June 1992.

[28] Charlet, B., J. Levine, and R. Marino, "Two sufficient conditions for dynamic feedback linearization," in *Analysis and Optimization of Systems, Lecture Notes in Control and Information Sciences*, vol. 111, A. Bensoussan and J.L. Lions, Eds. New York: Springer-Verlag, 1988.

[29] Charlet, B., J. Levine, and R. Marino, "On dynamic feedback linearization," *Systems and Control Letters*, 13, pp. 143-151, 1989

[30] Charlet, B., J. Levine, and R. Marino, "Sufficient conditions for dynamic feedback linearization," *SIAM Journal on Control and Optimization*, vol. 29, no. 1, pp. 38-57, January 1991.

[31] Akhrif, O., and G.L. Blankenship, "Symbolic computations in differential geometry applied to nonlinear control systems," in *Frontiers in Applied Mathematics, 5, Symbolic Computation Application to Scientific Computing, SIAM*, 1989.

[32] *MACSYMA Reference Manual*, Symbolics, Inc. 1988.

[33] Chiasson, J., "Dynamic feedback linearization of the induction motor," *IEEE Transactions on Automatic Control*, Vol. 38, No. 10, pp. 1588-1594, October 1993.

[34] Chiasson, J., and M. Bodson, "Nonlinear control of a shunt DC motor", *IEEE Transactions on Automatic Control*, Vol. 38, No. 11, pp. 1662-1665, November 1993.

[35] Chiasson, J., "A new approach to dynamic feedback linearization control of an induction motor," submitted to the *1995 CDC*, New Orleans LA.

[36] Novotnak, R.T., *Nonlinear Control of an Induction Motor by Input-Output Linearization: Theory and Experiment*, Ph.D. Thesis, University of Pittsburgh, 1995

[37] Martin, P., and Pierre Rouchon, Two remarks on induction motors, submitted to the *1995 CDC*, New Orleans LA.

[38] Ilic-Spong, M., R. Marino, S. Peresada, and D. Taylor, "Feedback linearizing control of switched reluctance motors", *IEEE Transactions on Automatic Control*, Vol. 32, No. 5, pp. 371-379, May 1987.

[39] Taylor, D.G., "Nonlinear control of electric machines: An overview, *IEEE Control Systems* magazine, Vol 14, No. 6, pp. 41-51, December 1994.

[40] Byrnes, C.I., A. Isidori, and J.C. Willems, "Passivity, feedback equivalence, and the global stabilization of minimum phase nonlinear systems", *IEEE Transactions on Automatic Control*, Vo. 36, No. 11, pp. 1228-1240, November 1991.

A CANONICAL FORM UNDER SIMILARITY OF A CLASS OF MINIMAL REALIZATIONS*

K. ÖZÇALDIRAN † § V. ELDEM §

*This research was supported by TÜBİTAK-DOPROG.

§ TÜBİTAK Marmara Research Center, Research Institute For Basic Sciences, Department of Mathematics, Gebze-Kocaeli, 41470, Turkey.

† Department of Electrical and Electronics Engineering, Boğaziçi University, 80815 Bebek, Istanbul, Turkey.

Abstract. In this paper, we consider minimal realizations of square, nonsingular and strictly proper transfer matrices. Equivalance classes of all such realizations under similarity transformations are characterized via a canonical form.

Résumé. On considère dans ce papier des réalizations minimales de matrices de transfert carrées, non singulières et strictement propres. On caractérise par une forme canonique les classes d'équivalence des ces réalizations sons l'action de transformations de similarité.

Key Words. infinite zeroes, canonical forms, linear systems

1. INTRODUCTION

Let Σ denote the set of all triples (C, A, B) where C is an $p \times n$, A is an $n \times n$ and B is an $n \times m$ real matrix with B and C having full rank.

In spite of the seminal results of Popov (1972), Brunovsky (1970), Morse (1973), Thorp (1973), and others (see for instance Guidorzi (1981), Eldem (1993) and references therein), study of canonical forms of linear systems is not complete yet. Indeed, classification of the similarity action of the general linear group $\mathcal{GL}(n, \mathcal{R})$ on Σ is an interesting problem in group analysis of linear systems which is still open. In this paper, we shall completely characterize, via a canonical form, the action of $\mathcal{GL}(n, \mathcal{R})$ on not Σ but on a suitably chosen subset Σ_{ns} of it. The reader should note that this paper is a brief version of Özçaldıran and Eldem (1995), and the results are exposed without proof.

Given $(C, A, B) \in \Sigma$, denote the supremal controlled invariant subspace contained in $\mathcal{K} := KerC$ by $\mathcal{V}^*$, and denote the infimal conditionally invariant subspace containing ImB by $\mathcal{T}^*$. Recall that $\mathcal{V}^*$ and $\mathcal{T}^*$ are given by the limits of the following subspace recursions (see Wonham (1979), Willems (1981)).

$$\mathcal{V}_{k+1} = \mathcal{K} \cap A^{-1}\{\mathcal{V}_k + ImB\} \quad \mathcal{V}_0 = \mathcal{K} \qquad (1)$$

$$\mathcal{T}_{k+1} = A\{\mathcal{T}_k \cap \mathcal{K}\} + ImB \quad \mathcal{T}_0 = 0 \qquad (2)$$

Now, denote the subset of Σ consisting of minimal and square (*i.e.*, $p = m$) realizations which satisfy $\mathcal{V}^* \oplus \mathcal{T}^* = \mathcal{R}^n$ by Σ_{ns}. Note that elements of Σ_{ns} are minimal realizations of square, nonsingular (over the field of real-rational functions) and strictly proper transfer matrices. Denote the subset of Σ_{ns} which consists of triples for which $\mathcal{V}^* = 0$ by Σ_{wp}. Elements of Σ_{wp} will be said to be *weakly prime*. A triple is weakly prime iff the transfer it realizes is square, nonsingular and has no finite zeroes. Also, denote the subset of Σ_{wp} which consists of *prime* triples (see Morse (1973)) by Σ_p, and denote the subset of Σ_{wp} which consists of triples whose duals are prime by Σ_p^*. Finally, denote $\Sigma_p \cap \Sigma_p^*$ by Σ_{sp}. Elements of Σ_p^* will be called *dually prime*, and elements of $\Sigma_p^* \cap \Sigma_p$ will be called *strongly prime*.

Given a realization (C, A, B), we shall denote the descendingly ordered list of its infinite zero orders $\{n_1, \ldots, n_m\}$ by $\mathcal{I}_n$ and the dual list $\{p_1, \ldots, p_{n_1}\}$ by $\mathcal{I}_p$. Note that $p_i = card\{n_j | n_j \geq i\}$, and it can be computed by the following formula (see Commault and Dion (1982)).

$$p_i = dim(\mathcal{V}_{i-1} \cap ImB / \mathcal{V}^* \cap ImB) \ ; \ i \geq 1 \qquad (3)$$

Now, if we denote the descendingly ordered lists of controllability and and observability indices of (C, A, B) by $\mathcal{I}_c$ and $\mathcal{I}_o$ respectively, then it is easy to show that $(C, A, B) \in \Sigma_{wp}$ is prime iff $\mathcal{I}_n = \mathcal{I}_c$, it is dually prime iff $\mathcal{I}_n = \mathcal{I}_o$, and it is strongly prime iff $\mathcal{I}_n = \mathcal{I}_o = \mathcal{I}_c$.

We close this section with a few remarks about notation. If B is a matrix, then $(B)_j$ will denote its jth column. If $\mathcal{L} = \{l_1, \ldots, l_q\}$ is a list then

$B_{\mathcal{L}}$ will denote the matrix $[(B)_{l_1} \dots (B)_{l_q}]$. $(\mathcal{L})^i$ will denote the sublist $\{j \in \mathcal{L} | j > i\}$ and $(\mathcal{L})_i$ will denote the sublist $\{j \in \mathcal{L} | j < i\}$. If n is a positive integer the $\mathbf{n}$ will denote the set $\{1, 2, \dots, n\}$.

2. A DIRECT SUM DECOMPOSITION OF ImB

In this section we shall consider a triple from Σ_{ns}. We shall establish the existence of a linearly independent family $\{ImB^i | i \in \mathbf{n_1}\}$ of subspaces of ImB satisfying

$$ImB^i \subset \mathcal{V}_{i-1} \quad ; \quad \sum_{j=1}^{i} ImB^j \cap \mathcal{V}_i = 0 \qquad (4)$$

and

$$ImB = ImB^1 \oplus ImB^2 \oplus \dots \oplus ImB^{n_1} \qquad (5)$$

To this end, we begin our discussion by defining a unique sublist $\mathcal{L}_1 := \{k_{11}, \dots, k_{1\sigma_1}\}$ of $\mathcal{I}_0 := \{1, \dots, m\}$ as follows. Let k_{11} be the least integer in $\mathcal{I}_0$ such that

$$(B)_{k_{11}} \notin \mathcal{V}_1 = \mathcal{K}$$

Let k_{12} be the least integer in $\mathcal{I}_0$ which is greater than k_{11} and for which

$$Im[\ (B)_{k_{11}} \quad (B)_{k_{12}}\ ] \cap \mathcal{V}_1 = 0$$

Proceeding as above and defining $\{k_{11}, k_{12}, \dots, \}$, let $k_{1\sigma_1}$ be the greatest integer in $\mathcal{I}_0$ such that

$$Im[\ (B)_{k_{11}} \quad \cdots \quad (B)_{k_{1\sigma_1}}\ ] \cap \mathcal{V}_1 = 0$$

Define another sublist $\mathcal{I}_1$ of $\mathcal{I}_0$ by $\mathcal{I}_1 = \mathcal{I}_0 / \mathcal{L}_1$. Note that given any $j \in \mathcal{I}_1$, it is by construction that there exist scalars β_{ij} and a vector $(B)_j'$ from $\mathcal{V}_1 \cap ImB$ such that

$$(B)_j' = (B)_j - \sum_{i \in (\mathcal{L}_1)_j} \beta_{ij} (B)_i$$

It is straightforward to show that β_{ij}'s, and consequently $(B)_j'$'s are unique. Note that the column operations given above can be realized by post-multiplying B by an upper triangular matrix G_1 with 1's on the diagonal.

Now, given $\mathcal{I}_1$, define a sublist $\mathcal{L}_2 = \{k_{21}, \dots, k_{2\sigma_2}\}$ of it as follows. Let k_{21} be the least integer in $\mathcal{I}_1$ such that

$$(BG_1)_{k_{21}} \notin \mathcal{V}_2$$

Let k_{22} be the least integer in $\mathcal{I}_1$ which is greater than k_{21} and for which we have

$$Im[\ (BG_1)_{k_{21}} \quad (BG_1)_{k_{22}}\ ] \cap \mathcal{V}_2 = 0$$

Proceeding as above and defining $\{k_{21}, k_{22}, \dots\}$, let $k_{2\sigma_2}$ be the greatest integer in $\mathcal{I}_1$ for which

$$Im[\ (BG_1)_{k_{21}} \quad \cdots \quad (BG_1)_{k_{2\sigma_2}}\ ] \cap \mathcal{V}_2 = 0$$

Given $\mathcal{L}_2$, define yet another sublist $\mathcal{I}_2$ of $\mathcal{I}_1$ by $\mathcal{I}_2 = \mathcal{I}_1 / \mathcal{L}_2$. Note that given any $j \in \mathcal{I}_2$, it follows from our construction that there exist unique scalars β_{ij} and a unique vector $(BG_1)_j'$ from $\mathcal{V}_2 \cap ImB$ so that

$$(BG_1)_j' = (BG_1)_j - \sum_{i \in (\mathcal{L}_2)_j} \beta_{ij} (BG_1)_j$$

Clearly, the column operations above can be realized by post-multiplying BG_1 by a unique upper triangular matrix G_2 with 1's on the diagonal.

Proceeding as above, suppose that $\mathcal{L}_1, \dots, \mathcal{L}_{s-1}$; $\mathcal{I}_1, \dots, \mathcal{I}_{s-1}$ and $G_1, \dots, G_{s-1}$ are defined. Define $\hat{G}_{s-1} = G_1 G_2 \cdots G_{s-1}$. Now, define a unique sublist $\mathcal{L}_s = \{k_{s1}, \dots, k_{s\sigma_s}\}$ of $\mathcal{I}_{s-1}$ as follows. Let k_{s1} be the least integer in $\mathcal{I}_{s-1}$ such that

$$(B\hat{G}_{s-1})_{k_{s1}} \notin \mathcal{V}_s$$

Let k_{s2} be the least integer in $\mathcal{I}_{s-1}$ which is greater than k_{s1} and for which we have

$$Im[\ (B\hat{G}_{s-1})_{k_{s1}} \quad (B\hat{G}_{s-1})_{k_{s2}}\ ] \cap \mathcal{V}_s = 0$$

Proceeding as above and defining $\{k_{s1}, k_{s2}, \dots\}$, let $k_{s\sigma_s}$ be the greatest integer in $\mathcal{I}_{s-1}$ for which we have

$$Im[(B\hat{G}_{s-1})_{k_{s1}} \quad \cdots \quad (B\hat{G}_{s-1})_{k_{s\sigma_s}}] \cap \mathcal{V}_s = 0$$

Having defined $\mathcal{L}_s$, define yet another sublist $\mathcal{I}_s$ of $\mathcal{I}_{s-1}$ by $\mathcal{I}_s = \mathcal{I}_{s-1} / \mathcal{L}_s$. For each $j \in \mathcal{I}_s$, there exist unique scalars β_{ij} and a unique vector $(B\hat{G}_{s-1})_j'$ from $\mathcal{V}_s \cap ImB$ so that

$$(B\hat{G}_{s-1})_j' = (B\hat{G}_{s-1})_j - \sum_{i \in (\mathcal{L}_s)_j} \beta_{ij} (B\hat{G}_{s-1})_i$$

Clearly, the column operations defined above can be realized by post-multiplying $B\hat{G}_{s-1}$ by the unique upper triangular matrix G_s with 1's on the diagonal.

At this point, we note the following properties which follow immediately from our construction.

$$ImB = \oplus_{j=1}^{s} Im(B\hat{G}_s)_{\mathcal{L}_j} \oplus Im(B\hat{G}_s)_{\mathcal{I}_s} \qquad (6)$$

$$Im(B\hat{G}_s)_{\mathcal{I}_s} = ImB \cap \mathcal{V}_s \qquad (7)$$

$$\mathcal{I}_s = \mathcal{L}_{s+1} \cup \mathcal{L}_{s+2} \cup \dots \cup \mathcal{L}_q \cup \mathcal{I}_{q+1} \qquad (8)$$

That n_1 is the greatest infinite zero order implies that $p_{n_1+1} = 0$. Then it follows from Equation (3) that $ImB \cap \mathcal{V}_{n_1} = ImB \cap \mathcal{V}^* = 0$. Thus, we conclude that $\mathcal{I}_{n_1} = \emptyset$ and consequently, the search for the lists $\mathcal{L}_j$'s stops at $j = n_1$. Therefore, it follows that $\{1,\ldots,m\} = \mathcal{L}_1 \cup \cdots \cup \mathcal{L}_{n_1}$. Defining

$$G = \hat{G}_{n_1} = G_1 \cdots G_{n_1} \tag{9}$$

we can now state the main result of this section.

Theorem 2.1 Given $(C, A, B) \in \Sigma_{ns}$, there exist unique lists $\mathcal{L}_i = \{k_{i1}, \ldots, k_{i\sigma_i}\}$, $i \in \mathbf{n_1}$, satisfying

1. $card\mathcal{L}_i = \sigma_i := p_i - p_{i+1}$ (where p_i's are as defined in Equation (3)),
2. $\cup_{i=1}^{n_1} \mathcal{L}_i = \{1, \ldots, m\}$,
3. $\mathcal{L}_j \cap \{\cup_{i \neq j} \mathcal{L}_i\} = \emptyset$,

and a unique upper triangular matrix G with 1's on the diagonal so that if

$$B^i := [\ (BG)_{k_{i1}} \quad \cdots \quad (BG)_{k_{i\sigma_i}}\ ]$$

then we have

$$ImB = ImB^1 \oplus \cdots \oplus ImB^{n_1}$$

In the decomposition above, $n \times \sigma_i$ matrices B^i's satisfy the following:

1. $ImB^i \subset \mathcal{V}_{i-1}$ but $\sum_{j=1}^{i} ImB^j \cap \mathcal{V}_i = 0$
2. $ImB \cap \mathcal{V}_i = \sum_{j=i+1}^{n_1} ImB^j$

Now, apply the theorem above to the dual triple (B^T, A^T, C^T). Denote the transpose of the unique upper triangular matrix associated by (A^T, C^T) by H. Then, it immediately follows from duality that:

Theorem 2.2 Given C, there exist n_1 unique lists $\hat{\mathcal{L}}_1, \ldots, \hat{\mathcal{L}}_{n_1}$ satisfying

1. $card\hat{\mathcal{L}}_i = card\mathcal{L}_i = \sigma_i$,
2. $\cup_{i=1}^{n_1} \hat{\mathcal{L}}_i = \{1, \ldots, m\}$,
3. $\hat{\mathcal{L}}_i \cap \{\cup_{j \neq i} \mathcal{L}_j\} = \emptyset$,

and a unique lower triangular matrix H with 1's on the diagonal so that if

$$C^i := \left[(C^T H^T)_{\hat{\mathcal{L}}_i}\right]^T$$

then

$$KerC = \bigcap_{i=1}^{n_1} KerC^i$$

In the above, C^i's are $\sigma_i \times n$ matrices which satisfy the following:

1. $KerC^i \supset \mathcal{T}_{i-1}$ and $\bigcap_{j=1}^{i} KerC^i + \mathcal{T}_i = \mathcal{R}^n$
2. $KerC + \mathcal{T}_i = \bigcap_{j=i+1}^{n_1} KerC^i$

3. A BASIS FOR $\mathcal{T}^*$

We resume our discussion by defining a list $\mathcal{I}_z = \{z_1, z_2, \ldots, z_m\}$ of positive integers as follows: $z_i = k$ iff $i \in \mathcal{L}_k$. Equivalently, z_i is the unique integer for which we have $(BG)_i \in B^{z_i}$. Note that the list $\mathcal{I}_z$ is nothing but a re-ordered version of the list $\mathcal{I}_n$ of infinite zero orders of (C, A, B). For notational convenience, we shall also define $s_1 = 0$ and $s_i = z_1 + \ldots + z_{i-1}$.

Now, define $\xi_i^1 = (BG)_i$. Note that $\xi_i^1 \in B^{z_i}$ and $B^{z_i} \subset \mathcal{V}_{z_i-1}$ together imply that

$$\begin{aligned} A\xi_i^1 \in A\mathcal{V}_{z_i-1} &\subset \mathcal{V}_{z_i-2} + ImB \\ &= \mathcal{V}_{z_i-2} + \sum_{j=1}^{z_i-2} ImB^j \end{aligned} \tag{10}$$

Noting that the sum $B^1 + \ldots + B^{z_i-2}$ above is equal to $Sp\{(BG)_p | \forall p \ni z_p \leq z_i - 2\}$, we conclude that there exist a $\xi_i^2 \in \mathcal{V}_{z_i-2}$ and scalars $\alpha_{i,j}$'s so that

$$A\xi_i^1 = \xi_i^2 + \sum_{\forall p \ni z_p \leq z_i-2} \alpha_{s_p+1, s_i+1}(BG)_p \tag{11}$$

Remark: Indexing of α's above may look unnatural at this point; however, we shall motivate it in the sequel.

Now, $\xi_i^2 \in \mathcal{V}_{z_i-2}$ implies

$$\begin{aligned} A\xi_i^2 \in A\mathcal{V}_{z_i-2} &\subset \mathcal{V}_{z_i-3} + ImB \\ &= \mathcal{V}_{z_i-3} + \sum_{j=1}^{z_i-3} ImB^j \end{aligned} \tag{12}$$

Again, noting that $B^1 + \ldots + B^{z_i-3}$ is equal to $Sp\{(BG)_p | \forall p \ni z_p \leq z_i - 3\}$, we conclude that there exist a $\xi_i^3 \in \mathcal{V}_{z_i-3}$ and scalars $\alpha_{i,j}$'s such that

$$A\xi_i^2 = \xi_i^3 + \sum_{\forall p \ni z_p \leq z_i-3} \alpha_{s_p+1, s_i+2}(BG)_p \tag{13}$$

Proceeding as above, we generate $\xi_i^3, \ldots, \xi_i^{z_i-1}$ where $\xi_i^j \in \mathcal{V}_{z_i-j}$, $j \in \mathbf{z_i} - 2$. Finally, $\xi_i^{z_i-1} \in \mathcal{V}_1$ implies

$$A\xi_i^{z_i-1} \in A\mathcal{V}_1 = \mathcal{V}_0 = \mathcal{R}^n \tag{14}$$

Thus, we simply define

$$\xi_i^{z_i} = A\xi_i^{z_i-1} \tag{15}$$

Thus, we have shown that for any $i \in \mathbf{m}$, there exists a chain of vectors satisfying the following recursion .

$$\begin{aligned} \xi_i^1 &= (BG)_i & (16) \\ \xi_i^j &= A\xi_i^{j-1} - \sum \alpha_{s_p+1, s_i+j-1}(BG)_p & (17) \end{aligned}$$

$$\xi_i^{z_i} = A\xi_i^{z_i-1} \tag{18}$$

In the ξ_i^j equation above, $j \in \mathbf{z_i} - 2$, and the summation is over all p such that $z_p \leq z_i - j$.

It is not hard to show that the chains defined above are unique in the sense that if $\{x_i^1, \ldots, x_i^{z_i}\}$ is another chain of vectors such that: (i) the recursion above is satisfied by x_i^j's, (ii) $x_i^1 = (BG)_i$, and (iii) $x_i^j \in \mathcal{V}_{z_i-j}$, then it necessarily follows that $x_i^j = \xi_i^j$, $j \in \mathbf{z_i}$.

Now, we can state the main result of this section.

Theorem 3.1 $\{\xi_1^1 \ldots \xi_1^{z_1}; \xi_2^1 \ldots \xi_2^{z_2}; \ldots; \xi_m^{z_m}\}$ is an ordered basis for $\mathcal{T}^*$.

Remark: Note that ξ_i^j is the $(s_i + j)^{th}$ vector in the basis $\{\xi_i^1, \ldots, \xi_i^{z_1}; \ldots; \xi_m^1, \ldots, \xi_m^{z_m}\}$. In specific, $\xi_p^1 = (BG)_p$ is the $(s_p + 1)^{st}$ vector in the basis. This should explain the reason why $\alpha_{i,j}$'s were indexed the way they were during the construction of the chains above.

4. CANONICAL FORM OF A WEAKLY PRIME TRIPLE

In this section, we shall consider a weakly prime triple (C, A, B). Recall that (C, A, B) is weakly prime iff it is a minimal realization with $rankB = m = p = rankC$, $\mathcal{V}^* = 0$ and $\mathcal{T}^* = \mathcal{R}^n$. We shall classify the orbits of the set Σ_{wp} of weakly prime triples under the similarity action of the general linear group $\mathcal{GL}(n, \mathcal{R})$ by explicitly defining a complete and independent orbital invariant. The canonical form defined by this complete orbital invariant will be presented.

Let T be the matrix whose columns are the (ordered) basis vectors given in Theorem 3.1. Note that since (C, A, B) is assumed to be weakly prime, we have $\mathcal{T}^* = \mathcal{R}^n$ and therefore T is nonsingular. For future reference, note that the jth column of BG is the $(s_j + 1)^{th}$ column of T, and that ξ_i^j is the $(s_i + j)^{th}$ column of T. Now, let $\{\alpha_{p,s_{i+1}} | \; p \in \mathbf{n}\}$ be the unique set of scalars defined by

$$A\xi_i^{z_i} = \sum_{p=1}^{n} \alpha_{p,s_{i+1}} (T)_p \tag{19}$$

Let $A_c = T^{-1}AT$. It follows by inspection that the $(s_i + j)^th$ column of A_c is given by:

1. $e_{s_i+j+1} + \sum \alpha_{s_{p+1},s_i+j} e_{s_p+1}$ when j is so that $s_i < j \leq s_{i+1} - 2$ (here, the summation is over all p such that $z_p \leq z_i - j - 1$),
2. e_{s_i+1} if $j = s_{i+1} - 2$,
3. $\sum_{p=1}^{n} \alpha_{p,s_{i+1}} e_p$ if $j = s_{i+1}$.

Now, considering $B_c := T^{-1}B$, we recall that $T^{-1}BG$ has its jth column equal to e_{s_j+1}. That is, we have

$$T^{-1}BG = [\; e_{s_1+1} \quad e_{s_2+1} \quad \ldots \quad e_{s_m+1} \;] \tag{20}$$

It is not trivial at all but nevertheless can be shown that

$$i^{th} \text{ row of } G^{-1} = e_i^T + \sum_{j \in (\mathcal{I}_{s_i})^i} \beta_{i,j} e_j^T \tag{21}$$

Therefore, we conclude that the jth row of B_c is given by:

1. identically zero if for no i, $j = s_i + 1$, and
2. $e_i^T + \sum_{p \in (\mathcal{I}_{s_i})^i} \beta_{i,p} e_p^T$ if $j = s_i + 1$ for some i.

Now, to describe the structure of the representation of C with respect to the basis chosen in Theorem 3.1, we first define a list $\mathcal{I}_x = \{x_1, \ldots, x_m\}$ as follows. If the ith row of C is in C^i then $x_i = j$. It is not hard to show that the list $\mathcal{I}_x$, if ordered descendingly, is equal to the list $\mathcal{I}_n$.

Note that the basis vector ξ_i^j lies in $\mathcal{K}$ unless $j = z_i$. Then,

$$C\xi_i^j = 0 \;\; ; \;\; j < z_i \tag{22}$$

and it can be shown that

$$C \, [\; \xi_1^{z_1} \quad \ldots \quad \xi_m^{z_m} \;] \tag{23}$$

is square and nonsingular. Also, it follows from the construction of H that

$$KerC^i \supset \mathcal{T}_{j-1} \tag{24}$$

As $i \in \hat{\mathcal{L}}_j$ iff $x_i = j$, the equation just above implies that

$$KerC^i \supset \mathcal{T}_{x_i-1} \tag{25}$$

Therefore, as $\xi_j^{z_j} \in \mathcal{T}_{z_j}$ (see Özçaldıran and Eldem (1995)) and as $\mathcal{T}_k \subset \mathcal{T}_{k+1}$, we have:

$$C^i \xi_j^{z_j} = 0 \;\; ; \;\; \text{if } z_j \leq x_i - 1 \tag{26}$$

Thus, the ith row of $\tilde{C} := HCT$ is given by:

$$\sum_{\forall j \ni z_j > x_i - 1} \theta_{i,s_{j+1}} e_{s_j+1}^T \tag{27}$$

Given the uniquely specified matrix $\tilde{C}$, define C_c by $C_c = H^{-1}\tilde{C} = CT$.

Now, fix an ordering for the parameters $\alpha_{i,j}$'s which appear in A_c. Do the same for the $\beta_{i,j}$'s which appear in B_c, for $\theta_{i,j}$'s which appear in $\tilde{C}$, and for $\gamma_{i,j}$'s which appear in H^{-1}. Define a subset $\mathcal{D}$ of $\mathcal{Z}^m \times \mathcal{Z}^m \times \mathcal{R}^{n_{A_c}} \times \mathcal{R}^{n_{B_c}} \times \mathcal{R}^{n_H} \times \mathcal{R}^{n_{\tilde{C}}}$ by the conditions:

1. If $p \in \mathcal{D}$ then the first m components $z_1, \ldots, z_m$ of p add up to n,

2. If $p \in \mathcal{D}$ then the second m components $x_1, \ldots, x_m$ of p add up to n,
3. The last $n_{\tilde{C}}$ components of p are so that the rank condition $rank\tilde{C} = m$ is satisfied.

Finally, define f so that it takes $(C, A, B) \in \Sigma_{wp}$ to the corresponding data point in $\mathcal{D}$.

Theorem 4.1 f is a complete and independent orbital invariant of the similarity action of the general linear group $\mathcal{GL}(n, \mathcal{R})$ on Σ_{wp}.

The following theorem is indeed a simple corollary to the theorem above.

Theorem 4.2 (C_c, A_c, B_c) is a canonical form of $(C, A, B) \in \Sigma_{wp}$ under similarity transformations.

We now restrict our attention to the set Σ_p of prime triples. Recall that the definition of primeness that we adopt from Morse (1973) implies that a triple from Σ_{wp} is prime iff $\mathcal{T}_i = \sum_{j=1}^{i-1} A^j ImB$. In this case, we have

$$\begin{aligned} A\mathcal{T}_i &= A(\textstyle\sum_{j=1}^{i-1} A^j ImB) \\ &\subset \textstyle\sum_{j=1}^{i} A^j ImB + ImB \\ &= \mathcal{T}_{i+1} \end{aligned}$$

Now, consider one of the chains $\{\xi_i^1, \ldots, \xi_i^{z_i}\}$ which appear in T. It can be shown that (see Özçaldıran and Eldem (1995)) $\xi_i^j \in \mathcal{T}_j$, $j \in \mathbf{z_i}$, and in specific, we have $\xi_i^{z_i} \in \mathcal{T}_{z_i}$. Then,

$$A\xi_i^{z_i} \in A\mathcal{T}_{z_i} \subset \mathcal{T}_{z_i+1}$$

implies that in the column of $T^{-1}AT$ which corresponds to $A\xi_i^{z_i}$, there will be $dim\mathcal{T}_{z_i+1}$, not n, α_{ij}'s. More specifically, those α_{ij}'s in the expansion of $A\xi_i^{z_i}$ which multiply ξ_j^k will be zero for all j, k such that $z_j > z_i$ and $k > z_i + 1$. Clearly, the parameters in B_c and C_c remain the same in this case.

Let $(C, A, B) \in \Sigma_{wp}$ be dually prime. Recall that $(C, A, B) \in \Sigma_{wp}$ is dually prime iff $\mathcal{V}_i = \bigcap_{j=0}^{i-1} A^{-j}\mathcal{K}$. Note that this condition immediately implies that the chains $\{\xi_i^1, \ldots, \xi_i^{z_i}\}$ which appear in T satisfy:

$$\xi_i^1 = (BG)_i \; ; \; \xi_i^j = A\xi_i^{j-1} \tag{28}$$

where $j = 2, \ldots, z_i$. Consequently, A_c will have no nonzero and nonunity parameters α_{ij}'s in column j unless $j = s_i$ for some i. The s_i^{th} columns of A_c will still have n parameters per column. In this case also, matrices B_c and C_c remain the same.

In case $(C, A, B) \in \Sigma_{wp}$ is strongly prime,the analyses of the previous two subsections can be combined to conclude that only columns $s_2, \ldots, s_{m+1}$ of A_c will have parameters α_{ij}^k's. The s_i^{th} column will have $dim\mathcal{T}_{z_i+1}$ such parameters. As it was the case with prime and dually prime triples, extra assumption of strong primeness does not change the B_c and C_c matrices.

5. CANONICAL FORM OF $(C, A, B) \in \Sigma_{ns}$

Finally, in this section, we shall consider a triple from Σ_{ns}. Recall that such a triple satisfies the following conditions:

1. (C, A, B) is a minimal realization,
2. $rankB = m = p = rankC$,
3. $\mathcal{V}^* \oplus \mathcal{T}^* = \mathcal{R}^n$.

Denote $dim\mathcal{V}^*$ by n_v and $dim\mathcal{T}^*$ by n_t. It follows from 3 above that $\mathcal{V}^*$ is a coasting subspace (see Willems (1981)), Then, there exists a unique Jordan form matrix J, a unique basis matrix $V = [v_1 \; \ldots \; v_{n_v}]$ and a unique matrix U so that (see Jaffe and Karcanias (1981))

$$AV = VJ + BU \tag{29}$$

Now, let T be the unique basis matrix for $\mathcal{T}^*$ which was defined in the previous sections. Let (A_c, B_c, C_c) be the canonical triple constructed in Section 4. Note that T is now an $n \times n_t$ matrix. Also note that we still have $B = TB_c$. Then, Equation (29) can be written as:

$$AV = \begin{bmatrix} T & V \end{bmatrix} \begin{bmatrix} B_cU \\ J \end{bmatrix} \tag{30}$$

Also note that:

$$AT = \begin{bmatrix} T & V \end{bmatrix} \begin{bmatrix} A_c \\ \overline{A}_{21} \end{bmatrix} \tag{31}$$

where the jth column of $\overline{A}_{21}$ is identically zero if $s_i < j < s_{i+1}$, $i \in \mathbf{m}$. If $j = s_i$ for some i, the the elements of the jth column are uniquely determined by the expansion

$$A\xi_i^{z_i} = \sum_{p=1}^{n_t} \alpha_{p,s_{i+1}} T_p + \sum_{p=n_t+1}^{n} \alpha_{p,s_{i+1}} V_{p-n_t} \tag{32}$$

Thus, it follows that $\overline{A}_{21}$ can be written as

$$\overline{A}_{21} = \tilde{A}_{21} \begin{bmatrix} e_{s_2}^T \\ \vdots \\ e_{s_{m+1}}^T \end{bmatrix} \tag{33}$$

where e_j above denotes the jth standard basis vector for $\mathcal{R}^{n_t}$, and $\tilde{A}_{21}$ is an $n_v \times m$ matrix. At this point, we note that there exists a unique nonsingular matrix $\overline{P}$ such that

$$\overline{P}C_c = \begin{bmatrix} e_{s_2}^T \\ \vdots \\ e_{s_{m+1}}^T \end{bmatrix} \tag{34}$$

Then, we conclude that

$$\overline{A}_{21} = PC_c \quad (35)$$

where P is defined by $P = \tilde{A}_{21}\overline{P}$.

Thus, representation $(\overline{C}, \overline{A}, \overline{B})$ of (C, A, B) with respect to the ordered basis $\overline{T} := [\, T \quad V \,]$ becomes:

$$([\, C_c \quad 0 \,], \begin{bmatrix} A_c & B_cU \\ PC_c & J \end{bmatrix}, \begin{bmatrix} B_c \\ 0 \end{bmatrix}) \quad (36)$$

Let $\overline{f}$ be the function from Σ_{ns} to $\mathcal{R}^{n_t \times n_t} \times \mathcal{R}^{n_t \times m} \times \mathcal{R}^{m \times n_t} \times \mathcal{C}^{n_v \times n_v} \times \mathcal{R}^{n_t \times n_v} \times \mathcal{R}^{n_v \times n_t}$ which takes $(C, A, B) \in \Sigma_{ns}$ to the corresponding six-tuple $(A_c, B_c, C_c, J, B_cU, PC_c)$. Then, we have:

Theorem 5.1 $\overline{f}$ is a complete and independent orbital invariant for the similarity action of $\mathcal{GL}(n, \mathcal{R})$ on Σ_{ns}.

As a corollary to the result above, we have:

Theorem 5.2 Let $(C, A, B) \in \Sigma_{ns}$. Let $\overline{A} = \overline{T}^{-1}A\overline{T}$, $\overline{B} = \overline{T}^{-1}B$ and $\overline{C} = C\overline{T}$ be as given in Equation (36). Then, $(\overline{C}, \overline{B}, \overline{A})$ is a canonical form of (C, A, B) under similarity transformations.

6. CONCLUSIONS

In this work, a complete characterization for the orbits of minimal systems with square, nonsingular transfer matrices under the action of similarity transformations is presented. The orbits which are equivalent under similarity are represented via a canonical form. This canonical form is first presented for a special subclass of square,nonsingular transfer matrices, namely weakly prime systems. Then its reduction to prime, dually prime and strongly prime systems are considered. Finally, it is extended to systems with square, nonsingular transfer matrices. It is believed that the canonical form depicted in this work will be helpful in studying a number of algeraic control problems. Indeed, it is shown somewhere else (see Özçaldıran *et al.* (1995)) that the canonical form constructed in this paper can be easily modified to yield a canonical form of $(C, A, B) \in \Sigma_{ns}$ under the action of the output feedback group $\mathcal{O}$ which is generated by similarity transformations and output feedback laws. Finally, we would like to pronounce our belief that the generalization of our results to nonsquare transfer matrices can be achieved with moderate effort.

7. REFERENCES

Brunovsky, P. (1970). A classification of controllable linear systems. *Kybernatika* **3**(6), 173–188.

Commault, C. and J. M. Dion (1982). Structure at infinity of linear multivariable systems:a geometric approach. *IEEE Transactions on Automatic Control* **27**, 112–117.

Eldem, V. (1993). Parametrization of mutivariable systems using output injections: Alpha canonical forms. *Automatica* **29**(4), 1127–1131.

Guidorzi, R. P. (1981). Invariants and canonical forms for systems structural and parametric identification. *Automatica* **11**, 117–133.

Jaffe, S. and N. Karcanias (1981). Matrix pencil characterization of almost (a, b)-invariant subspaces: A classification of geometric concepts. *International Journal of Control* **33**(1), 51–93.

Morse, A.S. (1973). A classification of controllable linear systems. *SIAM Journal on Control and Optimization* **11**(3), 446–465.

Özçaldıran, K. and V. Eldem (1995). *A complete classification of minimal realizations of nonsingular and strictly proper transfers under similarity transformations.* Submitted to *Automatica.*

Özçaldıran, K., V. Eldem and E.Tatar (1995). *A structural result on static output feedback.* Submitted to the 34th IEEE CDC.

Popov, V. M. (1972). Invariant description of linear, time-invariant, controllable systems. *SIAM Journal on Control and Optimization* **10**(2), 252–264.

Thorp, J. S. (1973). The singular pencil of a linear dynamical system. *International Journal of Control* **18**, 577–596.

Willems, J. C. (1981). Almost invariant subspaces: An approach to high gain feedback design. part i: Almost controlled invariant subspaces. *IEEE Transactions on Automatic Control* **26**, 235–252.

Wonham, W. M. (1979). *Linear Multivariable Control: A Geometric Approach, 2nd ed.*. Springer Verlag. New York.

A SEMI-CANONICAL FORM FOR A CLASS OF RIGHT INVERTIBLE LINEAR SYSTEMS.

A. HERRERA[*] J.F. LAFAY[**] P. ZAGALAK[***]

Sección de Control Automático Departamento de Ingeniería Eléctrica, Cinvestav IPN Av. IPN No. 2508, A.P. 14-740. 07000, México, D.F. e-mail: angel@ctrl.cinvestav.mx. MEXICO.

***LAN-ECN (CNRS URA 823), 1 rue de la Noë. 44072, CEDEX 03. NANTES, France. e-mail: lafay@lan.ec-nantes.fr*

**** UTIA, Academy of Sciences, 18208 Prague 8. Czech Republic.*

Abstract. We present a semi-canonical form for a class of right invertible systems under the action of the transformation group (T,F,G,Π), where Π is a permutation matrix acting on the outputs. Under certain additional conditions this form is canonical, in particular when the system is nonsingular and controllable.

Résumé. Nous présentons une forme semi-canonique pour une classe de systèmes linéaires inversibles à droite, sous l'action du groupe de transformations (T,F,G,Π), où Π désigne une permutation des sorties . Sous certaines conditions supplémentaires, cette forme devient canonique, en particulier lorsque le système est inversible et commandable.

Key Words. Linear systems, Semi-canonical form, Feedback group, Interactor matrix

1. INTRODUCTION

The solution of various control problems is closely related to the feedback invariants of linear multivariable systems, see (Morse, 1973) and references therein. Of particular interest in the study of feedback invariants are:

- The work of Brunovsky (Brunovsky, 1970), which shows that the controllability indices are the complete invariants of a controllable matrix pair (A, B) under the feedback group including state and input changes of coordinates (T, G) and state feedback (F).

- The work of Morse (Morse, 1973), which exhibits four lists of integers and a list of polynomials as the complete invariants of a triple (C, A, B) under the larger group of transformations (T, F, G, H, K), where H is a change of coordinates on the outputs, and K is an output injection. Elements of this group act on (C, A, B), as follows:

$$(T, F, G, H, K) \circ (C, A, B) =$$
$$\{HCT^{-1}, T(A + BF + KC)T^{-1}, TBG)\}$$

These invariants are the key of the solution of many difficult problems involving structural modifications of a linear system represented by the triple (C, A, B). As an example, let us mention the noninteracting control problem (Herrera, 91; Herrera and Lafay, 1993), the infinite structure placement (Loiseau, 1988), etc.

In the study of control problems like decoupling, we cannot use output injection, since it is not physically realizable, or a change of coordinates on the outputs. In this paper we consider the subgroup of transformations (T, F, G, Π), where Π denotes permutations of outputs. We establish a semi-canonical form, using the properties of the *row reduced interactor* defined in (Herrera *et al*, 1991). We shall show that this form is canonical for some class of linear systems (C, A, B), as those which are nonsingular and controllable.

2. PRELIMINARY RESULTS

Let the time invariant linear system described by

$$\dot{x}(t) = Ax(t) + Bu(t) \quad (1)$$

$$y(t) = Cx(t) \quad (2)$$

be given.

Throughout the paper $R[s]$ denotes the ring of polynomials over R, the field of reals. $R(s)$ stands for the field of rational functions over R, while the rings of proper and strictly proper functions are denoted, respectively, by $R_p(s)$ and $R_{sp}(s)$. Further, V^* denotes the supremal (A,B)-invariant subspace contained in $K := KerC$, and R^* the supremal controllability subspace contained in $K := KerC$ (Wonham, 1985); I_n denotes the $n \times n$ identity matrix, while J_m is the $m \times m$ nilpotent Jordan matrix. The $i-th$ column (row) degree of a given polynomial matrix M is denoted by $d_{ci}M$ ($d_{ri}M$). We will suppose (C,A,B) right invertible. Then there exists (Wolovich and Falb, 1976) a unique nonsingular lower triangular polynomial matrix $\Phi[s]$, and a *nonunique* biproper matrix $B_p(s)$ such that the transfer function $T(s) = C(sI-A)^{-1}B$ can be written as $T(s) = \left[\begin{array}{cc}\Phi_1^{-1}(s) & 0\end{array}\right] B_p(s)$. Matrix $\Phi_1[s]$ is called *the interactor* of (C,A,B) and satisfies:

1. $\Phi_{ii}[s] = s^{f_i}$, $i = 1,...,p$, for some positive integers f_i, and
2. $\Phi_{ji}[s]$ is zero or $\frac{\Phi_{ji}[s]}{\Phi_{ii}[s]}$ is a polynomial divisible by s.

A biproper matrix is a nonsingular proper matrix whose inverse is also proper. The degree of the $i-th$ column of the interactor is called the essential order of the $i-th$ output of $T(s)$, noted by n_{ie}, $i = 1,...,p$ (Commault *et al*, 1988).

Our main result is obtained from the following technical propositions (Herrera, 1991; Herrera *et al*, 1991)

Proposition 1 Let H be a $q \times q$ constant nonsingular matrix. Then there exist nonsingular matrices Π, L and R, where Π is a permutation matrix, such that $H = LR$ and

$$\Pi L = \begin{bmatrix} 1 & 0 & \ldots & 0 \\ l_{21} & 1 & \ldots & 0 \\ \ldots & \ldots & \ldots & \ldots \\ l_{q1} & l_{q2} & \ldots & 1 \end{bmatrix}$$

$$R = \begin{bmatrix} r_{11} & r_{12} & \ldots & r_{1q} \\ 0 & r_{22} & \ldots & r_{2q} \\ \ldots & \ldots & \ldots & \ldots \\ 0 & 0 & \ldots & r_{qq} \end{bmatrix}$$

Proposition 2 There exists a nonsingular matrix $T \in R^{q\times q}$ such that

$$T \left[\begin{array}{cc} \begin{bmatrix} a_0 \\ \ldots \\ a_{q-2} \\ 0 \end{bmatrix} & I_{q-1} \\ (0) & (0) \end{array}\right] T^{-1} =$$

$$\left[\begin{array}{cc} (0) & I_{q-1} \\ 0 & \begin{bmatrix} a_{q-2} & \ldots & a_0 \end{bmatrix} \end{array}\right]$$

Proposition 3 There exists a nonsingular matrix $T \in R^{q\times q}$ such that

$$T \left[\begin{array}{cc} \begin{bmatrix} * & I_{q_1-1} \\ 0 & (0) \end{bmatrix} & \begin{bmatrix} \begin{bmatrix} a_0 \\ \ldots \\ a_{q_1-2} \end{bmatrix} & (0) \\ 0 & (0) \end{bmatrix} \\ \begin{bmatrix} * & (0) \\ 0 & (0) \end{bmatrix} & \begin{bmatrix} (0) & I_{q_2-1} \\ 0 & (0) \end{bmatrix} \end{array}\right] T^{-1} =$$

$$\left[\begin{array}{cc} \begin{bmatrix} * & I_{q_1-1} \\ 0 & (0) \end{bmatrix} & \begin{bmatrix} (0) & (0) & (0) \\ 0 & M_1 & (0) \end{bmatrix} \\ \begin{bmatrix} * & (0) \\ 0 & (0) \end{bmatrix} & \begin{bmatrix} (0) & I_{q_2-1} \\ 0 & (0) \end{bmatrix} \end{array}\right]$$

where

$$M_1 = \begin{bmatrix} a_{q_1-2} & \ldots & a_0 \end{bmatrix}$$

and $q=q_1+q_2$, $q_1 \geq q_2$; $*$ denotes possible nonzero terms.

Proposition 4 There exists a nonsingular matrix $T \in R^{q\times q}$ such that

$$T \left[\begin{array}{cc} J_{q_1} & 0 \\ \begin{bmatrix} \begin{bmatrix} a_0 \\ \ldots \\ a_{q_2-q_1-2} \end{bmatrix} & (0) \\ (0) & M_2 \end{bmatrix} & J_{q_2} \end{array}\right] T^{-1} =$$

$$\left[\begin{array}{cc} J_{q_1} & 0 \\ \begin{bmatrix} \begin{bmatrix} a_0 \\ \ldots \\ a_{q-2} \\ 0 \end{bmatrix} & (0) \\ 0 & (0) \end{bmatrix} & J_{q_2} \end{array}\right]$$

where

$$M_2 = \left[\begin{array}{ccc} a_{q_2-2} & \cdots & a_{q_2-q_1} \end{array} \right]$$

$q = q_1 + q_2, q_2 \geq q_1$. If $q_2 = q_1$ and/or $2 \geq q_2 \geq 1$, then $q_{-j} = 0 \; \forall j > 0$.

3. MAIN RESULT

Suppose the Morse lists $I_4 = \left\{ n_i' \right\}_p$ and $I_2 = \{\sigma_i\}_{m-p}$ are nondecreasingly ordered. Then the main result of this paper is given by the following

Theorem 1 Let (C, A, B) be a right invertible system which satisfies $R^* = V^*$. There then exists a permutation matrix acting on the outputs of the system such that the interactor $\Phi_1[s]$ of (C, A, B) is row reduced (Lafay *et al*, 1992). Moreover, we can define $m - p$ fictitious outputs such that the extended system has the following row reduced extended interactor

$$\Phi_e[s] = \left[\begin{array}{cc} \Phi_1[s] & (0) \\ \Phi_2[s] & \Phi_3[s] \end{array} \right] \qquad (3)$$

where

$$\Phi_1[s] = \left[\begin{array}{ccc} s^{-n_1'} & \cdots & 0 \\ \cdots & \cdots & \cdots \\ \phi_{p1} & \cdots & s^{-n_p'} \end{array} \right] \qquad (4)$$

$$\Phi_3[s] = blockdiag \left\{ \begin{array}{ccc} s^{-\sigma_i} & \cdots & s^{-\sigma_{m-p}} \end{array} \right\} \qquad (5)$$

and

$$d_{r_i}\Phi_2[s] < \sigma_i \qquad (6)$$

A realization for the system $T(s) = \Phi_e^{-1}(s)$ has the following form

$$A_c = \left[\begin{array}{cc} A_4 & (0) \\ A_{21} & A_2 \end{array} \right] \quad B_c = \left[\begin{array}{cc} B_4 & (0) \\ (0) & B_2 \end{array} \right] \qquad (7)$$
$$C_c = \left[\begin{array}{cc} C_4 & (0) \end{array} \right]$$

where

$$A_4 = \left[\begin{array}{cccc} A_{11} & (0) & \cdots & (0) \\ A_{21} & A_{22} & \cdots & (0) \\ \cdots & \cdots & \cdots & \cdots \\ A_{p1} & A_{p1} & \cdots & A_{pp} \end{array} \right] \qquad (8)$$

$$A_2 = blockdiag \left\{ J_{\sigma_i} \right\} \qquad (9)$$

$$A_{ii} = J_{n_i'}$$

$$A_{ij} = \left[\begin{array}{cccc} -\alpha_1^{ij} & 0 & \cdots & 0 \\ \cdots & 0 & \cdots & 0 \\ -\alpha_{n_i'-n_j'-1}^{ij} & 0 & \cdots & 0 \\ 0 & 0 & \cdots & 0 \\ \cdots & \cdots & \cdots & \cdots \\ 0 & 0 & \cdots & 0 \end{array} \right] \forall i \neq j \qquad (10)$$

where $\alpha_k^{ij} \in R$ are the coefficients of the element, say ϕ_{ij}, of $\Phi_e[s]$ given by

$$\phi_{ij}[s] = \alpha_0^{ij} s^{n_i'} + \ldots + \alpha_{n_i'-n_j'}^{ij} s^{n_j'+1} \qquad (11)$$

$$B_4 = blockdiag \left\{ \left[\begin{array}{c} 0 \\ \cdots \\ 0 \\ 1 \end{array} \right]_{n_i' \times 1} \right\} \qquad (12)$$

$$B_2 = blockdiag \left\{ \left[\begin{array}{c} 0 \\ \cdots \\ 0 \\ 1 \end{array} \right]_{\sigma_i \times 1} \right\} \qquad (13)$$

Finally

$$C_4 = \left[\begin{array}{ccc} C_4^1 & \cdots & C_4^p \end{array} \right] \qquad (14)$$

where

$$C_4^1 = \left[\begin{array}{cccc} 1 & 0 & \cdots & 0 \\ -\alpha_0^{2\times 1} & 0 & \cdots & 0 \\ \cdots & \cdots & \cdots & \cdots \\ -\alpha_0^{p\times 1} & 0 & \cdots & 0 \end{array} \right]$$

$$C_4^2 = \left[\begin{array}{cccc} 0 & 0 & 0 & 0 \\ 1 & 0 & \cdots & 0 \\ -\alpha_0^{3\times 2} & 0 & \cdots & 0 \\ \cdots & \cdots & \cdots & \cdots \\ -\alpha_0^{p\times 2} & 0 & \cdots & 0 \end{array} \right]$$

.
.
.

$$C_4^p = \left[\begin{array}{cccc} 0 & 0 & \cdots & 0 \\ \cdots & \cdots & \cdots & \cdots \\ 1 & 0 & \cdots & 0 \end{array} \right]$$

Proof: (Herrera, 1991; Herrera *et al*, 1991; Herrera *et al*, 1993).

If (C,A,B) is right invertible with $R^* = V^*$, then under the action of the transformation group

(T, F, G), the system (C, A, B) is equivalent to the following triple (Morse, 1973):

$$\bar{A}_c = \begin{bmatrix} A_{n_1'} & .. & A_{1p} & (0) & .. & (0) \\ \cdots & .. & \cdots & \cdots & .. & \cdots \\ A_{p1} & .. & A_{n_p'} & (0) & .. & (0) \\ A_{(p+1)1} & .. & A_{(p+1)p} & A_{\sigma_1} & .. & (0) \\ \cdots & .. & \cdots & \cdots & .. & \cdots \\ A_{m1} & .. & A_{mp} & (0) & .. & A_{\sigma_{m-p}} \end{bmatrix}$$

where

$$A_{n_i'} = \begin{bmatrix} (*) & I_{n_i'-1} \\ 0 & (0) \end{bmatrix} \in R^{n_i' \times n_i'}, \quad i = 1, ..., p,$$

$$A_{\sigma_j} = \begin{bmatrix} (0) & I_{\sigma_j - 1} \\ 0 & (0) \end{bmatrix} \in R^{\sigma_j \times \sigma_j}, \quad j = 1, ..., m-p.$$

$$A_{ij} = \boxed{\begin{matrix} (*) & (0) \\ 0 & (0) \end{matrix}} \in \begin{cases} R^{n_i' \times n_j'}, \forall\, i \neq j; \\ i,\ j = 1, ..., p \\ R^{\sigma_j \times n_j'} \\ j = p+1, ..., m; \\ i = 1, ..., p \end{cases}$$

for $j = p+1, ..., m;\ i = 1, ..., p.$

$$\bar{B}_c = \begin{bmatrix} diag\left\{e_{n_i'}\right\} & (0) \\ (0) & diag\left\{e_{\sigma_j}\right\} \end{bmatrix}$$

$$\bar{C}_c = H \begin{bmatrix} diag\left\{e_1^T, ..., e_p^T\right\} & (0) \end{bmatrix}$$

where e_i^T is the i-th row of I_p, $e_{n_i'}$ and e_{σ_j} are the last columns of $I_{n_i'}$ and I_{σ_j}, respectively, $i = 1, ..., p; j = 1, ..., m-p$. H is a nonsingular matrix, and symbol $*$ denotes the position of possible non zero terms.

The similarity transformations of *Propositions 2 to 4* change some of the nonzero terms, indicated by $*$, in the columns of $\bar{A}_c$ into the rows of the same matrix associated with the nonzero terms in matrix $\bar{B}_c$. So, they can be cancelled by a state feedback. These transformations are characterized by the following:

Definition 1 A transformation T is said to be admissible if $C_c T^{-1} = C_c$ and if there exists a regular matrix G such that $T B_c G = B_c$. The property of these transformations is that they preserve the canonical structure of B_c and C_c. By extension (T, G, F) is admissible if so is T.

Now by *Proposition 1* and *Definition 1*, there exist a lower triangular matrix L, an uper triangular matrix R, and an admissible transformation T such that we get (14) ∎

Corollary 1 If the extended interactor $\Phi_e[s]$ is *strictly row reduced*, i.e.,

$$n_i' > \deg \phi_{ij}[s], \forall i \neq j, i = 1, ..., p; j = 1, ..., i-1$$

and

$$\sigma_k > \deg \phi_{kj}[s], \forall i = p+1, ..., m; j = 1, ..., p,$$

then the triple (C_c, A_c, B_c) is a canonical form of (C, A, B) under the action of the group of transformations (T, F, G, Π). The complete invariant is the extended interactor $\Phi_e[s]$.

Proof: If $n_i' = deg\phi_{ij}[s]$, $j < i$, then with Π_0 being the matrix which permutes columns i and j of $\Phi_1[s]$, we get that $\Phi_1[s]\Pi_0 = B_\Pi(s)\Phi_\Pi[s]$, where $B_\Pi(s)$ is a biproper matrix and the interactor $\Phi_\Pi[s]$ is still row reduced but not necessarilly equal to $\Phi_1[s]$. On the other hand, if $\Phi_1[s]$ is *strictly row reduced,* the interactor obtained after any permutation of two of its columns will give an interactor which is not longer row reduced, i.e., if Π_1 represents this permutation, there then exists a biproper matrix $B_{\Pi_1}(s)$ such that $\Phi_1[s]\Pi_1 = B_{\Pi_1}(s)\Phi_{\Pi_1}[s]$, where $\Phi_{\Pi_1}[s]$ is not row reduced. ∎

Corollary 2 Let $T(s) = C(sI - A)^{-1}B$ be square and nonsingular, (A, B) controllable, and the interactor $\Phi[s]$ of (C, A, B) *strictly row reduced.* Then, under the action of the group of transformations (T, F, G, Π), the triple $(C_c, A_c + B_c F_c, B_c)$ is a canonical form of (C, A, B), where $F_c \in R^{m \times n}$ is a unique state feedback satisfying

$$B_p^{-1}(s) = \left[I_n - F_c(sI - A)^{-1}B\right]^{-1} \tag{15}$$

with $B_p(s)$ being the biproper matrix appearing in the factorization:

$$\begin{aligned} T(s) &= C(sI - A)^{-1}B \\ &= \Phi^{-1}(s)\, B_p(s) \end{aligned}$$

Proof: Since $T(s)$ is nonsingular, the interactor $\Phi[s]$ and the biproper matrix $B_p(s)$ are unique. Note that if (N_1, D) is a right coprime factorization of $(sI - A)^{-1}B$, then $B_p^{-1}(s)$ satisfies the Heymann's conditions in order to be static realizable, i.e., $B_p^{-1}(s)$ is biproper and $B_p(s)D[s]$ is polynomial. Now if there exist F_1 and F_2 satisfying (15), then,

$$(F_1 - F_2)(sI - A)^{-1}B = 0$$

Now, since (A,B) is controllable, the left constant kernel of $(sI-A)^{-1}B$ is zero . Therefore $F_1 = F_2$, and so F_c is unique. ∎

4. CONCLUSIONS

We present a new semi-canonical form for a class of linear right invertible systems under the action of the feedback group completed by output permutations. This form is useful when we consider control problems with nonregular static state feedback($u = Fx + Gv, G$ only monic) like in the input-output decoupling. The values of the parameters appearing in (C_c, A_c, B_c) play a key role in the modification of controllability chains by nonregular feedback (Herrera, 1991; Herrera et al 1991). This form is canonical if the interactor of the system is *strictly row reduced*. In the case where the transfer function is square and nonsingular, (A, B) controllable, and the interactor $\Phi_1[s]$ is *strictly row reduced*, we get a canonical form under the action of the group of transformations (T, G, Π), whose complete invariants are the *strictly row reduced* interactor $\Phi_1[s]$ and a state feedback F_c.

5. REFERENCES

Brunovsky, P. (1970) *A Classification of Linear Multivariable Systems.* Kybernetika, vol. 6, no. 3, pp 173-188.

Commault, C.; Descusse, J.; Dion, J. M.; Lafay, J. F.; Malabre, M.(1988) *New Decoupling Invariants: The Essential Orders.* Int. Journal of Cont., vol. 44, no. 3, pp 689-700.

Herrera H., A. N. (1991) *Sur le découplage des systèmes linéaires par des lois statiques non régulières.* Thèse de Doctorat. Ecole Centrale de Nantes, France.

Herrera, A.N.; Lafay, J.F. (1993). *New Results about the Morgan's Problem.* IEEE Transactions on Automatic Control, Vol.38, No. 12, pp 1834-1838.

Zagalak, P.; Lafay, J.F.; Herrera H., A.N. (1993). *The row-by-row Decoupling Problem via State Feedback: A Polynomial Approach.* Automatica, Vol.29, No. 6, pp 1491-1499.

Herrera H., A. N.; Lafay, J. F.; Zagalak, P. (1991). *The extended Interactor in the Study of Nonregular Control Laws.* Proceedings of the 30th CDC, pp 273-278.

Herrera H., A. N.; Lafay, J. F.; Zagalak, P. (1993). *Invariants under the Feedback Group for a Class of Linear Systems.* Rapport Interne no. 93-04. Ecole Centrale de Nantes, France.

Heymann, M. (1976). *Controllability Subspaces and Feedback Simulation.* SIAM J. Contr. Optimiz., Vol. 14, pp 769-789.

Lafay, J.F.; Zagalak, P.; Herrera H., A.N. (1992). *Row Reduced Form of the Interactor Matrix.* IEEE Transactions on Automatic Control, Vol.37, No. 11, pp 1778-1782.

Loiseau, J.J. (1988) *Contribution à l'étude des sous espaces presque invariants.* SIAM J. Contr. Optimiz., Vol. 26, No. 3, pp 251-273.

Morse, A.S. (1973) *Structural Invariants of Linear Multivariable Systems.* SIAM J. Contr. Optimiz., Vol. 11, No. 3, pp 446-465.

Wolovich, W.A.; Falb, P.L.(1976) *Invariants and Canonical Forms under Dynamic Compensation.* SIAM J. Contr. Optimiz., Vol. 14, pp 996-1008.

Wonham, W.M. (1985) *Linear Multivariable Control: A Geometric Approach.* 3rd ed., New York. Appl. of Math., Springer-Verlag, Vol. 10.

POLE STRUCTURE ASSIGNMENT VIA NON-REGULAR STATIC STATE FEEDBACK

J. J. Loiseau *, P. Zagalak and V. Kučera **

* *Laboratoire d'Automatique de Nantes, URA CNRS 823, Ecole Centrale de Nantes - Université de Nantes, 1, Rue de la Noë, 44 072 Nantes Cedex 03, France.*

** *Institute of Information Theory and Automation, Academy of Sciences of the Czech Republic, P.O. Box 18, 182 08 Praha, Czech Republic.*

Abstract. The limits of static state feedback $u = Fx + Gv$, with G not necessarily nonsingular, in altering the transmission pole structure of a linear system $\dot{x} = Ax + Bu$ are studied. A necessary and sufficient condition for a list of polynomials to be the denominator polynomials of the Smith–McMillan form of the closed-loop transfer matrix $(sI - A - BF)^{-1}BG$ is established. The condition involves the calculation of the infimal set of the controllability indices that the system can attain through feedback while satisfying Rosenbrock's inequalities.

Résumé. La structure en pôles de transmissions d'un système linéaire est généralement modifié lorsque le système est bouclé par un retour d'état statique. Sont décrites des conditions nécessaires et suffisantes que doivent satisfaire les polynômes qui se trouvent au dénominateur dans la forme de Smith-MacMillan du transfert en boucle fermée. Ces conditions se basent sur le calcul d'une liste minimale d'indices de commandabilité que le système peut atteindre sous l'action du retour d'état, tout en satisfaisant les conditions de Rosenbrock.

Key Words. Linear Systems, Polynomial Approach, Feedback Simulation, Model Matching, Matrix Completion, Controllability Indices.

1. INTRODUCTION

We consider the linear, finite–dimensional, time–invariant system

$$\dot{x} = Ax + Bu \tag{1}$$

where $A \in \mathbb{R}^{n\times n}$ and $B \in \mathbb{R}^{n\times m}$ are real matrices. Applying to system (1) the control law

$$u = Fx + Gv \tag{2}$$

where $F \in \mathbb{R}^{m\times n}$ and $G \in \mathbb{R}^{m\times p}$, we get the closed–loop system

$$\dot{x} = (A + BF)x + BGv. \tag{3}$$

As far as control is concerned, we are interested in studying the properties of the closed–loop system (3), and how the choice of F and G can affect these properties. In particular one can be interested in studying:

- The eigenstructure of the closed–loop system, that is the set of the invariant factors, say $\alpha_i(s)$ for $i = 1, \ldots, n$, of the matrix $A + BF$. The eigenstructure plays a fundamental role when studying the internal stability of (3).
- The pole structure of the closed–loop system (3), that is the set of the invariant factors, say $\psi_i(s)$ for $i = 1, \ldots, p$, which appear at the denominators of the diagonal elements of the Smith–McMillan form of the closed–loop transfer matrix $(sI_n - A - BF)^{-1}BG$. The pole structure represents the transmission dynamics of the closed–loop system.

These two structures are sizably affected by the choice of F and G. In the case where the system (1) is controllable, that is if

$$\text{rank}[B \ \ AB \ \ldots \ A^{n-1}B] = n \tag{4}$$

the possible polynomials $\alpha_i(s)$ have been completely described by Rosenbrock. This famous result reads as follows.

Theorem 1. (Rosenbrock, 1970). Given the system (1) and a set of n monic polynomials arranged in non–increasing order

$$\alpha_1(s) \geq \alpha_2(s) \geq \ldots \geq \alpha_n(s)$$

where $\alpha_i(s) \geq \alpha_{i+1}(s)$ means that $\alpha_{i+1}(s)$ divides (without remainder) $\alpha_i(s)$. Assume that (1) is controllable, that is (4) holds. There exists a matrix $F \in \mathbb{R}^{m\times n}$ such that the invariant factors of $A + BF$ are $\alpha_1(s), \alpha_2(s), \ldots, \alpha_n(s)$ if and only if the following two conditions hold:

$$\sum_{j=i}^{n} \deg \alpha_j(s) \leq \sum_{j=i}^{n} c_j \qquad \text{for } i = 1, ..., n \quad (5)$$

$$\sum_{j=1}^{n} \deg \alpha_j(s) = \sum_{j=1}^{n} c_j = n \qquad (6)$$

where $c_1, c_2, ..., c_n$ is the ordered list of the controllability indices of (1), defined by

$$c_i = \text{card}\,\{j \geq 1 | c_j^* \geq i\} \qquad \text{for } i = 1, ..., n$$

and

$$\begin{aligned} c_1^* &= \text{rank} B \\ c_i^* &= \text{rank}[B \;\; AB \; ... \; A^{i-1}B] \\ &\quad -\text{rank}\;[B \;\; AB \; ... \; A^{i-2}B] \quad \text{for } i = 2, ..., n. \end{aligned}$$

Theorem 1 has been widely commented in the control literature. Alternative proofs have been proposed by Dickinson (1974) who used a state-space approach, Kučera (1981) who applied the theory of polynomial equations, Flamm (1980) and more recenlty Özçaldiran (1990) who studied the problem in a geometric framework. Generalizations have been given to the realm of implicit systems, see Kučera and Zagalak (1988), Zagalak and Loiseau (1992) and even to the case where (A, B) is not a controllable pair, see Zaballa (1987), Loiseau and Zagalak (1993). However, no such a strong result has been given concerning the assignment of the transmission invariant factors of the closed-loop system. Our present aim is to fill this gap.

2. NON-REGULAR STATIC STATE FEEDBACK

The difficulty of our problem clearly comes from the use of a possibly non-regular feedback: the matrix G in (2) can be singular and even non-square, if $p < m$. Indeed if the pair $(A+BF,\ BG)$ is controllable, that is in particular if (A, B) is controllable and G is square and nonsingular, then the pole structure coincides with the eigenstructure (Rosenbrock, 1970) and we have

$$\begin{aligned} \alpha_i(s) &= \psi_i(s) \quad \text{for} \quad i = 1, ..., m \\ \alpha_i(s) &= 1 \qquad\;\; \text{for} \quad i = m+1, ..., n. \end{aligned}$$

So in that case Theorem 1 answers our question.

The non-regular static state feedback has been shown useful in decoupling the linear system, that is in achieving non-interaction between specified sets of inputs and outputs of the sytem, see for instance Dion and Commault (1988). Then the controllability of the system (1) can be altered by the feedback (2) so that the eigenstructure and the pole structure of the closed-loop system can be different from each other, and Theorem 1 cannot be directly applied. However, partial picture can be obtained using well-known results on the effect of the feedback (2) on the controllability structure of the system (1).

Theorem 2. There exists an integer $q \leq n$ and a pair $A' \in \mathbb{R}^{q \times q}, B' \in \mathbb{R}^{q \times p}$ such that (A', B') is controllable and

$$(sI_n - A - BF)^{-1}BG = (sI_q - A')^{-1}B' \quad (7)$$

There are several ways, see Kailath (1980), how to construct this pair (A', B'), which is in fact the controllable part of $(A + BF,\ BG)$. Noting that the pole structure of the closed-loop system is given by the invariant factors of A', the application of Theorem 1 to the pair (A', B') immediately leads to the following corollary of Theorem 2.

Corollary 3. The following holds:

$$\sum_{j=i}^{n} \deg \psi_j(s) \leq \sum_{j=i}^{n} c'_j \qquad \text{for } i = 1, ..., n \quad (8)$$

$$\sum_{j=1}^{n} \deg \psi_j(s) = \sum_{j=1}^{n} c'_j = q \qquad (9)$$

where, by convention, $\deg \psi_i(s) = 0$ for $i = p + 1, .., n$ and where

$$c'_i = \text{card}\,\{j \geq 1 | c'^*_j \geq i\} \qquad \text{for } i = 1, ..., n$$

is the ordered list of integers defined by

$$\begin{aligned} c'^*_1 &= \text{rank}\, B' \\ c'^*_i &= \text{rank}\,[B' \;\; A'B' \; ... \; A'^{i-1}B'] \\ &\quad -\text{rank}\,[B' \;\; A'B' \; ... \; A'^{i-2}B'] \quad \text{for } i = 2, ..., n. \end{aligned}$$

This result will be useful together with that of Heymann, who described the possible values of the integers c'_i, the controllability indices of (A', B') and of $(A + BF,\ BG)$.

Theorem 4 (Heymann, 1976). Given a system (A, B) and a list of n nonnegative integers arranged in non-increasing order $c'_1 \geq c'_2 \geq ... \geq c'_n \geq 0$. There exist matrices $F \in \mathbb{R}^{m \times n}$ and $G \in \mathbb{R}^{m \times p}$ such that the controllability indices of $(A + BF,\ BG)$ are $c'_1, c'_2, ..., c'_n$ if and only if

$$\sum_{j | c'_j \leq i} c'_j \;\leq\; \sum_{j |\ c_j \leq i} c_j \qquad \text{for } i = 1, ..., n \quad (10)$$

where $c_1, c_2, ..., c_n$ are controllability indices of (A, B).

We are now able to state the main result of this section.

Corollary 5 Given the system (1) and a set of monic polynomials $\psi_1(s) \geq \psi_2(s) \geq ... \geq \psi_p(s)$, there exists a static state feedback (2) such that the set of ψ_is is the pole structure of the closed-loop system (3) if and only if there exists a list of integers c_i's which satisfy (8), (9) and (10).

The problem now comes down t0 the search for such a list of integers $c_1', c_2', ..., c_n'$. Let us note that, as n is finite, there is a finite number of lists $\{c_1', c_2', ..., c_n'\}$ of non-negative integers arranged in non-decreasing order

$$c_1' \geq c_2' \geq ... \geq c_n' \geq 0$$

and such that

$$\sum_{j=1}^{n} c_j' = q \leq n.$$

Thus Corollary 5 is implicit but one can ultimately compute, via a finite number of operations, whether or not there exists a list of integers satisfying the three requirements (8), (9) and (10).

We shall propose in the following a nicer way to solve the problem. It will require some developments of the structural properties of linear systems.

3. LATTICE PROPERTIES OF THE FINITE LISTS OF INTEGERS

Given an integer n, define the set $\mathcal{L}$ of all the lists of n integers, say $l = \{l_1, l_2, ..., l_n\}$, arranged in non-increasing order, $l_1 \geq l_2 \geq ... \geq l_n \geq 0$, and whose sum is bounded by n, $l_1 + l_2 + ... + l_n \leq n$.

The relation, say $\prec$, which appears in Theorem 1, called dominance (or majorization) and defined by

$$\forall \alpha, \beta \in \mathcal{L} : \alpha \prec \beta \iff \sum_{j=i}^{n} \alpha_j \leq \sum_{j=i}^{n} \beta_j \quad \text{for } i = 1, ..., n$$

is an order on $\mathcal{L}$, since it is obviously reflexive, transitive and antisymmetric.

To every list $l \in \mathcal{L}$, one can associate a conjugate denoted l^* which lies in $\mathcal{L}$, and is defined by

$$l_i^* := \text{card } \{j | l_j \geq i\} \qquad \text{for } i = 1, ..., n.$$

Conjugation is a biunivoque relation: indeed it is observed that $(l^*)^* = l$, see for instance Brylawsky (1973).

The conjugated order, denoted $*$ and defined by

$$\forall \alpha, \beta \in \mathcal{L} : \alpha * \beta \iff \sum_{j=1}^{i} \alpha_j \leq \sum_{j=1}^{i} \beta_j \quad \text{for } i = 1, ..., n$$

is also an order relation. We call it conjugated dominance and one can observe that β dominates α if and only if β^* conjugate-dominates α^*, see Loiseau (1988).

Theorem 6. $(\mathcal{L}, \prec)$ is a lattice.

Proof: With each list $l \in \mathcal{L}$, one can associate the function $\S_l : \{1, ..., n\} \to \{0, ..., n\}$, defined by

$$\S_l(i) = \sum_{j=i}^{n} l_j \qquad \text{for } i = 1, ..., n$$

as well as the function $\S_l^* : \{1, ..., n\} \to \{0, ..., n\}$, defined by

$$\S_l^*(i) = \sum_{j=1}^{i} l_j^* \qquad \text{for } i = 1, ..., n.$$

Then the following two operations are defined for every $x, y \in \mathcal{L}$, namely $\sup(x, y)$, the minimal element in $\mathcal{L}$ dominating both x and y, and $\inf(x, y)$, the maximal element in $\mathcal{L}$ dominated by both x and y. In fact, $z = \sup(x, y)$ is defined by

$$\S_z(i) = \max\{\S_x(i), \S_y(i)\} \qquad \text{for } i = 1, ..., n$$

and $w = \inf(x, y)$ is defined by

$$\S_w^*(i) = \min\{\S_x^*(i), \S_y^*(i)\} \qquad \text{for } i = 1, ..., n.$$

Then Brylawsky (1973) shows that $(\mathcal{L}, \prec)$ is a lattice with respect to the operations of sup and inf.

In view of Theorem 4 we are now interested in a relation denoted by H and defined by

$$\forall \alpha, \beta \in \mathcal{L} : \alpha H \beta \iff \sum_{j | \alpha_j \leq i} \alpha_j \leq \sum_{j | \beta_j \leq i} \beta_j .$$

for $i = 1, ..., n$. We shall refer to H as the Heymann ordering.

If only one of the two operations, sup and inf, is defined on $\mathcal{L}$, then $(\mathcal{L}, \prec)$ is called sup-semilattice or inf-semilattice, respectively. The following is a key result for our development.

Theorem 7. For each $c \in \mathcal{L}$, define the set $H_c = \{l \in \mathcal{L} \mid lHc\}$. Then $(H_c, \prec)$ is a sub-inf-semilattice of $(\mathcal{L}, \prec)$.

Theorem 7 will result from the two following lemmas.

Lemma 8. $\forall \alpha,\ \beta \in \mathcal{L},\ \alpha H \beta \implies \alpha \prec \beta$.

Proof: Define for $i = 1, ..., n$:

$$k_i = \max\left\{ k \in \mathbb{N}, 1 \leq k \leq n | \sum_{j=i}^{n} \alpha_j \leq \sum_{j=k}^{n} \beta_j \right\}$$

We observe that k_i is well defined if $\alpha H \beta$. In addition,

$$\sum_{j=i}^{n} \alpha_j \quad \leq \quad \sum_{j=k_i}^{n} \beta_j$$

by the definition of k_i and

$$\alpha_i \geq \beta_{k_i}, \qquad \text{if } \alpha H \beta.$$

Thus one can see that

$$\sum_{j=i+1}^{n} \alpha_j = \sum_{j=i}^{n} \alpha_j - \alpha_i \leq \sum_{j=k_i}^{n} \beta_j - \beta_{k_i} = \sum_{j=k_i+1}^{n} \beta_j$$

which shows that either $k_{i+1} \geq k_i + 1$ or $k_i = n$. It is then clear that $k_i \geq i$ for $i = 1, ..., n$ and so

$$\sum_{j=i}^{n} \alpha_i \leq \sum_{j=k_i}^{n} \beta_j \leq \sum_{j=i}^{n} \beta_j \qquad \text{for } i = 1, ..., n$$

which establishes the result.

Lemma 9 Let $x, y \in \mathcal{L}$ and $w = \inf(x, y)$. Then

$$\sum_{j|w_j \leq i} w_j \leq \max\left\{ \sum_{j|x_j \leq i} x_j, \sum_{j|y_j \leq i} y_j \right\}$$

for $i = 1, ..., n$.

Proof: As in the proof of Theorem 6 , w is defined by

$$\sum_{j=1}^{i} w_j^* = \min\left\{ \sum_{j=1}^{i} x_j^*, \sum_{j=1}^{i} y_j^* \right\}. \tag{11}$$

First observe that

$$\begin{aligned} w_i^* &= \sum_{j=1}^{i} w_j^* - \sum_{j=1}^{i-1} w_j^* \\ &= \min\left\{ \sum_{j=1}^{i} x_j^* - \min\left\{ \sum_{j=1}^{i-1} x_j^*, \sum_{j=1}^{i-1} y_j^* \right\}, \right. \\ &\qquad \left. \sum_{j=1}^{i} y_j^* - \min\left\{ \sum_{j=1}^{i-1} x_j^*, \sum_{j=1}^{i-1} y_j^* \right\} \right\} \\ &\geq \min\left\{ \sum_{j=1}^{i} x_j^* - \sum_{j=1}^{i-1} x_j^*, \sum_{j=1}^{i} y_j^* - \sum_{j=1}^{i-1} y_j^* \right\} \end{aligned}$$

and thus

$$w_i^* \geq \min\{x_i^*, y_i^*\}. \tag{12}$$

Now $w_i^* =$ card $\{j | w_j \geq i\}$ by definition. It appears that $w_i^* - w_{i+1}^* =$ card $\{j | w_j = i\}$. Thus we obtain

$$\sum_{j|w_j \leq i} w_j = \sum_{j=1}^{i} j(w_j^* - w_{j+1}^*) \qquad \text{for } i = 1, ..., n$$

where, by convention, we take $w_{n+1}^* = 0$, which can be rewriten as follows:

$$\sum_{j|z_j \leq i} w_j = \sum_{j=1}^{i} w_j^* - i w_{i+1}^* \quad \text{for } i = 1, ..., n. \tag{13}$$

Similarly we can establish that

$$\sum_{j|x_j \leq i} x_j = \sum_{j=1}^{i} x_j^* - i x_{i+1}^*, \quad \text{for } i = 1, ..., n \tag{14}$$

$$\sum_{j|y_j \leq i} y_j = \sum_{j=1}^{i} y_j^* - i y_{i+1}^* \quad \text{for } i = 1, ..., n. \tag{15}$$

The result clearly follows from (11) - (15).

Proof of Theorem 7: From Lemma 9, it is clear that $w = \inf(x, y)$ lies in H_c if x and y are themselves in H_c.

4. MAIN RESULTS

We shall now come back to the pole structure assignment problem. Using the notations introduced in Section 3, it should be clear that the list c' in Corollary 5 must satisfy

$$d \prec c' \tag{8$'$}$$

where

$$\begin{aligned} d_i &= \deg \psi_i(s) \quad &\text{for } i = 1, ..., p \\ d_i &= 0 \quad &\text{for } i = p+1, ..., n \end{aligned}$$

and

$$\sum_{j=1}^{n} c'_j = \sum_{j=1}^{n} d_j \tag{9$'$}$$

and

$$c' H\ c. \tag{10$'$}$$

Lemma 8 gives us a necessary condition for the existence of c', that is :

$$d \prec c. \tag{16}$$

Proposition 10. Under the condition (16), the set

$$\mathcal{H}_{c,d} = \{l \in \mathcal{L} \mid d \prec l \text{ and } lHc\}$$

is nonempty, for instance it contains c. In addition it is closed under the operation inf as in Theorem 7. Thus it posseses a minimal element

$$h = \inf \{l \in \mathcal{L} \mid d \prec l \text{ and } lHc\}.$$

There exists a feedback as in Corollary 5 if and only if

$$\sum_{j=1}^{n} h_j = \sum_{j=1}^{n} d_j.$$

The problem which now subsists is that of the effective computation of h. We shall for this purpose use the method which consists in chracterizing the inf of a semi-lattice as the minimal fix-point of some operator. see Wonham (1979) and Wonham and Ramadge (1987).

Given the list $c = \{c_1, c_2, ..., c_n\}$ of the controllability indices of the system (1) and the list $d = \{d_1, d_2, ... d_n\}$ of the degrees of the invariant polynomials the closed-loop system is to have, define the operator $\mathcal{H} : \mathcal{L} \longrightarrow \mathcal{L}$ as

$$l = \{l_1, l_2, ... l_n\} \longmapsto \mathcal{H}l = \sup\{l, l^1, l^2, ..., l^n\} \quad (17)$$

where sup is defined in Theorem 6 and, for $i = 1, ..., n$,

$$l_j^i = \begin{cases} \max\{l_i, c_{k_i}\} & \text{for } j = 1, ..., i \\ l_j & \text{for } j = i+1, ..., n \end{cases}$$

with k_i being defined as in the proof of Lemma 8 by

$$k_i = \max\left\{k \in \mathbb{N},\ \ 1 \leq k \leq n | \sum_{j=i}^{n} l_j \leq \sum_{j=k}^{n} c_j\right\}$$

for $i = 1, ..., n$. Without any limitation, we can assume at this level that $\sum_{j=1}^{n} c_j = n$, which ensures that k_i, and thus $\mathcal{H}$, is well-defined for every $l \in \mathcal{L}$.

Proposition 11.

(i) $\mathcal{H}$ is isotone with respect to dominance: $\forall x, y \in \mathcal{L},\ \ x \prec y \implies \mathcal{H}x \prec \mathcal{H}y$.

(ii) The lists which are dominated by c in the Heymann ordering are the fix-points of $\mathcal{H}$: $\forall l \in \mathcal{L},\ \ l\ H\ c \iff l = \mathcal{H}l$.

Proof:

(i) Assume that $x \prec y$, that is $\sum_{j=i}^{n} x_j \leq \sum_{j=i}^{n} y_j$ for $i = 1. ..., n$. It is then clear that

$$k_{x_i} \geq k_{y_i}$$

and that

$$c_{k_{x_i}} \geq c_{k_{y_i}}$$

where

$$k_{x_i} = \max\left\{k \in \mathbb{N},\ 1 \leq k \leq n | \sum_{j=i}^{n} x_j \leq \sum_{j=k}^{n} c_j\right\}$$

and

$$k_{y_i} = \max\left\{k \in \mathbb{N},\ 1 \leq k \leq n | \sum_{j=i}^{n} y_j \leq \sum_{j=k}^{n} c_j\right\},$$

for $i = 1, ..., n$. It implies that either

$$x^i = \{x_i, ..., x_i, x_{i+1}, ..., x_n\} \prec x \prec y,$$

if $x_i \geq c_{k_{x_i}}$, or

$$x^i = \{c_{k_{x_i}}, ..., c_{k_{x_i}}, x_{i+1}, ..., x_n\} \prec y^i,$$

if $x_i \leq c_{k_{x_i}}$. Anyway, in all the cases we have that

$$\sup\{x, x^1, ..., x^n\} \prec \sup\{y, y^1, ..., y^n\}$$

which establishes (i).

(ii) As it was pointed out in the proof of Lemma 8, $l\ H\ c \Longrightarrow l_i \geq c_{k_i}$ and $k_i \geq i$, for $i = 1, ..., n$, with k_i being defined as in (17). It appears that

$$l^i = \{l_i, ..., l_i, l_{i+1}, ..., l_n\} \prec l \qquad \text{for } i = 1, ..., n$$

and thus

$$\mathcal{H}l = \sup\{l, l^1, ..., l^n\} = l.$$

Conversely if $l = \mathcal{H}l$ then $l^i \prec l$ for $i = 1, ..., n$, that is $\sum_{j=k}^{n} l_j^i \leq \sum_{j=k}^{n} l_j$ for $k = 1, ..., n$. Taking $k = i$, it follows that $\max(l_i, c_{k_i}) + \sum_{j=i+1}^{n} l_j \leq l_i + \sum_{j=i+1}^{n} l_j$, which shows that

$$l_i \geq c_{k+i} \qquad \text{for } i = 1, ..., n.$$

By the definition of k_i, it means that

$$\sum_{j=i}^{n} l_j \leq \sum_{j | c_j \leq l_i} c_j \qquad \text{for } i = 1, ..., n$$

and, choosing $i = \min\{j | l_j \leq k\}$ for $k = 1, ..., n$, we finally obtain the result:

$$\sum_{j | l_j \leq k} l_j \leq \sum_{j | c_j \leq l_i} c_j \leq \sum_{j | c_j \leq k} c_j, \qquad \text{for } k = 1, ..., n.$$

Theorem 12. Define the sequence h^i on $\mathcal{L}$ as

$$\begin{aligned} h^0 &= 0 \\ h^i &= \sup\{d,\ \mathcal{H}h^{i-1}\} \quad \text{for } i \geq 1. \end{aligned}$$

The sequence h^i is non-decreasing and converges after a finite, less than n, number of steps to a limit denoted h^∞ which coincides with the minimal element defined in Proposition 10:

$$h^\infty = h = \inf \{l \in \mathcal{L} \mid d \prec l \quad \text{and} \quad l \; H \; c\}.$$

Proof: Clearly $h^0 = 0 \prec d = h^1$, and if $h^{i-1} \prec h^i$ then $\mathcal{H}h^{i-1} \prec \mathcal{H}h^i$, from assertion (i) of Proposition 11, and $h^i = \sup\{d, \mathcal{H}h^{i-1}\} \prec \sup\{d, \mathcal{H}h^i\} = h^{i+1}$. This recurrence establishes that h^i is non-decreasing. Further it follows that h^i is convergent, being non-decreasing and defined over a finite set.

The limit, say h^∞, satisfies

$$h^\infty = \sup\{d, \mathcal{H}h^\infty\} = \mathcal{H}h^\infty$$

since $d \prec h_\infty$ and $h^\infty \prec \mathcal{H}h^\infty$ by the definition of $\mathcal{H}$. Thus h^∞ is a fix-point of $\mathcal{H}$, and hence it is dominated by c with respect to H. It can be finally verified that $\forall f \in \mathcal{L}, \;\; d \prec f$ and $f H c \Longrightarrow$
$h^1 = d \prec f$, and
$h^i \prec f \Longrightarrow \mathcal{H}h^i \prec f$ and $d \prec f \Longrightarrow h^{i+1} \prec f$.
Thus $h^\infty \prec f$, and h^∞ coincides with the infimal element of Proposition 10.

Acknowledgement

P. Zagalak and V. Kučera gratefully acknowledge the support of the Grant Agency of the Academy of Sciences of the Czech Republic under Grant No. 275 102. J. J. Loiseau acknowledges the support of the European Union under Esprit basic research project No. 8924 (SESDIP).

References

Brylawski, T. (1973). The lattice of integer partitions. Discrete Mathematics, 6, 201–219.

Dickinson, B. W. (1974). On the fundamental theorem of linear state variable feedback. IEEE Trans. Automatic Control, AC–19, 577–579.

Dion, J. M. and C. Commault (1988). The minimal delay decoupling problem: Feedback implementation with stability. SIAM J. Control Optimiz., 26, 66–82.

Flamm, D. S. (1980). A new proof of Rosenbrock's theorem on pole assignment. IEEE Trans. Automatic Control, AC–25, 1128–1133.

Heymann, M. (1976). Controllability indices and feedback simulation. SIAM J. Control Optimiz., 14, 769–789.

Kailath, T. (1980). Linear Systems. Prentice Hall, Englewood Cliffs, N.J.

Kučera , V. (1981). Assigning the invariant factors by feedback. Kybernetika, 17, 118–127.

Kučera , V. and P. Zagalak (1978). Fundamental theorem of state feedback for singular systems. Automatica, 24, 653–658.

Loiseau, J. J. and P. Zagalak (1993). Eigenstructure assignment in linear and uncontrollable systems. Preprints MTNS'93 , Regensburg.

Loiseau, J. J. (1988). Sur la modification de la structure à l'infini par retour d'état statique. SIAM J. Control Optimiz., 26, 251–273.

Loiseau, J. J. and P. Zagalak (1994). On feedback simulation. Proc. 2nd IEEE Mediterranean Symposium, Chania, Crete, 275–282.

Rosenbrock, H. H. (1970). State–space and Multivariable Theory. Wiley, New York.

Özçaldiran, K. (1990). Fundamental theorem of linear state feedback for singular systems, Proc. 29th C. D. C., Honolulu.

Zaballa, I. (1984). Interlacing inequalities and control theory. Linear Algebra Appl., 87, 113–146.

Zagalak, P. and J. J. Loiseau (1992). Invariant factors assignment in linear systems. Preprints SINS'92, Ft. Worth, 197–204.

Wonham, W. M. (1979). Linear Multivariable Control: A Geometric Approach. Springer Verlag, New York.

Wonham, W. M. and P. J. Ramadge (1987). On the supremal controllable sublanguage of a given language. SIAM J. Control Optimiz., 25, 637–659.

ON THE FIXED POLES FOR DISTURBANCE REJECTION

M. MALABRE[1]* and J. C. MARTÍNEZ GARCÍA[2]**

* *Laboratoire d'Automatique de Nantes CNRS URA 823, Ecole Centrale de Nantes-Université de Nantes, 1, rue de la Noë, 44072 Cedex 03 Nantes FRANCE, Phone: (+33) 40 37 16 51, Fax: (33) 40 37 25 22, Email: malabre@lan.ec-nantes.fr*

** *Sección de Control Automático, Departamento de Ingeniería Eléctrica, CINVESTAV-IPN, Av. Instituto Politécnico Nacional No. 2508, 07300 México D.F. MÉXICO, Phone: (+ 52 5) 7 47 70 00 Ext. 3279, Fax: (+ 52 5) 7 47 70 89, Email: martinez@ctrl.cinvestav.mx*

Abstract. It is shown here that the (minimal) set of fixed poles of the closed-loop system, obtained after applying any disturbance rejecting static state feedback control law on a linear time invariant system with disturbances, is equal to the set of all the finite invariant zeros of the undisturbed system which are not finite invariant zeros of the combined system, obtained when considering the disturbance as a control input.

Key Words. Linear Systems; Geometric Control; Disturbance Rejection; Fixed Poles.

Résumé : Nous caractérisons pour la première fois les pôles fixes du Rejet de Perturbations par retour d'état, c'est à dire les pôles qui sont présents sur le système bouclé, quelle que soit la loi de commande solution choisie. Nous en donnons une caractérisation géométrique et nous montrons que ces pôles fixes correspondent aux zéros invariants du système sans perturbation qui ne sont pas des zéros invariants du système dans lequel la perturbation serait manipulée comme une entrée de commande.

1. Motivations

When a feedback control law is chosen in order to achieve a particular control objective, it usually exists some *fixed poles* in the closed-loop system, i.e. closed-loop poles that do not depend on the choice of the control law but precisely on the fact that this particular control problem is being solved. In the context of the (regular) *Decoupling Problem* (Morse and Wonham (1970) Descusse (1980), Grizzle and Isidori (1989)), it is well known that some poles of the decoupled system (after regular static state feedback) are fixed. These *Fixed Poles for Decoupling* have first been studied in geometric terms in Morse and Wonham (1970) and it has been established later (see Koussiouris (1980) for the first frequency domain approach characterization and Icart *et al.* (1990) for a geometric one) thay they are the finite invariant zeros of the open-loop system which are not finite invariant zeros of its row-subsystems.

The aim of this paper is to show (using a geometric approach) that as far as the *Disturbance Rejection Problem* (via static state feedback laws) is concerned, there also exist fixed poles and that these can also be characterized in terms of finite invariant zeros of the open-loop system.

2. Problem Formulation

Let us consider a linear time-invariant disturbed system (A, B, C, E) described by:

$$\left\{ \begin{array}{lcl} \dot{x}(t) & = & Ax(t) + Bu(t) + Ed(t), \\ y(t) & = & Cx(t), \end{array} \right. \qquad (1)$$

where: $x(\cdot) \in \mathcal{X} \approx \mathbb{R}^n$ denotes the state; $u(\cdot) \in \mathcal{U} \approx \mathbb{R}^m$ denotes the control input; $y(\cdot) \in \mathcal{Y} \approx \mathbb{R}^p$ denotes the output; $d(\cdot) \in \mathcal{D} \approx \mathbb{R}^q$ denotes the disturbance; and $A : \mathcal{X} \longrightarrow \mathcal{X}$, $B : \mathcal{U} \longrightarrow \mathcal{X}$, $C : \mathcal{X} \longrightarrow \mathcal{Y}$, and $E : \mathcal{D} \longrightarrow \mathcal{X}$ are linear maps represented in particular bases by real constant matrices.

We can associate to the disturbed system (A, B, C, E) the *undisturbed* system (A, B, C) (i.e. $d(\cdot) = 0$) and the *combined system* $(A, [\ B \mid E\ ], C)$ (i.e. $d(\cdot)$ is considered as a control input).

[1] Supported by ESPRIT Basic Research Project No. 8924 (SESDIP).

[2] This research was done when the author was at the Laboratoire d'Automatique de Nantes, supported by the National Council of Science and Technology of Mexico and by the Advanced Studies and Research Center of the IPN (CINVESTAV-IPN) of Mexico.

We can then formulate the *Disturbance Rejection Fixed Poles Problem* (**DRFPP**) as follows:

Definition 1 Given a disturbed system (A, B, C, E) as described by (1), find (if possible) a static state feedback control law:

$$u(t) \;=\; Fx(t),\; t \geq 0,$$

with $F : \mathcal{X} \longrightarrow \mathcal{U}$, such that the closed-loop system output $y(\cdot)$ does not depend on the disturbance $d(\cdot)$, i.e.:

$$C\left(sI_n - (A+BF)\right)^{-1} E \;\equiv\; 0, \qquad (2)$$

and to which extent can we choose F, with regards to the poles of $(A+BF)$: are there Fixed Poles for all F and how can we choose F in order to have the least possible number of Fixed Poles ?

3. Recalls

The solvability condition for the existence of solutions to the *Disturbance Rejection Problem* (2) is very well known (see for instance Wonham (1985) and Basile and Marro (1992)). Indeed, there exists a static state feedback control law $u(t) = Fx(t)$ such that $C\left(sI_n - (A+BF)\right)^{-1} E \equiv 0$ if and only if:

$$\mathrm{Im}E \;\subset\; \mathcal{V}^*, \qquad (3)$$

where $\mathcal{V}^* \subset \mathcal{X}$ is the supremal (A, ImB)-invariant subspace included in the kernel of C. $\mathcal{V}^*$ is obtained as the limit of the *Invariant Subspace Algorithm* (**ISA**):

$$\begin{array}{lll} \mathcal{V}^0 & := & \mathcal{X}; \\ \mathcal{V}^{\mu+1} & := & \mathrm{Ker}C \cap A^{-1}(\mathrm{Im}B + \mathcal{V}^{\mu}), \end{array} \qquad (4)$$

for all $\mu \geq 0$.

If (3) holds, it is always possible to reject the disturbance just picking a map F in $\mathbf{F}(\mathcal{V}^*) := \{F \mid F : \mathcal{X} \longrightarrow \mathcal{U}$ and such that $(A + BF)\mathcal{V}^* \subset \mathcal{V}^*\}$.

4. Geometric characterization of the Disturbance Rejection Fixed Poles

In what follows we shall call *Disturbance Rejection Fixed Poles* the minimal set of the closed-loop system fixed poles obtained via static state feedback, i.e. the intersection of all spectra $\sigma(A+BF_i)$, where F_i is any solution of (2). We shall prove that:

Theorem 1 Given a disturbed system (A, B, C, E), the Disturbance Rejection Fixed Poles imposed by any static state feedback control law $u(t) = Fx(t)$ are the finite invariant zeros of the undisturbed system (A, B, C) which are not finite invariant zeros of the combined system (A, $[\; B \mid E \;]$, C) (under the assumption that (A, B) is controllable).

Let us denote $\mathcal{R}^*$ the supremal (A, ImB)-controllability subspace included in $\mathcal{V}^*$. We can characterize the finite invariant zeros of (A, B, C) in geometric terms. Indeed, they are the eigenvalues of $\overline{A+BF_{\mathcal{V}^*}} \mid (\mathcal{V}^*/\mathcal{R}^*)$, the map induced by $A+BF_{\mathcal{V}^*}$, where $F_{\mathcal{V}^*} \in \mathbf{F}(\mathcal{V}^*)$, in the quotient space $\mathcal{V}^*/\mathcal{R}^*$ (see for instance Wonham (1985)). We can characterize the finite invariant zeros of (A, $[\; B \mid E \;]$, C) in a similar way.

Let us denote $\mathcal{V}_c^*$ the supremal (A, Im$[B \mid E]$)-invariant subspace included in Ker$C$. The finite invariant zeros of (A, $[\; B \mid E \;]$, C) are the eigenvalues of the map induced by $A + [B \mid E]F_{\mathcal{V}_c^*}$, with $F_{\mathcal{V}_c^*} \in \mathbf{F}(\mathcal{V}_c^*)$ (i.e. $A\mathcal{V}_c^* \subset \mathcal{V}_c^*$ + ImB + ImE), in the quotient space $\mathcal{V}_c^*/\mathcal{R}_c^*$, where $\mathcal{R}_c^*$ denotes the supremal (A, Im$[B \mid E]$)-controllability subspace included in $\mathcal{V}_c^*$.

If the Disturbance Rejection Problem has a solution (i.e. Im$E \subset \mathcal{V}^*$), it is well known that: $\mathcal{V}_c^* = \mathcal{V}^*$ (see Basile and Marro (1992), Malabre and Martínez García (1993)). Moreover, it will be shown that $\mathcal{R}_c^*$ is an (A, ImB)-invariant subspace. Then, the finite invariant zeros of (A, $[\; B \mid E \;]$, C) are the eigenvalues of $\overline{A+BF_{\mathcal{V}^*}} \mid (\mathcal{V}^*/\mathcal{R}_c^*)$, the map induced by $A+BF_{\mathcal{V}^*}$ in the quotient space $\mathcal{V}^*/\mathcal{R}_c^*$. Thus, if the Disturbance Rejection Problem has a solution, the finite invariant zeros of the undisturbed system (A, B, C) which are not finite invariant zeros of the combined system (A, $[\; B \mid E \;]$, C) are exactly the eigenvalues of:

$$\overline{A+BF_{\mathcal{V}^*}} \mid (\mathcal{R}_c^*/\mathcal{R}^*), \qquad (5)$$

the map induced by $A+BF_{\mathcal{V}^*}$ in the quotient space $\mathcal{R}_c^*/\mathcal{R}^*$. We shall establish in the sequel the connexion between these finite invariant zeros and the fixed poles. For that, let us first recall some basic results about eigenvalues placements when dealing with (A, ImB)-invariant subspaces.

Let $\mathcal{V} \subset \mathcal{X}$ be an (A, ImB)-invariant subspace. Suppose $F_{\mathcal{V}} \in \mathbf{F}(\mathcal{V})$ (i.e. $(A+BF_{\mathcal{V}})\; \mathcal{V} \subset \mathcal{V}$). Let us define $A_{\mathcal{V}} := A+BF_{\mathcal{V}}$. Then, $\sigma(A_{\mathcal{V}})$, the spectrum of $A_{\mathcal{V}}$, is given by (see Schumacher (1980)):

$$\sigma(A_{\mathcal{V}}) = \sigma_1(A_{\mathcal{V}}) \uplus \sigma_2(A_{\mathcal{V}}) \uplus \sigma_3(A_{\mathcal{V}}) \uplus \sigma_4(A_{\mathcal{V}}),$$

where $\uplus$ denotes union with all common elements repeated, with:

$$\sigma_1(A_{\mathcal{V}}) := \sigma\left(\overline{A_{\mathcal{V}}} \mid (\mathcal{X}/(\mathcal{V}+ < A \mid \mathrm{Im}B >))\right);$$

$$\sigma_2(A_{\mathcal{V}}) := \sigma\left(\overline{A_{\mathcal{V}}} \mid (\mathcal{V}+ < A \mid \mathrm{Im}B > /\mathcal{V})\right);$$

$$\sigma_3(A_\mathcal{V}) := \sigma\left(\overline{A_\mathcal{V}} \mid (\mathcal{V}/\mathcal{R}^*_\mathcal{V})\right);$$

$$\sigma_4(A_\mathcal{V}) := \sigma\left(\overline{A_\mathcal{V}} \mid \mathcal{R}^*_\mathcal{V}/0\right),$$

where $\sigma_1(A_\mathcal{V})$ and $\sigma_3(A_\mathcal{V})$ do not depend on the choice of $F_\mathcal{V}$ (so they are fixed) ($< A \mid \mathrm{Im}B >$ denotes the reachable subspace of $(A, \mathrm{Im}B)$). The map $F_\mathcal{V}$ can be chosen in order to place arbitrarily the elements of $\sigma_2(A_\mathcal{V})$ and $\sigma_3(A_\mathcal{V})$ in the complex plane (as symmetric sets on it). If (A, B) is controllable, $\sigma_1(A_\mathcal{V})$ does not exist and $\sigma_3(A_\mathcal{V})$ is the only fixed part of $\sigma(A_\mathcal{V})$. Since the number of elements of $\sigma_3(A_\mathcal{V})$ is equal to the dimension (noted $\dim(\cdot)$) of $\mathcal{V}$ minus the dimension of $\mathcal{R}^*_\mathcal{V}$, $\sigma_3(A_\mathcal{V})$ will be "small" if $\mathcal{V}$ is also "small" and $\mathcal{R}^*_\mathcal{V}$ is "big". Thus, in order to have the minimal quantity of fixed poles, we must select some "minimal" $(A, \mathrm{Im}B)$-invariant subspace which supremal $(A, \mathrm{Im}B)$-controllability subspace is as big as possible.

In what follows we shall consider, without loss of generality, that (A, B) is controllable.

From the previous discussion, we see that the Disturbance Rejection Problem Fixed Poles will actually be the finite invariant zeros of the undisturbed system (A, B, C) which are not finite invariant zeros of the combined system $(A, [\ B \mid E\ ], C)$ if and only if $\mathcal{R}^*_c$ can be chosen as a solution of the Disturbance Rejection Problem while giving rise to the least possible number of fixed poles. This is stated in the next result:

Theorem 2 Let a disturbed system (A, B, C, E) described by (1) be given. Suppose (A, B) controllable. If the Disturbance Rejection Problem has a solution (i.e. $\mathrm{Im}E \subset \mathcal{V}^*$), $\mathcal{R}^*_c$ can always be chosen as a support for the solutions and gives rise to the least possible number of fixed poles (i.e. taking F such that $(A+BF)\mathcal{R}^*_c \subset \mathcal{R}^*_c$).

Proof of Theorem 2. Let us first prove that if $\mathrm{Im}E \subset \mathcal{V}^*$, the subspace $\mathcal{R}^*_c$ is an $(A, \mathrm{Im}B)$-invariant subspace (included in $\mathrm{Ker}C$) containing $\mathrm{Im}E$.

If $\mathrm{Im}E \subset \mathcal{V}^*$, we have:

$$\begin{aligned}
\mathrm{Im}E &= \mathrm{Im}E \cap (\mathrm{Im}B + \mathrm{Im}E)\\
&\subset \mathcal{V}^* \cap (\mathrm{Im}B + \mathrm{Im}E)\\
&= \mathcal{V}^*_c \cap (\mathrm{Im}B + \mathrm{Im}E)\\
&= \mathcal{R}^*_c \cap (\mathrm{Im}B + \mathrm{Im}E),
\end{aligned}$$

i.e. $\mathrm{Im}E \subset \mathcal{R}^*_c$. Since $\mathcal{R}^*_c$ is by definition an $(A, \mathrm{Im}[B \mid E\ ])$-invariant subspace (included in $\mathrm{Ker}C$); the previous inclusion leads to:

$$A\mathcal{R}^*_c \subset \mathcal{R}^*_c + \mathrm{Im}B + \mathrm{Im}E \subset \mathcal{R}^*_c + \mathrm{Im}B,$$

i.e. $\mathcal{R}^*_c$ is an $(A, \mathrm{Im}B)$-invariant subspace included in $\mathrm{Ker}C$.

We can then choose $\mathcal{R}^*_c$ as a support for the solutions. It now remains to show that choosing $\mathcal{R}^*_c$ gives rise to the least possible number of fixed poles.

Let $\mathcal{V}$ be any geometric support for the solutions of the Disturbance Rejection Problem, and let us define, as in Basile and Marro (1992), $\overline{\mathcal{V}} := \mathcal{V} + \mathcal{R}^*$. One has:

i) $\overline{\mathcal{V}}$ is a geometric solution for the Disturbance Rejection Problem;

ii) $\overline{\mathcal{V}}$ is a self-bounded $(A, \mathrm{Im}B)$-invariant subspace (see Basile and Marro (1992)), i.e. $\mathrm{Im}B \cap \overline{\mathcal{V}} = \mathrm{Im}B \cap \mathcal{R}^*$;

iii) choosing $\overline{F} \in \mathbf{F}(\overline{\mathcal{V}})$ induces less fixed poles than choosing $F \in \mathbf{F}(\mathcal{V})$.

i) and *ii)* are obvious.

iii) As we have seen, the fixed poles associated with a given $(A, \mathrm{Im}B)$-invariant subspace, say $\mathcal{W}$, are the eigenvalues of $\overline{A+BF_\mathcal{W}} \mid (\mathcal{W}/\mathcal{R}^*_\mathcal{W})$, the map induced by $A+BF_\mathcal{W}$ in the quotient space $\mathcal{W}/\mathcal{R}^*_\mathcal{W}$ ($\mathcal{R}^*_\mathcal{W}$ denoting the supremal $(A, \mathrm{Im}B)$-controllability subspace contained in $\mathcal{W}$ and $F_\mathcal{W}$ being any map taken in $\mathbf{F}(\mathcal{W})$).

Then, choosing $\overline{\mathcal{V}} := \mathcal{V} + \mathcal{R}^*$ gives rise to fixed eigenvalues, namely those of:

$$\overline{A+BF_{\overline{\mathcal{V}}}} \mid \left(\overline{\mathcal{V}}/\mathcal{R}^*_{\overline{\mathcal{V}}}\right).$$

Note that $\mathcal{R}^*_{\overline{\mathcal{V}}}$ (the supremal $(A, \mathrm{Im}B)$-controllability subspace contained in $\overline{\mathcal{V}}$) is equal to $\mathcal{R}^*$. Then:

$$\begin{aligned}
\overline{\mathcal{V}}/\mathcal{R}^*_{\overline{\mathcal{V}}} &:= (\mathcal{V}+\mathcal{R}^*)/\mathcal{R}^*_{\overline{\mathcal{V}}}\\
&= (\mathcal{V}+\mathcal{R}^*)/\mathcal{R}^*\\
&\approx \mathcal{V}/(\mathcal{V}\cap\mathcal{R}^*).
\end{aligned}$$

Since $\mathcal{R}^*$ contains $\mathcal{R}^*_\mathcal{V}$, the supremal $(A, \mathrm{Im}B)$-controllability subspace contained in $\mathcal{V}$, we obviously have $\mathcal{V}\cap\mathcal{R}^*_\mathcal{V} \subset \mathcal{V}\cap\mathcal{R}^*$, and consequently:

$$\begin{aligned}
\dim(\mathcal{V}/\mathcal{R}^*_\mathcal{V}) &= \dim(\mathcal{V}/(\mathcal{V}\cap\mathcal{R}^*_\mathcal{V}))\\
&\geq \dim(\mathcal{V}/(\mathcal{V}\cap\mathcal{R}^*))\\
&= \dim\left(\overline{\mathcal{V}}/\mathcal{R}^*_{\overline{\mathcal{V}}}\right).
\end{aligned}$$

It means that choosing $\overline{F} \in \mathbf{F}(\overline{\mathcal{V}})$ induces less fixed poles than choosing $F \in \mathbf{F}(\mathcal{V})$.

We can now conclude the proof of Theorem 2.

Starting from any solution $\mathcal{V}$, using $\overline{\mathcal{V}} := \mathcal{V} + \mathcal{R}^*$ in place of $\mathcal{V}$ leads to solutions with less fixed poles. Such a $\overline{\mathcal{V}}$ is a self-bounded $(A, \text{Im}B)$-invariant subspace containing $\text{Im}E$: the infimal one is $\mathcal{R}_c^*$ (see Basile and Marro (1992)). Thus, using $\mathcal{R}_c^*$ as a support for the solutions of the Disturbance Rejection Problem (taking $F \in \mathbf{F}(\mathcal{R}_c^*)$) gives rise to the least possible number of fixed poles (i.e. $\dim(\mathcal{R}_c^*/\mathcal{R}^*)$). ∎

Proof of Theorem 1. From Theorem 2 we know that choosing a map $F_{\mathcal{R}_c^*} \in \mathbf{F}(\mathcal{R}_c^*)$, to solve the Disturbance Rejection Problem, gives rise to the least possible number of fixed closed-loop poles. These fixed poles are the eigenvalues of $\overline{A + BF_{\mathcal{R}_c^*}} \mid (\mathcal{R}_c^*/\mathcal{R}^*)$, the map induced by $A + BF_{\mathcal{R}_c^*}$ in the quotient space $\mathcal{R}_c^*/\mathcal{R}^*$. Now, from Morse (1973), we can choose a map $F_{\mathcal{V}^*} \in \mathbf{F}(\mathcal{V}^*)$ such that $F_{\mathcal{V}^*} \mid \mathcal{R}_c^* = F_{\mathcal{R}_c^*} \mid \mathcal{R}_c^*$. Then, we can rewrite (5), i.e. the geometric characterization of the finite invariant zeros of the undisturbed system (A, B, C) which are not finite invariant zeros of the combined system $(A, \left[\ B \mid E\ \right], C)$, as follows:

$$\overline{A + BF_{\mathcal{V}^*}} \mid (\mathcal{R}_c^*/\mathcal{R}^*)$$
$$= \overline{A + BF_{\mathcal{R}_c^*}} \mid (\mathcal{R}_c^*/\mathcal{R}^*),$$

which concludes the proof. ∎

5. Some Complements

- Suppose that the Disturbance Rejection Problem has a solution. It is clear that there exists stable solutions if and only if there are no unstable Disturbance Rejection Fixed Poles. As a natural consequence of Theorem 1 this should be the case if and only if (A, B, C) and $(A, \left[\ B \mid E\ \right], C)$ have the same unstable finite invariant zeros (see an alternative proof of this structural result in Malabre and Martínez García (1993)).
- If the disturbance $d(\cdot)$ is available to be incorporated in the control law $u(\cdot) = Fx(\cdot) + Gd(\cdot)$, all the previous results still hold without any modification (just replace $\text{Im}E \subset \mathcal{V}^*$ or $\text{Im}E \subset \mathcal{R}_c^*$ by $\text{Im}E \subset \mathcal{V}^* + \text{Im}B$ or $\text{Im}E \subset \mathcal{R}_c^* + \text{Im}B$).
- If the Disturbance Rejection Problem has a solution and $\mathcal{R}_c^* = \mathcal{R}^*$, there are no Disturbance Rejection Fixed Poles (since the quotient space $\mathcal{R}_c^*/\mathcal{R}^* = 0$). It is obvious that this is true if and only if $\text{Im}E \subset \mathcal{R}^*$, which is the famous geometric solvability condition for the Disturbance Rejection Problem with Pole Placement to have a solution (see Willems and Commault (1981)).
- Thanks to the geometric results of Icart *et al.* (1990), a natural extension of the present work will be the geometric characterization of the Simultaneous Disturbance Rejection and (regular) Decoupling Problem Fixed Poles (via static state feedback control laws) (see Malabre and Martínez García (1995)).

6. REFERENCES

Basile, G. and G. Marro (1992). *Controlled Invariants and Conditioned Invariants in Linear System Theory.* Prentice Hall. Englewood Cliffs, New Jersey, USA. A Simon & Schuster Company.

Descusse, J. (1980). *Contribution à la théorie du découplage des quadruplets linéaires invariants (A, B, C, D) (Thèse de Doctorat ès Sciences).* Université de Nantes. Nantes, France.

Grizzle, J. W. and A. Isidori (1989). Block non interacting control with stability via static state feedback. *Mathematics of Control, Signals and Systems* **2**, 315–341.

Icart, S., J. F. Lafay and M. Malabre (1990). A unified study of the fixed modes of systems decoupled via regular static state feedback. *Joint Conference on New Trends in System Theory, Gênes-Italie* pp. 425–432. Birkhauser Boston.

Koussiouris, T. (1980). A frequency domain approach to the block decoupling problem ii: Pole assignment while block decoupling a minimal system by state feedback and a constant non singular input transformation and the observability of the block decoupled system. *Int. J. Control* **32**, 443–464.

Malabre, M. and J. C. Martínez García (1993). The modified disturbance rejection problem with stability: A structural approach. *Second European Control Conference (ECC'93) (Groningen, The Netherlands)* pp. 1119–1124.

Malabre, M. and J. C. Martínez García (1995). Fixed poles of disturbance rejection and decoupling. *Submitted.*

Morse, A. S. (1973). Structure and design of linear model following systems. *IEEE Trans. on Auto. Control* **18**(8), 346–354.

Morse, A. S. and W. M. Wonham (1970). Decoupling and pole assignment in linear multivariable systems: a geometric approach. *SIAM J. of Control and Optimization* **8**(1), 1–18.

Schumacher, J. M. (1980). A complement on pole placement. *IEEE Trans. on Auto. Control* **25**(2), 281–282.

Willems, J. C. and C. Commault (1981). Disturbance decoupling by measurement feedback with stability and pole placement. *SIAM J. of Control and Optimization* **14**(4), 490–504.

Wonham, W. M. (1985). *Linear Multivariable Control: A Geometric Approach.* Springer Verlag. New York, USA. Third Edition, Applications of Mathematics No. 10.

DIAGONAL DECOUPLING PROBLEM WITH INTERNAL STABILITY : A SOLUTION BY DYNAMIC OUTPUT FEEDBACK AND CONSTANT PRECOMPENSATOR

Ulviye BAŞER* and Vasfi ELDEM**

** ITU, Faculty of Sciences and Letters, Department of Mathematics, Maslak, Istanbul, Turkey*
*** TUBITAK, Marmara Research Center, Division of Mathematics, Gebze, Kocaeli, Turkey*

Abstract: In this paper, the decoupling problems by dynamic output feedback and constant precompensator are considered under the constraint of weak or strong internal stability. This problem has been first considered by Hammer and Khargonekar (1984) for systems with square transfer matrices. Here, the problem is taken up in a more general setting and it is only assumed that transfer matrix has full row rank. In this respect, the problem being considered is the same as the one considered in Eldem (1994-a) except the constraint of internal stability. It is shown that the recently introduced notions of **diagonal stability and causality degree dominance** and **the joint essential stability functions** (Eldem (1994-b)) play a central role in the solution of the problem. Moreover, using the latter, one can obtain a characterization of all solutions if the constant precompensator part of the control is known.

Key Words: linear, time-invariant systems, diagonal decoupling, internal stability, essential orders.

Résumé: *Dans cet article les problémes de découplage par "sortie feedback" dynamique et précompensateur constant sont considérés sous la restriction de stabilité interne forte ou faible. Ce problème avait été traité pour la première fois par Hammer et Khargonekar (1948) pour des systémes aux matrices de transfert carrées. Dans cette communication le même problème est reconsideéré dans un contexte plus général. On suppose que la matrice de transfert a un rang de ligne complet, ce qui fait le probléme considéré pareil à celui considéré par (Eldem (1994-a)) à l'exception de restriction de stabilité interne. On montre que les notions récemment introduites comme la stabilité diagonale et la dominance de degré de causalité et ainsi que les fonctions joiuntes de stabilité essentielle (Eldem 1994-b) jouent un rôle capital dans la solution du problème. En plus, en utilisant ces notions nouvelles, on peut obtenir une charactérisation de tous les solutions si la partie précompensateur constant du contrôle est connue.*

1. INTRODUCTION:

Noninteracting control problems has been on the agenda of sytem theoreticians for almost three decades. The persistency of the interest in noninteracting control problems is, on one hand, due to their practical implications. The diagonal decoupling problem, for instance, enables the designer to control each output of the process independently. Thus , one can use the well established single input-single output design tecniques to achieve further design goals. From a theoretical point of view, on the other hand, these problems are quite complex and challenging in general. For instance, the constant state feedback version of the diagonal decoupling problem, which is also known as The Morgan's Problem (Morgan (1964)) and which has provided a strong stimulus for many reseaercherss over the years, is still unsolved in its full generality. Moreover, the investigation of noninteracting control problems has shed light to finer algebraic structures of linear time invariant systems. This yielded valuable information for the solution of problems such as eigen structure assignment, infinite zero modification, feedback realization etc. During the first two decades, the major research effort has been spended on solving the decoupling problems under various feedback schemes. The aspect of internal stability has been neglected on purpose, because the problems are quite complicated even without the constraint of internal stability. During the last decade, however,the results with internal stability has begun to appear in relatively simpler versions of the problem such as the ones which employ dynamic precompensator and or dynamic state feedback (See, for instance, Kucera (1983), Dion and Commault (1988) and Commault, Dion and Torres (1991)). Hautus and Heymann (1983) considered the block decoupling problem with internal stability. The most recent results along this line are due to Ozguler (1994) and Eldem (1994-b). The basic difference of the last two works from the previous results is that the controlled (decoupled) and the measured outputs are distinguished.

In the present work, the diagonal decoupling problem via dynamic output feedback and constant precompensator is taken up under the constraint of weak or strong internal stability. In this respect, the problem considered here is the same as the one considered in Eldem (1994a) except the constraint of internal stability.

The set of linear time invariant system under investigation in this study can be expressed as

$$y = Zu \tag{1}$$

where y and u take values from p and m dimensional vector spaces over R, respectively and Z is a full row rank, strictly proper transfer matrix. It is desired to find a control law of the following form

$$u = -Z_c y + Gv \tag{2}$$

where Z_c is a proper matrix, v is a p dimensional external input and G is a full column rank constant matrix, so that the closed-loop transfer matrix between y and v, i.e.

$$\Lambda := Z(I+Z_cZ)^{-1}G \tag{3}$$

is diagonal and nonsingular. It is further required that Z_c internally stabilizes the system. Informally, a system is said to be internally stable if all its modes, including the unreachable and unobservable ones, are stable. Various conditions for internal stability have been proposed in the literature. In our set up, it is suitable to use the conditions given by Hammer (1983). These can be briefly stated as follows: Z_c is an internally stabilizing compensator for Z iff $(I+Z_cZ)^{-1}$, $Z(I+Z_cZ)^{-1}$, $(I+Z_cZ)^{-1}Z_c$ and $Z(I+Z_cZ)^{-1}Z_c$ are stable, or equivalently, iff I-QZ, Z-ZQZ, ZQ and Q are stable, where $Q:=Z_c(I+ZZ_c)^{-1}$. Here, we can assume without loss of generality that Z is stable. Then, in view of this assumption, it is clear that the stability of Q is necessary and sufficient for internal stability. So, the problem being considered can be defined as follows:

Definition 1: (Decoupling Problem with Internal Stability (DPIS)) Given a stable, strictly proper, full row rank Z, find a proper Z_c and a full column rank, constant G such that
(i) $Z(I+Z_cZ)^{-1}G=\Lambda$ for some diagonal and nonsingular Λ,
(ii) $Z_c(I+ZZ_c)^{-1}$ is stable and proper.

Note that since both Z and $Z_c(I+ZZ_c)^{-1}$ are stable and proper, it follows that Λ is also stable and proper.

2. NOTATION and TERMINOLOGY

Through out the paper R denotes the field of real numbers. R(s) and R[s] are used to denote the field of rational functions and the ring of the polynomials in the indeterminant s with coefficients from R, respectively. The ring of proper rational functions and stable and proper rational functions are represented by $R_p(s)$ and $R_{ps}(s)$, respectively. A pxm matrix with entries from R(s) or its subrings will be denoted via superscripts ($R^{pxm}(s)$, $R_p^{pxm}(s)$, $R_{ps}^{pxm}(s)$).

Recall that any N(s) in $R^{pxm}(s)$ has a Laurent series expansion

$$N(s) = \sum_{i=-k}^{\infty} N_i s^{-i} \qquad (4)$$

where N_i's are pxm matrices over R.
For any N(s) in $R^{pxm}(s)$, we will use the notation N_a^+ to denote the strictly proper and antistable plus the strictly polynomial parts of N. With this notation N(s) can be written as

$$N = N_a^+ + N_s \qquad (5)$$

where N_a^+ denotes the antistable and anticausal part of N and N_s denotes the stable and proper part. Let p and q be polynomials and $\partial(.)$ denote the usual polynomial degree. Then, with the following degree functions $R_p(s)$ and $R_{ps}(s)$ become Euclidian domains:

$$\partial_p(p/q) := \partial(q) - \partial(p)$$
$$\partial_{ps}(p/q) := \partial(q) - \partial(p) + \text{no. unstable zeros of } p. \qquad (6)$$

Let π be defined as $\pi := s+a$, where $a>0$. Then, the π-Hermit form X of a full row rank transfer matrix Z over $R_{ps}(s)$ is a lower left triangular matrix defined as (Morse (1975))

$$Z := [X^{-1} \; : \; 0]B\;, \quad X := \{x_{ij}\}\;, \qquad (7)$$

$$x_{ij} := \begin{cases} a_{ij}/(\pi^{k_{ij}}) & \text{for } j \le i \\ 0 & \text{for } j > i \end{cases}$$

where B is biproper and bistable, a_{ii}'s are antistable polynomials and $k_{ij} < k_{ii}$.

For any diagonal matrix Λ, let Λ_v denote the vector obtained from the diagonal elements of Λ. Similarly, diag $\{\alpha\}$ denotes the diagonal matrix obtained from the elements of the vector α. Now, let H and L be pxp matrices. We denote by Mat(H,L), the following matrix:

$$Mat(H,L) := \left[M_1^T \; M_2^T \; \dots \; M_p^T\right]^T \qquad (8)$$
$$M_i := H \, diag\{L_{ci}\}$$

We denote, on the other hand, by Vec(H) the following vector:

$$Vec(H) := [\,(H)_{c1}^T \; (H)_{c2}^T \; \dots \; (H)_{cp}^T\,]^T \qquad (9)$$

Now, suppose that H and L are matrices over $R_{ps}(s)$ and let B_1 and B_2 be the biproper, bistable matrices such that

$$B_1 Mat(H,L)\, B_2 := \begin{bmatrix} \Gamma \\ 0 \end{bmatrix} \qquad (10)$$

where Γ is the Smith form of Mat(H,L) over $R_{ps}(s)$.

Definition 2. The joint essential stability functions $\gamma_i(H,L)$ of H and L are the least degree rational functions in $R_{ps}(s)$ such that $\gamma_i(H,L)\Gamma_{ii}$ is stable and proper (see Eldem (1994-b)). **The joint essential stability orders** $\{n_{jes,i}\}$ are the degrees of $\gamma_i(H,L)$, i.e., $n_{jes,i} := \partial_{ps}(\gamma_i(H,L))$.

In order to simplify the the notation we shall use γ_i, instead of $\gamma_i(H,L)$ in the rest of the paper. It is well known that $\gamma_i R_{ps}(s)$ is a subring of $R_{ps}(s)$. Let us denote the quotient ring $R_{ps}(s)/\gamma_i R_{ps}(s)$ by $R_{ps}(\gamma_i)$.

Definition 3. The joint essential stability spaces $JE_{si}(H,L)$ of H and L are defined as

$$JE_{si} := R\text{-}linear\,span\{\Gamma_{ii}\gamma_{ij}\}_a^+ \qquad (11)$$
$$j = 1,2,\dots,n_{jes,i}, \quad i = 1,2,\dots,p.$$

where γ_{ij}, $j=1,2,\dots,n_{jes,i}$ is a basis for the R-linear space $R_{ps}(\gamma_i)$, (see Eldem (1994-b)).

3. MAIN RESULTS

In this section we examine the solution of DPIS. First, consider the π-Hermit form X of Z given in (7). The first set of solvability conditions for DPIS is given by the following lemma:

Lemma 1. DPIS is solvable iff there exist a constant matrix G and a diagonal Λ over $R_{ps}(s)$ such that

(i) ZG is nonsingular,
(ii) $X - X\Lambda(ZG)^{-1}$ is proper and stable.

Proof: (Necessity) Assume that DPIS is solvable.

Thus, there exist diagonal and nonsingular Λ over $R_{ps}(s)$ and a constant matrix G such that

$$Z(I+Z_cZ)^{-1}G=(I+ZZ_c)^{-1}ZG=\Lambda \tag{12}$$

Since , both Λ and $(I+ZZ_c)$ are nonsingular, it turns out that ZG is nonsingular, i.e., (i) holds. Also note that

$$\begin{aligned}\Lambda &= Z(I+Z_cZ)^{-1}G=Z(I-Z_c(I+ZZ_c)^{-1}Z)G \\ &= ZG-ZQZG \\ \rightarrow \quad & X-X\Lambda(ZG)^{-1}=XZQ\end{aligned} \tag{13}$$

Since both $Q:=Z_c(I+ZZ_c)^{-1}$ and XZ are stable and proper, it follows that $X-X\Lambda(ZG)^{-1}$ is in $R_{ps}(s)$, i.e., (ii) also holds.

For the sufficiency, assume that there exists a constant matrix G and a diagonal,nonsingular Λ over $R_{ps}(s)$ such that ZG is nonsingular and $X-X\Lambda(ZG)^{-1}$ is proper and stable. Define

$$\begin{aligned}T &:= X-X\Lambda(ZG)^{-1} \\ Z_c &:= (XZ)^{-1}(I-TZ(XZ)^{-1})^{-1}T\end{aligned} \tag{14}$$

where $(XZ)^{-1}$ is stable, right biproper right inverse of XZ, which exists because XZ is right invertible over $R_{ps}(s)$. Since Z is strictly proper, $TZ(XZ)^{-1}$ is strictly proper and then $I-TZ(XZ)^{-1}$ is biproper. Thus, Z_c is proper. A simpler expression for Z_c can be obtained as follows:

$$\begin{aligned}Z_c &= (XZ)^{-1}[I-(X-X\Lambda(ZG)^{-1})Z(XZ)^{-1}]^{-1} \\ &\quad \times[X-X\Lambda(ZG)^{-1}] \\ &= (XZ)^{-1}[X\Lambda(ZG)^{-1}Z(XZ)^{-1}]^{-1} \\ &\quad \times[X-X\Lambda(ZG)^{-1}] \\ &= (XZ)^{-1}[Z(XZ)^{-1}]^{-1}[ZG\Lambda^{-1}-I]\end{aligned} \tag{15}$$

Then, using this expression we get,

$$\begin{aligned}Z(I+Z_cZ)^{-1}G &= (I+ZZ_c)^{-1}ZG \\ &= [I+Z(XZ)^{-1}(Z(XZ)^{-1})^{-1} \\ &\quad \times(ZG\Lambda^{-1}-I)]^{-1}ZG \\ &= [I+ZG\Lambda^{-1}-I]^{-1}ZG=\Lambda\end{aligned} \tag{16}$$

Thus Z_c and G decouples the system. In order to show internal stability, note that

$$\begin{aligned}Z_c(I+ZZ_c)^{-1} &= (XZ)^{-1}[Z(XZ)^{-1}]^{-1} \\ &\quad \times[ZG\Lambda^{-1}-I]\ [ZG\Lambda^{-1}]^{-1}\end{aligned} \tag{17}$$

$$\begin{aligned}\rightarrow Z_c(I+ZZ_c)^{-1} &= (XZ)^{-1}[Z(XZ)^{-1}]^{-1} \\ &\quad \times[I-\Lambda(ZG)^{-1}] \\ &= (XZ)^{-1}[X-X\Lambda(ZG)^{-1}) \\ &= (XZ)^{-1}T\end{aligned} \tag{18}$$

which is stable and proper by the definition of $(XZ)^{-1}$ and T. Consequently, (ii) holds.

Lemma 2. $X-X\Lambda(ZG)^{-1}$ is stable and proper iff $Vec(X)-Mat(X,(ZG)^{-1})\Lambda_v$ is stable and proper
Proof: A straight forward calculation yields the result.

Now let Θ_1 and Θ_2 be biproper and bistable matrices such that

$$\Theta_1 Mat(X,(ZG)^{-1})\Theta_2 := \begin{bmatrix}\Gamma \\ 0\end{bmatrix} \tag{19}$$

where Γ is a pxp diagonal matrix. Also, partition Θ_1 as

$$\Theta_1 := \begin{bmatrix}\Theta_{11} \\ \Theta_{12}\end{bmatrix} \tag{20}$$

where Θ_{11} consists of the first p rows of Θ_1.

Let us denote the joint essential stability functions of X and $(ZG)^{-1}$ by γ_i as in Definition 2. Recall that γ_i's are the least degree rational functions in $R_{ps}(s)$ such that $\gamma_i\Gamma_{ii}$ is stable and proper. Also let JE_{si} denote (for i=1,2,...,p.) the joint essential stability spaces of X and $(ZG)^{-1}$ as given in Definition 3. Then, we can state the following results:

Theorem 1. DPIS is solvable iff there exists a constant matrix G such that
(i) ZG is nonsingular,
(ii) Θ_{12} Vec(X) is stable and proper,
(iii) $[(\Theta_{11}Vec(X))_a{}^{+}]_i$ is in $JE_{si}(X,(ZG)^{-1})$.

Theorem 2. Given a stable, strictly proper and full row rank Z, let G be constant matrix such that ZG is nonsingular. Then,
(i) The set of all diagonal matrices Λ over R_{ps} such that $X-X\Lambda(ZG)^{-1}$ is stable and proper can be expressed as

$$\Lambda := diag\{\Theta_2^{-1}(\Delta^* + \varphi)\} \quad \textbf{(21)}$$

where Δ^* and φ are p-dimensional vectors. Moreover, φ_i's are in $\gamma_i R_{ps}(s)$ and Δ^*_i's are unique (up to the choice of joint essential stability functons γ_i of X and $(ZG)^{-1}$) and $(\Delta^*_i/\gamma_i)_s=0$.
(ii) For any diagonal matrix Λ over R_{ps} such that $X - X\Lambda(ZG)^{-1}$ is stable and proper, the set of all solutions Q of the equation $ZG-ZQZG=\Lambda$ can be written as

$$BQ := \begin{bmatrix} Q_1 \\ Q_2 \end{bmatrix} \quad \textbf{(22)}$$

where B is as given by equation (7), Q_2 is an arbitrary matrix over $R_{ps}(s)$ and

$$Q_1 := X - X\Lambda(ZG)^{-1} \quad \textbf{(23)}$$

Proof: The proofs of the above theorems are similar to the proofs of the theorem 2 and 3 in Eldem (1994) and therefore they will be ommitted here.

Remark 1. The above theorems give a set of solvability conditions and a characterization for the set of all solutions for DPIS. However, a procedure for the construction of the constant precompensator part of the controller (construction of G) is still missing. The recent work of Eldem (1994-a) gives a characterization for the set of all constant precompensators which solve the same problem without the constraint of internal stability. Using this characterization together with the above results, one can possibly find a finite procedure for the construction of G. But, that needs further study.

When Z is square and nonsingular, the solution of the decoupling problem by dynamic output feedback is based on the concept of **diagonal causality degree dominance** (DCDD) which was introduced by Eldem and Özgüler (1989). The extention of this concept to the cases with internal stability is **weak or strong diagonal stability and causality degree dominance** (DSCDD), which is investigared in Eldem (1994). Briefly, a square, nonsingular and strictly proper Z over $R_{ps}(s)$ is called weak DSCDD if there exists a diagonal matrix Λ over $R_{ps}(s)$ such that $Z^{-1}\Lambda Z^{-1}-Z^{-1}$ is stable and proper. Z is called strong DSCDD if there exists a diagonal Λ over $R_{ps}(s)$ such that both $Z^{-1}\Lambda Z^{-1}-Z^{-1}$ and $\Lambda^{-1}Z\Lambda^{-1}-\Lambda^{-1}$ are stable and proper. In the latter case, there exists a stable, proper and internally stabilizing compensator Z_c which decouples the system. In other words, decoupling is possible under strong stabilization. Note that the solvability condition (ii) of DPIS given in the Lemma 1 can be regarded as the generalization of weak DSCDD to non-square systems. Now we present a generalization for strong DSCDD:

Lemma 3. Let Z be a strictly proper and full row rank transfer matrix over $R_{ps}(s)$. Then, DPIS is solvable with a proper and stable compensator Z_c iff there exist a constant matrix G and a diagonal Λ over R_{ps} such that
(i) ZG is nonsingular,
(ii) $X - X\Lambda(ZG)^{-1}$ is stable and proper,
(iii) $X - XZG\Lambda^{-1}$ is stable and proper.

Proof: (Necessity) If the problem is solvable, then there exist Z_c and Λ over $R_{ps}(s)$ and a constant matrix G such that

$$Z(I+Z_cZ)^{-1}G=\Lambda.$$

This implies that $-XZZ_c = X - XZG\Lambda^{-1}$, which is stable and proper as both XZ and Z_c are stable and proper. Consequently, (iii) holds. The conditions (i) and (ii) obviously hold as they are necessary and sufficient for the solvability of DPIS.

Conversely, since $X - XZG\Lambda^{-1}$ is stable and proper for some diagonal Λ over $R_{ps}(s)$ and a constant G, then Z_c can be defined as

$$Z_c := (XZ)^{-1}(XZG\Lambda^{-1} - X) \quad \textbf{(24)}$$

using this expression we get

$$\begin{aligned} Z(I+Z_cZ)^{-1}G &= Z[I-(XZ)^{-1} \\ &\quad \times (X-XZG\Lambda^{-1})Z]^{-1}G \\ &= [I-Z(XZ)^{-1} \\ &\quad \times (X-XZG\Lambda^{-1})]^{-1}ZG \\ &= (I-I+ZG\Lambda^{-1})ZG=\Lambda \end{aligned} \quad \textbf{(25)}$$

Thus Z_c and G decouples the system. Clearly, Z_c is stable and proper. In order to show internal stability, note that

$$\begin{aligned} Z_c(I+ZZ_c)^{-1} &= -(XZ)^{-1}(X-XZG\Lambda^{-1}) \\ &\quad \times [I-Z(XZ)-1(X-XZG\Lambda^{-1})]^{-1} \\ &= -(XZ)^{-1}(X-XZG\Lambda^{-1})\Lambda(ZG)^{-1} \end{aligned} \quad \textbf{(26)}$$

$$\rightarrow Z_c(I+ZZ_c)^{-1} = -(XZ)^{-1}(X\Lambda(ZG)^{-1}-X)$$

which is stable by the condition (ii). Consequently, DPIS is solvable with a Z_c over $R_{ps}(s)$.

4. CONCLUSIONS

In this work the diagonal decoupling problem by dynamic output feedback and constant precompensator has been investigated under the constraint of weak or strong internal stability. It is shown that the notion of **diagonal stability and causality degree dominance**, which has been introduced by Eldem (1994-b) for systems with square transfer matrices, can be generalized to non-square cases. Moreover, **the joint essential stability functions**, which play a crucial role for the solution of the same problem for the square case, is also instrumental for the problem considered here. The construction procedure for the constant precompensator part of the controller is, however, still open. As mentioned previously, the characterization of the constant precompensators which achieve decoupling (given in Eldem (1994a)) can be used to resolve this problem. But, that needs furher study.

REFERENCES

Commault, C., J.M. Dion and J.A. Torres (1991). Minimal structure in the block decoupling problem with stability. *Automatica*, Vol.27, no.2., pp. 331-338.

Dion, J. M. and C. Commault, (1988). Minimal delay decoupling problem: Feedback implementation with stability. *SIAM J. Contr. Optimiz.*, vol. 20, no. 1, pp.66-82.

Eldem, V. and A. B. Özgüler (1989). A solution to the diagonalization problem by constant precompensator and dynamic output feedback. *IEEE Trans. Auto. Contr.*, Vol. AC-34, no.10, pp. 1061-1068.

Eldem, V. (1994a). The solution of diagonal decoupling problem by dynamic output feedback and constant precompesator: The general case. *IEEE Trans. Auto. Contr.*, Vol. 39, no.3, pp. 503-511.

Eldem, V. (1994b). Decoupling through specified input-output channels with internal stability. submitted to *Automatica.*

Hammer (1983). Linear dynamic output feedback: invariants and stability. *IEEE Trans. Auto. Contr.*, vol. AC-28, no.4, pp. 489-496.

Hammer, J. and P.P. Khargonekar (1984). Decoupling of linear systems by dynamic output feedback", *Math. System Theory,* Vol. 17, No.2, pp. 135-157.

Hautus, M. L.J. and M. Heymann (1983). Linear feeback decoupling-Transfer function analysis. *IEEE Trans. Auto. Contr.*, vol. AC-28, no.8, pp. 823-832.

Kucera, V., (1983). Block decoupling by dynamic compensation with internal stability and properness, *Problems of Control and Information Theory,* 12 (6), pp. 379-389.

Morgan, B. S., (1964). The synthesis of linear multivariable systems by state feedback" *J. A. C. C.*, vol. 64., pp. 468-472.

Morse, A.S. (1975). System invariants under feedback and cascade control. Proc. Int. Symp.,24 Udine Springer-Verlag, New York.

Özgüler, A.B. (1994). Linear multichannel control : A system matrix approach. *Prentice Hall.*

ON THE PROBLEM OF DECOUPLING

J. RUIZ†[1] , P. ZAGALAK†, and V. ELDEM‡

† *Institute of Information Theory and Automation, Czech Academy of Sciences. P.O. Box 18, 182 08 Prague 8, Czech Republic.*

‡ *Tubitak, Marmara Research Centre, Division of Mathematics. P.O. Box 21, Gebze, Kocaeli 41470, Turkey.*

Abstract. The row-by-row decoupling of linear multivariable systems with stability is considered. An implicit necessary and sufficient condition for this problem to have a solution is provided. The approach used here takes advantage of the properties of the ring of proper and stable rational functions.

Résumé. L'object de cet article est d'étudier le problème du découplage ligne par ligne des systèmes linéaires avec stabilité. Une condition explicite nécessaire et suffisante pour qu'il existe une solution á cette problème est présentée. Les propriétes de l'anneau des fontions propres et stables jouent un rô le-clé ici.

Keywords. Linear systems; state feedback; decoupling; stability; algebraic approach.

1. INTRODUCTION

We shall consider a linear time-invariant system (C, A, B) governed by

$$\dot{x}(t) = Ax(t) + Bu(t)$$
$$y(t) = Cx(t)$$

where $A \in \mathbb{R}^{n\times n}$, $B \in \mathbb{R}^{n\times m}$ and $C \in \mathbb{R}^{p\times n}$. All the matrices are over $\mathbb{R}$, the field of real numbers and, moreover, we assume that $rank\, B = m$, $rank\, C = p$, and that the system is controllable. As we are interested in the problem of decoupling, which will be precisely defined below, we also suppose that the system (C, A, B) is right invertible, i.e. $rank\, T(s) = p$, where

$$T(s) = C(sI - A)^{-1}B \tag{1}$$

is the transfer function of (C, A, B).
Let $T(s)$ be factorized as

$$T(s) = CN_1(s)D(s)^{-1} \tag{2}$$

where the polynomial matrices $N_1(s)$ and $D(s)$ form a normal external description (NED) of (C, A, B); see Malabre *et al.* (1990). The matrices $N(s) := CN_1(s)$ and $D(s)$ then form, not necessarily coprime, a matrix fraction description (MFD) of (C, A, B). Recall that $N_1(s)$ and $D(s)$ are right coprime with $D(s)$ being column reduced. The column degrees, $c_i := deg_{ci} D(s)$, $i = 1, 2, \ldots, m$, are the controllability indices of (C, A, B).

*This work was supported with a Grant from TUBITAK-DOPROG, TURKEY.

[1]Supported by the National Council of Science and Technology of Mexico.

It has already been mentioned that we are interested in the problem of decoupling, or more precisely, in the Morgan problem with stability (MPS), which is defined as follows.

MPS. Given a system (C, A, B), does there exist a (nonregular) state feedback

$$u(t) = Fx(t) + Gv(t) \tag{3}$$

where $F \in \mathbb{R}^{m\times n}$ and $G \in \mathbb{R}^{m\times p}$ with $rank\, G = p$, such that the transfer function $T_{F,G}(s)$ of the closed-loop system $(C, A + BF, BG)$ is of the form

$$\begin{aligned} T_{F,G}(s) &= C(sI_n - A - BF)^{-1}BG \\ &= T(s)[I_m - FN_1(s)D(s)^{-1}]^{-1}G \\ &= diag\{w_1(s), \ldots, w_p(s)\} =: W(s) \end{aligned} \tag{4}$$

where $w_i(s)$, $i = 1, 2, \ldots, p$, are strictly proper and stable rational functions, and $(C, A + BF, BG)$ is internally stable, i.e. $(A + BF)$ is stable?

The Morgan problem has a long history going back into sixtieths when Morgan (1964) gave the first precise definition of the problem. We do not want to review all results concerning the problem; the reader is referred to the work of Morse and Wonham (1971) for a more detailed historical background. But still it is necessary to mention at least the work of Falb and Wolovich (1967) where necessary and sufficient conditions for solvability of decoupling in the case of square systems are given. Morse and Wonham (1971) used the geometric approach to tackle the block decoupling problem, Descusse *et al.* (1988) established explicit necessary and sufficient conditions in the case of the so-called shifted systems and Zagalak *et al.* (1993) gave implicit necessary and sufficient conditions for there to exist a solution to the Morgan problem as defined above, where $w_i(s) = s^{-r_i}$ with $r_i \geq n_{ie}$, $i = 1, 2, \ldots, p$, n_{ie}'s being the essential orders of (C,A,B). Thus the Morgan problem remains unsolved in its most general case.

The main purpose of this paper is to tackle the Morgan problem with stability. To this end, we shall generalize the approach of Zagalak *et al.* (1993) including the stability issue and we shall also show why it is so difficult to establish explicit necessary and sufficient conditions of solvability in this case.

2. BACKGROUND

As we are going to deal with the stability issue, we repeat first some basic concepts concerning the ring of proper and stable rational functions over $\mathbb{R}$, hereafter denoted $\mathbb{R}_{ps}(s)$.

Let $\mathbb{R}(s)$ denote the field of rational functions over $\mathbb{R}$, i.e. the field of fractions $g(s)/f(s)$ where $g(s)$ and $f(s)$, $f(s) \neq 0$, belong to $\mathbb{R}[s]$, the ring of polynomials over $\mathbb{R}$. It is well-known that any element $h(s) \in \mathbb{R}(s)$ has as many zeros as poles if those at infinity are counted in, too. Let $\mathbb{R}_\Lambda(s)$ denote a subset of $\mathbb{R}(s)$ whose elements have poles in a symmetric set $\Lambda \subset \mathbb{C} \cup \{\infty\}$, i.e. Λ is such that it contains at least one point of the real axis and if $\alpha \in \Lambda$ then also α^* (complex conjugate) lies in Λ. As a special case, we shall also consider $\Lambda = \{\infty\}$.

Further, we define the degree of $h(s) \in \mathbb{R}_\Lambda(s)$, hereafter denoted $deg\, h(s)$, as the number of zeros of $h(s)$ lying outside of Λ (including those lying at infinity). With these definitions, $\mathbb{R}_\Lambda(s)$ becomes an Euclidean ring and hence, $\mathbb{R}_\Lambda(s)$ has similar properties as the well-known ring of polynomials $\mathbb{R}[s]$. If Λ is defined as the open left half complex plane, we get our ring $\mathbb{R}_{ps}(s)$. The following proposition summarizes some important facts about $\mathbb{R}_{ps}(s)$.

Proposition 1. Let the ring of proper and stable rational functions $\mathbb{R}_{ps}(s)$ be given. Then the following facts hold.

(P1) Given $h(s) \in \mathbb{R}_{ps}(s)$, $h(s) \neq 0$, and $deg\, h(s) = 0$, then $h(s)$ is a unit of $\mathbb{R}_{ps}(s)$.

(P2) Any $g(s) \in \mathbb{R}_{ps}(s)$ can be written in the form

$$g(s) = h(s)\, \frac{u(s)}{\pi^k}$$

where $h(s)$ is a unit of $\mathbb{R}_{ps}(s)$, $u(s) \in \mathbb{R}[s]$ is an antistable polynomial (i.e., all its roots are unstable), $\pi := (s + a)$, $a \in \mathbb{C}$, $Re\, a > 0$, and $k = deg\, g(s)$ (We shall say that $g(s)$ is monic if $h(s) = 1$.)

(P3) Given two elements $g(s)$, $f(s) \in \mathbb{R}_{ps}(s)$, $g(s) = h_g(s)\,\frac{u_g(s)}{\pi^{k_g}}$, and $f(s) = h_f(s)\,\frac{u_f(s)}{\pi^{k_f}}$, $g(s)$ divides $f(s)$, i.e. $f(s) = g(s)x(s)$, $x(s) \in \mathbb{R}_{ps}(s)$, (denoted $g(s) \leq f(s)$), if and only if $u_f(s) = u_g(s)w(s)$ for some $w(s) \in \mathbb{R}[s]$ and $k_f \geq k_g + deg\, w(s)$.

(P4) Given $g(s)$, $f(s) \in \mathbb{R}_{ps}(s)$, $g(s) \neq 0$, then there exist $q(s)$, $r(s) \in \mathbb{R}_{ps}(s)$ such that $f(s) = g(s)q(s) + r(s)$ and $deg\, r(s) < deg\, g(s)$, or $r(s) = 0$.

(P5) Any $v(s) \in \mathbb{R}(s)$ can be expressed as $v(s) = f(s)/g(s)$, for some $g(s)$, $f(s) \in \mathbb{R}_{ps}(s)$.

Notice that the division algorithm given by (P4) of Proposition 1 is defined for any two elements of $\mathbb{R}_{ps}(s)$.

When considering matrices over $\mathbb{R}_{ps}(s)$, it should now be clear that the well-known properties of polynomial matrices will hold, with more or less obvious changes, in this case, too. We shall briefly describe those facts that deserve a special attention.

A matrix $B(s) \in \mathbb{R}_{ps}^{m\times m}(s)$ will be called $\mathbb{R}_{ps}$-unimodular, or biproper and bistable if $B(s)^{-1}$ exists and $B(s)^{-1} \in \mathbb{R}_{ps}^{m\times m}(s)$, i.e., $det\, B(s)$ is a unit of $\mathbb{R}_{ps}(s)$. Similarly, as in the case of polynomial matrices, $\mathbb{R}_{ps}$-unimodular matrices are generated by a sequence of row (column) elementary operations performed on I_m over the ring $\mathbb{R}_{ps}(s)$.

Given $T(s) \in \mathbb{R}_{ps}^{p\times m}(s)$, $rank\, T(s) = p$, there exists a $\mathbb{R}_{ps}$-unimodular matrix $B(s)$ such that

$$T(s)B(s) = [\Phi_s(s)^{-1},\ 0] \tag{5}$$

where

$$\Phi_s(s)^{-1} = \begin{pmatrix} \varphi_{11}(s) & & (0) & \vdots & \\ \vdots & \ddots & & \vdots & (0) \\ \varphi_{p1} & \ldots & \varphi_{pp}(s) & \vdots & \end{pmatrix}$$

$\varphi_{ij}(s) \in \mathbb{R}_{ps}(s)$

$\varphi_{ij}(s) = \frac{a_{ij}(s)}{\pi^{k_{ij}}}, \quad i > j$

$\varphi_{ii}(s) = \frac{\epsilon_{ii}(s)}{\pi^{k_{ii}}}, \quad k_{ij} < k_{ii}, \quad j = 1, 2, \ldots, i-1$

and $\epsilon_{ii}(s)$'s are antistable polynomials.

The form just introduced is called the (generalized or π-)Hermite form of $T(s)$ and evidently, this form depends on π. For fixed π, the π-Hermite form is unique.

Using $\mathbb{R}_{ps}(s)$-unimodular matrices on both sides of $T(s)$, we obtain its (generalized, or π-)Smith form, which again parallels that known from the theory of polynomial matrices. Thus, given $P(s) \in \mathbb{R}_{ps}^{m\times n}(s)$ of rank r, there are $\mathbb{R}_{ps}(s)$-unimodular matrices $U(s)$ and $V(s)$ such that

$$P(s) = U(s) \begin{pmatrix} S_\pi(s) & 0 \\ 0 & 0 \end{pmatrix} V(s) \tag{6}$$

where $S_\pi(s) = diag[f_1(s), f_2(s), \ldots, f_r(s)]$ with $f_1(s) \leq f_2(s) \leq \cdots \leq f_r(s)$ being unique and monic elements of $\mathbb{R}_{ps}(s)$,

$$f_i(s) = \frac{a_i(s)}{\pi^{k_i}}, \qquad i = 1, 2, \ldots r \tag{7}$$

and $a_i(s)$'s are antistable polynomials.

Having defined $\mathbb{R}_\Lambda(s)$-unimodular matrices, the following equivalence relations of matrices over $\mathbb{R}(s)$ can be established. Let $P(s)$, $Q(s) \in \mathbb{R}^{m\times n}(s)$ and let $U(s) \in \mathbb{R}_\Lambda^{m\times m}(s)$ and $V(s) \in \mathbb{R}_\Lambda^{n\times n}(s)$ be $\mathbb{R}_\Lambda(s)$-unimodular. Then we shall say that $P(s)$ is left (right) $\mathbb{R}_\Lambda(s)$-equivalent to $Q(s)$ if $P(s) = U(s)Q(s)$ ($P(s) = Q(s)V(s)$). $P(s)$ is said to be $\mathbb{R}_\Lambda(s)$-equivalent to $Q(s)$ if it is left and right $\mathbb{R}_\Lambda(s)$-equivalent to $Q(s)$. For example, let $T(s)$ be the transfer function of (C, A, B) defined by (1). Define $\Lambda = \mathbb{C}$, i.e. $\mathbb{R}_\Lambda(s)$ is the ring of proper rational functions ($\mathbb{R}_p(s)$ will stand for this ring). Then $T(s)$ is $\mathbb{R}_p(s)$-equivalent to the well-known Smith-McMillan form of $T(s)$ at infinity, i.e.

$$T(s) \sim [S_\infty,\ 0], \qquad S_\infty := diag\{s^{-n_i}\}_{i=1}^{p} \qquad (8)$$

where the integers $0 < n_1 \leq n_2 \leq \cdots \leq n_p$ are the infinite zeros orders of $T(s)$.

Another example is the Smith-McMillan form of $T(s)$. Put $\Lambda = \{\infty\}$. Then $\mathbb{R}_\Lambda(s)$-unimodular matrices are just $\mathbb{R}[s]$-unimodular and we get

$$\begin{array}{l} T(s) \sim [S_M(s),\ 0], \\ S_M(s) := diag\{\frac{\theta_i(s)}{\psi_i(s)}\}_{i=1}^{p} \end{array} \qquad (9)$$

where all $\theta_i(s)$, $\psi_i(s) \in \mathbb{R}[s]$ and such that $\theta_1(s) \leq \theta_2(s) \leq \cdots \leq \theta_p(s)$ and $\psi_1(s) \geq \psi_2(s) \geq \cdots \geq \psi_p(s)$. Thus, $S_M(s) = \Theta(s)\Psi(s)^{-1}$ with $\Theta(s) = diag\{\theta_i(s)\}$ and $\Psi(s) = diag\{\psi_i(s)\}_{i=1}^{p}$.

Based on this form, we shall define now the finite zero structure of $T(s)$ as the set of the elementary divisors of $\Theta(s)$. Similarly, the finite pole structure of $T(s)$ is defined as the set of the elementary divisors of $\Psi(s)$. Decomposing the polynomials $\theta_i(s)$ ($\psi_i(s)$) as $\theta_i(s) = \theta_i^+(s)\theta_i^-(s)$ ($\psi_i(s) = \psi_i^+(s)\psi_i^-(s)$) where $\theta_i^+(s)$ is stable and $\theta_i^-(s)$ is antistable, we can define the stable zero (pole) structure of $T(s)$ in the same way as above.

As far as the infinite zero structure of $T(s)$ is concerned, we shall define it as the finite zero structure of $S_\infty(\frac{1}{\omega})$ at $\omega = 0$, that is to say, the integer n_i's in (8) are describing this structure.

3. MAIN RESULTS

The relationship (4) is not very convenient for getting the matrices F and G. Actually, this is not an equation as the functions $w_i(s)$'s are not known. However, we shall treat (4) as an equation and specify $w_i(s)$'s later on.

Thus, write (4) in the form

$$T(s)C(s) = W(s) \qquad (10)$$

where $C(s) := B(s)G$ with $B(s) := [I_m - FN_1(s)D(s)^{-1}]^{-1}$.

$B(s)$ is clearly biproper ($\mathbb{R}_p(s)$-unimodular), which means that $C(s)$ has a left proper inverse. Such matrices will be called column biproper (column (or right) $\mathbb{R}_p(s)$-unimodular). The main idea is to solve the equation (10) over the ring $\mathbb{R}_{ps}(s)$. This will guarantee the internal stability of $T(s)C(s)$ since all operations performed in this ring keep unstable pole and zero structures unchanged. There is no loss of generality if we suppose that (C, A, B) is a stable system. If not, a preliminary stabilizing feedback $u(t) = Fx(t) + v(t)$ can be used since (C, A, B) is controllable, to ensure its stability. Then, $T(s)$ is a stable matrix and hence, $C(s)$ will be right $\mathbb{R}_{ps}(s)$-unimodular, or column biproper and bistable. The stability of $T(s)$ also implies that $D(s)$ has no unstable zero structure, i.e. $D(s)^{-1} \in \mathbb{R}_{ps}^{m\times m}(s)$.

Now, as rank $N(s) = p$, there exists a right inverse of $N(s)$, say $\hat{N}(s)$, which can be written in the form

$$\hat{N}(s) = \hat{N}_0(s) + \hat{N}_k(s)T_p(s) \qquad (11)$$

where $\hat{N}_0(s)$ is a particular inverse and $\hat{N}_k(s)$ forms a basis for $ker\, N(s)$. $T_p(s)$ stands for a rational parameter. Using (11), the equation (10) can be rewritten into the form

$$M(s)[\hat{N}_0(s) + \hat{N}_k(s)T_p(s)]W(s) = Z \qquad (12)$$

where

$$M(s) = XD(s) + YN_1(s) \qquad (13)$$

with $X \in \mathbb{R}^{m\times m}$ nonsingular, $Y \in \mathbb{R}^{m\times n}$, and $Z \in \mathbb{R}^{m\times p}$ monic such that

$$F = -X^{-1}Y \quad and \quad G = X^{-1}Z \qquad (14)$$

The relationships (12), (13) and (14) represent a basic formulation of the Morgan problem. Once we find $W(s)$, a monic Z, and $M(s)$, $M(s)^{-1} \in \mathbb{R}_{ps}^{m\times m}(s)$, that is column reduced with the same column degrees as $D(s)$, we are able to compute F and G using (13) and (14); see Kučera and Zagalak (1991) for more details. The problem is how to find $M(s)$ and $W(s)$. To this end we use the concept of extended system, Zagalak *et al.* (1993)

Let $U(s)$ be $\mathbb{R}[s]$-unimodular such that

$$N(s)U(s) = [Q(s),\ 0] \qquad (15)$$

with a nonsingular $Q(s) \in \mathbb{R}^{p\times p}[s]$ and define

$$K(s) := \begin{pmatrix} Q(s) & 0 \\ 0 & I_{m-p} \end{pmatrix} U(s)^{-1} \qquad (16)$$

It was shown in Zagalak *et al.* (1993) that $U(s)$ can be chosen such that

$$T_e(s) = K(s)D(s)^{-1}$$

is strictly proper and, as $D(s)^{-1} \in \mathbb{R}_{ps}^{m\times m}(s)$, it follows that $T_e(s) \in \mathbb{R}_{ps}^{m\times m}(s)$, too. The matrix $T_e(s)$ is called an extension of $T(s)$ and a realization of $T_e(s)$, with the same order as (C, A, B), is called an extended system.

Lemma 1. Let $\Phi_e(s)^{-1}$ be the (right) π-Hermite form, see (5), of $T_e(s)$, i.e.

$$T_e(s)B_e(s) = \Phi_e(s)^{-1} \qquad (17)$$

for some $\mathbb{R}_{ps}$-unimodular matrix $B_e(s)$. Then the matrix $\Phi_e(s)$ is called the extended π-interactor of $T(s)$ and is of the form

$$\Phi_e(s) = \begin{pmatrix} \Phi_1(s) & 0 \\ \Phi_2(s) & \Phi_3(s) \end{pmatrix}$$

where $\Phi_1(s)$ is the π-interactor of $T(s)$ and $\Phi_3(s) = diag\{\pi^{\sigma_i}\}_{i=1}^{m-p}$ where the list $\{\sigma_i\}_{i=1}^{m-p}$ is the Morse list I_2, which is defined, in a polynomial setting, as the set of minimal column indices of the system matrix

$$P(s) := \begin{pmatrix} sI_n - A & -B \\ C & 0 \end{pmatrix} \qquad \blacksquare$$

Going back to the equation (12), the characterization of $M(s)$ seems to be much easier. Indeed, write first $\hat{N}(s)$ in the form

$$\hat{N}(s) = K(s)^{-1}[\bar{N}_0(s) + \bar{N}_k T_p(s)],$$

where

$$\bar{N}_0 = \begin{pmatrix} I_p \\ 0 \end{pmatrix}, \quad and \quad \bar{N}_k = \begin{pmatrix} 0 \\ I_{m-p} \end{pmatrix}$$

(see Zagalak *et al.* (1993) for details), to get

$$M(s)K(s)^{-1}[\bar{N}_0 + \bar{N}_k T_p(s)]W(s) = Z \qquad (18)$$

Notice further that

$$M(s) = [X + YN_1(s)D(s)^{-1}]D(s),$$

where $B_D(s) := [X + YN_1(s)D(s)^{-1}]$ is $\mathbb{R}_{ps}$-unimodular, we can write $M(s)$ in the form

$$M(s) = V(s)\Phi_e(s)K(s) \qquad (19)$$

with $V(s)$ biproper and bistable such that $M(s)$ is polynomial. Substituting (19) into (18) we arrive at the equation

$$V(s)\begin{pmatrix} \Phi_1(s)W(s) \\ \Phi_2(s)W(s) + \Phi_3(s)T_p(s) \end{pmatrix} = Z \qquad (20)$$

with the constraint

$$V(s)\Phi_e(s)K(s) \in \mathbb{R}^{m\times m}[s] \qquad (21)$$

The equation (20) and constraint (21) are essentially the same as those given in Zagalak *et al.* (1993) for the row-by-row and integrator decoupling. However, as can be readily seen, to work with matrices over $\mathbb{R}_{ps}(s)$ is much less comfortable.

The relationship (20) is nothing else than the Bezout identity for matrices over $\mathbb{R}_{ps}(s)$. Thus, if we find a diagonal matrix $W(s) \in \mathbb{R}_{ps}^{p\times p}(s)$ and a rational matrix $T_p(s)$ such that

$$Y(s) := \begin{pmatrix} Y_1(s) \\ Y_2(s) \end{pmatrix}$$

$Y_1(s) = \Phi_1(s)W(s)$, $Y_2(s) = \Phi_2(s)W(s) + \Phi_3(s)T_p(s)$ is right $\mathbb{R}_{ps}$-unimodular, then $V(s)$ will be $\mathbb{R}_{ps}$-unimodular such that (20) holds. (Without any loss of generality, we can put $Z = \begin{pmatrix} I_p \\ 0 \end{pmatrix}$.)

To achieve that, consider first the interactor $\Phi_1(s)$. This decomposes as

$$\Phi_1(s) = \Gamma(s)\,diag\{\frac{1}{h_i(s)}\}_{i=1}^p \qquad (22)$$

where $h_i(s) \in \mathbb{R}_{ps}(s)$, $i = 1, 2, \ldots, p$, and $\Gamma(s) \in \mathbb{R}_{ps}^{p\times p}(s)$. Let $h_i(s)$'s be of the lowest possible degrees satisfying (22). Certainly such $h_i(s)$'s exist and are unique up to units of $\mathbb{R}_{ps}(s)$. We shall call the degrees of $h_i(s)$ (which are indeed unique) the s-essential orders of (C, A, B) and they will be denoted as $n_{ie,s}$.

The s-essential orders determine the lower bound for the degrees of the functions $w_i(s)$ in $W(s)$ so that $\Phi_1(s)W(s) \in \mathbb{R}_{ps}^{p\times p}(s)$. Good candidates for such $w_i(s)$'s are just the functions $h_i(s)$'s, or the functions $w_i(s)$ satisfying

$$h_i(s) \le w_i(s), \quad i = 1, 2, \ldots, p \qquad (23)$$

We shall say that $W(s) = diag\{w_i(s)\}_{i=1}^p$ is an admissible matrix if $w_i(s)$'s satisfy the conditions (23). The problem now is if there exists a rational parameter $T_p(s)$ such that $Y(s)$ is right $\mathbb{R}_{ps}$-unimodular.

Lemma 2. There exists $T_p(s) \in \mathbb{R}^{(m-p)\times p}(s)$ such that $Y(s)$ is right $\mathbb{R}_{ps}$-unimodular if and only if there exists an admissible $W(s)$ such that the number of unit invariant factors of $\Phi_1(s)W(s)$, say k in number, satisfies

$$m \ge 2p - k \qquad (24)$$

■

Once the existence of $T_p(s)$ such that (20) holds is guaranteed, we must consider the constraint (21), too. The first observation we can do is that the matrix $\Phi_e(s)K(s)$ can be written in the form

$$\Phi_e(s)K(s) = B_0(\pi)diag\{\pi^{c_i}\}_{i=1}^m \qquad (25)$$

where c_i, $i = 1, 2, \ldots, m$ are the controllability indices of (C, A, B) and $\pi = s + a$, $Re\, a > 0$, is the stable term we used above. This follows from the fact that $\Phi_e(s)K(s)$ is a stable matrix having all its zeros at $s = -a$. $B_0(\pi)$ is then $\mathbb{R}_{ps}$-unimodular with entries of the form $\alpha_{ij}(\pi)/\pi^{k_{ij}}$ where $\alpha_{ij}(\pi) \in \mathbb{R}[\pi]$ if $a \in \mathbb{R}$, $a > 0$. Next, as $Y(s)$ is right $\mathbb{R}_{ps}$-unimodular, it implies that the entries can also be written in the π-form and hence so the entries of $V(\pi)$ do. Thus, the product $V'(\pi) := V(\pi)B_0(\pi)$ is $\mathbb{R}_{ps}$-unimodular having all its entries of the π-form, i.e. of the form $\frac{\beta_{ij}}{\pi^{d_{ij}}}$. Let $d_i := max\{d_{ij},\ j = 1, 2, \ldots, m\}$. Then, the polynomiality condition (21) is of the form

$$d_i \le c_i, \; i = 1, 2, \ldots, m \qquad (26)$$

which directly leads to the following.

Theorem 1. Given a right invertible, stable, and controllable system (C, A, B), then (with the above notation) there exists a solution to MPS iff there exists an admissible $W(s)$ such that (24) and (26) hold. ■

Now some remarks are in order.

Remark 1. Theorem 1 parallels that stated in Zagalak *et al.* (1993). This fact is not so surprising since the constraint of stability does not play, as we could see, so important role in the Morgan problem. The most difficult question here is the problem of a realization of a given precompensator $C(s)$, which is right $\mathbb{R}_{ps}$-unimodular (or right $\mathbb{R}_p$-unimodular if we consider the Morgan problem without stability), by state feedback (3). To this end, consider the polynomiality condition (21) again. This condition is equivalent to the existence of some constant matrices $\bar{X} \in \mathbb{R}^{m\times m}$, rank $\bar{X} = m$, and $\bar{Y} \in \mathbb{R}^{m\times n}$ such that

$$[\bar{X} + \bar{Y}N_1(s)D_\phi(s)^{-1}] = V(s) \tag{27}$$

where $D_\phi(s) = \Phi_e(s)K(s)$.
Substituting now (27) into (20), we get

$$[\bar{X} + \bar{Y}N_1(s)D_\phi(s)^{-1}]Y(s) = \begin{pmatrix} I_p \\ 0 \end{pmatrix}$$

which decomposes as

$$\bar{X}Y_0 = \begin{pmatrix} I_p \\ 0 \end{pmatrix} \tag{28}$$

and

$$[\bar{Y}N_1(s)D_\phi(s)^{-1}]Y(s) = -\bar{X}\hat{Y}(s) \tag{29}$$

where Y_0 denotes the constant part of $Y(s)$, i.e. $Y_0 := lim_{s\to\infty}Y(s)$, and $\hat{Y}(s)$ is the strictly proper part of $Y(s)$, $\hat{Y} = Y(s) - Y_0$.
The equation (29) describes just the situation in which the compensator $Y(s)$ is realizable by a static feedback (3) if we put $F = -X^{-1}Y$ and $G = X^{-1}\begin{pmatrix} I_p \\ 0 \end{pmatrix}$; see Herrera (1992) for details.

Remark 2. Lemma 2 guarantees for an admissible $W(s)$, the existence of $T_p(s)$ such that (20) holds if (24) is satisfied. But the construction of $V(s)$ satisfying (21) is not clear. What one can do is, first construct a right $\mathbb{R}_{ps}(s)$-unimodular $Y(s)$ and $\mathbb{R}_{ps}(s)$-unimodular $V(s)$ satisfying (20). Then, equation (29) can be used to check if $V(s)$ satisfies (21). If not, then $Y(s)$ and $V(s)$ should be reconstructed again. Unfortunately, no explicit procedure for obtaining $Y(s)$ and $V(s)$ satisfying both (20) and (21) is available yet, and this calls for further study.

Remark 3. As far as decoupling with stability by dynamic state feedback is concerned, the approach developed in this paper, which works within the ring $\mathbb{R}_{ps}(s)$, leads naturally to the so called "strong stability" case where, not only the feedback system is required to be stable, but also the feedback compensator $F(s)$. As a result, the feedback term $[I_m - F(s)N_1(s)D(s)^{-1}]^{-1}$ is biproper and bistable, since the system is supposed to be stable.
Necessary and sufficient conditions for a dynamic state feedback

$$u(s) = F(s)x(s) + Gv(s), \qquad F(s) \in \mathbb{R}_{ps}^{m\times n}(s)$$

to decouple the system (C, A, B) with stability are given by (24). The procedure to find such a feedback is as follows:
i) Complete $Y(s) := \begin{pmatrix} Y_1(s) \\ Y_2(s) \end{pmatrix}$ to be right $\mathbb{R}_{ps}(s)$-unimodular, and calculate a biproper and bistable $V(s)$ such that $V(s)\begin{pmatrix} Y_1(s) \\ Y_2(s) \end{pmatrix} = \begin{pmatrix} I_p \\ 0 \end{pmatrix}$
ii) Construct $M(s) = V(s)\Phi_e(s)K(s)$, which is a rational and stable matrix.
iii) Premultiply $M(s)$ by a constant nonsingular matrix $X \in \mathbb{R}^{m\times m}$ such that $(XM(s))_{hc} = D_{hc}$ where $(J)_{hc}$ denotes the highest column-degree coefficient matrix of $J(s)$.
iv) Find a proper and stable solution to the equation $\bar{Y}(s)N_1(s) = \hat{M}(s)$ where $\hat{M}(s) = XM(s) - D(s)$. Observe that there always exists such a solution. Write $\hat{M}(s) = M^+(s) + M^-(s)$, where $M^+(s)$ denotes the polynomial part of $\hat{M}(s)$ and $M^-(s)$ is the strictly proper and stable part of $\hat{M}(s)$.
One solution is given by

$$\bar{Y}(s) = Y_0 + [M^-(s)\Lambda(s),\ 0]B_1(s)$$

where Y_0 is a constant solution to the equation $Y_0N_1(s) = M^+(s)$ (Y_0 always exists since $deg_{ci}\,N_1(s) \geq deg_{ci}\,M^+(s)$ and $N_1(s)$ forms a polynomial basis), $\Lambda(s) \in \mathbb{R}_{ps}^{m\times m}(s)$ is a diagonal matrix such that $N_1(s)\Lambda(s)$ is column biproper and bistable, and $B_1(s)$ is a unit of $\mathbb{R}_{ps}^{n\times n}(s)$ such that $B_1(s)N_1(s)\Lambda(s) = \begin{pmatrix} I_m \\ 0 \end{pmatrix}$
v) Put $F(s) = \bar{Y}(s)$ and $G = X\begin{pmatrix} I_p \\ 0 \end{pmatrix}$

On the other hand, if only the feedback system is required to be stable, but noy necessarily $F(s)$, which is of course proper, then $[I + F(s)N_1(s)D(s)^{-1}]^{-1}$ is biproper and stable, but not bistable. The solution for decoupling by dynamic state feedback in this case ("weak stability") is the well known condition

$$m \geq 2p - k$$

where k is the rank at infinity of $\Phi(s)W(s)$, and $\Phi(s)$ is the system interactor (see Dion *et al.* (1988)).
The previously indicated procedure can also be applied here, with the difference that $V(s)$ is a unit of $\mathbb{R}_p^{m\times m}(s)$ and $\bar{Y}(s)$ in step iv) is a proper matrix (not necessarily stable).

Remark 4. The distinction between weak and strong stability is also evident when dynamic output feedback is used for decoupling. In this case, the control law is given by

$$u(s) = F(s)y(s) + Gv(s)$$

where $F(s)$ is proper and G is constant.
As it can be expected, the problem here is more involved than the case of decoupling by dynamic state feedback. In order to show that, let us consider equation (29), which under dynamic output feedback becomes

$$\bar{Y}(s)N(s)D_\phi(s)^{-1}Y(s) = -\bar{X}\hat{Y}(s)$$

If we express $Y(s)$ as $Y(s) = \Phi_e(s)\begin{pmatrix} I \\ T_p \end{pmatrix}W(s)$, then since $N(s)K(s)^{-1} = [I,\ 0]$, the above equation can be rewritten as

$$\bar{Y}(s) = -\bar{X}\hat{Y}(s)W(s)^{-1}$$

Since we are seeking for a proper $\bar{Y}(s)$ (or equivalently, a proper $F(s)$) the choice for $W(s)$ and $T_p(s)$ should be such that $\bar{X}\hat{Y}(s)W(s)^{-1}$ is proper. This condition is similar to the results given in Eldem (1993). In fact, it is equivalent to finding G and $W(s)$ such that

$$\Phi_1(s) - \Phi_1(s)T(s)GW(s)^{-1} \tag{30}$$

is proper. The details for the construction of G and $W(s)$ are given in Eldem (1994).

If the same problem is considered under the constraint of weak stability, then it is necessary and sufficient that (30) holds and $F(s)[I - T(s)F(s)]^{-1}$ is proper and stable. If we let $F(s) := [Y_0 - B_e(s)Y(s)]W(s)^{-1}$, then the latter condition becomes

$$F(s)[I - T(s)F(s)]^{-1} = [Y_0 - B_e(s)Y(s)][T(s)Y_0]^{-1}$$

as $T(s)B_e(s)Y(s) = W(s)$. Thus, $T_p(s)$ and $W(s)$ are to be chosen such that $[Y_0 - B_e(s)Y(s)][T(s)Y_0]^{-1}$ is stable and proper. It can be shown that this is equivalent to

$$\Phi_1(s) - \Phi_1(s)W(s)[T(s)G]^{-1}$$

being proper and stable (see Başer *et al.* (1995)).

If it is further required that $F(s)$ is stable (strong stability) then the necessary and sufficient conditions for the solvability of this problem is the existence of G and $W(s)$ such that both

$$\Phi_1(s) - \Phi_1(s)T(s)GW(s)^{-1}$$

and

$$\Phi_1(s) - \Phi_1(s)W(s)[T(s)G]^{-1}$$

are stable and proper. For the details the reader is again referred to Başer *et al.* (1995).

Remark 5. If the regular static state feedback $u(t) = Fx(t) + Gv(t)$, $G \in \mathbb{R}^{m\times m}$, $\mathrm{rank}\, G = m$, is used for decoupling, we can derive explicit necessary and sufficient conditions for the existence of a solution. To this end, notice that the nonsingularity of G implies that $\Phi_1(s)$ is $m \times m$, which means that $\Phi_1(s)W(s)$ must be $\mathbb{R}_{ps}$-unimodular. From that, $\Phi_1(s)$ is diagonal, we conclude. Thus, the proper and stable rational functions $h_i(s)$ in (22) are just the inverses of the diagonal terms of $\Phi_1(s)$, i.e.

$$h_i(s) = \frac{a_i(s)}{\pi^{n_{ie,s}}}, \; i = 1, 2, \ldots, m \qquad (31)$$

where $a_i(s) \in \mathbb{R}[s]$, $i = 1, 2, \ldots, m$, are some antistable polynomials. As the matrix $N(s) \in \mathbb{R}^{m\times m}[s]$, with $rank\, N(s) = m$, the product $\Phi_1(s)N(s) \in \mathbb{R}^{m\times m}[s]$, which means that all unstable zeros of $N(s)$ are canceled out. Hence, the polynomials $a_i(s)$ describe the same unstable zero structure as the unstable parts of the invariant factors of $N(s)$. We have proved the following.

Lemma 3. A square, invertible and stable system (C, A, B) is decouplable with stability by a regular static state feedback iff the π-interactor $\Phi_1(s)$ is diagonal. ■

This condition is completely equivalent to that given in Martinez Garcia *et al.* (1995), where the solution to the decoupling problem with stability (regular case) is given in terms of the so-called infinite and unstable contents.

4. CONCLUSIONS

Implicit necessary and sufficient conditions for the solution to the Morgan problem with stability are obtained. The approach used here takes advantage of the properties of $\mathbb{R}_{ps}(s)$. Connections with some other well known results are also mentioned.

5. REFERENCES

Başer, U., and V. Eldem (1995). Diagonal Decoupling Problem with Internal Stability: A Solution by Dynamic Output Feedback and Constant Precompensator. The General Case. Submitted to the 3rd. *IFAC Conference on System Structure and Control*, Nantes, France.

Descusse, J., J.F. Lafay, and M. Malabre (1988). Solution to Morgan's Problem. *IEEE Trans. Automat. Contr.*, No. 8, pp. 732-739.

Dion, J.M., and C. Commault (1988). The Minimal Delay Decoupling Problem: Feedback Implementation with Stability. *SIAM J. Contr. Optimiz.*, Vol. 26, No. 1, pp. 66-82.

Eldem, V. (1993). Feedback Realization of Open Loop Diagonalizers. *Kybernetika*, Vol. 29, No. 5, pp. 406-416.

Eldem, V. (1994). The Solution of Diagonal Decoupling Problem by Dynamic Output Feedback and Constant Precompensator: The General Case. *IEEE Trans. Automat. Contr.*, Vol. 39, No. 3, pp. 503-511.

Falb, P.L., and W.A. Wolovich (1967). Decoupling in the Design and Synthesis of Multivariable Control Systems. *IEEE Trans. Automat. Contr.*, Vol. AC-12, No. 6, pp. 651-659.

Herrera, A. (1992). Static Realization of Dynamic Precompensators. *IEEE Trans. Automat. Contr.*, Vol. 37, No. 9, pp. 1391-1394.

Kučera, V., and P. Zagalak (1991). Constant Solutions of Polynomial Equations. *Int. J. of Contr.*, Vol. 53, No. 2, pp. 495-502.

Malabre, M., V. Kučera, and P. Zagalak (1990). Reachability and Controllability Indices for Linear Descriptor Systems. *Systems&Control Letters*, Vol. 15, pp. 119-123.

Martinez Garcia, J.C., and M. Malabre (1995). The Row By Row Decoupling Problem with Stability. *IEEE Trans. Automat. Contr.* (To appear, January 1995).

Morgan Jr., B.S. (1964). The Synthesis of Linear Multivariable Systems by State-variable Feedback. *IEEE Trans. Automat. Contr.*, Vol. AC-9, pp. 405-411.

Morse, A.S., and W.M. Wonham (1971). Status of Noninteracting Control. *IEEE Trans. Automat. Contr.*, Vol. AC-16, No. 6, pp. 568-581.

Zagalak, P., J.F. Lafay, and A. Herrera (1993). The row-by-row Decoupling via State Feedback: A Polynomial Approach. *Automatica*, Vol. 29, No. 6, pp. 1491-1499.

PERFORMANCES AND LIMITATIONS OF SYMBOLIC COMPUTATION IN NONLINEAR ANALYSIS AND CONTROL: AN EXAMPLE USING MATHEMATICA

Gildas BESANÇON* and Guy BORNARD*

*Laboratoire d'Automatique de Grenoble, ENSIEG, BP 46, 38402 Saint Martin d'Hères, France.
e-mail: gildas@lag.grenet.fr

Abstract. This paper mainly presents the problem of bilinearization as an example of what can be done with symbolic computation in nonlinear control and design, and where are its weaknesses. It is a contribution to the development of symbolic computational tools for nonlinear control and design, and it results in a functional package computed with MATHEMATICA.

Key Words. Bilinearization, Symbolic Computation, Nonlinear Control Systems.

Résumé. Cet article présente le problème de bilinéarisation en guise d'illustration des performances du calcul formel appliqué à l'automatique non linéaire, en même temps que ses limites.
Il contribue ainsi au développement d'outils informatiques formels d'aide à l'étude des systèmes non linéaires et s'est concrétisé par la réalisation d'une bibliothèque de routines programmées avec Mathematica.

Mots Clefs. Bilinéarisation, Calcul Formel, Systèmes Non Linaires.

1. INTRODUCTION

Analyzing and designing nonlinear control systems involve a lot of differential calculus not easily done by hand. At a time when high-performance symbolic computing languages are available, it seems interesting to provide packages as an help for people dealing with nonlinear control. Attempts have already been proposed with Macsyma (Akhrif and Blankenship, 1991; Birk and Zeitz, 1991), and Maple (van Essen and de Jager, 1993). Our purpose is to give a contribution to the development of such tools, providing a package computed with Mathematica, as a continuation of results on linear MiMo systems (Benamara, 1993), covering the differential geometric approach (Isidori, 1989) as in (van Essen and de Jager, 1993), and deepening the specific issue of bilinearization (Hammouri and Gauthier, 1988) considered as an illustration of symbolic performances and limitations.
In section 2 are reviewed some standard control problems previously studied with MATHEMATICA (Besançon, 1993), mentioning the main symbolic limitations met; these limitations are highlighted in section 3, where a symbolically computable algorithm detecting bilinearizability is proposed. Section 4 generalizes the result to time-variable linearizability, and section 5 concludes.

2. SYMBOLIC COMPUTATION AND DIFFERENTIAL GEOMETRIC METHODS

Differential geometry applied to nonlinear control theory as presented in (Fossard and Normand-Cyrot, 1993),(Isidori, 1989), or (Bornard *et al.*, 1988) gives a framework well adapted to symbolic computation: it basically involves differential calculus such as Jacobian matrices, gradients, Lie brackets, Lie derivatives... easily computable in an environment like MATHEMATICA (Wolfram, 1988). This is why Akhrif and Blankenship (Akhrif and Blankenship, 1991) initiated the use of symbolic computation in nonlinear control and design based on differential geometry, with Macsyma, and more recently van Essen and de Jager with Maple (van Essen and de Jager, 1993), and ourselves with Mathematica (Besançon, 1993).
It is to be mentioned that some limitations are unavoidable when using symbolic computation, such as solving (sets of) nonlinear algebraic equations or (sets of) differential equations (either P.D.E. or O.D.E.): symbolic computing systems can indeed only solve problems which have symbolic solutions; this means, for instance, that a polynomial equation of degree higher than 5 cannot be solved by any symbolic computing language. Of course when dealing with equations "simple enough", some of these problems may be symbolically solved: this is the assumption made in (van

Essen and de Jager, 1993) in the problems of zero dynamics and linearization.
As to our previous work, it was restricted to exactly solvable problems in any case; in particular were computed tools to study controllability and observability, as well as linearizability, noninteracting control and disturbance decoupling. This work was the opportunity to review symbolic computations performances and limitations.
Based on a package of basic differential calculus, it was first inspected the information reachable by symbolic computation about nonlinear controllability and observability: this study is based on algorithms presented in (Isidori, 1989) to compute the smallest distribution (respectively codistribution) containing a given distribution Δ (respectively codistribution Ω), and invariant, with respect to Lie bracket (respectively Lie derivation), under some given vector fields $\tau_1, \ldots, \tau_n$ (respectively covector fields $\omega_1, \ldots, \omega_n$).
The controllability distribution is for instance given by

$$\begin{aligned} \Delta_0 &= Span\{f, g_1, \ldots, g_m\} \\ \Delta_k &= \Delta_{k-1} + [f, \Delta_{k-1}] \\ &\quad + \sum_{i=1}^{m} [g_i, \Delta_{k-1}] \\ \text{until } \Delta_k &= \Delta_{k-1} \end{aligned}$$

for a nonlinear system

$$\begin{aligned} \dot{x} &= f(x) + \sum_{i=1}^{m} g_i(x) u_i \\ y &= (h_1(x) \ldots h_p(x))^T, \qquad x \in I\!R^n. \end{aligned}$$

This kind of algorithm is valid provided all the computed distributions are nonsingular. With symbolic computation, the result can at best be generic, that is, true for almost all values of symbolic parameters: the algorithm is indeed based on rank computations, which requires to consider as not null all parameters not *exactly* (symbolically) null; for example, two vectors $\begin{pmatrix} x_1 \\ 1 \\ 0 \end{pmatrix}$ and $\begin{pmatrix} 1 \\ x_2 \\ 0 \end{pmatrix}$ will be considered as independent although this is only true for $x_1 x_2 \neq 1$.
Nevertheless, this drawback does not forbid the use of the algorithms mentioned above; one has just to keep in mind, when using them, that the results given are *generic*.

Some standard control problems for SiSo and MiMo nonlinear systems affine in the control were then inspected: feedback linearizability and maximal linearizable subsystem, based on rank computations of some specific distributions and on involutivity tests, linearization and disturbance decoupling based on relative degree computations etc, stopping computations as soon as solving nonlinear algebraic or differential equations was required.
Even though this study does not always provide complete solutions to the problems inspected, insofar as it does not symbolically compute the linearized system for instance, it answers the basic question of **existence of solutions.**
On the basis of this first approach, special attention is paid to the specific issue of bilinearization up to output injection, showing the symbolic computability of a bilinearizability test for nonlinear SiSo systems affine in the control, and computing the corresponding algorithm with MATHEMATICA. This study is presented in next section.

3. THE EXAMPLE OF BILINEARIZATION UP TO OUTPUT INJECTION

In this section is given an algorithm to recognize bilinearizable SiSo systems, only based on symbolically computable procedure.
Some notations and results on bilinearization (Hammouri and Gauthier, 1988) are first to be reminded:

given a nonlinear system:

$$(\Sigma)\left\{ \begin{array}{rcl} \dot{x} &=& f(x) + ug(x) = Z_u(x) \\ y &=& h(x), \quad x \in I\!R^n, y \in I\!R, u \in I\!R, \end{array} \right.$$

and a vector field X,
consider the following sequence of real vector spaces $\Omega_k^X(\Sigma)$:

$$\begin{aligned} &\Omega_1^X(\Sigma) = Span_{I\!R}\{(dL_{Z_u}(h)) \wedge dh, u \in I\!R\} \\ &\Omega_{k+1}^X(\Sigma) = Span_{I\!R}\{L_{Z_u}(i_X w) \wedge dh, u \in I\!R, \\ &\qquad w \in \Omega_k^X(\Sigma)\} \end{aligned}$$

where L_Z denotes the Lie derivative along Z, i_X the interior product by X, and $\wedge$ the exterior product, and set:

$$\Omega^X(\Sigma) = \textstyle\sum_{k=1}^{\infty} \Omega_k^X(\Sigma).$$

(Σ) is said to be locally bilinearizable at 0 up to output injection if there exists a change of coordinates preserving the origin steering (Σ) into:

$$\begin{aligned} \dot{z} &= Az + uBz + \Phi(u, y), \\ y &= Cz, \end{aligned}$$

Then the following result holds:

Theorem *3.1* A nonlinear system Σ, analytic and observable (in the sense of the observability rank condition) is locally (around $x = 0$) bilinearizable up to output injection if and only if:

1. $h(0) = 0$

2. $dh(0) \neq 0$

and there exists a vector field X analytic around 0 such that

3. $L_X(h) = 1$
4. $dim_{\mathbb{R}} \Omega^X(\Sigma) = n - 1$
5. $d(i_X \Omega^X(\Sigma)) = 0$

•

At this point it seems interesting to specify how the limitations mentioned in the previous section arise in such a problem as bilinearization: in this case the goal is twofold, on the one hand one has to check the conditions of theorem 3.1, and on the other hand, if the result is positive, the subsequent problem is to find a bilinearizing change of coordinates, and the bilinearized system.

The first part proved to be completely symbolically solvable by the algorithm proposed here; on the contrary, the second part a priori requires to solve sets of partial differential equations and non-linear algebraic equations: the change of coordinates is indeed given by

$$\begin{aligned} dz_1 &= dh \\ dz_2 &= i_X w_1 \\ &\vdots \\ dz_n &= i_X w_{n-1}, \end{aligned} \qquad (1)$$

where $w_1, \dots, w_{n-1}$ denotes a basis of $\Omega^X(\Sigma)$.

As soon as this diffeomorphism $z = \psi(x)$ is obtained, the bilinearized system is:

$$\begin{aligned} \dot{z} &= \left[\frac{\partial \psi}{\partial x}\right]_{x=\psi^{-1}(z)} Z_u(\psi^{-1}(z)) \\ y &= z_1 \end{aligned}$$

It is clear that inverting the diffeomorphism $\psi(x) = (\psi_1(x), \dots, \psi_n(x))$ is equivalent to solve in $x = (x_1, \dots, x_n)$ the following set of algebraic equations:

$$\begin{aligned} \psi_1(x_1, \dots, x_n) &= z_1 \\ &\vdots \\ \psi_n(x_1, \dots, x_n) &= z_n \end{aligned}$$

Both of these problems have in general no symbolic solution, and must either be considered from another point of view, or use numerical approximations.

Nevertheless, as for bilinearizability itself, our study results in an algorithm which avoids symbolic pitfalls previously mentioned: it consists in computing a vector field X such that $dim_{\mathcal{C}^\infty} \Omega^X(\Sigma) = n - 1$, and meeting with the other conditions of theorem 3.1; this computation involves some unspecified constants, and it is shown how their number can be reduced to $n - 1$. One is then able to check symbolically whether $dim_{\mathbb{R}} \Omega^X(\Sigma) = n - 1$ or not.

Remark *3.1* It can be seen from (Hammouri and Gauthier, 1992) that the vector field introduced in theorem 3.1 is defined up to $n - 1$ real constants: if a system is bilinearizable by some coordinate system $(z_1, \dots, z_n)$, every vector field X matching the conditions of theorem 3.1 is given by:

$$X = \frac{\partial}{\partial z_1} + \sum_{k=2}^{n} c_k \frac{\partial}{\partial z_k}, \text{ for some } c_2, \dots, c_n \in \mathbb{R}.$$

Our algorithm computes a family of such vector fields, parameterized by $n - 1$ real constants.

Let us now specify the procedure step by step:

Preliminary result:
Setting $w_{u_1} = dL_{Z_{u_1}}(h) \wedge dh$
and $w_{u_{k+1} \dots u_1} = L_{Z_{u_{k+1}}}(i_X w_{u_k \dots u_1}) \wedge dh$,
it comes

$$\Omega_k^X(\Sigma) = Span_{\mathbb{R}}\{w_{u_k \dots u_1} : (u_k, \dots, u_1) \in \mathbb{R}^k\}.$$

By induction, one can check the following:

Claim *3.1* $\forall (u_k, \dots, u_1) \in \mathbb{R}^k$
$w_{u_k \dots u_1} =$

$$\begin{aligned} & (-1)^{k+1} dL_{Z_{u_k}} \dots L_{Z_{u_1}}(h) \wedge dh \\ + \; & (-1)^k L_X L_{Z_{u_1}}(h) dL_{Z_{u_k}} \dots L_{Z_{u_2}}(h) \wedge dh \\ + \; & \sum_{l=2}^{k-1} \Bigg[\sum_{1<j(2)<\dots<j(l)\leq k} P^k_{1j(2)\dots j(l)} dL_{Z_{u_k}} \\ & \dots \hat{L}_{Z_{u_{j(l)}}} \dots \hat{L}_{Z_{u_{j(2)}}} \dots L_{Z_{u_2}}(h) \wedge dh \\ & \quad + \; (-1)^k L_X L_{Z_{u_l}} \dots L_{Z_{u_1}}(h) dL_{Z_{u_k}} \dots \\ & \quad L_{Z_{u_{l+1}}}(h) \wedge dh \Bigg] \end{aligned}$$

where the notation $\hat{L}_{Z_u}$ means omission of L_{Z_u}, and the coefficients $P^k_{1j(2)\dots j(l)}$ are in the algebra $\mathcal{A}_l$ generated by

$$\begin{aligned} &\{L_X L_{Z_{u_{i(\tau)}}} \dots L_{Z_{u_{i(1)}}}(h), \\ &u_{i(\tau)}, \dots, u_{i(1)} \in \{u_1, \dots, u_k\}, \\ &\text{and} \quad 1 \leq \tau \leq l - 1\} \end{aligned} \qquad (2)$$

and closed under L_{Z_u}, $\quad u \in \{u_1, \dots, u_k\}$.

They can be expressed by induction by:

- $P^k_{12\dots(k-1)} =$
$\sum_{l=2}^{k-2} \left[\sum_{1<j(2)<\dots<j(l)<k} P^{k-1}_{1j(2)\dots j(l)} \right.$
$\left. L_X L_{Z_{u_{k-1}}} \dots \hat{L}_{Z_{u_{j(l)}}} \dots \hat{L}_{Z_{u_{j(2)}}} \dots L_{Z_{u_2}}(h) \right]$
$+ \; (-1)^{k-1} \sum_{l=1}^{k-2} L_X L_{Z_{u_l}} \dots L_{Z_{u_1}}(h)$
$L_X L_{Z_{u_{k-1}}} \dots L_{Z_{u_{l+1}}}(h)$

- $P^k_{12\ldots(j-1)jk} = -L_{Z_{u_k}}(P^{k-1}_{12\ldots(j-1)j}) + (-1)^k L_{Z_{u_k}} L_X L_{Z_{u_j}} \ldots L_{Z_{u_1}}(h)$ for $1 \le j \le k-2$ and $P^{k-1}_1 = 0$.

- $P^k_{1j(2)\ldots j(l)} = -P^{k-1}_{1j(2)\ldots j(l)}$ for $2 \le l \le k-2$, $1 < j(2) < \ldots < j(l) < k$

- $P^k_{1j(2)\ldots j(l)k} = -L_{Z_{u_k}}(P^{k-1}_{1j(2)\ldots j(l)})$ for $2 \le l \le k-2$, $1 < j(2) < \ldots < j(l) < k$, $\{1, j(2), \ldots, j(l-1), j(l)\} \neq \{1, 2, \ldots, j-1, j\}$

It is the necessity of such calculations which motivates the use of symbolic computation for the problem of bilinearizability: their implementation will require some basic differential tools added to those already mentioned in previous section: Exterior product, Interior product, Exterior differentiation...

Algorithm:

1. **Computation of a basis of the observability codistribution:**
by the observability rank condition, it can be found sets of binary sequences $I_l = \{u_{i(l)} \ldots u_{i(1)} : u_{i(k)} \in \{0,1\}, 1 \le k \le l\}$ for $l = 1$ to p, such that
$$\{dh, dL_{Z_{u_{i(l)}}} \ldots L_{Z_{u_{i(1)}}}(h) : 1 \le l \le p\}$$
is a basis of the observability codistribution

2. **Computation of a basis of $\Omega^X(\Sigma)$ considered as a linear space on analytic functions** (instead of $\mathbb{R}$):
by claim 3.1, $\cup_{l=1}^p I_l$ provides indices for a basis $\mathcal{B}^X = \cup_{l=1}^p \{w_{u_{i(l)}\ldots u_{i(1)}} : u_{i(l)} \ldots u_{i(1)} \in I_l\}$ of $\Omega^X(\Sigma)$ considered as a linear space on C^∞.
At this point, it must be noted that:
for all $(v_1, \ldots, v_k) \in \mathbb{R}^k, 1 \le k \le p$,
$$dL_{Z_{v_k}} \ldots L_{Z_{v_1}}(h) \wedge dh = \sum_{l=1}^k \sum_{u_{i(l)}\ldots u_{i(1)} \in I_l} \phi^{v_k \ldots v_1}_{u_{i(l)}\ldots u_{i(1)}}(x) \quad (3)$$
$$dL_{Z_{u_{i(l)}}} \ldots L_{Z_{u_{i(1)}}}(h) \wedge dh$$
and
$$\mathcal{B}^2 = \cup_{l=1}^p \{dL_{Z_{u_{i(l)}}} \ldots L_{Z_{u_{i(1)}}}(h) \wedge dh, u_{i(l)} \ldots u_{i(1)} \in I_l\}$$
is a basis of the C^∞-space of 2-forms $d\gamma(x) \wedge dh$.

3. **Computation of an algebraic linear system in the unknowns $(X_1, \ldots, X_n)$** (components of the vector field X as defined in theorem 3.1), **up to $2^n - 2$ unspecified constants:**
given $u^0_{i(p)} \ldots u^0_{i(1)}$ an index chosen in I_p, the $L_X L_{Z_{u_{i(l)}}} \ldots L_{Z_{u_{i(1)}}}(h)$ can be computed by identifying $w_{u^0_{i(p)}\ldots u^0_{i(1)} u_{i(l)}\ldots u_{i(1)}}$ on $\mathcal{B}^X$ and $\mathcal{B}^2$ (inspection of claim 3.1): on the one hand, its coefficient can be obtained on $dL_{Z_{u^0_{i(p)}}} \ldots L_{Z_{u^0_{i(1)}}}(h) \wedge dh$, and on the other hand, this coefficient is its component (up to the sign) on $w_{u^0_{i(p)}\ldots u^0_{i(1)}}$, that is, must be real. Thus, it is forced to an unspecified constant $C^{u_{i(l)}\ldots u_{i(1)}}$.
It can be seen (still by claim 3.1) that at the *order* l, $L_X L_{Z_{u_{i(l)}}} \ldots L_{Z_{u_{i(1)}}}(h)$ only depends on terms of $\mathcal{A}_k, k < l$, say terms of the form $L_X L_{Z_{u_{j(k)}}} \ldots L_{Z_{u_{j(1)}}}(h), k < l$.
As a consequence, it comes by induction:
$$L_X(h) = 1$$
$$L_X L_{Z_{u_{i(1)}}}(h) = C^{u_{i(1)}} + \Phi^{u_{i(1)}}(x)$$
$$L_X L_{Z_{u_{i(2)}}} L_{Z_{u_{i(1)}}}(h) = -C^{u_{i(2)}u_{i(1)}} + \Phi^{u_{i(2)}u_{i(1)}}(x, C^{u_{i(1)}}, C^{u_{i(2)}})$$
$$\vdots$$
$$L_X L_{Z_{u_{i(p)}}} \ldots L_{Z_{u_{i(1)}}}(h) = (-1)^{p+1} C^{u_{i(p)}\ldots u_{i(1)}} + \Phi^{u_{i(p)}\ldots u_{i(1)}}(x, \ldots, C^{u_{i(p-1)}\ldots u_{i(1)}})$$

4. **Computation of X up to the $2^n - 2$ unspecified constants:**
the system previously obtained is clearly symbolically solvable in X since it is linear and the Jacobian matrix of the left member is invertible ($\mathcal{B}^2$ being a basis).

5. **Expression of X up to $n - 1$ constants:**
by equation (3), the $L_X L_{Z_{u_{j(k)}}} \ldots L_{Z_{u_{j(1)}}}(h)$ used to construct the linear system, but not directly involved, can be recurrently recomputed, and the $C^{u_{j(k)}\ldots u_{j(1)}}$ expressed with respect to the $C^{u_{i(l)}\ldots u_{i(1)}}$.
This finally gives
$X(C^{u_{i(l)}\ldots u_{i(1)}}, u_{i(l)} \ldots u_{i(1)} \in I_l, 1 \le l \le p)$.

At this point, it is to be noted that next results must be independent of the $n-1$ remaining unspecified constants (due to the fact that vector X is defined up to these $n - 1$ constants, as mentioned in remark 3.1).

6. **Verification of exact differentials** (condition 5 of theorem 3.1):
it only requires computations of interior products and exterior differentiations, and nullity tests; it is to be mentioned that these kind of tests within symbolic environment implies to simplify expressions before giving an answer (a priori a symbolic computing system consider $x_2(x_1 - 1) - x_1 x_2 + x_2 \neq 0$).

7. **Verification of the dimension of $\Omega^X(\Sigma)$ on $\mathbb{R}$** (condition 4 of theorem 3.1):

Denoting by $\{w_k, 1 \le k \le n-1\}$ the computed basis of $\Omega^X(\Sigma)$, it can be seen that $\mathcal{B}^X$ is an $(n-1)$-dimensional basis of $\Omega^X(\Sigma)$ on $\mathbb{R}$ if $L_{Z_u}(dh) \wedge dh$ and the $L_{Z_u}(i_X w_k) \wedge dh$ with $u \in \{0,1\}$ depends really on the basis. As a consequence, it is sufficient to express the coefficients on the basis of these $2n$ elements and compute their derivative with respect to the state variable x to check whether they are real or not.

Example:
The implementation of this algorithm with MATHEMATICA proved to be efficient in several examples; applied to the following system

$$
\begin{aligned}
\dot{x}_1 &= x_1^2 - x_3 \\
\dot{x}_2 &= \frac{2x_1^2(1-x_1x_2)+x_3(-3+2x_1x_2)}{1+x_3} \\
&\quad + x_2^2 + u\frac{x_1(-1+x_1x_2)}{1+x_3} \\
\dot{x}_3 &= 2x_1(x_1^2-x_3) - x_2(1+x_3) - ux_1^2 \\
y &= x_1
\end{aligned}
$$

it gives:

1. $dh,\ \ dL_{Z_0}(h),\ \ dL^2_{Z_0}(h)$ generates the observability codistribution.
2. $\Omega^X(\Sigma) = Span_{C^\infty}\{\omega_0, \omega_{00}\}$
3. linear system:

$$
\begin{aligned}
L_X(h) &= 1 \\
L_X L_{Z_0}(h) &= C^0 \\
L_X L^2_{Z_0}(h) &= (C^0)^2 - C^{00} + 3
\end{aligned}
$$

4. Solution:

$$
X = \begin{pmatrix} 1 \\ \frac{3+(C^0)^2-C^{00}+C^0x_2-2x_1x_2}{1+x_3} \\ 2x_1 - C^0 \end{pmatrix}
$$

5. $di_X\omega_0 = di_X\omega_{00} = 0$, since:

$$
\begin{aligned}
i_X\omega_0 &= (C^0 - 2x_1)dx_1 + dx_3 \\
i_X\omega_{00} &= (C^{00} - 3 - 2C^0x_1)dx_1 \\
&+ (1+x_3)dx_2 + (C^0+x_2)dx_3
\end{aligned}
$$

6. finally:

$$
\begin{aligned}
L_{Z_0}(dh) \wedge dh &= \omega_0 \\
L_{Z_1}(dh) \wedge dh &= \omega_0 \\
L_{Z_0}(i_X\omega_0) \wedge dh &= \omega_{00} \\
L_{Z_1}(i_X\omega_0) \wedge dh &= \omega_{00} \\
L_{Z_0}(i_X\omega_{00}) \wedge dh &= (C^{00} - (C^0)^2)\omega_0 \\
&\quad + C^0\omega_{00} \\
L_{Z_1}(i_X\omega_{00}) \wedge dh &= (C^{00} - (C^0)^2)\omega_0 \\
&\quad + C^0\omega_{00}
\end{aligned}
$$

The system is thus bilinearizable up to output injection.

Further computations give

$$
z = \begin{pmatrix} x_1 \\ C^0x_1 - x_1^2 + x_3 \\ (C^{00}-3)x_1 - C^0x_1^2 + x_2(1+x_3) + C^0x_3 \end{pmatrix}
$$

as a bilinearizing change of coordinates for any $C^0, C^{00} \in \mathbb{R}$, the bilinearized system being:

$$
\begin{aligned}
\dot{z} &= \\
&\begin{pmatrix} C^0 & -1 & 0 \\ C^{00}-3 & 0 & -1 \\ C^0(2C^{00}-3)-(C^0)^3 & (C^0)^2 - C^{00} & -C^0 \end{pmatrix} z \\
&+u\begin{pmatrix} 0 & 0 & 0 \\ 0 & 0 & 0 \\ -1 & 0 & 0 \end{pmatrix} z + \begin{pmatrix} 0 \\ -uy^2 \\ -y^2(1+uC^0) \end{pmatrix} \\
y &= (1 \quad 0 \quad 0)z
\end{aligned}
$$

4. TIME-VARIABLE LINEARIZATION UP TO OUTPUT INJECTION

In the case when the considered nonlinear system has the form

$$
\begin{aligned}
\dot{x} &= f(x,u) = Z_u(x) \\
y &= h(x)
\end{aligned} \tag{4}
$$

theorem (3.1) provides a Necessary and Sufficient Condition (Hammouri and Gauthier, 1992) for it to be locally diffeomorphic to

$$
\begin{aligned}
\dot{z} &= A(u)z + \varphi(u,y) \\
y &= Cz
\end{aligned} \tag{5}
$$

observable (in the sense of the observability rank condition).
The main difference in the algorithm proposed is that the control sequences $u_{i(l)} \dots u_{i(1)}$ determined at step 1 are supposed to be chosen in $\mathbb{R}^l$ instead of $\{0,1\}^l$, that is, in a symbolic computing environment, they are left as symbolic constants. In order to compute them, an algorithm similar to the one given in (Isidori, 1989) for control-affine systems observability can be used:

$$
\begin{aligned}
dO_0 &= Span\{dh\} \\
dO_{k+1} &= dO_k + \sum_{i=1}^{n_k} L_{f_{u_i^k}}(dO_k) \\
\text{until } dO_{k+1} &= dO_k = dO^*
\end{aligned}
$$

where u_i^k are symbolic constants for $1 \le i \le n_k$ and $1 \le k \le n-1$.
In fact, since $n-1$ elements are required, the computation will need $n-1$ constants.
As soon as the symbolic sequences of controls are obtained, the algorithm follows the procedure described in section 3.

Remark *4.1* Symbolic constants u_i^k will be involved in all computations, but the results must be independent of them; they only reflect the degree of freedom on the choice of the basis of Ω^X.

As an example, consider the following system:

$$\begin{aligned}\dot{x}_1 &= x_2u^2 + x_3(e^{-x_1} - u^2)\\ \dot{x}_2 &= x_1 + ux_1^2 + x_2e^{x1} + x_3(e^{u-x_1} - e^{x_1}\\ &\quad +u^2x_2) + x_3^2(e^{-x_1} - u^2)\\ \dot{x}_3 &= x_2(e^{x_1} + u^2x_3) - x_3e^{x_1}\\ &\quad +x_3^2(e^{-x_1} - u^2)\\ y &= x_1\end{aligned}$$

The algorithm concludes to time-variable linearizability, using symbolic constants u_1, u_2.
Further computations lead to a *linearizing* change of coordinates:

$$\begin{aligned}z_1 &= x_1\\ z_2 &= (C^{u_1} + \frac{(u_1u_2)^2e^{u_1} - 1}{u_2^2 - u_1^2})x_1 - u_1^2x_2\\ &\quad -(e^{-x_1} - u_1^2)x_3\\ z_3 &= (C^{u_2} + \frac{(u_1u_2)^2e^{u_2} - 1}{u_2^2 - u_1^2})x_1 - u_2^2x_2\\ &\quad -(e^{-x_1} - u_2^2)x_3\end{aligned}$$

for any u_1, u_2 defining a diffeomorphism, and any C^{u_1}, C^{u_2} in $\mathbb{R}$.
For instance, with $u_1 = 1, u_2 = 0$, and $C^{u_1} = C^{u_2} = 0$,the linearized system up to output injection, can be written as

$$\begin{aligned}\dot{z}_1 &= z_1 - u^2z_2 + (u^2 - 1)z_3\\ \dot{z}_2 &= (2 + e^u)z_1 - (1 + u^2)z_2\\ &\quad +(u^2 - e^u)z_3 + uz_1^2\\ \dot{z}_3 &= z_1 - (1 + u^2)z_2 + u^2z_3\\ y &= z_1\end{aligned}$$

5. CONCLUSION AND FURTHER DEVELOPMENTS

The purpose of this work was to clarify the possibilities and limitations of symbolic computation applied to nonlinear control and design, as well as to contribute to its development by providing a functional package computed with MATHEMATICA: this package gives elements of differential geometric methods for nonlinear SiSo and MiMo systems affine in the control, with a particular attention to bilinearizability study for SiSo systems.

Our next research will consist on the one hand, in extending the result on bilinearizability to MiMo systems (Hammouri and Gauthier, 1989), and on the other hand, in taking into account the limitations highlighted in this paper; the second point means two kinds of developments: constructing algorithms which avoid symbolically unsolvable problems (as in the case of bilinearizability study), and inspecting how to switch to numerical approximations when there is no other solution.

Acknowledgment: The authors would like to thank Prof Hassan Hammouri, (*Laboratoire d'Automatique et Génie des Procédés*, Lyon, France) for his support to this work.

6. REFERENCES

Akhrif, O. and G. L. Blankenship (1991). *Algebraic Computation for the Design of Nonlinear Control Systems.* Vol. 165 of *Lecture Notes in Cont. and Info. Sci., Algebraic Computing in Control.* Springer Verlag, Berlin. pp. 129–155.

Benamara, A. (1993). Systèmes Linéaires Multivariables, Etudes Algébriques, Calcul Formel et Optimisation H_∞. PhD thesis. Institut National Polytechnique de Grenoble.

Besançon, G. (1993). Utilisation du calcul formel pour l'etude des systèmes non linéaires. Master's thesis. Institut National Polytechnique de Grenoble. Laboratoire d'Automatique de Grenoble.

Birk, J. and M. Zeitz (1991). *Program for Symbolic and Rule-Based Analysis and Design of Nonlinear Systems.* Vol. 165 of *Lecture Notes in Cont. and Info. Sci., Algebraic Computing in Control.* Springer Verlag, Berlin. pp. 115–128.

Bornard, G., F. Celle-Couenne and H.Hammouri (1988). *Résultats et Perspectives en Automatique.* Hermès, Paris. chapter 3, pp. 116–149.

Fossard, A. J. and D. Normand-Cyrot (1993). *Systèmes Non Linéaires.* Masson, Paris.

Hammouri, H. and J.P. Gauthier (1988). 'Bilinearization up to output injection'. *Systems & Control Letters* **11**, 139–149.

Hammouri, H. and J.P. Gauthier (1989). The time-varying linearization up to output injection. In 'Proc. 28th IEEE Conf. on Decision and Control, Tampa, Florida'. pp. 1038–1039.

Hammouri, H. and J.P. Gauthier (1992). 'Global time-varying linearization up to output injection'. *Siam Journal on Control and Optimization* **30**(6), 1295–1310.

Isidori, A. (1989). *Nonlinear Control Systems.* Springer-Verlag, Berlin, 2nd edition.

van Essen, H. and B. de Jager (1993). Analysis and design of nonlinear control systems with the symbolic computation system maple. In 'Proc. 2nd European Control Conf., Groningen, The Netherlands'. pp. 2081–2085.

Wolfram, S. (1988). *Mathematica, a System for Doing Mathematics by Computer.* Addison-Wesley Publishing Company.

SOME ISSUES ON DEADBEAT CONTROL AND A RELATED SET OF CAD PROCEDURES

M.Budinsky * and L.Jetto **

* KAR Elektrotechnicka fakulta Ilkovicova 3, 81219 Bratislava, Slovakia.
** Dipartimento di Elettronica ed Automatica, Università di Ancona, via Brecce Bianche, 60131 Ancona, Italia.

Abstract. In this paper, a set of computer-aided procedures for the algebraic approach solution to some questions arising in deadbeat control is proposed. The analysis and design facilities of the proposed software package are highly interactive and are based on the recently proposed MAPLE toolbox. Both symbolic and numerical computations are allowed.

Résumé. Dans ce papier on propose un ensemble de procedures CAD pour l'analyse et le project des systèmes linéaires de réglage avec temp fini de réponse. Toutes les procédures sont basés sur la librarie MAPLE et sont interactives. On peut utiliser tant le calcul numérique que le calcul symbolique.

Keywords. Controller design; deadbeat control; polynomial methods; CAD procedures.

1 Introduction

One of the most extensively studied control problem is the output regulation in a finite-time (see e.g. Akashi and Imai (1978), Emami-Naeini and Franklin (1982), Kimura and Tanaka (1981), Kucera and Sebek (1984), Leden (1977), Sebakhy and Abdel-Moneim (1979), Wolovich (1983)). If the discrete-time plant originates from the sampling of a continuous time one, the absence of intersampling ripples in the steady-state , continuous-time response is also often required (Eichstaedt (1982), Jetto (1989, 1990, 1994) , Kaczorek (1982), Kucera (1979, 1980), Sebakhy (1989), Sirisena (1985), Urikura and Nagata (1987), Xiaotian et al. (1994)).

Another requirement that can be of interest is the simultaneous tracking of a reference variable from a generic but prespecified class. In fact, in various industrial applications some control situations arise where it is required that the output of a given plant be reset, in a finite time, to a new trajectory according to a reference input that may be changed from time to time.

The main purpose of this paper is to present a software package for the computation of the solution of the deadbeat tracking problem formulated according to the above considerations; namely under the assumption that the external reference may vary in a given set and that no ripple exist in-between the sampling intervals in the steady-state, continuous-time response . It is also assumed that both the state and the output of the continuous - time plant to be controlled are affected by some external disturbances. The chosen theoretical framework is the algebraic approach because this allows us to express solvability conditions in a concise form. To make it handy to deal with the above control problem, the algorithms for testing solvability conditions and for designing the compensator are implemented using the recently proposed computer algebra system Maple (Heck,1993). The package may also work in numerical mode (with arbitrary number of digits). This options allows us to save memory requirement and makes it possible to implement the algorithm on a small RAM memory PC.

For the sake of brevity, the proofs of theorems are not reported in this version.

2 Preliminaries and Problem Statement

Consider the following linear, continuous-time, time-invariant system $\Sigma_p \equiv (C_p, A_p, B_p)$

$$\dot{x}(t) = A_p x(t) + B_p u(t) + M_p d_1(t), \quad (2.1)$$
$$y(t) = C_p x(t) + N_p d_2(t), \quad (2.2)$$

where $x(t) \in \mathbb{R}^n$ is the state, $u(t) \in \mathbb{R}^q$ is the control input, $y(t) \in \mathbb{R}^p$ is the output, $d_1(t) \in \mathbb{R}^{p_1}$

and $d_2(t) \in \mathbb{R}^{p_2}$ are two external disturbances generated as the unforced output responses of the two following linear systems Σ_1 and Σ_2 respectively

$$\dot{\zeta}_1(t) = A_1\zeta_1(t), \quad (2.3)$$
$$d_1(t) = C_1\zeta_1(t), \quad (2.4)$$
$$\dot{\zeta}_2(t) = A_2\zeta_2(t), \quad (2.5)$$
$$d_2(t) = C_2\zeta_2(t), \quad (2.6)$$

where $\zeta_1(t) \in \mathbb{R}^{n_1}$ and $\zeta_2(t) \in \mathbb{R}^{n_2}$. The pairs (C_1, A_1) and (C_2, A_2) are assumed to be observable and $rank\, C_p = p$. The class of continuous-time references $r_i(t)(i = 1, ..., \nu)$ to be tracked is assumed to consist of the ensemble of the free output responses generated by the following linear, continuous-time, time-invariant reference models $\Sigma_{R,i}(i = 1, ..., \nu)$,

$$\dot{w}_i(t) = A_{r,i}w_i(t), \quad (2.7)$$
$$r_i(t) = C_{r,i}w_i(t), \quad (2.8)$$

where $w_i(t) \in \mathbb{R}^{n_{w_i}}$ is the state and $r_i(t) \in \mathbb{R}^p$ is the output. The pair $(A_{r,i}, C_{r,i})$ is assumed to be observable with $rank\, C_r, i = p$. Finally, all the eigenvalues of A_1, A_2 and $A_{r,i}$ are assumed to have nonnegative real part. The connection Σ_C of the plant Σ_p with the disturbance generators Σ_1 and Σ_2 is inserted between a zero-order holding device and a sampler in series in a classical error actuated servo-system structure. The sampling period and the hold interval are equal to T. The series connection of the holder, Σ_C and the sampler defines a sampled system Σ_D whose output response $y_i(kT), k \in Z^+$, has a d transform $\mathbf{y}_i(d)(i = 1, ..., \nu)$, which by (2.1)-(2.6), is given by

$$\begin{aligned}\mathbf{y}_i(d) &= A(d)^{-1}C(d)x_o + A(d)^{-1}B(d)\mathbf{u}_i(d) + \\ &\quad A(d)^{-1}B_1(d)F_1(d)^{-1}\zeta_{0,1} + \\ &\quad F_2(d)^{-1}G_2(d)\zeta_{0,2}, \end{aligned} \quad (2.9)$$

where $x_0, \zeta_{0,1}$ and $\zeta_{0,2}$ are the initial states of Σ_P, Σ_1 and Σ_2 respectively and $\mathbf{u}_i(d)$ is the d transform of $u_i(k)$ corresponding to the $i-th$ external reference $(i = 1, ..., \nu)$, and $A(d), C(d), B(d), M(d), F_1(d)$, $F_2(d)$ and $G_2(d)$ are polynomial matrices such that

$$\begin{aligned} C_p(I - dA_d)^{-1} &= A(d)^{-1}C(d), \\ C_p(I - dA_d)^{-1}B_d(d) &= A(d)^{-1}B(d), \\ C_p(I - dA_d)^{-1}\Phi d &= A(d)^{-1}B_1(d), \\ (I - dA_{1d}) &= F_1(d), \\ N_pC_2(I - dA_{2d})^{-1} &= F_2(d)^{-1}G_2(d), \end{aligned}$$

with

$$\begin{aligned} A_d &= exp(A_pT), \\ B_d &= \int_0^T exp(A_p\theta)B_p d\theta, \\ \Phi &= \int_0^T exp(A_p\theta) \\ &\quad M_pC_1 exp(A_1(T-\theta))d\theta, \\ A_{1d} &= exp(A_1T), \\ A_{2d} &= exp(A_2T), \end{aligned}$$

and with the pairs of matrices $(A(d), C(d)), (F_2(d), G_2(d))$ relatively left prime(rlp). Analogously, the d-transform $\mathbf{r}_i(d)$ of the sampled output $r_i(kT), k \in Z^+$, of the i-th reference generator is given by

$$\mathbf{r}_i(d) = R_i(d)^{-1}S_i(d)w_{0,i}, \quad (2.10)$$

where $w_{0,i}$ is the initial state of $\Sigma_{R,i}$ and $R_i(d)$ and $S_i(d)$ are rlp polynomial matrices such that

$$\begin{aligned} R_i(d)^{-1}S_i(d) &= C_{r,i}(I_{n_{w_i}} - dA_{r_d,i})^{-1}, \\ \text{with } A_{r_d,i} &= exp(A_{r,i}T), \end{aligned}$$

The sampling period is chosen according to the following assumption.

Assumption 2.1 For each $(i = 1, ..., \nu)$ there is not loss of observability due to sampling in the error system $\Sigma_{E,i}$ shown in figure 1, given by the parallel connection of $\Sigma_{R,i}$ and Σ_C. This means that

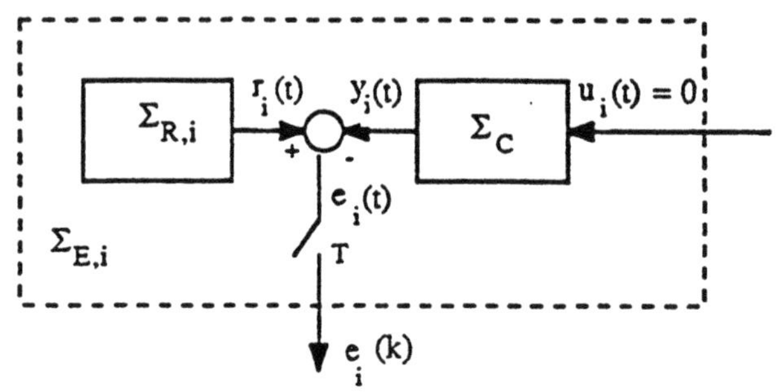

Figure 1: The error system

there is none nonzero free output response $e_i(t)$ of $\Sigma_{E,i}$ which is unobservable after sampling, namely $e_i(k) = 0$ for $k \geq \bar{k}$ ($\bar{k}$ being an arbitrary integer), if and only if $e_i(t) = 0 \quad (t \geq \bar{k}T)$.

Denoting with A_{Σ_c} and $A_{\Sigma_{E,i}}$ the dynamic matrices of Σ_C and $\Sigma_{E,i}$ respectively, given by

$$A_{\Sigma_C} = \begin{pmatrix} A_p & M_pC_1 & 0 \\ 0 & A_1 & 0 \\ 0 & 0 & A_2 \end{pmatrix},$$

$$A_{\Sigma_{E,i}} = \begin{pmatrix} A_{\Sigma_C} & 0 \\ 0 & A_{r,i} \end{pmatrix},$$

assumption 2.1 is satisfied by choosing T according to the condition (Chen (1984))

$$\lambda_f - \lambda_m \neq j2k\frac{\pi}{T}, \quad k = \pm1, \pm2, \cdots \quad (2.11)$$

Remark 2.1 Taking into account the structure of $A_{\Sigma_{E,i}}$ and A_{Σ_C} it is seen that the choice of T according to (2.11) implies $\grave{\lambda}_l - \grave{\lambda}_m \neq j2k\frac{\pi}{T}, \quad (k = \pm1, \pm2, \cdots)$ $\grave{\lambda}_l$ and $\grave{\lambda}_m$ being two distinct eigenvalues of A_p. Hence, condition (2.11) not only guarantees the fulfillment of assumption 2.1, but also that the subspaces S_{1P} and S_{2P} of the controllable and unobservable states respectively generated by the triplet (C_p, A_p, B_p) coincide with the analogous subspaces $\mathcal{S}_{1D}$ and $\mathcal{S}_2 D$ generated by the triplet (C_p, A_d, B_d) (Chen (1984)).

The deadbeat tracking problem considered in this paper can be now stated as follows.

Problem 2.1 Assuming condition (2.11) satisfied, find, if it exists, a linear discrete-time controller Σ_G of the following form

$$P(d)\mathbf{u}_i(d) = Q(d)\mathbf{e}_i(d) + T(d)\rho_0, \qquad (2.12)$$

where:

-$\rho_0$is the initial internal state of Σ_G,

- $P(d), Q(d)$ and $T(d)$ are polynomial matrices such that $detP(0) \neq 0$ and with the matrix $T(d)$ left prime,

- $\mathbf{u}_i(d)$ is the d-transform of the manipulated variable $u_i(k)$, input of the holder, and $\mathbf{e}_i(d)$ is the d-transform of the output error $e_i(k)$ defined as

$$e_i(d) = r_i(d) - y_i(d), \quad i = 1, \cdots, \nu, \qquad (2.13)$$

such that the following conditions hold

$$\begin{aligned} e_i(k) &= 0, \text{for } k \geq \bar{k}, i = 1, \cdots, \nu, && (2.14) \\ u_i(k) &= 0, \text{for } k \geq \bar{k}, i = 1, \cdots, \nu, && (2.15) \end{aligned}$$

for each initial state x_0 of the plant, ρ_0 of the controller, $w_{0,i}$ of the i-th $(i = 1, \cdots, \nu)$ external reference generators and $\zeta_{0,1}$ and $\zeta_{0,2}$ of the disturbance generators. □

Condition (2.14) is the classical deadbeat requirement, condition (2.15) guarantees the desired ripple-free behaviour. In fact, if (2.15) is fulfilled, then, for $k \geq \bar{k}$ the error system $\Sigma_{E,i}$ exhibits the unforced response. Such response is null at the sampling instants by (2.14), so that, by assumption 2.1, it is null even for each continuous-time instant $t \geq \bar{k}T$.

3 Conditions for problem solvability

3.1 Sampled plant conditions

Let $L(d)^{-1}M(d)$ be a rlp factorization of $A(d)^{-1}B_1(d)F_1(d)^{-1}$, hence from (2.9)

$$\begin{aligned} \mathbf{y}_i(d) &= A(d)^{-1}C(d)x_0 + A(d)^{-1}B(d)\mathbf{u}_i(d) + \\ &\quad L(d)^{-1}M(d)\zeta_{0,1} + \\ &\quad F_2(d)^{-1}G_2(d)\zeta_{0,2}, && (3.1) \end{aligned}$$

By equations (2.10), (2.12), (2.13) and (3.1) it is found that the d-transforms $\mathbf{e}_i(d)$ and $\mathbf{u}_i(d)$ of $e_i(k)$ and $u_i(k) (i = 1, \cdots, \nu)$ respectively are given by

$$\begin{aligned} \mathbf{e}_i(d) &= (A(d) + B(d)P(d)^{-1}Q(d))^{-1} \\ &\quad [A(d)R_i(d)^{-1}S_i(d)w_{0,i} - C(d)x_0 \\ &\quad -B(d)P(d)^{-1}T(d)\rho_0 - A(d)L(d)^{-1} \\ &\quad M(d)\zeta_{0,1} - A(d)F_2(d)^{-1} \\ &\quad G_2(d)\zeta_{0,2}], && (3.2) \\ \mathbf{u}_i(d) &= (P(d) + Q(d)A(d)^{-1}B(d))^{-1} \\ &\quad [Q(d)R_i(d)^{-1}S_i(d)w_{0,i} - Q(d)A(d)^{-1} \\ &\quad C(d)x_0 + T(d)\rho_0 - Q(d)L(d)^{-1} \\ &\quad M(d)\zeta_{0,1} - Q(d)F_2(d)^{-1} \\ &\quad G_2(d)\zeta_{0,2}], && (3.3) \end{aligned}$$

respectively. Conditions (2.14) and (2.15) impose that $\mathbf{e}_i(d)$ and $\mathbf{u}_i(d)$ defined by (3.2) and (3.3) be polynomial vectors. Denoting by $\bar{R}(d)$ the least common left multiple of the $R_i(d)$'s $(i = 1, \cdots, \nu)$ the following theorem can be now proved (Budinsky and Jetto,1994).

Theorem 3.1_Problem 2.1 can be solved if and only if the following conditions are satisfied_

(a_d) _$A(d)$ and $B(d)$ are rlp,_

(b_d) _$A(d)$ is a left multiple of $\bar{R}(d)$, $L(d)$ and $F_2(d)$. Namely $A(d) = H(d)\bar{R}(d) = H_1(d)L(d) = H_2(d)$ $F_2(d)$, for some polynomial matrices $H(d), H_1(d)$, $H_2(d)$._ □

Remark 3.1. The proof of this theorem follows from a result stated in (Jetto (1994)). The design procedure of the controller, reported in the sufficiency proof, requires the solution of the following diophantine equation

$$P(d)\bar{A}(d) + Q(d)\bar{B}(d) = I_q, \qquad (3.4)$$

where $\bar{A}(d)$ and $\bar{B}(d)$ are relatively right prime (rrp) polynomial matrices such that $\bar{B}(d)\bar{A}(d)^{-1} = A(d)^{-1}B(d)$.

3.2 Continuous-time plant conditions

As the original data of the problem are the continuous-time representations of $\Sigma_P, \Sigma_1, \Sigma_2$ and $\Sigma_{R,i}$, it seems useful to translate the conditions stated by theorem 3.1 on conditions relative to such continuous-time systems. This is done in this sub-section.

Denote by $\mathbf{y}_{c,i}(s), \mathbf{u}_{c,i}(s)$ and $\mathbf{r}_{c,i}(s)$ the Laplace transforms of $y_i(t), u_i(t)$and$r_i(t)$ respectively $(i = 1, \cdots, \nu)$. By (2.1)-(2.8) it is found that

$$\begin{aligned} \mathbf{y}_{c,i} &= A_c(s)^{-1}C_c(s)x_0 + A_c(s)^{-1}B_c(s)\mathbf{u}_{c,i}(s) \\ &\quad A_c(s)^{-1}B_{1c}(s)F_{1c}(s)^{-1}\zeta_{0,1} + \\ &\quad F_{2,c}(s)^{-1}G_{2,c}(s)\zeta_{0,2} && (3.5) \\ \mathbf{r}_{c,i}(s) &= F_{c,i}(s)^{-1}G_{c,i}(s)w_{0,i}, && (3.6) \end{aligned}$$

where: $A_c(s), C_c(s), B_c(s), B_{1c}(s), F_{1c}(s), F_{2c}(s)$ $G_{2c}(s), F_{c,i}(s)$ and $G_{c,i}(s)$ are polynomial matrices such that

$$\begin{aligned} A_c(s)^{-1}C_c(s) &= C_p(sI_n - A_p)^{-1}, \\ A_c(s)^{-1}B_c(s) &= C_p(sI_n - A_p)^{-1}B_p, \\ A_c(s)^{-1}B_{1c}(s) &= C_p(sI_n - A_p)^{-1}M_pC_1, \\ F_{1c}(s) &= (sI_{n_1} - A_1), \\ F_{2c}(s)^{-1}G_2c(s) &= N_pC_2(sI_{n_2} - A_2)^{-1}, \\ R_{c,i}(s)^{-1}S_{c,i}(s) &= C_{r,i}(sI_{n_{w_i}} - A_{r,i})^{-1}, \end{aligned}$$

with the pairs of matrices $(A_c(s), C_c(s)), (F_{2c}(s), G_{2c}(s))$ and $(R_{c,i}(s), S_{c,i}(s))$ rlp.

Let $L_c(s)^{-1}M_c(s)$ be a rlp factorization of $A_c(s)^{-1}B_{1c}(s)F_{1c}(s)^{-1}$ and let $\bar{R}_c(s)$ be the common least left multiple of the $R_{c,i}(s)'s$ $(i = 1, \cdots, \nu)$, the following theorem (Budinsky and Jetto) can be stated.

Theorem 3.2 *Problem 2.1 admits a solution if and only if the following conditions are satisfied:*
(a_c) $A_c(s)$ and $B_c(s)$ are rlp,
(b_c) $A_c(s)$ is a left multiple of $\bar{R}_c(s)$, $L_c(s)$ and $F_{2c}(s)$. Namely $A_c(s) = H_c(s)\bar{R}_c(s) = H_{1c}(s)L_c(s) = H_{2c}(s) F_{2c}(s)$ for some polynomial matrices $H_c(s), H_{1c}(s)$ and $H_{2c}(s)$. □

From theorem 3.2 it follows that, in general, the deadbeat control problem here considered does not admit a solution in terms of a simple discrete-time controller. In most cases a continuous-time precompensator Σ_{GC} must be introduced between the holder and Σ_P to guarantee the fulfillment of condition (b_c). The purpose of next section is to outline a possible procedure for designing such precompensator.

4 Introduction of the internal model

The procedure for designing the precompensator providing the required internal model is based on the notion of skew-prime polynomial matrices (Wolovich (1978)) and on the following assumption.

Assumption 4.1. Let $(\bar{R}_c(s), \bar{A}_c(s)), (\bar{F}_{1c}(s), \bar{M}_c(s))$ and $(\bar{F}_{2c}(s), \bar{A}_{2c}(s))$ be three pairs of rlp polynomial matrices such that

$$A_c(s)\bar{R}_{1c}(s)^{-1} = \bar{R}_c(s)^{-1}\bar{A}_c(s), \quad (4.1)$$
$$B_{1c}(s)F_{1c}(s)^{-1} = \bar{F}_{1c}(s)^{-1}\bar{B}_{1c}(s), \quad (4.2)$$
$$A_c(s)F_{2c}(s)^{-1} = \bar{F}_{2c}(s)^{-1}\bar{A}_{2c}(s), \quad (4.3)$$

and let $A_m(s)$ denote the least common multiple of $\bar{R}_c(s), \bar{F}_{1c}(s)$ and $\bar{F}_{2c}(s)$. Matrices $B_c(s)$ and $A_m(s)$ are assumed to be externally skew-prime. □

Assumption 4.1 states a necessary and sufficient condition for the applicability of the design procedure of Σ_{GC} described by the following steps.

1) Compute $\bar{R}_c(s), \bar{F}_{1c}(s)$ and $\bar{F}_{2c}(s)$ defined by (4.1)-(4.3).

2) Compute their lclm $A_m(s)$. Hence polynomial matrices $E(s), E_1(s)$ and $E_2(s)$ are obtained such that

$$\begin{aligned} A_m(s) &= E(s)\bar{R}_c(s) \\ &= E_1(s)\bar{F}_{1c}(s) \\ &= E_2(s)\bar{F}_{2c}(s). \end{aligned} \quad (4.4)$$

3) Compute two polynomial matrices $A_g(s)$ and $B_{2c}(s)$ such that

$$A_m(s)B_c(s) = B_{2c}(s)A_g(s), \quad (4.5)$$

with $A_m(s)$ and $B_{2c}(s)$ rlp, $B_c(s)$ and $A_g(s)$ rrp and $det(A_g(s)) = \alpha\, det(A_m(s))$ for some non zero scalar α.

Assumption 4.1 is necessary and sufficient for the existence of $B_{2c}(s)$ and $A_g(s)$ satisfying (4.5). Then the desired precompensator Σ_{GP} is given by $A_g(s)^{-1}B_g(s)$ where $B_g(s)$ is a polynomial matrix rlp with $A_g(s)$ (e.g. $B_g(s) = I_q$ may be assumed). In fact, the pairs $(A_c(s), B_c(s))$ and $(A_m(s), B_{2c}(s))$ are rlp, while $(A_g(s), B_c(s))$ is rrp, then no zero-pole cancellation occurs in the factorization

$$\begin{aligned} & A_c(s)^{-1}B_c(s)A_g(s)^{-1}B_g(s) = \\ & A_c(s)^{-1}A_m(s)^{-1}B_{2c}(s)B_g(s) = \tilde{A}_c(s)^{-1}\tilde{B}_c(s), \end{aligned}$$

with: $\tilde{A}_c(s) = A_m(s)A_c(s)$ and $\tilde{B}_c(s) = B_{2c}(s)B_g(s)$, so the pair $(\tilde{A}_c(s), B_c(s))$ is rlp and condition (a_c) is satisfied. To verify the fulfilment of condition (b_c), observe that $L_c(s)^{-1}M_c(s)$ is a rlp factorization of $A_c(s)^{-1}B_{1c}(s)F_{1c}(s)^{-1} = (\bar{F}_{1c}(s)A_c(s))^{-1}\bar{B}_{1c}(s)$, hence a polynomial matrix $D(s)$ exists such that

$$\bar{F}_{1c}(s)A_c(s) = D(s)L_c(s), \quad (4.6)$$

By (4.1)-(4.6) and definition of $\tilde{A}_c(s)$ it follows that

$$\begin{aligned} \tilde{A}_c(s)\bar{R}_c(s)^{-1} &= A_m(s)A_c(s)\bar{R}_c(s)^{-1}, \\ &= E_c(s)\bar{A}_c(s), \\ \tilde{A}_c(s)L_c(s)^{-1} &= A_m(s)A_c(s)((\bar{F}_{1c}(s)A_c(s))^{-1}D(s) \\ &= E_1(s)D(s), \\ \tilde{A}_c(s)F_{2c}(s)^{-1} &= A_m(s)A_c(s)F_{2c}(s)^{-1}, \\ &= E_2(s)\bar{A}_{2c}(s). \end{aligned}$$

Therefore, compensator Σ_{GC} guarantees the fulfilment of conditions (a_c) and (b_c) and hence the solvabitlity of problem 2.1.

5 Controller algebraic design using Maple

The aim of this section is to introduce a set of CAD procedures, related to the problem considered in the previous sections, based on Computer Algebra System: Maple V Release 2 (Heck, 1993). They include procedures providing solution of matrix diophantine equations (unilateral and bilateral), left and right coprime matrix fraction representation, skew-prime factorization, greatest common divisor and minimal common multiple of two polynomial matrices, Smith form, Smith-McMillan form, and other procedures necessary to manipulate polynomial and rational matrices. The user can choose between symbolic and numerical mode of computation. Symbolic approach avoids numerical difficulties and enables one to get exact calculus and to treat problems depending on one or more parameters (this is not possible with numerical computation). On the other hand, numerical way is useful to save memory and computing time when dealing with problems involving high degree polynomials. The possibility of choosing the number of digits of floating point arithmetic for numerical computation provides an ability to set any prescribed preciseness of calculation.

The name of the presented CAD procedures supporting algebraic design methods is "*Dioph Library Manual*" (Budinsky and Jetto,1994). They are based on Maple's basic facilities for computing with matrices presented by package Linalg (Linear Algebra package). Program system Maple V Release 2 is available on many computers platforms such as: McIntosh, NeXT, Amiga, IBM PC and compatibles. Package "*Dioph Library Manual*" was tested by authors using the version of Maple for IBM PC working under MS-Windows. A minimal recommended configuration for IBM PC (or compatible) is processor model 486 and 8MB of RAM.

The description of the most important procedures of the package is reported below.

Common divisors and multiples of two polynomial matrices

$$\underline{\text{Format}}: \begin{cases} gcdL(A,B,P,Q,R,S,G,L), \\ gcdR(A,B,P,Q,R,S,G,L) \end{cases}$$

The algorithm provides the greatest common left ($gcdL$) and right ($gcdR$) divisor G (in triangular form) and minimal right ($gcdL$) and left ($gcdR$) multiple of two given polynomial matrices A, B. It also gives four polynomial matrices P, Q, R, S satisfying

$$\begin{aligned} AP + BQ &= G, \quad AR + BS = 0, \\ PA + QB &= G, \quad RA + SB = 0. \end{aligned}$$

where the pairs (P, Q) and (R, S) are right and left coprime respectively. The algorithm is based on elementary column operations (Kucera,1979).

Left and right coprime matrix fraction representation

$$\underline{\text{Format}}: cmfr(G, Al, Bl, Ar, Br).$$

The procedure provides left and right coprime matrix fraction representation in the form $G = Al^{-1}$ $Bl = BrAr^{-1}$. The input parameter is the transfer function G, the output parameters are the polynomial matrices Al, Bl, Ar, Br. The algorithm uses the procedure for finding right and left greatest common divisors.

Smith and Smith-McMillan form of polynomial matrices

$$\underline{\text{Format}} = \begin{cases} S & := smith(A, U, V), \\ M & := mcmil(G, U, V) \end{cases}$$

The first subroutine returns the Smith form S of the polynomial matrix A togheter with two unimodular polynomial matrices U and V such that $S = UAV$. The second one returns the Smith-McMillan form of the rational matrix G togheter with two unimodular polynomial matrices U and V such that $G = UGV$. Both these subroutines are built-in functions of Maple.

Solution of unilateral matrix diophantine equations

$$\underline{Format}: \begin{cases} diophL(A, B, C, X0, X1, Y0, Y1, T), \\ diophR(A, B, C, X0, X1, Y0, Y1, T). \end{cases}$$

These subroutines provide the general solution (if it exists) of any polynomial matrix equation in the form

$$\begin{aligned} AX + BY &= C, \quad \text{diophL}, \\ XA + YB &= C, \quad \text{diophR}. \end{aligned}$$

The input parameters are given polynomial matrices A, B and C with the same row (respectively columns)dimensions. The output parameters are polynomial matrices $X0, X1, Y0, Y1, T$ defining the general solution : $X = X0 + X1T$, $Y = Y0 + Y1T$ (diophL) and $X = X0 + TX1$, $Y = Y0 + TY1$ (diophR), where T is a matrix with appropriate dimensions which represents any polynomial matrix as a parameter of the general solution. A preliminary test on equation solvability is performed. The solving procedure is based on the greatest common divisor algorithm and extraction of matrix divisors (Kucera,1979).

Solution of bilateral matrix diophantine equations

$$\underline{\text{Format}}: diophB(A, B, C, X, Y).$$

The procedure provides the solution (if it exists) of any polynomial matrix equation of the form

$$AX + YB = C.$$

The input parameters are given polynomial matrices A, B and C and the output parameters are the solution polynomial matrices X and Y. A preliminary solvability test based on the Roth (1952) condition is performed. The solving algorithm is based on transforming the bilateral equation in a set of scalar diophantine equations.

Skew-prime fraction representation

$$\underline{Format}: \begin{cases} skewI(A, B, P, Q), \\ skewE(A, B, P, Q). \end{cases}$$

The subroutine $skewI$ finds two internally skew-prime polynomial matrices Q and P , given two externally skew-prime polynomial matrices B and A. Analogously, $skewE$ finds two externally skew-prime polynomial matrices B and A given two internally skew-prime polynomial matrices Q and P.

More precisely, given the equation $AB = QP$, with A and Q rlp, B and P rrp (and with $det(P) = \alpha\, det(A)$ for some non zero scalar α when A is square and non singular), $skewl(skewE)$ finds the right (left) hand side QP (AB) when the left (right) hand side AB (QP) is known. The solving algorithm is based on the solution of the bilateral polynomial matrix equation $AX + YB = I$ and on the results of (Emre, 1982).

Other utilities of the package (e.g. test for left and right primeness of polynomial matrices, conversion from state-space form to transfer matrix description and viceversa, plant discretization, etc.) are not described here for the sake of brevity. For their description and for more details on the above procedures see Budinsky and Jetto (1994).

References

Akashi H., H. Imai (1978)"Output deadbeat controllers with stability for discrete-time multivariable linear systems" *IEEE Trans. Autom. Control*, vol. AC-23, pp. 1036-1043.

Budinsky, M. and L. Jetto"Dioph Library Manual "*Int. Rep. of Dept. of Electronics and Automatica for Tempus Project*, University of Ancona, italia.

Budinsky, M. and L. Jetto (1994) "Some issues on deadbeat control and a related set of CAD procedures" *Int. Rep. of Dept. of Electronics and Automatica for Tempus Project*, University of Ancona, italia.

Chen, C.T. (1984) "*Linear System Theory*" Holt, Rinehart and Winston, New York.

Eichstaedt, B., (1982)," Multivariable closed - loop deadbeat control: a polynomial approach" *Automatica*, vol. 18, pp. 589-593.

Emre, E. (1982) "Further results on skew-prime matrices for linear systems over rings" *Proceedings of Control and Decision Conference*, pp. 134-137.

Emami-Naeini, A., G.F. Franklin (1982) "Deadbeat control and tracking of discrete-time systems" *IEEE Trans. Autom. Control*, vol. AC-27, pp. 176-181.

Heck,A. (1993) "*Introduction to Maple*" Springer-Verlag, N.Y.

Jetto,L. (1989) "Ripple-free tracking problem" *Int. J. Control*, vol. 50, pp. 349-359.

Jetto,L. (1990) "Deadbeat controllers with ripple-free requirement for siso discrete systems" *IEE Proceedings Part D*, vol 137, pp. 323-328.

Jetto, L. (1994) "Deadbeat ripple-free control with disturbance rejection: a polynomial approach" *IEEE Trans. Autom. Control*, vol. AC- 39, pp. 1759-1764.

Kaczorek T. (1982) " Deadbeat control in multivariable linear systems" *Int. J. Control*, vol. 35, pp. 653 - 663.

Kimura T., Y. Tanaka (1981) "Minimal-time minimal order deadbeat regulator with internal stability" *IEEE Trans. Autom. Control* , vol. AC-26, pp. 1276 -1282.

Kucera, V. (1979) " *Discrete Linear Control- The Polynomial Equation Approach*" J. Wiley & Sons, Chicester.

Kucera, V., (1980), "A deadbeat servo problem" *Int. J. Control*, vol. 32, pp. 107-113.

Kucera, V, and M. Sebek (1984) "On deadbeat controllers", *IEEE Trans. Autom. Control*, vol. AC-29, pp. 719-722.

Leden, B. (1977) "Multivariable deadbeat control", *Automatica*, vol. 13, pp. 185-188.

Roth, W.E. (1952) "The equation $AX - YB = C$ and $AX - XB = C$ in matrices" *Proc. Amer. Math. Soc.*, vol. 3, pp. 392-396.

Sebakhy, O.A.,T.M. Abdel-Moneim (1979) "State-regulation in linear discrete-time systems in minimum time" *IEEE Trans. Autom. control*, vol. AC-24, pp. 84-88.

Sebakhy,O.A. (1989) "State-space design of deadbeat ripple-free deadbeat multivariable systems" *Int. J. Syst. Science* , vol. 20, pp. 2673-2694.

Sirisena, H.R. (1985) "Ripple-free deadbeat control of SISO discrete systems" *IEEE Trans. Autom Control*, vol. AC-30, pp.168-170.

Urikura S., and A. Nagata (1987) " Ripple - free deadbeat control for sampled data systems " *IEEE Trans. Autom. Control*, vol. AC- 31, pp. 474 - 482.

Wolovich, W.A. (1978) "Skew-prime polynomial matrices" *IEEE Trans. Autom Control*, vol. AC-23, pp. 880-887.

Wolovich, W.A. (1983) "Deadbeat error control of discrete multivariable systems" *Int. J. Control*, vol. 37, pp. 567-582.

Xiaotian C., L. Zhang, and J. C. Readle (1994) "The analytic formulas of the ripple - free tracking problem " *Automatica* , vol. 30, pp. 905 - 908.

IDENTIFICATION BASED ADAPTIVE CONTROL OF CHAOTIC CONTINUOUS-TIME SYSTEMS

ALEXANDER Yu. POGROMSKY

Institute of Problems of Mechanical Engineering, 61, Bolshoy, VO, St.Petersburg, 199178, RUSSIA

Abstract. The present paper further developes ideas of controlling chaos by the speed gradient method. It is shown how to design parameter identifier for chaotic systems ensuring achievement of the objective of parameter identification. The data obtained by the identifier can be used for adaptive control law design. Two examples are considered: identification of parameters of periodically perturbed Duffing equation and stabilizing the inherent unstable equilibrium point of the thermal convection loop model when parameters of the model are unknown.

Key Words. Chaotic systems, parameter identification, persistent excitation

1. INTRODUCTION

Recently there has been some increasing interest in the study of chaotic systems with potential application to the various fields of science and diverse approaches in control of chaos including open-loop (Hubler & Luscher, 1989) and feedback (Chen & Dong, 1993, a; Chen & Chou, 1994) controls have been successfully applied (see survey of Chen & Dong, 1993, b). Some attempts have been also made to use concept of adaptive control (Hubler, 1989; Vassiliadis, 1994) for controlling chaotic systems. Importance of the algorithms of parameter identification for adaptive control design is underlined in (Vassiliadis, 1994), therein several structures of parameter identifiers are proposed mostly for discrete-time systems. The present paper further extends the ideas of feedback control of chaotic continuous-time systems based on the speed gradient (SG) method. It has been already shown (Fradkov & Pogromsky, 1994,a) that this method can be successfully applied for purposes of stabilization and synchronization of chaotic continuous-time systems. The goal of the present consideration is to utilize the SG method for parameter identification of such system and to design adaptive control law based on the identification approach. As examples of the systems with chaotic behavior we use Duffing and Lorenz systems that has become the traditional objects in considerations regarding deterministic chaos.

2. SPEED GRADIENT ALGORITHM OF PARAMETER IDENTIFICATION

Assume that the model of plant depends on the vector of unknown parameters θ. Consider the tunable model based on the model reference approach and assume also that in this case the error equation has the following form:

$$\dot{e} = f_0(e,t) + f_1(t)(\theta - \theta_m), \qquad (2.1)$$

where $e = x - x_m$ is the error vector $e \in R^n$, x is the state vector of the observed model, x_m is the state vector of the tunable model, θ is the vector of unknown parameters, $\theta \in R^p$, θ_m are their estimates which are determined by an estimating algorithm.

Consider the SG algorithm of estimating which ensures achievement of the main objective:

$$Q(e,t) \to 0, \quad \text{as} \quad t \to \infty, \qquad (2.2)$$

where $Q(e,t) \geq 0$ is a scalar objective function. The SG algorithm provides the motion of the system in the direction along which the objective function decreases. Its differential form looks as follows (Fomin *et al.*, 1981; Fradkov, 1990)*) :

$$\dot{\theta}_m = -\Gamma \nabla_{\theta_m} \dot{Q}(e,t) = -\Gamma f_1^T(e,t) \nabla_e Q(e,t), \qquad (2.3)$$

where $\Gamma = \Gamma^T > 0$ is a $p \times p$ gain matrix.

All conditions under which this control law ensures achievement of the goal (2.2) are established in (Fomin *et al.*, 1981; Fradkov, 1990). Therein the conditions for which the additional goal

$$\theta_m \to \theta, \quad \text{as} \quad t \to \infty \qquad (2.4)$$

is achieved were also found.

*) Below $\nabla_x F$ stands for the vector of partial derivatives of F with respect to x

This property, usually called the identifying property of the SG algorithm, allows to construct parameter identifiers based on this algorithm. Establishment of this property depends on the presence of persistent excitation (PE), which is necessary to ensure asymptotic stability in various problems of adaptive control (Fomin *et al.*, 1981; Anderson et al. 1986).

Definition 1: Function $\omega(t)$: $R \to R^{m\times n}$ bounded for $t \geq 0$ is called persistently exciting ($\omega(t) \in PE$) as long as there exist $\alpha, \delta > 0$ such that for all $t_0 > 0$

$$\int_{t_0}^{t_0+\delta} \omega(t)\omega^T(t)dt - \alpha I_m \geq 0. \qquad (2.5)$$

In (Fradkov, 1990) six equivalent definitions of PE are given. Using this definition, we recall the standard stability theorem for identifying SG algorithms (Fomin *et al.*, 1981):

Theorem 1: Consider the nonlinear time-variant equation:

$$\dot{e} = f_0(e,t) + f_1(t)(\theta - \theta_m), \qquad (2.6)$$

where $e \in R^n$, $\theta, \theta_m \in R^p$ and the parameters are updated with the estimator:

$$\dot{\theta}_m = -\Gamma f_1^T(t)\nabla_e Q(e,t), \ \Gamma = \Gamma^T > 0, \qquad (2.7)$$

with the following assumptions:

A1. $f_0(0,t) \equiv 0$ for any $t > 0$.

A2. For $i = 0,1$ and $f_i \in C^1$, f_i and $\partial f_i / \partial t$ are uniformly bounded functions of t.

A3. $Q(e,t) \geq 0$, $Q(e,t) \in C^1$.

A4. $\dot{Q}(e,t)$ is convex in θ_m.

A5. (*attainability condition*) There exist $\theta_m^* \in R^p$ and scalar continuous function $\rho(Q)$, with property $\rho(Q) > 0$ when $Q > 0$, such that inequality

$$\dot{Q}(e,t)\Big|_{\theta_m=\theta_m^*} \leq -\rho(Q(e(t),t)) \qquad (2.8)$$

holds for any solutions $e(t)$ of the system (2.6).

A6. (*growth condition*) If $Q(e,t)$ is bounded then e is bounded as well.

A7. $f_1 \in PE$.

Under these conditions the equilibrium $e = 0$, $\theta_m = \theta$ of the system (2.6), (2.7) is globally uniformly asymptotically stable.

This theorem can be modified for functions $f_i(\cdot)$ dependent on the vector of tunable parameters (Ortega & Fradkov, 1993).

Of course the theorem presented above is an important tool for constructing identifying SG algorithms but it is possible to design identifying SG laws which do not satisfy the conditions of Theorem 1. Such example is presented in the next section. Certainly it means that Theorem 1 can be further generalized, however the author's desire is not to drive the reader crazy by the abundance of a new theorem conditions. Perhaps the best solution is to investigate the global behavior for the specific example and not to try to formulate more general theorem generalizing the way of proving. However, it is worth mentioning that establishment of the asymptotic stability for the SG parameter identifier considered in the next section depends on the condition that is stronger than A7 (persistent excitation).

In according to Lemma proved in (Fomin *et al.*, 1981) the persistent excitation ensures that for $\omega(t) \in PE, \theta(t) \in C^1$ and $\dot{\theta}(t) \to 0$, $\theta(t)$ tends to zero provided that $\theta^T(t)\omega(t) \to 0$. We define a subset PE_1 of the set PE by the following definition:

Definition 2: Function $\omega(t)$: $R \to R^{m\times n}$, $\omega(t) \in C^1$ is said to belong to PE_1 if $\omega(t) \in PE$ and $\dot{\omega}(t) \in PE$.

The following can be easily established

Proposition 1: Consider the vector functions $\omega(t)$: $R \to R^{m\times n}$ and $\theta(t)$: $R \to R^{n\times m}$. Suppose $\theta(t) \in C^{2}$ *), and $\dot{\theta}(t) \to 0$ when $t \to \infty$. Assume further that $\omega(t) \in PE_1$. Under these conditions $\theta(t) \to 0$, when $t \to \infty$ provided that $d\left(\theta^T(t)\omega(t)\right)/dt \to 0$, when $t \to \infty$.

This proposition is a consequence from the Lemma proved in (Fomin *et al.*, 1981) and to satisfy Definition 2 function $\omega(t)$ must demonstrate more complex behavior in comparison with the usual PE. This property can provide stability in the examples for which the standard PE is not sufficient to prove asymptotic stability.

3. PARAMETER IDENTIFICATION OF THE DUFFING EQUATION

Consider the problem of estimating of the coefficients of the Duffing equation written in the Cauchy form:

$$\begin{cases} \dot{x}_1 = x_2 \\ \dot{x}_2 = -p_1 x_1 - x_1^3 - p x_2 + q\cos(\omega t + \Delta) \end{cases} \qquad (3.1)$$

where p, p_1, q, Δ $(p > 0)$ are unknown parameters, which are to be estimated and ω is assumed to be

*) The result can be extended to the functions with picewise continuous derivative.

known. Without loss of generality we can put $q > 0$. In (3.1) parameter Δ expresses the phase uncertainty. Equations (3.1) can be rewritten in the following form:

$$\begin{cases} \dot{x}_1 = x_2 \\ \dot{x}_2 = -p_1 x_1 - x_1^3 - p x_2 + q_1 \cos \omega t + s_1 \sin \omega t \end{cases}, \quad (3.2)$$

where $q_1^2 + s_1^2 = q^2$ (3.3)

and Δ is a solution of the following equation:

$$q_1 \cos \Delta = s_1 \sin \Delta$$

It is well known that equations (3.1) posses complex behavior dependent on the value of parameters, including chaotic dynamics. We will study the problem of estimating of the coefficients when the observed system has the chaotic dynamics. The full order tunable model can be designed as

$$\begin{cases} \dot{x}_{1m} = x_{2m} \\ \dot{x}_{2m} = -p_{1m} x_{1m} - x_{1m}^3 - p_m x_{2m} + q_{1m} \cos \omega t + \\ \quad + s_{1m} \sin \omega t + K(x_1 - x_{1m}) - 3 x_1 x_{1m}(x_1 - x_{1m}), \end{cases} \quad (3.4)$$

$K > -p_1$,

where x_{1m}, x_{2m} are the states of the tunable model and $p_m, p_{1m}, q_{1m}, s_{1m}$ are the adjustable parameters.

Denoting $e_1 = x_1 - x_{1m}$, $e_2 = x_2 - x_{2m}$ one can obtain the error equations:

$$\begin{cases} \dot{e}_1 = e_2 \\ \dot{e}_2 = -(p_1 + K)e_1 + (p_{1m} - p_1)x_{1m} - e_1^3 - p e_2 + \\ \quad + (p_m - p)x_{2m} - (q_{1m} - q_1)\cos \omega t - (s_{1m} - s_1)\sin \omega t. \end{cases} \quad (3.5)$$

It is easy to notice that this system has the form (2.1). To estimate the unknown parameters consider the SG algorithm of estimating ensuring the achievement of the following main goal (Chen & Dong, 1993, a):

$$Q(e) = \left((K + p_1)e_1^2 + e_1^4 / 2 + e_2^2\right)/2 \to 0, \text{ when } t \to \infty \quad (3.6)$$

This goal is achieved when the error between the observed and tunable models vanishes as time increases. Clearly the function $Q(e)$ is nonnegative for $K + p_1 \geq 0$. Taking a derivative of $Q(e)$ along the trajectory of system (3.5) one can obtain

$$\dot{Q}(e) = -p e_2^2 + (p_{1m} - p_1)e_2 x_{1m} + (p_m - p)e_2 x_{2m} - \\ -(q_{1m} - q_1)e_2 \cos \omega t - (s_{1m} - s_1)e_2 \sin \omega t. \quad (3.7)$$

The SG algorithm of estimating of the unknown parameters can be obtained in the following differential form:

$$\dot{p}_{1m} = -\gamma_1 \partial \dot{Q}(e) / \partial p_{1m} = -\gamma_1 e_2 x_{1m}$$

$$\dot{p}_m = -\gamma_2 \partial \dot{Q}(e) / \partial p_m = -\gamma_2 e_2 x_{2m}$$

$$\dot{q}_{1m} = -\gamma_3 \partial \dot{Q}(e) / \partial q_{1m} = \gamma_3 e_2 \cos \omega t \quad (3.8)$$

$$\dot{s}_{1m} = -\gamma_4 \partial \dot{Q}(e) / \partial s_{1m} = \gamma_4 e_2 \sin \omega t$$

$$\gamma_1 > 0, \gamma_2 > 0, \gamma_3 > 0, \gamma_4 > 0.$$

Now, when the SG algorithm has been found, we check the achievement of the main and identifying objectives.

One can notice that all conditions of Theorem 1 hold except the *attainability condition*. Indeed, for $p_{1m} = p_1$, $p_m = p$, $q_{1m} = q_1$ and $s_{1m} = s_1$, we conclude from (3.7) that

$$\dot{Q}(e) = -p e_2^2, \text{ or } \dot{Q}(e) = -\rho(e), \quad (3.9)$$

where $\rho(e) = p e_2^2$ and clearly $\rho(\cdot)$ does not depend explicitly on $Q(e)$ as required in Eq.(2.8). However the Eq. (3.9) is sufficient to prove uniform convergence of e_2 to zero (Fradkov & Pogromsky, 1994,b).

Taking into account uniform convergence of e_2 to zero, from (3.5) one can obtain that $e_1 \to$ const and

$$f_1^T(t)\tilde{\theta}(t) \to \text{const}, \text{ as } t \to \infty, \quad (3.10)$$

where $f_1^T(t) = [x_{1m}\ x_{2m}\ \cos \omega t\ \sin \omega t]$,

$$\tilde{\theta} = [p_{1m} - p_1\ p_m - p\ q_{1m} - q_1\ s_{1m} - s_1]^T.$$

It is easy to obtain that (3.10) holds only for $\tilde{\theta}(t) \to 0$ if $f_1^T(t) \in PE_1$ (see Proposition 1), or, in other words, x_{1m} and x_{2m} must not tend to any constant values, that is, in turn, satisfied if the choice of the true parameters p, p_1, q_1, s_1 ensures the absence of stable equilibrium points in the system (3.2). Further from (3.5), using the fact that $\tilde{\theta}(t) \to 0$, from the uniform continuity of e_2 and $\dot{e}_2$, square integrability of e_2 and uniform boundedness of $\dot{e}_2$ one can obtain that $\dot{e}_2 \to 0$ as well, that in turn imlies that $e_1 \to 0$ In other words, equilibrium $e = 0, \tilde{\theta} = 0$ is uniformly asymptotically stable. It completes the proof of the following

Proposition 2: If the choice of true parameters of the system (3.2) ensures the absence of stable equilibrium points then the parameter identifier based on the tunable model (3.4) with the SG algorithm of estimating (3.8) provides convergence of the adjustable parameters $p_m, p_{1m}, q_{1m}, s_{1m}$ to their true values p, p_1, q_1, s_1.

Proposed identifier estimates values of the parameters p_1, p_{1m}. As an estimate of the value of q one can use the following expression:

$$q_m = \sqrt{q_{1m}^2 + s_{1m}^2}. \quad (3.11)$$

It is easy to notice that this estimate converges to the true value of parameter q as it appears in the system (3.1).

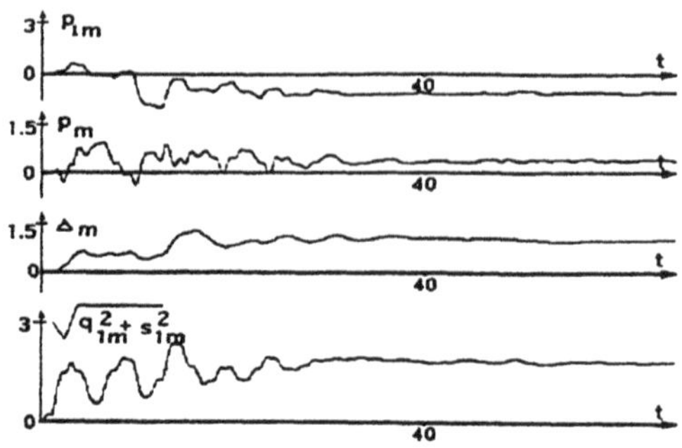

Fig 1. Convergence of the estimates to their true values.

To demonstrate the ability of proposed identifier to estimate the parameters of the Duffing's system we carried out computer simulation. Figure 1 shows the convergence of the estimated parameters to their true values when the observed system possesses chaotic dynamics (for the following parameters: $p = 0.4$, $p_1 = -1.1$, $q = 1.8$, $\omega = 1.8$. Notice that the value of parameter Δ does not influence the system behavior). Parameters of the identifier have been chosen as follows: $\gamma_1 = \gamma_2 = 1.9$, $\gamma_3 = \gamma_4 = 2$, $\gamma_5 = 0.6$, $K = 3$. One can notice that the objective of estimating is successfully achieved.

4. STABILIZING THE INHERENT EQUILIBRIUM POINT OF THE LORENZ SYSTEM

It has been discussed in the previous sections how to construct parameter identifier for systems with chaotic behavior. This section deals with the problem of control law design based on the considered identifier.

A great deal of considerations (Chen & Dong, 1993, a; Singer *et al.*, 1991; Wang & Abed, 1992) concerning controlling chaos regards stabilization of the inherent unstable equilibrium or periodic solution of the system with original chaotic behavior. Here it will be shown how this problem can be solved by the considered approach.

Singer, Wang and Bau (1991) study model of thermal convection loop of the following form:

$$\dot{x} = \sigma(y - x),\ \dot{y} = -y - xz,\ \dot{z} = -z + xy - r + u, \qquad (4.1)$$

where $\sigma > 0$ is the Prandtl number, $r > 0$ is the Raleigh number and u is a control action that is to be determined.

For $0 < r < 1$ the uncontrolled system has one stable globally attracting equilibrium - $(0,0,r)$ that corresponds to no-motion state of the thermal convection. For $1 < r < r_H$, where r_H is the value for which the Hopf bifurcation occurs, there are two asymptotically stable equilibria $C_+ : (x_+, y_+, -1)$ and $C_- : (-x_+, -y_+, -1)$, where $x_+ = y_+ = \sqrt{r-1}$. These two equilibrium points represent the states of steady convection in the counterclockwise or clockwise direction. For $r > r_H$ these equilibria lose their stability in Hopf bifurcation and for the greater values of r the system has no more equilibrium points.

Singer, Wang and Bau (1991) applied an on-off controller of the form $u = -\varepsilon \mathrm{sgn}(z+1)$ to stabilize one of the inherent unstable equilibria (C_+ or C_-) of the uncontrolled system (4.1) for $r > r_H$. It can be shown (Fradkov & Pogromsky, 1994,a) that this control law is the SG algorithm in the finite form and it provides convergence to the neighborhood of one of the inherent equilibria positions. It is worth mentioning that this controller can stabilize equilibrium C_+ or C_- and which of them will be stabilized depends on the initial conditions. We propose a continuous adaptive control law ensuring the same goal of stabilization based on the parameter identifier. Moreover, the researcher can choose which of the equilibria - C_+ or C_- is to be stabilized by the controller.

First we want to design the control law assuming that the state variables of the system (4.1) are measurable and $r > 1$ is known. To drive the system to the position corresponding to the inherent equilibrium point C_+, the following control is suggested:

$$u = -x_+(x - x_+) - \gamma_1(z+1),\ \gamma_1 > -1. \qquad (4.2)$$

Then the controlled system has two equilibrium points C_0 and C_+:

$$C_0 = (0,0,-1),\quad C_+ = (x_+, x_+, -1) = (\sqrt{r-1}, \sqrt{r-1}, -1) \quad (4.3)$$

and one of them (C_+) must be stabilized. Define the set

$$\Omega = \{x, y, z : (x - \sqrt{r-1})^2 / \sigma + (y - \sqrt{r-1})^2 + (z+1)^2 < (1 + 1/\sigma)(r-1)\} \qquad (4.4)$$

Proposition 3: All trajectories of the controlled system (4.1), (4.2) with initial conditions $x(0), y(0), z(0)$ from the set Ω converge to the equilibrium point C_+.

Proof: Consider the following Lyapunov function:

$$V(x,y,z) = \left((x - x_+)^2 / \sigma + (y - x_+)^2 + (z+1)^2\right)/2 \geq 0. \quad (4.5)$$

Its derivative along the trajectories of the system (4.1), (4.2) satisfied the inequality

$$\dot{V}(x,y,z) = -(x-y)^2 - (\gamma_1 + 1)(z+1)^2 \leq 0, \qquad (4.6)$$

and, hence, function $V(x,y,z)$ is nonincreasing. The set on which $\dot{V}(x,y,z) = 0$ contains two whole trajectories of the controlled system: equilibria C_0 and

C_+, and $\min_{x,y,z} V(x,y,z) = V(C_+)$. Therefore, for all initial conditions $x(0), y(0), z(0)$, such that $V(x(0), y(0), z(0)) < V(C_0)$, or

$$\{x(0), y(0), z(0)\} \in \Omega, \quad (4.7)$$

all trajectories uniformly converge to the equilibrium point C_+. □

Corollary: Equilibrium point C_0 of the controlled system (4.1), (4.2) is a nonrobust equilibrium.

Notice that proposed control algorithm (4.2) depends on the coordinates of the inherent equilibrium points of the uncontrolled system and the value of parameter r must be known to apply it. Assume further that the state variables of the system are measurable as was mentioned above and the value of r is unknown. Using the technique discussed in the second section consider the following parameter identifier:

$$\dot{z}_m = -z_m + xy - r_m + u, \quad \dot{r}_m = -\gamma_2(z - z_m), \quad \gamma_2 > 0. \quad (4.8)$$

Denoting $e = z - z_m$, one can obtain the error equations of the following form:

$$\dot{e} = -e - (r - r_m), \quad \dot{r}_m = -\gamma_2 e, \quad \gamma_2 > 0. \quad (4.9)$$

Choosing the objective goal in the scalar form $Q(e) = e^2/2$, it is easy to establish that the equations (4.9) are the SG parameter identifier and since all conditions of the Theorem 1 are satisfied, convergence $r_m(t) \to r$ and $z_m(t) \to z(t)$ as $t \to \infty$ is proved. Notice that this convergence is exponential.

Considered parameter identifier is suitable for estimating coordinates of the inherent equilibrium points of the uncontrolled system (4.1). To obtain the estimate of the coordinate x_+ one can calculate:

$$\tilde{x}_+ = \sqrt{|r_m - 1|}, \quad (4.10)$$

It may be seen that for any $r > 1$ this estimate converges to x_+ as time increases. Notice that this convergence does not depend on the control action.

Now one can design adaptive control law for stabilizing the inherent equilibrium point C_+ of system (4.1). The controller includes parameter estimator (4.8), (4.10) and nonadaptive regulator (4.2):

$$\begin{aligned} &\dot{z}_m = -z_m + xy - r_m + u, \quad \dot{r}_m = -\gamma_2(z - z_m), \\ &u = -\sqrt{|r_m - 1|}\,x + r_m - 1 - \gamma_1(z+1), \quad \gamma_1 > -1, \quad \gamma_2 > 0. \end{aligned} \quad (4.11)$$

First we establish the boundedness of trajectories of the adaptive control system. Consider the following Lyapunov function:

$$V(x,y,z) = x^2 + y^2 + (z - \sigma)^2, \quad (4.12)$$

whose derivative along the trajectory of the system for bounded r_m determines an ellipsoid outside which $\dot{V}(x,y,z) < 0$. It proves the boundedness of trajectories of the system (4.1), (4.11).

Since *exponential* stability of the solutions of the equations (4.9) is established one can conclude that

$$\int_0^\infty |r - r_m(t)|\,dt < \infty, \quad \int_0^\infty \left|\sqrt{r-1} - \sqrt{|r_m(t) - 1|}\right| dt < \infty. \quad (4.13)$$

And since all trajectories of the system (4.1), (4.11) are bounded the following inequalities hold:

$$\begin{aligned} &0 \le \mu_1 = \int_0^\infty |(r - r_m(t))(z(t)+1)|\,dt < \infty, \\ &0 \le \mu_2 = \int_0^\infty \left|\left(\sqrt{r-1} - \sqrt{|r_m(t) - 1|}\right)(z(t)+1)x(t)\right| dt < \infty. \end{aligned} \quad (4.14)$$

In order to analyze the system behavior consider the following Lyapunov functional:

$$\begin{aligned} V_t(x,y,z) = \frac{1}{2}\Big\{ &\left(x(t) - \sqrt{r-1}\right)^2/\sigma + \left(y(t) - \sqrt{r-1}\right)^2 + (z+1)^2 + \\ &+ \int_t^\infty |(r - r_m(t))(z(t)+1)|\,dt + \int_t^\infty \left|\left(\sqrt{r-1} - \sqrt{|r_m(t)-1|}\right)(z(t)+1)x(t)\right| dt \Big\}, \end{aligned} \quad (4.15)$$

whose derivative satisfies

$$\dot{V}_t(x,y,z) \le -(x(t) - y(t))^2 - (1+\gamma_1)(z(t)+1)^2. \quad (4.16)$$

As before, the set on which $\dot{V}_t = 0$ contains two equilibria (for $t \to \infty$): C_0 and C_+ and one can establish that all trajectories with the initial conditions from the set

$$\begin{aligned} \Omega_1 = \{x,y,z:\ &(x - \sqrt{r-1})^2/\sigma + (y - \sqrt{r-1})^2 + \\ &+ (z+1)^2 < (1 + 1/\sigma)(r-1) - \mu_1 - \mu_2\} \end{aligned} \quad (4.17)$$

uniformly converge to the point C_+. It completes the proof of the following

Proposition 4: All trajectories of the controlled system (4.1), (4.11) with initial conditions $x(0), y(0), z(0)$ from the set Ω_1 converge to the equilibrium point C_+.

Notice that the values μ_1 and μ_2 can be made arbitrary small by choosing appropriate values of the constant γ_2 and initial conditions $z_m(0), r_m(0)$. Hence the objective of stabilization will be successfully achieved if the control is applied at the instance when the initial data of the system (4.1) belong to the neighborhood (4.17) of the inherent equilibrium position that must be stabilized. However it is not necessary to design the additional control law which drives the system (4.1) to the position from the set (4.17) because of the chaotic behavior of the uncontrolled system (that was assumed). Indeed it can be shown (because of the page limit the proof is ommitted) that trajectory of the uncontrolled system gets into the set (4.4) sooner or later without any external force for almost all initial conditions. Figure 2 shows how the designed adaptive control affects the system for the following parameters: $r = 82, \sigma = 10, \gamma_1 = 0.6, \gamma_2 = 0.5$ and initial conditions $x(0) = y(0) = 12,\quad z(0) = z_m(0) = 6, r_m(0) = 20$. It is seen that the objective of stabilization is successfully achieved.

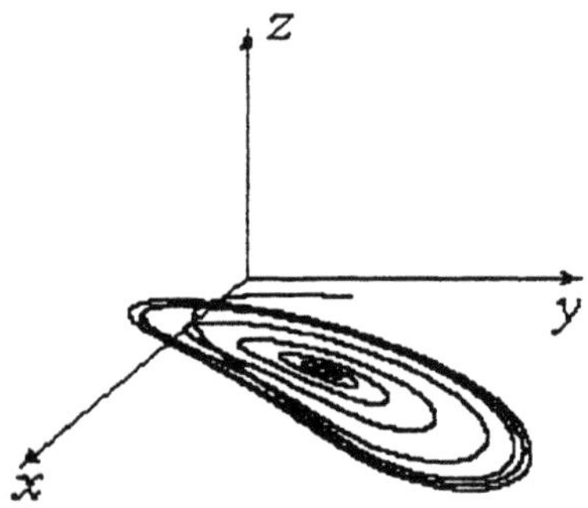

Fig. 2. Transient process in the adaptive system.

5. CONCLUSION

In the present paper it has been demonstrated how to design parameter identifier for chaotic systems and to utilize it for adaptive control of such systems. It is worth mentioning that the presence of chaotic behavior ensures appearance of the persistent excitation which facilitates the process of parameter identification. It has been also shown in the paper that the parameter identifier can be applied to organize adaptive control of the thermal convection loop model in order to stabilize the inherent unstable equilibrium points of the uncontrolled model. This problem can be solved by the proposed controller and all parameters of the model may be unknown. The most valuable feature of the controller is that the value of control vanishes as the objective of stabilization is achieved. To succeed in stabilizing the control must be applied at the moment when trajectory of the uncontrolled system gets into the determined neighborhood of the position of interest but chaotic behavior of the uncontrolled system ensures that it is not necessary to design additional controller which drives the system close enough to the target.

6. ACKNOWLEDGMENT

The author thanks Prof. A.L. Fradkov (IPME, Saint-Petersburg) for many helpful discussions during various stages of this work.

7. REFERENCES

Anderson B., Bitmead R., Johnson C., Kokotovic P., Kosut R., Mareels J., Praly L., Riedle B. (1986), *Stability of adaptive systems: passivity and averaging analisis*, (MIT Press, Cambridge, MA).

Chen G., Dong X., (1993,a) On feedback control of chaotic continuous-time systems. *IEEE Trans. Circuits Syst.*, **40**(9), pp. 591-601.

———, (1993,b) "From chaos to order- perspectives and methodologies in controlling chaotic nonlinear dynamical systems.", *Int. J. Bifurcation Chaos*, **3**(6), pp. 1363-1409.

Chen Y.H., Chou M.Y., (1994) Continuous feedback approach for controlling chaos. *Phys. Rev. E*, **50**, pp. 2331-2334.

Fomin V.N., Fradkov A.L., Yakubovich V.A., (1981) *Adaptive control of dynamic objects.*, (Moscow, Nauka), (in Russian).

Fradkov A.L. (1990), *Adaptive control in large-scale systems.*, (Moscow, Nauka), (in Russian).

Fradkov A.L., Pogromsky A.Yu. (1994, a) Speed-gradient adaptive control of chaotic continuous-time systems, *Preprint IPME*, **107**, pp. 29-48.

———,(1994, b) Speed gradient control of chaotic continuous-time systems (Submitted to *IEEE Trans. Cirq. Syst*).

Hubler A., (1989), Adaptive control of chaotic systems, *Helv. Phys. Acta*, **62**, pp.342-346.

Hubler A., Luscher E. (1989) Resonant stimulation and control of nonlinear oscillators, *Naturwissenschaften*, **76**, pp. 67-69.

Ortega R., Fradkov A., (1993) "Asymptotic stability of a class of adaptive systems." *Int. J. Adaptive Control and Signal Processing* , **7**, pp. 255-260.

Singer J., Wang Y.-Z., Bau H. (1991) "Controlling a chaotic system." *Phys. Rev. Let.*, **66**(9), pp. 1123-1125.

Vassiliadis D. (1994) Parametric adaptive control and parameter identification of low-dimensional chaotic systems, *Physica D*, **71**, pp. 319-341.

Wang H., Abed E.H., (1992) "Bifurcation control of chaotic dynamical system." *IFAC Symp. NOLCOS'92*, pp.283-288.

COMPLETE COMPUTATION OF THE RELATIONS FOR IMPLICIT DYNAMICAL SYSTEMS

F. BOULIER * , M. PETITOT **

* Université Lille I Villeneuve d'Ascq - France. *email:* boulier@lifl.lifl.fr

** L.S.S, Ecole SUPELEC Gif-sur-Yvette - France. *email:* petitot@lss.supelec.fr

Abstract. We give the mathematical foundations to compute *all* the differential algebraic relations satisfied by variables of some dynamical system defined by *implicit* polynomial equations. The resulting algorithm is implemented on UNIX workstations.

Résumé. *On donne les bases mathématiques permettant un calcul* complet *des relations algébro-différentielles qui lient les grandeurs d'un système dynamique donné* sous forme implicite *par des équations différentielles polynomiales. L'algorithme qui en découle est implanté sur station UNIX.*

Keywords. Differential algebra, nonlinear control, implicit system, observability

1. INTRODUCTION

We are concerned with *implicit* systems defined by some finite set of differential polynomial equations with coefficients in any differential ground field k of characteristic zero. We do not distinguish between the variables of the system (inputs, outputs, state variables, parameters). Such an approach, using differential algebra may be found in (Fliess 1989, Diop 1991, Fliess and Glad 1993, Forstman 1991, Michalik and Willems 1994).

We propose an algorithm which exhaustively computes the differential algebraic relations satisfied by the variables of some dynamical system. It directly solves some control theory problems (output-tracking, input-output inversion, observability test, local and global structural identifiability for parameters (Ollivier 1990)). It may also be very useful for the Disturbance Decoupling Problem by static or dynamic feedback.

The computation of the above cited relations is easier for *explicit* systems such as the following one:

$$\left\{\begin{array}{rcl} \dot{x}_1 & = & f_1(x,u) \\ \vdots & & \vdots \\ \dot{x}_n & = & f_n(x,u) \\ y & = & h(x) \end{array}\right. \qquad (1)$$

For such systems, the solutions described in the literature (geometrical approach, structure algorithm ...) are based on the *Church–Rosser* property of the rewriting system:

$$\mathcal{R}\left\{\begin{array}{rcl} \dot{x}_1 & \longrightarrow & f_1(x,u) \\ \vdots & & \vdots \\ \dot{x}_n & \longrightarrow & f_n(x,u) \\ y & \longrightarrow & h(x) \end{array}\right.$$

In other words, a differential polynomial relation $p = 0$ is a consequence of the system 1 if and only if p is reduced to zero by the rewriting system $\mathcal{R}$:

$$p(x,u,y) = 0 \bmod I \iff p \rightarrow_{\mathcal{R}} 0$$

It follows that the differential ideal I generated by the equations (1) is prime[1].

However, M. Fliess in (Fliess 1989) does not define a dynamical system as a set of equations but as a finitely generated differential field extension K of the ground field k. The set of the relations satisfied by the elements of a field is an differential ideal. It is easy to show that this ideal is always prime.

For implicit systems, the radical $\sqrt{I}$ of the ideal I generated by the equations is not necessarily prime[2]. We are thus concerned with all the *models* of Σ i.e. with all the differential field extensions of the ground field which contain at least a solution for the system Σ, i.e., with all the differential prime ideals containing I (see theorem 3.1).

We may find in (Ritt 1950), *Constructive methods*, p. 107, an algorithm which describes $\sqrt{I}$ as the intersection of some finite set of differential prime ideals given

[1] It may be shown that the differential algebra of the residue classes modulo I is isomorphic to the *free* algebra $k[x, u, \dot{u}, \ddot{u} \ldots]$. Since this latter has no zero-divisor, the ideal I is prime.

[2] it is an open problem to decide whether the radical of any finitely generated ideal is prime.

by *characteristic sets*. Up to our knowledge, this algorithm has never been implemented because it needs factorizations (Fliess and Glad 1993) of polynomials over towers of algebraic field extensions of the ground field k.

Our algorithm does not perform any factorization. It is based on the following ideas: Ritt's reduction algorithm, Seidenberg's elimination algorithm (Seidenberg 1956), Tarski's models theory, Rosenfeld's coherent autoreduced sets and some more recent techniques using Gröbner bases (Buchberger 1965). It exhaustively computes the relations satisfied by any subset V of the set of the variables of the system. More precisely:

- it describes the ideal $\sqrt{I}$ as the intersection of some finite set of differential radical ideals.

$$\sqrt{I} = J_0 \cap J_1 \cap \cdots \cap J_l$$

- each ideal $J_i, (i = 0..l)$ is given by a standard basis B_i which furnishes a membership test for J_i. The basis contains the "simplest" polynomials of the ideal $J_i \cap k\{V\}$ for some *admissible ranking* choosen by the user.
- when the ideal $\sqrt{I}$ is prime, we have:

$$\sqrt{I} = J_0$$

In this case, F. Boulier showed in (Boulier 1994) p. 76 how to extract some *Ritt-characteristic set* of $\sqrt{I}$ from B_0.

According to Seidenberg, our algorithm splits the models of the initial system. In practice, it appears that it performs much less splittings than Seidenberg's elimination methods. It carries out rankings which are not elimination ones. This allows engineers to compute the relations they really need.

Though most of nonlinear control theory problems may be expressed via *ordinary* differential algebra (the only derivation used is the the time–derivation $\frac{d}{dt}$) we implemented our algorithm more generally, for partial differential systems.

2. BASIC DEFINITIONS AND NOTATIONS

2.1 Differential fields

A *differential field* k is a commutative field endowed with a finite set of *derivations* $\Delta = (\delta_1, \delta_2, \cdots \delta_m)$ which commute together i.e. $\forall \delta \in \Delta, \forall \alpha, \beta \in k$,

$$\left\{ \begin{array}{rcl} \delta(\alpha + \beta) & = & \delta(\alpha) + \delta(\beta) \\ \delta(\alpha\beta) & = & \delta(\alpha)\beta + \alpha\delta(\beta) \\ \delta_i\delta_j & = & \delta_j\delta_i \quad (i, j = 1 \ldots m) \end{array} \right. \tag{2}$$

We denote Θ the free commutative monoid generated by Δ. Every operator $\theta \in \Theta$ can be written as:

$$\theta = \delta_1^{j_1}\delta_2^{j_2}\ldots\delta_m^{j_m}, \quad j_1, \ldots, j_m \in \mathrm{N}$$

2.2 Differential polynomials

Let X be a finite alphabet. The action of the derivations over X generates a new infinite alphabet ΘX which may be ordered such that:

$$\begin{array}{rl} \forall x \in \Theta X, \forall \theta \in \Theta, \theta \neq 1, & x < \theta x \\ \forall x, y \in \Theta X, & x < y \Rightarrow \theta x < \theta y \end{array}$$

The set of the differential polynomials with coefficients in the differential field k is denoted $k\{X\}$. An element $p \in k\{X\}$ is also a purely algebraic (i.e. nondifferential) polynomial in $k[\Theta X]$. The *leader* of p is the greatest element u of ΘX which appears in p. The *initial* of p is denoted I_p. Its *separant* $\partial p/\partial u$ is the initial of all the derivatives of p. It is denoted S_p.

Let A be any subset of $k\{X\}$.

(A)	denotes the algebraic ideal generated by A.
$[A]$	denotes the differential ideal generated by A.
$\sqrt{[A]}$	denotes the radical of the ideal $[A]$.
H_A	denotes the product of all the initials and all the separants of the elements of A.
$\mathcal{H}_A^\infty$	denotes the multiplicatively stable family generated by the initials and the separants of the polynomials of A.

2.3 Autoreduced and coherent sets

A polynomial $p \in k\{X\}$ is said to be *partially reduced* w.r.t. a polynomial $q \in k\{X\}$ if it does not contain any proper derivative of the leader of q.

A polynomial $p \in k\{X\}$ is said to be *reduced* w.r.t. a polynomial $q \in k\{X\}$ if it is partially reduced w.r.t. q and if its degree in the leader of q is less than the degree of q in its leader.

A polynomial $p \in k\{X\}$ is said to be reduced w.r.t. a set A of $k\{X\}$ if it is reduced w.r.t. each element of A.

A subset A of $k\{X\}$ is said to be *autoreduced* if each of its elements is reduced w.r.t. each other element of A. Kolchin showed that any autoreduced set is finite.

Given a polynomial $p \in k\{X\}$ and an autoreduced subset A of $k\{X\}$, there exists a reduction algorithm, due to Ritt (Ritt 1950), which computes a polynomial r, denoted p rem A such that:

$$\left\{ \begin{array}{l} \exists h \in \mathcal{H}_A^\infty, \quad h \cdot p \equiv r \mod [A] \\ \qquad r \text{ is reduced w.r.t. } A \end{array} \right. \tag{3}$$

Let $p_1, p_2 \in k\{X\}$ be two polynomials, whose leaders $u_1, u_2 \in \Theta X$ have some common derivative i.e.

$$\exists u \in \Theta X, \exists \theta_1, \theta_2 \in \Theta, \quad u = \theta_1 u_1 = \theta_2 u_2$$

We define the Δ-polynomial between p_1 and p_2 to be the polynomial $\Delta(p_1, p_2) = S_{p_2}\theta_1 p_1 - S_{p_1}\theta_2 p_2$ where θ_1 and θ_2 are the derivative operators of least order satisfying the formula above.

The set A is said (Rosenfeld 1959) to be *coherent* if all the Δ-polynomials, which may be formed with any set of two elements of A are reduced to zero by A, i.e.

$$\forall p_1, p_2 \in A, \quad \Delta(p_1, p_2) \text{ rem } A = 0$$

In ordinary differential algebra (only one derivation), any autoreduced set is coherent.

LEMMA 2.1 Let $F \subset k\{X\}$ be finite. It is possible to compute $A \subset k\{X\}$ autoreduced and coherent such that:

$$\begin{array}{l} \forall p \in F, \quad p \text{ rem } A = 0 \\ {[A]} \subset [F] \subset [A] : \mathcal{H}_A^{\infty} \end{array}$$

Proof: see (Boulier 1994) page 65. □

3. MODELS

Let Σ be a system of polynomial differential equations and inequations over the ground field k:

$$\Sigma \left\{ \begin{array}{lcl} a_1 & = & 0 \\ \vdots & & \\ a_n & = & 0 \\ q & \neq & 0 \end{array} \right.$$

An *algebraic model* (resp. *differential model*) of Σ is a field extension (resp. differential field extension) K of k where Σ has a solution. More precisely, a model Φ is a homomorphism of k-algebra (respectively differential k-algebra) $\Phi : k[\Theta X] \longrightarrow K$ (resp. $\Phi : k\{X\} \longrightarrow K$) such that

$$\left\{ \begin{array}{lcl} \Phi(a_1) & = & 0 \\ \vdots & & \\ \Phi(a_n) & = & 0 \\ \Phi(q) & \neq & 0 \end{array} \right. \tag{4}$$

Clearly, every differential model is also an algebraic model. The algebraic point of view consists in considering the variables $x \in X$ and their derivatives $\theta x \in \Theta X$ as unknowns which are completely independent.

THEOREM 3.1 (Nullstellensatz) The system Σ admits an algebraic model (resp. differential model) if and only if the polynomial q does not belong to the radical of the algebraic ideal $(A) = (a_1, a_2, \dots, a_n)$ (resp. of the differential ideal $[A] = [a_1, a_2, \dots, a_n]$) generated by the equations of Σ.

Proof: We only prove the implication from right to left. If q does not belong to the radical of the (differential) ideal generated by A, then there exists a (differential) prime ideal P which contains this ideal, but not q. The homomorphism which maps $k[\Theta X]$ on $k[\Theta X]/P$ ($k\{X\}$ on $k\{X\}/P$) furnishes a (differential) model of Σ. □

The relation $p = 0$ is thus true in all the (differential) models of A if and only if $p \in \sqrt{(A)}$ (resp. $p \in \sqrt{[A]}$).

4. REGULAR SYSTEMS

DEFINITION 4.1 (Regular systems) A system of differential equations and inequations which has the form:

$$\left\{ \begin{array}{lcl} a_1 & = & 0 \\ \vdots & & \\ a_n & = & 0 \\ H_A & \neq & 0 \end{array} \right. \tag{5}$$

(H_A denotes the product of the initials and of the separants of the a_i) is said to be *regular* if A is autoreduced and coherent.

LEMMA 4.1 (Rosenfeld) Let A be an autoreduced and coherent system of $k\{X\}$. Any polynomial of $[A] : \mathcal{H}_A^{\infty}$, partially reduced w.r.t. A, also belongs to $(A) : \mathcal{H}_A^{\infty}$.

Proof: See (Rosenfeld 1959) page 397. □

Then the following theorem may be easily proven:

THEOREM 4.1 A regular system has a differential model if and only if it has an algebraic model.

LEMMA 4.2 (D. Lazard) Let $A \subset k\{X\}$ be autoreduced and coherent. The ideal $(A) : \mathcal{H}_A^{\infty}$ is radical i.e.

$$\exists n \in \mathbb{N}, p^n \in (A) : \mathcal{H}_A^{\infty} \iff p \in (A) : \mathcal{H}_A^{\infty}$$

Proof: The quite difficult demonstration (Boulier *et al.* 1995) of this lemma uses the triangular presentation of the equations, whose multiple roots are eliminated, since the separants appear among the inequations. □

As consequence, the differential ideal $[A] : \mathcal{H}_A^{\infty}$ is as well radical.

4.1 Description of the models

Let J be the radical of the differential ideal generated by the polynomial $p \in k\{X\}$ such that $p = 0$ is true in all the differential models of some regular system Ω. Clearly, the algebraic models of Ω are not modified by transforming the inequation $H_A \neq 0$ as the equation $zH_A = 1$, where z is some new variable. We consider the nondifferential ideal $I = (a_1, \cdots, a_n, zH_A - 1)$ then we compute a *Gröbner basis* $B \subset k[\Theta X]$ such that $(B) = I \cap k[\Theta X] = (A) : \mathcal{H}_A^{\infty}$. The basis B is obtained from a Gröbner basis of I, for some elimination ordering $z > u, \forall u \in \Theta X$, by discarding all the polynomials which contain z. The following relations may be shown using lemmas 4.1 and 4.2 :

$$\begin{array}{rcll} p \in J & \iff & p \in [A] : \mathcal{H}_A^\infty & \\ & \iff & (p \text{ rem } A) \in (A) : \mathcal{H}_A^\infty & (6) \\ & \iff & (p \text{ rem } A) \to_B 0 & \end{array}$$

This gives an algorithm to decide on the membership to J. This test necessitates a reduction by A using Ritt's algorithm, then a normal form computation using the basis B.

5. PROPOSED ALGORITHM

THEOREM 5.1 Let Σ be any finite system of equations and inequations of $k\{X\}$. It is possible to compute a finite family of regular systems Ω_i such that every differential model of Σ is a differential model of some Ω_i and every differential model of every system Ω_i is a differential model of Σ i.e.

$$\begin{array}{rcl} \mathcal{M}od(\Sigma) & = & \bigcup_{i=0}^{l} \mathcal{M}od(\Omega_i) \\ \mathcal{M}od(\Omega_i) \cap \mathcal{M}od(\Omega_j) & = & \emptyset, \ \forall i \neq j \end{array}$$

Proof: Lemma 2.1 is used to build an autoreduced and coherent set A. The algorithm then splits on the initials and the separants of A as in Seidenberg's elimination algorithms (see (Boulier 1994) p. 65–67). □

The algorithm gathers as entry a system Σ and proceeds in the following way:

1. Compute the regular systems $\Omega_i, (i = 0..l)$ according to theorem 5.1.
2. For each system Ω_i, compute the Gröbner basis B_i according to section 4.1.
3. Discard the inconsistent systems Ω_i, whose Gröbner basis B_i contains 1.

5.1 Computation of the relations

Each system Ω_i defines a radical ideal J_i as in the section 4.1. Let I be the differential ideal generated by the equations of the initial system Σ. According to the theorem of zeros, we have $p \in \sqrt{I}$ if and only if the relation $p = 0$ is true for all the differential models of Σ:

$$\sqrt{I} \quad = \quad J_0 \cap J_1 \cap \cdots \cap J_l \qquad (7)$$

This furnishes an algorithm to decide on the membership to the ideal $\sqrt{I}$. More generally, the reader will convince himself that the relations, useful in control theory, may be read directly in the Gröbner bases B_i.

5.2 Principal component

The algorithm distinguishes among the Ω_i's a regular system $\Omega_0 = \{A_0 = 0, H_{A_0} \neq 0\}$ called *principal* such that:

$$[A_0] \subset I \subset [A_0] : \mathcal{H}_{A_0}^\infty$$

When the differential ideal $\sqrt{I}$ is prime, the following relation may then be easily established:

$$\sqrt{I} \quad = \quad J_0 = [A_0] : \mathcal{H}_{A_0}^\infty$$

It is shown in (Boulier *et al.* 1995) how to extract some *characteristic set* (Ritt 1950) of $\sqrt{I}$ from B_0.

6. APPLICATION TO OBSERVABILITY TEST

6.1 Principle of the test

A quantity x is said to be globally observable w.r.t. the quantities $u = (u_1, \ldots, u_m)$ and $y = (y_1, \ldots, y_p)$ if and only if there exists in $\sqrt{I}$ some polynomial of the form $b(u,y)x - a(u,y)$ where $a, b \in k\{X\}$, $b \notin \sqrt{I}$. This amounts to assume that a relation of the following form is true:

$$x \quad = \quad \frac{a(u,y)}{b(u,y)} \qquad (8)$$

We may look for such a polynomial in the Gröbner basis B_0 of Ω_0, for any ranking such that $x > y$ and $x > u$. We may choose for instance:

$$\cdots \ddot{x} > \dot{x} > x > \cdots \ddot{y} > \ddot{u} > \dot{y} > \dot{u} > y > u$$

If such a polynomial r exists, then a polynomial $\bar{r}$ of the same form, but partially reduced w.r.t. A_0 exists too. According to (6), we have $\bar{r} \to_{B_0} 0$. Because of the chosen ranking, the reduction is possible only if there exists in B_0 a polynomial of the form $b_1 x - a_1$. This latter is in some sense the "simplest" polynomial which permits to establish the observability of x.

Note that a relation of the form (8) may exist for the models of the principal component Ω_0 but may degenerate for some components Ω_i. We thus retrieve the notion of *universal input* as defined by E. Sontag.

6.2 Mini-example

Let us test the identifiability of the parameter p in the following implicit system, where u is the input.

$$\Sigma \left\{ \begin{array}{rcl} \dot{x}_1 & = & u x_1 + p x_2 \\ \dot{x}_2 & = & u x_3 \\ 0 & = & x_3^2 + x_1^2 - x_1 \\ \dot{p} & = & 0 \\ y & = & x_1 \\ u \neq 0 & ; & p \neq 0 \end{array} \right.$$

For the ranking :

$$\begin{array}{l} \cdots > \dot{x}_1 > \dot{x}_2 > \dot{x}_3 > x_1 > x_2 > x_3 > \cdots \\ \qquad \cdots > \dot{p} > p > \cdots > \dot{y} > \dot{u} > y > u \end{array}$$

the algorithm computes 3 regular systems (the initials and separants are not equal to 0) :

$$
\Omega_0 \left\{ \begin{array}{rcl}
2y^{(3)}y^2u - 2y^{(3)}yu - 2\ddot{y}\dot{y}yu + \ddot{y}\dot{y}u & & \\
-2\ddot{y}\dot{u}y^2 + 2\ddot{y}\dot{u}y - 2\ddot{y}y^2u^2 + 2\ddot{y}yu^2 & & \\
-2\ddot{u}y^3u + 2\ddot{u}y^2u + 2\dot{y}^2yu^2 - \dot{y}^2u^2 & & \\
+\dot{y}\dot{u}yu + 2\dot{u}^2y^3 - 2\dot{u}^2y^2 & = & 0 \\
(u^2y^2 - u^2y)p^2 + \ddot{y}^2 + (-2u\dot{y} - 2y\dot{u}) & & \\
\ddot{y} + u^2\dot{y}^2 + 2uy\dot{u}\dot{y} + y^2\dot{u}^2 & = & 0 \\
upx_3 - \ddot{y} + u\dot{y} + y\dot{u} & = & 0 \\
px_2 - \dot{y} + uy & = & 0 \\
x_1 - y & = & 0
\end{array} \right.
$$

$$
\Omega_1 \left\{ \begin{array}{rcl}
y - 1 & = & 0 \\
\dot{u} & = & 0 \\
x_3 & = & 0 \\
px_2 + u & = & 0 \\
x_1 - 1 & = & 0
\end{array} \right.
$$

$$
\Omega_2 \left\{ \begin{array}{rcl}
y & = & 0 \\
x_3 & = & 0 \\
x_2 & = & 0 \\
x_1 & = & 0
\end{array} \right.
$$

This computation *proves* that the parameter p is not globally identifiable. On the otherhand, p^2 may be identified in the principal component Ω_0, but not in the other components. The variables (x_1, x_2, x_3) are globally identifiable, as soon as the value of p is known.

7. CONCLUSION

We have shown that the computation of the relations which constrain the quantities defined by a implicit system may be solved using differential algebra.

Any admissible ranking of the differential variables may be chosen. Thus, we may either eliminate some variables, either look for the less differentiated relations.

The user is *sure* to find the relations he is looking for in the Gröbner bases of the regular systems Ω_i.

The choice of the ranking greatly modifies the time of the computations. In practice, elimination rankings lead to computations wich are far more difficult than the other ones.

When the initial system is linear, the algorithm computes a $\mathcal{D}$-module *standard basis* (Pankratev 1989). It does not perform any splitting.

REFERENCES

Boulier, F. (1994). Etude et implantation de quelques algorithmes en algèbre différentielle. PhD thesis. Univ. Lille I. 59655 Villeneuve–d'Ascq.

Boulier, F., D. Lazard, F. Ollivier and M. Petitot (1995). Representation for the radical of a finitely generated differential ideal. In 'proc. ISSAC'95'. submited.

Buchberger, B. (1965). An Algorithm for Finding a Basis for the Residue Class Ring of a Zero-Dimensional Polynomial Ideal (German). PhD thesis. Inst. Univ. of Innsbruck.

Diop, S. (1991). 'Elimination in control theory'. *Math. Control Signals Systems* **4**, 17–32.

Fliess, M. (1989). 'Automatique et corps différentiels'. *Forum Math.* **1**, 227–238.

Fliess, M. and S. Glad (1993). An algebraic approach to linear and nonlinear control. In H. Trentelman and J. Willems (Eds.). 'Essays on Control : Perspective in the Theory and its Applications'. Birkhäuser. Boston. pp. 223–267.

Forstman, K. (1991). 'Constructive commutative algebra in nonlinear control theory'. *Linköping studies in Science and Technology.*

Michalik, J. and J. Willems (1994). The elimination problem in differential algebric systems. In 'proc. MTNS'93 in Mathematical Research'. Vol. 79. pp. 357–360.

Ollivier, F. (1990). Le problème de l'identifiabilité structurelle globale: approche théorique, méthodes effectives et bornes de complexité. PhD thesis. Ecole polytechnique.

Pankratev, V. (1989). 'Computations in differential and difference modules'. *Applicandae Mathematicae* **16**, 167–189.

Ritt, J. F. (1950). *Differential Algebra.* Amer. Math. Soc. New York.

Rosenfeld, A. (1959). 'Specializations in differential algebra'. *Trans. Amer. Math. Soc.* **90**, 394–407.

Seidenberg, A. (1956). 'An elimination theory for differential algebra'. *Univ. California Publ. Math. (N.S.)* **90**, 31–38.

VARIABLE-STRUCTURE CONTROL OF NONLINEAR SYSTEMS

P. ZLATEVA

Institute of Control and System Research, Bulgarian Academy of Sciences
Acad. G. Bonchev Street, Bl.2, P.O.Box 79, Sofia 1113, Bulgaria

Abstract. A class of nonlinear systems expressed by nonlinear differential equations with uncertain parameters is investigated. The nonlinear model is transformed into a linear model with parameters varying in definite intervals. Variable structure control in sliding mode is designed. The sliding mode is realized with respect to an auxiliary variable. The theoretical results are proved by computer simulations of the designed variable structure control in sliding mode for nonlinear fermentation processes.

Resumé. Dans ce travail on presente une recherche sur une classe de systèmes non-linéaire aves paramètres mal définis. Le modèle non-linéaire se transforme en modèle linéaire avec paramètres variables dans des intervalles determinés. On présente la synthèse d'une commande avec structure variable en régime glissant. Le régime glissant est réalisé par rapport d'une variable auxiliaire. Les résultats théoriques sont approuvés par des simulations numériques d'un système aves structure variable en régime glissant pour commande d'un processus de fermentation non-lineaire.

Key Words. Variable structure systems; sliding mode; nonlinear systems; control design; nonlinear differential equations; fermentation processes

1. INTRODUCTION

In recent years, great attention has been given to control design for nonlinear and uncertain plants by using the theory of variable structure systems (Cortez *et al.*, 1989; Zinober, 1990; Utkin, 1992; Zlateva, 1994). Variable structure control is widely used in the control of robotics manipulators, power converters, chemical processes (Sira-Ramirez, 1993; Young, 1993).

A central feature of variable structure control is sliding mode, which occurs when the system state repeatedly crosses certain subspaces, or sliding surfaces, in the state or error vector space. The design of the sliding surface completely determines the performance of the closed loop system (Emelianov, 1967; Utkin, 1978). When a certain matching criterion holds, nonlinear variable structure control can yield invariance to the parameter uncertainties and external disturbances (Drazenovic, 1969).

A difficulty in applying the variable structure control approach is the need for the knowledge of the full-state information (Utkin, 1978). In practice, it is neither possible nor feasible to have full information regarding the state. Thus, when the state can not be measured, the output is measured instead. The use of state observers is a natural step towards the relaxation of this condition. However, this increases the complexity of system implementation, and only asymptotic tending of the state to the sliding surface can be achieved. An alternative to state feedback for simplification of system implementation is the use of output feedback (Sira-Ramirez, 1993; Young, 1993).

Fermentation processes involve living organisms. Their dynamics is often badly understood, strongly nonlinear and non stationary. The model parameters do not remain constant over long periods due to metabolic various and physiological modifications (Bastin and Dochain, 1991; Tzonkov *et al.*, 1992; Simeonov, 1993).

In this case, one class of systems which are able to perform high quality automatic control are variable structure systems.

It should be marked that a fermentation processes have been an area in which, traditionally, the variable structure control techniques could not be applied (Sira-Ramirez, 1993). The specific nature of these processes with the required smoothness of the input variables and the inherent limitations in today's actuators in fermentation industry, did not allow naturally for discontinuous feedback strategies. Moreover, the typical chattering responses in sliding mode are undesirable for the delicate and precise regulation of the fermentation processes.

However, there are single reports about variable structure control of fermentation processes (Cortez *et al.*, 1989; Sira-Ramirez, 1993; Zlateva et al., 1994; Zlateva, 1994). Two approaches to control design are used. The first one is comprised of two steps: first, nonlinear model linearization at one operation point, and second, variable structure control design using the obtained linear model. In this case, the stability of closed loop system is guaranteed only in the vicinity of the chosen operating point. The second approach uses directly the nonlinear model and full - biological state information.

In the present work, a class of the nonlinear systems with uncertain parameters (a class of the continuous fermentation processes) with only the input and output measurable is considered. A variable structure control in sliding mode is proposed. To avoid the chattering phenomenon the sliding mode is realized with respect to an auxiliary input variable.

2. PROBLEM STATEMENT

The kinetics of the class of continuous fermentation processes described by the following nonlinear differential equations is investigated (Bastin and Dochain, 1991):

$$\frac{dX}{dt} = \varphi - DX; \tag{1}$$

$$\frac{dS}{dt} = -k_3\varphi + D(S_{in} - S), \tag{2}$$

where $\varphi = \dfrac{k_1 S}{k_2 + S}X$ is the growth rate;

X and S are state variables - the biomass and substrate concentration, respectively;

D is the dilution rate;

S_{in} is the influent substrate concentration;

k_i (i=1,...3) are the characteristic coefficients of the fermentation process.

The control input is dilution rate ($u = D$) and the output - substrate concentration ($y = S$). State variables, dilution rate, influent substrate concentration and characteristic coefficients belong to given intervals, depending on technological requirements:

$$0 < X_{min} \le X \le X_{max}; \quad 0 < S_{min} \le S \le S_{max}; \quad 0 < S_{in}^{min} \le S_{in} \le S_{in}^{max}; \quad 0 < k_i^{min} \le k_i \le k_i^{max}; \tag{3}$$

$$0 \le D_{min} \le D \le D_{max}, \tag{4}$$

where X_{min}, X_{max}, S_{min}, S_{max}, D_{min}, D_{max}, S_{in}^{min}; S_{in}^{max}; k_i^{min} k_i^{max} are constants.

It is assumed that the state variable - biomass concentration is not directly measured, and that only the output - substrate concentration is measured.

The problem is to design a variable structure control law such that nonlinear and uncertain fermentation process to be stable and invariance to the parameter uncertainties and external disturbances.

3. VARIABLE STRUCTURE CONTROL DESIGN

3.1. Transformation of nonlinear fermentation process model

The system (1)-(2) is replaced by a second - order algebraic differential equation of the form:

$$\frac{d^2S}{dt^2} = -k_3 X\frac{d}{dt}\left(\frac{k_1 S}{k_2 + S}\right) - \frac{k_1 k_3 \varphi}{k_2 + S}S + \frac{k_1 k_3}{k_2 + S}DS - D\frac{dS}{dt} + (S_{in} - S)\frac{dD}{dt}. \tag{5}$$

In order to apply the main principles of the variable structure control theory (Emelianov, 1967), the nonlinear differential equation (5) is transformed into an input - output linear differential equation with uncertain parameters:

$$\frac{d^2y}{dt^2} + a_1\frac{dy}{dt} + a_0 y = b_1\frac{du}{dt} + b_0 u, \tag{6}$$

where the following substitutes are carried out:

$$a_1 = \frac{k_2[-k_1\varphi + D(S_{in} - S)]}{(k_2 + S)S};$$

$$b_1 = S_{in} - S;$$

$$a_0 = \frac{[-k_3\varphi + D(S_{in} - S)]k_1}{(k_2 + S)};$$

$$b_0 = -(S_{in} - S)\{[-k_3\varphi + D(S_{in} - S)]. \tag{7}$$

$$..[\frac{(k_2 - k_1 S}{(k_2 + S)S} - 1] + \varphi k_3\}$$

In this case, the parameters a_i and b_i, i = 0, 1 are represented as nonlinear and nonstationary functions of the fermentation process state variables and characteristic coefficients. Taking into account the conditions (1)-(4) these parameters are considered as uncertain parameters which vary in known intervals:

$$a_i^{min} \le a_i \le a_i^{max}; \qquad b_i^{min} \le b_i \le b_i^{max}, \tag{8}$$

where $a_i^{min}, a_i^{max}, b_i^{min}, b_i^{max}$ are constants.

The intervals boundaries are obtained as follows (Zlateva, 1994):

$$a_i^{max} \ge \sup(a_i); \qquad b_i^{max} \ge \sup(b_i);$$

$$a_i^{min} \le \inf(a_i); \qquad b_i^{min} \le \inf(b_i). \tag{9}$$

It should be marked that the parameters expressions (7) are not drawn in a uniquely possible way.

3.2. Control law

Suppose it is desired to regulate the substrate concentration ($S = y$) to its set-point ($S^* = g$). The substrate concentration error is:

$$e = S^* - S = g - y, \tag{10}$$

where g is positive constant.

Differentiating (10) and using (6), it is obtained the following second - order differential equation of the error:

$$\frac{d^2e}{dt^2} + a_1\frac{de}{dt} + a_0 e = b_1\frac{du}{dt} + b_0 u + a_0 g. \tag{11}$$

In this case, the set-point signal g is considered as an external disturbance.

According to the proposed transformation, the investigated nonlinear system (1)-(2) is considered as linear uncertain system with differentiated control input.

To quarantee the invariance of the variable structure system in sliding mode to the parameter uncertainies and external disturbances, it is required to hold the certain matching conditions. For this reason, the system is represented by the following generalized controller observability form (Fliess, 1989):

$$\frac{de_1}{dt} = e_2$$
$$\frac{de_2}{dt} = -a_0 e_1 - a_1 e_2 - b_0 u - b_1\frac{du}{dt} + a_0 g, \tag{12}$$

where $e_1 = e$.

Define the auxiliary variable:

$$v = \frac{du}{dt}. \tag{13}$$

Substituting (13) into (12), it is obtained the following extended system model:

$$\frac{d}{dt}\begin{vmatrix} e_1 \\ e_2 \\ u \end{vmatrix} = \begin{vmatrix} 0 & 1 & 0 \\ -a_0 & -a_1 & -b_0 \\ 0 & 0 & 0 \end{vmatrix} \begin{vmatrix} e_1 \\ e_2 \\ u \end{vmatrix} + \begin{vmatrix} 0 \\ -b_1 \\ 1 \end{vmatrix} . v + \begin{vmatrix} 0 \\ a_0 \\ 0 \end{vmatrix} . g \tag{14}$$

where v is an auxiliary input variable and the original input variable u is now an added state to the system.

To avoid the chattering phenomenon the sliding mode is realized with respect to the auxiliary input variable v. Thus, the dynamic variable structure feedback controller exhibiting smoothing features for the control input signal designed u is obtained.

On the basis of the variable structure system theory (Emelianov, 1967; Utkin, 1978), the control law is designed as follows:

$$v = \left(\psi_1|e_1| + \psi_2|e_2| + \psi_3|u| + \psi_4|g|\right).sng(\sigma(e)), \tag{15}$$

where ψ_i, i=1,...,4 are switching coefficients;

$\sigma(e)$ is sliding surface.

The sliding surface is defined as follows:

$$\sigma(e) = c.e_1 + e_2, \tag{16}$$

where c is a strictly positive constant.

The condition for the existence and stability of a sliding mode (Emelianov, 1967) is given by:

$$\sigma(e).\frac{d(\sigma(e))}{dt} < 0, \quad \text{for} \quad \sigma(e) \ne 0. \tag{17}$$

On the basis of this condition, it is defined the following inequalities which have to satisfy the switching coefficients:

$$\psi_1 \geq \sup\left|\frac{-a_0}{b_1}\right|; \qquad \psi_2 \geq \sup\left|\frac{c-a_1}{b_1}\right|;$$

$$\psi_3 \geq \sup\left|\frac{-b_0}{b_1}\right|; \qquad \psi_4 \geq \sup\left|\frac{a_0}{b_1}\right|. \qquad (18)$$

According to the method of the equivalent control (Utkin, 1978), the discontinuous motion on the sliding surface $\sigma(e)=0$ can be described in an idealized fashion, by invariance conditions $\sigma(e)=0$ and $d(\sigma(e))/dt=0$. Under ideal sliding condition $\sigma(e)=0$, at least one of the variables no longer qualifies as a state variable for the system since it is expressible as a linear combination of the remaining state. The function $sng(\sigma(e))$, is then ideally replaced by zero. The ideal (autonomous) closed loop dynamics may then be expressed in terms of a reduced state vector which only includes the remaining $n-1$ variable (in investigated case $n=2$). This leads to the following ideal sliding dynamics:

$$\frac{de_1}{dt} = -ce_1. \qquad (19)$$

The equivalent control is defined as a virtual feedback control action ideally achieving a smooth evolution of the system trajectories on the constraining sliding surface $\sigma(e)=0$, provided initial conditions are precisely set on such a switching surface. The equivalent control is formally obtained from the condition $d(\sigma(e))/dt=0$.

Taking into account the technological constraints (4) the dilution rate is submitted to following condition:

$$D = \begin{cases} D_{max} & when \quad u \geq D_{max} \\ u & when \quad D_{min} < u < D_{max} \\ D_{min} & when \quad u \leq D_{min} \end{cases} \qquad (20)$$

4. SIMULATION STUDIES

This section present some simulation results using the proposed variable structure control in sliding mode to regulate the substrate concentration of the continuous nonlinear fermentation process.

In this case, the biomass concentration, substrate concentration, dilution rate, influent substrate concentration and characteristic coefficients intervals have the following values (Bastin and Dochain, 1991):

$$0.01 \leq X \leq 0.2; \qquad 0.1 \leq S \leq 1.8;$$

$$0 \leq D \leq 0.39;$$

$$2 \leq S_{in} \leq 4; \qquad 0.28 \leq k_1 \leq 0.52;$$

$$0.28 \leq k_2 \leq 0.52; \qquad 19.1 \leq k_3 \leq 35.5$$

Taking into account the conditions (7)-(9), the following boundaries of the transformed model parameters are calculated:

$$-13 \leq a_1 \leq 20; \qquad 0.2 \leq b_1 \leq 4;$$

$$-1.3 \leq a_0 \leq 1.8; \qquad -73 \leq b_0 \leq 54$$

On the basis of the conditions (15)-(18), the auxiliary input variable is obtained as follows:

$$v = \left(15|e_1| + 0.2|e_2| + 3|u| + 3|g|\right).sng(\sigma(e)),$$

where the sliding surface is:

$$\sigma(e) = 0.5e_1 + e_2$$

It is investigated the performance of the designed variable structure control with respect to parameter uncertainties by varying the characteristic coefficients of the fermentation process and to external disturbances by varying the influent substrate concentration and the process set-point.

In Figure 1. is shown the evolution of the substrate concentration S, dilution rate D, auxiliary input variable v at step change of the set-point S^*. The good performance of the nonlinear system and the smooth nature of the control input - dilution rate D are evident.

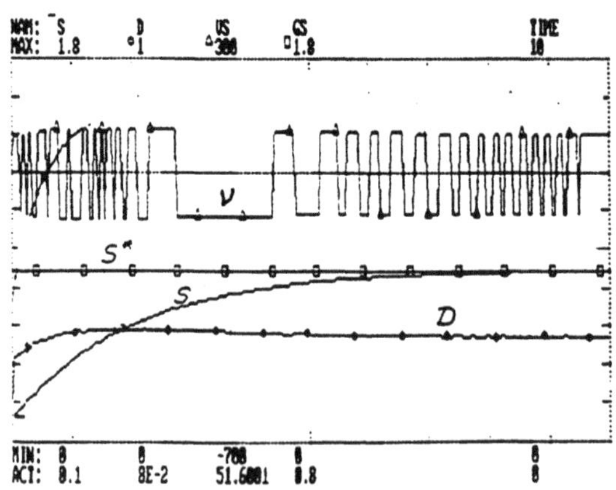

Fig. 1. Evolution of the substrate concentration S, dilution rate D, auxiliary input variable v at step change of the set-point S^* (from 0.1 to 0.8)

The simulation results corroborate that the closed nonlinear system is invariant to the step change of the influent substrate concentration. It is depicted in Fig.2. and Fig.3. where S_{in}=2 and S_{in}=3, respectively. It can be observed that dilution rate only takes extreme value (D_{min}=0) during a small time intervals. This time is just enough to the substrate concentration S reaches the boundary layer around set-point S^*.

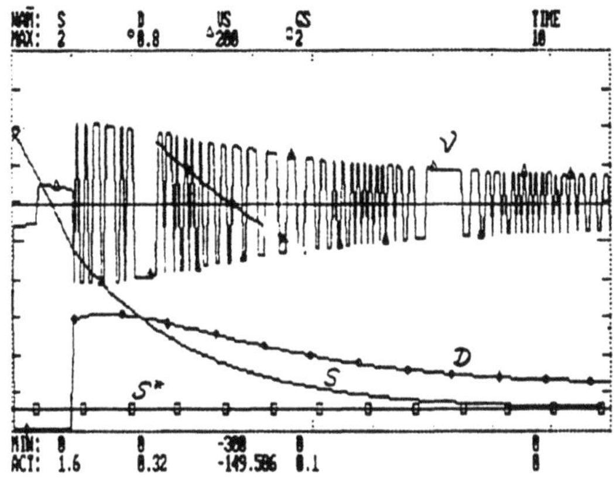

Fig. 2. Evolution of the substrate concentration S, dilution rate D, auxiliary variable v where S_{in} =2; S^*=0.1; $S(0)$=1.6

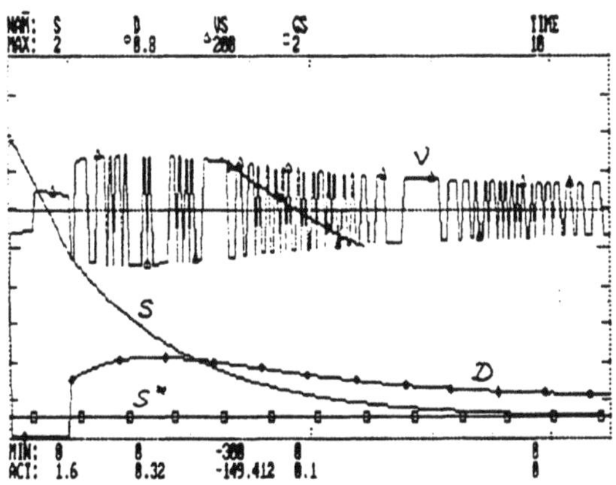

Fig. 3. Evolution of the substrate concentration S, dilution rate D, auxiliary variable v where S_{in} =3; S^*=0.1; $S(0)$=1.6

5. CONCLUSION

The design of a variable structure control in sliding mode for a class of nonlinear systems has been developed. This class of systems is described by nonlinear differential equations with uncertain parameters. For control design purpose, the nonlinear model is transformed into a linear model with parameters, varying in definite intervals. To avoid the chattering phenomenon the sliding mode is realized to respect to an auxiliary variable. It is shown that the nonlinear system is invariant to the parameters uncertainties and external disturbances. The obtained results proved that the designed variable structure control can be successfully used for set-point stabilization of the nonlinear and uncertain fermentation processes.

6. REFERENCES

Bastin, G. and D. Dochain (1991). *On-line Estimation and Adaptive Control of Bioreactors*, Elsevier, Amsterdam.

Drazenovic, B. (1969). The invariance conditions in variable structure systems. *Automatica*,**5**,3, 287-295.

Emelianov, S. V. (1967). *Variable structure controlsystems*. Nauka, Moscow. (in Russian)

Fliess, M. (1989). Nonlinear control theory and differential algebra. *Modelling and adaptive control*, (Ch. Byrnes and A. Khurzhansky, Ed.) Lecture Notes in Control and Information Sciences, vol. 105, Springer-Verlag, Berlin.

Simeonov, I. (1994). Modelling and Control of Anaerobic Digestion of Organic Waste, *Chem.Biochem.Eng.Q.*, **8**, 45-52.

Sira-Ramirez, H. (1993). On the dynamical sliding mode control of nonlinear systems. *Int. J. Control*, 57, 1039-1062.

Suarez-Cortez, R., J. Alvarez-Gallegos and E. Gonzalez-Mora (1989). Sliding conroller desing for a nonlinear fermentation system. *Biotechnol. Bioeng.*, .**33**, 377-385.

Tzonkov, S., D. Filev, I. Simeonov and L. Vaklev (1992). *Control of the biotechnological processes*. Technica, Sofia. (in Bulgarian)

Utkin, V. I. (1978). *Sliding modes and their applications in variable structure systems*. Mir, Moscow. (in Russian)

Utkin, V. I. (1992). *Sliding modes in optimization and control problems*. Springer-Verlag, New York.

Young, K.D., Ed. (1993). *Variable structure control for robotics and aerospace applications*. Elsevier Science, Amsterdam.

Zinober, A. S., Ed. (1990). *Deterministic control of uncertain systems*. Peter Peregrinus, London.

Zlateva, P., V. Lubenova, I. Simeonov, and M. Ignatova (1994). Sliding mode control of continuous fermentation processes with state estimation. In: Proc. IFAC Int. Workshop on New Trends in Design of Control Systems, Smolenice, Slovakia, September 1994, 181-184.

Zlateva, P. (1994). A method to fermentation processes control in the class of variable structure systems. *Automatica & Informatics*., **1**, 19-22. (in Bulgarian).

Analysis of Small Time Local Controllability of the Rigid Body Model

El Yazid KERAI*

*Ecole Normale Supérieure de Cachan, Centre de Mathématiques et de Leurs Applications, 61 av. Président Wilson, 94235 Cachan Cedex, France.

Abstract. This manuscript investigates the control system modelling a rigid spacecraft with gas-jet actuators. In the case of only one gas-jet this system is shown to never be small-time locally controllable. For the case of two linearly independent gas-jet actuators, necessary and sufficient conditions for small-time local controllability around a rest point are derived.

Résumé. L'objet de ce papier est l'étude de la contrôlabilité locale d'un système modélisant le mouvement d'un satellite artificiel commandé par des tuyères. Dans le cas d'un seul couple moteur, le système n'est pas localement contrôlable en temps petit. Pour le cas de deux couples moteurs, nous donnons une condition nécessaire et suffisante de contrôlabilité locale en temps petit.

Key Words. Small-time local controllability.

1. INTRODUCTION

In this article we study the small-time local controllability of the problem of rigid spacecraft. Let us mention that the question of global controllability of this problem has been completely solved by Crouch (1984), see also the previous works (Bonnard, 1982). However the Small-Time Local Controllability (STLC), which is not implied by the global controllability, does not seem to have been very studied.
Starting from the equations for the rigid-body model of a satellite controlled by gas-jet actuators -where gravitational forces have been neglected- using Euler angles $\eta = (\phi, \theta, \psi)$ representing rotations about the axes of the fixed frame. The equations describing the model on the state manifold $M = I\!R^3 \times I\!R^3$, have the following form

$$J\dot{\omega} = S(\omega)J\omega + \sum_{i=1}^{m} v_i b_i \qquad (1)$$

$$\dot{\eta} = A(\eta)\omega \qquad (2)$$

where $\omega = (\omega_1, \omega_2, \omega_3)$ is the angular velocity, $v_i b_i \in I\!R^3 \quad i = 1, \dots, m$ are the torques applied to the satellite, J is the inertia matrix which is symmetric, $S(\omega)$ is the matrix representation of the wedge-product, i.e.

$$S(\omega) = \begin{pmatrix} 0 & \omega_3 & -\omega_2 \\ -\omega_3 & 0 & \omega_1 \\ \omega_2 & -\omega_1 & 0 \end{pmatrix} \qquad (3)$$

and

$$A(\eta) = \begin{pmatrix} \cos\theta & 0 & \sin\theta \\ \sin\theta\tan\phi & 1 & -\cos\theta\tan\phi \\ -\sin\theta/cos\phi & 0 & \cos\theta/\cos\phi \end{pmatrix}. \qquad (4)$$

Let us recall the definition of STLC :

Definition 1. (Coron, 1992; Sussmann, 1987) : The system $\dot{x} = f(x, u)$ is small-time locally controllable at zero if starting from zero at $t = 0$ and using an arbitrary small control one can reach in an arbitrary small time a neighborhood of zero.

Let P be a subspace of $\mathbb{R}^3$, spanned by b_1 and b_2, and consider the condition :

$$\text{Span}\{b_1, b_2, S(\omega)J^{-1}\omega; \quad \omega \in P\} = \mathbb{R}^3$$

Clearly, this condition is equivalent to :

$$\text{Span}\{g_1, g_2, g_3(\gamma_1, \gamma_2); \ \gamma_1, \gamma_2 \in \mathbb{R}\} = \mathbb{R}^3 \quad (5)$$

where $g_1 = J^{-1}b_1$, $g_2 = J^{-1}b_2$ and

$$g_3(\gamma_1, \gamma_2) = 2J^{-1}S(\gamma_1 g_1 + \gamma_2 g_2)J(\gamma_1 g_1 + \gamma_2 g_2).$$

It is proved in (Crouch, 1984) that (5) holds if and only if the Lie algebra rank condition at zero holds. So by Sussmann and Jurdjevic's theorem, (5) is a necessary condition for STLC. Let us recall that in (Crouch, 1984) it is shown that (1)-(2) is globally controllable. Hashémi (1992), in his stage report, has studied the STLC question for this problem in the particular case $g_1 = (1,0,0)$ and $g_2 = (0,1,0)$, he has found that $a_1 a_2 a_3 \neq 0$ is a sufficient condition for STLC (we will see bellow what are the a_i).
The goal of this paper is to prove the following results:
Theorem 1. For $m = 1$ the system (1)-(2) cannot be STLC.

Theorem 2. For $m = 2$ small-time local controllability of (1)-(2) and (5) are equivalent properties.

2. PROOF OF THEOREM 1

First recall the Crouch condition (Crouch, 1984) which is necessary for the local or global controllability in the case $m = 1$:
Let P be a subspace of $\mathbb{R}^3$, spanned by $g_1 = J^{-1}b_1$ and $g_2 = J^{-1}S(g_1)Jg_1$, the Crouch condition is

$$\text{Span}\{g_1, \ g_2, \ J^{-1}S(\omega)J\omega \ ; \quad \omega \in P\} = \mathbb{R}^3. \quad (6)$$

Before, we come to conditions for STLC, we need the following notations : Let $L(f_1, f_0)$ be the Lie algebra generated by the set of vector fields $\{f_0, f_1\}$. Its homogeneous components $L^{(k,l)}(f_1, f_0)$ are spanned by all iterated brackets containing k factors f_1 and l factors f_0. We also have the following theorem which gives a necessary condition for STLC in the single-input case.

Theorem 3. (Kawski, 1987; Sussmann, 1983) : If the system $\dot{x} = f_0(x) + u_1 f_1(x)$ is STLC then

$$[f_1, [f_1, f_0]](0) \in S^1(f_1, f_0)(0).$$

Let us apply this theorem to system (1)-(2) : we have

$$f_0(x) = (J^{-1}S(\omega)J\omega, A(\eta)\omega) \ \text{ and}$$

$$f_1(x) = (g_1, 0) \text{ such that } x = (\omega, \eta).$$

Hence, with a simple calculation of iterated Lie brackets, we can immediately verify that

$$[f_1, [f_1, f_0]](0) \notin S^1(f_1, f_0)(0)$$

which ends the proof of Theorem 1.

3. PROOF OF THEOREM 2

In order to prove Theorem 2, we look for a sufficient condition for small-time local controllability which implies condition (5). There are various sufficient conditions for STLC, for example the conditions which appear in (Hermes, 1977; Sussmann, 1987; Kawski, 1990; Bianchini-Stefani, 1986; ...), we choose Sussmann's condition.

Definition 2. (Sussmann, 1987; section 7) : Consider the control system $\dot{x} = f_0(x) + \sum_{i=1}^{m} u_i f_i(x)$, $x \in \mathbb{R}^n$. Zero is the equilibrium point. Let $Br(f)$ be the set of iterated Lie brackets of $\{f_0, f_1, \ldots, f_m\}$. For such h in $Br(f)$ let $\delta_i(h)$ the number of times that f_i appears in h, let $\theta \in [0, +\infty)$, then $\dot{x} = f_0(x) + \sum_{i=1}^{m} u_i f_i(x)$ satisfies Sussmann's condition $s(\theta)$ if whenever

$h \in Br(f)$ with $\delta_0(h)$ odd and $\delta_i(h)$ even for all i in $[1, m]$, then

$$h(0) \in D_\theta(h) = \text{Span}\{g(0); g \in Br(f)$$
$$\theta\delta_0(g) + \sum_{i=1}^{m} \delta_i(g) < \theta\delta_0(h) + \sum_{i=1}^{m} \delta_i(h)\}. \quad (7)$$

With this definition, we have the following sufficient condition, due to Sussmann ,

Theorem 4. (Sussmann, 1987) : Assume that $f_0(0) = 0$, $Br(f)(0)$ spans $I\!R^6$ and $s(\theta)$ holds for some $\theta \in [0, 1]$, then $\dot{x} = f_0(x) + \sum_{i=1}^{m} u_i f_i(x)$ satisfies the STLC condition.

After this theorem, we now turn to the proof of Theorem 2. The equations, in their present form (1)-(2) do not allow us to do further calculation in order to verify condition (7). Hence we have to change variables for state and for control and then use condition (5) to bring this system into a form more amenable to do further calculation -for more details, see the Appendix of this paper-: After a change of control variables, (1) can be replaced by

$$\dot{\omega_1} = Q(\omega) + \omega_1 L_1(\omega), \quad \dot{\omega_2} = V_1, \quad \dot{\omega_3} = V_2, \quad (8)$$

with

$$\begin{aligned} L_1\omega &= D_1\omega_1 + E_1\omega_2 + F_1\omega_3 && (9) \\ Q(\omega) &= A\omega_2^2 + B\omega_2\omega_3 + C\omega_3^2. && (10) \end{aligned}$$

For system (8) the controls are V_1 and V_2. It is proved in the Appendix that Q changes sign if and only if condition (5) holds -which is a necessary condition for local controllability (see Crouch, 1984; Jurdjevic-Quinn, 1978)-.

From now on we assume that Q changes sign, hence after a suitable change of coordinates :

$$\omega = P\tilde{\omega} = \begin{pmatrix} 1 & 0 & 0 \\ 0 & a_p & b_p \\ 0 & c_p & d_p \end{pmatrix} \tilde{\omega} \quad (11)$$

such that $c := a_p d_p - b_p c_p > 0$. Then, (2) and (8) can be written

$$\begin{cases} \dot{\tilde{\omega}}_1 = \tilde{\omega}_2\tilde{\omega}_3 + \tilde{\omega}_1 L_2(\tilde{\omega}), \quad \dot{\tilde{\omega}}_2 = u_1, \quad \dot{\tilde{\omega}}_3 = u_2 \\ \dot{\eta} = A(\eta)P\tilde{\omega} \end{cases} \quad (12)$$

with $L_2\tilde{\omega} = D_2\tilde{\omega}_1 + E_2\tilde{\omega}_2 + F_2\tilde{\omega}_3$.
System (12) is an affine control system, i.e. $\dot{x} = f_0(x) + u_1 f_1(x) + u_2 f_2(x)$ such that the vector field f_0 and the control vector fields f_i are given by

$$f_0(x) = (\tilde{\omega}_2\tilde{\omega}_3 + \tilde{\omega}_1 L_2(\tilde{\omega}), 0, 0, A(\eta)P\tilde{\omega}), \quad (13)$$

$$f_1(x) = (0, 1, 0, 0, 0, 0), \quad (14)$$

$$f_2(x) = (0, 0, 1, 0, 0, 0). \quad (15)$$

where $x := (\tilde{\omega}, \eta)$ is the variable of state and $u = (u_1, u_2)$ is the variable of control.

We now show that system (12) verifies condition (7). First, we call $h \in Br(f)$ a **bad bracket** if it verifies $\delta_0(h)$ odd and $\delta_i(h)$ even for $i \in [1, m]$, and a **good bracket** if it does not verify these properties. We denote also

$$\delta(h) = \delta_0(h) + \sum_{i=1} \delta_i(h).$$

Now we try to find six good brackets which span $I\!R^6$. Here we have

$$[f_1, f_0](0) = (0, 0, 0, a_p, 0, c_p), \quad (16)$$

$$[f_2, f_0](0) = (0, 0, 0, b_p, 0, d_p). \quad (17)$$

Computing the Lie bracket $[f_1, f_0]$ with f_2, we obtain

$$[f_2, [f_1, f_0]](0) = (1, 0, 0, 0, 0, 0). \quad (18)$$

Computing this bracket with f_0 we obtain

$$[[f_2, [f_1, f_0]], f_0](0) = (0, 0, 0, 0, 1, 0). \quad (19)$$

Hence the good brackets :
$e_1 = [f_2, [f_1, f_0]](x)$, $e_2 = f_1(0)$, $e_3 = f_2(0)$,

$e_4 = [[f_2,[f_1,f_0]],f_0](0)$, $e_5 = [f_1,f_0](0)$ and $e_6 = [f_2,f_0](0)$ span $I\!R^6$ and

$$\text{for each } j \in [1,6], \quad \delta(e_j) \leq 4.$$

Let h be a bad bracket in $Br(f)$,

- if $\delta(h) > 4$, then $h(0)$ is a linear combination of $(e_j)_{j=1}^6$ and therefore $h(0) \in D_1(h)$.
- if $\delta(h) \leq 4$, then we have two possibilities:
  1. $\delta_0(h) = 1$ and $\delta_i(h) = 0$, $i = 1,2$, in which case $h = f_0$ and $h(0) = 0 \in D_1(h)$.
  2. $\delta_0(h) = 1$ and $\delta_i(h) = 2$, for some $i \in [1,2]$, in which case $h = [[f_0,f_i],f_i](0) = 0$.

Then condition $s(\theta)$ holds for $\theta = 1$ which, by Theorem 4, ends the proof of Theorem 2.

Remark 1. Remark that if $m = 1$ or 2 the system (1)-(2) does not verify the Brockett condition (Brockett, 1983); so it cannot be asymptotically stabilized by means of a continuous feedback law, see also (Byrnes-Isidori, 1991), but in the case where (5) holds, the system (1)-(2) is locally asymptotically stabilizable by means of a continuous time-varying feedback law. This result is a consequence of (the proof) of Theorem 2, (Coron, 1992; 1993). An explicit stabilizing time-varying feedback law has been given recently in (Morin-Samson-Pomet and Jiang, 1994) for a special case and in (Coron-Keraï, 1994) for the general case.

APPENDIX

The constant control-vector fields g_1 and g_2 of the Euler equations (1) satisfy (5). If g_1 and g_2 are not orthogonal, we take

$$k_1 = g_1 \text{ and } k_2 = \alpha_1 g_1 + \alpha_2 g_2$$

such that we have k_2 orthogonal to g_1, $(\alpha_1,\alpha_2) \in I\!R^2$, $\alpha_2 \neq 0$.

One may take

$$\tilde{v}_1 = \parallel k_1 \parallel (v_1 - \frac{\alpha_1}{\alpha_2} v_2) \tag{20}$$

$$\tilde{v}_2 = \frac{1}{\alpha_2} \parallel k_2 \parallel v_2 \tag{21}$$

Then system (1) becomes

$$\dot{\omega} = J^{-1}S(\omega)J\omega + G_1\tilde{v}_1 + G_2\tilde{v}_2 \tag{22}$$

where $G_i = k_i / \parallel k_i \parallel$, $i = 1, 2$. G_1 and G_2 are ortho-normal, so we can take $G_1 = (0,1,0)^T$ and $G_2 = (0,0,1)^T$ without loss of generality. Using these new controls (20)-(21), system (22) can be rewritten in the same form as (8) by taking as controls

$$V_1 = (J^{-1}S(\omega)J\omega)G_1 + \tilde{v}_1,$$
$$V_2 = (J^{-1}S(\omega)J\omega)G_2 + \tilde{v}_2.$$

We are interested in the quadratic form (10) : an explicit calculation gives

$$A = \det(JG_1, JG_2, S(G_1)JG_1),$$
$$B = \det(JG_1, JG_2, S(G_2)JG_1 + S(G_1)JG_2),$$
$$C = \det(JG_1, JG_2, S(G_2)JG_2).$$

The quadratic form (10) changes sign if

$$\triangle = B^2 - 4AC > 0.$$

Knowing that J is symmetric and invertible, there exists a matrix $R \in SO(3)$ such that

$$J = R^T \tilde{J} R,$$

where

$$\tilde{J} = \text{diag}\{J_1, J_2, J_3\}, \quad J_i > 0 \text{ for } i \in [1,3]$$

we note also that

$$RG_j = \tilde{g}_j = (x_j, y_j, z_j)^T, \quad j \in [1,2].$$

Then

$$\triangle = (\det\tilde{J})^2 \tilde{\triangle}$$

where

$$\begin{aligned}\tilde{\Delta} &= a_1^2X^4 + a_2^2Y^4 + a_3Z^4 - 2a_1a_2(XY)^2 \\ &- 2a_1a_3(XZ)^2 - 2a_2a_3(YZ)^2\end{aligned}$$

such that $a_1 = (J_2 - J_3)/J_1$, $a_2 = (J_3 - J_1)/J_2$, $a_3 = (J_1 - J_2)/J_3$ and

$$\begin{aligned}X &= y_1z_2 - y_2z_1, \\ Y &= x_2z_1 - x_1z_2, \\ Z &= x_1y_2 - x_2y_1.\end{aligned}$$

Since a_1, a_2 and a_3 cannot have the same sign, condition (5) implies $\tilde{\Delta} > 0$. More precisely, we can write condition (5) -after changing coordinates- like follows :

$$a_1HLX + a_2KLY + a_3HKZ \neq 0$$

where $(K, H, L)^T = \gamma_1\tilde{g}_1 + \gamma_2\tilde{g}_2$. Hence,

(i) for example if $a_1 \geq 0$, $a_2 \leq 0$ and $a_3 = 0$:

$$\tilde{\Delta} = (a_1X^2 - a_2Y^2)^2 \geq 0$$

$\tilde{\Delta} = 0$ if and only if $a_1 = a_2 = 0$ or $X = Y = 0$. Therefore condition (5) fails.

(ii) If $a_1 \geq 0$, $a_2 \geq 0$ and $a_3 \leq 0$,

$$\begin{aligned}\tilde{\Delta} &= (a_1X^2 - a_2Y^2)^2 + a_3Z^2(a_3Z^2 \\ &- 2a_2Y^2 - 2a_1X^2) \geq 0.\end{aligned}$$

$\tilde{\Delta} = 0$ implies that $a_i = 0$ for $i \in \{1, 2, 3\}$, or $X = Y = Z = 0$, or $Z = 0$ and $a_1X^2 = a_2Y^2$ in this last case if a_1 or a_2 vanishes then Y or X vanishes,
otherwise, having $Z = 0$ and for instance $x_1 \neq 0$, we obtain $XY = (-y_1/x_1)Y^2$, and so $Y = (-y_1/x_1)(a_1/a_2)X$.
Observing that

$$\begin{aligned}&a_1HX + a_2KY = \\ &a_1HX + a_2K(-y_1/x_1)(a_1/a_2)X = \\ &(a_1/x_1)X(x_1H - y_1K) = 0.\end{aligned}$$

Therefore condition (5) fails.

Since $\tilde{\Delta} > 0$, we can write (10) like :

$$Q(\tilde{\omega}) = \tilde{\omega}_2\tilde{\omega}_3$$

after changing coordinates with a linear invertible matrix P (see (11)). Taking :

$$(0, u_1, u_2)^T = P^{-1}(0, V_1, V_2)^T,$$

system (8)-(2) takes the form of (12).

Acknowledgment. I thank Professor J.M. Coron who helped and encouraged me to write this paper.

4. REFERENCES

Aeyels, D. (1985). Stabilization by smooth feedback of the angular velocity of a rigid body, *Systems and Control Letters*, pp. 59-64.

Bianchini, R.M., and G. Stefani (1986). Sufficient conditions of local controllability, *Proceedings 25th Conference on Decision and Control*, Athènes, Grèce, Décembre, pp. 967-970.

Bonnard, B. (1982). Controllability of mechanical control systems on Lie groups, *Laboratoire d'Automatique de Grenoble, Ecole Nationale Superieure d'Ingenieurs*, France, Note 82-08, Jan.

Brockett, R.W. (1983). Asymptotic stability and feedback stabilization, in : R.W. Brockett, R.S. Millman and H.J. Sussmann eds.. *Differential Geometric Control Theory*; Birkhäuser, Basel-Boston.

Byrnes, I. and Isidori (1991). On the attitude stabilization of rigid spacecraft, *Automatica vol. 1*, pp. 87-95.

Coron, J.-M. (1992). Links between local controllability and local continuous stabilization, *Université Paris-Sud et ETH Zurich*, Octobre 1991 et NOLCOS, Bordeaux, june.

Coron, J.-M. (1993). On the stabilization in finite time of locally controllable systems by means of continuous time-varying feedback laws, *preprint CMLA*, ENS Cachan.

Coron, J.-M., and E.-Y. Keraï (1994). Explicit feedbacks stabilizing the attitude of a rigid spacecraft with two control torques, *preprint CMLA*, ENS Cachan.

Crouch, P.E. (1984). Spacecraft Attitude Control and Stabilization : Applications of Geometric Control Theory to Rigid Body Models, *IEEE Transactions on Automatic control*, vol. AC-29,$N^0$4, April.

Hashémi, H. (1992). Commandabilité et stabilisabilité des satellites artificiels, *Rapport d'option automatisme*, Ecole Polytechnique.

Hermes, H.G. (1977). Controlled Stability, *Annali di Matematica pura ed applicata 4*, CXIV.

Jurdjevic, V., and J.P. Quinn (1978). Controllability and stability, *J. Diff. Equ.* 28, pp. 381-399.

Kawski, M. (1990). High-order small-time local controllability, *monograph and texbooks in pure an applied mathematics*, Marcel Dekker, vol. 133, pp. 431-467.

Meyer, G. (1966). On the use of Euler's theorem on rotations for the synthesis of attitude control systems, *Ames Res. Cen.*, Moffet Field, CA, NASA TN. D-3643.

Morin, P., C. Samson, J.-B. Pomet, and Z.P. Jiang (1994). Time-varying feedback stabilization of the attitude of a rigid spacecraft with two controls,*preprint INRIA*, 2275, May.

Sussmann, H.S. (1983). Lie brackets and local controllability : a sufficient condition for scalar input systems, *SIAM J.Control Optim. 21*, pp. 686-713.

Sussmann, H.J. (1987). A general theorem on local controllability, *SIAM J. Control Optim. 25*, pp. 158-194.

AUTHOR INDEX

www.ingramcontent.com/pod-product-compliance
Ingram Content Group UK Ltd.
Pitfield, Milton Keynes, MK11 3LW, UK
UKHW052301280926
15471UKWH00006B/1039

9 780080 423562